Conversion Factors

Length

1 in. = 2.54 cm

1 ft = 0.3048 m

1 mi = 5280 ft = 1.609 km

1 m = 3.281 ft

1 km = 0.6214 mi

1 angstrom (Å) = 10^{-10} m

Mass

1 slug = 14.59 kg

1 kg = 1000 grams = 6.852×10^{-2} slug

1 atomic mass unit (u) = 1.6605×10^{-27} kg

(1 kg has a weight of 2.205 lb where the gravitational acceleration is 32.174 ft/s^2)

Time

1 day = 24 h = 1.44×10^3 min = 8.64×10^4 s

1 yr = 365.24 days = 3.156×10^7 s

Speed

1 mi/h = 1.609 km/h = 1.467 ft/s = 0.4470 m/s

1 km/h = 0.6214 mi/h = 0.2778 m/s = 0.9113 ft/s

Force

1 lb = 4.448 N

1 N = 10^5 dynes = 0.2248 lb

Work and Energy

1 J = 0.7376 ft $\cdot$ lb = 10^7 ergs

1 kcal = 4186 J

1 Btu = 1055 J

1 kWh = 3.600×10^6 J

1 eV = 1.602×10^{-19} J

Power

1 hp = 550 ft $\cdot$ lb/s = 745.7 W

1 W = 0.7376 ft $\cdot$ lb/s

Pressure

1 Pa = 1 N/m^2 = 1.450×10^{-4} lb/in.^{2}

1 lb/in.^{2} = 6.895×10^3 Pa

1 atm = 1.013×10^5 Pa = 1.013 bar = 14.70 lb/in.^{2} = 760 mm Hg

Volume

1 liter = 10^{-3} m^3 = 1000 cm^3 = 0.03531 ft^3

1 ft^3 = 0.02832 m^3 = 7.481 U.S. gallons

1 U.S. gallon = 3.785×10^{-3} m^3 = 0.1337 ft^3

Angle

1 radian = 57.30°

1° = 0.01745 radians

2π radians = 360°

Standard Prefixes Used to Denote Multiples of Ten

Prefix	Symbol	Factor
Tera	T	10^{12}
Giga	G	10^{9}
Mega	M	10^{6}
Kilo	k	10^{3}
Hecto	h	10^{2}
Deka	da	10^{1}
Deci	d	10^{-1}
Centi	c	10^{-2}
Milli	m	10^{-3}
Micro	μ	10^{-6}
Nano	n	10^{-9}
Pico	p	10^{-12}
Femto	f	10^{-15}

Basic Mathematical Formulae

Area of a circle = πr^2

Circumference of a circle = $2\pi r$

Surface area of a sphere = $4\pi r^2$

Volume of a sphere = $\frac{4}{3}\pi r^3$

Pythagorean theorem: = HYP^2 = OPP^2 + ADJ^2

Sine of an angle: $\sin \theta$ = OPP/HYP

Cosine of an angle: $\cos \theta$ = ADJ/HYP

Tangent of an angle: $\tan \theta$ = OPP/ADJ

Law of cosines: $c^2 = a^2 + b^2 - 2ab \cos \gamma$

Law of sines: $a/\sin\alpha = b/\sin\beta = c/\sin\gamma$

Quadratic formula:

If $ax^2 + bx + c = 0$, then, $x = (-b \pm \sqrt{b^2 - 4ac})/(2a)$

eGrade Plus

www.wiley.com/college/touger
Based on the Activities You Do Every Day

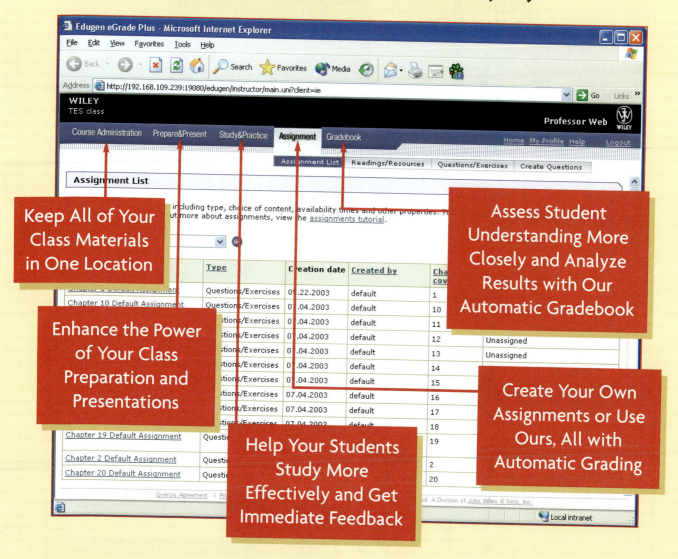

Keep All of Your Class Materials in One Location

Enhance the Power of Your Class Preparation and Presentations

Help Your Students Study More Effectively and Get Immediate Feedback

Assess Student Understanding More Closely and Analyze Results with Our Automatic Gradebook

Create Your Own Assignments or Use Ours, All with Automatic Grading

All the content and tools you need, all in one location, in an easy-to-use browser format. Choose the resources you need, or rely on the arrangement supplied by us.

Now, many of Wiley's textbooks are available with eGrade Plus, a powerful online tool that provides a completely integrated suite of teaching and learning resources in one easy-to-use website. eGrade Plus integrates Wiley's world-renowned content with media, including a multimedia version of the text, PowerPoint slides, and more. Upon adoption of eGrade Plus, you can begin to customize your course with the resources shown here.

See for yourself!
Go to www.wiley.com/college/egradeplus for an online demonstration of this powerful new software.

Students,
eGrade Plus Allows You to:

Study More Effectively

Get Immediate Feedback When You Practice on Your Own

Our website links directly to **electronic book content,** so that you can review the text while you study and complete homework online. Additional resources include **self-assessment quizzing** with detailed feedback, **Interactive Learningware** with step by step problem solving tutorials, and **interactive simulations** to help you review key topics.

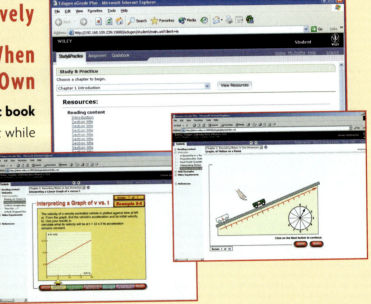

Complete Assignments / Get Help with Problem Solving

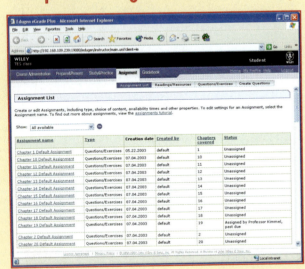

An **"Assignment"** area keeps all your assigned work in one location, making it easy for you to stay on task. In addition, many homework problems contain a **link** to the relevant section of the **electronic book,** providing you with a text explanation to help you conquer problem-solving obstacles as they arise.

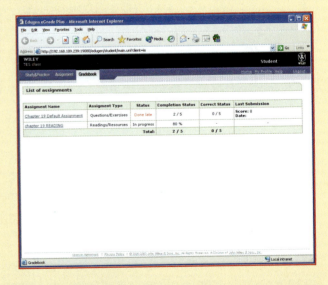

Keep Track of How You're Doing

A **Personal Gradebook** allows you to view your results from past assignments at any time.

INTRODUCTORY

Physics

Building Understanding

Jerold Touger

Curry College
Milton, Massachusetts

JOHN WILEY & SONS, INC.

For my wife Hallie,
my daughters Molly and Naomi,
and in loving memory of my parents,
Morris and Freda Touger

ACQUISITIONS EDITOR	Stuart Johnson
SENIOR DEVELOPMENT EDITOR	Ellen Ford
PRODUCTION EDITOR	Barbara Russiello
MARKETING MANAGER	Amanda Wygal
DESIGN DIRECTOR	Harry Nolan
TEXT DESIGNER	Dawn L. Stanley
COVER DESIGN	Norm Christensen
ILLUSTRATION EDITOR	Sigmund Malinowski
ART MANAGEMENT	Edward T. Starr & Associates
ELECTRONIC ILLUSTRATIONS	Precision Graphics
CHAPTER OPENING CARTOON ART	Jerold Touger
PHOTO EDITOR	Hilary Newman
PHOTO RESEARCHER	Ramón Rivera Moret
COVER PHOTOGRAPH	©Stuart Westmorland/The Image Bank/Getty Images

This book was set in 10/12 ITC Garamond by *The GTS Companies*/York, PA Campus, and printed and bound by Von Hoffmann Press. The cover was printed by Von Hoffmann.

This book is printed on acid free paper. ∞

Library of Congress Cataloging in Publication Data

Touger, Jerold.
  Introductory physics : building understanding / by Jerold Touger.—1st ed.
    p. cm.
  Includes index.
  ISBN 0-471-41873-0 (cloth)
    1. Physics. I. Title.

QC21.3.T68 2006
530—dc22

                                      2004059614

Printed in the United States of America

10 9 8 7 6 5 4 3 2 1

Jerold Touger is an experienced teacher as well as an active researcher in cognitive aspects of physics teaching and learning. He received his own physics education at Cornell University (B.A., 1966) and the City University of New York (Ph.D., 1974). That education has continued through 28 years of full-time teaching at Curry College in Milton, Massachusetts, where he is now Professor of Physics. At Curry he has developed his own courses and course materials, some with NSF support, and has reported on this work in the *American Journal of Physics* and the *Journal of College Science Teaching*. He has also been involved with the physics teaching profession more broadly. He has chaired the Committee on Research in Physics Education of the American Association of Physics Teachers (AAPT), is a past president of the AAPT's New England Section, and has done curriculum development for TERC in Cambridge, Massachusetts.

For over fifteen years, he has maintained an active research association with the Physics Education Research Group at the University of Massachusetts at Amherst, supported at various times by three successive NSF Research Opportunity Awards and a Visiting Research Professorship. His research findings have appeared in several journal articles and in the proceedings of four international conferences. His own research and that of his collaborators, together with a broad and deep exposure to the outcomes and applications of physics education research, establish much of the perspective from which he has written this textbook.

Preface

PURPOSE AND GOALS

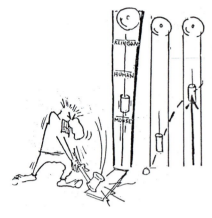

This Preface describes a textbook with a difference. It is intended for the standard two-semester, non-calculus course. But unlike traditional books, it draws extensively on a vast body of physics education research to develop pedagogical approaches that directly address students' needs, and in particular it makes extensive use of on-line technology to make the textbook an environment for active learning.

For over two decades, physics education research has been enriching our understanding of what works and what doesn't in physics teaching. At a broad range of institutions, from high schools to Ivy League universities, these new insights have been transforming the nature of physics teaching.

The purpose of this book is to meet the needs of faculty and students and to support the changes that an ever-increasing number of faculty have already made in their teaching. It addresses a range of teaching concerns raised by the research, with special focus on these:

- Students learn better in properly guided active learning environments. Most students do not spontaneously use textbooks in an active style, and for that reason many of them use their textbooks minimally if at all.

- Students can't be treated as empty vessels passively receiving the knowledge we bestow on them. They enter our classes with their own conceptions about the physical world and try to reconcile new ideas with their existing knowledge. Their prior conceptions and strategies are resistant to change, even in the face of clear and meticulous explanations. Merely explaining why they are wrong, or how it "really" is, just doesn't do it for most students. They need to be engaged in active learning; they need to be guided in discovering new understanding for themselves.

- Most of our students have to construct abstract ideas as extrapolations from a prior body of concrete experience. They need to engage in a conceptual exploration of the physical phenomena before the introduction of mathematical formalisms can be meaningful for them.

- The connections among ideas in a physics course are not self-evident to students. Many students need a "story line" to scaffold the connectedness that physicists take for granted. For these students, we need to motivate the approaches that we take. We need to introduce phenomena and raise questions before offering explanations. We need to motivate and construct concepts before naming them. They must also see the ideas represented in multiple ways (verbal, pictorial, diagrammatic, graphical, algebraic, even kinesthetic) that ultimately reinforce one another.

- The "plug and chug" approach that many beginning students use to solve quantitative problems shows that they fail to connect qualitative ("conceptual") and quantitative reasoning. Students need explicit guidance in using qualitative thinking about physics concepts in the service of quantitative problem solving.

- They often find conventional usage obscure or misleading; they require language that is direct, explicit, and self-consistent.

- Many of the students entering introductory physics courses are inadequately prepared. For these students to succeed, courses and the textbooks they use must address the challenge of lowering thresholds to meet students where they are while maintaining appropriate goals for outcomes.

How the proposed book deals with the concerns raised here—pedagogical devices that draw the reader into active dialogue with the ideas, compelling situations that motivate inquiry or challenge students' preconceptions, and many other approaches—will be described more fully in the next section.

FEATURES OF THE TEXT

Students *construct* understanding of the physical world. To do so, they must be actively engaged in making sense of the phenomena and the ideas. Explicit recognition of this is a major thrust of the proposed text. The features of the text outlined here are keyed to the specific aspects of student cognition that they are intended to address:

Students' Conceptions: Students construct knowledge by assimilating new material to their existing knowledge structures. Introduction of new ideas must therefore address existing conceptions. Widely held students' conceptions will be addressed through specific **STOP&Think** situations and **On-The-Spot Activities**, using readily available materials, that acquaint students with the phenomena and at times place their conceptions in conflict with the evidence. Like the on-line features, these features are aimed at promoting active learning rather than passive reading.

Story Line: Whenever possible, the rationale for introducing new ideas is provided by a coherent narrative thread or story line. The overall story line is that physics is an ongoing search for (or attempt to model) the fundamental rules by which the "game of nature" is played. More local story lines inform individual chapters. For example, Chapter 3 explores the general notion that complex motions can be viewed as composites of simpler motions, in order to motivate the introduction of vectors so that motion in two or three dimensions can be treated as a composite of one-dimensional components. Chapter 6 first explores qualitatively the idea that many situations in nature involve something being gained as something else is lost. This provides a context in which to consider energy and momentum as conserved quantities.

Conceptual before Quantitative: An extensive conceptual exploration of the phenomena introduces each new major area or concept before it is addressed quantitatively. An entire introductory qualitative chapter is devoted to each of the two most central and conceptually difficult areas, Newton's Laws (Chapter 4) and Electrical Phenomena (Chapter 18).

Role of Mathematics: Mathematics is treated as a way of modeling the behavior of measurable quantities. New mathematics gets "invented" or constructed as needed to model physical phenomena, *after* the phenomena are explored conceptually. For example, in Chapter 3 the introduction of vectors is motivated by the idea, first explored qualitatively, that simpler motions can be combined to produce more complex motions, and that in particular mutually perpendicular one-dimensional motions can be combined to produce two-dimensional motions. In this spirit also, formulas get discussed in terms of how they encapsulate meaning. When appropriate, examples show solutions from basic definitions as well as from derived formulas (*e.g.,* in Example 3-8 in Chapter 3).

Some of the beginning material may seem very basic to some instructors, but the level of sophistication increases through each chapter, reflecting the "start low, aim high" approach of meeting students where they are, but providing a pathway to where they should appropriately end up.

Design of Experiments: Insofar as it is possible, where the experimental basis for physical understanding is treated, students are encouraged to think through the construction of the experimental design themselves (*e.g.,* to consider how one might try to produce a two-source pattern for light analogous to what one can observe in a ripple tank, and thus to re-invent the double slit experiment).

Analogical Thinking: Researchers have noted that experienced physicists draw heavily on analogical thinking in attacking new problems. Some researchers have found considerable value in using "bridging analogies" which allow students to see how concepts may be applied in a range of situations analogous to their application in a concrete "anchor" situation in which the concept's application is easily grasped. The text will call the student's attention repeatedly to the power of analogical thinking. Instances of this include:

- The bridging between forces exerted by springs and springy surfaces and the upward force a table exerts on a book in Chapter 4
- The donut situation in Chapter 14, an analog to the Doppler effect
- The variation of the shadow cast by a rotating radius (in Chapter 13) as an analogy for all sinusoidal cycles, and the idea that angle, which tells you where the radius is in its cycle, becomes an analog to *phase,* which tells you where you are in a cycle (like that of a block on a spring) in which no physical angle is involved.

Visual Imagery: Visualization—literally "getting the picture"—is an important aspect of a physicist's thinking. The WebLinks are one feature that encourages this. Where appropriate, print figures will iterate the theme of evolving an idea in several steps (as in Figure 3-17, which is developed more fully in WebLink 3-5), because students often don't know where to begin when they encounter only a single completed figure with a great deal of information. In addition, the book will have a repertoire of easily remembered pictorial situations. Rather like the anchors of bridging analogies, these serve students as concrete departure points for constructing concepts that can then be applied to less readily visualized situations.

Deep Structure/Governing Concepts and Hierarchical Thinking: The tendency of students to resort to formulas rather than thinking about applicable principles is well known to both researchers and experienced teachers. Heavy emphasis in both in-chapter examples and end-of-chapter question-and-problem sets will therefore be placed on identifying applicable principles prior to (or even in lieu of) doing any calculation. Cumulative Review question-and-problem sets are included for mechanics and for electricity and magnetism. Students will not have available as clues to solutions the fact that a problem appears at the end of a particular chapter, let alone under a specific section heading. Rather, summary lists of principles appear at the beginning of each cumulative review set, and students are explicitly instructed to identify which principles are relevant before writing down formulas or doing algebra. In the chapters and in the end-of-chapter summaries, principles and the scope of their applicability are stressed.

Qualitative and Functional Reasoning Integrated with Quantitative Reasoning: An extensive body of research provides irrefutable evidence that students who perform well on exams by cranking out the right numbers often have little understanding of how the system they have considered actually behaves. For that reason, qualitative and quantitative questions and problems at the ends of chapters are not separated; often a problem will have both qualitative and quantitative parts. There will be a variety of question types that require students to do qualitative and functional, but not explicitly calculational or algebraic, reasoning. These will include:

- "What if" questions requiring students to determine what changes in behavior will result if a change is made in a set-up and even what variables to address (*e.g.,* if a block slides down a ramp onto a spring, how is the motion of the block changed if the set-up is operated on the moon rather than on Earth?).
- Comparisons
- "Examining an argument" questions that ask students to critique an argument (including arguments containing typical student misconceptions)

- Ranking questions, such as "list the bulbs in this circuit in rank order from brightest to dimmest, indicating any equalities"
- Always, Sometimes, or Never questions
- Discussion questions (suitable for in-class discussion)
- "How would you test . . . ?" or "design an experiment to . . ." questions
- A cumulative set of Review Problems covering mechanics, electricity and magnetism can be found in the back of the book. These are referenced at the end of Chapters 7, 9, 10, 21, and 23.

Attention to Use of Language: The author's own research[1] has shown that the locutions of conventional physics usage (e.g., "a force acts") can miscommunicate the shape of the ideas (what acts is an agent, not an action) to someone not already acculturated to the physicist's language. At other times students fail to adopt usage ("an object has energy" rather than "a situation is energy") that properly conveys the structure of the ideas. The proposed text will explicitly call students' attention both to the pitfalls of some usages and the conceptual appropriateness of others, as well as being meticulous in its own use of language.

Encouragement of Active Reflection: Both the text and on-line material seek to engage the student in a dialogue about his/her own thinking, thus encouraging the student to reflect actively and critically on how he/she is approaching the subject matter.

Cases and Examples: Cases are concrete situations in which a line of reasoning is explored or which brings out the need for a line of reasoning. Often they precede the generalization of the ideas that they involved, so that students can go from concrete to abstract, rather than the other way around. This is in contrast to the Examples, which illustrate the use of concepts or principles that have already been presented. Cases also contrast with Examples in that their format is not one of posing a problem, then working out a solution; the Cases are rather more like explorations. Cases may serve to provide "**memorable instances,**" which act as a concrete "anchor" for a concept (Case 3-1 and WebLink 3-4 do this for projectile motion as a composite of two one-dimensional motions).

Implicit Made Explicit: Example problems in textbooks are too often careful expositions of algebra, leaving the student to ask, "How did you know to do that?" Examples within chapters will usually begin by addressing the *Choice of approach*, and in many cases end with comments on *Making sense of the solution*. These features may address which principles are applicable (and how we know), why other principles are not applicable (*e.g.*, why energy principles are applicable to motion on a roller coaster but constant acceleration equations of motion are not), any strategic considerations, interconnections between various representations (*e.g.*, diagrammatic, graphic, and algebraic representations of motion), the significance of the result, etc.

On-line Features: The on-line features have been fully integrated into the book—they are in no way "peripherals." Material is delivered electronically rather than in print when it is more appropriate and serves active-engagement objectives. The overall Web environment in which this material will be found is labeled by an icon in the text —basically a cue to the student to go on line. Within this environment, there are three main types of on-line features:

- ***Web Examples:*** These are worked out examples, just as worked out examples appear within the body of each chapter in all texts. But on-line the student participates actively in the solution. In particular the student is often required to make preliminary decisions (and then receives immediate "just-in-time"

[1]Touger, J.S., When Words Fail Us, *The Physics Teacher,* **29,** 90–95 (1991)

feedback) about which concepts to apply. See, for instance, Web Example 5-7, in which students must sort out ideas about Newton's second and third laws, and Web Example 7-8, in which students must think about which conservation laws are applicable under which conditions. These examples underscore the importance of qualitative thinking in quantitative problem solving.

- **WebLinks:** WebLinks provide an on-line alternative to static figures that compress a great deal of information. Students will turn to these brief sequences, which will be available on the Web, as they now turn to ordinary figures. But their frame-by-frame and continuous animations will allow students to control the pace at which they work through the development of an idea. Some WebLinks (*e.g.,* WebLink 3-8) raise questions along the way that require students to confront their own thinking, while others (*e.g.,* WebLink 3-2) are simple animations to help students "get the picture."

- **End-of-chapter problems:** Both Web Examples and WebLinks have immediate follow-up questions with which students can test their understanding. But in addition, most of the end-of-chapter problems will be on-line (see description of eGrade Plus below). On-line assignments are widely used and make possible more timely feedback than when assignments are hand-submitted and hand-graded. However, not all questions and problems lend themselves to being answered on-line—for example, discussion questions, or questions that ask for an explanation, graph, or diagram. These will remain in print form.

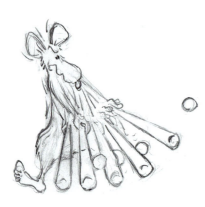

SUPPLEMENTS

An extensive package of supplements to accompany *Introductory Physics* is available to assist both the teacher and the student.

Instructor Supplements *HELPING TEACHERS TEACH*

Instructor's Resource Guide provides many useful suggestions for how the text material can be used to promote active learning in the classroom. References to additional resources and articles are also provided.

Instructor's Solution Manual, written by Sen-Ben Liao of the Massachusetts Institute of Technology, is an excellent resource for instructors. This manual provides worked-out solutions to all of the end-of-chapter Qualitative and Quantitative Problems. Each solution is thoroughly stepped through and numerous original diagrams have been created to enhance the presentation. Volume 1 contains the solutions for Chapters 1–14. Volume 2 contains the solutions for Chapters 15–28.

Test Bank, written by Bruce Libby of Manhattan College, contains a range of conceptual and quantitative questions. The questions are sorted by chapter sections and have varying difficulty levels. These items are also available in the *Computerized Test Bank* (*see below*).

Instructor Resource CD-ROM This CD contains:

- **Instructor's Solution Manual** in both Microsoft Word and PDF files
- **Computerized Test Bank** program with full editing features allows you to customize exams
- **Text Art** suitable for classroom projection and printing

Wiley Physics Demonstration DVD contains over 80 classic physics demonstrations that will engage and instruct your students. An accompanying instructor's guide is available.

Wiley Physics Simulations CD-ROM includes 50 interactive simulations that can be used for classroom demonstrations.

Web Resources for Classroom Management include Web CT, Blackboard, Web Assign, and eGrade Plus.

eGrade Plus is a powerful online tool that provides instructors with an integrated suite of teaching and learning resources, including an online version of the text, in one easy-to-use website. Organized around the essential activities you perform in class, eGrade Plus allows you to create class presentations, assign homework and quizzes that will be automatically graded, and track your students' progress. The system links homework problems to the relevant section of the online text, providing students with context-sensitive help. View a demo and learn more about eGrade Plus by visiting www.wiley.com/college/egradeplus.

WebAssign and WebAssign Plus WebAssign is an on-line homework management program that gives instructors the ability to deliver and grade homework and quizzes via the Web. WebAssign Plus includes links to an on-line version of the text.

WebCT and Blackboard Many of the instructor and student supplements have been coded for easy integration into WebCT and Blackboard, which are course management programs that allow instructors to set up a complete on-line course with chat rooms, bulletin boards, quizzing, as well as built-in student tracking tools, etc.

Student Supplements *HELPING STUDENTS LEARN*

Student Solutions Manual, written by Sen-Ben Liao of the Massachusetts Institute of Technology, provides students with complete worked-out solutions for selected odd-numbered end-of-chapter problems. These problems are indicated in the text with the **SSM** icon. The problem-solving method emphasizes mathematical clarity and also makes use of visual representations.

Student Workbook, written by Todd Zimmerman of Madison Area Technical College, is an essential supplement for students. The skills students develop completing the workbook exercises will aid them in solving more quantitative problems in the textbook. Students will learn to interpret the various representations used, such a sketches, graphs, formulas, and verbal descriptions and translate a problem between the representations. Exercises in the workbook draw on many years of research of the most common conceptual difficulties students encounter.

Student Web Site www.wiley.com/college/touger is designed to assist students further in their study of physics. At this site students can access the following resources:

- Solutions to selected end-of-chapter problems that are identified with a **WWW** icon.
- *WebLinks:* These are on-line interactive versions of some of the illustrations in the text. Please see the more complete description above.

- *Web Examples:* These are on-line interactive versions of some of the worked examples in the text. Please see the more complete description above.
- *eGrade Plus* includes all of the companion website's assets and more. In classes using eGrade Plus, students are able to complete homework and receive immediate feedback on their progress, as well as have their work tracked by their instructor.

ACKNOWLEDGMENTS

The author of a textbook is like the tip of an iceberg. The reader sees the author's name on the cover, but the book is a project to which large numbers of people have made essential contributions in a variety of ways, not only those who have been directly involved in its development and publication, but the professional colleagues whose ideas and encouragement are the foundation on which the entire project rests. In this section, I would like to express my appreciation to as

many of these individuals as possible, both at John Wiley & Sons and in the broader world of physics teaching and physics education research.

At Wiley I especially want to thank physics editor Stuart Johnson, whose sympathetic understanding and support of what I wanted to accomplish were vital both to getting the project launched and keeping it on track. Special thanks also to my developmental editor Ellen Ford and my production editor Barbara Russiello for suggestions, support, and ongoing meticulous scrutiny throughout the project, and for keeping the entire enterprise on task and on time. The look of the book is thanks to Dawn Stanley, designer.

I must express my appreciation as well for the very substantial contributions of developmental editor Anne Scanlan-Rohrer; media editor Martin Batey; photo research manager Hilary Newman; photo researcher Ramón Rivera-Moret; project editor Geraldine Osnato; media project manager Bridget O'Lavin; illustration editor Sigmund Malinowski; art coordinator, Ed Starr; and Russell Roy, who checked the final manuscript for accuracy. Their supportiveness and professionalism have been invaluable.

I must also express my appreciation to Sen-ben Liao, Massachusetts Institute of Technology, for his work on the Instructor and Student Solutions Manuals, and to Bruce Libby, Manhattan College, for creating the Test Bank.

This book has also benefited immeasurably from the contributions of an outstanding group of external reviewers, both those who read chapters and provided critical feedback and those who participated in focus groups. In addition, a superb group of individuals have worked on reviewing and revising the electronic media pieces that are an integral part of this project.

My sincere gratitude to all of the following:

Reviewers

B. N. Narahari Achar, *University of Memphis*

Rhett Allain, *Southeastern Louisiana State University*

Martina Arndt, *Bridgewater State College*

Steven Barnes, *California State University, San Bernardino*

Scott Bonham, *Western Kentucky University*

Robert Boughton, *Bowling Green State University*

Robert Buckley, *Hudson Valley Community College*

Charles Burkhardt, *St. Louis Community College*

Richard Cannon, *Southeast Missouri State University*

Tom Carter, *College of Dupage*

Keith Clay, *Green River Community College*

Michael Crescimanno, *Youngstown State University*

Dennis Crossley, *University of Wisconsin, Sheboygan*

Renee Diehl, *Pennsylvania State University*

Andrew Duffy, *Boston University*

Robert Dufresne, *University of Massachusetts, Amherst*

John Dykla, *Loyola University, Chicago*

Zbigniew Dziembowski, *Temple University*

Simon George, *California State University, Long Beach*

David Gerdes, *University of Michigan*

Barry Gilbert, *Rhode Island College*

Ron Greene, *University of New Orleans*

Kastro Hamed, *University of Texas, El Paso*

David Hammer, *University of Maryland*

Athula Herat, *Northern Kentucky University*

Yu-Kuang Hu, *University of Akron*

Tom Johannesmeyer, *Otero Junior College*

Adam Johnston, *Weber State University*

Stephen Kaback, *University of Maine*

Paul Kramer, *Farmingdale State University*

Mark Lattery, *University of Wisconsin, Oshkosh*

Mark Lucas, *Ohio University*

David Meltzer, *Iowa State University*

Jose Mestre, *University of Massachusetts, Amherst*

Marina Milner-Bolotin, *Rutgers University*

Peter Moeck, *Portland State University*

Rod Nave, *Georgia State University*

Anthony Nicastro, *West Chester University*

Scott Nutter, *Northern Kentucky University*

N. Sanjay Rebello, *Kansas State University*

Charles Robertson, *University of Washington*

Larry Rowan, *University of North Carolina, Chapel Hill*

Russell Roy, *Santa Fe Community College*

Roy Rubins, University of Texas at Arlington

Chandralekha Singh, *University of Pittsburgh*

Tim Slater, *Montana State University*

Timothy Slater, *University of Arizona*

Daniel Smith, *South Carolina State University*

William Smith, *Boise State University*

David Sokoloff, *University of Oregon*

Ronald Stoner, *Bowling Green State University*

Michael Strauss, *University of Oklahoma*

Nilgun Sungar, *California Polytechnic University, San Luis Obispo*

Aaron Titus, *North Carolina Agricultural and Technical University*
Beth Ann Thacker, *Texas Technical University*
Mary Urquhart, *University of Texas, Dallas*
Lee Widmer, *University of Cincinnati*
Paul Williams, *Aims Community College*
Susan Wyckoff, *Arizona State University*

William McNairy, *Duke University*
Arthur Mittler, *University of Massachusetts, Lowell*
Chris Pearson, *University of Michigan, Flint*
David Reid, *Eastern Michigan University*
N. Sanjay Rebello, *Kansas State University*
Aaron Titus, *High Point University*

Media Reviewers

Doyle Davis, *New Hampshire Community Technical College, Berlin*
Karim Diff, *Santa Fe Community College*
Nancy Donaldson, *Rockhurst University*
Vern Lindberg, *Rochester Institute of Technology*

Focus Group Participants

Nancy Donaldson, *Rockhurst University*
Robert Endorf, *University of Cincinnati*
William McNairy, *Duke University*
Arthur Mittler, *University of Massachusetts, Lowell*
Steve Shropshire, *Idaho State University*

The existence of this work is due in no small part to the people who gave me encouragement and direction at critical stages in my career. In chronological order I would like to thank:

My older brother Larry, whose enthusiasm for his high school physics course first brought physics to my attention,

Harry Lustig, who as chair of The City College (New York City) physics department awarded me my first teaching assistantship, which opened my eyes to the possibilities of teaching,

Naomi Sager, of the NYU Linguistics String Project, for involving me in a project that whetted my lifelong interest in questions of cognition and language,

Myriam Sarachik, my dissertation adviser at the City University of New York, whose substantial research projects never precluded taking teaching seriously,

John Hovorka, my first chairperson at Curry College, and a friend and mentor by personal example,

Lillian McDermott, whose uniquely insistent style of advice and encouragement when I first connected with the world of physics education research (PER) in 1983 was pivotal in determining the direction my career would follow,

The late Roman Sexl, whose interest in my early PER efforts also provided critical encouragement,

Jose Mestre and William (Bill) Gerace, who brought me in as a collaborator in the University of Massachusetts Physics Education Research Group, and the other colleagues in the group with whom it has been my pleasure to work over many years: Bob Dufresne, Bill Leonard, and Pamela T. Hardiman (and thanks to Jack Lochhead for introducing me to Jose and Bill),

Colleagues at TERC in Cambridge, MA, with whom I collaborated on curriculum development projects, especially David Crismond, June Foster, S. Riley Hart, Nancy Ishihara, and Irene Baker,

Ron Pullins, who first suggested that I write an introductory physics textbook informed by my PER perspective,

And Ron Thornton, for first putting me in contact with Stuart Johnson. The rest is history.

Although it is an impossible task to list everyone in the PER and physics teaching communities whose work has influenced mine, I wish to acknowledge the following people who (in addition to those thanked above), by their papers, presentations, and/or personal conversation or correspondence have directly influenced some of the content of this book: Arnold Arons, Bob Beichner, Ruth Chabay, John Clement, Dewey Dykstra, Igal Galili, Uri Ganiel, Fred Goldberg, Edith Guesne, Uri Haber-Schaim, Ibrahim Halloun, Paula Heron, David Hestenes, Curt Hieggelke, Priscilla Laws, David Maloney, Jim Minstrell, Tom O'Kuma, Edward (Joe) Redish, Peter Shaffer, Rachel Scherr, Bruce Sherwood, Josip Slisko, David Sokoloff, Mel Steinberg, Ron Thornton, Andrée Tiberghien, Alan Van

Heuvelen, Laurence Viennot, Stamatis Vokos. There are no doubt others who ought to be on this list but whom I have inadvertently excluded; to them, my sincere apologies.

I also wish to thank the three decades of physics students at Curry College whom I have had the pleasure of teaching, and who have provided me with an ongoing reality check.

Finally, I would like to thank my wife Hallie and my daughters Molly and Naomi for their support and for their patience over the years that this project has been a persistent and obtrusive presence in their lives.

Brief Contents

Contents

A Note to the Student

INTRODUCTORY PHYSICS: Building Understanding was written with you, the student, in mind:

- We believe you will learn better in a properly guided active learning environment and by using your textbook in an active style.

- We believe you enter our classes with your own conceptions about the physical world and try to reconcile new ideas with existing knowledge.

- We believe you ordinarily construct abstract ideas as extrapolations from a prior body of concrete experience and that you need to engage in a conceptual exploration of the physical phenomena before the mathematical treatment of ideas can be meaningful for you.

- We believe the connections among ideas in a physics course are not always self-evident to you. We provide a "story line" to scaffold the connectedness that physicists take for granted.

- We believe we need to motivate the approaches that we take. We need to introduce phenomena and raise questions before offering explanations. We need to motivate and construct concepts before naming them. You must be able to see the ideas represented in multiple ways (*verbal, pictorial, diagrammatic, graphical, algebraic, even kinesthetic*) that ultimately reinforce one another.

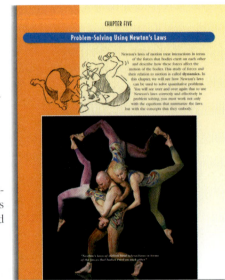

This text was written with these beliefs in mind and with careful attention to ways of making the learning experience more successful. If you take full advantage of the unique features and elements of the textbook, we believe your experience in physics will be both fulfilling and enjoyable.

VISUALIZATION

Getting the picture is an important aspect of thinking like a physicist. Be sure to pay close attention to the numerous in-text figures and photos. In physics, you need to be able to visualize what is happening in real-world situations. Developing concepts to explain what is happening also involves forming mental images. Real understanding is reached when you "see" the connections between the real-world actions and the concepts. The illustrative material is designed to help you do that.

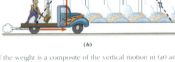

Figure 3-5 The ring-the-bell carnival game in time and space. (*a*) A sequence in time: Here the carnival game remains stationary. Spreading a sequence of stop-action pictures of the weight is a composite of the vertical motion in (*a*) and the horizontal motion of the truck. The truck's motion spreads that succes...

ON-THE-SPOT ACTIVITIES

These brief and easily do-able assignments encourage "active learning." By exploring a concept in a hands-on way, you are more likely to remember and understand. These can be done in class or outside of the classroom and often allow you to work with other class members.

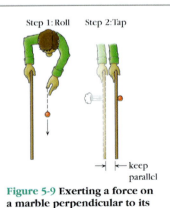

Step 1: Roll   Step 2: Tap

keep parallel

Figure 5-9 Exerting a force on a marble perpendicular to its motion.

On-The-Spot Activity 5-1

Lay a meter stick or other long straight object on the floor. Roll a marble or small ball along a straight-line path parallel to and a few centimeters to one side of the meter stick (Figure 5-9). Keeping the meter stick parallel to the ball's path, move it sideways until it taps the ball gently from the side while it is rolling. What is the path of the ball after being tapped? Does it just go in the direction of the tap? Repeat this several times, tapping the ball slightly harder each time and observing the consequent final direction. As long as the direction of the tap is perpendicular to the initial direction of motion, can the ball ever end up in a direction *perfectly* perpendicular to its initial direction? Experiment with your set-up to determine how hard and in what direction the tap must be if the ball's final direction *is* to be perpendicular to its initial direction.

STOP & THINK SITUATIONS

These "resting points" allow you to pause and think about the implications of what you are reading. Carefully consider these questions to check your understanding.

Typically, in dissipation, the energy becomes less concentrated as more and more objects get progressively smaller shares. Let's consider how this happens for a chair dropped to the ground from a height. When the chair is about to strike the ground, all its PE has been converted into KE. **STOP&Think** When the chair hits the ground, what happens to its KE? Does it convert back to PE? ◆

Experience tells us the chair is unlikely to bounce very high. Because grav-

CASES AND EXAMPLES

Cases are explorations typically involving an easily remembered instance or analogy that can serve as a concrete anchor for a concept we wish to introduce. Examples pose a specific problem using concepts already presented and then work through the solution. Both of these serve to enhance your understanding of the subject matter.

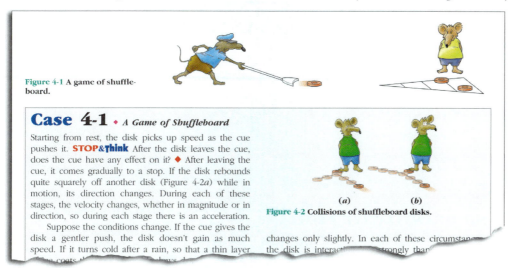

Figure 4-1 A game of shuffle-board.

Case 4-1 ◆ *A Game of Shuffleboard*

Starting from rest, the disk picks up speed as the cue pushes it. **STOP&Think** After the disk leaves the cue, does the cue have any effect on it? ◆ After leaving the cue, it comes gradually to a stop. If the disk rebounds quite squarely off another disk (Figure 4-2*a*) while in motion, its direction changes. During each of these stages, the velocity changes, whether in magnitude or in direction, so during each stage there is an acceleration.

Suppose the conditions change. If the cue gives the disk a gentler push, the disk doesn't gain as much speed. If it turns cold after a rain, so that a thin layer

(*a*)   (*b*)
Figure 4-2 Collisions of shuffleboard disks.

changes only slightly. In each of these circumstan[ces] the disk is interact[ing] [more] strongly than

WEBLINKS AND WEB EXAMPLES

The text website, www.wiley.com/college/touger, provides on-line features that are completely integrated with the book material and will further enhance your visualization skills and promote active learning. A web/book icon in the text points to material to be found on the website. *WebLinks* focus on individual concepts. They are an on-line alternative to the static figures found in the text. Frame-by-frame animation allows you to work through an idea at your own pace. *Web Examples* are guided, interactive, step-by-step versions of the Examples in the text that allow you to participate actively in the solution. You are encouraged to make decisions and receive immediate feedback on your choices.

For **WebLink 2-1** A Spaceship in a Resistive Medium, go to www.wiley.com/college/touger

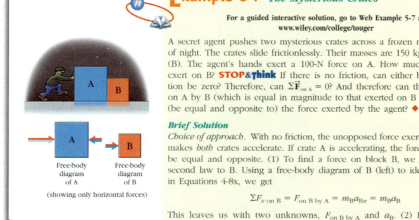

Example 5-7 *The Mysterious Crates*

For a guided interactive solution, go to Web Example 5-7 at www.wiley.com/college/touger

A secret agent pushes two mysterious crates across a frozen river in the da[rk] of night. The crates slide frictionlessly. Their masses are 150 kg (A) and 50 [kg] (B). The agent's hands exert a 100-N force on A. How much force does [A] exert on B? **STOP&Think** If there is no friction, can either block's accelera-tion be zero? Therefore, can $\Sigma \vec{F}_{\text{on A}} = 0$? And therefore can the force exerted on A by B (which is equal in magnitude to that exerted on B by A) "balanc[e]" (be equal and opposite to) the force exerted by the agent? ◆

Free-body diagram of A   **Free-body diagram of B**

(showing only horizontal forces)

Brief Solution

Choice of approach. With no friction, the unopposed force exerted by the age[nt] makes *both* crates accelerate. If crate A is accelerating, the forces on A cann[ot] be equal and opposite. (1) To find a force on block B, we apply Newto[n's] second law to B. Using a free-body diagram of B (left) to identify the forc[es] in Equations 4-8x, we get

$$\Sigma F_{x \text{ on B}} = F_{\text{on B by A}} = m_B a_{Bx} = m_B a_B \qquad (5-11)$$

This leaves us with two unknowns, $F_{\text{on B by A}}$ and a_B. (2) However, if we apply the second law to block A as well (see free-body diagram of A at left), we ge[t]

MARGIN NOTES

These reminders, cautions, and notations call attention to important terminology, hints, and tips related to the discussion.

➡ **A note on language:** It is usual in physics to speak of "the work done by a force," whereas in everyday usage, wo⋯ need to re⋯ a person ⋯ saying tha⋯ we are rea⋯ whatever ⋯ transferring⋯ doing so. ⋯

➡ **Note:** For beginning studen⋯ usually more instructive to d⋯ tions starting from fundam⋯ whenever possible, because it⋯ you thinking more about the ⋯ (what is happening and wh⋯ various quantities mean) than⋯ the algebra.

Warning: In contrast to modern everyday usage, Newton's use of the word *reaction* does *not* imply that the reaction comes after the action. On the con⋯ two inseparable si⋯ occur together in ⋯ *force* and *the oth⋯* which you call ⋯ reaction.

➡ **Caution:** Do not confuse equal and opposite forces *on the same body*, which produce equilibrium, with the equal and opposite inter-action forces that two bodies exert *on each other*.

CHAPTER SUMMARY

All the major points of the chapter are summarized at the end of each chapter for easy study and review. The Summary will have limited value, however, unless you have worked through the chapter.

◆ SUMMARY ◆

In keeping track of energy-related quantities, it is generally valuable to focus on a **system**, a set of objects that we group together mentally for convenience. We defined the *work* done on an object by a force $\vec{F}$ as

Work

$$W \equiv \frac{\text{amount of force exerted}}{\text{in the direction of motion}} \times \frac{\text{distance over which}}{\text{the force is exerted}} \quad (6\text{-}1)$$

or

$$W \equiv F \Delta s \cos \theta \quad (6\text{-}2)$$

F is the *magnitude* of the force vector. Δs is the distance (the *magnitude* of the displacement $\Delta \vec{s}$). The sign of work depends only on whether $\cos \theta$ is + or −.

(SI units: newtons · meters = **joules[J]**)

This work is an *energy input* (or *output* if negative) to the system containing the object.

• *the total work done by this force around any closed path (one that returns to its starting place) is zero.*

Conservative forces include gravitational forces (so that changes in gravitational PE depend only on changes in height, irrespective of the path; see Case 6-4) and elastic or Hooke's law forces.

Hooke's Law $\qquad F = -kx \qquad (6\text{-}17)$
(F = force exerted *by* spring; x = spring's displacement)

In contrast, the work done by **nonconservative forces**, such as sliding *frictional forces* on an object (which change direction as the object's path changes), depends on path length. The effect of such forces is frequently to convert energy from energies of whole objects to **internal** or **distributed energies**, such as **thermal energy** or **chemical energy.** Such energy is said to be **dissipated** and is readily transferable piecewise to the surroundings of a system ⋯

◆ QUALITATIVE AND QUANTITATIVE PROBLEMS ◆
Hands-On Activities and Discussion Questions

The questions and activities in this group are particularly suitable for in-class use.

6-1. Discussion Question. For the sequence in Figure 6-11, discuss the additional energy inputs, transfers, and conversions necessary so that the energy stored in carbohydrate molecules formed by wheat growing in a field in Kansas and peanut plants growing in a field in Georgia becomes available to a softball player in Boston in the form of a peanut butter sandwich.

6-2. Discussion Question. Two identical blocks are simultaneously released from the same height above a level floor. Block A reaches the floor by dropping straight down. Block B reaches the floor by sliding down a frictionless ramp.

a. A student argues as follows: "At any time during their fall, the blocks should have the same kinetic energy and therefore the same velocity. If one is never going faster than the other and both are released from the same height at the same time, both hit the table top at the same instant." Do you agree with this student's conclusions? Explain why or why not.

b. Before you knew any physics, if someone described this situation to you, which block would you have expected to reach the floor first? Do your conclusions based on physics agree with this expectation? If not, try to identify the fallacy in the one you think is wrong. Explain your reasoning.

Review and Practice

Section 6-1 An Intuitive Introduction to Energy Ideas

In each of Problems 6-3 to 6-10 below, (1) identify a suitable system, and (2) describe the energy conversions within the system and the ⋯

other until the athlete's feet cease touching the floor. (Neglect microscopic effects.)
a. Does the floor do work on the athlete? Brie⋯

END-OF-CHAPTER PROBLEMS

We encourage you to understand the concepts behind the problems, as opposed to just looking for the "right formula" or "right equation." The end of chapter problems are a mix of qualitative and quantitative, and include a variety of question types, such as hands-on and discussion questions, and follow-up questions on WebLinks. Working through these problems will give you a deeper understanding of the physics principles, not "just the math."

REVIEW PROBLEM SETS

Two sets of cumulative problems at the end of the text offer review and practice on concepts across several chapters. These serve to show the integration of concepts and require you to identify and apply principles before doing any calculations.

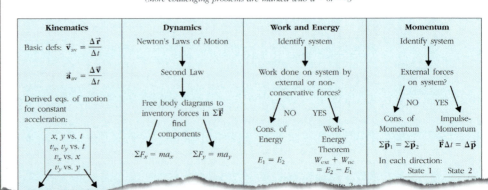

◆ REVIEW PROBLEM SET I ◆
Picking the Principles: Mechanics
(More challenging problems are marked with a • or ••.)

Kinematics	Dynamics	Work and Energy	Momentum
Basic defs: $\vec{v}_{av} = \dfrac{\Delta \vec{r}}{\Delta t}$	Newton's Laws of Motion	Identify system	Identify system
$\vec{a}_{av} = \dfrac{\Delta \vec{v}}{\Delta t}$	Second Law	Work done on system by external or non-conservative forces?	External forces on system?
Derived eqs. of motion for constant acceleration:	Free body diagrams to inventory forces in $\Sigma \vec{F}$ find components	NO YES	NO YES
x, y vs. t	$\Sigma F_x = ma_x \quad \Sigma F_y = ma_y$	Cons. of Energy Work-Energy Theorem	Cons. of Momentum Impulse-Momentum
v_x, v_y vs. t		$E_1 = E_2$ $W_{ext} + W_{nc} = E_2 - E_1$	$\Sigma \vec{p}_1 = \Sigma \vec{p}_2$ $\vec{F} \Delta t = \Delta \vec{p}$
v_x vs. x			In each direction:
v_y vs. y			State 1 State 2

CHAPTER ONE

Physics, Mathematics, and the Real World

Physics is a fundamentally human activity. It is a collective expression of the sense of wonder we feel before the rich diversity of the natural universe.

> *"The most beautiful experience we can have is the mysterious. It is the fundamental emotion that stands at the cradle of true art and true science."*
> —ALBERT EINSTEIN

Your gut-level "Oh, wow!" response when you witness a spectacular sunset is an expression of wonder. But wonder isn't *just* "Oh, wow!" and certainly not "Oh, wow! Now let's go to dinner." Wonder couples the "Oh, wow!" response with curiosity, with the urge to explore what you're seeing. In this sense, science and the arts are somewhat alike.

> *"There is no science without fancy, no art without facts."*
> —VLADIMIR NABOKOV

But physics, like all true science, goes an important step further by aspiring to *collective* understanding—not just how *I* understand something but reaching agreement on how *we* understand it.

> *"Art is I. Science is we."*
> —CLAUDE BERNARD, *nineteenth-century physiologist*

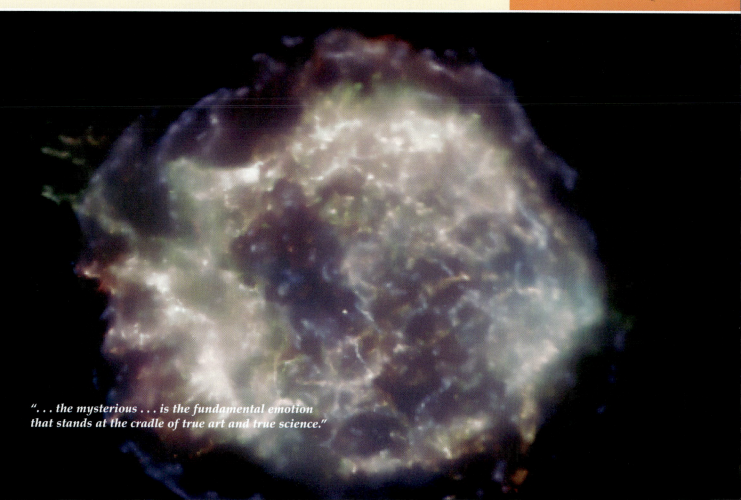

"... the mysterious ... is the fundamental emotion that stands at the cradle of true art and true science."

Reaching agreement is not merely a matter of majority rule. Physicists must exchange ideas and verify one another's observations and reasoning, testing whether behavior implied by the reasoning is borne out by further observations. Thus, they carry on the very human activity of consensus building in careful and refined ways. Physics is a consensual body of knowledge to which many individuals have contributed. Humans are social animals, and physics is very fundamentally the activity of a social species.

1-1 What Is Physics?

The activity of trying to understand is much the same in any area of knowledge. Imagine yourself as a small child, watching an older sibling playing soccer or baseball. Eventually, by careful observation and reasoning, you begin to figure out the rules of the game. In much the same way, we are all observers of the great game of nature, and after much observation we may find that certain rules appear to be followed without exception: the sun always rises in the east and sets in the west, a rock released in midair always falls down, hot and cold objects mixed together always reach a common in-between temperature. *Physics is the activity of trying to find the rules by which nature plays.*

Underlying physics, therefore, or any science for that matter, is the belief that there *are* rules, that nature is in some sense orderly. But that doesn't mean that the rules are easy to figure out.

Case 1-1 ◆ *Inferring the Rules of Baseball from an Obstructed Viewpoint*

Imagine yourself always watching baseball from the same overpriced but lousy seat in the ballpark. Your seat has an obstructed view (see Figure 1-1); you can see the batter, but not the catcher, the home plate umpire, or the path by which either takes or leaves the field. You are also too far away to hear what is happening on the field. Even if you start out knowing absolutely nothing about the game, you might quickly *infer* that there is a catcher from the way the ball keeps coming back from the unseen region. **STOP&Think** To **infer** something is to think that it is implied by the evidence. On what basis might you infer the existence of a home plate umpire? ◆

Figuring out the whole conceptual structure of balls and strikes is more difficult, but at length you think you have it worked out. Although you cannot see the scoreboard, you know whether the home crowd reacts positively or negatively after each pitch. You become more confident about your understanding each time a batter walks off the field after a third strike. Then one day a batter swings and misses at a third strike and runs to first base. This is an *anomalous* event—one that doesn't fit the previously established pattern. Baseball, it turns out, has a dropped third strike rule: If the catcher fails to catch the pitch on a third strike, the batter can attempt to advance to first base, and must be thrown out like any other base runner.

If you have never seen the catcher, though you've inferred that one exists, it is more difficult to figure out what is happening. You must build a mental picture of what the catcher is doing that is consistent with what you *can* see. But that picture is tentative, always subject to revision—literally picturing anew—if it turns out to conflict with some later observation.

Your understanding of the rules governing balls and strikes must also be considered tentative. You can never be sure you have seen all there is to see and that there will not be some unexpected event like a dropped third strike to make you reconsider. When that does happen, it is not the rules of baseball that have changed, only your understanding of what those rules are.

Figure 1-1 An obstructed view of a baseball game. The batter is visible, but not the umpire or catcher.

(a)

(b)

Figure 1-2 Flyers old and new. The study of motion and forces can help us understand the aerodynamics of (a) the archaeopteryx, a prehistoric ancestor of today's birds, or (b) the next-generation supersonic passenger jet as envisioned by NASA.

Our view of nature is also obstructed. There are aspects of the game of nature that our senses cannot detect. For example, humans never see ultraviolet radiation or X rays, but we infer from the exposure of photographic film that they exist. The details of the exposure in turn let us draw inferences about the objects that emit them, even objects thousands of light years away.

As we do for the baseball game, we assume that the rules of nature exist and are unchanging. But as in Case 1-1, the activity of reaching an understanding of those rules is endless. Our present understanding must always be viewed as tentative. It is a **model,** a mental picture in which the pieces of the picture obey the rules we've deduced. It is a good model to the extent that the objects in the real world behave as our mental picture would lead us to expect. In other words, we consider a model or theory valid, and potentially useful and productive, if it fits with all the evidence *so far*.

➡**Models:** Although hobbyists, architects, and others may build models out of wood, Plexiglas, and so on, that people can see and hold, physicists build models in their heads. The model is the idea itself; it is a theory.

◆**THE SCOPE OF PHYSICS** Because the behaviors we encounter in the natural universe are so overwhelmingly diverse, physics must cover a broad range of topics, though it turns out that underlying much of that diversity are the rules in a few fundamental areas. You will find that the areas listed here overlap and interconnect in surprising ways.

The study of *motion and forces* encompasses the orbits of planets and the paths of comets, the spin on a curve ball or the hovering of a Frisbee, and the aerodynamics of a supersonic jet or of the prehistoric forerunners of present-day birds (Figure 1-2).

The study of *electricity and magnetism* provides insight into phenomena as diverse as:

- Atmospheric effects, such as lightning and the aurora borealis, or northern lights.
- The technology underlying television, personal computers, use of solar energy, and a vast array of basic and not-so-basic appliances and instrumentation.
- The details of chemical reactions, including those (collectively called your *metabolism*) that occur in your body.

Nature's sound and light show. The study of electricity and magnetism provides insight into some of nature's most dramatic "special effects."

(a)

(b)

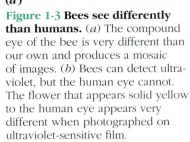

Figure 1-3 Bees see differently than humans. (*a*) The compound eye of the bee is very different than our own and produces a mosaic of images. (*b*) Bees can detect ultraviolet, but the human eye cannot. The flower that appears solid yellow to the human eye appears very different when photographed on ultraviolet-sensitive film.

- Properties of new materials, such as high-temperature superconductors.
- The transmission of signals in your own nervous system by means of charged particles moving in electric fields.
- The *electrophoresis* technique used in genetics and in blood identification in criminal and paternity cases to compare samples of DNA by looking at the different flow rates of its component parts in an electric field.

Optics, the study of light and related emissions (from X rays to radio waves), relates to:

- The working of your eyeglasses or microscope.
- The diverse ways in which different living things see (e.g., bees view the world through compound eyes and can see ultraviolet though we cannot; see Figure 1-3).
- Why things look as they do (a rainbow, say, or a blue sky or a red sunset).
- The use of X rays to analyze the submicroscopic structure of crystals and large molecules, a technique that has been of interest to geologists, to solid state or condensed matter physicists, and to biologists trying to determine the structure of DNA.
- The development and use of lasers and holography.
- The identification of substances by the light and related emissions that they give off, and the application of this procedure (called **spectroscopy**) to emissions from distant stars, quasars, and so on. This has enabled us to develop a considerable understanding, all of it inferred, of the composition of the objects that populate outer space and of the origins of our universe. In the late twentieth century the study of those origins, called **cosmology,** evolved rapidly from philosophical speculation to hard science.

Studies of *heat and temperature* apply equally to matters of auto engine efficiency, the risks of hot tubs, home insulation, determining whether dinosaurs were warm- or cold-blooded, the risks to global climate of increasing carbon dioxide and other gases in the atmosphere, or the heat generation processes in the interiors of stars.

Physicists at the cutting edge of physics still need a thorough understanding of these basic areas, called **classical physics,** whether they are studying the smallest known components of the universe in elementary particle physics or the largest in astrophysics and cosmology. Whether working on one of the world's largest high-energy particle accelerators (Figure 1-4*a*), or on the Hubble Space Telescope (Figure 1-4*b*), they are constantly considering how the basic underlying rules play out in new and unexpected contexts. But they are also asking where our understanding breaks down and where we may have to infer new or revised rules.

(a) **(b)**

Figure 1-4 At the frontiers of human knowledge. (*a*) The 3-km-long high-energy particle accelerator at Stanford probes the tiniest constituents of matter. (*b*) The Hubble Space Telescope probes the depths of interstellar space.

1-2 Measurement and Units

"The supreme task of the physicist is to arrive at those universal elementary laws from which the cosmos can be built up by pure deduction. There is no logical path to these laws; only intuition, resting on sympathetic understanding, can lead to them."
—ALBERT EINSTEIN

Before you can figure out the laws or rules things follow, you must first observe them. To develop a "sympathetic understanding," you need to become familiar with how things behave. Physics necessarily begins with what we detect by means of our senses. But then, to develop a shared understanding, we must be able to agree in detail on what we have observed and we must be able to communicate what we see without risk of being misunderstood. It is too vague, for example, to say that a ball player is large. Which athlete in Figure 1-5 is larger? By what standard? It would be clearer to give each athlete's height, weight, or shoulder width. We could agree on how the measurements compare, no matter which athlete we call larger.

In physics, therefore, observations are generally **quantitative,** that is, they are expressed in terms like *height* and *weight* that can have numerical values. Something that can have a numerical value is called a **quantity.** Speed, area, and the price per pound of potatoes are all examples of quantities. For the remainder of this chapter, we will focus on some aspects of how we treat quantities. As we do so, bear in mind Einstein's emphasis on intuition and a sympathetic understanding of how things behave. Physics is not just mathematics, or even primarily mathematics. Your use of mathematics has to be guided by thinking about how things behave and what rules or physical principles govern their behavior.

Figure 1-5 Comparing the "largeness" of athletes. Who is "larger," the tall, thin basketball player or the shorter but much broader football player?

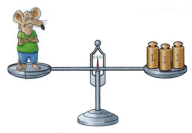

Figure 1-6 Balancing a laboratory rat. The three standard 1-kg masses just balance a (very large) 3-kg laboratory rat.

Measurements are quantitative observations made in comparison with a **standard,** which we call a **unit of measurement.** For many years, the distance between two fine lines engraved on a bar of platinum-iridium alloy kept at the International Bureau of Weight and Measures outside Paris was the internationally recognized standard meter. The standard **meter** is now defined as the length of the path traveled by light in a vacuum during a time interval of 1/299 792 458 of a second. When we say a soccer field is 100 meters long, we mean it is 100 times as long as the carefully marked-off unit called a meter.

Likewise, we can establish a unit of mass by choosing a particular block of metal: a cylinder of platinum-iridium alloy serves this purpose at the International Bureau of Weights and Measures. Another mass is equal to this mass if it just balances it on an equal arm balance in a uniform gravitational environment. If we call our standard unit of mass a **kilogram,** a rat will have a mass of 3 kg if it just balances three of these units on a balance scale (Figure 1-6). For now, what we will mean by the mass of an object is the number of standard units that it can counterbalance on an equal arm balance. Physicists call this an **operational definition,** because we are defining mass by what we do (the "operation" we perform) to measure it.

Crudely speaking, mass measured in this way gives us a feel for "how much stuff" we have. But mass is different from weight. Placed on the moon, the contents of each pan in Figure 1-6 will weigh less, but the rat still counterbalances the three standard kilograms and thus still has a mass of 3 kg.

To measure **time duration,** we must choose the duration of some particular happening as our standard or unit. But we cannot pick something that happens just once, because we could never go back and check it. How would we know if our clock has sped up if the standard is gone? We therefore have to pick a happening that keeps repeating itself, such as the back-and-forth swing of a pendulum. Such occurrences are said to be **cyclical** or **periodic.** Two well-known examples of periodic occurrences or cycles used as standard units of time duration are the duration of a complete rotation of Earth on its axis, which we call a day, and the duration of one complete orbit of Earth about the sun, which we call a year. These and other units are now taken as multiples of the **second,** which is itself defined as a multiple of a cycle characteristic of a particular type of radiation emitted by cesium atoms.

The units most commonly used in physics are the units of the Système Internationale (**SI units**), a current version of the metric system generally agreed on by the international scientific community and in extensive everyday use in nearly every country in the world except the United States. The basic SI units for fundamental quantities, including those we have considered so far, are listed in Table 1-1.

SI units are sometimes called *mks* (meter-kilogram-second) units. Other larger or smaller units of these quantities are expressed as multiples of basic units by a system of prefixes. These prefixes, which represent multiplication by different powers, are summarized in Table 1-2. For instance, 1 nanosecond is 1×10^{-9} seconds, and 5 kilograms is 5×10^3 grams. In the latter case, it is the kilogram that we take as basic, not the gram.

By basic units, we mean that units of all other quantities can be defined in terms of these. In contrast to basic units, those are called *derived units.* For instance, the unit of *two-dimensional space* or **area** is a square 1 m by 1 m, called a **square meter.** A rectangle measuring 3 m by 2 m (Figure 1-7*a*) thus has an area equal to length × width = 3 m × 2 m = 6 m^2 because there are three rows of two square-meter squares in this rectangle. When we multiply units as well as numbers, we get m^2 as units of area. To make this meaningful, we choose to identify 1 "m^2" as a square meter.

We can similarly derive units of three-dimensional space (**volume**). If we have a block measuring 3 m by 2 m by 4 m (Figure 1-7*b*), we can picture it as made up of cubes 1 m on a side, which we call **cubic meters.** As the figure shows,

Table 1-1 Basic SI units for Fundamental Quantities

Quantity	SI Unit
Distance	meter (m)
Mass	kilogram (kg)
Time duration	second (s)
Electric current	ampere (A)
Temperature	kelvin (K)
Amount of substance	mole (mol)
Luminous intensity	candela (cd)

Table 1-2 Prefixes for SI (or Metric) Units

The Prefix . . .	Is Abbreviated . . .	And Means . . .	The Prefix . . .	Is Abbreviated . . .	And Means . . .
yetta-	Y	10^{24}	centi-	c	10^{-2}
zetta-	Z	10^{21}	milli-	m	10^{-3}
exa-	E	10^{18}	micro-	μ (mu)	10^{-6}
peta-	P	10^{15}	nano-	n	10^{-9}
tera-	T	10^{12}	pico- or	p	10^{-12}
giga-	G	10^{9}	micromicro-	or $\mu\mu$	
mega-	M	10^{6}	femto-	f	10^{-15}
kilo-	k	10^{3}	atto-	a	10^{-18}
hecto-	h	10^{2}	zepto-	z	10^{-21}
deka-	da	10	yocto-	y	10^{-24}
deci-	d	10^{-1}			

each layer has three rows of two cubes (six cubes in all), and there are four layers, so in all there are $3 \times 2 \times 4 = 24$ cubic meters. In effect, we have multiplied length by width by height to get volume ($V = lwh$). Multiplying units as well as numbers gives $V = 3 \text{ m} \times 2 \text{ m} \times 4 \text{ m} = 24 \text{ m}^3$. This is meaningful only if we identify m^3 as a cubic meter.

To get a volume in m^3, length, width, and height must all be in meters. This is always the case: To get a derived quantity in standard units, the quantities you use to calculate it must be in standard units. Some derived units have names that obscure their derivation. It will turn out, for example, that the SI unit of energy is $1\frac{\text{kg} \times \text{m}^2}{\text{s}^2}$ which is called a joule. In energy calculations, your energy will not come out in joules unless you are working with mass in kilograms, length in meters, and time duration in seconds.

The basic quantities involved in the definition of a derived quantity are called its *dimensions*. If we represent the basic quantities mass, length, and time duration by the bracketed symbols [M], [L], and [T], then the dimensions of energy are $\left[\frac{ML}{T^2}\right]$ or $[MLT^{-2}]$. Appendix F provides a fuller treatment of dimensions and of a method called **dimensional analysis** for checking dimensions to see whether there is an error in a mathematical relationship among physical quantities.

◆**CONVERTING UNITS** When you do a calculation, the available values of quantities are not always in the units you want. In that case, you have to convert units. This is often true outside of physics as well. A change machine, for example, is a device that converts from dollars to quarters. You end up with the *same value*, but expressed in different units.

To convert units, you first need a conversion relationship, such as "one dollar equals four quarters" or 1 min = 60 s. Dividing both sides of the equation by the same thing, you can arrive at either

$$\frac{1 \text{ min}}{60 \text{ s}} = \frac{60 \text{ s}}{60 \text{ s}} = 1 \quad \text{or} \quad 1 = \frac{1 \text{ min}}{1 \text{ min}} = \frac{60 \text{ s}}{1 \text{ min}}$$

Thus you can write one (1) as either $\frac{1 \text{ min}}{60 \text{ s}}$ or $\frac{60 \text{ s}}{1 \text{ min}}$. Multiplying by one never changes the value of something. When we convert, we want to change the units in which a value is expressed without changing the value. We can do that by multiplying by one (1) written in suitable form:

To convert 5 min to seconds: $5 \text{ min} = 5 \text{ min} \times \dfrac{60 \text{ s}}{1 \text{ min}} = 300 \text{ s}$

To convert 300 s to min: $300 \text{ s} = 300 \text{ s} \times \dfrac{1 \text{ min}}{60 \text{ s}} = 5 \text{ min}$

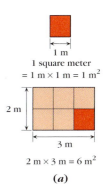

1 m
1 square meter
$= 1 \text{ m} \times 1 \text{ m} = 1 \text{ m}^2$

2 m

3 m

$2 \text{ m} \times 3 \text{ m} = 6 \text{ m}^2$

(*a*)

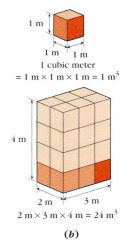

1 m

1 m   1 m
1 cubic meter
$= 1 \text{ m} \times 1 \text{ m} \times 1 \text{ m} = 1 \text{ m}^3$

4 m

2 m   3 m

$2 \text{ m} \times 3 \text{ m} \times 4 \text{ m} = 24 \text{ m}^3$

(*b*)

Figure 1-7 Units of area and volume.

Like one dollar and four quarters, 300 s is the same value as 5 min, but expressed in different units. In each case, we pick the form of one so that when we multiply, the units we don't want "cancel out," and we are left with the units we do want.

PROCEDURE 1-1

➥Caution: Notice that the numerical value is different in different units. It is therefore meaningless to give a numerical value for a quantity without giving its units as well.

Converting a Quantity to Different Units

1. Rewrite your conversion relationship (x first units = y second units) as either

$$\frac{x \text{ first units}}{y \text{ second units}} = 1 \quad \text{or} \quad \frac{y \text{ second units}}{x \text{ first units}} = 1$$

2. Multiply the quantity by 1 in whichever of these two forms cancels out the units you don't want and leaves the units you do want in the right place (numerator or denominator).
3. Check to make sure your result makes sense: You should always get more of the smaller unit, fewer of the larger unit.

Example 1-1 *Converting Speed*

For a guided interactive solution, go to Web Example 1-1 at www.wiley.com/college/touger

A bus travels 110 km/h (kilometers per hour) on open highway. What is this speed in standard SI units?

Brief Solution

1. Identify the units you want for your answer. In SI, distances are in meters (m) and time durations are in seconds (s).
2. 110 km/h is really a fraction $110 \frac{km}{h}$ or $\frac{110 \text{ km}}{1 \text{ h}}$, and means 110 kilometers are traveled *in each* hour or *per* hour.
3. Write the conversion relations between the units you start out with and those you want. In this case, it may be easier to convert time in two steps, first from hours to minutes, then from minutes to seconds.

$$1 \text{ km} = 1000 \text{ m} \qquad 1 \text{ h} = 60 \text{ min} \qquad 1 \text{ min} = 60 \text{ s}$$

4. As fractions, these relations become

$$\frac{1 \text{ km}}{1000 \text{ m}} = \frac{1000 \text{ m}}{1 \text{ km}} = 1 \qquad \frac{1 \text{ h}}{60 \text{ min}} = \frac{60 \text{ min}}{1 \text{ h}} = 1 \qquad \frac{1 \text{ min}}{60 \text{ s}} = \frac{60 \text{ s}}{1 \text{ min}} = 1$$

5. Multiply by 1 as many times as necessary to get the units you want:

$$110 \frac{\cancel{km}}{\cancel{h}} \times \frac{1000 \text{ m}}{1 \cancel{km}} \times \frac{1 \cancel{h}}{60 \cancel{min}} \times \frac{1 \cancel{min}}{60 \text{ s}} = \frac{110\,000 \text{ m}}{3600 \text{ s}} = \textbf{30.6 m/s}$$

"cancels out" km in num., puts m in num.   "cancels out" h in denom., puts min in denom.   "cancels out" min in denom., puts s in denom.

Alternative method. You can also do unit conversion by substitution. For instance, in Example 1-1 you can substitute 1000 m for 1 km, 60 min for 1 h, and 60 s for 1 min. Thus,

$$110 \frac{km}{h} = \frac{110(1000 \text{ m})}{60 \text{ min}} = \frac{110(1000 \text{ m})}{60(60 \text{ s})} = 30.6 \text{ m/s or } \textbf{30.6 m/s}$$

◆ Related homework: Problems 1-5, 1-10, 1-11, 1-12, and 1-20.

Example 1-2 *Buying a Carpet*

You want to carpet a 12 ft × 15 ft room. You can readily calculate that the floor area is 180 ft², but carpeting is sold by the square yard (yd²). How many square yards do you need? **STOP&Think** Since 1 yd = 3 ft, should you just divide by 3? ◆

Solution

1. We have the conversion relation 1 yd = 3 ft, which we can rewrite as $\frac{1 \text{ yd}}{3 \text{ ft}} = 1$ or $\frac{3 \text{ ft}}{1 \text{ yd}} = 1$.

2. Remember that 1 ft² = 1 ft × 1 ft. Thus, 180 ft² = 180 ft × ft, and we have to end up with yd² = yd × yd. We therefore have to multiply twice by $\frac{1 \text{ yd}}{3 \text{ ft}}$:

$$180 \text{ ft}^2 = 180 \text{ ft} \times \text{ft} = 180 \, \cancel{\text{ft}} \times \cancel{\text{ft}} \times \frac{1 \text{ yd}}{3 \, \cancel{\text{ft}}} \times \frac{1 \text{ yd}}{3 \, \cancel{\text{ft}}}$$

$$= 20 \text{ yd} \times \text{yd} = \mathbf{20 \ yd^2}$$

◆ Related homework: Problems 1-6 and 1-9.

◆**SIGNIFICANT FIGURES** No measurement is completely precise. You cannot read distances much smaller than 0.001 m (1 mm) on a meter stick, nor can you read more than a certain number of places on any instrument that has a numerical readout, be it an electronic balance calibrated in units of mass or a multimeter that measures electric current and voltage. The number of places that you can legitimately read with your measuring instrument is called the number of significant figures. A numerical value should always be written to show the number of significant figures. Suppose you measure "exactly" two meters on a tape measure that has 0.001 m accuracy. The measured value is not really exact, but it is closer to 2.000 m than to 2.001 m or to 1.999 m. Therefore, you must write 2.000 m, not 2 m, to represent your measurement. If you converted to kilometers (1 m = 10^{-3} km), you would have to write 2.000×10^{-3} m, not 2×10^{-3} m.

STOP&Think Does a mass of 5000 kg represent one significant figure? Two? Three? Four? More? ◆

Ordinary writing of numbers is sometimes ambiguous, but in scientific notation we can distinguish readily the number of significant figures in 5×10^3 kg (one), 5.000×10^3 kg (four), or 5.000000×10^3 kg (seven, that is, more).

When you use your measured values to calculate a result, *you cannot claim greater accuracy (more significant figures) for your result than for the measurements from which it came.* Suppose $A = 2.000$ m and $B = 3.000$ m are the measured lengths of the two legs of a right triangle. You wish to calculate the length of the hypotenuse using the Pythagorean theorem: $A^2 + B^2 = C^2$. Using your calculator, you obtain the value $C = 3.605551725$ m. The last six places of this calculator readout are meaningless because your measurements could give you only four significant figures. Because your calculator readout is closer to 3.606 than to 3.605, you must write that $C = 3.606$ m.

If the measurements you had were $A = 2.000$ m and $B = 3.0$ m because B was measured by a less precise instrument, you would have to write your result as $C = 3.6$ m. Your result cannot have more significant figures than *any* of the values you used to find it.

When you *estimate*, you can sometimes be more flexible, because you are basing your calculations on numbers that you either guess at based on experience or round off for convenience. For example, Mrs. Wang knows she can get carpeting for $8.79 a square yard. She eyeballs her children's playroom and says,

"This looks to be about 10 feet by 15 feet. That's about 150 square feet. A square yard is around 10 square feet (actually it is 9), so that's about 15 square yards. It's probably a bit more, but if I figure $10 a square yard, that will compensate, so I should budget about $150 ($10 a square yard × 15 square yards) to carpet the room."

In pursuing its quest of the rules by which nature plays, physics must adhere to the rules and tools of careful logical reasoning. In addressing measurement and units, we have taken a few small steps toward building the rich and varied toolkit that we will need.

◆SUMMARY◆

We have discussed physics as a collective human activity that involves observing diverse phenomena in the natural world and trying to figure out the rules that govern their behavior. In practice, observation generally means *measurement*. Measurements are always made in comparison to carefully defined standards called *units of measurement*. Values of physical **quantities,** things like time duration and mass, which may have numerical values, must be expressed in terms of such units; it matters whether you tell your friend you will meet her in 2 minutes or 2 weeks. In physics, we ordinarily express values in **SI units** (see Table 1-1), but sometimes we need to convert from other units. See **Procedure 1-1** for converting units.

In writing numerical values, you need to be aware of the number of significant figures (e.g., 5×10^3 has one, 5.000×10^3 has four). You cannot claim greater accuracy (more significant figures) for your result than for the measurements from which it came.

◆QUALITATIVE AND QUANTITATIVE PROBLEMS◆
Review and Practice

Section 1-1 What Is Physics?

1-1.

a. In Case 1-1, suppose you have been to 20 baseball games, and in all these games you have only seen batters walk off the field after taking a third strike. Can you conclude from this that a batter can *never* advance to first base after a third strike?

b. You have asked 50 children if they like lollipops. They have all said yes. Does this prove that all children like lollipops?

c. You hang 10 different weights, one at a time, on a spring. The first two weights are equal. Each weight after that is double the previous weight. Thus, the total weight doubles each time an additional weight is hung. With each added weight, the spring stretches twice as much as previously. Does this *prove* that doubling the weight suspended from the spring always doubles the distance the spring is stretched?

d. Can you ever prove a law of nature; that is, can you prove that nature *always* behaves in a particular way?

Section 1-2 Measurement and Units

1-2. Below are comparative figures for the Empire State Building in New York and the Pentagon in Washington.

	Empire State Building	Pentagon
Number of stories	102	5
Height in feet	1250	71
Acres of ground covered	about 2	29
Sq. ft. of office space	2.1 million	3.7 million
Volume in cubic ft.	37 million	77 million

Is it meaningful to ask which is the larger building? Explain why or why not.

1-3. Does the duration of time between sunrise and sunset make a good unit of time? Briefly explain.

1-4. The National Institute of Standards and Technology in Gaithersburg, Maryland, declares that the distance between two fine parallel lines on a particular metal rod is a standard unit of length. The rod is then shipped to a research station in Antarctica to be used as a standard for some high-precision measurements. Is there a problem with this? Briefly explain.

1-5. The speed limit on many U.S. highways is 55 miles/hr. What is this speed ***a.*** in ft/s? ***b.*** in SI units?

1-6.

a. How many meters is 20 feet? How many feet is 20 meters?

b. How many square meters is 20 square feet? How many square feet is 20 square meters?

c. How many cubic centimeters are there in 20 m^3? How many cubic meters are there in 20 cm^3?

1-7. SSM WWW Judith Jamison, long the principal dancer of the Alvin Ailey Dance Theater, was a striking presence on the stage in part because she was 5′10″, or 70 inches tall, a height that had traditionally been considered too tall for a ballet dancer.

a. What is Ms. Jamison's height in cm?

b. What is Ms. Jamison's height in meters?

c. A mischievous publicist for the dance company decides to report Ms. Jamison's height in fictional units as 5 pseudometers tall. For this to be correct, how many inches must there in a pseudometer?

d. If instead, the publicist wishes to report Ms. Jamison's height correctly as 20 pseudometers tall, how many inches must there in a pseudometer?

e. For her height to be a larger number of units, should each unit be larger or smaller?

1-8. Memory on electronic storage devices is measured in bytes, or in larger units such as kilobytes (kB), megabytes (MB), or gigabytes (GB). Suppose a floppy disk has a memory capacity of 1440 kB.

a. If a CD (compact disc) has a memory capacity of 700 MB, then compared to the floppy, how many times as much memory capacity does the CD have?

b. If a DVD (digital video disc or digital versatile disc) has a memory capacity of 4.7 GB, then compared to the CD, how many times as much memory capacity does the DVD have?

1-9. Using the data in Problem 1-2, find **a.** the total area of office space in the Empire State building in SI units. **b.** the total volume of the Pentagon in SI units.

1-10. A runner has just completed a 4-minute mile. What was his average speed (total distance divided by total time) in m/s?

1-11. A top Major League fastball pitcher can throw a baseball 95 mi/h (miles per hour). What is this speed in m/s?

1-12. A runner is entered in the 5000-meter event. She wishes to know how many miles she is running. Do the conversion for her.

Going Further

The questions and problems in this group are not organized by section heading, so you must determine for yourself which ideas apply. Some of them will be more challenging than the Review and Practice questions and problems (especially those marked with a • or ••).

1-13. In 1959, members of an MIT fraternity measured the nearby Massachusetts Avenue Bridge by rolling fellow student Oliver Smoot end over end across it and painting a mark after each length. The paint marks have been kept fresh ever since, and you can still read "364.4 SMOOTS" at the MIT end of the bridge. Assuming a reasonable height for Oliver, estimate the length of the bridge in meters.

1-14. A small sample of water from Sludgeport Harbor contains 0.002 g/cm^3 (0.002 g in each cubic centimeter) of a certain pollutant. How many kilograms of this pollutant are contained in each cubic meter of water from the harbor?

1-15.

a. Estimate the speed (total distance divided by total time) in m/s at which your hair grows. State the assumptions on which you base your calculation.

b. According to our present understanding of continental drift, a continental mass will typically drift a distance of about 3 m in a century. How does the speed at which the continents move compare with the speed at which your hair grows?

1-16. How many hours would it take a person walking at a speed of 1.4 m/s to complete Boston's 20-mile Walk for Hunger?

1-17. In one reference, you read that the average human brain at birth has a mass of about 0.390 kg. In another reference, you find that the mass of the typical adult human brain is about 1350 g. Typically, how many times as massive as a newborn brain is an adult brain?

1-18. Professional basketball player Yao Ming is 7'6″ tall. What minimum height in meters must a doorway have for him to be able to go through the doorway barefoot without having to bend at all?

1-19. A liter (L) is equal to 1000 cm^3. Allergists are concerned with the volume of air their patients' lungs can hold. An allergist determines that his patient has a lung capacity of 3.9 L. What is this patient's lung capacity in cubic meters?

1-20. An American training schooner puts in at a Caribbean port to get some replacement rope. The captain knows that the rope they buy at home weighs 0.13 pounds per foot (lb/ft). In the islands, weights (actually masses) are in kilograms and lengths in meters. What weight per unit length in kg/m should the captain look for?

1-21. **SSM** When dealing with thin sheet metal, you might be interested in the mass per unit area rather than the mass per unit volume, or density. The units would then be units of mass divided by units of area. Here are four possibilities:

$$\frac{kg}{m^2} \qquad \frac{kg}{cm^2} \qquad \frac{g}{m^2} \qquad \frac{g}{cm^2}$$

The numerical value of the mass per unit area for a particular kind of sheet metal would depend on the units in which it was expressed. Rank the four possible units in order of the numerical value the mass per unit area would have when expressed in each of these units. Order them from least to greatest, making sure to indicate any equalities.

1-22. The label on a paint can says that the coverage is 450 square feet per gallon. What would be the SI units for coverage? See Appendix C for conversion factors. Simplify your answer if possible.

CHAPTER TWO

Describing Motion in One Dimension

The aim of physics is to understand the rules by which nature plays. As participant-observers, we try to deduce the rules from our observations. The success of our efforts will depend not only on our powers of observation but on what questions we ask and how carefully we formulate them.

What might early humans have asked about the points of light they saw in the night sky? They might have asked why a few looked redder than the rest. But we know now that the red ones include the planet Mars and several distant stars. For that reason, a question that lumped them together would have been unlikely to advance their understanding. They might instead have wondered about their motions,

"We can often describe the changes that we see in terms of . . . matter in motion."

noticed that like the sun and moon, they all rise and set, and left it at that. Or they might have inquired about the motions in greater detail and noticed that although most stars move in formation, a few do not. Like a drum major moving up and down a formation of marchers in a parade, they change their positions against the fixed patterns of the constellations. The ancient Greeks called these stars *planētēs,* meaning wanderers. We call them planets. Before telescopes, it was the planets' motions alone that distinguished them from other stars.

At each stage, then, we try to pose questions that advance our understanding. But there is no rule for how to do this, no fixed "scientific method." What should we ask about in dealing with all of nature? What aspects should we focus on? A dominant aspect of what happens in the physical world is change. We can often describe the changes that we see in terms of rearrangements of the matter or "stuff" that makes up everything in our world. This in turn requires us to think of *matter in motion.*

2-1 Matter in Motion

The focus on matter and motion has been central to physics for hundreds of years. It happens at many levels. Our understanding of distant objects in the vastness of space began with careful observers tracking their motions. On a less obvious level, we often try to understand a complex system by thinking about the tiniest particles that make up the system—how they move and how they affect one another's motions. For example, we may think about the motions of the atoms or molecules that make up the atmosphere or the seas or that participate in the furious internal activity of our own sun or a distant supernova.

Even the study of life lends itself to this approach. The atoms that make up a living being come together in precise and specialized configurations during the individual's lifetime, are rearranged as they participate in the body's *chemistry,* and are eventually dispersed again after the individual's death.

In modern times, indirect evidence about the motion of distant stars has helped guide us to the idea of an expanding universe originating in a "Big Bang." Detection devices that show the tracks of subatomic particles passing through them have provided insights into the most fundamental constituents of matter.

To deal scientifically with matter and motion, physicists had to do two things. First, they had to develop a precise quantitative vocabulary to describe how things move. But objects affect one another as they move; in other words, they interact—they exert forces on one another. So physicists also had to look for rules governing these forces or interactions and their effect on motion. The detailed quantitative description of motion will be the focus of this chapter and the next; the concept of force and its relation to motion will be developed in Chapters 4 and 5. In a sense, we will investigate how the universe works by studying the motions and interactions of its parts, much as if it were a great machine. For that reason, the study of forces and motion in nature is called **mechanics.**

2-2 A Vocabulary for Describing Motion

We often use terms like *speed, distance,* and *time* when describing motion. But even these terms are not sufficiently clear for our purposes. In ordinary language, for example, we use the word *time* in two distinct ways. If I ask, "What time is it?," you might reply "a quarter to two" or "9:54 AM." If I ask, "How much time

For some of nature's creatures, one-dimensional motion comes naturally.

do you need?," you might respond with something like "2 hours." No proper response to the first question legitimately answers the second and vice versa; yet in both cases I am asking for "the time." We need to distinguish between specific **instants** or *clock readings,* on the one hand, and **time intervals** or *durations* on the other.

◆**INSTANTS AND INTERVALS** If I ask you for an *instant* ("At what time . . . ?"), you need look at your watch only once to respond. If I ask you for a time interval ("How long does it take to do X?"), you must look at your watch twice, when you start and when you finish. If your readings were 3 o'clock and 5 o'clock, you would then conclude that it took you 5 hr − 3 hr = 2 hr. We will denote an **instant,** a *single clock or watch reading,* by t. To find an **interval,** which we denote by Δt, we need to obtain two readings, t_1 and t_2, and subtract the earlier reading from the later one:

$$\Delta t \equiv t_2 - t_1$$

Interval ≡ difference between two **instants.**

The triple-bar equal sign tells you this is a definition.

◆**POINT OBJECTS** When an object moves, it changes its position during some time interval. If we are not careful, however, this claim can get us into some silly discussions, such as:

You: But I can wave my arms around while standing in one place.

I: But then your arms are changing *their* positions.

You: Yes, but my hands are moving further than my biceps. Which change in position are we talking about?
   etc.

We avoid all this by thinking first about an object small enough so that we can treat it mentally as though it were no more than a point. We call such an object a **point object** or **point particle.**

There is no such thing as a point object in nature. But it can be a useful way of thinking about a larger object when the object's size doesn't matter. For example, in describing the motion of a car on a cross-country drive, the length of your car is such a tiny fraction of the distance traveled that it might as well be a point. Physicists say its size is *negligible* in this situation. But if your car goes into a skid on an icy turn, you are very concerned that the tail of the car may be moving differently than the front end; in that case, a point object approximation would cut out much of what is of interest in analyzing the situation. We could, however, treat each tiny piece of the car as a separate point object. In that way, developing a description of the motion of point objects lays a foundation for describing the motions of all objects. We will start by describing one-dimensional (straight-line) motion. We'll then extend our ideas to two-dimensional motion (Figure 2-1) in Chapter 3.

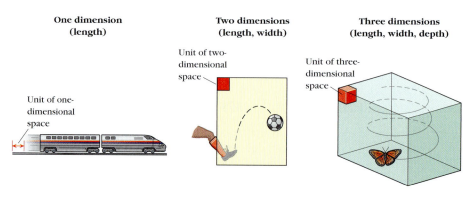

| One dimension (length) | Two dimensions (length, width) | Three dimensions (length, width, depth) |

Unit of two-dimensional space

Unit of three-dimensional space

Unit of one-dimensional space

Figure 2-1 Motion in one, two, and three dimensions.

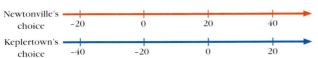

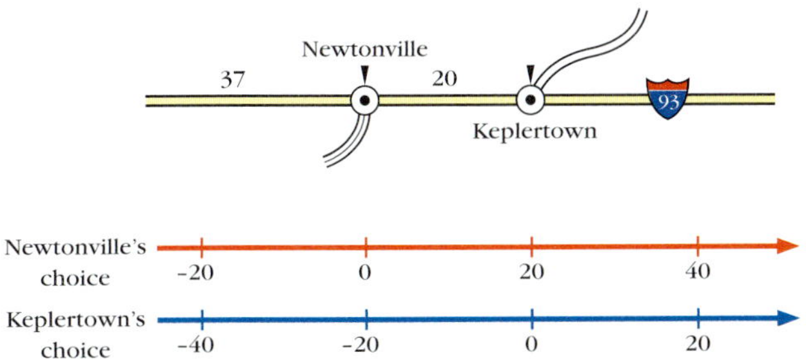

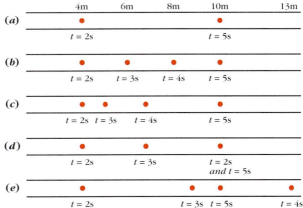

Figure 2-2 Number lines or axes to indicate position.

◆**POSITIONS AND DISTANCES** We can describe the **position** of any point or point object by giving its coordinates on some set of coordinate axes. Each axis is basically a real number line. For one-dimensional motion, position means location along a single real number line, on which consecutive integers are one unit (such as a meter) apart.

We can picture such a number line stretched along a long, straight road (Figure 2-2). But where should the zero go? The choice is yours; it depends on what question you wish to address. If you are interested in how far things are from Newtonville in Figure 2-2, you would choose the red number line, with its zero fixed at Newtonville. If you care about distances from Keplertown, on the other hand, the blue number line's zero is better located. Once we choose a particular number line, we can measure an object's position at various instants during its motion. In Figure 2-3a, for example, the ball's position has been measured at $t = 2$ s and again at $t = 5$ s. **STOP&Think** Based on these two measurements, what can we say about the total distance the ball traveled during that 3-s interval? ◆

In fact, those two measurements provide no details about how the ball moved during the interval. Any of the scenarios shown (Figures 2-3b to 2-3e) is possible; the number of possibilities is infinite. If (as in Figure 2-3e) the motion is not all in one direction, the total distance traveled is *not* just 10 m − 4 m = 6 m. Even in Figure 2-3b, we cannot say for certain that the ball traveled 6 m, because there might have been back-and-forth motion *in between* clock readings (say, between $t = 3$ s and $t = 4$ s). All we know for sure is that in each of the scenarios the *net* or *resulting* change in position was 6 m to the right during the 3-s interval depicted. We call this change in position the **displacement** Δx in one dimension:

Figure 2-3 Different ways of achieving the same displacement and average velocity.

Definition: In one dimension,

$$\textit{displacement} \qquad \Delta x \equiv x_2 - x_1 \qquad (2\text{-}1)$$

- The displacement is a difference between two positions in space, just as an interval is a difference between two instants in time. It can have both positive and negative values, as you will see in Example 2-1.

- If an object ends up at x_2 and if x_1 is the zero point or origin, then $\Delta x = x_2$; in other words, an object's position is its displacement from the origin.

The displacement tells you how far x_2 is from x_1, but not how far an object has traveled to get to x_2 from x_1. In contrast, distance is the length of the actual path traveled. Each piece of the overall length is positive, so the longer you travel, the greater the distance you've gone.

For each straight line path segment where there is no reversal of travel direction,

$$\textit{distance} = |\Delta x| = |x_2 - x_1| \qquad (2\text{-}2)$$

$|\Delta x|$ denotes "absolute value of Δx"; it is always positive.

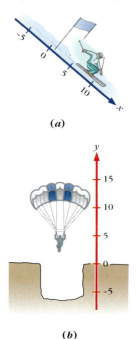

(a)

(b)

Figure 2-4 Axes indicating position needn't be horizontal.

Table 2-1 Basic Quantities Used in Describing Motion

Term	Position	Displacement	Instant	Time Interval
Meaning	Number line or coordinate axis reading, point in space	Space interval	Clock reading, point in time	Time duration, time elapsed
Symbol	x	$\Delta x = x_2 - x_1$	t	$\Delta t = t_2 - t_1$
Question addressed	Where . . . ?	How far . . . ? (and sign tells, Which way . . . ?)	When . . . ? At what time . . . ?	How long . . . ? How much time . . . ?

Table 2-1 summarizes the relationships among these quantities and the type of question each quantity addresses.

◆**NOTATION** A straight line path may be horizontal, diagonal, or vertical (Figure 2-4). If we use y instead of x for position along a vertical path, then the displacement between two points would be $\Delta y = y_2 - y_1$.

◆**NEGATIVE VALUES** When we use a real number line to identify position along a straight line, half of all possible positions will have negative values. In Figure 2-2, a position 3 m to the left of the origin is expressed as $x = -3$ m.

Example 2-1 *Negative and Positive Displacements*

Find the displacement of a cyclist who rides
a. from a position of 500 m to a position of 300 m.
b. from a position of –500 m to a position of –300 m.
c. Find the distance traveled by the cyclist in each case (assuming no backtracking).

Solution

Sketch the situation. It is important to have a clear picture of what is happening. Figure 2-5 shows the displacements asked for in **a** and **b.**

Figure 2-5 Positions and displacements for Example 2-1.

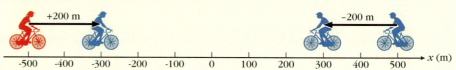

Use the definition. Displacement is defined as $\Delta x = x_2 - x_1$, where x_1 is the earlier and x_2 the later position. In **a,** $x_1 = 500$ m and $x_2 = 300$ m, so

$$\Delta x = 300 \text{ m} - 500 \text{ m} = \textbf{–200 m}$$

and is negative. In **b,** however, $x_1 = -500$ m and $x_2 = -300$ m, so

$$\Delta x = -300 \text{ m} - (-500 \text{ m}) = -300 \text{ m} + 500 \text{ m} = \textbf{+200 m}$$

and is positive.
c. The absolute value of either displacement is 200 m, so the *distance* traveled is 200 m in each case. Distances are always positive.

Making sense of the results. In all cases,

if the motion is to the right along a left-to-right real number line, the displacement is positive and if it is to the left, the displacement is negative.

Figure 2-5 illustrates this point.

◆ Related homework: Problems 2-7, 2-8, and 2-9.

In one sense, the displacement provides more information than the distance: It answers not only "How far?" but "Which way?" as well. In the same spirit, we wish to address "How fast?" and "Which way?" together. To do so, we define a quantity $\bar{v}$ called the *average velocity during a time interval* Δt (the bar over the v denotes *average*):

Definition: During the time interval Δt, the

$$\text{\textit{average velocity}} \qquad \bar{v} \equiv \frac{\text{displacement}}{\text{time interval}} \equiv \frac{\Delta x}{\Delta t} \equiv \frac{x_2 - x_1}{t_2 - t_1} \qquad (2\text{-}3)$$

You can find the average velocity of an object using a meter stick and a clock. The definition is equivalent to the following procedure:

PROCEDURE 2-1

Determining Average Velocity in One Dimension

1. Measure the object's coordinate x_1 at instant t_1.
2. Measure the object's coordinate x_2 at instant t_2 (later than t_1).
3. Calculate the increments $\Delta x = x_2 - x_1$ and $\Delta t = t_2 - t_1$.
4. Divide Δx by Δt to obtain $\bar{v}$.

With t_2 as the later instant, Δt is positive. Then $\bar{v}$ has the same sign as the displacement Δx, and thus is in the same direction. We still need to clarify two aspects of average velocity: (1) why *average?* and (2) how is velocity different than our everyday notion of speed?

◆**WHY AVERAGE?** Consider Figure 2-3. The ball in Figure 2-3*a* is at position $x_1 = 4$ m when the clock reads $t_1 = 2$ s and at $x_2 = 10$ m when $t_2 = 5$ s. The ball's displacement is therefore $\Delta x = 10$ m $- 4$ m $= +6$ m (to the right) during a time interval $\Delta t = 5$ s $- 2$ s $= 3$ s. During this time interval, the average velocity is $\bar{v} = \frac{\Delta x}{\Delta t} = \frac{+6 \text{ m}}{3 \text{ s}} = +2$ m/s (also to the right). But Figure 2-3*a* provides no information about how the ball moved between the two instants $t = 2$ s and $t = 5$ s. The possible scenarios in Figures 2-3*b* to 2-3*e* show that the ball may have traveled at uniform speed, sped up, gotten to $x = 10$ m quickly and then stopped there until $t = 5$ s, and so on. We can't say which scenario the ball followed or how fast it was going at any particular instant during the interval, only that its velocity averaged 2 m/s for the whole 3-s interval.

◆**HOW IS AVERAGE VELOCITY DIFFERENT THAN SPEED (OR AVERAGE SPEED)?** Physicists are basically stating the everyday notion that most people have of speed when they define *average speed* as the total *distance* traveled divided by the time spent traveling:

$$\text{\textit{average speed}} \equiv \frac{\text{total distance traveled}}{\text{total time interval}} \qquad (2\text{-}4)$$

For each segment with no direction reversals, the key difference is this: *The average velocity can be positive or negative; the sign indicates its direction. The average speed is always positive; it provides no information about direction.* When there are direction reversals, the average speed and average velocity can have numerically different values, as in the following example.

An albatross gliding on prevailing winds. In the Southern Hemisphere, albatrosses have been known to go completely around the Earth in this way. At 40° south latitude, what information would you need to look up to calculate how long this trip would take?

Example 2-2 *Average Speed versus Average Velocity*

For the motion of the ball in Figure 2-3e during the time interval from $t = 2$ s to $t = 5$ s, find **a.** the total distance, **b.** the total displacement, **c.** the average speed, and **d.** the average velocity. Assume the only change in direction occurs at $t = 4$ s.

Solution

Choice of approach. We can find the average speed and average velocity from their definitions (Equations 2-4 and 2-3). Because distance equals $|x_2 - x_1|$ only for path segments in which the object doesn't reverse direction, we must find separately the distances traveled from $t = 2$ s to $t = 4$ s and from $t = 4$ s to $t = 5$ s, and then add the two distances. We will get the same displacement, however, whether we add the displacements for the two shorter intervals or just subtract the position at $t = 2$ s from the position at $t = 5$ s.

The mathematical solution. To make comparison easier, we will arrange the calculations in a table.

Distance			Displacement Δx			Average Speed	Average Velocity
Between $t = 2$ s and $t = 4$ s	Between $t = 4$ s and $t = 5$ s	Total	Between $t = 2$ s and $t = 4$ s	Between $t = 4$ s and $t = 5$ s	Total	Between $t = 2$ s and $t = 5$ s	$\left(\dfrac{\text{total distance}}{\Delta t}\right)$	$\left(\dfrac{\Delta x}{\Delta t}\right)$
$\|13\,\text{m} - 4\,\text{m}\|$ $= 9$ m	$\|10\,\text{m} - 13\,\text{m}\|$ $= 3$ m	$9\,\text{m} + 3\,\text{m}$ $= \mathbf{12\ m}$	$13\,\text{m} - 4\,\text{m}$ $= 9$ m	$10\,\text{m} - 13\,\text{m}$ $= -3$ m	$9\,\text{m} + (-3\,\text{m})$ $= \mathbf{6\ m}$	$10\,\text{m} - 4\,\text{m}$ $= 6$ m	$\dfrac{12\,\text{m}}{3\,\text{s}}$ $= \mathbf{4\ m/s}$	$\dfrac{6\,\text{m}}{3\,\text{s}}$ $= \mathbf{2\ m/s}$

A critical feature of the calculations is that unlike the distance, the displacement from $t = 4$ s to $t = 5$ s is negative. This leads to an average velocity that is different than the average speed.

◆ Related homework: Problems 2-10, 2-11, and 2-12.

For **WebLink 2-1:** A Spaceship in a Resistive Medium, go to www.wiley.com/college/touger

✦**INSTANTANEOUS VELOCITY** To find an average velocity experimentally, you must take position readings at two instants. These measurements cannot tell you how fast something is going at any *single* instant. If you want to know the velocity at a particular instant t, you must approach it by taking readings for smaller and smaller time intervals containing that instant. Using these measurements, you find the average velocity for each interval. These average velocity values close in on a particular value (called a *limiting value*) as Δt shrinks closer and closer to zero ($\Delta t \rightarrow 0$). To see in detail how this works, work through WebLink 2-1. This limiting value is what we call the *instantaneous* velocity at t.

Definition: In one dimension,

$$\textit{instantaneous velocity} \qquad v \equiv \lim_{\Delta t \to 0} \bar{v} \equiv \lim_{\Delta t \to 0} \frac{\Delta x}{\Delta t} \qquad (2\text{-}5)$$

The symbol v with no bar above it denotes *instantaneous velocity*. Just as average velocity only has meaning over a particular time interval, instantaneous velocity only has meaning at a particular point in time—a particular *instant t*.

- When the time interval containing an instant is reduced enough to exclude all direction changes, distance $= |\Delta x|$ and average speed $= \frac{|\Delta x|}{\Delta t}$. Then, shrinking Δt to zero, we can take the absolute value of the instantaneous velocity to be the *instantaneous speed*.

- If an object does reverse direction, in slow motion it would seem to stop and then begin moving in the opposite direction. There is an instant of reversal. If the velocity is positive before this instant, then after, it is negative. In changing sign, it must pass through zero. At this instant, then, the instantaneous velocity is zero.

We've defined the average velocity as $\bar{v} = \frac{\Delta x}{\Delta t} = \frac{x_2 - x_1}{t_2 - t_1}$. Suppose we choose the first instant t_1 to be $t = 0$. This is like resetting our stopwatch to $t = 0$ when we start our observations. We can let x_o denote the object's position at this instant, so that x_o and 0 become the "values" of x_1 and t_1. If x represents its position at any later instant t (so that these become the "values" of x_2 and t_2), then

$$\bar{v} = \frac{\Delta x}{\Delta t} = \frac{x - x_o}{t - 0} = \frac{x - x_o}{t}$$

Solving for x in this equation gives us

$$x = x_o + \bar{v}t \qquad (2\text{-}6)$$

In words,

position at time t = initial position + additional displacement traveled
during the interval $\Delta t = t - 0$

Both x_o and $\bar{v}t$ are lengths. Adding them gives a total length.

◆**UNIFORM MOTION** When an object's velocity is **uniform** (the same at every instant during the time interval being analyzed), we do not have to distinguish between average velocity $\bar{v}$ and instantaneous velocity v. Then Equation 2-6 becomes

$$x = x_o + vt \qquad \text{(uniform motion only)} \qquad (2\text{-}6a)$$

2-3 Representing Motion Graphically

When we plot *position* versus *clock reading,* the slope is

$$\frac{\Delta(\text{variable plotted vertically})}{\Delta(\text{variable plotted horizontally})} = \frac{\Delta x}{\Delta t}$$

and therefore tells us the object's *average velocity.* Equation 2-6 has the standard form of a linear or straight line equation: v in Equation 2-6 has the same role as m in $y = b + mx$, the standard equation for a straight-line graph of y versus x.

To review basic ideas about linear equations and for more on how this math applies to real-world situations, go to WebLinks 2-2 and 2-3. Even if you are confident about the math, these WebLinks will help you think about the math in ways that are useful for application to physics.

For **WebLink 2-2: Proportionality, Rates, Slope, and Straight Lines** and
WebLink 2-3: Real-World Quantities and Units in Linear Equations, go to
www.wiley.com/college/touger

◆**VERTICAL INTERCEPTS AND THEIR MEANING** The vertically plotted variable is now x rather than y, so its initial value x_o is the vertical intercept. The following example treats the graphing of a uniform motion situation in detail.

Example 2-3 *Driving on Cruise Control*

For a guided interactive solution, go to Web Example 2-3 at
www.wiley.com/college/touger

Cruise control keeps your car automatically at constant speed. An SI-oriented teenager has been heading east for 4000 s with the cruise control set at 31 m/s (almost 70 miles/hour). He is now 150 000 m east of Ridgemont.
a. How far east of Ridgemont did he start out?
b. Sketch a graph of x versus t for this motion.

Brief Solution
a. *Assumptions.* If we think of the instant he started out as $t = 0$, the given information then tells us that his position is $x = 150\,000$ m at $t = 4000$ s.

EXAMPLE 2-3 continued

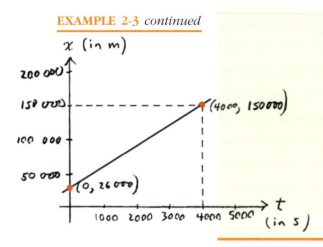

Mathematical solution. Solving for x_0 in Equation 2-6 gives us

$$x_0 = x - vt = 150\,000 \text{ m} - (31 \text{ m/s})(4000 \text{ s})$$

$$= 150\,000 \text{ m} - 124\,000 \text{ m}$$

$$= \mathbf{26\,000 \text{ m}}$$

b. We need two points to determine a straight line. We now know x at two instants t because x_0 is by definition the value of x at $t = 0$. We can therefore plot the two points $(t = 0, x = 26\,000 \text{ m})$ and $(t = 4000 \text{ s}, x = 150\,000 \text{ m})$ and draw the straight line connecting them. The resulting graph is shown at left.

◆ Related homework: Problems 2-18 and 2-19.

◆**INTERPRETING SLOPE** Positions and velocities, and also the slopes that represent velocities, may be either positive or negative. To reinforce your understanding of how the signs work in various situations, consider Figure 2-6. Parts *a* and *b* show the positions of four cars at two different instants ($t = 0$ and $t = 4$ s). We shall assume that each car is being driven at uniform velocity. Note that when the cars are to the left of the origin they have negative positions.

In Figure 2-6*c*, we use two positions x and the corresponding instants t to calculate the average velocity of each car. For each car, the two pairs of values give us two points on a plot of x versus t (Figure 2-6*d*). The two points determine the straight line graph for that car. Each average velocity in *c* is the slope

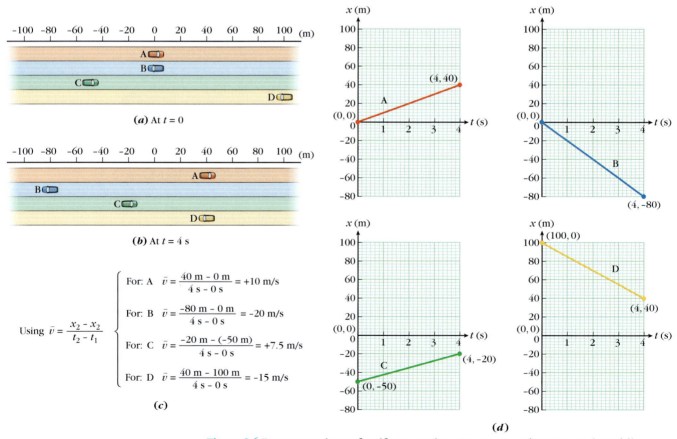

Figure 2-6 Representations of uniform motion. We can use either pictures (*a* and *b*), equations (*c*), or graphs (*d*) to communicate aspects of the motion. As you come to understand them better, these different ways of representing the motion should reinforce each other.

of the corresponding graph in *d*. The graphs have constant slope—that's what makes them straight lines—because the cars are moving at constant velocity.

Let's see what some of the features of the graphs in *d* represent. (*Note:* If possible, you should try to explore these motions with a motion detector.)

Feature of Graph(s)	Aspect of Motion Represented
The graphs for cars A and C slant upward to the right (their slopes are positive).	Cars A and C are moving toward the right. Their displacements as time advances are positive; thus, their velocities are positive.
The graphs for cars B and D slant downward to the right (their slopes are negative).	Cars B and D are moving toward the left, giving negative displacements and therefore negative velocities.
The graphs for cars B and C lie below the horizontal axis.	Cars B and C are to the left of the origin throughout the interval, so their positions *x* are always negative.
The graph for car C lies *below* the *x* axis, but the slope is *positive*.	C is *positioned to the left* of the origin during this interval but is *moving toward the right*.
The graph for car D lies *above* the *x* axis, but the slope is *negative*.	D is *positioned to the right* of the origin during this interval but is *moving toward the left*.
A's graph is steeper than C's (the scale is the same).	A is going at greater speed than C (also at greater velocity).
B's graph is steeper (more nearly vertical) than D's. Its slope is more negative, so the absolute value of its slope is greater.	B is going at greater speed than D (but *not* at greater velocity—a more negative number is not greater than a less negative number).

The idea of slope, a property of a straight line, can also be applied to nonuniform motion, such as the motion graphed in Figure 2-7*a*. If you draw a straight line segment or *secant* connecting the points P_1 and P_2 on the graph, its slope is $\frac{\Delta x}{\Delta t} = \frac{x_2 - x_1}{t_2 - t_1}$, which is the average velocity $\bar{v}$. The straight line segment connecting points P_2 and P_3 has a greater slope, so the average velocity is greater over this interval. Interval by interval, the slopes of the straight secants tell us about the motion represented by the curve.

If we want to know about instantaneous velocity, we can zoom in until we are looking at a segment of the curve that still includes the instant in question, but is so tiny that it is essentially straight (Figure 2-7*b*). In the same way, as Figure 2-7*c* shows, a small stretch of ocean surface on a windless day may appear flat, even though the Earth is round. To help see how the segments are sloping, we extend them (dotted in Figure 2-7*b*). Each extended line is very nearly a

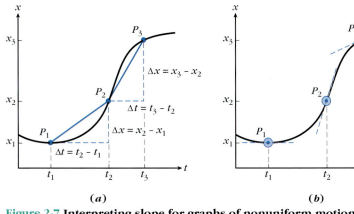

(a) (b) (c) Over a small enough distance, the Earth seems flat.

Figure 2-7 **Interpreting slope for graphs of nonuniform motion.**

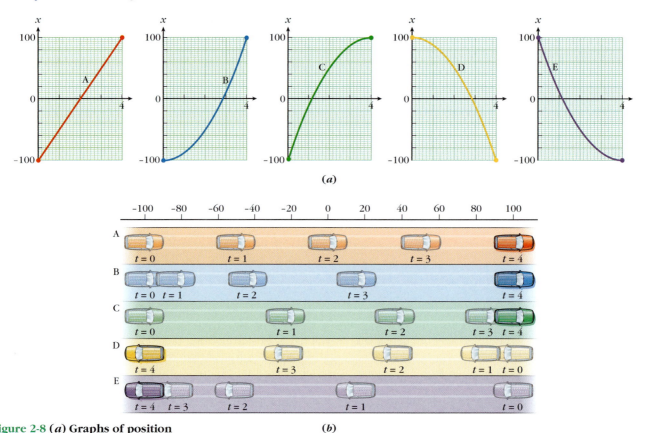

Figure 2-8 (*a*) **Graphs of position versus clock reading for cars with constant (A) and non-constant (B–E) velocities.** See WebLink 2-4 for more detail. (*b*) Picturing the motions described by these graphs. The centers of the cars are at the graphed positions.

For **WebLink 2-4:**
Interpreting Motion Graphs, go to
www.wiley.com/college/touger

tangent, a line that just grazes the graph at a single point. So we can pick a tiny close-up segment of the curve that includes a certain instant *t*, and we can take the slope of this segment to be a very good approximation to the instantaneous velocity at *t*. In other words, we find the instantaneous velocity approximately by applying Procedure 2-1 to this tiny segment.

Graphs are a concise way of communicating a great deal of information about the motion of an object, whether uniform (constant velocity) or not. It is therefore important for you to learn to read their features. WebLink 2-4 provides an opportunity to develop some experience with this. Figure 2-8 summarizes the motions treated in the WebLink and their graphs.

◆**GRAPHS OF *v* VERSUS *t*** We have already seen how to use the idea of slope to find the instantaneous velocity *v* at each instant *t*. Because we can find a value of *v* for each value of *t*, we can also graph *v* versus *t*. Figure 2-9*a* shows *x* versus *t* (from Figure 2-7*a*) and *v* versus *t* graphs for the same motion. In the following table, note how the features of the *v* versus *t* graph correspond to those of the *x* versus *t* graph.

Figure 2-9 Graphs of *x* versus *t* and *v* versus *t* for the same nonuniform motion. (*a*) In the *x* versus *t* plot, the slopes of the tangents give the velocities. (*b*) In the *v* versus *t* plot, those slopes are plotted against *t*. Note that the horizontal tangent at t_1 has zero slope.

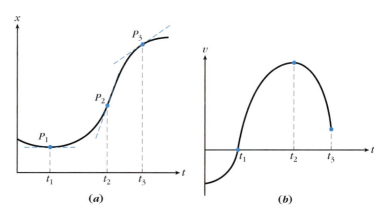

	Graph of x versus t	Graph of v versus t	What Object Is Doing
From t_1 to t_2	Graph gets steeper; slope increases	v increases in value; graph rises	Speeding up
At t_2	Graph is steepest; slope of tangent is at maximum value	v has maximum value; graph peaks	Going fastest
From t_2 to t_3	Graph "levels off"; slope decreases	v decreases; graph goes back down	Slowing down
For entire period covered by graph	Graph slants upward toward the right throughout	Values of v are always positive; thus, all points are above horizontal axis	Always moving toward the right

Example 2-4 *Sketching v versus t Graphs*

Figure 2-10 displays x versus t plots for two different moving objects. Sketch a graph of instantaneous velocity v versus clock reading t for each of these motions.

Solution

Get the picture. We first pick tiny segments of the curve at some typical values of t, because the slopes of these segments are roughly the instantaneous velocities at those values of t. This is done in Figure 2-11.

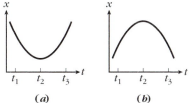

Figure 2-10 x versus t graphs for two different moving objects.

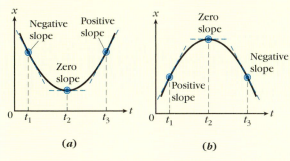

Figure 2-11 The slopes of tiny segments of the graph give us the instantaneous velocities at different values of t.

See how v changes with t. We see in Figure 2-11 that the slope of graph a is first negative (at t_1), then zero (at t_2), and then positive (at t_3). Thus, the value of v goes from negative to zero to positive; like the slope that represents it, v is always increasing. The graph of v versus t must show v behaving in this way. The v versus t plot sketched in Figure 2-12a shows the required behavior.

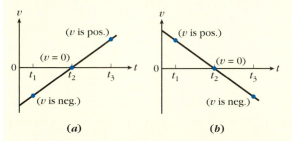

Figure 2-12 These graphs of velocity versus t are for the same motions as the x versus t graphs in Figures 2-10 and 2-11.

In contrast, the slope of graph b in Figure 2-11 is first positive (at t_1), then zero (at t_2), and then negative (at t_3). The value of v likewise goes from positive to zero to negative; it is always *decreasing*. This behavior is displayed by the v versus t plot sketched in Figure 2-12b.[1]

◆ Related homework: Problems 2-22, 2-23, and 2-24.

---

[1]We have not yet shown that the graphs in Figure 2-12 must be straight lines, but Section 2-5 will show that when the x versus t graphs are parabolas, the v versus t graphs do turn out to be straight lines.

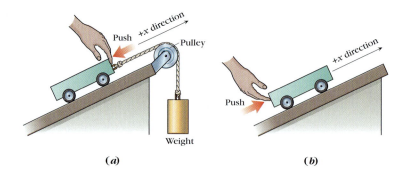

Figure 2-13 These situations exhibit the motions that are graphed in Figures 2-10 through 2-12. In (*a*), the hand gives the car an initial velocity downhill. The net force on the car is uphill. In (*b*), the initial velocity is uphill, the net force downhill.

(*a*)   (*b*)

For **WebLink 2-5: Graphs of Motion on a Ramp,** go to www.wiley.com/college/touger

Figures 2-13*a* and 2-13*b* show real-world objects whose velocities behave in the manner described by the graphs in Figure 2-12 when set in motion by an initial push. You can see the graphs develop as the motions are animated in WebLink 2-5.

2-4 Acceleration and Graphs of Accelerated Motion

We have looked at several situations in which an object's velocity changes over time. We know that an object, in everyday language, can pick up or lose speed. A powerful sports car picks up speed more quickly than an economy car; a braking car loses speed more quickly on dry pavement than on an icy road. It is therefore useful to talk about the *rate* at which it happens. But we want to do so in language consistent with describing motion in terms of *position, displacement,* and *velocity*. We therefore will define *acceleration* as the rate at which velocity changes with time. If an object's velocity is v_1 at instant t_1 and v_2 at instant t_2, then we define *average acceleration* as follows:

Definition Over the interval $\Delta t = t_2 - t_1$,

$$\text{average acceleration} \qquad \bar{a} \equiv \frac{\Delta v}{\Delta t} \equiv \frac{v_2 - v_1}{t_2 - t_1} \qquad (2\text{-}7)$$

for one-dimensional motion.

This definition is an abbreviated statement of the following procedure.

PROCEDURE 2-2

Determining Average Acceleration in One Dimension

1. Determine an object's instantaneous velocity v_1 at time t_1.
2. Determine its instantaneous velocity v_2 at later time t_2.
3. Calculate the increments $\Delta v = v_2 - v_1$ and $\Delta t = t_2 - t_1$.
4. Use Equation 2-7 to calculate $\bar{a}$.

Let's see how this definition works in the case of the cart in Figure 2-13*a*. The cart leaves the hand in the downhill direction at $t = 0$, but soon reverses direction because of the suspended weight. Look at its v versus t graph (Figure 2-12*a*). At t_1 its velocity is negative; it is moving to the left. At t_2 its velocity is zero. Thus, over the interval $\Delta t = t_2 - t_1$,

$$\bar{a} \equiv \frac{\Delta v}{\Delta t} \equiv \frac{v_2 - v_1}{t_2 - t_1} = \frac{0 - (-\text{value of } v)}{\text{positive time interval}} = \frac{\text{positive numerator}}{\text{positive denominator}}$$

$$= \text{positive value of average acceleration.}$$

Likewise, during the interval $\Delta t = t_3 - t_2$,

$$\bar{a} \equiv \frac{\Delta v}{\Delta t} \equiv \frac{v_3 - v_2}{t_3 - t_2} = \frac{(+\text{value of } v) - 0}{\text{positive time interval}} = \frac{\text{positive numerator}}{\text{positive denominator}}$$

$$= \text{positive value of average acceleration.}$$

The cart's acceleration is positive during both intervals! That is because its velocity increases when it goes from a negative value (cart going left) to zero as well as when it goes from zero to a positive value (cart going right). If the velocity is increasing, its rate of change must be positive.

By similar arguments, we can conclude that the average acceleration of the cart in Figure 2-13b (which goes right immediately after leaving the hand at $t = 0$, but later goes left) is negative throughout. We can now summarize our reasoning:

	Time Interval Δt	Change in Velocity Δv	$\bar{a} = \dfrac{\Delta v}{\Delta t}$	What Object Is Doing
Cart in Figure 2-13a	$t_2 - t_1 = +\text{value}$	$v_2 - v_1 = 0 - (-\text{value})$ to left $= +\text{value}$	$+$	Moving left but slowing down
	$t_3 - t_2 = +\text{value}$	$v_3 - v_2 = (+\text{value}) - 0$ to right $= +\text{value}$	$+$	Moving right and speeding up
Cart in Figure 2-13b	$t_2 - t_1 = +\text{value}$	$v_2 - v_1 = 0 - (+\text{value})$ to right $= -\text{value}$	$-$	Moving right but slowing down
	$t_3 - t_2 = +\text{value}$	$v_3 - v_2 = (-\text{value}) - 0$ to left $= -\text{value}$	$-$	Moving left and speeding up

◆**UNITS OF AVERAGE ACCELERATION** Average acceleration is the *rate of change* of velocity. For instance, if a car went from 0 to 60 mi/hr in 5 seconds, it would be gaining an average of 12 mi/hr each second, or per second. Its acceleration in mixed units would be 12 (mi/hr)/s In SI, the units of average acceleration are m/s (meters per second) divided by s, giving us (m/s)/s (meters per second per second). If, for example, your velocity changes from 15 m/s to 35 m/s as your stopwatch advances from 3 s to 7 s, your velocity increases on the average by 5 m/s during each second of the 4-s time interval. Then

$$\bar{a} = \frac{\Delta v}{\Delta t} = \frac{20\,\frac{m}{s} - 5\,\frac{m}{s}}{4\,s} = \frac{5\,\frac{m}{s}}{s} \quad \text{or} \quad 5\left(\frac{m}{s}\right)/s$$

Note that mathematically, (m/s)/s means

$$\frac{\frac{m}{s}}{s} = \frac{\frac{m}{s}}{\frac{s}{1}} = \frac{\frac{m}{s} \times \frac{1}{s}}{\frac{s}{1} \times \frac{1}{s}} = \frac{\frac{m}{s \times s}}{1} = \frac{m}{s^2}$$

so you will also see (m/s)/s written as m/s^2 (meters per second squared) or as ms^{-2}. No matter—it still means the same thing.

◆**AVERAGE ACCELERATION AS SLOPE OF v VERSUS t GRAPH** If we apply the general definition of slope

$$\text{slope} = \frac{\Delta\ (\textit{quantity plotted vertically})}{\Delta\ (\textit{quantity plotted horizontally})}$$

to v versus t plots such as those in Figure 2-12, we find that the slope is $\frac{\Delta v}{\Delta t}$. But $\frac{\Delta v}{\Delta t}$ is our definition of average acceleration.

> The **slope of a v versus t graph** between t_1 and t_2 is equal to the **average acceleration** over that interval.

If the graph is not a straight line between t_1 and t_2, then the slope is understood to mean the slope of the secant connecting (t_1, v_1) and (t_2, v_2).

The graph in Figure 2-12a rises left to right—its slope is always positive. The descending graph in Figure 2-12b always has a negative slope. But the slope is the average acceleration. So as we concluded previously, the acceleration is always positive for the motion graphed in Figure 2-12a and always negative for the motion graphed in Figure 2-12b.

◆**INSTANTANEOUS ACCELERATION** Like average velocity, average acceleration is defined over an interval. As we did for average velocity in WebLink 2-1, we can find average accelerations for progressively smaller time intervals. The value approached as the interval closes in on a particular instant is defined to be the acceleration at that instant.

Definition In one dimension,

$$\textit{instantaneous acceleration} \qquad a \equiv \lim_{\Delta t \to 0} \bar{a} \equiv \lim_{\Delta t \to 0} \frac{\Delta v}{\Delta t} \qquad (2\text{-}8)$$

Without a bar, a denotes instantaneous acceleration.

When the v versus t graph is not a straight line, we can "zoom in" on a segment that includes any instant t that interests us. If we zoom in close enough, the segment is essentially straight, and its slope gives the instantaneous acceleration at that instant.

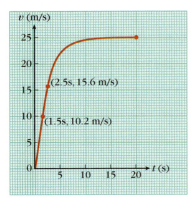

Figure 2-14 Graph of v versus t for Example 2-5.

Example 2-5 *Determining Instantaneous Accelerations by Finding Slopes*

The motion of a certain object is represented by the v versus t graph in Figure 2-14. Find approximate values of the instantaneous acceleration
a. at $t = 2.0$ s   **b.** at $t = 20.0$ s

Solution
Choice of approach. Pick a small segment of the graph that includes each instant, then find the slope of each segment.

The mathematical solution.
a. To include $t = 2.0$ s, we arbitrarily pick the interval from $t_1 = 1.5$ s to $t_2 = 2.5$ s. (You can redo this choosing an even smaller interval to see how much it changes your result—see Problem 2-100 at the end of the chapter.) From the graph, we find that the velocity at $t_1 = 1.5$ s has the value $v_1 = 10.2$ m/s, and at $t_2 = 2.5$ s, its value is $v_2 = 15.6$ m/s. The slope of the chosen segment is

$$\frac{\Delta v}{\Delta t} = \frac{v_2 - v_1}{t_2 - t_1} = \frac{15.6 \text{ m/s} - 10.2 \text{ m/s}}{2.5 \text{ s} - 1.5 \text{ s}} = 5.4 \text{ (m/s)/s}$$

Then at $t = 2.0$ s, the instantaneous acceleration $a \approx$ **5.4 (m/s)/s** or **5.4 m/s²**.
b. As we go beyond $t = 10$ s, we see the graph becoming more nearly horizontal. Because the slope of any horizontal line is zero, the slope of a small enough segment including $t = 20.0$ s is approximately zero, and so $a \approx$ **0**.

Making sense of the results. The value obtained in **b** reflects the fact that by $t = 20$ s, the velocity is no longer changing.

◆ Related homework: Problems 2-28, 2-29, and 2-100.

2-5 Constant Acceleration and Equations of Motion

◆**CONSTANT ACCELERATION** If an object's acceleration is constant, or *uniform,* its value never changes, so the instantaneous and average accelerations are the same. In that case, as in Figure 2-12, a graph of v versus t is a straight line, that is, its slope is the same everywhere.

◆**GRAPHING UNIFORMLY ACCELERATED MOTION** For constant acceleration, Equation 2-7 becomes

$$a = \frac{\Delta v}{\Delta t} = \frac{v_2 - v_1}{t_2 - t_1}$$

Proceeding as we did to obtain Equation 2-6, we call the velocity v_o at $t = 0$, and let v be its velocity at any later instant t. Then

$$a = \frac{\Delta v}{\Delta t} = \frac{v - v_o}{t - 0} = \frac{v - v_o}{t}$$

Solving for v in $a = \frac{v - v_o}{t}$ gives us

$$v = v_o + at \tag{2-9}$$

In words: *velocity at time t = initial velocity + change in velocity due to acceleration during the interval $(t - 0)$.*

This has the general form $y = b + mx$ or $y = y_o + mx$ of a straight line equation; t and v are the horizontally and vertically plotted variables, v_o is the initial value of the latter—that is, the *vertical intercept*—and a is the rate of change or *slope*. To see how you can use this to interpret a graph of v versus t, let's work through Example 2-6.

Example 2-6 *Interpreting a Linear Graph of v versus t*

For a guided interactive solution, go to Web Example 2-6 at
www.wiley.com/college/touger

The velocity of a remote-controlled vehicle is plotted against time below.

a. From the graph, find the vehicle's acceleration and its initial velocity.

b. Use your results to calculate what the velocity will be at $t = 32$ s if the acceleration remains uniform.

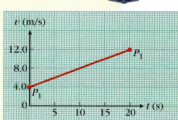

Brief Solution

Choice of approach. Note that $v = v_o + at$ has the same form as $y = y_o + mx$.

a. *Slope and intercept:* We can read the vertical intercept directly off the graph (point P_1). This is the initial velocity: $v_o = \textbf{4 m/s}$.

The acceleration is the slope of the graph, which we can find from any two points on the line. In the given graph, we arbitrarily select points

$$P_1(t_1 = 0, v_1 = 4.0 \text{ m/s}) \quad \text{and} \quad P_2(t_2 = 20 \text{ s}, v_2 = 12.0 \text{ m/s}).$$

Thus, $\quad$ slope $= a = \dfrac{v_2 - v_1}{t_2 - t_1} = \dfrac{12.0 \text{ m/s} - 4.0 \text{ m/s}}{20 \text{ s} - 0} = \textbf{0.4 m/s/s}$

b. *Equation of motion:* Once we know v_o and a, we can use Equation 2-9 to find v at any value of t. Thus,

$$v = v_o + at = 4.0 \text{ m/s} + (0.4[\text{m/s}]/\text{s})(32 \text{ s})$$
$$= 4.0 \text{ m/s} + 12.8 \text{ m/s} = \textbf{16.8 m/s}$$

Note that both terms, v_o and at, have units of velocity.

◆ Related homework: Problem 2-30.

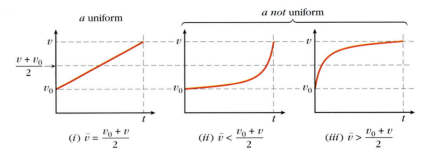

Figure 2-15 Average velocities for cases of uniform and non-uniform motion.

◆**AVERAGE VELOCITY** Suppose an object accelerates uniformly from a velocity v_o at $t = 0$ to a velocity v at some later instant t. The average velocity over this interval will then be midway between the values of v_o and v. In other words, it is the numerical average of these two values:

$$\text{condition for constant acceleration} \qquad \bar{v} = \frac{v_o + v}{2} \qquad (2\text{-}10)$$

It is important to realize that this is generally not true when the acceleration is not *uniform*. Figure 2-15 shows this: in graph *ii*, the velocity is below $\frac{v_o + v}{2}$ for most of the interval from 0 to t, so its time-averaged value is lower. For graph *iii*, by similar reasoning, the average velocity is higher than $\frac{v_o + v}{2}$.

Before solving problems about uniformly accelerated bodies, we should first ask where in the real world do we find bodies experiencing constant acceleration? We will have more to say about the conditions needed for constant acceleration in Chapters 4 and 5, when we discuss how objects affect one another's motions. For now, we won't try to generalize. We simply acknowledge that there is constant acceleration when measurements show that there is, and we give examples of this occurring.

◆**RANGE FINDER MEASUREMENTS** One way of doing the necessary measurements is with a *sonic range finder* (or *motion detector*) connected to a computer (Figure 2-16). Software loaded into the computer enables it to do calculations and plot graphs using the input from the range finder. The range finder emits ultrasound pulses that travel at constant speed. It determines how far away a body is by sending out a pulse, detecting the pulse reflected back from the body, and timing the duration of the round trip. One common model makes 15 such measurements each second, so that for a moving body, displacements Δx can be calculated for tiny intervals Δt, which are measured by the computer's internal clock. Then, using $\bar{v} = \frac{\Delta x}{\Delta t}$, the software directs the computer to calculate values of average velocities that are approximately instantaneous because the intervals are so small. From these values, the computer can therefore use $\bar{a} = \frac{\Delta v}{\Delta t}$ to calculate acceleration values that are likewise approximately instantaneous.

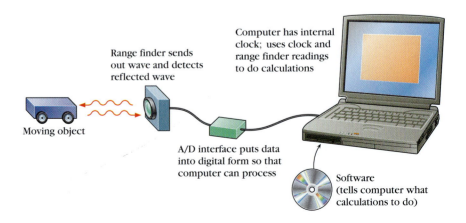

Figure 2-16 Range finder set-up for motion measurements.

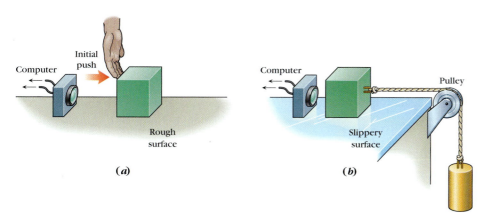

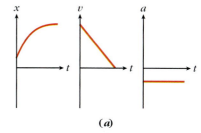

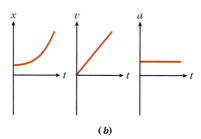

Figure 2-17 **Making measurements on the motion of a block with a range finder.**

Using these simple equations (which do *not* require *a* to be uniform) to do hundreds of calculations each second, the computer can find *x*, *v*, and *a* at successive clock readings *t*, and it can display the results as graphs.

Figure 2-17 shows two situations for which these measurements are easily done. In 2-17a, the hand gives the block a shove to the right and the detector is turned on ($t = 0$) as soon as the block leaves the hand. In 2-17b, the detector is turned on when the weight strung over the pulley begins to fall. Figure 2-18 shows the resulting graphs. (Compare the *x* versus *t* plots with the graphs for cars B and C in Figure 2-8.) The *v* versus *t* graphs have constant slope, so the acceleration $\frac{\Delta v}{\Delta t}$ is uniform. The *a* versus *t* plots reinforce this point: They are horizontal—the value of *a* is not going up or down.

Recall that the *v* versus *t* graphs (in Figure 2-12) for the carts in Figure 2-13 are also straight lines, so the carts' accelerations are also constant. In short, there are a variety of situations that we can treat as having constant or roughly constant acceleration.

Figure 2-18 **Motion graphs for the blocks in Figure 2-17.**

Example 2-7 *A Racing Car Speeds Up*

A racing car goes from 30 m/s to 50 m/s over a 5.0-s interval. If the acceleration is constant, how far does it go during this time?

Solution

Restating the problem. The question asks for a *distance* traveled during a *time interval:* What is $|\Delta x|$ during a particular Δt of 5 s if *v* changes from $v_1 = 30$ m/s to $v_2 = 50$ m/s during this time interval?

What we know/what we don't.

$$v_1 = 30 \text{ m/s} \qquad v_2 = 50 \text{ m/s} \qquad \Delta t = 5.0 \text{ s} \qquad \Delta x = ?$$

Choice of approach. (1) Because the acceleration is constant, the average velocity meets the condition that $\bar{v} = \frac{v_o + v}{2}$, which we can use to find $\bar{v}$. (2) Once we know $\bar{v}$, we can use the definition of average velocity ($\bar{v} = \frac{\Delta x}{\Delta t}$) to find Δx.

The mathematical solution.

1. $\bar{v} = \left(\dfrac{v_o + v}{2}\right) = \dfrac{30 \text{ m/s} + 50 \text{ m/s}}{2} = 40$ m/s

2. Since $\bar{v} = \frac{\Delta x}{\Delta t}$, $\Delta x = \bar{v}\Delta t = (40 \text{ m/s})(5 \text{ s}) = $ **200 m**.

◆ Related homework: Problems 2-35, 2-36, and 2-37.

◆COMPLETELY DESCRIBING MOTION FROM INITIAL CONDITIONS

Suppose an object starts out at $t = 0$ with a certain initial velocity v_o, and it has an acceleration *a*, which we know is constant. From the definition of acceleration, we obtained Equation 2-9, which can tell us the velocity *v* of the object at each subsequent instant t (see Example 2-6*b*). To describe the object's motion completely, we would need to know not only its velocity but its position x (or its displacement $x - x_o$) at each instant. To find an expression that tells us this, we can reason algebraically from definitions and the condition for constant acceleration.

The definition of average velocity (Equation 2-3) now becomes $\bar{v} = \frac{x - x_o}{t - 0}$, so that

$$x - x_o = \bar{v}t$$

But $\bar{v} = \frac{v_o + v}{2}$ (Equation 2-10), so

$$x - x_o = \left(\frac{v_o + v}{2}\right)t$$

Because we have previously found that $v = v_o + at$, we can use this to substitute for v:

$$x - x_o = \left(\frac{v_o + v_o + at}{2}\right)t$$

Simplifying the right-hand side, we get

$$x - x_o = v_o t + \tfrac{1}{2}at^2 \qquad (a \text{ constant}) \qquad (2\text{-}11)$$

If we know the initial values (x_o, v_o, and an a that doesn't change), we can use this equation to find the object's position x at any later instant t.

Compare Equation 2-11 with $x - x_o = vt$, which is valid whenever v is constant. If v does not change from its initial value v_o, then a, its *rate* of change, is zero, and Equation 2-11 reduces to $x - x_o = v_o t$. Otherwise, the last term on the right-hand side of Equation 2-11 represents an additional contribution to the position. This contribution is needed because the velocity is changing.

In describing a body's motion, you may also wish to find its velocity at each position x, without having to know t. To obtain an equation that does this, we start with what we already know and do some algebra:

Start with

$$x - x_o = \bar{v}t$$

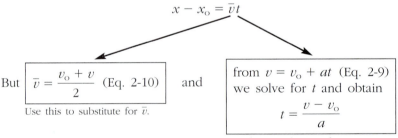

But $\boxed{\bar{v} = \dfrac{v_o + v}{2} \quad (\text{Eq. 2-10})}$ and $\boxed{\begin{array}{c}\text{from } v = v_o + at \ (\text{Eq. 2-9}) \\ \text{we solve for } t \text{ and obtain} \\[4pt] t = \dfrac{v - v_o}{a}\end{array}}$

Use this to substitute for $\bar{v}$.   Use this to substitute for t.

With these substitutions, we get

$$x - x_o = \left(\frac{v_o + v}{2}\right)\left(\frac{v - v_o}{a}\right) = \frac{v^2 - v_o^2}{2a}$$

For simplicity, we next solve for v^2 rather than v, obtaining

$$v^2 = v_o^2 + 2a(x - x_o) \qquad (2\text{-}12)$$

Note that solving for v would then give you

$$v = \pm\sqrt{v_o^2 + 2a(x - x_o)}$$

that is, you get both positive and negative values of v. For a situation like the one in Figure 2-13*b*, both values are meaningful because the cart will pass a position x once on its way up and again on its way down.

2-6 Solving Kinematics Problems I: Uniform Acceleration

We have begun doing **kinematics,** the mathematical description of motion. Equations 2-9, 2-11, and 2-12 are called the **equations of motion** or the **kinematic equations** for a uniformly accelerated body. They describe various aspects of the body's motion—for example, they tell where the object is at each instant, or what its velocity is at each position.

The first step in solving a problem in physics is to determine which physical principles and definitions are relevant to the problem, then let that guide your selection of equations. In kinematics problems, the most basic relationships are the definitions of average velocity and average acceleration. In addition, if the acceleration is uniform, we can write that the average velocity $\bar{v}$ is $\frac{v_o + v}{2}$. Equations 2-9 to 2-12 follow algebraically from these more fundamental relationships. You cannot solve any problem with them that cannot be solved using the more fundamental relationships. Sometimes, in fact, it is easier to start from fundamentals (see Example 2-7). Other times, it may be more convenient to use the equations of motion.

Example 2-8 *Example 2-7 Revisited*

Repeat Example 2-7 using the equations of motion.

Solution

Restating the problem. Although the question asks for a distance traveled during a time interval, it is often convenient to assume that the object in question is starting out at $x = 0$ and $t = 0$. (This is just a matter of deciding where to put your origin and when to start your stopwatch.) With this assumption, $\Delta x = x - x_o = x - 0 = x$ and $\Delta t = t - 0 = t$. The question then becomes $x = ?$ when $t = 5$ s.

What we know/what we don't.

$$v_o = 30 \text{ m/s at } t = 0 \qquad v = 50 \text{ m/s at } t = 5 \text{ s} \quad a = ?$$

$$x = x_o = 0 \text{ at } t = 0 \qquad x = ? \text{ at } t = 5 \text{ s}$$

Choice of approach. Because the acceleration a is unknown, we first solve Equation 2-9 for a. Once a is known, you can use Equation 2-11 to find x.

The mathematical solution. Solving Equation 2-9 for a gives

$$a = \frac{v - v_o}{t} = \frac{50 \text{ m/s} - 30 \text{ m/s}}{5 \text{ s}} = 4 \, (\text{m/s})/\text{s}$$

(Notice that this step is equivalent to using the definition of average acceleration.) Equation 2-11 then gives us

$$x - x_o = x = v_o t + \tfrac{1}{2} a t^2 = (30 \text{ m/s})(5 \text{ s}) + \tfrac{1}{2}(4[\text{m/s}]/\text{s})(5\text{s})^2$$

$$= 150 \text{ m} + 50 \text{ m} = \textbf{200 m, as before.}$$

◆ Related homework: Problem 2-38.

In general, we can use the kinematic equations to generate tables of numbers. With the numbers, we plot graphs. From the graphs, we "read" how the object moves. It is therefore very important for you to get used to reading and interpreting graphs. To see how the kinematic equations generate motion graphs for the ball in Figure 2-19a, work through Example 2-9.

Example 2-9 *Uniform Acceleration Pinball*

For a guided interactive solution, go to Web Example 2-9 at
www.wiley.com/college/touger

A pinball machine slopes slightly downward toward the player (Figure 2-19). Dastardly Dude, local pinball champ, gets the feel of the plunger by first shooting the ball gently enough so that it rolls back to the plunger. When he does so, the ball leaves the plunger with an initial velocity of 0.60 m/s. Its velocity 1.0 s later is 0.30 m/s.

a. Find the ball's acceleration.
b. Find its velocity every half second from $t = 0$ to $t = 4$ s.
c. Find its distance from the plunger every half second from $t = 0$ to $t = 4$ s.
d. Sketch graphs of x versus t, v versus t, and a versus t for the time period from $t = 0$ to $t = 4$ s.

Brief Solution

Restating the question. If $x_o = 0$, then in part **c** we need only find x.

What we know/what we don't.

$v = 0.60$ m/s at $t = 0$ $x = x_o = 0$ at $t = 0$

$v = ?$ at $t = 0.5$ s $a = ?$ $x = ?$ at $t = 0.5$ s, 1.0 s, . . . 4.0 s

$v = 0.30$ m/s at $t = 1.0$ s

$v = ?$ at $t = 1.5$ s, 1.0 s, . . . 4.0 s

Choice of approach. When the acceleration is uniform, we can use the definition of average acceleration to find a. Once a is known, we can use Equation 2-9 to find v at other values of t and Equation 2-11 to find x at each value of t.

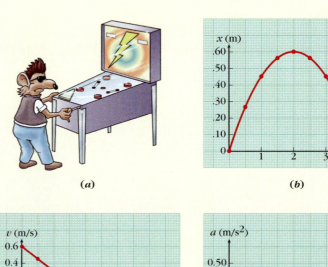

(a)   (b)

Figure 2-19 Describing a ball's motion graphically (see Web Example 2-9). (*a*) Dastardly Dude checks out the pinball machine by firing the ball gently enough so that it rolls uphill a short distance then returns to the plunger. The graphs that follow describe the ball's motion: (*b*) its position versus time, (*c*) its velocity versus time, and (*d*) its acceleration versus time.

(c)   (d)

Mathematical solution to parts a–c. The definition gives us

$$a = \frac{\Delta v}{\Delta t} = \frac{v - v_o}{t - 0} = \frac{0.30 \text{ m/s} - 0.60 \text{ m/s}}{1.0 \text{ s}} = -0.30\,(\text{m/s})/\text{s}$$

Note that the velocity has decreased, so the acceleration is *negative*. For parts b and c we will do a sample calculation for $t = 1.5$ s. We can do the same for each value of t and make a table of the results, then use the table entries as coordinates to plot points

Sample calculation. When $t = 1.5$ s,

$$v = v_o + at = 0.60 \text{ m/s} + (-0.30 \,[\text{m/s}]/\text{s})(1.5 \text{ s}) = 0.15 \text{ m/s}$$

and

$$x - x_o = x = v_o t + \tfrac{1}{2}at^2 = (0.60 \text{ m/s})(1.5 \text{ s}) + \tfrac{1}{2}(-0.30[\text{m/s}]/\text{s})(1.5 \text{ s})^2$$

$$= 0.90 \text{ m} - 0.34 \text{ m} = 0.56 \text{ m}$$

Graphing the Results. The resulting graphs are summarized in Figure 2-19b–d. To obtain the *a* versus *t* graph, recall that *a* is uniform; its value is −0.30 (m/s)/s at every value of *t*.

◆ Related homework: Problems 2-40 and 2-41.

◆**ASKING QUESTIONS THAT EQUATIONS CAN ANSWER** In Example 2-9 it was not sufficient to ask "what is the value of *x* (or *v*)?" Because the values were changing, we always had to ask, "what is the value of *x* at a particular value of *t*?" The equations we use are always relationships between two or more variables. For an equation to be useful in solving for the answer to a question, the question cannot simply be, "What is the value of some quantity Q_1?" ($Q_1 = ?$) It must have a more complete form, such as

$$Q_1 = \underline{\;?\;} \text{ when } Q_2 = \underline{\;\;\;\;} \qquad\qquad \text{(Form 2-1)}$$

(**In words:** *What is the value of Quantity 1 when Quantity 2 = a known value?*)

or "$Q_1 = \underline{\;?\;}$ when $Q_2 = \underline{\;\;\;\;}$ and $Q_3 = \underline{\;\;\;\;}$ and so on?" When a question of Form 2-1 is not completely stated, you need to state the rest of it for yourself. You must do this even for questions like "how far does an object go?" or "how long does it take to get there?," as in the following example.

Example 2-10 *"How Far . . . ?"*

For a guided interactive solution, go to Web Example 2-9 at www.wiley.com/college/touger

The hand in Figure 2-13a gives the cart an initial downhill shove. The cart leaves the hand with a speed of 3.0 m/s. While traveling downhill, it loses speed at a rate of 5 (m/s)/s. How far down the ramp does it go before coming back up?

Restating the question. The *speed* is the absolute value of the velocity. The rate at which the velocity changes is the *acceleration*. If we take the downhill direction to be positive, $v_1 = 3.0$ m/s but is decreasing, so the acceleration must be negative: $a = -5.0$ (m/s)/s or −5.0 m/s each second.

The turnaround point is the position where $v = 0$. The question "how far?" is really asking: $\Delta x = ?$ as v changes from $v_1 = 3.0$ m/s to $v_2 = 0$.

EXAMPLE 2-10 *continued*

➡**Note:** For beginning students, it is usually more instructive to do solutions starting from fundamentals whenever possible, because it keeps you thinking more about the physics (what is happening and what the various quantities mean) than about the algebra.

Brief Solution 1 (reasoning from fundamentals)

Choice of approach. Knowing both values of v, we can (1) find $\bar{v}$ and (2) find Δv. Knowing Δv, we can use the definition of acceleration to find Δt. Then, knowing $\bar{v}$ and Δt, we can use the definition of average velocity to find Δx.

What we know/what we don't.

$$\Delta x = ? \text{ as } v \text{ changes from } v_1 = 3.0 \text{ m/s to } v_2 = 0 \qquad a = -5.0 \text{ (m/s)/s}$$

$$\Delta t = ? \text{ as } v \text{ changes from } v_1 = 3.0 \text{ m/s to } v_2 = 0 \qquad \bar{v} = ?$$

The mathematical solution. Over the interval in question, the velocity changes by

$$\Delta v = v_2 - v_1 = 0 - 3.0 \text{ m/s} = -3.0 \text{ m/s}$$

The average velocity $\bar{v} = \dfrac{v_1 + v_2}{2} = \dfrac{3.0 \text{ m/s} + 0}{2} = 1.5 \text{ m/s}$

Since $a = \dfrac{\Delta v}{\Delta t}$ when the acceleration is constant, we can solve for Δt:

$$\Delta t = \frac{\Delta v}{a} = \frac{-3.0 \text{ m/s}}{-5.0 \text{(m/s)/s}} = 0.60 \text{ s}$$

Now, because $\bar{v} = \dfrac{\Delta x}{\Delta t}$, $\Delta x = \bar{v}\Delta t = (1.5 \text{ m/s})(0.60 \text{ s}) = \textbf{0.90 m}$

Brief Solution 2 (using equations of motion)

The One-Step Solution. Solving for x in Equation 2-12 (with $x_\text{o} = 0$) gives us

$$x = \frac{v^2 - v_\text{o}^2}{2a} = \frac{0^2 - (3.0 \text{ m/s})^2}{2(-5.0 \text{ [m/s]/s})} = +0.90 \frac{\text{m}^2/\text{s}^2}{\text{m/s}^2} = \textbf{0.90 m}$$

The Two-Step Solution. From Equation 2-9,

$$t = \frac{v - v_\text{o}}{a} = \frac{0 - 3.0 \text{ m/s}}{-5.0 \text{ (m/s)/s}} = +0.60 \frac{\text{m/s}}{\text{m/s}^2} = 0.60 \text{ s}$$

With this value of t, Equation 2-11 gives us

$$x - x_\text{o} = x = v_\text{o}t + \tfrac{1}{2}at^2 = (3.0 \text{ m/s})(0.60 \text{ s}) + \tfrac{1}{2}(-5.0 \text{ [m/s]/s})(0.60 \text{ s})^2$$

$$= 1.80 \text{ m} - 0.90 \text{ m} = \textbf{0.90 m}$$

◆ Related homework: Problem 2-42.

2-7 Gravitational Acceleration and Free Fall

When an object's fall to Earth is not helped or opposed by anything else, not even air resistance, we say the object is *freely falling* or in **free fall.** If you drop a dense, heavy object from rest and monitor its fall with a sonic range finder, you will find that the object has uniform acceleration. Moreover, you will find the value of the acceleration to be very nearly 9.8 (m/s)/s or 9.8 m/s². This value is called its **gravitational acceleration.**

If you drop this textbook, the acceleration will have approximately that value. If you drop a single sheet of paper, in contrast, you will find that the acceleration is not uniform and is on average much less than 9.8 m/s², so that the sheet of paper takes much longer to reach the ground. If you crumple the sheet of paper into a ball before dropping it, you reduce the effect of air resistance, and the paper falls more nearly like the textbook.

On-The-Spot Activity 2-1

Take a smooth sheet of paper, small enough so that it doesn't extend beyond the edges of the cover of this book, and hold it to the underside of the book, as in

Figure 2-20*a*. Hold the book and paper horizontally and release them together. The book and the sheet of the paper should hit the ground together. Under these circumstances, their accelerations are the same.

"Big deal!" you say. The book is *making* the paper move with it. But what if you hold the sheet of paper flat *on top* of the book, as in Figure 2-20*b*, and drop it again? First decide what you think will happen to the paper, then try it. Does the paper do what you expected? If not, why not?

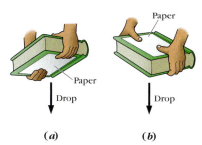

Figure 2-20 Comparing the accelerations of falling objects when air resistance is negligible.

When air resistance is prevented from affecting the sheet of paper, the paper falls with the same acceleration as the book, roughly 9.8 m/s².

In free fall, all bodies have the same acceleration.

Dropped simultaneously from the same height, bodies in free fall will hit level ground at the same time. A hammer and a feather did exactly that when they were dropped in the airless conditions on the moon by a member of the 1971 Apollo 15 mission. In another dramatic demonstration in place for many years at the Boston Museum of Science, feathers released inside a two-story glass column would drop to the bottom like the proverbial ton of bricks when the air was pumped out. Over short distances above or below Earth's surface (a few hundred meters or less), the acceleration of a freely falling body will vary with height by less than one part in a thousand. To two-place accuracy, it remains 9.8 m/s². We use *g* as a symbol for this special value.

Magnitude of **gravitational acceleration** near Earth's surface:

$$g = 9.8 \text{ m/s}^2$$

Important: The symbol *g* stands for this *positive* number, no matter which direction we call positive. When we take the positive direction to be upward, the acceleration $a = -g$; that is, $a = -9.8 \text{ m/s}^2$.

Applying the Constant Acceleration Equations of Motion to Free Fall

1. If you choose the positive direction to be upward, let $a = -g$ in all the equations (let $a = +g$ if you let downward be positive).
2. Because we tend to label the vertical axis *y*, it is common—though not necessary—to replace *x* with *y* in all the equations to indicate vertical position.

By these two steps, you should be able to *show* whenever needed (rather than memorize more equations) that Equations 2-9, 2-11, and 2-12 become

$$v = v_o - gt$$
$$y - y_o = v_o t - \tfrac{1}{2}gt^2$$
$$v^2 = v_o^2 - 2g(y - y_o)$$

Try working with these ideas now in the following example.

Example 2-11 *A Rock Dropped Downward*

For a guided interactive solution, go to Web Example 2-11 at
www.wiley.com/college/touger

A rock is dropped into a 200-m-deep mine shaft. How long does it take
a. to fall halfway to the bottom?   **b.** to hit bottom?

EXAMPLE 2-11 *continued*

➡**A note on language:** Downward is not the same direction for someone in New York as for someone in Hong Kong. Wherever you are, downward means toward the center of the Earth.

Brief Solution

Assumptions. The rock is dropped, not thrown, so it starts from rest ($v_o = 0$). We choose the rock's starting position to be the origin ($y = y_o = 0$), the starting instant to be $t = 0$, and the downward direction to be negative. Then the bottom of the shaft is at $y = -200$ m, and halfway down is $y = -100$ m.

Restating the question. In Form 2-1, the questions become
a. $t = ?$ when $y = -100$ m   **b.** $t = ?$ when $y = -200$ m.

What we know/what we don't.

$$v_o = 0 \text{ (starts from rest)} \qquad y_o = 0$$

$$a = -g = -9.8 \text{ (m/s)/s} \quad t = ? \text{ when } y = -100 \text{ m} \quad t = ? \text{ when } y = -200 \text{ m}$$

Choice of approach. The constant acceleration equations of motion are convenient to use here.

The mathematical solution. With $v_o = 0$, $y_o = 0$, and $a = -g$, Equation 2-11 becomes $y = -\frac{1}{2}gt^2$. Solving for t and substituting the values for part **a** gives

$$t = \pm\sqrt{\frac{-2y}{g}} = \pm\sqrt{\frac{-2(-100 \text{ m})}{(9.8[\text{m/s}]/s)}} = \pm 4.52 \text{ s}$$

We choose the positive root because the rock is falling only when $t \geq 0$. Thus,

$$y = -100 \text{ m at } t = \textbf{4.52 s}$$

STOP&Think Will it take twice as long to fall twice as far? ◆

Doing the calculation for **b** in the same way, we get

$$t = \pm\sqrt{\frac{-2y}{g}} = \pm\sqrt{\frac{-2(-200 \text{ m})}{(9.8 \text{ [m/s]}/s)}} = \pm 6.39 \text{ s}$$

so $$y = -200 \text{ m at } t = \textbf{6.39 s}$$

The position y is proportional to t^2, not t, so it takes less than twice the time to fall twice the distance.

◆ Related homework: Problems 2-44, 2-47, and 2-48.

Example 2-11 can also be done starting with basic definitions (see Problem 2-96).

◆**OTHER VALUES OF g** Objects near the surface of the moon or another heavenly body (e.g., a planet or one of *its* moons) may also be freely falling, but not with the same acceleration. The value of g is different at each body's surface. For example, $g_{\text{moon}} \approx \frac{1}{6}g_{\text{Earth}}$, and $g_{\text{Jupiter}} \approx 2.5g_{\text{Earth}}$. The value of g will also vary from location to location on Earth's surface. For example, for cities at different elevations, $g_{\text{Denver}} = 9.796$ m/s^2, but $g_{\text{New York}} = 9.803$ m/s^2. For sites at different latitudes, $g_{\text{Equator}} = 9.78$ m/s^2, but $g_{\text{North Pole}} = 9.83$ m/s^2. These are *observed* values—they come from measurements. We will not be able to address reasons for these values until we treat gravitational forces. At this point, we are only describing how objects move, not asking why they move that way.

We have said that an object is in free fall when only the gravitational pull of Earth is affecting its motion. By this definition, a ball thrown straight up is in free fall from the instant it leaves the thrower's hand, *even while it is on its way up* (carefully examine Figure 2-21). A sonic range finder would show that its acceleration remains constant at -9.8 m/s^2 from that instant until it lands. Because the acceleration is negative, the velocity first becomes less and less positive, then more and more negative, just as was true for the ball in Example 2-8.

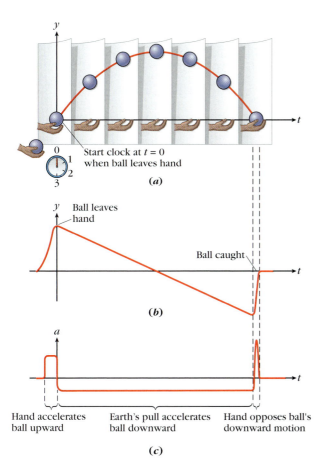

Figure 2-21 **Motion graphs for the ball thrown upward in Example 2-12.** The ball is in free fall, so that the constant acceleration equations are applicable, only between the instant it leaves the hand and the instant when it is again makes contact with the hand.

Example 2-12 *A Ball Thrown Upward*

A softball player throws a ball straight upward with a velocity of 17 m/s, and catches it exactly where it left her hand.

a. How long does the ball remain in the air?
b. How high does it go?

Solution

Assumptions. We take the origin to be the point where the ball leaves her hand. We assume that effects like air resistance are negligible, so that the ball is in *free fall*.

Restating the problem. The ball remains in the air until it is back where it started—at $y = 0$. Part **a** therefore asks

$$t = ? \quad \text{when} \quad y = 0$$

Part **b,** similarly to Example 2-9, asks

$$y = ? \text{ when } v = 0$$

What we know/what we don't. In addition to this restatement of the question of the question, we know that

$$y_o = 0 \text{ (by assumption)} \qquad v_o = 17 \text{ m/s}$$

and because the ball is falling freely,

$$a = -g = -9.8 \, (\text{m/s})/\text{s}$$

Choice of approach. Again, we may apply the constant acceleration equations of motion. For part **a,** the equation relating y and t is Equation 2-11, written as $y - y_o = v_o t + \frac{1}{2} a t^2$. If y is known, t will be the only unknown.

EXAMPLE 2-12 *continued*

For part **b**, we apply Equation 2-12, written as $v^2 = v_o^2 + 2a(y - y_o)$ for vertical motion, to obtain a one-step solution.

The mathematical solution.

a. When $y = y_o = 0$ and $a = -g$, Equation 2-11 becomes

$$0 = v_o t - \tfrac{1}{2}gt^2$$

This is a quadratic equation that can be solved for t by factoring:

$$0 = t(v_o - \tfrac{1}{2}gt)$$

so that either one or the other factor must be 0:

$$0 = t \quad \text{or} \quad 0 = v_o - \tfrac{1}{2}gt$$

in which case, solving for t gives

$$t = 2\frac{v_o}{g} = 2\left(\frac{17 \text{ m/s}}{9.8[\text{m/s}]/\text{s}}\right) = \textbf{3.5 s}$$

There are two solutions for t, and both have meaning in the actual situation. The ball is at $y = 0$ first at $t = 0$ when it leaves the player's hand, then again at $t = 3.5$ s when it returns to her glove. The second solution answers the question of how long the ball remains in the air. (In fact, for *any* value of y, Equation 2-11 yields two solutions for t because as Figure 2-21*a* shows, the ball passes each value of y, other than the maximum value, both on the way up and on the way down.)

b. With $y_o = 0$ and $a = -g$, $v^2 = v_o^2 + 2a(y - y_o)$ becomes $v^2 = v_o^2 - 2gy$. Proceeding much as in Example 2-9, when $v = 0$ we get

$$y = \frac{v^2 - v_o^2}{-2g} = \frac{v_o^2}{2g}$$

$$= \frac{(17 \text{ m/s})^2}{2(9.8[\text{m/s}]/\text{s})} = +14.7\frac{\text{m}^2/\text{s}^2}{\text{m/s}^2} = \textbf{14.7 m}$$

◆ Related homework: Problems 2-49 and 2-102.

◆SUMMARY◆

In this chapter we defined the terms needed to describe quantitatively the motion of **point objects** (or objects that can be treated as point objects) in one dimension (see also Table 2-1).

◆DEFINITIONS

- **Instant** $t \equiv$ a single clock reading (a single point on a time line or axis)

- **Time interval** $\Delta t \equiv t_2 - t_1$ (distance between two points on a time line or axis)

In one dimension

- **Position** $(x$ or $y)$ is a coordinate along an arbitrarily placed real number line with successive integers (in SI) 1 m apart.

- **Displacement** $\Delta x \equiv x_2 - x_1$ (2-1)

- **Distance** $|\Delta x| \equiv |x_2 - x_1|$ (2-2) for path segments where there is no reversal of travel direction. Distances are always positive.

- **Average* velocity** $\equiv \bar{v} \equiv \dfrac{displacement}{time\ interval} = \dfrac{\Delta x}{\Delta t} = \dfrac{x_2 - x_1}{t_2 - t_1}$ (2-3)

- **Average* speed** $\equiv \dfrac{total\ distance\ traveled}{total\ time\ interval}$ (2-4)

- **Instantaneous velocity** $v \equiv \underset{\Delta t \to 0}{\text{limit}}\ \bar{v} \equiv \underset{\Delta t \to 0}{\text{limit}}\ \dfrac{\Delta x}{\Delta t}$ (2-5)

- **Average* acceleration** $\bar{a} \equiv \dfrac{\Delta v}{\Delta t} \equiv \dfrac{v_2 - v_1}{t_2 - t_1}$ (2-7)

- **Instantaneous acceleration** $a \equiv \underset{\Delta t \to 0}{\text{limit}}\ \bar{a} \equiv \underset{\Delta t \to 0}{\text{limit}}\ \dfrac{\Delta v}{\Delta t}$ (2-8)

(**averaged over Δt*)

Table 2-1 in Section 2-2 summarizes the relationships among these quantities and the type of question each quantity addresses.

Graphs of x, v, and a versus t can provide meaningful descriptions of an object's motion. *Slopes* and *vertical intercepts* of these graphs have particular meaning. Because the *slope* always gives a *rate of change* and the vertical intercept always give an initial value, it follows that

For a graph of . . .	The slope of a (secant/tangent) gives . . .	And the vertical intercept gives . . .
x versus t	(Avg./instantaneous) velocity	Initial position
v versus t	(Avg./instantaneous) acceleration	Initial velocity

For straight lines (uniform slope), the average and instantaneous values are the same. Sections 2-3 and 2-4 provide further guidelines for interpreting graphs of x versus t and v versus t. The tables in those sections summarize key points.

From the definition of average velocity, it follows that

$$x - x_{\mathrm{o}} = \bar{v}t \tag{2-6}$$

which for constant velocity situations becomes

$$x = x_{\mathrm{o}} + vt \qquad \text{(uniform motion only)} \tag{2-6u}$$

Whenever the *acceleration* is constant,

$$\begin{array}{c}\text{condition for}\\ \text{constant acceleration:}\end{array} \qquad \bar{v} = \frac{v_o + v}{2} \tag{2-10}$$

From the definitions, we can use algebraic reasoning to obtain *equations of motion* that describe the motion of an object under particular conditions. When the condition is constant acceleration,

Equations of motion (kinematic equations) for constant (uniform) a only:

$$v = v_{\mathrm{o}} + at \tag{2-9}$$

$$x - x_{\mathrm{o}} = v_{\mathrm{o}}t + \tfrac{1}{2}at^2 \tag{2-11}$$

$$v^2 = v_{\mathrm{o}}^2 + 2a(x - x_{\mathrm{o}}) \tag{2-12}$$

Things to remember when solving problems:

- Make sure you get the picture—sketch the situation.
- Know how the relevant quantities are defined and be able to use the definitions.
- Think first about what is happening—what physical concepts

or principles apply to the situation—and let that guide your choice of equations.

- To further guide your choice in deciding which definitions or principles or other equations relating known and unknown quantities to use, it is generally useful to restate the question in a form that directly connects known and unknown quantities, such as

$$Q_1 = \underline{\ ?\ } \text{ when } Q_2 = \underline{\qquad} \tag{Form 2-1}$$

In words: *What is the value of Quantity 1 when Quantity 2 = a known value?*

Measurements made with a sonic range finder demonstrate that there are a range of situations where a is very nearly constant, so that Equations 2-9 through 2-12 are applicable. They can be used, for example, when an object is in **free fall,** affected only by a gravitational pull. If the upward direction is taken as positive, objects falling freely to Earth have an acceleration $-g$, where g itself is always positive.

Magnitude of **gravitational acceleration** near Earth's surface

$$g = 9.8 \ (\text{m/s})/\text{s} = 9.8 \ \text{m/s}^2$$

To apply the constant acceleration equations of motion to *free fall,*

1. If you choose the positive direction to be upward, let $a = -g$ in all the equations (let $a = +g$ if downward is positive).
2. Because we tend to label the vertical axis y, it is common—though not necessary—to replace x with y in all the equations to indicate vertical position.

◆ QUALITATIVE AND QUANTITATIVE PROBLEMS ◆
Hands-On Activities and Discussion Questions

The questions and activities in this group are particularly suitable for in-class use.

2-1. Hands-On Activity.

a. Throw a ball vertically up in the air so that it rises about 1 m after leaving your hand. Does it slow down as it rises? Does it speed up as it falls? Can you see this happening, or is it too difficult to judge because things happen so fast?

b. Now find a table top (or a board or other flat surface) that is at least 1 m long and that you can tilt by propping up one end. Tilt it very slightly, just barely enough so that a

ball rolled uphill on the tilting surface will roll back down. Roll the ball so that it goes about 1 m up the tilting surface before turning around. Now can you see whether the ball slows down on its way uphill? Can you see whether the ball speeds up on its way downhill? Comment on your observations.

2-2. Discussion Question. A car manufacturer launches a new advertising campaign claiming that its cars are capable of accelerations of 30 000 miles/hr². Is this an outrageous claim?

Review and Practice

Section 2-1 Matter in Motion

2-3. How can you think of each of the following as involving matter in motion? (You are not expected to give a full explanation here. Simply think about what in each situation must have moved or been rearranged, and indicate why. In some of these situations, it will be helpful to think about the atoms or molecules. Also indicate what in each situation is unclear to you.)

a. About a week after you see a half moon on a particular night, the moon appears full.

b. A beaker of ammonia spills in a chemistry lab. The students closer to it smell it before those who are further away.

c. When some cold milk is added to a child's too-hot cocoa, everything reaches a drinkable in-between temperature.

d. Green plants form glucose, a simple sugar, from water and carbon dioxide.

Section 2-2 A Vocabulary for Describing Motion

2-4. Can you treat a baseball as a point object when you are analyzing *a.* the path of a batted ball hit over the right field wall? *b.* how a pitcher makes the ball curve?

2-5. **SSM** Assume that the given values in Figure 2-2 are in km (10^3 m).

a. What is the position of a car according to a Newtonville observer when a Keplertown observer reports that it is −40 km? When a Keplertown observer reports that it is zero?

b. If the car travels from the first of these positions to the second, what is its displacement according to the Newtonville observer? According to the Keplertown observer?

2-6. In the room where you are now working, identify the wall closest to you.

a. If the origin is at the point on this wall nearest to you, what is your approximate position?

b. If the origin is at the point on the opposite wall nearest to you, and you choose the *same direction* as positive, what is your approximate position? Draw a sketch that shows your reasoning for both *a* and *b*.

c. Walk 1 m toward the opposite wall. How would you now answer *a*? *b*? What is your displacement for each choice of origin?

d. What distance have you gone for each choice of origin?

2-7. Find the displacement of a jogger who follows a straight path

a. from a position of 100 m to a position of 600 m.

b. from a position of 600 m to a position of 100 m.

c. from a position of −100 m to a position of −600 m.

d. from a position of −600 m to a position of −100 m.

2-8. Find the distance gone by the jogger in each part of Problem 2-7.

2-9. For each of the following straight-line motions, show the positions on a real number line and sketch the displacement, showing both its direction and its value. Assume east is the positive direction in each case and that the origin is at the stoplight.

a. A car starts out 250 m east of the light and ends up 350 m east of the light.

b. A bicyclist starts out 250 m west of the light and ends up 350 m east of the light.

c. A delivery truck starts out 250 m east of the light and ends up 350 m west of the light.

d. A pedestrian starts out 250 m west of the light and ends up 350 m west of the light.

2-10. In Problem 2-7, assume in each part that the initial position occurs when the jogger's stopwatch reads 30 s and the final position occurs when the watch reads 280 s.

a. Find the jogger's average velocity and average speed in Problem 2-7*a*.

b. Find the jogger's average velocity and average speed in Problem 2-7*b*.

c. Find the jogger's average velocity and average speed in Problem 2-7*c*.

d. Find the jogger's average velocity and average speed in Problem 2-7*d*.

e. The jogger's speed when the watch reads 100 s (is 0.5 m/s; is −0.5 m/s; is 2 m/s; is −2 m/s; is 5 m/s; is −5 m/s; or cannot be calculated).

2-11. A walker follows the path from A to E (Figure 2-22) over an 80-s time interval.

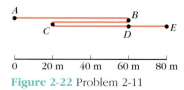

Figure 2-22 Problem 2-11

a. What total distance does the walker cover between A and E?

b. What is the walker's total displacement between A and E?

c. What is the walker's average velocity between A and E?

d. What is the walker's average speed between A and E?

e. What is the walker's average velocity between points B and D?

f. What is the walker's average speed between B and D?

2-12. A test vehicle is being developed that can react almost instantaneously to objects in its path. Its position x during one trial is graphed against clock reading t in Figure 2-23. What is the vehicle's average

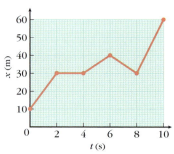

Figure 2-23 Problem 2-12

a. velocity between $t = 4$ s and $t = 6$ s?

b. velocity between $t = 6$ s and $t = 8$ s?

c. velocity between $t = 4$ s and $t = 8$ s?

d. speed for the interval in *a*?

e. speed for the interval in *b*?

f. speed for the interval in *c*?

g. velocity between $t = 0$ and $t = 10$ s?

h. speed during this interval?

2-13. **SSM WWW** The position of a test vehicle is plotted against clock reading t in Figure 2-23. Consider the vehicle's average velocity during each of the following time intervals: from $t = 6$ s to $t = 8$ s; from $t = 2$ s to $t = 4$ s; from $t = 4$ s to $t = 8$ s; from $t = 0$ to $t = 2$ s. Put the intervals in rank order according to average velocity, starting with the value (lowest or most negative) that would occur farthest to the left on a real number line. Make sure to indicate equalities if there are any.

2-14. In Figure 2-24, the positions of three travelers A, B, and C are plotted against clock reading t over a 25-s time interval.

a. List these travelers in rank order of their *velocity* during this time interval, starting with the value (lowest or most negative) that would occur furthest to the left on a real number line. Make sure to indicate equalities if there are any.

b. List these travelers in rank order of their *speed* during this time interval, starting with the value (lowest or most negative) that would occur furthest to the left

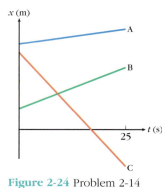

Figure 2-24 Problem 2-14

on a real number line. Make sure to indicate equalities if there are any.

2-15. SSM In Figure 2-23, which records the motion of a test vehicle, what is the instantaneous velocity of the vehicle
a. at $t = 1$ s?   **b.** at $t = 7$ s?   **c.** at $t = 9$ s?

2-16. A commuting student rushing to get home before her children get home from school has left campus and is 200 m beyond the college gate at exactly 2:19:20 PM (in hours, minutes, seconds). The bus stop is 1100 m beyond the gate in the same direction. The last bus that will get her home in time departs promptly at 2:23:00 PM. She runs the remaining distance at a constant speed of 3.9 m/s.
a. Determine by a calculation whether she is running fast enough to catch her bus. (Also, does she have exact change when she gets there?)
b. Suppose that the 3.9 m/s had been an average speed rather than a constant speed. Would that have any effect on your answer to **a**? Explain.

2-17. A college student is driving home for spring break along a straight highway. When she is 20 000 m from the college, she sets her cruise control at 25 m/s and starts a stopwatch that she has borrowed from her physics lab. What does the watch read when she is 50 000 m from the college in the same direction?

Section 2-3 Representing Motion Graphically

2-18. A storm front advances eastward at a constant speed of 9.2 km/h. A meteorologist notes from satellite photos that at $t = 2.0$ h, the storm front has reached a point 13.8 km east of his weather station.
a. Where was the storm front at $t = 0$?
b. If the weather station is at $x = 0$, what is the value of the vertical intercept of a graph of x versus t?
c. Sketch the graph.

2-19. A student in a high-rise residence hall is disgruntled because an elevator descending at constant speed fails to stop at his floor. When he pushes the button, the elevator is 6.4 m above him; 5.0 s later, it is 9.6 m below him.
a. If the upward direction is positive, what is the velocity of the elevator?
b. What is the slope of a graph of the elevator's position (plotted vertically) versus t?
c. Sketch the graph.

2-20. Consider Figure 2-8b. For which 1-s interval (0 to 1 s, 1 s to 2 s, 2 s to 3 s, or 3 s to 4 s) is the average velocity
a. of car C a maximum?   **c.** of car D a maximum?
b. of car C a minimum?   **d.** of car D a minimum?

2-21. Figure 2-8a shows the position versus time graphs of five race cars.
a. Suppose you wanted to know during which 1-s interval (0 to 1 s, 1 s to 2 s, etc.) car C had its maximum average velocity. How is this information communicated by the graph for car C?
b. Repeat **a** for car D.

2-22. In Figure 2-25, match each graph of velocity v versus t with the graph that plots position x versus t for the same motion.

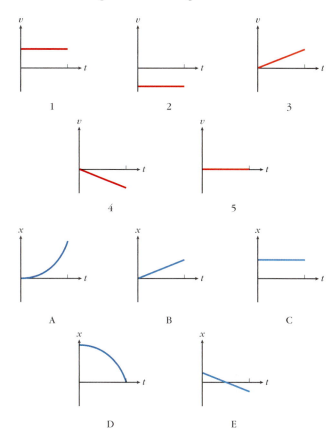

Figure 2-25 Problem 2-22

2-23. In the v versus t graph in Figure 2-9b, the velocity is negative between $t = 0$ and $t = t_1$. What feature of Figure 2-9a (the x versus t graph of the same object) tells you that the velocity is negative between these two instants?

2-24. Sketch the v versus t graph for each of the x versus t graphs in Figure 2-26.

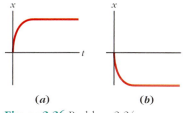

Figure 2-26 Problem 2-24

Section 2-4 Acceleration and Graphing Uniformly Accelerated Motion

2-25. *Converting units.*
a. A car manufacturer launches a new advertising campaign claiming that its cars are capable of accelerations as great as 20 000 miles/hr². Is this an outrageous claim?
b. The specs on a 2000 Volkswagen Jetta indicate that it can go from zero to 60 miles/hr in 7.6 s. Find its average acceleration in units of (miles/hr)/s.
c. Convert the acceleration that you found in **b** to (miles/hr)/hr (or miles/hr²). Think about how this compares to the car in **a**, and whether you would still give the same answer to part **a**.

2-26. Consider the equations $x = x_o + vt$ and $v = v_o + at$.
a. Do a feature-by-feature comparison of the graphs of these two equations.
b. Under what conditions can each of these equations be used? Can the two equations be used together under the same conditions?

2-27. SSM WWW Sketch a graph of acceleration a versus t for each of the following cars.

a. Car A starts from rest and speeds up uniformly toward the east.

b. Car B travels westward at a constant speed of 10 m/s.

c. Car C starts out at a speed of 10 m/s due east and slows down uniformly.

d. Car D starts from rest and speeds up uniformly toward the west.

2-28. In the motion graphed in Figure 2-14, does the acceleration increase, decrease, or remain the same as t increases?

2-29. SSM For the v versus t graph in Figure 2-14, find the approximate instantaneous acceleration at $t = 7.5$ s.

Section 2-5 Constant Acceleration and Equations of Motion

2-30. In the graph in Figure 2-27, the velocity of a ball rolling on a ramp is plotted against time.

a. Find the ball's initial velocity.

b. Find the ball's acceleration.

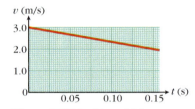

Figure 2-27 Problem 2-30

c. Use your results to determine the time at which the velocity will be -4.0 m/s if the acceleration remains uniform.

2-31. What happens to Equations 2-9 through 2-12 when $a = 0$? Write the resulting equations and interpret them.

2-32.

a. Is it possible for an object with constant acceleration to slow down and then speed up again? Explain.

b. Is it possible for an object with constant acceleration to speed up and then slow down again? Explain.

2-33. A child rolls a ball up a playground slide. The ball has constant acceleration once it leaves the child's hand, and it passes a certain spot on the slide both on its way up and on its way down.

a. Does the ball have the same velocity each time it passes the spot? Explain.

b. Does the ball have the same speed each time it passes the spot? Explain.

2-34. A car starts at rest at the origin when the driver's stopwatch reads $t = 0$ and accelerates uniformly in the positive x direction. When the car has gone a certain distance d_1, it has a speed v_1 and the stopwatch reading is t_1. When the car has gone twice this distance, it has a speed v_2 and the stopwatch reading is t_2.

a. Will v_2 be equal to twice v_1, greater than twice v_1, or less than twice v_1? Explain.

b. Will t_2 be equal to twice t_1, greater than twice t_1, or less than twice t_1? Explain.

2-35. A driver begins to brake when her car is traveling at 15.0 m/s, and the car comes to a stop 4.0 s later. How much further does the car go after she begins to brake, assuming the acceleration is constant?

2-36. A truck starting from rest along a straight road has an acceleration of 2.0 m/s².

a. What is its speed after 1 s? 2 s? 3 s?

b. How far has it gone after 1 s? 2 s? 3 s?

2-37. SSM WWW A subway train starting from rest along a straight track has a uniform acceleration of 1.8 m/s² for the first 20 m it travels.

a. Calculate its speed when it has traveled 10 m.

b. When it has traveled 20 m, will its speed be less than double, double, or more than double its speed at 10 m? (Answer this *before* doing a calculation.)

c. Now *calculate* its speed when it has traveled 20 m. Does this agree with your answer to **b**?

Section 2-6 Solving Kinematics Problems I: Uniform Acceleration

2-38. A certain vehicle requires a distance of 30 m to stop when it is traveling at 25 m/s.

a. What acceleration is produced by braking under these conditions?

b. If a jaywalker is 28 m in front of the vehicle when the driver first applies the brakes, how much time does the pedestrian have to get out of the way?

2-39. A car starting from rest is given a constant acceleration of 4 m/s².

a. What is its speed when it has gone a distance of 50 m?

b. How long does it take to go this distance?

2-40. A ball is rolled up a long, sloping driveway with an initial velocity of 4 m/s. The absolute value of the ball's acceleration is 2 m/s². Assume the ball's initial position is zero and the uphill direction is positive.

a. Find the ball's velocity every second from $t = 0$ to $t = 6$ s.

b. Find the ball's position every second from $t = 0$ to $t = 6$ s.

c. During this time period, at how many different instants does the ball have a speed of 2 m/s?

d. During this time period, at how many different instants does the ball have a speed of 6 m/s?

2-41.

a. Sketch a graph of velocity versus time using the values from Problem 2-40**a**.

b. Sketch a graph of position versus time using the values from Problem 2-40**b**.

2-42. A ball is rolled up a ramp with an initial velocity of 10 m/s. The absolute value of the ball's acceleration is 2 m/s².

a. How far up the ramp does the ball go?

b. How long does it take the ball to reach this point?

c. If the ball starts out at $t = 0$, at how many different instants t will it be at a point 9 m from its starting position—zero, one, or two? (Answer this before doing a calculation.)

d. If the ball starts out at $t = 0$, when will it be 9 m from its starting position? What will its velocity be at each instant? Answer this by doing a calculation, and see if your result agrees with your answer to **c**.

2-43. A test driver starts her timer and begins to accelerate uniformly as she passes the zero marker on a posted straightaway. She passes the 40-m mark when her timer reads 2 s, and the 60-m mark when her timer reads 4 s. Find her velocity at the zero marker and her acceleration.

2-44. Identical balls A and B are released at the openings of the hollow vertical shafts in Figure 2-28. The balls strike the bottoms of the two shafts at speeds v_A and v_B, respectively. Suppose we know that $v_B = 2\,v_A$, and air resistance is negligible.

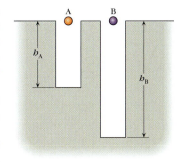

a. If h_A and h_B are the depths of the two shafts, h_B is then equal to ____ ($\frac{1}{4}h_A$; between $\frac{1}{4}h_A$ and $\frac{1}{2}h_A$; $\frac{1}{2}h_A$; h_A; between h_A and $2h_A$; $2h_A$; $4h_A$).

Figure 2-28 Problem 2-44

b. If t_A and t_B are the descent times of the two balls, t_B is then equal to ____ ($\frac{1}{4}t_A$; between $\frac{1}{4}t_A$ and $\frac{1}{2}t_A$; $\frac{1}{2}t_A$; t_A; between t_A and $2t_A$; $2t_A$; $4t_A$).

Section 2-7 Gravitational Acceleration and Free Fall

2-45. Which of the highlighted objects shown in Figure 2-29 are freely falling (or very nearly so)?

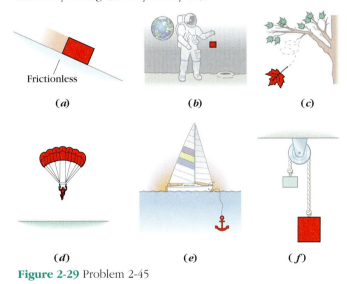

(a)   *(b)*   *(c)*

Frictionless

(d)   *(e)*   *(f)*

Figure 2-29 Problem 2-45

2-46. A child's toy rocket is propelled straight upward by releasing a compressed spring. At which (one or more) of the following instants is the rocket's acceleration equal to -9.8 m/s^2 (assuming the upward direction is positive and neglecting air resistance)? Explain.

a. t_1: just before the rocket loses contact with the spring.

b. t_2: just after the rocket loses contact with the spring.

c. t_3: when the rocket reaches the highest point of its ascent.

d. t_4: when the rocket is on its way back down to the ground.

2-47. Two rocks are dropped from rest and fall freely from the same horizontal railing. If the second falls for twice as much time as the first, how many times as far does it fall?

2-48. A penny dropped from the top of the No-Trump Tower would hit the pavement below in 6.8 s. How tall is the No-Trump Tower?

2-49. **SSM** A stunt man in a movie action scene is propelled straight up in the air with a velocity of 14.4 m/s.

a. How high does he go?

b. How long does he remain in the air, assuming he is in free fall until he reaches the ground again?

2-50. A rocket is launched upward from atop the edge of a cliff at $t = 0$. It has an initial velocity of 39.2 m/s. It just misses the edge of the cliff on the way down.

a. Find its position and velocity at 2-s intervals from $t = 0$ to $t = 10$ s.

b. How high above or below the edge of the cliff is the rocket when its velocity is zero?

2-51. For the motion described in Problem 2-50, sketch a graph of the rocket's

a. velocity versus time from $t = 0$ to $t = 10$ s.

b. position versus time from $t = 0$ to $t = 10$ s.

c. acceleration versus time from $t = 0$ to $t = 10$ s.

2-52. The gravitational acceleration on the moon is about one-sixth what it is at Earth's surface ($g_{moon} = \frac{1}{6}g_{Earth}$). Let t_{moon} be the time it takes a rock dropped from rest to hit the moon's surface, and t_{Earth} be the time for a rock dropped from rest from the same height to hit Earth's surface. Find the numerical value of the ratio $\frac{t_{moon}}{t_{Earth}}$.

Going Further

The questions and problems in this group are not organized by section heading, so you must determine for yourself which ideas apply. Some of them will be more challenging than the Review and Practice questions and problems (especially those marked with a • or ••).

2-53.

a. A lens mounted on a straight calibrated track is moved from a position of -0.30 m to a position of $+0.25$ m. Find its displacement.

b. Find the displacement of the lens when it is moved from a position of $+0.30$ m to a position of -0.25 m.

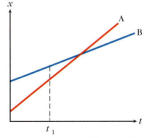

Figure 2-30 Problem 2-54

2-54. The positions of two bodies A and B are plotted over the same time interval in Figure 2-30. Is A or B moving faster at time t_1? Briefly explain. (This and the following problem are adapted from materials by Lillian C. McDermott and the University of Washington physics education group.)

2-55. Two bodies P and Q move along the same straight line. Their velocities are plotted over the same time interval in Figure 2-31. What information does this graph provide about whether P and Q collide? (See note on previous problem.)

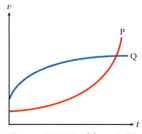

Figure 2-31 Problem 2-55

2-56. According to coordinate frame A (Figure 2-32), a sparrow flies from $x = 2$ m to $x = 5$ m at constant speed. Which of the following quantities change if we consider things from the point of view of coordinate frame B? The sparrow's initial position, the sparrow's final position, the sparrow's displacement, the sparrow's velocity. Explain.

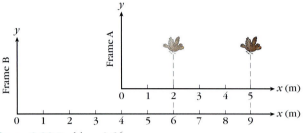

Figure 2-32 Problem 2-56

2-57. *Converting units.* Many U.S. highways have a speed limit of 65 miles/hour. Using the fact that 1 mile = 1609 m, what is this speed limit in SI units of m/s?

2-58. American baseball does not use SI units. A typical Major League pitcher throws a 90 mi/hr fastball and an 80 mi/hr change-up. The distance from the point where the ball leaves his hand to home plate is 55 feet. (Based on data from Robert Beck/ Allsport USA as published in G. F. Will, *Men at Work*, Macmillan, NY 1990.)

a. How much more time (in seconds) does a batter have to react to a change-up than to a fastball?

b. Suppose that the ball can be struck by the bat over about 2 feet of the pitched ball's path. How long a time interval does the bat have to make contact with a 90 mi/hr fastball?

2-59. Albatrosses, large seabirds of the Southern Hemisphere, have been known to circumnavigate the globe by gliding on prevailing winds. At 40° south latitude, the distance around Earth is about 3×10^7 m. Riding winds averaging 5 m/s (about 11 knots, actually light for this region), how long would it take an albatross to complete a trip around the world? Convert your result to days.

2-60. A marine biologist shooting a video of a rare species of fish uses the video data to produce the graph of the fish's position x plotted against clock reading t in Figure 2-33.

a. What is the fish's average velocity between $t = 0$ and $t = 8$ s?

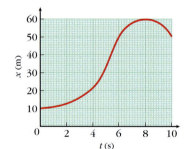

Figure 2-33 Problem 2-60

b. between $t = 8$ s and $t = 10$ s?

c. between $t = 0$ and $t = 10$ s?

d. Is your answer to **c** the average of your answers to **a** and **b**? Explain.

•**e.** What is the instantaneous velocity at $t = 8$ s?

•**f.** Estimate the instantaneous velocity at $t = 5$ s.

2-61. **SSM** A driver pulls out of his driveway, drives east at constant speed to pick up his daughter at school, and stops to chat with his daughter's teacher for a few minutes. Returning home, he must drive at a slower constant speed because westbound traffic is heavier. He finally pulls back into his driveway at time T. Which graph of velocity versus t in Figure 2-34 most accurately represents this round trip?

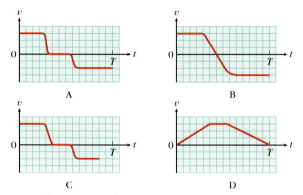

Figure 2-34 Problem 2-61

2-62. An assembly line worker runs to get a replacement part (100 m away) in 15 s and then takes 20 s to run back again. Find his average velocity **a.** for the first 100 m. **b.** for the second 100 m. **c.** for the total run. **d.** Do you get the answer to **c** if you just average your answers to **a** and **b**? Explain why, or why not.

•**2-63.** A dog starts at its owner's feet, runs to fetch a stick 20 m away, then returns the stick to its owner at a more leisurely pace. If the dog's average speed is 8.8 m/s on its way to the stick, and 6.6 m/s on its way back, what is its average speed for the entire round trip? (Assume that the time to pick up the stick and turn around is built into the two legs of the trip, so there's no additional time to account for.)

2-64. In Figure 2-35, the positions of three ants A, B, and C are plotted against clock reading t over a 25-s time interval.

a. List these ants in rank order according to their average velocity during this time interval. Start with the value (lowest or most negative) that would occur furthest to the left on a real number line. Make sure to indicate equalities if there are any.

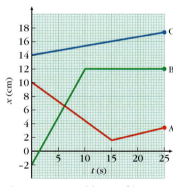

Figure 2-35 Problem 2-64

b. List these ants in rank order according to their average speed during this time interval. See directions in part **a.**

2-65.

a. Graph a in Figure 2-36a shows an object's velocity plotted against t. Sketch a graph of its position x plotted against t.

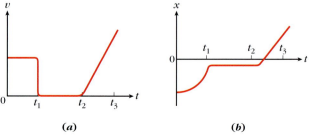

Figure 2-36 Problem 2-65

b. Repeat **a,** but this time assume the object is to the right of the origin at $t = 0$. (Did you make any assumption about where the object was at $t = 0$ when you answered part **a** the first time?)

c. Graph *b* in Figure 2-36*b* shows an object's position plotted against *t*. Sketch a graph of its *velocity* against *t*.

2-66. The velocity of a uniformly accelerated body increases from $v_1 = 8$ m/s to $v_2 = 14$ m/s over the interval from $t_1 = 2$ s to $t_2 = 4$ s. **a.** Find the body's acceleration. **b.** Find the body's initial velocity at $t = 0$.

2-67. SSM A laboratory cart starts out traveling to the right. Sketch a graph of its velocity *v* versus *t*
a. if its acceleration is as shown in Figure 2-37*a*.
b. if its acceleration is as shown in Figure 2-37*b*.

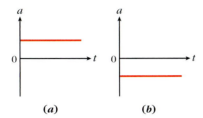

Figure 2-37 Problem 2-67

2-68. For each of the graphs shown in Figure 2-38, tell whether Equations 2-9 through 2-12 can be used to treat the motion that the graph represents and briefly explain your answer.

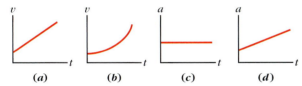

Figure 2-38 Problem 2-68

2-69. The two cars in Figure 2-39 start from rest at $t = 0$ and each accelerates at a constant rate until it crosses the finish line. If car A takes 20% longer than car B to cross the finish line,
a. how do the speeds at which the two cars cross the finish line compare? Express your answer as $v_A = \underline{\quad} v_B$. (Write the correct numerical multiplier in the blank.)
b. how do the accelerations of the two cars compare? Express your answer as $a_A = \underline{\quad} a_B$. (Write the correct numerical multiplier in the blank.)

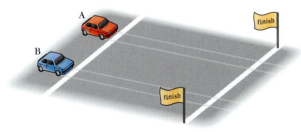

Figure 2-39 Problem 2-69

2-70.
a. When you take a spoonful of honey and then let the honey run off in an unbroken stream, the stream is narrower at the bottom. Why?
b. If you allow the stream to descend a great enough distance from the spoon, why do you get a break in the stream at that distance?

2-71. In each of the following cases, tell whether the object's position *x* doubles as the clock reading *t* doubles. Justify all your answers.
a. The object passes the origin at $t = 0$ traveling along the *x* axis at constant speed.
b. The object passes the point (3,0) at $t = 0$ traveling along the *x* axis at constant speed.
c. The object passes the point (0,3) at $t = 0$ traveling parallel to the *x* axis at constant speed.
d. The object starts from rest at the origin at $t = 0$ and has a constant acceleration in the positive *x* direction.
e. The object passes the origin at $t = 0$ traveling at constant speed along a straight line making an angle of 30° with the *x* axis.
f. The object is released from rest at the origin at $t = 0$ and is allowed to fall freely.
g. The object starts from rest at the origin at $t = 0$ and travels in the positive *x* direction. Its speed is proportional to its distance from the origin.

2-72. In each of the following cases, tell whether the object's velocity *v* doubles as the clock reading *t* doubles. Justify all your answers.
a. The object starts from rest at the origin at $t = 0$ and has a constant acceleration in the negative *x* direction.
b. The object passes the origin at a speed of 5 m/s at $t = 0$ and has a constant acceleration in the positive *x* direction.
c. The object starts from the point (5,0) at $t = 0$ and has a constant acceleration in the positive *x* direction.
d. The object is released from rest at the origin at $t = 0$ and is allowed to fall freely.
e. The object is thrown straight up from the origin with an initial speed (at $t = 0$) of 10 m/s. (Neglect air resistance.)

2-73. A car accelerates uniformly from rest to a speed of 20 m/s over a 6-s time interval, starting at $t = 0$.
a. What is the car's average (time-averaged) speed during this time interval?
b. At what instant is the car's instantaneous velocity equal to this average velocity?
c. Over the entire 6 s, the car travels a certain distance. At the instant described in **b,** will the car have traveled half of this distance, more than half this distance, or less half this distance? Answer this without doing any calculations, but briefly explain your reasoning.
d. Now calculate the distance the car will have traveled at the instant described in **b,** and also at the instant $t = 6$ s. Use these results to check your answer to **c,** and if your answer to **c** was not correct, try to figure out why.

2-74. A bus starts from rest and comes to a stop at a destination 30 miles away 1 hour later.
a. What is the minimum number of instants (zero? one? two? more?) at which the bus must have had an instantaneous speed of 30 mi/hr during this trip?
b. Is it possible to say when these instants occurred? Explain.
c. What is the minimum number of instants (zero? one? two? more?) at which the bus must have had an instantaneous velocity of 30 mi/hr during this trip?
d. Is it possible to say when these instants occurred? Explain.

• **2-75.** An auto race is held on a straight course. All motion is in the same direction. Race car A passes the starting line at $t = 0$ at a speed of 20 m/s.

a. If it is capable of an acceleration of 7.0 m/s^2, how far from the starting line will it be after 6 s?

b. At the distance found in **a,** race car B overtakes car A. If car B started from rest from the starting line just as car A passed it, what acceleration (assumed constant) did it maintain?

c. At a certain instant, the two cars had the same velocity. Was this instant before $t = 6$ s, just at $t = 6$ s, or after $t = 6$ s? (Try to answer this without doing any calculation.)

d. Now calculate the instant at which both cars had the same velocity. See if your result agrees with your answer to **c.**

2-76. Race car A passes the starting line at a speed of 24 m/s and over the next 6 s has a constant acceleration of 6.0 m/s^2. Race car B starts from rest from the starting line just as car A passes it and accelerates uniformly. At an instant 8 s later, the two cars have the same velocity.

a. On the same set of axes, sketch a graph of each car's velocity versus t.

b. Is there a feature of this graph that tells you at what time car B passes car A? If so, tell what this feature is.

c. Is there a feature of this graph that tells you at what time car B has the same velocity as car A? If so, tell what this feature is.

2-77. A proposed new unit of time is defined as the amount of time required for a sphere of solid lead 10^{-2} m in diameter to fall a distance of 1 m. How satisfactory is this as a unit of time? Explain.

2-78.

a. If a raindrop was freely falling from a cloud at an altitude of 3000 m, with what speed would it strike the ground? (Neglect variations in g.)

b. Is the free fall approximation a good one for raindrops? Explain.

2-79.

a. How far does a rock dropped from rest fall in 4 s?

b. What is its velocity after 4 s?

c. What is its average velocity over the 4 s interval?

d. At what distance below its starting point does the rock's instantaneous velocity have the value you found in **c**? Compare this to the distance you found in **a.**

2-80. In 1991, Carl Lewis set a new world record of 9.86 s for the 100-m dash. His "split times" for various parts of the 100 m distance were as follows (Data based on R. Myers, *The Physics Teacher,* **30,** 89, 1992):

0 to 10 m	1.88 s	50 to 60 m	0.84 s
10 to 20 m	1.08 s	60 to 70 m	0.84 s
20 to 30 m	0.92 s	70 to 80 m	0.83 s
30 to 40 m	0.89 s	80 to 100 m	1.73 s
40 to 50 m	0.84 s		

a. Why did the first 10 m take the longest time?

b. During which 10-m stretch was his average *acceleration* greatest? Estimate its value, and state any assumptions you make in doing your calculation.

2-81. Problem 2-80 gives data for Carl Lewis's record run of the 100-m dash.

a. During which 10 m stretch was his average speed greatest? What was its value?

b. What was his average speed between 50 and 60 m?

c. Estimate his average acceleration over this stretch. (Why can't you be sure it is an exact value?)

2-82. The form of an equation can itself provide information. Answer the following *without* doing any calculations. Two cars A and B are stationary until $t = 0$. From $t = 0$ forward, the positions are given by $x_A = 300 - 50t$ and $x_B = 800 + 40t$, when x is in meters and t is in seconds.

a. How do the speeds of the two cars compare? (*Choose the best answer.*)

A's speed is greater   B's speed is greater
The two speeds are equal   Not enough information to tell

b. Does either car ever pass the other? (*Choose the best answer and explain.*) Yes   No   Not enough information to tell

2-83. **SSM WWW** Answer the following *without* doing any calculations. Two cars A and B are stationary until $t = 0$. From $t = 0$ forward, their velocities are given by $v_A = 30 - 4t$ and $v_B = 20 + 4t$, when v is in m/s and t is in seconds.

a. How do the accelerations of the two cars compare? (*Choose the best answer.*)

A's acceleration is greater   B's acceleration is greater
The two accelerations are   Not enough information to
equal   tell

b. Does either car ever pass the other? (*Choose the best answer and explain.*) Yes   No   Not enough information to tell

2-84. Answer the following *without* doing any calculations. The position of a certain bus traveling north on a straight road is given by $x = 2t^2 + 15t - 300$, where x is in meters and t is in seconds. The origin is at the entrance to the bus depot.

a. What is the bus's acceleration?

b. Where is the bus at $t = 0$?

c. How fast is it going at that instant?

d. Which is the positive direction, north or south? How can you tell?

e. Repeat **b** for the case when the position of a bus traveling north is given by $x = 2t^2 - 15t - 300$.

f. Repeat **c** for the case when the position of a bus traveling north is given by $x = 2t^2 - 15t - 300$.

2-85. The positions of several objects are shown at 1-s intervals in Figure 2-40. Choose the (one or more) objects that can possibly have constant nonzero accelerations over the entire time period shown. Explain.

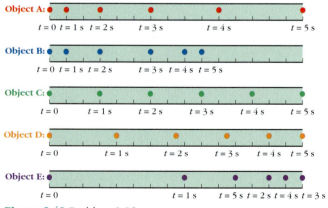

Figure 2-40 Problem 2-85

•**2-86.** Find the initial velocity and the acceleration of each of the objects in Figure 2-38 that has a constant acceleration.

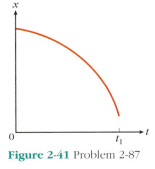
Figure 2-41 Problem 2-87

2-87. The motion of a laboratory cart over a certain interval of time is recorded on the position versus time graph in Figure 2-41. Is the cart speeding up or slowing down during this interval? Explain.

2-88. The motion of an airplane is represented by the position versus time graph in Figure 2-42. During which time interval is the plane's average velocity greater, the interval from $t = 2$ s to $t = 10$ s or the interval from $t = 2$ s to $t = 20$ s? Explain.

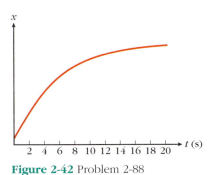
Figure 2-42 Problem 2-88

2-89. Graphs A and B in Figure 2-43 represent the motions of two runners along the same straightaway. Which of these runners has the greater acceleration? Explain your reasoning.

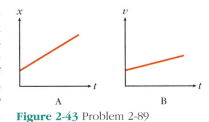
Figure 2-43 Problem 2-89

2-90. Figure 2-44 shows the trajectory of a ball thrown straight up in the air from near the bottom of a pit. The upward and downward parts of the trajectory actually are along the same straight line but are slightly shifted sideways in the diagram for readability. Fill in each space in the table below with +, 0, or − to indicate whether the position, velocity, and acceleration are positive, zero, or negative for the indicated parts of the trajectory.

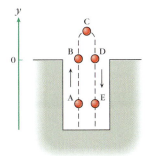
Figure 2-44 Problem 2-90

	y	**v**	**a**
While ball is going from A to B			
While ball is going from B to C			
When ball is at C			
While ball is going from C to D			
While ball is going from D to E			

2-91. Three used cars are available from a dealer. When you floor the accelerator of each, car A achieves an acceleration of 29 mi/hr², car B achieves an acceleration of 290 mi/hr², and car C achieves an acceleration of 29 000 mi/hr². Which car should you buy for normal highway driving? Explain.

2-92. Parts **a–e** are separate, unconnected situations. For the car in each situation, determine whether it is possible to calculate

• the average velocity over a 5-s interval
• the instantaneous velocity at $t = 5$ s
• both
• neither

a. Car A passes the 200-m mark on a test track at $t = 2$ s and the 260-m mark at $t = 7$ s.
b. Car B passes the 200-m mark at $t = 0$ and experiences a uniform acceleration of 15 m/s² for the next 5 s.
c. Car C passes the starting line at a speed of 20 m/s. It has a uniform acceleration of 2.4 m/s² for the next 5 s.
d. Car E passes the starting line at a speed of 20 m/s. It has an average acceleration of 2.4 m/s² for the next 5 s.
e. The driver of car E steps on the brake as he sees himself approaching the edge of a very high cliff. The car reaches a velocity negligibly greater than zero just as the car passes the cliff's edge. At this instant, the timer's stopwatch is started.

2-93. A rocket is launched upward from atop the edge of a cliff at $t = 0$. It has an initial velocity of 39.2 m/s. It just misses the edge of the cliff on the way down.
a. Find the two instants when it is 34.3 m above the cliff top.
b. What is its velocity at each of these instants?
c. The rocket lands at the base of the cliff 10 s after launching. How high is the cliff?

2-94. The correct time is 2:35 PM. Your watch reads 2:40 PM.
a. In ordinary language, would you say your watch is fast?
b. Is your watch necessarily running any faster than a watch that reads the correct time? How could you test it to see whether it is?

2-95. A newspaper article reports, "Federal earthquake relief in 1994 is credited with accelerating the state's financial recovery by a full fiscal quarter." Is the writer of this article using the idea of acceleration in a way that would be appropriate in physics? In particular, is the writer speaking of a change (an increment) or a rate of change? In physics, does acceleration refer to a change or a rate of change?

2-96. Redo Example 2-11 *without* using Equations 2-9 to 2-12, the constant acceleration equations of motion (or kinematic equations). (That means you cannot use Equations 2-13 to 2-16 either.) Instead, start using the definitions of average velocity and average acceleration ($\bar{v} = \frac{\Delta x}{\Delta t}$ and $\bar{a} = \frac{\Delta v}{\Delta t}$) and the fact that $\bar{v} = \frac{v_1 + v_2}{2}$ when the acceleration is constant.

2-97. The hand in Figure 2-45 rolls the ball up the ramp, releasing it at point A at $t = 0$. The ball gets as far as point B before coming back down.
a. Sketch a graph of the ball's position plotted against clock reading t.
b. Sketch a graph of the ball's velocity plotted against t.
c. Sketch a graph of the ball's acceleration plotted against t.

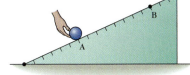
Figure 2-45 Problem 2-97

2-98. The hand in Figure 2-45 rolls the ball up the ramp, releasing it at point A at $t = 0$. The ball gets as far as point B before coming back down. If the marks on the ramp are 0.20 m apart,

a. find the ball's acceleration.

b. find the velocity with which the ball reaches the bottom of the ramp.

••**2-99.** A team of spelunkers (cave explorers) find an opening in the ground that seems to go straight down, but they cannot see how far. They need to find this out before risking a descent. So they drop a rock, and 3.2 s later they hear it hit bottom. Assuming that sound travels at a constant speed of 340 m/s, how far is the vertical drop?

2-100. Repeat Example 2-5, using the time interval from $t_1 = 1.75$ s to $t_2 = 2.25$ s.

2-101. Each of the phrases below describes one of the two graphs in Figure 2-46: negative slope, positive slope, quantity increasing with time, quantity decreasing with time,

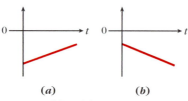

(a)   **(b)**

Figure 2-46 Problem 2-101

quantity becoming more negative over time, quantity becoming less negative over time. Tell which of the graphs each phrase describes.

2-102. On planet Zork, which has no atmosphere, space explorers discover a jet of water shooting out of the ground with a speed of 20.0 m/s. If the jet of water reaches a maximum height of 16.67 m, what is the gravitational acceleration near the surface of Zork?

2-103. A Web site on pest control warns that Norway rats can jump vertically more than 77 cm. To do so, the speed with which they propel themselves off the ground would have to exceed what value?

2-104. A bored engineer rigs his toaster to pop the toast up much higher than usual. If a slice of toast falls back down to the toaster 1.5 s later, how high does the toaster pop up? (Neglect air resistance and low ceilings.)

2-105. A certain vehicle requires a distance of 30 m to stop when it is traveling at 25 m/s.

a. What acceleration is produced by braking under these conditions?

b. If a jaywalker is 28 m in front of the vehicle when the driver first applies the brakes, how much time does the pedestrian have to get out of the way?

Problems on WebLinks

2-106. For the data in WebLink 2-1, we can also arrive at the instantaneous velocity at $t = 3.00$ s by considering progressively smaller intervals starting at this instant. Use the data from WebLink 2-1 to complete the following table, and from the table find the approximate instantaneous velocity at $t = 3.00$ s.

Interval between …	Δx (m)	Δt (s)	$\bar{v} = \dfrac{\Delta x}{\Delta t}$ (m/s)
$t = 3.00$ s and $t = 6$ s			
$t = 3.00$ s and $t = 4$ s			
$t = 3.00$ s and $t = 3.1$ s			
$t = 3.00$ s and $t = 3.01$ s			

2-107. Use the data from WebLink 2-1 to find the approximate instantaneous velocity at $t = 2.995$ s.

2-108. In the last graph shown on WebLink 2-2, ***a.*** is y proportional to x? ***b.*** is Δy proportional to Δx?

2-109. In the last graph shown on WebLink 2-2, ***a.*** is y proportional to x? ***b.*** is Δy proportional to Δx?

2-110. A table presented in WebLink 2-3 gives data for a beaker of water placed on a balance. For each volume of water in the beaker, the table gives the total mass on the balance.

a. What is the mass of the beaker?

b. Based on the data, calculate the density of water.

2-111. In the table discussed in Problem 2-109,

a. is the total mass m on the balance proportional to the volume V of the water in the beaker?

b. is Δm proportional to ΔV?

2-112.

a. In WebLink 2-4, is car C moving to the left or to the right over the 4-s interval shown?

b. Is the speed of car C increasing, decreasing, or not changing during this interval?

2-113.

a. In WebLink 2-4, is car D moving to the left or to the right over the 4-s interval shown?

b. Is the speed of car D increasing, decreasing, or not changing during this interval?

2-114. WebLink 2-5 shows the motion of a cart in each of the two situations shown in Figure 2-13. When the cart is in the situation shown in Figure 2-13b,

a. its velocity at $t = 2$ s is _____.
    positive and increasing;
    positive and decreasing;
    negative and increasing (becoming less negative);
    negative and decreasing (becoming more negative).

b. its velocity at $t = 6$ s is _____.
    positive and increasing;
    positive and decreasing;
    negative and increasing (becoming less negative);
    negative and decreasing (becoming more negative).

c. its acceleration at $t = 2$ s is _____ and its acceleration at 6 s is _____.
    positive . . . positive;
    positive . . . negative;
    negative . . . positive;
    negative . . . negative.

2-115. Repeat Problem 2-114 for the case when the cart is in the situation shown in Figure 2-13a.

CHAPTER THREE

Constructing Two-Dimensional Motion from One-Dimensional Motions

This chapter deals with a more varied selection of the motions found in the physical universe—from the graceful arc of a leaping dolphin to the corkscrew path of the tip of a boat propeller churning through the water, from the orbit of a telecommunications satellite or a planet to the dizzying gyrations you experience on a carnival ride.

In the previous chapter, we developed vocabulary and equations for describing motion along a straight line. In this chapter, building on the one-dimensional description, we develop the conceptual and mathematical tools you need for describing and analyzing the richer array of motions that occur in two or more dimensions.

"... From the graceful arc of a leaping dolphin ... to the dizzying gyrations you experience on a carnival ride."

3-1 Constructing Complex Motions from Simpler Motions

Complicated motions can be built up from two or more simpler motions. In the activity that follows, you will produce a two-dimensional motion by combining two one-dimensional motions.

On-The-Spot Activity 3-1

Part A Follow steps 1–3 of the figure. The marks left by your pen point show where the pen point has been, but the later marks cover over the earlier ones.

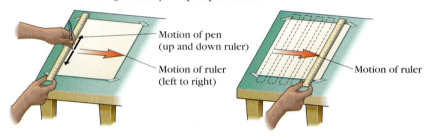

Part A
1. Fasten paper to flat surface (table, book, etc.)
2. Line up ruler with left edge of paper.
3. Run your pen up and down length of ruler several times.

Part B
4. Move ruler to the right *while* you repeat pen motion.

Motion of pen (up and down ruler)

Motion of ruler (left to right)

Motion of ruler

For **WebLink 3-1:**
Combining Two Motions, go to
www.wiley.com/college/touger

Part B Now repeat your pen motion while moving your ruler slowly to the right (see step 4 of the figure). This time the mark left by your pen provides a clearer record of the pen's motion. It is a composite motion, combining the purely vertical motion of your pen along the ruler with the purely horizontal motion of the ruler. In the language of physics, we say the actual motion has *a horizontal and a vertical component.* Go to WebLink 3-1 to explore these ideas further.

When motions combine, we sometimes say they have been **superimposed,** or that the resulting complex motion is a **superposition** of the simpler motions. We may also find it useful to speak of the complex motion as the **resultant** of two or more simpler **component** motions. Many kinds of complex motions can be built from simpler component motions.

Here are some real-world examples. You may have seen devices, in hospitals, laboratories, or elsewhere, in which an electronically controlled pen moves up and down a track, just as your pen moves along the ruler in Activity 3-1, to keep a record of some quantity that is varying over time (Figure 3-1). The height of the pen above its lowest point is proportional to the quantity being measured. While the pen goes up and down, the paper scrolls to the left, so that the record or *trace* left by the pen advances toward the right end of the paper. (This is accomplished by sliding your ruler to the right in Activity 3-1.)

Such devices are called *xy*-recorders because the trace is like a graph drawn on an *xy* coordinate frame. In a device called an *oscilloscope* (see Figure 3-2), a

(a)   (b)

Figure 3-6 **Why do these graphs look similar?** Compare (a) a plot of y versus x for the weight in Figure 3-5 and (b) a graph of y versus t for an object thrown vertically upward (reproduced from Figure 2-21a).

STOP&Think Compare this trajectory to the graph of y versus t for an object thrown vertically upward (Figure 3-6). Why do they look so similar? ◆

The motion graphed in Figure 3-6b, which we considered in Chapter 2, is purely vertical. But in plotting y against t, we represent values of t as displacements along the horizontal axis. For the flatbed driven at constant speed, on the other hand, the x axis of the graph of y versus x (Figure 3-6a) shows the truck's actual horizontal displacement. If we simply relabel the horizontal axis, replacing each horizontal position reading with the clock reading at which it occurs (we can do this when x and t are proportional), we end up with a graph of y versus t.

To explore the importance of Case 3-1 further, turn to WebLink 3-4.

Suppose a lit bulb was mounted on the weight as it followed the path in Figure 3-5b. Figure 3-7 shows what you would see if you watched the motion of the bulb on a dark night when only the bulb was visible. If you had no knowledge of the elaborate set-up producing its motion, what would you assume you were seeing? Probably just a glowing object—a flare or a firework perhaps—that someone had thrown or fired through the air (Figure 3-8). In short, it would look like just like some kind of *projectile*.

A **projectile** is an object thrown, launched, or otherwise projected so that once released, if air resistance is negligible, its path is affected only by Earth's gravitational attraction. A pop fly in baseball is an example of projectile motion.
STOP&Think Once the ball leaves the bat, does the bat have any influence on the subsequent motion of the ball? ◆

When we say a projectile is affected *only* by Earth's gravitational attraction, it means that the ball becomes a projectile only when it ceases to be in contact with the bat. We'll return to this point in Chapter 4.

For **WebLink 3-4:**
Carnival Game on Flatbed, go to
www.wiley.com/college/touger

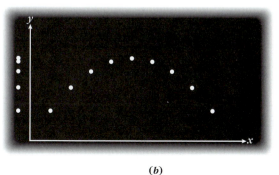

Figure 3-8 **Trajectories of fireworks.** A fireworks display shows a tracery of parabolic paths that have been altered somewhat by air resistance.

Figure 3-7 **Key ideas from Weblink 3-4.** (a) Combining the components of the weight's motion. (b) The trajectory of the illuminated weight as observed at night.

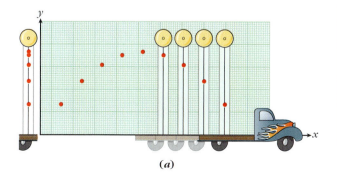

(a)   (b)

3-2   Breaking Down Two-Dimensional Motions into One-Dimensional Components: Projectile Motion

In Case 3-1 and also in WebLink 3-4, a two-dimensional motion was constructed from one-dimensional component motions. We could as easily have worked backward, starting with the resultant motion and breaking it down into its two component motions. It turns out that a powerful technique for analyzing *any* two-dimensional motion is to reduce it to two one-dimensional component motions. To see how to do this, we again focus our attention on projectile motion.

Picture a ball thrown into the air, but not straight up as in the previous chapter. The trajectory (Figure 3-9) of the ball, shown as it appears when air resistance is negligible, is two-dimensional. Watching the ball travel, you do not actually see the component parts of the motion separately, but it is possible to separate them *mentally*. Consider the following.

Case 3-2 ◆ *The Moving "Shadows" of a Thrown Ball*

Suppose that, as in Figure 3-9, you are standing in a room with your back to a wall as you throw the ball. Imagine that as the ball travels, it casts two idealized "shadows," one directly below it on the ground, the other directly behind it (and you) on the wall. (The shadows are idealized because real shadows wouldn't always be directly below and directly behind the ball.) If coordinate axes are drawn on the floor and wall, the two shadows mark the x and y coordinates of the ball itself. Mathematicians sometimes speak of the coordinates of a point as the *projections* of the point onto the x and y axes, just as, using two distant spotlights, we might *project* our two shadows. It follows that if you know the motions of the shadows, which are one-dimensional, you will know how the ball's coordinates change; that is, you will know the motion of the ball.

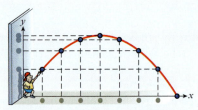

Figure 3-9 Analyzing the trajectory of a thrown ball. The ball's motion is tracked by the sequence of positions of its two shadows. The positions of the ball—and therefore of the shadows it casts on the wall and on the ground—are separated by equal time intervals.

In the next situation, we compare measured and calculated values for the coordinates. Figure 3-10 shows a multiple exposure flash photograph of a golf ball—the yellow one—that has been launched horizontally. Coordinate axes have been added to the photo, as have the ball's horizontal and vertical "shadows" for each exposure. On the photo, we can measure the positions x and y of these shadows. Table 3-1 records these measured values. The table shows that the measured y coordinates have the same values that we obtain by calculation using $y = -\frac{1}{2}gt^2$ for a body dropped vertically from rest. This is corroborated by the red ball in the figure, which was dropped vertically at the instant the yellow ball was launched, and has exactly the same vertical motion as the yellow ball. The table also shows that if we neglect a very small decrease due to air resistance, the shadow on the ground advances the same distance every $\frac{1}{30}$ s—in other words, it moves *at constant speed*.

These are the same component motions that we had for the flatbed—the motion of the weight along the slide wire affected only by gravitational acceleration, the horizontal motion provided only by the truck's constant speed. For the carnival game on the flatbed, you can run either component alone. It is harder to separate the golf ball's component motions in reality, but the shadows let us picture or *conceptualize* the two component motions separately. As Figures 3-7 and 3-9 show, the resulting two-dimensional trajectories are the same in

Table 3-1 Comparison of Calculated (Theoretical) and Measured Positions of Golf Ball in Figure 3-10

t (s)	Measured x (cm) (position of horizontal shadow)	Measured y (cm) (position of vertical shadow)	y coordinate of body dropped vertically from rest (calculated by $y = -\frac{1}{2}gt^2$ (cm))
0	0	0	0
first exposures of ball at $\frac{1}{30}$-s intervals not visible			
$\frac{4}{30}$	12.3	−8.8	−8.7
$\frac{5}{30}$	14.9	−13.2	−13.6
$\frac{6}{30}$	17.5	−18.6	−19.6
$\frac{7}{30}$	20.4	−26.0	−26.7
$\frac{8}{30}$	23.0	−33.4	−34.8
$\frac{9}{30}$	25.6	−42.8	−44.1
$\frac{10}{30}$	28.4	−53.0	−54.4
$\frac{11}{30}$	31.3	−64.6	−65.9
$\frac{12}{30}$	34.1	−77.2	−78.4

(Measured values are ±0.3 cm)

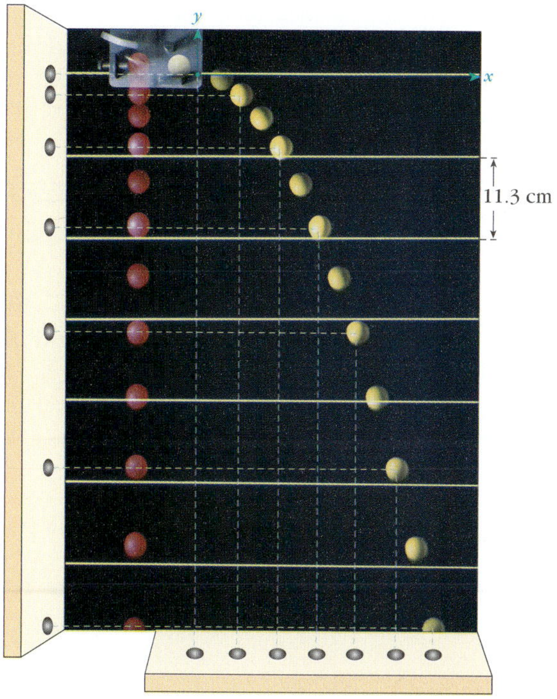

11.3 cm

Figure 3-10 Measuring the horizontal and vertical "shadow" motions of a ball that is launched horizontally. In the photo, the red ball is given no horizontal motion, but is released from rest at the same time that the yellow ball is launched horizontally. Note that both balls have the same vertical "shadow" motion. (The balls are shown at $\frac{1}{30}$ s intervals.)

the two cases. Because the object occupies the same sequence of positions in the two cases, *their mathematical descriptions are the same.*

In principle, *any* two-dimensional motion can be described by its two shadow motions. All we need now is a way of translating back and forth between the two shadow descriptions and the composite two-dimensional description. For this, we turn to the mathematics of vectors.

3-3 Vectors

◆**VECTORS IN ONE DIMENSION** In Figure 2-4, the displacements of two cyclists were shown as arrows. In Figure 3-11, we expand on this idea and use arrows to represent the cyclists' positions as well as their displacements. Each cyclist in the figure has an earlier position x_1 and a later position x_2. Because a position is a displacement from the origin ($\Delta x = x - 0 = x$), it also has a direction. For each of the cyclists, $x_2 = x_1 + \Delta x$, whether the values are positive (as for the cyclist going right) or negative (as for the cyclist going left). For each cyclist, we can draw a new arrow (x_2) from the *tail* of the first (x_1) to the *tip* of the second (Δx) to represent this addition.

A **vector** is a quantity that, like these arrows, must be described by giving its *direction* as well as the *magnitude* (numerical size). The arrows are one way of representing vectors. We denote a vector by placing an arrow above the symbol for the quantity. In our cyclist example, when we draw the arrow from the tail of the first vector to the tip of the second, we can write the addition that it represents as

$$\vec{\mathbf{x}}_2 = \vec{\mathbf{x}}_1 + \Delta\vec{\mathbf{x}} \qquad (3\text{-}1)$$

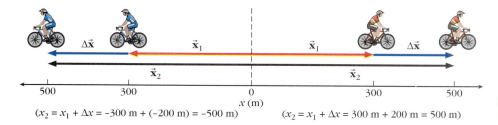

$(x_2 = x_1 + \Delta x = -300 \text{ m} + (-200 \text{ m}) = -500 \text{ m})$ $(x_2 = x_1 + \Delta x = 300 \text{ m} + 200 \text{ m} = 500 \text{ m})$

Figure 3-11 Positions and displacements of two cyclists.

➡**Notation:** Writing longhand, you will always use the arrow. In most printed material, the vector is printed in **bold** and the arrow omitted. For ease of recognition, this text will use both the arrow and boldface to identify vectors.

The procedure for combining any two vectors $\vec{A}$ and $\vec{B}$ to obtain a *vector sum* or *resultant* vector $\vec{R}$ is called **vector addition.** We always write $\vec{A} + \vec{B} = \vec{R}$ to represent this procedure, although in two dimensions it will be different than ordinary addition of numbers.

PROCEDURE 3-1

Picturing Vector Addition: The "Tail-to-Head" Method of Finding $\vec{R} = \vec{A} + \vec{B}$

1. Draw $\vec{A}$ to some scale (say, 1 cm represents 1 m) in the proper direction.
2. Draw $\vec{B}$ to the same scale in *its* proper direction, starting its tail at the tip of $\vec{A}$. This is called adding $\vec{B}$ to $\vec{A}$.
3. Draw the resultant $\vec{R}$ from the tail of $\vec{A}$ to the tip of $\vec{B}$.

Procedure 3-1 tells us how to draw the arrow representation of $\vec{R}$. Figure 3-12 shows this procedure for several examples. To find the magnitude or absolute value of $\vec{R}$, you would measure its length and use your scale to convert back to the original units.

Vector representation $(\vec{A} + \vec{B} = \vec{R})$		Representation in words	Arithmetic representation $(A + B = R)$
(a) 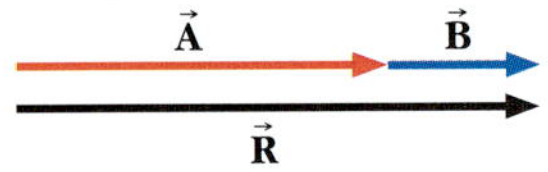	$\vec{A}$ $\vec{B}$ $\vec{R}$	A: Walk 5 m east B: ...then walk another 2 m east. R: End up 7 m east	5 m + 2 m = 7 m
(b) 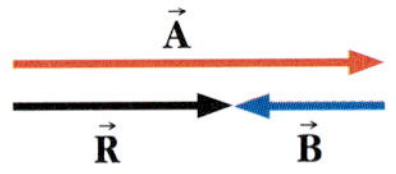	$\vec{A}$ $\vec{R}$ $\vec{B}$	A: Walk 5 m east B: ...then walk 2 m west. R: End up 3 m east	5 m + (-2 m) = 3 m
(c) 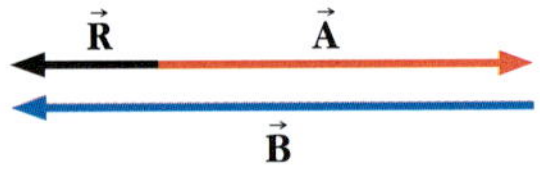	$\vec{R}$ $\vec{A}$ $\vec{B}$	A: Walk 5 m east B: ...then walk 7 m west. R: End up 2 m west	5 m + (-7 m) = -2 m
(d) 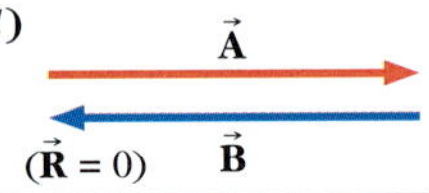	$\vec{A}$ $(\vec{R} = 0)$ $\vec{B}$	A: Walk 5 m east B: ...then walk 5 m west. R: End up back at origin	5 m + (-5 m) = 0

Figure 3-12 Examples of $\vec{A} + \vec{B} = \vec{R}$ in one dimension.

◆**NEGATIVES OF VECTORS** Note that in Figure 3-12*d*, $\vec{A} + \vec{B} = 0$. We can rewrite this as $\vec{B} = -\vec{A}$. Figure 3-12*d* then shows that the negative of a vector is a vector equal to it in magnitude but opposite in direction. As an example of this reasoning, we can rewrite Equation 3-1 (depicted in Figure 3-13*a*) as

$$\vec{x}_2 + (-\vec{x}_1) = \Delta\vec{x} \qquad \text{or simply as} \qquad \vec{x}_2 - \vec{x}_1 = \Delta\vec{x}$$

Subtracting a vector simply means *adding the negative of that vector.* This is shown in Figure 3-13*b*.

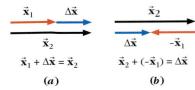

$\vec{x}_1 + \Delta\vec{x} = \vec{x}_2$ $\vec{x}_2 + (-\vec{x}_1) = \Delta\vec{x}$

(*a*)   (*b*)

Figure 3-13 Picturing two equivalent vector equations. In (*a*) we add $\Delta\vec{x}$ tail-to-head to $\vec{x}_1$. In (*b*) we take $-\vec{x}_1$, which is equal and opposite to $\vec{x}_1$, and add it tail-to-head to $\vec{x}_2$.

◆**VECTORS AND SCALARS** We have concentrated on positions and displacements as examples of vectors because they are easy to visualize. But velocities also must be described by both a magnitude and a direction in space, and this is also true of many other quantities we shall encounter. Such quantities are called **vector quantities,** and all can be treated mathematically by the same rules (such as Procedure 3-1). In contrast, quantities that do not have a spatial direction (such as mass or time) and can be described fully by a single number are often called **scalars** or **scalar quantities.**

◆**VECTORS IN TWO DIMENSIONS** In two dimensions, as in one, we can draw an arrow representing the magnitude and direction of a vector quantity. For example, the vector in Figure 3-14 indicates that Dallas is 3.6×10^5 m (224 miles) from Houston in a direction 22° W of N (west of north). Implicitly, we have picked the origin (Houston) and orientation (fixing the directions N, S, E, and W) of a coordinate system. The magnitude is 3.6×10^5 m; the direction is stated as an angle. Similarly, a pilot who reports "heading 30° N of E at 268 m/s [600 miles/hour]" is providing a magnitude and direction description of the plane's velocity.

We use the symbol $\vec{r}$ to denote a two- or three-dimensional position vector, and we use either r or $|\vec{r}|$ to denote its magnitude. As the absolute value bars indicate, the magnitude of a vector is always taken to be positive. In two dimensions, a vector's direction is given by its directional angle (Figure 3-15), so there is no need to use sign to indicate direction. As the figure shows, vectors in opposite directions are described by angles that differ by 180°.

Figure 3-14 A position vector for Dallas, if the origin is at Houston.

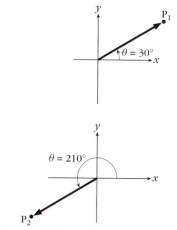

Figure 3-15 Position vectors with equal magnitudes but opposite directions.

PROCEDURE 3-2

Determining the Magnitude and Direction of the Position Vector of a Point in Two-Dimensional Space

1. Choose an xy coordinate frame (origin, orientation, and scale).
2. Draw the vector $\vec{r}$ as an arrow from the origin to the point in question.
3. Measure its length to find the magnitude ($|\vec{r}|$ or r).
4. Measure the angle the vector makes with the $+x$ axis. We designate this directional angle by θ (the Greek letter *theta*).

If you stand at the origin facing along the $+x$ axis, and then rotate counterclockwise (↺) until you are facing along the position vector, the angle through which you have rotated is θ. If you rotate clockwise (↻), the value of θ is negative.

◆**ABOUT COORDINATE FRAMES** Note that the first step in Procedure 3-2 is to choose a coordinate frame. As in one dimension (Figure 2-1), the description of a position depends on the choice of origin. In two dimensions, you are also free to choose the *orientation* of the x and y axes, as long as the axes remain perpendicular to each other. In Figure 3-16, if we choose coordinate frame A, the motions of the targets on the horizontal walkway have no component motion in the y direction. If we choose coordinate frame B, the targets on the ramp have no component motion in the y direction. If we want to analyze the motion of a particular object, we can often simplify our analysis by picking the frame in which the object has just one component motion. In making our choice, the axes need not be in the horizontal and vertical directions.

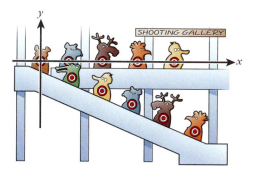

(**a**) A good choice for describing the targets on the horizontal walkway

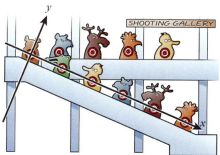

(**b**) A good choice for describing the targets on the ramp

Figure 3-16 Choosing a coordinate frame in two dimensions.
(*a*) A good choice for describing the targets on the horizontal walkway.
(*b*) A good choice for describing the targets on the ramp.

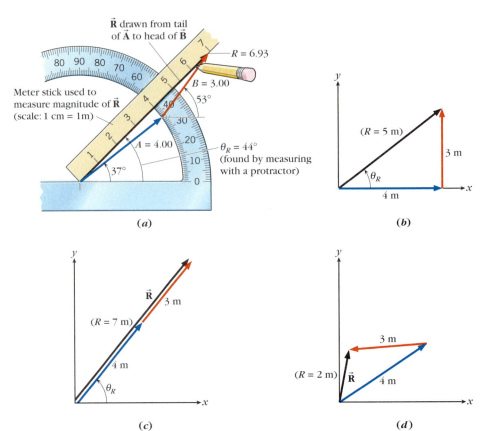

Figure 3-17 The resultant $\vec{R}$ of two vectors with magnitudes $A = 4$ and $B = 3$ is obtained by the tail-to-head method for different directions of $\vec{A}$ and $\vec{B}$.
(*a*) Shows how the resultant may be found for one such pair of vectors. (*b*)–(*d*) Show how the resultant may vary when $\vec{A}$ and $\vec{B}$ are in different directions.

◆**ADDING VECTORS IN TWO DIMENSIONS** Suppose you walk 4 m in a particular direction, then 3 m in another direction (Figure 3-17*a*). How far are you from your original starting point? in what direction? The figure shows the two successive displacements $\vec{A}$ and $\vec{B}$ and the resulting displacement $\vec{R}$ from the starting point. As in one dimension, $\vec{R}$ is the vector sum (or resultant) of $\vec{A}$ and $\vec{B}$ ($\vec{R} = \vec{A} + \vec{B}$). As Figure 3-17*a* illustrates, the resultant can be found by the drawing method of Procedure 3-1. WebLinks 3-5 and 3-6 will help you picture how this works in two dimensions.

Before you begin calculating the sums of vectors in the next section, you need a good mental picture of what vector addition means. You should be able to sketch roughly what the sum of any two vectors looks like. Even if the magnitudes of two vectors remain the same, both the magnitude and direction of their vector sum will change if the directions of the individual vectors change (Figures 3-17*b*–*d*). Note that you can get the magnitude of the resultant by doing ordinary arithmetic addition of the individual magnitudes only (as in Figure 3-17*c*) if $\vec{A}$ and $\vec{B}$ are in the same direction.

On-The-Spot Activity 3-2

In each step below, you will combine the two motions shown in the figure: (1) moving a ruler 6 cm to the right while keeping it parallel to the edge of the paper, and (2) moving your pen 8 cm up the ruler. What will differ from one step to the next is the order in which you execute these motions. But first, draw a line 6 cm in from the left edge of the paper, as shown, so you will know when your ruler has moved 6 cm.

a. Line the ruler up top-to-bottom on a sheet of paper, as you did in Activity 3-1. Place your pencil point against the ruler at a point near the bottom and label that point O. Run your pencil point 8 cm up the ruler, then, keeping the pencil point

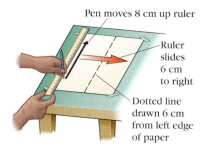

Pen moves 8 cm up ruler

Ruler slides 6 cm to right

Dotted line drawn 6 cm from left edge of paper

pressed against the ruler, slide the ruler 6 cm to the right. Mark the point on the paper where your pencil point ends up.

b. Now start at O again. This time, with the pencil point pressed against the ruler, you should *first* slide the ruler 6 cm to the right and *then* run your pencil point 8 cm up the ruler. Now where does the pencil point end up?

c. Finally, start at O and try to run your pencil point 8 cm up the ruler *while* you are sliding the ruler 6 cm to the right. Where does the pencil point end up this time?

In Activity 3-2, you should find that the pencil point ends up in the same place each time. If "8 cm upward" is $\vec{A}$ and "6 cm to the right" is $\vec{B}$, the first two trials illustrate that $\vec{A} + \vec{B}$ is the same as $\vec{B} + \vec{A}$ (see Figure 3-18). When $\vec{A}$ and $\vec{B}$ are concurrent (one occurs *while* the other does), you still end up in the same place. In that case, you can draw them originating from the same point (see Figure 3-18c). If you then complete the parallelogram (in this case a rectangle), you can again get the same resultant $\vec{R}$ by drawing the diagonal. It doesn't matter whether you call this $\vec{A} + \vec{B}$ or $\vec{B} + \vec{A}$; whether the vectors occur in one order or the other or concurrently, you get the same result.

Adding three vectors simply means adding a third vector to the resultant of the first two, and so on for additional vectors. Each time you add another vector to a previous resultant, you follow the same procedure.

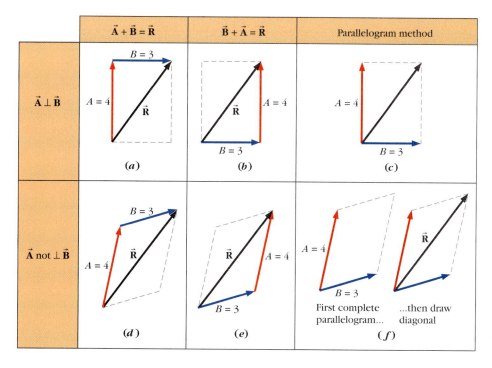

Figure 3-18 **Three ways of adding two vectors $\vec{A}$ and $\vec{B}$.**

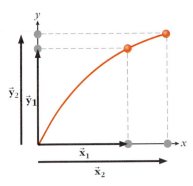

(a) Vector component description

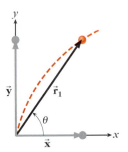

(b) Magnitude-and-direction description

Figure 3-19 The position vector changes along the trajectory of a projectile. (*a*) Vector component description. (*b*) Magnitude-and-direction description.

3-4 Working with Vector Components

◆**COMPONENTS OF A VECTOR** In Figure 3-7, you saw the one-dimensional component motions for the two-dimensional motion of a projectile. Figure 3-19*a* shows the position vectors for these two component motions. Now we draw in the first two-dimensional position vector $\vec{r}_1$ for the actual motion (Figure 3-19*b*). We then see that the position vectors for the component motions are actually the "shadows" or projections of the two-dimensional vector, just as the object itself casts idealized "shadows" on the two axes. They are called the **component vectors** of the two-dimensional vector. As an object moves, its position vector changes, and the two component vectors $\vec{x}$ and $\vec{y}$ change correspondingly.

The numerical values x and y (the coordinates of the objects) associated with the component position vector are called its **scalar components.** When we just speak of **components,** we will mean scalar components. The sign of a scalar component indicates the direction of the associated vector component. For example, in Figure 3-20*a*, where $\vec{x}$ is to the right and $\vec{y}$ upward, x and y are both positive. In Figure 3-20*b*, where $\vec{x}$ is to the left and $\vec{y}$ downward, x and y are both negative.

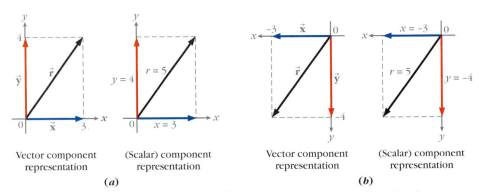

| Vector component representation | (Scalar) component representation | Vector component representation | (Scalar) component representation |

(a)   **(b)**

Figure 3-20 Component vectors $\vec{x}$ and $\vec{y}$, and scalar components (or simply components) x and y.

On-The-Spot Activity 3-3

Try drawing a position vector that has a negative x component and a positive y component. Repeat for a positive x component and a negative y component.

Example 3-1   *Finding the Position Vector of a Shipwreck*

The map in Figure 3-21 shows the location of the sunken fishing vessel *Maki.* For the coordinate frame in Figure 3-21*a*, use a protractor and a cm ruler to

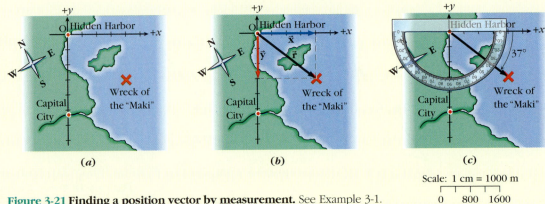

(a)   **(b)**   **(c)**

Scale: 1 cm = 1000 m

0   800   1600

Figure 3-21 Finding a position vector by measurement. See Example 3-1.

find (a) the magnitude and direction of the wreck's position vector and (b) the (scalar) components of its position vector.

Solution

a. Following Procedure 3-1, we draw an arrow (see Figure 3-21b) from the origin at Hidden Harbor to the wreck and measure its length to be 2.0 cm. Using the scale (1 cm represents 1000 m), the magnitude $r =$ **2000 m**. Remember—vector magnitudes are always positive.

If we place a protractor as shown in Figure 3-21c, we get a reading of 37°. But because we are measuring clockwise from the $+x$ axis, we must take this as negative, so that $\theta = $ **−37°**. As the figure shows, it is equally correct to say that $\theta = $ **+323°**.

b. The component vectors $\vec{\mathbf{x}}$ and $\vec{\mathbf{y}}$ have been drawn in Figure 3-21b. Their directions must be conveyed by the *signs* of the components x and y: x positive because $\vec{\mathbf{x}}$ is to the right, y negative because $\vec{\mathbf{y}}$ is downward. Thus the (scalar) components are $x = $ **1600 m** and $y = $ **−1200 m** (again using 1 cm to represent 1000 m, so that the marks on the axes are 400 m apart).

You should pay careful attention to the signs on all results in this example.

◆ Related homework: Problems 3-25 and 3-29.

Notice that a different choice of coordinate frame (Figure 3-22) changes both the wreck's two-dimensional position vector and its components.

Most of what we have said about position vectors holds true for any kind of vector, such as velocity, force, or other vector quantities that you will encounter later. To completely describe any vector $\vec{\mathbf{V}}$ in two dimensions, you must either

1. give its magnitude V (or $|\vec{\mathbf{V}}|$) and an angle θ representing orientation or direction, or

2. give the values of its components in the x and y directions. We speak of these as the **x and y components** of the vector, and we label them V_x and V_y. We call (1) the **magnitude and direction description** of the vector, and we call (2) the **component description**. In either case, the fact that the vector is two-dimensional means it requires two numbers for its complete description. In contrast, a *scalar* quantity such as time or density is fully described by a single number.

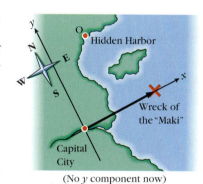
(No y component now)

Figure 3-22 A different choice of coordinate frame. Compare with Figure 3-21.

Summary of Notation for Vectors

- Vectors in general are represented in **bold** type (in printed texts) or with an arrow above. In this text we use both: $\vec{\mathbf{r}}$ for position, $\vec{\mathbf{F}}$ for force, etc. When writing longhand, use the arrow.

- The magnitude of a vector $\vec{\mathbf{V}}$ is represented by V (*not* in bold type) or $|\vec{\mathbf{V}}|$. The absolute value bars remind you that the magnitude is always positive.

- The directional angle is represented by θ. If a situation involves several vectors $\vec{\mathbf{A}}$, $\vec{\mathbf{B}}$, etc., their directional angles will be represented by θ_A, θ_B, etc.

- The component vectors of $\vec{\mathbf{V}}$ in the x and y directions are represented by $\vec{\mathbf{V}}_x$ and $\vec{\mathbf{V}}_y$.

- The x and y (scalar) components of a vector $\vec{\mathbf{V}}$ are represented by V_x and V_y. The x and y components may be positive or negative, depending on the vector's direction.

- The components of a point's position vector $\vec{\mathbf{r}}$ are usually written simply as x and y to remind you that these particular components are just the point's coordinates.

Magnitude-and-
direction description

Component
description

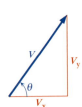

$$\sin \theta = \frac{\text{opp}}{\text{hyp}} = \frac{V_y}{V}$$

$$\cos \theta = \frac{\text{adj}}{\text{hyp}} = \frac{V_x}{V}$$

$$\tan \theta = \frac{\text{opp}}{\text{adj}} = \frac{V_y}{V_x}$$

$$V_x^2 + V_y^2 = V^2$$

**Figure 3-23 A right triangle
connects the magnitude-and-
direction description and the
component description.**

◆**CONVERTING BETWEEN DESCRIPTIONS** The next step is for you to learn to convert back and forth between the two vector descriptions. In Figure 3-23, the features of both the *magnitude and direction* and *component* descriptions are shown together. The resulting diagram is simply a right triangle with sides V_x and V_y and hypotenuse V (the magnitude or size of $\vec{V}$). We now apply the relationships that hold true for all right triangles. For vectors, the Pythagorean theorem becomes

$$V_x^2 + V_y^2 = V^2 \tag{3-2}$$

and the ratios of sides (see figure) become

$$\cos \theta = \frac{V_x}{V} \tag{3-3}$$

$$\sin \theta = \frac{V_y}{V} \tag{3-4}$$

$$\tan \theta = \frac{V_y}{V_x} \tag{3-5}$$

Multiplying both sides by V in Equations 3-3 and 3-4, you get expressions for V_x and V_y:

Finding the component description (V_x and V_y) of a vector from the magnitude-and-direction description:

$$V_x = V \cos \theta \tag{3-6}$$

$$V_y = V \sin \theta \tag{3-7}$$

(θ must be the angle the vector makes with the $+x$ axis; if a different angle is given, you must first find θ.)

➥**Using your calculator:** For the material that follows, make sure you are familiar with using the *sin* and *cos* keys of your calculator. Suppose you want to find cos 140°. Making sure that your calculator is dealing with angles in degrees (a typical calculator may have a *DRG* key that you can push until *DEG,* for degrees, appears on your readout display), key in the number 140 and then press the *cos* key. (The order of these last two steps is reversed on some calculators.) You should get cos 140° = −0.766.

◆**FINDING COMPONENTS BY INSPECTION** When a vector of known magnitude is purely in the (±) *x* or *y* direction, it is easier to find its components just by looking at it (called finding components *by inspection*) than by using the equations. For instance, a vector $\vec{V}$ of magnitude 5 directed toward the left points entirely in the negative *x* direction and has *no y* component, so $V_x = -5$ and $V_y = 0$.

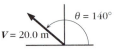

$\theta = 140°$

$V = 20.0$ m

(a) The given vector

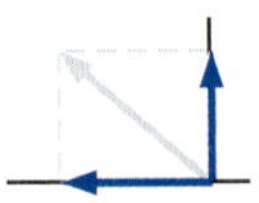

(b) its components

**Figure 3-24 Vector for Example
3-2.** (*a*) The given vector. (*b*) Its
components.

Example 3-2 *From Magnitude and Direction to Components*

The vector in Figure 3-24*a* is shown with its magnitude and direction description. Find its component description and draw it.

Solution
Calculating the components. Use your calculator to obtain values of cos 140° and sin 140°, and insert them in Equations 3-6 and 3-7:

$$V_x = V \cos \theta = (20.0 \text{ m}) \cos 140° = (20.0 \text{ m})(-0.766) = \mathbf{-15.3 \text{ m}}$$

and

$$V_y = V \sin \theta = (20.0 \text{ m}) \sin 140° = (20.0 \text{ m})(0.643) = \mathbf{12.9 \text{ m}}$$

Note that when you draw the component vectors (Figure 3-24*b*), V_x is to the left and V_y is upward, because the vector in Figure 3-25*a* is simultaneously to the left and upward.

◆ Related homework: Problem 3-30.

With Equations 3-2 and 3-5, you can calculate V and $\tan\theta$ (and thus find θ) if you know V_x and V_y. In other words,

to go *from a component description to a magnitude and direction description,* use

$$V^2 = V_x^2 + V_y^2 \qquad (3\text{-}2)$$

and

$$\tan\theta = \frac{V_y}{V_x} \qquad (3\text{-}5)$$

From the signs of V_x and V_y, decide what quadrant θ is in, then find θ from $\tan\theta$—that is, take the inverse function $\theta = \tan^{-1}\left(\frac{V_y}{V_x}\right)$.

Example 3-3 *From Components to Magnitude and Direction*

In Figure 3-25a, Ms. Mulkeen's window is 5.2 m to the left and 14.0 m up from the base of Firefighter Towle's ladder. In what direction should the ladder be angled and how far along it must Towle climb to rescue Ms. Mulkeen (assuming the truck must remain stationary)?

Solution

Interpreting the question. Taking the base of the ladder as the origin, the components of Ms. Mulkeen's position vector are $x = -5.2$ m and $y = 14.0$ m. The question is answered by finding the magnitude and direction of this vector. For position vectors, V_x becomes x, V_y becomes y, and V becomes r.

Magnitude. From Equation 3-1, it follows that

$$r = \sqrt{x^2 + y^2}$$

Substituting the given values, you get

$$r = \sqrt{(-5.2 \text{ m})^2 + (14.0 \text{ m})^2} = \sqrt{223.04 \text{ m}^2} = \mathbf{14.9\ m}$$

Direction. Applying Equation 3-4, we get

$$\tan\theta = \frac{14.0 \text{ m}}{-5.2 \text{ m}} = -2.69$$

To find θ, first key in -0.371, then press the tan^{-1} key (or the *inv* and *tan* keys; this varies from one calculator to another). Your calculator will tell you that the angle that has this tangent is $\theta = -69.6°$. Does this (Figure 3-25b) look right? Your calculator always gives an angle between $-90°$ and $+90°$ for the given tangent value. But the negative tangent value can represent either $\frac{\text{negative } y \text{ component}}{\text{positive } x \text{ component}}$, as in Figure 3-25b, or $\frac{\text{positive } y \text{ component}}{\text{negative } x \text{ component}}$, as in Figure 3-25c. The angles for the two situations are 180° apart. As Figure 3-25c shows, we want the situation where the ladder points upward to the left—a positive y component and a negative x component—so

$$\theta = -69.6° + 180° = \mathbf{110.4°}$$

♦ Related homework: Problem 3-31.

$V_y = 14.0$ m

θ

$V_x = 5.2$ m

(a)

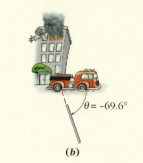

$\theta = -69.6°$

(b)

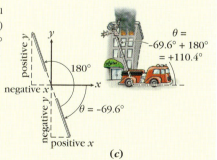

$\theta = -69.6° + 180° = +110.4°$

$180°$

positive y

negative y

negative x

positive x

$\theta = -69.6°$

(c)

Figure 3-25 Picturing Example 3-3. (*a*) The set-up. (*b*) A first attempt at a solution. (*c*) The corrected solution.

♦**CALCULATING VECTOR SUMS** In Section 3-3 we developed a tail-to-head picture of what vector addition means. We now develop a procedure for

For **WebLink 3-7:**
**Vector Addition
(Component Description),**
go to
www.wiley.com/college/touger

calculating the resultant or vector sum of two or more vectors. Figure 3-26 shows how the component description of vectors is involved in this procedure. To develop the idea in the figure in fuller detail, work through WebLink 3-7. The figure shows that the lengths of $\vec{A}_x$ and $\vec{B}_x$, placed end on end, add up to the length of $\vec{R}_x$. In other words, $A_x + B_x = R_x$. In the same way, $A_y + B_y = R_y$. To add scalar components, including those that are zero or negative, we do ordinary arithmetic addition.

Adding more than two vectors just means adding each new vector to the previous resultant. As Figure 3-27 shows, this just means arithmetically adding more components in each direction, so that in general,

$$A_x + B_x + C_x + \cdots = R_x \tag{3-8x}$$

and

$$A_y + B_y + C_y + \cdots = R_y \tag{3-8y}$$

(Individual components may be positive, zero, or negative.)

By doing these additions, we are in fact finding the vector $\vec{R}$, as given by its component description. We can think of the vector addition equation

$$\vec{A} + \vec{B} + \vec{C} + \cdots = \vec{R} \tag{3-8}$$

as a shorthand summary of the two arithmetic addition equations (3-8x, 3-8y).

But what if you want a magnitude-and-direction description of $\vec{R}$? As we saw earlier, once you know either description of $\vec{R}$, you can find the other. The component description, however, is better suited for calculating a resultant, because the actual steps of combining vectors (Equations 3-8x and 3-8y) require only simple arithmetic addition.

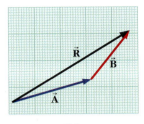

(a)

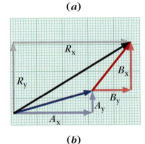

(b)

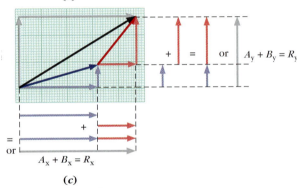

(c)

Figure 3-26 Developing a component description for the addition of two vectors. (*a*) Adding $\vec{A}$ and $\vec{B}$ by the tail-to-head method to obtain the resultant $\vec{R}$. (*b*) The *x* and *y* components are shown for each vector. (*c*) Adding the two vectors means we can add the components in each direction to get the components of the resultant.

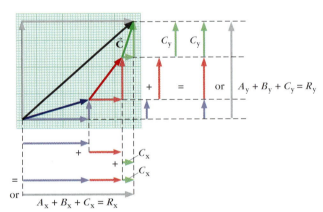

Figure 3-27 Adding three vectors. By the same method as in Figure 3-26, we add the third vector to the resultant of the first two. Again we find the components are additive.

If you start out with magnitude-and-direction descriptions of individual vectors, and you would like to calculate a magnitude and vector description of their resultant, there are three main steps:

PROCEDURE 3-3

Calculating the Resultant of Two or More Vectors by Components

1. Use

$$V_x = V \cos \theta \tag{3-6}$$

and

$$V_y = V \sin \theta \tag{3-7}$$

to obtain a component description of each given vector (or find components by inspection if the vector is purely in the $\pm x$ or $\pm y$ direction). (θ must be the angle the vector makes with the $+x$ axis; if a different angle is given, you must first find θ.)

2. Add components in each direction (Equations 3-8x and 3-8y) to obtain R_x and R_y, the components of the resultant.

3. Once you know R_x and R_y use

$$R_x^2 + R_y^2 = R^2 \qquad (3\text{-}2 \text{ with } \vec{V} = \vec{R})$$

and

$$\tan\theta = \frac{\text{opposite}}{\text{adjacent}} = \frac{R_y}{R_x} \qquad (3\text{-}5 \text{ with } \vec{V} = \vec{R})$$

to find the magnitude R and the directional angle θ_R of the resultant.

Example 3-4 *Calculating the Resultant of Two Vectors*

For a guided interactive solution, go to Web Example 3-4 at www.wiley.com/college/touger

A fishing trawler radios that after leaving harbor, it sailed 15 km in a direction $37°$ N of E, then headed due south for 21 km before its engines died. How far and in what direction must a Coast Guard vessel based in the harbor go to provide assistance?

(*a*) The given vectors   (*b*) Sketching $\vec{A} + \vec{B}$

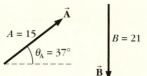

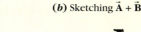

Brief Solution

Sketch a diagram. Sketch the two legs of the trawler's trip as vectors ($\vec{A}$ and $\vec{B}$ shown above). The course followed by the Coast Guard vessel should be their vector sum $\vec{R}$. Sketch the vector addition (shown above in *b*) carefully enough so you can anticipate what $\vec{R}$ should look like.

Now follow Procedure 3-3.

Convert to component description. Because $\vec{B}$ is downward, we can find its components by inspection. $\vec{B}$ has no x component and a negative y component, so $B_x = 0$ and $B_y = -21$. We verify this when we find the components of both vectors by applying Equations 3-6 and 3-7:

$A_x = A\cos\theta_A = 15\cos37° \;\; = (15)(0.80) = 12$	$A_y = A\sin\theta_A = 15\sin37° \;\; = (15)(0.60) = 9$
$B_x = B\cos\theta_B = 21\cos270° = (20)(0) \quad\;\; = \;\; 0$	$B_y = B\sin\theta_B = 21\sin270° = (21)(-1) \; = -21$
Add components in each direction $\qquad R_x = 12$	$R_y = -12$

Adding gives the resultant vector in component form.

Convert back to magnitude and direction description. Applying Equations 3-2 and 3-5 gives

$$R = \sqrt{R_x^2 + R_y^2} = \sqrt{(12)^2 + (-12)^2} = \mathbf{17}$$

to two significant figures, and $\tan\theta_R = \frac{R_y}{R_x} = \frac{-12}{12} = -1$, so that $\theta_R = \mathbf{-45°}$. (The signs on R_x and R_y tell you that $\vec{R}$ is indeed in the fourth quadrant.)

◆ Related homework: Problem 3-33.

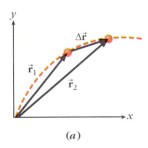

(a)

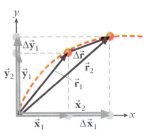

(b)

Figure 3-28 $\vec{r}_1 + \Delta\vec{r} = \vec{r}_2$.

◆**DISPLACEMENT VECTORS** Although it is perfectly correct to use $\vec{A}$, $\vec{B}$, and $\vec{R}$ to represent any two vectors and their resultant, we more commonly choose symbols that tell us what physical quantities the vectors represent—$\vec{v}$ for velocity, $\vec{r}$ for position, and so on. In two dimensions, Equation 3-1 becomes $\vec{r}_1 + \Delta\vec{r} = \vec{r}_2$. In Figure 3-28a, as in Figure 2-4, an object moves from the tip of the earlier position vector $\vec{r}_1$ to the tip of the later position vector $\vec{r}_2$. So $\Delta\vec{r}$ is a *change in position* vector; that is, a **displacement vector.**

$$\vec{r}_1 + \Delta\vec{r} = \vec{r}_2 \tag{3-9}$$

In words: earlier position + displacement = later position
                vector   vector   vector

The component vectors of a position vector $\vec{r}$ are simply labeled $\vec{x}$ and $\vec{y}$. If we draw perpendiculars to the axes to locate the "shadows", we see that

$$\vec{x}_1 + \Delta\vec{x} = \vec{x}_2 \quad \text{and} \quad \vec{y}_1 + \Delta\vec{y} = \vec{y}_2$$

Vector addition applies equally for the one-dimensional component motions that make up our two-dimensional motion. The component equations $A_x + B_x = R_x$ and $A_y + B_y = R_y$ now become

$$x_1 + \Delta x = x_2 \tag{3-9x}$$

$$y_1 + \Delta y = y_2 \tag{3-9y}$$

Example 3-5 *Calculating Another Resultant of Two Vectors*

For a guided interactive solution, go to Web Example 3-4 at www.wiley.com/college/touger

NASA scientists are giving the Mars rover *Sojourner* its first test run on the Martian surface. They set it in motion, sending it 40.0 m in a direction 15° N of E (north of east), and then another 20.0 m in a direction 70° N of W before stopping. Taking its initial deployment position as the origin, find the magnitude and direction of *Sojourner*'s position vector when it stops.

Brief Solution

Sketch a diagram. You should get roughly the diagram in Figure 3-29. Note that 70° N of W is 110° counterclockwise from the $+x$ direction. Now follow Procedure 3-3.

Convert to component description. For this we use Equations 3-6 and 3-7. $\vec{r}_1$ has components

$$x_1 = r_1 \cos\theta = (40.0 \text{ m}) \cos 15° = (40.0 \text{ m})(0.966) = 38.64 \text{ m}$$

$$y_1 = r_1 \sin\theta = (40.0 \text{ m}) \sin 15° = (40.0 \text{ m})(0.259) = 10.36 \text{ m}$$

$\Delta\vec{r}$ has components

$$\Delta x = \Delta r \cos\theta = (20.0 \text{ m}) \cos 110° = (20.0 \text{ m})(-0.342) = -6.84 \text{ m}$$

$$\Delta y = \Delta r \sin\theta = (20.0 \text{ m}) \sin 110° = (20.0 \text{ m})(+0.940) = +18.8 \text{ m}.$$

Add components in each direction. We apply Equations 3-9x and 3-9y.

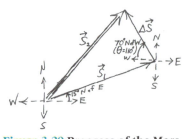

Figure 3-29 Progress of the Mars rover *Sojourner*. A vector addition diagram sketched for Example 3-5, and a NASA artist's rendition of *Sojourner* rumbling across the Martian landscape.

	Adding x components	*Adding y components*
components of $\vec{r}_1$	$x_1 = 38.64$ m	$y_1 = 10.36$ m
components of $\Delta\vec{r}$	$\Delta x = \underline{-6.84 \text{ m}}$	$\Delta y = \underline{18.8 \text{ m}}$
components of resultant $\vec{r}_2$	$x_2 = 31.8$ m	$y_2 = 29.2$ m

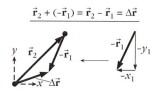

Convert back to magnitude and direction description. (by Equations 3-7 and 3-10)

$$r_2 = \sqrt{x_2^2 + y_2^2} = \sqrt{(31.8 \text{ m})^2 + (29.2 \text{ m})^2} = \textbf{43.2 m}$$

and

$$\tan \theta = \frac{y_2}{x_2} = \frac{29.2 \text{ m}}{31.8 \text{ m}} = 0.918 \quad \text{so that } \theta = \textbf{42.6°}$$

◆ Related homework: Problems 3-34 and 3-35.

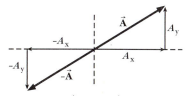

◆DISPLACEMENT VECTORS AND VECTOR SUBTRACTION As noted earlier, subtracting a vector means adding the negative of the vector (a vector with the same magnitude but opposite direction). From Equations 3-9x and 3-9y, it follows that

$$\Delta x = x_2 + (-x_1) = x_2 - x_1 \qquad (3\text{-}10\text{x})$$

and

$$\Delta y = y_2 + (-y_1) = y_2 - y_1. \qquad (3\text{-}10\text{y})$$

Just as Equation 3-9 summarizes Equations 3-9x and 3-9y, Equations 3-10x and 3-10y can be summarized by

$$\Delta \vec{\mathbf{r}} = \vec{\mathbf{r}}_2 + (-\vec{\mathbf{r}}_1) = \vec{\mathbf{r}}_2 - \vec{\mathbf{r}}_1. \qquad (3\text{-}10)$$

Figure 3-30 compares the vector addition diagrams for Equations 3-9 and 3-10. The vector $-\vec{\mathbf{r}}_1$ has components $-x_1$ and $-y_1$.

Some general rules that vectors obey are summarized below (see Figures 3-31 and 3-32):

Figure 3-30 Visualizing two equivalent vector equations.

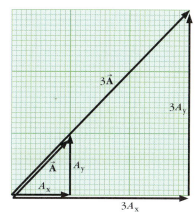

Figure 3-31 $\vec{\mathbf{A}}$ and $-\vec{\mathbf{A}}$.

Any vector $\vec{\mathbf{A}}$ obeys these rules

- The magnitudes of $\vec{\mathbf{A}}$ and $(-\vec{\mathbf{A}})$ are equal.
- The directions of $\vec{\mathbf{A}}$ and $(-\vec{\mathbf{A}})$ are opposite.
- The scalar components of $\vec{\mathbf{A}}$ and $(-\vec{\mathbf{A}})$ are equal in absolute value but have opposite signs.
- When you multiply or divide $\vec{\mathbf{A}}$ by a scalar c,
  —its components will also get multiplied or divided by c
  —its magnitude will be multiplied or divided by $|c|$, the absolute value of c.
  —the resulting vector will be in the same direction as the original vector if c is positive, and opposite if c is negative.

To develop a fuller understanding of these rules, work through WebLink 3-8.

It follows from these rules that multiplying a vector by a positive scalar only changes the *scale* of the vector (hence the term *scalar*).

Figure 3-32 Multiplying the vector $\vec{\mathbf{A}}$ by a scalar. The multiplication is depicted here for the scalar $c = 3$.

3-5 Velocity and Acceleration Vectors

In two dimensions, as in one, we can define an average velocity vector $\vec{\mathbf{v}}$ over a time interval Δt. In equation form, the

$$\textbf{average velocity} \quad \vec{\overline{\mathbf{v}}} \equiv \frac{\Delta \vec{\mathbf{r}}}{\Delta t} \qquad (3\text{-}11)$$

with **average velocity components** $\quad \bar{v}_x \equiv \dfrac{\Delta x}{\Delta t} \quad$ and $\quad \bar{v}_y \equiv \dfrac{\Delta y}{\Delta t} \quad \left(\begin{matrix} 3\text{-}11\text{x} \\ 3\text{-}11\text{y} \end{matrix}\right)$

You can find the average velocity of an object using a meter stick and a clock by following Procedure 2-1 to determine each of its components.

For **WebLink 3-8:**
Some Basic Rules for Vectors, go to
www.wiley.com/college/touger

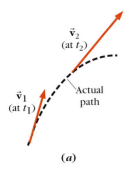

$\vec{v}_2$
(at t_2)

$\vec{v}_1$
(at t_1)

Actual
path

(a)

$\vec{v}_1 + \Delta \vec{v} = \vec{v}_2$
(equivalent to $\Delta \vec{v} = \vec{v}_2 - \vec{v}_1$)

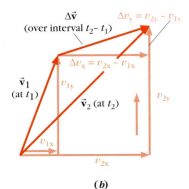

$\Delta \vec{v}$
(over interval $t_2 - t_1$)

$\Delta v_y = v_{2y} - v_{1y}$

$\Delta v_x = v_{2x} - v_{1x}$

$\vec{v}_1$
(at t_1)

v_{1y}

v_{2y}

$\vec{v}_2$ (at t_2)

v_{1x}

v_{2x}

(b)

Figure 3-33 Finding the change $\Delta \vec{v}$ in a velocity vector.

◆**INSTANTANEOUS VELOCITY** As in the one-dimensional case, the average components $\bar{v}_x$ and $\bar{v}_y$ close in on *instantaneous* values v_x and v_y as Δt is shrunk down to a single instant t. In two dimensions, **instantaneous velocity** is the vector $\vec{v}$ that has v_x and v_y as its components. If velocity vectors are not specifically labeled average, they are instantaneous.

Notice that v_x and v_y are the rates at which the coordinates of a point object are changing. In Case 3-2, they would be the velocities of the two "shadows" of the thrown ball. **STOP&Think** Look at the shadows in Figure 3-10. Is v_x increasing, decreasing, or remaining about the same? What about v_y? ◆

The **speed** of an object (also instantaneous unless otherwise stated) is defined to be the magnitude of its velocity vector. You calculate it much as you calculate any vector magnitude:

$$\textbf{speed} \qquad v \text{ or } |\vec{v}| = \sqrt{v_x^2 + v_y^2} \qquad (3\text{-}12)$$

By this definition, *two objects may be traveling at the same speed, but if they are going in different directions, their velocities are different.*

If an object's instantaneous velocity changes from $\vec{v}_1$ at instant t_1 to $\vec{v}_2$ at instant t_2 (Figure 3-33), the change in velocity $\Delta \vec{v} = \vec{v}_2 - \vec{v}_1$ is a vector as well. As always, the vector equations are summaries of the component equations in each direction:

$$\Delta \vec{v} = \vec{v}_2 - \vec{v}_1 \qquad (3\text{-}13)$$

$$\Delta v_x = v_{2x} - v_{1x} \qquad (3\text{-}13x)$$

$$\Delta v_y = v_{2y} - v_{1y} \qquad (3\text{-}13y)$$

Example 3-6 *Finding Your Change in Velocity*

**For a guided interactive solution, go to Web Example 3-6 at
www.wiley.com/college/touger**

On an amusement park ride, your velocity changes from $\vec{v}_1$ to $\vec{v}_2$ (see part *a* of figure). Find the magnitude and direction of your change in velocity $\Delta \vec{v}$.

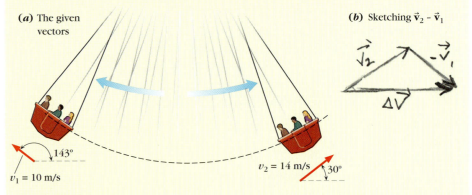

(a) The given vectors

(b) Sketching $\vec{v}_2 - \vec{v}_1$

$v_1 = 10$ m/s   143°

$v_2 = 14$ m/s   30°

Brief Solution
Doing the subtraction $\Delta \vec{v} = \vec{v}_2 - \vec{v}_1$ means adding $-\vec{v}_1$ to $\vec{v}_2$. We sketch this addition $\vec{v}_2$ (part *b* of figure) to anticipate what $\Delta \vec{v}$ should look like. Then we apply Procedure 3-3 to calculate its magnitude and direction.

Convert to component description. By Equations 3-6 and 3-7, $\vec{v}_1$ has components

$$v_{1x} = v_1 \cos \theta = (10 \text{ m/s}) \cos 143° = (10 \text{ m/s})(-0.80) = -8.0 \text{ m/s}$$

$$v_{1y} = v_1 \sin \theta = (10 \text{ m/s}) \sin 143° = (10 \text{ m/s})(0.60) = 6.0 \text{ m/s}$$

and $\vec{v}_2$ has components

$$v_{2x} = v_2 \cos \theta = (18 \text{ m/s}) \cos 30° = (18 \text{ m/s})(0.87) = 15.6 \text{ m/s}$$

$$v_{2y} = v_2 \sin \theta = (18 \text{ m/s}) \sin 30° = (18 \text{ m/s})(0.50) = 9.0 \text{ m/s}$$

Because Equations 3-12x and 3-12y are equivalent to Equations 3-13x and 3-13y, step 2 of Procedure 3-3 becomes . . .

Subtract components in each direction.

Subtracting x components:	Subtracting y components:
$v_{2x} = \quad 15.6 \text{ m/s}$	$v_{2y} = \quad 9.0 \text{ m/s}$
$-v_{1x} = \underline{-(-8.0 \text{ m/s})}$	$-v_{1y} = \underline{-6.0 \text{ m/s}}$
$\Delta v_x = v_{2x} - v_{1x} = \quad 23.6 \text{ m/s}$	$\Delta v_y = v_{2y} - v_{1y} = \quad 3.0 \text{ m/s}$

This gives us the components of $\Delta\vec{v}$.

Convert back to magnitude and direction description. (Equations 3-7 and 3-10)

$$\Delta v = \sqrt{(\Delta v_x)^2 + (\Delta v_y)^2} = \sqrt{(23.6 \text{ m/s})^2 + (3.0 \text{ m/s})^2} = \textbf{23.8 m/s}$$

and

$$\tan \theta = \frac{\Delta v_y}{\Delta v_x} = \frac{3.0 \text{ m/s}}{23.6 \text{ m/s}} = 0.13, \text{ so that } \theta = \textbf{7.4°}$$

◆ Related homework: Problem 3-45.

STOP&Think If you experienced the velocity change that we calculated in Example 3-6, how would it feel? Would it feel different if it occurs over a 5-s interval than over a 0.5-s interval? ◆

  The following case addresses this question.

For **WebLink 3-9: Circle,**
go to
www.wiley.com/college/touger

Case 3-3 ◆ *Two Direction Reversals*

The blue car in Figure 3-34a travels halfway around a traffic circle at a speed of 10 m/s (about 22 miles/hour). The red car in Figure 3-34b is traveling at 10 m/s, and then is propelled backward at 10 m/s after colliding with an oncoming truck. In both cases, the direction is reversed, so the velocity has changed even though the initial and final speeds are equal. Since $\vec{v}_2 = -\vec{v}_1$ (Figure 3-34c), $\vec{v}_1 + \Delta\vec{v} = -\vec{v}_1$. Solving for $\Delta\vec{v}$ then gives $\Delta\vec{v} = -\vec{v}_1 - \vec{v}_1 = -2\vec{v}_1$. In both cases $\Delta\vec{v}$ has the same magnitude, 20 m/s. But for the blue car, the change is gradual, occurring over a few seconds; for the red car it is sudden, occurring over a small fraction of a second. What you experience differently in the two cases is the *rate* of change $\frac{\Delta\vec{v}}{\Delta t}$, which—as in one dimension—we will call acceleration.

  To understand more fully what happens to the velocity of an object going around a circle at constant speed, work through WebLink 3-9.

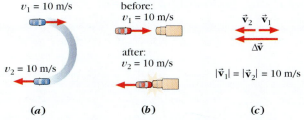

Figure 3-34 Changes in velocity during reversals of direction.

(a) $v_1 = 10$ m/s, $v_2 = 10$ m/s

(b) before: $v_1 = 10$ m/s; after: $v_2 = 10$ m/s

(c) $\vec{v}_2$ $\vec{v}_1$, $\Delta\vec{v}$, $|\vec{v}_1| = |\vec{v}_2| = 10$ m/s

(a)   (b)

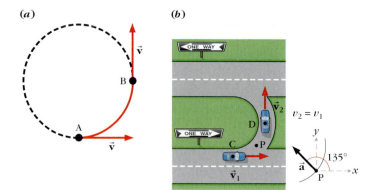

Figure 3-35 Traveling in a circle at constant speed. Is the velocity changing? Is the car accelerating? (Point P is halfway along the arc connecting positions C and D.)

◆**ACCELERATION IN TWO DIMENSIONS** When the velocity $\vec{v}$ of a body changes (Equation 3-12) over a time interval Δt, its components in each direction correspondingly change (Equations 3-12x and 3-12y). Then by Procedure 2-2, we can find an *average acceleration* in each direction:

$$\bar{a}_x \equiv \frac{\Delta v_x}{\Delta t} \quad \text{and} \quad \bar{a}_y \equiv \frac{\Delta v_y}{\Delta t} \quad \text{(3-14x, 3-14y)}$$

This procedure defines the

average acceleration vector $\qquad \vec{\bar{a}} \equiv \frac{\Delta \vec{v}}{\Delta t} \qquad$ (3-14)

by prescribing the steps necessary to find its components. As in one dimension, when you calculate $\Delta \vec{v} = \vec{v}_2 - \vec{v}_1$, $\vec{v}_1$ and $\vec{v}_2$ must be *instantaneous* velocities. **STOP&Think** By this definition, is an object accelerating if it travels along a circular path at constant speed? Jot down your answer and your reasoning.

Did you recall that constant speed means only that the *magnitude* of the velocity is constant, not the direction? Does that alter your answer?

Now consider Figure 3-35a, showing the path of this object. The velocity vectors of the object at points A and B are shown equal in magnitude, as required. Is the horizontal component v_x the same for these two vectors? Is Δv_x equal to zero? And thus, is $\bar{a}_x = \frac{\Delta v_x}{\Delta t}$ equal to zero? What about Δv_y and $\bar{a}_y$? Once more: Is an object accelerating if it travels along a circular path at constant speed? If this is still unclear, it may help to look again at WebLink 3-9. ◆

If $\Delta \vec{v} \neq 0$, the average acceleration vector $\vec{\bar{a}} = \frac{\Delta \vec{v}}{\Delta t}$ is also nonzero and is in the same direction as $\Delta \vec{v}$. There is a change $\Delta \vec{v}$ in the velocity as long as either the magnitude or the direction of the velocity changes, and the direction continuously changes when an object travels in a circle at constant speed.

Example 3-7 *Calculating the Average Acceleration of a Car Following a Circular Path*

Figure 3-35b shows a car taking a semicircular U-turn at constant speed. Its velocity vectors at points C and D, 90° apart, are also shown. Suppose its speed is 10.0 m/s and it takes 4.0 s to get from C to D. Find the components of the average acceleration over this time interval, and then find the magnitude and direction of the average acceleration for the same interval.

Solution

Choice of approach. You cannot be sure you are calculating acceleration correctly unless you are clear on its definition. Once you find its components from the definition, you can convert to a magnitude and direction description.

What you know/what you don't.

$v_{1x} = 10.0$ m/s and $v_{1y} = 0$ (because at point C, the velocity is entirely in the $+x$ direction)

$v_{2x} = 0$ and $v_{2y} = 10.0$ m/s   (because at point D, the velocity is entirely in the $+y$ direction)

$\Delta t = 4.0$ s (so no need to know t_1 and t_2) $\Delta v_x = ?$ $\Delta v_y = ?$

$\bar{a}_x = ?$ $\bar{a}_y = ?$ $\bar{a} = ?$ $\theta = ?$

The mathematical solution. We can now follow Procedure 2-2:

$$\Delta v_x = v_{2x} - v_{1x} = 0 - 10.0 \text{ m/s} = -10.0 \text{ m/s} \qquad (\Delta \vec{v}_x \text{ is to the left})$$

and

$$\Delta v_y = v_{2y} - v_{1y} = 10.0 \text{ m/s} - 0 = 10.0 \text{ m/s} \qquad (\Delta \vec{v}_y \text{ is upward})$$

So

$$\bar{a}_x = \frac{\Delta v_x}{\Delta t} = -\frac{10.0 \text{ m/s}}{4.0 \text{ s}} = \mathbf{-2.5 \text{ m/s}^2}$$

and

$$\bar{a}_y = \frac{\Delta v_y}{\Delta t} = \frac{10.0 \text{ m/s}}{4.0 \text{ s}} = \mathbf{-2.5 \text{ m/s}^2}$$

Making sense of the result. Like $\Delta \vec{v}_x$, $\vec{a}_x$ is to the left, because multiplying or dividing a vector by a positive scalar such as Δt doesn't change its direction. Likewise, $\vec{a}_y$ must be upward because $\Delta \vec{v}_y$ is upward.

Now apply step 3 of Procedure 3-3 to find $\bar{a}$ and θ:

$$\bar{a} = \sqrt{\bar{a}_x^2 + \bar{a}_y^2} = \sqrt{(-2.50 \text{ m/s}^2)^2 + (2.50 \text{ m/s}^2)^2}$$

$$= \mathbf{3.54 \text{ m/s}^2}$$

Important: Although the speed does not change, the velocity does because the direction is changing, so the value we get for the acceleration is *not* zero.

$$\tan \theta = \frac{\bar{a}_y}{\bar{a}_x} = \frac{2.50 \text{ m/s}^2}{-2.50 \text{ m/s}^2} = -1.00$$

This corresponds to $\boldsymbol{\theta = 135°}$

in the second quadrant (and *not* $-45°$) because the signs on its components tell us that $\vec{a}$ points up and to the left. The instantaneous acceleration $\vec{a}$ has its average value $\vec{a}$ at the midpoint P of the car's path; at this point (see inset in Figure 3-35), the direction $\theta = 135°$ is *toward the center of the circle.*

◆ Related homework: Problems 3-44 and 3-47.

STOP&Think Without doing any further calculation, what is the direction of $\Delta \vec{v}$ in Example 3-7? How do you know? If you are not sure, try this On-The-Spot Activity: From the head of $\vec{v}_1$ in Figure 3-36, draw the vector $\Delta \vec{v}$ that, when added to $\vec{v}_1$ in this way, gives you the correct $\vec{v}_2$ as the vector sum. ◆

◆**INSTANTANEOUS ACCELERATION** As was true in one dimension, $\bar{a}_x$ and $\bar{a}_y$ (averages over an interval Δt) approach limiting values—a_x and a_y—when the interval shrinks to a single instant t. The **instantaneous acceleration** $\vec{a}$ is the vector that has a_x and a_y as components at that instant. (In the previous example, the instantaneous values a_x and a_y at point P are equal to the average values $\bar{a}_x$ and $\bar{a}_y$ for the trip from C to D.)

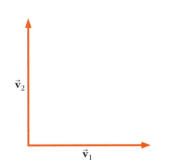

Figure 3-36 Complete the vector diagram so that it shows the vector addition $\vec{v}_1 + \Delta \vec{v} = \vec{v}_2$. Draw the vector $\Delta \vec{v}$ so that the resultant $\vec{v}_2$ goes from the tail of the first vector $\vec{v}_1$ to the head of the second vector $\Delta \vec{v}$.

3-6   Solving Motion Problems in Two Dimensions: Projectile Motion Revisited

The mathematics of vectors lets us separate a two-dimensional motion problem into problems about one-dimensional component or "shadow" motions, to which we can apply the methods of Chapter 2.

	In one dimension we used . . .	In two dimensions we now use . . .	
Definition of average velocity	$\bar{v} \equiv \dfrac{\Delta x}{\Delta t}$	$\bar{v}_x \equiv \dfrac{\Delta x}{\Delta t}$ and $\bar{v}_y \equiv \dfrac{\Delta y}{\Delta t}$	(3-11x, y)
Definition of average acceleration	$\bar{a} \equiv \dfrac{\Delta v}{\Delta t}$	$\bar{a}_x \equiv \dfrac{\Delta v_x}{\Delta t}$ and $\bar{a}_y \equiv \dfrac{\Delta v_y}{\Delta t}$	(3-14x, y)
Condition for uniform motion	$\bar{v} = \dfrac{v_{\mathrm{o}} + v}{2}$	$\bar{v}_x = \dfrac{v_{\mathrm{ox}} + v_x}{2}$ and $\bar{v}_y = \dfrac{v_{\mathrm{oy}} + v_y}{2}$	(3-15x, y)

Alternatively, when the acceleration is constant, we can use the set of equations (2-9, 2-11, and 2-12) that we derived for that special case (as in Solution 2 of Example 2-11). However, now we need a set of equations for each of the two one-dimensional component motions.

Equations of Motion in Two Dimensions for Constant Acceleration

$$x - x_{\mathrm{o}} = v_{\mathrm{ox}}t + \tfrac{1}{2}a_x t^2 \qquad \text{(3-16x)}$$
$$v_x = v_{\mathrm{ox}} + a_x t \qquad \text{(3-17x)}$$
$$v_x^2 = v_{\mathrm{ox}}^2 + 2a_x(x - x_{\mathrm{o}}) \qquad \text{(3-18x)}$$

at the same instant t as

$$y - y_{\mathrm{o}} = v_{\mathrm{oy}}t + \tfrac{1}{2}a_y t^2 \qquad \text{(3-16y)}$$
$$v_y = v_{\mathrm{oy}} + a_y t \qquad \text{(3-17y)}$$
$$v_y^2 = v_{\mathrm{oy}}^2 + 2a_y(y - y_{\mathrm{o}}) \qquad \text{(3-18y)}$$

The equations of motion in each direction involve only the components in that direction. They become simpler when you can assume the object is at the origin at $t = 0$, so that $x_{\mathrm{o}} = 0$ and $y_{\mathrm{o}} = 0$. They are also simpler if you choose your coordinate system so that one axis is in the direction of the acceleration. **STOP&Think** Why? How do the equations become simpler? We do this now for projectile motion. ◆

◆**PROJECTILE MOTION** Figures 3-7 and 3-9 (and also WebLink 3-4) showed that a projectile (such as a thrown ball) follows the same kind of trajectory as the weight in the truck-mounted carnival game, assuming that we can neglect air resistance. Like the constant horizontal velocity component provided by the truck, the projectile's horizontal component v_x is constant (recall Figure 3-10), so $a_x = 0$. In both cases, $a_y = -g$ (the $+y$ direction is upward). So the constant acceleration equations become

---

horizontal motion		vertical motion
$x - x_{\mathrm{o}} = v_{\mathrm{ox}}t$	at the same instant t as	$y - y_{\mathrm{o}} = v_{\mathrm{oy}}t - \tfrac{1}{2}gt^2$
$\cancel{v_x = v_{\mathrm{ox}}}$		$v_y = v_{\mathrm{oy}} - gt$
$\cancel{v_x^2 = v_{\mathrm{ox}}^2}$		$v_y^2 = v_{\mathrm{oy}}^2 - 2g(y - y_{\mathrm{o}})$

When $a_x = 0$, the terms with a_x disappear. The equations for v_x and v_x^2 then just say that the horizontal velocity v_x at any instant remains equal to the initial horizontal velocity v_{ox}. (We cross these out to indicate they provide no new information.)

If you solved for t in the first of the horizontal motion equations, giving you $t = \dfrac{x - x_{\mathrm{o}}}{v_{\mathrm{ox}}}$, and substituted this into the first of the vertical motions, you would find that y is equal to a quadratic expression in x, such as $ax^2 + bx + c$, where

a, b, and c are constants. The graph of a quadratic expression is a *parabola*. A graph of y versus x shows the two-dimensional path that the object actually follows. If you are familiar with parabolas from your math courses, you should recognize that the paths followed by the objects in Figures 3-5b and 3-9 are parabolic.

As an object travels along a two-dimensional trajectory, its position and velocity vectors $\vec{r}$ and $\vec{v}$ vary. As they do so, their x and y components likewise vary. Each of Equations 3-16x through 3-18y is a relationship among the vector components in a single direction. To use these equations, you must first have a component description of each vector.

PROCEDURE 3-4

A Strategy for Solving Motion Problems

1. In identifying what you know and what you don't, give a component description of each vector. Use Equations 3-6 and 3-7 to find components when necessary.
2. In your solution, use separate equations for the motions in the x and y directions. This lets you solve for the components of a vector quantity.
3. [*if necessary*] You can find the magnitude and direction of a vector quantity by applying Equations 3-2 and 3-5 once you know its components.

Example 3-8 *A Symmetrical Trajectory*

For a guided interactive solution, go to Web Example 3-5 at www.wiley.com/college/touger

A test rocket, due to technical failure, experiences no further thrust after leaving a ground-level launching pad at a speed of 44.0 m/s and at an angle of 63° above the ground. Find x, y, v_x, v_y, and the speed v at 1-s intervals until the rocket lands. Then plot x, y, v_x, and v_y against time to see graphically the symmetry in the numerical results.

Brief Solution
The mathematical solution. First you must find the components of v_o.

$$v_{ox} = v_o \cos 63° = (44.0 \text{ m/s})(0.454) = 20.0 \text{ m/s}$$

$$v_{oy} = v_o \sin 63° = (44.0 \text{ m/s})(0.891) = 39.2 \text{ m/s}$$

The equations and values are presented in tabular form to make patterns in the numbers more evident and to facilitate graphing. The numerical calculations that lead to the results are displayed for one representative instant ($t = 2$ s).

t (s)	$x = v_{ox}t$ (m)	$y = v_{oy}t - \frac{1}{2}gt^2$ (m)	$v_x = v_{ox}$ (m/s)	$v_y = v_{oy} - gt$ (m/s)	$v = \sqrt{v_x^2 + v_y^2}$ (m/s)
0	0	0	20.0	39.2	44.0
1	20.0	34.3	20.0	29.4	35.6
2	20.0×2 = 40.0	$(39.2)(2) - \frac{1}{2}(9.8)2^2$ = 58.8	20.0	$39.2 - (9.8)(2)$ = 19.6	$\sqrt{(20.0)^2 + (19.6)^2}$ = 28.0
3	60.0	73.5	20.0	9.8	22.3
4	80.0	78.4	20.0	0	20.0
5	100.	73.5	20.0	−9.8	22.3
6	120.	58.8	20.0	−19.6	28.0
7	140.	34.3	20.0	−29.4	35.6
8	160.	0	20.0	−39.2	44.0

EXAMPLE 3-8 *continued*

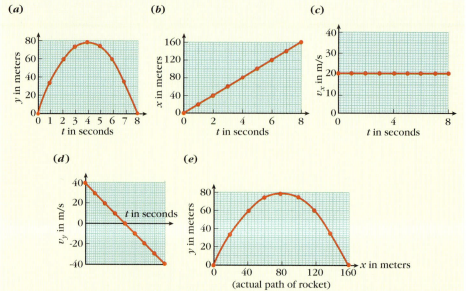

Figure 3-37 Graphs of quantities describing the motion of the rocket in Example 3-8. Notice what is plotted in each graph, and make sure you understand why each graph looks the way it does.

The quantities x, y, v_x, and v_y are plotted against time in Figure 3-37, as well as a graph of y versus x that displays the actual shape of the trajectory. **STOP&Think** Is the velocity zero at the rocket's highest point? (Why, or why not?) Does $v_y = 0$ at this point? (Why, or why not?) What about v_x? In what direction is the rocket moving at its highest point? ◆

◆ Related homework: Problems 3-52, 3-53, 3-54, and 3-55.

Figure 3-38 shows what happens to the velocity vector and its components during the rocket's symmetrical trajectory. Figure 3-38*d* illustrates the fact that we get the velocity vector at the end of each like time interval by adding (in a vector sense) the same $\Delta\vec{v}$ to the velocity vector at the beginning of the interval. Each $\Delta\vec{v} = \vec{a}\,\Delta t$ is the same because $\vec{a}$ does not change (Figure 3-38*e*).

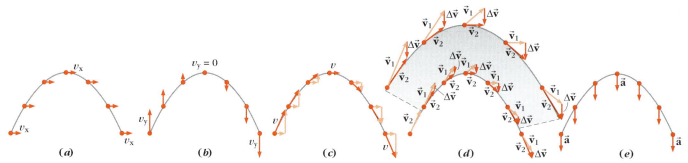

Figure 3-38 Vector aspects of a symmetrical parabolic trajectory. The following vectors are shown at equal time intervals Δt during the course of the trajectory. (*a*) The horizontal vector component of the velocity. (*b*) The vertical vector component of the velocity. (*c*) The resultant velocity (with its horizontal and vertical components shown in a lighter shade). (*d*) Each resultant velocity $\vec{v}_2$ is obtained by adding the same $\Delta\vec{v}$ to the previous velocity $\vec{v}_1$. (*e*) The acceleration vector. Note that because $\Delta\vec{v} = \vec{a}\,\Delta t$, each $\Delta\vec{v}$ is the same in (*d*) and is directed vertically downward because the acceleration $\vec{a}$ is a constant downward vector.

Example 3-9 *Finding the Final Velocity of a Projectile*

For a guided interactive solution, go to Web Example 3-9 at
www.wiley.com/college/touger

A car in a movie chase scene is traveling on level ground at a speed of 45 m/s when it hurtles off a 20-m-high sea cliff. With what speed and in what direction does the car hit the water below?

Brief Solution

Restating the problem. If we let the diver's starting position be $x_o = y_o = 0$, the water's surface is at $y = -20$ m. "Speed" is the *magnitude* of the velocity vector. The quantities in the equations of motion are components of vectors, but the question is asking for the magnitude and direction of the final velocity vector:

$$v = ? \text{ and } \theta = ? \text{ when } y = -20 \text{ m.}$$

Choice of approach. We will work from basic definitions and the condition for uniform acceleration (Equations 3-15x, y). It is also possible (see Web version) to use the derived equations of motion to obtain the same result.

What you know/what you don't. State as components

$$x_o = y_o = 0 \qquad a_x = 0 \qquad a_y = -g = -9.8 \text{ m/s}^2$$

$$v_{ox} = 45 \text{ m/s} \qquad v_{oy} = 0 \text{ (the road is level where the car leaves the cliff)}$$

$$v = ? \text{ and } \theta = ? \text{ when } y = -20 \text{ m (or when } \Delta y = -20 \text{ m)}$$

The mathematical solution. First find the components of $v(v_x = ?, v_y = ?)$. In the x direction, $v_x = v_{ox} = 45$ m/s because $a_x = 0$. In the y direction, the condition for uniform acceleration is $\bar{v}_y = \frac{v_{oy} + v_y}{2} = \frac{v_y}{2}$ because $v_{oy} = 0$. Then the definition $\bar{v}_y \equiv \frac{\Delta y}{\Delta t}$ gives us

$$\bar{v}_y \Delta t = \frac{v_y}{2} \Delta t = \Delta y$$

The definition of average acceleration $\bar{a}_y \equiv \frac{\Delta v_y}{\Delta t}$ gives us $-g = \frac{v_y - 0}{\Delta t}$, so that $\Delta t = \frac{v_y}{-g}$. Then, $\Delta y = \frac{v_y}{2} \Delta t = \frac{v_y}{2} \left(\frac{v_y}{-g} \right)$. Thus $v_y^2 = -2g \Delta y$ and $v_y = \sqrt{-2gy} = \pm \sqrt{-2(9.8 \text{ m/s}^2)(-20 \text{ m})} = \pm 20$m/s. Because the vertical velocity is *downward*, we must choose the *negative* solution: $v_y = -20$ m/s.

Now we find the magnitude (speed) and direction of the final velocity from its components: $v^2 = v_x^2 + v_y^2 = (45 \text{ m/s})^2 + (-20 \text{ m/s})^2 = 2425 \text{ m}^2/\text{s}^2$, so that the speed

$$v = \textbf{49 m/s}$$

and $\tan \theta = \frac{v_y}{v_x} = \frac{-20 \text{ m/s}}{45 \text{ m/s}} = -0.44$, so that $\theta = \textbf{−24°}$ (or $\theta = \textbf{336°}$)

◆ Related homework: Problems 3-56 and 3-57.

In Example 3-9, we got two solutions for v_y, but only the negative one was correct for the descending car. When you get two solutions for v_y or for t at a given value of the height y, both have meaning for the real-world situation only if the projectile actually passes that height during both ascent and descent. This may not happen if the projectile's initial and final y coordinates are different, as when a cannonball is fired from a cliff. In contrast to the cannonball in

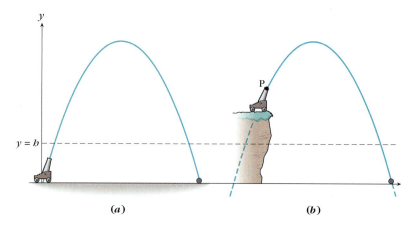

Figure 3-39 Trajectories of two cannonballs. One cannonball is fired from ground level, the other from a height. The dashed line in (*b*) is an unbounded parabola—one that has neither beginning nor end. This is the trajectory represented by the equations of motion. The trajectories of the two cannonballs are both bounded segments of this parabola.

Figure 3-39*a*, the cannonball in Figure 3-39*b* passes $y = b$ only while descending, even though the two trajectories are identical beyond point P. Both trajectories, in fact, are segments of a trajectory (the dotted curve) that has no beginning or end. The equations of motion give information about the whole dotted trajectory from $t = -\infty$ to $t = +\infty$, but the information is meaningful only during the time interval when the projectile is actually in flight. You must reject solutions that occur outside this interval.

When—but only when—the trajectory is symmetrical, the velocity has the same magnitude for any point on the upward path and the corresponding point on the downward path. In that case, the upward and downward paths must take the same amount of time. You could calculate the instant at which $v_y = 0$ (the peak of this trajectory), and double it to obtain the time duration of the entire trajectory. Symmetry arguments will simplify the solution in the next example.

Example 3-10 *Another Projectile*

For a guided interactive solution, go to Web Example 3-9 at
www.wiley.com/college/touger

A kicked soccer ball leaves the soccer player's foot at a speed of 19.21 m/s in a direction 38.66° above the horizontal. How high does the ball rise? How far from the point where it was kicked does the ball land?

Brief Solution
Choice of approach. See Example 3-8. Assume the ball was kicked at the origin.

Restating the problem. Now the vertical "shadow" motion has zero velocity at the highest point, so "how high?" becomes: $y = ?$ when $v_y = 0$. How far is asking for the final horizontal position x when . . . when *what?* Because y is again zero at this point, the question then becomes: $x = ?$ when $y = 0$.

Your equations don't give you a direct relationship between y and x, but knowing y you can find t, and then knowing t you can find x. *The time t provides the link between the horizontal and vertical situations* (see Figure 3-40), because the two component or "shadow" motions are occurring together in time.

What you know/what you don't.

$$x_o = y_o = 0 \qquad a_x = 0,\ a_y = -g = -9.8\ \text{m/s}^2 \qquad v_o = 19.21\ \text{m/s} \qquad \theta_o = 38.66°$$

$$y = ?\ \text{when}\ v_y = 0 \qquad x = ?\ \text{when}\ y = 0 \qquad (t = ?\ \text{when}\ y = 0)$$

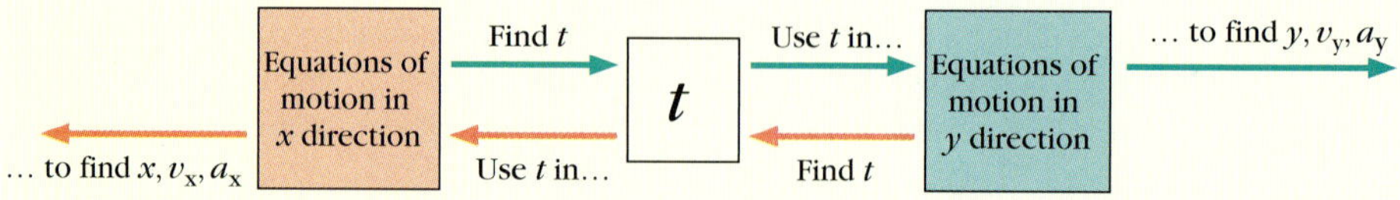

Figure 3-40 A strategy for approaching problems involving motion in both the *x* and *y* directions.

Equations of motion in *x* direction —Find *t*→ *t* —Use *t* in...→ Equations of motion in *y* direction —... to find y, v_y, a_y→

... to find x, v_x, a_x ←Use *t* in...— Find *t*

The mathematical solution.

- (Step 1 of Procedure 3-4) Obtain the component description. For velocity,

$$v_{ox} = v_o \cos 38.66° = 19.21 \text{ m/s} (0.7809) = 15.00 \text{ m/s}$$

$$v_{oy} = v_o \sin 38.66° = 19.21 \text{ m/s} (0.6247) = 12.00 \text{ m/s}$$

- (Step 2 of Procedure 3-4) Once you know v_{oy}, Equation 3-20y permits you to solve for y when $v_y = 0$:

$$v_y^2 = v_{oy}^2 - 2gy$$

becomes

$$0^2 = v_{oy}^2 - 2gy$$

so that the maximum height

$$y = \frac{v_{oy}^2}{2g} = \frac{(12.00 \text{ m/s})^2}{2(9.8 \text{ m/s}^2)} = \textbf{7.35 m}$$

Next, find the time t at which the ball returns to $y = 0$. Finding t is simplified if you realize from symmetry that $v_y = -v_{oy}$ at this time. You can then substitute $-v_{oy}$ for v_y in Equation 3-17y

$$-v_{oy} = v_{oy} - gt$$

and solve for t, getting

$$t = \frac{2v_{oy}}{g} = \frac{2(12.00 \text{ m/s})}{9.8 \text{ m/s}^2} = 2.44 \text{ s}$$

Thus, the ball lands at

$$x = v_{ox}t = (15.00 \text{ m/s})(2.44/\text{s}) = \textbf{36.6 m}$$

A second symmetry argument. Because the trajectory is symmetrical when the ball starts and lands at the same height, you could also find the time t at which the ball reaches maximum height—that is, when half the trajectory is completed—and then double it. Using the fact that $v_y = 0$ at this point, Equation 3-17y gives you $0 = 12.00 \text{ m/s} - (9.8 \text{ m/s})t$. Solving for t, you get $t = 1.22 \text{ s}$ at the halfway point, so that $t = 2.44$ seconds when the ball lands, just as before.

◆ Related homework: Problems 3-58 and 3-59.

An equally valid approach to Example 3-10 is to use Equation 3-16y to find the time t at which the ball lands, then use Equation 3-17x as before. The calculation is more difficult in this approach because you generally need to apply the quadratic formula to solve for t in Equation 3-16y. However, if the initial and final heights are different, it is necessary to do it this way because you cannot make use of symmetry arguments. (To check your understanding of this second approach, try Problem 3-71.)

◆ SUMMARY ◆

This chapter developed the conceptual and mathematical tools for treating motion in two dimensions. We began with the idea that simple motions can be combined or **superimposed** to produce more complex **resultant** motions. Reversing this, you can break down or **analyze** or **resolve** a complex motion into its **component** parts.

In particular, you can break a two-dimensional motion down into two one-dimensional "shadow" motions, and do the same things that we did in one dimension for the components in each direction. We could solve problems (such as Examples 3-6 and 3-7) using the component definitions of average velocity and average acceleration:

$$\bar{v}_x \equiv \frac{\Delta x}{\Delta t} \quad \text{and} \quad \bar{v}_y \equiv \frac{\Delta y}{\Delta t} \qquad (3\text{-}11x, \ 3\text{-}11y)$$

$$\bar{a}_x \equiv \frac{\Delta v_x}{\Delta t} \quad \text{and} \quad \bar{a}_y \equiv \frac{\Delta v_y}{\Delta t} \qquad (3\text{-}14x, \ 3\text{-}14y)$$

When the motion is *uniformly* accelerated (Section 3-6), we need to supplement these with the condition for uniform acceleration

$$\bar{v}_x = \frac{v_{ox} + v_x}{2} \quad \text{and} \quad \bar{v}_y = \frac{v_{oy} + v_y}{2} \quad (3\text{-}15x, \ 3\text{-}15y)$$

or use the derived equations of motion for uniform acceleration in each direction:

$$
\begin{array}{l}
x - x_o = v_{ox}t + \frac{1}{2}a_x t^2 \\
\hspace{3cm} (3\text{-}16x) \\
v_x = v_{ox} + a_x t \quad (3\text{-}17x) \\
v_x^2 = v_{ox}^2 + 2a_x(x - x_o) \\
\hspace{3cm} (3\text{-}18x)
\end{array}
\left.
\begin{array}{l}
\text{at} \\
\text{the} \\
\text{same} \\
\text{instant} \\
t \text{ as}
\end{array}
\right\{
\begin{array}{l}
y - y_o = v_{oy}t + \frac{1}{2}a_y t^2 \\
\hspace{3cm} (3\text{-}16y) \\
v_y = v_{oy} + a_y t \quad (3\text{-}17y) \\
v_y^2 = v_{oy}^2 + 2a_y(y - y_o) \\
\hspace{3cm} (3\text{-}18y)
\end{array}
$$

Projectile motion (when air resistance is negligible) is a particular instance of uniform acceleration. In this case, we take $a_x = 0$ and $a_y = -g$ in the above equations. (In particular, 3-16x becomes $x - x_o = v_{ox}t$ and 3-17x and 3-18x reduce to $v_x = v_{ox}$.)

Vectors provide a framework for treating motion in two dimensions mathematically. Vectors can be specified by a **magnitude and direction description** or by a **component description.**

and

Any vector equation is a summary of equations involving components in each direction. For example, $\vec{R} = \vec{A} + \vec{B}$ summarizes the two equations $R_x = A_x + B_x$ and $R_y = A_y + B_y$. It is useful to express vector equations in component form because the components add by ordinary arithmetic. But you should also be able to form a picture (such as Figure 3-17) by Procedure 3-1 of what the sum of any two vectors looks like.

Subtracting vectors simply means adding the negative of one vector to another: $\vec{A} - \vec{B} = \vec{A} + (-\vec{B})$. If $\vec{B}$ has components B_x and B_y, $-\vec{B}$ has components $-B_x$ and $-B_y$ and is opposite in direction to $\vec{B}$. In particular, we use this when finding changes in a vector such as

$$\Delta\vec{r} = \vec{r}_2 - \vec{r}_1 \quad (3\text{-}10) \quad \text{or} \quad \Delta\vec{v} = \vec{v}_2 - \vec{v}_1 \quad (3\text{-}13)$$

Resolving vectors into components lets you solve two-dimensional motion problems using one-dimensional equations of motion, each involving components of position, velocity, and acceleration vectors (see Procedure 3-4).

Because **speed** is defined as the *magnitude* of the velocity vector,

$$v \text{ or } |\vec{v}| = \sqrt{v_x^2 + v_y^2} \quad (3\text{-}12)$$

an object changing direction can change velocity without changing speed. Because the acceleration vector $\vec{a} = \frac{\Delta\vec{v}}{\Delta t}$ is nonzero whenever there is a change $\Delta\vec{v}$ in the velocity, an object is accelerating if its velocity vector is changing in any way at all, even if only in direction (see Example 3-7).

◆ QUALITATIVE AND QUANTITATIVE PROBLEMS ◆
Hands-On Activities and Discussion Questions

The questions and activities in this group are particularly suitable for in-class use.

3-1. Hands-On Activity. Remove the labels from a large soda bottle so that it is transparent. Cut a circular hole in a stiff piece of cardboard, just large enough to let the bottle slide freely through it (Figure 3-41). Have a friend press down on the bottle to hold it still. With the cardboard held stationary at one height, make a circle on the bottle by tracing along the inner edge of the cardboard with a marking pen. Next, keeping the cardboard horizontal, lower it along the bottle at a slow, steady speed. As you do this, continue to mark the bottle by running your marker along the circular inner edge of the cardboard. As the figure shows, you should end up with a spiral trace on the bottle. Describe the two component motions that combine to produce the spiral motion for the marker tip.

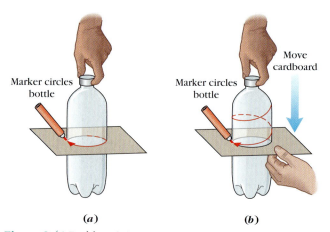

Marker circles bottle

Move cardboard

Marker circles bottle

(a)   *(b)*

Figure 3-41 Problem 3-1

3-2. Hands-On Activity.

a. Draw a circle at least 0.2 m (about 8 in.) in diameter on a large sheet of paper (a double page of newspaper will do nicely). Move a drinking glass or beaker at steady speed along this circle as though it were a circular track, making certain to keep the center of the glass's bottom on the track. As the glass circles the track, trace around the bottom with a pencil so that a continuous path is marked out on the paper (Figure 3-42).

b. How do the two component motions in this activity compare to the two component motions in Figure 3-4?

c. How does the resultant motion of your pencil point compare with the actual path of the Octopus rider in Figure 3-4*b*?

Motion of marker around glass   Motion of glass around circle

Figure 3-42 Problem 3-2

3-3. Discussion Question. At the end of Section 3-1, a pop fly in baseball was given as an example of projectile motion. A baseball player can also hit a line drive. How is this different than a pop fly? How is it the same? Is the motion of a bullet fired from a gun projectile motion? In the example of the carnival game on the back of a truck (Case 3-2), how can you vary the situation to make the trajectory of the weight either more like a pop fly or more like a line drive?

•• **3-4. Discussion Question.** A ball can go from point A to point B by rolling down either of the tracks in Figure 3-43. If it is released from rest at A, will it get to B faster by rolling down track I or track II? Explain. (Adapted from W. J. Leonard and W. J. Gerace, *The Physics Teacher*, **34**, 280–283 [1996].)

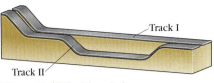

Track I

Track II

Figure 3-43 Problem 3-4

3-5. Discussion Question. The set-up described in Figure 3-50 (see Problem 3-16) is altered so that now there is only one screen (Figure 3-44), but there are two trains traveling on separate tracks between points X and Y. Suppose both trains travel at constant speed, but not the same constant speed. However, both leave point X at the same time and both arrive at point Y at the same time. If you watch only the shadows that the two balls cast on the screen, how would the motions of these two shadows compare?

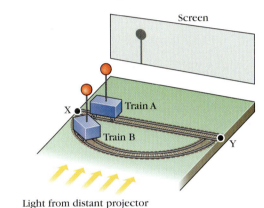

Screen

X   Train A

Train B   Y

Light from distant projector

Figure 3-44 Problem 3-5

Review and Practice

Section 3-1 Constructing Complex Motions from Simpler Motions

3-6. *Analysis of complex motion into simpler motions*: The "sawtooth" pattern in Figure 3-45 is traced on an *xy* recorder. The motion of the pen that left the trace can be analyzed into two one-dimensional component motions. Describe these two motions. (To check whether your idea works, move a piece of paper by hand and try it yourself.)

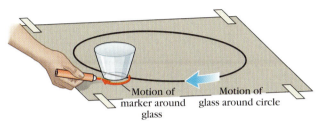

Figure 3-45 Problem 3-6

3-7. *Synthesis of complex motions from simpler motions*: A cartoon figure is drawn on a disk (Figure 3-46*a*). A pencil poked through the disk's center (Figure 3-46*b*) serves as an axis of rotation. To animate the figure, you can rotate the disk about the pencil at the same time you move the pencil in a straight line to the right, as shown. (You can trace or photocopy the disk and cut it out to try this for yourself.) What does the

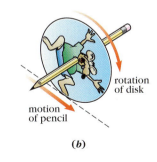

rotation of disk

motion of pencil

(*a*)   (*b*)

Figure 3-46 Problem 3-7

cartoon figure appear to be doing when both of these motions are executed at once?

3-8. As the paper of an *xy*-recorder scrolls, the pen remains at height b_1 for 1 s, drops abruptly to height b_2 (2 cm lower) where it remains for the next second, returns to height b_1 for the third second, to b_2 for a fourth second, and continues alternating in this way. Sketch the trace left by the pen on the paper.

3-9. *Synthesis of complex motions from simpler motions:* In Figure 3-47, a heavy metal ball hangs from a long rope over a sandbox on wheels. A slender metal rod attached to the bottom of the ball can leave a track in the sand when the ball swings. Suppose that the sandbox is pulled to the right at a slow constant speed. Meanwhile, the ball is given a gentle push so that it swings toward side A of the sandbox. Sketch an overhead view of the track traced in the sand as the ball swings forward and back several times.

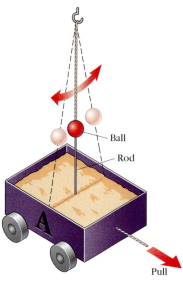

Figure 3-47 Problem 3-9

3-10. The pen of an *xy*-recorder moves between two heights b_1 and b_2 (see Figure 3-48a) while the paper scrolls to the right. Below are five descriptions of the pen's motion. Figure 3-48b shows five traces left by the pen on the paper. For each description of motion, write the number of the trace that that motion produces.

a. Pen drops from b_1 to b_2 at constant speed.

b. Pen drops suddenly from b_1 to b_2, pauses for a second, goes up suddenly from b_2 to b_1, pauses for a second, then repeats the same sequence of steps.

c. Pen goes up and down between b_1 and b_2 at constant speed (except while quickly reversing direction).

d. Pen speeds up as it goes from b_1 to b_2.

e. Pen slows down as it goes from b_1 to b_2.

3-11. A principal step in manufacturing screws is to cut the thread that winds around the shaft. Describe two motions that are necessary for the process of cutting the thread.

3-12. A straight track placed on a table top is seen from above in Figure 3-49. It is shown at five equally spaced instants as it is slid across the table at constant speed v_{track}. As it begins sliding (at $t_o = 0$), a cart at the left end of the track is given a velocity to the right. The cart rolls along the track as the track is slid, but the cart slows down at a constant rate because of friction. It just reaches the other end of the track at $t_4 = 1.00$ s.

a. The dot in the middle of the cart in the figure indicates its initial position. Draw dots on the figure showing roughly where the cart is when the track is at each of its later positions.

b. Now draw a smooth line connecting the five dots that represent the cart's position at the five depicted instants. Is this line straight or curved?

c. The smooth line that you have drawn shows the path of the cart as seen by a stationary observer looking down at the table. Can the motion along this path be understood as a superposition of straight line motions? Explain.

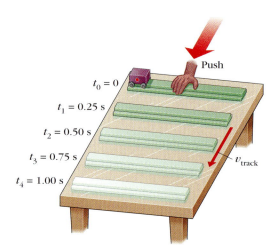

Figure 3-49 Problem 3-12

3-13. SSM In Case 3-1, the weight goes up and down the slide wire of the carnival game while the truck is moving (see Figure 3-7a). Assume friction in negligible. Indicate whether each item below is a true or false statement about the resulting two-dimensional motion of the weight.

a. It has constant velocity in both the vertical and horizontal directions.

b. It has the same acceleration as a ball thrown vertically into the air.

c. It has a constant upward velocity and then a constant downward velocity.

d. It is accelerating in both the vertical and horizontal directions.

e. It has zero acceleration at the highest point in its path.

f. It has zero velocity at the highest point in its path.

g. Its horizontal velocity doesn't change.

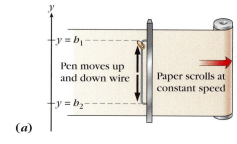

(a)

Trace left by pen

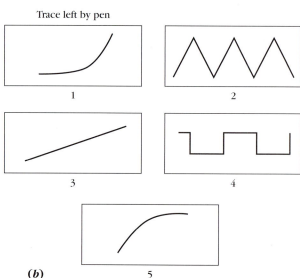

(b)

Figure 3-48 Problem 3-10

Section 3-2 Breaking Down Two-Dimensional Motions into One-Dimensional Components: Projectile Motion

3-14. A basketball player shoots a foul shot in an outdoor playground where the sun is directly overhead. As the ball travels from the player's hands to the basket, the shadow that the ball casts on the ground ____. (*accelerates; slows down; slows down then speeds up; speeds up then slows down; moves at constant speed*)

3-15. Using the data in Figure 3-10 (Section 3-2) and making any necessary measurements, calculate the speed of the golf ball's horizontal shadow motion.

3-16. A toy electric train moves along a curved track in the shadow box in Figure 3-50. A ball on a post is mounted on the train. When light beams from distant slide projectors (or strong narrow-beam flashlights) X and Y are directed at the ball, the ball casts shadows on screens X and Y as shown.

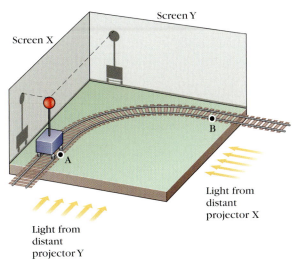

Figure 3-50 Problems 3-16, 3-80, and 3-81

a. Suppose the train moves along the track at constant speed from A to B in the figure. Does the shadow that the ball casts on screen X speed up, slow down, or move at constant speed? Explain your answer.

b. Does the shadow that the ball casts on screen Y speed up, slow down, or move at constant speed? Explain your answer.

c. Can a constant speed motion in two dimensions have one-dimensional component motions that are accelerated? Explain.

d. Suppose the track were laid in a straight line from A to B instead of being curved. In this case, would the shadow that the ball casts on screen X speed up, slow down, or move at constant speed? Explain your answer.

e. In **d,** would the shadow that the ball casts on screen Y speed up, slow down, or move at constant speed? Explain your answer.

f. The answers to **a** and **b** are ____.

    A. the same as the answers to **d** and **e.**

    B. different from the answers to **d** and **e** because the train's speed is changing in **a** and **b.**

    C. different from the answers to **d** and **e** because the train's direction is changing in **a** and **b.**

g. As the train follows different possible paths from A to B but always at constant speed, is it ever possible that one of the two shadows cast by the ball will be accelerating but not the other? Briefly explain.

Section 3-3 Vectors

3-17. Which of the following are vector quantities and which are scalars? Explain your reasoning.

speed	length	time
displacement	velocity	angle
v_x	acceleration	

3-18.

a. Turn to any page of your physics textbook. If the origin is at the lower left-hand corner of this page (and the axes oriented in the usual way), determine by measurement the magnitude and direction of the position vector for the upper right-hand corner of the page.

b. If the origin is at the upper left-hand corner of this page, determine by measurement the magnitude and direction of the position vector for the lower right-hand corner of the page.

3-19. **SSM WWW** Sketch the vector $\vec{A}$ and find its components if its magnitude A

a. is 10 and its direction is 37° above (counterclockwise from) the $+x$ axis.

b. is 10 and its direction is 135° counterclockwise from the $+x$ axis.

c. is 8 and its direction is 60° below (clockwise from) the $+x$ axis.

d. is 5 and its direction is 37° below the negative x axis.

3-20. The aerial surveillance picture in Figure 3-51 show the location X of a proposed statue, but on each copy of the picture, a different coordinate frame has been drawn for determining positions. For each choice of coordinate frame, sketch the position vector for X, then use a cm ruler to obtain the components of this vector. Assume the scale is 1 cm = 50 m.

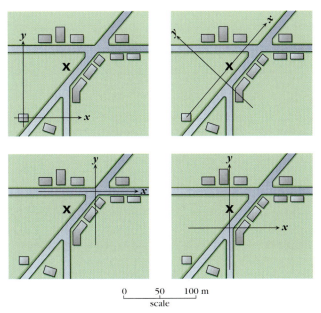

Figure 3-51 Problem 3-20

3-21. Use Procedure 3-1 to draw the vector sum or resultant of each pair of one-dimensional vectors $\vec{A}$ and $\vec{B}$ shown in Figure 3-52. For each resultant you obtain, state the value of its magnitude and a single word (left, right, up, or down) to indicate its direction.

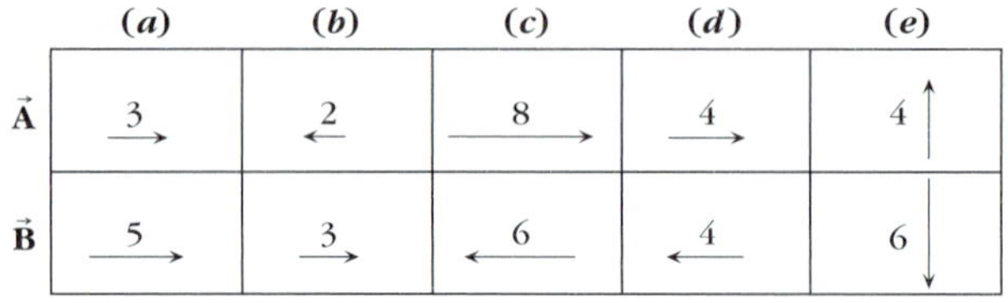

	(a)	(b)	(c)	(d)	(e)
$\vec{A}$	3 →	2 ←	8 →	4 →	4 ↑
$\vec{B}$	5 →	3 →	6 ←	4 ←	6 ↓

(Numerical values are for $|\vec{A}|$ and $|\vec{B}|$)

Figure 3-52 Problem 3-21

3-22. Use Procedure 3-1 to draw the vector sum or resultant of each pair of vectors $\vec{A}$ and $\vec{B}$ shown in Figure 3-53. Measure the magnitude and directional angle for each resultant, and give their values.

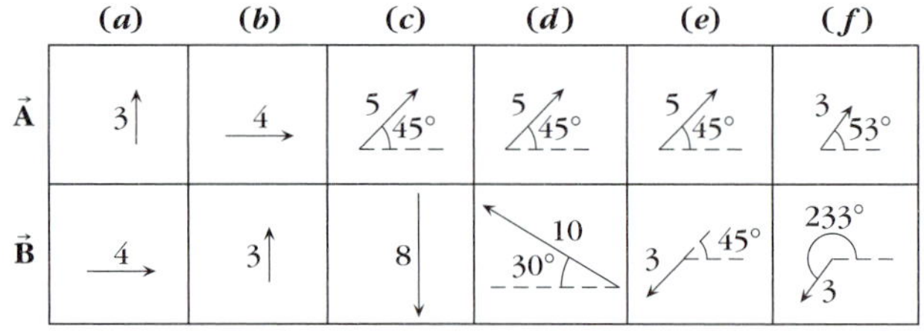

	(a)	(b)	(c)	(d)	(e)	(f)
$\vec{A}$	3 ↑	4 →	5 / 45°	5 / 45°	5 / 45°	3 / 53°
$\vec{B}$	4 →	3 ↑	8 ↓	10 / 30°	3 / 45°	233° / 3

(Numerical values are for $|\vec{A}|$ and $|\vec{B}|$)

Figure 3-53 Problems 3-22, 3-33, 3-36, 3-37, and 3-78

3-23. SSM A hiking trail goes to Vernal Falls. The sign posted at the beginning of the trail indicates that it follows a river for 1450 m in a direction 15° S of E (south of east), and then cuts through woodland for 900 m in a direction 65° N of E to reach the falls. Taking the beginning of the trail as your origin, use a ruler and protractor to find the position vector (magnitude and direction) of the falls.

3-24.
a. The flight plan of a plane leaving O'Hare Airport first takes it 3.6×10^4 m in a direction 20° N of E (north of east), and then 4.8×10^4 m in a direction 70° N of W. Taking O'Hare as your origin, use a ruler and protractor to find the plane's position vector (magnitude and direction) at the end of the flight.
b. A second plane out of O'Hare follows the two stages of the flight plan in reverse order. Sketch the vector addition and the final position vector for this plane's flight. How do the two planes' final position vectors compare?

3-25. Figure 3-54 shows four different pairs of vectors $\vec{A}$ and $\vec{B}$. All the vectors $\vec{A}$ are equal in magnitude. So are all the vectors $\vec{B}$. Rank these four pairs in order of the magnitude of the resultant vector $\vec{A} + \vec{B}$, starting with the smallest. Indicate which values, if any, are equal. Briefly explain your reasoning.

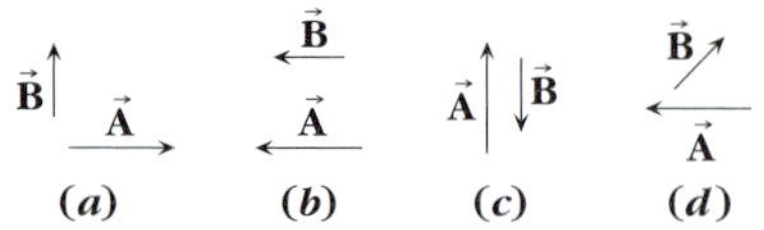

Figure 3-54 Problem 3-25

3-26. Vectors $\vec{A}$ and $\vec{B}$ are to be added to obtain a resultant $\vec{R}$. Vector $\vec{A}$ has magnitude $A = 3$ and is directed toward the right. Vector $\vec{B}$ has magnitude $B = 4$; you are free to choose its direction. Below are descriptions of six different possible resultants $\vec{R}$. For each description, sketch what the vector $\vec{B}$ must look like, and sketch how you would add it to $\vec{A}$ to produce that resultant.
a. Its magnitude is $R = 7$.
b. Its magnitude is $R = 1$.
c. Its magnitude is $R = 5$.
d. Its magnitude is $R = 5$, but its direction is different than in c.
e. Its x component is $R_x = -1$.
f. Its y component is $R_y = -4$.

3-27. SSM Vectors $\vec{A}$ and $\vec{B}$ each have a magnitude of 10. Sketch the vector sum $\vec{A} + \vec{B}$, choosing the directions of $\vec{A}$ and $\vec{B}$ so that the resultant $\vec{R}$
a. has a magnitude of 20.
b. is zero.
c. has a magnitude of 10.
d. has a magnitude of 2.

Section 3-4 Working with Vector Components

3-28. A jogger goes 300 m east, then 225 m north.
a. Find the magnitude and direction of the jogger's displacement by drawing and measuring (to anticipate what a calculation should give you).
b. Calculate the magnitude and direction, and see if it agrees with your measurements.

3-29. Using the same scale as in Example 3-1, use a protractor and a centimeter ruler to find
a. the magnitude and direction of the position vector of Hidden Harbor based on the coordinate frame in Figure 3-22.
b. the x and y components of this position vector.

3-30.
a–e. Find the x and y components of each vector in Figure 3-55. Where possible, find the components both by inspection and by calculation to see if you can get the same results both ways. You may wish to answer parts f and g first.
f. In each of parts **a–e**, what value of θ (measured counterclockwise from the $+x$ axis) could you use to calculate the components?
g. In which parts could you find the components by inspection?

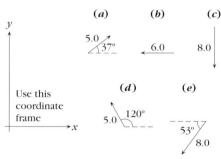

Figure 3-55 Problem 3-30

3-31. Find the magnitude and direction of the velocity $\vec{v}$ if
a. $v_x = 6$ m/s and $v_y = 8$ m/s.
b. $v_x = -6$ m/s and $v_y = 0$.
c. $v_x = 6$ m/s and $v_y = -8$ m/s.
d. $v_x = -6$ m/s and $v_y = -8$ m/s.

3-32. *Connection with one-dimensional motion:* Suppose Equation 3-6 is applied to the velocity vector of an object on a straight track directed along the x axis.
a. What is the value of θ when the object is traveling to the right? What is the value of $\sin \theta$?
b. What is the value of θ when the object is traveling to the left? What is now the value of $\sin \theta$?
c. Are there any other possible values of θ for one-dimensional motion?
d. What does Equation 3-6 tell you about the relationship between the velocity and the speed for one-dimensional motion in the x direction? Does this agree with the way we accounted for direction in Chapter 2?

3-33. Use Procedure 3-3 to calculate the magnitude and direction of the vector sum $\vec{A} + \vec{B}$ for each pair of vectors shown in Figure 3-53.

3-34. A primate research station attaches a tiny radio transmitter to a chimpanzee born in captivity before releasing it into the wild. One day, the station picks up a signal indicating the chimp is 4000 m from the station in a direction 15° S of W. Over the next day, the chimp wanders 2500 m in a direction 40° N of E. At this point, how far from the station and in what direction is the chimp?

3-35. **SSM WWW** An aircraft surveillance post reports an unidentified aircraft 800 km from them in a direction 45° S of E and traveling due north. If it continues in this direction for 600 km, at what distance from the surveillance post and in what direction will the aircraft be observed?

3-36. By drawing and measuring, find the vector difference $\vec{A} - \vec{B}$ for each pair of vectors in Figure 3-53.

3-37. By calculation, find the magnitude and direction of the vector difference $\vec{A} - \vec{B}$ for each pair of vectors in Figure 3-53.

3-38. Vector $\vec{A}$ has components $A_x = -2$ and $A_y = 3$.
    Vector $\vec{B}$ has components $B_x = 4$ and $B_y = -2$.
    Vector $\vec{C}$ has components $C_x = -5$ and $C_y = -4$.
Find
a. a component description of $\vec{A} + \vec{B} + \vec{C}$.
b. a magnitude and direction description of $\vec{A} + \vec{B} + \vec{C}$.
c. a magnitude and direction description of the vector $3\vec{A}$.
d. a magnitude and direction description of the vector $-\frac{1}{2}\vec{B}$.

3-39. Vectors $\vec{A}$ and $\vec{B}$ each have a magnitude of 5.
a. What is the range of possible values for the magnitude of $\vec{A} + \vec{B}$?
b. On what does the actual value depend?
c. What is the range of possible values for the magnitude of $\vec{A} - \vec{B}$?
d. If the directions of $\vec{A}$ and $\vec{B}$ differ by 60°, what is the magnitude of $\vec{A} + \vec{B}$?

3-40. A vector makes an angle of 60° with the positive x axis. When you multiply the vector by 0.5, does the angle increase, decrease, or remain the same?

Section 3-5 Velocity and Acceleration Vectors

3-41. **SSM WWW** An ornithologist studying raptors releases three red-tailed hawks after fitting them with small radio transmitters. Figure 3-56a, shows hawk A's position $\vec{r}_1$ at a certain instant, and its position $\vec{r}_2$ at an instant 80.0 s later.
a. To help picture what you will be calculating, first sketch $\Delta\vec{r}$, the change in hawk A's position.
b. Calculate the components of $\Delta\vec{r}$.
c. Find the components of hawk A's average velocity $\vec{\overline{v}}$ over the 80.0-s interval.
d. Find the magnitude and direction of $\vec{\overline{v}}$.
e. How does the direction of $\vec{\overline{v}}$ compare to the direction of $\Delta\vec{r}$?

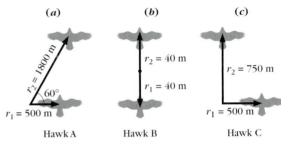

(a)   (b)   (c)

$r_2 = 1800$ m $r_2 = 40$ m $r_2 = 750$ m
60° $r_1 = 40$ m $r_1 = 500$ m
$r_1 = 500$ m

Hawk A   Hawk B   Hawk C

Figure 3-56 Problems 3-41, 3-42, and 3-43

3-42. Repeat Problem 3-41 for Figure 3-56b.

3-43. Repeat Problem 3-41 for Figure 3-56c.

3-44. An object's velocity vector is shown at an instant t_1 and a later instant t_2 in Figure 3-57.
a. Give a component description of each vector.
b. Give a magnitude and direction description of each vector.
c. Find the magnitude and direction of $\Delta\vec{v}$.
d. Has the object's velocity changed?
e. Has the object's speed changed?

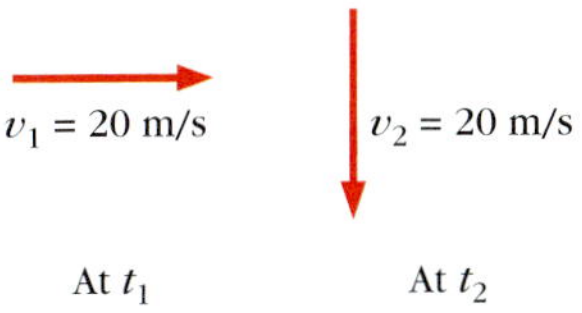

$v_1 = 20$ m/s $v_2 = 20$ m/s

At t_1 At t_2

Figure 3-57 Problem 3-44

3-45. A cheetah has velocity vector $\vec{v}_1$ (Figure 3-58) at $t = 3$ s and $\vec{v}_2$ at $t = 8$ s. Find the magnitude and direction of the cheetah's change in velocity $\Delta\vec{v}$ over this interval.

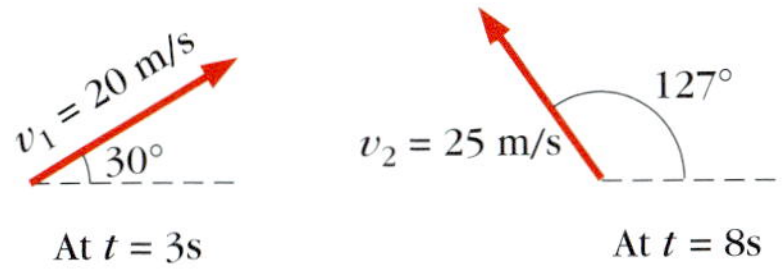

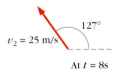

$v_1 = 20$ m/s   127°
30° $v_2 = 25$ m/s

At $t = 3$s   At $t = 8$s

Figure 3-58 Problems 3-45 and 3-46

3-46. Using the information from Problem 3-45, find the cheetah's average acceleration (both the component description and the magnitude and direction description) between $t = 3$ s and $t = 8$ s.

3-47. A motorcycle goes clockwise around a circular track at a constant speed of 20 m/s.

a. Is the motorcycle accelerating?

b. How do the magnitude and direction of its velocity vectors at points A and B (figure) compare with $\vec{v}_1$ and $\vec{v}_2$ in Problem 3-44?

c. As the motorcycle goes from A to B, how do the magnitude and direction of $\Delta\vec{v}$ for the motorcycle during this period compare with the solution in Problem 3-44?

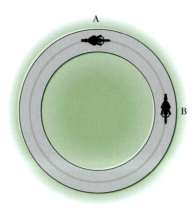

Figure 3-59 Problem 3-47b

d. If it takes the motorcycle 4 s to get from A to B, use the solution to Problem 3-44 to calculate the motorcycle's average acceleration (magnitude and direction) between A and B. Compare this with your answer to **a**.

Section 3-6 Solving Motion Problems in Two Dimensions: Projectile Motion Revisited

3-48. In Figure 3-60, a kicked soccer ball is shown in several stop-action pictures taken at half-second intervals during its trajectory. The x and y shadows at each of these instants are also shown.

a. Is the horizontal shadow motion uniform? Is the vertical shadow motion uniform?

b. Using the coordinate frame in the figure to determine positions, find the *average* horizontal and vertical velocities for each half-second time interval (0 to 0.5 s, 0.5 s to 1.0 s, etc.). Do your calculations support your answer to part **a**?

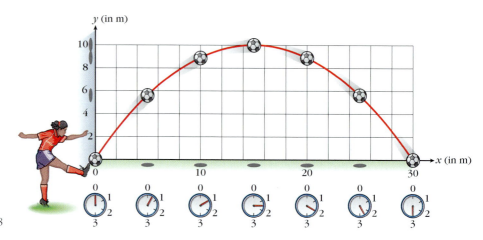

Figure 3-60 Problem 3-48

3-49. Figure 3-61 shows a cannonball at five points during its flight. The coordinate frame shown applies to all parts of the question. Without doing any calculations, answer each of the following.

a. Indicate whether the vertical acceleration is positive, negative, or zero at each of these points.

b. Indicate whether the speed is positive, negative, or zero at each of these points.

c. At which of these positions is the speed a minimum?

d. At which of these positions is the speed zero?

e. Which of these quantities change sign during the cannonball's flight? x, y, v_x, v_y, v, a_x, a_y, a.

3-50. When you shoot a free throw in basketball, does the ball have zero velocity at its highest point? Briefly explain.

3-51. Figure 3-62 shows the trajectory of a ball thrown up in the air at an angle from near the bottom of a pit. Fill in each space in the table below with +, 0, or − to indicate whether the components of position, velocity, and acceleration are positive, negative, or zero for the indicated parts of the trajectory.

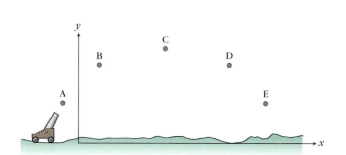

Figure 3-61 Problem 3-49

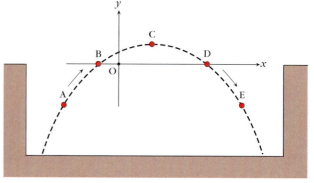

Figure 3-62 Problem 3-51

	x	v_x	a_x	y	v_y	a_y
While ball is going from A to B						
While ball is going from B to C						
When ball is at C						
While ball is going from C to D						
While ball is going from D to E						

3-52. The theater at Einstein Community College has a sloping stage (Figure 3-63). Because of the slope, a ball released from rest on the stage would roll toward the edge of the stage with a downhill acceleration of magnitude 2.4 m/s². A group of physics students map out coordinate axes on the stage, and then give a ball an initial velocity as shown in the figure. Find the x and y coordinates of the ball at each of the following instants: $t = 0$, $t = 1$ s, $t = 2$ s, $t = 3$ s, $t = 4$ s.

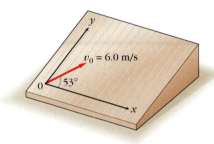

Figure 3-63 Problem 3-52

3-53. Use the values obtained in Problem 3-52 to plot graphs of x versus t and y versus t. How do these graphs compare with the graphs of x versus t and y versus t obtained in Example 3-8?

3-54. From Problem 3-52, find the components v_x and v_y of the ball's velocity at each of the following instants: $t = 0$ $t = 1$ s, $t = 2$ s, $t = 3$ s, $t = 4$ s.

3-55. Use the values obtained in Problem 3-54 to plot graphs of v_x versus t and v_y versus t. How do these graphs compare with the graphs of v_x versus t and v_y versus t obtained in Example 3-8?

3-56. An artillery shell is fired horizontally from the top of a cliff 78.4 m high. It leaves the gun with a velocity of 1200 m/s.
a. How long does it take to reach the ground?
b. How far from the base of the cliff does it land?
c. Find the components of its velocity at the instant it strikes the ground.
d. Find its speed at that instant.

3-57. Redo Example 3-9 taking the downward direction as positive.

3-58. A golf ball is given an initial velocity of 49.0 m/s at an angle of 37° above the horizontal. Ignoring air resistance,
a. how long does it take the ball to reach its highest point?
b. how far has the ball gone horizontally when it is at its highest point?
c. how long does the ball remain in flight?

3-59. **SSM WWW** A soccer ball leaves a player's foot at an angle of 22° to the ground. With negligible air resistance, it takes 0.833 s to reach the highest point in its trajectory.
a. With what speed did it leave the player's foot?
b. What is its speed at its highest point?
c. How far away does it land?

Going Further

The questions and problems in this group are not organized by section heading, so you must determine for yourself which ideas apply. Some of them will be more challenging than the Review and Practice questions and problems (especially those marked with a • or ••).

3-60. In Figure 3-64, points P and Q are shown on a graph paper background where each box represents one unit of distance. Copy the diagram four times, so that you can do each part below on a separate diagram. Now sketch a coordinate frame (one for each part) in which the position vector for
a. point P has components $x = 1$ and $y = 0$.
b. point P has components $x = 0$ and $y = -2$.
c. point P has components $x = -1$ and $y = 3$.
d. point Q has components $x = 5$ and $y = 0$.

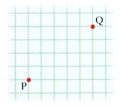

Figure 3-64 Problem 3-60

3-61.
a. Figure 3-65 shows some unusually trained laboratory animals. Sketch the coordinate frame (superimpose it on the figure) that would be best suited for describing the motion of animal A after the rope burns through. Give reasons for your choice.
b. Repeat for animal B.
c. Repeat for animal C.

Figure 3-65 Problem 3-61

3-62. While a truck creeps along in traffic at a speed of 5 m/s, a man walks diagonally across the back of the truck at 2 m/s. To a traffic helicopter hovering in a stationary position above

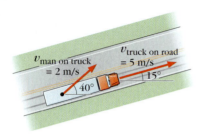

Figure 3-66 Problem 3-62

the road, the man's apparent motion is the superposition of these two motions; that is, his apparent velocity is the vector sum of the two given velocities. If the directions are as shown in Figure 3-66, find the magnitude and direction of the man's velocity as it appears to an observer in the helicopter.

3-63. SSM Although a diver's body rotates in space in a complicated way during a dive, the diver's center of mass (we'll define this carefully in Chapter 5) moves like a projectile (see Figure 3-67). A diver leaves a 10-m-high diving board with a speed of 3.3 m/s at an angle 68° above the horizontal. Find the approximate speed with which she hits the water below. (*Hint:* The solution will be approximate when you make the following simplifying assumptions: Ignore the difference in height between the diving board (where her feet are before the dive) and her center of mass; ignore the time difference between the instant her hands make contact with the water and the instant her center of mass is at the water's surface; ignore air resistance.)

Figure 3-67 Problem 3-63

3-64. A vector $\vec{F}$ has a magnitude of 20 and is directed vertically downward. Find its components F_x and F_y in each of the coordinate frames shown in Figure 3-68.

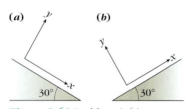

Figure 3-68 Problem 3-64

3-65. SSM A radio signal is relayed from station A to station B to station C to station D. The distances and directions for these three stages are shown in Figure 3-69. How far and in what direction would a signal have to go if it were to be sent directly from station A to station D?

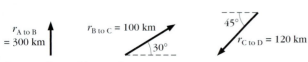

Figure 3-69 Problem 3-65

3-66. Show that for any vector $\vec{V}$, tan θ is the same for both $\vec{V}$ and $-\vec{V}$.

3-67. For each of the following situations, determine whether the velocity of the object *cannot* be constant, *might* be constant, or *must* be constant.

a. A billiard ball rebounds off the cushion of a billiards table.

b. A car travels down a straight road at an average speed of 20 m/s.

c. A car maintains a steady speed of 10 m/s around a traffic circle.

3-68. The velocity components of a ball are plotted against time in Figure 3-70. What would you observe the ball doing during the time period graphed? Sketch its trajectory.

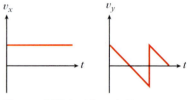

Figure 3-70 Problem 3-68

•**3-69.** A plane is in the situation depicted in Figure 3-71 when it runs out of fuel. If it clears the mountain, it can glide to a safe landing. The mountain's elevation is 4450 m. By a suitable calculation, determine whether the plane will make it over. Neglect air resistance.

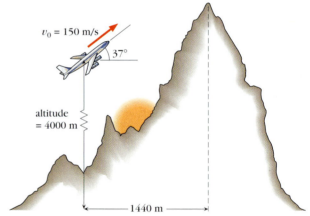

Figure 3-71 Problem 3-69

•**3-70.** Starting with Equations 3-17x and 3-17y, show by a series of mathematical steps that $v^2 = v_0^2 - 2gy$.

•**3-71.** A cannonball is fired from a hilltop bunker as shown in Figure 3-72.

a. If it is to land at the foot of the hill (P), what must be its initial speed?

b. If a truck at P requires 11 s to move far enough out of the way, will it escape damage?

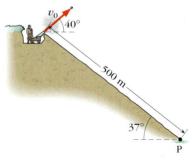

Figure 3-72 Problem 3-71

•**3-72.** A projectile at ground level is given an initial velocity of 100 m/s at an angle of 30° with the horizontal. At what two horizontal distances does the projectile have a height of 81.4 m?

•**3-73.** A kicked football leaves the kicker's foot at a speed of 20 m/s. At what angle above the horizontal was the ball kicked if it reaches a maximum height of 14.4 m?

•**3-74.** A model rocket carries no fuel; it is set in motion entirely by its ground-level launcher. 8.0 s after being launched, the rocket lands on the ground 410 m from the launcher.
a. Find a component description of the rocket's initial velocity.
b. Find the speed and directional angle at which the rocket was launched.
c. If the launch occurs at $t = 0$, at what two instants t does the rocket have a speed of 55 m/s?

••**3-75.** In 1918, the Paris Gun was state-of-the-art German artillery, with a muzzle velocity of 1.61×10^3 m/s (1 mile/s). It was set up at Laon, France, with the intent of shelling Paris, a horizontal distance of 1.08×10^5 m (108 km) away (slightly further by land because of Earth's curvature).
a. To strike a target at this distance, at what angle did the gun have to be fired? (Neglect air resistance.)
b. How long would the shell then take to reach its target?

3-76. In each of the following cases, what is the direction of the object's *average* acceleration vector during the interval described? (Draw sketches if necessary to clarify your answer).
a. A golf ball in a mini-golf game moves to the right, hits an obstacle head on, and bounces straight back at the same speed.
b. The golf ball hits the obstacle at a 30° angle to the perpendicular and bounces off it at the same angle and speed (see Figure 3-73).
c. A race car is driven halfway around a circular track at constant speed (see Figure 3-73).
d. The race car is driven two full laps around the circular track at constant speed.

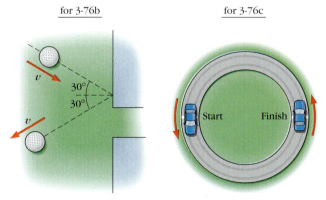

Figure 3-73 (for Problem 3-76)

3-77. In Figure 3-37 in Section 3-6, the choice of origin is shown in the graph of y versus x. Suppose we move the origin to the right so that $x = 0$ is now directly under the highest point of the trajectory. Which of the other graphs in Figure 3-37 would change?

3-78. Find the x and y components of each of the vectors in Figure 3-53 if the coordinate frame is oriented as in Figure 3-74.

3-79. Give an example of two vectors that *cannot* be added together.

Figure 3-74 Problem 3-78

3-80. Suppose the track in Figure 3-50 (described more fully in Problem 3-16) is extremely flexible, so that it can be arranged to form a path of any shape from A to B. (Do not worry about difficulties the train may have in making turns that are too sharp.) Sketch a path from A to B such that if the train follows that path at constant speed from A to B, the ball's shadow on
a. screen X reverses direction once, but its shadow on screen Y does not change directions at all;
b. screen Y reverses direction twice, but its shadow on screen X does not change directions at all;
c. screen Y reverses direction twice, but its shadow on screen X reverses direction only once.

3-81. Suppose the track in Figure 3-50 (see Problem 3-16) is extremely flexible. (Do not worry about difficulties the train may have in making turns that are too sharp.) Tell whether it is possible to bend the track to do each of the following. If it is possible, sketch one way of doing it. If it is not possible, tell why not.
a. Arrange the flexible track so that when the ball moves at constant speed, its shadow on screen X speeds up while its shadow on screen Y slows down.
b. Arrange the flexible track so that when the ball moves at constant speed, its shadow on screen X speeds up while its shadow on screen Y also speeds up.
c. Arrange the flexible track so that when the ball moves at constant speed, its shadow on screen X speeds up while its shadow on screen Y moves at constant speed.
d. Arrange the flexible track so that when the ball speeds up, its shadow on screen X speeds up while its shadow on screen Y slows down.

3-82. An object follows the path shown in red in Figure 3-75. Its position along the path is shown at 1-s intervals.
a. Is v_x, the x component of the object's velocity, changing as the object travels? Explain.
b. Is v_y, the y component of the object's velocity, changing as the object travels? Explain.

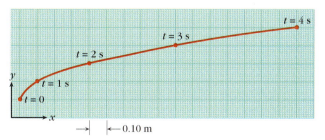

Figure 3-75 Problems 3-82 and 3-83

•**3-83.** Find the position of the object in Figure 3-75 at $t = 5$ s.

3-84. An object follows the path shown in red in Figure 3-76. Its position along the path is shown at 1-s intervals.

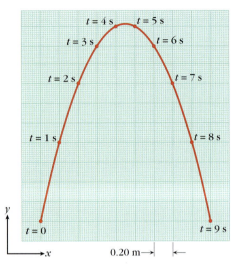

Figure 3-76 Problem 3-84

a. Complete the following table.

Time Interval	Δx	Δy
from 0 to $t = 1$ s		
from $t = 1$ s to $t = 2$ s		
from $t = 2$ s to $t = 3$ s		
from $t = 3$ s to $t = 4$ s		
from $t = 4$ s to $t = 5$ s		
from $t = 5$ s to $t = 6$ s		
from $t = 6$ s to $t = 7$ s		
from $t = 7$ s to $t = 8$ s		
from $t = 8$ s to $t = 9$ s		

b. Use the values from the table in a to find the average velocity components $\bar{v}_x$ and $\bar{v}_y$ required to complete the following table.

Time Interval	$\bar{v}_x$	$\bar{v}_y$
from 0 to $t = 1$ s		
from $t = 1$ s to $t = 2$ s		
from $t = 2$ s to $t = 3$ s		
from $t = 3$ s to $t = 4$ s		
from $t = 4$ s to $t = 5$ s		
from $t = 5$ s to $t = 6$ s		
from $t = 6$ s to $t = 7$ s		
from $t = 7$ s to $t = 8$ s		
from $t = 8$ s to $t = 9$ s		

c. Do either of the component motions have zero acceleration? Do either of the component motions have nonzero constant acceleration? Explain.

d. If either component motion has a nonzero constant acceleration, calculate its value.

3-85.
a. Can the magnitude of a vector remain constant while one of its components is increasing? Explain.

b. Can the magnitude of a vector remain constant while both of its components are increasing? Explain.

3-86. A rock is thrown horizontally with an initial speed v_o from a vertical cliff of height h. It lands on the flat plain below at a distance R from the base of the cliff. This happens somewhere on Earth. If the identical scenario is repeated on the moon,
a. how is the horizontal component of the velocity with which the rock hits the ground affected? Is it greater than, equal to, or less than it was on Earth? Explain. (Ignore differences in air resistance.)
•***b.*** how is the horizontal distance R traveled by the rock affected? Is it greater than, equal to, or less than it was on Earth? Explain. (Again, ignore differences in air resistance.)

3-87. SSM A small craft pilot in A-ville wishes to get to C-town, which is located 480 km from A-ville in a direction 37° N of E. But first she must make a stop in B-burg, 600 miles due east of A-ville. How far and in what direction must she fly to get from B-burg to C-town?

3-88. Would you prefer to have cruise control keep your car at constant speed or constant velocity? Briefly explain why.

3-89. Vectors $\vec{A}$ and $\vec{B}$ make angles θ_A and θ_B with the $+x$ axis. $\vec{R} = \vec{A} + \vec{B}$ is the vector sum of $\vec{A}$ and $\vec{B}$. The angle that $\vec{R}$ makes with the $+x$ axis is (*always, sometimes,* or *never*) equal to $\theta_A + \theta_B$. Briefly explain your answer.

3-90.
a. Sketch two vectors $\vec{A}_1$ and $\vec{A}_2$ that satisfy the following condition: $\vec{A}_1$ and $\vec{A}_2$ are equal in magnitude, and their difference $\Delta\vec{A} = \vec{A}_2 - \vec{A}_1$ is zero.
b. Now sketch two vectors $\vec{A}_1$ and $\vec{A}_2$ that satisfy this condition: $\vec{A}_1$ and $\vec{A}_2$ are equal in magnitude, but their difference $\Delta\vec{A} = \vec{A}_2 - \vec{A}_1$ is *not* zero. How do the vectors that you've drawn this time differ from the vectors you drew in part ***a***?
c. Suppose an object's speed is the same at instants t_1 and t_2. Is the change in velocity $\Delta\vec{v}$ necessarily zero over the time interval $\Delta t = t_2 - t_1$? Explain.
d. Is the average acceleration of the object in part ***c*** necessarily zero over the time interval $\Delta t = t_2 - t_1$? Explain.

3-91.
a. An object moves from position $\vec{r}_1$ at instant t_1 to position $\vec{r}_2$ at instant t_2, so that its position changes by $\Delta\vec{r} = \vec{r}_2 - \vec{r}_1$ over the time interval $\Delta t = t_2 - t_1$. If the object has an average velocity $\vec{\bar{v}}$ over this same interval, how do the directions of the vectors $\Delta\vec{r}$ and $\vec{\bar{v}}$ compare? Can you say anything about how their magnitudes compare? Explain.
b. An object's velocity changes from $\vec{v}_1$ at instant t_1 to $\vec{v}_2$ at instant t_2, so that $\Delta\vec{v} = \vec{v}_2 - \vec{v}_1$ over the time interval $\Delta t = t_2 - t_1$. If the object has an average acceleration $\vec{a}_{av}$ over this same interval, how do the directions of the vectors $\Delta\vec{v}$ and $\vec{a}_{av}$ compare? Can you say anything about how their magnitudes compare? Explain.

3-92. A team of engineering students has designed a radio-controlled model car. On its first test run, a team member stands with the control box while the car travels 7.8 m in a direction 60° S of E, and then 5.2 m due north, and then

15.0 m in a direction 15° N of W. If the car sends a signal back to the control box at that point, how far and in what direction must the signal be sent?

3-93. SSM In Example 3-7, we calculated the magnitude and the direction of the car's average acceleration as it traveled around a quarter of a circle at constant speed. If the car continued for another quarter-circle at the same speed,
a. would the magnitude of the average acceleration be the same as for the first quarter-circle? Explain.
b. would the direction of the average acceleration be the same as for the first quarter-circle? Explain.
c. are Equations 3-14x and 3-14y applicable to a car traveling in a circle at constant speed? Explain.
d. are Equations 3-16x through 3-18y applicable to a car traveling in a circle at constant speed? Explain.

3-94. A table tennis ball thrown from a window is strongly affected by air resistance, so it doesn't keep speeding up indefinitely as it falls. Instead, its speed levels off to some final value. Can Equations 3-16x through 3-18y be applied to the motion of this ball? Briefly explain.

3-95. Two students are arguing about playing soccer on the moon, where g has about $\frac{1}{6}$ of its value on Earth. Student A claims that with the same kick, the ball would rise higher and land farther down-field on the moon than on Earth. Student B agrees that the ball will rise higher, but argues that because the gravitational acceleration only affects the vertical motion, the ball will not land farther downfield than on Earth. With which student would you side? Briefly explain.

3-96. A runner doing a complete lap around the track in Figure 3-77 at constant speed (*always, sometimes,* or *never*) has a nonzero acceleration. Choose the correct answer, and briefly explain.

Figure 3-77 Problem 3-96

3-97. A car circles the track in Figure 3-78 at constant speed. Consider its horizontal velocity component v_x. Rank the labeled points on the track in order according to the value of v_x at each point, starting with the lowest (or most negative) value. Indicate where (if at all) the values are equal.

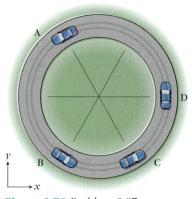

Figure 3-78 Problem 3-97

3-98. Suppose you have a camcorder and viewer that let you produce and view videos frame by frame. You decide to make a video of a child's ball bouncing along a sidewalk in front of a picket fence. How can you determine whether the ball's horizontal velocity component is constant or not? What complications can you think of in this approach, and how would you deal with them?

3-99. Describe a coordinate system (location of origin, direction of axes) that you could use in Figure 3-16 if you wanted the targets moving up the ramp to have positive velocities with no y components.

•• **3-100.** A ball is thrown from the point $x = 0$, $y = 0$ with an initial velocity of 50 m/s at an angle of 30° above the horizontal. Using the equations of motion for a projectile, obtain an equation that expresses the vertical position y in terms of the horizontal position x. It is possible to write this equation in the form $y = ax^2 + bx + c$.
a. Find the numerical values of a, b, and c.
b. Which of these constants is unitless?

3-101. Suppose that in Activity 3-2 you start with the pen toward the top of the ruler. You first move the ruler 10.0 cm to the right, and then move the pen 5.0 cm along the ruler toward the bottom of the paper.
a. What are the x- and y-components of the resultant vector?
b. Find the magnitude and direction of the resultant vector.

3-102. Both situations in Figure 3-79 take place on horizontal surfaces. In situation 1, the ball is rolling around the circular track at a constant speed of 10 m/s. In situation 2, the ball strikes a wall traveling at a constant speed of 10 m/s and bounces off in the direction shown at the same speed. For each situation, consider the ball's average acceleration over the interval that the ball takes to get from A to B.
a. In which situation, if either, is the magnitude of this average acceleration greater?
b. In which of these situations (1, 2, both, or neither) is the direction of this average acceleration towards the center of the circle?

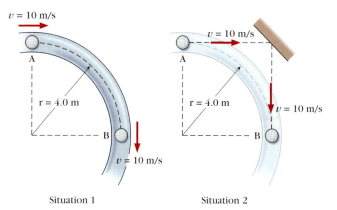

Situation 1   Situation 2

Figure 3-79 Problems 3-102 and 3-103

• **3-103.** Consider the two situations described in Problem 3-102.
a. In situation 1, find the magnitude of the ball's average acceleration over the interval that the ball takes to get from A to B.
b. Repeat for situation 2.

3-104. Suppose an object starts out traveling to the right (the $+x$ direction) at a speed of 10 m/s. Suppose its velocity then changes by $\Delta \vec{v}$, and $\Delta \vec{v}$ has a magnitude of 10 m/s. In what two directions could $\Delta \vec{v}$ be for the object to end up with a speed of 3 m/s? Express each answer by giving the angle that $\Delta \vec{v}$ makes with the $+x$ axis.

Problems on WebLinks

3-105. Suppose that in Figure 3-80, points 1 and 2 are the endpoints of the path originally traced by the camper's lantern in WebLink 3-1. Which of the paths in Figure 3-80 would have been traced by the lantern if the camper had not changed his motion along the raft but the river's flow had gained speed?

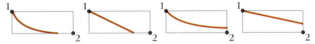

Figure 3-80 Problem 3-105

3-106. In WebLink 3-2, a three-dimensional motion is produced by combining ____

a. a uniformly accelerated straight-line motion and a constant-speed straight-line motion.

b. a constant-speed straight-line motion and a circular motion.

c. a uniformly accelerated straight-line motion and a circular motion.

d. two circular motions.

3-107. In WebLink 3-3, a motion that one might experience on an amusement-park ride is produced by combining ____.

a. a uniformly accelerated straight-line motion and a constant-speed straight-line motion.

b. a constant-speed straight-line motion and a circular motion.

c. a uniformly accelerated straight-line motion and a circular motion.

d. two circular motions.

3-108. In WebLink 3-4, the weight that moves up and down the slide wire has a light on it that lets it be seen at night. The motion observed at night that is actually shown in the WebLink is a composite motion. Its two component motions are ____.

a. uniformly accelerated upward motion and uniformly accelerated downward motion.

b. uniformly accelerated vertical motion and uniformly accelerated horizontal motion.

c. uniformly accelerated vertical motion and constant speed horizontal motion.

d. constant speed upward motion and uniformly accelerated downward motion.

3-109. WebLink 3-5 shows that the sum of two vectors (*always, sometimes,* or *never*) depends on the order in which you add them.

3-110. In WebLink 3-6, suppose that instead of having equal magnitudes, vector $\vec{A}$ has a magnitude of 5 and vector $\vec{B}$ has a magnitude of 3. Then as vector $\vec{B}$ rotates, the resultant $\vec{R}$ can have a magnitude anywhere in the range from ____ to ____.

3-111. In Problem 3-110, the resultant $\vec{R}$ can have an *x* component anywhere in the range from ____ to ____ as vector $\vec{B}$ rotates.

3-112. In Problem 3-110, the resultant $\vec{R}$ can have a *y* component anywhere in the range from ____ to ____ as vector $\vec{B}$ rotates.

3-113. According to WebLink 3-7, which (one or more) of the following must you do when you add two vectors?

a. Add their magnitudes to get the magnitude of the resultant.

b. Add their *x* components to get the *x* component of the resultant.

c. Add their *y* components to get the *y* component of the resultant.

d. Add their directional angles to get the directional angle of the resultant.

3-114. Figure 3-81 shows a portion of the circular motion treated in WebLink 3-9. During the interval in which the object travels from A to B at constant speed, ____.

a. Δv_x is positive and Δv_y is negative.

b. Δv_x is negative and Δv_y is positive.

c. Δv_x and Δv_y are both positive.

d. Δv_x and Δv_y are both negative.

e. Δv_x and Δv_y are both zero.

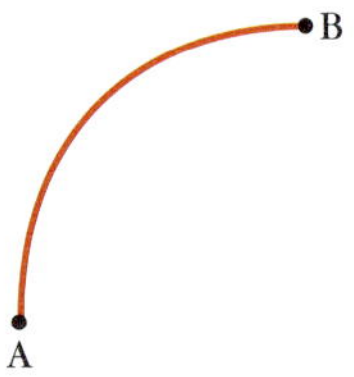

Figure 3-81 Problem 3-112

CHAPTER FOUR

Interactions and Newton's Laws of Motion

The quantitative tools we developed in Chapters 2 and 3 *describe* motion, but they don't tell us *why* objects move as they do. We must now ask under what conditions does an object speed up, slow down, change direction, or just keep moving the same way?

In 1665, another student began seriously contemplating these same questions. He was home from Trinity College of Cambridge University because the university had been forced to close by the Great Plague that was again sweeping England. Over the next 18 months, the young Isaac Newton, then 22 years old, generated many of the ideas that have been at the very foundations of physics ever since. Among other things, Newton developed an interconnected framework of ideas, which came to be known as **Newton's laws of motion,** for dealing with the question of why things move as they do.

In this chapter you will explore the concepts involved in Newton's laws. In Chapter 5, we will use these concepts in applying

"Newton's laws of motion . . . deal . . . with the question of why things move as they do."

Newton's laws to a range of quantitative problems. As you encounter Newton's ideas, you need to consider how they match with your own. Do they conflict? Is one set of ideas more useful than the other? Which ideas are more consistent with the way things really behave? Do things always behave as your ideas would lead you to expect? These are questions you will need to consider over and over as you progress through the chapter.

4-1   Newton's First Law: Inertia and the Concept of Force

Let's begin with *your* ideas about why things move the way they do. **STOP**&**Think** Under what conditions does an object speed up, slow down, change direction, or just keep moving the same way? Which of these will happen only if something else does something to the object? Write down your reasoning. It is important that you take the time to articulate and examine your own ideas. They represent a lifetime of impressions formed from both your own direct experience of the physical world and your ongoing exposure to what other people say and think. ◆

In developing his ideas, Newton was engaging in that most basic of physics activities: observing how things behave in the natural world and trying to figure out the rules by which the great game of nature is played. To get meaningful answers in this kind of inquiry, you have to ask good questions. This can itself be a very difficult task. Newton had to develop a suitable and well-defined vocabulary with which to word the questions and in which the answers could take shape.

If we look at a moving object and the ways in which its motion may be affected, we can perhaps reason somewhat as Newton did. To do so, let's follow the progress of a disk in a game of shuffleboard (Figure 4-1) played in its traditional shipboard setting.

Figure 4-1 A game of shuffle-board.

Case 4-1 ◆ *A Game of Shuffleboard*

Starting from rest, the disk picks up speed as the cue pushes it. **STOP**&**Think** After the disk leaves the cue, does the cue have any effect on it? ◆ After leaving the cue, it comes gradually to a stop. If the disk rebounds quite squarely off another disk (Figure 4-2a) while in motion, its direction changes. During each of these stages, the velocity changes, whether in magnitude or in direction, so during each stage there is an acceleration.

Suppose the conditions change. If the cue gives the disk a gentler push, the disk doesn't gain as much speed. If it turns cold after a rain, so that a thin layer of ice coats the deck, the disk slows down more gradually and travels further before stopping. If the first disk just grazes the second one (Figure 4-2b), its direction

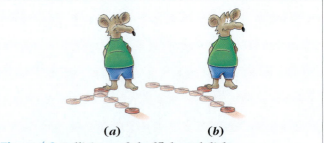

         (a)   *(b)*
Figure 4-2 Collisions of shuffleboard disks.

changes only slightly. In each of these circumstances, the disk is interacting less strongly than before—with the cue, with the ship's deck, with the second disk. In each case, its velocity changes less, whether in magnitude or direction, so that the acceleration is less.

Here we see the usefulness of talking about the vector quantity *velocity*. To find a rule that covers all circumstances, we have to associate interactions with changes of velocity rather than speed. For example, in every one of the situations in Case 4-1, the less the disk interacted with its surroundings, the more nearly it continued moving at the same velocity. Newton reasoned that if an object were moving free of *all* interaction with its surroundings then the object would keep going at the same velocity forever. Such total isolation can never be completely achieved in the real world, but it is nevertheless imaginable, and imagining what happens in extreme cases can be a valuable guide to our thinking.

This line of reasoning did not originate with Newton. Galileo had been the first to propose it, and it was revolutionary because of its implication that if an object is moving in isolation, you don't have to do anything to keep it going. This is counter to our ordinary experience, because in our everyday world objects do *not* exist in isolation, and they *do* slow down when we do nothing. But conditions like the iced-over deck point the way to thinking about what happens when an object's interactions with surrounding objects are progressively reduced. It was Galileo's genius to extend this reasoning to an idealized situation free of all interactions, and it was Newton's genius to recognize the power of this idea and incorporate it into the interconnected framework of his laws of motion.

Newton's expression of Galileo's idea is called Newton's first law of motion, here translated from the original Latin: *"Every body perseveres in its state of rest, or of uniform motion in a right line, unless it is compelled to change that state by forces impressed thereon."* (In modern English, a *right line* is a straight line.) This statement is also called the **law of inertia,** because inertia means the tendency of an object to resist change and continue doing exactly what it was doing. With regard to motion, it means the tendency to keep going at the same speed in the same direction; in other words, at the same velocity.

➥ Italian scientist Galileo Galilei (1564–1642), who made major contributions to physics and astronomy, died the year Newton was born.

Newton's first law of motion, restated in modern English:

A body will continue to move forever at the same velocity (same speed and direction) unless a nonzero total outside force is exerted on it.

4-2 Exploring the Meaning of Force

Newton's first law introduces the concept of *force* by describing what happens in its absence. But what *is* a force? Because force is central to the vocabulary in which Newton's ideas took shape, you must develop a feel for what force means in the Newtonian picture. **STOP&Think** What associations do *you* have with the word *force?* What *is* force (or a force)? What examples of force or forces can you think of? Again, it is important that you write down your ideas. ◆

STOP&Think In your view, which of the following words are synonyms for force?

Power   Energy   Push   Action

Review your answers from time to time as you progress through this and subsequent chapters, and revise them as necessary. ◆

◆**TYPES OF FORCES** Forces were Newton's way of dealing with interactions. In some interactions, the interacting objects press or rub against each other. When the interacting objects actually touch each other, like the disk and the cue, or the disk and the deck, we speak of the forces associated with these interactions as

Examples of contact (touching) interactions or forces	Examples of action-at-a-distance interactions or forces
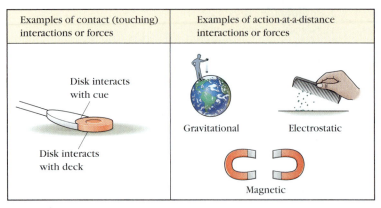 Disk interacts with cue / Disk interacts with deck	Gravitational Electrostatic / Magnetic

Figure 4-3 Types of interactions or forces.

contact (touching) **forces** (Figure 4-3). But objects can also interact when they do not actually touch. You see instances of this when you release two magnets separated by a small distance and they come together, or when you run a comb through your hair and then hold it above some tiny scraps of tissue paper (try this now if you've never done it), so that the scraps "jump" to the comb. A less obvious instance occurs when you release a rock from a height and it falls to the ground. **STOP**&**Think** What is the rock interacting *with?* ◆ The interaction between the paper scraps and the comb is called **electrostatic,** the interaction between Earth and the falling rock is called **gravitational.** Note that the scraps are *not* attracted to a magnet; electrostatic and magnetic forces are *not the same.* Gravitational, electrostatic, and magnetic forces all fall into the category of *noncontact* forces, which are commonly called **action-at-a-distance forces.** The two categories of forces are summarized in Figure 4-3.

◆**A FORCE AS ONE SIDE OF AN INTERACTION** In some limited contexts, as when speaking of the forces that occur in nature, physicists do not distinguish between forces and interactions. But interactions are two-way; ordinary use of the word *force* focuses on one of the two "directions." For example, as the disk slides across the deck, we may say it is slowed down by *friction,* which is not a thing but rather the type of interaction that occurs when the disk and the deck rub or scrape against each other (as evidence of this, both deck and disk become scuffed after extensive play). We can describe the two-way interaction in terms of two component one-way actions:

(1) { The deck rubs against the disk.
     The deck acts on the disk.
     The deck exerts a force on the disk.

and

(2) { The disk rubs against the deck.
     The disk acts on the deck.
     The disk exerts a force on the deck.

It is important for you to recognize that action here does not necessarily mean *willful* action. It simply means that one body or object affects the other. If you willfully smash your fist into a brick wall, the wall's response is totally passive, but the wall certainly has an effect on your fist and its motion. In this Newtonian sense, the wall acts on your hand. A force refers to this action of one body or object (A) on another (B). To describe a particular force, you must be able to identify A and B in a sentence of the form,

A exerts a force on B.   (Form 4-1)

Warning: In contrast to modern everyday usage, Newton's use of the word *reaction* does *not* imply that the reaction comes after the action. On the contrary, action and reaction, as two inseparable sides of one interaction, must occur together in time. Thus, like saying *one force* and *the other force,* it doesn't matter which you call the action and which the reaction.

The reciprocal aspect of the interaction, or what Newton called the *reaction,* can then be described by a similar sentence with the roles of A and B reversed:

B exerts a force on A.   (Form 4-2)

In Newton's terms, it is the action or force *on B* that affects the motion of B. More broadly, Newton's first law says that the speed and direction of an object such as B remain unchanged "unless a nonzero total outside force is exerted on it." Why does the law speak of a *nonzero total* force? Why does it speak of an *outside* force? What does it mean by these?

Consider an equally matched game of tug-of-war. Each team exerts a force that by itself would set the rope in motion. But because the forces are in opposite directions, the combination of forces (the *total force*) on the rope produces no motion. We can anticipate that because the forces have directions, we can treat them as vectors, and that the vector sum of equal and opposite forces on a single object is zero. In other words, there is *not* a *nonzero total* force, so the rope's motion remains unchanged.

And why do we speak of an *outside* force? Consider the critter on the skateboard in Figure 4-4a. **STOP&Think** By pressing on its own nose, as shown, can the critter set itself in motion? Speed up? Slow down? Change directions? Can you ever affect your own motion by giving yourself a push? ◆ We will have more to say about this when we discuss Newton's third law of motion, but your own experience should tell you that forces a body exerts on itself cannot affect the motion of the body as a whole. The critter in Figure 4-4b can flail his limbs, but lacking something outside himself with which to interact, he can go nowhere.

◆**FORCES AND OUR INTUITIVE IDEAS ABOUT PUSHES AND PULLS** As these examples suggest, it is often useful, even though it is an oversimplification, to think of a force as a *push* or a *pull,* because you have well-developed intuitions about how it feels when *you* push or are pushed and how things move in such circumstances. But be careful. Your intuitions about pushes and pulls may not mesh completely with the physicist's view of forces within the carefully structured context of Newton's laws of motion. For example, if you picture yourself giving a child on a swing a push, the push has duration. The force you exert may increase and then ease up over the whole time interval of the push; it has different instantaneous values at different instants during the push. So it is more accurate to think of a force as the instantaneous strength of a push or pull. We can only have meaningful discussion about the effect of forces on motion if we share the same understanding of what forces are and can agree on what forces are involved in a given situation.

For instance, the shuffleboard cue clearly pushes the disk, but does the deck push on the disk in the opposite direction while the disk is traveling, or does it merely resist the disk's motion? Is that a meaningful distinction? Different readers may be led by their intuitions to answer these questions differently. But no matter which way you answer, the observable behavior is that when the deck is in contact with the disk, it acts on it; that is, it has an effect on its motion. For that reason, a physicist would say the deck exerts a force on the disk.

What about the interaction between the cue and the disk? The two sides of this interaction, put into sentences of Forms 4-1 and 4-2, are

Cue exerts force on disk.

Disk exerts force on cue.

Does this make intuitive sense to you? It is usual to say that the cue pushes on the disk. But does the disk "push on" the cue, or in any event exert a force on it?

What happens in general when two bodies interact in this way? For example, if you put your physics book on a table, does the table exert an upward force on the book? The following discussion may help you develop a fit between your intuitions and Newton's ideas.[1] For a fuller and more interactive version of this discussion, see WebLink 4-1.

[1]This treatment is based on ideas developed by J. Minstrell (*Physics Teacher,* **20,** p. 10, 1982) and J. Clement (Proc. 2nd Int. Seminar: Misconceptions and Educational Strategies in Science and Math, Vol. 3, p. 84, 1987)

➡**Watch your language:** We say that one body exerts a force on or acts on another. Thus it is correct to say that *a force is exerted* on a body but it can be misleading to say that *a force acts* on a body. Strictly speaking, a *body* acts on a body, and the action itself is what we call a force. Nevertheless, you will find *a force acts* in common usage, even by physicists, when *a force is exerted* (that is, exerted *by some body*) is meant.

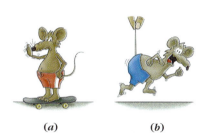

(a)   (b)

Figure 4-4 An object's motion cannot change without an *external* force on it. (*a*) You cannot set yourself in motion by pushing on yourself. (*b*) You cannot set yourself in motion with nothing to push off from.

For **WebLink 4-1:** **Does the Table Exert a Force on the Book?,** go to www.wiley.com/college/touger

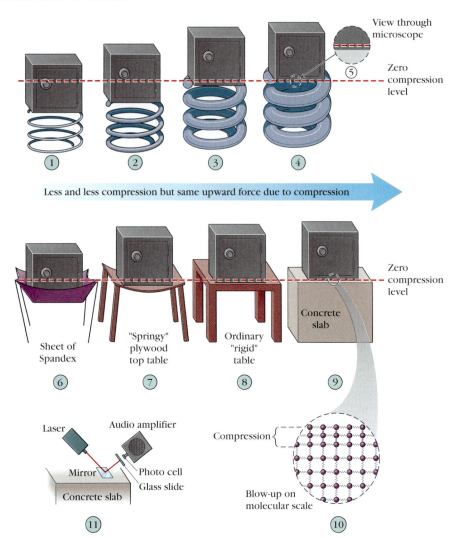

Figure 4-5 Picturing how an upward force is exerted on an object when it rests on a flat surface.

People generally agree that when you push on a spring to compress it, it pushes back. In parts ①–④ of Figure 4-5, the spring pushes back up on the safe sitting on it. The tighter and thicker the spring, the less it has to be compressed for it to push up hard enough to support the safe. We can extend that reasoning even to part ④, where the compression is microscopically small ⑤, but is still responsible for an upward force on the safe. Objects other than springs also have varying degrees of springiness. The objects in parts ⑥–⑨ follow the same progression from more to less springy. The blow-up ⑩ shows that in ⑨ there is still springiness at the atomic or molecular level. There is always some deformation and some consequent springback of any body that supports the weight of another; that is the **mechanism** by which the supporting body exerts a force on its load. For a concrete slab, the extremely small deformations can be observed by the detection set-up in ⑪. When the mirror is displaced microscopically and then restored to its original position, the laser beam's intensity varies correspondingly (the details of why this happens depend on certain properties of lasers). When the variation is turned into a sound signal by the audio amplifier, you actually hear the diminishing *SPROI-oi-oing* of the concrete slab's surface (on which the mirror rests) returning to its undisturbed position. You can see that this happens when a body deforms *visibly,* and you *hear* the same thing with the concrete slab. So it makes sense to envision the intermolecular or interatomic bonds—that is, the forces that hold the molecules or atoms together—as springlike ⑩. We can picture the microscopic deformation of all these little springs, so that the concrete slab, like the table, exerts an upward force on the book.

These little springs are, of course, not literally there. They are a model, a useful imaginative metaphor for the workings of interatomic forces that are actually rather more complicated. The atoms or molecules are separated by empty space, and the interatomic forces are actually action-at-a-distance forces. What is happening is more like pushing together a pair of mutually repelling magnets, though the forces involved here are not magnetic but electrostatic. The contact forces that we experience on a macroscopic level are always the consequence of more fundamental action-at-a-distance forces exerted on an atomic scale. We have *not* explained how action-at-a-distance forces work. Newton did not explain that either, stating that he did not make hypotheses but simply inferred the existence of these forces from how objects were observed to behave.

We have established that within Newton's framework of ideas,

a force is an action or effect of one of two interacting bodies (A) on the other (B), which, in the absence of other bodies, would *change the motion of body B.*

This is the real Newtonian criterion for whether a force exists. In the case of some contact forces, our ability to describe an underlying mechanism resembling compressed springs can help us understand why two bodies exert forces on each other. Where we can make connections to human effort, our intuitive ideas about pushes and pulls may give us a feel for *some* of these contact forces. For action-at-a-distance forces, arguments based on mechanism or intuition are not so readily at hand, but we can still see that the above criterion for a force is met. In Figure 4-5 ⑧, for example, Earth exerts an action-at-a-distance force on the safe, and if you quickly pulled the table out from under the safe, this force would clearly change the motion of the safe.

4-3　Newton's Second and Third Laws

Although Newton's first law is stated in the negative, it implies that the velocity *will* change, in magnitude or direction or both, and that therefore the body will accelerate when a nonzero total force is exerted on it. But it does not tell us how to measure or in any way attach numbers to the notion of force. It does not tell us how much force or how strong an interaction it takes to produce a particular amount of acceleration, or whether that depends on what's being accelerated. So our concept of force is not yet fully developed. The way we address these questions must be consistent with what Newton's first law says about forces. To do so, we will develop the ideas expressed in Newton's third and second laws, in that order. As you will see, the third law makes more specific the interactive nature of forces. The second law then clarifies how one side of the interaction—a force—affects the motion of a single body.

In one common type of situation, two bodies that are both free to move "push off" from each other—a swimmer jumping off a rowboat, a child jumping forward from a wagon, a bullet fired from a recoiling rifle. The bodies exert forces on each other and are set in motion—they accelerate—in opposite directions. But the accelerations are not equal in magnitude. Common experience should tell you that in each case the heavier or more massive body will not end up going as fast;[2] it will accelerate less. If you doubt this assertion, get together with someone of substantially different weight than yourself, and while both of you are on skateboards or roller skates or ice skates, push off from each other and see what happens.

If the two bodies interact more strongly—they push off harder from each other—each accelerates more, but measurements (such as those described in

---

[2]We will later draw a distinction between mass and weight, but under the same conditions, the more massive object will still be the heavier one.

Figure 4-6 Measuring how the accelerations of two interacting objects vary with time. (*a*) Magnets Ⓜ are mounted on two air-track gliders Ⓖ so that they repel each other. The gliders are moved close together on the air track Ⓐ and then released. Reflectors Ⓡ serve as targets for the range finders Ⓕ, which plot the accelerations of the gliders against time. (*b*) The accelerations are plotted for a pair of gliders with masses m_1 and m_2. The table shows values of the accelerations a_1 and a_2 at equal time intervals, as well as their ratio $\frac{a_2}{a_1}$. Note that as a_1 and a_2 vary, this ratio remains constant. (*c*) The accelerations are plotted here for a second pair of gliders having a different ratio of masses ($\frac{m_3}{m_1} \neq \frac{m_2}{m_1}$). Note that the ratio $\frac{a_3}{a_1}$ also remains constant, but it has a different constant value than $\frac{a_2}{a_1}$. The data shown here were produced by a computer simulation, not by actual measurement. Values of t are in s and a in m/s^2.

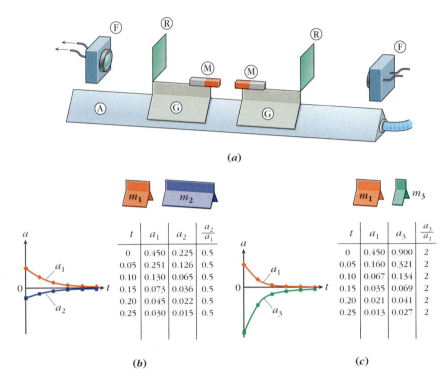

(*a*)

t	a_1	a_2	$\frac{a_2}{a_1}$
0	0.450	0.225	0.5
0.05	0.251	0.126	0.5
0.10	0.130	0.065	0.5
0.15	0.073	0.036	0.5
0.20	0.045	0.022	0.5
0.25	0.030	0.015	0.5

(*b*)

t	a_1	a_3	$\frac{a_3}{a_1}$
0	0.450	0.900	2
0.05	0.160	0.321	2
0.10	0.067	0.134	2
0.15	0.035	0.069	2
0.20	0.021	0.041	2
0.25	0.013	0.027	2

(*c*)

Figure 4-6) invariably show that the *ratio* of the magnitudes of the accelerations remains the same, so long as the same two bodies are interacting, and the combined effect of all other interactions on these bodies is negligible. We express this relationship as $\frac{a_B}{a_A} = c_{AB}$, where the constant c_{AB} is a property of the two bodies.

♦**NEWTON'S THIRD LAW AND THE CONCEPT OF MASS** We observe the behavior of interacting bodies and we see two patterns being followed without exception: (1) $\frac{a_B}{a_A} = c_{AB}$, for a given pair of interacting bodies, and (2) the more massive body's motion is less changed, or, in the language we introduced earlier, it has greater *inertia*. We will use these patterns to give precise quantitative meaning to the concept of **mass.** We will take mass to be a measure of a body's inertia, and we will speak of it as the **inertial mass** of the body.

We can express both patterns at once if we *define* mass so that the constant c_{AB} for a given pair of bodies will be the ratio of their masses m_A and m_B; that is, we set $c_{AB} = \frac{m_A}{m_B}$, so that

$$\frac{a_B}{a_A} = \frac{m_A}{m_B} \qquad (4\text{-}1)$$

Note that we have set up the ratio so that if one body's acceleration is the numerator on the left, the other body's mass is the numerator on the right. That is precisely because if m_A is larger, the other body's acceleration a_B will be larger (and both ratios will be greater than one). Note also that the actual values of m_A and m_B don't matter. For instance, whether $\frac{m_A}{m_B} = \frac{2\text{ kg}}{1\text{ kg}}$ or $\frac{2000\text{ g}}{1000\text{ g}}$, the ratio stays the same. Picking the value of m_B is equivalent to choosing units: Once m_B is fixed, m_A will be c_{AB} times as massive. The standard unit of mass in SI is the kilogram (kg). After defining force, we will discuss how mass differs from weight.

♦**DEFINING FORCE** Equation 4-1 can be rewritten as

$$m_A a_A = m_B a_B \qquad (4\text{-}2)$$

or in vector form as

$$m_A \vec{a}_A = -m_B \vec{a}_B \qquad (4\text{-}3)$$

The minus sign tells us that the acceleration vectors are in opposite directions, as they must be to match our observations. In this form, only quantities describing body A are on the left, and only quantities describing body B are on the

➤**A note on language:** Many physics books will speak of masses when they mean bodies. Strictly speaking, mass is a property or characteristic of a body, not the body itself.

right. We have separated the interaction out into two sides, each side describing the action on one of the participating bodies. But the action on one of the two bodies is what we have called the *force* on that body. We can make the math consistent with this idea of force by *defining* $m_A \vec{a}_A$ to be the force exerted on body A

$$\vec{F}_{\text{on A (by B)}} \equiv m_A \vec{a}_A \qquad (4\text{-}4a)$$

and $m_B \vec{a}_B$ to be the force exerted on body B

$$\vec{F}_{\text{on B (by A)}} \equiv m_B \vec{a}_B \qquad (4\text{-}4b)$$

when bodies A and B interact only with each other (or equivalently, when the combined effect of all other interactions is negligible).

The vectors on both sides of each equation must have the same direction, so the **force vector** exerted on a body must be in the same direction as the body's acceleration. Also, because we can speak of either average or instantaneous acceleration, we correspondingly can speak of either an average or an instantaneous force.

With the forces defined by Equations 4-4a and 4-4b, Equation 4-3 says that

$$\vec{F}_{\text{on A (by B)}} = -\vec{F}_{\text{on B (by A)}} \qquad (4\text{-}5)$$

Equation 4-5 is a concise statement of **Newton's third law of motion.** In Newton's own words (translated from his Latin), the third law says, *To every action there is always opposed an equal reaction: or the mutual actions of two bodies upon each other are always equal, and directed in contrary parts.*

Newton's third law of motion, restated in modern English: *The forces that two interacting bodies exert on each other are always equal in magnitude and opposite in direction.*

In Figure 4-7, the forces on the gliders in Figure 4-6 are shown at three successive separations. Note that although the forces weaken as the gliders move apart, Equation 4-5 remains instantaneously true at each separation. The graph in Figure 4-7 underscores this point.

The two forces in Newton's third law, commonly called an *action-reaction pair* but better called an **interaction pair,** are the two one-way actions that constitute the interaction. Therefore,

- *they can never be on the same body,* but are always exerted by two bodies *on each other.* **STOP**&**Think** If your hand and the doorknob exert equal and opposite forces on each other, does that mean you can't open the door? ◆

- they must both be the *same kind of force* (normal, gravitational, frictional, etc.) because they are two aspects of a single interaction.

- they are *instantaneously* equal and opposite, never one after the other. A pair of equal and opposite forces that does not meet all these criteria cannot be an interaction pair.

PROCEDURE 4-1

Identifying Interaction (Action-Reaction) Pairs

1. Identify the type of interaction.
2. Identify the two "sides" of the interaction:
   a. Describe one of the two forces by a sentence of Form 4-1.
   b. To describe the other force, interchange A and B in Form 4-1 to obtain a sentence of Form 4-2.
3. Because the two forces are vectors, give their directions, making sure they are opposite.

Not all pairs of equal and opposite forces are interaction pairs.

What does Newton's third law tell us? Is the force that the child is exerting on the father less than, equal to, or greater than the force that the father exerts on the child?

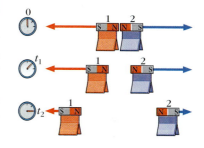

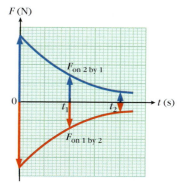

Figure 4-7 Time dependence of the forces that two objects exert on each other. Here are the two gliders with repelling magnets from Figure 4-6 at three successive separations. Note that although the magnets exert less force on each other as they separate, the forces they exert on each other remain equal and opposite.

◆**INTERNAL FORCES** The idea of interaction pairs sheds further light on Figure 4-4a. When the critter's finger exerts a force on its nose, its nose must exert an equal and opposite force on its finger. Both of these are forces on the critter, so the sum of the forces on the critter taken as a whole is zero. Thus, they cannot affect the critter's motion. Because forces exerted by parts of a body on each other—**internal forces**—necessarily occur in interaction pairs, their net effect on a body's motion is always zero.

Example 4-1 *A Gravitational Interaction*

For a guided interactive solution, go to Web Example 4-1 at www.wiley.com/college/touger

A hammer dropped by a roofer speeds up as it falls to the ground. Assume that air resistance is negligible. As completely as you can, describe the force responsible for the acceleration of the hammer as it falls, and then describe the reaction force. Write down your answer before reading the solution.

Brief Solution

We follow the steps of Procedure 4-1.

1. *Type of interaction: gravitational* interaction between the hammer and Earth.
2a. *"Earth exerts a gravitational force on the hammer"* (sentence of Form 4-1 with A = Earth, B = hammer). The type of interaction—*gravitational*—necessarily describes each side of the interaction.
2b. Interchange A and B: *"The hammer exerts a gravitational force on Earth."*
3. The gravitational force on the hammer is toward Earth, so the gravitational force on Earth is toward the hammer.

Making sense of the results. Does the hammer really exert a gravitational force on Earth, and if so, why doesn't Earth come up part of the way to meet it, or does it? From Equation 4-3, it follows that $\vec{\mathbf{a}}_2$ will be smaller in magnitude than $\vec{\mathbf{a}}_1$ if m_2 is larger than m_1. If the hammer's mass is about 1 kg, Earth's mass is about 6×10^{24} times as great, so its acceleration will be only $\frac{1}{6 \times 10^{24}}$ as great. We *can* in fact think of the two bodies as being pulled together, but Earth's share of the motion is negligibly small.

◆ Related homework: Problems 4-15 and 4-18.

Example 4-1 deals with gravitational forces. The gravitational force exerted on an object by Earth or another large heavenly body is what we commonly call the object's **weight**. (We will have more to say about this in Chapter 5.)

Mass Is Not the Same Thing as Weight

- An object's weight is proportional to its mass (we'll treat this in more detail in Chapters 5 and 8), so more massive objects weigh more under the same gravitational conditions. For that reason, either quantity can be used, say, to measure how much flour you are buying. The English system uses pounds (units of weight); the metric system uses kilograms (units of mass). However . . .

- How much a given mass weighs depends on what is attracting it gravitationally and how far away it is. You weigh less on the moon than on Earth, and even less a million kilometers from either, but your mass, a measure of your inertia, is the same everywhere. Therefore . . .

- When two bodies "push off" from each other in deep space, away from all large gravitational attractors, the more massive of the two still accelerates less even though both bodies are nearly weightless.

- Weight, the gravitational force on an object, is a vector; mass is a scalar.

◆**UNITS OF FORCE** Because Equations 4-4a and 4-4b define force, a unit of force must be a unit of mass (kg) times a unit of acceleration (m/s^2), so that one unit of mass times one unit of acceleration equals $1\ \text{kg} \cdot \text{m/s}^2$. The resulting SI unit of force is called a **newton** (N):

$$1\ \text{N} \equiv (1\ \text{kg})(1\ \text{m/s}^2) \qquad (4\text{-}6)$$

A newton is the amount of force that will cause a 1 kg mass to experience an acceleration of 1 m/s^2. $1\ \text{N} \approx 0.225\ \text{lb}$ (a little less than a quarter of a pound). **STOP&Think** Estimate your own weight in newtons. ◆

In the English system, pounds (lb) are units of force, not mass. Because acceleration in English units is in ft/s^2, the corresponding units of mass must be units of force divided by units of acceleration: $\frac{\text{lb}}{\text{ft/s}^2}$. The amount of mass that has an acceleration of 1 ft/s^2 when 1 lb of force is exerted on it is called a **slug**:

$$1\ \text{slug} \equiv \frac{1\ \text{lb}}{1\ \text{ft/s}^2} \qquad (4\text{-}7)$$

STOP&Think The label on a ketchup bottle reads "net wt. one pound (454 g)." Is one pound the same thing as 454 g? ◆

◆**NEWTON'S SECOND LAW OF MOTION** From Equations 4-4a and 4-4b, we can generalize that if a body (call it A) accelerates due to the combined effect of its interactions with bodies 1, 2, 3, etc., then

$$\Sigma \vec{\mathbf{F}}_{\text{on A}} = \vec{\mathbf{F}}_{\text{on A by 1}} + \vec{\mathbf{F}}_{\text{on A by 2}} + \vec{\mathbf{F}}_{\text{on A by 3}} + \cdots = m_A \vec{\mathbf{a}}_A \qquad (4\text{-}8)$$

This is **Newton's second law of motion,** stated in equation form.

Note that the sum $\Sigma \vec{\mathbf{F}}_{\text{on A}}$ is a vector sum (see Figure 4-8). It is sometimes called the net force on A. This resultant force, rather than a force exerted on A by any individual body (1, 2, 3, etc.), is responsible for A's acceleration.

In Newton's own words (again translated), the second law of motion says, *The alteration of motion is ever proportional to the motive force impressed; and is made in the direction of the right line in which that force is impressed.* In modern English, an *alteration of motion* always involves a change in the velocity vector of an object of fixed mass. *Impressed* means exerted, and a *right line* is a straight line. Newton's words correspond to rewriting Equation 4-8 as $\vec{\mathbf{a}}_A = \frac{1}{m_A} \Sigma \vec{\mathbf{F}}_{\text{on A}}$ (the constant of proportionality is $\frac{1}{m}$).

Newton's second law of motion, restated in modern English: *The acceleration vector of a body is proportional to the vector sum of all forces exerted on that body by other bodies. The constant of proportionality is the inverse of the body's **inertial mass.***

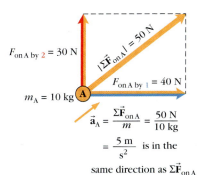

$F_{\text{on A by 2}} = 30\ \text{N}$

$|\Sigma \vec{\mathbf{F}}_{\text{on A}}| = 50\ \text{N}$

$F_{\text{on A by 1}} = 40\ \text{N}$

$m_A = 10\ \text{kg}$

$$\vec{\mathbf{a}}_A = \frac{\Sigma \vec{\mathbf{F}}_{\text{on A}}}{m} = \frac{50\ \text{N}}{10\ \text{kg}}$$

$$= \frac{5\ \text{m}}{\text{s}^2} \ \text{is in the}$$

same direction as $\Sigma \vec{\mathbf{F}}_{\text{on A}}$

Figure 4-8 Newton's second law involves a *vector* sum of forces.

Example 4-2 *A Qualitative Application of the Second and Third Laws*

In a supermarket parking lot that has become dangerously iced over, a large delivery truck goes into a skid and collides with an abandoned shopping cart.
a. Of the two colliding objects, which one affects the other's motion more?
b. Which of the two colliding objects exerts a greater force on the other during the collision?
c. Is there any disagreement between the answers to **a** and **b**? Explain.
STOP&Think Try to answer this as much of this as you can before you read the solution. ◆

EXAMPLE 4-2 *continued*

Solution

Choice of approach. Part **a** can be answered on the basis of your common-sense impressions or intuitions. In parts **b** and **c,** you need to apply the physics reasoning of Newton's second and third laws. In particular, in part **c,** you should resolve any conflicts between your "commonsense" answers to **a** and your physics answer to **b.**

The details.

a. The shopping cart may get knocked across the lot, while the cart would do little to alter the truck's skid. Most of us would agree that the shopping cart's motion is affected more. Common experience is likely to tell us that the cart would be more affected in other ways as well.

b. The collision is an interaction. The forces that the truck and the cart exert on each other are the two "sides" of the interaction. Newton's third law tells us that these two forces are not only opposite but equal.

c. Newton's second law can be rewritten as $\vec{a}_A = \frac{1}{m_A}\Sigma\vec{F}_{\text{on A}}$. Assume the only unbalanced force (and therefore the total force) on each of the colliding objects is the force that the other object exerts on it, so $\Sigma\vec{F}_{\text{on cart}}$ and $\Sigma\vec{F}_{\text{on truck}}$ are equal. But m_{cart} and m_{truck} are not equal. Because the mass is in the denominator, the less massive cart will have the greater acceleration. That is the effect on its motion. The second law tells us that equal forces may affect the motion of two interacting objects unequally.

◆ Related homework: Problems 4-19 and 4-20.

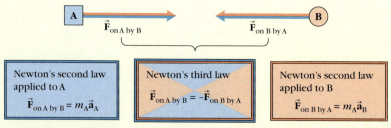

Example 4-3 *Attracting Magnets: A Quantitative Application of the Second and Third Laws*

For a guided interactive solution, go to Web Example 4-3 at www.wiley.com/college/touger

Two magnets, aligned to attract each other, are held apart in an environment of weightlessness and simultaneously released. At the instant of release, magnet A exerts a force of 0.60 N on magnet B. Magnet B has a mass of 0.060 kg.
a. What is the acceleration of magnet B at the instant of release?
b. If at that instant, the acceleration of magnet A in the opposite direction is 15 m/s², what is the mass of magnet A?

Brief Solution

Choice of approach. We follow the approach outlined in Figure 4-9.

| A | $\vec{F}_{\text{on A by B}}$ | $\vec{F}_{\text{on B by A}}$ | B |

| Newton's second law applied to A $\vec{F}_{\text{on A by B}} = m_A\vec{a}_A$ | Newton's third law $\vec{F}_{\text{on A by B}} = -\vec{F}_{\text{on B by A}}$ | Newton's second law applied to B $\vec{F}_{\text{on B by A}} = m_A\vec{a}_B$ |

Figure 4-9 Newton's third law is the link between applications of the second law to each of two interacting bodies A and B.

Anticipating and checking results. It follows from the second law that if two bodies are subject to the same force, the body with less inertial mass will experience a greater acceleration. You should make sure that your numerical results satisfy this condition.

What we know/what we don't.

For 2nd Law Applied to A	For 2nd Law Applied to B
$\Sigma F_{\text{on A}} = F_{\text{on A by B}} = ?$	$\Sigma F_{\text{on B}} = F_{\text{on B by A}} = 0.60$ N
$a_{\text{A}} = 15$ m/s^2	$a_{\text{B}} = ?$
$m_{\text{A}} = ?$	$m_{\text{B}} = 0.060$ kg

The mathematical solution.

a. The second law applied to B says $\Sigma F_{\text{on B}} = F_{\text{on B by A}} = m_{\text{B}}a_{\text{B}}$. Solving for a_{B} gives

$$a_{\text{B}} = \frac{F_{\text{on B by A}}}{m_{\text{B}}} = \frac{0.60 \text{ N}}{0.060 \text{ kg}} = 10 \text{ N/kg}$$

Because 1 N = 1 kg $\times$ m/s^2, B's acceleration toward A is

$$a_{\text{B}} = 10 \frac{\text{kg} \times \text{m/s}^2}{\text{kg}} = \textbf{10 m/s}^2$$

and is toward A.

b. The *third* law tells you that the magnitude of $\vec{\mathbf{F}}_{\text{on A by B}}$ equals that of $\vec{\mathbf{F}}_{\text{on B by A}}$, so that $F_{\text{on A by B}} = 0.60$ N. The second law as applied to magnet A is

$$\Sigma F_{\text{on A}} = F_{\text{on A by B}} = m_{\text{A}}a_{\text{A}}$$

Solving for m_{A} gives

$$m_{\text{A}} = \frac{F_{\text{on A by B}}}{a_{\text{B}}} = \frac{0.60 \text{ N}}{15 \text{ m/s}^2} = 0.040 \frac{\text{N}}{\text{m/s}^2} = 0.040 \frac{\text{kg} \times \text{m/s}^2}{\text{m/s}^2} = \textbf{0.040 kg}$$

As expected, because A's acceleration is greater than B's, its inertial mass is smaller.

◆ Related homework: Problems 4-24, 4-25, 4-29, and 4-32.

◆**TRANSLATIONAL EQUILIBRIUM** A corollary of Newton's second law is that unless $\Sigma\vec{\mathbf{F}}_{\text{on A}}$ is *not* zero, A's acceleration *is* zero and its velocity remains constant, which is the essence of the first law. We can thus restate the first law as

$$\vec{\mathbf{v}}_{\text{A}} = constant \ unless \ \Sigma\vec{\mathbf{F}}_{\text{on A}} \neq 0$$

If $\vec{\mathbf{a}}_{\text{A}} = 0$, then $\Sigma\vec{\mathbf{F}}_{\text{on A}} = 0$, that is, the net force on A is zero. This condition is called **translational equilibrium.** Remember: Zero acceleration does not imply zero velocity. In Chapter 5, you will see the usefulness of the idea of equilibrium in solving problems.

↪**A note on language:** Motion along a straight line is called *translational* motion, in contrast to *rotational* motion. For point objects, we can ignore the distinction and just speak of *equilibrium*. That is because rotation involves the parts of a body going around some central point, but for a point object, the central point is all there is.

Case **4-2** ◆ *The Accelerating Suitcase*

Suppose you are checking out of a hotel room. You are talking to your friend as you get into the elevator, and you realize you've left your suitcase on the landing as your elevator starts from rest and accelerates downward. To someone still on the landing, your suitcase is at rest, and also has zero acceleration. This satisfies the first-law condition that $\vec{\mathbf{v}}_{\text{A}}$ is constant unless $\Sigma\vec{\mathbf{F}}_{\text{on A}} \neq 0$. But what happens in a coordinate frame where *you* are the origin? As this origin accelerates downward relative to the landing, the suitcase accelerates upward relative to this origin. In other words, in this coordinate frame the suitcase's velocity is not remaining constant, even though the total force on it is zero. Newton's first law does not hold true in all reference frames. **STOP&Think** In that case, can the second law hold true in all reference frames? ◆

Based on Case 4-2, we might wish to say that the first law doesn't hold true if the coordinate frame is accelerating. But can you always say which one is accelerating? Imagine people on two spacecraft affected only by gravitational forces. According to their own reference frames, they each see the other's spacecraft accelerating. With only the backdrop of space, how can they tell which is "really" accelerating? Newton would reply that the coordinate frame that is not accelerating is the one for which $\vec{\mathbf{v}}_A$ is constant unless $\Sigma\vec{\mathbf{F}}_{\text{on A}} \neq 0$. Such a frame is called an *inertial reference frame*.

> **Inertial reference frame:** *A coordinate frame in which Newton's first law holds true.*

This is an arbitrary choice but is necessary to be able to use Newton's laws consistently in reasoning about forces and motion. For this reason, it is sometimes said that the real importance of the first law is that it defines the reference frame in which Newton's laws are valid. The first law does not hold true in reference frames that accelerate relative to inertial frames. Those are called **noninertial reference frames.**

4-4   Reexamining Your Own Ideas about Forces

Do the implications of Newton's second law make sense in terms of *your* understanding of forces and *your* expectations of how things behave? Review your own ideas about the following. **STOP&Think** Must an unopposed force (so that $\Sigma\vec{\mathbf{F}} \neq 0$) be exerted on a body to set it in motion? To keep it going in a straight line at constant speed? To keep it going along a curved path at constant speed? ◆

What does Newton's second law say about whether an unopposed force is needed in these situations? Does it agree with your understanding? Which view agrees with how things actually behave? To test this, you first need a way to tell experimentally whether there *is* a nonzero force.

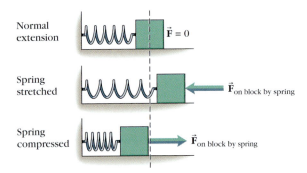

Figure 4-10 A spring's displacement is an indicator of the force it exerts.

◆**MEASURING FORCES** Strictly speaking, a force can be measured by finding the mass and acceleration of a body on which it is exerted in the absence of other forces. This is difficult to do in practice. A more practical and intuitive means of measuring a force makes use of the situation in which the force $\vec{\mathbf{F}}$ is opposed by the action of a stretched or compressed spring (see Figure 4-10). The more a spring is stretched or compressed, the harder it pulls or pushes back. At rest, the force it exerts equals the other force on the body. The distance the spring stretches or compresses is a measure of the force it exerts and is therefore a measure as well of the magnitude of the equal and opposite force $\vec{\mathbf{F}}$.

The displacement of a spring, or of any other length of elastic material, is best suited as a measure of the force when it is directly proportional to the force.[3] Some materials satisfy this condition for a substantial range of forces, although many of you found out as children that if you stretch a Slinky too far it doesn't spring back, and if you stretch a rubber band too far it breaks. For our purposes, however, it will suffice if the displacement is zero when the force is zero and if the displacement increases as the force does, whether linearly or not. A length of easily stretchable or compressible material can serve as a practical tool for detecting whether a nonzero $\vec{\mathbf{F}}$ is being exerted and for comparing different forces. You will use this in the activities that follow.

---

[3]If the relationship between force and displacement is not linear, the length of material can still be calibrated as long as the same force repeatedly produces the same displacement.

On-The-Spot Activities

In these activities you will need:

1. The loosest, stretchiest rubber band you can find. Cut through it in one place to turn it from a loop into a length of elastic string.

2. An object of reasonable size that can move across a surface with as little friction as possible. The ideal movable object, if you have access to one in your physics lab or elsewhere, would be a glider on an air track or air table. Next best would be any object on freely turning wheels—a roller skate, a skateboard, a child's toy car or truck (oil the wheels if you have to), and so on.

3. A good-sized flat surface that you can tilt slightly. Sliding books or magazines under two legs of any rectangular table or desk will accomplish this nicely. But make sure it is *very* flat; any warping will complicate your observations. The air track or air table, if you are using one, will serve as this surface.

4. If available, a wind-up toy (the heavier the better) that travels in a straight line.

On-The-Spot Activity 4-1

Compensating for Frictional Forces Set your movable object on the flat surface. Then tilt the surface just enough so that if you give the object a small downhill velocity, it keeps moving at that velocity. Judge this by eye as best you can. It may be a bit tricky to do because strictly speaking variations in friction would require corresponding variations in the compensating tilt of the surface. Try varying somewhat the velocity that you give the object. Does this alter the constancy of its velocity? Try putting the object in one place on the surface, thus giving it zero velocity. Does it stay there; that is, does the velocity remain constant at a value of zero? Is there an unopposed force on the object under these conditions?

The motion in Activities 4-2 to 4-4 should be up or down the slight slope you produced in Activity 4-1.

On-The-Spot Activity 4-2

Is an Unopposed Force Needed to Set a Body in Motion? Tie the elastic string to the movable object so that you can pull the object by the string. Pull on the string until it is just taut, so that you know how far the string extends when it is taut but not stretched. Now put the movable object on the surface and pull with the string in the forward direction to set the object in motion. Does the string stretch as motion begins, and therefore does the string exert a force on the object while setting it in motion?

On-The-Spot Activity 4-3

How Does a Force on a Body Affect Its Acceleration? Repeat Activity 4-1, but now try to maintain the string at the same amount of stretch as the object moves. Does the object move at a constant speed, or does it speed up? What happens if you increase the stretch of the elastic string? When you decrease it? When you reduce it to zero? When the string is stretched so that it exerts a force, is there an acceleration? Is there more acceleration when the force is greater?

On-The-Spot Activity 4-4

Is an Unopposed Force Needed to Keep a Body in Motion? Now set the object in motion as before, but once it is moving, pull no harder than necessary to keep the object going at constant speed (judged by eye). While the object is moving at constant speed, is the string just extended to its taut length or is it extended beyond that length? When there is no acceleration, in other words, is the string exerting a force?

On-The-Spot Activity 4-5

Is an Unopposed Force Required to Keep a Body Going Along a *Curved* Path at Constant Speed? (As an alternative to this activity, you can do Problem 4-4.) At-tach the elastic string as near to the front of the wind-up toy as is practical. Wind the toy up and let it travel across a horizontal surface at constant velocity. While it is traveling, keep the string loosely pulled to one side (perpendicular to the direction of motion; see Figure 4-11a). Then, well before the toy winds down, pull harder to the side so that the string stretches slightly. Until the toy travels a short distance fur-ther, try to keep the string pulled so that it always has the same amount of stretch and is always perpendicular to the toy's direction of motion, even as the direction of motion changes (see Figure 4-11b). After the toy has traveled a short distance un-der these conditions, let the string go slack again and observe the subsequent mo-tion of the toy.

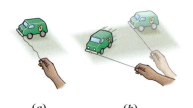

(a)   **(b)**

Figure 4-11 Exerting a force per-pendicular to the velocity of a toy car. In (a), the string is kept perpen-dicular to the motion but loosely enough not to exert a force. In (b), the string is pulled taut to exert a force on the car.

STOP&Think Must an unopposed force be exerted on a body to set it in motion? To keep it in motion at constant velocity? To accelerate it? To keep it moving at constant speed along a curved path? How would you answer these questions on the basis of Newton's second law of motion? How would you answer the same questions on the basis of your observations in Activities 4-1 to 4-5? Do the expectations arising from the second law consistently agree with what you observe? (The chapter summary will touch briefly on these points.) ◆

STOP&Think In Activity 4-5, during what part of the motion was an unop-posed force exerted on the toy? What happened to the motion of the toy before the string exerted a force on it? While it exerted a force? After it stopped exerting a force? ◆

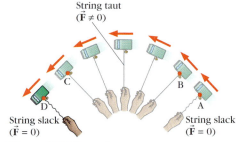

String taut
($\vec{F} \neq 0$)

C

B

D

String slack
($\vec{F} = 0$)

A

String slack
($\vec{F} = 0$)

Figure 4-12 How the toy car moves.

In case you were unable to find a suitable wind-up toy for Activity 4-5, Figure 4-12 summarizes the observations you would likely have made. Between points A and B, while the string is slack, there is zero total force exerted on the toy. The toy moves at constant velocity: constant speed in a fixed direction. Over this stretch, Newton's first law (or the second law with $\Sigma \vec{F} = 0$) applies. Between B and C, the string exerts a force on the toy. $\Sigma \vec{F} \neq 0$, and there is a change in velocity, not in magnitude but in direction. Thus, as expected from the second law, there is an acceleration. (These observations are consistent with Example 3-7 and Figure 3-35.) At point C, the string stops exerting a force. At that instant, the vector $\vec{v}$ stops changing, and therefore, as you would expect from the first law—the law of inertia—the toy continues to move at the velocity it had at that instant.

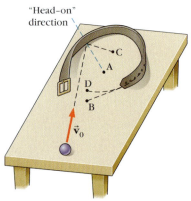

"Head-on"
direction

C

A

D

B

$\vec{v}_0$

Figure 4-13 Exploring the impli-cations of Newton's second law.

On-The-Spot Activity 4-6

To make sure that you understand this point, take a belt made of leather or other stiff material, and set it up on edge to form a curved barrier, as in Figure 4-13. You are going to roll a marble or other small ball toward the belt, giving it a substantial velocity $\vec{\mathbf{v}}_o$ at a considerable angle to the head-on direction (see figure). Before you do so, think about the path the marble will follow after striking the belt. Will it go through point A, B, C, D, or some other point? Why? Now roll the marble as directed. How did the marble actually travel after striking the belt? After leaving the belt? At what instant did the belt cease to exert a contact force on the marble? Was the velocity (direction as well as magnitude) changing until that instant? After it?

If the water skier lets go of the tow rope, will she continue going in a curved path or will she go straight?

➡**A note on language:** You sometimes hear people say that an object *has* a force, or that it travels *with* great force. In light of Newton's laws of motion, this is incorrect usage. What is meant is that the object is traveling with great *speed* or *velocity*. As the two straightaway stretches in Figure 4-12 show, a body at constant velocity has *no* net force exerted on it, and it is meaningless, within the framework of Newton's ideas, to say that a body *has* force or has *a* force. A force is not a property of a single object, but an *action of one body on another*. By the same token, you cannot give something a force or impart a force to it (although you can give it a push).

By now, you should be ready to say with some confidence that for an object to speed up, slow down, or change direction, an unopposed force must be exerted on it by another body. These points are summed up in Newton's second law. But "to just keep doing the same thing," changing neither speed nor direction, requires no action by another body. The total external force on the object must be zero; that is the essence of the first law.

◆ SUMMARY ◆

In this chapter, you were introduced to Newton's laws of motion, an interconnected and self-consistent framework of ideas dealing with how *interactions* between bodies affect their motions.

Newton's laws of motion in modern English and in equation form:

1. *A body will continue to move indefinitely at the same velocity (same speed and direction) unless a net outside force is exerted on it.*

$$\vec{\mathbf{v}}_A = \text{constant } unless \ \Sigma\vec{\mathbf{F}}_{\text{on A}} \neq 0$$

2. *The acceleration vector of a body is proportional to the vector sum of all forces exerted on that body by other bodies. The constant of proportionality is the inverse of the body's **inertial mass.***

$$\vec{\mathbf{a}}_A = \frac{1}{m}\Sigma\vec{\mathbf{F}}_{\text{on A}}$$

or

$$\Sigma\vec{\mathbf{F}}_{\text{on A}} = \vec{\mathbf{F}}_{\text{on A by 1}} + \vec{\mathbf{F}}_{\text{on A by 2}} + \vec{\mathbf{F}}_{\text{on A by 3}} + \cdots$$

$$= m_A \vec{\mathbf{a}}_A \qquad (4\text{-}8)$$

3. *The forces that two interacting bodies exert on each other are always equal in magnitude and opposite in direction.*

$$\vec{\mathbf{F}}_{\text{on 2 by 1}} = -\vec{\mathbf{F}}_{\text{on 1 by 2}} \qquad (4\text{-}5)$$

An **inertial reference frame** is a coordinate frame in which the first law holds true.

An object's **weight** is the gravitational force (a vector) on an object. It is proportional to the mass under constant gravitational conditions, but can vary according to what is attracting the object gravitationally and how far away it is. The object's **mass** or **inertial mass** is a measure of its **inertia**—its tendency to continue at the same velocity. The mass is the same everywhere. For differences between mass and weight, see Section 4-3.

To understand Newton's laws, you must understand the concept of a **force** and how a body is affected when a force is or is not exerted on it. A force is one side of an interaction. An *interaction* between A and B means that A acts on B *and* B acts on A, or, restated,

$$A \text{ exerts a force on } B. \qquad \text{(Form 4-1)}$$

and $\quad B \text{ exerts a force on } A. \qquad \text{(Form 4-2)}$

Identifying an interaction fully requires following the steps of Procedure 4-1.

We have stressed the importance of examining your own ideas about forces and reconciling them with the Newtonian view. To the question raised at the beginning of Section 4-2, you should be able to respond that the best synonym of a force is an action. It is sometimes helpful to think of a force as a push or pull, but ultimately you must determine whether or not A exerts a force on B on the basis that

a force is an action or effect of one of two interacting bodies (A) on the other (B), which, in the absence of other bodies, would change the motion of body B.

In the first and second laws, it is the *total outside* (or **external**) force that affects a body's motion. If the total outside force on a body is zero, the body is said to be in **translational equilibrium.**

The forces and accelerations in a situation can vary over time (see Figures 4-6 and 4-7). Newton's laws hold true at each instant; they are statements about *instantaneous* quantities.

We have identified both **contact** (touching) **forces** and **action-at-a-distance forces.** Picturing a mechanism of exertion for how contact forces are exerted (such as the deformation and springback of bodies pressed against each other) may help you accept the idea that a force *is* being exerted, even by an inanimate and seemingly rigid object (see Figure 4-5). The existence of action-at-a distance forces is inferred from observations of how objects move, but at this point, like Newton, we have not provided an explanation of how they are exerted.

The stretching of a spring or other elastic material provides a way of measuring a force and therefore of determining experimentally whether a force is being exerted. This is done in Activities 4-1 to 4-5. Consistent with Newton's laws, the elastic stretches when you set the object in motion (a force is required to accelerate it from rest to a nonzero velocity) but not when you keep it moving in a straight line at constant speed (no acceleration, no unbalanced force). But keeping the object moving at constant speed along a circular path requires that the elastic stay stretched and always perpendicular to the object's instantaneous direction of motion (see Figures 4-11 and 4-12). If the velocity *vector* is changing in any way (magnitude or direction), there is a nonzero acceleration so there must be a nonzero force.

◆QUALITATIVE AND QUANTITATIVE PROBLEMS◆
Hands-On Activities and Discussion Questions

The questions and activities in this group are particularly suitable for in-class use.

4-1. Discussion Question. In the midst of a fiercely contested Ping-Pong game, you slam the ball at your opponent.
a. Is the force that your paddle exerts on the ball greater than, equal to, or less than the force that the ball exerts on the paddle?
b. What does Newton's third law of motion tell us about how these two forces compare?
c. What does Newton's second law of motion tell us about how the paddle and the ball are affected by these two forces? Is one object affected more than the other? Why?
d. In ordinary experience, there appears to be something unequal about the interaction between the ball and the paddle. What quantities are unequal in the interaction? What quantities are not unequal?

4-2. Discussion Question. After a bowl of Jello has firmly set, a grape is placed gently atop its surface.
a. Does the grape exert a force on the Jello? Does the Jello exert a force on the grape? Answer this part first, and state your reasons. The subsequent parts will then provide you with opportunities to rethink your answer.
b. *Argument based on mechanism.* When it is sufficiently set to support the grape on its surface, the Jello is quite springy. In that sense, the grape on the Jello is like the safe on the spring. As the Jello sets more firmly, its springiness lessens, so like a tighter spring, it pushes back as hard with less give. Taking this into account, would you still give the same answer to part *a*? Explain.
c. *Arguing from the effect of force on motion.* If the grape is released from the same position and there is *no* Jello in the bowl, the grape falls freely to the bottom of the bowl. If there is Jello in the bowl but it has not yet begun to set, the grape will sink, but it will not hit bottom as quickly as in the absence of Jello. The Jello is acting to slow down the grape. The grape is meanwhile pushing liquid Jello out of its way as it falls. How would you answer part *a* now? Incorporate the ideas of part *c* into your explanation.
d. Suppose the substance that you gradually allow to set is poured concrete rather than Jello. If a grape is lying on the pavement, does the grape exert a force on the pavement? Does the pavement exert a force on the grape?

4-3. Discussion Question. Your textbook is lying flat on a desk. Someone places an unabridged dictionary on top of your textbook. Does the dictionary exert a force on the desk? (Ignore any forces that are too small to measure.)

4-4. Hands-On Activity. In this activity, you will explore how the force that a rope exerts on you affects your motion. You will need a rope, a secure place to tie it to, and a friend. An inanimate object can substitute for the friend.
a. Tie one end of a rope around your waist, and the other end around something (such as a post or a doorknob) that will not move even if you pull hard on it.
b. Start out in the arrangement shown in Figure 4-14a. In this arrangement, (1) P is the point where the rope is tied, (2) you must be far enough from point P so that the rope is pulled tight, (3) you are facing your friend along line AC, and (4) the angle θ is about 30°.
c. Now begin to walk toward your friend along line AC. Can you keep going in a straight line until you reach your friend?
d. Suppose you have difficulty continuing along the straight line once you reach point B in Figure 4-14b. If you try as hard as you can to keep walking toward your friend, what happens to the direction of your motion? Sketch the path that you end up following.
e. As you walk from point A to point B, do you feel the rope pressing into you at all? When you continue beyond point B, do you feel the rope pressing into you? Where do you feel it, and in what direction do you feel it pressing?
f. Once you go beyond point B, what is the shape of the path that you find yourself following? How can you describe the direction of the force that makes you follow this path?

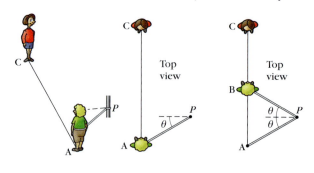

(*a*) Starting position   (*b*) Location of point B

Figure 4-14 Problem 4-4

Review and Practice

Section 4-1 Newton's First Law: Inertia and the Concept of Force

4-5.

a. A trunk is pushed across a level floor at constant speed. Is the force exerted by the pusher the only horizontal force on the trunk? Briefly explain.

b. A methane molecule is drifting in deep space at constant velocity. Must there be any forces on the molecule for this to happen? Briefly explain.

4-6. The multi-exposure sequence in Figure 4-15 shows a bowler releasing a ball. In each exposure, fully identify and sketch the direction of each force exerted on the ball. Other than frictional forces or air resistance, is there any horizontal force on the ball once the ball is released? If so, what is exerting the force, or, if not, why does the ball keep going?

Figure 4-15 Problem 4-6

4-7. In *Genius: The Life and Times of Richard Feynman* (Pantheon, NY, 1992), James Gleick describes inertia as "a particle's memory of its past velocity." Explain what he means by this.

4-8. To convey the notion that a new electric car prototype is quite powerful, a popular press article reports, "The force of acceleration pushes you back against the seat." From a physics point of view, what is wrong with this statement?

4-9. **SSM** You are riding in the front passenger seat of a car with a somewhat reckless driver. When the driver does each of the following, what happens to you (or what do you feel physically), and why?

a. The driver floors the accelerator to make a light before it turns red.

b. The driver sees a squirrel in the road and brakes suddenly.

c. The driver takes a sharp right turn a bit too quickly.

d. The driver maintains a constant very high speed (at least 75 mi/hr) over a long, straight stretch of highway.

Section 4-2 Exploring the Meaning of Force

4-10. In each blank, fill in *always, sometimes,* or *never,* then briefly explain your answer.

a. An object traveling at constant velocity in the horizontal direction _____ has horizontal forces exerted on it.

b. When the only two forces on a certain object are equal in magnitude, the object _____ accelerates.

4-11. When a book is lying on a shelf, is there any interaction between the book and the shelf? If there is an interaction, describe the forces that constitute the interaction. If there is no interaction, explain why not.

4-12. The wooden boards of a bench seat are nailed in place. The head of one nail sticks up just enough so that when you sit on it, it causes some discomfort. When you are sitting still on this bench, _____

a. both the boards and the nail exert forces on your bottom.

b. only the boards exert a force on your bottom.

c. only the nail exerts a force on your bottom.

d. neither the boards nor the nail exert a force on your bottom.

4-13. When a food parcel dropped from a Red Cross helicopter is falling to the ground, is it interacting with anything besides the air? If it is, briefly describe the forces involved in each interaction. If it isn't, briefly explain why not.

4-14. Push against a wall. **a.** Does the wall exert a force on your hand? **b.** If so, by what mechanism does it do this?

4-15. **SSM** Two students have a disagreement about action and reaction forces. Student A says, "If I hit a board with a karate chop, my hand exerts a force on the board. That's an action. Then the board exerts a force back on my hand. That's the *reaction*." Student B says, "Wait a minute. If I come down with my hand, my hand keeps going until the board stops it. That's the action. Then the force that my hand exerts on the board is the *reaction*." How do you resolve the disagreement?

4-16. Your body is made up of a vast number of atoms. Because these atoms stay together rather than flying off in all directions, we must assume that they exert forces on one another. In applying Newton's laws to your own motion, why isn't it necessary to take these forces into account? In particular, why isn't necessary to include them in the sum of forces $\Sigma\vec{F}_{\text{on you}}$ if you apply Newton's second law to yourself?

4-17. As a figure of speech, people sometimes speak of "pulling yourself up by your own bootstraps." Is it physically possible to do this? Briefly explain your answer.

Section 4-3 Newton's Second and Third Laws

4-18. An ice skater skates across a rough patch of ice. For each force in the first column, give the number of the force in the second column that goes with it to make up an interaction (or action-reaction) pair.

a. the backward force that the skater exerts on the ice to propel herself forward

b. the downward gravitational force that Earth exerts on the skater

c. the upward force with which the ice supports the skater

1. the downward force that the skater exerts on the ice

2. the upward force that the ice exerts on the skater

3. the backward frictional force that the ice exerts on the skater, tending to slow her down

4. the downward gravitational force that Earth exerts on the skater

5. the upward force that the skater exerts on Earth

6. the forward force that the ice exerts on the skater

4-19. A brick from an upper story of a condemned building falls on an ant walking across the sidewalk. At any instant during the tiny interval of contact, is the force that the brick exerts on the ant less than, equal to, or greater than the force that the ant exerts on the brick? Briefly explain your choice.

4-20. Suppose that a piece of space debris collides with an unmanned satellite in space. If we compare a_{debris} and $a_{satellite}$, the magnitudes of the accelerations that the two objects experience due to the collision, we find that $a_{satellite} = 2.5 \, a_{debris}$. Then if F_{debris} and $F_{satellite}$ are the magnitudes of the forces the two objects exert on each other during the collision, $F_{debris} = \underline{\quad ? \quad} F_{satellite}$. What is the value of the numerical multiplier that goes in the blank?

4-21. SSM A swimmer pushes off horizontally from a boat. He then repeats the same action, pushing off exactly as hard as before, but meanwhile a load with twice the mass of the boat has been placed in the boat.
a. Is the swimmer's acceleration on the second try less than, equal to, or greater than his acceleration on the first?
b. Is the boat's acceleration on the second try less than, equal to, or greater than its acceleration on the first?

4-22. Choose the word that best completes the sentence: When a heavy passenger bus collides with a lightweight bicycle, the force that the bus exerts on the bicycle is (*always, sometimes,* or *never*) greater than the force that the bicycle exerts on the bus. Briefly explain.

4-23. A skater jumps horizontally from a skateboard, so that she and the skateboard go in opposite directions. She then repeats the action with a second skateboard, which has twice the mass of the first. If she pushes off equally as hard as before, how does the change affect her motion? How does it affect the skateboard's motion?

4-24. A 30-kg child dives from the front of a 120-kg boat. When the child's acceleration is 22 m/s² in the forward direction, what are the magnitude and direction of the boat's acceleration?

4-25. Two figure skaters push off from each other. When A's acceleration is 35 m/s², B's acceleration is 25 m/s² in the opposite direction. **a.** Which skater has the greater mass? **b.** What is the ratio $\left(\frac{greater\ mass}{smaller\ mass}\right)$ of their masses?

4-26. Two astronauts are floating in space. Astronaut A's mass is 1.25 times as great as that of astronaut B. If astronaut A pushes astronaut B, how will the motion of each astronaut be affected? Compare the effects of the push on the two astronauts, and be as quantitative as possible in your answer.

4-27. While a constant total force of 10 N is exerted on a cart, the cart's acceleration is 6 m/s². Find the mass of the cart.

4-28.
a. An astronaut pushes off from a space vessel. The vessel's mass is 10 times as great as the astronaut's. At any instant during the push, is the vessel's acceleration also 10 times that of the astronaut? If so, why? If not, why not?
b. A swimmer pushes off horizontally from a boat. The boat's mass is 10 times the swimmer's mass. At no time during the push is the boat's acceleration one-tenth that of the swimmer. Why not?

4-29. SSM WWW A two-part aerial fireworks device is designed so that before they explode, the two parts A and B break apart by mechanically propelling each other in opposite directions. A has a mass of 0.12 kg and B has a mass of 0.08 kg. If B exerts a force of 1.2 N on A during the break-up, what is **a.** A's acceleration? **b.** B's acceleration?

4-30. A boater proposes to propel his sailboat by standing on board and blowing on the sail with a powerful bellows. Use Newton's laws to assess his prospects of success. Also, is there any way he can use the bellows more effectively?

4-31. SSM WWW The hammer in Example 4-1 falls 4.9 m before striking the ground. If the hammer's mass is 1.0 kg, and the mass of Earth is 6×10^{24} kg, **a.** find the acceleration of Earth resulting from this interaction. **b.** find the resulting distance traveled by Earth.

4-32. As the magnets in Example 4-3 move closer together, the acceleration of magnet B changes. At a certain instant (a tiny fraction of a second later), it has increased to 16 m/s². At this later instant, what is the acceleration of magnet A, and what force does each magnet exert on the other? (Use given data and results from Example 4-3 as needed.)

4-33. If the hand in Figure 4-16 is causing blocks A and B to accelerate, the force that A exerts on B is (*always, sometimes,* or *never*) equal and opposite to the force that B exerts on A.

Figure 4-16
Problem 4-33

4-34. What is *your* approximate weight in newtons?

4-35. The label on a ketchup bottle reads "net wt. one pound (454 g)." Is one pound the same thing as 454 g?

4-36. A bottle of ketchup falling from a shelf (with only Earth's gravitational force exerted on while it is falling) has an acceleration of 9.80 m/s². What is the weight in newtons of 454 g of ketchup?

4-37. The forces shown in Figure 4-17 are exerted on a 60-kg block. **a.** What total force (magnitude and direction) does the block experience? **b.** What is the block's acceleration (magnitude and direction)?

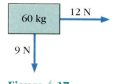

Figure 4-17
Problem 4-37

4-38. Figure 4-18 shows two minivans passing each other. Each has a ball suspended on a string from its roof. Suppose you are observing both of these vans in an inertial reference frame.

Figure 4-18 Problem 4-38

a. Which van, if either, is accelerating? Is it A, B, both, neither, or is it impossible to determine from the given information?
b. Which van is going at greater speed? Is it A or B, or are both going at the same speed, or is it impossible to determine from the given information?

4-39. Is it possible to have more than one inertial reference frame? Briefly explain.

Section 4-4 Reexamining Your Own Ideas about Forces

4-40. A row of ducks wait near a coiled water hose to be sprayed with water. Clamps keep the hose anchored in place.

Assuming the hose shoots a narrow stream of water, which duck in the figure will get wet when the valve is first opened?

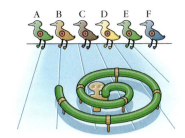

Figure 4-19 Problem 4-40

4-41. A spacecraft is initially drifting in the direction shown in Figure 4-20. Exhaust jets on the rocket (A, B, C, and D) may be fired singly or in combination to alter the rocket's motion.

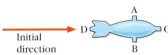

Figure 4-20 Problems 4-41, 4-42, and 4-43

a. Jet A is turned on at time t_1 and turned off at a slightly later time t_2. Sketch the path of the spacecraft over the time interval from shortly before t_1 to shortly after t_2.

b. Repeat **a** for the case when jets A and C are turned on in combination during the interval from t_1 to t_2.

c. Repeat **a** for the case when jets A and D are turned on in combination during the interval from t_1 to t_2.

d. Starting with the original motion, what jet or combination of jets would you have to turn on if you wish the rocket's velocity to end up directed toward the left side of the page? toward the top of the page?

4-42. A spacecraft viewed from above is initially drifting in the direction shown in the figure. Exhaust jets on the rocket (A, B, C, and D) may be fired singly or in combination to alter the rocket's motion. Assume the firings are of short duration.

a. If the rocket is to reverse direction, which exhaust jet(s) must be fired?

b. If the rocket's final direction is to be toward the lower left corner of the page (as viewed from above), which exhaust jet(s) must be fired?

c. If the rocket's final direction is to be straight down toward the bottom of the page (as viewed from above), which exhaust jet(s) must be fired?

4-43. Again consider the spacecraft initially drifting in the direction shown in Figure 4-20. If exhaust jet A is turned on at instant t_1 and turned off at instant t_2, which diagram in Figure 4-21 most accurately shows the spacecraft's path?

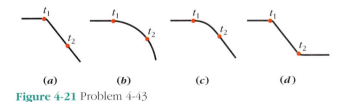

Figure 4-21 Problem 4-43

4-44. According to Newton's second law of motion, a body traveling in a circular path (*always, sometimes,* or *never*) has a nonzero total force exerted on it. Briefly explain.

4-45. In this problem, consider only forces exerted in the plane of the table, which is shown viewed from above in the diagrams. Neglect forces directed upward or downward from the table.

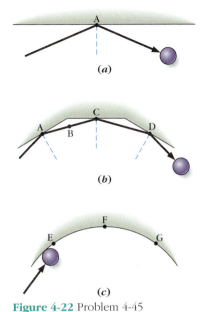

Figure 4-22 Problem 4-45

a. A ball rebounds off the side of a billiards table (Figure 4-22a). What is the direction of the force on the ball at the instant it is at point A?

b. A ball rolling on a table with a many-sided raised edge undergoes a series of rebounds (Figure 4-22b). What is the direction of the force on the ball at point A? B? C? D? Use a carefully drawn arrow to answer each part.

c. A ball is rolled on a table on which a barrier shaped like an arc of a circle has been set up (Figure 4-22c). What is the direction of the force on the ball at point E? F? G? When the ball has gone a short distance beyond G? What is the direction of the ball's *velocity* at each of these points?

4-46. A bowling ball is rolling at a fairly slow speed in a straight line across a gym floor. A student with a broom is allowed to hit the ball once with the broom. In doing so, he must make the ball's direction change by 20° to the right, without speeding it up or slowing it down. (Assume the ball loses speed *very* slowly if we just let it roll.)

a. Sketch a labeled diagram showing the ball's directions before and after being hit with the broom, and the direction in which the broom must exert a force on the ball to cause this change in direction.

b. A second student with a broom stands 2 m away from the first in the direction the ball is traveling after the first student hit it. Her task is to hit the ball once and change the ball's direction another 20° to the right, again without speeding it up or slowing it down. Add the following to the diagram that you sketched in **a**: the force exerted on the ball by the second student's broom and the direction the ball travels after she has hit it. Continue to label carefully.

c. Sixteen more students, each 2 m from the previous student, do the same thing in turn. Roughly, what is the overall shape of the path followed by the ball as it passes all eighteen students? What can you say in general about the direction(s) of the force(s) needed to keep the ball following this path at constant speed?

Going Further

The questions and problems in this group are not organized by section heading, so you must determine for yourself which ideas apply.

Some of them will be more challenging than the Review and Practice questions and problems (especially those marked with a • or ••).

4-47. If you could throw a ball from a space probe in a remote region of space, how far would it go? Explain.

4-48. As the moon travels around Earth, is there any force that keeps it going or keeps it from slowing down? Explain.

4-49. **SSM** A child's toy consists of a car pulled by a rubber band. Two children start their cars (A and B) from rest at the same time. The velocities of the two cars are graphed against *t* in Figure 4-23.

a. At instant t_1, is A's rubber band stretched more, is B's rubber band is stretched more, or are the rubber bands are stretched equally? Briefly explain.

b. Repeat part *a* for instant t_2.

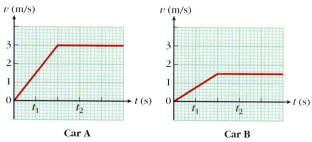

Car A   **Car B**

Figure 4-23 Problem 4-49

••**4-50.** A group of students wishes to show that when the mass of an object is constant, the object's acceleration is proportional to the total force exerted on it. They set up the following arrangement. A low-friction cart is allowed to move along a horizontal track on a table. A string at the front of the cart passes over a pulley at the end of the table. A hanger to which weights can be added is suspended from the dangling end of the string. The students reason that they can vary the force on the cart by varying the amount of weight hanging from the string.

a. Sketch the set-up. Why isn't this a satisfactory arrangement for showing what the students want to show?

b. How can they improve the set-up?

4-51. Two objects A and B are interacting. They are isolated from other objects. Object B has twice the mass of object A.

a. In Figure 4-24a, the first graph on the left shows the force that B exerts on A, plotted against time. Which of the

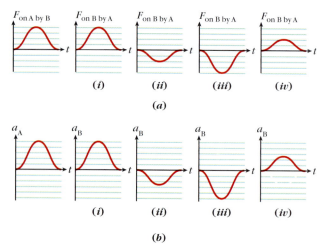

(a)

(b)

Figure 4-24 Problem 4-51

graphs to the right of it (i, ii, iii, or iv) correctly shows the force that A exerts on B? Briefly explain.

b. In Figure 4-24b, the first graph on the left shows the acceleration of A during the interaction. Which of the graphs to the right of it (i, ii, iii, or iv) correctly shows the corresponding acceleration of B? Briefly explain.

4-52. Figure 4-25 shows a sequence of stop-action pictures of a bumper car that travels to the left, collides with a spring, and bounces back to the right (the +*x* direction).

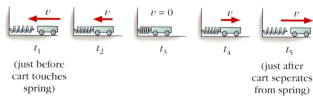

t_1 t_2 t_3 t_4 t_5

(just before cart touches spring)   (just after cart seperates from spring)

Figure 4-25 Problems 4-52 and 4-53

a. At which (one or more) of the instants shown is the force that the spring exerts on the car positive? Briefly explain.

b. At which instant(s) is the force that the spring exerts on the car negative? Briefly explain.

c. At which instant(s) is the force zero? Briefly explain.

d. At which instant(s) does the force that the spring exerts on the car have the greatest magnitude? Briefly explain.

4-53. For the time sequence shown in Figure 4-25, sketch a graph of $F_{\text{on cart by spring}}$ plotted against time from $t = t_1$ to $t = t_5$.

4-54. When a parachutist who has jumped from a plane opens his parachute, his speed continues to increase for a bit, but not uniformly. Instead, it gradually levels off to a final speed, so that the speed for the last part of his descent is constant. From this we can conclude that the upward force that the air exerts on his parachute _____.

a. increases as his velocity increases.

b. is always equal and opposite to his weight.

c. is always greater in magnitude than his weight.

d. increases until it is greater in magnitude than his weight.

4-55.

a. How could you use the device shown in Figure 4-26 to measure the acceleration of your car going down a straight street?

b. What would you do differently than in *a* if you wanted to use this device to test whether the car is accelerating when it goes around a corner at constant speed?

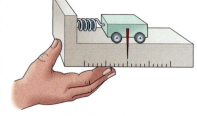

Figure 4-26 Problem 4-55

4-56. As a car is driven around the track in Figure 4-27 at constant speed, the net force on the car is (*always, sometimes,* or *never*) zero.

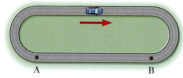

Figure 4-27 Problems 4-56 and 4-57

4-57. Suppose a car travels clockwise at constant speed from point A to point B on the track in Figure 4-27. Over this interval, the average net force on the car is _____. (*toward the top of the page; toward the bottom of the page; zero; toward the right side of the page*)

4-58. An exploratory spacecraft is traveling at a height of 50 m above the surface of a newly discovered planet. A crew member drops a 0.20-kg rock from the spacecraft and finds it takes 4.0 s to reach the planet's surface. Assume the planet's atmosphere has a negligible effect.
a. How much gravitational force does the planet exert on the rock?
b. How much gravitational force does the rock exert on the planet?

4-59. SSM WWW Two pucks glide without friction on an air table. When they collide head on, the change in velocity $\Delta\vec{v}$ that puck A experiences is twice as great in magnitude as the change $\Delta\vec{v}$ that puck B experiences but is opposite in direction.
a. Compare the masses m_A and m_B of the two pucks by filling in the numerical value in the blank in the following equation: $m_A = \underline{\quad} m_B$.
b. Compare the magnitudes of the forces $F_{\text{on A by B}}$ and $F_{\text{on B by A}}$ that the two pucks exert on each other during the collision by filling in the numerical value in the following equation: $F_{\text{on A by B}} = \underline{\quad} F_{\text{on B by A}}$.

4-60.
a. For how long must a net force of 20 N be exerted on a 4-kg crate to increase its speed by 10 m/s?
b. For how long must the same net force of 20 N be exerted on a 4-g paperclip to increase its speed by 10 m/s?

Problems on WebLinks

4-61. Suppose you press down with your thumb on a sturdy wooden tabletop, as in WebLink 4-1. If you then remove your thumb, the region of the tabletop that was in contact with your thumb will _____. (*move upward visibly; move upward a microscopic distance; not move at all; move downward a microscopic distance.*)

4-62. [*Choose the best answer.*] When you press down with your thumb on a sturdy wooden tabletop as in WebLink 4-1, the atoms in (*your thumb; the table; both; neither*) that are nearest to the surfaces in contact are pressed closer together.

CHAPTER FIVE

Problem-Solving Using Newton's Laws

Newton's laws of motion treat interactions in terms of the forces that bodies exert on each other and describe how these forces affect the motion of the bodies. This study of forces and their relation to motion is called **dynamics.** In this chapter, we will see how Newton's laws can be used to solve quantitative problems. You will see over and over again that to use Newton's laws correctly and effectively in problem solving, you must work not only with the equations that summarize the laws but with the concepts that they embody.

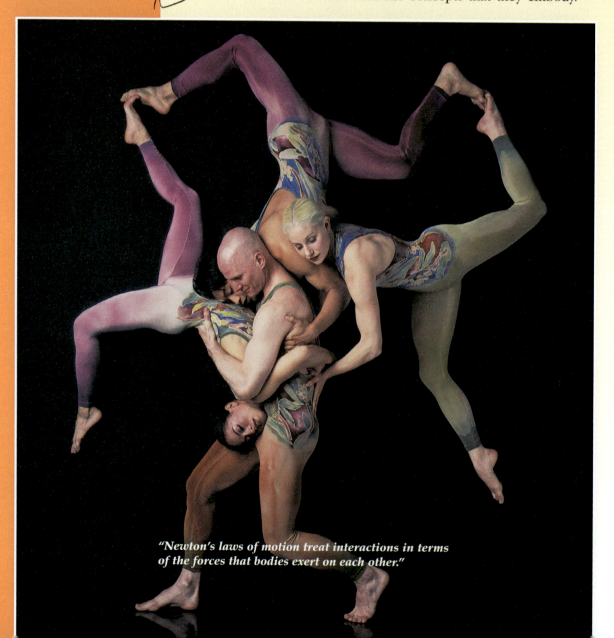

"Newton's laws of motion treat interactions in terms of the forces that bodies exert on each other."

5-1 Inventorying Forces: Applying Newton's Second Law

Newton's second law (Equation 4-5) addresses the motion of a single body (call it A). To apply it, you must be able to identify all the external forces—the forces that *other* bodies exert on A. In combination ($\Sigma\vec{F}_{\text{on A}}$), these forces produce A's acceleration and thus alter A's motion. When you treat the motions of interacting objects, the second law applies separately to each object.

◆**REPRESENTING FORCES IN DIAGRAMS** Systematically drawing diagrams is of great value in using Newton's laws. We will proceed in the following way:

PROCEDURE 5-1

Representing Forces by Arrows

1. Represent each force on a body by an arrow pointing in the direction in which the force is exerted. The body would move in this direction if it started from rest with just this one force exerted on it.
2. If you have information about the magnitudes of the force vectors, try to keep the arrow lengths at least roughly proportional to the magnitudes.
3. Place the tail of the arrow at the point of application of the force.
4. Label each arrow $\vec{F}_{\text{on}\_\_\text{by}\_\_}$. Because each force is "one side" of an interaction between two bodies, fill in the blanks with the two interacting bodies. (For example, if the interaction is a dog straining at its leash, the two forces are $\vec{F}_{\text{on leash by dog}}$ and $\vec{F}_{\text{on dog by leash}}$.)
5. (Optional) If each object is drawn in a different color, you can color-code each arrow in the colors of both interacting bodies (see Figure 5-1).

When multiple forces are involved, drawing carefully labeled diagrams will help you keep them sorted out, as in the following situations.

As we noted in Chapter 4, if a body is not accelerating, the vector sum of the external forces exerted on it must be zero, and the body is said to be in translational equilibrium (or simply in equilibrium if we can treat it as a point object). If the body is in equilibrium when *two* forces are exerted on it, the forces must be equal in magnitude and opposite in direction. It is important to distinguish these forces from the equal and opposite forces in an interaction (action-reaction) pair. **STOP&Think** Can the two forces in an interaction pair ever, by themselves, result in the equilibrium of a single body? ◆

A critical difference is that interaction forces are always exerted on two different bodies by each other, whereas equilibrium of a single body results from forces exerted on that one body. To explore this distinction further, consider the following case. For a more detailed and interactive version of this case, go to WebLink 5-1.

➡**Caution:** Do not confuse equal and opposite forces *on the same body,* which produce equilibrium, with the equal and opposite interaction forces that two bodies exert *on each other.*

For **WebLink 5-1: Forces on a Rocket,** go to www.wiley.com/college/touger

Case 5-1 ◆ *Forces on a Rocket*

Figure 5-1 shows a rocket adrift in space. As it drifts away from the moon and toward Earth, the moon's pull on it weakens and Earth's pull on it grows stronger. In Figure 5-1*a*, the rocket is passing a point where the gravitational forces that Earth and the moon exert on the rocket are equal in magnitude; at this instant the rocket is in equilibrium. Figure 5-1*a* shows the two interaction pairs involved in the situation and describes them. The color-coding follows Step 5 of Procedure 5-1. Figure 5-1*b* reproduces just the part of 5-1*a* showing the rocket and the forces *on it.* Those two forces, though equal and opposite, belong to different interaction pairs.

Figure 5-1*c* shows the rocket when it is closer to Earth, where Earth's pull is stronger and the moon's is weaker. Figure 5-1*c* shows that there are still two interaction pairs, each with equal and opposite forces, but

Case 5-1 ◆ *Forces on a Rocket (continued)*

5-1*d* makes clear that the two forces on the rocket are no longer equal and opposite; a rocket adrift at this point is *not* in equilibrium. The force on a single object may be equal and opposite at one instant, producing equilibrium, but not at another. But the forces making up an interaction pair always remain equal and opposite, even as the forces grow stronger or weaker.

Figure 5-1 Distinguishing between force pairs producing equilibrium and interaction pairs. The rocket is drifting because its thrust is turned off. The black-and-yellow force vectors show the two sides of the interaction between the black rocket and the yellow moon; the black-and-blue force vectors show the two sides of the interaction between the black rocket and the blue Earth. (*a*) The situation at instant t_1. (*b*) Free-body diagram of the rocket at t_1: At this location the moon and Earth pull equally hard on the rocket. The rocket is in equilibrium. (*c*) The situation at later instant t_2. (*d*) Free-body diagram of the rocket at t_2: At this location Earth pulls harder on the rocket than the moon does. The rocket is no longer in equilibrium.

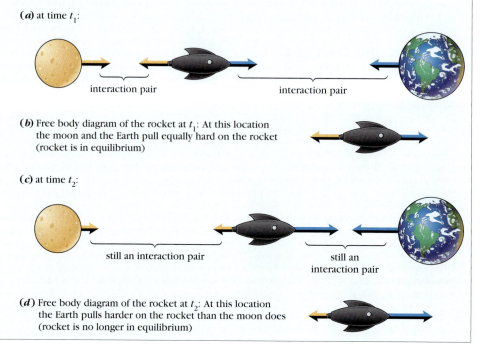

(*a*) at time t_1:

interaction pair   interaction pair

(*b*) Free body diagram of the rocket at t_1: At this location the moon and the Earth pull equally hard on the rocket (rocket is in equilibrium)

(*c*) at time t_2:

still an interaction pair   still an interaction pair

(*d*) Free body diagram of the rocket at t_2: At this location the Earth pulls harder on the rocket than the moon does (rocket is no longer in equilibrium)

◆**FREE-BODY DIAGRAMS** Figures 5-1*b* and 5-1*d* isolate one body, a drifting rocket, at two different instants. *Drifting* means there is no thrust. They show all the forces exerted on it (and *not* the forces that it exerts on other bodies). When you apply the second law to the rocket at one of these instants, the diagram serves as a visual inventory of all the forces you need to include in the vector sum $\Sigma \vec{\mathbf{F}}$ on rocket. Visual inventories of this type are called **free-body diagrams** or **force diagrams.**

PROCEDURE 5-2

Drawing a Free-Body Diagram

1. Isolate a single body to which you wish to apply the second law.
2. Draw each non-negligible force on that body as a vector or arrow according to Procedure 5-1.
   a. Draw all contact forces exerted by other bodies touching this body.
   b. Draw in any action-at-a-distance forces.

Free-body diagrams are especially valuable when several forces are exerted on a body. Newton's second law remains an unusable abstraction until you know what specific forces make up the vector sum on the left side of Equation 4-5. Therefore, an important first step is to take inventory of the forces that are exerted on the object of interest. The free-body diagram of the object gives us a picture summary of the inventory. In the following paragraphs, we will expand on the ideas in Figure 4-3 and identify some of the kinds of forces that you should look for in your inventories.

◆**CONTACT OR TOUCHING FORCES** The most intuitively evident forces are those that bodies exert on each other when they actually touch, or make contact. (*Contact* comes from *tactus,* a form of the Latin verb *tangere,* to touch, from

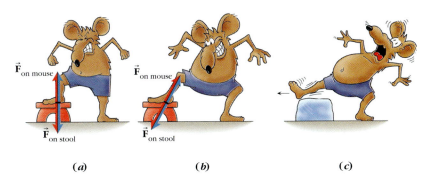

Figure 5-2 Exerting a contact force.

which *tangent* also comes.) The simplest such forces occur when bodies press directly against each other in a direction perpendicular to the contact surfaces, as in Figure 5-2*a*. Forces of this type, which we explained in Chapter 4 in terms of deformation and springback, are called **normal forces.** In this context, the word *normal* means *perpendicular*. Notice that for every normal force, there is a reaction normal force.

The critter in the figure can alter the direction of the net force by pushing, as in Figure 5-2*b*. Now the net force exerted by the critter on the stool is the vector sum of two forces; there is a force *along* the contact surface as well as one perpendicular to it. What is the nature of the force along the surface? It could move the stool sideways if the stool were free to move. But if the critter tried doing the same thing to a block of ice (Figure 5-2*c*), the ice wouldn't move and there would be little resistance to his foot. The force along the surface is thus a resistive force, and is called a **frictional force.** Frictional forces also occur as action-reaction pairs. The frictional force that the critter exerts on the stool may move the stool sideways; the frictional force that the stool exerts on the critter provides the resistance that is lacking in Figure 5-2*c*.

When one body presses against another, look for

- *normal* forces *perpendicular* to the surface of contact, and
- *frictional* forces *along* the surface of contact.

These two forces, normal and frictional, are not independent of each other. The harder two bodies press against each other, the more difficult it becomes to make one of them slide across the other. In other words, the greater the normal force on a body at the contact surface, the greater the frictional force on it that you must overcome to set or keep it in motion. Further details of the relationship between frictional and normal forces are presented in Section 5-5 (optional).

Another kind of contact force is exerted when flexible objects such as strings or chains pull on other bodies. If a string is connected to a block and you pull the string taut (Figure 5-3), the string aligns in the direction you are pulling, and

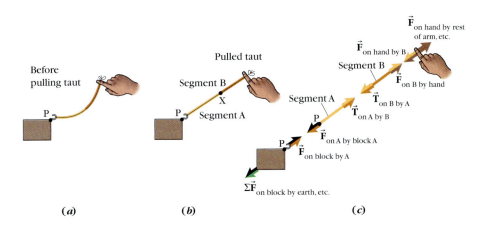

Figure 5-3 Tension forces. Tension forces are the forces that connected parts of a flexible object like a rope or string exert on each other. They are always directed along the flexible object. This is easiest to picture when the object is pulled taut so that the forces are in the same direction all along its length.

For **WebLink 5-2:**
Tension Forces, go to
www.wiley.com/college/touger

$\vec{\mathbf{F}}_{\text{on mouse by earth}}$

$\vec{\mathbf{F}}_{\text{on earth by mouse}}$

Figure 5-4 Free-body diagram of a body in free fall.

For **WebLink 5-3:**
Weighing In, go to
www.wiley.com/college/touger

the string in turn pulls in the same direction on the block. Moreover, if we pick any point X along the string, thereby dividing the string mentally into two segments A and B, the two segments pull on each other. The pulling forces that they exert on each other are called **tension forces** (denoted by $\vec{\mathbf{T}}$). They are always directed *along the string*. To see in greater detail how thinking about tension forces is useful when we apply Newton's second and third laws, work through WebLink 5-2.

When the string's mass is negligibly small, the total force on either segment is zero because $\Sigma F = ma = 0$ when m is zero as well as when $a = 0$. It then follows from Figure 5-3c that the tension forces $\vec{\mathbf{T}}_{\text{on segment A by segment B}}$ and $\vec{\mathbf{T}}_{\text{on segment B by segment A}}$ are equal in magnitude to (and along the same line as) the forces that the hand and rope and the rope and block exert on each other. (This is explained more fully in WebLink 5-2.) But these forces are *not* all equal if the string or other flexible object has a significant mass (see Problem 5-14).

As Figure 5-3 and WebLink 5-2 illustrate, you can mentally isolate any object, part of an object, or combination of objects and represent it by a free-body diagram. Whatever you isolate, you treat as a single "lump." You identify the forces *external* to this "lump," which you will in turn relate to its motion.

Figure 5-3 illustrates another important point, namely, *in dealing with a situation involving several bodies, you should draw a separate free-body diagram of each body.* Separating the pictures, even of bodies that are actually touching, allows you to see clearly which forces are exerted on which body. For example, it allows you to see that an action-reaction pair of contact forces such as $\vec{\mathbf{F}}_{\text{on rope by hand}}$ and $\vec{\mathbf{F}}_{\text{on hand by rope}}$ have the same point of application but are exerted on different bodies.

◆**ACTION-AT-A-DISTANCE FORCES, GRAVITATIONAL** We now examine the role of the gravitational force on an object—assumed to be a point object—as it falls freely to Earth (Figure 5-4) and when it lands on a scale (Figure 5-5). In both figures, free-body diagrams identify the forces on the objects involved and thus enable us to apply the following second and third law reasoning. (For a more complete treatment of the reasoning, see WebLink 5-3.)

- The only force on an object in free fall (Figure 5-4) is the gravitational force that Earth exerts on it. Also, we know that a freely falling object's acceleration has magnitude g. So $\Sigma\vec{\mathbf{F}} = m\vec{\mathbf{a}}$ reduces to

$$\vec{\mathbf{F}}_{\text{grav}} = m\vec{\mathbf{g}} \qquad \text{or} \qquad F_{\text{grav}} = mg \qquad (5\text{-}1)$$

[$\vec{\mathbf{g}}$, the gravitational acceleration vector, has magnitude g and is directed toward Earth's (or other attracting body's) center.]

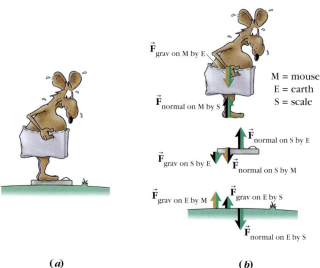

M = mouse
E = earth
S = scale

$\vec{\mathbf{F}}_{\text{grav on M by E}}$

$\vec{\mathbf{F}}_{\text{normal on M by S}}$

$\vec{\mathbf{F}}_{\text{normal on S by E}}$

$\vec{\mathbf{F}}_{\text{grav on S by E}}$

$\vec{\mathbf{F}}_{\text{normal on S by M}}$

$\vec{\mathbf{F}}_{\text{grav on E by M}}$

$\vec{\mathbf{F}}_{\text{grav on E by S}}$

$\vec{\mathbf{F}}_{\text{normal on E by S}}$

Figure 5-5 Free-body diagrams of bodies in equilibrium.

(a)   *(b)*

- It is equivalent to say that the object falls because it has weight. Most physicists define your weight to be the gravitational force exerted on you, ordinarily by Earth.[1]

Definition: $$\textbf{Weight} \equiv \vec{\textbf{F}}_{grav} = m\vec{\textbf{g}} \qquad (5\text{-}2)$$

Your weight is determined by your interaction with another body; unlike mass, it is not a property of you alone. On the moon, for example, the value of g is smaller, so that you would weigh less, and g varies with both elevation and latitude even on Earth's surface (see Section 2-7).

- In Figure 5-5, the same gravitational force is exerted on the object, but the object is now in equilibrium ($\Sigma\vec{\textbf{F}} = 0$). Note that there is also a *normal* force exerted upward on the object by the scale:

$$\vec{\textbf{F}}_{normal\ on\ obj.\ by\ scale} + \vec{\textbf{F}}_{grav\ on\ obj.\ by\ Earth} = 0 \qquad (5\text{-}3)$$

The two forces are equal in magnitude but opposite in direction.

- A simple scale exerts a normal force because its springs are compressed and push back up. The dial reading increases when the springs compress further—in other words, if there is an object on the scale, the dial reads the normal force that the scale exerts on the object.

- By Equation 5-3, the scale reading (normal force) is equal to the object's weight (gravitational force), provided that the object is in equilibrium ($\Sigma\vec{\textbf{F}} = 0$) and these are the only forces on the object. (In two-dimensional situations, these conditions need only apply in the vertical direction.) If these conditions are not met, the normal force will not equal the object's weight.

According to our definition of weight, the scale will not read your correct weight if you are holding a heavy package. The forces that add to zero are now (1) the normal force the scale exerts on you (read by the scale), (2) your weight, *and* (3) the force the package exerts on you, so (1) and (2) cannot be equal and opposite. Even when your weight and the normal force exerted by the scale are the only two forces on you, they are equal and opposite only when you (like the critter) are in equilibrium (i.e., when $\vec{\textbf{a}} = 0$). Otherwise, in contrast to Equation 5-3, $\Sigma\vec{\textbf{F}} \neq 0$, so that $\vec{\textbf{F}}_{normal\ by\ scale\ on\ you} \neq \vec{\textbf{F}}_{grav\ on\ you}$ (the scale's reading is not equal to your weight). **STOP&Think** Will a scale on a ramp read correct weights? ◆

Several of the forces we have discussed are involved in the situation depicted in Figure 5-6. A free-body diagram identifies the forces on the anvil, and Newton's second law is written out to include these forces explicitly. **STOP&Think** See if you can identify all the forces in the free-body diagram. Why is the weight of the hanging mass not included among the forces in the free-body diagram? ◆ In Example 5-1, you will see that a correct inventory of forces is a necessary first step when you apply Newton's second and third laws in solving quantitative problems.

[1]We choose this definition because it is much more commonly used in the United States. However, it is also possible to define the weight of a body to be the elastic contact force (normal force) that the body exerts on whatever is supporting it (such as $\vec{\textbf{N}}_{on\ M\ by\ S}$ in Figure 5-5b).

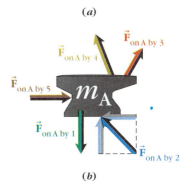

(a)

(b)

$$\Sigma\vec{\textbf{F}} = \vec{\textbf{F}}_{on\ A\ by\ 1} + \vec{\textbf{F}}_{on\ A\ by\ 2} + \vec{\textbf{F}}_{on\ A\ by\ 3}$$
$$+ \vec{\textbf{F}}_{on\ A\ by\ 4} + \vec{\textbf{F}}_{on\ A\ by\ 5} = m\vec{\textbf{a}}$$

(c)

Figure 5-6 Inventorying forces. (a) The objects in this real-world picture are color-coded. (b) A free-body diagram of the anvil. Each force vector bears the colors of the interacting objects responsible for that force. (c) $\Sigma\vec{\textbf{F}}$ is written out in detail for the anvil subject to the forces inventoried in (b).

Example 5-1 *Mouse on a Scale in Equilibrium*

For a guided interactive solution, go to Web Example 5-1 at www.wiley.com/college/touger

Suppose that in Figure 5-5, the mouse is a large stage mouse designed for a performance of the *Nutcracker* ballet. Its mass is 3.0 kg, and the scale's mass is 1.5 kg. Find

EXAMPLE 5-1 *continued*

a. the upward normal force that the scale exerts on the mouse and

b. the upward normal force that Earth exerts on the scale.

Brief Solution

Choice of approach. We apply Newton's second law separately to each body. Also, because the mouse and the scale interact with each other, the third law requires that the forces they exert on each other must be equal.

Free-body diagrams. To apply Newton's second law to a body, we must draw a free-body diagram to inventory the forces on the body. Separate free-body diagrams of the mouse and scale are shown in Figure 5-5b.

Anticipating results. Earth is supporting both the mouse and the scale. Therefore, we should expect the upward force exerted by Earth to equal the total of their two weights.

The mathematical solution. (M = mouse, S = scale, E = Earth)

a. To find a force on the mouse, we apply the second law to the mouse, with $a = 0$. We sum the forces in the free-body diagram of the mouse. Taking the downward direction as negative, we write the sum in terms of magnitudes as

$$F_{\text{normal on M by S}} - F_{\text{grav on M by E}} = 0$$

Because the mouse's weight is $m_M g$, $F_{\text{normal on M by S}} - m_M g = 0$, or

$$F_{\text{normal on M by S}} = m_M g = (3.0 \text{ kg})(9.8 \text{ m/s}^2) = \textbf{29.4 N}$$

(The normal force exerted by the scale equals the mouse's weight.)

b. Now we sum the forces in the free-body diagram of the scale:

$$\Sigma \vec{F}_{\text{on S}} = \vec{F}_{\text{normal on S by E}} + \vec{F}_{\text{normal on S by M}} + \vec{F}_{\text{grav on S by E}}$$

In terms of magnitudes, the second law applied to the scale is

$$\Sigma F_{\text{on S}} = F_{\text{normal on S by E}} - F_{\text{normal on S by M}} - F_{\text{grav on S by E}} = 0 \quad (5\text{-}4)$$

But by Newton's third law, $F_{\text{normal on S by M}} = F_{\text{normal on M by S}}$. Thus,

$$F_{\text{normal on S by E}} = F_{\text{normal on M by S}} + F_{\text{grav on S by E}}$$

$$= \underset{\text{from a}}{m_M g} + \underset{\text{by Eq. 5-3}}{m_S g} \quad (5\text{-}5)$$

$$= (3.0 \text{ kg})(9.8 \text{ m/s}^2) + (1.5 \text{ kg})(9.8 \text{ m/s}^2) = \textbf{44.1 N}$$

◆ Related homework: Problems 5-10, 5-11, and 5-12.

In Example 5-1, we conclude from Newton's laws that the floor exerts an upward normal force just sufficient to support the combined weight of the mouse and the scale. This should not be a surprising result. What is interesting is how we get it. We are applying a very spare set of rules—Newton's laws—involving a restricted concept of force that is free of the associations (active effort, etc.) we sometimes make with force in nonscientific language. Yet when we deduce what happens from these laws, what we get is the way the world actually behaves. (Strictly speaking, they are not really laws but a widely accepted theory or model—a human attempt to state what the laws of nature really are. You will discover later that Newton's has not always been the last word.)

STOP&Think Suppose the mouse and scale in Example 5-1 had been on an escalator moving at constant speed. How would the solution have changed (other than letting E = escalator rather than Earth)? ◆

Constant speed in a straight line means there is no acceleration. Thus Equation 5-3 does not change, nor does any part of the solution that follows. A constant velocity situation is an equilibrium situation whether $v = 0$ or not, and as long as there is equilibrium, we get the same results. In the next two examples, we look at cases where the velocity is not constant.

Example 5-2 *Apparent and Actual Weights in an Elevator*

An executive weighing 441 N (how heavy is this in pounds?) gets on the elevator of an office building. The elevator takes 1.60 s to go from rest to a steady upward speed of 2.00 m/s. It takes an equal amount of time to come to a stop when it reaches the floor where her office is located. While she is riding, she is standing on a scale calibrated in newtons (she is *heavily* into fitness).
a. What is the average scale reading while the elevator is starting up?
b. What is the average scale reading while the elevator is coming to a stop?

Solution

Restating the question. The scale reading measures the deformation responsible for the normal force $\vec{F}_N$ that the scale exerts on the executive. Part **a** asks you to find this normal force. You are asked for its average value over the intervals when the elevator is going from rest to a positive velocity and vice versa, that is, when its acceleration is positive and vice versa. Average acceleration values can be found for the two 1.6-s intervals. The question in each part is thus

$$F_N = ? \text{ when } a = \underline{\quad}$$

but you must first find a.

Choice of approach. Our approach is sketched out in Figure 5-7. After clarifying what the problem is asking, your next step should always be to identify the main *principle(s)* that will allow you to solve it—see (1) in the figure. As soon as you decide to apply Newton's second law to a body, you should draw a free-body diagram of that body (2). The free-body diagram (except for the identity of the passenger) is identical to that of the mouse in Figure 5-5b. The diagram of the executive shows the forces contributing to $\Sigma F_{on\ EX}$, so that the second law becomes $F_N - F_{grav} = m\bar{a}$. After input from steps 3 and 4, you can then solve for F_N.

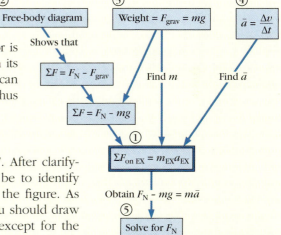

Figure 5-7 Schematic diagram of problem-solving approach for Example 5-2 (EX = executive).

Anticipating results. During the intervals described, the executive's acceleration is the same as the elevator's and is not zero, so the total force on her cannot be zero. By our definition of weight, her actual weight has a constant value mg and is directed downward. When her acceleration is upward (positive), the total force on her must be upward, so the normal force exerted by the scale must be greater than her weight. In fact, you *feel* momentarily heavier when an elevator starts upward, and you might expect the scale reading to be indicative of this *apparent* (in contrast to actual) weight. Likewise, when the acceleration is downward (negative), the total force must be downward. So the normal force must be *less* than her weight. (You feel momentarily lighter when an ascending elevator comes to a stop.)

➡ The apparent weight that the scale reads is an actual (real) force—the normal force exerted on the scale. But when the person on the scale is accelerating, the normal force is not equal to the gravitational force on the person, and the gravitational force is what we have defined to be the actual weight.

What we know/what we don't. (EX = executive, E = Earth, S = scale)

For definition of average acceleration	For second law
$\Delta t = 1.60$ s	weight = $F_{grav\ on\ EX\ by\ E} = mg = 441.$ N
	$g = 9.80$ m/s^2 $m = ?$

a. $v_1 = 0$ (rest) $v_2 = 2.00$ m/s $a = ?$ $\longrightarrow a = ?$ $F_N = ?$

b. $v_1 = 2.00$ m/s $v_2 = 0$ (rest) $a = ?$ $\longrightarrow a = ?$ $F_N = ?$

EXAMPLE 5-2 *continued*

The mathematical solution.

a. The situation is one-dimensional, so we needn't write Newton's second law in vector form. Thus, from the free-body diagram,

$$\Sigma F_{\text{on EX}} = F_{\text{N on EX by S}} - F_{\text{grav on EX by E}}$$

and $\Sigma F = ma$ becomes

$$F_{\text{N on EX by S}} - F_{\text{grav on EX by E}} = ma$$

or

$$F_{\text{N on EX by S}} = F_{\text{grav on EX by E}} + ma \qquad (5\text{-}6)$$

From step 3 of Figure 5-7,

$$m = \frac{F_{\text{grav on EX by E}}}{g} = \frac{441.\ \text{N}}{9.80\ \text{m/s}^2} = 45.0\ \text{kg}$$

Step 4 of Figure 5-7 is

$$a_{\text{EX}} = \frac{\Delta v}{\Delta t} = \frac{v_2 - v_1}{\Delta t} = \frac{2.00\ \text{m/s} - 0}{1.60\ \text{s}} = 1.25\ \text{m/s}^2$$

With these values, Equation 5-6 becomes

$$F_{\text{N on EX by S}} = 441\ \text{N} + (45.0\ \text{kg})(1.25\ \text{m/s}^2) = 441\ \text{N} + 56.3\ \text{N} = \mathbf{497\ N}$$

The executive's apparent weight, as read by the scale, is in fact greater than the actual weight of 441 N.

b. Here the values of v_1 and v_2 are reversed, so that

$$a = \frac{v_2 - v_1}{\Delta t} = \frac{0 - 2.00\ \text{m/s}}{1.60\ \text{s}} = -1.25\ \text{m/s}^2$$

Now the acceleration is negative, because the elevator and passenger are slowing down in the upward direction. If you use *this* value in Equation 5-7,

$$F_{\text{N on EX by S}} = 441.\ \text{N} + (45.0\ \text{kg})(-1.25\ \text{m/s}^2) = 441\ \text{N} - 56.3\ \text{N} = \mathbf{385\ N}$$

The executive's apparent weight is now less than the actual weight.

◆ Related homework: Problems 5-19 and 5-20.

Reminder: Results never have more significant figures than the values from which they are calculated.

In situations where ΣF has a constant, nonzero value, the resulting constantly accelerated motion is describable by the usual equations of motion: If both forces and motion are in a single direction, these will be either Equations 3-16x to 3-18x or Equations 3-16y to 3-18y for constant acceleration. The acceleration is the link between the *kinematics,* which is purely the description of motion without reference to forces, and the *dynamics,* which deals with the effect of forces on motion. The kinematics is expressed by the equations of motion and the dynamics by Newton's second law. This suggests the plan of attack summarized in Figure 5-8 for dealing with problems involving forces and motion. Example 5-3 shows this plan of attack in action.

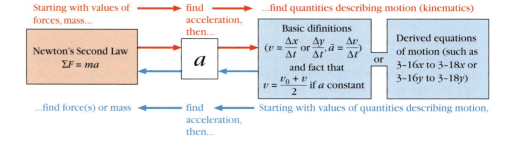

Figure 5-8 Plan of attack for problems involving forces and motion in one dimension.

Example 5-3 *Motion of a Block Subject to Two Forces*

For a guided interactive solution, go to Web Example 5-3 at
www.wiley.com/college/touger

As a block weighing 39.2 N is raised from rest by a rope, the force exerted
by the rope is maintained at a constant 40.0 N.
a. Find the upward acceleration of the rope.
b. How high is the block raised in the first two seconds of lifting?

Brief Solution

Choice of approach. Following the plan of attack in Figure 5-8 from left to
right, we apply Newton's second law to find $\vec{a}$, then use $\vec{a}$ to do kinematics.
Sketch the free-body diagram for yourself. The forces on the block are exerted
by the rope (a force equal to the tension T) and by Earth (the weight mg).

What we know/what we don't.

For second law	For equation of motion
weight $F_{grav} = mg = 39.2$ N	$y_o = 0$ $y = ?$
$m = ?$ $g = 9.8$ m/s^2	$t = 2.0$ s
$T = 40$ N $a = ?$ $\longrightarrow a = ?$	

The mathematical solution.
a. The second law gives us

$$\Sigma F = T - F_{grav} = ma \tag{5-7}$$

From Equation 5-3, $\qquad m = \dfrac{F_{grav}}{g}$

which we substitute for m in
Equation 5-7 yielding $\qquad T - F_{grav} = \dfrac{F_{grav}}{g} a$

Solving for a, $\quad a = \dfrac{g(T - F_{grav})}{F_{grav}} = \dfrac{(9.8\ \text{m/s}^2)(40.0\ \text{N} - 39.2\ \text{N})}{39.2\ \text{N}} = \mathbf{0.20\ m/s^2}$

b. From the definition of
acceleration $a_y = \dfrac{\Delta v_y}{\Delta t} \qquad \Delta v_y = v_y - v_{oy} = a_y \Delta t$

so $\qquad v_y = v_{oy} + a_y \Delta t = 0 + (0.2\ \text{m/s}^2)(2.0\ \text{s}) = 0.40$ s

Because the acceleration is constant, $\bar{v}_y = \dfrac{v_{oy} + v_y}{2} = \dfrac{0 + 0.40\ \text{m/s}}{2}$

$$= 0.20\ \text{m/s}$$

From the definition of
velocity $v_y = \dfrac{\Delta y}{\Delta t} \qquad \Delta y = \bar{v}\Delta t = (0.20\ \text{m/s})(2.0\ \text{s}) = \mathbf{0.40\ m}$

Or

Use Equation 3-15y to find y:
Since $v_{oy} = 0$ and $y_o = 0$ $\qquad y = v_{oy}t + \dfrac{1}{2}at^2 = \dfrac{1}{2}at^2$

Thus, $\qquad y = \dfrac{1}{2}(0.20\ \text{m/s}^2)(2.0\ \text{s})^2 = \mathbf{0.40\ m}$

◆ Related homework: Problems 5-21, 5-22, 5-23, and 5-24.

5-2 Applying Newton's Second and Third Laws in Two Dimensions

In this section, we will apply Newton's second law to two-dimensional situations. As we saw in Case 4-1, interactions may affect the magnitude and/or direction of a body's velocity. In Newtonian terms, a body's velocity changes when a net force is exerted on it. In the following activities, you will examine this change in more detail.

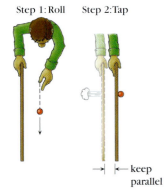

Step 1: Roll   Step 2: Tap

← keep parallel

Figure 5-9 Exerting a force on a marble perpendicular to its motion.

On-The-Spot Activity 5-1

Lay a meter stick or other long straight object on the floor. Roll a marble or small ball along a straight-line path parallel to and a few centimeters to one side of the meter stick (Figure 5-9). Keeping the meter stick parallel to the ball's path, move it sideways until it taps the ball gently from the side while it is rolling. What is the path of the ball after being tapped? Does it just go in the direction of the tap? Repeat this several times, tapping the ball slightly harder each time and observing the consequent final direction. As long as the direction of the tap is perpendicular to the initial direction of motion, can the ball ever end up in a direction *perfectly* perpendicular to its initial direction? Experiment with your set-up to determine how hard and in what direction the tap must be if the ball's final direction *is* to be perpendicular to its initial direction.

The meter stick exerts an average force $\vec{F}$ during the time interval Δt when it is in contact with the ball. $\vec{F}$ is in the direction of the tap. So is $\Delta\vec{v}$, because $\Delta\vec{v} = \vec{a}\Delta t = \frac{\vec{F}}{m}\Delta t$ is a scalar times $\vec{F}$ and is therefore in the same direction as $\vec{F}$. Figure 5-10 shows $\Delta\vec{v}$ and the final velocity $\vec{v}_2 = \vec{v}_1 + \Delta\vec{v}$ that result from exerting each of several different forces $\vec{F}$. Notice that in Figure 5-10a, the y component of the velocity doesn't change. A force with only an x component can affect only the x-direction "shadow motion"—it results in an acceleration $\vec{a}$ and a velocity

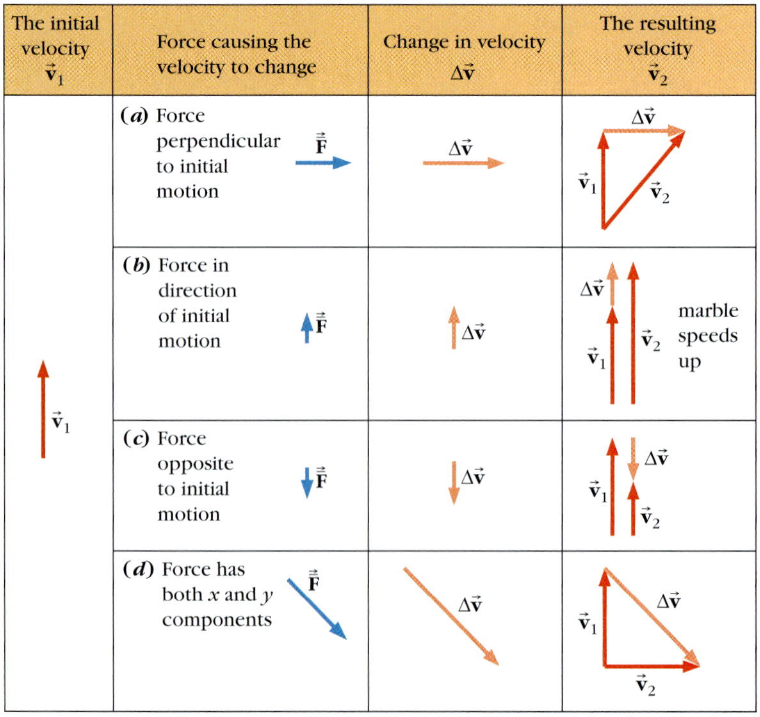

Figure 5-10 Vectors representing $\Delta\vec{v}$. Each vector diagram shows a different change of velocity $\Delta\vec{v}$. Each $\Delta\vec{v}$ is in the direction of the average force $\vec{F}$ causing the change.

The initial velocity $\vec{v}_1$	Force causing the velocity to change	Change in velocity $\Delta\vec{v}$	The resulting velocity $\vec{v}_2$
$\vec{v}_1$	(*a*) Force perpendicular to initial motion $\vec{F}$	$\Delta\vec{v}$	$\Delta\vec{v}$ $\vec{v}_1$ $\vec{v}_2$
	(*b*) Force in direction of initial motion $\vec{F}$	$\Delta\vec{v}$	$\Delta\vec{v}$ $\vec{v}_1$ $\vec{v}_2$ marble speeds up
	(*c*) Force opposite to initial motion $\vec{F}$	$\Delta\vec{v}$	$\vec{v}_1$ $\Delta\vec{v}$ $\vec{v}_2$
	(*d*) Force has both x and y components $\vec{F}$	$\Delta\vec{v}$	$\vec{v}_1$ $\Delta\vec{v}$ $\vec{v}_2$

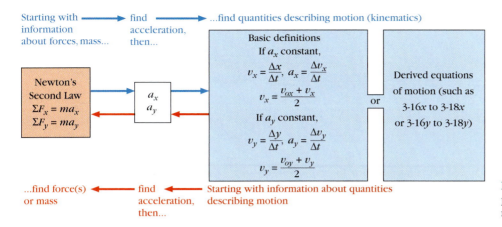

change $\Delta\vec{v}$ that are purely in the x direction. In Figure 5-10d the y-direction shadow motion is affected because the force has a component in the $-y$ direction. For the marble to end up moving perpendicular to its initial direction, its velocity must lose its y component and gain an x component. To change the velocity components in both directions, the force must have components in both directions.

More formally, we can say that because

$$\Sigma\vec{F} = m\vec{a} \tag{4-8}$$

is a vector equation, it summarizes the two one-dimensional equations

$$\Sigma F_x = ma_x \tag{4-8x}$$

and

$$\Sigma F_y = ma_y \tag{4-8y}$$

which separately relate the total force in each direction to changes in motion in that direction. Like kinematics, dynamics is best treated using the component description of vectors. Review Procedure 3-3 on how to do this. We now add Equations 4-8x and 4-8y to the equations that we can use in each direction. Again the kinematics and dynamics are linked by the acceleration. The strategy summarized in Figure 5-8 for one dimension takes the form shown in Figure 5-11 for the two-dimensional case.

In Example 5-3, we followed the approach in Figure 5-8 left to right. In Example 5-4, we follow the approach in Figure 5-11 from right to left, starting with a kinematic treatment of motion to find the acceleration, and then using the result in Newton's second law to calculate unknown forces.

Example 5-4 *Towing a Log*

For a guided interactive solution, go to Web Example 5-4 at www.wiley.com/college/touger

Tow ropes are used to pull a 400 kg log. Starting from rest, the log travels a distance of 1.50 m in its first 3.00 s of motion. During this time, the tow ropes exert a steady combined horizontal force of 2000 N on the log.
a. Find the normal and frictional forces exerted by the ground on the log.
b. Find the force exerted by each rope individually.

Brief Solution

Choice of approach. See paragraph preceding this example. The forces to be included in applying the second law are shown in the free-body diagrams in the figure below. Note that in part **b**, each of the forces exerted separately by the ropes has components in two directions.

EXAMPLE 5-4 *continued*

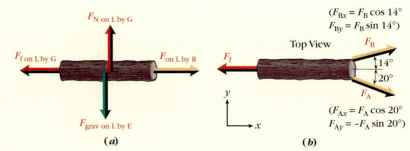

Free-body diagrams for Example 5-4. L = log, G= ground, E = Earth, R = ropes pulling together, A and B are the separate ropes.

What you know/what you don't. (L = log, G = ground, E = Earth, R = ropes pulling together, A and B are the two separate ropes)

for equation of motion	for second law
$x_o = 0$ $x = 1.50$ m	$m = 400$ kg $g = 9.8$ m/s^2
$v_{ox} = 0$ $t = 3.00$ s	$F_{\text{on L by R}} = 2000$ N $F_{\text{f on L by G}} = ?$
	$F_{\text{N on L by G}} = ?$ $F_A = ?$ $F_B = ?$

$$a_x = ? \longrightarrow a_x = ?\quad a_y = 0$$

The mathematical solution.

a. Equation 3-15x with $v_{ox} = 0$ is

$$x = \frac{1}{2}a_x t^2$$

Solving for a_x gives us

$$a_x = \frac{2x}{t^2} = \frac{2(1.50\text{ m})}{(3.00\text{ s})^2} = 0.333\text{ m/s}^2$$

Now we can apply Newton's second law in component form.

$\Sigma F_y = ma_y = 0$ leads to $F_{\text{N on L by G}} = mg = (400\text{ kg})(9.8\text{ m/s}^2) = \textbf{3920 N}$

$\Sigma F_x = ma_x$ for $F_{\text{f on L by G}}$ gives $F_{\text{f on L by G}} = F_{\text{on L by R}} - ma$

$= 2000\text{ N} - (400\text{ kg})(0.333\text{ m/s}^2) = 2000\text{ N} - 133\text{ N} = \textbf{1870 N}$

b. The horizontal components of the two forces must add up to the combined horizontal force of 2000 N:

$$F_A \cos 20° + F_B \cos 14° = 2000\text{ N}$$

Applying the second law with the forces in the second free-body diagram, we get

$$\Sigma F_y = F_A \sin 20° - F_B \sin 14° = ma_y = 0$$

From the second of these equations, we can get $F_A = F_B \frac{\sin 14°}{\sin 20°}$, which we can substitute for F_A in the first equation, yielding

$$F_B \frac{\sin 14°}{\sin 20°}\cos 20° + F_B \cos 14° = \textbf{2000 N}$$

Solving for F_B gives us $F_B = 1250$ N

Then $F_A = F_B \dfrac{\sin 14°}{\sin 20°} = \textbf{885 N}$

◆ Related homework: Problem 5-31.

Part **b** of Example 5-4 emphasizes the need to consider components of forces when working with Newton's second law. We encounter that need again in the following example.

Example 5-5 *Pulling a Glider on an Air Track: Normal Force ≠ Weight*

A 0.20-kg glider is pulled along an air track, for which frictional forces are negligible, by a string kept taut at an angle 30° above the horizontal.
a. What must be the force exerted by the string if the block is to have an acceleration of 5.0 m/s²?
b. What normal force will the track then exert on the glider?

Solution
Choice of approach. (1) Again we apply Newton's second law in component form. (2) We inventory the forces on the glider, and their components, in a free-body diagram (Figure 5-12).

What we know/what we don't.

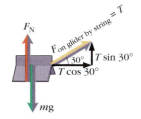

Figure 5-12 Free-body diagram for Example 5-5.

$a_x = 5.0 \text{ m/s}^2$	$a_y = 0$	
Forces in x direction	Forces in y direction	$g = 9.8 \text{ m/s}^2$
$T\cos 30°$	N	$m = 0.20$ kg
	$-mg$	
	$T\sin 30°$	

Note: $\Sigma F_y = 0$ because the motion is entirely horizontal. Then F_N cannot equal mg because unlike in the previous two examples, these are no longer the only two forces in the y direction. In everyday terms, the weight of the block is only partly supported by the track; the rest of the support is provided by the upward pull of the string.

The mathematical solution. Equations 4-8x and 4-8y become

$$\Sigma F_x = T\cos 30° = ma_x \tag{5-8x}$$

and
$$\Sigma F_y = F_N + T\sin 30° - mg = 0 \tag{5-8y}$$

When you solve for T in Equation 5-8x and substitute values, you get

$$T = \frac{ma_x}{\cos 30°} = \frac{(0.20 \text{ kg})(5.0 \text{ m/s}^2)}{0.866} = \textbf{1.2 N}$$

Solving for F_N in Equation 5-8y and substituting values, including the previous result, gives you

$$F_N = mg - T\sin 30° = (0.2 \text{ kg})(9.8 \text{ m/s}^2) - (1.15 \text{ N})(0.500)$$
$$= 1.96 \text{ N} - 0.58 \text{ N} = \textbf{1.4 N}$$

In this calculation, 1.96 N is the weight; the normal force, as we anticipated, is less.

◆ Related homework: Problem 5-32.

What are the forces on the child at the instant captured here? In what direction is the total force on the child at this instant? In what direction is the child's acceleration at this instant? Does the direction of the acceleration change as the child swings?

Recall (see Figure 3-19b) that in some situations it is more useful to choose a coordinate system where the x and y axes are not horizontal and vertical, but instead line up along and perpendicular to an object's motion along an incline. This is true of the following example.

Example 5-6 *The Water Slide*

Water slides are designed to minimize friction. The critter in Figure 5-13 has found the ideal water slide, one that exerts *no* frictional force on the critter. If the water slide makes an angle of 35° with the horizontal, find the critter's acceleration.

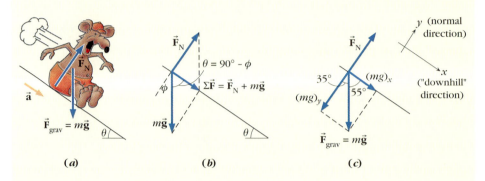

Figure 5-13 Free-body diagram for Example 5-6.

(*a*)   (*b*)   (*c*)

Solution

Choice of approach. (1) To apply Newton's second law, we first inventory the forces involved in $\Sigma\vec{F}$ by drawing a free-body diagram (Figure 5-13*a*). Forces are exerted on the critter only by Earth (the critter's weight $\vec{F}_{grav} = m\vec{g}$) and the slide $\vec{F}_{on\ critter\ by\ slide}$ (purely a normal force $\vec{F}_N$, because there is no frictional force). Neither of these forces is directed down the slide, but their total must be (Figure 5-13*b*), or the direction of motion would change. Note also that $\vec{F}_{grav}$ is always downward (toward Earth's center), whereas normal forces are always perpendicular to the contact surfaces. So if the contact surfaces aren't horizontal, $\vec{F}_N$ cannot be opposite to $\vec{F}_{grav}$, and usually is not equal to it in magnitude. (2) If we choose the coordinate system in Figure 5-13*c*, all motion is in the *x* direction, so $a_y = 0$ and $a = a_x$. (3) We can now identify components of the forces in the free-body diagram that make up ΣF_x and ΣF_y. *Important:* In the coordinate system we've chosen, $\vec{F}_{grav}$ has both *x* and *y* components (see Figure 5-13*c*). As Figure 5-13*b* shows, ϕ, the angle of inclination, is also the angle that $\vec{F}_{grav}$ makes with the new *y* (not *x*) direction, so we must find components as in Figure 5-14.

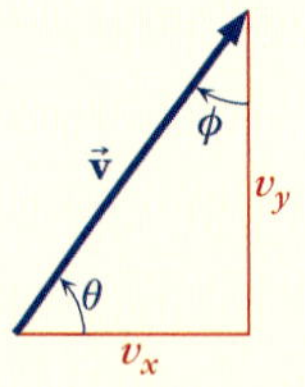

Figure 5-14 Components of a vector expressed in terms of θ and in terms of ϕ. θ is the angle the vector makes with the +*x* axis and ϕ is the angle it makes with the +*y* axis.

$$\sin\theta = \frac{\text{side opposite to }\theta}{v} = \frac{v_y}{v}, \text{so } v_y = v\sin\theta$$

$$\cos\phi = \frac{\text{side adjacent to }\phi}{v} = \frac{v_y}{v}, \text{so } v_y = v\cos\phi$$

$$\cos\theta = \frac{\text{side adjacent to }\theta}{v} = \frac{v_x}{v}, \text{so } v_x = v\cos\theta$$

$$\sin\phi = \frac{\text{side opposite to }\phi}{v} = \frac{v_x}{v}, \text{so } v_x = v\sin\phi$$

What we know/what we don't.

Second law in *x* direction	Second law in *y* direction	
$a_x = a = ?$	$a_y = 0$	

Forces in *x* direction	Forces in *y* direction	
$(F_{grav})_x = mg\sin\phi\ (\phi = 35°)$	$(F_{grav})_y = -mg\cos\phi$	$g = 9.8\ \text{m/s}^2$
	$F_N = ?$	$m = ?$

The mathematical solution. We first write out the force components that make up the sums in Newton's second law:

$$\Sigma F_x = mg \sin \phi = ma_x = ma \tag{5-9x}$$

$$\Sigma F_y = F_N - mg \cos \phi = ma_y = m(0) = 0 \tag{5-9y}$$

We don't know the mass, but if we do the algebra before substituting values, this turns out not to be a problem. Dividing both sides of Equation 5-9x by m gives

$$g \sin \phi = a \tag{5-10x}$$

so $\quad\quad a = (9.8 \text{ m/s}^2)(\sin 35°) = (9.8 \text{ m/s}^2)(0.574) = \textbf{5.6 m/s}^2$

Making sense of the results. The fact that m dropped out of Equation 5-10x shows that the acceleration doesn't depend on the value of the mass or the weight mg. Think of the value of $\sin \phi$ as a fraction. "$g \sin \phi$" means the critter experiences this fraction of the free-fall acceleration g—a fraction that depends only on the tilt angle ϕ of the slide. From Equation 5-9y we could show that

$$\frac{F_N}{mg} = \cos \phi \tag{5-10y}$$

so $\cos \phi$ is the fraction of the critter's weight that presses "normally" against the slide.

◆ Related homework: Problems 5-33, 5-34, and 5-35.

STOP&Think Use Equations 5-10x and 5-10y to see what happens to a and to $\frac{F_N}{mg}$ when $\phi = 0$ and when $\phi = 90°$. Do these results agree with your intuitions about what should happen? ◆

Is the normal force that the slope exerts on the skier less than, equal to, or greater than the weight of the skier in this photo?

5-3 Bodies with Linked Motions

Sometimes the motions of two bodies are linked. This is true of blocks A and B in each of the situations in Figure 5-15. In *a*, A and B clearly move together; their velocities and accelerations must therefore be the same at every instant. In *b* and *c*, if the string remains taut and doesn't stretch, the magnitudes of A's velocity and acceleration must be the same as B's, even though the directions are different.

In all three situations, there is a force exerted on B equal and opposite to one that is exerted on A. In *a*, this is because Newton's third law tells us that the normal forces that A and B exert on each other must be equal and opposite

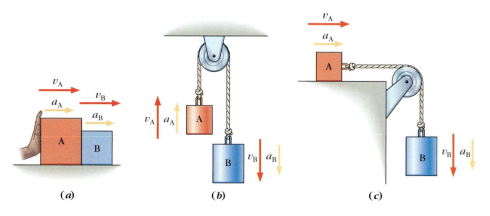

(*a*) (*b*) (*c*)

Figure 5-15 Examples of situations involving two bodies with linked motions. The set-up in (*b*) is sometimes called an **Atwood's machine** and the set-up in (*c*) a **modified Atwood's machine.**

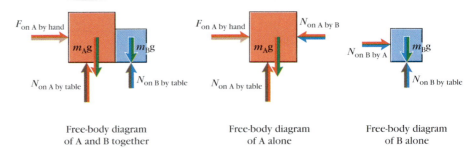

Figure 5-16 Free-body diagrams for Figure 5-15a. For legibility, we have placed the heads rather than the tails of the normal force vectors at their points of application.

(see Figure 5-16). In *b* and *c*, blocks A and B do not make direct contact with each other. But as long as (1) the string connecting them and the pulley are light enough to be treated as massless and (2) there is no slippage between the string and the pulley,[2] the forces that the string exerts at its two ends will both be equal in magnitude to the tension in the string.

To see how we use this third law reasoning when two objects move together, work through the next two examples.

[2]If the pulley's mass is negligible, the frictional force that the string must exert on it to keep string and pulley moving together is negligible, so that the frictional force the pulley exerts on the string will likewise be negligible. Under these conditions, the situation is tantamount to the string moving frictionlessly over the axle of the pulley. The pulley serves as light wheels or bearings might in reducing the frictional force on a body in linear motion across a surface.

Example 5-7 *The Mysterious Crates*

For a guided interactive solution, go to Web Example 5-7 at www.wiley.com/college/touger

A secret agent pushes two mysterious crates across a frozen river in the dark of night. The crates slide frictionlessly. Their masses are 150 kg (A) and 50 kg (B). The agent's hands exert a 100-N force on A. How much force does A exert on B? **STOP&Think** If there is no friction, can either block's acceleration be zero? Therefore, can $\Sigma \vec{F}_{on\,A} = 0$? And therefore can the force exerted on A by B (which is equal in magnitude to that exerted on B by A) "balance" (be equal and opposite to) the force exerted by the agent? ◆

Brief Solution
Choice of approach. With no friction, the unopposed force exerted by the agent makes *both* crates accelerate. If crate A is accelerating, the forces on A cannot be equal and opposite. (1) To find a force on block B, we apply Newton's second law to B. Using a free-body diagram of B (left) to identify the forces in Equations 4-8x, we get

$$\Sigma F_{x\,on\,B} = F_{on\,B\,by\,A} = m_B a_{Bx} = m_B a_B \qquad (5\text{-}11)$$

This leaves us with two unknowns, $F_{on\,B\,by\,A}$ and a_B. (2) However, if we apply the second law to block A as well (see free-body diagram of A at left), we get

$$\Sigma F_{x\,on\,A} = F_{on\,A\,by\,agent} - F_{on\,A\,by\,B} = m_A a_{Ax} = m_A a_A \qquad (5\text{-}12)$$

But $F_{on\,A\,by\,B} = F_{on\,B\,by\,A}$. Also, $a_A = a_B$ (label this common acceleration *a*). Thus, from Equations 5-11 and 5-12, we get

$$F_{on\,B\,by\,A} = m_B a \qquad (5\text{-}13B)$$

and

$$F_{on\,A\,by\,agent} - F_{on\,B\,by\,A} = m_A a \qquad (5\text{-}13A)$$

which we can solve for the two unknowns $F_{on\,B\,by\,A}$ and *a*.

Free-body
diagram
of A

Free-body
diagram
of B

(showing only horizontal forces)

The mathematical solution. We use Equation 5-13B to substitute for $F_{\text{on B by A}}$ in Equation 5-13A:

$$F_{\text{on A by agent}} - m_B a = m_A a$$

Then
$$F_{\text{on A by agent}} = m_B a + m_A a = (m_B + m_A)a$$

and
$$a = \frac{F_{\text{on A by agent}}}{m_B + m_A} = \frac{100 \text{ N}}{50 \text{ kg} + 150 \text{ kg}} = \mathbf{0.50 \text{ m/s}^2} \qquad (5\text{-}14)$$

Substituting this value and $m_A = 50$ kg into Equation 5-13B gives us

$$F_{\text{on B by A}} = (50 \text{ kg})(0.50 \text{ m/s}^2) = \mathbf{25 \text{ N}}$$

Making sense of the results. We have found that $F_{\text{on B by A}}$ is one quarter of the force exerted by the hand on A. Likewise, m_B is one quarter of the combined mass $m_A + m_B$ pushed by the hand. As you would expect from Newton's second law, to impart the same acceleration to a mass one-fourth as great requires one fourth the force.

◆ Related homework: Problems 5-36, 5-41, 5-44, and 5-94.

If you are riding on one car of this train, how does the force that the car in front of you exerts on your car compare to the force that your car exerts on the car behind you? Is it greater, equal, or smaller? How do you know?

Example 5-8 *The Modified Atwood's Machine*

In Figure 5-15c the masses are again $m_A = 15.0$ kg and $m_B = 5.0$ kg, and the string and pulley are light enough to be considered massless. If A moves frictionlessly across the table, find the acceleration of block B.

Solution I

Choice of approach. (1) As in Example 5-7, we can apply Newton's second law. So we must draw free-body diagrams to inventory the forces in $\Sigma \vec{F}$. (2) Because the string is massless, $F_{\text{on A by string}} = F_{\text{on B by string}} = T$, the tension in the string. (3) To find a, we can apply the second law separately to each of the two blocks and keep the usual x and y directions. For block A, $(a_A)_x = a$, and for block B, which is descending, $(a_B)_y = -a$. With these accelerations and the forces in the free-body diagrams (Figures 5-17a and b),

$$\Sigma F_{x \text{ on A}} = m_A (a_A)_x \qquad \text{becomes} \qquad T = m_A a \qquad (5\text{-}15\text{A})$$

and
$$\Sigma F_{y \text{ on B}} = m_B (a_B)_y \qquad \text{becomes} \qquad T - m_B g = m_B(-a) \qquad (5\text{-}15\text{B})$$

The mathematical solution. We must solve these two simultaneous equations for T and a. We begin by using $T = m_A a$ (Equation 5-15A) to substitute for T in Equation 5-15B:

$$m_A a - m_B g = m_B(-a)$$

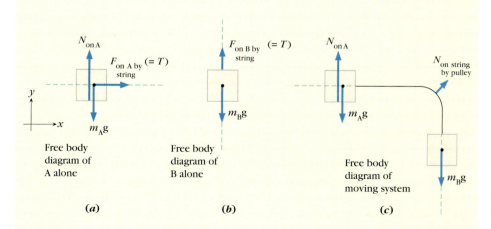

Free body diagram of A alone

Free body diagram of B alone

Free body diagram of moving system

(a)

(b)

(c)

Figure 5-17 Free-body diagram for Example 5-8.

EXAMPLE 5-8 *continued*

so that
$$m_A a + m_B a = m_B g$$

and
$$a = \frac{m_B g}{m_A + m_B} = \frac{(5.0 \text{ kg})(9.8 \text{ m/s}^2)}{15.0 \text{ kg} + 5.0 \text{ kg}} = \textbf{2.45 m/s}^2 \qquad (5\text{-}16)$$

(If we also wanted to know the tension force, we could readily substitute this value of a into Equation 5-15A to find T.)

Solution II

We can also apply the second law to the *combined* system if we resolve all vectors into components along and perpendicular to the path of the actual motion (dotted in Figure 5-17*c*) even though the path is not entirely along a single straight line. The only force on the combined system that acts along the path (and thus the *total* force, because the vector sum of the other forces is zero) is $m_B g$. Thus, $\Sigma F_{\text{on (A+B)}} = m_{\text{(A+B)}} a$ becomes

$$m_B g = (m_A + m_B)a$$

from which Equation 5-16 immediately follows.

Making sense of the result. Because B is not freely falling, we expect its downward acceleration to be smaller in magnitude than g, and it is. When only a fraction of the system's mass hangs over the edge, you get only that same fraction ($\frac{1}{4}$ in this case) of the acceleration that you would get when the *whole* system is over the edge, and therefore freely falling with acceleration g.

◆ Related homework: Problems 5-40 and 5-45.

STOP&Think In Example 5-7, where the secret agent pushed crates A and B together, we found that $a = \frac{F_{\text{on A by agent}}}{m_A + m_B}$. Why is this so similar to Equation 5-16 in Example 5-8? What do the two situations have in common? ◆

5-4 Static Equilibrium, Rigid Bodies, and the Concept of Torque

In general, $\Sigma \vec{F}_A = 0$ when $\vec{a}_A = 0$. So an object is in translational equilibrium when its velocity has *any* constant value, not just when it is constant at zero. However, the equilibrium of stationary bodies is of frequent interest, as in structural situations where you wish to know how much force a floor beam or a bridge support must be able to sustain. It thus has a special name, **static equilibrium,** but you apply Newton's second law the same way.

Example 5-9 *On the Luggage Chute*

For a guided interactive solution, go to Web Example 5-9 at www.wiley.com/college/touger

A duffle bag and a suitcase, each weighing 39.2 N, are on a luggage chute making an angle of 25° with the horizontal. While the duffle bag remains stationary on the chute, the suitcase slides past it at a constant speed of 0.3 m/s.
a. Find the frictional and normal forces that the chute exerts on each piece of luggage.
b. Find the total force that the chute exerts on each piece of luggage.

Brief Solution

Choice of approach. Each piece of luggage moves at constant velocity (whether at 0 or at 0.3 m/s). Because neither one is accelerating, $\Sigma \vec{F} = 0$ on each. The free-body diagram (below) is thus the same for the duffle bag and the suit-case. Note the directions of the normal force and the weight; they are not equal and opposite. Because $\vec{a} = 0$, there must also be an uphill frictional force on each piece of luggage equal and opposite to the resultant of the other two forces exerted on it. We choose a coordinate frame oriented so that the $+x$ direction is downhill and identify components along the axes of this coordinate frame to use in $\Sigma F_x = 0$ and $\Sigma F_y = 0$.

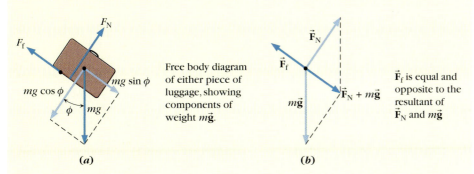

(a)   **(b)**

What we know/what we don't.

$$a_x = 0 \qquad\qquad a_y = 0$$

Forces in x direction on either one	Forces in y direction on either one	
$mg \sin \phi$	F_N	$m = 39.2$ N for either
$-F_f$	$-mg \cos \phi$	$\phi = 25°$

The mathematical solution.

a. Putting these specifics into Equations 4-8x and 4-8y, we get

$$\Sigma F_x = mg \sin \phi - F_f = 0$$

$$\Sigma F_y = F_N - mg \cos \phi = 0$$

from which it readily follows that

$$F_f = mg \sin 25° = (39.2 \text{ N})(0.423) = \textbf{16.6 N}$$

$$F_N = mg \cos 25° = (39.2 \text{ N})(0.906) = \textbf{35.6 N}$$

where F_f and F_N mean $f_{\text{on A or B}}$ and $N_{\text{on A or B}}$.

b. $-F_f$ and F_N are the x and y components of the total force $\vec{F}_{\text{by ramp}}$ on either piece of luggage. Thus $F_{\text{by ramp}} = \sqrt{(-16.6 \text{ N})^2 + (35.6 \text{ N})^2} = 39.2$ N, that is, $\vec{F}_{\text{by ramp}}$ is equal in magnitude to the luggage's weight. (Show for yourself that it is also opposite in direction, as equilibrium requires.)

◆ Related homework: Problems 5-47 and 5-48.

So far, we have treated all bodies as though the fact that they weren't really point objects didn't matter. Often this is not the case. In Figure 5-18, the block X (*a*) is in static equilibrium—it remains stationary when the two hands exert equal and opposite forces on it. We are free to regard the block as a point object (*b*). But in *c*, the baton does *not* remain stationary when the two hands pull equally in opposite directions. It rotates until it is horizontal (*d*), at which time the hands pulling left and right produce no further rotation. Treating the baton as a point object (*e*) gives us an adequate picture of what the hands are doing to the baton when the baton is oriented as in *d*, but *not* when it is oriented as in *c*.

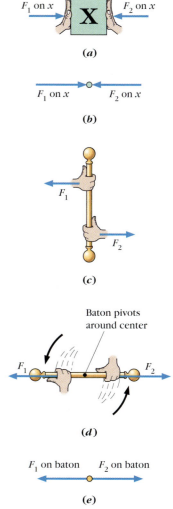

Figure 5-18 Objects in translational equilibrium.

The center of the baton in *d* stays in the same place throughout the rotation. If we ignore the extent of the baton beyond its center, the baton goes nowhere; it is in translational equilibrium. But to get the whole picture, we must consider **rotational equilibrium** as well as translational equilibrium. Even to speak of rotation, we must consider the baton as an extended body. Rotation means that the parts of the extended body change their orientations (directions) relative to some pivot point. In *c*, the baton is in translational equilibrium but not in rotational equilibrium. In *d*, it is in both.

The boom in Figure 5-19 is likewise in both translational and rotational equilibrium. In this case, we see that if rope 1 broke, $\vec{T}_2$ and the weight $\vec{F}_{grav}$ of the boom would together make the boom rotate clockwise about the hinge. The effect of force $\vec{T}_1$ on the boom's rotation must therefore be equal and opposite to the combined effect of $\vec{F}_{grav}$ and $\vec{T}_2$. In the following paragraphs, we will identify more precisely what the effect of a force is on the rotation of a body. In doing so, we will confine our attention to bodies that do not stretch, compress, bend, or otherwise deform—in short, bodies in which no part of the body ever changes its distance from any other part. We call these **rigid bodies.**

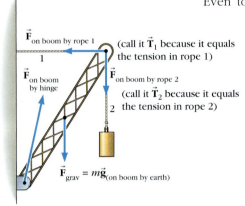

Figure 5-19 A boom in both translational and rotational equilibrium.

Case 5-2 ◆ *Two for the Seesaw: The Law of the Lever*

You probably recall from your own childhood that when two children of unequal weights sit on a seesaw (the author grew up in Brooklyn, where this is *never* a "teeter-totter"), the heavier child must sit closer to the pivot to make the seesaw balance (Figure 5-20). For example, suppose Tom weighs 60 lb and Maria weighs 45 lb. If Tom sits 3 ft from the pivot, Maria must sit 4 ft from the pivot. Notice that 3 ft × 60 lb = 4 ft × 45 lb. It is easy to verify experimentally that when balance is achieved, this kind of relationship always holds true. In general, if d_1 and d_2 are their distances from the

pivot, and $F_1 = m_1g$ and $F_2 = m_2g$ are the magnitudes of their weights (any other kind of downward force would do as well), the values of these quantities satisfy the condition

$$d_1F_1 = d_2F_2 \qquad (5\text{-}17)$$

This condition is sometimes called the **law of the lever,** because we can think of one child's weight as the force applied to pry up the other (inset in Figure 5-20). The equality of the two products requires that if $F_2 > F_1$, then $d_2 < d_1$.

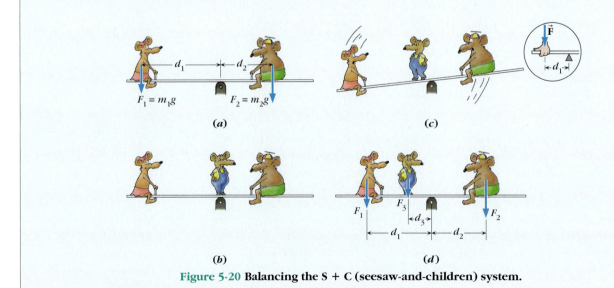

Figure 5-20 Balancing the S + C (seesaw-and-children) system.

On-The-Spot Activity 5-2

If there is a children's playground nearby, you and two friends can use the seesaw there to test out the various steps in this and the following case. Or you can simulate these cases using any balanced board and three unequal weights. In either event, you will have to substitute your own measured values for those given here.

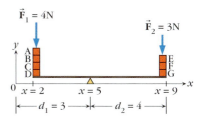

Figure 5-21 Center of mass. When the seesaw balances, the average position of a unit of mass is at the pivot (see calculation).

It turns out that to balance the seesaw, the children have distributed themselves so that the average position of one unit of weight is right at the pivot point. Figure 5-21 gives an example of how this works. Each block weighs exactly 1 N, so each block is a unit of weight. Blocks A–D (a total of 4 N) are 3 m to the left of the pivot. Blocks E–G (a total of 3 N) are 4 m to the right of the pivot. The law of the lever tells us they balance because 4 N × 3 m = 3 N × 4 m. To find the average position of one unit of weight, we add up positions of all the blocks (x_A, x_B, etc.) and divide by seven:

$$\frac{x_A + x_B + x_C + x_D + x_E + x_F + x_G}{7}$$

$$= \frac{2\,m + 2\,m + 2\,m + 2\,m + 9\,m + 9\,m + 9\,m}{7} = \frac{35\,m}{7} = 5\,m$$

which you see in the figure is the pivot point.

The average position of a unit of weight in a system is called the **center of gravity** of the system. If we average positions of the units of mass, we call it the **center of mass.** Not all center of gravity or center of mass calculations are as simple as the one we just did, but all are based on the same idea.

Like the seesaw-and-rider system, any body will balance at its center of mass (see Figure 5-22*a*), unless it is so large that *g* varies significantly from one point on the body to another. You can then find the center of mass experimentally by seeing where it balances. Under these conditions,

the weight of the body thus affects the body as though it were a force applied totally at the body's center of mass.

Otherwise (Figure 5-22*b*), the weight and the normal force would not be exerted along the same line and would make the body rotate just as the two forces in Figure 5-18*c* make the baton rotate.

➥Imagine the object to be subdivided into tiny pieces having equal mass. You get the *x* coordinate of the center of mass by averaging the *x* coordinates of all such pieces. A like calculation gives you the *y* coordinate. The center of mass of an object is thus the average position of a unit of its mass. Calculating the center of mass for all but a few simple shapes requires mathematics beyond the level used in this book.

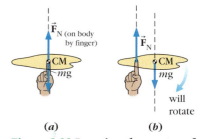

Figure 5-22 Locating the center of mass experimentally. (*a*) $\vec{F}_N$ *and* $m\vec{g}$ are along the same line of action. The object is balanced. (*b*) $\vec{F}_N$ and $m\vec{g}$ are along different lines of action. The object is unbalanced.

Case 5-3 ◆ *One More for the Seesaw*

Note that Figure 5-20 is essentially a free-body diagram, and that the combination of seesaw and children (S + C) is in equilibrium. A third child now joins the first two. When he stands right on the pivot point (Figure 5-20*b*), he has no effect on the balance. His weight is now added to the downward forces, so that the upward force exerted by the pivot must be correspondingly greater. But when the third child walks to the left of the pivot, the S + C system is no longer in static equilibrium (Figure 5-20*c*): It is thrown off balance and rotates counterclockwise.

The first child can remedy this imbalance if she moves in toward the pivot as the third child moves outward. Measurements confirm that static equilibrium is again achieved (Figure 5-20*d*) when

$$d_1F_1 + d_3F_3 = d_2F_2 \qquad (5\text{-}18)$$

As in the law of the lever, we can think of each child's contribution *dF* as the amount of leverage associated with that child's weight, much as if a hand were exerting a force *F* perpendicular to the end of a lever of length *d* (inset in Figure 5-20). We may call this length the **lever arm.**

We can rewrite Equation 5-18 as $d_1F_1 + d_3F_3 - d_2F_2 = 0$, or more concisely as

$$\Sigma(\pm)dF = 0 \qquad (5\text{-}19)$$

where the sign (+ or −) for each child's contribution indicates whether that child's weight alone would make the seesaw rotate counterclockwise (+) or clockwise (−). This choice of sign is consistent with our convention that you rotate through a positive angle as you go counterclockwise from the +*x* axis.

Equation 5-19 is valid in a limited context: The seesaw is horizontal and the forces are vertical and therefore perpendicular to the lever arms. In the following case, we explore how the ideas summarized in Equation 5-19 can be made more general, and introduce the concept of *torque*.

Case 5-4 ◆ *A Pivoting Pegboard: Arriving at a Definition of Torque*

The pegboard in Figure 5-23*a* is free to pivot about its center. The red line on the pegboard is like the seesaw in Cases 5-2 and 5-3. The vertical extent of the pegboard provides options for applying forces at points above and below this line and in various directions. Row and column numbers enable you to identify easily the hole at which a particular force is applied. We will see how various such forces affect the rotation of the board.

For a more extended and interactive presentation of this case, go to WebLink 5-4. You may also find it helpful to treat this case as an On-the-Spot Activity by testing what happens on a set-up of your own. A piece of stiff cardboard with holes punched in it can substitute for the pegboard. Make sure the pegboard rotates easily on the pivot. Strings with knots tied at one end can be passed through holes to support weights, or bolts and nuts of various sizes can serve as the weights (Figure 5-23*b*).

1. Hang a weight from any hole in column 5. You should find that no rotation occurs.

2. Now try positioning the weight at any hole in column 1 or 9. The pegboard should rotate vigorously. But does the vigorousness of rotation depend on which of these holes you choose? You should be able to verify that it does not.

3. To check more carefully whether moving a weight up or down a column changes its effect on rotation, we do the following. Position a weight F_1 at hole C-1 and position twice that weight ($F_2 = 2F_1$) at C-7, half as far from the pivot, so that, like the seesaw, the pegboard balances; it is in *rotational equilibrium*. If you now move F_2 to any other hole in column 7, you should still find that no rotation occurs. In contrast, you cannot achieve rotational equilibrium by positioning F_2 at any hole in a different column (try it).

Equation 5-17 agrees with the observations in (3) if we take *d* to mean the distance measured along the red line from the pivot to the *column,* and *not* the distance from the pivot to the particular hole that serves as the point of application of the force. The vertical force applied at each hole is directed *along* the column. The column is a segment of the blue dashed line, so the force at each hole is along this line. We call this line **the line of action of the force.** As you see in Figures 5-23*d* and *e,* if *r* is the actual distance from the pivot to a particular point of application, then *d* is just the perpendicular component of *r*, that is, *the perpendicular distance* $r_\perp$ *from the pivot to the line of action of the force.* If θ is the angle between *r* and the line of action, then $\sin\theta = \frac{r_\perp}{r}$ or $\frac{d}{r}$, so that

$$d = r_\perp = r\sin\theta \tag{5-20}$$

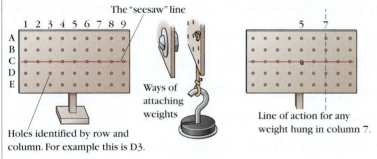

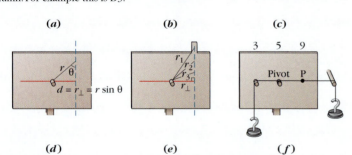

(a)   (b)   (c)

(d)   (e)   (f)

Figure 5-23 Set-up in Case 5-4 for exploring torque.

For **WebLink 5-4:**
Arriving at a Definition of Torque, go to
www.wiley.com/college/touger

We see from Case 5-4 that if we substitute $r_\perp = r\sin\theta$ for *d,* and *F* is *any* force (whether a weight or not), then Equation 5-19, rewritten as

$$\Sigma(\pm)r_\perp F = \Sigma(\pm)rF\sin\theta = 0 \tag{5-21}$$

remains the condition for rotational equilibrium. Each contribution $(\pm)rF\sin\theta$ is, loosely speaking, the amount of leverage associated with one force. We call this the **torque.** Denoting the torque by τ, the Greek letter *tau,* we can rewrite Equation 5-21 and summarize its significance:

$$\Sigma \tau = 0 \qquad (5\text{-}22a)$$

is **the condition for rotational equilibrium,**

where
$$\tau \equiv (\pm) r_\perp F = (\pm) r F \sin \theta \qquad (5\text{-}22b)$$

is the **torque** about a given pivot point ($r_\perp$ will be different for different pivot points) due to the force $\vec{F}$.

In the language of physics, the lever arm $r_\perp = r \sin \theta$ is more usually called the **moment arm;** in this context, the word *moment* is a synonym of *torque.* In Figure 5-24, the same force is exerted at the same distance r from the pivot point as we go from *a* to *c.* But as the direction of the force changes, its line of action passes further from the pivot point. The moment arm $r_\perp$ tells you how far from the pivot point the line of action passes. Parts *d* to *f* of Figure 5-24 show the analogous situation for a hand pushing on a wrench. **STOP&Think** If you position the pencil as in Figure 5-23*f,* what is the value of $r_\perp$ for the force exerted by the string attached at point P? In an idealized set-up with a frictionless pivot, could you achieve balance if you sufficiently increased the force exerted by this string? In practice, could you prevent the pegboard from rotating if you could pull hard enough to the right on this string? Explain your reasoning. ◆

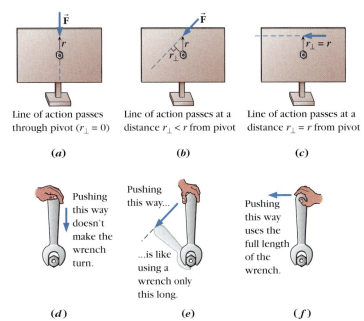

Line of action passes through pivot ($r_\perp = 0$)

(a)

Line of action passes at a distance $r_\perp < r$ from pivot

(b)

Line of action passes at a distance $r_\perp = r$ from pivot

(c)

Pushing this way doesn't make the wrench turn.

(d)

Pushing this way... ...is like using a wrench only this long.

(e)

Pushing this way uses the full length of the wrench.

(f)

Figure 5-24 Moment arms and "leverage." Having a larger moment arm $r_\perp$ is like using a longer lever or wrench.

◆**UNITS** Units of torque are units of $r_\perp$ times units of F and thus are meters × newtons (m · N) in SI units.

◆**CHOOSING A PIVOT POINT** If an object is stationary, then no matter what point we pick, the object doesn't rotate about it. So the sum of all torques is zero, and Equation 5-22a holds true about *any* choice of pivot point. The choice is made for convenience. In Case 5-4 it was convenient to choose the point where the pegboard is supported, because otherwise we would have to consider the torque due to the upward force exerted by the support. (That torque is zero when the support point *is* the pivot point because the moment arm from the support point to itself is zero.) In Example 5-12, you will see what happens when you choose different pivot points.

Let's now summarize how to find the torques that contribute to the sum in Equation 5-22a.

PROCEDURE 5-3

Determining the Torques Exerted on a Body

1. Draw the free-body diagram of the body (Procedure 5-2). Be sure to indicate clearly the *direction* and *point of application* of each force on the body.
2. Draw the line of action of each force.
3. Pick a pivot point. (Any point can be a pivot point, but some choices are more convenient than others.)
4. For each force, draw the moment arm $r_\perp$ (the perpendicular from the pivot point to the line of action of that force).

5. Find each moment arm either by measuring or by using $r_\perp = r\sin\theta$.
6. Calculate the magnitude of the torque due to each force by $|\tau_1| = r_{1\perp}F_1$, $|\tau_2| = r_{2\perp}F_2$, etc.
7. If the torque tends to produce counterclockwise motion, make it positive; if it tends to produce clockwise motion, make it negative.

The sign convention in step 7 is arbitrary; it is simply a way of distinguishing between two opposite directions. Example 5-10 provides practice in applying this procedure.

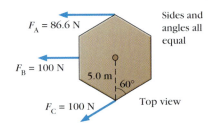

$F_A = 86.6$ N
$F_B = 100$ N
5.0 m
60°
$F_C = 100$ N
Sides and angles all equal
Top view

Example 5-10 *Torques on a Rotating Platform*

For a guided interactive solution, go to Web Example 5-10 at www.wiley.com/college/touger

A platform that rotates on a shaft (see figure) is built for a student drama production. The stage crew controls its motion by ropes attached at various points. At a particular instant, three of the ropes are pulled. The forces they exert are shown in the top view. Find the total torque about the shaft.

Brief Solution

Choice of approach. We follow the steps of Procedure 5-3.

What we know/what we don't. The figure shows the forces and enough about the geometry to find the moment arms.

The mathematical solution.

Force F	Moment arm $r_\perp$	Sign	$\tau = (\pm)r_\perp F$
$F_A = 86.6$ N	5.0 m	+ (counter-clockwise)	$+(5.0 \text{ m})(86.6 \text{ N})$ $= +433$ m·N
$F_B = 100$ N	0 because force's line of action passes through pivot	none	0
$F_C = 100$ N	5.0 m sin 60° = 4.33 m (see figure)	− (clockwise)	$-(4.33 \text{ m})(100 \text{ N})$ $= -433$ m·N

Adding the three torques, we get

$$\Sigma\tau = \Sigma(\pm)r_\perp F = \mathbf{0}$$

◆ Related homework: Problems 5-52 and 5-53.

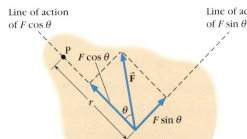

Line of action of $F\cos\theta$

Line of action of $F\sin\theta$

P
$F\cos\theta$
$\vec{F}$
r
θ
$F\sin\theta$

Torque = $r(F\sin\theta) + 0(F\cos\theta) = rF\sin\theta$

Figure 5-25 The torque due to a force is the sum of the torques due to its components.

Because $|\tau| = r_\perp F = (r\sin\theta)F$, we can reorder the terms on the right side and write $|\tau| = r(F\sin\theta)$. But $F\sin\theta$ is $F_\perp$, the component of $\vec{F}$ perpendicular to r. As Figure 5-25 shows, if $\vec{F}$ is replaced by its two components $F\cos\theta$ and $F\sin\theta$, the moment arm of $F\sin\theta$ is r, but that of $F\cos\theta$ is zero (its line of action passes through the pivot). The total torque due to both component forces is therefore $rF\sin\theta$, which is just the torque due to $\vec{F}$ itself.

◆**BODIES IN BOTH ROTATIONAL AND TRANSLATIONAL EQUILIBRIUM** An extended body in static equilibrium (no rotational or straight line motion) is in both rotational equilibrium and translational equilibrium. For any such body (call it A), the requirements that $\Sigma\tau_{\text{on A}} = 0$ and $\Sigma\vec{F}_{\text{on A}} = 0$ must both be met.

Example 5-11 *A Painter on a Scaffold*

The painter in Figure 5-26 weighs 800 N and is working on a very light 3.2-m-long scaffold suspended by ropes at each end. Find the total tension forces F_L and F_R exerted by the ropes at the left and right ends of the scaffold when the painter is standing 0.8 m from the left end.

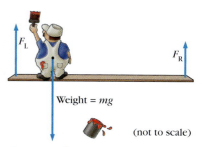

Weight = mg

(not to scale)

Figure 5-26 Free-body diagram for the painter + scaffold system.

Solution

Assumption. We take "very light" to mean we can neglect the weight of the scaffold.

Choice of approach.

1. The scaffold is in both rotational and translational equilibrium:

$$\Sigma\tau = 0 \text{ and } \Sigma\vec{F} = 0$$

2. To apply either equation, we must draw a free-body diagram to identify the terms in the sum (Figure 5-26).

3. $\Sigma\vec{F} = 0$ by itself turns out to be inadequate, because there are no forces in the x direction, and $\Sigma F_y = 0$ becomes

$$F_L + F_R - mg = 0 \qquad (5\text{-}23)$$

   which has two unknowns. A second equation is needed to solve for both.

4. The second equation is provided by applying $\Sigma\tau = 0$, but about what pivot point? Because the scaffold is stationary, there is no rotation about *any* point, so $\Sigma\tau$ will be zero about any point we pick. The following table shows what Procedure 5-3 gives us for three different pivot choices, points A, B, and C (see Figure 5-26). You should verify all the entries in the table for yourself.

	Moment arm measured from A	Sign of torque about A	Moment arm measured from B	Sign of torque about B	Moment arm measured from C	Sign of torque about C
F_L	0	none	0.8 m	−	3.2 m	−
mg	0.8 m	−	0	none	2.4 m	+
F_R	3.2 m	+	2.4 m	+	0	none

From first two columns: $\quad \Sigma\tau_{\text{about A}} = -(0.8 \text{ m})mg + (3.2 \text{ m})F_R = 0 \quad (5\text{-}24\text{A})$

From middle two columns: $\quad \Sigma\tau_{\text{about B}} = -(0.8 \text{ m})F_L + (2.4 \text{ m})F_R = 0 \quad (5\text{-}24\text{B})$

From last two columns: $\quad \Sigma\tau_{\text{about C}} = -(3.2 \text{ m})F_L + (2.4 \text{ m})mg = 0 \quad (5\text{-}24\text{C})$

Since the weight mg is known, any two of Equations 5-23, 5-24A, 5-24B, and 5-24C are sufficient to solve for F_L and F_R.

The mathematical solution. It is easiest in this case to use Equations 5-24A and 5-24C to solve for F_L and F_R, because each has only one unknown. From Equation 5-24A,

$$(-0.8 \text{ m})(800 \text{ N}) + (3.2 \text{ m})T_R = 0$$

so $\quad F_R = \dfrac{(0.8 \text{ m})(800 \text{ N})}{3 \text{ m}} = \textbf{200 N}$

From Equation 5-24C,

$$(-3.2 \text{ m})F_L + (2.4 \text{ m})(800 \text{ N}) = 0$$

so $\quad F_L = \dfrac{(2.4 \text{ m})(800 \text{ N})}{3 \text{ m}} = \textbf{600 N}$

EXAMPLE 5-11 *continued*

Making sense of the results. As you might expect, F_L and F_R together equal the painter's weight of 800 N. The values thus satisfy Equation 5-23. (Check that they also satisfy Equation 5-24B.) Also, because the painter is three-fourths of the way to the left, the left ropes support three-fourths (600 N) of his weight. Indeed, we could show (see Problem 5-84) that if he goes all the way to the left, his weight is entirely supported by the left ropes, and the force exerted by the right ropes is zero.

◆ Related homework: Problems 5-57, 5-58, 5-59, and 5-84.

5-5 Frictional Forces

It is a common misconception that the rougher surfaces are, the greater the frictional forces that they exert on each other. We commonly think of glass as smooth, but if you try to slide one pane of glass across another, you will find that the frictional forces they exert on each other are considerable. Machinists are familiar with a similar situation with polished metal surfaces. If you polish very rough metal surfaces, you will reduce the frictional forces they exert on each other. But if you polish them finely enough, the frictional forces they exert on each other will increase.

The reason is as follows. On a microscopic scale, even flat, smooth surfaces appear jagged and make contact at relatively few tiny areas (see Figure 5-27). Where contact does occur, bonding occurs between the atoms of the pressed-together surfaces, causing a "cold weld." Because these welds must be broken to set one object in motion or keep it moving across another, we feel the resistance to our push that we call friction. Frictional forces are the macroscopic sum of the binding forces that occur at the micro-areas of contact.

We will concern ourselves here not with the microscopic details but with the effects that are observable on the scale of everyday human experience. One effect is that when we increase the normal forces that two surfaces exert on each other, we also increase the frictional forces they exert on each other. Let's look at what happens in more detail.

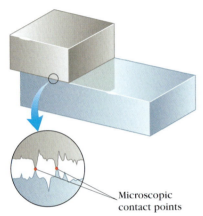

Microscopic contact points

Figure 5-27 On a microscopic scale, even highly polished surfaces appear jagged and make contact at relatively few tiny areas.

ON-THE-SPOT ACTIVITY 5-3

1. Take a heavy, flat-bottomed object such as an unabridged dictionary and push it from one side (Figure 5-28), but gently enough so that it remains stationary. (If you do this with friends, a stack of physics books can substitute for the dictionary, as long as you always push on the bottom book so that the stack doesn't topple.) Next, very gradually build up the force that you are exerting until the dictionary (or similar object) starts to move.

2. Once the dictionary is set in motion, try to keep it sliding at constant speed. Notice that once the dictionary "breaks loose" from its stationary situation (indeed, some microscopic surface-to-surface bonds are being broken), there is a tendency for the dictionary to lurch forward (accelerate). You in fact have to ease up and push less hard to keep the dictionary going at constant speed than you did to set it in motion.

3. Try repeating the previous steps with a substantial weight sitting on the dictionary, taking care not to let the weight slip when you set the dictionary in motion. Notice that the force you must exert at each step is now correspondingly greater.

Push

Figure 5-28 Step 1 of On-the-Spot Activity 5-3.

"No motion" and "constant speed" are both equilibrium situations; in both cases $\Sigma\vec{F} = 0$. Thus the free-body diagrams in Figures 5-29*a–d* show that as you exert more force (short of setting the dictionary in motion), the frictional force

Small F by you   Increased F "Breaking point"   In motion

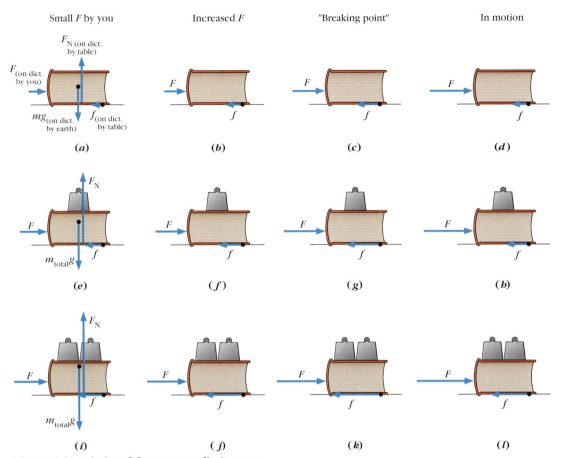

Figure 5-29 Frictional forces on a dictionary.

exerted by the table must correspondingly increase. Likewise, if you have to exert less force to keep the dictionary in motion at constant speed than you did to set it in motion, the frictional force must also be less for $\Sigma\vec{F}$ to remain zero. In the microscopic picture, we can think of welds constantly forming and breaking as one surface slides over the other, but never being quite as extensive as when the surfaces are stationary but hitting the "breaking point," that is, the onset of motion, when the "break" occurs.

Figures 5-29e–h repeat these free-body diagrams for the case when the weight is placed on the dictionary. Figures 5-29i–l repeat them for a larger weight sitting on the dictionary. In any single column of free-body diagrams in Figure 5-29, you see that the frictional force is proportional to the normal force, that is, $\frac{f}{F_N}$ = constant. The value of this constant is greatest in the third column, labeled *breaking point,* and less in the fourth column, labeled *in motion* (the constant speed situation).

The constant for the *breaking point* column, where the frictional force has reached a maximum, is called the **coefficient of static friction,** and we denote it by μ_s (μ is the lower case Greek letter *mu*).

➤**Notation:** Up until now, we have used notation like $F_{f\ on\ dictionary\ by\ table}$ to reinforce the point that frictional forces are always exerted on one object by another; they are *not* exerted "by friction." However, it will be more convenient from here on simply to represent the frictional force by a lower case f.

Coefficient of Static Friction

$$\mu_s \equiv \frac{f_{max}}{F_N} \tag{5-25}$$

at breaking loose or onset of motion.

The constant for the *in motion* column is called the **coefficient of kinetic or sliding friction** and we denote it by μ_k (we've used s for *static,* so s for *sliding* wouldn't do).

Coefficient of Kinetic Friction

$$\mu_k \equiv \frac{f}{F_N} \tag{5-26}$$

when sliding.

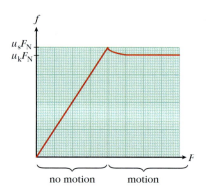

Figure 5-30 Graph of frictional force f versus the force F that you exert in the opposite direction.

Both coefficients of friction depend on the materials making up the two surfaces in contact. It is easier, say, to pull a sled with metal runners over hard-packed snow than over pavement. In either case, if the surfaces and the pull are horizontal, the normal force equals the weight of the sled. On snow, we must match a smaller f to keep the sled sliding and a smaller f_{max} to set it in motion in the first place. It thus follows from the definitions (Equations 5-25 and 5-26) that μ_k and μ_s are both smaller for metal on snow than for metal on pavement.

Warning: If an object is pushed too gently to set it in motion (as in columns 1 and 2 of Figure 5-29), then the frictional force on the object is usually less than f_{max}, and therefore f is not $\mu_s F_N$.

The graph in Figure 5-30 summarizes what happens to the opposing frictional force as you increase the force you exert on a body. When the body is set in motion, f drops in value from $\mu_s F_N$ to $\mu_k F_N$. It remains at that value as you increase the force that you exert. The net force $\Sigma F = F_{\text{on body by you}} - f$ therefore continues to increase, and the body's acceleration increases.

Example 5-12 *Statics on a Luggage Chute, Revisited*

Consider the duffle bag in Example 5-9.
a. From the data given in that example (the duffle bag weighs 39.2 N and is stationary on a chute inclined at a 25° angle), can you find μ_s?
b. Suppose that $\mu_s = 0.90$. To what maximum angle can you tilt the chute and still have the duffle bag remain motionless?

Solution

Choice of approach.
1. We know the values of f and F_N from Example 5-11. But Equation 5-25 relates f_{max} and F_N. However, we don't know whether the chute can be tilted even more steeply without setting the duffle bag in motion, so we don't know if the f we found is f_{max}. Thus we **cannot** use Equation 5-25 to find μ_s in part **a.**

2. In part **b**, the condition of maximum tilt tells us we can apply Equation 5-25. However, because ϕ is changed, f and F_N are changed as well, and to find them we must apply Newton's second law anew.

What we know/what we don't.

Second law in x direction	Second law in y direction		
$a_x = 0$	$a_y = 0$		

Forces in x direction	Forces in y direction	$F_N = ?$	$f = ?$
$mg \sin \phi$	F_N	$g = 9.8 \text{ m/s}^2$	$mg = 39.2$ N
$-f$	$-mg \cos \phi$	$\phi = ?$	$\mu_s = 0.90$

*The mathematical solution to **b**.* Just as in Example 5-11, it follows from Equations 4-8x and 4-8y that $f = mg \sin \phi$ and $F_N = mg \cos \phi$ (though clearly with ϕ changed), so that the values of f and F_N are different than in **a**. Dividing the first of these equations by the second, we get

$$\frac{f}{F_N} = \frac{mg \sin \phi}{mg \cos \phi} = \tan \phi$$

But now $\frac{f}{F_N}$ is $\frac{f_{max}}{F_N}$, which is μ_s by definition. Thus,

$$\mu_s = \tan\phi \qquad (5\text{-}27)$$

so
$$\phi = \tan^{-1}\mu_s = \tan^{-1}0.90 = 42°$$

◆ Related homework: Problems 5-61, 5-64, and 5-65.

Equation 5-27 provides a straightforward way of determining μ_s experimentally, because you can readily measure the angle at which the block starts to slide.

✦ SUMMARY ✦

When you solve quantitative problems using Newton's second law of motion, you must generally apply the law in component form,

$$\Sigma F_x = ma_x \quad \text{and} \quad \Sigma F_y = ma_y \qquad (4\text{-}8x, 4\text{-}8y)$$

to each body of interest.

Free-body diagrams are an important technique for inventorying the forces that contribute to $\Sigma\vec{F}$ on each body. **Procedure 5-1** describes the drawing and labeling of arrows representing the forces on a body.

Carefully drawn diagrams help you to distinguish (see Case 5-1) between pairs of equal and opposite forces arising from two conceptually distinct circumstances:

Two forces producing equilibrium exerted on a single body A by two other bodies — satisfy second law: $\Sigma\vec{F}_{on\ A} = \vec{F}_{on\ A\ by\ 1} + \vec{F}_{on\ A\ by\ 2} = 0$

Inter-reaction pair exerted by two bodies A and B on each other — satisfy third law: $\vec{F}_{on\ A\ by\ B} = -\vec{F}_{on\ B\ by\ A}$

The inventory of forces on a body A must include **contact forces** exerted by all bodies actually touching (in *contact* with) A. These include:

• **normal forces** $(\vec{F}_N)$ — pressing forces *perpendicular* to the contact surfaces.

• **frictional forces** $(\vec{f})$ — resistive forces *along* the contact surfaces and always directed so as to resist A's slipping along the surface of the other body.

• **tension forces** $(\vec{T})$ — exerted by connected parts of a flexible body (rope, string, etc.) on each other, and always directed along the length of the flexible body.

The frictional force a body exerts on A depends on the normal force that the body exerts on A. The degree of frictional interaction depends on the materials that are in contact, and is characterized by the ratio of f to F_N.

Coefficient of Static Friction $\mu_s = \dfrac{f_{max}}{F_N}$ (5-25)

at "breaking loose" or onset of motion

Coefficient of Kinetic Friction $\mu_k = \dfrac{f}{F_N}$ (5-26)

when sliding

A useful feature of the *tension* force under certain conditions is that its magnitude is the same anywhere along its length and is also equal to the force the rope exerts on an object on which it pulls. These conditions are: the rope is massless, is kept taut, and no external forces are exerted along the rope's length other than at its ends.

The inventory of forces in $\Sigma\vec{F}$ must also include all **action-at-a-distance** forces, including **gravitational forces.** The gravitational force exerted on a body (by Earth or another planet, moon, etc.) is called its weight:

$$\text{Weight} \equiv F_{grav} = mg$$

A scale measures the normal force exerted on its platform (or hook); this equals the correct weight only when these two forces alone produce equilibrium in one direction (say, $\Sigma F_y = 0$), so that $F_N - mg = 0$. Otherwise, whether $\vec{a} = 0$ (Example 5-9) or not (Example 5-2), the scale reading F_N does *not* equal mg.

When treating bodies that move together (remain the same distance apart, as measured along the path of motion), we make use of the fact that their accelerations are equal in magnitude (though not necessarily in direction; see Example 5-8), and that there is a force on one body that equals a force on the other, either because the forces are an interaction pair (Example 5-7) or because a third body (such as a rope) exerts equal forces on both of them (Example 5-8).

The forces in Newton's second law can result in **translational equilibrium** $(\vec{a} = 0)$. But forces on an *extended body* (in contrast to a point mass) can contribute to its *rotation* as well, so for these bodies we must also consider **rotational equilibrium.** Building from the **law of the lever,**

$$d_1F_1 = d_2F_2 \qquad (5\text{-}17)$$

for downward forces (such as weights) on a horizontal see-saw, we developed the notion of each weight's leverage ($\pm dF$) into the more general concept of **torque,** defined by

$$\tau = (\pm)r_{\perp}F = rF \sin \theta \qquad (5\text{-}22b)$$

To find individual torques using the definition, you must follow **Procedure 5-3** in Section 5-4.

In regions of constant g, the point of application of a gravitational force on a body (its weight) is the body's **center of mass,** the average position of a unit of mass in the body.

Conditions for a body A in static equilibrium (no motion):

$$\Sigma F_x = 0 \qquad \Sigma F_y = 0 \qquad \Sigma \tau_{\text{about any pivot point P}} = 0$$

Together, these conditions enable you to calculate individual forces on the body (Example 5-11).

✦QUALITATIVE AND QUANTITATIVE PROBLEMS✦

General Reminders

In problems involving Newton's laws, you should always

a. identify the object(s) to which you will apply the laws,

b. draw free-body diagrams,

c. use a suitably colored two-tone arrow to represent each force vector (optional),

d. identify each force with an identifying label of the form $\vec{F}_{\text{on\_by\_}}$, with the appropriate pair of interacting bodies filling in the two blanks.

Hands-On Activities and Discussion Questions

The questions and activities in this group are particularly suitable for in-class use.

5-1. *Discussion Question.*

a. While you are standing on a bathroom scale, a friend hands you a bowling ball. Does the bowling ball exert a force on the scale? Ignore any forces that are too small to measure.

b. Which of the following forces change when the ball is handed to you? The force that the scale exerts on you, the force that you exert on the scale, the force that the floor exerts on the scale, the force that the scale exerts on the floor.

c. Harry weighs as much as you and the bowling ball together. Suppose you stepped off the scale and Harry got on. How

does the force that the scale exerts on Harry's feet compare to the force it exerted on your feet when you were on it holding the ball? How does the force that Harry's feet exert on the scale compare to the force that your feet exerted on the scale when you were on it holding the ball? If need be, reconsider your answer to **a.**

5-2. *Discussion Question.* You are driving a subcompact car and have just begun a long, steep single-lane downgrade (with guard rails on either side). There is a humongous 18-wheeler truck close behind you. The truck driver suddenly contacts you by CB to let you know his brakes have just given out completely and he can't slow down. Is it possible for you to do anything to avoid having the truck smash into you? On what condition(s), if any, does your answer depend?

Review and Practice

Section 5-1 Inventorying Forces: Applying Newton's Second Law

5-3. Each part of this question asks you to identify one or more forces. Each force should be identified by telling what kind of force it is, what object is exerting the force, and what object the force is exerted on.

a. When you are standing on a stool, what forces are exerted on you, and by what bodies?

b. Your weight is one of two forces in an interaction pair. What is the other force?

c. Besides the force you gave in answer to **b,** what other force is equal and opposite to your weight?

d. The force you identified in **c** is one of two forces in an interaction pair. What is the other force?

5-4. When you walk or run, what exerts the force on you that propels you forward? What *kind* of force is it?

5-5.

a. Draw a free-body diagram of the baseball in Figure 5-31 at each depicted point (A to E) in its trajectory. Assume the ball is about to lose contact with the hand at A and is just making contact with the glove at E.

b. Tell what is exerting each of the forces shown in your free-body diagrams.

Figure 5-31 Problems 5-5 and 5-6

5-6.

a. At each of points B to D in Figure 5-31, determine the direction of $\Sigma\vec{F}_{on\ ball}$.

b. In what direction is the ball's acceleration at each of these points?

c. Is $\vec{a}$ in the same direction as $\Sigma\vec{F}$ at each point? Must it be?

d. At points B, C, and D, does the direction of your acceleration agree with Figure 3-38e? Briefly explain why or why not.

e. At points B, C, and D, is anything besides Earth exerting a force on the ball?

5-7. SSM Your neighbor's kid is caught in a stop-action picture as he is jumping off his skateboard (Figure 5-32). Draw separate free-body diagrams of the kid and the skateboard and carefully identify *all* forces on each.

Figure 5-32 Problem 5-7

5-8.

a. Draw a free-body diagram of the fish suspended from the fishing line in Figure 5-33.

b. When you determine which forces to include in the free-body diagram, does it matter whether the fish is hanging stationary or being reeled in?

c. How do the sizes of the arrows in your free-body diagram compare if the fish is hanging stationary (state 1)? If it is being reeled upward at constant speed (state 2)? If it is changing gradually from state 1 to state 2?

Figure 5-33 Problem 5-8

5-9. Figures 5-34a–c show a shipping clerk moving a parcel around the mail room at constant speed. The parcel is always pressed against the surface (floor, wall, ceiling) on which it is being slid. Draw a free-body diagram of the parcel for each of the three depicted situations. (Suggested by a paper by Alan van Heuvelen, *AAPT Announcer,* **24** [4], 46 [1994].)

(**a**) Sliding along floor (**b**) Sliding up wall (**c**) Sliding along ceiling

Figure 5-34 Problem 5-9

5-10. When a specimen with a mass of 0.050 kg is placed in a tray and put on a laboratory scale, the scale reads a weight of 0.686 N.

a. What upward normal force does the scale exert on the tray?

b. What upward normal force does the tray exert on the specimen?

c. What is the mass of the tray?

5-11. In Problem 5-10, what upward normal force does the scale exert on the specimen?

5-12. A crate sits on a scale in a warehouse. The scale exerts an upward normal force of 60 N on the crate. The warehouse floor exerts an upward normal force of 75 N on the crate. **a.** What is the mass of the crate? **b.** What is the mass of the scale?

5-13. The glider in Figure 5-35 is on a tilted air track (frictionless).

a. Draw a free-body diagram of the glider.

Figure 5-35 Problem 5-13

b. Graphically find the vector sum (resultant) of the force vectors in your free-body diagram. (You can only sketch this, because you have no numerical values.)

c. In what direction should the resultant point? Why? Does the resultant you obtained graphically in **b** point at least roughly in this direction? If not, try to determine why the discrepancy has occurred.

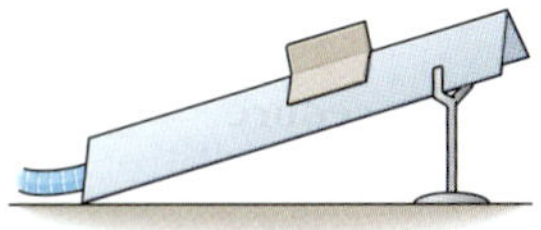

5-14. *Settle the argument:* The weight of the suspended cylinder in Figure 5-36 is 5 N. Two students attempt to predict the scale reading. Student A argues, "The cylinder is in equilibrium. The tension in the rope, exerted upward on the cylinder, must be equal and opposite to the 5 N weight. Since the tension is also the force exerted downward on the scale, the scale must read 5 N." Student B argues, "The scale is supporting both the cylinder and the big, fat rope. The scale must therefore read more than 5 N." Tell which student is right, and then briefly explain what is wrong with the other argument.

Figure 5-36 Problem 5-14

5-15. Explain in your own words the difference between weight and (inertial) mass.

5-16. A boxer trains at a ski resort near Denver and then flies to New York for the flight. When he weighs in, he is disqualified because he exceeds the 600 N weight limit for his class by 0.4 N. The boxer claims he was just at the limit when he weighed himself before the flight and had nothing to eat or drink on the plane. Explain the discrepancy.

5-17. When a wagon is stationary on a horizontal road, the normal force that the road exerts on the wagon is (*always, sometimes, never*) equal and opposite to the wagon's weight.

5-18. A 1000-kg safe sits on a reinforced floor.

a. What normal force does the floor exert on the safe?

b. Without reinforcement, the floor could sustain a weight of 2000 N. Would reinforcement be required if the office were on the moon?

5-19.

a. What is your weight in pounds? (Estimate if you don't know.)

b. Based on your answer to **a,** what is your weight in newtons?

c. Based on your answer to **a,** what is your mass in kilograms?

d. If you were standing on a bathroom scale (calibrated in newtons) in an elevator with an upward acceleration of 2 m/s², what would the scale read?

e. Repeat **d** for a downward acceleration of 2 m/s².

5-20. Is the normal force that the floor exerts on you greater
a. in an up elevator that is coming to a stop or a down elevator that is coming to a stop? Briefly explain.
b. in a down elevator that is starting from rest or a down elevator that is coming to a stop? Briefly explain.

5-21. SSM A lifeguard exerts a horizontal force of 80 N to pull an injured swimmer whose mass is 60 kg. The resistive force exerted by the water is 75 N. **a.** What is the swimmer's acceleration? **b.** If they start from rest, how far will they have gone after 3 s?

5-22. An 800-kg car goes from a speed of 20 m/s to a speed of 10 m/s in 4 s. What average force is exerted by the brakes over this interval?

5-23. SSM WWW A mover exerts a force of 200 N to push a 180-kg file cabinet across a floor at constant velocity.
a. What frictional force does the floor exert on the file cabinet?
b. The mover suddenly encounters a rough area where the frictional force increases by 50 N. Find the file cabinet's acceleration across this area.

5-24. A truck with a mass of 30 000 kg speeds up from 30 m/s to 40 m/s over a distance of 175 m. What average total force must be exerted on the truck over this stretch?

5-25. Figure 5-37 shows a moving flatcar with a crate on it at two different instants t_1 and t_2. Between t_1 and t_2, the crate slides to the back of the flatcar. The flatcar exerts a frictional force on the crate as it slides.

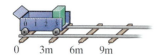
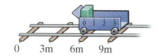

Figure 5-37 Problem 5-25

a. Does the crate move to the right or the left between t_1 and t_2?
b. Is the frictional force that the flatcar exerts on the crate toward the right or toward the left?
c. Does this frictional force oppose the motion of the crate?
d. Now consider the coordinates marked on the side of the flatcar. These coordinates show the position of the crate *relative to the flatcar*. Does the crate move to the right or the left relative to the flatcar?
e. Does the frictional force oppose the *relative* motion of crate and flatcar?
f. Which of the following statements is correct?
   I. A sliding frictional force on an object always opposes the object's motion.
   II. Sliding frictional forces always oppose the relative motion of two objects.

Section 5-2 Applying Newton's Second and Third Laws in Two Dimensions

5-26. The crate in Figure 5-38a is subjected to a number of forces, and is stationary. Figure 5-38b shows a free-body diagram of the crate under these conditions. Fully identify (what type of force, exerted on what *by* what) each of the five forces in the free-body diagram.

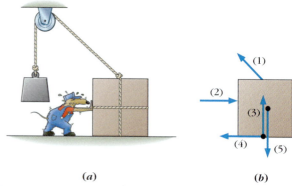

Figure 5-38 Problems 5-26 and 5-27

5-27. Redraw the free-body diagram in Figure 5-38b using the color coding of step 5 of Procedure 5-1.

5-28. The crate in Figure 5-39a is subjected to a number of forces and is stationary. Figure 5-39b shows a free-body diagram of the crate under these conditions. Fully identify (what type of force, exerted on what by what) each of the three forces in the free-body diagram.

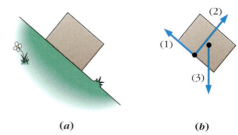

Figure 5-39 Problems 5-28 and 5-29

5-29. Redraw the free-body diagram in Figure 5-39b using the color coding of step 5 of Procedure 5-1.

5-30. In Figure 5-40, three different free-body diagrams of a block are shown. For each of these,
a. sketch a real-world physical situation that gives rise to that free-body diagram. Make each object in the sketch a different color, then
b. color-code and fully label the forces in the diagram.

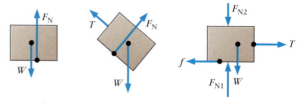

Figure 5-40 Problem 5-30

5-31. A father pushes his baby daughter in a carriage (Figure 5-41). Baby and carriage together have a mass of 9.0 kg. When

Figure 5-41 Problem 5-31

the father exerts a force on the carriage at an angle of 25° to the forward direction, a sideways frictional force keeps the carriage from changing its direction, and the carriage's speed increases from 0.70 m/s to 1.30 m/s over 3.0 s. Neglect frictional forces along the direction of motion. Find the magnitude of
a. the force that the father exerts on the carriage.
b. the sideways frictional force.

5-32. A stock clerk pushes a carton on a dolly (Figure 5-42). The carton and dolly have a combined mass of 25 kg. The clerk exerts a force on the carton in the direction shown. Assume friction is negligible. If the carton and dolly undergo an acceleration of 0.40 m/s²,

Figure 5-42 Problem 5-32

a. find the magnitude of the force exerted by the clerk if the carton and dolly undergo an acceleration of 0.40 m/s²;
b. find the normal force that the floor exerts on the dolly;
c. find the normal force that the dolly exerts on the floor.

5-33. Find the acceleration of a block sliding down a frictionless 20° incline.

5-34. The block in Figure 5-39 is at rest. Suppose that the angle the slope makes with the horizontal is 30°, and that the frictional force on the block is 40 N. Find the weight of the block.

5-35. A skier encounters virtually no friction over a 20-m stretch of slope making an angle of 15° with the horizontal. If the skier comes to the top of this stretch at a speed of 5 m/s, what is her speed at the bottom?

Section 5-3 Bodies with Linked Motions

5-36. Figure 5-43 shows eight situations in which a mover pushes two crates along a horizontal surface. In each case, the crates move together with a constant acceleration.
a. Focus on the forces that crate A and crate B exert on each other. Below are four possible statements about these two forces. After each statement, write the Roman numeral of each of the depicted situations that it correctly describes.

Block A exerts a greater force. ____

Block B exerts a greater force. ____

Blocks A and B exert equal forces on each other. ____

Blocks A and B exert no forces on each other. ____

Briefly explain your reasoning.

b. Which of the situations shown are only possible if one or more frictional forces are involved? Fully identify those forces (on what by what).

5-37
a. Sketch a free-body diagram of crate B in situation VI of Figure 5-43. What force is responsible for its acceleration? Is it in the same direction as the acceleration?
•**b.** Sketch separate free-body diagrams of crates A and B in situation VIII of Figure 5-43.

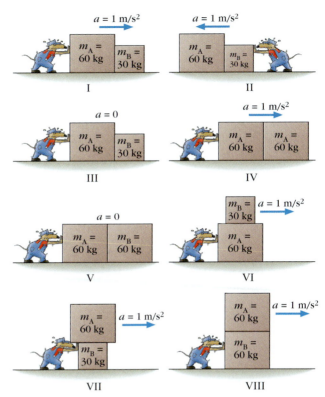

Figure 5-43 Problems 5-36, 5-37, and 5-38

5-38. Suppose that in situation IV of Figure 5-43, the mover is pushing the two crates together on a frictionless surface. Determine which of the following goes in each blank below: *twice; equal to; half of; between one and two times; between 0.5 and 1.0 times.*
a. The acceleration of crate A is ____ the acceleration of crate B.
b. In magnitude, the force that crate B exerts on crate A is ____ the force that the mover exerts on crate B.

5-39. Two blocks, A and B, remain stationary in the positions shown in Figure 5-44. If the pulley is frictionless, which block, if either, is heavier? Explain your answer.

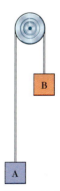

Figure 5-44 Problem 5-39

5-40. The set-ups on the two identical light, frictionless pulleys in Figure 5-45 are simultaneously released from rest. Does B hit the ground before, after, or at the same time as C? Explain your reasoning. (Adapted from an interview question described by Mark Somers, *AAPT Announcer,* 21 [4], paper FG4, 1991.)

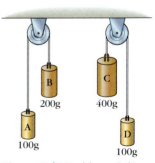

Figure 5-45 Problems 5-40 and 5-45

5-41. A mover pushes two crates of unequal mass across a frictionless surface (Figure 5-46).

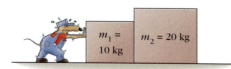

Figure 5-46 Problems 5-41 and 5-44

a. Do the two crates exert equal forces on each other? Explain.

b. Is the total horizontal force on each of the crates the same? Explain.

c. Are the accelerations of the two blocks equal? Explain.

d. Use Newton's second law to check whether your answers to **b** and **c** are consistent. Should you change your answer to **b**? Should you change your answer to **c**?

e. Is the total vertical force on each of the blocks the same? Explain.

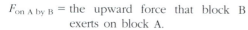

5-42. Blocks A and B in Figure 5-47 each have a mass of 2 kg. When block A slides, the table exerts a frictional force of 11.6 N on it. Find the acceleration of the two blocks.

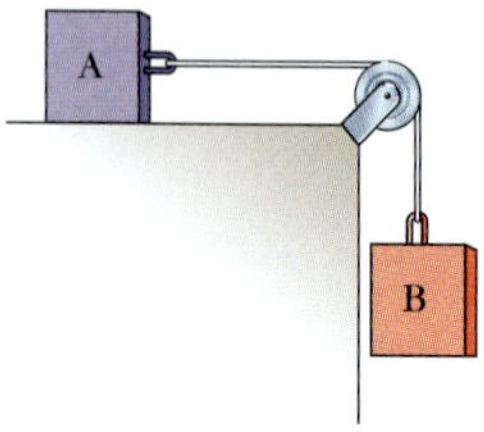

Figure 5-47 Problem 5-42

5-43. SSM WWW In Figure 5-48, block B is hollow; its mass is negligible compared to that of block A.

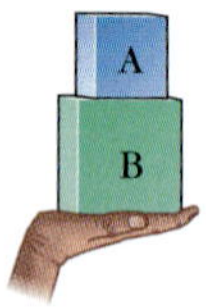

$F_{\text{on A by B}}$ = the upward force that block B exerts on block A.

$F_{\text{on B by A}}$ = the downward force that block A exerts on block B.

Figure 5-48 Problem 5-43

$F_{\text{on B by h}}$ = the upward force that the hand exerts on block B.

a. Suppose the hand and blocks are at rest. Rank these three forces in order of magnitude, starting with the value (lowest or most negative) that would occur furthest to the left on a real number line. Make sure to indicate equalities if there are any.

b. Repeat **a** for the case when the hand and blocks are accelerating downward together.

c. Repeat **a** for the case when the acceleration of the hand and blocks is zero.

5-44. If the mover in Figure 5-46 exerts a 6.0 N force and the floor is frictionless, find **a.** the force that block 1 exerts on block 2, and **b.** the force that block 2 exerts on block 1.

5-45. In Figure 5-45, find
a. the acceleration of cylinder B.
b. the tension force exerted by the rope connecting cylinders A and B.
c. the acceleration of cylinder C.
d. the tension force exerted by the rope connecting cylinders C and D.

Section 5-4 Static Equilibrium, Rigid Bodies, and the Concept of Torque

5-46. An airplane in flight experiences a total resistive force equal to the total thrust force propelling the plane forward. Should the passengers be concerned? Explain your reasoning.

5-47. SSM A cyclist is pedaling up an upgrade at constant speed. The upgrade makes an angle of 8.5° with the horizontal. The cyclist and bicycle have a combined mass of 92 kg.
a. Find the frictional force that the road exerts on the bicycle. (Think carefully: In what direction must the frictional force be?)
b. Find the normal force that the road exerts on the bicycle.

5-48. In Figure 5-49, the mass of block A is 20 kg and that of cylinder B is 5 kg. If both objects are at rest,
a. calculate the weight you would read on the scale.
b. find the frictional force that the scale exerts on block A.

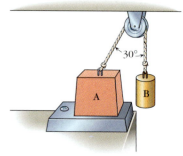

Figure 5-49 Problem 5-48

5-49. Two children sit on a 4.4-m-long seesaw pivoted at its center. Child A weighs 300 N; child B weighs 96 N.
a. If child A sits 0.8 m from the pivot, can child B sit far enough from the pivot to balance the seesaw? Briefly explain.
b. If child B sits at the very end of the seesaw, where must child A sit to achieve balance?

5-50. For the situation in Problem 5-49**b**,
a. Calculate the torques about the seesaw's center due to each child's weight.
b. Are they equal and opposite?
c. Repeat **a** for a pivot point located at the end where child B sits.
d. Are the torques due to the two children's weights equal and opposite about *this* pivot point? Explain.

5-51. Consider the pegboard in Figure 5-23a, which moves freely on the pivot at hole C-5. Suppose a string is connected between hole 9-E and the table top directly below it, so that the string is just barely taut. A weight is then hung from hole D-3. How is the tension in the string affected—is it increased, decreased, or unchanged—if the hanging weight is moved
a. from hole D-3 to hole E-5?
b. from hole D-3 to hole A-3?
c. from hole D-3 to hole A-4?
Briefly explain each of your answers.

5-52. Figure 5-50 shows several forces exerted on a square board. The tail of each force vector is placed at the force's point of application. All the forces have the same magnitude except $\vec{F}_5$, which has twice the magnitude of the others.
a. List the forces in rank order of the torque that each force produces about point A, starting with the torque value (lowest or most negative) that would occur furthest to the left on a real number

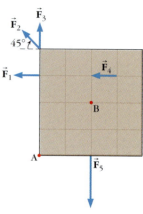

Figure 5-50 Problem 5-52

line. Take the counterclockwise direction to be positive. Make sure to indicate equalities if there are any.

b. Now list the forces in rank order (from most negative to most positive) of the torque that each force produces about point B. Again take the counterclockwise direction to be positive.

5-53. **SSM** A square board, 4 m on a side, pivots about its center point P (Figure 5-51). Six forces, each with a magnitude of 5 N, are exerted at the points and in the directions shown. Find the torque about P due to each force at the instant shown.

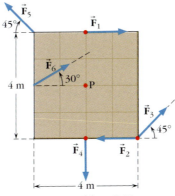

Figure 5-51 Problem 5-53

5-54. You are having a bowl of Sweet 'n' Krispies for breakfast and are reading the special offer on the back of the box. To catch the light better, you push against the box near the top to keep it tilted backward. Then you tilt it a little more, but still not so far that it falls over. (Based on a situation proposed by A. diSessa, *Cognition and Instruction*, **10**, 105–225 [1993].)

a. At which of these two tilts must you push harder to keep the box in its tilted position?

b. How much force must you exert to keep the box tilted to where it is just on the verge of falling over?

5-55. **SSM WWW** Acrobats A and B hold a 5-m pole while acrobat C, who weighs 500 N, balances on the pole 3 m from one end (Figure 5-52).

Figure 5-52 Problem 5-55

a. How much upward force must each of the other two acrobats exert if the pole is to remain stationary?

b. How must these forces change as acrobat C moves toward one end of the pole?

5-56. An acrobat stands at one end of a springboard that is held in place at the other end by a rope (Figure 5-53). The acrobat weighs 570 N. The tension force exerted by the rope is 900 N. Find the angle ϕ that the rope makes with the vertical direction.

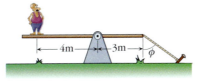

Figure 5-53 Problem 5-56

5-57.

a. In Example 5-11, what does Equation 5-23 tell us about how the total of the forces exerted by the ropes at both ends compares to the weight of the painter?

b. Does Equation 5-23 tell you what fraction of the painter's weight is supported by each rope? Is it possible to figure this out using only Newton's second law ($\Sigma \vec{F} = 0$ when $\vec{a} = 0$)? Explain.

c. Does your answer to either *a* or *b* change if the mass of the scaffold is no longer negligible? If so, how? If not, why not?

5-58. In Example 5-11,

a. use Equations 5-23 and 5-24A to solve for F_L and F_R;

b. use Equations 5-23 and 5-24B to solve for F_L and F_R.

In each case, check that your results agree with those obtained in the example.

5-59. Redo Example 5-11 for the case when the scaffold's weight, instead of being negligible, is 50.0 N and uniformly distributed.

Section 5-5 Frictional Forces

5-60. When an 8-N force is exerted horizontally on a 10-kg block, the block's acceleration across a level surface is 0.2 m/s^2.

a. Find the frictional force on the block.

b. Find the coefficient of kinetic friction μ_k between the block and the surface.

5-61. In Example 5-9,

a. can you find the coefficient of kinetic friction from the given data? Briefly explain.

b. can you find the coefficient of static friction from the given data? Briefly explain.

5-62. Is there enough information to do each of the following problems? If so, do the problem. If not, what additional information do you need?

a. A child pulls a friend on a sled. The child maintains a constant tension of 29.4 N in the rope. The combined mass of friend and sled is 30 kg, and the coefficient of sliding friction between the sled and snow is 0.1. How far can the child pull her friend in 6 s?

b. A box slides down a 20° incline at constant speed. What is the coefficient of sliding friction between box and incline?

5-63. **SSM** When a board is held at an angle of 30° to the ground, a brick sliding down it has a constant speed of 4 m/s. Find the coefficient of kinetic friction between brick and board.

5-64. A small block sits at the midpoint of a 2.0-m plank lying on a level floor. A student slowly raises one end of the plank until the block starts to slide. At that instant, the raised end is 0.55 m above the floor. Find the coefficient of static friction between block and plank.

5-65. The coefficient of static friction between a parcel and a loading chute is 0.20. To what maximum angle can the chute be tilted before the parcel begins to slide?

5-66. A spring is connected to a block on a table (Figure 5-54). A person pulling horizontally on the spring gradually stretches it until the block is set in motion. Once the block begins sliding, the table exerts a kinetic frictional force of magnitude f_k on the block. Before the block begins to slide, the force exerted on the block by the spring is (*always, sometimes,* or *never*) greater than f_k. Explain.

Figure 5-54 Problem 5-66

Going Further

The questions and problems in this group are not organized by section heading, so you must determine for yourself which ideas apply. Some of them will be more challenging than the Review and Practice questions and problems (especially those marked with a • or ••).

5-67. Figure 5-55 shows three situations involving a wooden block on a rough-surfaced ramp. In each case the indicated motion is produced when a rope (not shown) is exerting a tension force on the block. Exerting the tension force does not change the normal force on the block. For each of these situations, complete the free-body diagram, making sure that the direction and approximate size of the additional force vectors that you draw are consistent with the indicated motion.

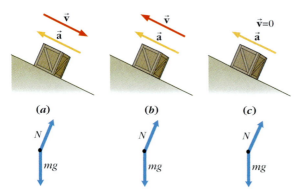

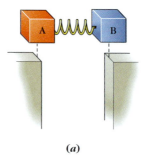

Figure 5-55 Problems 5-67 and 5-68

5-68. Suppose that in part *c* of Problem 5-67 (that is, in Figure 5-55*c*), the normal force on the block is reduced when the rope exerts a tension force. Redraw the free-body diagram for this change in conditions.

5-69. An assembly consisting of two blocks connected by a spring (Figure 5-56*a*) is wider than the opening of a hollow shaft when the spring is uncompressed. The assembly is then squeezed into the opening of the shaft, so that it remains securely in place (Figure 5-56*b*).

a. Draw a free-body diagram of the complete assembly, showing and fully identifying all forces exerted on it.

b. In your free-body diagram, what force(s) are exerted oppositely to the assembly's weight to keep the assembly from falling?

c. What is the reaction force to each force in your free-body diagram? Sketch the color-coded arrow (recall step 5 of Procedure 5-1) for each of these reaction forces.

d. Now sketch a free-body diagram of block A only. Is there any force represented in this diagram that does not appear in your free-body diagram of the complete assembly? Why, or why not?

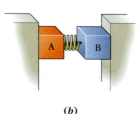

(a)

(b)

Figure 5-56 Problem 5-69

5-70. *Settle the argument:* A block is pulled along a table top at a constant speed of 5 m/s by a string kept at constant tension. Student A argues, "If the block doesn't slow down, the contact between block and table must be frictionless." Student B argues, "There has to be a frictional force equal to the tension because the block doesn't speed up." Student C argues, "There can be friction, but the frictional force must be *less* than the tension for the block to keep going." Determine which student is correct, and then tell what is wrong with each of the other two students' arguments.

5-71. In each of the four situations in Figure 5-57, a hanging weight produces a torque about the pivot P. Rank the situations in order of the torque about P in each situation. Rank them from least to greatest, considering only the magnitude of each torque, not its sign. Be sure to indicate any equalities.

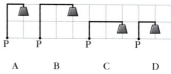

A   B   C   D

Figure 5-57 Problem 5-71

5-72. A worker exerts a 200-N horizontal force on a 40-kg crate. The crate's acceleration across a level floor is 0.5 m/s^2. Find the frictional force that the floor exerts on the crate.

5-73. **SSM** A 2.0-kg sled is dragged across an ice slick (presumed frictionless) by a rope exerting a 6.5-N tension force in the direction shown (Figure 5-58). Find the sled's acceleration.

Figure 5-58 Problem 5-73

5-74. The block in Figure 5-59 weighs 98 N. If the slope is frictionless, find the force exerted by the rope.

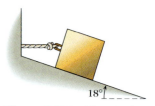

Figure 5-59 Problem 5-74

5-75. In each of the situations in Figure 5-60, the yo-yo is kept moving in a horizontal circle. For each case, sketch a free-body diagram of the yo-yo, and on the basis of your diagram decide whether the depicted motion is possible.

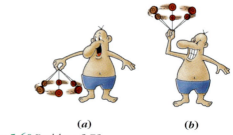

(a)   (b)

Figure 5-60 Problem 5-75

5-76.

a. Draw a separate free-body diagram of each of the two acrobats in Figure 5-61. (In this part, assume that the safety cable connected to the upper acrobat is slack and exerts no force on her.)

b. If the upper acrobat were suddenly lifted by the safety cable, what would happen to the lower acrobat? Use your free-body diagrams and Newton's laws to guide your reasoning.

Safety cable

Figure 5-61 Problem 5-76

•5-77. In the set-up in Figure 5-62, the pulley and ramp are frictionless. If objects A and B remain stationary, what is the ratio $\frac{m_A}{m_B}$ of their masses?

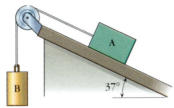

Figure 5-62 Problem 5-77

5-78. Draw a free-body diagram of the rope in Figure 5-15c. Assume the rope's weight is negligible.

5-79. Using results of Example 5-8 as needed, find the normal force that the pulley exerts on the string if the string is effectively massless. (*Hint:* Think carefully about your free-body diagram.)

••5-80. During a winter festival, a Zamboni ice-resurfacing vehicle is used to pull six identical ice boats along a frozen canal (Figure 5-63). The force F_1 exerted by the rope between the Zamboni and the first boat is 600 N. Calculate the forces exerted by the other five ropes connecting the boats. (*Hint:* Assume the ice exerts no frictional force on the boats. Under these conditions, can their acceleration be zero?)

$\vec{F}_6$ $\vec{F}_5$ $\vec{F}_4$ $\vec{F}_3$ $\vec{F}_2$ $\vec{F}_1$

Figure 5-63 Problem 5-80

5-81. A garden hose is wound around a large spool when not in use, as in Figure 5-64. If the gardener pulls on the hose hand over hand to unwind it, and exerts the same force with each pull by either hand, how does the torque that the gardener exerts on the hose change as the hose unwinds—does it increase, decrease, or remain the same? Briefly explain.

Figure 5-64 Problem 5-81

•5-82. [*Choose the best answer.*] Two blocks of equal mass are connected by a light string. The blocks are held stationary on a frictionless ramp with the string just barely taut, as shown in Figure 5-65. They are

Figure 5-65 Problem 5-82

then simultaneously released. As the blocks slide down the ramp, the tension force that the string exerts on each block (*remains zero (or nearly so); remains constant, but clearly above zero; increases; decreases*).

5-83. In Example 5-11, if the painter is a distance x from the left end of the scaffold, solve for F_L and F_R in terms of x.

5-84. In Example 5-11, find the values of F_L and F_R when the painter is *at* the left end of the scaffold.

•5-85. When two children are balanced on a seesaw, the total force on the system (children + seesaw) is zero. If one child then slides toward the other, does the sum of the forces remain zero? Justify your answer, and if it is no, tell which force(s) change. (*Hint:* Think about what happens to the center of mass.)

5-86. Three men were being interrogated by FBI agents. The agents believed that one of them had left a federal office with 20 lb of top-secret documents under his coat, gone down the freight elevator, and slipped into a waiting getaway car. All three had admitted to having stood on a scale kept in the freight elevator to weigh incoming shipments.

Suspect A said, "The scale read my normal body weight only while the elevator was just starting to descend."

Suspect B said, "I only got on the scale for a few seconds about halfway down. It seemed to read my normal body weight."

Suspect C said, "When the elevator was coming to a stop, the scale read about 20 lb more than my normal body weight."

If all three men were telling the truth, who was the likeliest suspect? Briefly explain your choice.

5-87. In Figure 5-66, the rope is fixed at its upper end and then connected through two frictionless pulleys, one of which is free to ride up and down on the rope. The lower end of the rope is connected to block B. Block A is suspended from the movable pulley.

a. How do the accelerations of the two blocks compare ($a_A = \frac{?}{} a_B$) when the mass of block B is greater than that of block A? (Answer by giving the correct numerical multiplier to replace the question mark.)

b. How should the mass of block B compare to that of block A ($m_B = \frac{?}{} m_A$) if there is to be *no* acceleration?

m_B m_A

Figure 5-66
Problems 5-87 and 5-88

c. Will block A move up or down when the system is released from rest if $m_B = \frac{3}{4} m_A$? Briefly justify your answer.

••5-88. In Figure 5-66, if $m_B = \frac{3}{4} m_A$, find the tension in the rope and the acceleration of each block in terms of g and m_A.

5-89. A wheelbarrow and its load together weigh 260 N, and you are pushing it over perfectly level horizontal ground.

a. If the motion of the wheelbarrow is not opposed by any resistive forces, what constant force must you exert on the wheelbarrow at an angle of 20° to the horizontal if you wish to bring the wheelbarrow from rest to a speed of 3.2 m/s within 2.0 s?

b. Find the horizontal component of the force that *you* must exert on the ground to accomplish this, and tell what kind of force it must be.

c. What additional information would you need to know to find the total force that you must exert on the ground?

5-90. Figure 5-67 shows a detail of a rope that is tied around a stationary post and then pulled horizontally until it is taut.

Figure 5-67 Problems 5-90 and 5-91

a. Draw a free-body diagram of the part of the rope (and *only* that part) that is highlighted in yellow. Make sure that you fully identify all forces in the diagram, indicating what type of force each is and what exerts it.

b. Fully identify the reaction force to each of the forces identified in **a.** Which of these are tension forces?

c. Suppose you were now going to draw a free-body diagram of the post. When the way in which the rope is attached to the post is not shown in detail, it would be common practice for this free-body diagram to show a tension force $\vec{T}$ exerted by the rope on the post. Strictly speaking, is this correct? Explain.

d. Besides the post itself, should anything else be included in the free-body on which $\vec{T}$ is exerted? Explain.

5-91. Can the person pulling on the other end of the rope shown in Figure 5-67 keep the rope taut in the absence of all frictional forces? Explain your answer.

5-92. Does it make sense to speak about the coefficient of friction of the linoleum on the floor of your aunt's kitchen? Explain.

•**5-93.** In a stunt aviation performance, the co-pilot of a plane hangs by a rope from the tail section while the plane is in flight (Figure 5-68). When the pilot subjects the plane to a constant acceleration, the rope hangs at an angle ϕ with the vertical.

a. Starting with $\Sigma F_x = ma_x$ and $\Sigma F_y = ma_y$, and neglecting the very real effects of air resistance (so that what we get is a

crude approximation), show that $\tan \phi = \frac{a}{g}$. The significance of this result is that the value of ϕ depends only on the acceleration, so that a body suspended by a cord or rope from an accelerating vehicle can serve as an *accelerometer*—a device to measure acceleration—if the body is heavy enough to keep the rope taut. Our dangling co-pilot is a human accelerometer!

b. Find the value of ϕ when the plane's acceleration is 7.35 m/s².

c. Try this in a car, *but with someone else driving:* Have a friend accelerate as uniformly as possible from 0 to 30 mph (speedometer readings) in 10 s. Find the value of this acceleration in m/s². Measure the angle that a string with a weight on the end makes with the vertical during this interval. Calculate the acceleration using $\tan \phi = \frac{a}{g}$, and compare with the value calculated from the speedometer readings.

5-94. In Example 5-7, we could have tried applying Newton's second law to the combined mass of the two crates. The second law deals only with external forces on a body, so only external forces on the combined mass are inventoried in this free-body diagram.

a. Could you have used this approach to find the acceleration of the blocks? Briefly explain.

b. Could you have used this approach to find the force that block A exerts on block B? Briefly explain.

••**5-95.** A painter weighing 600 N climbs a very light ladder propped against the side of a house at an angle of 60° with the horizontal (Figure 5-69). The ladder is 5.00 m long. The coefficient of static friction between ladder and ground is 0.40. Assume friction between the ladder and the wall is negligible.

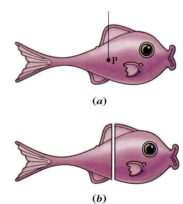

Figure 5-69 Problems 5-95, 5-96, and 5-97

a. How high up the ladder (*d* in figure) can the painter go before the ladder slips?

b. Could the painter climb at all without the ladder slipping if the contact between the ladder and the ground were frictionless? Briefly explain.

•**5-96.** Redo part **a** of Problem 5-95 for the case when the ladder's weight, instead of being negligible, is 60.0 N.

5-97. Show that if the ladder in Problems 5-95 and 5-96 has a non-negligible weight W_L, and the contact between ladder and ground is frictionless $(\mu_s = 0)$, there will be slipping even when the painter's distance *d* up the ladder is zero. Why must this be so?

5-98. A carved fish remains stationary when suspended in the window of a craft shop as shown in Figure 5-70a.

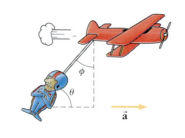

Figure 5-68 Problem 5-93

Figure 5-70 Problem 5-98

a. Is the fish's center of mass to the left of point P, right at point P, or to the right of point P?

b. The fish has been assembled from the two pieces in Figure 5-70*b*, which are joined at point P. For the fish to balance at P, is the weight of the piece with the tail less than, equal to, or greater than the weight of the piece with the head? (Adapted from an interview question described by Paula Heron, *AAPT Announcer,* 33 [2], 160, 1991.)

5-99. SSM WWW Identical twins Cara and Sara are seated on a seesaw at equal distances from the pivot. Their mother raises Cara's end so that the seesaw slants downward from Cara to Sara, but neither child's feet reach the ground. When the mother releases the seesaw, _____.

a. Cara moves upward and Sara moves downward.

b. Cara moves downward and Sara moves upward.

c. both twins remain stationary.

Briefly explain. (Adapted from an interview question described by Paula Heron, *AAPT Announcer,* 33 [2], 160, 1991.)

5-100. As an object travels along a straight line, there is a force on it in the direction of motion and another force on it in the opposite direction. At a particular instant, these forces are equal and opposite.

a. Describe a specific situation of this type in which these two forces could remain equal and opposite as the object travels.

b. Now describe a specific situation of this type in which these two forces could *not* remain equal and opposite as the object travels.

Problems on WebLinks

5-101. In WebLink 5-1, a drifting rocket starts out at a point where Earth and the moon pull equally hard on it. At this point, which (one or more) of the following pairs of forces are interaction pairs? The forces that Earth and the moon exert on the rocket, the forces that Earth and the rocket exert on each other, the forces that the rocket and the moon exert on each other, the forces that the rocket exerts on Earth and on the moon.

5-102.

a. In WebLink 5-1, the drifting rocket starts out at a point where Earth and the moon pull equally hard on it. As the rocket drifts closer to Earth, which of the pairs of forces listed in Problem 5-101 remain equal and opposite?

5-103. Problems 5-100 and 5-101 illustrate the following points.

a. The interaction forces that two objects exert on each other (*always, sometimes,* or *never*) remain equal and opposite as the object travels.

b. If two forces on the same object are equal and opposite at a particular instant, so that they produce equilibrium, they (*always, sometimes,* or *never*) remain equal and opposite as the object travels.

5-104. In WebLink 5-2, a block is being pulled by a rope, and we focus on one segment (segment A) of the rope. When the rope is taut, the forces that the other parts of the rope exert on this segment are equal and opposite only when _____.

a. the rope can be considered massless

b. the block is stationary

c. the block is moving at constant speed

d. the block is moving at constant velocity.

5-105. In WebLink 5-3, which of the following is always equal in magnitude to the mouse's weight? The normal force that the scale exerts on the mouse, the gravitational force that Earth exerts on the mouse, the scale reading, or the normal force that the mouse exerts on the scale.

5-106. Figure 5-71 shows a set-up from WebLink 5-4. If the contact between the pencil and the string that passes over it is frictionless, which of the following distances must be equal to the distance PC for the set-up to be in equilibrium? AB, BP, AP, AD, AE.

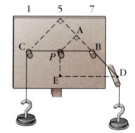

$r_\perp$ now equals distance from pivot to C7

Figure 5-71 Problem 5-106

CHAPTER SIX

Bookkeeping on Physical Systems: The Concept of Energy

In Chapters 2 through 5, we chose to view the natural world as a collection of objects in motion, in which objects' motions are affected when they interact with one another (exert forces on one another). From that perspective, we arrived at Newton's laws of motion as one way of expressing the rules by which the game of nature is played. In Chapters 6 and 7, we will view nature from a rather different perspective—one that says you never get something for nothing, that there is a kind of overall balance in nature. This perspective first developed in physics

"In nature as in industry, nothing is gained without something else being lost."

during the time of the Industrial Revolution. In nature as in industry, nothing is gained without something else being lost. Indeed, industry is not something apart from nature—the creations of human beings are a part of the natural world and are bound by the same laws.

In describing this balance, we will draw extensively on our prior notions of motion and forces, but the balance itself will be expressed in terms of two new concepts—the concept of *energy*, which we will develop at length in this chapter, and the concept of *momentum*, which we will develop in Chapter 7.

6-1 An Intuitive Introduction to Energy Ideas

What *is* energy? You use the word frequently in a range of everyday situations, but can you define it? This, it turns out, is no simple task. A somewhat simpler task is to think of some of the things that you associate with energy.

On-The-Spot Activity 6-1

Before reading further, briefly list the things or circumstances that *you* associate with the word *energy*. Try to group the items on your list into any categories you can think of. This will help you to compare your own ideas with those presented in the chapter.

A typical list might include the following:

> *heat*   *light*   *the sun* (or solar energy)
>
> *batteries*   *fusion*   *atoms* (or atomic or nuclear energy)
>
> *food* (or specific foods, such as sugars, carbohydrates, fats)
>
> *fuel* (or specific fuels, such as oil or gasoline)
>
> *motion or activity* (people who do more or move around more are more energetic)

Different categories might occur to you for sorting out the items in the list:

- *sources* of energy: the sun, atoms (Does food belong here?)
- *kinds* of energy: heat? (perhaps as hard to define as energy), light (Does motion or activity belong here?)
- things that *store* energy: batteries (Does food belong *here?* How about fats?)

We are not yet prepared to answer all the questions we are raising. In ordinary (imprecise) speech, wood in a fireplace is a source of energy to heat a room. Is the energy in the wood before you light it? Does that mean it is stored? Do you have to add energy to the wood by lighting it to get energy back? Why? As we refine our ideas, we will be able to eliminate some of the fuzziness in the ways we talk about energy.

Let's look now at some everyday situations (Figure 6-1) and see what they have in common. We will then think about how to express their common aspects in energy terms. In Figure 6-1a, as the height of the ball increases, its speed decreases. In Figure 6-1b, as the spring's compression increases, the car slows down. In each case, there is a kind of balance: something is gained, something else lost. In each of these cases, this exchange is reversible: As the ball comes back down, it speeds up again; as the spring decompresses, the car regains its speed.

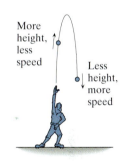

More height, less speed

Less height, more speed

(a)

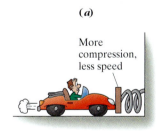

More compression, less speed

(b)

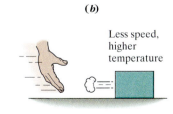

Less speed, higher temperature

(c)

Figure 6-1 Reversible and irreversible exchanges.

The situation in Figure 6-1c, where a block has been given a push, also involves a loss and a gain: As frictional forces slow the block down, the temperature increases where the rubbing occurs (like when you rub your hands together). Here again there is a kind of balance, but the exchange is *not* reversible: The block does *not* gain speed as the rubbed surfaces cool down.

Let's take a stab at putting this discussion into energy terms. If an object has energy because it is moving, this energy of motion decreases as the object slows down. The car in Figure 6-1b gets back its energy of motion as the spring recoils. What happened to the energy in between? One way of accounting for it is to say that it was *stored* in the spring as the spring compressed, and then returned to the car as the spring expanded back to its uncompressed length. In the language of physics, energy of motion is called **kinetic energy** (from the Greek *kinētikos*, meaning *of motion*), and stored energy is called **potential energy.** Using these terms, we can say that as the spring compresses, kinetic energy is lost and an equal amount of potential energy is gained; there is a balance because the *total* of the two kinds of energy remains the same. *For the remainder of this and subsequent discussions, we will abbreviate potential energy and kinetic energy as PE and KE.*

Can we talk about the situation in Figure 6-1a in the same way? To see that we can, consider the situations in Figures 6-2a–c. In each of these situations, objects (the two blocks in 6-2a, a chair and Earth in 6-2b, a just-used comb and pieces of tissue paper in 6-2c) are pulled apart against forces (elastic in 6-2a, gravitational in 6-2b, electrical in 6-2c) that tend to bring them back together. We will call these forces **restoring forces.** In each situation, we will speak of the objects that have been pulled apart as a *system*. The word **system** simply means one or more objects that we mentally group together for convenience. We pick the system so that the restoring forces are exerted by objects within the system on one another—they are *internal* forces.

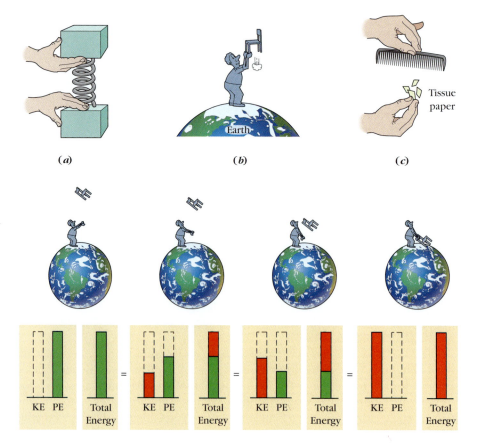

Figure 6-2 **Storing energy by separating objects against forces tending to bring them back together.** The objects are shown separated against (a) a spring or elastic force, (b) a gravitational force, and (c) an electrical (electrostatic) force. The graphs in (d) keep track of the energy in the system in (b) once the system is released. With no further energy input to the system, the total energy stays the same. But as the chair and Earth come back together, PE decreases and KE increases.

In each case, energy is stored by *positioning* the objects in the system so that the forces they exert on each other will increase their speed—and thus their KE—when they are released. The bar graphs in Figure 6-2d show that when no further energy is put into the system, the total energy stays the same, but as the objects in the system come back together, PE decreases and KE increases. Note that this occurs bit by bit, not all at once. To work through these ideas in more step-by-step detail, go to WebLink 6-1.

In the examples in Figure 6-2, to position the objects so that the forces they exert on each other will increase their KE when they are released, the objects are pulled apart against attractive forces between them. But it can also be done by pushing objects together against a force tending to separate them, for example, when there is a compressed spring between them or they have like electric charges. The amount of energy stored depends on *where* the objects are placed; in other words, the PE depends on the objects' *positions*. (The KE, in contrast, depends on their *speed*.) Depending on what type of internal restoring force is exerted when the system is released, the PE is called *elastic potential energy, gravitational potential energy, electrical potential energy,* and so on.

Energy storage involving both gravitation and springs is found in early clocks. In the earliest pendulum clocks (Figure 6-3), a weight was suspended from a cord wrapped around a spool connected to gears. As the weight descended, the turning gears moved the hands. By winding the clock, you turned the spool to raise the weight, storing the energy needed for the clock's continued operation. In later clocks, the spool was connected to a coiled spring rather than a weight; the uncoiling spring drove the clock.

◆**DISTRIBUTED ENERGY/"INTERNAL" ENERGY AND "HEAT"** We return now to the situation in Figure 6-1c. As the block slides across the table, it slows down—the KE decreases. **STOP&Think** Does this mean the PE increases? What is your criterion for judging? Is the block moving to a position where forces within the system can cause it to speed up? ◆

Because the answer to the last question is no, there is no gain in PE. Then what happens to the lost KE? You know from experience that objects exerting frictional forces on each other will increase in temperature. In this case both block and table experience temperature increases. If they become warmer than the surrounding air, the block and table will cool back down.

Until now we have only concerned ourselves with whole objects. The energies we discussed (PE and KE) actually involved the positions and velocities of their *centers of mass*. But in this case, energy is becoming *distributed;* that is, small amounts of it are being passed from atom to atom within these objects and then getting passed on across the boundaries of the system to atoms in the surroundings. Picture the atoms of an object as though they were connected by springs. When surfaces scrape across each other, for example, atoms at the surfaces can be shifted or caused to jiggle faster. The energy changes that occur have to do with the changes in the positions and velocities of atoms or molecules or their constituent parts. They are changes within the objects, independent of any changes in position or velocity that the objects experience as a whole.

- When energy of the objects as a whole is changed to energy of the particles that make up the objects, physicists speak of a rise in the **internal energy** of the system.

- In particular, the internal energy associated with a body being at a particular temperature is called **thermal energy.**

- In contrast, physicists use the word **heat** only to refer to the amount of energy passing into or out of a system because of a temperature difference between the system and its surroundings. In many common situations, this energy passes from a hotter to a colder region—say, from a radiator to a cool room. But

For **WebLink 6-1:**
Restoring Forces and Potential Energy, go to
www.wiley.com/college/touger

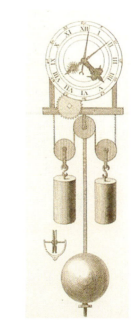

Figure 6-3 Weight-driven and spring-driven pendulum clocks.

Figure 6-4 An object that dissipates energy. The weights on this cart are suspended on rubber bands allowing energy to be transferred within the combined object (the cart plus its contents). **An object that does not dissipate energy.** Here the contents of the cart are rigidly fixed in place. (Photos from U. Ganiel, *The Physics Teacher*, 30, 18–19, 1992. Courtesy of the author.)

sometimes energy can pass from a colder to a hotter region, as when an air conditioner cools an inside space while it greets a bypasser outside with a blast of hot air. (Chapter 10 treats heat, temperature, and thermal energy more fully and investigates their basis in microscopic scale behaviors.)

- Another type of internal energy that is distributed on a microscopic scale is **chemical energy.** We will discuss this shortly.

- In general, when the "whole object" KE and PE of objects in a system is converted to energies of many smaller *parts* of objects (usually on an atomic scale, but see Figure 6-4), both in and out of the system, we say the energy is **dissipated.**

Typically, in dissipation, the energy becomes less concentrated as more and more objects get progressively smaller shares. Let's consider how this happens for a chair dropped to the ground from a height. When the chair is about to strike the ground, all its PE has been converted into KE. **STOP&Think** When the chair hits the ground, what happens to its KE? Does it convert back to PE? ◆

Experience tells us the chair is unlikely to bounce very high. Because gravitational potential energy depends on how high the chair is raised, the KE does not change back into gravitational PE. There is also no heating to speak of. What then? The cartoonist's conventional way of picturing a crash (Figure 6-5) can be instructive here. As the picture shows, there is some compression where contact occurs, and the surroundings are set to vibrating. Loosely speaking, there is some elastic PE associated with the compression. There is also energy connected with the vibrations. Some of this vibrational energy is kinetic, because vibration involves motion. Vibration also involves position changes, so some of the vibrational energy is potential energy. The energy connected with the vibrations is distributed, or *dissipated,* throughout the surroundings, and so is lost from the system. In fact, you know the energy is being dissipated when you hear the chair hit the ground. Sound involves the transmission of vibrations (and therefore of KE and PE) to your ear, which in turn sets your eardrum to vibrating. The thump can be heard anywhere nearby, so we know the sound energy is being transferred away from the system in all directions.

Figure 6-5 A system losing energy by dissipation.

The frictional force on a block sliding along a table and the upward force that the ground exerts on a falling chair are **dissipative forces.** When you release an object after you have moved it against a frictional force, the frictional force will not cause the object to speed up (gain KE). The energy you've put into the system has not been stored as PE. Likewise, when KE is lost because of such forces, PE is not gained. The total of all whole-object PE and KE in a system is called the **mechanical energy** of the system. In systems with dissipative forces, the total does not stay the same. Some of this energy becomes distributed, and part of the distributed energy is in turn lost to the system's surroundings.

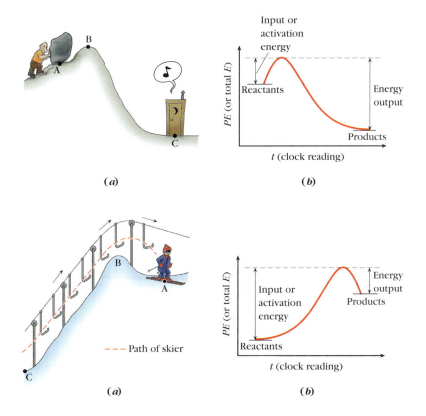

Figure 6-6 A gravitational analogy to an exergonic (also called exothermic) reaction. (*a*) The "real-world" picture (*b*) The corresponding energy graph.

Figure 6-7 A gravitational analogy to an endergonic (also called endothermic) reaction (*a*) The "real-world" picture (*b*) The corresponding energy graph.

When the total amount of mechanical energy in a system stays the same, we say that mechanical energy is **conserved.** For that reason, dissipative forces are also called **nonconservative forces.**

◆**AN ANALOGY APPROACH TO CHEMICAL ENERGY** The pictures in Figures 6-2*b* and *c* involving gravitational and electrical potential energy are similar or analogous. Work through WebLink 6-2 to see how this analogy can be extended to the breaking of old bonds and the formation of new bonds in chemical reactions. Figures 6-6 and 6-7 summarize some of the ideas of WebLink 6-2. It discusses **fuels** as molecules in which energy has been stored as a consequence of endergonic reactions. (Reactions are called **endergonic** or **exergonic** depending on whether the system of atoms involved in the reaction ends up with more or less energy after the reaction than before.) The energy is released in an exergonic reaction. This is a reversal of the storage process, just as moving the rock from A to C is the reverse of moving it from C to A. This is equally true of the gasoline that you burn in your automobile engine and the sugars, such as glucose, that you "burn" (oxidize) in the cells of your body (see Figure 6-8).

For **WebLink 6-2:**
An Analogy Approach to Chemical Energy, go to
www.wiley.com/college/touger

➡**A note on language:** In endergonic reactions, there is a net energy input to the system of atoms; in exergonic reactions, there is a net energy output. Because the inputs and outputs are commonly heat, they are also called endothermic and exothermic reactions.

Figure 6-8 Photosynthesis and cell respiration. Photosynthesis is an example of an endergonic reaction in which a fuel is formed and much of the energy input is stored. In this case, the fuel is glucose and the energy input is radiated energy from the sun. Cell respiration is an example of an exergonic reaction in which the fuel—here it is again glucose—is then oxidized or "burned," releasing stored energy. In this case, the cell can then use the energy output. (Note: Each of these "reactions" is really a long sequence of reactions. The overview shown here gives the starting reactants and the final products, but omits the in-between details.)

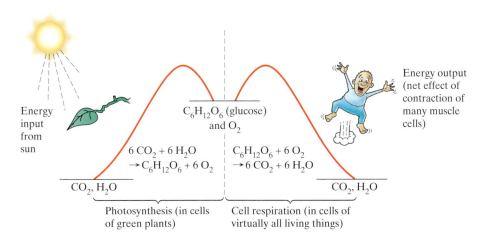

◆**ENERGY TRANSFERS AND CONVERSIONS** We've discussed several situations where one quantity increases as another decreases. We can analyze a broad range of such situations by keeping track of energy, or what is sometimes called **energy bookkeeping.** Indeed, we may keep track of energy in various energy "transactions," much as bookkeeping keeps track of money and other assets in business transactions:

Physicists
Bookkeepers keep track of each kind of *energy* / *asset* held by each *object or system* / *individual or organization*

Energy "transactions" are typically either

- **energy transfers,** in which energy is given by or transferred from one object to another, or

- **energy conversions** or **transformations,** in which one kind of energy is converted or changed into another.

Sentences about energy transfers take different kinds of subjects and objects than sentences about energy conversions, and you should distinguish carefully between them. Note the following examples.

	Typical descriptive sentence	Energy example	Analogous example for money
Energy transfer	(*Object A*) transfers or gives energy to (*object B*).	A bat transfers energy to a ball.	I give you a dollar.
Energy conversion	(*One kind of energy*) is converted into (*another kind of energy*).	Kinetic energy is converted into potential energy (during the upward part of a jump).	You exchange a dollar bill for 100 pennies, or convert yen into pesos.

Case 6-1 ◆ *Energy Transfers and Conversions in an Interconnected System*

We can trace the flow of energy through the system in Figure 6-9 by describing some of the energy transfers and conversions that take place (we will not attempt to cover every detail). ① As the river flows downstream toward lower elevations, PE is converted into KE. ② The river then hurtles over a precipice, becoming a waterfall, where more of its PE is converted into KE as it falls. A hydroelectric plant is set up at the falls. ③ Here KE is transferred from the water to the turbine (rotating part) of an electric generator, which converts some of this KE into electrical energy. (We will treat the electrical aspects in later chapters.) ④ Much of this energy is transferred to the electrical system of your home via the transmission lines. The rest is dissipated en route. ⑤ When you plug an electric fan into an outlet and switch it on, you provide a path along which the electrical PE of charged particles can be converted into KE. Within the electric fan, there must be further energy transfers and conversions, because the motion of the fan blades tells us the energy is ultimately transferred to the blades of the fan, where it has again been converted into KE. ⑥ Some of that KE is in turn transferred to the surrounding air, producing wind.

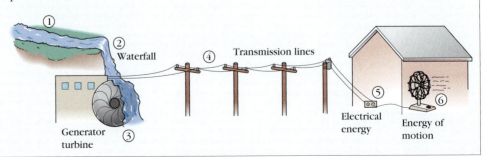

Figure 6-9 The Flow of energy through an interconnected system.

(*a*)   (*b*)   (*c*)   (*d*)

Figure 6-10 Radiated energy.
(*a*) The radii of a circle radiate from the center C. (*b*) The spokes of a wheel radiate from its hub. (*c*) Light radiates from a bulb. (*d*) "Heat" radiates from a hot stove.

STOP&**Think** What was the origin of the energy that we traced through the system in Case 6-1? ◆ For water droplets to form clouds and fall as rain upriver, surface water must have absorbed energy from the *sun* and evaporated. Although we are not yet ready to discuss details of light and heat, we may call this energy from the sun **radiated energy** (or *radiant energy*). Radiated energy travels outward in all directions from its source; that is, it follows all the possible *radii* of a sphere with the source at its center (Figure 6-10). We know it as energy because like PE and KE, radiated energy may be converted into other kinds of energy when it is transferred to an object. This conversion can also take place in the other direction. When a light bulb is turned on, the energy it receives is converted into radiated energy. By radiating outward in all directions, this energy is transferred to the surroundings of the system and thus is lost from the system. Note that because some energy is always being dissipated and lost by the system, the system requires a continued input of energy from outside to go on functioning in the same way.

➡**How thin is thin?** Although the biosphere has a thickness of several kilometers, this is thin compared to Earth's radius. The relative thickness is comparable to the thickness of a piece of tissue paper used to wrap an apple or pear relative to the radius of the fruit itself.

Case 6-2 ◆ *A Global Energy System: The Biosphere*

In the system shown in Figure 6-11, the initial energy input is again radiated energy from the sun. Here this energy is converted into chemical energy stored in carbohydrate molecules by plants, in this case wheat and peanut plants. Further energy inputs and conversions are necessary (see Problem 6-1) so that those molecules, with their stored energy, become available to our softball player in the form of a peanut butter sandwich. She then comes to bat, gets a base hit, and quickly finds herself sliding into second base. All the while her body is making use of the energy stored in the carbohydrate molecules. Through the detailed processes of energy flowing through a living organism, collectively called **metabolism,** some of that energy is converted into KE as she races around the base path, and some of it is radiated to her surroundings as she gets "heated up." When she slides into second base, her KE is dissipated because of frictional forces. The objects rubbing together are warmed, and the internal energy they gain is in turn radiated to the surroundings.

The plants and the human that eats them typify what goes on in the **biosphere,** the thin life-supporting layer at or near Earth's surface.

The source of energy for very nearly all living things—the plants that carry on photosynthesis, the herbivores that eat the plants, and the meat-eaters (carnivores) that eat the herbivores—is ultimately the sun. This energy enters the system—the biosphere—as radiated energy. Ultimately, through a variety of dissipative mechanisms, most of it gets radiated back into space and is lost to the system. For life to exist, some energy must be retained long enough so that all the energy transfers and conversions that constitute life processes can take place before the energy is irretrievably reradiated into space.

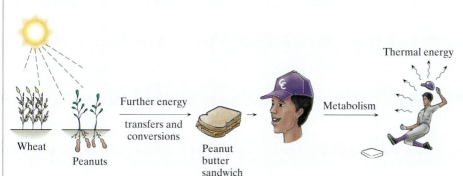

Figure 6-11 Energy transfers and conversions in the biosphere.

Fossil Fuels

Humans have been able to inhabit many otherwise uninhabitable regions of Earth by burning fuels to provide heat for warmth, cooking, and eventually other more sophisticated purposes. Learning to use fire meant learning to control exergonic reactions by which energy that had been stored in fuel molecules could be released. Over the history of humankind, almost all the fuel molecules that humans have used in extensive quantities were originally formed by plants in the process of photosynthesis. At first, in wood fires, living or recently living plants were burned directly. Later on, coal and petroleum became our major fuels, and they remain so today. But these materials are essentially the fossil remains of plant matter (coal) and sea life (petroleum) that have been compressed for millions of years, clearly with some chemical changes, but still retaining the stored energy of molecules that were formed by photosynthesis millions of years ago. Fossil fuel energy is thus energy from the sun that nature has been able to keep in storage for millions of years. But when it is burned, the energy is quickly radiated back into space and lost to further human use. A critical issue for humankind at the start of the twenty-first century is that we have become heavily dependent on this finite supply of stored energy—indeed we have used most of the supply within the past century—and that when this supply is gone it will take millions of years to replace.

6-2 Making Energy Concepts Quantitative

The qualitative ideas in Section 6-1 provide a framework for talking about a kind of balance in nature. To bear up under the scrutiny of measurement, however, the ideas must have precise quantitative meaning. With this as our goal, we first define one kind of energy input, to which physicists assign the name *work*. We then show that under carefully chosen conditions, this energy input may be equal to the increase in one of two quantities. We can identify these quantities as kinetic energy and potential energy. (Under other conditions, both of these quantities may change, but we pick situations that permit us to focus on one or the other.)

➥ **A note on language:** It is usual in physics to speak of "the work done by a force," whereas in everyday usage, work is done by people. You need to remember that a force is not a person or thing; it is an action. By saying that the force is doing work, we are really saying that whoever or whatever is exerting this force is transferring energy to the object by doing so.

◆ **THE CONCEPT OF WORK** A simple way to put energy into a system is to exert forces on one or more objects in the system to increase their KE and/or PE. This adds energy only when you exert a force over a distance, because only then do you change either an object's speed (and thus its KE) or its position (and thus its PE). This kind of energy input is called the **work** done by a force, and is defined as follows:

$$\text{Work} \equiv \frac{\textit{amount of force exerted in}}{\textit{the direction of motion}} \times \frac{\textit{distance over which}}{\textit{the force is exerted}} \qquad (6\text{-}1)$$

You are probably unaccustomed to using the word *work* in this way. You may feel that if you spend 10 minutes trying to push up a jammed window without budging it, you have worked very hard, and indeed you have. But your effort has borne no result; the energy you have expended has been dissipated. It has not been transferred to the window (which has gained neither KE nor PE) and so represents zero energy input into that system. By the physicist's definition, then, you have done no work on the window (compare Figures 6-12a and b).

In Figure 6-12c, both movers are exerting forces on the piano bench, but only the force exerted by the larger mover is responsible for giving the bench kinetic energy. Suppose we consider the effect of the total force exerted by the two movers (Figure 6-12d). The total force's component in the direction of motion is just the force exerted by the larger mover. In general, only *components of forces in the direction of motion* of an object contribute to the work done on that object. If $\Delta \vec{s}$ represents the object's displacement while a force $\vec{F}$ (assumed to be uniform

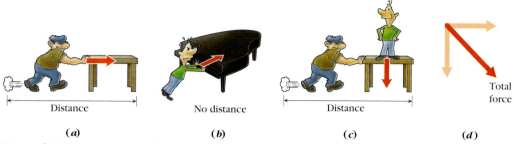

Figure 6-12 Forces that result in work/forces that don't. (*a*) Work is done by the large mover. (*b*) No work is done by the small mover. (*c*) Here too, no work is done by the small mover. (*d*) The forces the two movers exert are the components of the total force on the piano stool in (*c*).

or average) is exerted on it, then the component of $\vec{F}$ in the direction of $\Delta\vec{s}$ is $F\cos\theta$ (Figure 6-13*a*). The definition of work therefore can be rewritten as

$$\textbf{Work:} \qquad W \equiv F\Delta s \cos\theta \qquad\qquad (6\text{-}2)$$

When the force is entirely in the direction of the displacement (Figure 6-13*b*), $\cos\theta = \cos 0 = 1$, so that $W = F\,\Delta s$. When the force is perpendicular to the displacement (Figure 6-13*c*), $\cos\theta = \cos 90° = 0$. Then $W = F\Delta s\,(0) = 0$, consistent with our claim that *the perpendicular force provides no energy input*. The SI units of work are units of force (N) times units of displacement (m) and are called **joules** (J):

$$1\text{ joule} \equiv 1\text{ newton} \times 1\text{ meter} \qquad\qquad (6\text{-}3)$$

$$(1\text{ J} = 1\text{ N}\cdot\text{m})$$

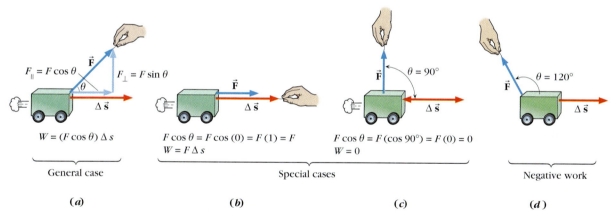

Figure 6-13 Understanding the equation *Work = FΔs cos θ*.

Example 6-1 *Negative Work*

A cart rolls 0.60 m to the right. Throughout its roll, a force of 40 N is exerted on the cart in the direction shown in Figure 6-13*d*. How much work is done on the cart by this force?

Solution
We apply the definition of work (Equation 6-2) here:

$$W \equiv F\Delta s \cos\theta = (40\text{ N})(0.60\text{ m})\cos 120° = (40\text{ N})(0.60\text{ m})(-0.50) = \textbf{−12 J}$$

Making sense of the result. The work done here is negative because the cart is being pulled from behind to slow it down. The force component parallel to the displacement is directed opposite to the displacement. It therefore reduces the cart's energy and slows the cart down. Thus work, because it is the energy input, is *negative*.

◆ Related homework: Problems 6-14, 6-15, and 6-17.

Note: F is the magnitude of the force vector and Δs is the distance (the magnitude of the displacement $\Delta \vec{s}$), so both are positive. Then the sign on work depends only on whether $\cos \theta$ is positive or negative.

In one dimension,

Work is + when $\vec{F}$ and $\Delta \vec{s}$ in same direction ($\longrightarrow$ or $\longleftarrow$) ($\cos 0 = +1$)

Work is − when $\vec{F}$ and $\Delta \vec{s}$ in opposite directions ($\longrightarrow$) ($\cos 180° = -1$)

By the reasoning in Example 6-1, a force that opposes the motion of an object over its entire path, as sliding frictional forces commonly do, always does negative work on the object, thus reducing its energy. This is what you would expect of a dissipative force.

For **WebLink 6-3:**
Obtaining a Quantitative
Definition of Kinetic energy, go to
www.wiley.com/college/touger

♦**TOWARD A QUANTITATIVE DEFINITION OF KINETIC ENERGY** Let's see what happens when work is done on a system consisting of a single point object. For simplicity, we'll consider a one-dimensional situation, such as the one in Figure 6-13b. For a point object there are no internal forces, so there can be no build-up of potential (stored) energy. The only effect of exerting a force on the object is to accelerate it from speed v_1 to speed v_2. Restated in energy terms, the only effect of doing work on this object is to increase its kinetic energy. The force is exerted over a distance Δs during a time interval Δt. The work done (with $\cos \theta = \cos 0 = 1$) is then $W = F\Delta s$. By working through WebLink 6-3 (and perhaps gaining some insight into how a mathematical derivation works), you will see that this work equals the change in a quantity $\frac{1}{2}mv^2$. In brief summary, the derivation is done for a constant total force F for simplicity, but the result is valid more generally. It uses the fact that $F = ma = m\frac{\Delta v}{\Delta t} = m\frac{v_2 - v_1}{\Delta t}$, and also that the average velocity $\bar{v} = \frac{\Delta s}{\Delta t}$ but also is equal to $\frac{v_2 + v_1}{2}$ when F and a are constant. Then

$$W = F\Delta s = ma\Delta s = m\left(\frac{v_2 - v_1}{\Delta t}\right)\bar{v}\,\Delta t = m(v_2 - v_1)\left(\frac{v_2 + v_1}{2}\right) \qquad (6\text{-}4)$$

Simplifying, we get

$$W = \tfrac{1}{2}mv_2^2 - \tfrac{1}{2}mv_1^2 \qquad (6\text{-}5)$$

or

$$W = (\tfrac{1}{2}mv^2)_2 - (\tfrac{1}{2}mv^2)_1 = \Delta(\tfrac{1}{2}mv^2)$$

But we've already said that the work done equals the change in KE. The quantity $\frac{1}{2}mv^2$ simply depends on the speed of the object (note that v^2 is positive whether v is plus or minus) and on the amount of mass m moving at this speed—that is, it meets our qualitative requirement for kinetic energy. Thus we *define* the kinetic energy as

$$KE \equiv \tfrac{1}{2}mv^2 \qquad (6\text{-}6)$$

Note that $KE \geq 0$ at all times. Because the energy gained equals the work input, the SI units of energy are also joules.

♦**WHY *KE* DEPENDS ON *m*** We would expect that to change the speed of two identical blocks in this way would require twice as much total energy input, and each block would end up with the same KE as the one. Because the distance between the blocks should not matter, we can now mentally reduce the distance between them to zero, so that we effectively have one block with twice the mass. It is thus reasonable to expect that having twice the mass at the same speed doubles the KE, that is, that an object's KE is proportional to its mass, as Equation 6-6 implies.

◆TOWARD A QUANTITATIVE DEFINITION OF GRAVITATIONAL POTENTIAL ENERGY Now let's see what happens when we do work on a system consisting of a rock and Earth, say, by lifting the rock. We include Earth in the system so that we can deal with the gravitational force as an internal force, one that objects within the system exert on each other. But because Earth is so much more massive than ordinary objects, the interaction has negligible effect on the position and speed of Earth (recall Example 4-1). Likewise, if we exert equal and opposite forces on the rock and Earth to separate them, the effect on Earth is negligible and we can concentrate on what happens to the rock.

The work we do is an energy input to this system. In general, some of this input will become PE and some will become KE. To simplify matters, imagine lifting the rock vertically from a point A (at height y_A) to a point B (at height y_B) by exerting a force equal and opposite to the rock's weight (Figure 6-14a). This way the rock doesn't accelerate and doesn't gain any KE. You will actually have to exert a force slightly larger than the weight to start the rock upward and slightly less than the weight to bring it to a stop at B. But in all, you are exerting a force that has very nearly a constant magnitude mg through a displacement $y_B - y_A$. (In fact, the *average* force can be kept *exactly* equal to mg.) The work you do on the rock is

$$W = mg(y_B - y_A) = mgy_B - mgy_A = \Delta(mgy)$$

When you lift the rock in this way, none of the work you do on the rock ends up as KE at B. All the energy input W is stored. As the algebra shows, this stored energy equals the change in a quantity mgy. This quantity depends on the rock's *position,* and thus meets our qualitative criteria for PE. (More generally, PE depends on the separation between the interacting objects in a system.) We therefore *define* this quantity to be the gravitational PE:

$$PE_{grav} \equiv mgy \qquad (6\text{-}7)$$

when g constant (e.g., over short distances from a planet's surface).

Note that raising a greater weight mg requires more work—the resulting stored energy is thus greater if either m or g is greater. Note also that our reasoning holds true only if g remains constant over the entire distance; we cannot apply it, say, to a rocket launched to a height so great that Earth's gravitational pull is significantly weaker.

Suppose the same rock is now released from rest from a point above B (Figure 6-14b). For the part of its descent between B and A, the displacement has

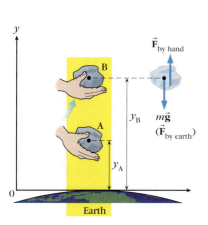

(a) Hand exerts external force and does external work an energy input to system which is stored as *PE*

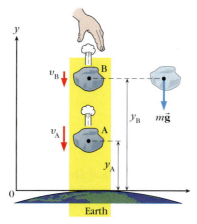

(b) Earth exerts internal force and does internal work. Equivalently we say that much *PE* is converted to *KE*

Figure 6-14 Energy aspects of raising and dropping a rock. (In this and subsequent figures, the system is highlighted in yellow.)

magnitude $\Delta s = y_B - y_A$ and is directed downward. The force on it has magnitude mg and is also directed downward, so that $\cos \theta = \cos 0 = 1$. Then

$$W_{\text{done by Earth on rock}} = F\Delta s \cos \theta = mg(y_B - y_A) = mgy_B - mgy_A \quad (6\text{-}8)$$

But between B and A, the gravitational force is unopposed, and Equation 6-5 tells us that when *that* is the case, the work equals the change in KE. Thus,

$$W_{\text{done by Earth on rock}} = mgy_B - mgy_A = \tfrac{1}{2}mv_2^2 - \tfrac{1}{2}mv_1^2$$

For the *descent*, y_B is the earlier position (y_1) and y_A the later position (y_2), so that

$$mgy_1 - mgy_2 = \tfrac{1}{2}mv_2^2 - \tfrac{1}{2}mv_1^2$$

This can be rearranged to give us

$$mgy_1 + \tfrac{1}{2}mv_1^2 = mgy_2 + \tfrac{1}{2}mv_2^2 \quad (6\text{-}9)$$

By the definitions (Equations 6-6 and 6-7) that we have just established, Equation 6-9 says

$$PE_1 + KE_1 = PE_2 + KE_2 \quad (6\text{-}10)$$

In words, this says that if there is no energy input from outside the system, and the objects in the system can be treated as point objects (internal energies not relevant), then

$$(Total\ Mechanical\ Energy)_1 = (Total\ Mechanical\ Energy)_2$$

This is an instance of the **law of conservation of energy.** When there is no transfer of energy into or out of the system, the total energy of a system remains constant and is said to be *conserved*.

We have not yet fully set out the *conditions* required for the law of conservation of energy to apply, but when it does apply, it is a powerful calculational tool. A little algebra, for example, takes us from Equation 6-10 to

$$KE_2 - KE_1 = -(PE_2 - PE_1)$$

which we can write more concisely as

$$\Delta KE = -\Delta PE \quad (6\text{-}11)$$

If the total of two kinds of energy remains constant, then the changes in the two must be equal but opposite in sign: A gain in one must be matched by a corresponding loss in the other. Thus, a falling rock *gains* KE as it *loses* PE.

A necessary condition for energy conservation, which held true for the rock–Earth system in Figure 6-14b but not in Figure 6-14a, is that the total work done on the system by external forces must be zero. When this condition is satisfied, there is no energy input from outside the system. We are not saying that no work is done at all; *within* the system, objects attracting each other gravitationally do work on each other. But the equivalence of Equations 6-8 through 6-11 tells us that this *internal work is equivalent to an energy conversion within the system.* Rather than talk about internal work, it is usually more convenient to talk about an increase in one kind of energy and a loss in another.

Case **6-3** ♦ *Three Observers Watch a Ball Drop*

In Figure 6-15, suppose each observer decides, "I am at $y = 0$." Then as the ball drops from point 1 to point 2,

Observer A says $y_1 = 8$ m and $y_2 = 5$ m, so $\Delta y = 8$ m $- 5$ m $= 3$ m.

Observer B says $y_1 = 2$ m and $y_2 = -1$ m, so $\Delta y = 2$ m $- (-1$ m$) = 3$ m.

Observer C says $y_1 = -4$ m and $y_2 = -7$ m, so $\Delta y = (-4$ m$) - (-7$ m$) = 3$ m.

The three observers measure different values of y_1 and y_2. Therefore,

the gravitational PE will have different values in different coordinate frames.

In other words, the observers disagree on the values of *PE*, which depend on y. But they agree on the value of Δy, so that the *change* in PE ($\Delta PE = mg\,\Delta y$) does *not* vary with the coordinate frame. It is the change, ΔPE, that has physical meaning. For instance, Equation 6-11 tells us that a change in PE results in a change in KE, and therefore in the change in motion, which is the observable behavior. Switching to another coordinate frame does not affect what the ball does; it only changes the way we describe it.

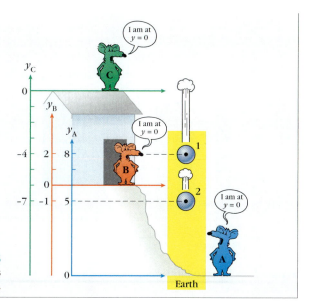

Figure 6-15
Gravitational PE depends on the choice of origin.

It is important to keep in mind that the language of energy is another way of describing observable physical behavior, which in some cases we have already described in the language of kinematics and dynamics (summarized in Table 6-1).

Table 6-1 Ways of Describing Motion

Observed Behavior	Dynamic Description	Energy Description
As A pushes B over a distance, B speeds up.	As A exerts a force on B, B accelerates.	As A does work on B, B gains KE.
A pulls on B to keep B moving in a circle around A at constant speed (but *not* constant direction).	As A exerts a force on B, B accelerates.	A does *no* work on B, so B's KE does not change.*
As an object falls toward Earth, it gains speed.	Earth exerts a downward force on the object, so it accelerates downward.	The object loses gravitational PE and gains KE.

*The dynamic description is a vector description and depends on direction; the energy description is a scalar description and does not.

However, the law of conservation of energy can be a powerful calculational tool.

Example 6-2 *A Ball Thrown Upward (Again)*

As in Example 2-12, a softball player throws a softball straight upward with a velocity of 17.0 m/s.
a. How high does the ball go?
b. How fast is it traveling when it is at a height of 10.0 m and on its way down?

Solution

Choice of approach. Although **a** was solved using kinematics in Example 2-12, here we will solve both **a** and **b** by applying the law of conservation of energy (Equation 6-10 or 6-9) to the system of softball and Earth. Equation 6-9 involves the object's position and velocity at two different times. But time does *not* appear in the equation (we therefore could *not* use conservation of energy to

EXAMPLE 6-2 *continued*

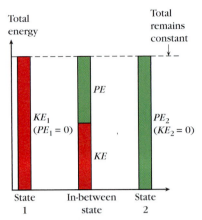

Total
energy

Total
remains
constant

PE

KE_1
$(PE_1 = 0)$

PE_2
$(KE_2 = 0)$

KE

State
1

In-between
state

State
2

Figure 6-16 Energy Bar Graph for Example 6-2. (adapted from work of A. Van Heuvelen and A. Zou; see for example *Am. J. Phys.* **69.** 184 (2001))

find t in part **a** of Example 2-12), so we say instead that Equation 6-9 equates the total energy of the ball + Earth system in two different **states,** an initial or earlier state 1 and a final or later state 2. Each state is characterized by a position (on which the ball's PE depends) and a velocity (on which the KE depends). We can draw a bar graph showing the KE and PE in each state (Figure 6-16) as an aid in seeing that the total energy is the same in the two states, as well as in any in-between states, as the KE and PE bars gradually change height.

We will assume $y = 0$ at ground level.

Restating the problem. As in Examples 2-10 and 2-12, **a** can be restated as $y = ?$ when $v = 0$. For part **a,**

What we know/what we don't. For energy problems, it is useful to organize your data according to what you know about each *state*.

State 1	State 2	Constants
$y_1 = 0$ $v_1 = 17.0$ m/s	$y_2 = ?$ $v_2 = 0$	$m = ?$
$PE_1 = 0$ $KE_1 = \frac{1}{2}mv_1^2$	$PE_2 = mgy_2$ $KE_2 = 0$	$g = 9.80$ m/s^2

The mathematical solution.

$$(Total\ Energy)_1 = (Total\ Energy)_2$$

or

$$PE_1 + KE_1 = PE_2 + KE_2 \qquad (6\text{-}10)$$

becomes

$$0 + \tfrac{1}{2}mv_1^2 = mgy_2 + 0$$

Then

$$\tfrac{1}{2}v_1^2 = gy_2$$

Solving for y_2 gives

$$y_2 = \frac{v_1^2}{2g} = \frac{(17.0\ \text{m/s})^2}{2(9.80\ \text{m/s}^2)} = \mathbf{14.7\ m}$$

For part **b,**

What we know/what we don't.

State 1	State 2	Constants
$y_1 = 0$ $v_1 = 17.0$ m/s	$y_2 = 10.0$ m $v_2 = ?$	$m = ?$
$PE_1 = 0$ $KE_1 = \frac{1}{2}mv_1^2$	$PE_2 = mgy_2$ $KE_2 = \frac{1}{2}mv_2^2$	$g = 9.80$ m/s^2

The mathematical solution.

$$PE_1 + KE_1 = PE_2 + KE_2 \qquad (6\text{-}10)$$

becomes

$$0 + \tfrac{1}{2}mv_1^2 = mgy_2 + \tfrac{1}{2}mv_2^2$$

or

$$0 + \tfrac{1}{2}v_1^2 = gy_2 + \tfrac{1}{2}v_2^2$$

(Again, the result will not depend on the ball's mass.)
Solving for v_2 in steps, we get

$$v_2^2 = v_1^2 - 2gy_2$$
$$v_2 = \pm\sqrt{v_1^2 - 2gy_2}$$
$$= \pm\sqrt{(17.0\ \text{m/s})^2 - 2(9.80\ \text{m/s}^2)(10.0\ \text{m})} = \pm 9.64\ \text{m/s}$$

The negative solution gives the value of the downward velocity, but the speed (magnitude of velocity) is the same in either direction at this height:

$$|v| = \mathbf{9.64\ m/s}$$

◆ Related homework: Problems 6-23 and 6-24.

Note: As we saw in Example 6-2, KE depends only on speed, not velocity; the direction doesn't matter. KE is a *scalar* quantity, and so are all energies. Adding energies (as in Equation 6-10) thus requires only ordinary arithmetic and is much easier than adding vectors. This is a strong argument for applying energy concepts rather than Newton's laws where possible.

Case 6-4 ◆ *Two Ways of Getting to the Top of the Slope*

Jean-Claude Fastueux (nicknamed Fast), an Olympic skier, can get to the top of Slalom Slope, a vertical height h above the base, either by an elevator that rises through a vertical shaft in the mountain or by a chairlift that makes an angle θ with the horizontal (Figure 6-17). The weights of the elevator and chair are identical; the combined weight of Fast and either conveyance is mg. Both do the lifting at a constant very slow speed and stop at the top, so that essentially all the energy added to the Earth + skier + conveyance system is stored as PE (assuming negligible friction).

As the free-body diagrams show, the elevator cable must exert an upward force of magnitude mg through a displacement h. It does an amount of work $W = F\Delta s = mgh$. For constant velocity ($\Sigma \vec{F} = 0$), the force exerted on the chair support is $mg \sin \theta$, but this force must be exerted through the longer distance d, so that the work done here is $F\Delta s = (mg \sin \theta)d$.

But (see inset in Figure 6-17) $d \sin \theta = h$, so that again the work done (and the PE gained) is mgh. The gravitational PE gained is the same in either case; it depends only on the *vertical* displacement and not on the specific path by which that displacement is achieved.

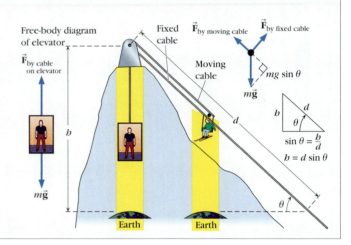

Figure 6-17 Gravitational PE depends only on the height (part I). This means it depends on the separation between the object and Earth.

Free-body diagram of chair support that rides along fixed cable

STOP&Think In the above case, does the *normal* force $\vec{F}_N$ exerted by the fixed cable on the chair support do any work on it? ◆

We can extend the reasoning of Case 6-4 piecewise to motion along a path of continuously varying steepness (Figure 6-18), so for a path of any degree of complexity, the change in gravitational PE depends only on the vertical displacement. Here is an example of how this is useful.

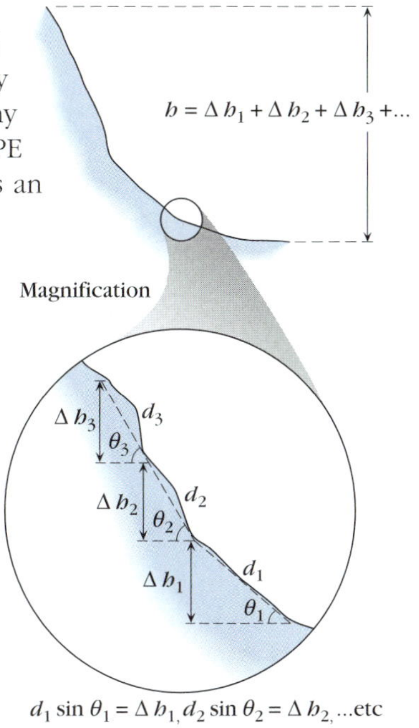

$$h = \Delta h_1 + \Delta h_2 + \Delta h_3 + \dots$$

Magnification

Figure 6-18 Gravitational PE depends only on the height (part II). Since it is true for each approximately straight piece of the curved slope (compare with Figure 6-17), it is true for the sum of these pieces, that is, the entire curved slope.

$$d_1 \sin \theta_1 = \Delta h_1, \ d_2 \sin \theta_2 = \Delta h_2, \dots \text{etc}$$

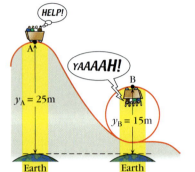

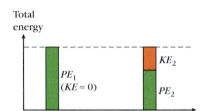

Total energy

Example 6-3 *The Holy Horror*

For a guided interactive solution, go to Web Example 6-3 at www.wiley.com/college/touger

On the Holy Horror, a roller coaster with a complete loop-the-loop (see figure), car and riders begin their descent from the highest point (A) at negligible speed. Assuming negligible friction effects, what is their speed at the top of the loop-the-loop (point B)?

Brief Solution

Choice of approach. We know enough about the car's position and velocity at points A and B so that we can use an energy approach. In the absence of friction, energy is conserved by the Earth + roller coaster system.

What we know/what we don't. We assume $y = 0$ at ground level.

State A	State B	Constants
$y_A = 25$ m $v_A = 0$	$y_2 = 15$ m $v_2 = ?$	$m = ?$
$PE_A = mgy_A$ $KE_A = \frac{1}{2}mv_A^2$	$PE_B = mgy_B$ $KE_B = \frac{1}{2}mv_B^2$	$g = 9.8$ m/s^2

The mathematical solution.

$$(Total\ Energy)_A = (Total\ Energy)_B$$

or

$$PE_A + KE_A = PE_B + KE_B \qquad (6\text{-}10)$$

becomes

$$mgy_A + 0 = mgy_B + \tfrac{1}{2}mv_B^2$$

or simply

$$gy_A + 0 = gy_B + \tfrac{1}{2}v_B^2$$

Solving for v_B^2, we get

$$v_B^2 = 2g(y_A - y_B)$$

so that

$$v_B = \sqrt{2g(y_A - y_B)} = \sqrt{2(9.8\ \text{m/s}^2)(25\ \text{m} - 15\ \text{m})} = \mathbf{14\ m/s}$$

◆ Related homework: Problems 6-26 and 6-27.

6-3 The Law of Conservation of Energy

In Section 6-2, we gave certain energy terms quantitative definitions, and we saw that in limited circumstances we could account for an increase in one kind of energy in a system either by a corresponding loss of another kind of energy (an energy conversion) or by an input of energy from outside the system (an energy transfer). The underlying assumption, which cannot be proven by mathematical derivations, is that we can keep track of energy because it is never simply created or destroyed. Rather, we have hit on something that all human experience and observation to date tells us exists in a fixed total amount in the universe as a whole. It is one of those things we believe to be a law of nature, that is, a fundamental rule by which nature behaves, because no one has come up with any verifiable evidence to the contrary.

The Law of Conservation of Energy The total amount of energy in the universe remains constant: Energy can change form but it can never be created or destroyed.

It follows that in an isolated system—one that cannot exchange energy with its surroundings—energy will also be conserved. Real systems are never *completely* isolated, though under certain conditions they may be isolated to a high degree of approximation. More generally, energy may be transferred to a system either as

- **heat** (denoted as Q): energy transferred by any mechanism from one region to another *because of* a temperature difference between the two regions, or
- *work done by an external force* (denoted by W_{ext}).

These quantities may be either positive (energy inputs) or negative (energy outputs). For example, energy may be transferred into our Earth + atmosphere system from the sun by *radiation* or out of a poorly insulated home by *conduction* through the walls, as well as by infrared *radiation*, which your eyes cannot detect but will expose infrared film. If you pull on a moving toy car from behind, gradually bringing it to a stop, you are doing negative work on that one-body system and reducing its energy.

In general, then, if E_1 and E_2 denote the total energy of a system in earlier and later states 1 and 2, then

$$E_1 + Q + W_{ext} = E_2 \qquad (6\text{-}12)$$

The total energy E includes the *mechanical energy* of the system (the sum of the whole-object potential and kinetic energies of all objects in the system) as well as the distributed or internal energies (thermal and chemical) of which we spoke earlier. Thus,

$$PE_1 + KE_1 + (E_{internal})_1 + Q + W_{ext} = PE_2 + KE_2 + (E_{internal})_2$$

<table>
<tr><td>Mechanical energy in state 1</td><td>Chemical and thermal energies in state 1</td><td>Mechanical energy in state 2</td><td>Chemical and thermal energies in state 2</td></tr>
</table>

Total energy of system in state 1	Total of energy inputs and outputs	Total energy of system in state 2	

$$(6\text{-}13)$$

If the system consists of more than one object, each energy term must be considered for each object in the system, as Figure 6-19 shows.

In general, when applying Equation 6-13, you need only include the inputs and outputs and those mechanical and internal energies that *change* from the earlier to the later state. For point objects, concerns about internal energy and temperature differences that cause heat exchange do not apply, and Equation 6-13 reduces to

$$PE_1 + KE_1 + W_{ext} = PE_2 + KE_2$$

or

$$W_{ext} = (PE_2 + KE_2) - (PE_1 + KE_1) = (E_{mechanical})_2 - (E_{mechanical}) \qquad (6\text{-}14)$$

Equation 6-14 is the **work-energy theorem** for point objects.

→**A note on language:** In ordinary language, when we talk about heat (in its everyday usage) and light from the sun, we are just responding to the fact that two different sense systems—touch and sight—are detecting the same radiation.

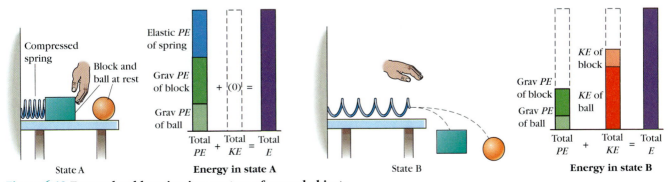

Figure 6-19 Energy bookkeeping in a system of several objects.

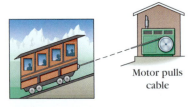

Motor pulls cable

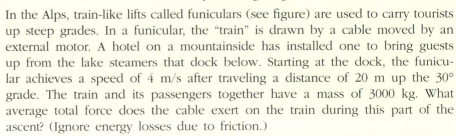

Example 6-4 *Work-Energy Applied to Alpine Lifts*

For a guided interactive solution, go to Web Example 6-4 at
www.wiley.com/college/touger

In the Alps, train-like lifts called funiculars (see figure) are used to carry tourists up steep grades. In a funicular, the "train" is drawn by a cable moved by an external motor. A hotel on a mountainside has installed one to bring guests up from the lake steamers that dock below. Starting at the dock, the funicular achieves a speed of 4 m/s after traveling a distance of 20 m up the 30° grade. The train and its passengers together have a mass of 3000 kg. What average total force does the cable exert on the train during this part of the ascent? (Ignore energy losses due to friction.)

Brief Solution

Choice of approach. We consider the work done by to this force and apply the work-energy theorem (Equation 6-14). Remember that PE depends only with on *vertical* position, which we can calculate as in Figure 6-17.

What we know/what we don't.

Work	State 1		State 2	Constants
$F_{ext} = ?$	$y_1 = 0$	$v_1 = 0$	$y_2 = (20\ m)(\sin 30°) = 10\ m$	$m = 3000\ kg$
$\Delta s = 20\ m$	$PE_1 = 0$	$KE_1 = 0$	$v_2 = 4\ m/s$	$g = 9.8\ m/s^2$
			$PE_2 = mgy_2 \quad KE_2 = \frac{1}{2}mv_2^2$	

The mathematical solution. By Equation 6-14,

$$W_{ext} = (PE_2 + KE_2) - (PE_1 + KE_1) = PE_2 + KE_2$$

$$F_{ext}\Delta s = mgy_2 + \frac{1}{2}mv_2^2$$

$$F_{ext} = \frac{m(gy_2 + \frac{1}{2}v_2^2)}{\Delta s}$$

$$= \frac{(3000\ kg)[(9.8\ m/s^2)(10\ m) + \frac{1}{2}(4\ m/s)^2]}{20\ m} = \textbf{16 000 N}$$

◆ Related homework: Problems 6-30 and 6-31.

◆**CONSERVATIVE AND NONCONSERVATIVE FORCES** We have seen that it is important to identify the system to which you apply energy principles such as conservation of energy or the work-energy theorem. In deciding which principle is applicable, it is necessary to determine whether there is any energy entering or leaving the system—for example, whether work is being done by objects outside the system exerting external forces on objects within the system.

Suppose that in Figure 6-20, an object ascends from point 1 to point 2 by any of the paths shown. The work done on the object by the gravitational force is the same for each path because Δy is always the same (recall Case 6-4). If the object returns to point 1 by the same path, the direction of each little displacement

Figure 6-20 Changes in gravitational PE depend only on Δy. Here Δy remains the same no matter which path is followed, so the change in gravitational PE from 1 to 2 is the same by any of these paths.

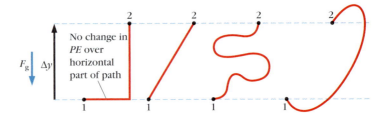

$\Delta \vec{s}$ is reversed, but the gravitational force is still downward, so the work done during the descent is the negative of the work done on the upward trip. Then the total work over a round trip via any one path is zero. Not only that, because the work was the same for each upward path, the object can go up by any path and descend by any other, and the total work done by the gravitational force (and hence the total change in PE) will still be zero. Forces like the gravitational force that do zero work on an object over any round trip are called **conservative forces.**

> A force is a **conservative force** if *the work done by this force is the same along any path connecting the same two points* or equivalently, if *the total work done by this force around any closed path (one that returns to its starting place) is zero.*

This condition is *not* met when sliding frictional forces are exerted. A sliding frictional force is always directed opposite to the relative motion (relative to the surface exerting the force), so the work done by a frictional force is always negative. Because the contributions ΔW for the different path segments are all negative, they cannot add up to zero. Frictional forces are therefore **nonconservative forces.**

◆**SITUATIONS INVOLVING FRICTIONAL FORCES** When a block sliding across a table slows to a stop, mechanical energy is converted into distributed forms of energy. Both block and table get warmer; both gain thermal energy. Then to keep track of the energy involved, we need to choose a system that includes the block *and* the table. Suppose the system is insulated from its surroundings, so that the heat exchange $Q = 0$. Then Equation 6-13 becomes

$$PE_1 + KE_1 + W_{ext} = PE_2 + KE_2 + \Delta E_{internal} \qquad (6\text{-}15)$$

where

$$\Delta E_{internal} = (E_{internal})_2 - (E_{internal})_1$$

$\Delta E_{internal}$ will be positive (an increase) when, for example, the block slows down on the table, that is, when the table exerts a frictional force on the block *opposite* to the block's motion, so that the work the table does on the block is negative. (The block exerts an equal and opposite frictional force on the table, but the table is stationary—its $\Delta s = 0$—so no work is done on it.) Thus, we can think of the change in internal energy as equivalent to an *equal and opposite* amount of work[1] done by nonconservative (e.g., frictional) forces F_{nc} within the system:

$$\Delta E_{int} = -(-f\Delta s) = f\Delta s$$

or, more generally,

$$\Delta E_{int} = \Sigma F_{nc}\Delta s$$

Expressed in terms of this work-like quantity, Equation 6-15 becomes the following.

> If the system chosen so that
>
> **1.** the nonconservative forces F_{nc} are *internal* forces, and
>
> **2.** the heat transfer into or out of the system is negligible ($Q = 0$),
>
> the **Work-Energy Theorem is**
>
> $$PE_1 + KE_1 - \Sigma F_{nc}\Delta s + W_{ext} = PE_2 + KE_2 \qquad (6\text{-}16)$$
>
> | Mechanical energy in state 1 | Changes in mechanical energy | Mechanical energy in state 2 |
> |:---:|:---:|:---:|

[1]This is not work in the strict sense of an energy input into a system, but is calculated as ordinary work is.

Example 6-5 *A Cliffhanger*

Star Cadet Kim is tackled by two hostile extraterrestrials 3.5 m from the edge of a sheer precipice. Because Kim's briefcase contains secret documents that mustn't fall into hostile hands, she desperately shoves the briefcase along the ground toward the edge. The ground exerts a sliding frictional force of 8.0 N on the briefcase (mass 2.8 kg, including contents). What minimum initial velocity must Kim give the briefcase to keep it from being captured?

Solution

Choice of approach. If we choose our system to include the briefcase and the ground, we can apply the work-energy theorem.

Restating the question. The velocity of the briefcase must be negligibly more than zero when it reaches the edge of the cliff. The question is thus $v_1 = ?$ if $v_2 \approx 0$ when $\Delta s = 3.5$ m.

What we know/what we don't.

State 1	Changes in Mech. Energy	State 2	Constants
$y_1 = 0 \quad v_1 = ?$	$W_{ext} = 0$	$y_2 = 0 \quad v_2 \approx 0$	$m = 2.8$ kg
$PE_1 = 0 \quad KE_1 = \frac{1}{2}mv_1^2$	$-\Sigma F_{nc}\Delta s = -f\Delta s$	$PE_2 = 0 \quad KE_2 = 0$	$g = 9.8$ m/s^2
	$f = 8.0$ N $\quad \Delta s = 3.5$ m		

The mathematical solution. Equation 6-16 becomes

$$KE_1 - f\Delta s = KE_2$$

or

$$\tfrac{1}{2}mv_1^2 - f\Delta s = 0$$

so that

$$v_1 = \sqrt{\frac{2f\Delta s}{m}} = \sqrt{\frac{2(8.0 \text{ N})(3.5 \text{ m})}{2.8 \text{ kg}}} = \textbf{4.5 m/s}$$

◆ Related homework: Problems 6-31, 6-32, and 6-33.

6-4   Another Conservative Force: The Elastic Force

In Section 4-4, we introduced the idea of using the displacement (stretching or compression) of a length of elastic material, such as a spring, as a measure of the force it exerts. If we suspend an object from a vertical spring, the object's weight downward will be equal in magnitude to the upward elastic force that the spring exerts on the object. Given any spring, we can hang various weights on it, measure the distance the spring stretches, and tabulate our measurements (Figure 6-21). We pick the x axis to be along the spring, so that the displacement x is one-dimensional. Often (but *not* always), as in Figure 6-21, the ratio of x to the suspended weight will remain constant over a considerable range of stretches. In such cases, we can write the proportionality relationship as $mg = kx$, where k is the proportionality constant. Because the elastic force F exerted *by* the spring is opposite in direction to the weight, and to the displacement that it causes, we can write:

$$F = -kx. \tag{6-17}$$

(F = force exerted *by* spring; x = spring's displacement)

This is called **Hooke's law.**

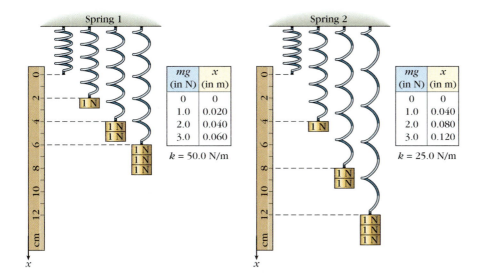

mg (in N)	x (in m)
0	0
1.0	0.020
2.0	0.040
3.0	0.060

$k = 50.0$ N/m

mg (in N)	x (in m)
0	0
1.0	0.040
2.0	0.080
3.0	0.120

$k = 25.0$ N/m

Figure 6-21 Displacements of springs with different spring constants.

Hooke's law is not really a law because it is not a universally observed behavior. Rather, it is observed to hold true for *some* lengths of elastic material for limited ranges of x. If too much weight is suspended, for example, the spring will permanently distort or break. At the other extreme, a certain minimum weight may be necessary to stretch the spring at all.

On-The-Spot Activity 6-2

See how much weight you can put on your bathroom scale or on a produce scale in a supermarket without producing any movement of the dial.

The proportionality constant k is called the *elastic constant* or *spring constant* and has units of newtons per meter (N/m) in SI units. We see in Figure 6-21 that k is greater for Spring 1, which is harder to stretch. The tables tell us that 2 N of weight are required to stretch Spring 1 a distance of 0.04 m, but only 1 N is required to stretch Spring 2 the same amount. Thus Spring 1 is tighter; more force is required for each cm that the spring is stretched. It is this property of the spring that is communicated by k.

Example 6-6 *The One that Got Away*

For a guided interactive solution, go to Web Example 6-6 at www.wiley.com/college/touger

Uncle Harry fishes with 20-N test monofilament fishing line, a type of line that has a little bit of give. He lets out 15 m of line and sets his bail, so that if a fish strikes, no further line unwinds. Under these conditions, if a fish exerts a force of 8 N on the line, the line will stretch 4×10^{-4} m.
a. What is the elastic constant of the 15 m length of line?
b. How much will this length of line stretch when a larger fish exerts a 16-N force on it? When the granddaddy of all fishes in the lake exerts a 200-N force on it?
c. If 30 m of line are let out, how much will the line stretch when a fish exerts a 16-N force on it?

Brief Solution
Choice of approach. This is an exercise in reasoning with Hooke's law. We don't care about direction here, so we choose Equation 6-17b.

EXAMPLE 6-6 *continued*

Restating the question. Part **a** asks, what is k if $x = 4 \times 10^{-4}$ m when $F = 8$ N? Part **b** asks, What is x when $F = 16$ N (or 200 N) if k = value found in **a**? In part **c**, you no longer have the same length of string, so that the value of k is changed. But each half of the 30-m length is a 15-m length of string. The tension force is the same anywhere along the length of the string, so the same 16-N force is exerted on each of the two 15-m lengths. We need only answer the question we asked in **b**, and then double our result. Notice that reasoning about the physics of the situation solves **c** more quickly than using the Hooke's law "formula."

The mathematical solution.

a. $$F = kx$$

or $$k = \frac{F}{x} = \frac{8 \text{ N}}{4 \times 10^{-4} \text{ m}} = \mathbf{2 \times 10^4 \text{ N/m}}$$

b. $$x = \frac{F}{k} = \frac{16 \text{ N}}{2 \times 10^{-4} \text{ N/m}} = \mathbf{8 \times 10^{-4} \text{ m}}$$

If $F = 200$ N, a similar calculation gives $x = \dfrac{200 \text{ N}}{2 \times 10^4 \text{ N/m}} = \mathbf{1 \times 10^{-2} \text{ m}}$

STOP&Think Does this result have any real-world meaning for the given situation? ◆

c. Twice the result from **b** is $2(8 \times 10^{-4} \text{ m}) = \mathbf{1.6 \times 10^{-3} \text{ m}}$

◆ Related homework: Problems 6-36, 6-37, 6-38, and 6-39.

◆**ELASTIC POTENTIAL ENERGY** To stretch a spring so slowly that it always has negligible KE, the force you exert must always be equal and opposite to the elastic force. This means you must exert a force close to zero when the spring is barely stretched at all and gradually build up to a force kx when the spring is at full stretch x. Assuming the spring obeys Hooke's law, the *average* force (averaged over distance, not time) that you exert is $\frac{0 + kx}{2} = \frac{1}{2}kx$, and thus the work you do is

$$W = F\Delta s = \left(\tfrac{1}{2}kx\right)x = \tfrac{1}{2}kx^2$$

Because x^2 is always positive, this result does not depend on the sign of x.

Suppose you stretch a vertical spring by adding weights to its lower end one at a time, then *remove* those same weights one at a time. If the spring shows the same stretch when it is supporting the same amount of weight whether it is being loaded or unloaded, the spring is genuinely *elastic*. The elastic force is the same at the same position x whether we measure it during expansion or contraction. But because the displacements are opposite during expansion and contraction, the total work the spring does as it is stretched from and returns to a given position is zero. The elastic force that the spring exerts is thus a conservative force. The work done *on* the spring each time an additional weight is added is stored in the system as potential energy. When you remove a weight, this work is in turn done *by* the spring on the remaining weights, converting the PE into KE. This connection with work leads to the following definition.

Elastic Potential Energy $PE_{\text{elas}} \equiv \tfrac{1}{2}kx^2$ (6-18)

Whether we apply conservation of energy (Equation 6-10) or the work-energy theorem (Equation 6-16) to a system, we must now keep in mind as we solve problems that the PE in each state can be the sum of more than one kind of PE:

$$PE_1 = (PE_{grav})_1 + (PE_{elas})_1 + \cdots \qquad (6\text{-}19a)$$

$$PE_2 = (PE_{grav})_2 + (PE_{elas})_2 + \cdots \qquad (6\text{-}19b)$$

Example 6-7 *The Spring Launcher*

For a guided interactive solution, go to Web Example 6-7 at
www.wiley.com/college/touger

A beginning physics student designs a model rocket (see figure) of mass 0.15 kg to be launched by compressing a very light spring with a spring constant of 450 N/m. Because the spring is wider than the opening of its enclosure, contact between rocket and spring terminates when the spring hits the rim of the opening (the spring's equilibrium position). If the spring is initially compressed 0.10 m, what is the launch velocity of the rocket?

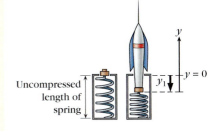

Brief Solution

Choice of approach. In applying conservation of energy, we identify a system (rocket + spring + Earth) that has both gravitational and elastic PE. We then identify two states of the system that can be well specified: In State 1, the spring is at maximum compression and the rocket at zero velocity. In State 2, the spring is back to zero compression as the rocket reaches its launch velocity.

What we know/what we don't.

State 1	State 2	Constants
$y_1 = -0.10$ m $\quad v_1 = 0$	$y_2 = 0 \quad v_2 = ?$	$m = 0.15$ kg
$(PE_{grav})_1 = mgy_1 \quad KE_1 = 0$	$(PE_{grav})_2 = 0 \quad KE_2 = \frac{1}{2}mv_2^2$	$g = 9.8$ m/s^2
$(PE_{elas})_1 = \frac{1}{2}ky_1^2$	$(PE_{elas})_2 = 0$	$k = 450$ N/m

The mathematical solution. Combining Equation 6-10,

$$PE_1 + KE_1 = PE_2 + KE_2$$

with Equations 6-19,

$$PE_1 = (PE_{grav})_1 + (PE_{elas})_1 = mgy_1 + \tfrac{1}{2}ky_1^2$$

$$PE_2 = (PE_{grav})_2 + (PE_{elas})_2 = 0 + 0 = 0$$

we get $\qquad mgy_1 + \tfrac{1}{2}ky_1^2 + 0 = 0 + \tfrac{1}{2}mv_2^2$

Solving for v_2, we get $v_2 = \sqrt{2gy_1 + \dfrac{ky_1^2}{m}}$

$$= \sqrt{2(9.8 \text{ m/s}^2)(-0.10 \text{ m}) + \frac{(450 \text{ N/m})(-0.10 \text{ m})^2}{0.15 \text{ kg}}}$$

$$= \mathbf{5.3 \text{ m/s}}$$

◆ Related homework: Problems 6-40 and 6-41.

6-5 Energy Rates: Power and Intensity

In many real-world situations, energy transfers happen gradually over time: the longer the time interval, the greater the transfer. For example if you leave the light on in your room for five hours, it will use five times as much energy as it would in one hour. The electric company must provide you and thousands of other consumers with five times as much electrical energy in five hours as in one, assuming consumer use remains constant over time. In five hours, five times as

➡**Consumer use of energy:** To simplify the discussion, we are assuming that consumer use remains constant throughout the day. In reality, an electric company must deal with different levels of usage at different times of day and have the capacity to generate energy at a rate sufficient to meet peak usage.

much water will pour through the company's hydroelectric plant transferring KE to the turbines to turn them.

When energy production or delivery is uniform, changes in the energy totals are proportional to the time elapsed; that is, ΔE is proportional to Δt. Thus, $\frac{\Delta E}{\Delta t}$ = a constant, and that constant can be interpreted as a *rate of energy input, change, or output*. This rate is of sufficiently widespread interest that it has its own name: **power.** Because ΔE is in joules (J) and Δt in seconds in SI units, power is in joules per second (J/s). A J/s is called a **watt** (W).

A warning about language: In physics, Equation 6-20 defines *power*. It has *no* other meaning. *Power* and *energy* are *not* interchangeable. *Power* is a rate, *energy* a total; *power* tells you how much each second, *energy* tells you how much in all.

Power	Definition:	$P \equiv \dfrac{\Delta E}{\Delta t}$	(6-20)
	Units:	$\text{watts} = \dfrac{\text{joules}}{\text{seconds}} \quad (\text{W} = \text{J/s})$	

Strictly speaking, this is a definition of *average power*. This definition suffices when the rate is constant. (We could otherwise define an instantaneous value of power much as we did for velocity and acceleration.)

Working through Examples 6-8 and 6-9 will help you connect power with other energy-related concepts.

Example 6-8 *An Uplifting Experience*

For a guided interactive solution, go to Web Example 6-8 at www.wiley.com/college/touger

A motor-driven lift raises parcels to a delivery truck's load platform 1.4 m above the ground. The motor is capable of delivering 50 W of power. Assuming no energy is dissipated, what minimum time is required for the lift to raise a 200-N parcel from the ground to the platform?

Brief Solution

Choice of approach. The lift does *work,* which is an energy input, so $\Delta E = W$. The total work that must be done is fixed. The power rating of the motor tells us the maximum *rate* at which it can provide energy. If you take more time to do the work, you won't need as large an input rate to produce the same total.

What we know/what we don't.

$$F = 200 \text{ N} \qquad \Delta s = 1.4 \text{ m}$$
$$P = 50 \text{ W} \qquad \Delta t = ?$$

The mathematical solution. Because the energy input equals the work done when no energy is dissipated, $\Delta E = W = F\Delta s$. Then the power $P = \frac{\Delta E}{\Delta t} = \frac{F\Delta s}{\Delta t}$. Solving for Δt gives us

$$\Delta t = \frac{F\Delta s}{P} = \frac{(200 \text{ N})(1.4 \text{ m})}{50 \text{ W}} = \textbf{5.6 s}$$

◆ Related homework: Problems 6-44 and 6-45.

Example 6-9 *Not a Bright Idea*

You have left a 100-W bulb burning in an unoccupied room for 24 hours.
a. How much energy has it used?
b. If the electric power company charges $0.1375 per kwh (kilowatt-hour)—they charge you for energy, not power—what is the cost of your carelessness?

Solution

Choice of approach. This is an exercise in using the definition of power. You also need to know that the **kilowatt-hour** is a unit of energy commonly used by electric power companies in the United States (check your electric bill). It is the total amount of energy used when you use energy at a rate of 1000 watts (or 1 kilowatt) for a 1-hour (h) interval; it is 1 kw × 1 h. Both parts require unit conversion.

The mathematical solution.

a. Because $P = \dfrac{\Delta E}{\Delta t}$, $\Delta E = P\Delta t$

$$= (100 \text{ W})(24 \text{ h}) = (100 \text{ J/s})(24 \text{ h})\left(\frac{3600 \text{ s}}{1 \text{ h}}\right) = \mathbf{8.64 \times 10^6 \text{ J}}$$

b. $(100 \text{ W})(24 \text{ h}) = (100 \text{ W})(24 \text{ h})\left(\dfrac{1 \text{ kw}}{1000 \text{ W}}\right) = 2.4 \text{ kWh}$

$$\text{Cost} = 2.4 \text{ kWh} \times \$0.1375/\text{kWh} = \mathbf{\$0.33}$$

◆ Related homework: Problems 6-43 and 6-46.

Another unit of power commonly used in the United States is the **horsepower** (hp):

$$1 \text{ hp} = 746 \text{ W}$$

A 200-hp internal combustion automobile engine converts 200(746 W)(1 s) = 149 200 J of chemical energy to other forms of energy each second.

When work is done on or by a point object due to a force in the direction of its motion, the object's energy change ΔE is just $\Delta W = F\Delta s$. In such cases, the power may be written as

$$P = \frac{\Delta W}{\Delta t} = \frac{F\Delta s}{\Delta t} = Fv \tag{6-21}$$

because $\frac{\Delta s}{\Delta t}$ is the speed v (uniform or average).

◆**ENERGY FLUX DENSITY** Suppose a solar panel is set out to receive energy from the sun. The solar panel will receive more energy if either we (1) increase its area or (2) expose it to the sun for a longer time interval. What remains constant is the amount of energy that reaches each unit of solar panel area each second. This rate is called the **energy flux density** (*flux* means flow). Figure 6-22 is an aid to picturing an energy flux density of 2 (J/m²)/s (2 Joules per square meter per second) or 2 W/s. We see that 2 J (the joules are represented by red dots) have fallen on each square (each m²) when the clock reads 1 s. A total of 8 J have fallen on the entire area of 4 m². If 2 J fall on each square during each subsequent second as well, then after 3 s, 6 J will "have" fallen on each square,

Solar panels are set up to capture radiated energy from the sun. Because the total power input is proportional to the area, large areas are required to provide sufficient power for applications such as heating water or living spaces.

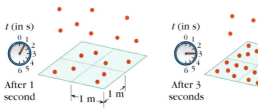

Figure 6-22 Visualizing an energy flux density of 2 J/m²/s.

t (in s)   After 1 second   ←1 m→ 1 m

t (in s)   After 3 seconds   (Each red dot represents 1 joule) for a total of 24 J on the entire area:

$$(2[J/m^2]/s)(4 m^2)(3 s) = 24 J$$

$$\underset{\substack{\text{Energy flux} \\ \text{density}}}{} \times \underset{\substack{\text{Area struck "head on"} \\ \text{by the flow}}}{} \times \underset{\substack{\text{Time} \\ \text{interval}}}{} = \underset{\substack{\text{Total} \\ \text{energy}}}{}$$ (6-22)

◆**INTENSITY, IRRADIANCE, INSOLATION** When the energy arriving at a surface is radiated energy, the energy flux density is commonly called **intensity** or **irradiance.** If it is arriving from the sun (*sol* in Latin), it is called **solar irradiance** or **insolation** (a different word from ins*u*lation). Because of the curvature of Earth's surface, the same amount of irradiated energy from the sun gets spread over a larger area near the poles than near the equator (see figure in Example 6-10), and so each unit of area gets less energy near the poles. That is why the polar regions are much colder. To visualize how this happens, examine WebLink 6-4.

The idea of solar irradiance is also used in the following example.

For **WebLink 6-4:**
Solar Irradiance, go to
www.wiley.com/college/touger

Example 6-10 *The Solar Constant*

For a guided interactive solution, go to Web Example 6-8 at
www.wiley.com/college/touger

If a disk equal in radius to Earth were placed just above the atmosphere facing the sun (see figure), that disk would, on average, be subjected to an irradiance of 1353 $(J/m^2)/s$ or 1353 W/m^2. This is the amount of solar energy intercepted by each square meter of Earth's cross-section each second and is called the **solar constant.** Not all this energy is absorbed at Earth's surface; an average of 34% is reflected back by clouds, water surfaces, snowy areas, and so on. How much energy is absorbed at Earth's surface each day?

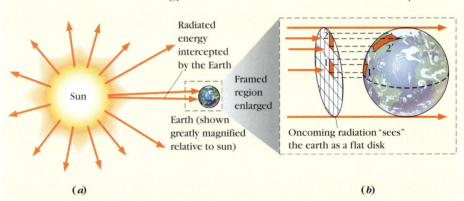

Radiated energy intercepted by the Earth

Framed region enlarged

Earth (shown greatly magnified relative to sun)

Oncoming radiation "sees" the earth as a flat disk

(a)   (b)

Solution

Choice of approach. Make use of Equation 6-22, noting that the area of a circular disk is πr^2, and that Earth's radius is 6.371×10^6 m.

The mathematical solution. By Equation 6-22, the energy reaching Earth's "disk" each day is

$$\text{solar constant} \times \pi r^2 \times 24 \text{ h} = (1353 \text{ J/m}^2/s)(3.14)(6.371 \times 10^6 \text{ m})^2(24 \text{ h})\left(\frac{3600 \text{ s}}{1 \text{ h}}\right)$$

$$= 1.49 \times 10^{22} \text{ J}$$

If 34% is reflected back into space, the remaining 66% is absorbed:

$$0.66(1.49 \times 10^{22} \text{ J}) = \mathbf{9.8 \times 10^{21} \text{ J}}$$

◆ Related homework: Problem 6-50.

The rain forest is a vast energy storage facility for living things. By using the solar energy input to carry out photosynthesis, green plants produce molecules that store the energy required by living cells. This is the beginning of the food chain for almost all life on Earth.

◆ SUMMARY ◆

In keeping track of energy-related quantities, it is generally valuable to focus on a **system,** a set of objects that we group together mentally for convenience. We defined the *work* done on an object by a force $\vec{F}$ as

---

Work

$$W \equiv \begin{array}{c} \textit{amount of force exerted} \\ \textit{in the direction of motion} \end{array} \times \begin{array}{c} \textit{distance over which} \\ \textit{the force is exerted} \end{array} \quad (6\text{-}1)$$

or $$W \equiv F\,\Delta s \cos\theta \qquad (6\text{-}2)$$

F is the *magnitude* of the force vector. Δs is the distance (the *magnitude* of the displacement $\Delta \vec{s}$). The sign of work depends only on whether $\cos\theta$ is $+$ or $-$.

(SI units: newtons · meters = **joules[J]**)

This work is an *energy input* (or *output* if negative) to the system containing the object.

In one dimension,

Work is $+$ when $\vec{F}$ and $\Delta \vec{s}$ in same direction (⟶)

(cos 0 = +1)

Work is $-$ when $\vec{F}$ and $\Delta \vec{s}$ in opposite directions (⇄)

(cos 180° = −1)

In suitable circumstances, work can be converted to other quantities that we can likewise define as types of energy of the object or system.

Kinetic Energy

$$KE \equiv \frac{1}{2}mv^2 \qquad (6\text{-}6)$$

Gravitational Potential Energy

$$PE_{\text{grav}} \equiv mgy \qquad (6\text{-}7)$$

when g is constant (e.g., over short distances from a planet's surface)

Elastic Potential Energy

$$PE_{\text{elas}} \equiv \frac{1}{2}kx^2 \qquad (6\text{-}18)$$

These may be understood as energies of *point objects,* or objects taken as a whole. You sum all of these whole-object energies to get the total **mechanical energy.**

Work and energy are scalar quantities.

PE is *stored energy.* It always depends on *position.* External work on a system is converted into PE when it is done against *conservative forces* that objects *within the system* exert on one another. As a result, the objects may end up positioned where the internal forces can set them in motion (turning PE into KE) when released from rest (as in Figures 6-1*a* and *b*).

A force is a **conservative force** if

- *the work done by this force is the same along any path connecting the same two points,* or equivalently,

- *the total work done by this force around any closed path (one that returns to its starting place) is zero.*

Conservative forces include gravitational forces (so that changes in gravitational PE depend only on changes in height, irrespective of the path; see Case 6-4) and elastic or Hooke's law forces.

Hooke's Law $\qquad F = -kx \qquad (6\text{-}17)$

(F = force exerted *by* spring; x = spring's displacement)

In contrast, the work done by **nonconservative forces,** such as sliding *frictional forces* on an object (which change direction as the object's path changes), depends on path length. The effect of such forces is frequently to convert energy from energies of whole objects to **internal** or **distributed energies,** such as **thermal energy** or **chemical energy.** Such energy is said to be **dissipated** and is readily transferable piecewise to the surroundings of a system. **Heat** refers solely to the energy passing into or out of a system because of a temperature difference between the system and its surroundings. It always represents a *change*—an *input* or *output*—in the energy of a system.

Once we define these quantities, we find, *not* by deriving it from prior knowledge of mechanics but as a rule or pattern that is found to hold true for all behavior thus far observed in the universe, that *energy is a conserved quantity.*

The Law of Conservation of Energy The total amount of energy in the universe remains constant: Energy can change form but it can never be created or destroyed.

Because energy is conserved in the universe as a whole, and remains constant in smaller *systems* insofar as they are isolated, energy concepts provide us with a way of analyzing the behavior of objects or systems of objects: We do so by keeping track of the energy involved in **energy conversions** from one kind of energy to another or in **energy transfers** between one object or system and another.

- For a system consisting of a single point object (so that there is no PE arising from forces objects in a system exert *on one another*),

$$W = \frac{1}{2}mv_2^2 - \frac{1}{2}mv_1^2 \qquad (6\text{-}5)$$

- When there is no energy input from outside the system,

$$(\textit{Total Energy})_1 = (\textit{Total Energy})_2$$

- If the objects in such a system can be treated as point objects (internal energies not relevant), then *mechanical energy is conserved.*

$$PE_1 + KE_1 = PE_2 + KE_2 \qquad (6\text{-}10)$$

or $$\Delta KE = -\Delta PE \qquad (6\text{-}11)$$

where $\quad \Delta KE = KE_2 - KE_1 \quad$ and $\quad \Delta PE = PE_2 - PE_1$

- For a system in which internal energies *do* matter, and where inputs and outputs *do* occur,

$$E_1 + Q + W_{\text{ext}} = E_2 \qquad (6\text{-}12)$$

or

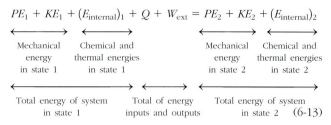

$$PE_1 + KE_1 + (E_{internal})_1 + Q + W_{ext} = PE_2 + KE_2 + (E_{internal})_2$$

| Mechanical energy in state 1 | Chemical and thermal energies in state 1 | | Mechanical energy in state 2 | Chemical and thermal energies in state 2 |

| Total energy of system in state 1 | Total of energy inputs and outputs | Total energy of system in state 2 | (6-13) |

If the system consists of more than one object, each energy term must be a total for all the objects in the system.

• For point objects (for which concerns about internal energy and temperature differences that cause heat exchange do not apply), Equation 6-13 reduces to *the work-energy theorem for point objects:*

$$PE_1 + KE_1 + W_{ext} = PE_2 + KE_2$$

or
$$W_{ext} = (PE_2 + KE_2) - (PE_1 + KE_1)$$

$$= (E_{mechanical})_2 - (E_{mechanical})_1 \qquad (6\text{-}14)$$

• In dealing with situations involving changes in *internal energy* due to **nonconservative forces,** such as *frictional forces,* it is wise to choose a system in which the frictional forces, say, are *internal forces* (i.e., such that the objects exerting frictional forces on each other are both in the system). You can then apply the work-energy theorem.

The **work-energy theorem** for a system chosen so that

1. the nonconservative forces F_{nc} are *internal* forces, and

2. the heat transfer into or out of the system is negligible ($Q = 0$):

$$PE_1 + KE_1 - \Sigma F_{nc}\Delta s + W_{ext} = PE_2 + KE_2 \qquad (6\text{-}16)$$

| Mechanical energy in state 1 | Changes in mechanical energy | Mechanical energy in state 2 |

In approaching energy problems and deciding which circumstances apply, follow Procedure 6-1.

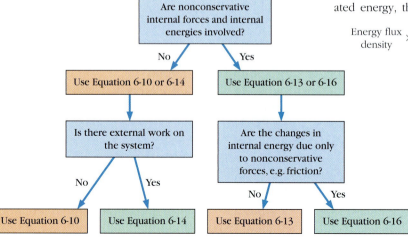

PROCEDURE 6-1 _____

An Approach to Energy Problems

1. Identify the *system* that you wish to consider, and decide what is *internal* to the system and what is *external*. If two objects exert restoring forces on each other, then
   a. if you want to consider the potential energy of the system due to those forces, include both objects in the system (so that the forces are internal forces)
   b. if you want to consider the work done by one of the two objects on the system, don't include that object in the system (so that the force it exerts is an external force).
2. Identify initial (or earlier) and final (or later) *states* of the system, characterized by positions (and thus by PEs) and velocities (and thus by KEs).
3. Decide whether you need to consider nonconservative forces and internal energies.
4. Decide whether external work is done on the system between its initial and final states.

You can then select equations by the decision-making scheme in Figure 6-23.

Sometimes it is useful to speak of *rates* of energy input, use, or output. One such rate is

Power   Definition of *average* power:

$$P = \frac{\Delta E}{\Delta t} \qquad (6\text{-}20)$$

SI units: **watts** $= \dfrac{\text{joules}}{\text{seconds}}$ $(W = J/s)$

1 **horsepower** or 1 hp = 746 W

When work is done on or by a point object due to a force in its direction of motion, the power is

$$P = \frac{\Delta W}{\Delta t} = \frac{F\Delta s}{\Delta t} = Fv \qquad (6\text{-}21)$$

Also useful in many contexts is the power per unit of area—the energy falling on each unit of area each second—which is called the **energy flux density,** or, in cases of radiated energy, the **intensity** or the **irradiance:**

$$\begin{array}{ccccc} \text{Energy flux} \\ \text{density} \end{array} \times \begin{array}{c} \text{Area struck "head on"} \\ \text{by the flow} \end{array} \times \begin{array}{c} \text{Time} \\ \text{interval} \end{array} = \begin{array}{c} \text{Total} \\ \text{energy} \end{array} \qquad (6\text{-}22)$$

Figure 6-23 Selecting energy equations. This chart shows the kinds of questions you must learn to ask on your own. Do not keep using this chart, but make sure you understand the reasoning in it.

◆QUALITATIVE AND QUANTITATIVE PROBLEMS◆
Hands-On Activities and Discussion Questions

The questions and activities in this group are particularly suitable for in-class use.

6-1. *Discussion Question.* For the sequence in Figure 6-11, discuss the additional energy inputs, transfers, and conversions necessary so that the energy stored in carbohydrate molecules formed by wheat growing in a field in Kansas and peanut plants growing in a field in Georgia becomes available to a softball player in Boston in the form of a peanut butter sandwich.

6-2. *Discussion Question.* Two identical blocks are simultaneously released from the same height above a level floor. Block A reaches the floor by dropping straight down. Block B reaches the floor by sliding down a frictionless ramp.

a. A student argues as follows: "At any time during their fall, the blocks should have the same kinetic energy and therefore the same velocity. If one is never going faster than the other and both are released from the same height at the same time, both hit the table top at the same instant." Do you agree with this student's conclusions? Explain why or why not.

b. Before you knew any physics, if someone described this situation to you, which block would you have expected to reach the floor first? Do your conclusions based on physics agree with this expectation? If not, try to identify the fallacy in the one you think is wrong. Explain your reasoning.

Review and Practice

Section 6-1 An Intuitive Introduction to Energy Ideas

In each of Problems 6-3 to 6-10 below, (1) identify a suitable system, and (2) describe the energy conversions within the system and the energy transfers into or out of the system that occur during the action(s) described.

6-3. You launch a toy airplane upward with a rubber band. It gets stuck in a tree on the way back down.

6-4. You jump up and down on a trampoline.

6-5. Your portly uncle flops down on a soft sofa.

6-6. You drop an egg. It splatters when it hits the floor.

6-7. You catch your friend's fastest pitch in your catcher's mitt.

6-8. You unplug an electric fan that has been running on high. The fan gradually slows to a stop.

6-9. **SSM** You push your little cousin on a swing.

6-10. You strike a match and it bursts into flame.

Section 6-2 Making Energy Concepts Quantitative

6-11. The definition of work (Equation 6-2) is interpreted by Student A to mean

    Work ≡ force exerted × component of distance
                             in the direction of the force

Student B says, "That's wrong; it disagrees with the word equation in the chapter [Equation 6-1]," but A replies, "No it doesn't. The two word equations are equivalent."
a. Which student is right? Briefly justify your answer.
b. After you have committed yourself on part **a**, look at Case 6-4 and decide whether it supports either student's position.

6-12. When an athlete performs a high jump, the gym floor and the athlete's feet exert equal and opposite forces on each other until the athlete's feet cease touching the floor. (Neglect microscopic effects.)
a. Does the floor do work on the athlete? Briefly explain.
b. Does the athlete do work on the floor? Briefly explain.
c. The athlete has a certain amount of kinetic energy at the instant of liftoff. Where does this energy come from?

6-13. **SSM** Push as hard as you can against the wall of your room for 30 s. (Neglect microscopic effects.)
a. Have you expended any energy?
b. Have you done any work?
c. Are your answers to **a** and **b** the same or different? Briefly explain why.

6-14. In the following table, for each force vector $\vec{F}$ and corresponding displacement $\Delta\vec{s}$, put a +, −, or 0 in the last column to indicate whether the work done is positive, negative, or zero.

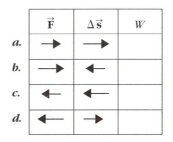 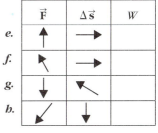

6-15. How would your answers to Problem 6-14 change if, instead of the usual coordinate frame, you choose a coordinate frame in which the positive *x* direction is toward the right but the positive *y* direction is downward?

6-16. A block subjected to the combination of forces shown in Figure 6-24 moves 2 m to the right.

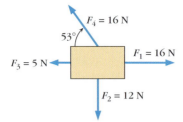

a. How much work is done on the block by each individual force?

b. What is the total work done on the block?

Figure 6-24 Problem 6-16

c. If the block moves along a frictionless surface, how much kinetic energy does it gain?

6-17. SSM WWW An airplane ascends in an eastward direction at a constant angle of 25° above the horizontal. As it does so, a wind blowing due west exerts a constant horizontal force of 550 N on it. How much work is done on the plane by this force for each 100 m the plane travels during its ascent?

6-18. Complete the following additional entries to Table 6-1.

Observed Behavior	Dynamic Description	Energy Description
As a thrown object rises, it loses speed.	?	?
As a shuffleboard disk slides across a wooden floor, it loses speed.	?	?

6-19. You are exerting a 10-N force on a block as it moves a distance 5 m to the right. In what direction(s) should you exert this force if the work that you do is

a. to be a maximum?

b. to have half its maximum possible value?

c. to equal zero?

d. to have the most negative value possible?

6-20. A ball of mass 0.2 kg is thrown straight upward with a velocity of 18.4 m/s. How much kinetic energy is lost as the ball rises from a height of 3.0 m to a height of 5.1 m?

6-21. Two rocks are dropped from a height of 2 m, one on the moon and one on Earth. (Assume the air resistance on Earth is negligible.)

a. Which rock strikes the ground at greater speed? Briefly justify your answer in energy terms.

b. Does your answer to **a** depend on whether the two rocks have equal masses? Briefly explain.

6-22. By what factor do you multiply the kinetic energy of a moving body if you

a. double its mass? **c.** reverse its direction?

b. double its speed? **d.** change its direction by 45°?

6-23. A spring mechanism launches a toy rocket from the ground. If the rocket reaches a maximum height of 7.2 m, with what velocity did the rocket leave the spring? Do not use kinematics to solve.

6-24. A ball is dropped from a window 19.6 m above street level.

a. What is its speed when it hits the sidewalk?

b. At what height does it achieve half this speed?

c. What is its speed when it has fallen half the distance?

6-25. SSM A 2-kg flowerpot falls from a 10-m-high window ledge.

a. How much energy is dissipated when the flower pot strikes the pavement?

b. How much work does the pavement do on the flowerpot?

6-26. A skier starts from rest at the top of a 50-m-long slope. The slope makes an angle of 20° with the horizontal. If friction is negligible, what is the skier's speed on reaching the bottom of the slope?

6-27. SSM A roller coaster car passes point A (Figure 6-25) at a speed of 2.0 m/s. If frictional forces are negligible, what is the speed of the roller coaster car at point B?

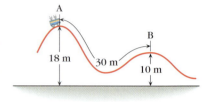

Figure 6-25 Problems 6-27 and 6-30

6-28. Two identical blocks are simultaneously released from the same height above a level floor. Block A reaches the floor by dropping straight down. Block B reaches the floor by sliding down a frictionless ramp.

a. Which block reaches the floor first? Briefly explain.

b. Which block arrives at the table with greater speed? Briefly explain.

Section 6-3 The Law of Conservation of Energy

6-29. A basketball dropped from a height of 2.0 m bounces back up to a height of 1.1 m. If the ball's mass is 0.6 kg,

a. how much energy is dissipated when it hits the floor?

b. with what speed does it bounce back from the floor?

6-30. Suppose now that the friction force on the roller coaster car in Problem 6-27 is no longer negligible. If riders fill the car to a total mass of 1100 kg, what is the maximum average frictional force to which the car could be subjected between A and B and still make it over the hump at B?

6-31. SSM WWW A standard hockey puck has a mass of 0.17 kg. In a hockey game on a frozen lake, the puck is hit into an area where the ice is rough. In this area, the ice exerts a sliding frictional force of 0.40 N on the puck. If the puck enters the area at a speed of 5.0 m/s, how far does it go before coming to a stop?

6-32. How fast would the puck in Problem 6-31 have to be going when it reaches the rough area in order to travel 11.0 m before coming to a stop?

6-33. A 0.006-kg bullet fired through a door enters at 800 m/s and leaves at 600 m/s. If the door material is known to exert an average resistive force of 5600 N on bullets of this type at usual speeds, find the thickness of the door.

•6-34. A test driver and her car have a combined weight of 5500 N. While driving on a straightaway at 32 m/s, she brakes suddenly to send her car into an intentional skid. After she brakes, the car travels another 14 m before stopping. How far would the car skid if she were traveling at half this speed when she braked?

6-35. A missile fired straight upward from ground level will, on return, hit the ground at the same speed with which it was fired.

a. Does this remain true even if the upward distance traveled is large compared to the radius of Earth? Briefly explain.

b. Does it remain true for small upward distances if the effect of air resistance is not negligible? Briefly explain.

Section 6-4 Another Conservative Force: The Elastic Force

6-36. An artificial Achilles' tendon for a prosthetic leg is made up of coupled springs that simulate the properties of the living tendon. One of these springs is reported to have a spring constant of 63 900 N/m. Suppose that in testing its properties, a lab technician suspends a block with a mass of 30 kg from this spring. How far does it stretch? (Data from G. K. Klute, J. M. Czerniecki, and B. Hannaford, *Chicago 2000 World Congress on Medical Physics and Biomedical Engineering.*)

6-37. A uniform spring has a spring constant of 90 N/m. If the spring is cut in half,

a. will the spring constant for each half be half as much, the same, or twice as much as for the original spring?

b. will the upward force that each half would exert on a 5-kg block that is hung from it be half as much, the same, or twice as much as the original spring would exert on the same block?

c. will the distance that each half would stretch when a 5-kg block is hung from it be half as much, the same, or twice as much as the original spring would stretch when the same block is hung from it?

6-38. A uniform spring is cut into two pieces. Piece A is longer than piece B. If identical weights are hung from each piece, will length A stretch more than, the same amount as, or less than length B?

6-39. Two lengths of stretch fabric, A and B, have spring constants $k_A = 400$ N/m and $k_B = 600$ N/m, respectively. If one end of A is sewn to an end of B and then the two lengths of fabric are pulled in opposite directions by their free ends, how much will B be stretched when A is stretched 0.09 m?

6-40. When the test dummy in Figure 6-26 is dropped from rest from the position shown, the spring compresses. The dummy comes to rest when the spring is exactly level with the ground. Find the elastic constant of the spring.

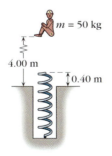

Figure 6-26
Problem 6-40

6-41. SSM WWW A 5.0-kg cart on a level surface is connected to a wall at one end by a spring with a spring constant of 500 N/m. The cart is pushed toward the wall, compressing the spring a distance of 0.10 m, and then released. What is the speed of the cart when the spring is at half its initial compression? Assume frictional forces are negligible.

Section 6-5 Energy Rates: Power and Intensity

6-42. When an appliance is running steadily, is it correct to speak about the appliance using a certain number of "watts per second"? Briefly explain.

6-43. How much energy (in SI units) is used by a 500-W hair dryer in 2 minutes?

6-44. A piston exerts an average force of 8500 N on the axle of a vehicle during a single stroke (Figure 6-27) of length 0.24 m. A stroke takes 0.06 s. How much power does the piston deliver to the axle?

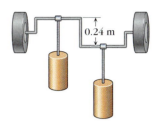

Figure 6-27 Problem 6-44

6-45. SSM A lift has a power rating of 480 W. How high can it raise a crate in 2.0 s if the crate weighs 1200 N?

6-46.

a. How much energy do you expend when you lift a barbell with a weight of 200 N (a little less than 50 lb) from the floor to above your head? Assume the total distance you raise it is 2.0 m.

b. How long would it take a 60-W light bulb to use the same amount of energy?

6-47. A popular press article reports that the new Inter-City Express of the German Federal Railways is "propelled forward by 13 000 horsepower" and "can reach speeds of up to 250 kilometers an hour."

a. What total force is exerted on the train in the forward direction when it is going at peak speed (assumed constant)?

b. What is the total resistive force on the train when it is traveling at this speed?

6-48. The light intensity reaching a wall 4 m wide by 3 m high is 5 W/m². How much energy reaches the wall in 20 s?

6-49. A 2 m by 3 m solar panel provides 1200 J of energy to the heating system of a solar home over a 10-s interval. The panel is 10% efficient (it converts 10% of the energy input into useful output). What is the average irradiance striking the panel over that interval?

6-50. Professor Touger's home, in the greater Boston area, was billed by the electric company for 516 kWh of energy for the month of December 2003. The average (over all hours of *day and night*) solar irradiance falling on a horizontal surface in Boston in December is 58 W/m². (*Note:* December is the worst-case scenario. The average irradiance for June and July in Boston is over four times as great.)

a. How large an area of solar panel, laid out flat and assumed 100% efficient, would Professor Touger need to provide all his household electrical energy for December?

b. Repeat *a* for 20% efficiency.

c. Which calculation is more realistic?

Going Further

The questions and problems in this group are not organized by section heading, so you must determine for yourself which ideas apply. Some of them will be more challenging than the Review and Practice questions and problems (especially those marked with a • or ••).

6-51. In the set-up in Figure 6-28, if the magnifying glass focuses the sunlight on the rope long enough, the rope will burn through. Everything else is stationary until this happens. Describe all the energy transfers and energy conversions that then take place.

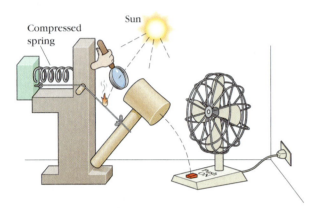

Figure 6-28 Problem 6-51

6-52. We can speak of both average and instantaneous forces. Can we also speak of average and instantaneous work? Briefly explain.

6-53. The table below shows the net force on an object (assumed constant) and the object's displacement for six situations (*a–f*). The values give the magnitudes of the vectors. Rank the situations in order of how much work is done on the object, starting with the value (lowest or most negative) that would occur furthest to the left on a real number line. Make sure to indicate equalities if there are any.

	$\vec{\mathbf{F}}$	$\Delta\vec{\mathbf{s}}$		$\vec{\mathbf{F}}$	$\Delta\vec{\mathbf{s}}$
a.	1 N →	2 m →	*d.*	2 N ↑	1 m ↓
b.	2 N ↙	1 m →	*e.*	1 N ↖	2 m ↓
c.	2 N ↑	2 m →	*f.*	2 N ←	2 m ←

6-54. Does it require the same amount of work to reduce your speed by 1 m/s no matter how fast you are going? Briefly explain.

6-55. In each of the situations graphed in Figure 6-29, a force F_x (in N) varies as it is exerted on an object moving from $x = -5$ m to $x = 5$ m. For each situation, find the total work done on the object by the force F_x.

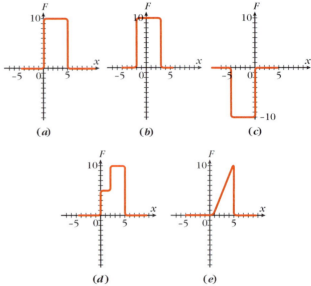

Figure 6-29 Problem 6-55

6-56. Two objects of unequal mass start out at rest. Equal, unopposed forces of magnitude F are then exerted on the two objects over the same time interval Δt. Student A claims, "The lighter [less massive] object gains more kinetic energy because a force has a greater effect on a lighter object." Student B argues, "How can that be? Kinetic energy is $\frac{1}{2}mv^2$, so if an object has less mass, it should have less kinetic energy, not more." Which of these students is right? What is wrong with the other student's argument?

6-57.
a. How can a moving object change its velocity without changing its kinetic energy?
b. Can the same object change its kinetic energy without changing its velocity? Explain.

6-58. A bottle falls from a window ledge a height h above the sidewalk. When it is at a height $\frac{1}{2}h$,
a. what fraction of its maximum kinetic energy has it achieved? (Assume air resistance is negligible.)
b. what fraction of its maximum speed has it achieved?

6-59. **SSM WWW** A student's physics textbook has twice as much mass as his psychology textbook.
a. Suppose the two books are initially at rest on a frictionless horizontal surface. Compare the kinetic energies of the two books after equal horizontal forces have been exerted on the two books for one second. Express the comparison by an equation $KE_{physics} = \underline{\ ?\ } \times KE_{psych}$, with the correct numerical multiplier in the blank.
b. Suppose the student, after a frustrating evening of study, releases both books from rest at the same instant from an upper story window of a high-rise dorm. Assume air resistance is negligible. Compare the kinetic energies of the two books after both have fallen for one second ($KE_{physics} = \underline{\ ?\ } \times KE_{psych}$).
c. Comment on any similarity or difference between the situations in *a* and *b*.

6-60. A student's physics textbook has twice as much mass as his psychology textbook. Think about whether the kinetic energies of the two books are equal or not after 1 second in situations A and B below, and comment on how the two situations are similar or different in this regard.

Situation A: The two books are initially at rest on a frictionless horizontal surface. Equal horizontal forces are then exerted on the two books for 1 second.

Situation B: The student, after a frustrating evening of study, releases both books from rest at the same instant from an upper-story window of a high-rise dorm. Assume air resistance is negligible, and the books fall freely for 1 second.

6-61. A projectile is fired straight upward at a speed of 50 m/s from the edge of a cliff 20 m above a river. It just misses the cliff on the way down. With what speed does it land in the river?

6-62. A rock climber on the face of a sheer escarpment fires a flare directly upward at a speed of 20 m/s. It reaches the bottom of the escarpment at a speed of 28 m/s. How high up was the climber when the flare was fired?

6-63. Suppose you raise a book of mass m from the floor to the top of a bureau of height b. As you do so, Earth exerts a gravitational force on the book and does work on it. What is the relationship between this work and the change in gravitational potential energy of the book + Earth system? Identify any assumptions in your reasoning.

6-64. A house cat brushes against a can of soup on a high shelf just hard enough to topple the can off the shelf so that it falls to the kitchen floor below. The can has considerable kinetic energy just before it hits the ground.

a. How does the amount of kinetic energy gained by the can compare with the amount of work that the cat does on the can?

b. What other object does work on the can during this scenario?

c. How does the amount of work done by this object compare with the amount of work that the cat does on the can?

•**6-65.** What must be the velocity of the swinger in Figure 6-30 at the bottom of her swing, to reach a maximum angle of 45° with the vertical, as shown?

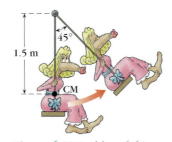

Figure 6-30 Problem 6-65

6-66. A cable raises an elevator in a high-rise building from the ground floor. The mass of the elevator and its passengers is 600 kg. Over the first 10 m of the elevator's ascent, the cable exerts an average force of 6800 N on it. During this stretch, what percent of the work done by the cable is used to raise the elevator and what percent is used to accelerate it from rest?

6-67.

a. A certain force on an object always has the same magnitude and direction no matter which way the object moves; this force is (*always, sometimes,* or *never*) a conservative force.

b. Some springs remain slightly stretched after very large weights that have been suspended from them are removed; the elastic forces these springs exert are (*always, sometimes,* or *never*) conservative forces.

6-68. A block bobs up and down on a vertical spring. As the block descends from its highest to its lowest point,

a. the total upward force on the block is (*always, sometimes,* or *never*) increasing.

b. the total potential energy is (*always, sometimes,* or *never*) increasing.

6-69. An ice skater gliding frictionlessly across the ice at a speed of 10.0 m/s encounters a 4.0-m-long rough spot and is slowed down to a speed of 9 m/s. (If you learned about coefficients of friction, do Version 2; if not, do Version 1.)

Version 1: If the skater's mass is 70 kg, find the frictional force that the rough spot exerts on the skater's skates.

Version 2: Assume the skater's mass is unknown. Find the coefficient of kinetic friction between the rough spot and the skater's skates.

6-70. At the rate charged in Example 6-9 ($0.1375/kWh), what is the cost of 1 J of energy?

6-71. Estimate the kinetic energy of

a. a 9-month-old baby crawling across a living room floor.

b. the moon ($m_{moon} = 7.35 \times 10^{22}$ kg) in its orbit around Earth.

6-72.

a. Imagine a 40-W bulb to be at the center of a transparent sphere of radius 1 m. Recalling that the surface area of a sphere is $4\pi r^2$, what is the irradiance at (or crossing) the sphere?

b. Repeat **a** for a sphere of radius 2 m, then for a sphere of radius 3 m.

c. If the irradiance is proportional to r^p, where r is the distance from the bulb, what is the value of the exponent p?

6-73. **SSM** When radiated energy reaches an object, the area that the object blocks (like the disk in WebLink 6-4 and Example 6-11) is called the object's *cross section*. If the irradiance falling on Earth's cross section from the sun is 1353 W/m², and the mean radius of Earth's orbit around the sun is 1.5×10^{11} m, what is the sun's total power output?

6-74. A team of space explorers has boldly gone to a planet where no one has gone before. One of them finds a rock and throws it upward with an initial speed of 10 m/s. If the rock rises to a height of 2.5 m, what is the gravitational acceleration at this planet's surface? Neglect air resistance.

6-75. A block weighing 98.0 N is placed on a level surface, pushed horizontally against a spring with a spring constant of 1200 N/m, and released.

a. If the surface exerts a sliding frictional force of 40 N on the block when it is in motion, how much must the spring be compressed to give the block a maximum acceleration of 2 m/s²?

b. What is the acceleration of the block when the spring is at half its maximum compression?

c. During the block's motion, is there any spring displacement at which the block is in equilibrium? If so, find it.

•**d.** At what spring displacement does the block come to a stop?

•**6-76.** To fire a projectile of mass m upward from a spring gun, you push the projectile down against a spring of elastic constant k and then let it go. In terms of k and m, how far down must you compress the spring so that on release the projectile will go further upward than the end of the spring?

6-77. In the specs on an electric jigsaw, its maximum mechanical energy output is given as "Max. HP: $\frac{1}{7}$" (1 horsepower = 1 HP = 746 W). The higher of its two "speeds" is given as 3000 spm (strokes per minute).
a. How much work is done by the saw each minute when operating at maximum output?
b. If the blade moves 1 cm during each stroke, what average force does it exert under these conditions?

6-78. A speeding driver slams on the brake and comes to a screeching halt just in time to avoid hitting an oblivious jogger.
a. At the instant the car stops, has all the energy that was initially the car's kinetic energy been transferred to the car's surroundings?
b. Does any part of this energy get transferred *after* the car stops? Briefly explain.

6-79. The glider in Figure 6-31 is connected to the left end of a horizontal air track, on which it moves frictionlessly, by a light spring with an elastic constant $k = 11.103$ N/m. The glider's mass is 0.28125 kg. The glider is pulled until it is centered at point B, 0.02000 m to the right of its equilibrium position at point A, and is then released. What is the glider's velocity when its center reaches point A?

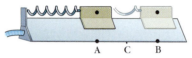

Figure 6-31 Problems 6-79 and 6-85

•6-80. To understand energy policy issues, you often have to make sense of energy information provided by a variety of sources in a variety of units, as in the following case. A century-old book on steam power gives the following specifications (in deplorably un-SI units) for the U.S.S. *Brooklyn,* which saw service during the Spanish-American War:

> Length: 400'6"
> Displacement: 8150 tons
> Total mean hp of machinery: 18 769.62
> Speed in knots: 21.912
> Typical coal use per hp: 2–2.5 lb./hour

a. To maintain the *Brooklyn* at steady speed, what resistive force did its machinery have to overcome, assuming that the total power output served this purpose?
b. If the energy content of bituminous coal is about 30×10^6 J/kg, approximately how efficient was the *Brooklyn's* machinery in converting chemical energy to mechanical energy? Express your answer as a percent.

6-81. A company manufactures square solar panels to gather energy from the sun. They provide customers with a table (first two columns below) for determining the maximum amount (M) of electrical energy per second that can be provided by a panel having a side of length l.

$M\left(\text{in } \dfrac{\text{J}}{\text{m}^2 \cdot \text{s}}\right)$	l (in m)	l^2 (in m^2)
17	0.25	
70	0.50	
157	0.75	
280	1.00	

a. Complete the third column of the table.
b. Plot graphs of M versus l and M versus l^2, and find the slope of the one that is a straight line.
c. To what power of l is M directly proportional?
••d. Can you suggest a model or picture to explain why M depends on this particular power of l?

6-82. The room in Figure 6-32 is lit only by a single suspended light bulb. How does the light intensity on the wall at point A compare with the light intensity on the opposite wall at point B? Answer by finding the numerical multiplier in $I_A = \underline{\ ?\ } \times I_B$.

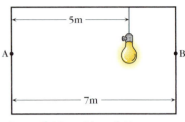

Figure 6-32 Problem 6-82

6-83. Look up the average distances of Earth and Mars from the sun. Based on this information, how does the intensity of solar radiation striking Earth compare with the intensity of solar radiation striking Mars? Answer by finding the numerical multiplier in $I_E = \underline{\ ?\ } \times I_M$.

6-84. A ball is thrown up into the air and allowed to fall back to the ground.
a. Is the work that Earth does on the ball positive or negative over the time interval when the ball is rising?
b. Is it positive or negative over the time interval when the ball is falling?

6-85. In Figure 6-31, a glider on a frictionless air track is attached to a very light spring and is in equilibrium when its center is at point A. It is then pulled to the right until its midpoint is at B, and released. If point C is halfway between A and B,
a. how does the elastic potential energy when it is at C compare with the elastic potential energy when it was at B? Answer by finding the numerical multiplier in $PE_C = \underline{\ ?\ } \times PE_B$.
b. how does the glider's kinetic energy when it is reaches A compare with the kinetic energy it had at C? Answer by finding the numerical multiplier in $KE_A = \underline{\ ?\ } \times KE_C$.

6-86. Tell whether each of the following can increase the potential energy of a system.
a. moving a positively charged object and a negatively charged object apart
b. moving two positively charged objects apart
c. work done by frictional forces
d. pulling down on a spring suspended from a ceiling

6-87. A 32-kg child on a sled is going at a speed of 7 m/s when he passes point P in Figure 6-33. If he is still going at the same speed when he reaches the bottom of the hill, how much work is done on him (and the sled) by the resistive forces opposing his motion as he goes from P to the bottom?

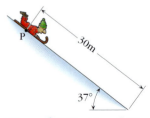

Figure 6-33 Problem 6-87

6-88. When you ignite a pile of newspapers with a match, far more energy is given off than you would get from the match alone. Where did the rest of the energy come from?

6-89. If there is energy stored in a tank of propane, a common fuel, why do you need to put energy in by igniting it to get the stored energy out?

6-90. A 1600-kg truck weighs about 2 tons.
a. Is it possible for a lift with a power rating of 0.5 W to raise this truck 1 m above the ground? If so, briefly explain how; if not, briefly explain why not.
b. How long would it take this lift to raise the truck 1 m above the ground?
c. Convert your answer to **b** into hours.

6-91. Suppose that in Figure 6-34, radiated energy arrives in the positive x direction and that the irradiance is uniform. Consider the three rectangles ABCD (rectangle I),

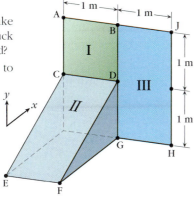

Figure 6-34 Problem 6-91

CDEF (rectangle II), and BGHJ (rectangle III). Rank these rectangles in order of how much radiated energy they receive each second. Make sure to indicate any equalities.

6-92. A delivery truck facing east pulls straight forward a distance of 5.0 m. As it does so, three constant forces are exerted on it. The first is in the forward direction and has a magnitude of 800 N, the second is in the backward direction and has a magnitude of 500 N, and the third is directed due west and has a magnitude of 300 N. How much total energy is transferred to the truck by these forces as it pulls forward?

6-93. SSM A kiddie ride in an amusement park consists of a miniature fire truck traveling around a circular track with a radius of 7.5 m. The truck is kept on the track by a cable connected to a post in the center of the circle. As the truck makes one complete circle, the cable exerts a force of average magnitude 90 N on the truck, and the track exerts a frictional force of constant magnitude 12 N opposing the truck's motion. During the completion of this one lap, how much work is done on the truck by the
a. frictional force?
b. force exerted by the cable?

6-94. A child does 200 J of work pushing her dog in a baby carriage. As this happens, how much work does the carriage do on the child?

Problems on WebLinks

6-95.
a. In WebLink 6-1, if the positively charged sphere is raised toward the hand holding the negatively charged sphere, is the potential energy of the system increased, decreased, or unchanged?
b. If the hand releases the positively charged sphere from this new position, will it gain more KE than before, less KE than before, or the same amount of KE as before?
c. In this new situation, is the amount of PE lost more than, equal to, or less than the amount of KE gained?

6-96. The energy of a system made up of Earth and a rock changes when the rock is pushed up a hill and allowed to lodge in a shallow depression at the top of the hill. This is analogous to the changes that occur in the system of atoms involved in which of the following chemical reactions in WebLink 6-2?
a. $2H_2 + C \rightarrow CH_4$ **c.** both
b. $CH_4 + O_2 \rightarrow CO_2 + 2H_2O$ **d.** neither

6-97. Consider the system of carbon (C), hydrogen (H), and oxygen (O) atoms involved in the chemical reaction in WebLink 6-2 in which methane (CH_4) is burned. Is the energy of this system greatest when the atoms are joined to form CH_4 and O_2, when they are joined to form CO_2 and $2H_2O$, or when they are separated from one another?

6-98. In the derivation in WebLink 6-3, $\bar{v}\Delta v$ is equal to which of the following expressions?
a. $v_1 v_2$ **c.** $v_2^2 - 2v_1v_2 + v_1^2$
b. $v_2^2 + v_1^2$ **d.** $v_2^2 - v_1^2$

6-99. In the derivation of $W = \frac{1}{2}mv_2^2 - \frac{1}{2}mv_1^2$ in WebLink 6-3, which of the following is *not* used?
a. the definition of work
b. the definition of kinetic energy
c. the definition of acceleration
d. the definition of average velocity
e. Newton's second law

6-100. WebLink 6-4 shows both Earth's surface and a flat disk representing what radiated energy reaching Earth "sees." Is the irradiance the same for every square unit of area on both of these, on Earth's surface only, on the disk only, or on neither?

6-101. In Figure 6-35, light traveling to the left passes through identical square openings A_1 and B_1. The light passing through A_1 then falls on region A_2 on the surface of a model of Earth. The light passing through B_1 falls on region B_2. According to the reasoning developed in WebLink 6-4,

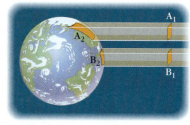

Figure 6-35 Problem 6-101

a. is the irradiance over region B_2 less than, equal to, or greater than the irradiance over region A_2?
b. is the total amount of energy reaching region B_2 each second less than, equal to, or greater than the total amount of energy reaching region A_2 each second?

CHAPTER SEVEN

More Bookkeeping: Collisions and the Concept of Momentum

The events of the physical universe are played out in an arena of time and space. When a force is exerted on an object over some distance, that is, an *interval of space,* there is an input of a quantity we call energy, which turns out to exist in an unchanging amount in the universe as a whole or in an isolated system. We can also consider the exertion of a force over an *interval of time,* such as the brief interval for which objects remain in contact during a collision. In doing so, we shall see that there is an input of another quantity, called *momentum,* which also exists in an unchanging amount in the universe. Momentum, like energy, is a conserved quantity. As we saw for energy, we will see in this chapter that keeping track of momentum as systems go from initial to final states is a powerful calculational and reasoning tool.

"When we consider the exertion of a force over an interval of time, there is an input of another quantity, called momentum."

7-1 Impulse and Momentum

◆**IMPULSE** Much as we defined work to be $(F\cos\theta)\Delta s$, the product of a space interval and the force exerted in the direction of that interval, we define **impulse** to be the product of a time interval and a force exerted over that time interval:

$$\text{Impulse} \equiv \vec{F}\Delta t \qquad (7\text{-}1)$$

Impulse, as the product of a vector and a scalar, is itself a vector quantity. The impulse associated with a force $\vec{F}$ is in the same direction as $\vec{F}$.

If a force is exerted on an object for a longer time interval, it has more effect on the object. Impulse is a way of quantifying this effect. In fact, our intuitive understanding of forces is really an understanding of impulses: We do not directly see or experience the effect of a force at an instant but rather its cumulative effect over an interval (see Problem 7-4). As Figures 7-1 and 7-2 show, the effect of large forces can be very considerable, even over one or a few milliseconds when a golf club remains in contact with a ball or a foot with a fully inflated football. Indeed, these effects are more than meets the eye, because the human visual system is incapable of responding to such short time exposures.

◆**IMPULSE AND NEWTON'S SECOND LAW: INTRODUCING MOMENTUM**
Using Newton's second law and the definition of (average) acceleration, we can rewrite Equation 7-1. Assuming that the mass m remains constant,

$$\text{Impulse} = \vec{F}\Delta t = m\vec{a}\Delta t = m\frac{\Delta\vec{v}}{\Delta t}\Delta t = m\Delta\vec{v} = m\vec{v}_2 - m\vec{v}_1 = \Delta(m\vec{v}) \qquad (7\text{-}2)$$

This result is actually more general than the conditions for which we derived it. $\Delta(m\vec{v}) = m\Delta\vec{v}$ only when the mass is constant; more generally, it allows for the possibility that m can change. Equation 7-2 says that the impulse equals the change in a quantity $m\vec{v}$. This quantity depends not only on an object's velocity but also on how much mass is moving at that velocity. For that reason, it was at one time called the *quantity of motion*. In the language of physics today, we call it **momentum**, and denote it by $\vec{p}$.

Definition: **Momentum**

$$\vec{p} \equiv m\vec{v} \qquad (7\text{-}3)$$

Figure 7-1 A collision may be "quicker than a wink"—much quicker. A well-timed strobe flash lasting only 1/100 000 s captures the kicker's foot deforming the football as the human eye can never see it, because our eye and brain cannot detect anything with a duration much less than about 1/15 s.

Figure 7-2 An impulse occurring over a millisecond time interval. The duration of this sequence captured by strobe photography is a couple of thousandths of a second. (*a*) A golf ball struck by a golf club is visibly compressed while in contact with the club head. (*b*) The ball is momentarily elongated as it springs back beyond its normal outline.

Equation 7-2 then says

Impulse on a body = the change in the body's momentum

$$\vec{F}\Delta t = \Delta \vec{p} \tag{7-4}$$

This is called the **impulse-momentum theorem.**

➡Because $\vec{F}\Delta t = \Delta(m\vec{v})$ is more general than $\vec{F}\Delta t = m\Delta\vec{v}$, a more general form of Newton's second law is $\Sigma\vec{F} = \frac{\Delta(m\vec{v})}{\Delta t}$ or $\Sigma\vec{F} = \frac{\Delta\vec{p}}{\Delta t}$.

In component form, this says:

$$F_x\Delta t = \Delta p_x \qquad F_y\Delta t = \Delta p_y \tag{7-4x, 7-4y}$$

For a point object in one dimension (so that we can drop subscripts), $F\Delta t = \Delta p$ is analogous to the work-energy theorem, which says that the work on a point object equals the change in the body's energy ($F\Delta s = \Delta E$).

Example 7-1 *How Hard Is Hard?*

For a guided interactive solution, go to Web Example 7-1 at www.wiley.com/college/touger

A Major League pitcher's fastball often exceeds 40 m/s (about 90 mi/h). A strong fastball hitter can line the ball back at the pitcher as fast as it was pitched. Assuming the ball both reaches and leaves the bat with a speed of 40 m/s, and the bat remains in contact with the ball for 1.0 ms (1.0×10^{-3} s), what average force does the bat exert on the ball during this contact (i.e., how hard is the ball hit)? A baseball has a mass of about 0.15 kg.

Brief Solution

Choice of approach. We apply the impulse-momentum theorem. Note that although the speed is the same before and after, the velocity reverses direction. We need to choose opposite signs for the two directions. For convenience, we will take the "after" direction to be positive.

What we know/what we don't.

$$m = 0.150 \text{ kg} \qquad v_1 = -40 \text{ m/s} \qquad v_2 = +40 \text{ m/s} \qquad \Delta t = 0.0010 \text{ s} \qquad F = ?$$

The mathematical solution. In one dimension, we can write Equation 7-2 as

$$F\Delta t = m\Delta v = m(v_2 - v_1)$$

so that

$$F = \frac{m(v_2 - v_1)}{\Delta t} = \frac{(0.15 \text{ kg})(40 \text{ m/s} - |-40 \text{ m/s}|)}{0.0010 \text{ s}}$$

$$= \textbf{12 000 N} \text{ or } \textbf{1.2} \times \textbf{10}^{\textbf{4}} \textbf{ N}$$

Making sense of the result. Because a newton is a little less than a quarter of a pound, this is well over a ton!

◆ Related homework: Problems 7-6 and 7-7.

◆**COLLISIONS AND NEWTON'S THIRD LAW** When two objects interact, the forces that they exert on each other are equal at every instant. This is Newton's third law. Collisions are interactions that usually involve contact forces,[1] as is

---

[1]But when two objects that repel each other, such as the two magnets in Figure 4-6, are given initial motions toward each other, they might "bounce" back from each other without ever actually touching.

Figure 7-3 **Table tennis is all about collisions.**

emphatically the case in Figure 7-1. At each instant during the interval of contact, the contact forces that two objects A and B exert on each other are equal and opposite. Review Example 4-2 to satisfy yourself that this is really true. This also means that the paddle and the ball in Figure 7-3 exert equal forces on each other, although the motion of the much less massive ball is affected (accelerated) far more in the collision. **STOP&Think** In Figure 7-4, how did the forces that the comet and Jupiter exerted on each other compare? How did their accelerations compare? ◆

Because A and B exert equal and opposite forces on each other throughout any time interval Δt (necessarily the same interval for both forces), the impulses on A and B must also be equal and opposite:

$$\vec{\mathbf{F}}_{\text{on A by B}}\Delta t = -\vec{\mathbf{F}}_{\text{on B by A}}\Delta t$$

But because $\quad \vec{\mathbf{F}}_{\text{on A by B}}\Delta t = \Delta\vec{\mathbf{p}}_{A} \quad$ and $\quad \vec{\mathbf{F}}_{\text{on B by A}}\Delta t = \Delta\vec{\mathbf{p}}_{B}$

$$\Delta\vec{\mathbf{p}}_{A} = -\Delta\vec{\mathbf{p}}_{B} \tag{7-5}$$

or $\qquad \vec{\mathbf{p}}_{A2} - \vec{\mathbf{p}}_{A1} = -(\vec{\mathbf{p}}_{B2} - \vec{\mathbf{p}}_{B1})$

(letter subscripts denote bodies, number subscripts denote instants or states)

which you can show by a little algebra is equivalent to

$$\vec{\mathbf{p}}_{A1} + \vec{\mathbf{p}}_{B1} = \vec{\mathbf{p}}_{A2} + \vec{\mathbf{p}}_{B2}$$

or $\qquad\qquad \Sigma\vec{\mathbf{p}}_{1} = \Sigma\vec{\mathbf{p}}_{2} \tag{7-6}$

On-The-Spot Activity 7-1

Do the above-mentioned algebra for yourself.

Figure 7-4 **Comet Shoemaker-Levy collision with Jupiter.** From July 16 through July 22, 1994, pieces of Comet Shoemaker-Levy 9 collided with Jupiter. This is the first collision of two solar system bodies ever to be observed.

Equation 7-5 says that the amount of momentum that A gains is the negative of what B gains—it's what B loses. So the total momentum of the two objects is the same before and after the collision. In other words, the total momentum of the system of two objects is conserved, and Equation 7-6 is a statement of **conservation of momentum.**

As we did for energy, we must think of momentum as conserved by a system. If we think of a system with just a single object, the impulse-momentum theorem (Equation 7-4) tells us that when an external force is exerted on the system, the system's momentum will change. This will also be true for systems of two or more objects. But when two objects within a system exert forces *on each other,* Equation 7-5 tells us that the changes in their momenta (plural of momentum) are equal and opposite. In that case, the sum of their momenta, and thus the total momentum of the system, remains unchanged, in agreement with Equation 7-6. Thus we can write a general statement:

Law of Conservation of Momentum
When there is zero total external force on a system, the system conserves momentum, that is, the sum of all momenta in the system (a vector sum) remains unchanged:

$$\vec{p}_{A1} + \vec{p}_{B1} + \vec{p}_{C1} + \cdots = \vec{p}_{A2} + \vec{p}_{B2} + \vec{p}_{C2} + \cdots$$

or
$$\Sigma\vec{p}_1 = \Sigma\vec{p}_2 \tag{7-6}$$

(letter subscripts denote objects, number subscripts denote instants or states)

Note that the states here are velocity or momentum states. Although positions must be specified for energy states, we need not know anything about positions to apply momentum concepts to a situation.

7-2 Conservation of Momentum Applied to Problem Solving

◆**COLLISIONS** Momentum is conserved when bodies within a system exert forces on each other. This is always the case when they collide. Whenever two bodies collide, you should immediately think of them as a system to which the law of conservation of momentum applies. We will first examine how this works when all motion is along a single straight line, so that Equation 7-6 becomes

$$\Sigma p_1 = \Sigma p_2 \quad \text{(in one dimension)} \tag{7-7}$$

or
$$m_A v_{A1} + m_B v_{B1} + \cdots = m_A v_{A2} + m_B v_{B2} + \cdots \tag{7-8}$$

Example 7-2 *Scene from a Mall*

On an icy winter day, a shopper ($m = 50$ kg) leaves her new hatchback car ($m = 860$ kg) in neutral in a mall parking lot while she mails a letter ($m = 0.015$ kg). She forgets to use her emergency brake. Several students from the local college attempt to pull up behind her in a big station wagon ($m = 1940$ kg, including passengers), but they hit an ice slick and, unable to stop, they plow into her car at a speed of 6.0 m/s. With what speed is the hatchback propelled forward if the station wagon's speed is reduced to 2.0 m/s by the collision?

Solution
Choice of approach. A collision should immediately set you to thinking about conservation of momentum. Moreover, time is not explicitly involved, only masses and velocities, so Equation 7-7 should readily suggest itself.

What we know/what we don't. As in the case of energy problems, it is helpful to organize your data and unknowns by state. We choose State 1 to be immediately before the vehicles touch and State 2 to be immediately after they cease to touch.

State 1	State 2	Constants
$v_{sw1} = 6.0$ m/s $v_{hb1} = 0$	$v_{sw2} = 2.0$ m/s $v_{hb2} = ?$	$m_{sw} = 1940$ kg
		$m_{hb} = 860$ kg
(hb = hatchback, sw = station wagon)		Other masses irrelevant

The mathematical solution. $\Sigma p_1 = \Sigma p_2$ becomes

$$m_{sw}v_{sw1} + m_{hb}v_{hb1} = m_{sw}v_{sw2} + m_{hb}v_{hb2}$$

$$(1940 \text{ kg})(6.0 \text{ m/s}) + 0 = (1940 \text{ kg})(2.0 \text{ m/s}) + (860 \text{ kg})v_{hb2}$$

Proceeding from either of the above lines, with $v_{hb1} = 0$, we get

$$v_{hb2} = \frac{m_{sw}v_{sw1} - m_{sw}v_{sw2}}{m_{hb}}$$

or $$v_{hb2} = \frac{(1940 \text{ kg})(6.0 \text{ m/s} - 2.0 \text{ m/s})}{860 \text{ kg}} = \textbf{9.0 m/s}$$

Making sense of your results. See the next Stop&Think.

◆ Related homework: Problems 7-18, 7-19, 7-20, and 7-21.

STOP&Think In Example 7-2, suppose you had made an algebraic error in calculation and had ended up with $v_{hb2} = 1.5$ m/s. Why should you have suspected that your answer was in error? Why would you have suspected an error if instead you had gotten $v_{hb2} = -3$ m/s? Can the lighter vehicle's speed change less than the heavier one's? In which direction must its final velocity be? ◆

The next example shows a somewhat different application.

Example 7-3 *Recoil*

For a guided interactive solution, go to Web Example 7-3 at www.wiley.com/college/touger

When a bullet is fired from a gun, the gun kicks back or *recoils* in the opposite direction. A 110-grain (0.00713-kg) bullet fired from a 0.30-caliber M-1 carbine rifle with a mass of 3.632 kg has a muzzle velocity of 607 m/s. What is the "free recoil" velocity of the M-1, that is, the velocity the rifle would be given if not restrained by the shooter? (Data based on E. Matunas, *American Ammunition and Ballistics,* Winchester Press, Tulsa, OK, 1979.)

Brief Solution

Choice of approach. The conservation of momentum situation is simplified by the fact that the initial momenta of *both* objects are zero. Then the total momentum of the system in Equation 7-7 must be zero in the final state as well.

Anticipating the result. How should the signs of the two final velocities compare? Which object should have the greater final speed? Why?

What we know/what we don't. (b = bullet, r = rifle)

State 1	State 2	Constants
$v_{b1} = 0$ $v_{r1} = 0$	$v_{b2} = 607$ m/s $v_{r2} = ?$	$m_b = 0.00713$ kg
		$m_r = 3.632$ kg

EXAMPLE 7-3 *continued*

Figure 7-5 An inelastic collision. When a spring is compressed, the energy input is stored and is returned to KE when it springs back. What happens to the energy input here?

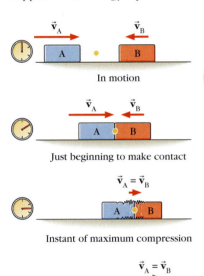

In motion

Just beginning to make contact

Instant of maximum compression

A subsequent instant
(bodies continue to stick together)

Figure 7-6 A totally inelastic collision of two deformable bodies. Note that the center of mass, shown as a yellow dot, moves the same distance to the right during each like time interval. Its velocity remains constant because the total momentum of the system—the two bodies taken together—does not change.

For **WebLink 7-1:**
Inelastic Collisions,
go to
www.wiley.com/college/touger

The mathematical solution. Now $\Sigma p_1 = \Sigma p_2$ becomes

$$m_b v_{b1} + m_r v_{r1} = 0 = m_b v_{b2} + m_r v_{r2}$$

so that $v_{r2} = -\dfrac{m_b v_{b2}}{m_r} = -\dfrac{(0.00713 \text{ kg})(607 \text{ m/s})}{3.632 \text{ kg}} = \mathbf{-1.19 \ m/s}$

Making sense of the result. As we should expect, the minus sign indicates that the rifle's velocity is in the opposite direction, and that the larger mass—the rifle—experiences a smaller change in velocity than the bullet.

◆ Related homework: Problems 7-23, 7-24, and 7-25.

STOP&Think In what way(s) is an explosion like the recoil situation in Example 7-3, or like a pair of ice dancers who push off from each other, or like the firing of a spring gun? In what way(s) is it different? ◆

In all of these cases, the parts of the system are initially together, whether they are moving or not. Stored energy, whether chemical energy in explosives or food or the potential energy of a loaded spring, is converted into kinetic energy of the parts, and they separate. Mechanical energy is therefore not conserved in the cases where chemical energy is involved, but momentum is conserved in all of these cases. The total momentum of all the parts after the separation must equal the initial momentum of the whole.

7-3 Elastic and Inelastic Collisions

When two bodies are in contact with each other during a collision, the forces that they exert on each other may in general be both normal and frictional forces. A table tennis paddle may strike a ball "head on," exerting only a normal force, but an experienced table tennis player can also use the paddle to exert a force *along* the surface of the ball to put a spin on it. A bullet fired through a block of wood on a table will drag the block along: The bullet and block will exert both normal and frictional forces on each other until the bullet emerges on the other side. In either case, we can think of the interval for which these forces are exerted as the duration of the collision.

For now, let's focus on normal forces. As we saw in Chapter 4, we can think of normal forces as composites of elastic forces exerted on a microscopic scale. Objects undergo deformations, either microscopically (as in Figure 4-5) or macroscopically (Figures 7-1 and 7-2) or both, when they exert normal forces on each other during collisions. In some cases, as Figure 7-2 shows, the bodies "spring back" to their original uncompressed shapes as they bounce back from each other without any energy being dissipated. In other cases (Figure 7-5), the deformation is permanent. These are the extreme situations. Between these extremes, the bodies might spring back but not completely, or they might spring back completely but with some dissipation of energy. We speak of the two extremes as **elastic collisions** (which implies totally elastic) and **totally inelastic collisions.**

In a totally inelastic collision (Figure 7-6), there is no springback whatsoever. The two bodies therefore remain together after the collision and thus have the same final velocity.

A **totally inelastic collision** is one in which the colliding bodies have the same final velocity after the collision:

$$\vec{v}_{A2} = \vec{v}_{B2}$$

Ideas about inelastic collisions are developed more fully in WebLink 7-1.

In a totally elastic collision, all the kinetic energy that is converted to elastic potential energy as the colliding objects compress is converted back into kinetic energy as the objects spring apart. However, while the colliding objects are in contact, the total kinetic energy is less than before or after. This point is more fully illustrated in WebLink 7-2, in which you can work through the details of some simple elastic collisions.

For **WebLink 7-2:**
Elastic Collisions,
go to
www.wiley.com/college/touger

An **elastic collision** is one in which kinetic energy is conserved. This means that the total kinetic energy of the system is the same immediately after the collision as before . . . but it is *not* the same *during* the collision.

$$\tfrac{1}{2} m_A v_{A1}^2 + \tfrac{1}{2} m_B v_{B1}^2 = \tfrac{1}{2} m_A v_{A2}^2 + \tfrac{1}{2} m_B v_{B2}^2 \qquad (7\text{-}9)$$

(letter subscripts denote objects, number subscripts denote states *before* and *after* collision)

No energy is stored or dissipated as a result of an elastic collision.

If the objects in Figure 7-6 were of equal mass and started out with equal and opposite velocities, the total momentum of that system would also be zero. Because those objects have a shared final velocity, their final velocity would then have to be zero as well. Thus, in contrast to the elastic situation, the objects are brought to a halt and all kinetic energy is lost. In Figure 7-6, the shared final velocity is not zero; the KE is not *all* lost, but the total KE is still reduced.

Example 7-4 *High Jinks at the Ice Show: An Inelastic Collision*

Groucho ($m_G = 80.0$ kg) and Harpo ($m_H = 60.0$ kg) are two clowns in the ice show. As Groucho skates across the ice with a velocity of magnitude 8.00 m/s, Harpo leaps into his arms (see figure). Groucho, clutching Harpo, continues forward at 4.00 m/s.

a. What is Harpo's airborne velocity (assumed horizontal) as he collides with Groucho?

b. How much kinetic energy is lost by the pair of clowns in the collision?

Solution
Choice of approach. After the collision, Groucho has the same velocity as Harpo; we apply conservation of momentum to a totally inelastic collision.

What we know/what we don't.

State 1	State 2	Constants
$v_{G1} = 8.00$ m/s $v_{H1} = ?$	$v_{G2} = v_{H2} = 4.00$ m/s (call this common value v_2)	$m_G = 80.0$ kg $m_H = 60.0$ kg

The mathematical solution.

a. Here $\Sigma p_1 = \Sigma p_2$ becomes

$$m_G v_{G1} + m_H v_{H1} = m_G v_{G2} + m_H v_{H2} = (m_G + m_H)v_2$$

so that $v_{H1} = \dfrac{(m_G + m_H)v_2 - m_G v_{G1}}{m_H}$

$$= \frac{(80.0 \text{ kg} + 60.0 \text{ kg})(4.00 \text{ m/s}) - (80.0 \text{ kg})(8.00 \text{ m/s})}{60.0 \text{ kg}}$$

$$= \mathbf{-1.33 \ m/s}$$

that is, 1.33 m/s in a direction *opposite* to Groucho's initial direction.

EXAMPLE 7-4 *continued*

b. We total the kinetic energies in each state

$$\tfrac{1}{2}m_G v_{G1}^2 = \tfrac{1}{2}(80.0 \text{ kg})(8.00 \text{ m/s})^2 = 2560 \text{ J}$$

$$\tfrac{1}{2}m_H v_{H1}^2 = \tfrac{1}{2}(60.0 \text{ kg})(-1.33 \text{ m/s})^2 = \underline{53 \text{ J}}$$

$$\text{Total initial KE} = 2613 \text{ J}$$

$$= 2610 \text{ J to three significant figures}$$

Because the clowns move together as a combined mass after the collision,

$$\text{Total final KE} = \tfrac{1}{2}(m_G + m_H)v_2^2$$

$$= \tfrac{1}{2}(80 \text{ kg} + 60 \text{ kg})(4.00 \text{ m/s})^2 = 1120 \text{ J}$$

The change in the system's total KE is thus

$$\text{Total final KE} - \text{Total initial KE} = 1120 \text{ J} - 2610 \text{ J} = \mathbf{-1390 \text{ J}}$$

(The minus sign tells us that the change is a decrease or loss.)

◆ Related homework: Problems 7-29, 7-30, and 7-31.

In Example 7-4, although we didn't describe it that way, we actually solved for a second unknown, Harpo's velocity after the collision. By saying it was the same as Groucho's, we were actually solving a second equation, $v_{G2} = v_{H2}$, the condition for a totally inelastic condition. In general, to solve for two unknowns, we must have a second equation in addition to the conservation of momentum equation (Equation 7-7). The second equation states a further condition that the system must satisfy in addition to conservation of momentum. If the second condition is that the collision is elastic, the second equation is the more cumbersome Equation 7-9. If, for example, we knew the masses and initial velocities of two bodies involved in a one-dimensional elastic collision, and we wished to find the two final velocities v_{A2} and v_{B2}, we would have to solve Equations 7-7 and 7-9 simultaneously for these two unknowns. Because Equation 7-9 involves the squares of the unknowns, this is a difficult task. However, starting with Equations 7-7b and 7-9, it is possible to show that for elastic collisions, Equation 7-9 is equivalent to the following simpler condition, which we can use in its place:

$$v_{B2} - v_{A2} = -(v_{B1} - v_{A1}) \tag{7-10a}$$

for elastic collisions.

For **WebLink 7-3:** **Deriving Another Condition for Elastic Collisions**, go to www.wiley.com/college/touger

To see how we obtain this result, work through WebLink 7-3 as an example of "if . . . then" reasoning that uses both the logic of algebra and our knowledge of physics.

Figure 7-7 shows how to interpret the differences between velocities that occur in Equation 7-10a. In one hour, car A goes 40 miles, car B goes 50 miles, and car C goes 30 miles. But if we keep the origin of a coordinate frame fixed on car A, then in that coordinate frame, car B has gone from 0 to 10 mi in 1 h and car C has dropped back from 0 to −10 mi in the same hour. B is gaining 10 mi each hour on A and C is losing 10 miles each hour. We can say the same thing by saying B's velocity relative to A is 10 mi/h and C's velocity relative to A is −10 mi/h. In each case, this **relative velocity** is just the difference between the two velocities that would be observed by an observer standing beside the road: $v_{B \text{ relative to A}} = v_B - v_A = 50 \text{ mi/h} - 40 \text{ mi/h} = 10 \text{ mi/h}$, and $v_{C \text{ relative to A}} = v_C - v_A = 30 \text{ mi/h} - 40 \text{ mi/h} = -10 \text{ mi/h}$. In general, this is a vector difference:

The velocity of B relative to A $\vec{v}_{B \text{ relative to A}} = \vec{v}_B - \vec{v}_A$ (7-10b)

Equation 7-10a then says that *the relative velocity $v_B - v_A$ is the same in magnitude after an elastic collision as before but is reversed in direction.* This means

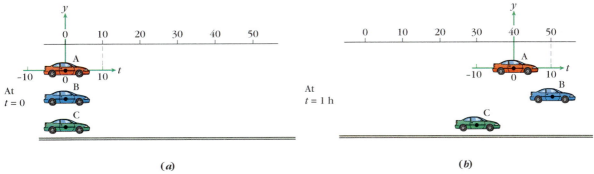

Figure 7-7 Velocities relative to a moving coordinate frame. (*a*) The cars start out together at $t = 0$. Car A carries its coordinate frame with it. (*b*) Positions after 1 hour if the cars have been traveling at speeds $v_A = 40$ mi/hr, $v_B = 50$ mi/hr, and $v_C = 30$ mi/hr. A's coordinate frame shows that B has gained 10 m in 1 hr, a velocity of 10 mi/hr relative to A, and C has lost 10 m in 1 hr, a velocity of -10 mi/hr relative to A.

that if before the collision two bodies are coming closer by so many meters each second, after the collision they will be moving apart by the same number of meters each second. For a fuller step-by-step introduction to relative velocities, go to WebLink 7-4.

For **WebLink 7-4:**
Relative Velocities,
go to
www.wiley.com/college/touger

Example 7-5 *An Off-the-Wall Example*

A ball thrown to the right is traveling horizontally at 20 m/s when it hits a wall and bounces straight back in an elastic collision. What is the ball's velocity after the collision?

Solution

Choice of approach. It is easiest to think of the ball's velocity relative to Earth, that is, to a coordinate frame that is fixed to Earth. Because we know that the collision is elastic, Equation 7-9 applies, as does Equation 7-10a. We can use Equation 7-9 in this case, because the mathematics is simplified by the fact that in our chosen coordinate frame one of the colliding bodies, the wall, remains stationary.

What we know/what we don't. (let W = wall, B = ball)

State 1		State 2		Constants
$v_{W1} = 0$	$v_{B1} = 20$ m/s	$v_{W2} = 0$	$v_{B2} = ?$	$m_W = ?$
				$m_B = ?$

The mathematical solution. Because $v_{A1} = v_{A2} = 0$, Equation 7-9 is reduced to $\frac{1}{2} m_B v_{B1}^2 = \frac{1}{2} m_B v_{B2}^2$, so that $v_{B2} = \pm v_{B1}$. We know the ball bounces back to the left, so we must choose the minus sign. Then

$$v_{B2} = -20 \text{ m/s} = \mathbf{-20 \text{ m/s}}$$

Making sense of the results. The ball's velocity relative to the wall is

$$v_{B1} - v_{W1} = 20 \text{ m/s} - 0 = \mathbf{20 \text{ m/s}}$$

After collision, its velocity relative to the wall is

$$v_{B2} - v_{W2} = (-20 \text{ m/s}) - 0 = \mathbf{-20 \text{ m/s}}$$

Beginning with the assumption that KE is conserved, we find that the relative velocity reverses after the collision, satisfying the condition (Equation 7-10a) for elastic collisions.

♦ Related homework: Problems 7-32 and 7-33.

Example 7-6 provides more practice with elastic collisions.

Example 7-6 *While the Chemistry Prof's Away: An Elastic Collision*

**For a guided interactive solution, go to Web Example 7-6 at
www.wiley.com/college/touger**

Spheres of various sizes representing atoms are used to build molecular models in chemistry courses. While Professor Tramondozzi is out of the room, two students are using the spheres to simulate atomic collisions on the surface of their lab bench. At a certain point, a sphere of mass 0.060 kg moving at a speed of 10.0 m/s collides head on with a sphere of mass 0.040 kg coming in the opposite direction at 2.5 m/s. Find the velocities of both spheres after the collision if the collision is elastic.

Brief Solution

Choice of approach. Because there will be two unknowns (the two final velocities) in the conservation of momentum equation, we need a second equation involving the two unknowns. Because the collision is elastic, we can use Equation 7-10a. Because the spheres start out in opposite directions, their initial velocities must have opposite signs.

What we know/what we don't. (Call the two spheres A and B.)

State 1	State 2	Constants
$v_{A1} = 10.0$ m/s	$v_{A2} = ?$	$m_A = 0.060$ kg
$v_{B1} = -2.5$ m/s	$v_{B2} = ?$	$m_B = 0.040$ kg

Anticipating results. Considering the comparative masses and speeds of the two spheres, in which direction would you expect sphere B to be moving after the collision?

The mathematical solution. Equations 7-7a ($\Sigma p_1 = \Sigma p_2$) and 7-10a (the condition for elastic collisions) become

$$m_A v_{A1} + m_B v_{B1} = m_A v_{A2} + m_B v_{B2} \tag{7-11}$$

and
$$v_{B2} - v_{A2} = -(v_{B1} - v_{A1}) \tag{7-10a}$$

From Equation 7-12,

$$v_{B2} = v_{A2} - (v_{B1} - v_{A1}) = v_{A2} - v_{B1} + v_{A1}$$

We use this to substitute for v_{B2} in Equation 7-11 to obtain an equation with only v_{A2} as an unknown:

$$m_A v_{A1} + m_B v_{B1} = m_A v_{A2} + m_B(v_{A2} - v_{B1} + v_{A1})$$

Lumping terms with v_{A2} together on one side and then solving for v_{A2}, we first get

$$m_A v_{A2} + m_B v_{A2} = m_A v_{A1} + 2m_B v_{B1} - m_B v_{A1}$$

and then

$$v_{A2} = \frac{m_A v_{A1} + 2m_B v_{B1} - m_B v_{A1}}{m_A + m_B}$$

$$= \frac{(0.060 \text{ kg})(10.0 \text{ m/s}) + 2(0.040 \text{ kg})(-2.5 \text{ m/s}) - (0.040 \text{ kg})(10.0 \text{ m/s})}{0.060 \text{ kg} + 0.040 \text{ kg}}$$

that is,
$$v_{A2} = \mathbf{0}$$

Substituting this result into Equation 7-10a yields

$$v_{B2} = -(v_{B1} - v_{A1}) = -(-2.5 \text{ m/s} - 10.0 \text{ m/s}) = \mathbf{12.5 \text{ m/s}}$$

Making sense of the results. We could reasonably anticipate that because A, the heavier sphere, was moving to the right much faster than sphere B was moving to the left, sphere B would end up moving to the right after the collision.

◆ Related homework: Problems 7-37 and 7-39.

In general, you will have to think about whether collisions are elastic, totally inelastic, or somewhere between these two extremes, and what each implies about the collision.

Example 7-7 *Colliding Freight Cars*

For a guided interactive solution, go to Web Example 7-7 at www.wiley.com/college/touger

Freight cars A and B roll toward each other on the same track. A's mass is 6000 kg and B's mass is 4000 kg. Before they collide, A is rolling eastward at a speed of 1.5 m/s, and B is rolling westward at a speed of 2.5 m/s. If the collision is totally inelastic, with what speed and in what direction does each car move immediately after the collision?

Brief Solution
Choice of approach. Here we have only one unknown in the conservation of momentum equation because in a totally inelastic collision, the two final velocities are equal. The trains must move together after the collision.

What we know/what we don't. (Call the two spheres A and B.)

State 1	State 2	Constants
$v_{A1} = 1.5$ m/s	$v_{A2} = v_{B2} = ?$	$m_A = 6000$ kg
$v_{B1} = -2.5$ m/s		$m_B = 4000$ kg

The mathematical solution. We drop the letter subscripts on the two final velocities because they are equal, and $\Sigma p_1 = \Sigma p_2$ becomes

$$m_A v_{A1} + m_B v_{B1} = m_A v_2 + m_B v_2$$

Solve for v_2:

$$v_2 = \frac{m_A v_{A1} + m_B v_{B1}}{m_A + m_B}$$

$$= \frac{(6000 \text{ kg})(1.5 \text{ m/s}) + (4000 \text{ kg})(-2.5 \text{ m/s})}{6000 \text{ kg} + 4000 \text{ kg}} = \mathbf{-0.10 \text{ m/s}}$$

◆ Related homework: Problems 7-42 and 7-32.

7-4 Conservation of Mechanical Energy, Conservation of Momentum: Which Do We Use When?

Objects conserve momentum as long as the total force exerted on them by other objects, such as the vector sum of normal and gravitational forces, is zero. In particular this is true of a system of two colliding objects. We have seen that the two-object system does *not* necessarily conserve mechanical energy. Mechanical energy may be conserved in elastic collisions, where the total KE *is* conserved and the total PE often changes negligibly over the interval of contact. But during inelastic collisions, the two objects do work on each other by exerting forces that are nonconservative, so the work-energy theorem (Equation 6-16) tells us that mechanical energy is *not* conserved.

In general, you must remain aware that

- momentum can only be conserved by a system when the total outside force on the system is zero.

- mechanical energy can be conserved by a system even if the total outside force on the system is not zero, as long as the total *work* due to outside forces is zero. It *will* be conserved if there are no heating effects and there is no conversion of internal energy (e.g., chemical energy) to mechanical energy.

It will be important to consider which condition applies over each time interval in the following case.

Case 7-1 ♦ *The Ballistic Pendulum*

A ballistic pendulum (Figure 7-8) is a device for measuring the speed of a horizontally traveling projectile. If the projectile (of mass m) remains lodged in the suspended block (of much greater mass M), the vertical height h to which the block rises provides a measure of the entry speed v_1 of the projectile. The analysis involves the application of both energy and momentum conservation.

Identifying a system. Whenever we apply a conservation law, we must first identify a system to which it is applicable. In considering conservation of momentum, it is simplest to include only the two colliding bodies (block and projectile) in our system (call this system I). To deal with gravitational PE when the block rises, we must include Earth in our system as well (call this system II), because PE is in general a way of dealing with work done by *internal* conservative forces. With either choice, the ropes by which the block is suspended,

which we will take to be effectively massless, are *not* part of the system. The forces they exert on the block are *external* forces.

Identifying states of the system. As things happen, we focus on the states at those instants that stand out. The duration of the collision is from the instant the leading edge of the projectile makes contact with the block to the instant the projectile ceases to penetrate any further into the block. We will refer to the states of the system at these two instants as states 1 and 2. The block travels a negligibly small distance while the projectile is entering it. Then its upward motion during this interval of collision is likewise approximately zero. Another landmark state occurs at the instant when the block reaches the highest point in its upward swing. Call this state 3.

Which law applies when? Between states 1 and 2, an *inelastic* collision occurs. KE is lost, and according to our approximation, no gravitational PE is gained. Then mechanical energy is not conserved; it is dissipated in the collision. But the external forces on system I (see free-body diagram in Figure 7-8b)—the upward forces exerted by the ropes and the gravitational force—remain equal and opposite, so that the total external force on the system is zero. Therefore, *system I conserves momentum between states 1 and 2.*

Between states 2 and 3, the only nonvertical external forces on system I are the forces exerted by the ropes. The total external force is therefore not zero, and system I does not conserve momentum between states 2 and 3. But the ropes are always directed along the radius of the arc followed by the block as it swings upward. The forces they exert are therefore always perpendicular to the motion, so no work is done by these forces, and *system II conserves mechanical energy between states 2 and 3.* System II includes Earth, so in this system the gravitational force is an internal force, and the work it does is accounted for as a change in gravitational PE.

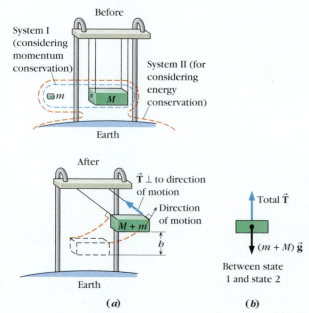

Figure 7-8 A ballistic pendulum. See Case 7-1.

Now we can state our conclusions in equation form. Because the collision is inelastic and the block's velocity is zero in state 1, we can write *system I conserves momentum between states 1 and 2* as

$$mv_1 = (m + M)v_2$$

Because the gravitational PE is zero in state 2 and the KE is zero in state 3, we can write *system II conserves mechanical energy between states 2 and 3* as

$$\tfrac{1}{2}(m + M)v_2^2 = (m + M)gh$$

which simplifies to $\quad v_2 = \sqrt{2gh}$

Substituting this into the momentum equation, we get

$$v_1 = \frac{m + M}{m}\sqrt{2gh} \qquad (7\text{-}13)$$

enabling us to calculate the initial velocity of the projectile from the maximum height of the swing. (The change in height is generally small and difficult to measure, so commonly one measures either the horizontal displacement or the angular displacement, and then uses geometry to calculate the height.)

We apply the ideas in Case 7-1 in the following example.

Example 7-8 *Tarzan and Jane*

For a guided interactive solution, go to Web Example 7-8 at www.wiley.com/college/touger

Tarzan ($m_T = 80$ kg) wishes to swing down from his tree house at a height of 15 m, grab Jane ($m_J = 50$ kg) at the bottom of the swing, and end up with the two of them on the limb of an opposite tree (see figure). If the target limb is 10 m above the ground, will Tarzan and Jane reach it?

15 m

10 m

Not drawn to scale

Brief Solution

Choice of approach. We use the same reasoning that we applied to the two bodies in Case 7-1. To keep our states numbered as in Case 7-1, we take the state at the instant when Tarzan begins his *downward* swing to be state 0. Energy is conserved between states 0 and 1 just as it is between states 2 and 3. We then (1) apply conservation of mechanical energy between states 0 and 1 to find v_1; (2) follow the same reasoning as in Case 7-1: momentum is conserved between states 1 and 2, so $m_T v_1 = (m_T + m_J)v_2$ and mechanical energy is conserved between states 2 and 3, so $v_2 = \sqrt{2gh_3}$; and (3) use these two equations to solve for h_3 to see if they go high enough.

What we know/what we don't.

State 0	State 1	State 3	Constants
$v_0 = 0$	$v_1 = ?$	$v_3 = 0$	$m_T = 80$ kg
$h_0 = 15$ m	$h_1 = 0$	$h_3 = ?$	$m_J = 50$ kg

EXAMPLE 7-8 *continued*

(Because we did the reasoning involving state 2 to obtain Equation 7-13, we do not have to consider state 2 further.)

The mathematical solution. Step (1) above is

$$\tfrac{1}{2}m_T v_0^2 + m_T g h_0 = \tfrac{1}{2}m_T v_1^2 + m_T g h_1$$

Since KE_0 and PE_1 are both zero, solving for v_1 gives us

$$v_1 = \sqrt{2gh_0} = \sqrt{2(9.8 \text{ m/s}^2)(15 \text{ m})} = 17.1 \text{ m/s}$$

Step (2): Solving for h_3 in $v_2 = \sqrt{2gh_3}$ gives us

$$h_3 = \frac{v_2^2}{2g}$$

We then write $\Sigma_{p1} = \Sigma_{p2}$ as $m_T v_1 = (m_T + m_J)v_2$ to solve for v_2 in terms of v_1:

$$v_2 = \left(\frac{m_T}{m_T + m_J}\right)v_1$$

and substitute this into our expression for h_3:

$$h_3 = \frac{v_2^2}{2g} = \frac{v_1^2 m_T^2}{2g(m_T + m_J)^2} = \frac{(17.1 \text{ m/s})^2(80 \text{ kg})^2}{2(9.8 \text{ m/s}^2)(80 \text{ kg} + 50 \text{ kg})^2} = \textbf{5.6 m}$$

Tough break, big guy!

◆ Related homework: Problems 7-43, 7-45, and 7-46.

7-5 Momentum and Center of Mass

Suppose the baton in Figure 7-9 is sent spinning across a frictionless icy surface. Because it is spinning, not all parts of the baton are moving at the same speed or even in the same direction, and the speed and direction of an individual part, such as A or B, changes as the baton travels. But what about its center of mass?

We introduced the idea of center of mass in Section 5-4: When the mass of an object or system is spread out rather than concentrated at a single point, the center of mass tells you the average position of a unit of mass. Then you would get the x coordinate $\bar{x}$ of the center of mass by averaging the x coordinates of all the kilograms of mass in the system. For instance, if there are 4 kg at $x = 2$ m and 3 kg at $x = 9$ m, we would take the average of 2 m four times and 9 m three times:

$$\bar{x} = \frac{2 \text{ m} + 2 \text{ m} + 2 \text{ m} + 2 \text{ m} + 9 \text{ m} + 9 \text{ m} + 9 \text{ m}}{7}$$

$$= \frac{(4 \times 2 \text{ m}) + (3 \times 9 \text{ m})}{7} = 5 \text{ m}$$

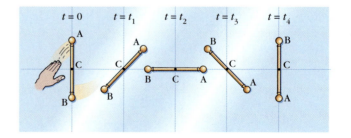

Figure 7-9 Center of mass of a baton spinning across an icy surface. The baton is shown at equally spaced instants. Point C is the baton's center of mass. Which of points A, B, and C are traveling at constant velocity?

Note that this result is closer to 2 m than to 9 m because there is more mass at 2 m. We can multiply the numerator and denominator of the above fraction by kg, giving us

$$\bar{x} = \frac{(4 \text{ kg} \times 2 \text{ m}) + (3 \text{ kg} \times 9 \text{ m})}{7 \text{ kg}} = \frac{(4 \text{ kg} \times 2 \text{ m}) + (3 \text{ kg} \times 9 \text{ m})}{4 \text{ kg} + 3 \text{ kg}}$$

In this last form, each x coordinate is "weighted" by the number of units of mass at that coordinate, and we then divide by the total number of units of mass. For a system of two parts, we can write this symbolically as

$$\bar{x} = \frac{m_1 x_1 + m_2 x_2}{m_1 + m_2} \qquad (7\text{-}14)$$

(For more parts, we would write $\bar{x} = \frac{m_1 x_1 + m_2 x_2 + m_3 x_3 + \cdots}{m_1 + m_2 + m_3 + \cdots}$.)

As the baton in Figure 7-9 moves across the ice, its center of mass travels at a speed

$$v_{CM} = \frac{\Delta \bar{x}}{\Delta t} = \frac{m_1 \dfrac{\Delta x_1}{\Delta t} + m_2 \dfrac{\Delta x_2}{\Delta t}}{m_1 + m_2}$$

But $\frac{\Delta x_1}{\Delta t} = v_1$ and $\frac{\Delta x_2}{\Delta t} = v_2$, where v_1 and v_2 are the velocities of the two parts.

Hence,

$$v_{CM} = \frac{m_1 v_1 + m_2 v_2}{m_1 + m_2}$$

We can now recognize the numerator as the total momentum of the two-part system. Conservation of momentum tells us this total is constant when there is no net outside force on the system. Then under these conditions, even if the parts of the system are moving differently and nonuniformly, the center of mass of the system moves at constant velocity. In Figure 7-9, you can verify that point C is moving at constant velocity although points A and B are not. We also see the center of mass in Figure 7-6 moving at constant velocity.

✦ SUMMARY ✦

Whenever a force is exerted on an object over a time interval, there is an impulse.

$$\textbf{Impulse} \equiv \vec{\mathbf{F}} \Delta t \qquad (7\text{-}1)$$

As a consequence of Newton's second law,

Impulse on a body = the change in the body's momentum

$$\vec{\mathbf{F}} \Delta t = \Delta \vec{\mathbf{p}} \qquad (7\text{-}4)$$

(impulse-momentum theorem)

where **momentum** is defined as

$$\vec{\mathbf{p}} \equiv m \vec{\mathbf{v}} \qquad (7\text{-}3)$$

Because the impulses, like the forces, that two bodies A and B exert on each other must be equal and opposite (by Newton's third law), the changes in momenta that the objects undergo during a collision must likewise be equal and opposite,

$$\Delta \vec{\mathbf{p}}_A = -\Delta \vec{\mathbf{p}}_B \qquad (7\text{-}5)$$

An immediate consequence of this is momentum conservation:

Law of Conservation of Momentum

When there is zero total external force on a system, the system conserves momentum, that is, the sum of all momenta in the system (a vector sum) remains unchanged:

$$\vec{\mathbf{p}}_{A1} + \vec{\mathbf{p}}_{B1} + \vec{\mathbf{p}}_{C1} + \cdots = \vec{\mathbf{p}}_{A2} + \vec{\mathbf{p}}_{B2} + \vec{\mathbf{p}}_{C2} + \cdots$$

or
$$\Sigma \vec{\mathbf{p}}_1 = \Sigma \vec{\mathbf{p}}_2 \qquad (7\text{-}6)$$

(letter subscripts denote objects, number subscripts denote instants or states)

This law is a powerful calculational tool for dealing with the states of motion of objects immediately before and immediately after collisions. When there are two unknowns in Equation 7-6, you must look for additional information so that you can write a second equation and solve simultaneously for the two unknowns. Often, the additional information may be that the collision is either **elastic** or **totally inelastic.**

A **totally inelastic collision** is one in which the colliding objects have the same final velocity after the collision:

$$\vec{v}_{A2} = \vec{v}_{B2}$$

An **elastic collision** is one in which kinetic energy is conserved. This means that the total kinetic energy of the system is the same immediately after the collision as before . . . but it is *not* the same *during* the collision.

$$\frac{1}{2}m_A v_{A1}^2 + \frac{1}{2}m_B v_{B1}^2 = \frac{1}{2}m_A v_{A2}^2 + \frac{1}{2}m_B v_{B2}^2 \qquad (7\text{-}9)$$

(letter subscripts denote objects, number subscripts denote states *before* and *after* collision)

For calculations, it is simpler to use a condition equivalent to the previous one:

In an **elastic collision,** *the relative velocity* of the two colliding objects is reversed. In one dimension,

$$v_{B2} - v_{A2} = -(v_{B1} - v_{A1}) \qquad (7\text{-}10a)$$

In more complicated situations, you may need to apply both conservation of momentum and conservation of mechanical energy. In such cases (see Case 7-1), you must be aware of when (i.e., between *which states of the system*) each quantity is conserved:

• Momentum can only be conserved by a system when the total outside force on the system is zero.

• Mechanical energy can be conserved by a system (in the absence of heating effects) if the total *work* done by exerting outside forces is zero.

If a system conserves momentum, its center of mass moves at constant velocity.

◆QUALITATIVE AND QUANTITATIVE PROBLEMS◆
Hands-On Activities and Discussion Questions

The questions and activities in this group are particularly suitable for in-class use.

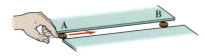

7-1. *Hands-On Activity.* Line up two pennies A and B between two straight-edges on a smooth flat surface (see

figure). Flick penny A quickly toward penny B with your finger, as shown.

a. What happens to penny A when it collides with penny B?

b. What happens to the kinetic energy of penny A after the collision?

c. What conclusion can you draw, at least approximately, about the velocity of penny B immediately after the collision?

Review and Practice

Section 7-1 Impulse and Momentum

7-2. In what ways are *impulse* and *work* alike? In what ways are they different?

7-3. Forces and many other quantities can have instantaneous values. Is it meaningful to speak of an instantaneous impulse? Briefly explain.

7-4. Ask a friend who isn't taking physics to show you (by acting it out) a "push." In a few words, write down a description of what your friend did. Did it have a beginning and an end? In that case, did it have some duration, or was it instantaneous? Can your friend's "push" be better described as a force or as an impulse? Why?

7-5. SSM

a. What does Newton's first law of motion tell you about the momentum of a moving object?

b. What does Newton's second law of motion tell you about the momentum of a moving object?

7-6. A truck jackknifes, blocking two lanes on a highway. In the right lane a car doing 40 mi/hr has no time to brake. It hits the truck at that speed, and is stopped within a few

thousandths of a second. In the left lane, the nearest car is doing 60 mi/hr when the driver sees the truck, but because this car is farther away, the driver can bring the car to a screeching halt over a 6-s interval, stopping just short of the truck. Assuming the cars have equal masses, which car, if either, is brought to a stop by a greater impulse? Briefly explain.

7-7. A regulation tennis ball has a mass of 0.059 kg. A ball reaches a player's racket at a speed of 40 m/s and is returned at a speed of 35 m/s in the opposite direction. If the racket exerts an average force of 2200 N while it is in contact with the ball, how long do the racket and ball remain in contact?

7-8. A disgruntled basketball player flings a ball straight downward. The ball strikes the floor at a speed of 5.6 m/s. The ball has a mass 0.59 kg. If the floor exerts an average upward force of 3000 N on the ball over the 0.002 s that floor and ball remain in contact, with what upward speed does the ball bounce back from the floor?

7-9. SSM In a crash test, identical vehicles carry identical test-dummy "drivers." Both vehicles are going at 15 m/s (a little over 30 mi/hr) when they crash head-on into a brick wall and are brought to a stop. Dummy A's vehicle has no safety devices. Dummy B's vehicle has an air bag that inflates on impact.

a. Compare the total impulses exerted on the two drivers as they are brought to a stop: The impulse on dummy B is (*greater than, equal to,* or *less than*) the impulse on A. (Choose the phrase that correctly completes the sentence.)

b. Which of the following quantities differ in the two situations?

$$F \quad \Delta t \quad m \quad v_i \quad v_f \quad \Delta v \quad impulse \quad p_i \quad p_f \quad \Delta p$$

7-10. In a medical journal article on "Injuries Due to Falling Coconuts," (P. Barss, *Journal of Trauma,* **24** [11], 990–991 [1984]), the author writes, "A 4-year review of trauma admissions to the Provincial Hospital, Alotau, Milne Bay Province, reveals that 2.5% of such admissions were due to being struck by falling coconuts. Since mature coconut palms may have a height of 24 up to 35 meters and an unhusked coconut may weigh 1 to 4 kg, blows to the head of a force exceeding 1 metric ton are possible." (A metric ton is the weight of 1000 kg. All "weights" or "forces" given here should be multiplied by g to obtain weights in SI units.) Based on this information, estimate the time interval for which the coconut remains in contact with the head of its victim. Assume average values for the tree height and coconut weight.

7-11. A ball is thrown straight up in the air and falls back down to its starting point, that is, the point where it left the hand. Assume air resistance is negligible.

a. Keeping in mind that impulse is a vector, compare the total impulse exerted on the ball during its ascent after leaving the hand with the total impulse exerted on it during its descent. Consider both magnitude and direction.

b. What exerts these impulses?

7-12. Two movers push bureaus the same distance across a room with a rough wooden floor. The movers exert equal constant forces over the entire distance. The bureaus are also identical, except that mover A's bureau is on casters so that it can be rolled, but mover B's bureau must be slid.

a. Compare the amounts of *work* done by the two movers to get their bureaus across the room. Briefly explain your reasoning.

b. Compare the *impulses* the two movers deliver to their bureaus to get them across the room. Briefly explain your reasoning.

7-13. (In this problem you will have to look up some values in the tables on the inside cover of the book.) Find the momentum *in SI units* of

a. a proton traveling in a particle accelerator at a hundredth of the speed of light.

b. Earth traveling in its orbit about the sun.

c. the world's smallest hummingbird (mass ≈ 2 grams) when it is flying at its peak speed of about 60 km/hr.

d. a 5-ton elephant barely shuffling along, about 3 inches each second.

7-14. How fast would *you* have to be moving to have the same momentum as ***a.*** the proton in 7-13***a***? ***b.*** Earth in 7-13***b***? ***c.*** the hummingbird in 7-13***c***? ***d.*** the elephant in 7-13***d***?

7-15. A circular hoop having a mass of 0.20 kg and a circumference of 0.80 m is kept spinning at a constant rate of 4 complete revolutions each second. What is the total momentum of the hoop?

7-16.

a. What is the magnitude of the instantaneous momentum of any 1 cm segment of the spinning hoop in Problem 7-15?

b. What is the *average* momentum of this segment as the hoop spins?

7-17. A pitched baseball ($m = 0.15\,kg$) reaches home plate at a speed of 35 m/s. The batter hits it back toward the pitcher at a speed of 25 m/s. If the bat exerts an average force of 6000 N on the ball, how long are the bat and ball in contact?

Section 7-2 Conservation of Momentum Applied to Problem Solving

7-18. The mass of a proton is 1840 times that of an electron. When a certain proton and electron collide in a particle accelerator, the proton's velocity is reduced by $2 \times 10^3\,m/s$. Assume all motion is along one straight line.

a. By how much does the electron's velocity change? (Try to answer this *without* using any numerical values not given in the problem.)

b. What can you say about how the electron's *speed* changes?

7-19. **SSM** Two ice cubes are slid toward each other on a wet glass tabletop. Cube A, with a mass of 0.050 kg, moves to the right with a speed of 4.0 m/s. Cube B, which has a mass of 0.070 kg, moves toward the left with a speed of 3.0 m/s. When they collide, cube A bounces back with a speed of 2.0 m/s. Find the velocity of cube B after the collision.

7-20. Two air track gliders have masses of 0.30 kg and 0.80 kg. As both gliders move toward the right on the air track, the lighter glider bumps the heavier one from behind. Before the bump, the lighter glider has a speed of 6.00 m/s and the heavier glider has a speed of 1.50 m/s. Find the speed and direction of the lighter glider immediately after the collision if the heavier glider's speed has increased to

a. 3.95 m/s.　　***b.*** 3.20 m/s.

7-21. In Problem 7-20, what is the minimum speed the heavier glider can have after the collision?

7-22. A link-up is made in space between a passenger module drifting at a speed of 3.0 m/s in one direction and an equipment module drifting at a speed of 4.0 m/s in the opposite direction. If the mass of the passenger module (including passengers) is 380 kg and the mass of the equipment module is 120 kg, what is the speed of the two modules after link-up?

7-23. An exploding piñata is part of a fireworks display for Cinco de Mayo, an important Mexican holiday. The piñata, which hangs stationary before it explodes, has a mass of 0.45 kg. It breaks into eight pieces when it explodes. What is the total momentum of these pieces immediately after the explosion?

7-24. A child with a mass of 30 kg is initially at rest on a skateboard with a mass of 1.5 kg. If the child jumps directly forward off the skateboard with a speed of 6.0 m/s,

a. with what speed does the skateboard move?

b. in what direction does the skateboard move?

7-25. SSM WWW A child winds up a mechanical toy alien and places it on a small toy wagon (Figure 7-10). The alien has a mass of 25 g and the wagon has a mass of 70 g. When the alien is placed on the wagon with its wheels already turning, the wagon starts rolling to the child's right at a speed of 0.05 m/s. If we assume all wheels turn without slipping, then relative to the stationary child,

Figure 7-10 Problem 7-25

a. in what direction is the alien moving?

b. with what speed is the alien moving?

7-26. Tell whether each of the following one-dimensional collisions is possible. (Assume that changes in speed are due only to the objects mentioned in the descriptions.) For each possible collision, tell how it can happen. For each collision that is not possible, explain why not.

a. Objects A and B both travel at a speed of 5 m/s before colliding. After the two objects collide, object B has a speed of 25 m/s.

b. Two objects both have speeds of 10 m/s before they collide with each other. After the collision, both speeds are reduced to 2 m/s.

c. Two objects of equal mass are traveling to the right. The faster one bumps the slower one from behind. The object that was behind slows down from a speed of 8 m/s to a speed of 6 m/s. The object that was ahead speeds up from a speed of 4 m/s to a speed of 7 m/s.

d. Two objects of equal mass are traveling at the same speed before they collide. After the collision, one of the two objects has a greater speed than the other.

e. Two objects of equal mass are traveling at the same speed before they collide. After the collision, one of the two objects has a greater velocity than the other.

Section 7-3 Elastic and Inelastic Collisions

7-27. What happens to the mechanical energy lost in an inelastic collision?

7-28. While your back is turned, two objects moving on an air track collide with a loud crash. Was the collision perfectly elastic? Briefly explain how you can tell.

7-29. A group of in-line skaters are playing basketball on skates. The ball is passed toward the left at a speed of 12 m/s to a player who is skating toward the right. The player's mass is 70 kg and the ball's mass is 0.60 kg. If the player and ball have a speed of 2.0 m/s immediately after the ball is caught, what was the player's speed just before catching the ball?

7-30. Experienced drivers know that collisions between insects and their windshields are often totally inelastic.

a. A 900-kg compact car traveling at 20 m/s hits a fly with a mass of 4×10^{-6} kg coming toward it at 5 m/s. How much is the car slowed down by the collision?

b. Is the impulse that the car delivers to the fly greater than, equal to, or less than the impulse that the fly delivers to the car? Briefly explain your answer.

7-31. SSM As a child is skating at a speed of 4 m/s, her father comes up behind her at a speed of 12 m/s and picks her up. The father weighs 680 N, the child weighs 170 N. What percent of their total kinetic energy is lost when he picks her up?

7-32. Light metal gliders can move virtually without friction on an air track. A 0.20-kg glider traveling from the left at 0.60 m/s and a 0.40-kg glider traveling from the right at 0.25 m/s collide at the center of the track. If the more massive glider moves back toward the right at 0.06 m/s after the collision,

a. find the speed and direction of the other glider after the collision.

b. determine whether the collision is *elastic, totally inelastic,* or *neither.*

7-33. A ball of mass 0.20 kg is dropped from a height of 1.3 m. To what maximum height will it bounce after hitting the floor if its collision with the floor is totally elastic?

Note: Do either *Version 1* or *Version 2* of Problem 7-34.

7-34. (*Version 1*) A soldier finds herself on a railroad handcar on which a machine gun is permanently mounted. The handcar and gun have a total mass of 500 kg. The soldier in her gear has a mass of 60 kg. The handcar is rolling toward a bombed-out bridge at a speed of 4 m/s. The machine gun fires 0.016-kg cartridges at a speed of 1000 m/s. The soldier has 145 cartridges. Does she have enough cartridges to save both herself and the handcar by firing the gun in the direction of motion, or must she abandon the handcar and save herself? (Do a calculation to decide.)

7-34. (*Version 2*) A soldier finds herself on a railroad handcar on which a machine gun is permanently mounted. The handcar and gun have a total mass of 500 kg. The soldier in her gear has a mass of 60 kg. The handcar is rolling toward a bombed-out bridge at a speed of 4 m/s. The machine gun fires 0.016 kg cartridges at a speed of 1000 m/s. She begins firing the gun to save herself.

a. In which direction must she fire the gun to save herself?

b. How many cartridges must she fire to save herself?

7-35. SSM A passenger train is traveling east at a speed of 32 m/s. Assume that the eastward direction is positive. Relative to this train, what is the relative velocity of a freight train going

a. east at 15 m/s?   **b.** west at 20 m/s?

7-36. A weather helicopter is flying west at a speed of 50 m/s. Assume that the eastward direction is positive. With what speed and in what direction would the wind be blowing, according to a stationary observer, if the pilot reports that the wind velocity relative to the helicopter is

a. 60 m/s?   **b.** 44 m/s?

7-37. Two identical billiard balls approach each other along an x axis drawn on the pool table. Before they collide, ball A is traveling to the right at a speed of 2.4 m/s and ball B to the left at a speed of 1.7 m/s.

a. What is the relative velocity of ball B relative to ball A?

b. What is the relative velocity of ball A relative to ball B?

c. The two balls then collide elastically. What are their velocities after the collision?

d. What is the relative velocity of ball B relative to ball A after the collision?

e. How do the relative velocities in **d** and **a** compare? Consider both magnitude and sign.

7-38. Suppose that in Problem 7-37, we set up a new set of coordinate axes, with its origin fixed in ball A (so that the origin moves as ball A does—see Figure 7-11).

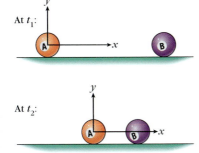

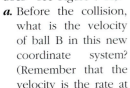

Figure 7-11 Problem 7-38

a. Before the collision, what is the velocity of ball B in this new coordinate system? (Remember that the velocity is the rate at which the position vector changes, and the position vector is always from the origin to the position in question.)

b. How does the magnitude of this velocity compare to the value you obtained in part **a** of Problem 7-37?

c. What is the velocity of ball A in this new coordinate system before the collision?

d. How does your answer to **c** compare to the way you usually describe the velocity of a rock sitting on Earth's surface as Earth travels through space?

e. If the balls collide elastically, what are their velocities in the new coordinate system after the collision?

f. How does the magnitude of the velocity of ball B compare to the value you obtained in part **c** of Problem 7-37?

g. How does the magnitude of the velocity of ball B compare to the value you obtained in part **d** of Problem 7-37?

7-39. Suppose that in Problem 7-37, ball B is replaced by a trick ball. This ball looks like the original ball B, but it has twice as much mass. The initial conditions of the two balls are otherwise the same. Repeat all parts of Problem 7-37 for this new situation.

7-40. Repeat all parts of Problem 7-38 for the new situation described in Problem 7-39.

7-41. Cart A, with a mass of 2 kg, collides with cart B, which has a mass of 4 kg. Before the two carts collide, cart B is moving to the right at a speed of 3 m/s and cart A is stationary. After the collision, cart B is stationary and cart A moves to the right at 6 m/s.

a. Check whether this collision satisfies conservation of momentum.

b. Is this collision possible? Briefly explain.

7-42. Two freight cars move toward each other along a straight track. Car A moves to right with a speed of 4.5 m/s and car

B moves to the left with a speed of 3.0 m/s. If car B has twice as much mass as car A, with what speed and in what direction does each car move after they collide if the collision is totally inelastic?

Section 7-4 Conservation of Mechanical Energy, Conservation of Momentum: Which Do We Use When?

7-43. In a ballistic pendulum (Case 7-1), suppose the ropes reach less than halfway to the ground. What is the effect on the maximum height to which the block and projectile rise if we double the length of the ropes from which the block is suspended (assuming the projectile's initial velocity is unchanged)?

7-44. In Case 7-1, the conservation of momentum equation is unchanged if we apply it to system II instead of system I, because Earth's velocity is zero before and after the collision. The ropes still exert an upward external force on the system, but the gravitational force on the block is now an internal force. Is the requirement for conservation of momentum—that the total external force on the system must be zero—met by system II? Briefly explain.

7-45. Suppose the bullet in Example 7-3 is fired horizontally into a wooden block suspended by ropes. If the block has a mass of 0.993 kg, how high does it rise when the bullet lodges in it?

7-46. A bullet with a mass of 0.0080 kg is fired at a velocity of 600 m/s at a 1.9920-kg wooden block backed by a spring with a spring constant of 5000 N/m. If the bullet remains embedded in the block, find the maximum compression of the spring.

7-47. Two identical hockey pucks slide across the same rough area of ice. Both travel the same distance across this area, and both are slowed down. But puck A enters the area 2 m/s faster than puck B enters the area. It leaves the rough area 3 m/s faster than B leaves it. Explain how this can happen

a. by reasoning about impulse–momentum relationships.

b. by reasoning about work–energy relationships.

Section 7-5 Momentum and Center of Mass

7-48. What is the velocity of the center of mass in the three-car system in Figure 7-7, assuming the cars have equal masses?

7-49. A 2000-kg SUV and a 1200-kg compact car are traveling in the same direction along a highway. The SUV passes the 316-km marker at a speed of 80 km/hr. At the same instant, the compact, trailing behind the SUV, passes the 308-km marker at a speed of 70 km/hr.

a. Find the center of mass of the two-car system at this instant.

b. Find the velocity of the center of mass of the two-car system.

Going Further

The questions and problems in this group are not organized by section heading, so you must determine for yourself which ideas apply. Some of them will be more challenging than the Review and Practice questions and problems (especially those marked with a • or ••).

7-50. *Extending conservation of momentum to two dimensions:* A 1200-kg passenger car runs a stop sign at a speed of 14 m/s. A 4000-kg truck, unable to brake in time, strikes the car at a speed of 4.0 m/s. The initial directions are shown in Figure 7-12.

a. Find the x and y components of each vehicle's initial velocity.

b. This is a two-dimensional problem. In two dimensions, momentum is separately conserved in each direction:

$\Sigma(p_x)_i = \Sigma(p_x)_f$ and $\Sigma(p_y)_i = \Sigma(p_y)_f$. Suppose the truck and car do not separate after impact. Write the details of the two momentum-conservation equations and use them to find the speed (v) and direction (θ) of the vehicles immediately after colliding.

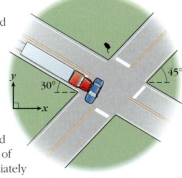

Figure 7-12 Problem 7-50

7-51. SSM WWW *Extending conservation of momentum to two dimensions:* A red shuffleboard disk traveling at 8.00 m/s makes a glancing strike on an opponent's green disk, identical except for color. The subsequent directions of the two disks are shown in Figure 7-13. This is a two-dimensional problem. In two dimensions, momentum is separately conserved in each direction: $\Sigma(p_x)_i = \Sigma(p_x)_f$ and $\Sigma(p_y)_i = \Sigma(p_y)_f$. Write the details of the two momentum-conservation equations for this collision and use them to find the speeds of the two disks after collision.

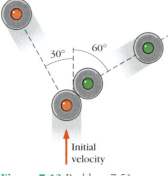

Figure 7-13 Problem 7-51

7-52. *Extending the impulse-momentum theorem to two dimensions:* In two dimensions, the impulse-momentum theorem must be written as two component equations: $F_x\Delta t = \Delta p_x$ and $F_y\Delta t = \Delta p_y$. Use these two component equations to solve the following problem. A 500-kg car goes around a circular track with a radius of 20 m. If it travels the 90° arc between points A and B (Figure 7-14) at a constant speed of 8 m/s, find the magnitude and direction of the impulse that the track delivers to the car during this interval.

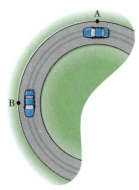

Figure 7-14 Problem 7-52

•**7-53.** A toy cart is set in motion by pulling it with a rubber band stretched by a constant amount *d*. A second cart, identical to the first, is set in motion by pulling it the same distance with *two* such rubber bands, *each* stretched an amount *d*. (The rubber bands are attached side by side, not end on end.) A student concludes that the total impulse on the second cart will be twice as great, giving it double the final speed. *a.* Is the student's conclusion correct? Why or why not? *b.* Use energy conservation to check your answer to *a.*

7-54. Super Balls are popular toys because they conserve energy much better than ordinary rubber balls and therefore bounce exceptionally high. A Super Ball and a tennis ball of identical mass and volume are dropped from the same height. Which exerts a greater impulse on the ground? Briefly explain how you know.

7-55. SSM A popular carnival game involves trying to knock over milk bottles with a thrown ball. A variation of this game uses exceptionally large and heavy milk bottles. Suppose you step up to the booth and are given a choice of two balls of equal mass and volume to throw at the bottles. You drop both of them. One bounces high; the other barely bounces at all. Which ball would give you the best chance of knocking over the bottles? Briefly explain why.

7-56. *When worlds collide:* When a meteorite strikes Earth, the meteorite or its pieces remain on or near Earth's surface.

Suppose a giant object from space enters the solar system and collides head on with Earth as Earth travels in its path around the sun.
a. If the object's speed is twice Earth's speed, how massive must it be to reduce Earth's orbital velocity by even a hundredth of a percent?
b. Compare its mass to the mass of a body that you consider to be a useful basis for comparison, that is, make a comparison like "it's about the same as" or "it's ten (or some other number) times as great as."

7-57. The swing in Figure 7-15 is suspended from the ceiling of a lightweight car mounted on tracks. The wheels of the car turn easily. Tell what else happens in this situation as the child swings back and forth. Describe the expected behavior as precisely as you can, and explain why you expect this behavior.

Figure 7-15 Problem 7-57

7-58. *Dirty pool?* In a standard game of billiards, the cue ball is identical in size and mass to all the other balls. The judges in a billiards tournament are reviewing a videotape of a shot by a player who used his own cue ball. The ball is observed to approach the eight ball, which is stationary, at a speed of 8.000 m/s. After a head-on collision, the cue ball moves forward at a speed of 0.027 m/s while the eight ball is propelled forward at 8.727 m/s in the same direction.
a. What can you conclude about the mass of the cue ball?
b. What should the velocities of the two balls be after the collision if it is completely elastic?

•**7-59.** Two springs, such as those in Weblink 7-3, collide head on in an elastic collision. Spring A has a mass of 0.080 kg, and spring B has a mass of 0.060 kg. Their speeds just before contact are 6.0 m/s and 4.0 m/s, respectively.
a. Find the speeds of the two springs when they are *at maximum compression.*
b. Find their speeds at the instant they cease to touch each other.

•**7-60.** A juggler is invited to perform at a casino on the opposite bank of a river. But the only boat available to bring him across can carry a maximum weight of 600 N. The juggler weighs 590 N. The three dumbbells that he plans to juggle weigh 5 N each. "That's no problem," boasts the juggler. "I can keep juggling all the way across so that one of the dumbbells is always in the air. That way I can bring all my dumbbells with me and still keep to the weight limit." Will the juggler's strategy succeed? Use physics principles to prove your answer.

7-61. Two balls A and B collide and bounce back from each other. After the collision, each ball's velocity is equal in magnitude and opposite in direction to what it was before.
a. Assuming no external forces are exerted on this two-ball system, find the system's total momentum.
b. Is the collision elastic? Briefly explain.

7-62. Carefully describe what must be happening in a collision that satisfies this description: two objects of equal mass are traveling at the same speed before they collide; after the collision, the two objects have equal velocities.

•**7-63.** Two objects A and B collide elastically. Just before the collision, object A, which has mass m_A, is moving at constant speed, but object B, which has mass m_B, is stationary. In this problem, you will be asked to compare A's final velocity to its initial velocity.

a. Before doing any math, write down what you would expect to happen to A's velocity as a result of the collision (i) if m_B is infinitesimally small compared to m_A, and (ii) if m_B is practically infinitely large compared to m_A. Describe a situation involving real-world objects that meets each of these conditions.

b. Now, for the collision described, use conservation of momentum and the condition for elastic collisions (Equation 7-10) to derive an equation that expresses the final velocity of A in terms of its initial velocity and the two masses.

c. Test the validity of your equation by seeing if it gives you the results you expected for the two extreme conditions described in **a.**

••**7-64.** A car goes around a circular track of radius R at constant speed.

a. Use the impulse-momentum theorem (in combination with other principles, if needed) to show that the time interval required for the car to change its direction by 90° is inversely proportional to the car's speed.

b. Is there a simpler way to show this? Justify your answer.

•**7-65.** Suppose all the contacts between surfaces in Figure 7-16 are frictionless, including the contact between the wedge and the table top.

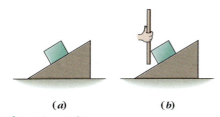

(a)   (b)

Figure 7-16 Problem 7-65

a. If the block is released from rest in the position shown in Figure 7-16a, what happens? Why?

b. Is momentum conserved in either the x or the y direction in this situation? Explain.

c. Is mechanical energy conserved in this situation? Explain.

d. The situation described in **a** is repeated while a meter stick is held motionless, as shown in Figure 7-16b. What happens now when the rest of the set-up is released from rest?

e. Repeat parts **b** and **c** for the situation involving the meter stick.

7-66. When a bumper car bounces off a large spring, the spring first compresses and then springs back. Assuming the collision is elastic, how does the impulse exerted by the spring over the interval when it is compressing compare to the impulse exerted by the spring over the interval when it is springing back?

a. The two impulses are equal in both magnitude and direction.

b. The impulse exerted during compression has a somewhat greater magnitude than the impulse exerted while the spring is springing back.

c. The two impulses are equal in magnitude but opposite in direction.

d. The impulse exerted during compression has a much greater magnitude than the impulse exerted while the spring is springing back.

7-67.

a. Find the total impulse delivered by the force that is plotted against time in Figure 7-17.

b. What feature of the graph represents the total impulse?

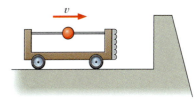

Figure 7-17 Problem 7-67

7-68. A cart (Figure 7-18) has a long pole running front to back. The pole passes through the center of a large, heavy ball that is fixed in place so that it cannot slide along the pole. The cart, initially traveling at a speed v, is brought to a stop when it crashes into the wall. The wall exerts a force of average magnitude F over an interval Δt in bringing the cart to a stop. What is the effect on the details of the crash if the ball, instead of being fixed in place, is free to slide frictionlessly along the pole? Does it matter whether or not the ball collides elastically with the end(s) of the cart?

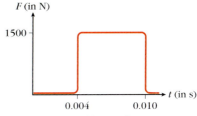

Figure 7-18 Problem 7-68

7-69. Is the collision of the marble with the belt in Figure 4-13 elastic or inelastic? How do you know?

7-70. Two objects A and B with equal masses collide elastically.

a. How does the velocity of A after the collision compare to the velocity of B before the collision?

b. How does the velocity of B after the collision compare to the velocity of A before the collision?

c. Write a sentence in your own words that summarizes your answers to **a** and **b.**

7-71. An astronaut drifting in space holds two bar magnets together, north pole to north pole (like poles repel), and then releases them. Magnet A has a mass of 0.020 kg and magnet B has a mass of 0.030 kg. When magnet A has a speed of 0.24 m/s, what will be the speed of magnet B?

7-72. A collision can happen _____ (only over a time interval; only at precisely one instant or point in time; or either over a time interval or at precisely one instant).

7-73.

a. Give an example of a situation in which an object or system conserves momentum but not kinetic energy from start to finish.

b. Give an example of a situation in which an object or system conserves kinetic energy but not momentum from start to finish.

•7-74. A system consists of two identical springs. The springs travel toward each other at the same speed and collide end on end. Before colliding, the total kinetic energy of the two springs is 8.0 J. If the collision is elastic, what is the elastic potential energy of the system when the springs are at maximum compression?

7-75. A bus and a truck are traveling toward each other on a two-lane highway. The bus is traveling at a constant speed of 30 m/s in the eastbound lane, and the truck is traveling at a constant speed of 20 m/s in the westbound lane. They continue at the same speeds after passing each other. Which of the following correctly describes the velocity of the bus relative to the truck after the two vehicles pass each other?

a. It has the same magnitude and direction as it did before they passed each other.

b. It has the same magnitude as before but is in the opposite direction.

c. It has the same direction as before but not the same magnitude.

d. Its magnitude is different than before, and its direction is opposite.

7-76. A board hangs by ropes from a tree limb. A soccer ball kicked at the board makes the board swing upward. In applying conservation laws to this situation, a student decides to consider the state of the system at the instant the ball collides with the board. Is this a good choice? Briefly tell why or why not.

7-77. **SSM** Two identical bumper cars A and B collide three times. The initial state of the two cars (Figure 7-19) is the same before all three collisions.

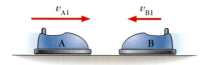
Figure 7-19 Problems 7-77 and 7-78

• Collision I resulted in car A having zero velocity after the collision.

• Collision II is elastic.

• Collision III is totally inelastic.

Rank the three collisions in order of how much car B's momentum is changed by the collision, starting with the smallest change.

7-78. For the information given in Problem 7-77, rank the three collisions in order of how much impulse is delivered to car B in the collision, starting with the smallest impulse.

•7-79. A softball coach decides to suspend a heavy box with a hole in it from 2.5-m ropes to measure pitching speed. She would like to set it up so that when a softball is thrown horizontally into the hole at a speed of 45 m/s and lodges in the box, the ropes swing through an angle of 30°, as in Figure 7-20. If a softball has a mass of 0.18 kg, what should be the mass of the box? (Neglect the mass of the ropes.)

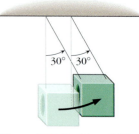
Figure 7-20 Problem 7-79

•7-80. A Maserati rear-ends an old Yugo convertible on a highway. Including their loads, the Maserati's mass is 2000 kg and the Yugo's mass is 1000 kg. The collision is totally inelastic, and the coupled vehicles go into a prolonged skid. The length of the skid marks is found to be 73 m. The coefficient of sliding friction between the asphalt road and the vehicles' tires is 0.70. The Yugo driver reports that he was going at 28 m/s just before the collision. How fast was the Maserati going at that instant?

•7-81. A model rocket is made up of two pieces: Piece A has a mass of 1.5 kg and piece B has a mass of 0.5 kg. The rocket is fired from ground level with a launch speed of 60 m/s at an angle of 37° above the horizontal. When the rocket is at its highest point, piece B is propelled straight backward with a speed of 20 m/s relative to the ground. How far from its initial launch point does piece A land?

7-82. A pellet with a mass of 0.08 kg is fired horizontally with a speed of 14 m/s and strikes the back of a small toy locomotive. The locomotive is initially stationary and has a mass of 0.32 kg. If the system of pellet + locomotive loses 20% of its total kinetic energy in the collision, find the velocities of the two objects immediately after the collision. Let the initial direction of the pellet be the positive direction.

7-83. Jack and Naomi are bowling, using identical bowling balls. In a mischievous mood, Jack bowls his ball while Naomi's ball is still going down the lane, and it strikes Naomi's ball elastically from behind. After the collision, Naomi's ball has a velocity of 11 m/s relative to the floor, and 4 m/s relative to Jack's ball. What were the velocities of the two balls before the collision?

Problems on WebLinks

7-84. WebLink 7-1 discusses a situation in which two objects move apart after a collision, but less quickly than they came together. Which (one or more) of the following must be true of such a situation?

a. The total energy of the system is conserved.

b. The collision is inelastic.

c. The total momentum of the system is conserved.

d. The total kinetic energy of the system is conserved.

e. One of the objects remains stationary throughout the collision.

7-85. (*Choose the best answer.*) WebLink 7-2 considers a head-on elastic collision of two identical (except for color) billiard balls, such that before the collision, one of the balls is moving at twice the speed of the other. In this situation, at the instant when the compression of the balls is a maximum, the total kinetic energy of the system is ____ (*zero; a minimum but not zero; the same as before the balls collided; a maximum; negative*).

7-86. In the situation in Problem 7-85, at the instant when the two balls first make contact, is the force that the faster ball exerts on the slower ball greater than, equal to, or less than the force that the slower ball exerts on the faster ball?

7-87. Suppose the balls in Problem 7-85 collide elastically again, but this time ball A is stationary before the collision, and ball B has a speed of 7 m/s. What is the speed of ball A at the instant during the collision when the compression of the two balls is maximum? (*Hint:* A similar situation is considered in the follow-up questions to WebLink 7-2.)

7-88. In the derivation in WebLink 7-3, we used the fact that $m_A(v_{A2} - v_{A1}) = m_B(v_{B1} - v_{B2})$. These two expressions are equal because ____
a. the velocity of A relative to B is reversed by the collision
b. the collision conserves momentum

c. the collision conserves kinetic energy
d. whenever you interchange A and B you automatically interchange 1 and 2.

7-89. In WebLink 7-4, what would be the velocity relative to car A of a second car traveling to the *a.* right at 60 mi/hr? *b.* right at 20 mi/hr? *c.* left at 20 mi/hr?

7-90. In WebLink 7-4, what would be the velocity of car A relative to a second car traveling to the *a.* right at 60 mi/hr? *b.* right at 40 mi/hr? *c.* left at 40 mi/hr?

For a cumulative review, see Review Problem Set I on page 825.

CHAPTER EIGHT

Circular Motion, Central Forces, and Gravitation

Throughout the universe, instances of motion in a circle occur in an endless variety of contexts. Planets orbit the sun; satellites orbit Earth; electrons orbit nuclei in simple models of the atom; stars sweep out circumgalactic orbits in spiral galaxies; masses of air whirl violently in tornadoes, hurricanes, or the spinning formation of ice, clouds, and fog shown here. Riders on amusement park rides, a yo-yo whirled on a string, the head of a golf club in full swing, heavy ions hurtling through the 2.4-mile ring of a particle accelerator (Figure 8-1)—all follow circular or nearly circular paths. Every point of a rotating object also traces a circular path—the tread of an automobile tire, the tip of a propeller, the outstretched hand of a pirouetting ballerina or a break dancer spinning, each cubic centimeter of Earth spinning on its axis. We can also think of certain

"...instances of motion in a circle occur in an endless variety of contexts... masses of air whirling violently...a break dancer spinning"

more complex motions as compounded of circular motions—the motion of the rider on the Octopus (Figure 3-4), for example, or of a point on the rim of a spinning top or on Earth as it precesses (Figure 8-2).

In dealing with two-dimensional motion, we have so far treated circular motion only minimally. Unlike the component motions of projectile motion, the one-dimensional component or "shadow" motions of circular motions are not uniformly accelerated (see Problem 8-82 or 8-83), so we cannot treat them using the equations of motion we developed in Chapter 3. In this chapter we will develop methods for treating circular motion quantitatively. As always, your use of the equations we develop must be guided by a good conceptual grasp of the physics principles involved in circular motion.

Figure 8-1 The Relativistic Heavy Ion Collider (RHIC). Bunches of ions travel both clockwise and counterclockwise around the ring of this facility at Brookhaven National Laboratory at speeds approaching that of light. Collisions between the ion beams provide crucial data about the fundamental constituents of matter. A view along the tunnel shows how gradual the curve is along the 2.4-mile circumference of the ring.

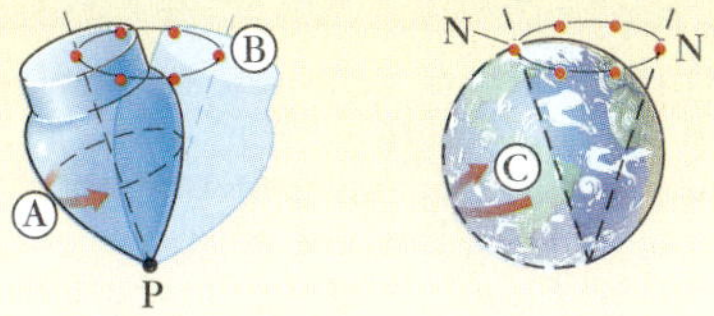

Figure 8-2 A spinning top and Earth's precession. Each point on the body of the top (A) travels in a circle about the top's axis, while the top of the axis (B) moves in a circle about a vertical line passing through point P. This motion is called **precession.** Earth's axis of rotation (C) precesses in the same way.

8-1 Circular Motion Is Accelerated Motion

Let's review some points from previous chapters. In Example 3-7 we calculated the acceleration of a car following a circular path at constant speed. We found that in uniform circular motion, the acceleration is *not zero*. Velocity is by definition a vector. In the magnitude-and-direction description of $\vec{v}$, the magnitude v—the speed—does not change in uniform circular motion, but the direction changes continuously. This means that in the component description, v_x and v_y are continuously changing (Figure 8-3), so that over the time interval Δt required to travel a small fraction of the circle, neither Δv_x nor Δv_y is zero. Then over this interval the average acceleration components $\bar{a}_x = \frac{\Delta v_x}{\Delta t}$ and $\bar{a}_y = \frac{\Delta v_y}{\Delta t}$ cannot be zero either, and the acceleration $\vec{a}$ has a nonzero magnitude $a = \sqrt{(\bar{a}_x)^2 + (\bar{a}_y)^2}$.

In Example 3-7, we further found that the direction of the car's acceleration was toward the center of the circle. WebLink 3-9 examined why this must be true for any uniform circular motion. Consistent with this, we saw in On-the-Spot Activity 4-5 (Figures 4-11 and 4-12) and Problem 4-45 (Figure 4-22) that to bend a would-be straight line path into a circular path, at each instant a force must be exerted perpendicular to the object's motion at that instant; that is, toward the center of the circle. Figure 8-4 reiterates the point. To get a sense of what this

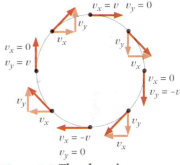

Figure 8-3 The changing components of the velocity vector in uniform circular motion.

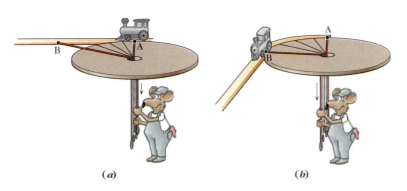

(*a*)   (*b*)

Figure 8-4 The "bending" of a straight-line path by a radial force. A toy train rides on a flexible rubber track. Whenever the train reaches a point on the track, the "conductor" pulls down on the string connected to the track at this point. By this means, the would-be straight line path of the train is progressively "bent" into a circular path (the track is wrapped around the circular hub) as each string in turn exerts a force toward the center of the circle.

Tornadoes are dramatic examples of circular motion. The overall motion combines circular motion with motion along the path of advance.

For **WebLink 8-1: Circular Motion and Radial Forces,** go to
www.wiley.com/college/touger

➥**Caution:** Labeling vectors as radial does not imply that they are in a fixed direction. Their direction changes, because they are not always along the same radius.

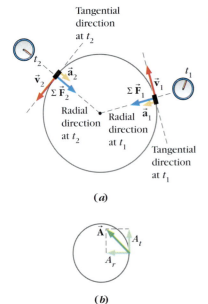

(a)

(b)

Figure 8-5 The radial and tangential directions.

feels like—literally—do the hands-on activity in Problem 8-1 at the end of the chapter. To see in step-by-step detail what the velocity changes that occur during uniform circular motion tell us about the net force that causes them, work through WebLink 8-1.

Remember that Newton's second law, $\Sigma \vec{F}_{on\,A} = m_A \vec{a}_A$, is a vector relationship: The vectors on both sides of the equation must be in the same direction.

> *In uniform circular motion, both the total force on an object and the object's acceleration must be perpendicular to $\vec{v}$ and toward the center of the circle.*

Newton's second law is a relationship between instantaneous quantities. In Figure 4-12, at the instant when the force toward the center of the circle becomes zero, the car's acceleration correspondingly becomes zero. Newton's first law tells us that an object will change neither its speed *nor its direction* when no outside force is exerted on it, so the car continues straight in the direction it was heading at the instant the force on it was terminated. Note that this direction is *tangent* to the circle at that point. By inertia, the car would keep going in the same direction. In uniform circular motion, the direction keeps changing, so the car cannot maintain this motion simply by inertia. Similarly, if you whirl a yo-yo on a string and the string breaks, the yo-yo does *not* keep going in a circle; it literally "flies off on a tangent."

◆**RADIAL ACCELERATIONS AND FORCES** We have established that in uniform circular motion, the direction of the velocity vector $\vec{v}$ is always "along" the circle, but this "along" direction is continuously changing. Because the direction of the acceleration and of the total force is instantaneously perpendicular to the velocity, it too is continuously changing. At any instant, the acceleration and the total force point inward along some radius of the circle. It is useful to call this the **radial direction,** and to speak of the acceleration and forces in this direction as **radial acceleration** and **radial forces** (also called *central forces*).

Acceleration toward the center of the circle is also commonly called *centripetal acceleration.* The total radial force, which is responsible for this acceleration, is often called the *centripetal force;* the word *centripetal* comes from the Latin for "center-seeking."

> **Caution:** Centripetal forces do not exist *in addition to* other forces on an object. In Figures 4-12 and 8-5, there are not both a centripetal force and a tension force on the object. Rather, the tension force *is* the centripetal force, meaning it is toward the center of the circle. *Centripetal* is just a term describing the direction of an already existing force. For that reason, this book prefers the term *total radial force,* which more self-evidently describes a combination of forces or force components in a particular direction.

In contrast to the radial direction, the instantaneous direction of a body moving along a circular path is called the **tangential direction** and is perpendicular to the radial direction. Like the radial direction, this does not imply a fixed direction (see Figure 8-5a). *In uniform circular motion, the velocity vector is always in the tangential direction.*

In general, any vector $\vec{A}$ may have radial and tangential components A_r and A_t (Figure 8-5b). It may also have a component perpendicular to the plane of the circle (not shown).

The geometry of a circle makes it possible to find a useful relationship between the radial acceleration a_r and the speed v, which is tangential, of a body traveling in a circle of radius r. To see in detail how this relationship is obtained,

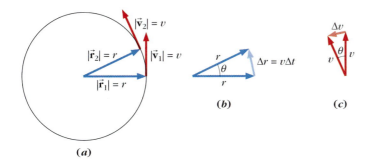

(a)

(b)

(c)

Figure 8-6 **Deriving an expression for radial acceleration.** As an object follows a circular path at constant speed (*a*), both *v* and *r* remain constant, but the directions of $\vec{v}$ and $\vec{r}$ change. Because $\vec{v}$ remains perpendicular to $\vec{r}$ as the object travels, their directions change by the same angle θ (*b* and *c*). The vector diagrams (*b* and *c*) showing the changes in the two vectors are similar triangles, so $\frac{v\Delta t}{r} = \frac{\Delta v}{v}$, or $\frac{v^2}{r} = \frac{\Delta v}{\Delta t}$; that is, $\frac{v^2}{r} = a$.

work through WebLink 8-2. The derivation in the WebLink is summarized in Figure 8-6, and yields this important result:

For **WebLink 8-2:**
Radial Acceleration,
go to
www.wiley.com/college/touger

Radial acceleration (also called *centripetal acceleration*)

$$a_r = \frac{v^2}{r} \tag{8-1}$$

(scalar component or magnitude of acceleration vector)

If we let the time interval Δt in the derivation shrink toward a single instant t, Equation 8-1 becomes a completely accurate equation for the *instantaneous* radial acceleration.

By Newton's second law, the total force required to give a body of mass m this acceleration is the total radial force.

Total radial force (also called the *centripetal force*)

$$\Sigma F_r = ma_r = \frac{mv^2}{r} \tag{8-2}$$

(scalar component or magnitude of total force vector)

It is the sum of all forces or force components in the radial direction.

On-The-Spot Activity 8-1

Walk quickly down the street and without slowing down, go around the corner (a 90° corner). Do this twice. The first time make a wide, gradual arc to make the turn; the second time make the turn very sharply. Notice how the force you feel on your feet compares in the two cases. How does the radius of curvature differ in the two cases? How does the force that you feel relate to the radius of curvature?

Now repeat the sharp right turn more slowly. How does the force you feel relate to your speed?

What does Equation 8-2 say about the dependence of ΣF_r on the speed v and the radius of curvature r? Does it agree with your observations?

Think of Equation 8-2 in terms of pulling inward to bend a would-be straight path into a circular one. The greater the body's tangential velocity, the harder you would have to push or pull to bend the path into a circle of any given radius. Thus, as Equation 8-2 indicates, F_r must increase as v increases. If, on the other hand, you wish to bend the path into a circle of much larger radius, the curvature will be much more gradual, and you need not push or pull so hard to produce it. It then makes sense that r should be in the denominator in Equation 8-2, indicating that if r is greater, the required F_r is smaller.

Example 8-1 *Achieving Two g's*

An amusement park ride whirls you in a horizontal circle of radius 7.00 m. How fast must you be going to experience a horizontal acceleration of "two g's" (twice the gravitational acceleration)?

Solution

Choice of approach. The horizontal acceleration is radial, so Equation 8-1 applies: $a_r = \frac{v^2}{r}$.

What we know/what we don't.

$$a = 2g = 2(9.8 \text{ m/s}^2) = 19.6 \text{ m/s}^2 \qquad r = 7.00 \text{ m} \qquad v = ?$$

The mathematical solution. From Equation 8-1, $v^2 = a_r r$, so

$$v = \sqrt{a_r r} = \sqrt{(19.6 \text{ m/s}^2)(7.00 \text{ m})} = \textbf{11.7 m/s}$$

Making sense of the results. If you don't have a good intuitive feel for SI units, determine how fast this is in mi/h. How large a force on *you* would be necessary to produce this acceleration? Think of amusement park rides you are familiar with. What would be exerting the force in each case?

◆ Related homework: Problems 8-7, 8-8, and 8-10.

Figure 8-7 Figure skaters experiencing radial forces. On which skater is the force inward toward the center of the circle? On which skater is the force outward from the center of the circle?

8-2 Examples of Radial Forces

Any type of contact or action-at-a-distance force can be exerted in the radial direction. In other words, any kind of force may be part or all of the total radial force constraining a body to follow a circular path. **STOP&Think** In each of the following cases, what type of force is keeping the object moving in a circle, and what is exerting this force? (Recall Section 5-1: Inventorying Forces.) ◆

- The car following a circular stretch of road in Example 3-7.
- A rider on the Turkish Twist (Figure 8-10*a*), an amusement park ride in which people are pressed against the outer wall as it spins, so that at sufficient rotational speeds the bottom can drop out and people remain pressed against the wall.
- A figure skater being whirled by her partner (Figure 8-7).

As a lead-in to Case 8-1, try the following.

On-The-Spot Activity 8-2

Place a heavy object on a table and push on it firmly, but not so hard as to set it in motion (Figure 8-8). What force keeps it from sliding? In what direction is it? Next, push on the object as before, but from a different angle. In what direction now is the force that prevents sliding? If you keep changing the direction of your push, what other force keeps changing its direction? What exerts this force?

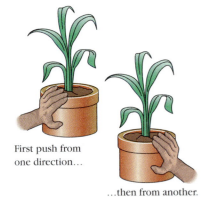

First push from one direction...

...then from another.

Figure 8-8 Changing the directions of forces exerted on a stationary object.

Case 8-1 ◆ *A Vehicle on a Circular Roadway*

The motion of the car in Example 3-7 is governed by Equation 8-2, which is a specific instance of Newton's second law. In applying the second law, we always draw a free-body diagram (Figure 8-9*a*) to inventory the forces on the object of interest. But F_r in the diagram is not yet fully identified. What is exerting it?

What kind of force is it? Newton's second law requires that it be an external force, so nothing within the truck can be exerting it. What is left?

Now think about what happened in On-the-Spot Activity 8-2. As you push from different directions, the static frictional force that the table exerts on the object will always be equal and *opposite* to your push. Its direction keeps changing to prevent the sliding that would occur in its absence.

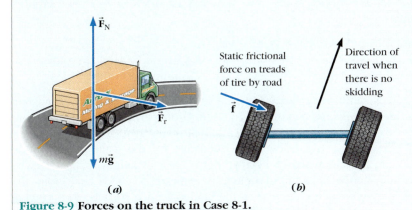

Figure 8-9 Forces on the truck in Case 8-1.

A similar thing happens when you angle your front tires to negotiate a curve. If your car or truck "skidded out," it would go off on a tangent to the turn. In doing so, its radial distance from the center of curvature must increase, so there must be a *static* frictional force inward in the radial direction to prevent this sliding. As you go around the curve, the direction of the would-be slide changes, so the frictional force changes direction to remain perpendicular to your motion. The road exerts this force across your tire treads (Figure 8-9*b*); it is the total force in the *radial* direction.

What happens when this force is reduced? This happens on a rainy or icy day, when the coefficient of static friction is lower. If the car is moving too fast, the radial frictional force cannot bend the car's path enough to keep it on the road (Equation 8-2 requires a larger total force when v is larger). The car's path remains straighter than the driver would wish: It skids out. Once skidding begins, the force opposing the skid is further reduced because the *sliding* frictional force is always less than the maximum *static* frictional force.

For experience using these ideas in problem solving, work through Example 8-2.

Example 8-2 *Slippery When Wet*

For a guided interactive solution, go to Web Example 8-2 at www.wiley.com/college/touger

You have learned from experience that on dry days the maximum speed at which you can drive your car around the corner near your house without skidding is 9.0 m/s. Suppose that when the road is wet, the frictional force that the road exerts on your tire treads is reduced 25%. How fast can you safely take the turn in the rain?

Brief Solution

Choice of approach. We apply Newton's second law in the form of Equation 8-2, with the frictional force as the total radial force.

Restating the question.

If $v_1 = 9.0$ m/s *when* $F_r = f_1$, *then* $v_2 = ?$ *when* $F_r = f_2 = f_1 - 0.25f_1 = 0.75f_1$

The mathematical solution. Because the radial force $F_r = f = \frac{mv^2}{r}$, v_2 must also be reduced 25%. Solving for v gives us

$$v = \sqrt{\frac{r}{m}} \sqrt{f}$$

For a *given* car on a *given* turn, m and r are both constant, so the ratio of v to $\sqrt{f}$ stays the same:

$$\frac{v_1}{\sqrt{f_1}} = \frac{v_2}{\sqrt{f_2}}$$

EXAMPLE 8-2 *continued*

Thus, $v_2 = \dfrac{v_1}{\sqrt{f_1}}\sqrt{f_2} = (9.0 \text{ m/s})\dfrac{\sqrt{0.75 f_1}}{\sqrt{f_1}} = (9.0 \text{ m/s})\sqrt{0.75} = \textbf{7.8 m/s}$

Making sense of the results. Make sure you understand why v_2 is not simply 25% less than v_1.

◆ Related homework: Problems 8-14, 8-15, and 8-16.

For **WebLink 8-3: Bouncing Around,** go to www.wiley.com/college/touger

Normal forces can also be radial forces. These are the forces that occur when bodies press directly against each other in a direction perpendicular to the contact surfaces. By working through WebLink 8-3, you can see the connection between the forces on a ball bouncing sequentially off the walls of a polygon and a ball rolling around within a circular rim. This will help you picture why the force keeping the ball moving in a circle is a normal force. At each point the normal force is exerted perpendicular to the circumference (that is, to a tangent to the circumference) and therefore *toward the center of the circle.*

At full rotational speed, the rider on the Turkish Twist (Figure 8-10*a*), like the marble going around the belt in Figure 4-13, is constrained to a circular path by a normal force directed inward toward the center of the circle. The forces exerted on the rider when the floor drops out are shown in the free-body diagram (Figure 8-10*b*). Note that when the floor drops out, the only upward force on the rider is the frictional force exerted by the wall. The frictional force is proportional to the normal force, which in turn, by Equation 8-2, is proportional to v^2. Hence, the frictional force can only keep the rider from falling when the Twist is at sufficient speed. (We'll return to the forces on the rider later in the chapter.)

STOP&Think Is it a good idea to dress in smooth, shiny fabrics on this ride? ◆

If the normal force is suddenly removed (Figure 8-10*c*), the rider will fly off in a tangential direction just as a yo-yo whirled on a string does when you suddenly eliminate the tension force by letting go of the string.

Radial force problems are still problems involving accelerated motion due to forces, and therefore involve Newton's second law, which is a vector relationship ($\Sigma F_x = ma_x$ and $\Sigma F_y = ma_y$). Sometimes only a component of a force, not the entire force, is in the radial direction, as in the following example.

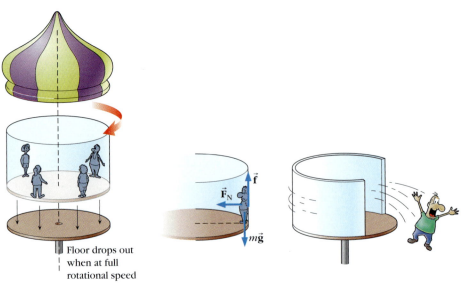

Floor drops out when at full rotational speed

Figure 8-10 The Turkish Twist.   (*a*)   (*b*)   (*c*)

Example 8-3 *Radial Acceleration in the Context of Newton's Second Law*

A tetherball (Figure 8-11*a*) with a mass of 0.30 kg hangs by a 2.0-m cord from the top of its post. At what speed must you keep the tetherball moving if you want the cord to maintain a 40° angle with the vertical?

Solution

Choice of approach. The ball's path is shown by the dotted line in Figure 8-11*a*. Note that the center of the circle is point C, *not* the top of the post. The radius of the circle can be found from the trigonometry of triangle ABC. When applying the second law in two dimensions, an important first step is to draw a free-body diagram (Figure 8-11*b*). Recall that when a vector $\vec{V}$ makes an angle ϕ with the vertical rather than θ with the horizontal, the *x* component is *opposite* ϕ, and the *y* component is *adjacent* to ϕ, so that $V_x = V\sin\phi$ and $V_y = V\cos\phi$. In this example, the perpendicular to the *y* direction is labeled *r* (radial) rather than *x*.

What we know/what we don't. We use Newton's second law to organize our known and unknown values, as we did in Chapter 5. The radial direction replaces the *x* direction. The force exerted by the cord is labeled *T* because it is equal in magnitude to the tension in the cord.

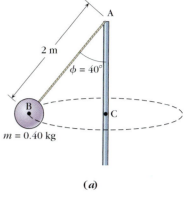

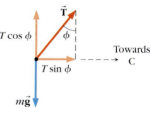

Figure 8-11 Diagrams for Example 8-3. (*a*) The physical set-up. (*b*) A free-body diagram of the ball.

2nd Law in *r* Direction	2nd Law in *y* Direction	Constants
$a_r = \dfrac{v^2}{r}\ (v = ?)$	$a_y = 0$	$g = 9.8 \text{ m/s}^2$
Forces in *r* Direction	**Forces in *y* Direction**	$m = 0.30 \text{ kg}$
$T\sin\phi\ (T = ?, \phi = 40°)$	$T\cos\phi$	$r = (2.0 \text{ m})\sin 40°$
	$-mg$	

The mathematical solution. $\Sigma F_r = ma_r$ becomes

$$T\sin\phi = m\frac{v^2}{r} \tag{8-3}$$

and $\Sigma F_y = ma_y$ becomes

$$T\cos\phi - mg = 0 \tag{8-4}$$

We can solve these two equations for the two unknowns v and T. Dividing the first by the second, we get

$$\frac{T\sin\phi}{T\cos\phi} = \frac{m\dfrac{v^2}{r}}{mg} = \frac{v^2}{gr}$$

or

$$\tan\phi = \frac{v^2}{gr}$$

so that

$$v = \sqrt{gr\tan\phi} = \sqrt{(9.8 \text{ m/s}^2)(2.0 \text{ m})\sin 40° \tan 40°} = \textbf{3.1 m/s}$$

Alternate approach. We can solve Equation 8-4 for *T*. Substituting that value into Equation 8-3 leaves *v* as the only unknown.

◆ Related homework: Problems 8-17 and 8-19.

In some situations, more than one force will have a component in the radial direction. Curved roadways and tracks are often banked (Figure 8-12) so that a

Figure 8-12 Banking on a normal force.

component of the normal force will be directed toward the center of the path of motion. To explore how normal and frictional forces contribute to the total radial force on a car negotiating a banked turn, work through Example 8-4.

Example 8-4 *Identifying Radial Forces*

For a guided interactive solution, go to Web Example 8-4 at www.wiley.com/college/touger

A test driver is taking a new model car around a steeply banked icy curve.
a. First he drives the car as fast as it will go without skidding out on the curve.
b. Then he drives the car as slowly as it will go without sliding downhill. Identify the radial forces on the car as fully as you can (kind of force, direction, comparative size) in each of these two situations.

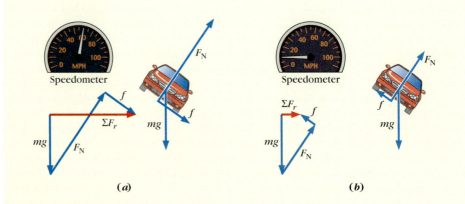

(a)   (b)

Brief Solution

Choice of approach. We draw free-body diagrams (see figure above) for the two situations. The frictional force must oppose impending motion uphill in **a** and downhill in **b**.

The solution. Although the road is banked, if the car doesn't skid up or down the bank, it moves in a horizontal circle. Both the frictional force f and normal force F_N that the road exerts on the tires have *components* along the radius of this circle. These components are the radial forces. In **a** they are both toward the center of the circle; in **b** they oppose each other. The total radial force (in red) must be greater in **a** because the car is going faster and $\Sigma F_r = \frac{mv^2}{r}$. As the online version shows in greater detail, both f and F_N must therefore be greater in **a**.

◆ Related homework: Problems 8-18, 8-24, and 8-25.

Figure 8-13 Motion in a vertical circle.

When an object moves in a vertical circle, the magnitude of the total radial force on it does *not* usually remain constant. In Figure 8-13, the radial force exerted by the string is always perpendicular to the ball's motion, so it does no work on the ball. Because no work is done, mechanical energy is conserved. As the ball drops from point A to point B, it loses PE; consequently it gains KE. This means it moves at greater speed v at the bottom of the circle, and the total radial force $m\frac{v^2}{r}$ on it will also be greater. **STOP&Think** How must the force the string exerts at B then compare with the force it exerts at A? ◆ To explore this question further, work through Example 8-5.

Example 8-5 *Motion in a Vertical Circle*

For a guided interactive solution, go to Web Example 8-5 at
www.wiley.com/college/touger

On the Holy Horror roller coaster, riders experience a complete loop-the-loop. In the figure, the diagram at upper left is a free-body diagram of the car and riders (treated as a single object) at point A. Which of the free-body diagrams below it is the correct free-body diagram of the car and riders at point B? Neglect friction and air resistance.

Brief Solution

Choice of approach. Newton's second law applied to circular motion tells us $\Sigma F_r = \frac{mv^2}{r}$ at any point on the circle. But because the car conserves energy and loses PE from A to B, it gains KE, so v must be greater at B than at A.

The solution. It follows that if v is greater at B, $\Sigma F_r = \frac{mv^2}{r}$ must be greater in magnitude at B than at A. At A, $\Sigma F_r = 5000 \text{ N} + 15\,000 \text{ N} = 20\,000$ N (units in the diagrams are thousands of newtons), so at B the net radial force, now upward, must be greater than 20 000 N. Only choice **V**, in which $\Sigma F_r = 45\,000 \text{ N} - 5000 \text{ N} = 40\,000$ N, satisfies this.

Making sense of the results. Why should the net radial force be 40 000 N? By assuming a value for the radius r, you can use $\Sigma F_r = \frac{mv^2}{r}$ to find the speed the car has at A and at B. You can then show that for these values, the increase in KE is just equal to the loss of PE from the top of the circle to the bottom, so energy is conserved.

◆ Related homework: Problems 8-27, 8-28, and 8-29.

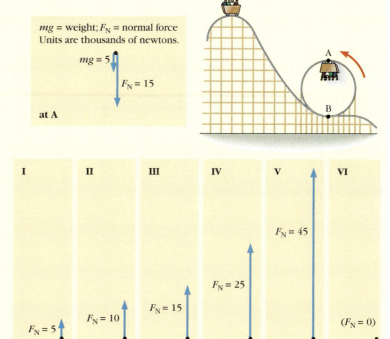

8-3 The Universal Law of Gravitation

We have said that in physics we try to find the rules by which nature plays. In this section we will look at Isaac Newton's reasoning in deducing one such rule: the rule governing gravitational forces. We will see that two points were critical in Newton's reasoning: (1) He recognized that the gravitational force is the radial force that constrains bodies in the heavens to their nearly circular paths, and (2) he understood that in Newton's second law, when $\Sigma \vec{F}$ on a body is radial, the body's acceleration will be $\frac{v^2}{r}$.

To follow Newton's work, we must be aware of what Newton knew about the work of his predecessors, especially that of Galileo and Kepler. From Galileo came a great wealth of discoveries and insights, including the understanding that projectile motion "is compounded of two other motions . . . one which is uniform and horizontal and . . . another which is vertical and naturally accelerated,"[1] as we saw in Chapter 3.

---

[1]From the Crew and deSalvio translation of Galileo's *Dialogue Concerning Two New Sciences*, (McGraw-Hill, New York, 1963). Many other editions exist.

Table 8-1 Data for Orbits of Planets Visible to Naked Eye

Planet	T (Earth years)	R (km $\times 10^6$)	T^2 (Earth yr^2)	R^3 (km$^3 \times 10^{24}$)	$\frac{T^2}{R^3}$ $\left(\times 10^{-25} \frac{\text{yr}^2}{\text{km}^3}\right)$
Mercury	0.2409	57.9	0.05803	0.1941	2.989
Venus	0.6152	108.2	0.3785	1.267	2.988
Earth	1.0000	149.6	1.000	3.348	2.987
Mars	1.8809	227.9	3.538	11.84	2.989
Jupiter	11.8622	778.3	140.7	471.5	2.985
Saturn	29.4577	1427.0	867.8	2905.	2.986

The entries in these two columns are proportional.

➡Danish astronomer Tycho Brahe (1546–1601) spent much of his life recording data about the planets' paths with unprecedented precision. After his death, Johannes Kepler (1571–1630), a German, applied his mathematical skills to seeking pattern in this voluminous data.

Kepler had examined Tycho's extensive data on the motion of the planets, looking for mathematical patterns. His years of trial and error led him to three surprising patterns. We will limit ourselves to considering one of these—a precise relationship between T, the time that it takes each planet to complete one orbit around the sun, and R, the planet's average distance from the sun. T is called the planet's *period* or *year,* and R is so labeled because it is radius-like. As Table 8-1 shows, the ratio $\frac{T^2}{R^3}$ has the same value for every planet. If we represent this constant value as k, we can write the pattern as $\frac{T^2}{R^3} = k$. This pattern came to be known as Kepler's third law of planetary motion.

Galileo's work on projectile motion addressed the behavior of bodies close to Earth. Kepler's laws described the behavior of bodies in the heavens. Galileo, too, had directed his attention to the heavens. The first to use a telescope in this endeavor, he was the discoverer of Jupiter's four largest moons. It was later found (see Problem 8-77 and Table 8-2) that the ratio $\frac{T^2}{R^3}$ had the same value for the orbit of each of Jupiter's moons about Jupiter. But Galileo made no connection between projectile motion and the motions of planets and moons. The ancient Greeks had believed that one set of rules was appropriate to bodies on Earth and another to bodies in the heavens. When Newton was born in 1643, within a year of Galileo's death, this notion persisted. It was Newton's great insight to look for a single rule governing bodies in *both* domains—a *universal* law.

In Galileo's analysis of projectile motion, one component motion "is vertical and naturally accelerated" (with acceleration g). Newton realized that *vertical* or *downward* is not the same direction in Tokyo as it is in New York or Buenos Aires, but rather it always means *toward the center of Earth*. The path of a projectile fired horizontally from a height (Figure 8-14b) is bent because it is pulled radially; otherwise, by inertia it would keep going straight along a tangent. It is fine to think in terms of horizontal and vertical when distances are short enough so that a "flat Earth" approximation will suffice, but not if a projectile is fired with so much speed that its landing point is significantly affected by Earth's curvature. In that case, the radial and tangential directions change from one point along the projectile's path to the next. By pursuing this reasoning, Newton realized that projectiles fired from a height at progressively greater horizontal speeds would come closer and closer to circumnavigating, or orbiting, Earth. To follow Newton's reasoning in detail, work through WebLink 8-4. His ideas are summarized briefly in Figure 8-14.

It now becomes possible for Newton to consider the moon, a body that is actually orbiting Earth, and argue that by doing so it is continuously falling toward Earth from its would-be tangential path, just as ordinary projectiles do. Therefore, the "causes" that keep the moon in its orbit and that make a thrown rock fall back to Earth must be one and the same. It is the cause to which we ordinarily attach the name **gravitation.**

For **WebLink 8-4:**
Gravitation Is Universal, go to
www.wiley.com/college/touger

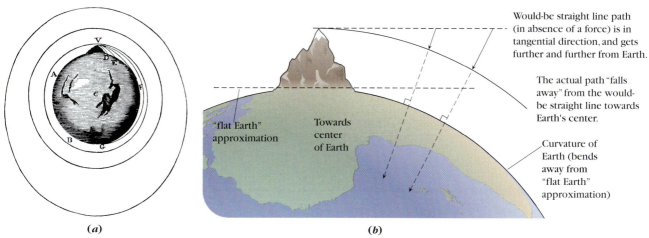

Would-be straight line path (in absence of a force) is in tangential direction, and gets further and further from Earth.

The actual path "falls away" from the would-be straight line towards Earth's center.

Curvature of Earth (bends away from "flat Earth" approximation)

"flat Earth" approximation

Towards center of Earth

(a)   (b)

Figure 8-14 From projectiles to orbiting bodies. (*a*) Newton's diagram of the changing paths of projectiles as they are fired at progressively greater horizontal speeds from a mountain top and then from greater heights so that they would orbit Earth much as the planets orbit the sun. (*b*) The path of a projectile fired fast enough to circle Earth.

Naming something does not mean that we understand it. Newton made no attempt to propose an actual mechanism to explain how gravitation worked. "I do not make hypotheses," he wrote. But he identified this "cause" as a *force* that Earth exerts both on the falling rock and the orbiting moon. Because the motions of these bodies could be described quantitatively, Newton set out to find a quantitative rule governing the gravitational force.

On what, he asked, must the gravitational force depend? He thought in terms of a gravitational *interaction:* two interacting bodies exerting equal and opposite forces *on each other.* He supposed that matter possessed some property responsible for gravitational forces, and for an interaction to occur, that property must be present in both interacting bodies. If its presence is increased in either body, the force will be stronger. Newton called this property *gravitational mass.* It is important to recognize that as a concept, gravitational mass is totally distinct from inertial mass, but as we discuss later, this does not mean we can distinguish between their measured values.

Newton also supposed that this action-at-a-distance force would weaken as the interacting objects moved further apart. He therefore proposed that the force should be proportional to $\frac{1}{r^2}$, where r is the separation between the centers of mass of the two bodies. It was already known that the intensity of light from a point source (see Problem 6-72) drops off with distance as $\frac{1}{r^2}$, so it is plausible to assume that awareness of the geometrical reason for this might have inspired Newton to anticipate a similar rule for gravitation.

Combining these two ideas, Newton proposed that the *magnitude* of the gravitational force each of two bodies exerts on the other be given by

$$F = G\frac{M_1 M_2}{r^2} \tag{8-5}$$

(the *direction* of the force on each body is *toward* the other). The gravitational masses of the two bodies are denoted by capital letters M_1 and M_2 to contrast with their inertial masses m_1 and m_2, and G is a constant, indicating that F is proportional to $\frac{M_1 M_2}{r^2}$. As we would expect of the force, this fraction gets bigger when either gravitational mass increases and gets smaller when r increases, increasing the denominator. The value of G, however, remains the same for any two masses at any separation: It is therefore called the **universal gravitational constant.**

Equation 8-5 is called **Newton's universal law of gravitation.** This was a *proposed* rule. To see if it was valid, it had to be tested against observation. Would the behaviors that followed from it as logical consequences actually be observed in nature? Because Newton was proposing it as a *universal* rule, it had to be

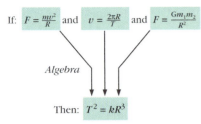

If: $F = \frac{mv^2}{R}$ and $v = \frac{2\pi R}{T}$ and $F = \frac{Gm_1 m_2}{R^2}$

Algebra

Then: $T^2 = kR^3$

Newton's mathematics

If: | Dana is a police officer | and | All police officers own guns |

Propositional logic

Then: Dana owns a gun

"Ordinary" reasoning

Figure 8-15 Algebra is a tool of deductive logic, just as ordinary if-then reasoning is. Both provide a way of reasoning from initial premises to conclusions.

tested against the observed behavior of objects in the heavens as well as on Earth. Kepler's laws of planetary motion were patterns observed in the behavior of heavenly bodies. We will check to see if Newton's "universal law" is consistent with the pattern described by Kepler's third law: $\frac{T^2}{R^3} = k$. To simplify the math, we will treat the planets' orbits around the sun as circles. (In fact, they are ellipses, as Figure 8-21 will show, but they are only slightly off from being circles.)

As Figure 8-15 shows, we can do deductive reasoning and draw conclusions using algebra just as we can by ordinary if-then logic. We follow the mathematical argument outlined in the figure, beginning with three premises:

1. Each planet is constrained to its approximately circular orbit by a radial force of magnitude $F_r = \frac{m_p v^2}{R}$, where m_p is the planet's inertial mass, and $r = R$ is the radius of the planet's orbit and hence its distance from the sun.

2. The planet travels one orbit circumference in the time interval T; hence its speed $v = \frac{2\pi R}{T}$.

3. The total radial force is a gravitational force of the form $F = G\frac{M_p M_s}{R^2}$, where M_p and M_s are the gravitational masses of the planet and the sun. This is the test premise.

From these premises, let's see what conclusion we can draw using the step-by-step logic of algebra. If the gravitational force in its proposed form is the total radial force, the forces in **1** and **3** are equal, and we can use **2** to substitute for v. Thus,

$$\boxed{G\frac{M_p M_s}{R^2}} = \frac{m_p v^2}{R} = \frac{m_p \left(\frac{2\pi R}{T}\right)^2}{R} = \boxed{m_p\frac{4\pi^2 R}{T^2}}$$

Equating the two boxed expressions and rearranging, we get

$$\frac{T^2}{R^3} = \frac{4\pi^2 m_p}{GM_p M_s} \tag{8-6}$$

This agrees with Kepler's third law *if* the right side of the equation is a constant, that is, if its value does not change from planet to planet, so we could just call it k. The only part of the right side that depends on the choice of planet is $\frac{m_p}{M_p}$, the ratio of the planet's inertial mass to its gravitational mass. Then *if* this ratio is the same for all planets, Newton's universal law of gravitation would indeed predict the pattern of behavior described by Kepler's third law.

Remarkably, although gravitational mass and inertial mass are conceptually distinct ideas, no one has ever been able to devise an experiment that reveals any measurable difference between them. So far as we know, with our usual choice of units, they are exactly equal (with *any* choice of units, they are proportional), though there is no theoretical reason why that should be true. Because they are always numerically the same, we will hereafter denote both kinds of mass by m, and Equation 8-6 becomes

$$\frac{T^2}{R^3} = \frac{4\pi^2}{Gm_s} \tag{8-7}$$

where the right-hand side is Kepler's k. In the same vein, we will hereafter write Equation 8-5 as

$$F_g = G\frac{m_1 m_2}{r^2} \tag{8-5a}$$

♦**BIG *G*, LITTLE *g*, NEITHER ONE IS "GRAVITY"** We have said that your weight on a planet is the gravitational force that the planet exerts on you. At Earth's surface (a distance R_E from its center of mass), it follows that

$$G\frac{mM_E}{R_E^2} = mg$$

where M_E is Earth's mass, m is the mass of you or some other Earthbound object, and g is, as it has always been, the object's *gravitational acceleration*. We thus see that at Earth's surface

$$g = G\frac{M_E}{R_E^2} \qquad (8\text{-}8)$$

and likewise, at a greater distance r from Earth's center

$$a = G\frac{M_E}{r^2} \qquad (8\text{-}9)$$

Note that g and G are not equal—and neither one is simply "gravity." Little g is an *acceleration*, and by Equation 8-8 has different values at the surfaces of other planets, where different masses and radii substitute for M_E and R_E. G, in contrast, is a proportionality *constant* between the gravitational force and the combination of variables $\frac{m_1 m_2}{r^2}$ on which it depends. It is a *universal* gravitational constant. Calling either one gravity strips it of its physical meaning and obscures these important distinctions.

◆**FINDING G** By calling his law *universal*, Newton was saying that there is nothing special about the "stuff" that makes up Earth or the sun; there is just a lot of it. The gravitational force described by Equation 8-5a describes a force that exists between *any* two bodies of nonzero mass. But the forces that objects of everyday size exert on one another are too small to be readily detectable. To find a value for G, however, one must detect, and indeed *measure*, such a force. Once F is measured for a pair of objects of known mass at a known separation, we could use Equation 8-5a to calculate G.

A set-up (Figure 8-16) of sufficient precision to measure such a force was finally devised by Henry Cavendish (1731–1810) about a century after Newton. When the large lead spheres are moved into Position 2 in the figure, the force they exert on the small spheres causes the latter to rotate and eventually come to rest at an angle at which the suspension wire exerts an equal and opposite torque. If the torque necessary to twist the wire through a certain angle has been previously calibrated, the gravitational force can be determined. The value that Cavendish could then calculate for G was

$$G = 6.67 \times 10^{-11} \text{ N} \cdot \text{m}^2/\text{kg}^2$$

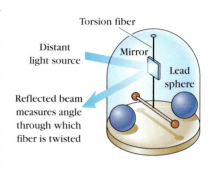

Torsion fiber

Distant light source

Mirror

Lead sphere

Reflected beam measures angle through which fiber is twisted

TOP VIEW

Large spheres in position 1

Large spheres in position 2

BEFORE (no net torque on rod with small spheres)

AFTER (net torque due to gravitational forces; rotation of small spheres exaggerated)

Figure 8-16 The Cavendish balance. A device to measure the universal gravitational constant g.

Example 8-6 *"Weighing" Earth*

Find Earth's mass based on the observations that the moon's average distance from Earth is 3.84×10^8 m, and it completes its orbit in 27.3 days.

Solution

Choice of approach. We again apply Newton's second law in the form of Equation 8-2: $\Sigma F_r = ma_r = \frac{mv^2}{r}$. The radial force ΣF_r that keeps the moon in orbit consists solely of the gravitational force that Earth exerts on the moon, which by Equation 8-5a is $G\frac{m_M m_E}{r^2}$. From the given orbital data, we can find the moon's speed v, which we need to determine $a_r = \frac{v^2}{r}$.

What we know/what we don't. (M = moon, E = Earth)

$$r = 3.84 \times 10^8 \text{ m} \qquad m_M = ? \qquad m_E = ? \qquad G = 6.67 \times 10^{-11} \text{ N} \cdot \text{m}^2/\text{kg}^2$$

Distance for a complete orbit $\Delta s = 2\pi r = 2\pi(3.84 \times 10^8 \text{ m}) = 2.41 \times 10^9$ m

Time for a complete orbit $\Delta t = (27.3 \text{ days})\left(\dfrac{24 \text{ h}}{1 \text{ day}}\right)\left(\dfrac{3600 \text{ s}}{1 \text{ h}}\right) = 2.36 \times 10^6$ s

EXAMPLE 8-6 *continued*

The mathematical solution. Combining Equations 8-2 and 8-5a,

$$G\frac{m_M m_E}{r^2} = \frac{m_M v^2}{r}$$

Because both sides are divisible by m_M, the solution does not depend on the mass of the moon. Solving for m_E, we get

$$m_E = \frac{v^2 r}{G}$$

But

$$v = \frac{\Delta s}{\Delta t} = \frac{2.41 \times 10^9 \text{ m}}{2.36 \times 10^6 \text{ s}} = 1.02 \times 10^3 \text{ m/s}$$

so

$$m_E = \frac{(1.02 \times 10^3 \text{ m/s})^2 (3.84 \times 10^8 \text{ m})}{6.67 \times 10^{-11} \text{ N} \cdot \text{m}^2/\text{kg}^2} = \mathbf{5.99 \times 10^{24} \text{ kg}}$$

This differs in the last decimal place from the value in standard tables because of rounding errors.

◆ Related homework: Problems 8-40 and 8-41.

Example 8-7 provides more practice reasoning with Newton's universal law of gravitation.

Example 8-7 *How Constant Is g?*

For a guided interactive solution, go to Web Example 8-4 at www.wiley.com/college/touger

If the "standard" value of g is at sea level, at what altitude would your weight be reduced by 1%?

Brief Solution

Choice of approach. Your weight is the gravitational force that Earth exerts on you. If your weight mg is reduced by 1%, it means the gravitational acceleration is $0.99g$, or $\frac{a}{g} = 0.99$. At sea level (Equation 8-8), $g = G\frac{M_E}{R_E^2}$, and at a greater distance r from Earth's center (Equation 8-9), $a = G\frac{M_E}{r^2}$, so

$$\frac{a}{g} = \frac{G\dfrac{M_E}{r^2}}{G\dfrac{M_E}{R_E^2}} = \frac{R_E^2}{r^2}$$

What we know/what we don't.

$$\frac{a}{g} = 0.99 \qquad R_E = 6.38 \times 10^6 \text{ m} \qquad r = ? \qquad \text{Altitude } h = r - R_E$$

The mathematical solution. Because $\frac{a}{g} = \frac{R_E^2}{r^2}$, we can solve for r to get

$$r = R_E\sqrt{\frac{g}{a}} = (6.38 \times 10^6 \text{ m})\sqrt{\frac{1}{0.99}} = 6.41 \times 10^6 \text{ m}$$

Then

$$h = r - R_E = 6.41 \times 10^6 \text{ m} - 6.38 \times 10^6 \text{ m} = \mathbf{3 \times 10^4 \text{ m}}$$

◆ Related homework: Problems 8-36 and 8-37.

STOP&Think Look at the result we just calculated. Can you get this high above sea level anywhere on Earth without being airborne? Over vertical distances of, say, 1000 m or less, is it a reasonable approximation to assume that g is constant? ◆

◆**APPARENT WEIGHTLESSNESS IN SPACE** We can think of an object orbiting Earth as continually falling from its would-be tangential path toward Earth's center (Figure 8-14). In fact, it is *freely falling,* because its acceleration is just the gravitational acceleration. As is true of any object orbiting Earth, the acceleration of a "manned" satellite orbiting Earth in an orbit of radius r is $a = G\frac{M_E}{r^2}$, and so is the acceleration of the astronaut inside it. So just as a freely falling elevator exerts no normal force on its occupant (recall Example 5-2), the satellite exerts no normal force on the astronaut, and her *apparent* weight is zero. There is still a gravitational force on her, so she is not weightless in that sense. But the gravitational force doesn't pull her up or down relative to the cabin, so she can drift freely within it—up, down, sideways. The normal forces that parts of your body ordinarily exert on one another are likewise absent, as is gravitation's effect on the flow of blood and other fluids relative to the solid parts of your body. The physiological effects that this may have on your body are a subject of ongoing research in major space programs.

To explore this idea further, go to **WebLink 8-5: Free Fall on Earth and in Space** at www.wiley.com/college/touger

8-4 Gravitational Potential Energy Revisited

In Chapter 6, our derivation of the expression mgy for gravitational PE rested on the assumption that g, and the gravitational force mg, remained constant over the vertical distances of interest. Then $PE_{grav} = mgy$ is valid only when y is small enough for that to be a good approximation of reality.

In general, as in Figure 6-18, the total change of position in the direction of the gravitational force, which we may now call the radial rather than the vertical direction, is $b = \Delta b_1 + \Delta b_2 + \Delta b_3 + \cdots$, or equivalently, $\Delta r = \Delta r_1 + \Delta r_2 + \Delta r_3 + \cdots$ But now we must accommodate the fact that the value of the gravitational force $G\frac{m_1 m_2}{r^2}$ varies from one Δr to the next. Under these circumstances, the increase in gravitational PE—that is, the work that we must do to move a body from r_1 to r_2 against the gravitational force at constant speed (so that $\Delta KE = 0$)—is

$$\Delta PE = \Sigma F_r \Delta r = G\frac{m_1 m_2}{r_1^2}\Delta r_1 + G\frac{m_1 m_2}{r_2^2}\Delta r_2 + G\frac{m_1 m_2}{r_3^2}\Delta r_3 + \cdots$$

In the sum, the value of r is understood to be different in each denominator.

Because finding this sum exactly requires calculus, we must omit some steps in the mathematical reasoning. But by doing the summation, it is possible to show that

$$\Delta PE = \left(-G\frac{m_1 m_2}{r_2}\right) - \left(-G\frac{m_1 m_2}{r_1}\right) = \Delta\left(-G\frac{m_1 m_2}{r}\right)$$

that is, it equals a change in a quantity which depends, as it should, on the position r. We thus take that quantity to be the gravitational PE.

When dealing with distances large enough so that the gravitational force must be treated as variable, the **gravitational PE** is

$$PE_{grav} = -G\frac{m_1 m_2}{r} \qquad (8\text{-}10)$$

On June 18, 1973, Sally Ride, a 32-year-old physicist with four degrees, became the first American woman in space.

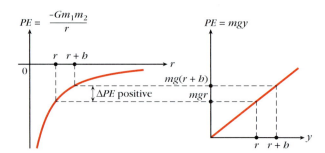

Figure 8-17 How gravitational potential energy varies with r.

◆**Caution:** Showing that something "makes sense" is not in itself a guarantee that it is correct. A plausibility argument is *not* a proof.

Let's look at some reasons why this result "makes sense" in terms of our previous understanding:

- *It has units of work or energy:* Because the force varies as $\frac{1}{r^2}$, its units are $\frac{\lfloor \text{units of } Gm_1m_2 \rfloor}{\text{m}^2}$. If we multiply this by a distance in meters to get a unit of work, we get

$$\frac{\lfloor \text{units of } Gm_1m_2 \rfloor}{\text{m}^2} \times \text{m} = \frac{\lfloor \text{units of } Gm_1m_2 \rfloor}{\text{m}}$$

which are the units of PE_{grav} in Equation 8-10.

- *It increases as the height increases:* Figure 8-17 shows a graph of PE_{grav} against r. Notice that although it is always negative, it becomes less negative as the denominator r increases. It thus goes up as we go from a smaller distance r to a greater distance $r + b$, just as mgy, the more limited expression for PE_{grav}, increases when y increases by an amount b.

- *For very small increases in r, Equation 8-10 and $PE_{\text{grav}} = mgy$ give us the same change in potential energy:* This can be shown mathematically (see Problem 8-79).

Because it keeps getting less negative as r increases, $-G\frac{m_1m_2}{r}$ eventually approaches zero. By taking this as our expression for PE_{grav}, we are in effect saying that the zero level of PE is at $r = \infty$. Although this may at first seem like a counterintuitive choice, remember that only changes in PE have physical meaning, and as we have seen above, $-G\frac{m_1m_2}{r}$ does change in the expected ways. Also, it reinforces the idea that a system has potential energy because of the interaction(s) between the objects in the system, and the interactions are reduced to zero when the objects are infinitely far apart.

Case 8-2 ◆ *Bound and Unbound Systems*

For an isolated system in which the only internal forces are gravitational, an advantage of having a negative expression for the PE is that the total mechanical energy—$E = KE + PE_{\text{grav}}$—consists of one term that is always positive and one that is always negative. It is interesting to see what happens when the total is positive and what happens when it is negative.

Figure 8-18 distinguishes between two instances in which a rocket blasts off from Earth's surface. In one instance the rocket escapes from Earth; in the other, it falls back. We will assume that the thrust continues for only a short distance above the launching pad. After that, the rocket gains no further energy

from the fuel, and the total energy $E = KE + PE_{\text{grav}}$ remains constant. Then the rocket slows down as it rises—it loses KE as it gains gravitational PE, just like a ball thrown in the air. The KE becomes less positive as the increasing PE becomes less negative, but the total energy—the sum of the two—remains the same.

In Instance 1, the total energy is positive. When the rocket is so far away from Earth that the gravitational PE is essentially zero, the KE is still positive. The gravitational force at this distance is effectively reduced to zero as well, so the rocket keeps going; it *escapes* from Earth's gravitational pull.

In Instance 2, where the total energy is negative, the KE reaches zero from above *before* the PE reaches zero from below (state 3 in the figure). The rocket's speed is reduced to zero while it is still at a finite distance r from Earth and still experiences Earth's gravitational pull. It therefore falls back to Earth, just as any ordinary projectile does when its PE reaches a maximum and its speed drops to zero. As it falls back (states 4 and 5), speeding up in the downward direction, its KE becomes increasingly positive and its PE increasingly negative.

We see that

when the total energy is . . .	the rocket . . .
positive	escapes
negative	falls back

STOP&Think What happens to the rocket when the total energy is zero? ◆

In the first instance, the Earth + rocket system is an **unbound system,** meaning that the separation between the rocket and Earth is unlimited. In the second instance it is a **bound system:** The rocket can get no further than a certain maximum distance from Earth.

This language is also used to describe systems in which the PE is electrical, not gravitational: The electrons + nucleus system that comprises an atom, or the system of atoms making up a molecule. If you have taken a chemistry course, you may already have encountered chemists speaking of *binding energies,* which are negative because the total energy of a bound system is negative.

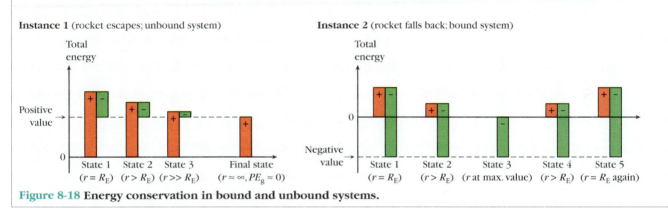

Instance 1 (rocket escapes; unbound system)

Instance 2 (rocket falls back; bound system)

Figure 8-18 Energy conservation in bound and unbound systems.

The threshold between a rocket escaping ($E > 0$) and falling back ($E < 0$) occurs when $E = 0$. The initial velocity required for this borderline condition when there is no further thrust after liftoff is called the **escape velocity,** denoted by v_e. For a rocket of mass m to escape from a planet or moon of mass M, it must have a total energy

$$\tfrac{1}{2}mv_e^2 + \left(-G\frac{mM}{R}\right) = 0$$

Solving for v_e, we get

$$v_e = \sqrt{\frac{2GM}{R}} \qquad\qquad (8\text{-}11)$$

Note that this general result (R and M can be the radius and mass of any astronomical body) does not depend on the escaping object's mass m.

Example 8-8 *Anticipating Black Holes*

Light travels in a vacuum at a speed of 3.00×10^8 m/s (this value is generally denoted as c) and at very nearly this speed in air. In 1796, P. S. Laplace supposed (incorrectly) that light is made up of particles of very small mass. He then raised the following question about the velocity these particles must have to escape from a star: Into how small a sphere must the star's mass be compressed for the escape velocity to equal the speed of light?

EXAMPLE 8-8 *continued*

Solution

Restating the question. A measure of the size of a sphere is its radius. The question is therefore: *What is R when $v_e = c$?*

Choice of approach. Conservation of energy has already led us to Equation 8-11 for the escape velocity. Although Einstein's theory of special relativity later altered this equation for particles traveling at or near the speed of light, we can still use it as Laplace did to get a rough estimate of R.

What we know/what we don't.

$$G = 6.67 \times 10^{-11} \, \frac{\text{N} \cdot \text{m}^2}{\text{kg}^2} \qquad v_e = c = 3.00 \times 10^8 \, \text{m/s}$$

$$M = M_{\text{star}} = ? \qquad\qquad R = ?$$

Because we are not dealing with a specific star, our result will not be a numerical value but must be expressed in terms of M_{star}.

The mathematical solution. Solving Equation 8-11 for R gives us

$$R = \frac{2GM_{\text{star}}}{v_e^2} = \frac{2\left(6.67 \times 10^{-11} \, \dfrac{\text{N} \cdot \text{m}^2}{\text{kg}^2}\right) M_{\text{star}}}{(3.00 \times 10^8 \, \text{m/s})^2} \qquad (8\text{-}12)$$

$$= \left(\mathbf{1.48 \times 10^{-27} \, \frac{m}{kg}}\right) \boldsymbol{M_{\text{star}}}$$

➥**Evidence for black holes:** We cannot "see" black holes, either by visible light or by other radiated energy, which travels at the same speed. But their gravitational pull acts as the inward radial force on bodies whose motions we can detect, for example, there are stars orbiting black holes, and hot gas pulled from neighboring stars forms whirling disks about black holes, somewhat as scattered matter forms rings around Saturn.

Making sense of the results. We see from Equation 8-11 that as R decreases, v_e increases. But light always travels through space at the same speed, never faster. Thus, if a star were compressed into a sphere of less than this radius, then by Laplace's reasoning no light could escape from its surface and it would therefore appear black. Laplace's ideas were abandoned at the beginning of the nineteenth century when new evidence led to the rejection of the idea that light had mass. But in the twentieth century, Einstein's theory of general relativity revived the idea that large masses have a significant gravitational effect on light. Furthermore, it was found that some very massive stars do begin to collapse inward on themselves late in their lifetimes, and indirect astronomical evidence of these "collapsed stars" has accumulated. In 1968, John Wheeler introduced the term **black holes** to describe them.

The mass of our own sun is about $M_{\text{sun}} = 2 \times 10^{30}$ kg. Using this value for M_{star} in Equation 8-12 gives us $R = (1.48 \times 10^{-27} \, \text{m/kg})(2 \times 10^{30}) \approx 3 \times 10^3$ m or 3 km. Our own sun would have to be compacted into a sphere with a radius of just 3 km to become a black hole!

Expressed in terms of M_{sun}, Equation 8-12 becomes

$$R = \left(1.48 \times 10^{-27} \, \frac{\text{m}}{\text{kg}}\right) M_{\text{star}} \times \frac{M_{\text{sun}}}{M_{\text{sun}}} = \left(1.48 \times 10^{-27} \, \frac{\text{m}}{\text{kg}}\right) M_{\text{star}} \times \frac{2 \times 10^{30} \, \text{kg}}{M_{\text{sun}}}$$

or

$$R_S \approx (3 \times 10^3 \, \text{m}) \frac{M_{\text{star}}}{M_{\text{sun}}} \qquad (8\text{-}13)$$

The S is because this radius is commonly called the Schwarzschild radius.

◆ Related homework: Problems 8-46, 8-47, and 8-48.

✦ SUMMARY ✦

Moving objects deviate from straight line paths when and only when they are subjected to forces or force components perpendicular to the direction of motion. For a body traveling in a circle, the perpendicular direction at each point is the **radial** direction, and the direction of motion is the **tangential** direction.

For an object traveling in a circle *at constant speed,* the acceleration and the total force have *only* a radial component, that is, they are toward the center of the circle. We obtain this radial acceleration from the geometry of circular motion:

Radial acceleration (also called *centripetal acceleration*)

$$a_r = \frac{v^2}{r} \qquad (8\text{-}1)$$

(scalar component or magnitude of acceleration vector)

By Newton's second law, we find the total force required to give a body of mass *m* this acceleration:

Total radial force (also called the *centripetal force*)

$$\Sigma F_r = ma_r = \frac{mv^2}{r} \qquad (8\text{-}2)$$

(scalar component or magnitude of total force vector)

The total radial (or centripetal) force is not a *kind* of force in addition to other forces—it is made up of existing forces such as friction forces (Case 8-1 and Example 8-2), tension-like forces (Example 8-3), normal forces (Example 8-4), or gravitational forces. These are the forces that should appear in free-body diagrams.

For *vertical* circular paths, velocity is *not* ordinarily constant but may be related to height by applying conservation of mechanical energy (Example 8-5).

Newton recognized that the radial force keeping heavenly bodies in their orbits was the same type of force that causes projectiles to fall back to Earth. He formulated his universal law of gravitation:

Newton's Universal Law of Gravitation

$$F = G\frac{m_1 m_2}{r^2} \qquad (8\text{-}5a)$$

Universal gravitational constant $G = 6.67 \times 10^{-11}\,\text{N} \cdot \text{m}^2/\text{kg}^2$

This gives the magnitude of the force that *any* two objects with mass exert on each other. The *direction* of the force on each object is toward the other.

Using Equation 8-5a, Newton showed that for all objects in orbit around the same central body (planets around the sun, moons around Jupiter), the ratio $\frac{T^2}{R^3}$ should have the same value, thus explaining why the pattern known as Kepler's third law of planetary motion held true.

From Equation 8-5a, it follows that a body *at or near Earth's surface* (and only there) experiences an acceleration

$$g = G\frac{M_E}{R_E^2} \qquad (8\text{-}8)$$

Conservation of energy may also be applied to situations where objects travel such large distances that *g* cannot be taken as constant.

When dealing with distances large enough so that the gravitational force must be treated as variable, the **gravitational PE** is

$$PE_{\text{grav}} = -G\frac{m_1 m_2}{r} \qquad (8\text{-}10)$$

In **bound systems,** where the separation between bodies is limited, *their total mechanical energy E = PE + KE is negative.* In **unbound systems,** where one bodies can escape completely from each other, *their total mechanical energy is positive.* From the *threshold* condition (*E* = 0), we find that the **escape velocity** is

$$v_e = \sqrt{\frac{2GM}{R}} \qquad (8\text{-}11)$$

◆QUALITATIVE AND QUANTITATIVE PROBLEMS◆
Hands-On Activities and Discussion Questions

The questions and activities in this group are particularly suitable for in-class use.

8-1. *Hands-On Activity.* Take a long length of rope. Tie one end of the rope around a post or other fixed object that will not be moved or damaged if you pull fairly hard on it. Tie the other end around your waist so that it won't slip when you pull on it. The length of rope between you and the post

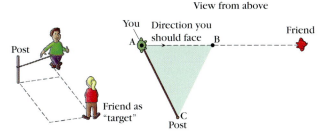

Figure 8-19 Problem 8-1

should be roughly 5 to 10 feet. Have a friend join you, and arrange yourselves as shown in the top view in Figure 8-19. If you imagine an equilateral triangle ABC, the post should be at C, you should be at A, and your friend should be lined up with point B, but further away from you. Now start walking toward your friend. Try to follow as straight a path as possible.

a. How easy is this to do before you reach point B? After you reach point B?

b. What happens after you reach point B?

c. Do you feel the rope pushing or pulling on you or digging into you before you reach point B? What about after you reach point B? If so, where on your body do you feel it? In what direction does it seem to be pushing or pulling?

d. What is the connection between your motion after you pass point B and the force that the rope exerts on you?

8-2. *Discussion Question.* Compare the role of the frictional force in Case 8-1 with the role of the force exerted on you by the rope as you try to go beyond point B in Problem 8-1.

Review and Practice

Section 8-1 Circular Motion Is Accelerated Motion

8-3. Student A argues, "When I whirl a block on a string in a nearly horizontal circle, the forces on the block in the horizontal plane are the force exerted by the string and the centripetal force." "No," responds Student B, "the block stays at a fixed distance from the center because the string's force is balanced by a centrifugal force, that is, a force that tends to cause the block to fly out from the center." How should you respond to each of these students?

8-4. In Problem 8-3, if Student A loses his grip on the string, eliminating the tension-like force on the block, does the block fly out from the center as Student B suggests? Sketch a diagram showing how the block moves before and after the string is released.

8-5. A centrifugal force is a force directed outward from the center of a circle. (The word comes from the Latin for "fleeing from the center"; it has the same root *fuge* from the Latin "to flee" as does *refugee*.) Is the operation of a centrifuge based on centrifugal force? Briefly explain.

8-6. Consider On-the-Spot Activity 4-6, as depicted in Figure 4-13.
a. If the marble rolls along the belt at constant speed, is it accelerating? If so, in what direction?
b. As the marble rolls along the belt, what object exerts the total radial force on the marble and what kind of force is it?
c. When the marble passes the tip of the belt, does this radial force continue to be exerted?
d. What can you say about the marble's acceleration once it passes the tip of the belt?
e. Based on your answer to **d,** will the marble's direction continue to change? Therefore, will the marble continue to point B or point D in Figure 4-13?

8-7. SSM Suppose you drive your car around a curve at a constant speed of 10.0 m/s.
a. What must be the radius of curvature of the curve if the car is to have a radial acceleration equal in magnitude to the gravitational acceleration *g*?
b. Repeat **a** for a radial acceleration of "two *g*'s."
c. Which is the sharper turn?

8-8. Suppose you drive your car around a curve with a radial acceleration equal in magnitude to the gravitational acceleration *g*. What happens to you as the driver? (What do you feel? What do you experience?) Why? How would it be different if you take the curve fast enough so that your radial acceleration is "two *g*'s"?

8-9. An object traveling along a circular path has a speed of 30 m/s at $t = 0$. If its radial acceleration remains constant at 4 m/s^2, what is its speed at $t = 2$ s?

8-10.
a. Compare the radial accelerations of two cars going around the same curve, car A at 10.0 m/s, and car B at 20.0 m/s. Express your comparison as a ratio $\frac{a_A}{a_B}$ of the two accelerations.

b. If the curve is unbanked, compare the frictional forces that must be exerted on the tire treads to keep the two cars from skidding. Express your comparison as a ratio $\frac{f_A}{f_B}$.

8-11. A softball player wishing to stretch a single into a double makes a wide turn as she rounds first base toward second. A sharp turn would reduce the distance that she runs. Why doesn't she do that? Use Equation 8-2 in explaining how the relevant quantities vary.

8-12. Radial forces operate on both the tiniest and the grandest scales in our universe. Find the magnitude of the total radial force that keeps each of the following objects in its circular orbit under the given conditions.
a. An electron: In the Bohr model of the hydrogen atom, when the electron is most tightly bound to the single-proton nucleus, it travels about the nucleus in an orbit of radius 5×10^{-11} m at a speed of about 2×10^6 m/s.
b. The sun: Our sun is 2.6×10^{20} m from the center of the great spiral galaxy, which, from our position within it, we see as the Milky Way. It is moving in a roughly circular path about the center at a speed of 2.2×10^5 m/s.

Section 8-2 Examples of Radial Forces

8-13. In Figure 8-7, what kind(s) of force make up the total radial force that the figure skater's partner exerts on her? Why do figure skaters sometimes put rosin on their hands?

8-14. Professor Samuelson's car can safely take a certain bend in the road at a maximum speed of 15 m/s on a dry day. If the maximum frictional force between her tires and the road is reduced by 20% when it is raining, at what maximum speed can she go around the bend on rainy days?

8-15. SSM WWW A truck on a level semicircular exit ramp with a radius of 100 m begins to skid out when it exceeds a speed of 19 m/s. What is the coefficient of static friction between the road and the truck's tire treads?

8-16. A truck driver knows she can take a certain turn at a maximum safe speed of 12 m/s when her truck is empty. At what maximum speed can she safely take the turn when she is carrying a load equal to the weight of her truck?

8-17. Fully describe the forces or force components that comprise the total radial force on Tarzan in Example 7-8 when he is halfway through the downward part of his swing.

8-18.
a. Draw a carefully labeled free-body diagram of a car of mass *m* parked on a banked road, assuming the road makes an angle *θ* with the horizontal.
b. Now compare your free-body diagram with free-body diagram *a* in Example 8-4. That diagram is for a car traveling in a horizontal circle around a banked curve. Do any of the forces have different directions in the two diagrams? Briefly state the reasons for any differences.
c. A free-body diagram inventories all the forces contributing to the total force on an object. How does the total force on the car compare in the two situations? Justify your answer.

d. Suppose that the car that was parked begins to move very slowly around the banked curve, without sliding up or down the bank. How does the direction of the frictional force on the car compare to the direction of $\vec{f}$ shown in diagram **a** in Example 8-4? Explain.

8-19. **SSM WWW** The block in Figure 8-20 is whirled by two ropes. The tension in rope 1 is just equal to the object's weight, that is, $T_1 = mg$. Find the block's radial acceleration.

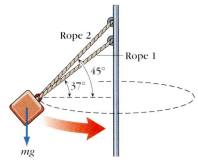

Figure 8-20 Problems 8-19 and 8-54

8-20. A block is connected to a post by a spring. When the block is whirled in a horizontal circle around the post, the spring stretches. Why?

8-21. A block of mass 0.40 kg is connected to a post by a spring with an unstretched length of 0.18 m. When the block is whirled in a horizontal circle around the post at a speed of 8.0 m/s, the spring stretches to a length of 0.20 m. Find the spring constant k of the spring.

8-22. A car goes around a turn at a speed of 9.2 m/s. The radius of curvature of the turn is 8.3 m. If the upholstery of the passenger seat is so slippery as to exert negligible frictional force, how much force will a passenger with a mass of 75.0 kg exert on the car door during the turn assuming the passenger is on the outside of the turn?

8-23. Repeat Problem 8-22 for the case where the frictional force that the seat exerts on the passenger is reduced not to zero but to 15% of the normal force.

8-24. A vehicle is going around a banked curve making an angle θ with the horizontal. F_N, f, and mg are the magnitudes of the normal, frictional, and gravitational forces on the vehicle. Consider the following force components:

F_N $\quad$ $F_N \cos\theta$ $\quad$ $F_N \sin\theta$ $\quad$ mg $\quad$ f $\quad$ $f\cos\theta$ $\quad$ $f\sin\theta$

a. Which (one or more) of these components are in the radial direction?

b. Which (one or more) of these components are in the vertical direction?

8-25. You are about to drive around a curve with a radius of curvature of 50.0 m. The road is so iced over that the frictional force it can exert on your car is virtually zero. But the curve has a 20° bank. At what maximum speed can you negotiate this curve without skidding? (*Hint:* Think in terms of the components you identified in Problem 8-24.)

8-26. When a child rides on an ordinary playground swing, is the combined force that the ropes exert at the highest point of the child's path greater than, equal to, or less than the force that they exert at the lowest point? Briefly explain your answer.

8-27. **SSM** A ball attached to a string is thrown horizontally with an initial speed v_0 (Figure 8-21) and is constrained by the string to move in a vertical circle.

a. At point C, is the force that the string exerts on the ball less than, equal to, or greater than at point A?

b. Is the horizontal component of the force that the string exerts on the ball greatest at point A, point B, or point C?

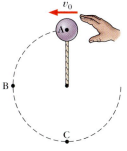

Figure 8-21
Problem 8-27

8-28. Suppose the circle in Figure 8-21 has a radius of 0.80 m. The ball has a mass of 0.30 kg and its speed at point A is 2.9 m/s. Find the tension force that the string exerts on the ball at point A and at point B.

8-29. Verify the claims made in the *Making sense of the results* section of Example 8-5. In particular, begin by assuming that, as shown in the correct free-body diagrams, the roller-coaster car has a weight of 5000 N, and the normal force on it is 15 000 N and directed downward when it is at A, and 45 000 N and directed upward when it is at B. Assume further that the loop has a radius of 8 m.

a. Find the speed of the car when it is at A.

b. Find the speed of the car when it is at B.

c. Find the gain in KE as the car descends from A to B.

d. Use $\Delta PE = mg\Delta y$ to find the loss in PE as the car descends from A to B.

e. Is the amount of PE lost equal to the amount of KE gained? (Your answer to this question should end up the same no matter what value you used for r. You could verify this by repeating your calculations with any other value of r.)

Section 8-3 The Universal Law of Gravitation

8-30. Explain the difference between *gravitational mass* and *inertial mass*.

8-31. Estimate the gravitational attraction that you and a close (adult human) friend or relative of your choice exert on each other when you are hugging. Remember that the distance in the denominator of Newton's universal law of gravitation is the distance between the centers of mass of the two bodies, not between their surfaces.

8-32. By what numerical factor does the gravitational force that two objects exert on each other get multiplied if the distance between the objects is **a.** tripled? **b.** reduced by two-thirds of its original value? **c.** multiplied by $\sqrt{2}$? **d.** 0.9 times its original value?

8-33. **SSM WWW** Two objects exert a gravitational force of magnitude F_g on each other. By what numerical factor must you multiply the distance between two objects to **a.** reduce F_g to one-ninth of its original value? **b.** increase F_g to double its original value?

8-34. Calculate the gravitational force that the sun exerts on Earth (assume Earth's orbit is approximately a circle).

8-35.

a. Calculate the gravitational force that the sun exerts on you, and compare it to your weight. Express your comparison as a ratio $\dfrac{F_{\text{by sun on you}}}{\text{your weight}}$.

b. What effect does this force have on you?
i. Almost none. **ii.** It keeps you going around Earth's axis as Earth rotates. **iii.** It keeps you orbiting with Earth around the sun. **iv.** It very gradually brings you closer to the sun.

8-36. At what altitude above Earth's surface would the value $g = 9.8$ m/s^2 differ from the actual gravitational acceleration by 5%?

8-37.

a. Use Newton's universal law of gravitation to find the gravitational force of a 40-kg dancer at a point in a leap when her center of mass is 1.5 m above Earth's surface.

b. Compare this result to her weight mg at Earth's surface. Express your comparison as a ratio $\frac{\text{answer to } \boldsymbol{a}}{\text{weight at surface}}$.

8-38. What is the approximate gravitational force that the rest of Earth exerts on a 1-kg chunk of matter at the center of Earth's core?

8-39. A weather satellite is in a circular orbit about Earth. By what factor will the time required for a single orbit be multiplied if **a.** the radius of its orbit is doubled? **b.** its mass is tripled? **c.** it is instead placed in an orbit of the same radius about Mars ($M_{\text{Mars}} = 6.58 \times 10^{23}$ kg)?

8-40.

a. If a telecommunications satellite in a circular orbit with a radius of 2.66×10^7 m takes one day to complete an orbit about Earth, how long does it take a satellite with a radius of 3.99×10^7 m to circle Earth?

b. Which of these orbits is more desirable for a telecommunications satellite? Briefly explain why.

8-41. **SSM** Europa, one of Jupiter's moons, orbits Jupiter in 3.551 days. The radius of the orbit is 6.709×10^8 m. Which (one or more) of the following could you calculate without any additional data about Jupiter or Europa? The mass of Europa; the mass of Jupiter; the force that Jupiter exerts on Europa; how much you would weigh at the surface of Jupiter.

8-42. The centers of masses of two objects are a distance R apart. How does the gravitational force that one of these object exerts on the other change (how many times its original value is it) when this distance is

a. increased to $4R$?   **b.** reduced to $\frac{R}{4}$?

Section 8-4 Gravitational Potential Energy Revisited

8-43. Earth's mass is 5.98×10^{24} kg; the moon's mass is 7.35×10^{22} kg. The average distance from Earth to the moon (center to center) is 3.84×10^8 m.

a. Calculate the gravitational potential energy of the Earth-moon system.

b. Would you expect the kinetic energy of the moon in its orbit around Earth to be less than, equal to, or greater than the absolute value of the kinetic energy you calculated in **a**? Briefly explain why. Don't do part **c** until after you answer **b**.

c. Do a calculation to estimate the kinetic energy of the moon as it orbits Earth, and see if the result you obtain agrees with your answer to **b**.

8-44. Equation 8-10 tells us that the gravitational PE of a system consisting of Earth and an object of mass m is $PE_{\text{grav}} = -G\frac{mM_E}{r}$.

a. At what value of r would the gravitational PE be zero? Explain.

b. Does the PE increase or decrease as the object falls toward Earth?

c. What would happen to the value of PE_{grav} at $r = 0$? Is it physically possible for r to have this value? Explain.

d. Equation 8-10 is based on the universal law of gravitation for point objects. Would you expect it to hold true for an object that has bored deep into Earth's core? Explain your answer by talking about what happens physically.

8-45. The paths of the planets around the sun are elliptical (Figure 8-22). Although most of the planetary orbits are fairly close to being circles, this is not the case for Mercury or Pluto. At **perihelion** (P), its point of nearest approach to the sun, Mercury is 4.59×10^{10} m from the sun. At **aphelion** (A), its most distant point, it is 6.97×10^{10} m from the sun.

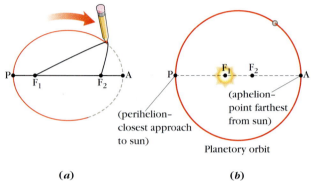

(*a*)   (*b*)

Figure 8-22 Kepler's second law of planetary motion: The path of each planet is an ellipse with the sun at one focus. For Problem 8-45. (*a*) An ellipse can be drawn by moving the pencil while keeping the string taut. The points where the string's ends are fixed (F_1 and F_2) are called the foci (plural of focus) of the ellipse. (What happens to the ellipse as the foci are moved closer together? Try it for yourself.) (*b*) The orbit of Mercury (with the sun's size exaggerated). The sun's center is at one focus (F_1). For some planets, such as Earth, the second focus F_2 is so close to F_1 that the planet's orbit is almost a perfect circle.

a. Is Mercury's speed at aphelion greater than, equal to, or less than its speed at perihelion? Briefly explain why. Don't do part **b** until after you answer **a**.

b. If Mercury's orbital speed is 3.9×10^4 m/s at aphelion, calculate its orbital speed at perihelion.

8-46. The moon has a mass of 7.35×10^{22} kg and a radius of 1.74×10^6 m. At what speed would a rocket have to be launched from the moon's surface (assuming no further thrust after liftoff) in order not to fall back to the moon?

8-47. In considering a rocket's escape velocity, why can we no longer simply apply the equation $E = KE + PE_{\text{grav}}$ if after the rocket is off the ground, it continues to burn fuel and expel the products of combustion to maintain thrust?

8-48. Into how small a sphere would the mass of our sun have to be compacted to prevent particles traveling at half the speed of light from escaping it? Answer by giving the value of the sphere's radius.

Going Further

The questions and problems in this group are not organized by section heading, so you must determine for yourself which ideas apply. Some of them will be more challenging than the Review and Practice questions and problems (especially those marked with a • or ••).

8-49. SSM The pendulum bob in Figure 8-23 is released from rest at point A and swings through point B. Of the following locations, choose the one that correctly completes each statement below about the pendulum bob's acceleration: *only at point A; only at point B; at both point A and point B; at no point; at a point in between A and B.*

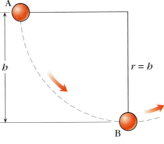

Figure 8-23 Problem 8-49

a. Its acceleration has a zero *radial* component ____.

b. Its acceleration has a zero *tangential* component ____.

c. Its acceleration has a zero *vertical* component ____.

8-50. The three identical blocks A, B, and C in Figure 8-24 are simultaneously released from rest at $y = b$. Assume air resistance and friction forces are negligible in all three situations. Rank the blocks in order of each of the following quantities. Rank them from least to greatest, indicating any equalities.

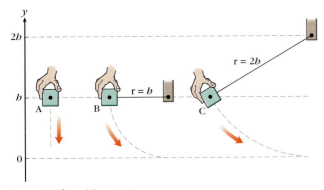

Figure 8-24 Problem 8-50

a. The vertical velocity component v_y that each block has at the instant it reaches $y = 0$.

b. The horizontal velocity component v_x that each block has at the instant it reaches $y = 0$.

c. The speed that each block has at the instant it reaches $y = 0$.

d. The time that each block takes to go from $y = b$ to $y = 0$.

e. The total force on each block at the instant it reaches $y = 0$.

f. The horizontal acceleration component a_x that each block has at the instant it reaches $y = 0$.

g. The vertical acceleration component a_y that each block has at the instant it reaches $y = 0$.

h. The magnitude of the acceleration that each block has at the instant it reaches $y = 0$.

i. The magnitude of the acceleration that each block has as soon as it is released at $y = b$.

8-51. When you drive along a curved stretch of road, can the radial force that keeps the car in its curved path be any of the forces that

a. you exert on the steering wheel or steering column? Briefly explain.

b. the steering column exerts on another part of the car? Briefly explain.

8-52. What does $F = \frac{mv^2}{r}$ tell you about the total radial force of a train moving at high speed along a *straight* track? (*Hint:* What is the radius of curvature of this track?)

8-53.

a. In laying railroad tracks, a straight length of track is never connected directly to a length of track that forms an arc of a circle. Instead, there must be a length of track that builds gradually from being straight to having the desired radius of curvature. Why?

b. In driving a car, what action with the steering wheel would produce an equivalent experience to traveling directly from a straight stretch of track to one that is an arc of a circle? How is this different than what you ordinarily do?

•8-54.

a. If the block in Figure 8-20 moves in a circle of radius 0.40 m, for what range of values of the block's speed will the tensions in both ropes be nonzero?

b. Briefly tell what happens when the value of the block's speed is *above* the range of values in **a.**

c. Briefly tell what happens when the value of the block's speed is *below* the range of values in **a.**

8-55. A marble is placed in a bowl shaped like a hemisphere, and the bowl is swirled around, setting the marble in motion. What happens to the marble if the swirling becomes faster? Why?

8-56. An in-line skater goes around the inside of a circular park fountain in which the water has been shut off because of a water shortage. The floor of the fountain slopes down toward the center at an angle of 18° to the horizontal. Going in a circle of radius 8.5 m, the skater encounters a short stretch of slick surface.

a. Assuming the slick surface exerts a negligibly small frictional force on the skates, how fast must the skater pass over this area to avoid slipping sideways, that is, up or down the slope?

b. If the skater crosses the slick area at a speed greater than the value found in **a,** does she end up on a path that is further up the slope, on a path that is further down the slope, or on the same path but with a different speed?

c. Answer the question in **b** if the skater crosses the slick area at a speed less than the value found in **a.**

•8-57. Again consider the in-line skater in Problem 8-56. Away from the small area of slick surface, the maximum static frictional force that the fountain surface can exert on the skates is 0.30 times the normal force that it exerts. Between what minimum and what maximum speed can the skater continue to go around in a circle of radius 8.5 m without slipping sideways either down or up the sloped fountain floor?

8-58. A 0.12-kg ball is attached to a length of rubber band and whirled in a horizontal circle fast enough so that the rubber band is essentially horizontal as well. The rubber band has an elastic constant of 90 N/m.

a. When the ball is whirled at a speed of 8.5 m/s, the rubber band is stretched to a total length of 0.50 m. At this speed, what is the tension in the rubber band?

•*b.* By an appropriate calculation, check to see if our assumption that the rubber band is very nearly horizontal at this speed is a good one.

c. How much further does the rubber band stretch when the ball's speed is increased to 9.5 m/s?

8-59. Suppose you are riding the Turkish Twist in Figure 8-10, and it has a radius of 9.0 m. As you ride, the wall exerts a normal force on you. Suppose the floor drops out when you are moving at a speed of 18 m/s. If the maximum static frictional force that the wall can exert on you is 30% of the normal force, is the Twist spinning fast enough to prevent you from sliding downward? Support your answer with a suitable calculation.

8-60. Pick the word that best completes each of the following sentences about a cyclist riding a bicycle (Figure 8-25) around a curve in a road banked at an angle of 15° to the horizontal. Give a brief reason for each.

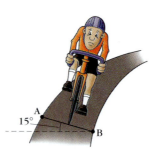

Figure 8-25 Problem 8-60

a. The total normal force that the road exerts on the bicycle is (*always, sometimes,* or *never*) equal and opposite to the total weight of bicycle + cyclist as they go around the turn.

b. The total radial force on the cyclist is (*always, sometimes,* or *never*) directed outward from the center of the circular path followed by the bicycle and cyclist.

c. The frictional force that the road exerts on the tires (*always, sometimes,* or *never*) has a component that is directed up the slope in the road (in the direction from B toward A in the figure) as they go around the turn.

•**8-61.** Show that for a set-up like that in Figure 8-13, if the ball is given a horizontal velocity at the top of the circle, the tension in the string when the ball is at the bottom of the circle will exceed the tension when the ball is at the top by six times the weight of the ball.

•**8-62.** A 20-kg child is swinging on a playground swing suspended from 3.5-m ropes. At the extremes of the child's swing, the ropes make an angle of 40° with the vertical. Find the total force exerted by the ropes at *a.* either extreme of the swing. *b.* the lowest point of the swing.

8-63. Calculate the gravitational force that the moon exerts on you, and compare it to your weight. Express your comparison as a ratio $\frac{F_{\text{by moon on you}}}{\text{your weight}}$.

8-64. Applying Equation 8-5a tells you that as you travel outward from Earth's surface, the gravitational force on you (your weight) decreases. To go the other way, imagine a tunnel drilled to the center of Earth. Why wouldn't your weight increase as you travel inward from Earth's surface?

8-65. How high above Earth's surface should a satellite orbit if it is always to remain directly above the same point on Earth's equator? (*Hint:* What else can you say about the satellite's orbit if it meets this condition?)

8-66.

a. Use the data in Table 8-2 (see Problem 8-77) to find the mass of Jupiter.

b. Now use the data in Table 8-1 to find Jupiter's acceleration due to the force that the sun exerts on Jupiter.

c. Combine the results of *a* and *b* to obtain the magnitude of the force that the sun exerts on Jupiter. Compare this to the force that the sun exerts on Earth (Problem 8-34).

d. Now use the universal law of gravitation to find the gravitational force that the sun exerts on Jupiter. Is this greater than, equal to, or less than the value you obtained in *c*?

8-67. The masses of Uranus and Neptune are $M_{\text{U}} = 8.66 \times 10^{25}$ kg and $M_{\text{N}} = 1.03 \times 10^{26}$ kg. Their average distances from the sun are $R_{\text{U}} = 2.87 \times 10^{12}$ m and $R_{\text{N}} = 4.50 \times 10^{12}$ m. Compare the gravitational force that the sun exerts on Uranus with the force that Neptune exerts on Uranus when the two planets are closest. Express your comparison as a ratio $\frac{F_{\text{on U by N}}}{F_{\text{on U by sun}}}$. Does the latter force seem large enough to affect the orbit of Uranus about the sun? (In fact, observed perturbations in the orbit of Uranus led to Neptune's discovery.)

8-68. Find the escape velocity required by a spacecraft attempting to return to Earth from the moon. (Neglect the effect of Earth.)

8-69.

a. Verify that the units of the constant that multiplies M_{star} in Equation 8-12 work out to be m/kg, and therefore can be expressed as some distance in meters divided by the mass of the sun in kg.

b. Show that that distance is $\approx 3 \times 10^3$ m, as indicated in Equation 8-13.

8-70. Although no one has measured any difference between gravitational mass and inertial mass, we know of no theoretical reason why they must be equal. Consider the following equations:

i. $\Sigma \vec{F}_{\text{on A}} = m_{\text{A}} \vec{a}_{\text{A}}$ **ii.** $F = G\frac{m_1 m_2}{r^2}$ **iii.** $g = G\frac{M_{\text{E}}}{R_{\text{E}}^2}$

a. Which of these equations is a statement purely about gravitational mass?

b. Which of these equations is a statement purely about inertial mass?

c. Which of these equations only holds true if the gravitational mass and inertial mass are always equal? Briefly explain.

8-71. Jupiter, with a radius 11.1 times as great as Earth, has a mass that is 318 times Earth's mass. How would the velocity required for a rocket to escape from Jupiter compare with the velocity required for it to escape from Earth? Express your comparison as a ratio $\frac{v_{\text{esc from Jupiter}}}{v_{\text{esc from Earth}}}$.

8-72. The amusement park ride in Figure 8-26 has cars (one shown) suspended from a rotating disk. Which of the distances shown in the figure should you use for r when calculating the radial acceleration $a_r = \frac{v^2}{r}$ of the car?
a. r_{a}; *b.* r_{b}; *c.* r_{c}; *d.* r_{d}; *e.* other.

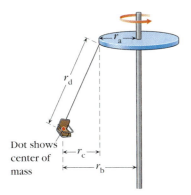

Figure 8-26 Problem 8-72

8-73. **SSM** The diameter of planet A is 1.5 times the diameter of planet B. How would your weight on the surfaces of the two planets compare if
a. the two planets have equal masses? Express your comparison as a ratio $\frac{\text{weight on A}}{\text{weight on B}}$.
•**b.** the two planets have equal densities? Express your comparison as a ratio $\frac{\text{weight on A}}{\text{weight on B}}$.

8-74. Suppose the mass of the sun were compressed into a ball with a radius of 10 km (the distance of the 10 000-m run in the Olympics). How much would you weigh at the surface of such a star? (Neutron stars, formed as certain stars collapse in on themselves near the ends of their lifetimes, may have dimensions similar to those described here.)

•**8-75.** When the space probe *Galileo* approached Ganymede, Jupiter's largest moon, in 1996, it detected a magnetic field. Because physicists knew that Ganymede had a low density, they believed it unlikely that Ganymede would have an iron core like Earth and therefore inferred that the field was probably due to an iced-over ocean of liquid water. But how did they know that Ganymede had a low density?
a. Explain how it is possible to determine the density of Ganymede based only on observations made from Earth. What quantities would you have to measure? What calculations would you have to do?
b. Describe a method by which equipment on the space probe could be used to verify the result obtained in **a.**

8-76.
a. Write down Newton's universal law of gravitation (Equation 8-5) and solve for the quantity Fr^2 (i.e., write the equation as $Fr^2 = $ _____, with the right side filled in).
b. Does anything on the right side of the equation change as the two masses move closer together or further apart?
c. What does your answer to **b** tell you about the quantity Fr^2?
d. Based on your answers to **b** and **c**, how does the magnitude F of the force change as the distance between the masses doubles? Explain.

8-77. The average orbital radii R and times T for a complete orbit are given in Table 8-2 for the four Galilean moons of Jupiter.
a. From the table, find the ratio $\frac{T^2}{R^3}$ for each of the four moons.

Table 8-2 Data for Orbits of the Galilean Moons of Jupiter

Moon	T (Earth days)	R (measured from Jupiter, km $\times 10^6$)	T^2 (Earth day^2)	R^3 (km$^3 \times 10^{18}$)
Io	1.769	0.422	3.129	0.0752
Europa	3.551	0.671	12.61	0.3021
Ganymede	7.155	1.070	51.19	1.225
Callisto	16.69	1.880	278.6	6.645

b. Is there a relationship among the values that you found? If so, briefly explain how this relationship compares with Kepler's third law of planetary motion.

8-78. When two objects are at a distance r_1 from each other, each exerts a gravitational force of magnitude F_1 on the other. When the distance between them changes to r_2, the magnitude of the gravitational force changes to F_2. How does F_2 compare to F_1
a. if $r_2 = 2r_1$? Express your comparison as a ratio $\frac{F_2}{F_1}$.
b. if $r_2 = \frac{1}{2}r_1$? Express your comparison as a ratio $\frac{F_2}{F_1}$.

•**8-79.** Suppose a body of mass m is raised a small ($\ll R_E$) distance h above Earth's surface, so that r goes from R_E to $R_E + h$. Equation 8-10 tells us that the change in PE is $\Delta PE_{\text{grav}} = \left(-G\frac{mM_E}{R_E + h}\right) - \left(-G\frac{mM_E}{R_E}\right)$. Using this and the fact that $g = G\frac{M_E}{R_E^2}$, show that $\Delta PE_{\text{grav}} \approx mgh$. In other words, if h is small compared to R_E, it gives the same change in PE that we would get from $PE_{\text{grav}} = mgy$ when $\Delta y = h$.

8-80. Indicate which of the descriptions below correctly describe the quantities that head the columns by checking the appropriate boxes in the table.

Description	g	G	$\frac{T^2}{R^3}$
Varies with altitude above Earth's surface			
Has very nearly the same value for an object in any circular orbit having Earth at its center			
Has the same value when you treat planets in orbit around the sun as when you treat moons in orbit around Jupiter			

8-81.
•**a.** **SSM** Find the gravitational potential energy of a 500-kg rocket at the point between Earth and the moon where the total gravitational force on the rocket is zero. (Ignore any effect of the sun.)
b. Is the gravitational potential energy zero at this point? If so, explain why, *or* if not, tell where, if anyplace, it *would* be zero.

8-82. A satellite travels at a constant speed of 500 m/s around the circular orbit in Figure 8-27. At this speed, it takes the satellite 50 000 s to go halfway around the circle.

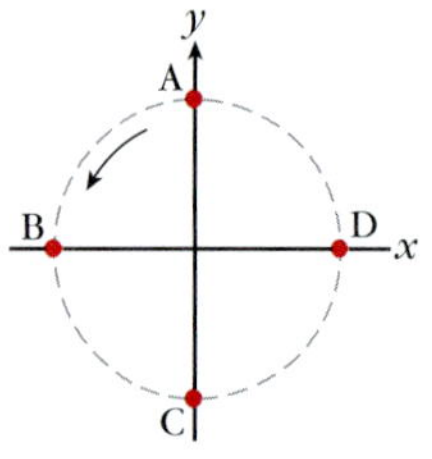

a. What are the x and y components of its velocity at point A? At point B? At point C?

b. Find $a_x = \dfrac{\Delta v_x}{\Delta t}$ and $a_y = \dfrac{\Delta v_y}{\Delta t}$, the x and y components of its average acceleration, as it goes from point A to point C.

Figure 8-27 Problems 8-82 and 8-83

c. Now find the x and y components of its average acceleration as it goes from point B to point D.

d. Are the components of the acceleration the same in **b** as they are in **c**?

8-83. (*Choose the best answer.*) A satellite travels at a constant speed around the circular orbit in Figure 8-27.

a. Consider its velocity components v_x and v_y as it travels from point A to point C. At point C, _____

   i. v_x is the same as at point A, but v_y is opposite in sign.

   ii. v_y is the same as at point A, but v_x is opposite in sign.

   iii. v_x and v_y are both the same as at point A.

   iv. v_x and v_y are both opposite in sign to their values at point A.

b. Over the semicircle from A to C, the satellite's average acceleration has _____

   i. a zero x component and a nonzero y component.

   ii. a zero y component and a nonzero x component.

   iii. zero components in both the x and y directions.

   iv. nonzero components in both the x and y directions.

c. Now consider the satellite's velocity components v_x and v_y as it travels from point B to point D. At point D, _____

   i. v_x is the same as at point B, but v_y is opposite in sign.

   ii. v_y is the same as at point B, but v_x is opposite in sign.

   iii. v_x and v_y are both the same as at point B.

   iv. v_x and v_y are both opposite in sign to their values at point B.

d. Over the semicircle from B to D, the satellite's average acceleration has _____

   i. a zero x component and a nonzero y component.

   ii. a zero y component and a nonzero x component.

   iii. zero components in both the x and y directions.

   iv. nonzero components in both the x and y directions.

e. Based on your answers to **b** and **d**, do the components of the satellite's acceleration remain constant as it travels in a circular path?

8-84. At low enough temperatures and pressures, ice sublimates; that is, it turns directly to water vapor without going through a liquid water stage. Suppose a block of ice is put in a circular orbit around Earth. As the ice sublimates, will the radius of the orbit of the remaining block of ice increase, decrease, or remain the same?

•8-85. The child's swing in Figure 8-28 is held in position A and released from rest. If the swing has a mass of 0.70 kg, what is the total force exerted on the swing by the ropes at the instant it passes position B?

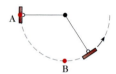

Figure 8-28
Problems 8-85, 8-86 and 8-87

8-86. The child's swing in Figure 8-28 is held in position A and released from rest. At the instant the swing passes position B, is the total force that the ropes exert on the swing greater than, equal to, or less than the weight of the swing?

8-87. In Problem 8-86, as the swing goes from position A to position B, does its radial acceleration increase, decrease, or remain the same?

Problems on WebLinks

8-88. In WebLink 8-1, an object travels in a horizontal circle at speed v_0. Suppose $v_0 = 3$ m/s.

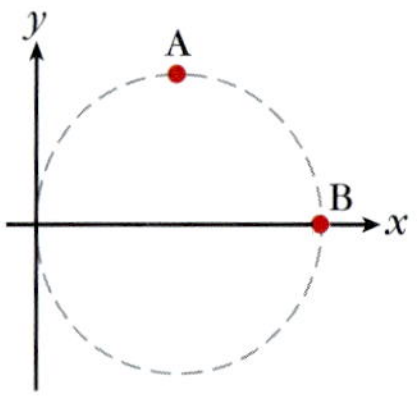

a. As the object goes from point A to point B (Figure 8-29), what is the change Δv_x in the x component of its velocity?

Figure 8-29
Problem 8-88

b. Over the same interval, what is the change Δv_y in the y component of its velocity?

c. To produce the change you found in **a**, should the x component of the average force on the object be in the positive x direction, in the negative x direction, or zero during this interval?

d. To produce the change you found in **b**, should the y component of the average force on the object be in the positive y direction, in the negative y direction, or zero during this interval?

e. In what direction is the resultant average force during this interval?

8-89. In WebLink 8-2, we rotate a sheet of paper and watch what happens to two adjacent edges of the sheet. The purpose of this is to show that which of the following is true as an object goes around a circle at constant speed? (*Note:* More than one of the choices is true, but only one is the real point of the demonstration.)

a. $\vec{v}$ remains perpendicular to $\vec{r}$;

b. The direction of $\vec{v}$ always changes by the same amount as the direction of $\vec{r}$;

c. The magnitudes of $\vec{v}$ and $\vec{r}$ do not change;

d. When $\vec{r}$ changes by an amount $\Delta\vec{r}$, $\vec{v}$ changes by an equal amount $\Delta\vec{v}$.

8-90. In WebLink 8-2, suppose the object travels just far enough around the circle so that the direction of $\vec{r}$ changes by 40°.

a. By what angle will the direction of $\vec{v}$ change?

b. By how much will the magnitude of $\vec{v}$ change?

8-91. (*Choose the best answer.*) WebLink 8-3 shows that each time a billiard ball rebounds elastically from the cushion of a pool table, the change $\Delta \vec{v}$ in its velocity is ____ (*along the cushion; normal to the cushion; in the direction the ball follows after the bounce; opposite to the direction the ball had before the bounce*).

8-92. In WebLink 8-3, a ball is made to follow a path in the shape of a regular polygon by rebounding elastically off one wall after another. What main point about circular motion is demonstrated by increasing the number of sides in the polygon?

8-93. If there were no gravitational force on the cannonball that is fired horizontally in WebLink 8-4, would its radial distance from Earth's center increase, decrease, or remain the same?

8-94. (*Choose the best answer.*) In WebLink 8-4, the faster the cannonball is fired horizontally ____ (*the more nearly it follows a circular path; the further from Earth it gets; the more potential energy it gains as it loses speed; the more it slows down*).

8-95. (*Choose the pair of answers that go in the two blanks.*) In WebLink 8-5, the magnitude of the astronaut's apparent weight is given by the ____ force that ____ exerts on the astronaut.
a. normal . . . the satellite's outer wall
b. gravitational . . . Earth
c. gravitational . . . the whole satellite
d. normal . . . the satellite's inner wall

CHAPTER NINE

Rotational Kinematics and Dynamics

In Chapter 8 we looked mainly at *uniform* circular motion. But objects may speed up or slow down as they move along curved paths (as in Example 8-5 or Figure 9-1). The rotational speed of spinning objects can also increase or decrease. Rotational speeds change when virtually any machine is started up and its rotating parts are brought up to speed, when the winds circling the eye of a hurricane slow down as the storm weakens, when a ballet dancer goes into or comes out of a spin, or

"Rotational speeds change . . . when the winds circling the eye of a hurricane slow down as the storm weakens . . ."

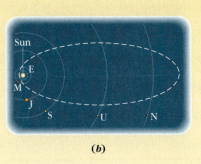

(a)   (b)

Figure 9-1 Halley's Comet. (*a*) As a comet draws closer to the sun, it loses PE and gains KE. Thus it travels fastest when closest to the sun. (*b*) The diagram shows the elliptical orbit of Halley's comet (dashed line), at its near point coming closer to the sun than Earth and at its far point extending beyond the orbit of Neptune. The solid curves are the orbits of Earth (E), Mars (M), Jupiter (J), Saturn (S), Uranus (U), and Neptune (N).

when whirling aggregates of matter in space coalesce because of mutual gravitational attraction to form stars and planets (Figure 9-2).

To deal more fully with *nonuniform* circular or rotational motion, we will first develop a more complete way of quantitatively describing rotational motion—that is, we will develop a **rotational kinematics.** This will let us do further reasoning about rotational motion much as translational kinematics did for translational or straight-line motion.

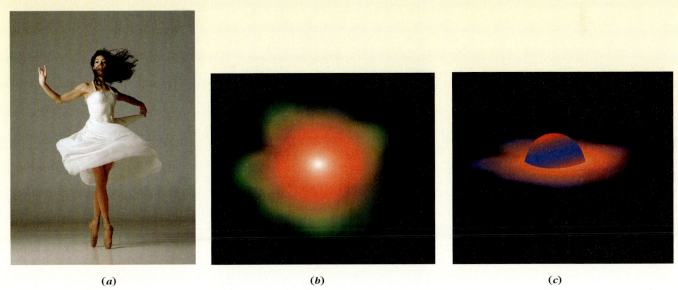

(a)   (b)   (c)

Figure 9-2 Aggregations of matter changing rotational speed. (*a*) A dancer slows down as she comes out of a spin. (*b*) The solar system is believed to have formed from a rapidly spinning cloud of interstellar gas. (*c*) As the matter at the center contracted into a sun, a disk of matter, from which the planets would eventually form, extended further from it. The disk slowed down as it spread out, much as the dancer does.

9-1 Measures of Rotation

When we describe a vector's direction by an angle, we are really telling how much it is rotated from some reference direction, such as the direction of the $+x$ axis. An angle is a measure of rotation. Very likely the units of angle, and thus of rotation, that you are most familiar with are *degrees*. But dividing a complete circle or rotation (also called a **revolution** or **cycle**) into 360 equal subdivisions called degrees is a totally arbitrary choice. Why not divide the circle into a hundred or a thousand units to make it consistent with other SI units, or into some other number of units for some other reason? Any choice would be equally correct. The question is, which is most useful?

◆**RADIAN MEASURE** It turns out that for doing calculations about rotational motion, degrees are *not* the most convenient choice of units. We will now develop another measure of rotation that will prove more useful. This measure is rooted in the fact that in all circles, the circumference is 2π times as big as the radius: $C = 2\pi r$, or $\frac{C}{r} = 2\pi$. We can think of the circumference as the arc that subtends a central angle of 360° (Figure 9-3a). If the arc length s is only half of a circumference ($s = \frac{1}{2}C$—see Figure 9-3b), it subtends only 180° and $\frac{s}{r} = \pi$. Notice that the angle is the same no matter how big r is, and $\frac{s}{r}$ is the same no matter how big r is. Figures 9-3c and d extend this reasoning to a quarter of a circle and a twelfth of a circle. The ratio $\frac{s}{r}$ remains proportional to the number of degrees, and therefore can serve just as well as a measure of the angle θ. It is equivalent to saying that we are subdividing a complete rotation into 2π divisions rather than 360 divisions.

By choosing this measure (called **radian measure**), we are choosing to let

$$\frac{s}{r} = \theta \qquad (9\text{-}1)$$

➡**Caution:** The relationship $\frac{s}{r} = \theta$ holds true *only* when θ is given in radian measure.

Although the choice may at first seem peculiar, it has this advantage: If a point is a distance r from the center of a spinning object (Figure 9-4), Equation 9-1 provides an immediate simple relationship between the distance s the point travels and the angle θ through which the body rotates.

Notice that in Equation 9-1, the measure of the angle is a ratio of distances. This means that in SI units, θ is in $\frac{\text{meters}}{\text{meters}}$ (m/m). The meters "cancel," so θ is in fact *dimensionless*—it is a pure number. Nevertheless, because of the way we ordinarily speak, if we say a complete rotation is 2π rather than 360°, people

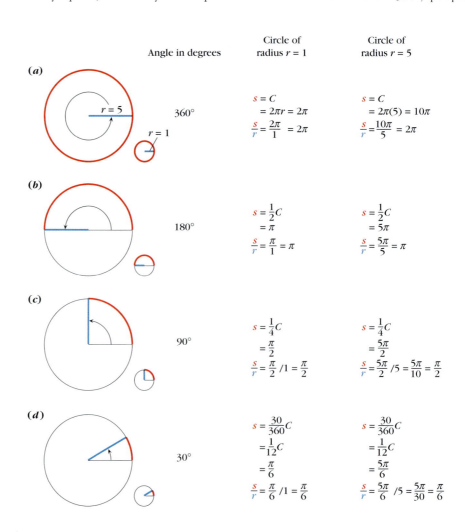

Figure 9-3 The ratio $\frac{s}{r}$ as a measure of angle or rotation (radian measure) (s = arc length; r = radius; C = circumference).

tend to ask "2π what?" To provide the filler word that people seem to need, mathematicians and physicists respond by saying "2π *radians.*" The **radian** (abbreviated *rad*) is considered a supplementary unit in SI. But you should remember that because radians are dimensionless, if we rewrite Equation 9-1 as

$$s = r\theta \qquad (9\text{-}2)$$

a radius in meters times an angle in radians gives you an arc length in meters (for example, if $r = 5$ m and $\theta = \pi$ radians, $s = 5\pi$ m $\times$ radians $= 5\pi$ m $\times$ m/m $= 5\pi$ m). In other words, without any loss of information, we can equate one complete rotation in degrees and in radian measure by writing either

$$360° \equiv 2\pi \text{ rad}$$

or simply

$$360° \equiv 2\pi \qquad (9\text{-}3)$$

In **radian measure,** we take the dimensionless ratio $\frac{s}{r}$ as our measure of angle or rotation. For one complete rotation or revolution or cycle, this measure has the value 2π.

The number of complete revolutions or cycles, n, is also dimensionless, because, like $\frac{s}{r}$ it is also the ratio of two distances—the actual distance traveled and the circumference of the particular circle:

$$\text{Number of revolutions or cycles } n = \frac{s}{C} = \frac{s}{2\pi r} \qquad (9\text{-}4)$$

From this it follows that

$$\frac{s}{r} = 2\pi n = \text{amount of rotation in radian measure} = \theta \qquad (9\text{-}5)$$

➥**Note:** We can omit the word *radians* only when we express all angles as multiples of π. If we see $\theta = 2\pi$, we know it is the same angle as $\theta = 360°$ because the π enables us to recognize that the angle is in radian measure. But because $\pi \approx 3.14$, we can also write $\theta \approx 6.28$. There is no π to cue us, so we are forced to write "$\theta \approx 6.28$ rad."

Example 9-1 *Converting from Degrees to Radian Measure*

Convert each of the following angles from degrees to radian measure: 60°, $-45°$, 720°.

Solution

Choice of approach. Equation 9-3 is the needed equivalence relationship between units. Converting units as in Section 1-2, we can rewrite Equation 9-3 as either $\frac{2\pi}{360°} = 1$ or $\frac{360°}{2\pi} = 1$, giving us two ways of writing the number 1. To convert from degrees to radians, we multiply the given values by one written in the first of these two ways.

The mathematical solution.

$$60° \times \frac{2\pi}{360°} = \frac{\pi}{3} \qquad -45° \times \frac{2\pi}{360°} = -\frac{\pi}{4} \qquad 720° \times \frac{2\pi}{360°} = 4\pi$$

Note: Angles expressed in radians are commonly left as multiples of π, so that they can be easily compared to π (half a rotation) or 2π (a full rotation). For instance, $4\pi = 2\pi \times 2$ represents four half rotations or two full rotations ($n = 2$).

◆ Related homework: Problems 9-3 and 9-4.

On-The-Spot Activity 9-1

Find $\sin(\pi/3)$ on your calculator, and make sure you get the same value as you do for $\sin 60°$.

➥**Using your calculator:** Your calculator should have a key (sometimes labeled DRG or MODE) that allows you to switch back and forth between degrees and radian measure.

Example 9-2 *Converting from Radians to Degrees*

Each of the following angles is given in radian measure:

$$-3\pi, \frac{3\pi}{4}, 1.60 \text{ rad}$$

Convert each of these to degrees.

Solution

Choice of approach. To convert the other way, we multiply each value by one in the form $\frac{360°}{2\pi}$.

The mathematical solution.

$$-3\pi \times \frac{360°}{2\pi} = \mathbf{540°}$$

$$\frac{3\pi}{4} \times \frac{360°}{2\pi} = \mathbf{135°}$$

$$1.60 \text{ rad} \times \frac{360°}{2\pi} = \frac{288°}{\pi} = \frac{288°}{3.14} = \mathbf{91.7°}$$

◆ Related homework: Problems 9-5 and 9-6.

9-2   Angular Velocity and Angular Acceleration

If a rotating object is rigid, the amount of rotation is the same for all of its parts. When the turntable in Figure 9-4 rotates, each of the mice lined up along a radius of the turntable will rotate through the same angle $\Delta\theta$ in a given time interval Δt, and each takes the same time to go through one full rotation. But as the dotted paths show, they do not all travel the same distance in that time interval. A mouse further out from the axis moves in a larger circle than one closer in. It therefore travels a greater distance during the time interval required for a complete rotation and is thus traveling at greater speed. Likewise, point B on the rim of the turntable travels at a much greater tangential speed than point A on the surface of the axle.

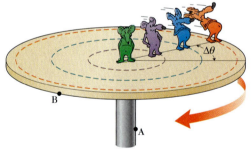

Figure 9-4 Speed increases with distance *r* from the axis of rotation.

Nevertheless, because all the mice are going around together, we would like somehow to say that they are rotating at the same rate. What is true is that during a time interval Δt, each mouse goes through the same increase Δn in the number of rotations or the same increase $\Delta\theta = 2\pi\Delta n$ (from Equation 9-5) in the amount of rotation in radian measure. Then the average *rate of change* of n or θ is the same for each mouse. These rates of change are thus useful quantities for describing rotational motion, and we give each a name and a symbol:

$$\textbf{Frequency } f \equiv \frac{\Delta n}{\Delta t} \tag{9-6}$$

$$\textbf{Angular velocity } \omega \equiv \frac{\Delta \theta}{\Delta t} \tag{9-7}$$

(ω is the lower case Greek letter *omega*) From $\Delta\theta = 2\pi\Delta n$, it follows that

$$\omega = 2\pi f \tag{9-8}$$

Strictly speaking, Equations 9-6 and 9-7 define *average* frequency and angular velocity. As always, we can obtain the corresponding instantaneous quantities by finding the average values over progressively smaller intervals Δt shrinking toward a single instant t.

The frequency tells the number of revolutions or cycles that occur per second. Because n is dimensionless, the units of f are $\frac{1}{s}$ or s^{-1}. The SI unit of frequency, called the hertz (Hz), is defined to be $1\ s^{-1}$. The SI units of angular velocity ω are most commonly written as rad/s (radians per second).

◆**A FURTHER NOTE ON UNITS** Because θ in radian measure is also dimensionless, ω is sometimes expressed in $\frac{1}{s}$ or s^{-1} as well. For instance, suppose a rotating object completes 20 revolutions in 4 s. Then its frequency is $\frac{\Delta n}{\Delta t} = \frac{20}{4\ s} = \frac{5}{s}$ (which we read as "five per second") or $5\ s^{-1}$ (also written as $5\ \frac{rev}{s}$ or $5\ \frac{cycles}{s}$ or 5 Hz). Because 20 revolutions $= 20(2\pi) = 40\pi$ in radian measure, its *angular velocity* is $\frac{\Delta\theta}{\Delta t} = \frac{40\pi}{4\ s} = \frac{10\pi}{s}$ (read as "10π per second") or $10\pi\ s^{-1}$. However, the units here are *not* hertz, because the π cues us that this is $\pi\ s^{-1}$ in radian measure. It is generally simpler to write 10π rad/s to avoid this confusion.

Because $s = r\theta$, it follows that the distance traveled by an object or point tracing a circular path (so that r is fixed) is

$$\Delta s = s_2 - s_1 = r\theta_2 - r\theta_1 = r(\theta_2 - \theta_1) = r\Delta\theta$$

If it travels this distance in a time interval Δt, its average speed is

$$v = \frac{\Delta s}{\Delta t} = r\frac{\Delta\theta}{\Delta t}$$

so that by the definition of ω (Equation 9-7),

$$v = r\omega \tag{9-9}$$

We have treated v and ω as average values, but Equation 9-9 also holds true at each instant. When we take v and ω to be instantaneous values, the speed v is the magnitude of the instantaneous *tangential* velocity.

We can think of s and θ as the linear position and the angular position of the object along its circular path. If we do so, we see that the same relationship holds between linear and angular position (Equation 9-2) as between linear and angular velocity.

An object traveling along a curved path experiences a radial acceleration, even if its tangential velocity is constant. We showed in Chapter 8 that this acceleration $a_r = \frac{v^2}{r}$. From Equation 9-9, it follows that

$$a_r = \frac{(r\omega)^2}{r} = r\omega^2 \tag{9-10}$$

Now work through Example 9-3 to see how these ideas are used to produce "artificial gravity."

Example 9-3 *Simulating the Experience of Weight in a Remote Space Station*

For a guided interactive solution, go to Web Example 9-3 at www.wiley.com/college/touger

Many futurists have envisioned space stations as donut shapes in which crew members walk on the inside of the outer walls (figure on next page). The experience of weight is caused by the rotation of the station about the central axis CC'. Crew members are constrained to a circular path by the normal forces exerted on them by the space station's outer wall. If such a space station has

EXAMPLE 9-3 *continued*

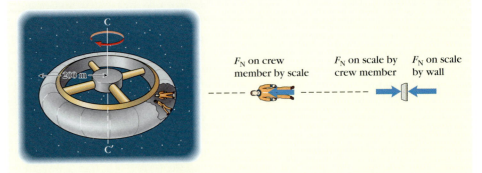

F_N on crew member by scale F_N on scale by crew member F_N on scale by wall

a radius of 200 m, find the angular velocity necessary if a scale placed on the outer wall is to read a crew member's correct Earth weight. (Assume the space station is in interstellar space, so that actual gravitational forces are negligible.)

Brief Solution

Choice of approach. The scale reads the normal force that the crew member exerts on it. The correct Earth value of this "weight" is mg. By Newton's third law, this force is equal and opposite to the normal force that the scale exerts on the crew member. When we apply Newton's second law to the crew member, this normal force is the sole radial force. By Equation 9-10, the second law becomes $\Sigma F_r = ma_r = mr\omega^2$, and in this case $\Sigma F_r = F_N = mg$.

What we know/what we don't.

2nd law in r direction	Constants
$a_r = r\omega^2 \quad (\omega = ?)$	$r = 200.$ m
$F_r = F_N = mg$	$g = 9.8$ m/s^2
	$m = ?$

The mathematical solution. $\Sigma F_r = ma_r$ becomes

$$F_N = mg = mr\omega^2$$

Then $\omega = \sqrt{\dfrac{mg}{mr}} = \sqrt{\dfrac{g}{r}} = \sqrt{\dfrac{9.8 \text{ m/s}^2}{200. \text{ m}}} = \textbf{0.221 s}^{-1}$ (in radian measure) (or **0.221 rad/s** or 0.0703π/s)

Making sense of the results. Fortunately, the necessary angular velocity ω does not depend on the crew member's mass. It would certainly not be practical if different rotational speeds were required for different crew members.

◆ Related homework: Problems 9-15 and 9-16.

◆**ANGULAR ACCELERATION** Not all circular motion is uniform. An object can speed up or slow down along its circular path. As its tangential velocity v changes, its angular velocity $\omega = \frac{v}{r}$ correspondingly changes. Recall that there is radial acceleration even if the tangential velocity is constant, but if the tangential velocity changes, there is *tangential* acceleration as well:

$$a_t = \frac{\Delta v}{\Delta t} = \frac{v_2 - v_1}{\Delta t} = \frac{r\omega_2 - r\omega_1}{\Delta t} = r\frac{\Delta\omega}{\Delta t}$$

We can rewrite this as $a_t = r\alpha$ (9-11)

where $\alpha \equiv \dfrac{\Delta\omega}{\Delta t}$ (9-12)

is called the average **angular acceleration,** just as the rate of change of linear velocity is called the linear acceleration. Units of a are units of ω (rad/s) over units of t (s), and thus are rad/s^2 or rad $\times$ s^{-2}. Just as ω requires that θ be in radian measure, so does α. For instance, if the angular velocity increases from 6π s^{-1} to 21π s^{-1} over an interval of 3 s, the angular acceleration would be

$$\alpha = \frac{21\pi \text{ s}^{-1} - 6\pi \text{ s}^{-1}}{3 \text{ s}} = 5\pi \text{ s}^{-2} \text{ or } \frac{5\pi}{\text{s}^2}$$

("5π per second squared", or "5π per second per second"). As a number you would have to write 5(3.14) rad $\times$ s^{-2} or 15.7 rad/s^2.

STOP&Think When the radial acceleration of an object traveling in a circle is constant, what can you say about the object's tangential acceleration? Record your reasoning before you continue reading. ◆

A circle is a path of fixed radius. Because $v = r\omega$, if the tangential speed is constant, so is the angular velocity, and we have *uniform* circular motion. For *any* circular motion, there must be a radial acceleration $a_r = \frac{v^2}{r} = r\omega^2$. But if the radial acceleration is constant, v and ω must be constant. And if the angular velocity ω is constant, its rate of change, that is, the angular acceleration, must be zero. Then by Equation 9-11, the tangential acceleration $a_t = 0$. Now work through Example 9-4 to see what happens when v is *not* constant.

Example 9-4 *Speeding Up on a Curve*

For a guided interactive solution, go to Web Example 9-4 at www.wiley.com/college/touger

Part *a* of the figure shows the speed of a locomotive at three different points along a stretch of track with radius of curvature 30 m. Assuming the increase in speed to be uniform over the time interval depicted, find the magnitude and direction of the locomotive's linear acceleration at point 2.

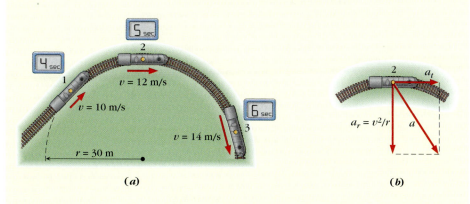

(a)   (b)

Brief Solution
Choice of approach. Because the locomotive is traveling along a circular path, it must have a radial acceleration, which depends on its speed (Equation 8-1). But because its speed along the track is increasing, it also has a *tangential* acceleration $a_t = \frac{\Delta v}{\Delta t}$. The radial and tangential accelerations are the two components of the *total* linear acceleration (part *b* of the figure). We can then convert to a *magnitude and direction* description (a, θ_a) as we do for any vector.

What we know/what we don't.

$v_1 = 10$ m/s	$v_2 = 12$ m/s	$v_3 = 14$ m/s	
$t_1 = 4.0$ s	$t_2 = 5.0$ s	$t_3 = 6.0$ s	$r = 30$ m
$a_r = ?$	$a_t = ?$	$a = ?$	$\theta_a = ?$

EXAMPLE 9-4 *continued*

The mathematical solution. At point 2, the radial acceleration is

$$a_r = \frac{v_2^2}{r} = \frac{(12 \text{ m/s})^2}{30 \text{ m}} = 4.8 \text{ m/s}^2$$

Because it is uniform, the tangential acceleration may be found from the data for any two of points 1, 2, and 3:

$$a_t = \frac{\Delta v}{\Delta t} = \frac{v_3 - v_1}{t_3 - t_1} = \frac{14 \text{ m/s} - 10 \text{ m/s}}{6.0 \text{ s} - 4.0 \text{ s}} = 2.0 \text{ m/s}^2$$

Then $a = \sqrt{a_r^2 + a_t^2} = \sqrt{(4.8 \text{ m/s}^2)^2 + (2.0 \text{ m/s}^2)^2} = \textbf{5.2 m/s}^2$

Because a_r is in the $-y$ direction at point 2, $a_y = -a_r$, and

$$\tan\theta_a = \frac{a_y}{a_x} = \frac{-a_r}{a_t} = \frac{-4.8 \text{ m/s}^2}{2.0 \text{ m/s}^2} = -2.4$$

so $\theta_a = \textbf{−67°}, \text{ or } -67°\left(\dfrac{2\pi}{360°}\right) = \textbf{−0.37}\boldsymbol{\pi}$

◆ Related homework: Problems 9-17, 9-18, 9-19, and 9-20.

For **WebLink 9-1:**
**Tarzan's Swing
Revisited,** go to
www.wiley.com/college/touger

To see how some of these ideas apply to motion along a vertical circle, and to help connect tangential and radial accelerations into a broader framework of ideas, work through WebLink 9-1.

9-3 Torque and Angular Acceleration

In Example 9-4, we considered an object that underwent both a radial and a tangential acceleration. Then by Newton's second law, there must be both radial and tangential force components exerted on the object, and the total force on the object must be in the direction of the total acceleration. The object will experience an *angular acceleration* only if the force exerted on it has a tangential component, because

$$F_t = ma_t = mr\alpha \qquad (9\text{-}13)$$

From this equation, we also see that if an object travels on a path of greater radius r, a greater tangential force will be required to give it the same angular acceleration as before (see Figure 9-5).

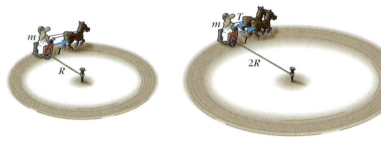

$F_1 = T$	Tangential force	$F_2 = 2T$
$r_1 = R$	Radius	$r_2 = 2R$
$a_1 = \frac{T}{m} = a$	Tangential acceleration	$a_2 = \frac{2T}{m} = 2a$
$\alpha_1 = \frac{a_1}{r_1} = \frac{a}{r}$	Angular acceleration	$\alpha_2 = \frac{a_2}{r_2} = \frac{2a}{2r} = \frac{a}{r}$

Figure 9-5 Tangential forces and angular acceleration.

STOP&Think In Figure 9-5, both the radius and the total tangential force are twice as great in situation 2 as in situation 1. If the angular acceleration is the same in both situations, how will the change in angular velocity during a time interval Δt compare in the two situations? Taking into account the difference between the radii, how will the change in tangential velocity during Δt compare in the situations? Then to produce the same radial acceleration, how must the force compare in the two situations? ◆

If we multiply both sides of Equation 9-13 by r, we get

$$rF_t = mr^2\alpha$$

Because F_t is perpendicular to the radius r, which serves as a moment arm (compare with Figure 5-24c), rF_t is the *torque* τ. If the quantity mr^2 is increased, the same torque would produce less angular acceleration α. Hence, this quantity is a measure of an object's tendency *not* to change its angular velocity, that is, to keep executing the same rotational motion, just as the inertial *mass* is a measure of the object's tendency to keep executing the same linear motion. It is therefore called the object's **moment of inertia,** and is denoted by I (but I isn't always mr^2—see later). The equation then reduces to $\tau = I\alpha$. More generally,

$$\Sigma\tau = I\alpha \qquad (9\text{-}14)$$

This is analogous to Newton's second law of motion in one dimension:

	Reason for acceleration		Inertial property		Type of acceleration
Linear motion	Total external force ΣF	=	Inertial mass m	×	Linear acceleration a
Rotational motion	Total torque $\Sigma\tau$	=	Moment of inertia I	×	Rotational acceleration α

✦**MOMENTS OF INERTIA OF RIGID BODIES** The moment of inertia is mr^2 only when all the mass is located at the same distance r from the axis of rotation. This is true exactly for point objects and approximately for rings or hollow cylinders of negligible thickness rotating about their central axes. It is *not* true for extended rigid bodies, which have parts at various distances from any axis of rotation.

We can see how to treat such a body by first considering a rigid body (Figure 9-6a) made up of two parts having masses m_1 and m_2 concentrated at distances r_1 and r_2 from pivot point P. An external force F_{ext} exerted on part 2, as shown, results in a torque τ_{ext}. At their point of contact, the equal and opposite forces the parts exert on each other (Figure 9-6b) produce equal and opposite torques $\tau_{on\ 2\ by\ 1}$ and $\tau_{on\ 1\ by\ 2}$ about point C. Because they move together, the two parts must have same angular acceleration α. We now apply Equation 9-14 to each part:

$$\Sigma\tau_{on\ 1} = I_1\alpha \qquad \text{and} \qquad \Sigma\tau_{on\ 2} = I_2\alpha$$

With the torques and moments of inertia written out in detail, the first of these becomes

$$\tau_{on\ 1\ by\ 2} = mr_1^2\alpha$$

and the second becomes

$$\tau_{ext} + \tau_{on\ 2\ by\ 1} = mr_2^2\alpha$$

But because $\tau_{on\ 2\ by\ 1} = -\tau_{on\ 1\ by\ 2}$, summing these two equations gives us

$$\tau_{ext} = (m_1r_1^2 + m_2r_2^2)\alpha$$

or

$$\tau_{ext} = (\Sigma mr^2)\alpha$$

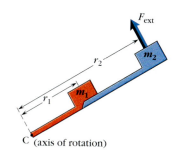

(a)

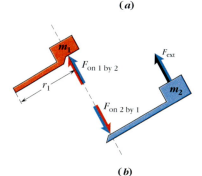

(b)

Figure 9-6 Forces giving rise to torques on a two-part rigid body.

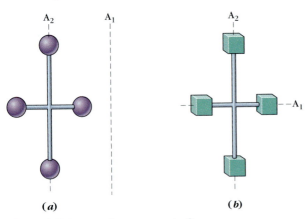

(a)

(b)

Figure 9-7 Comparing moment of inertia about different axes. The moment of inertia is less about axis A2 than about A1 in each of the cases shown.

The quantity in parentheses is the *total moment of inertia* of the rigid body.

Moment of inertia	$I_A = \Sigma mr^2$	(9-15)

for a rigid body having parts of various masses m at different distances r *from a given axis of rotation* A. *Important:* The distances r depend on the axis of rotation you choose, so the body's moment of inertia does as well.

(In SI, the units of I_A are kg · m²)

In the cases shown in Figures 9-7a and b, the pieces of mass are on average closer to axis A2 than to axis A1, so in these cases the moment of inertia is smaller about axis A2. In particular, if two axes are parallel, as they are in Figure 9-7a, the moment of inertia will always be less about the axis that passes through the object's center of mass.

Work through Example 9-5 to see in more detail how the choice of axis affects an object's moment of inertia.

Example 9-5   *The Moment of Inertia of a Barbell*

For a guided interactive solution, go to Web Example 9-5 at www.wiley.com/college/touger

An unevenly weighted barbell has a single weight of mass M placed at one end of a rod of length l, and two weights of mass M at the other. The mass of the rod is negligible compared to the masses at either end. Find the moment of inertia of the barbell about each of the axes shown in the figure (axis A2 is through the rod's center) and express your results in terms of M and l.

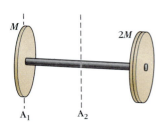

Brief Solution
Choice of approach. This is an exercise in applying Equation 9-15. We must find the distances r *from each axis.*

What we know/what we don't.

	About axis A1		About axis A2
$m_1 = M$	$r_1 = 0$		$r_1 = \dfrac{l}{2}$
$m_2 = 2M$	$r_2 = l$		$r_2 = \dfrac{l}{2}$
	$I_{A1} = ?$		$I_{A2} = ?$

The mathematical solution. About axis A1,

$$I_{A1} = m_1 r_1^2 + m_2 r_2^2 = M(0)^2 + 2M(l)^2 = \mathbf{2Ml^2}$$

About axis A2,

$$I_{A2} = m_1 r_1^2 + m_2 r_2^2 = M\left(\frac{l}{2}\right)^2 + 2M\left(\frac{l}{2}\right)^2 = \frac{3}{4}Ml^2$$

Making sense of the results. In general, the further the component masses are from the axis, the more rotational inertia the object has about that axis.

◆ Related homework: Problem 9-21.

In the following example, the moment of inertia must be found in order to apply the rotational equivalent of Newton's second law.

Example 9-6 *The Double Wheel*

The two bicycle wheels depicted here rotate on a common axle. The axle and the spokes may be assumed to have a negligible mass. If the wheels are initially rotating at an angular velocity of 4.0π s^{-1}, how much frictional force must the brake exert on the smaller wheel to bring the entire assembly to a stop in 2.0 s?

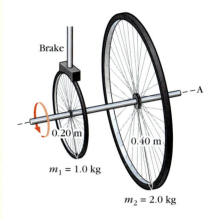

Brake

0.20 m   0.40 m

$m_1 = 1.0$ kg

$m_2 = 2.0$ kg

A

Solution

Choice of approach. We apply the rotational equivalent of Newton's second law: $\Sigma\tau = I\alpha$. We can use the definition of moment of inertia to find I and the definition of angular acceleration to find α. We can then find the torque resulting from the frictional force f. Finally, we can use $\tau = rF \sin\theta$ to find the magnitude of the force. Because the spokes of bicycle wheels are very light, virtually all of the mass of each wheel is at approximately the same distance from the axis. So each wheel's moment of inertia is mr^2, rather than a sum of contributions due to parts at different radial distances.

What we know/what we don't.

To apply definition of α	For applying $I_A = \Sigma mr^2$	For applying $\tau = rF \sin\theta$
$\omega = 4.0\pi$ s^{-1} at $t = 0$	$m_1 = 1.0$ kg $m_2 = 2.0$ kg	$\theta = 90°$ ($\sin\theta = 1$)
$\omega = 0$ at $t = 2.0$ s	$r_1 = 0.20$ m $r_2 = 0.40$ m	$r_1 = 0.20$ m
$\alpha = ?$ ————————	$I_A = ?$ ————→	$\tau = ?$ $F = f = ?$

The mathematical solution. Solving for α in Equation 9-12, we get

$$\alpha = \frac{\Delta\omega}{\Delta t} = \frac{\omega_2 - \omega_1}{t_2 - t_1} = \frac{0 - 4.0\pi \text{ s}^{-1}}{2.0 \text{ s} - 0} = -2.0\pi \text{ s}^{-2}$$

We care only about the magnitude of the frictional force, so we will omit minus signs. Applying $I_A = \Sigma mr^2$ (with the axle as the axis of rotation), we get

$$I_A = m_1 r_1^2 + m_2 r_2^2 = (1.0 \text{ kg})(0.20 \text{ m})^2 + (2.0 \text{ kg})(0.40 \text{ m})^2 = 0.36 \text{ kg}\cdot\text{m}^2$$

Now we can apply Equation 9-14

$$\Sigma\tau = r_1 f \sin\theta = r_1 f = I\alpha$$

and solve for the magnitude f of the frictional force:

$$f = \frac{I\alpha}{r_1} = \frac{(0.36 \text{ kg}\cdot\text{m}^2)(2\pi \text{ s}^{-2})}{0.20 \text{ m}} = \textbf{11 N}$$

◆ Related homework: Problems 9-25 and 9-26.

r_1
m_1
r_2
m_2

Figure 9-8 Finding the moment of inertia of an extended rotating body. We must first break up the object mentally into pieces so small that each piece is effectively at a single distance from the axis, like the two pieces of the dancer shown here. We would then have to add up the moments of inertia of all such pieces to get the dancer's total moment of inertia.

The mass of most solid objects, like the dancer in Figure 9-8, is spread out continuously over a range of distances from their axes, rather than being concentrated at specific distances. In such cases, doing the summation required by Equation 9-15 is more difficult. Figure 9-8 suggests the approach. We omit the mathematical details because they require calculus. But we can look at the results that the math yields for the moments of inertia of common solid shapes

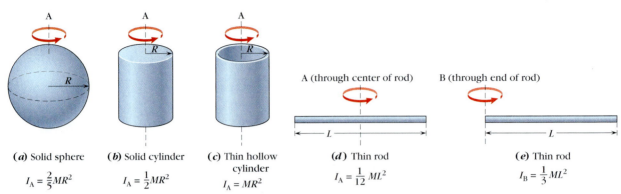

(a) Solid sphere

$I_A = \frac{2}{5}MR^2$

(b) Solid cylinder

$I_A = \frac{1}{2}MR^2$

(c) Thin hollow cylinder

$I_A = MR^2$

(d) Thin rod

$I_A = \frac{1}{12}ML^2$

(e) Thin rod

$I_B = \frac{1}{3}ML^2$

Figure 9-9 Moments of inertia of simple solids with uniform density. The solids here all have mass M. The expression for I is about the dashed axis in each case.

(Figure 9-9) and see that they display features that we must expect of *all* moments of inertia:

1. The moment of inertia of the same solid is in general different about different axes.
2. The moment of inertia has units of (mass) × (distance)^2.
3. If objects of equal mass rotate about the same axis, the moment of inertia will be greater for the object whose mass is distributed farther from the axis.

STOP&Think Can you see how the order in which the shapes are presented in Figure 9-9 illustrates point 3? ◆

9-4 Rotational Kinetic Energy

Because the tangential velocity $v = r\omega$ for a point object traveling in a circle, the kinetic energy of such an object is

$$KE = \tfrac{1}{2}mv^2 = \tfrac{1}{2}m(r\omega)^2 = \tfrac{1}{2}mr^2\omega^2 = \tfrac{1}{2}I\omega^2$$

We can think of a rigid body as being made up of many tiny pieces, each approximately a point mass, rotating together, so they have the same angular velocity ω. The total kinetic energy of these masses is then

$$\tfrac{1}{2}m_1 r_1^2 \omega^2 + \tfrac{1}{2}m_2 r_2^2 \omega^2 + \cdots = \tfrac{1}{2}(\Sigma mr^2)\omega^2 = \tfrac{1}{2}I\omega^2$$

just as it is for a point object.

The **rotational kinetic energy** of a rigid body is

$$KE_{\text{rot}} = \tfrac{1}{2}I\omega^2 \tag{9-16}$$

The roles of I and ω in rotational motion are analogous to those of m and v for linear motion, so the form of the rotational kinetic energy is analogous to $\tfrac{1}{2}mv^2$.

In Chapter 3, we looked at motions (Figures 3-3 and 3-4) that involved the superposition of (1) *the motion of the center of mass along some path* and (2) *a rotational motion of the object about its center of mass*. For any object *rolling* in a straight line (see figure), the linear motion is perpendicular to the axis of rotation.

It follows that if two objects of equal mass are traveling across a surface at the same speed, but one is sliding and one is rolling, the one that is rolling has more kinetic energy. In addition to the linear motion of the sliding object, it has a rotational motion about its axis. Its total energy is the sum of the kinetic energies associated with the two component motions:

$$KE_{\text{total}} = KE_{\text{lin}} + KE_{\text{rot}} = \tfrac{1}{2}mv_{\text{CM}}^2 + \tfrac{1}{2}I\omega^2 \tag{9-17}$$

If an object of radius r rolls without slipping (as you hope your tires will do on a wet road), the tangential speed $v_t = r\omega$ is just equal to v_{CM}, as the following activity demonstrates.

On-The-Spot Activity 9-2

Take a roll of toilet paper and tape one end of it to an inclined surface, as shown in Figure 9-10a. Then roll the paper back up again until the free end is just where the roll makes contact with the surface. Mark this point with a dot (see Figure 9-10b). Now allow the toilet paper to unroll until the dot is once again where the roll makes contact with the surface (Figure 9-10c). The length l of the paper that has unrolled is then equal to the distance s that the dot has gone around (you can check this by rolling the paper back up), traveling at a speed v_t. But as Figure 9-10c shows, l is also the distance that the center of mass moves, traveling at a speed v_{CM}. Because these equal distances are traveled during the same time interval, the two speeds must be equal.

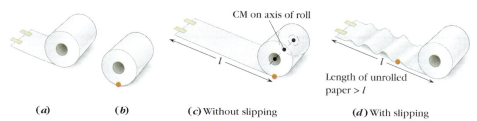

CM on axis of roll

Length of unrolled
paper > l

(a)   (b)   (c) Without slipping   (d) With slipping

Figure 9-10 Comparing v_{CM} and v_t for a rolling object.

In contrast, if the roll of paper slips as it rotates, it would have to rotate more to go the same distance down the incline. The length of paper unrolled would then be greater than the downhill distance the roll has traveled (Figure 9-10d).

Example 9-7 *Downhill Racers*

For a guided interactive solution, go to Web Example 9-7 at www.wiley.com/college/touger

The stock clerks at Mendoza's Market race canned goods when they get bored. Two identical cans A and B are released from the same height on identical ramps. Can A's ramp is iced over, so can A slides or skids down it. But there is enough frictional interaction between can B and its ramp so that can B *rolls* down *without slipping*. How do the speeds with which the two cans reach bottom compare? (Treat the filled cans as solid cylinders.)

Brief Solution

Restating the question. Sliding means no rolling whatsoever: $\omega = 0$. For rolling without slipping, $\omega = \frac{v_t}{r} = \frac{v_{CM}}{r}$. To compare the center of mass speeds in state 2 (see below), we want to know their ratio: $\frac{v_{A2}}{v_{B2}} = ?$ *if* $\omega_A = 0$ and $\omega_B = \frac{v_{B2}}{r}$.

Choice of approach. We apply conservation of energy, taking into account rotational KE. With no numerical values available, we do symbolic algebra.

What we know/what we don't. We note where quantities are the same for both cylinders:

	State 1	State 2
$m_A = m_B = m$	$h_{A1} = h_{B1} = h$	$h_{A2} = h_{B2} = 0$
$I = \frac{1}{2}mr^2$ for a solid cylinder (Figure 9-9)	$v_{A1} = v_{B1} = 0$	$v_{A2} = ?$ $v_{B2} = ?$ $\omega_{B2} = \frac{v_{B2}}{r}$
	$PE_{A1} = PE_{B1} = mgh$	$PE_{A2} = PE_{B2} = 0$
	$KE_{A1} = KE_{B1} = 0$	$KE_{A2} = \frac{1}{2}mv_{A2}^2$ $KE_{B2} = \frac{1}{2}mv_{B2}^2 + \frac{1}{2}I\omega_{B2}^2$

9-7 continued

The mathematical solution. For cylinder A,

$$PE_{A1} + KE_{A1} = PE_{A2} + KE_{A2}$$

becomes

$$mgh + 0 = 0 + \tfrac{1}{2}mv_{A2}^2$$

Then

$$v_{A2}^2 = 2gh$$

For cylinder B,

$$PE_{B1} + KE_{B1} = PE_{B2} + KE_{B2}$$

becomes

$$mgh + 0 = 0 + \tfrac{1}{2}mv_{B2}^2 + \tfrac{1}{2}I\omega_{B2}^2$$

So

$$mgh = \tfrac{1}{2}mv_{B2}^2 + \tfrac{1}{2}(\tfrac{1}{2}mr^2)\left(\frac{v_{B2}}{r}\right)^2 = \tfrac{1}{2}mv_{B2}^2 + \tfrac{1}{4}mv_{B2}^2 = \tfrac{3}{4}mv_{B2}^2$$

Then

$$v_{B2}^2 = \tfrac{4}{3}gh$$

The ratio

$$\frac{v_{B2}^2}{v_{A2}^2} = \frac{(4/3)gh}{2gh} = \tfrac{2}{3}, \text{ so } \frac{v_{B2}}{v_{A2}} = \sqrt{\frac{2}{3}} = \mathbf{0.816}$$

Making sense of the results. The center of mass velocity of the rolling object is less because part of its PE is converted into rotational KE instead of translational KE.

◆ Related homework: Problems 9-29, 9-32, and 9-34.

9-5 Another Basis for Bookkeeping: Angular Momentum

We have seen that when we use the law of conservation of energy as a basis for balancing the books for systems in the physical world, we must take into account rotational as well as translational motion. What about when we use momentum conservation for our bookkeeping? Momentum exchanges can occur on a circular track as well as on a straightaway. Can we draw any generalizations about such exchanges that can provide us with greater analytical power in treating rotating bodies?

Consider the collision at point P_1 in Figure 9-11. Over the duration of the collision, the two bodies exert equal and opposite impulses of magnitude $F\Delta t$ on each other, so the two bodies experience equal and opposite momentum changes of magnitude $\Delta(mv)$ (recall Equation 7-2). If we picture an axis perpendicular to the page at point C, the two bodies also exert equal and opposite torques on each other, since the moment arm $r_\perp$ is the same for the two equal and opposite forces. This motivates us to multiply both sides of Equation 7-2 by $r_\perp$:

$$r_\perp F\Delta t = r_\perp \Delta(mv) = \Delta(r_\perp mv) \tag{9-18}$$

Because the torque $\tau = r_\perp F$ is analogous to force in linear motion, we can extend the analogy by defining a quantity called *angular momentum*, which we denote by L.

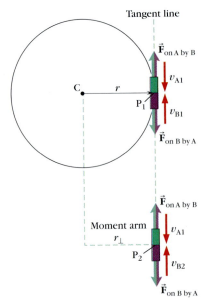

Tangent line

$\vec{F}_{\text{on A by B}}$

v_{A1}

C $\xrightarrow{\quad r \quad}$

P_1

v_{B1}

$\vec{F}_{\text{on B by A}}$

$\vec{F}_{\text{on A by B}}$

Moment arm $r_\perp$

v_{A1}

P_2

v_{B2}

$\vec{F}_{\text{on B by A}}$

Figure 9-11 A collision involves the same exchange of angular momentum anywhere along a tangent line.

> Definition: **Angular momentum:**
>
> $$L \equiv r_\perp mv \tag{9-19}$$
>
> about a given axis.

Equation 9-18 then becomes

$$\tau\Delta t = \Delta L \tag{9-20}$$

in analogy to $F\Delta t = \Delta p$.

The two bodies in Figure 9-11 exert equal and opposite torques on each other, so the total torque on the system is zero. Then

$$\Sigma \tau \Delta t = \Sigma \Delta L = 0$$

and therefore, if the system goes from state 1 to state 2,

$$\Sigma L_1 = \Sigma L_2 \qquad (9\text{-}21)$$

In words, this says that the *total angular momentum of the system is conserved if the total external torque on the system is zero*. This statement of the **law of conservation of angular momentum** is completely analogous to the law of conservation of linear momentum, which says linear momentum is conserved in the absence of a total external force on the system. The conservation law for angular momentum applies equally to a system of many bodies, whether they are separate point objects or the tiny pieces, each approximately a point mass, that make up an extended rigid body. **STOP&Think** Do Equations 9-20 and 9-21 also apply to the bodies colliding at point P_2 in Figure 9-11? ◆

The forces that the two bodies exert on each other are along the same line of action (the tangent line) whether the collision occurs at P_1 or at P_2. Then the moment arm $r_\perp$ about C is the same in either case. Therefore, nothing in our arguments changes if we apply them to the collision at P_2; the results are equally valid for that case.

For a point object in circular motion, because $v = r\omega$ and $r_\perp = r$, Equation 9-19 becomes

$$L = r_\perp m v = r m (r\omega) = m r^2 \omega$$

We can think of an extended rigid body as made up of tiny point-like parts all rotating at the same angular velocity ω, so the body's total angular momentum is

$$L = (\Sigma m r^2)\omega = I\omega \qquad (9\text{-}22)$$

Equation 9-21 then takes the following form:

Angular momentum in the hammer throw. In the hammer throw event, the athlete whirls with the hammer before releasing it. Does the angular momentum of the hammer change at the instant it leaves her hand?

Law of Conservation of Angular Momentum:
When there is zero total external torque on a system, the system conserves angular momentum, that is, the sum of all angular momenta in the system remains unchanged. For rigid bodies rotating around a fixed axis,

$$\Sigma I_1 \omega_1 = \Sigma I_2 \omega_2$$

or $\quad\quad I_{A1}\omega_{A1} + I_{B1}\omega_{B1} + \cdots = I_{A2}\omega_{A2} + I_{B2}\omega_{B2} + \cdots \qquad (9\text{-}23)$

(letter subscripts denote bodies, number subscripts denote instants or states)

The moments of inertia are subscripted 1 and 2 because Equation 9-23 applies also to objects that shift some of their mass in the radial direction, as when the dancer in Figure 9-8 pulls in his extended leg. In that case, the moments of inertia as well as the angular velocities may change.

Example 9-8 *Dropping a Load at the Casino*

For a guided interactive solution, go to Web Example 9-8 at www.wiley.com/college/touger

A disgruntled casino patron drops a 2.0-kg sack of quarters onto the outer edge of a 12.0-kg roulette wheel from a height of 0.30 m. The wheel is rotating initially with an angular velocity of 0.10π s^{-1}. If the sack does not slip when it lands, find the angular velocity of the roulette wheel after the sack is dropped on it.

EXAMPLE 9-8 *continued*

Brief Solution

Choice of approach. This is a collision-like situation for rotating bodies, so we apply conservation of angular momentum. In choosing expressions for moments of inertia, we treat the wheel as a solid cylinder (Figure 9-9*b*) and the sack as a point mass.

What we know/what we don't. (S = sack, W = wheel)

State 1	State 2	Constants
$\omega_{W1} = 0.10\pi \text{ s}^{-1}$ $\omega_{S1} = 0$	$\omega_{W2} = \omega_{S2} = ?$ (call this common value ω_2)	$m_W = 12.0 \text{ kg}$ $m_S = 2.0 \text{ kg}$
$I_{W1} = I_{W2} = \frac{1}{2}m_W R^2$	$I_{S1} = I_{S2} = m_S r^2$	$R = ?$

The mathematical solution. Equation 9-23 for these bodies,

$$I_W\omega_{W1} + I_S\omega_{S1} = I_W\omega_{W2} + I_S\omega_{S2}$$

becomes $\qquad (\frac{1}{2}m_W R^2)\omega_{W1} + 0 = (\frac{1}{2}m_W R^2)\omega_2 + (m_S R^2)\omega_2$

When we solve for ω_2, R drops out of the calculation, so its value doesn't matter:

$$\omega_2 = \frac{\frac{1}{2}m_W}{\frac{1}{2}m_W + m_S}\omega_{W1} = \frac{\frac{1}{2}(12.0 \text{ kg})}{\frac{1}{2}(12.0 \text{ kg}) + 2.0 \text{ kg}}(0.10\pi \text{ s}^{-1}) = \mathbf{0.075\pi \text{ s}^{-1}}$$

◆ Related homework: Problems 9-38, 9-39, and 9-61.

A body's moment of inertia may be changed by the exertion of either internal or external forces. For example, the ice skater in Figure 9-12*a* has exerted an internal force to pull her arms in and thus decrease her moment of inertia Σmr^2. Because there are negligible torques on her due to external forces, her angular momentum is conserved. For a single body, the conservation equation (9-30) reduces to

$$I_1\omega_1 = I_2\omega_2 \qquad\qquad (9-24)$$

If $I_2 < I_1$, then $\omega_2 > \omega_1$, and she will spin faster. Conversely, to slow herself down, she will come out of her spin by extending her arms and a leg (Figure 9-12*b*).

When a planet in its orbit (Figure 8-22) changes its distance from the sun, its moment of inertia about the sun changes. The planet does have an external

Figure 9-12 Conserving angular momentum. Ice skater Michelle Kwan's angular momentum is roughly the same in both photos, but her moment of inertia is not. In which position is her moment of inertia greater? In which position does she spin more rapidly?

force on it—the gravitational force exerted by the sun—but because that force is directed toward the sun, its moment arm is zero, and it results in zero external torque. The planet's angular momentum about the sun therefore remains constant, and Equation 9-24 applies to the planet in any two positions in its orbit.

Example 9-9 *The Eccentric Orbit of Mercury*

The elliptical paths of most planets around the sun are fairly close to being circles, but the orbits of Mercury and Pluto are more elongated or *eccentric*. At its point of nearest approach, or **perihelion** (see Figure 8-22), Mercury is 4.59×10^{10} m from the sun. At its furthest point, or **aphelion,** it is 6.97×10^{10} m from the sun. If its orbital speed is 3.9×10^4 m/s at aphelion, what is its orbital speed at perihelion?

Solution

Choice of approach. There is no external torque on the planet, so we can apply conservation of angular momentum. For a single body, this is stated by Equation 9-24. Mercury's moment of inertia is well represented by that of a point object: $I = mr^2$. For an elliptical path, a line drawn from the sun to a point in the orbit is not necessarily perpendicular to the path, as it is for circular paths. In general, then, the tangential speed would be $v = r_\perp \omega$, but at aphelion and perihelion $r_\perp = r$, so the orbital speed $v = r\omega$ at these two points.

What we know/what we don't.

State 1 (aphelion)	State 2 (perihelion)	Constants
$v_1 = 3.9 \times 10^4$ m/s	$v_2 = ?$	$m = ?$
$r_1 = 6.97 \times 10^{10}$ m	$r_2 = 4.59 \times 10^{10}$ m	
$\omega_1 = \dfrac{v_1}{r_1}$	$\omega_2 = \dfrac{v_2}{r_2}$	
$I_1 = mr_1^2$	$I_2 = mr_2^2$	

The mathematical solution. $I_1\omega_1 = I_2\omega_2$ becomes

$$mr_1^2\left(\frac{v_1}{r_1}\right) = mr_2^2\left(\frac{v_2}{r_2}\right)$$

or
$$r_1 v_1 = r_2 v_2$$

and we need only solve for v_2:

$$v_2 = \frac{r_1 v_1}{r_2} = \frac{(6.97 \times 10^{10}\ \text{m})(3.9 \times 10^4\ \text{m/s})}{4.59 \times 10^{10}\ \text{m}} = \mathbf{5.92 \times 10^4\ m/s}$$

Making sense of the results. The planet speeds up as it moves inward toward the sun, just as the figure skater spins more quickly as she draws her arms and legs in toward her axis of rotation.

This problem appeared previously as Problem 8-45, where the expectation was that you would solve it using conservation of energy, with $PE_{\text{grav}} = -G\frac{m_{\text{Merc}} m_{\text{sun}}}{r}$. Do you get the same result by either approach? Can you show that you must?

The more eccentric or elongated an elliptical orbit is, the greater the percent increase in the object's velocity will be as it goes from aphelion to perihelion. This change is most extreme for the highly elongated orbits of comets about the sun (Figure 9-1).

◆ Related homework: Problem 9-41.

◆A FINAL NOTE ABOUT ANALOGIES Recognizing analogies between linear and rotational motion will help you understand rotational motion better. In attempting to explain an unfamiliar situation, we often begin by asking what it has in common with a situation we understand better. Even if the elements in the two situations are different, the more familiar situation can offer us guidance in approaching the less familiar one if we can identify *relationships* among the elements that are the same in both cases. For example, there is the same relationship between torque and angular momentum as there is between force and linear momentum. It is also possible to develop

• equations for *rotational kinematics* analogous to the kinematics equations for constant acceleration that we developed in Chapter 2 (see Problems 9-63 and 9-68 to 9-71).

• a rotational form of the work-energy theorem (see Problems 9-75 and 9-76).

Use of these analogies is a powerful technique in the physicist's enterprise of trying to learn the rules by which nature plays.

As a final application of the use of analogy, WebLink 9-2 uses the power of analogy to explain why the precession motion shown in Figure 8-2 occurs.

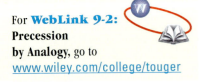

For **WebLink 9-2:**
Precession
by Analogy, go to
www.wiley.com/college/touger

◆ SUMMARY ◆

We have defined quantities describing rotational motion analogous to those that describe linear or translational motion. In many contexts, the amount of rotation is best expressed in radians.

> **Radian measure:** We take the dimensionless ratio $\frac{s}{r}$ as our measure of angle or rotation θ. For one *complete* rotation (*revolution* or *cycle*), this measure has the value 2π.

Thus
$$2\pi = 360° \qquad (9\text{-}3)$$
and
$$\theta = 2\pi n$$

where n is the number of revolutions or cycles. We can then define the following average quantities (and their instantaneous counterparts):

$$\textbf{Frequency } f \equiv \frac{\Delta n}{\Delta t} \qquad (9\text{-}6)$$

$$\textbf{Angular velocity } \omega \equiv \frac{\Delta \theta}{\Delta t} \qquad (9\text{-}7)$$
$$(\theta \text{ must be in radian measure})$$

and $\qquad$ $\textbf{Angular acceleration } \alpha \equiv \frac{\Delta \omega}{\Delta t} \qquad (9\text{-}12)$

From $\theta = 2\pi n$ and Equations 9-6 and 9-7, we get this further relationship:

$$\omega = 2\pi f \qquad (9\text{-}8)$$

From the definition of radian measure follows a simple set of relationships between quantities describing rotational motion and the analogous quantities describing the related linear motion *in the tangential direction:*

$$s = r\theta \qquad (9\text{-}2)$$
$$v_t = r\omega \qquad (9\text{-}9)$$
$$a_t = r\alpha \qquad (9\text{-}11)$$
$$(\theta \text{ must be in radian measure})$$

Equation 9-9 allows us to express the *radial* (centripetal) acceleration as

$$a_r = r\omega^2 \qquad (9\text{-}10)$$

The relationships among θ, ω, and α are identical to the relationships among s, v, and a. In dynamics, we can draw further analogies between the quantities involved in linear and rotational motion.

Linear (translational) motion	Rotational motion
A body undergoes **linear acceleration** a when there is a nonzero total external **force** ΣF on it.	A body undergoes **angular acceleration** α when there is a nonzero total external **torque** $\Sigma \tau$ on it.
This **change in linear motion** is less if the **inertial mass** m is greater.	This **change in angular motion** is less if the **moment of inertia** I is greater.
If the total external **force** on a system is zero, its **linear momentum** $\Sigma p = \Sigma mv$ is **conserved.**	If the total external **torque** on a system is zero, its **angular momentum** $\Sigma L = \Sigma I\omega$ is **conserved.**

> The **moment of inertia** $\qquad I_A = \Sigma mr^2 \qquad (9\text{-}15)$
>
> for a rigid body having parts of various masses m at different distances r *from a given axis of rotation* A. Important: The distances r depend on the axis of rotation you choose, so the body's moment of inertia does as well.
>
> In SI, the units of I_A are kg · m².

Figure 9-9 shows the resulting moments of inertia for specific geometric shapes about particular axes.

The motion of a *rolling* object is a *superposition* of linear motion of the center of mass and rotational motion around it. The total kinetic energy is thus

$$KE_{\text{total}} = KE_{\text{lin}} + KE_{\text{rot}} = \tfrac{1}{2}mv_{CM}^2 + \tfrac{1}{2}I\omega^2 \qquad (9\text{-}17)$$

For a cylindrically symmetrical object *rolling without slipping,*

$$v_t = v_{CM}$$

We define angular momentum as follows:

Angular momentum of a point object:

$$L \equiv r_\perp mv \qquad (9\text{-}19)$$

about a given axis.

By dividing an extended rotating body into many tiny pieces, each approximately a point mass, and summing their angular momenta, we can show it to have a total angular momentum

$$L = (\Sigma mr^2)\omega = I\omega \qquad (9\text{-}22)$$

We can summarize the connections between the treatment of dynamics and conservation laws for (one-dimensional) linear and rotational motion:

Linear relationships	Analogies	Rotational relationships
Newton's second law $\Sigma F = ma$	$v \leftrightarrow \omega$	$\Sigma \tau = I\alpha$ $\qquad$ (9-14)
$p = mv$	$a \leftrightarrow \alpha$	$L = I\omega$ $\qquad$ (9-22)
Law of conservation of (linear) momentum: $\Sigma m_1 v_1 = \Sigma m_2 v_2$	$m \leftrightarrow I$ $F \leftrightarrow \tau$ $p \leftrightarrow L$	Law of conservation of angular momentum for rigid bodies about a fixed axis: $\Sigma I_1\omega_1 = \Sigma I_2\omega_2$ $\qquad$ (9-23)
$KE_{lin} = \frac{1}{2}mv^2$		$KE_{rot} = \frac{1}{2}I\omega^2$

✦ QUALITATIVE AND QUANTITATIVE PROBLEMS ✦

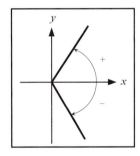

Reminder: The same sign convention that holds true for the angle θ applies to arc length *s*, angular velocity ω, angular acceleration α, angular momentum *L*, and torque τ. **Counterclockwise positive, clockwise negative.**

Hands-On Activities and Discussion Questions

The questions and activities in this group are particularly suitable for in-class use.

9-1. Hands-On Activity. Among Miranda's many interests are astronomy and juggling. Her folding telescope is close at hand, so she decides to balance it on her fingertip by the eyepiece end (keeping the lens cap on to protect it).

a. Will it be easier for her to balance when it is folded into itself (Figure 9-13*a*) or when it is fully extended (Figure 9-13*b*)? Try to use ideas from this chapter to explain your reasoning.

b. Now try balancing some objects (broomsticks, pens, rulers, etc.) of different lengths to see which are easier to balance. How does ease of balancing seem to depend on the length of the object? Is this what you expected?

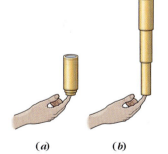

Figure 9-13 Problem 9-1 $\qquad$ **(a)** $\qquad$ **(b)**

Review and Practice

Section 9-1 Measures of Rotation

9-2. As a rotating machine part goes through an angle of $\frac{\pi}{3}$, a point on its outer rim travels a distance of 0.5 m.
a. Find the radius of the machine part.
b. During the same time interval, what distance is traveled by a point halfway between the axis of rotation and the outer rim?

9-3. Convert each of the following to radian measure: 53°, −90°, 210°.

9-4. Convert each of the following to radian measure: 1.4 revolutions, 60 cycles.

9-5. Convert each of the following angles (given in radian measure) to degrees: π, $\frac{5\pi}{6}$, $-\frac{\pi}{6}$, $\frac{2\pi}{3}$, $\frac{9\pi}{4}$, $\frac{11\pi}{12}$, 1.4π.

9-6. Convert each of the following angles (given in radian measure) to degrees: $\frac{3\pi}{2}$, 3.14, 3π, 3.0.

9-7. Find numerical values for each of the following: $\sin\left(\frac{5\pi}{6}\right)$, $\cos 1.4\pi$, $\tan \pi$, $\cos\left(\frac{11\pi}{12}\right)$.

9-8. A dog is on a 2.5-m leash tied to a post in your neighbor's yard. The dog runs in circles, keeping the leash taut. How far does the dog run if the leash rotates through an angle of
a. $\pi/3$. **b.** 3π. **c.** $-\pi/5$.

9-9. As a merry-go-round rotates through an angle of $5\pi/6$, a point on its outermost edge travels 9.25 m. Find the radius of the merry-go-round.

9-10. As a turntable rotates $40°$, a point on its outermost edge travels 0.10 m. Find the radius of the turntable.

Section 9-2 Angular Velocity and Angular Acceleration

9-11. Until the early 1950s, most phonograph records were made to play at 78 rpm (revolutions per minute). What was the angular velocity of these records? What was their frequency in SI units?

9-12. Find Earth's angular velocity in rad/s.

9-13. Six amusement park employees are working on a carousel that has a radius of 5.0 m. One is standing right at the center of the carousel, one is right on the outer edge, and the others are positioned between them at 1-m intervals. When the carousel is taking 10 seconds to complete a single revolution, what is the
a. frequency of the carousel?
b. angular velocity of each employee?
c. linear velocity of each employee?

9-14. We read on a newspaper science page (never respectful of SI units) that the world's smallest motor, fabricated atom by atom at Sandia Laboratories in New Mexico, has a diameter of 50 micrometers and rotates at 500 000 rpm, "faster than any other known device."
a. Which do you think has a greater linear speed, an atom on the outer rim of the rotating motor or a person standing on the equator of a rotating Earth? Try to give a brief reason for your choice without doing a complete calculation.
b. Now calculate the two linear speeds to check your prediction. If the results disagree with your prediction, explain why. (For example, was there anything you neglected to take into account?)

9-15. **SSM** The passenger car of a certain amusement park ride travels in horizontal circles about a central axis to which it is connected by a 5.0-m cable. If the cable is in danger of breaking when the force on it exceeds 10 000 N, what is the
a. maximum safe angular velocity when the car is filled to a total car + passenger mass of 500 kg?
b. car's linear velocity at this maximum?

9-16. Mechanics are installing a large round rotating platform or carousel to be used as a luggage return. In testing its endurance, they try running it much faster than would be allowed when passengers were present. Suppose they operate it at an angular velocity of $\frac{\pi}{3}$ rad/s. When a loaded suitcase weighing 100 N is placed on the platform, the maximum static frictional force that the platform can exert on it is 66 N. How far out from the center can the suitcase be placed on the platform without slipping when the platform turns at this angular velocity?

9-17.
a. Can a train on a circular track ever have a radial acceleration without any tangential acceleration? If so, when? If not, why not?
b. Can it ever have a tangential acceleration without any radial acceleration? If so, when? If not, why not?
c. If it starts from rest and maintains a constant tangential acceleration, is the magnitude of its radial acceleration *zero, nonzero but constant, increasing,* or *decreasing?*

9-18. Find the angular acceleration of the locomotive between points 1 and 3 in Example 9-4.

9-19. **SSM WWW** A motor part at a distance of 0.15 m from the motor's axis of rotation experiences a constant angular acceleration of 0.25 rad/s^2. What is the magnitude of its linear acceleration at the instant when its angular velocity is 0.5 rad/s?

9-20. A driver brakes, reducing the speed of her vehicle at a constant rate as she drives along a semicircular exit ramp. Sketch a vector diagram showing the car's tangential acceleration, its radial acceleration, and its total linear acceleration at a typical point along the ramp.

Section 9-3 Torque and Angular Acceleration

9-21. Four small balls, each of mass 0.40 kg, are fixed at the corners of a very light square frame, which is 0.20 m on each side. Find the moment of inertia of this rigid arrangement about an axis perpendicular to the square if the axis passes through
a. the center of the square. **b.** one corner of the square.

9-22. Calculate the moment of inertia of Earth in SI units about
a. its own axis of rotation.
b. the sun (or more precisely, a line through the sun perpendicular to the plane in which the planets orbit).

9-23.
a. The moment of inertia of the long, thin cylindrical rod in Figure 9-14a is not the same about every axis. List the axes by letter in rank order according to the rod's moment of inertia about the axis. List them from least to greatest, indicating any equalities.
b. Repeat **a** for the thin rectangular plate in Figure 9-14b.

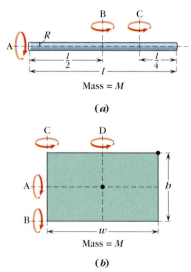

Figure 9-14 Problems 9-23 and 9-24

9-24. Make use of the information in Figure 9-9 to answer the following:
a. Determine the moment of inertia of the long, thin cylindrical rod in Figure 9-14a about each of the axes shown. (You should compare your results with the ordering you did in Problem 9-16**a** and resolve any inconsistencies.)
b. Of the various expressions for moments of inertia given in Figure 9-9, which, if any, could also be the moment of

inertia for a thin rectangular plate rotated about one edge? Briefly explain your reasoning. Assume the plate has uniform density.

c. Determine the moment of inertia of the thin rectangular plate in Figure 9-14*b* about each of the axes shown. (You should compare your results with the ordering you did in Problem 9-23*b* and resolve any inconsistencies.)

9-25. Of two hula hoops, hoop A has twice the mass but half the radius of hoop B. Is the moment of inertia of hoop A less than, equal to, or greater than that of hoop B? Briefly explain.

9-26. A propeller mounted on an axle has a moment of inertia of $0.050 \text{ kg} \cdot \text{m}^2$ about the axle. On average, how much torque must the axle exert on the propeller to bring it from rest to an angular speed of 30π rad/s in 4.0 s?

9-27. A solid disk-shaped machine part with a radius of 5.0×10^{-2} m and a mass of 0.32 kg rotates at 20 rev/s. It can be brought to a stop by a brake mechanism that exerts a frictional force on the outer rim. How much frictional force must the brake exert to bring the machine part to a stop in 0.75 s?

9-28. The pulley in Figure 9-15 is a uniform solid disk of mass 3.0 kg and radius 0.12 m. The grappling hook suspended from the rope has a mass of 1.1 kg. The rope itself is light enough for us to treat as massless. When the hook is allowed to fall, the pulley turns frictionlessly on its axis.

a. Find the force that the rope exerts on the hook as it falls.

b. What is the hook's acceleration as it falls?

c. What is the pulley's angular acceleration as the hook falls?

Figure 9-15
Problem 9-28

Section 9-4 Rotational Kinetic Energy

9-29. **SSM** A lightweight metal hoop used in a magician's act has a mass of 0.12 kg and a radius of 0.25 m. If the hoop is rolling with an angular velocity of $3\pi/s$ (3π rad/s), how much kinetic energy is lost from the system observed by the audience when the hoop suddenly disappears?

9-30. A baseball (not necessarily regulation) has a mass of 0.15 kg and a radius of 0.045 m. Calculate the total kinetic energy of the baseball when it is rolling (without slipping) at a speed of 6.0 m/s.

9-31. Figure 9-10 illustrates an argument showing that when a body rolls without slipping, the tangential velocity of its outer edge is equal to v_{CM}. Suppose the mass of the rolling object is not symmetrically distributed, so that its center of mass is off to one side of its central axis. Is the argument illustrated in Figure 9-10 still valid? Briefly explain.

9-32. A light metal hoop is released from rest at the top of a 20 m incline making an angle of 37° with the horizontal. The hoop has a mass of 0.24 kg and a radius of 0.45 m. If the hoop rolls down the incline without slipping, what is its speed when it gets to the bottom?

9-33. An acrobat can roll down a hill in either of the two positions shown in Figure 9-16. In which position will he get to the bottom faster? Briefly explain why.

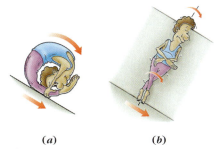

(a)   *(b)*

Figure 9-16 Problem 9-33

9-34. Two identical spheres A and B are released from the same height h on identical ramps. But the first ramp is iced over, so that contact between sphere A and its ramp is frictionless, and sphere A slides (or skids) down it. But there is enough frictional interaction between sphere B and its ramp so that sphere B rolls down without slipping. Compare the speeds with which the two spheres reach bottom. State your answer as a ratio of speeds $\frac{v_B}{v_A}$.

9-35. Consider the first three solids in Figure 9-9. Assuming they have equal masses and radii, rank these solids in order (from least to greatest, indicating any equalities) of

a. the kinetic energy that each solid has if all three solids are rotating at the same angular velocity.

b. the kinetic energy that each solid acquires if all three solids start at rest and have the same constant external torque applied to them for 5 s.

Section 9-5 Another Basis for Bookkeeping: Angular Momentum

9-36. Consider the rider thrown from the Turkish Twist in Figure 8-10*c*. The rider's direction is tangential to the circular perimeter of the ride. When the rider is thrown from the ride, does the angular velocity of the ride increase, decrease, or stay the same? Briefly explain.

9-37. Suppose you sit down on a rotating piano stool with your arms fully outstretched and an exercise weight clutched in each hand, and you spin yourself around on the stool. What will happen if, while you are spinning, you pull the weights in close to your chest? Why?

9-38. While a small playground merry-go-round is rotating at an angular speed of $\frac{2}{5}\pi$ rad/s, a child with a mass of 20 kg standing next to it jumps onto the outer edge (Figure 9-17*a*). The merry-go-round, which is roughly a solid disk or flat cylinder, has a mass of 32 kg and a radius of 1.0 m. Assuming that the child's jump is completely in the radial direction, what is the angular velocity of the merry-go-round after the child lands?

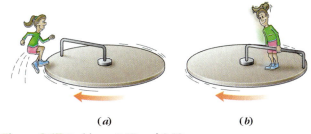

(a)   *(b)*

Figure 9-17 Problems 9-38 and 9-39

9-39. SSM After landing on the outer edge of the merry-go-round, the child in Problem 9-38 walks halfway from the outer edge to the merry-go-round's center (Figure 9-17*b*).

a. What is the angular velocity of the merry-go-round when the child reaches this point?

b. Is it possible to answer part *a* of this problem without first solving Problem 9-38?

9-40. A figure skater with his arms and a leg maximally extended (position 1) has a moment of inertia of 5.5 kg · m². By pulling his arms and legs in as close as possible to the central axis of his body (position 2) he is able to reduce his moment of inertia to 1.1 kg · m². If he begins spinning in position 1 with an angular velocity of 0.30π/s (in radian measure), what will be his angular velocity if he then switches to position 2?

9-41. SSM WWW At its perihelion (Figure 8-22), Pluto is 4.45×10^{12} m from the sun and is traveling at 6.1×10^3 m/s. At its aphelion, it is 7.38×10^{12} m from the sun. Find its speed at aphelion two ways,

a. by applying conservation of angular momentum, and

b. by applying conservation of mechanical energy.

c. Do you get the same value both ways?

Going Further

The questions and problems in this group are not organized by section heading, so you must determine for yourself which ideas apply. Some of them will be more challenging than the Review and Practice questions and problems (especially those marked with a • or ••).

9-42. The disks in Figure 9-18 have rubber rims that make pressure contact with each other, so that when one turns on its axis, it turns the other without slipping. A weight hangs from a string wrapped around the hub of the smaller disk, so that when the weight is allowed to drop, it turns the disk. At any instant while the weight is dropping, compare the

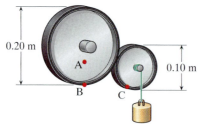

Figure 9-18 Problem 9-42

a. linear velocity of dots B and C.

b. linear velocity of dots B and A.

c. angular velocity of dots B and A.

d. angular velocity of dots B and C.

e. linear acceleration of dots B and C.

f. linear acceleration of dots B and A.

(Adapted from a demonstration interview problem by G. Francis and R. Herman, *AAPT Announcer*, 22 (4), p. 83, 1992)

9-43. Figure 9-19 shows the working parts of a rotational motion machine commonly used for laboratory demonstrations. (Ask your instructor if there is one that you can examine.) Both disks are made of hard rubber, and there is a pressure contact between them. Therefore, when disk A turns in a vertical plane, it turns disk B in a horizontal plane without slipping. Disk B can be slid

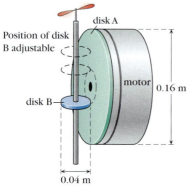

Figure 9-19 Problem 9-43

up or down its axis and then locked in position making contact at any point along the diameter of disk A. Suppose the motor turns disk A at a frequency of 10 revolutions/s.

a. How should disk B be moved along its axis to increase its angular velocity? To decrease its angular velocity? In each case, choose the best answer: *always upward; always toward the center of disk A; always downward; always away from the center of disk A.*

b. What is the maximum tangential velocity that can be achieved by the outer rim of disk B? What is the frequency of disk B when this is the case?

c. What is the frequency of disk B when it makes contact with disk A at A's center?

d. At what contact point along the diameter of disk A do the rims of the two disks have the same tangential velocity? Answer by telling how far this point must be from the center of A.

e. At what point along the diameter of disk A do the two disks have the same angular velocity? Answer by telling how far this point must be from the center of A.

9-44. A mechanical car with front wheel drive consistently veers to the right as it rolls forward. One front wheel turns out to have a slightly greater diameter than the other. Which is it, the left or the right wheel? Explain.

9-45. A large circular platform rotates at a constant rate of 3 revolutions/second. Several people stand on the platform at different distances r from the center while it rotates. Student A finds the equation $F_r = mr\omega^2$ in the textbook and concludes, "The radial force increases as r increases, so the platform must exert a greater frictional force on the people who are further out to keep them from slipping." Student B finds the equation $F_r = m\frac{v^2}{r}$ in the textbook and concludes, "Because r is in the denominator, the radial force increases as r decreases, so the platform must exert a greater frictional force on the people who are closer to the center to keep them from slipping." Determine which student is correct, and explain why the other student's argument is incorrect.

•9-46. In Example 9-3, if the crew member has a mass of 70.0 kg and the scale has a mass of 2.00 kg, find the normal force that the wall exerts on the scale when the scale is reading the crew member's correct Earth weight.

•**9-47.** A yo-yo (Figure 9-20) on a string of length *l* is whirled in a horizontal circle at an angular velocity *ω*. How would *ω* change if you wanted to keep the yo-yo rotating on a longer string making the same angle *φ* with the vertical?

•**9-48.** *Regulating a steam engine:* A fly-ball governor (Figure 9-21) is a device to prevent sudden speed changes in a steam engine. As the shaft rotates more rapidly, the balls rise, lifting the collar. This in turn connects to a valve control, so that when it is lifted, it reduces the steam pressure that is causing the speeding up.

a. Make the simplifying assumptions that the force exerted on each ball by a rod of length *l* is purely a tension force and that the connecting pieces are very light compared to the balls. Under these assumptions, show that the relationship between the angle *φ* (see figure) and the shaft's angular velocity *ω* is given by $\cos\phi = \frac{g}{l\omega^2}$.

b. Does this relationship indicate that the angle *φ* is increasing or decreasing with *ω*? Explain.

9-49. Use analogies between linear and rotational quantities to obtain an equation for the power input to an axle needed to keep the axle rotating at a constant angular speed *ω*. What other quantity would you need to know to calculate this power input?

9-50. In Figure 9-9, why does the hollow cylinder have a greater moment of inertia about its central axis than the solid cylinder having the same *R* and *M*? Why does the thin rod have a greater moment of inertia about axis B than about axis A?

9-51. Measurements on a large, roughly spherical molecule of total mass *M* and radius *R* show that it has a moment of inertia of roughly $\frac{1}{12}MR^2$ about one particular axis. What might this possibly imply about the arrangement of atoms within this molecule?

9-52. A light, stiff straight wire is set to rotating about one end. A bead of mass 0.15 kg is fixed 0.10 m from the pivot end and a bead of mass 0.05 kg is fixed 0.30 m from the pivot end. What is the moment of inertia of this entire arrangement?

•**9-53.** **SSM** Of two hula hoops with the same radius, hoop A has twice as much mass per unit length (kg/m) as hoop B and twice B's radius. By what factor must you multiply the moment of inertia of hoop B to get the moment of inertia of hoop A?

9-54. Do the moment of inertia values in Problem 9-40 make sense? To answer this, estimate your own moment of inertia in position 2, making measurements on yourself as necessary.

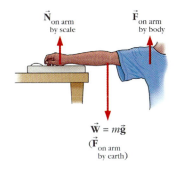

Figure 9-20
Problem 9-47

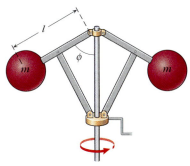

Figure 9-21 A ball governor for a steam engine, for Problem 9-48

You can approximate your body by a solid cylinder of suitable mass and radius.

••**9-55.** Estimate your own moment of inertia when you are balanced on one foot with your other leg and both arms extended straight out (a common position in figure skating or ballet). In this position, you will need to estimate the masses of your arms and legs and treat them as rods. You can get an approximate weight of your arm by holding it out horizontally and supporting the outstretched end on a scale (Figure 9-22*a*). You must keep your arm relaxed to let the scale do its share. Then to the extent that your arm weight is uniformly distributed, each upward force will support half of it (why?), so the scale reads half of your arm weight. Figure 9-9 gives you the moment of inertia of a rod about one end. But your arm ends at your shoulder, not at your central axis. To find the moment of inertia of your outstretched arm about your central axis A (Figure 9-22*b*), you can mentally extend your arm to the central axis by adding a piece of length *r* (the radius of your body cylinder). Because its length is a fraction $\frac{r}{L}$ of the arm's length *L*, its mass will be the same fraction of the arm's mass *m*, that is, it will be $\frac{r}{L}m$. You can then find the moment of inertia of the arm + extension about A, find the moment of inertia of the extension alone about A, and subtract the latter from the former to get the moment of inertia of the arm itself. Follow a similar procedure to find the moment of inertia of the extended leg.

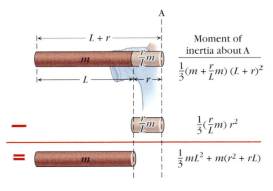

(*a*) Weighing your arm

	Moment of inertia about A
	$\frac{1}{3}(m + \frac{r}{L}m)(L + r)^2$
	$\frac{1}{3}(\frac{r}{L}m)r^2$
	$\frac{1}{3}mL^2 + m(r^2 + rL)$

(*b*) Finding your arm's moment of inertia

Figure 9-22 Problem 9-55

•**9-56.** Two cans of soup have the same weight and dimensions. One is chicken noodle (thin, with lots of noodles); the other is split pea (very thick, especially before adding water). The two cans are simultaneously released from rest at the top of a long ramp and allowed to roll to the bottom. The can of split pea soup reaches the bottom first. Why?

9-57. A mouse begins running clockwise along the outer rim of a turntable that is free to rotate on its axis.

a. What happens to the turntable? On what principle do you base your conclusion?

b. Can you suggest an analogous situation involving straight-line motion? What principle governs that situation?

9-58. A turntable with an unknown moment of inertia is rotating at a measurable angular speed ω_1. How can you find the moment of inertia of the turntable by dropping a small, heavy object on it? In your answer, tell what quantities you would have to measure and how you would use these quantities to calculate the turntable's moment of inertia.

9-59. Example 9-6 deals with an assembly consisting of two different size wheels on an axle. When the assembly is initially rotating with an angular velocity ω_o, the brake must exert a certain frictional force on the smaller wheel to bring it to a stop in 2.0 s. Would the magnitude of the required frictional force be affected if the larger wheel were removed from the axis before the brake was applied to the smaller wheel? Explain your answer.

9-60.

a. For the system consisting of the roulette wheel and the sack of quarters in Example 9-8, find the total mechanical energy before and after the collision.

b. Is energy lost? Briefly explain why or why not.

9-61. Consider the system consisting of the roulette wheel and the sack of quarters in Example 9-8.

a. If there is no slipping, is kinetic energy lost between the initial and final states? If so, what happens to it? If not, why not?

b. Does this situation have anything in common with the situation in Example 7-4? Explain.

9-62. For which of the following equations must the angular quantities always be in radian measure, and for which equations is conversion from degrees to radians unnecessary? Briefly explain each answer.

a. $s = r\theta$.   ***d.*** $\Sigma\tau = I\alpha$.

b. $\omega = \alpha t$ when $\omega_o = 0$.   ***e.*** $KE_{\text{total}} = \frac{1}{2}mv^2 + \frac{1}{2}I\omega^2$.

c. $\omega = \Delta\theta/\Delta t$.

9-63. The angular velocity $\omega = \frac{\Delta\theta}{\Delta t}$ is analogous to the linear velocity $v = \frac{\Delta x}{\Delta t}$. For each graph of ω versus t in Figure 9-23, indicate which graph of θ versus t corresponds to the same motion.

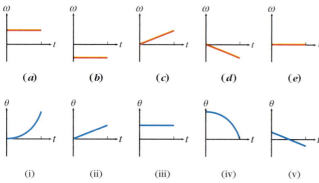

Figure 9-23 Problem 9-63

9-64. The angles through which two bodies A and B rotate are plotted over the same time interval in Figure 9-24. Does A or B have a greater angular velocity at time t_1? Briefly explain. (This and the following problem are adapted from materials by Lillian C. McDermott and the University of Washington physics education group.)

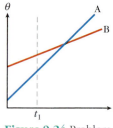

Figure 9-24 Problem 9-64

9-65. Two hands P and Q of an oddly behaving clock rotate around the same central axis. Their angular velocities are plotted over the same time interval in Figure 9-25. What information does this graph provide about whether one of the two hands overtakes the other during this interval?

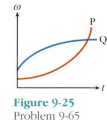

Figure 9-25 Problem 9-65

9-66. A turntable has a 30-cm diameter. An ant walks along a diameter of this turntable from one end to the other at a constant speed of 2.0 cm/s (relative to the turntable). Just at the instant the ant starts out, the turntable, which was at rest, begins to rotate with a constant angular velocity. Because the turntable has now rotated, the ant arrives at the other end of the diameter to discover it is exactly where it started out.

a. What is the smallest angular velocity the turntable can have that would result in the ant ending up where it started?

b. What is the next smallest angular velocity that would lead to this result?

9-67. SSM

a. In Figure 9-26a, the three axes shown are perpendicular to the triangular metal plate (that is, they are directed into and out of the picture). The plate has uniform thickness and density. Rank the axes in order of the moment of inertia that the triangular plate would have about each axis. Rank them from least to greatest, indicating any equalities.

b. The three axes shown in Figure 9-26b run along the edges of the triangular metal plate. Rank these axes in order, from least to greatest, of the moment of inertia that the triangular plate would have about each axis. Again, indicate any equalities.

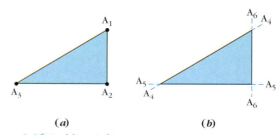

(a)   *(b)*

Figure 9-26 Problem 9-67

9-68. *Rotational kinematics:* It is often useful to think about the analogies between translational motion and rotational motion. The following equations held true for linear motion when there is constant linear acceleration.

$$v = v_o + at$$

$$x - x_o = \bar{v}t = \left(\frac{v_o + v}{2}\right)t$$

$$x - x_o = v_o t + \frac{1}{2}at^2$$

$$v^2 = v_o^2 + 2a(x - x_o)$$

Transform these equations into equations for rotational motion with constant angular acceleration by replacing the variables x, v, and a wherever they occur by their appropriate angular analogs. Choose replacement variables from among the following: ω, r, α, θ. You will have to decide whether you need every one.

Use the equations that you found in Problem 9-68 for rotational motion with constant angular acceleration to solve the next three problems.

9-69. (See Problem 9-68) A turntable starting from rest is given an angular acceleration of $0.2\pi/s^2$.
a. Through what total angle does it rotate in the first 3.0 s?
b. How far from the center must a bug be standing on the turntable in order to reach a speed of 2.0 m/s after 3.0 s?

9-70. (See Problem 9-68) A certain electric window fan ordinarily rotates at 8.0 rev/s. When suddenly switched from the "in" to the "out" setting, it takes 4.0 s to reach operating speed in the reverse direction.
a. What is its angular acceleration during this interval?
b. If the fan has a radius of 0.20 m, what is the tangential acceleration of the tips of the blades?
c. What is its total angular displacement during this interval?

9-71. (See Problem 9-68) You have been giving your tea three stirs per second. When you stop, your tea continues to swirl for another five seconds.
a. What is the angular acceleration of the tea over this interval?
b. Through what angle (in radians) does the tea swirl during this interval?

9-72. Three students calculate the moment of inertia of the same object. Each student gets a different value, and each student's value is correct. If the object does not change in any way, and the students are all working in the same units, how is this possible?

9-73. An object drops vertically downward onto a rotating turntable. Will the turntable slow down more if the object lands near the center or near the rim, or doesn't it matter? Briefly explain.

9-74. A solid grinding wheel with a radius of 0.20 m is rotating with an angular velocity of 6π rad/s. If a knife edge is held against the wheel just hard enough to exert a constant frictional force of 3.8 N, it brings the wheel to a stop in 10.0 s. Find the mass of the wheel.

9-75. *Rotational work-energy theorem:* The work-energy theorem for motion along a straight line can be written as $F\Delta s = \frac{1}{2}mv_2^2 - \frac{1}{2}mv_1^2$. By substituting the analogous rotational quantity for each quantity in this equation, find the equation that gives the work-energy theorem for purely rotational motion.

9-76.
a. Find the work done by the knife edge in Problem 9-74.
b. Use the equation you found in Problem 9-75 to solve Problem 9-74 by the rotational work-energy theorem.

Problems on WebLinks

9-77. As Tarzan swings in WebLink 9-1, he passes through the positions A, B, and C identified in the WebLink. At which of these positions (A, B, C, all, none) is his tangential acceleration in the same direction as his velocity?

9-78. At which of the positions in Problem 9-77 (A, B, C, all, none) is Tarzan's radial acceleration perpendicular to his velocity?

9-79.
a. At which of the positions in Problem 9-77 (A, B, C, all, none) is Tarzan's total acceleration perpendicular to his velocity?
b. At which of the positions in Problem 9-77 is his total acceleration parallel to his velocity?

9-80. At which of the positions in Problem 9-77 does the rope exert a nonzero force on Tarzan that is neither completely parallel nor completely perpendicular to his velocity? If there is more than one correct answer, identify each one. If there are no correct answers, write *none*.

9-81. The tilted top in Figure 9-27 is rotating about axis A in the direction shown. At the instant depicted, axis A is in the xy plane (the plane of the page). (The z axis is perpendicular to the page.) Based on the reasoning in WebLink 9-2, determine which of these axes (axis A, the x axis, the y axis, the z axis) goes in each of the blanks below.
a. The top's angular momentum vector is directed along ____.
b. The torque vector due to the weight of the top is directed along ____.
c. The change $\Delta\vec{L}$ in the angular momentum vector is directed along (or parallel to) ____.

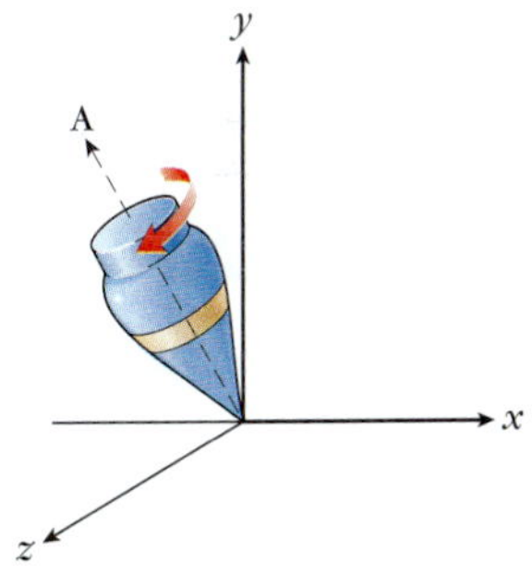

Figure 9-27 Problem 9-81

For a cumulative review, see Review Problem Set I on page 825.

CHAPTER TEN

Statics and Dynamics of Fluids

Any form of matter that cannot maintain its own shape, and therefore flows, is called a **fluid.** Liquids and gases are both fluids. Our direct experience of fluids is quite different than our experience of typical solids. Push on a wooden block and it seems impenetrable; you can feel it pushing back or resisting. Move your hand through air or water and your hand readily penetrates it; it flows around your hand as your hand moves. Yet you see evidence that fluids exert forces when a sailing ship stays afloat or when the wind (moving air) propels it forward. You see evidence that forces in turn affect the motion of fluids when you open a faucet or squeeze the air out of a rubber raft. Fluids are subject to the same laws of motion as solids, but because the shape of a body of fluid can change,

"... you see evidence that fluids exert forces when a sailing ship stays afloat or when the wind propels it forward."

its motion and its response to forces may be more complicated than that of a point object or rigid body. More ingenuity is then required in applying the basic laws of mechanics to fluids. In this chapter we will look at the mechanics of fluids.

- In Sections 10-1 and 10-2 we will consider effects by and on stationary bodies of fluid, including such common phenomena as air and water pressure, floating and sinking. The study of these situations is called the **statics** of fluids.

- In Sections 10-3 through 10-5 we will consider fluids in motion: what affects their motion and what effects moving fluids have on other objects. In connecting forces and motion, we will be treating the **dynamics** of fluids.

10-1   The Statics of Fluids: Pressure

When you buy a backpack, you look for one with wide straps because skinny straps cut into your shoulders. Round pads under the legs of your living room furniture keep them from digging into your carpeting. The multiple nails in Figure 10-1b cause less pain than the single nail in Figure 10-1a. In all of these cases, the effect of the force is lessened by distributing it over a greater area.

The quantity in physics that tells how much of the total force pressing against a surface is exerted on each square unit of area is **pressure** (note that the root word is *press*). Finding pressure is simplest when a total force is distributed uniformly over a surface and is exerted perpendicular to, or normal to, the surface. We just divide the magnitude of the force by the total surface area to determine what share of the force is "felt" by each *unit* of surface area. The pressure P is then $P = \frac{F}{A}$.

If the force is not perpendicular to the surface, only its normal component $F_\perp$ (or F_N) presses against the surface. **STOP&Think** Can a frictional force ever contribute to pressure?

If the force is unevenly distributed, dividing it by the area gives the *average* pressure:

The **average pressure on a surface of area A** is

$$P = \frac{F_\perp}{A} \qquad (10\text{-}1)$$

($F_\perp$ = the *total* normal force exerted on the entire surface)

Because it involves only a scalar component of force, pressure is a scalar.

◆**UNITS** Units of pressure are newtons per square meter ($\frac{N}{m^2}$) in metric units and pounds per square inch ($\frac{lb}{in^2}$, commonly abbreviated **psi**) in the English system. A $\frac{N}{m^2}$ is called a **pascal** ($1\ \text{Pa} \equiv 1\ \frac{N}{m^2}$) in SI.

We often think of pressure being exerted by continuous fluids like air or water. The water in the fish tank in Figure 10-2 exerts pressure on all the surfaces it makes contact with. The domino does not have to be secured because the water exerts equal pressure on both sides of it, so the resulting forces are equal and opposite. This is not so for the walls of the tank, so the mountings that hold each wall in place must exert a total force equal and opposite to the force exerted by the water. It follows from Equation 10-1 that the force has magnitude $F_\perp = PA$. It is usually easiest to determine the pressure exerted by a fluid when it results in a force on only one side of a body, as in Examples 10-1 and 10-2.

(a)

(b)

Figure 10-1 The effect of a force is different when concentrated at one point than when distributed over a large area.

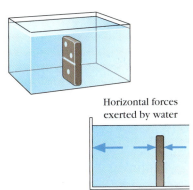

Horizontal forces exerted by water

Figure 10-2 Objects under equal and unequal pressure. The water exerts equal pressure on both sides of the domino but not on both sides of the fish tank walls.

Example 10-1 *Measuring Air Pressure*

For a guided interactive solution, go to Web Example 10-1 at www.wiley.com/college/touger

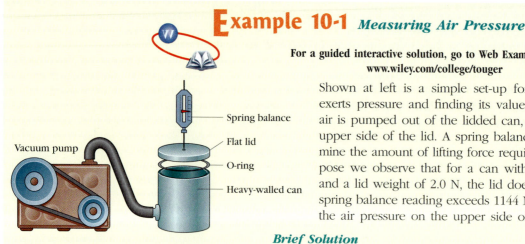

Spring balance
Flat lid
O-ring
Heavy-walled can
Vacuum pump

Shown at left is a simple set-up for demonstrating that air exerts pressure and finding its value at sea level. When the air is pumped out of the lidded can, air remains only on the upper side of the lid. A spring balance is then used to determine the amount of lifting force required to raise the lid. Suppose we observe that for a can with a diameter of 12.0 cm and a lid weight of 2.0 N, the lid does not come off until the spring balance reading exceeds 1144 N (a realistic value). Find the air pressure on the upper side of the lid.

$F_{\text{on lid by sb}}$

$m_{\text{lid}}g$

$F_{\text{on lid by air}}$

Free-body diagram of lid
(vacuum on underside)

Brief Solution

Choice of approach. When there is air on both sides of the lid, there is an upward force due to the air pressure below the lid as well as a downward force due to the air pressure above. When the air within the can is pumped out, the upward force is eliminated. We draw a free-body diagram of the lid (see figure) when it is on the verge of coming off. We then apply Newton's second law with $\vec{a} = 0$ and solve for the force exerted by the air. After finding the area of the lid, we can then find the pressure using the definition (Equation 10-1).

What we know/what we don't. (See free-body diagram; sb = spring balance):

Forces in y direction	
$F_{\text{on lid by sb}} = 1144$ N	$d = 12$ cm $= 0.12$ m $= 2r$
$-F_{\text{on lid by air}} = ?$	$A = ?$ $P = ?$
$-m_{\text{lid}}g$ $(m_{\text{lid}}g = 2.0$ N$)$	

The mathematical solution. $\Sigma F_y = 0$ becomes

$$F_{\text{on lid by sb}} - F_{\text{on lid by air}} - m_{\text{lid}}g = 0$$

Then $F_{\text{on lid by air}} = F_{\text{on lid by sb}} - m_{\text{lid}}g = 1144$ N $- 2.0$ N $= 1142$ N

This is exerted on an area $A = \pi r^2 = (3.1416)\left(\dfrac{0.12 \text{ m}}{2}\right)^2 = 0.0113$ m^2

so $P_{\text{by air}} = \dfrac{F_{\text{on lid by air}}}{A} = \dfrac{1142 \text{ N}}{0.0113 \text{ m}^2} = \mathbf{1.01 \times 10^5 \text{ N/m}^2}$ or

$$\mathbf{1.01 \times 10^5 \text{ Pa}}$$

This is the generally accepted value. It is equivalent in the English system to 14.7 lb/in^2 (14.7 psi).

◆ Related homework: Problem 10-4.

Example 10-2 *Keeping the Pressure On*

For a guided interactive solution, go to Web Example 10-2 at www.wiley.com/college/touger

Suppose the experiment in Example 10-1 is repeated using a can and lid with a diameter of 24.0 cm. Again, the upward pull on the lid is gradually increased. When the spring balance reads 1144 N, how much downward force does the air exert on the lid?

Brief Solution

Choice of approach. Once you know the pressure, use the definition of pressure with the area of the new lid to find the new total force. **STOP&Think** Do you need the spring balance reading? ◆

What we know/what we don't.

$$P = 1.01 \times 10^5 \text{ N/m}^2 \text{ (from Example 10-1)}$$

$$d = 24 \text{ cm} = 0.24 \text{ m} = 2r \qquad A = ? \qquad \text{new } F_{\text{on lid by air}} = ?$$

The mathematical solution. The new area

$$A = \pi r^2 = (3.1416)\left(\frac{0.24 \text{ m}}{2}\right)^2 = 0.0452 \text{ m}^2$$

so now $F_{\text{on lid by air}} = P_{\text{by air}}A = (1.01 \times 10^5 \text{ N/m}^2)(0.0452 \text{ m}^2) = $ **4565 N**

which, allowing for rounding off errors, is *four* times the original force.

Making sense of the results. When the radius or diameter is multiplied by n (here $n = 2$), the area is multiplied by n^2. The total force increases proportionally to the area, so it must also be multiplied by n^2.

◆ Related homework: Problems 10-5, 10-6, and 10-10.

We will frequently need the value we found in Example 10-1:

Atmospheric pressure at sea level

$$1 \text{ atmosphere (atm)} = 1.01 \times 10^5 \text{ N/m}^2 = 1.01 \times 10^5 \text{ Pa} = 14.7 \text{ psi}$$

We can reason about the behavior of fluids at rest by mentally isolating some part of the fluid and treating it as a free body to which we can apply Newton's second law. Much of this section and the next is based on this approach; it enables us to draw important conclusions about fluids. As you work through these sections, keep in mind that knowing how we draw on basic physics ideas in the reasoning is as important as the conclusions.

Now follow the reasoning in Figure 10-3. Isolating a wedge-shaped chunk of water lets us consider the forces due to water pressure on its three sides. See how the equilibrium condition enables us to conclude that

at a given depth, a fluid exerts the same pressure against surfaces on all sides of it.

To follow this reasoning in guided, stepwise detail, go to WebLink 10-1.

Considering the chunk of water in Figure 10-3 enabled us to compare the pressure that a fluid exerts on different surfaces at the same depth; considering

For **WebLink 10-1:
Newton II
Underwater,** go to
www.wiley.com/college/touger

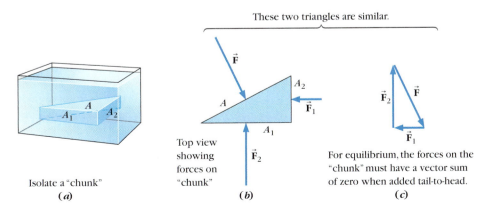

These two triangles are similar.

Top view showing forces on "chunk"

For equilibrium, the forces on the "chunk" must have a vector sum of zero when added tail-to-head.

Isolate a "chunk"

(*a*)   (*b*)   (*c*)

Figure 10-3 Reasoning about an isolated "chunk" of water.
Because the two triangles are similar, their sides are in the same ratio: $\frac{F_1}{A_1} = \frac{F_2}{A_2} = \frac{F}{A}$. But this says that $P_1 = P_2 = P$, so we conclude that the pressure is the same on all three vertical surfaces of the chunk. ◆

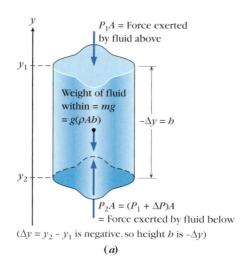

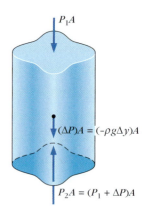

Figure 10-4 More reasoning about an isolated "chunk" of water. A column of water mentally isolated as a free body.

Table 10-1 Densities of Selected Materials

Material	Density (kg/m³)
Solids	
Gold	19.3×10^3
Iron	7.8×10^3
Cement	$2.7–3 \times 10^3$
Aluminum	2.7×10^3
Bone	$1.7–2.0 \times 10^3$
Ice	0.92×10^3
Oak	$0.6–0.9 \times 10^3$
Pine	0.5×10^3
Liquids	
Mercury	13.6×10^3
Whole blood	1.05×10^3
Sea water	1.025×10^3
Water (3.98°C)	1.00×10^3
Ethyl alcohol	0.81×10^3
Liquid nitrogen	0.81×10^3
Gasoline	$0.66–0.69 \times 10^3$
Gases	
Air (dry, 0°C)	1.29
Air (dry, 20°C)	1.21
Helium	0.178
Hydrogen	0.090

Values are at 0°C and 1 atm of pressure unless otherwise stated.

For **WebLink 10-2: Newton II Goes Deeper,** go to www.wiley.com/college/touger

the chunk of water in Figure 10-4 will enable us to compare the pressure in the fluid at different depths. But first we must introduce the concept of density. At various times, it may be useful to picture different "chunks" of the same body of fluid. It is then important to realize that the mass of the isolated chunk will be proportional to its volume:

$$m = \rho V \qquad (10\text{-}2)$$

The proportionality constant ρ (mass per unit of volume) is called the **density** of the material. It is a property of the material that doesn't depend on the size of the chunk. It is a basis for comparing how "heavy" different materials are. You can have a ton of Styrofoam or an ounce of lead, but if we compare equal volumes of each, lead is heavier. Density, which tells us how much mass is in a *unit* of volume, is a basis for comparing equal volumes. Table 10-1 gives densities for some common materials (not just fluids).

If the height of a column-shaped chunk of fluid that we have mentally isolated is not negligible (Figure 10-4), then the pressure on the bottom of the column is greater than the pressure on the top because the fluid below must also support the weight of the fluid within the column. For this reason, water pressure increases with depth and air pressure increases as you descend toward sea level. Again we apply Newton's second law. By identifying the forces on the column and setting the sum equal to zero, we arrive at a quantitative relationship between the change in pressure, ΔP, and the change in depth, Δy. We see that if the column has a height $b = -\Delta y$ (y *drops* by b), the pressure increases from its value P_1 at the top by an amount ΔP. The volume of the column is Ab, so the fluid in the column has a weight $\rho g b A = -\rho g \Delta y A$. Because the column is in equilibrium, the sum of the forces on it is 0. Then from Figure 10-4b,

$$(P_1 + \Delta P)A - P_1 A - (-\rho g \Delta y)A = 0$$

Simplifying then gives us

$$\Delta P = -\rho g \Delta y \qquad (10\text{-}3)$$

To follow this reasoning in guided, stepwise detail, go to WebLink 10-2.

When you go downward from y_1 to y_2, Δy is negative. The minus sign in Equation 10-3 indicates that the change ΔP is an increase in pressure when there is a decrease in y, because a decrease in y means the depth is increasing.

STOP&Think If you were going upward in air rather than downward in water, how would the air pressure change? Would Δy be positive or negative? Would ΔP be positive or negative? ◆

Because $\Delta P = P_2 - P_1$, it is often convenient to write Equation 10-3 as

$$P_2 = P_1 + \Delta P = P_1 - \rho g \Delta y \qquad (10\text{-}4)$$

Example 10-3 *Two Divers*

The expert diving team of Dimaggio and Lopez is exploring an underwater wreck. The density of sea water is 1025 kg/m³.

a. Dimaggio is at the surface (see figure). At what depth will he experience double the pressure that is on him now?

b. Lopez is 20 m below the surface. At what depth will she experience the same increase in pressure as Dimaggio experiences in **a**?

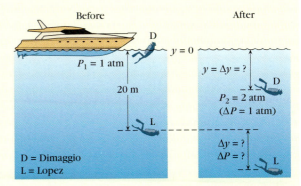

Before — After
D
$y = 0$
$P_1 = 1$ atm
$y = \Delta y = ?$
D
20 m
$P_2 = 2$ atm
($\Delta P = 1$ atm)
L
$\Delta y = ?$
$\Delta P = ?$
L
D = Dimaggio
L = Lopez

Solution

Assumptions. We treat each diver as being all at one depth. Equivalently, we can choose a reference point on each diver, perhaps a pressure-sensitive one like the eardrums or the sinuses.

Restating the problem. **STOP&Think** Is there any point to the question in **a** if we assume the pressure at the surface is zero? ◆

If the upper face of the column is at the water's surface, it is experiencing atmospheric pressure (1 atm = 1.01×10^5 Pa). So **a** asks, at what depth will Dimaggio experience 2 atm of pressure? Then in part **b,** Lopez also has to experience a pressure increase of 1 atm.

Choice of approach. Newton's second law takes the form of Equation 10-4 when the column in Figure 10-4 is in equilibrium ($\Sigma F_y = 0$). The column's top and bottom surfaces now correspond to the diver's initial and final positions.

We do not need to find the pressure experienced by Lopez either at 20 m or at her new depth. Equation 10-3 shows that the change in pressure depends only on the change in depth, so to experience the same change in pressure she must increase her depth the same amount that Dimaggio does.

What we know/what we don't.

a. $\qquad P_1 = 1 \text{ atm} = 1.01 \times 10^5 \text{ Pa} \qquad P_2 = 2 \text{ atm} = 2.02 \times 10^5 \text{ Pa}$

$$\rho = 1025 \text{ kg/m}^3 \qquad g = 9.80 \text{ m/s}^2 \qquad \Delta y = ?$$

b. $\qquad \Delta y = \text{same as in } \mathbf{a} \qquad \text{New } y = -20 \text{ m} + \Delta y = ?$

The mathematical solution.

a. Since $P_2 = P_1 + \Delta P$,

$$\Delta P = P_2 - P_1 = 2.02 \times 10^5 \text{ Pa} - 1.01 \times 10^5 \text{ Pa} = 1.01 \times 10^5 \text{ Pa}$$

But since $\Delta P = -\rho g \Delta y$,

$$\Delta y = -\frac{\Delta P}{\rho g} = -\frac{1.01 \times 10^5 \text{ Pa}}{(1025 \text{ kg/m}^3)(9.80 \text{ m/s}^2)} = \mathbf{-10 \text{ m}} \quad \text{(to two places)}$$

The minus sign indicates *down* from the surface.

b. New $y = -20 \text{ m} + \Delta y = -20 \text{ m} + (-10 \text{ m}) = \mathbf{-30 \text{ m}}$

The pressure increases by 1 atm for each additional 10 m that you descend, no matter what your starting point.

◆ Related homework: Problems 10-7, 10-8, 10-11, and 10-12.

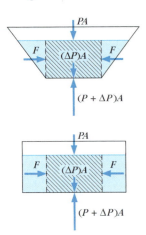

Figure 10-5 The pressure at a given depth is the same no matter what the shape of the tank. F is the magnitude of the force exerted on each vertical wall of the column by the adjacent water outside the column.

Their ability to multiply forces makes hydraulic lifts indispensable in auto repair shops.

Picturing an isolated vertical column of water and looking at the forces on it also enables us to compare the pressure at a given depth in containers of different shape. In the two tanks of water in Figure 10-5, we have isolated identical columns of water. Both columns have the same pressure P and total force PA on their upper surfaces. Because $\Sigma F_y = 0$, the upward force at the bottom equals PA plus the weight of the column in each case because the surrounding water exerts only horizontal forces on the columns. Therefore the forces on the bottoms of the two columns are equal, and because the areas are the same, the pressures there are also equal. We conclude that

the pressure at a given depth does not depend on the shape of the container.

STOP&Think Are the total forces exerted by the two tank bottoms equal? Do the sides of the tanks contribute to supporting the weight of the water in either case? ◆

By similar static equilibrium reasoning, we can explain how a **hydraulic press** works. A hydraulic press is a device that produces an output force much greater than the force exerted by the user. Consider the two tanks of water in Figure 10-6. A piston pushes down with a force of magnitude F_1 on the left side of each tank, causing the right side to push up on a second piston. The pressure is the same at the surface on both sides of tank I. So if the force on the left establishes a pressure $\frac{F_1}{A}$ on the left side, the water on the right exerts equal pressure at its surface, and therefore exerts a force F_1 on the right piston. The same argument applies to the part of the water in tank II that we have shaded in Figure 10-6b. So this "chunk" of the water exerts an upward force F_1 on the left half of the right piston. But by the same reasoning again, the "chunk" of water shaded in Figure 10-6c exerts an upward force F_1 on the *right* half of the right piston. Then the total upward force on the right piston of tank II is $F_2 = F_1 + F_1 = 2F_1$. (So the pressure on the right side of tank II is $\frac{2F_1}{2A} = \frac{F_1}{A}$, the same as on the left.)

The net effect is that if the cross-sectional area on the right side is doubled, the force on the right piston is twice that exerted by the piston on the left. If the cross-sectional area is multiplied by n, the output force is n times the input force. Hydraulic lifts in dentist's chairs and hydraulic brakes in cars and trucks are adaptations of this basic device.

By isolating a vertical column of liquid as a free body, we can demonstrate a more general principle as well. Suppose the shaded volume of water in Figure 10-7 is in equilibrium before the piston pushes down to add external pressure at the top surface. We assume the liquid is incompressible. The free-body diagrams before and after external pressure is applied show that for the shaded body of water to remain in equilibrium ($\Sigma F_y = 0$), the pressure at any arbitrary depth Δy must increase by the same amount P_{ext} as the pressure at the top. We can therefore think of this additional pressure as being passed on or transmitted through the fluid without being diminished or otherwise changed. French philosopher-mathematician Blaise Pascal articulated this idea in 1652.

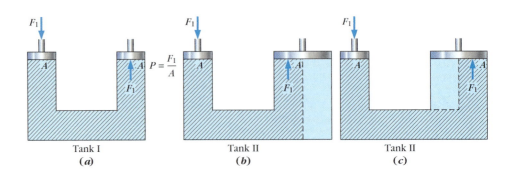

Figure 10-6 Multiplication of force in a hydraulic press.

Tank I	Tank II	Tank II
(a)	(b)	(c)

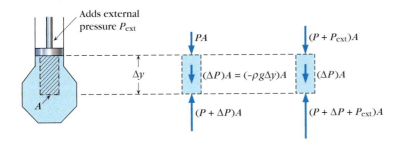

Figure 10-7 The transmission of external pressure through a fluid (Pascal's principle).

Pascal's principle: A change in pressure occurring in one part of a static, enclosed body of fluid occurs equally (undiminished) throughout that body of fluid.

Pascal's principle tells us, for example, that a drop in pressure in the water mains that bring water to your home results in the same drop in pressure in your kitchen sink, or that squeezing one end of a water balloon will provide the added pressure to produce a bulge everywhere else on the balloon.

10-2 The Statics of Fluids: Buoyant Forces

We next compare the free-body diagrams of a submerged box (Figure 10-8a) and the box-shaped region of fluid that it displaces (Figure 10-8b). This "chunk" of fluid would be in equilibrium, so $\Sigma F_y = 0$. More specifically,

$$F_{\substack{\text{on bottom of region} \\ \text{by fluid beneath it}}} - m_{\substack{\text{fluid within} \\ \text{region}}}g - F_{\substack{\text{on top of region} \\ \text{by fluid above it}}} = 0$$

or

$$F_{\substack{\text{on bottom of region} \\ \text{by fluid beneath it}}} - F_{\substack{\text{on top of region} \\ \text{by fluid above it}}} = m_{\substack{\text{fluid within} \\ \text{region}}}g \qquad (10\text{-}5)$$

The left side of this equation is the total upward force that the fluid outside the "chunk" exerts on the fluid within it.

Now we substitute the submerged box (Figure 10-8a) for the "chunk" of fluid. Because the box is the same size and shape as this region of fluid, the surrounding fluid is not moved or rearranged; it continues to exert the same pressure on the top and bottom of the region—that is, on the box that now fills it. Thus it still exerts the same total upward force $m_{\text{fluid within region}}g$. With a change in wording, then, we can restate Equation 10-5 in a form known as *Archimedes' principle*.

Archimedes' principle: The total upward force exerted by a fluid on an object submerged in it (called the **buoyant force $\vec{B}$**) is equal in magnitude to the weight of the fluid the object displaces. In equation form,

$$F_{\substack{\text{total on object} \\ \text{by fluid surrounding it}}} \equiv B = m_{\substack{\text{fluid} \\ \text{displaced}}}g \qquad (10\text{-}5a)$$

or if the volume V of fluid displaced has uniform density ρ_{fluid},

$$B = \rho_{\text{fluid}}V_{\substack{\text{fluid} \\ \text{displaced}}}g \qquad (10\text{-}5b)$$

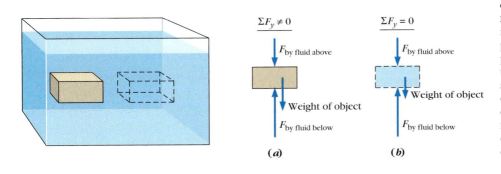

(a)   (b)

Figure 10-8 Free-body diagrams of a submerged object (a) and the fluid it displaces (b). The fluid is in equilibrium before being displaced. After it is displaced, the pressure at the top and bottom are unchanged because they are at the same depth as before. So the forces exerted by the outside fluid don't change. If the object's density differs from the fluid's, the only force that differs is the weight, so the object cannot be in equilibrium.

Based on what you see in this photo, can you estimate the weight of the water that this eight-person shell is displacing? Would the estimate be easier if you could neglect the weight of the oars and the shell itself?

This principle is named for the ancient Greek philosopher who inferred it from a series of careful observations, rather than deducing it from Newton's second law as we have done here.

Let's see what Archimedes' principle tells us about how objects behave. An object displaces its own *volume* of water when it is completely submerged. If it is denser than water, it is heavier than the water it displaces. Because the surrounding water continues to exert an upward buoyant force just sufficient to keep the displaced water in equilibrium, the heavier-than-water object will not be in equilibrium. It will sink at increasing speed until a speed-dependent drag force builds up sufficiently to produce equilibrium at some terminal speed.

If an object less dense than water is released in a submerged position, the upward buoyant force is still equal in magnitude to the weight of the displaced water and is therefore greater than the weight of the object. The total force is therefore upward, and the object rises until it is partly above the water's surface. It then displaces less water (Figure 10-9), so the buoyant force is less. The boat and passengers in Figure 10-9 are still totally surrounded by fluid, but the fluid

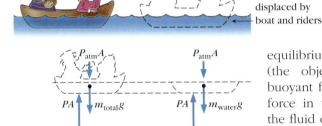

they are displacing is no longer uniform; above the water's surface, the displaced fluid is air. When a floating body such as the boat is in equilibrium, the force downward (the object's weight) equals the buoyant force upward. The buoyant force in turn equals the weight of the fluid displaced. Because air is so much lighter than water, the total fluid weight is only negligibly more

Figure 10-9 A floating body displaces its own weight of liquid.

than the weight of the displaced water. (It follows from the densities in Table 10-1 that a volume of air has only one-thousandth the weight of an equal volume of water.) To that degree of approximation, we can say

a floating body displaces its own weight of liquid.

For WebLink 10-3: Buoyant Forces, Floating, and Sinking, go to www.wiley.com/college/touger

To help you picture how objects are affected by buoyant forces, work through WebLink 10-3.

Case 10-1 ◆ *Crazy as a Loon*

A heavier floating body must displace more water to float, so heavier bodies ride lower in the water. The loon, one of very few birds with heavy bones, has an average density only slightly less than that of water, so it has more of its body underwater when it floats (Figure 10-10) than most aquatic birds. The loon can raise its average

Figure 10-10 The denser loon sits lower in the water than other aquatic birds.

density to slightly above that of water by expelling air from its body and under its feathers.[1] As it does so, the loon effortlessly sinks under water with hardly any disturbance of the water's surface. Loons are called *divers* in Great Britain and Ireland because this trait makes them one of the most graceful diving birds. The trait, in short, is the ability to slip across the threshold from being less dense than water to denser than water, and thus to change from a floating object to a sinking object.

[1]O. L. Austin Jr. *Water and Marsh Birds of the World.* Golden Press, NY, 1967.

STOP&Think How might a submarine accomplish a similar transition from floating to sinking? ◆

Example 10-4 *The Floating Soap*

A bar of Ivory brand soap floating in a tub of water displaces two-thirds of its own volume of water. **STOP&Think** Write an equation relating the two volumes—that of the bar of soap and that of the displaced water—before checking below. How does the density of the soap compare to that of water? ◆

Solution

Choice of approach. Because $\Sigma F_y = 0$, the buoyant force upward equals the weight of the soap downward. Archimedes' principle tells us that the buoyant force must equal the total weight of displaced water, because the air weight is negligible. So the weight of the soap (or any floating object) equals the weight of the water it displaces. If the weights are equal, so are the masses. Because the mass $m = \rho V$ for each object, the one with a larger volume V must have a smaller density ρ (Figure 10-11).

What we know/what we don't. $V_{\text{displaced water}} = \frac{2}{3} V_{\text{soap}}$

$$\rho_{\text{soap}} = \underset{=}{?} \rho_{\text{water}}$$

The mathematical solution. $m_{\text{soap}} = m_{\text{water}}$

Because $m = \rho V$,

$$\rho_{\text{soap}} V_{\text{soap}} = \rho_{\text{water}} V_{\text{displaced water}} = \rho_{\text{water}} \left(\frac{2}{3} V_{\text{soap}} \right)$$

Dividing the left- and rightmost expressions by V_{soap} gives us

$$\boldsymbol{\rho_{\text{soap}} = \frac{2}{3} \rho_{\text{water}}}$$

Making sense of the results. As the example shows, if an object's average density is a fraction of that of the liquid in which it is floating, the volume of the liquid it displaces is that same fraction of its own volume. Because an object cannot displace more than its own volume of liquid, it follows that an object denser than a given liquid cannot float in it.

◆ Related homework: Problems 10-21 and 10-23.

Figure 10-11 offers another way of thinking about Archimedes' principle. Imagine that parts *b–e* of the figure all show the same object, stretched or compressed vertically to occupy different volumes. Suppose it has the same mass as

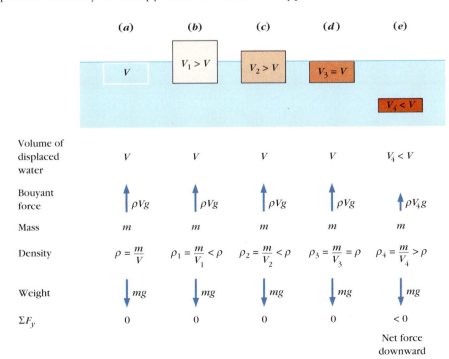

	(a)	(b)	(c)	(d)	(e)
Volume of displaced water	V	V	V	V	$V_4 < V$
Buoyant force	$\rho V g$	$\rho V g$	$\rho V g$	$\rho V g$	$\rho V_4 g$
Mass	m	m	m	m	m
Density	$\rho = \dfrac{m}{V}$	$\rho_1 = \dfrac{m}{V_1} < \rho$	$\rho_2 = \dfrac{m}{V_2} < \rho$	$\rho_3 = \dfrac{m}{V_3} = \rho$	$\rho_4 = \dfrac{m}{V_4} > \rho$
Weight	mg	mg	mg	mg	mg
ΣF_y	0	0	0	0	< 0
					Net force downward

Figure 10-11 Archimedes' principle applied to an object as it shrinks and stretches to occupy different volumes. The object shown in **b** through **e** has the same mass m as the "chunk" of water in **a**. But in each part, the object's volume is different. The volumes in the four different cases are compared to the volume V of the chunk.

the volume of water outlined in white in *a*. This "chunk" of water is in equilibrium. The object replacing it in *d* has the same volume as well as the same mass, so it too must be in equilibrium. Suppose the object expands so that its lower surface remains at the same depth while its top surface rises out of the water (*b* and *c*). Its weight doesn't change, and the buoyant force that the water exerts on its lower surface doesn't change, so it remains in equilibrium (it floats). However, if the object shrinks while its lower surface remains at the same depth, its top becomes submerged. The water that pours in on top of it adds to the weight supported by the water at the lower surface. Because the pressure at this depth doesn't change, the buoyant force doesn't increase, so there is now a net force downward on the object, and it sinks.

For practice in reasoning with buoyant forces, work through Example 10-5.

Example 10-5 *Romancing the Stone*

**For a guided interactive solution, go to Web Example 10-5 at
www.wiley.com/college/touger**

An archaeologist has found a mysterious, heavy stone artifact and lowers it into a beaker of water to rinse it off. The beaker rests on an electronic balance calibrated in newtons. Before the artifact is lowered into the water, the balance reads 6.50 N. When the artifact is totally submerged (but still suspended by a string), will the balance reading be greater than, equal to, or less than 6.50 N?

Brief Solution

Choice of approach. We apply Newton's second law and identify the forces making up $\Sigma\vec{F}$ on the water and on the artifact when they are in equilibrium. Newton's third law connects the forces that the water and the artifact exert on each other.

The solution. When the artifact is out of the water, the only forces on it are the upward force by the string and the weight downward. The force that the string must exert is reduced when it is submerged because the water exerts a buoyant force that partly supports the weight. (Sketch a free-body diagram for yourself to picture this.) Because the water exerts an upward force on the artifact, the third law tells us that the artifact exerts an equal downward force on the water. With an additional downward force on the water, the balance must exert a greater upward force on the beaker. This balance reading is equal to this force, and is therefore **greater than 6.50 N**.

◆ Related homework: Problem 10-24.

We now apply these ideas to a calculation.

Example 10-6 *The Diving Bell*

A ship's cable is used to raise and lower a spherical diving bell. The diving bell has a diameter of 1.5 m and a weight (including contents) of 18 000 N. If the cable always raises and lowers the diving bell with only very gradual changes in speed, how much tension must the cable sustain as long as the diving bell remains completely submerged?

Solution

Choice of approach. We will assume that the speed at which the bell is raised or lowered changes gradually enough so that the acceleration is negligible. Then Newton's second law reduces to $\Sigma F_y = 0$. We draw a free-body diagram (Figure 10-12) to inventory the forces making up the total force ΣF_y. Among

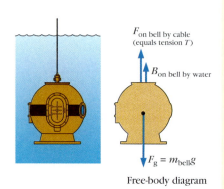

$F_{\text{on bell by cable}}$
(equals tension T)

$B_{\text{on bell by water}}$

$F_g = m_{\text{bell}}g$

Free-body diagram

Figure 10-12

these are the buoyant force $B = \rho_{water}V_{water\ displaced}\,g$ (Equation 10-5b) and the force that the cable exerts on the bell, which equals the tension. The volume of the water displaced is the volume $\frac{4}{3}\pi r^3$ of the sphere, where the radius r is half the diameter.

What we know/what we don't.

$$\frac{\Sigma F_y = 0}{}$$

$m_{bell}g = 18\ 000\ N$ $\qquad g = 9.80\ m/s^2$

$B = ?$ $\qquad r = \frac{1}{2}(1.50\ m) = 0.75\ m$ $V = ?$

$F_c = T = ?$ $\qquad \rho_{sea\ water} = 1025\ kg/m^3$ (Table 10-1)

The mathematical solution. From the free-body diagram,

$$\Sigma F_y = T + B - m_{bell}g = 0$$

But $\qquad B = \rho_{water}V_{water\ displaced}\,g = \rho_{water}g\left(\frac{4}{3}\pi r^3\right)$

so $\qquad T = m_{bell}g - B = m_{bell}g - \rho_{water}g\left(\frac{4}{3}\pi r^3\right)$

$$= 18\ 000\ N - (1025\ kg/m^3)(9.80\ m/s^2)\left(\frac{4}{3}\pi\left[0.75\ m\right]^3\right)$$

$$= 18\ 000\ N - 17\ 800\ N = \textbf{200 N}$$

STOP&Think 200 N is less than 50 lb. Is this enough force to pull the diving bell out of the water? ◆

Making sense of the results. The force needed to raise the diving bell can be very small as long as the bell is totally submerged, because the buoyant force exerted by the water provides most of the support. But the buoyant force starts to drop as soon as the top of the bell begins to emerge from the water (so that the bell's weight must increasingly be supported by the cable), because the part of the bell that is above the surface is displacing air, and the buoyant force now depends on the combined total mass of displaced water and air. As the bell lifts completely out of the water, the tension in the cable must equal the bell's weight.

◆ Related homework: Problems 10-25, 10-28, and 10-29.

10-3 Introduction to Fluid Dynamics: Flow Rate and Continuity

In Sections 10-1 and 10-2 we looked at stationary fluids. Even when water was moved by lowering an object into it, we focused on the stationary before-and-after states. Now we will focus on the motion or flow of fluids.

◆**TYPES OF FLOW** Intuitively we can distinguish between different kinds of flow. The smooth, steady flow of a great river like the Nile or the Mississippi on a pleasant day is quite different from the violent churning motion of whitewater rapids (Figure 10-13). Physicists likewise find it useful to distinguish between these two broad categories of motion, although intermediate stages occur as well.

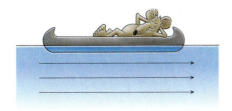
(a)

(b)

Figure 10-13 Streamline flow (*a*) versus turbulent flow (*b*).

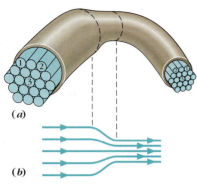

Figure 10-14 Laminar or stream-line flow.

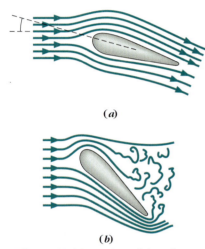

Figure 10-15 Laminar (*a*) and turbulent (*b*) flow of air past an airplane wing.

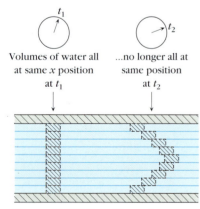

Figure 10-16 The effect of friction-like forces on laminar flow. The layers closest to the walls are slowed down most.

We can picture a fluid's path of flow as though it were made up of a series of imaginary tubes or channels extending along the direction of flow (Figure 10-14*a*). In the simplest type of flow to analyze, fluid entering by a given tube (e.g., 1, 2, or 3 in the figure) remains in (and leaves by) the same tube. The molecules of the fluid move much as cars on a multilane highway would if they never changed lanes. This kind of flow is called **laminar flow** (*laminar* means layered). The paths followed by the molecules under these conditions (Figure 10-14*b*) are often called **streamlines,** so laminar flow is also called **streamline flow.**

Contrasting with this is the churning and disordered motion that physicists call **turbulent flow.** The whirlpool-like eddies that form around rocks sticking out of a rushing stream of water are examples of turbulence; so is the spitting effect when you just barely open the nozzle of a garden hose connected to a wide-open valve. Figure 10-15 shows that the flow of air past an airplane wing (depicted in cross-section) can be laminar or turbulent, depending on the upward tilt, or *angle of attack,* of the wing. (Laminar flow must be maintained for proper lift.) In turbulent flow, the molecules do not "stay in lane."

◆**FRICTION-LIKE EFFECTS (QUALITATIVE)** When an ideal fluid travels through a straight length of pipe, all of it travels at the same speed. The molecules traveling in one tube or layer are unaffected by the molecules traveling in any other or by the walls of the pipe. The behavior of real-world fluids differs significantly from this ideal, even in laminar flow. The molecules in the layer of a real-world fluid nearest the wall of the pipe may stick somewhat to the walls of pipe, especially where there are irregularities. This has the effect of exerting a friction-like resistive force on this outermost layer. In addition, the molecules in each layer of the fluid stick somewhat to the molecules of the adjacent layers, so each layer in turn exerts a resistive force on the next layer toward the center.

The overall effect on the body of fluid flowing through the pipe is that the layers nearest the walls of the pipe are slowed down most, and the innermost layers are slowed down least (Figure 10-16). The property of a fluid that describes the extent to which each layer of the fluid has a friction-like effect on the next is called the **viscosity** of the fluid. An ideal fluid has no viscosity at all. Gases come close to this ideal, but liquids do not and exhibit a great range of viscosities. A bottle of "white" vinegar and a bottle of clear corn syrup may look nearly identical sitting on the supermarket shelf, but the corn syrup (basically pancake syrup without the coloring) pours much more slowly than the vinegar. It is more **viscous.** "Slow as molasses" is a phrase we use to compare people to one very viscous liquid.

> *Superfluids:* Fluids with zero viscosity exhibit surprising behavior, and are therefore called *superfluids.* Liquid helium is a superfluid below 2.172 K (about −271°C). It flows freely through even the tiniest capillary tubes, and it will flow as a thin film over the walls of an open-topped container holding it.

Turbulence also contributes to the friction-like forces to which a fluid is subjected. To the extent that molecules shift randomly from a faster "lane" to a slower one, it has the effect of increasing the average speed in the slower lane and decreasing the average speed in the faster lane. The result is similar to what happens when two *solid* layers traveling at different speeds exert sliding frictional forces on each other. The sliding frictional forces oppose the relative motion of the two layers, speeding up the slower one and slowing down the faster one.

Small-scale turbulence effects produced by bumps, depressions, or other irregularities on the walls of the pipe or pipe-like carrier can slow down the layer of fluid they immediately affect. This slowing down gets passed inward from layer to layer. Examples of these irregularities include outcroppings along a stream or riverbank, rust or corrosion on the interior walls of a water or oil pipe, or accumulations of matter on the inner walls of blood vessels in the human body. (This condition is called *atherosclerosis;* an advanced stage of it is the disease *arteriosclerosis,* or "hardening of the arteries.") All of these have a friction-like effect in slowing the fluid down. When the individual occurrences of turbulence are small but have a significant collective effect on the overall flow of a fluid, we can treat the motion as laminar flow subjected to an overall resistive force.

◆**PRESSURE DIFFERENCES CENTRAL TO FLUID DYNAMICS** Although you can roll a ball uphill, the ball's general tendency is to roll downhill. Likewise, a fluid can be given a velocity in the direction of increasing pressure, but its general tendency is to flow from regions of high pressure to regions of low pressure: from the high pressure in a water main to the lowered pressure near your open tap and from the high pressure in the chambers of your heart that pump blood into your arteries, to the low pressure in the chambers that receive the blood back from your veins. Gases are also fluids, so the same is true of gas flow. Air flows from the high pressure within a service station air pump to the lower pressure in a slightly flattened tire. On a much grander scale, the movement of air masses from higher to lower pressure regions (the "highs" and "lows" of weather forecasters) is a dominant factor in the shaping of global weather patterns.

Pressure was a key concept when we described the forces on stationary bodies of fluids. As we proceed from the statics of fluids to considering how forces affect fluids in motion (the dynamics of fluids), pressure remains a key concept. Before considering how pressure affects the flow of a fluid, we must define suitable quantities for describing the flow.

◆**FLOW RATE** The most important feature of an oil pipeline is how much oil (how many barrels or gallons or liters) it can deliver each second to any receiving point along the line. (We care about how much volume in each unit of time, not how much in all—you can get the same total at any rate of delivery if you wait long enough, but if it takes 1000 years to accumulate the total that you need, it does you no good.) Likewise, the important property of a blood vessel is what volume of oxygen- and nutrient-carrying blood it can transport each second past any cell along its delivery route. This property is called the **flow rate.**

We define the flow rate as follows. Suppose a large volume of fluid is flowing past point P in Figure 10-17. The shaded region shows the additional volume ΔV that will pass point P during the next tiny time interval Δt. The flow *rate* (denoted by Q) is then defined as follows:

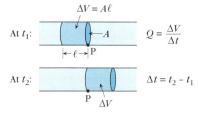

Figure 10-17 Defining flow rate.

Flow rate	$Q \equiv \dfrac{\Delta V}{\Delta t}$	(10-6)

If the shaded region has a length l and cross-sectional area A, its volume is $\Delta V = Al$, and $Q = \frac{Al}{\Delta t}$. But l is just the distance the shaded region of fluid travels between t_1 and t_2 (Figure 10-17), so $\frac{l}{\Delta t}$ is just the average speed $\bar{v}$ of the fluid. Therefore the flow rate is

$$Q = A\bar{v} \qquad (10\text{-}7)$$

Equation 10-7 tells us that the flow rate depends on

• the speed of the fluid's flow;

• the cross-sectional area of the channel through which it flows.

For example, in atherosclerosis, blood flow through the arteries is diminished because the accumulation of cell debris, fibrin, fatty material, and so on, on the arterial walls to form layers called *plaques* affects both of these factors. The build-up of material on the walls reduces the diameter, and thus the area, of the opening

➡Even if we consider the speed at one instant, we use the average speed $\bar{v}$ in the flow rate. Because the speed of a fluid through a tube is typically greater in midstream than near the walls, we must still average over all the molecules at different distances from the center.

the blood can pass through. As we have already noted, it also gives rise to increased resistive forces, reducing the steady-state speed even more. A further danger is that a plaque can be dislodged and block a smaller blood vessel downstream.

◆**PRINCIPLE OF CONTINUITY** If there is no build-up or loss of fluid along a path of flow, the flow rate at any two points 1 and 2 along the path must be the same. This basic idea is called the *continuity principle:*

Continuity principle $Q_1 = Q_2$ (10-8a)

if 1 and 2 are points along the path of flow, and no fluid is added to or removed from the flow.

Example 10-7 applies this principle along with what you've already learned about flow rate.

Example 10-7 *Down the Pipes: Part I*

For a guided interactive solution, go to Web Example 10-7 at www.wiley.com/college/touger

Jamal opens the faucet of his kitchen sink just enough so that he can fill a 1.00-liter (1 liter = 1×10^{-3} m³) bottle in 10.0 s. The faucet is fed by water pipes under the sink that have an inner diameter of 1.27 cm. Find the speed at which water flows through the pipes under the sink while Jamal is filling the bottle.

Brief Solution

Choice of approach. Because we know what volume of water comes out of the pipes in a given time interval, we can find what volume flows out each second; this is the flow rate Q. The rate at which water flows out of the pipe is equal to the flow rate within the pipe. After finding the cross-sectional area of the pipe from its diameter, we can then use the fact that $Q = A\bar{v}$ to find $\bar{v}$.

What we know/what we don't. $\Delta V = 1.00 \times 10^{-3}$ m³ $\Delta t = 10.0$ s $Q = ?$

$$d = 1.27 \text{ cm} = 0.0127 \text{ m} \quad A = ? \quad \bar{v} = ?$$

The mathematical solution. The flow rate (by Equation 10-6) is

$$Q = \frac{\Delta V}{\Delta t} = \frac{1.00 \times 10^{-3} \text{ m}^3}{10.0 \text{ s}} = 1.00 \times 10^{-4} \text{ m}^3/\text{s}$$

The cross-sectional area

$$A = \pi r^2 = \pi \left(\frac{d}{2}\right)^2 = \pi \frac{d^2}{4} = \pi \frac{(0.0127 \text{ m})^2}{4} = 1.27 \times 10^{-4} \text{ m}^2$$

Then since $Q = A\bar{v}$, the flow speed

$$\bar{v} = \frac{Q}{A} = \frac{1.00 \times 10^{-4} \text{ m}^3/\text{s}}{1.27 \times 10^{-4} \text{ m}^2} = \textbf{0.787 m/s}$$

◆ Related homework: Problems 10-32, 10-33, and 10-2.

To a high degree of accuracy, a typical liquid in a pipe or other enclosed channel is incompressible, so its density remains constant. If we consider the length of pipe between points X and Y in Figure 10-18, the continuity principle tells us that the volume of fluid that enters at X during a time interval Δt must equal the volume of fluid that leaves at Y during the time interval. These two volumes are shaded in the figure. Because the pipe has a smaller cross-sectional area at Y, a body of liquid with this cross-sectional area must be correspondingly

If a fluid's density doesn't change during a time interval Δt, then...

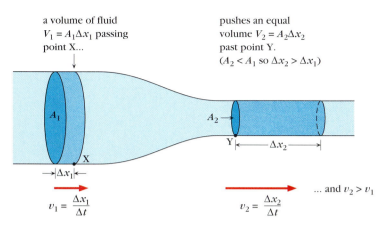

a volume of fluid
$V_1 = A_1\Delta x_1$ passing
point X...

pushes an equal
volume $V_2 = A_2\Delta x_2$
past point Y.
($A_2 < A_1$ so $\Delta x_2 > \Delta x_1$)

$v_1 = \dfrac{\Delta x_1}{\Delta t}$

$v_2 = \dfrac{\Delta x_2}{\Delta t}$

... and $v_2 > v_1$

Figure 10-18 The continuity principle.

longer to have the same volume. During the time interval Δt, each of the shaded bodies of fluid travels its own length. The volume passing Y travels a greater distance ($\Delta x_2 > \Delta x_1$) in this interval and therefore must be traveling faster. (The two velocities are $\bar{v}_1 = \frac{\Delta x_1}{\Delta t}$ and $\bar{v}_2 = \frac{\Delta x_2}{\Delta t}$.) Thus, *the fluid speeds up in the region where the pipe narrows.* **STOP**&**Think** Intuitively, would you have expected the fluid to be slowed down by the narrowing of the pipe? Does this result make sense? ◆ We will address this question shortly.

Saying that the same volume passes X and Y in the same time interval is the same as saying that the flow rate $Q = \frac{\Delta V}{\Delta t}$ is the same past both points, and because $Q = A\bar{v}$, we can write the continuity principle mathematically as follows:

Continuity equation for incompressible fluids:

$$A_1\bar{v}_1 = A_2\bar{v}_2 \tag{10-8b}$$

Now we can address the predicament we raised. If part of the pipe is constricted or narrowed, it does in fact slow down the flow. This should be obvious if you carry the narrowing to an extreme: When you reduce the area of the opening to zero at Y, you have no flow; the flow speed is zero. But then you have no flow at X either. When you narrow the pipe at Y, whether you narrow it to zero or not, you get a back-up effect. Slowing down the flow at Y in turn slows down the flow at X. Working through WebLink 10-4 will help you develop a mental picture of how continuity works. This WebLink examines an analogous situation in which two lanes of traffic must merge into one.

This reasoning will help make sense of the result we calculate using the continuity principle in Example 10-8.

For **WebLink 10-4:**
**Continuity
on the Highway,** go to
www.wiley.com/college/touger

Example 10-8 *Down the Pipes: Part II*

**For a guided interactive solution, go to Web Example 10-8 at
www.wiley.com/college/touger**

The water pipes under Jamal's sink in Example 10-7 are in turn connected to water pipes in his basement that have an inner diameter of 2.54 cm. Find the speed at which water flows through the pipes in the basement while Jamal is filling the bottle.

Brief Solution

Choice of approach. Because the water is incompressible, we can apply the principle of *continuity:* The flow rate through the pipes in the basement must equal the flow rate through the pipes under Jamal's sink. We will use the data and results of Example 10-7 in applying Equation 10-8b.

EXAMPLE 10-8 *continued*

What we know/what we don't.

$$A_1 = 1.27 \times 10^{-4} \text{ m}^2 \qquad \bar{v}_1 = 0.787 \text{ m/s}$$

$$d_2 = 2.54 \text{ cm} = 0.0254 \text{ m} \qquad A_2 = ? \qquad \bar{v}_2 = ?$$

The mathematical solution. The cross-sectional area of the basement pipes is

$$A_2 = \pi r_2^2 = \pi \left(\frac{d_2}{2}\right)^2 = \pi \frac{d_2^2}{4} = \pi \frac{(0.0254 \text{ m})^2}{4} = 5.07 \times 10^{-4} \text{ m}^2$$

Because $A_1 \bar{v}_1 = A_2 \bar{v}_2$,

$$\bar{v}_2 = \frac{A_1 \bar{v}_1}{A_2} = \frac{(1.27 \times 10^{-4} \text{ m}^2)(0.787 \text{ m/s})}{5.07 \times 10^{-4} \text{ m}^2} = \textbf{0.197 m/s}$$

Making sense of the results. The flow rates are equal, but the flow *speeds* are not. Because the basement pipes have twice the diameter of the pipes under the sink, their cross-sectional area is four times as great, so the flow speed in these pipes is only one-fourth as great.

◆ Related homework: Problems 10-34 and 10-36.

10-4 Fluid Dynamics: Work-Energy Considerations and Bernoulli's Equation

We now use work-energy considerations to deduce a relationship between pressures, heights, and speeds of flow at different points along the path of an ideal incompressible fluid, that is, one with laminar flow and negligible viscosity. The relationship that we will obtain is called Bernoulli's equation. It was first published by Swiss physicist-mathematician Daniel Bernoulli in 1738, well before the formulation of the work-energy theorem. But we can derive it as an important consequence of the work-energy theorem, bearing on a range of situations from the lift on an airplane wing to roofs blowing off houses in hurricanes to problems in the human circulatory system.

As a preliminary step, we first consider one stretch of a longer pipe (Figure 10-19) that is transporting an ideal incompressible fluid (laminar flow and negligible viscosity) and has differing pressures and cross-sectional areas at its two ends. When a volume of fluid $A_1 \Delta x_1$ enters the pipe (Figure 10-19a), an equal volume of fluid $A_2 \Delta x_2$ gets pushed out the other end (Figure 10-19b).

How much work is done in moving the entire shaded body of fluid through the pipe? The force $P_1 A_1$ is exerted through a displacement Δx_1. The other end undergoes a displacement Δx_2 (also to the right), but the force on it, directed toward the left, is $-P_2 A_2$. Then the total work done is

$$W = P_1 A_1 \Delta x_1 - P_2 A_2 \Delta x_2$$

But because the entering volume $A_1 \Delta x_1$ equals the exiting volume $A_2 \Delta x_2$, we can use the same symbol ΔV to represent either of these volumes:

$$W = P_1 \Delta V - P_2 \Delta V$$

Now suppose that our length of pipe is not perfectly horizontal but travels uphill or downhill, as real conduits for fluids generally do: for example, pipes that carry water down from mountain sources to city reservoirs, or from basement water heaters to upstairs bathrooms, and vessels that carry blood up to your brain, down to your feet, and back to your heart. In that case, the work that is done equals the total change in mechanical energy (kinetic energy plus gravitational potential energy). This is the work-energy theorem (Equation 6-14). If the

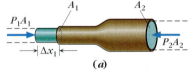

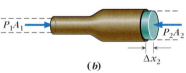

Figure 10-19 Forces and displacements involved in finding the work done on a fixed volume of fluid.

fluid enters with speed v_1 at height y_1, and leaves with speed v_2 at height y_2 (Figure 10-20), then

$$W = \Delta PE_{\text{grav}} + \Delta KE$$

becomes
$$P_1 \Delta V - P_2 \Delta V = (mgy_2 - mgy_1) + (\tfrac{1}{2}mv_2^2 - \tfrac{1}{2}mv_1^2)$$

But because the mass that is relocated from the lower to the upper end of the pipe is equal to its density times the volume ($m = \rho \Delta V$), we can substitute this expression for each occurrence of m, giving us

$$P_1 \Delta V - P_2 \Delta V = (\rho \Delta V g y_2 - \rho \Delta V g y_1) + (\tfrac{1}{2}\rho \Delta V v_2^2 - \tfrac{1}{2}\rho \Delta V v_1^2)$$

Now we can divide all terms by the common factor ΔV:

$$P_1 - P_2 = \rho g y_2 - \rho g y_1 + \tfrac{1}{2}\rho v_2^2 - \tfrac{1}{2}\rho v_1^2$$

By rearranging this, we get one form of Bernoulli's equation:

Bernoulli's equation:

$$P_1 + \rho g y_1 + \tfrac{1}{2}\rho v_1^2 = P_2 + \rho g y_2 + \tfrac{1}{2}\rho v_2^2 \qquad (10\text{-}9a)$$

Note: $\rho g y$ = potential energy of each unit of volume

$\tfrac{1}{2}\rho v^2$ = kinetic energy of each unit of volume

Points 1 and 2 can be any two points along the pipe, so this says that the quantity $P + \rho g y + \tfrac{1}{2}\rho v^2$ is the same at any point, or in other words,

$$P + \rho g y + \tfrac{1}{2}\rho v^2 = \text{constant} \qquad (10\text{-}9b)$$

(Pressure + PE/volume + KE/volume = constant)

The work-energy theorem as we have stated it assumes no conversion of macroscopic mechanical energy to internal energy. For a fluid, this means no dissipation of energy due to viscosity or turbulence, so Bernoulli's equation applies only under these conditions. Otherwise,

$$P_1 + \rho g y_1 + \tfrac{1}{2}\rho v_1^2 = P_2 + \rho g y_2 + \tfrac{1}{2}\rho v_2^2 + \begin{smallmatrix}\text{internal energy}\\\text{per unit of volume}\end{smallmatrix}$$

We have already seen that to maintain continuity, an incompressible fluid must speed up where the cross-sectional area of the tube in which it flows decreases. We can now use Bernoulli's equation to see how the narrowing of the tube affects the pressure. **STOP&Think** What do you expect to happen to the pressure? Jot down your expectation before you read on, so you can check it against the conclusions that we draw. ◆

We can first apply Bernoulli's equation (in the form of Equation 10-9b) qualitatively to a horizontal tube. The potential energy per unit of volume doesn't change if the tube is horizontal, so the pressure must decrease if the kinetic energy per unit of volume increases in order for the total of pressure + PE/volume + KE/volume to remain constant. Thus, contrary to common expectations, *the quicker flowing fluid through a constricted segment of the tube exerts less pressure.* **STOP&Think** Exerts less pressure *on what?*

Combining Bernoulli's equation (10-9a) and the continuity equation (10-8b), we can calculate the change in pressure. For a horizontal pipe ($y_1 = y_2$), Equation 10-9a becomes

$$P_2 - P_1 = \tfrac{1}{2}\rho v_1^2 - \tfrac{1}{2}\rho v_2^2 = -(\tfrac{1}{2}\rho v_2^2 - \tfrac{1}{2}\rho v_1^2) \qquad (10\text{-}10)$$

(or $\Delta P = -\Delta KE/\text{volume}$)

The minus sign indicates that an increase in KE, due to increased speed, results in a decrease in pressure. Example 10-9 provides another application of Bernoulli's equation.

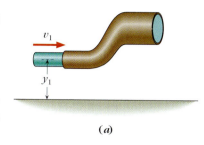

(a)

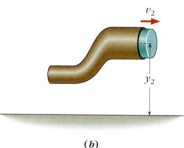

(b)

Figure 10-20 Two states of a fixed volume of fluid. The shaded volume has moved to the other end of the pipe segment in the second state. Applying the work-energy theorem to these two states leads to Bernoulli's equation.

Example 10-9 *Watering the Garden*

For a guided interactive solution, go to Web Example 10-8 at
www.wiley.com/college/touger

A gardener wishes the water from his hose to be able to reach flowers 3.0 m away when he holds the hose nozzle horizontally at a height of 1 m. Using kinematics, he figures out that the water must leave the hose at a speed of 6.64 m/s to do this (see Problem 10-73). The nozzle of his hose has an inner diameter of 1.5 cm. The other end of his hose (which we will assume is horizontal) is connected to an exit pipe in the side of his house with an inner diameter of 2.0 cm. What must the water pressure in the exit pipe be for the water from his hose to reach the flowers?

Brief Solution

Choice of approach. (1) We can use the continuity principle $v_1 A_1 = v_2 A_2$ to determine the speed at the exit pipe. (2) We can then use Bernoulli's equation and the continuity principle to find the pressure difference necessary to produce this nozzle speed. In the pressure difference, we take P_2 to be the pressure where the water emerges from the nozzle. This is just atmospheric pressure. Then we can solve for P_1, the pressure in the exit pipe.

What we know/what we don't.

Step 1: $r_1 = \dfrac{2.0 \text{ cm}}{2} = 1.0 \text{ cm} = 0.010 \text{ m}$ $\qquad r_2 = \dfrac{1.5 \text{ cm}}{2} = 0.75 \text{ cm} = 0.0075 \text{ m}$

$A_1 = ? \qquad A_2 = ? \qquad v_2 = 6.64 \text{ m/s} \qquad v_1 = ?$

Step 2: $\qquad \rho_{\text{H}_2\text{O}} = 1000 \text{ kg/m}^3$ (from tables) $\qquad y_2 = y_1 = 1.0 \text{ m}$

$P_2 = 1 \text{ atm} = 1.0 \times 10^5 \text{ N/m}^2 \qquad P_1 = ?$

The mathematical solution. Step 1: From the continuity equation, we get $v_1 = \frac{v_2 A_2}{A_1}$. The ratio of cross-sectional areas is $A_2/A_1 = \pi r_2^2 / \pi r_1^2 = r_2^2/r_1^2$, so

$$v_1 = \frac{v_1 r_2^2}{r_1^2} = \frac{(6.64 \text{ m/s})(0.0075 \text{ m})^2}{(0.010 \text{ m})^2} = 3.74 \text{ m/s}$$

Step 2: Since $y_1 = y_2$, Bernoulli's equation (Equation 10-9a) simplifies to

$$P_1 + \frac{1}{2} \rho v_1^2 = P_2 + \frac{1}{2} \rho v_2^2$$

Solving for P_1 gives us

$$P_1 = P_2 + \frac{1}{2} \rho (v_2^2 - v_1^2)$$

$$= 1.0 \times 10^5 \text{ N/m}^2 + \frac{1}{2} (1000 \text{ kg/m}^3)([6.64 \text{ m/s}]^2 - [3.74 \text{ m/s}]^2)$$

$$= \mathbf{1.15 \times 10^5 \text{ N/m}^2}$$

Making sense of the results. This agrees with the conclusion we reached previously that the pressure is greater where the speed is lower, because the speed is lower where the cross-sectional area is greater.

◆ Related homework: Problems 10-38, 10-42, and 10-73.

To observe another kind of behavior that Bernoulli's equation can explain, try the following activity.

On-The-Spot Activity 10-1

Take a tall plastic bottle (such as a two-liter soft drink bottle or a gallon milk container) and with a sharp object make two puncture holes one above the other at different heights as shown in Figure 10-21. Try to keep the holes quite small and round, as much alike as possible, and reasonably far apart. Now fill the bottle to the top with water and watch how it streams out of the two holes. On the basis of your observation, can you tell which stream of water emerges from its hole at greater speed? Why do you suppose the difference in the two streams occurs?

Open

Hole B

Hole A

Figure 10-21 A behavior explained by Bernoulli's equation. This is the set-up for On-the-Spot Activity 10-1.

   In this situation, water meets air at the top surface of the water and at the hole, so the pressure at both of these positions is just the atmospheric pressure. When we apply Bernoulli's equation (Equation 10-9a) to two points where the pressures are equal, it simplifies to $\rho g y_1 + \frac{1}{2}\rho v_1^2 = \rho g y_2 + \frac{1}{2}\rho v_2^2$.
   In addition, the area of the hole is much smaller than the top surface area of the water, so the water level in the bottle drops at a speed v_1 that is very small compared to the speed v_2 of the water streaming out of the hole. (This follows from the continuity principle $A_1 v_1 = A_2 v_2$. If A_2 is much smaller than A_1, v_1 must be much smaller than v_2.) If we take v_1 to be approximately zero, Bernoulli's equation further simplifies to $\rho g y_1 = \rho g y_2 + \frac{1}{2}\rho v_2^2$. Dividing both sides by the density ρ, we get

$$g y_1 = g y_2 + \tfrac{1}{2} v_2^2 \qquad \text{or} \qquad g(y_1 - y_2) = \tfrac{1}{2} v_2^2$$

which we then solve for v_2:

$$v_2 = \sqrt{2g(y_1 - y_2)}$$

   This is like the familiar $v = \sqrt{2gh}$ that we get when we apply conservation of energy to an object (Figure 10-22a) that starts at rest and descends without friction by any path through a vertical distance $h = y_1 - y_2$. The overall energy change of the water is as though each little bit of water that leaves the hole had followed a path from the upper surface of the water to the hole (Figure 10-22b) to build up speed. Though the actual movement of water in the bottle is more complicated than this, this is the net effect.
   In more general circumstances, the pressure and the height may both differ between two points along the path of a fluid, and no pair of corresponding terms on the two sides of Bernoulli's equation (Equation 10-9a) will be equal.

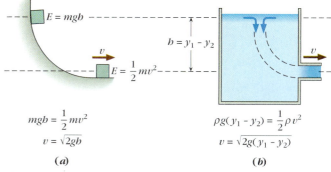

Figure 10-22 Energy changes for an ideal fluid. The overall change from PE to KE for an ideal fluid streaming from a hole in the side of an open tank is the same as if that liquid (shaded) had come from the upper surface and the rest of the liquid had closed over it.

10-5 Real-World Applications of Fluid Dynamics

Applications of fluid dynamics in the real world are many and varied, but certain fluids are of great enough interest so that studies of these fluids are major areas of scientific specialization.

- **Aerodynamics** focuses on the air around us. Aerodynamicists study the motion of air and its flow around moving objects, and the forces that the air and the objects it encounters exert on one another. This bears on a range of matters as diverse as the lift on airplane wings, the action of a curveball in baseball, and the reason why windows have popped out of improperly designed skyscrapers.
- **Hemodynamics** is the study of blood flow in living things, and, by extension, the characteristics of the circulatory system that carries the blood and affects its flow.

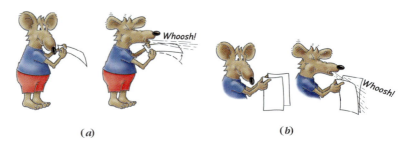

Figure 10-23 An important consequence of Bernoulli's equation. The pressure exerted by a fluid decreases as its speed increases.

(a)   (b)

Office workers on New York City's famous Wall Street often complain about how windy the street is. Based on the photo, why would you expect this street to be especially windy?

Figure 10-24 A seriously important consequence of Bernoulli's equation. The pressure exerted by high-speed air in a hurricane may be so much less than the pressure exerted by the stationary air within a house that there is a net upward force sufficient to lift the roof off.

In aerodynamics, Bernoulli's equation finds important applications. Bernoulli's equation tells us that $P + \rho gy + \frac{1}{2}\rho v^2$ remains constant (Equation 10-9b) apart from dissipative effects. For this total to have the same value at different points along a level path, *air pressure must decrease as air speed increases.*

On-The-Spot Activity 10-2

You can demonstrate this consequence with one or two sheets of paper by following the steps illustrated in Figure 10-23. Hold a sheet of paper just under your lips as in Figure 10-23a and blow hard over the top of it (a single explosive burst of air works best, just as it does for blowing out birthday candles). What happens to the sheet of paper? Now hold two sheets of paper suspended below your lips as in Figure 10-23b, and blow an explosive burst of air between the two sheets of paper. What happens? In both cases, what is the direction of the resultant force on a sheet of paper when the air on one side of it is in rapid motion? To produce this resultant force, how must the pressure exerted by the rapidly moving air compare to the pressure exerted by the stationary air on the other side?

In Figure 10-23a, there is a net upward force because the pressure on the upper surface of the paper by the moving air is less than the pressure on the lower surface of the paper by the stationary air. As a result, the paper is pushed upward. For the same reason, in Figure 10-23b, there is a net inward force on each sheet of paper and the two sheets come together while the air between them is in rapid motion.

In hurricane conditions, the pressure exerted by high-speed air on the upper surface of a roof may be so much less than the pressure exerted by the stationary air within the house that the roof will lift off (Figure 10-24). Tragically, this occurs most commonly where cost-cutting in home construction has not permitted adequate protective countermeasures. Buildings at the upper end of the economic scale are by no means invulnerable, however. When the Hancock Tower, a glass-sheathed skyscraper in Boston, was first constructed, glass windows from the upper stories popped out and fell to the pavement below because the designers had not adequately compensated for the pressure difference between the stationary air within the building and the lower pressure to which the outer surfaces are subjected when there are high winds. (The situation has since been remedied.)

A more positive but also more complex consequence of this pressure difference is the upward resultant force (**lift**) that is produced on an airplane wing because of the shape of its cross-section. The figure in the margin shows laminar flow around a typical airplane wing cross-section. (Because *relative* velocity is what matters, moving air and a stationary cross-section will result in the same flow pattern as a moving cross-section through still air.) If all the air approaching from the left continued straight ahead, no air would end up in the region blocked by the wing. But because the air is a fluid, pressure exerted by one

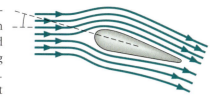

portion of the air on another forces air into this region. The net effect of a complex balancing process is that the air flows along the surface of the cross-section and "fills in" behind it. The characteristics of the flow pattern can be observed experimentally. One way is to let trails of smoke be carried along the entering streamlines and see which way they flow when they approach the airplane wing. The streamlines are equally spaced until they approach the wing. The figure at right shows two important features of the resulting flow pattern:

• The streamlines above the wing have moved closer together than those below.

• Because of the tilt of the wing, the air that has traveled closest to the wing, following the surface contour, has a downward velocity component as it leaves the rear edge of the wing.

Where the streamlines are closer together, the air is traveling through a smaller cross-sectional area, and therefore by the continuity principle must be traveling faster. Then the pressure is less on the upper surface of the wing than on the lower surface, and there is a resultant upward force on the wing (the lift force).

The second observation suggests we can also account for the lift force by Newton's third law. If the air that has approached the wing horizontally is given a downward velocity component by its encounter with the wing, it means that the total force that the wing exerted on the air had to have a downward vertical component. By the third law, then, the force that the air exerts on the wing must have an equal and opposite—that is, upward—vertical component.

This force is sometimes called a *dynamic lift* force because it does not occur if the wing (or the plane to which it is attached) is stationary relative to the air. If a plane's engines conk out and it loses horizontal speed, it is in deep trouble. That is why conventional aircraft, such as jetliners, which are much heavier than air, must reach sufficient speed on a runway before they lift off.

The most studied *hemodynamic* system is the human circulatory system. Briefly summarized, the system consists of an elaborate network of tubes—blood vessels—of varying diameters that carry blood to every cell in the body, and a pump—the heart—that drives the flow. Figure 10-25 shows a simplified schematic of the system, which is also called the *cardiovascular* system (*cardio* refers to the heart, and *vascular* refers to blood vessels). Hemodynamics is also called *cardiovascular dynamics*.

The heart, like any pump, moves the blood from lower to higher pressure, and the blood, a fluid, then flows from higher to lower pressure. Blood pressure is highest where it is pumped out of the heart into the **aorta**, the largest artery, and lowest in the **venae cavae** (singular, *vena cava*), the large veins that return it to the heart. Blood pressures are commonly expressed in units of mm of mercury (mm Hg) because a device called a mercury *manometer* (described in Problem 10-15) was traditionally used to measure blood pressure.

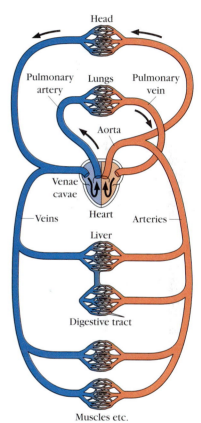

Figure 10-25 Schematic diagram of human circulatory system.

Pressure unit equivalencies:

$$1 \text{ atm} = 1.01 \times 10^5 \text{ N/m}^2 \text{ (or Pa)} = 760 \text{ mm Hg (or torr)}$$

A mm of mercury is also called a torr.

Blood pressures are relative pressures; that is, a blood pressure reading at a particular location is the amount by which the blood pressure in that location exceeds atmospheric pressure. (These relative pressures are also called *gauge pressures*.) It is usual in medical practice to state blood pressures in this way, so we will do so throughout our discussion of the circulatory system. What matters for blood flow is the difference between the pressures at two locations in the circulatory system; this difference will be the same whether or not we add the atmospheric pressure to each value.

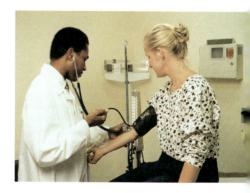

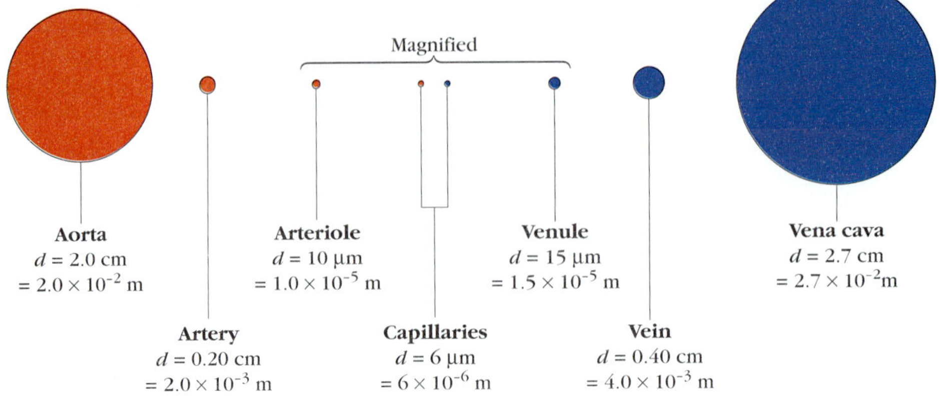

Figure 10-26 Typical inner diameters of blood vessels in the human circulatory system. The larger vessels are shown actual size; the smaller vessels are greatly magnified.

Aorta
$d = 2.0$ cm
$= 2.0 \times 10^{-2}$ m

Artery
$d = 0.20$ cm
$= 2.0 \times 10^{-3}$ m

Magnified

Arteriole
$d = 10$ μm
$= 1.0 \times 10^{-5}$ m

Capillaries
$d = 6$ μm
$= 6 \times 10^{-6}$ m

Venule
$d = 15$ μm
$= 1.5 \times 10^{-5}$ m

Vein
$d = 0.40$ cm
$= 4.0 \times 10^{-3}$ m

Vena cava
$d = 2.7$ cm
$= 2.7 \times 10^{-2}$ m

Blood pressures vary typically from an average of about 90–95 mm Hg in the arteries to an average of about 7–8 mm Hg in the veins nearest the heart. (To sustain the greater pressure, artery walls are much thicker than the walls of veins of comparable inner diameter.) Because the heart pumps rhythmically, your blood pressure varies cyclically between a maximum value when your heart is fully contracted and a minimum value when your heart is most relaxed. The maximum value is called the **systolic** blood pressure and the minimum is called the **diastolic** blood pressure. Standard medical blood pressure readings give the systolic and diastolic pressures in an artery of the arm. Thus a reading of 110/80 (read as "110 over 80") means an arterial systolic pressure of 110 mm Hg and an arterial diastolic pressure of 80 mm Hg. In comparison, a typical systolic pressure of blood emerging from the heart into the pulmonary system that provides blood to your lungs may only be about 22 mm Hg. Readings are generally obtained when you are lying down. If you are standing with your arm raised, the average arterial pressure near your heart will remain about 90 mm Hg, but it may be as low as 40 mm Hg in the arteries of your raised wrist and as high as 170 mm Hg in the arteries of your feet.[2] **STOP&Think** Why this difference? Also, why might you get dizzy if you sit up from a lying-down position too quickly?

The complexity of the circulatory system is due to the elaborate branching, as a small number of large diameter blood vessels—the **arteries,** which carry blood away from the heart—branch out into many more smaller-diameter blood vessels, which in turn branch out into even more vessels with still smaller diameters. The most numerous and smallest diameter vessels, the **capillaries,** are the local delivery system, bringing blood to every cell, dropping off some of its cargo of oxygen and nutrients and carrying away waste products. On returning to the heart, the blood is gathered into progressively larger-diameter and numerically fewer vessels and finally returns to the heart via large-diameter **veins.** Figure 10-26 shows the comparative diameters of typical human blood vessels. The flow patterns along these multiple paths are far more complex than for a single channel.

The complexity is compounded by two significant factors. First, blood is not an ideal fluid. Normal whole blood at body temperature is about five times as viscous as water at the same temperature and, as Figure 10-26 shows, must pass through capillaries with microscopically small diameters. Second, unlike water pipes, blood vessels are not rigid. They may expand somewhat in response to increased pressure, so there is feedback between pressure and blood vessel diameter: To some extent, increasing pressure gives rise to an increasing of diameter, which in turn eases up on the pressure.

[2]R. F. Rushmer. *Cardiovascular Dynamics,* 4th ed. Saunders, Philadelphia, 1976, pp. 220–221.

Any form of matter that cannot maintain its own shape, and therefore flows, is called a **fluid.** Newton's laws of motion also apply to fluids, but additional concepts are of value in treating the *statics and dynamics of fluids.*

Pressure describes how a total force is distributed over an area.

The *average pressure on a surface of area* A is

$$P = \frac{F_\perp}{A} \qquad (10\text{-}1)$$

SI units: 1 pascal (Pa) = 1 N/m^2

Atmospheric pressure at sea level:

1 atm = 1.01×10^5 N/m^2 (or Pa) = 14.7 lb/in^2 (or psi)

= 760 mm Hg (or torr)

Focusing on an isolated "chunk" of fluid and applying free-body diagram reasoning allows us to arrive at some important conclusions about fluids:

- A fluid exerts the same pressure against surfaces on all sides of it (at same height).
- When there is a change in position Δy in a uniform fluid of **density** ρ ($m = \rho V$), the pressure changes by

$$\Delta P = -\rho g \Delta y \qquad (10\text{-}3)$$

(minus because a decrease in y represents an increase in depth)

- The pressure at a given depth does not depend on the shape of the container.
- **Pascal's principle:** A change in pressure occurring in one part of a static, enclosed body of fluid occurs equally (undiminished) throughout that body of fluid.
- In a **hydraulic press** (Figure 10-8), if the cross-sectional area of the output column $A_{\text{output}} = nA_{\text{input}}$, then the output force $F_{\text{output}} = nF_{\text{input}}$.
- **Archimedes' principle**

Archimedes' principle: The total upward force exerted by a fluid on an object submerged in it (called the **buoyant force $\vec{B}$**) is equal in magnitude to the weight of the fluid the object displaces. In equation form,

$$F_{\substack{\text{total on object} \\ \text{by fluid surrounding it}}} \equiv B = m_{\substack{\text{fluid} \\ \text{displaced}}} \, g \qquad (10\text{-}5a)$$

or if the volume V of fluid displaced has uniform density ρ_{fluid},

$$B = \rho_{\text{fluid}} V_{\substack{\text{fluid} \\ \text{displaced}}} \, g \qquad (10\text{-}5b)$$

In nonstatic situations, fluids flow from higher to lower pressure regions. The flow can be

- **laminar flow** (or **streamline flow**), in which molecules follow characteristic paths called **streamlines,** or

- **turbulent flow,** in which the motion of the molecules is churning and disordered.

The property of a fluid that describes the extent to which each layer of the fluid has a friction-like effect on the next is called the **viscosity** of the fluid.

Quantities which describe the flow of a fluid are the

$$\textbf{flow rate} \qquad Q \equiv \frac{\Delta V}{\Delta t} \qquad (10\text{-}6)$$

(meaning a volume of fluid ΔV passes any point P along the path of flow in each time interval Δt)

and the fluid's average **speed** $\bar{v}$ of flow. (*Important:* Speed is still *distance* per unit time and should not be confused with the flow rate.) They are related by

$$Q = A\bar{v} \qquad (10\text{-}7)$$

(A = cross-sectional area of the channel or pipe)

Continuity equation for incompressible fluids:

$$Q_1 = Q_2 \qquad (10\text{-}8a)$$

or
$$A_1 \bar{v}_1 = A_2 \bar{v}_2 \qquad (10\text{-}8b)$$

for any two points 1 and 2 in a single pipe

so *a fluid flows at greater speed through the narrower stretches of a pipe.*

The work-energy theorem applied to an ideal fluid results in Bernoulli's equation:

Bernoulli's equation:

$$P_1 + \rho g y_1 + \frac{1}{2}\rho v_1^2 = P_2 + \rho g y_2 + \frac{1}{2}\rho v_2^2 \qquad (10\text{-}9a)$$

Note: $\rho g y$ = potential energy of each unit of volume

$\frac{1}{2}\rho v^2$ = kinetic energy of each unit of volume

or
$$P + \rho g y + \frac{1}{2}\rho v^2 = \text{constant} \qquad (10\text{-}9b)$$

(pressure + PE/volume + KE/volume = constant)

One consequence of Bernoulli's equation is that *the quicker-flowing fluid through a more constricted segment of a tube exerts less pressure.*

Areas in which fluid dynamics is applied include **aerodynamics,** which deals with the motion of air and its flow around moving objects and the forces that the air and the objects it encounters exert on one another, and **hemodynamics,** the study of blood flow in living things.

In aerodynamics, Bernoulli's equation helps explain the upward resultant force (**lift**) that is produced on an airplane wing because of the shape of its cross-section and explains why roofs may be lifted off houses in hurricanes.

Hemodynamics is concerned with pressures and resulting flow rates in the human circulatory system. Quantities of interest include the **systolic blood pressure** (the maximum value, which occurs when the heart is most contracted) and **diastolic blood pressure** (the minimum value, which occurs when the heart is most relaxed) in the arteries.

QUALITATIVE AND QUANTITATIVE PROBLEMS
Hands-On Activities and Discussion Questions

The questions and activities in this group are particularly suitable for in-class use.

10-1. *Hands-On Activity.* For this exercise, you will need a 1-L (one liter) or 2-L bottle ($1 \text{ L} = 1 \times 10^{-3} \text{ m}^3$), some sort of ruler or tape measure, a watch or clock with a second hand, and access to a sink where the pipes leading to it are visible.

a. Using only the hot or cold water, let the water run at a steady rate slow enough so that you can fill the bottle without any water missing the mouth of the bottle. Time how long it takes to fill the bottle in this way.

b. Use this result to calculate the flow rate of water out of the tap, in m^3/s. (Assume viscosity has a negligible effect, so you can ignore velocity differences between water in the middle of the pipe and water close to the pipe walls.)

c. Measure the diameter of the pipe that leads to the faucet. Compensate for the thickness of the pipe walls and estimate the *inner* diameter of the pipe.

d. Calculate the speed at which water flowed through this pipe while you were filling the bottle.

10-2. *Hands-On Activity.* For this exercise you will need two glasses or bottles; a clean, flexible, narrow tube (two flexible straws taped end on end will do); and some water. The tube should be long enough to extend well below the middle of both glasses (Figure 10-27a). Begin with one glass nearly full and the other empty. Insert one end of the tube below the middle of the full glass, and draw on the other end with your mouth (like sucking on a straw) just until water begins to reach your mouth (Figure 10-27b). Then pinch the mouth end closed, lower that end into the empty glass so that it reaches well below the middle, and unpinch the end so that the water can

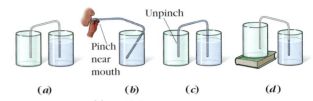

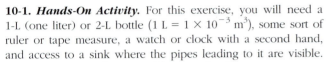

Figure 10-27 Problem 10-2

flow out of it (Figure 10-27c). A tube used in this way is called a *siphon*.

a. Describe what happens when you unpinch the end of the tube. Does the water flow? (If so, why, *or* if not, why not?)

b. When the water stops flowing, how do the water levels in the two glasses compare?

c. Now repeat the experiment, but this time with a book or other object about an inch thick under the empty glass (Figure 10-27d). This time, when the water stops flowing, how do the water levels in the two glasses compare if you measure from the bottom of each glass? How do the water levels in the two glasses compare if you measure from the table top?

d. What do your observations in **b** and **c** have in common? Use physics principles to explain why this shared behavior occurs.

e. What would happen if you repeated the experiment with the book placed under the glass that starts out full? Why?

f. Would your results be different depending on whether the two glasses have the same or different diameters? Explain.

g. Try testing part **f** experimentally. Briefly describe your procedure and your results.

Review and Practice

Section 10-1 The Statics of Fluids: Pressure
*Unless otherwise indicated, use the value $\rho_{water} = 1000 \text{ kg/m}^3$ for the **density of water.***

10-3. In English units, the atmospheric pressure at sea level is about 14.7 lb/in^2. What total force is exerted by the atmosphere on a square foot of pavement?

10-4. A diver at the bottom of a lake has to exert a force of 130 N to open a small vacuum-sealed specimen jar she has brought down with her. The lid of the jar has a diameter of 3.0 cm, and its weight is negligible. Assuming the canister was completely evacuated, find the water pressure at the bottom of the lake.

10-5. Your eardrum has an area of about 60 mm^2. Sounds become painful to your ear when the pressure variations that are involved in sound reach 200 Pa above and below normal air pressure. At this level, called the *threshold of pain,* how much force is exerted on your eardrum?

10-6. A totally evacuated metal canister is suspended in air by a cord attached to the lid (Figure 10-28). The canister and lid

each have a radius of 0.025 m. The lid weighs 2.5 N.

a. If the only thing keeping the canister closed is air pressure, what is the maximum weight the canister can have without falling from the lid?

b. Find the upward force exerted by the cord if the canister has the weight calculated in **a.**

Figure 10-28
Problem 10-6

10-7. **SSM** The atmospheric pressure on the surface of Venus is $9.0 \times 10^6 \text{ N/m}^2$. Taking the density of sea water to be 1025 kg/m^3, find the oceanic depth on Earth at which this same pressure is reached.

10-8. How much pressure is experienced by a whale swimming 200 m below the ocean's surface?

10-9. Would it be easier to open a vacuum-sealed jar on Earth or on the moon? Briefly explain.

10-10. Two jars are vacuum-sealed. The lid of jar A has triple the diameter of the lid of jar B. Compare the forces required to lift the two lids off the jars by finding a numerical value of their ratio (force to lift lid A)/(force to lift lid B).

10-11. The density of light crude oil is 7.97×10^2 kg/m³. A tank of light crude is open to the air at the top. At what depth below the surface is the pressure 25% greater than at the surface?

10-12. A tank of unidentified liquid is open to the air at the top. At a depth of 1.0 m below the surface, the pressure is 1.08×10^5 Pa. What is the pressure at a depth of 2.0 m below the surface?

10-13. **SSM** Rank the containers in Figure 10-29 in order (from least to greatest, indicating any equalities) according to how much

a. pressure the water exerts on the bottom of the container.

b. force the water exerts on the bottom of the container.

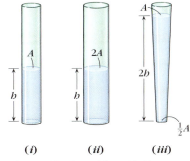
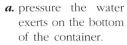

Figure 10-29 Problem 10-13

10-14. The mercury barometer in Figure 10-30*a* is used to measure atmospheric pressure. (The density of mercury is 1.36×10^4 kg/m³.) Because the space above the mercury in the closed column contains only mercury vapor, which exerts negligible pressure, the atmospheric pressure on the exposed surface of mercury forces mercury up the column.

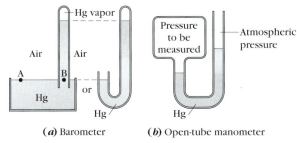

Figure 10-30 Problems 10-14 and 10-15

a. How does the pressure compare at points A and B?

b. How high will the mercury rise in the column when the atmospheric pressure is 1.01×10^5 N/m²?

c. What change in atmospheric pressure (from the value given in **b**) will cause the mercury in the column to drop 1 cm?

d. What pressure (much smaller than atmospheric pressure) would cause the mercury in the column to rise to a height of just one millimeter? This amount of pressure is sometimes called "1 mm of mercury" (1 mm Hg).

10-15. Blood pressures are commonly expressed in units of mm Hg because a device much like the barometer in the previous problem, called a mercury manometer (Figure 10-30*b*), was traditionally used to measure blood pressure. (The barometer is in fact a manometer; it just gets a special name when it is used for measuring atmospheric pressure.) In the open-tube manometer, the *difference* in the heights of the two columns is a measure of the *difference* between the pressure to be measured and atmospheric pressure. In other words, it gives a pressure reading *relative to atmospheric pressure*. Blood pressure readings are readings of this kind. Suppose a patient's blood pressure is reported to be 120/80 (read as "120 over

80"), meaning that the blood pressure, which goes up and down as the pumping heart contracts and expands, varies cyclically between 120 mm Hg and 80 mm Hg. Find the equivalent absolute pressures in N/m² or Pa.

10-16. Draw a free-body diagram of each of the following objects, and describe all forces in each free-body diagram as fully as possible:

a. the crumpled piece of paper in Figure 10-31*a* immediately after the vacuum cleaner is turned on;

b. the suction cup dart adhering to the dartboard in Figure 10-31*b*.

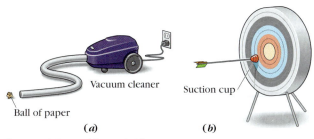

Figure 10-31 Problem 10-16

10-17. The box in Figure 10-32 is filled with air and sealed. Rank the three surfaces A, B, and C in order from least to greatest according to the magnitude of the force that the air within the box exerts on the surface. Indicate any equalities.

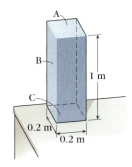

10-18. When a high dam such as the Hoover Dam in Nevada or the Aswan Dam in Egypt is built to block a river, why must the dam be thicker near the base than near the top?

Figure 10-32 Problem 10-17

10-19. An empty, inverted drinking glass is lowered into a sink full of room temperature water (Figure 10-33) until the bottom of the glass is level with the surface of the water in the sink. The lowering is done slowly and carefully so that the sides of the glass are kept perfectly vertical throughout. The figure shows four choices (*a–d*) for completing the diagram to show what has happened inside the glass. Tell which of these choices most accurately shows how far up the inside of the glass the water will go, and briefly explain your reasoning.

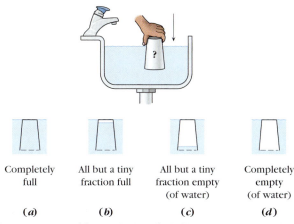

Figure 10-33 Problems 10-19 and 10-58

10-20.

a. The cross-sectional area at the output end of a hydraulic lift used to raise a dentist's chair is 12 times the cross-sectional area at the input end. The movable part of the chair weighs 170 N. When a patient weighing 670 N is seated in the chair, what force must the dentist exert to lift the patient?

• **b.** Through what total distance must the dentist exert this force to raise the patient 0.05 m?

Section 10-2 The Statics of Fluids: Buoyant Forces

10-21. Without looking up the values of their densities, determine which is denser, ice or water. State the reasoning or evidence on which you base your answer.

10-22. A light wooden cylinder floats in water. Will more of the cylinder be submerged if we place it in the water with orientation A or with orientation B (Figure 10-34)? Briefly explain.

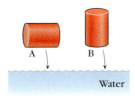

Figure 10-34
Problem 10-22

10-23. SSM WWW The density of water is 1.00×10^3 kg/m^3, and that of benzene is 0.90×10^3 kg/m^3. A certain body floats in water with one third of its volume extending above the surface. What fraction of its volume would extend above the liquid's surface if it were floating in benzene?

10-24. The properties of a new ceramic material to be used in artificial hips are being tested. A scale with a beaker of water on it sits on a small lift (Figure 10-35). A block of the material is suspended above it from a spring balance. As the lift is raised from position A to position C in the figure,

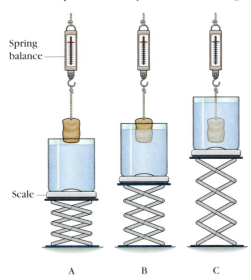

Figure 10-35 Problem 10-24

a. does the reading of the scale under the beaker increase, decrease, or remain the same?

b. does the reading of the spring balance increase, decrease, or remain the same?

10-25. SSM A solid metal cube 0.10 m on a side is released from rest. Draw a free-body diagram fully identifying all forces on the cube a millisecond after its release if it is released with its bottom side

a. 0.05 m below the surface of a deep tank of water.

b. 0.50 m below the surface of a deep tank of water.

10-26. A steel (density 7.8×10^3 kg/m^3) cube with edges 0.10 m long is released from rest with its bottom 5.0 m below the surface in a deep, still freshwater lake. Find the acceleration of the cube at the instant it is released. (It will help to draw a free-body diagram showing all forces on the cube. In particular, identify all vertical forces as fully as you can.)

10-27. A boat with its present cargo displaces 9.0 m^3 of water.

a. What is the present combined weight of boat and cargo?

b. If the total volume of the boat's hull is 13.5 m^3, what additional weight of cargo can it take on before the hull becomes submerged? (Assume the boat doesn't tip.)

10-28. Calculate the buoyant force and the total force on each of the following objects a couple of seconds after being released at the surface of a deep tank of water.

a. A solid oak cube with edges 0.100 m long and a mass of 0.700 kg

b. A solid aluminum cube with edges 0.064 m long and the same mass as the oak cube

10-29. In Example 10-6, what must be the tension in the cable in order to support the diving bell when it is half out of the water?

10-30. Cylinders A and B in Figure 10-36 have the same height but different diameters. Both are floating half submerged. If cylinder A is hollow and cylinder B solid, is it possible for their weights to be equal? Briefly explain.

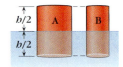

Figure 10-36
Problem 10-30

Section 10-3 Introduction to Fluid Dynamics: Flow Rate and Continuity

10-31. Explain the difference between a fluid's flow rate Q and the average speed $\bar{v}$ of the fluid. Use a concrete example to illustrate the difference.

10-32. At the gas pump at your neighborhood service station, putting 11.0 gallons of gasoline (1 gallon = 3.79×10^{-3} m^3) into your gas tank takes 2.2 minutes. Find the flow rate into your tank in m^3/s.

10-33. Assuming the hose and nozzle on the gas tank in Problem 10-32 have an inner diameter of 2.0 cm, find the speed at which gasoline flows out of the nozzle.

10-34. Two successive lengths of artery, A and B, are identical except that whereas A is clear, the accumulation of plaque on the walls of B has reduced its effective diameter to 80% of its original value.

a. How does the flow rate in length B compare to the flow rate in length A? ($Q_B = \underline{\ ?\ } Q_A$)

b. How does the speed at which blood travels through B compare to the speed at which it travels through A? ($\bar{v}_B = \underline{\ ?\ } \bar{v}_A$)

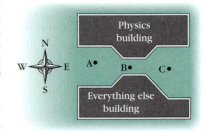

10-35. Figure 10-37 shows two high-rise buildings on a campus

Figure 10-37 Problem 10-35

map and the momentary positions of three students, A, B, and C. A steady wind blows from west to east. Rank the three students A, B, and C in order, from least to greatest, of the wind speeds they feel. Indicate any equalities.

10-36. Water flows into the storage tank in Figure 10-38 through a cylindrical pipeline with an inner diameter of 0.10 m. When the water level in the tank is rising at a speed of 5.0×10^{-4} m/s (half a millimeter per second),

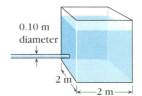

Figure 10-38
Problem 10-36

a. at what speed is the water flowing through the pipeline?

b. how many liters of water are entering the tank each second?

Section 10-4 Fluid Dynamics: Work-Energy Considerations and Bernoulli's Equation

10-37. **SSM** The viscosity of blood depends substantially on the concentration of red blood cells in the blood. A person suffering from anemia has lower blood viscosity than a person with a normal red cell count. Suppose the speed of blood flow changes from v_1 to v_2 as the cross-sectional area changes from A_1 to A_2 along a horizontal stretch of blood vessel.

a. If the change from A_1 to A_2 is a decrease in cross-sectional area, is the corresponding decrease in pressure greater in absolute value for an anemic's blood or for normal blood? Briefly explain.

b. If the change from A_1 to A_2 is an increase in cross-sectional area, is the corresponding increase in pressure greater for an anemic's blood or for normal blood? Briefly explain.

10-38. The liquid in Figure 10-39 flows through a long horizontal pipe of variable diameter. Three very tall, narrow standpipes, A, B, and C, connect to different

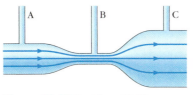

Figure 10-39 Problem 10-38

sections of the horizontal pipe. These standpipes are all open at the top, and water rises partway to the top of each. The tops of the three columns of water are not shown, so you cannot see how high the water rises in each standpipe. Rank the three standpipes A, B, and C in order from least to greatest of the height to which the water rises in each. Indicate any equalities.

10-39. A plastic soft-drink bottle is filled to a height of 0.200 m with water. Two holes, A and B, are made one above the other in the side of the bottle, as in Figure 10-21. Hole A is made at a height of 0.040 m above the bottom of the bottle. At what height above the bottom of the bottle should hole B be made if the water is to leave hole B at half the speed with which it leaves hole A?

10-40. In the human circulatory system, blood flows from the arteries to the veins. Is the blood pressure in your arteries greater than, equal to, or less than the blood pressure in your veins? Briefly explain.

10-41.

a. A common medical practice is to administer medication or nutrients *intravenously;* that is, dissolved in a fluid that is injected directly into a vein. Why is a vein preferable to an artery?

b. A collapsible plastic bag containing the fluid that is being administered is commonly kept in an elevated position on a post at the patient's bedside (Figure 10-40 below). What is the reason for this elevation?

c. A nurse finds that a patient is overreacting to intravenously administered medication and concludes that the rate at which the medication is being introduced into the patient's vein should be reduced. How can the nurse most simply adjust the intravenous set-up to reduce this rate?

10-42. A horizontal pipeline narrows at a certain point from a cross-sectional area of 2.5×10^{-3} m^2 to 1.0×10^{-3} m^2. If the water flowing in the wider portion has a speed of 0.10 m/s^2 at a pressure of 6.0×10^4 N/m^2, what is the

a. speed at which water flows through the narrower portion?

b. pressure in the narrower portion?

10-43. An intravenous feed much like the one in Figure 10-40 is used to administer blood to a patient. The needle is inserted into a vein in the patient's forearm. For blood to flow through the needle at a rate sufficient for the patient to receive a liter of blood in 20 minutes, the pressure in the hose just before the blood enters the needle must be 6750 N/m^2.

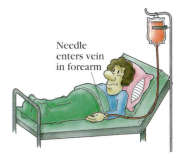

Figure 10-40 Problems 10-41 and 10-43

a. Find the flow rate through the hose in m^3/s.

b. If the hose has an inner radius of 5.0×10^{-3} m (half a cm), at what speed does blood flow through the hose?

c. If the bag containing the blood supply has an average inner radius of 4.0×10^{-2} m, at what average speed does the blood level drop in the bag?

d. Because the bag starts out with no air in it, the pressure on the upper surface of the blood as it descends in the bag is negligible. How high above the patient's forearm must the bag be raised for blood to flow into the patient's vein at the desired rate?

10-44. When a certain fire hydrant is open (Figure 10-41), the flow rate out of it is 3.0 L/second, or 3.0×10^{-3} m^3/s. The radius of the opening is 0.020 m. The hydrant is fed by an underground water main 5.0 m below the level of the hydrant opening. The water main has an inner radius of 0.075 m.

a. Find the speed at which the water flows out of the hydrant.

b. Find the speed at which the water flows through the main, assuming it all flows to the hydrant.

c. What must be the water pressure in the main to sustain the flow rate out of the hydrant?

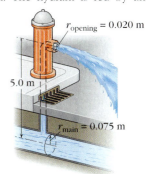

Figure 10-41
Problem 10-44

10-45. SSM A downspout on an old house is totally blocked at ground level. After a rainstorm, a narrow stream of water squirts horizontally out a small hole in the spout and lands 0.8 m from the base of the downspout. The hole is 1.2 m above the ground.

a. At what speed is the water leaving the hole?

b. To what height above ground level is the water backed up in the downspout?

Section 10-5 Real-World Applications of Fluid Dynamics

10-46. Figure 10-42 shows two cabins in different stages of construction. The cabins will eventually be identical, and their

A	B

Figure 10-42 Problem 10-46

flat roofs are already attached in the same way. But cabin A has just been framed, whereas cabin B has already been sided and sealed up. Which cabin's roof is more likely to blow off in a hurricane? Briefly explain.

10-47. The molecule hemoglobin binds loosely and reversibly with oxygen. The presence of hemoglobin in red blood cells enables them to serve as oxygen carriers. Because hemoglobin is a very heavy molecule, the density of blood increases as the concentration of hemoglobin increases. Blood is accepted from donors only if it has a certain minimum concentration of hemoglobin. Use what you know about the statics of fluids to design a simple test that blood drive personnel can use to check a drop of blood drawn from a would-be donor and see if it meets this minimum standard. (If you are already aware of such a test, explain the principle(s) by which it works.)

10-48. It takes about 15 minutes to draw a liter (1.0×10^{-3} m^3) of blood from a fairly typical blood donor. If the needle through which the blood is drawn has an inner radius of 2.5×10^{-4} m, at what speed does blood travel through the needle?

Going Further

The questions and problems in this group are not organized by section heading, so you must determine for yourself which ideas apply. Some of them will be more challenging than the Review and Practice questions and problems (especially those marked with a • or ••).

10-49. Spheres A and B have equal masses, but the radius of sphere A is double that of sphere B. How do the densities of the two spheres compare? (Express in form $\rho_A = \underline{\;?\;} \rho_B$.)

10-50. In Figure 10-1, assume that you are the principal individual in each part of the figure. Assume the bed of nails has one nail for each cm^2 of area. Estimate the pressure exerted by

a. the thumbtack in Figure 10-1*a;*

b. each nail (on the average) in Figure 10-1*b.* Briefly state any assumptions you make in your estimate.

10-51.

a. In Example 10-1, convert the force that the spring balance must exert in part *a* into pounds.

b. Does this result seem very large to you?

c. Would you have to use that much force if you pried the lid off with a "church key" type can opener? Briefly explain your answer.

10-52. An aquarium director calls in an engineer to determine the water pressure on the glass windows of a particular large tank. He tells the engineer the dimensions of the glass. The engineer informs him that he needs more information. Why?

10-53. Devise a procedure by which you could determine the density of air at 20°C and 1 atm of pressure if you knew the density of water. Briefly describe your procedure and the reasoning on which it is based.

10-54. A square slab of material of thickness 2.0 cm floats in water with 0.6 cm of its thickness extending above water. The same sample of material can be remolded into a smaller, thicker square so that 1.0 cm of its thickness will extend above water when it floats. What fraction of a side of the old square must a side of the new square be to accomplish this?

10-55. The density of ice water is 1000 kg/m^3, that of ice is 917 kg/m^3. A ship's captain spots an iceberg in the distance and estimates that the part of the iceberg visible above the surface of the water has a volume of 600 m^3. What volume of ice should the captain assume remains hidden below the surface of the water?

10-56. Use what you know about pressure in a fluid to explain why fluids assume the shape of their containers.

•10-57. If energy is conserved, a hydraulic press cannot have more energy as output than you put into it. Prove that the work done by the upward force $2F_1$ on the right in Figure 10-6 is equal to the work done on the left by the downward force F_1. Justify any assumptions you make by discussing what happens to the water.

•10-58. Suppose the inverted glass in Figure 10-33 is lowered into a deep pond rather than a sink.

a. Explain how the inverted glass can act as a depth gauge.

b. At approximately what depth will the water have risen halfway up the inverted glass?

c. Identify the factor(s) that make your answer to *b* approximate rather than exact.

10-59. SSM WWW A small swimming pool in a municipal playground is 13 m long by 8.3 m wide, and the average water depth is 1.0 m. At the start of the summer, it takes $5\frac{1}{2}$ days (24 hours/day) to fill the pool to this depth. If the pool is fed by a single 1-inch pipe (approximately 0.025 m inner diameter), at what average speed does the water flow through the pipe while the pool is filling?

10-60. Huck Finn is floating downriver on a raft. Along a certain stretch of river, he notices that the distance between the banks remains the same, and there is no change in elevation, yet his raft is slowing down. What might he conclude about the river along this stretch?

10-61. You can readily observe that when you run water from a faucet at a moderate steady flow rate, the stream of water narrows as it descends. Does this imply that the speed of the water at point B (in Figure 10-43) is less than, equal to, or greater than the speed at A? Briefly explain.

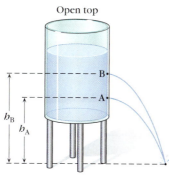

Figure 10-43
Problem 10-61

10-62. Figure 10-44 shows a water tank that has two holes in the tank wall. The two holes, A and B, are at different heights but are otherwise identical. Water streams out of both holes. Is it possible for there to be a water level in the tank (above both holes) such that when the water is at this level, the streams from both holes will land at the same distance from the base of the tank? (In other words, is it possible for the streams of water from the two holes to appear as they do in the figure?) Briefly explain.

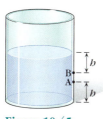

Figure 10-44 Problem 10-62

••10-63. Figure 10-45 shows another open water tank that has two holes in the tank wall. The two holes, A and B, are at different heights but are otherwise identical. Hole A is at a height h above ground level. The surface of the water in the tank is at a height h above hole B. Water streams out of both holes. Each of the two streams of water lands a certain horizontal distance from the base of the tank.

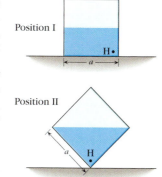

Figure 10-45
Problem 10-63

a. By a suitable (non-numerical) calculation, determine how these two distances compare.
b. Does your result depend on the vertical distance between holes A and B?

10-64. A small hole H (Figure 10-46) has appeared in the bottom of a cubic Plexiglas fish tank of side A. When the tank is *half full*,
a. is the initial flow rate of water out of the hole greater when the tank is held in position I or position II? Briefly explain.
b. Find the ratio $Q_{\text{position I}}/Q_{\text{position II}}$ of the flow rates in the two positions.

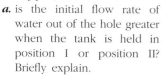
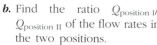

Figure 10-46 Problem 10-64

10-65. A syringe needle is used to draw blood from blood donors. How would the amount of time required to draw a pint of blood change—by what factor would it be multiplied—if the inner diameter of the needle were doubled?

10-66. Suppose this textbook is lying closed on your desk.
a. Is the total force that the air exerts on the book upward, downward, or zero? Explain your reasoning before going on to the next part of the question.
b. Is there any air pressure on the underside of the book? Explain your reasoning before going on to the next part of the question.
c. Estimate the total force that the air exerts on the top surface of the book. How much force would you have to exert to lift the book off the table if there were no air pressure on the underside of the book? Can you exert that much force? Can you lift the book?
d. After answering **c,** do you need to revise your answer to **a**? Comment.

10-67. When a truck passes you on a highway, it affects the pressure on the side of your car closest to the truck. Does this change in pressure tend to move your car toward or away from the truck? Briefly explain.

10-68. The four blocks floating in water in Figure 10-47 have the same horizontal dimensions (length and width) but different heights. They may also be made of different materials. Rank the blocks in order of their weights, from least to greatest, indicating any equalities. Give brief reasons for your order.

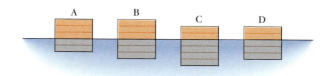

Figure 10-47 Problem 10-68

•10-69. The two plastic soft-drink bottles in Figure 10-48 are connected by a length of clear hose, so that water can flow freely and without leakage between the two bottles. A small puncture hole is made in the side of bottle II at point P.
a. Just after the hole is made, at what speed does water stream out of the hole at P?

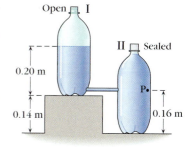

Figure 10-48 Problem 10-69

b. At what distance above the bottom of bottle I should another hole be made if we want water to leave this hole at the same speed with which it leaves the hole in bottle II?

10-70.
a. An underwater surveillance camera is submerged in a body of water. The density of the water is uniform throughout. As the camera is lowered deeper into the water, does the buoyant force on it increase, decrease, or remain the same? Briefly explain your answer.
b. As one moves into the upper atmosphere, the air becomes less dense. As a weather balloon rises, does the buoyant force on it increase, decrease, or remain the same? Briefly explain your answer.

10-71. A beaker of water sits on an electronic balance (Figure 10-49a) that reads its mass in kilograms. A solid metal fishing sinker is also in the beaker, resting on the bottom. A fishhook on the end of a length of line is then lowered into the beaker and used to raise the sinker slightly

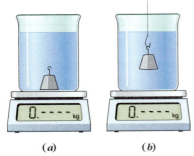

(a)   **(b)**

Figure 10-49 Problems 10-71 and 10-72

(Figure 10-49b). When the sinker is raised to this new position, by how much does the balance reading decrease? Choose the best answer: *by less than the mass of the sinker; by an amount equal to the mass of the sinker; by more than the mass of the sinker; not at all.* Briefly explain.

• **10-72.** A beaker of water sits on an electronic balance. A solid metal fishing sinker is lowered into the water on the end of a length of fishing line and held suspended below the water's surface (Figure 10-49b). In this position, the balance reads 0.810 kg. Before the sinker was lowered into the water, the balance reading was 0.780 kg.

a. Find the volume of the sinker.

b. Is it possible to find the mass of the sinker from the given information? If so, find its value. If not, tell what additional information you need.

10-73. In Example 10-9, show by a calculation that the water leaves the hose at a speed of 6.64 m/s.

10-74. A tank of liquid nitrogen (see Table 10.1) is open at the top. At a certain point P in the liquid, the pressure is 1.01 atm.

a. How far below the surface of the liquid is this point?

b. If a hole is tapped in the side of the tank at this depth, at what speed will the liquid spurt out of the hole?

10-75. A very deep water well is open to the air at ground level. At a certain point P below the upper surface of the water, the pressure is 2 atm, that is, twice what it is at the surface. The well is then capped and the air pressure at the surface is doubled. The pressure at point P is then ____ (*still 2 atm; greater than 2 atm but less than 4 atm; 4 atm; greater than 4 atm*).

10-76. Suppose a giant human had linear dimensions (height, width, thickness) that were 10 times those of a normal human.

a. How would the giant's volume compare to that of a normal human? Express the comparison as a ratio $\frac{V_{\text{giant}}}{V_{\text{normal}}}$.

b. How would the giant's density compare to that of a normal human? Express the comparison as a ratio $\frac{\rho_{\text{giant}}}{\rho_{\text{normal}}}$.

c. How would the giant's weight compare to that of a normal human? Express the comparison as a ratio $\frac{(mg)_{\text{giant}}}{(mg)_{\text{normal}}}$.

10-77. SSM WWW *Biomechanics:* The compressive stress on a bone is the pressure exerted on a cross-section perpendicular to the bone's length. When you stand, there is compressive stress on your femur, or thighbone. When the pressure on the

cross-section reaches about 10 times the value it has when you are standing, the femur will fracture. Based on this information, determine whether the giant in Problem 10-76 would be able to stand.

a. How would the cross-sectional area of the giant's femur compare to that of a normal human? Express the comparison as a ratio $\frac{A_{\text{giant}}}{A_{\text{normal}}}$.

b. How would the compressive stress on the giant's femur compare to that of a normal human? Express the comparison as a ratio of pressures $\frac{P_{\text{giant}}}{P_{\text{normal}}}$.

c. Could the giant's femur safely support his weight when he is standing? Briefly explain.

10-78. The block in Figure 10-50 is in equilibrium. Liquid A has a density of 1.00×10^3 kg/m³, liquid B has a density of 0.85×10^3 kg/m³, and the block has a density of 0.90×10^3 kg/m³. If the block's height is 10 cm, how many centimeters of this height is submerged in liquid A?

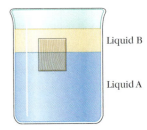

Liquid B

Liquid A

Figure 10-50 Problems 10-78 and 10-79

10-79. Suppose that in Problem 10-78, all of liquid B is drained off. After this happens, will more, less, or the same amount of the block be submerged in liquid A? Briefly explain.

10-80. In Figure 10-18, the flow rate is the same at points X and Y. From this, can we conclude that the acceleration of the average water molecule traveling from X to Y is positive, negative, or zero? Briefly explain.

10-81. If you want the water from your garden hose to spray further, should you increase or decrease the diameter of the nozzle opening? Briefly explain.

10-82. A city has had problems with all sorts of materials being dumped into a canal running through it. The canal was constructed with uniform width and depth. The city suspects that a local company has been dumping cement somewhere in the canal. However, they don't know where, and the water is too murky for anyone to see the bottom. However, a town engineer announces, "We don't need to see to the bottom. We can tell if large amounts of cement have sunk to the bottom somewhere if we watch the debris that floats along the surface." What is the engineer going to watch for, and how will that tell her where the cement has been dumped? Use principles discussed in this chapter to explain your answer.

• **10-83.** For a public fountain display, a pump 10 m below ground maintains a pressure of 4.04×10^5 Pa. Water driven by this pump passes through an opening at ground level, so that a jet of water shoots vertically upward. Neglecting the effect of air resistance, how high does this jet rise?

10-84. A spherical balloon with a radius of 0.050 m is filled with helium to a pressure of 2 atm, so that the density of helium in it is twice what it would be at 1 atm. When the balloon is released into the air at sea level, what is the total vertical force on it?

Problems on WebLinks

10-85. Consider the wedge-shaped "chunk" of water in WebLink 10-1. If this chunk were replaced by a submerged wooden block, which (one or more) of the following would be equal: the forces on all three sides; the pressure on all three sides, the forces on top and bottom, the pressure on top and bottom?

10-86. In WebLink 10-1, we consider the forces $\vec{F}_1$, $\vec{F}_2$, and $\vec{F}$ that the rest of the water exerts on the three sides of the wedge-shaped "chunk" of water. Suppose that $\vec{F}_3 = \vec{F} + \vec{F}_1$ is the vector sum of $\vec{F}$ and $\vec{F}_1$. Consider the magnitudes F_1, F_2, and F_3 of the three numbered forces. Rank these three magnitudes in order from least to greatest, indicating any equalities.

10-87. Because the chunk of water that we isolated in WebLink 10-2 is in equilibrium, should we conclude that the magnitude of the force that the rest of the water exerts on the top of this chunk is less than, equal to, or greater than the magnitude of the force that the rest of the water exerts on the bottom of the chunk?

10-88. In WebLink 10-3, we consider a totally submerged treasure chest. As the weight of treasure in the chest increases, does the buoyant force that the water exerts on the chest increase, decrease, or remain the same?

10-89. Suppose we have two treasure chests identical to the one in WebLink 10-3. At a particular instant, chest A is totally submerged and sinking toward the bottom, whereas chest B has so little treasure in it that it is floating at the top, only partially submerged. At this instant, is the buoyant force that the water exerts on chest A less than, equal to, or greater than the buoyant force that it exerts on chest B?

10-90. Suppose the treasure chest in WebLink 10-3 goes from position A to position B to position C in Figure 10-51.

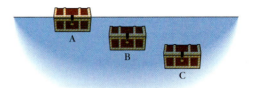

Figure 10-51 Problem 10-90

a. As it goes from A to B, does the buoyant force that the water exerts on it increase, decrease, or remain the same?

b. As it goes from B to C, does the buoyant force that the water exerts on it increase, decrease, or remain the same?

10-91. In WebLink 10-4,
a. how many cars have passed point Y when 4 cars have passed point X?

b. is the flow rate in cars/second of cars approaching point Y from the left less than, equal to, or greater than the flow rate of cars approaching point X?

c. is the speed of cars approaching point Y from the left less than, equal to, or greater than the speed of cars approaching point X?

For a cumulative review, see Review Problem Set I on page 825.

CHAPTER ELEVEN

Thermal Properties of Matter

A blazing sun provides life-sustaining energy to an otherwise frigid Earth. Energy released in combustion warms your home, powers your car, and drives the generators in many of the power plants that provide electricity to entire communities or regions. On a more personal scale, the release of energy stored in large molecules keeps your body temperature at 37.0°C (98.6°F), well above room temperature. Elaborate body mechanisms carefully regulate this temperature. Too high or too low and you are a goner.

"At all levels of importance, from life-and-death to trivial, transfers of energy between warmer and cooler objects are among the conspicuous behaviors that we encounter in nature."

At all levels of importance, from life-and-death to trivial, transfers of energy between warmer and cooler objects are among the conspicuous behaviors that we encounter in nature. To transfer energy quickly where winters are cold, we may use the water in heating pipes to carry energy from the basement furnace to an upstairs bedroom, but to slow down its transfer out of our homes we pack our walls with insulation. In doing so, we take advantage of the properties of water and of insulating materials.

In this chapter, we will look at conditions involving temperature, warming, and cooling. We will concentrate on a few **thermal** (related to temperature) properties of materials that affect their ability to store and transfer energy under these conditions. Identifying properties expressible as numerical values will allow us to calculate the effects of using particular materials.

Because most of us know that heating our homes and cooking our food requires energy, we will start out assuming that when a hotter object warms a cooler one, it does so by transferring energy to the cooler one. We will later have to describe observations and measurements that justify this assumption.

11-1 Temperature

"It must be 100° in the shade." "It's going below zero tonight." All of us are accustomed to associating our experiences of hot and cold with thermometer readings or temperature (Figure 11-1). In fact, we can start out *defining* temperature as what we read on a traditional liquid thermometer. In other words, because the liquid in the bulb of a thermometer expands as it gets warmer, and is therefore forced up a very narrow tube, the height of the liquid in the tube can serve as an indicator measuring how warm the thermometer's surroundings are. Later on, though, we will have to ask, "At the level of atoms or molecules, what makes the column of liquid respond to its surroundings in this way?" In addressing that question, we will arrive at a deeper understanding of what temperature means.

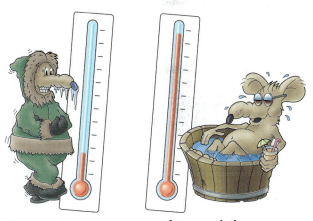

Figure 11-1 We are accustomed to associating our experiences of hot and cold with thermometer readings.

◆**ESTABLISHING A TEMPERATURE SCALE** The enclosed column of liquid in Figure 11-2 will always rise to the same height in the same conditions of

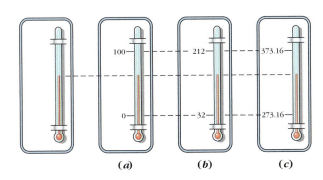

(*a*)   (*b*)   (*c*)

Figure 11-2 Establishing a temperature scale. The column of liquid stands at the same height no matter what numbers we write on the backing.

warmness or coolness, whether we put numbers on the backing or not. If we change the numbers on the backing, the height doesn't change: The liquid "reads" its surroundings, not our numerical labels. But labeling allows us to keep track of how high the column is and compare its height in different conditions. The choice of numbers that we place on the backing establishes a **temperature scale.**

We can take the height of the column when water freezes, assign to it whatever number we choose, and let that be "the temperature at which water freezes." We can assign another number to the height of the column when water boils, and let that be "the temperature at which water boils." If the two assigned numbers are further apart, there are more degrees (or units of temperature) between the two conditions, and each degree must be smaller.

Different choices of assigned numbers, both of which were widely adopted, were made by two early European scientists, Gabriel Fahrenheit (German, worked in the Netherlands, 1686–1736) and Anders Celsius (Swedish, 1701–1744).

	Temperature at which water freezes	Temperature at which water boils
Fahrenheit	32	212
Celsius	0	100*

*More precisely, 99.975 on the modern Celsius scale.

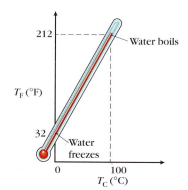

Figure 11-3 Comparing temperatures on the Fahrenheit and Celsius scales.

The Celsius scale was formerly called the centigrade ("hundred-degree") scale because there are a hundred degrees between the boiling and freezing temperatures that Celsius picked. **STOP&Think** Which is smaller, a "degree Fahrenheit" or a "degree Centigrade"? ◆ The Celsius scale is now officially used in almost every country except the United States.

The thermometer in Figure 11-3 has markings for both the Celsius and Fahrenheit scales. In the figure, the thermometer itself serves as the straight line relating Fahrenheit temperature (T_F, plotted vertically, is like y) and Celsius temperature (T_C, plotted horizontally, is like x). The freezing and boiling points have a coordinate on each scale. From these two points, (0°C, 32°F) and (100°C, 212°F), we determine that the slope is

$$\frac{\Delta T_F}{\Delta T_C} = \frac{212°F - 32°F}{100°C - 0°C} = \frac{9}{5} \qquad (11\text{-}1)$$

and the vertical intercept is 32°F. We can therefore write the equation of the straight line, which enables us to convert from °C to °F:

To convert from °C to °F $T_F = \frac{9}{5}T_C + 32°$ $\qquad (11\text{-}2)$

Solving this for T_C gives us an equation we can use to convert the other way:

To convert from °F to °C $T_C = \frac{5}{9}(T_F - 32°)$ $\qquad (11\text{-}3)$

Once you know Equation 11-2, you can obtain 11-1 by finding the slope and 11-3 by solving for T_C. If you learn one equation and the reasoning that gets you to the others, you don't need to memorize three separate equations.

Example 11-1 *An American Abroad*

Calvina Hobbes from Brooklyn is visiting a French beach resort where they post the water temperature. She knows that at home she doesn't like to swim unless the water temperature is at least 72°. What minimum posted temperature should she look for before taking the plunge?

Solution

Choice of approach. Calvina needs to convert from the temperature she knows on the Fahrenheit scale to the Celsius scale used in France.

What we know/what we don't.

$$T_F = 72° \qquad \text{The equivalent } T_C = ?$$

The mathematical solution. Applying the conversion equation (11-2) and solving for T_C (so that it takes the form of Equation 11-3),

$$T_C = \tfrac{5}{9}(72° - 32°) = \tfrac{5}{9}(40°) = \mathbf{22°}$$

◆ Related homework: Problems 11-6 and 11-7.

Caution: Equations 11-2 and 11-3 convert between T_C and T_F, which represent individual thermometer readings, *not* changes in temperature. But the slope of a graph (Equation 11-1) relates *changes* in temperature, as in the following example.

Example 11-2 *Different Differences*

For a guided interactive solution, go to Web Example 11-2 at www.wiley.com/college/touger

You are hiking with an Asian exchange student on a trail that takes you up a mountainside from the trailhead. You know it will be about 15° cooler (in °F) at the trail's upper end than at its start. As a gracious host, what temperature drop in °C should you tell your guest to anticipate?

Brief Solution

Choice of approach. Because you are converting a temperature *difference*, you should use Equation 11-1, not Equation 11-2 or 11-3.

What we know/what we don't.

$$\Delta T_F = 15° \qquad \text{The equivalent } \Delta T_C = ?$$

The mathematical solution. We want to solve for ΔT_C. Inverting Equation 11-1 gives us $\Delta T_C/\Delta T_F = \tfrac{5}{9}$. Then

$$\Delta T_C = \tfrac{5}{9}(\Delta T_F) = \tfrac{5}{9}(15°F) = \mathbf{8.3°C}$$

◆ Related homework: Problems 11-9, 11-10, and 11-12.

We have seen that by assigning numbers to two fixed points, we can establish a temperature scale. In the cases we have considered so far, the fixed points were the freezing and boiling points of water, but there is nothing sacred about these choices. Another fixed point, which leads to a temperature scale widely used by physicists, is arrived at in the following way.

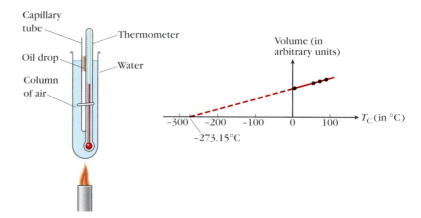

Figure 11-4 The volume of a column of air kept at constant pressure varies linearly with temperature. The solid part of the line is through actual data points; the dotted part is an extension.

Consider the experimental set-ups in Figures 11-4 and 11-5. In the first of these, a column of air is trapped under a drop of oil in a glass capillary tube open at the top. The weight of the oil drop sitting on the column of air keeps it at fixed pressure. Under these conditions, we would find that the height of the column of air, which is proportional to its volume, varies linearly with temperature. If we extend the straight line that passes through the data points, it will cross the temperature axis at a value in the vicinity of $-273.15°C$ (the value we would obtain with a more precise experimental set-up). This is the temperature at which the height of the column would be reduced to zero if it continued to shrink at the same rate. Because the height can go no lower, it may be that whatever accounts for temperature on an atomic or molecular scale has also reached a lower limit.

This idea of a lowest possible temperature is reinforced by the data in Figure 11-5. Here the air is enclosed in a copper ball. Its volume is kept fixed, but the pressure P is allowed to vary with temperature. Plotting pressure against temperature, we again find that the data points lie along the straight line. If we extend this line to where it crosses the temperature axis (at $P = 0$), we find that the temperature at which the *pressure* would be reduced to zero is the same ($273.15°C$) to whatever degree of precision our experimental set-up permits.

The two behaviors we have described here can be observed for a suitable range of temperatures and densities in any gas. The temperature at which the pressure (at constant volume) or the volume (at constant pressure) appears to reach zero is the same for each gas.

We need to be aware of what this line of reasoning ignores. For example, all real gases condense to liquids at sufficiently low temperature if there is sufficient pressure. Also, although the liquids occupy only a tiny fraction of the volume that the gases did (even if the liquids never solidified), it would seem impossible for the volume to shrink to exactly zero, as long as matter occupies some space. But we are doing an "even if" kind of reasoning here. *Even if* the tiny volume to which the material could be reduced were exactly zero, and *even if*

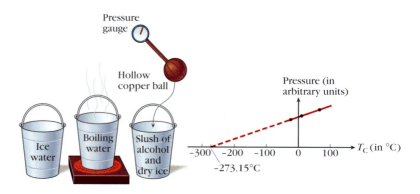

Figure 11-5 The air pressure in an enclosed container of fixed volume varies linearly with temperature.

the gases never condensed to form liquids, we could still get no lower than this temperature. Physicists call this lowest possible temperature **absolute zero.**

A scale that assigns the number 0 to absolute zero is called an **absolute scale.** A simple way to transform the Celsius scale into an absolute scale (Figure 11-6) is to add 273.15 to every Celsius temperature:

Temperature on absolute or Kelvin scale: $T_K = T_C + 273.15$ (11-4)

The resulting scale is called the **Kelvin scale** after William Thomson, Lord Kelvin (1824–1907), who proposed it. Each one-unit interval on the Kelvin scale is the same size as a °C, but instead of being called a degree Kelvin, it is simply called a **kelvin** (abbreviated K). **STOP&Think** If the temperature increases by 10°C, by how many kelvins does it increase? ◆

Although it is theoretically impossible to achieve temperatures below absolute zero, experimentalists have been able to produce temperatures within a tiny fraction of a kelvin of absolute zero. At MIT in 2003, a sample of sodium gas was cooled to 500 picokelvins, half-a-billionth of a kelvin above absolute zero.

The straight-line graphs in Figure 11-6 go through the origin and therefore show proportionalities. The experimental data indicate that *at fixed pressure, the volume of a gas is proportional to its temperature in kelvins,* and that *at fixed volume, the pressure exerted by a gas is proportional to its temperature in kelvins.* The experiments show that these patterns occur; we have not yet explained why.

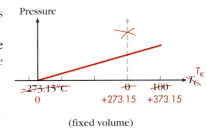

(fixed volume)

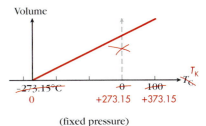

(fixed pressure)

Figure 11-6 Transforming the Celsius scale into an absolute scale. When 273.15 is added to each Celsius temperature, the new "lowest possible temperature" is zero. The straight lines then pass through (0,0) and therefore represent proportionalities.

11-2 Temperature Differences and Heat Transfer (Qualitative)

In many familiar situations, objects or substances at different temperatures are brought together. Leave your hot cup of coffee on the kitchen counter and it cools to room temperature. Take a gallon of milk out of the refrigerator; it, too, reaches room temperature, but much more slowly. In a Styrofoam cup, the coffee also cools more slowly, but it still reaches room temperature eventually. Add cold milk to your hot coffee and the mixture quickly reaches a temperature somewhere between the original temperatures of the two liquids. What temperature? It depends on how much of each liquid we started with, as well as the initial temperatures of the liquids.

There are several questions we can address when objects or substances at different temperatures are brought together. How much does the temperature of each change? How quickly do the changes occur? What properties of the objects or substances affect the outcomes? Why is it that things at different temperatures reach a common in-between temperature when they are brought together?

We can start to address the last question if we assume that as the temperature of a thing rises or drops, it is gaining or losing something. Then whatever the initially hotter thing loses, the initially cooler thing gains. Does this answer the last question fully? No, it doesn't. It doesn't tell us why the two things always reach the *same* temperature. It doesn't even tell us why the warmer object cools down and the cooler object warms up rather than the other way around. But that's the way we see nature behave. So we build these observed patterns of behavior into our assumption.

We assume that there is an amount of something being transferred from the warmer to the cooler object. We cannot yet say what this something is, but we give the amount of it that is being transferred a name: We call it **heat.** The reaching of a common temperature is called **thermal equilibrium.** It may happen quickly or slowly, depending on the extent of **thermal contact** between the two bodies. This refers to how readily heat can be transferred from the warmer to the cooler body. Hot soup in an open metal bowl transfers heat to the surrounding air much more quickly than if it were in a closed Styrofoam container.

➡ **A note on language:** While in ordinary language we talk about the air cooling the soup, this does not mean the air is giving something to the soup. Our model requires us to talk about heat transferring from the hotter to the cooler body. Something warms when its surroundings transfer heat to it; it cools when it transfers heat to its surroundings. (In each case, we are talking about the *total* or *net* amount of heat transferred.)

But the soup in the Styrofoam container doesn't stay hot indefinitely. The Styrofoam, a good **thermal insulator,** doesn't totally prevent heat transfer; it just slows it down.

Transferring an amount of heat to an object may raise its temperature. **STOP&Think** Are heat and temperature the same thing? ◆

Consider this situation. Two identical stove burners are set on high. A pot of water is placed on each. The pots are identical and just cover the burners. In each, the water starts out at room temperature, but one pot is nearly full whereas the other has water just a cm deep at the bottom. **STOP&Think** Which water boils first? When the water first starts to boil in one pot, will the water in the other pot be at or even near boiling temperature? Is the burner transferring the same amount of heat to each pot? ◆

Experience should tell us that the temperature of the fuller pot of water will have risen much less. But the burners don't know what's in the pots. At the same setting, both burners start out transferring heat through the pot bottoms at the same rate. When the smaller quantity of water first begins to boil, the same amount of heat has raised its temperature more, so we must conclude:

> *Heat cannot be the same thing as temperature.*

Nor can heat be the same thing as a change in temperature.

> *The amount of heat we need to transfer depends not only on the temperature increase we want but on how much mass we are warming.*

Does it depend on anything else? Let's look at a less familiar situation. We enclose equal masses of water and ethyl alcohol in two well-insulated containers. The insulation ensures that heat transfers to the surroundings take place extremely slowly compared to heat transfers within the container. Over short periods of time we can assume transfers to the surroundings are negligible, and we can concentrate on the transfers within. A container that meets these conditions is called a **calorimeter** (from the Latin words for *heat* and *measure*), because it provides the controlled conditions in which easy-to-interpret heat measurements can be done.

We now boil a bunch of glass marbles in water long enough so they are all at the temperature of the boiling water. With a slotted spoon, we transfer equal numbers of hot glass marbles to the calorimeters containing water and alcohol. **STOP&Think** Should the temperature of the water and the alcohol necessarily increase by the same amount? ◆

What we observe is that the temperature of the alcohol rises substantially more than that of the water. To get the water temperature to rise as much, we have to add several more hot marbles. Because each marble transfers heat to the cooler liquid, we conclude that more heat has to be transferred to water than to alcohol to raise it to the same temperature.

> The amount of heat that must be transferred to produce a given temperature increase depends on the *material* to which it is transferred.

Example 11-3 *Your Cocoa's Getting Cold*

When you pour yourself a cup of hot cocoa and forget about it for a while, it cools down to room temperature. So far we have talked about heat transfers in which a warm object transfers heat to a cooler object, so that as the one cools down, the other warms up. But now everything is leveling off to room temperature. Is anything warming up here? Or is something else happening? Explain your answer.

Solution

Choice of approach. We have shown that the amount of heat that must be transferred to produce a given temperature increase depends on the mass and the kind of material being heated. We will focus on the mass dependence.

The qualitative solution. The heat is being transferred from the cocoa. What is it being transferred to? At the very least, to all the air in the room. But the room is not isolated. It transfers heat quickly through open doors or windows, and more slowly through the walls. So in all there is a vast mass of stuff to which the heat is transferred. It would take a very large amount of heat to raise the temperature of all this stuff by even a substantial fraction of a degree. The heat transferred from the cocoa does raise the temperature of its surroundings, but the surroundings are so vast that the temperature increase is too small to be measurable.

Making sense of the results. We will do actual calculations for this problem in Section 11-5. To reinforce the basic idea, do On-the-Spot Activity 11-1.

◆ Related homework: Problem 11-14.

In the following activity, your hand will provide a rough qualitative measure of temperature change.

On-The-Spot Activity 11-1

Boil a pot or kettle of water, and pour enough to fill a china or other ceramic cup. *Caution:* Always be careful when working with boiling water! Quickly put your hand on the outside of the cup as soon as it is poured. Is the outside of the cup as hot as the boiling water? What happens when you keep your hand on the outside of the cup for a half a minute or a minute? (Move your hand away sooner if it gets too hot!) Is heat being transferred through the sides of the cup? Does this happen instantaneously or does it take time? How do you know?

Pour out the water and cool down the cup. Now place the cup upright in a small flat-bottomed bowl partly filled with room-temperature water, as in Figure 11-7a. Then *carefully* fill the cup with boiling water. Keep your hand in the water *in the bowl*. What happens to the temperature of this water over time?

Finally, empty and cool the cup again, and arrange the cup and bowl in a tub or other large body of room-temperature water as in Figure 11-7b. Fill the cup once more with boiling water, and monitor the temperature of the tub water with your hand. Do you feel any detectable change? (Does your answer depend on how far from the cup you place your hand?)

How do you suppose the tub temperature would be affected if you took a cup of boiling water, spilled into the tub, and then stirred? What would be the effect if you spilled the same cup of boiling water into a swimming pool? Into the ocean?

Figure 11-7 Steps of On-the-Spot Activity 11-1.

We can conclude that the larger the mass to which an amount of heat is transferred, the smaller the temperature increase it produces. For the entire environment surrounding a small object, the increase will be too small to detect.

11-3 Heat Transfer and Energy (Qualitative)

We have proposed that when two objects at different temperatures are brought together and reach a common in-between temperature, one loses something and the other gains something in equal amounts. We have given the name *heat* to the amount being transferred. Used in this way, heat never refers to the amount contained in an object, only to the amount going from one object to another. Now we must ask: an amount *of what?*

Well into the nineteenth century, many scientists viewed heat as a substance, a fluid that gets transferred from one body to another. In 1787, the French chemist Antoine Lavoisier named this fluid *caloric*. Caloric, in the eighteenth-century view, was a conserved substance. It existed in two forms: "sensible" (detectable by our senses) caloric flowed between objects of different temperature, and "latent" caloric occurred in combination with matter, and was absorbed or released during changes of state and during endothermic or exothermic chemical reactions. Each of the two forms of caloric could be converted into the other. In all, the caloric theory was quite successful at explaining the range of observations that we've discussed.

But well before Lavoisier there were already dissenters from this theory of heat as a fluid substance. As early as 1620, Francis Bacon suggested that "the very essence of heat . . . is motion and nothing else," a view amplified by Newton in 1704: "Heat consists in a minute vibratory motion of the particles of bodies." From a modern perspective, this suggests that heat involves the mechanical energy of particles.

In seeking evidence to enable us to choose between these two theories, we turn to situations where the temperature of one or more bodies increases without a decrease in the temperature of something else. This can happen, for example, when there is friction.

Sometimes these temperature changes can be dramatic. One such instance was brought to the world's attention in 1798 by Benjamin Thompson (1753–1814), a native of Massachusetts. He pursued his scientific career in Europe, gaining the title of Count Rumford. While superintending the boring of brass cannons for the military arsenal in Munich, Rumford was struck by the high temperatures reached by the brass of the guns and the even higher temperatures (well above the boiling point of water) of the brass chips that flew off the gunstocks in the boring process. When he tried submerging the boring tool and the part of the cannon being bored in a body of water, he found that the water temperature rose steadily during the boring operation and after $2\frac{1}{2}$ hours was brought to a boil. As he reported, "It would be difficult to describe the surprise and astonishment . . . of the bystanders on seeing so large a body of water heated, and actually made to boil, without any fire."

On a smaller scale, Rumford found he could bring the water in a test tube to a boil by rotating the test tube between two paddles making frictional contact with it (Figure 11-8). After other careful observations and measurements to rule out alternate explanations, Rumford concluded that the caloric theory could not account for his observations. If caloric was a conserved substance, there had to be a finite supply of it. But the heating of a cannon could go on without limit so long as the horse powering the boring mechanism kept moving, and one could continue to boil water in the test tube as long as one kept turning the crank. Thus, he wrote,

> *anything which any insulated body, or system of bodies, can continue to furnish without limitation, cannot possibly be a material substance; and it seems to me to be extremely difficult, if not quite impossible, to form any distinct idea of anything capable of being excited and communicated in the manner in which heat was excited and communicated, except it be MOTION.*[1]

But what aspect of motion? The experiments of Englishman James Prescott Joule (1818–1889), a brewer by profession and a scientist by avocation, provided a substantial part of the answer. Figure 11-9 shows an experimental arrangement

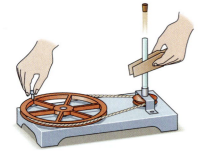

Figure 11-8 One of Rumford's demonstrations that a mechanical energy input produces a temperature increase. (Based on A. Privat-Deschanel, *Traité Elementaire de Physique,* reproduced in M. Wilson, *American Science and Invention,* New York: Simon & Schuster, 1959, p. 29)

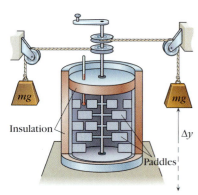

Figure 11-9 Joule's apparatus for measuring the heat equivalent of mechanical energy.

[1] Quoted in F. T. Bonner and M. Phillips, *Principles of Physical Science,* (Addison-Wesley, Reading, MA, 1957), p. 232.

with the same principal features as one set up by Joule. By means of gears, the descending weights make the paddlewheel rotate horizontally in the water. Because of the resistive forces exerted by the water on the paddles, the weights quickly level off to a rather slow terminal velocity, so that the descending weights gain very little kinetic energy. Fixed paddles break up the water's flow to keep the water from gaining much rotational kinetic energy. As the paddles turn against the friction-like resistance, the water temperature rises. (We will address this experiment quantitatively in Example 11-6.)

By this and related experiments, Joule was able to show that the same amount of heating—increasing the temperature of a given mass of water by a fixed number of degrees—was always associated with the same number of pounds falling a certain distance. But a force exerted through a distance gives us work—an energy input. In modern language, Joule had found that there is a fixed equivalence between the amount of work done by Earth (i.e., the gravitational potential energy that is lost) and the amount of heat transferred to the water. This is called the **mechanical equivalent of heat:** This much energy input always means that much heating. So once we establish units of heat, we can convert them to units of energy, much as we can convert meters to centimeters because both are units of distance. In short, *a heat transfer is an energy transfer,* and an amount of heat is an amount of energy being transferred.

Important note on language: In physics, *heat* refers only to the energy being transferred, never to the energy when it is in one object or another, which can be energy of various kinds. Therefore, we never speak of an object as containing heat.

11-4 Units

We saw in Section 11-2 that the amount of heat transferred to a body depends on

- the size of the temperature change the body undergoes,
- the mass of the body, and
- the material of which the body is made.

To take all three of these factors into account, we define a unit of heat as the amount of heat required to produce a particular increase in a particular mass of a standard material. One common unit defined in this way is the calorie:

A **calorie** (cal) is the amount of heat that must be transferred to 1 gram of water to raise its temperature by 1°C (or by 1 K).

This amount of heat actually differs slightly at different temperatures; our definition neglects those differences. The calorie is not an SI unit. But because an amount of heat is an amount of energy, the calorie can be converted to joules, the SI units for energy:

$$1 \text{ calorie} = 4.186 \text{ joules} \tag{11-5}$$

However, in heating contexts it remains common to express energy in calories or in other similarly defined units. Another such unit, the Calorie with a capital C (and abbreviated C) is actually a kilocalorie,

$$1 \text{ Calorie (C)} = 1 \text{ kcal} = 10^3 \text{ calories (cal)} \tag{11-6}$$

and is the unit commonly used in discussing the energy value of foods. (By the way, you don't really "burn calories"; you release calories of energy in the slow oxidation or "burning" of food molecules.) In Great Britain, the **British thermal unit (BTU)** was defined as the amount of energy that must be transferred to one pound of water to raise its temperature by 1°F (1 BTU = 252 cal = 1060 J). No longer used in Great Britain, the BTU remains the preferred unit in the SI-resistant world of heating contractors and home improvement centers in the United States.

Example 11-4 *Can a Little Breakfast Go a Long Way?*

The nutritional information on a box of Krispy Flakes, a typical unsweetened breakfast cereal, tells us that a single serving with skim milk has 150 Calories. What if your body were 100% efficient in metabolizing this breakfast and directing the output energy into pushing a wheelbarrow? How far (in meters) would it let you push the wheelbarrow, assuming you need to exert an average force of 200 N to keep it in motion?

Solution

Choice of approach. 100% efficiency tells us that the energy input (from the food) equals the energy output—the work you do to push the wheelbarrow. We must convert Calories to joules (N·m) to obtain a distance in meters.

What we know/what we don't.

$$W = 150 \text{ Calories} = ? \text{ joules}$$

$$F = 200 \text{ N} \qquad \Delta s = ?$$

The mathematical solution. We first convert units using the equivalences in Equations 11-5 and 11-6.

$$W = 150 \text{ Calories} = 150 \times 10^3 \text{ calories} \times \frac{4.186 \text{ joules}}{1 \text{ calorie}} = 6.28 \times 10^5 \text{ J}$$

Because $W = F\Delta s$,

$$\Delta s = \frac{W}{F} = \frac{6.28 \times 10^5 \text{ J}}{200 \text{ N}} = \textbf{3140 m} \text{ (about 2 miles!)}$$

Making sense of the results. If that seems like a lot to get out of one measly bowl of cereal, it is. Your body would expend energy to walk that distance even if you were not pushing anything. All the while, your body is also expending energy to maintain your body temperature. A person walking at a moderate speed of about 1.4 m/s (roughly 3 mi/h) has a total metabolic rate of about 280 J/s. Then in one second the person uses up about 280 J and walks 1.4 m. The amount of energy used to walk each meter is $\frac{280 \text{ J}}{1.4 \text{ m}} = 200$ J/m. To walk 3140 m using energy at this rate, you would use 3140 m × 200 J/m = 6.28×10^5 J, all the energy you had gotten from the cereal. There would be no energy left over to push the wheelbarrow!

So what was wrong with our calculation? We did the math correctly, but based it on an unreal assumption. In fact, you *cannot* direct all the energy from the cereal into work on the wheelbarrow. In physics, being able to manipulate equations isn't enough. You need to know when they apply.

◆ Related homework: Problems 11-20, 11-21, and 11-22.

11-5 Temperature Differences and Heat Transfer (Quantitative)

We define a unit of heat as the amount of energy needed to raise the temperature of each unit of mass of standard material by one degree. Then the total number of units of heat required to raise a body's temperature must be proportional to the number of grams (the mass m) being warmed and the number of degrees by which the temperature changes ($\Delta T = T_f - T_i$). If Q represents the amount of energy transferred to the body, then we can state this relationship as follows:

the **heat transfer associated with a temperature change** is

$$Q = cm\,\Delta T = cm(T_f - T_i) \qquad (11\text{-}7)$$

Heat transferred depends on $\begin{cases} \text{the material of which the body is made }(c) \\ \text{the mass of the body }(m) \\ \text{the size of the temperature change }(\Delta T) \end{cases}$

The constant of proportionality c depends on the material constituting the body. So the three factors that together determine the amount of heat transferred to a body are all incorporated into this mathematical statement of dependence.

The quantity c tells us how much energy must be transferred to each unit of mass for each degree the temperature is raised. It is called the **specific heat.** (Sometimes the term *specific heat capacity* is used but is misleading because it seems to suggest an ability of a material to hold heat, but heat is only the energy in transit from one object to another, never the energy within an object.) The word *specific* here indicates that c is a property of the *species* of material itself, and does not depend on the mass of the sample—it is the same for each kg of this material no matter how many kg there are. The SI units of specific heat are joules per kilogram per kelvin ($[\text{J/kg}]/\text{K} = \text{J}/[\text{kg}\cdot\text{K}]$). Because we are dealing only with changes in temperature, K and °C may be used interchangeably. Table 11-1 gives the specific heats of some common materials.

Table 11-1 Specific Heats of Common Materials at or Near Room Temperature at Constant Pressure

Material	Specific Heat (in J/[kg · K] or J/[kg · °C])	Specific Heat (in cal/[g · °C] or kcal/[kg · K])
Solids		
Ice (at −5°C)	2090	0.53
Aluminum	903	0.215
Rocks, bricks, dry dirt	800	0.2
Glass	400–800	0.1–0.2
Iron	450	0.11
Copper	386	0.0923
Silver	236	0.0564
Tungsten	134	0.0321
Gold	129	0.031
Liquids		
Water	4186	1.00
Gasoline	2100	0.5
Gases (at constant pressure)		
Hydrogen	14 175	3.42
Water vapor	2000	0.48
Dry air	1000	0.24

Because specific heats vary slightly with temperature, we list their values at 25°C where possible. (Obviously, the value for ice will have to be at a lower temperature, but not too much below the freezing point.) For the calculations required in this book, you may neglect changes from the values listed here unless you are explicitly told to do otherwise. The amount of heat required to raise the temperature of a sample of material depends on whether you maintain the sample at constant volume or at constant pressure during the transfer. The values here assume constant pressure.

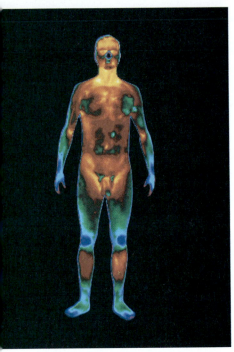

Thermograms or thermographs are infrared photographs. Because the infrared radiation that an object gives off depends on its temperature, a photograph like this is a temperature map of the body, and can alert medical personnel to abnormalities resulting in locally elevated temperatures.

Example 11-5 *Warm Bodies*

Forensic pathologists, who investigate medical evidence for the cause and circumstances of death, know that when an adult dies it may take about $1\frac{1}{2}$ hours (depending on body type, clothing, etc.) for the body to cool down to room temperature. In a living person, the energy that would be lost in cooling is continually replaced by energy released in the chemical reactions that we collectively call **metabolism.** Estimate the total amount of energy that must be released by the chemical reactions in all the cells of your body in $1\frac{1}{2}$ hours to maintain your normal body temperature in a 22°C room.

Solution

Choice of approach. Apart from sign, the input from metabolism must be equal in absolute value to the output from your body. We can use Equation 11-7 to find the energy output, the heat transferred from your body to its surroundings. (Remember that $T_f - T_i$ is the change in *temperature;* although we know the time interval $\Delta t = 1\frac{1}{2}$ hours, we have no need for it in our calculation.) Because your body mass is mostly water, the specific heat of water (Table 11-1) provides a good approximation to the specific heat of your body. (Water has the highest specific heat in the table, so you might reasonably guess that the non-water parts of the body lower its overall specific heat. In fact, a more accurate figure for the body's specific heat is 3470 J/(kg·°C)—lower, but still in the same ballpark.) For our sample calculation, we'll assume a body weight of about 150 lb.

What we know/what we don't.

$$T_i = \text{normal body temperature} = 37°C \qquad T_f = 22°C$$

$$1 \text{ kg weighs about } 2.2 \text{ lb,}$$

$$\text{so } m \approx 150 \text{ lb.} \times 1 \text{ kg}/2.2 \text{ lb} = 68 \text{ kg}$$

$$c_{\text{water}} = 4190 \text{ J/(kg·°C)}$$

The mathematical solution. The heat loss to the surrounding air is

$$Q = cm(T_f - T_i) \approx (4190 \text{ J/(kg·°C)})(68 \text{ kg})(22°C - 37°C) = -4.3 \times 10^6 \text{ J}$$

so the compensating heat input would have to be **+4.3 × 10⁶ J**

Making sense of the results. The body's energy output in $1\frac{1}{2}$ hours is over 4 million joules. For a comparison, what would be the total energy output in $1\frac{1}{2}$ hours from a 100 W bulb? From a 900 W microwave oven?

◆ Related homework: Problems 11-25, 11-26, and 11-27.

Now that we can treat heat transfers quantitatively, work through Example 11-6: Joule's Apparatus.

Example 11-6 *Joule's Apparatus*

For a guided interactive solution, go to Web Example 11-6 at
www.wiley.com/college/touger

In Joule's apparatus (refer to Figure 11-9), each of the weights weighed 4 lb and descended at a speed of about 1 ft/s. The total descent was 36 ft. After each descent, the apparatus was rewound and the weights were permitted to descend again. Joule measured the cumulative temperature increase after 16 descents. Let us assume that his container held 7 lb of water.

a. What fraction of the work done on one of the weights is turned into kinetic energy during each descent?

b. What increase in water temperature would Joule have detected?

Brief Solution

Choice of approach. Here we work as Joule did in British units. In part **a**, the work done on one weight is the product of the gravitational force mg on it and the distance Δy that it falls. Kinetic energy is gained as the weight quickly goes from rest to a roughly constant terminal speed of 1 ft/s.

In **b** we multiply by $2 \times 16 = 32$ to get the total work done on two such weights during 16 descents. This work equals the heat Q transferred to the water. We can then use Equation 11-7 to find the temperature *change* $T_f - T_i$. In the British system the specific heat of water, 1 BTU/(°F · lb) is expressed per pound, so we must substitute the weight in pounds for the mass in Equation 11-7.

What we know/what we don't.

a. $mg = 4$ lb $\qquad g = 32$ ft/s^2 $\qquad y = 36$ ft

$\qquad v_i = 0 \qquad v_f = 1$ ft/s $\qquad \dfrac{\Delta KE}{W} = ?$

b. $m_{water} = 7$ lb (actually a weight) $\qquad N = 2 \times 16 = 32$

$\qquad c_{water} = 1 \dfrac{BTU}{°F \cdot lb} = \dfrac{778.26 \text{ ft} \cdot lb}{°F \cdot lb} \qquad T_f - T_i = ?$

The mathematical solution.

a. The work $W = mg\Delta y = (4 \text{ lb})(36 \text{ ft}) = 144$ ft · lb.

The mass we need to calculate the kinetic energy is

$$m = \frac{mg}{g} = \frac{4 \text{ lb}}{32 \text{ ft/s}^2} = 0.125 \frac{lb}{ft/s^2} \text{ (called 0.125 } \textit{slugs} \text{ in the British system)}$$

The increase in kinetic energy (since $v_i = 0$) is

$$\Delta KE = KE_f - KE_i = KE_f = \tfrac{1}{2} mv_f^2 = \tfrac{1}{2}\left(0.125 \frac{lb}{ft/s^2}\right)(1 \text{ ft/s})^2 = 0.06 \text{ ft} \cdot lb.$$

Then $\qquad \dfrac{\Delta KE}{W} = \dfrac{0.06 \text{ ft} \cdot lb}{144 \text{ ft} \cdot lb} = \mathbf{4 \times 10^{-4}}$ (or $4 \times 10^{-4} \times 100\% = 0.04\%$)

As we assumed, the fraction of the work that goes into the KE of the weight is negligible.

b. The total work done on two weights descending 16 times is

$$NW = 32(144 \text{ ft} \cdot lb) = 4608 \text{ ft} \cdot lb$$

which we set equal to Q, the equivalent heat transfer, in $Q = cm(T_f - T_i)$. Solving for the temperature increase, we get

$$T_f - T_i = \frac{Q}{c_{water} m_{water}} = \frac{4608 \text{ ft} \cdot lb}{(778.26 \text{ ft} \cdot lb/[°F \cdot lb])(7 \text{ lb})} = \mathbf{0.84°F}$$

a small but measurable increase.

♦ Related homework: Problem 11-28.

♦**CALORIMETRY** When heat exchanges take place in a well-insulated calorimeter over a short period of time, we can neglect heat transfer into or out of the calorimeter. Conservation of energy then tells us that any energy lost by something in the calorimeter must be gained by something else in the calorimeter, and the total of all such gains (+) and losses (−) must be zero.

Within a perfectly insulated environment,

$$\Sigma Q = 0 \qquad (11\text{-}8)$$

When all the heat transfers are associated with temperature changes, this becomes (combining with Equation 11-7)

$$\Sigma cm(T_f - T_i) = 0$$

$$\Sigma \left(\begin{array}{c} \text{Heat gains} \\ = \text{energy inputs} \\ Q \text{ positive} \\ T_f > T_i \end{array} \quad \text{and} \quad \begin{array}{c} \text{Heat losses} \\ = \text{energy outputs} \\ Q \text{ negative} \\ T_f < T_i \end{array} \right) = 0$$

The temperature within the calorimeter can be monitored by a liquid thermometer squeezed through a tight-fitting hole in the lid or by an electronic temperature-measuring device, such as a thermocouple or a thermistor that would require only fine wires passing through the lid. A thin coating on the wires insulates them electrically but not thermally.

Example 11-7 *Finding the Specific Heat of Aluminum*

The following somewhat inefficient method can be used to find the specific heat of aluminum. Boil a load of aluminum beads in water until everything reaches the same temperature. Use a wire mesh strainer to quickly transfer some of the aluminum beads into a calorimeter containing 0.200 kg of water at 20.0°C. The final temperature to which the calorimeter levels off depends on the total mass of the beads you transfer. Suppose you keep repeating this with different masses of aluminum beads until you manage to get a final temperature of 60.0°C, so that the temperature drop of the aluminum exactly equals the temperature gain of the water. You measure the mass of aluminum beads that give you this result, and get a value of 0.944 kg. What is the specific heat of the aluminum?

Solution

Choice of approach. The aluminum beads start out at the boiling temperature of water, 100°C. The heat they lose in the calorimeter must be gained by the water. On the left side of Equation 11-8, you must list a heat loss or gain for each substance or object in the calorimeter.

What we know/what we don't.

Water: $c_{water} = 4190 \text{ J/kg} \cdot °C$ $m_{water} = 0.200 \text{ kg}$ $T_f = 60.0°C$ $T_i = 20.0°C$

Al beads: $c_{Al} = ?$ $m_{Al} = 0.944 \text{ kg}$ $T_f = 60.0°C$ $T_i = 100.0°C$

The mathematical solution. Applying Equation 11-8,

$$c_{water} m_{water} \Delta T_{water} + c_{Al} m_{Al} \Delta T_{Al} = 0$$

where $\quad \Delta T_{water} = 60.0°C - 20.0°C = 40.0°C$ (a gain)

and $\quad \Delta T_{Al} = 60.0°C - 100.0°C = -40.0°C$ (a loss)

so $\quad \Delta T_{water} = -\Delta T_{Al}$

Substituting this into the calorimetry equation and solving for c_{Al}, we get

$$c_{Al} = \frac{m_{water}}{m_{Al}} c_{water} = \frac{0.200 \text{ kg}}{0.944 \text{ kg}} (4190 \text{ J/kg} \cdot °C) = \textbf{888 J/kg} \cdot °C$$

Making sense of the results. Our calculation shows that the specific heat of aluminum is less than a quarter of that of water. This explains why it took more than four times as much aluminum as water to lose the amount of heat that the water gained. But we also get a result that is lower than the value of c_{Al} in Table 11-1. **STOP&Think** Why? ◆

No calorimeter is a perfect insulator. So some of the heat lost by the aluminum is transferred to the container rather than the water. We therefore have to increase the mass of aluminum somewhat for the water to gain enough heat to warm up to 60.0°C. Because we used $c_{Al} = \frac{m_{water}}{m_{Al}} c_{water}$ for our calculation, a larger mass of aluminum than in ideal circumstances would give us a smaller value of c_{Al}.

◆ Related homework: Problems 11-29, 11-30, 11-31, and 11-32.

The next two examples provide further practice in applying the ideas of calorimetry.

Example 11-8 *A Simple Calorimetry Experiment*

For a guided interactive solution, go to Web Example 11-8 at www.wiley.com/college/touger

Suppose you again boil a load of aluminum beads in water until everything reaches the same temperature. You now transfer some of the aluminum beads into a calorimeter containing 0.200 kg of water at 20.4°C. You observe that the temperature of the calorimeter levels off to a final temperature of 27.8°C. Use your data to determine the total mass of the beads you transferred.

Brief Solution

Choice of approach. As in Example 11-7, the heat the aluminum beads lose in the calorimeter as they cool from 100°C must be gained by the water. Again the sum of the gain (+) and the loss (−) must be zero (Equation 11-8).

What we know/what we don't. (c values from Table 11-1)

Water: $c_{water} = 4190 \text{ J/kg} \cdot \text{°C}$ $m_{water} = 0.200 \text{ kg}$ $T_f = 27.8\text{°C}$ $T_i = 20.4\text{°C}$

Al beads: $c_{Al} = 903 \text{ J/kg} \cdot \text{°C}$ $m_{Al} = ?$ $T_f = 27.8\text{°C}$ $T_i = 100\text{°C}$

The mathematical solution. Applying Equation 11-8, we get

$$c_{water}m_{water}(T_f - T_i)_{water} + c_{Al}m_{Al}(T_f - T_i)_{Al} = 0$$

and we solve for m_{Al}.

$$c_{Al}m_{Al}(T_f - T_i)_{Al} = -c_{water}m_{water}(T_f - T_i)_{water}$$

$$m_{Al} = -\frac{c_{water}m_{water}(T_f - T_i)_{water}}{c_{Al}(T_f - T_i)_{Al}}$$

$$= -\frac{(4190 \text{ J/kg} \cdot \text{°C})(0.200 \text{ kg})(27.8\text{°C} - 20.4\text{°C})}{(903 \text{ J/[kg} \cdot \text{°C]})(27.8\text{°C} - 100\text{°C})}$$

$$= \mathbf{9.51 \times 10^{-2} \text{ kg}} \text{ (or 95.1 g)}$$

Making sense of the results. In this experimental situation, you can easily check the result by measuring the mass of the aluminum beads. More usually, a known mass of material is added to the water, and the set-up is used to determine the specific heat of the material.

◆ Related homework: Problem 11-33.

Example 11-9 *Your Cocoa's Getting Cold Again*

For a guided interactive solution, go to Web Example 11-9 at www.wiley.com/college/touger

We now return to the cup of cocoa cooling down to room temperature, and attempt to consider it quantitatively. Suppose that after you pour 0.25 kg of cocoa into a paper hot-drink cup of negligible mass, it has a temperature of 90°C. You then put the cup into an enclosed, well-insulated space filled with nothing but air at a room temperature of 20°C. Estimate the final, common temperature of the cocoa and air if the enclosure is

a. a thick-walled and lidded Styrofoam picnic carrier having dimensions 0.200 m × 0.200 m × 0.300 m;

b. a square room of your home, 4.00 m on a side with a 2.50 m high ceiling. The density of dry air at 20°C is 1.29 kg/m³.

Brief Solution

Choice of approach. With perfect insulation and no other objects present, conservation of energy tells us that the heat loss of the cocoa and the heat input to the air are equal but opposite in sign, so their sum is zero. We multiply density by volume to find the mass of air in each case.

What we know/what we don't.

cocoa: $c_{cocoa} \approx c_{water} = 4190$ J/kg·°C $m_{cocoa} = 0.250$ kg $T_f = ?$ $T_i = 90$°C

air: $c_{air} \approx 1000$ J/kg·°C $\rho_{air} = 1.29$ kg/m³ $m_{air} = ?$ $T_f = ?$ $T_i = 20$°C

a. $V_{air} = 0.200$ m × 0.200 m × 0.300 m = 0.0120 m³
b. $V_{air} = 4.00$ m × 4.00 m × 2.50 m = 40.0 m³

Anticipating the results. Because there is a much greater mass of air in **b,** we should expect that the air temperature will undergo a much smaller increase.

The mathematical solution.

In **a,** $m_{air} = \rho_{air}V_{air} = (1.29$ kg/m³$)(0.012$ m³$) = 0.0155$ kg

In **b,** $m_{air} = \rho_{air}V_{air} = (1.29$ kg/m³$)(40.0$ m³$) = 51.6$ kg

Applying Equation 11-8,

$$c_{cocoa}m_{cocoa}(T_f - T_i)_{cocoa} + c_{air}m_{air}(T_f - T_i)_{air} = 0$$

To solve, first we multiply out, then we group together the terms involving the unknown T_f:

$$c_{cocoa}m_{cocoa}T_f - c_{cocoa}m_{cocoa}T_{i,\,cocoa} + c_{air}m_{air}T_f - c_{air}m_{air}T_{i,\,air} = 0$$

$$T_f(c_{cocoa}m_{cocoa} + c_{air}m_{air}) - c_{cocoa}m_{cocoa}T_{i,\,cocoa} - c_{air}m_{air}T_{i,\,air} = 0$$

Then $$T_f = \frac{(c_{cocoa}m_{cocoa}T_{i,\,cocoa} + c_{air}m_{air}T_{i,\,air})}{c_{cocoa}m_{cocoa} + c_{air}m_{air}}$$

For **a,**
$$T_f = \frac{(4190\text{ J/[kg·°C]})(0.250\text{ kg})(90°C) + (1000\text{ J/[kg·°C]})(0.0155\text{ kg})(20°C)}{(4190\text{ J/[kg·°C]})(0.250\text{ kg}) + (1000\text{ J/[kg·°C]})(0.0155\text{ kg})}$$

= **88.4°C**

For **b,** the calculation is identical except that it uses the value $m_{air} = 51.6$ kg for the mass of air in the room:

$$T_f = \frac{(4190\text{ J/[kg·°C]})(0.250\text{ kg})(90°C) + (1000\text{ J/[kg·°C]})(51.6\text{ kg})(20°C)}{(4190\text{ J/[kg·°C]})(0.250\text{ kg}) + (1000\text{ J/[kg·°C]})(51.6\text{ kg})}$$

= **22.2°C**

Making sense of the results. How realistic are the values we've obtained? As expected, the temperature of the cocoa drops much closer to the initial room temperature when put into a whole roomful of air. Even so, we would not expect the cocoa to raise the room temperature by over 2°C, as it did according to our calculation. It would if all the energy lost by the cocoa were transferred to the air in the room, but is that really what happens? In fact, most of this energy is transferred to or through the walls, floor, and ceiling, and to the objects in the room. See the Web version of this example for a fuller discussion.

◆ Related homework: Problems 11-34 and 11-35.

11-6 Changes of State and Heat Transfer

Case 11-1 ◆ *The Case of the Melted Mothballs*

Let's consider a simple experiment. We heat a test tube containing para-dichlorobenzene (mothballs) in a beaker of water (Figure 11-10a) until everything reaches a temperature just below the boiling point of water. We then separate the test tube and beaker (with their respective contents), place a thermometer in each, and take temperature readings as they cool down to room temperature. The results are graphed in Figures 11-10b and 11-10c.

Not surprisingly, the water temperature drops continuously, sharply at first, then leveling off as it approaches room temperature (20°C). The mothballs also level off at room temperature, but their temperature drop is *not* continuous. They first level off at 53°C, plausible for a midsummer day in Death Valley but hardly room temperature. There is a *plateau* on the graph as the mothballs remain at this elevated temperature for several minutes before starting to drop again. Along with this difference between the two substances, we observe another difference. Above 53°C, para-

dichlorobenzene is a clear liquid. At 53°C it crystallizes or solidifies. During solidification there is no temperature change. Water, in contrast, remains liquid down to room temperature.

Both materials continuously transfer energy to their surroundings as they cool. But during the plateau, when the mothballs aren't cooling, are they losing energy? We would expect an energy transfer from the warmer mothballs to their cooler surroundings, but there is no drop in temperature providing evidence of such an energy loss.

We resolve the dilemma with a measurement. We repeat the experiment, this time putting the test tube of mothballs in a calorimeter containing room-temperature water. We can now monitor the temperature of the surrounding water as well as the mothballs. With careful measurement, we can observe that the water temperature in fact continues to rise even after the mothball temperature levels off. While the para-dichlorobenzene temperature remains constant, the water is gaining energy. From where? The mothballs are the only possibility. Therefore the mothballs must be *losing energy* to the surrounding water.

In discussing chemical energy in Section 6-1, we noted that an energy input is required to break bonds, and that energy is released as new bonds form. Intermolecular bonds are broken when a substance melts and form again as the substance solidifies. Thus an energy input from the surroundings of a substance is necessary for a change of state from solid to liquid, and the same amount of energy is lost to the surroundings during the reverse change of state. In this case, the solidifying of the mothballs is itself evidence of an energy loss.

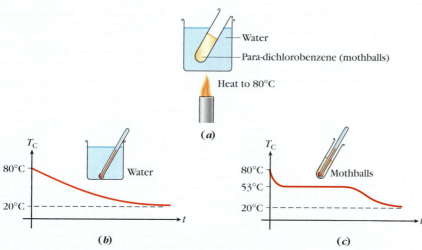

Figure 11-10 **When water and mothballs are warmed to 80°C, their cooling curves as they drop back to room temperature are strikingly different.**

In general, there is a heat transfer associated not only with each change in temperature but *with each change of state* as well (between solid and liquid or between liquid and gas). Just as for temperature changes, the amount of energy involved depends both on the kind of material changing state and its mass. The amount of heat gained or lost by each unit of mass when a body undergoes a change of state is called its **heat of transformation.** We denote it by L because historically, it was called the *latent heat.* Because L is really the amount of energy transferred as a unit of mass changes state, it might better be called the energy of transformation, but this is not usual.

> In general, the **heat transfer (energy transfer) associated with a change of state** is
>
> $$Q = (\pm)Lm \qquad (11\text{-}9)$$
>
> $+$ if transferred to a body, $-$ if transferred from a body

We determine the sign of Q and the specific labeling of L as follows:

Change of State	Change of Energy per Unit of Mass (Q/m, in joules/kilogram)	Sign of Q
Melting: solid to liquid	L_{SL}	$+$
Freezing or solidifying: liquid to solid	L_{SL}	$-$
Boiling: liquid to gas	L_{LG}	$+$
Condensation: gas to liquid	L_{LG}	$-$

L_{SL} and L_{LG} have historically been called the *latent heat of fusion* and the *latent heat of vaporization,* respectively. These terms are still in fairly common use by physicists and engineers, although *fusion* (when it means solidifying) and *vaporization* (when it means evaporation) are no longer everyday language. The term *latent heat* is a carryover from the rejected caloric theory. For this reason we prefer to speak of *heat of transformation.* Here we shall label the heat of transformation by the two states between which the transformation occurs (e.g., LG for "between liquid and gas").

Table 11-2 lists the heats of transformation associated with changes of state of some representative materials, and the temperatures at which the changes occur.

Table 11-2 Heats of Transformation of Common Materials

Material	Melting or Freezing		Boiling or Condensation	
	Melting Temperature (°C)	Heat of Transformation L_{SL} (J/kg)	Boiling Temperature (°C)	Heat of Transformation L_{LG} (J/kg)
Iron	1530	2.721×10^5	2500	6.364×10^6
Copper	1083	1.34×10^5	2600	
Silver	962	1.05×10^5	2050	2.336×10^6
Aluminum	658.7	3.984×10^5	2300	10.5×10^6
Water	0.00	3.335×10^5	100.0	2.256×10^6
Mercury	-39	0.118×10^5	357	0.272×10^6
Ethyl alcohol	-114	1.04×10^5	78	0.854×10^6
Nitrogen	-210.1	0.257×10^5	-195.81	0.201×10^6
Oxygen	-218.8	0.138×10^5	-182.97	0.213×10^6
Hydrogen	-259.3	0.586×10^5	-252.89	0.452×10^6
Helium	(He has no solid state at standard pressure.)		-268.93	0.02×10^6

Note: Some materials go directly from a solid to a gaseous state without becoming liquid. Such materials are said to *sublimate.* Dry ice, the solid form of carbon dioxide, sublimates to become carbon dioxide gas at about $-79°C$. The heat of transformation, which we would denote by L_{SG}, is sometimes called the *heat of sublimation.*

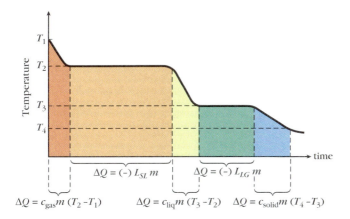

Figure 11-11 Cooling curve of a gas that first condenses (liquefies) then freezes solid. The heat transfer terms that would appear on the left side of Equation 11-10 or 11-11 are shown for each part of the curve.

In a perfectly insulated environment, Equation 11-8 now becomes

$$\Sigma cm(T_f - T_i) + \Sigma(\pm)Lm = 0 \qquad (11\text{-}10)$$

terms associated with temperature changes
positive when $T_f > T_i$ (warming)
negative when $T_f < T_i$ (cooling)

terms associated with phase changes
+ for melting, evaporating
− for freezing, condensing

(See Figure 11-11.) When there is imperfect insulation or none at all, the total amount of heat that a system gains from (or loses to, if negative) its environment,

$$\Sigma cm(T_f - T_i) + \Sigma(\pm)Lm = \Sigma Q \qquad (11\text{-}11)$$

is *not* zero (see Figure 11-12).

Figure 11-12 Cooling by evaporation. Because of the high heat of transformation L_{LG} of water, the evaporation of sweat requires a transfer of energy from your body surface, which cools your body. Because dogs don't sweat, they must hang out their tongues in hot weather to produce the same effect.

Example 11-10 *A Steamy Situation*

4.00 kg of hot iron fragments are plunged into a well-insulated vat containing 2.50 kg of warm water, initially at 45°C. The mixture is stirred regularly to keep the heating as uniform as possible. If the water reaches boiling temperature and 0.20 kg of the water turns to steam, what was the initial temperature of the iron?

Solution

Choice of approach. There is no heat transfer into or out of a perfectly insulated system. Under these conditions, and only under these conditions, Equation 11-10 tells us that the total of all heat transfers to and from the materials within the system must be zero. These conditions hold true approximately for a well-insulated vat. Applying Equation 11-10 means identifying the terms in the sum. For each material present, there must be a term for each temperature change and each change of state:

- The energy the water must gain to raise its temperature from 45°C to 100°C.
- The additional energy gain required to convert part of the 100°C water to steam at 100°C (recall from Case 11-1 that there is no change in temperature during a phase change).
- The energy the iron loses as it drops to 100°C.

Everything must finally reach a common temperature. To the extent that the materials are well mixed so that heating is uniform, all the water must evaporate before its temperature as steam starts rising above 100°C. Then the fact that only part of the water has evaporated tells you that the temperature rise has stopped at 100°C.

EXAMPLE 11-10 *continued*

What we know/what we don't. (from Tables 11-1 and 11-2)

$$c_{Fe} = 450 \text{ J/(kg} \cdot {}°C) \quad c_{water} = 4186 \text{ J/(kg} \cdot {}°C) \quad L_{LG, water} = 2.256 \times 10^6 \text{ J/kg}$$

$$m_{Fe} = 4.00 \text{ kg} \quad m_{water} = 2.50 \text{ kg} \quad m_{steam} = 0.20 \text{ kg}$$

$$T_{i, water} = 45°C \quad T_{f, water} = 100°C \quad T_{i, Fe} = ? \quad T_{f, Fe} = 100°C$$

The mathematical solution. We write out the terms in Equation 11-10 for the three energy transfers listed in *choice of approach*:

$$m_{water}c_{water}(T_{f, water} - T_{i, water}) + m_{steam}L_{LG, water} + m_{Fe}c_{Fe}(T_{f, Fe} - T_{i, Fe}) = 0$$

Solving for $T_{i, Fe}$ gives us

$$T_{f, Fe} - T_{i, Fe} = -\frac{m_{water}c_{water}(T_{f, water} - T_{i, water}) + m_{steam}L_{LG, water}}{m_{Fe}c_{Fe}} \quad \text{or}$$

$$T_{i, Fe} = T_{f, Fe} + \frac{m_{water}c_{water}(T_{f, water} - T_{i, water}) + m_{steam}L_{LG, water}}{m_{Fe}c_{Fe}}$$

$$= 100°C + \frac{(2.50 \text{ kg})(4186 \text{ J/[kg} \cdot {}°C])(100°C - 45°C) + (0.20 \text{ kg})(2.256 \times 10^6 \text{ J/kg})}{(4.00 \text{ kg})(450 \text{ J/[kg} \cdot {}°C])}$$

$$= \mathbf{672°C}$$

Making sense of the results. We should check to make sure that iron is solid at this very high temperature. (It is; the melting point of iron is 1350°C.)

◆ Related homework: Problems 11-41, 11-42, 11-44, and 11-45.

Strictly speaking, the specific heats of materials are not the same at all temperatures. For example, it requires much more energy to raise the temperature of a kilogram of a typical metal by one degree from a warm 50°C to 51°C than to raise it by one degree from a frigid −250°C to −249°C. But from room temperature up, the specific heats of many metals, including iron, vary little with temperature, so the calculation in the last example is reasonably correct. The details of this temperature dependence are due to the atomic and electronic structure of metals. Knowing little more than that, we can readily surmise that the specific heat of a substance will be very different for the solid state than for the liquid state, because the structures are so different in the two states. For example, the specific heat of ice at temperatures slightly below freezing is only about half the specific heat of water at temperatures slightly above. If liquid H_2O in a bucket drops to a subfreezing temperature, its energy loss down to the freezing point (as a liquid) and from the freezing point on down (as a solid) must be separately calculated.

Example 11-11 *A Tale of a Tub*

For a guided interactive solution, go to Web Example 11-11 at www.wiley.com/college/touger

A large washtub containing 1.60 kg of water is left outdoors while the air temperature drops from a daytime high of 8.00°C to a low of −10.00°C. The tub is large enough so that the water forms a shallow layer, and there is little temperature variation between its top surface and the bottom of the tub. How much energy does the water gain or lose during this temperature change?

Brief Solution

Choice of approach. The total ΣQ is no longer zero because the water is not insulated from its surroundings. In Equation 11-11, the sum must now include:

• the energy the water loses in dropping from 8.00°C to the freezing point (exactly 0°C);

- the energy lost by the H_2O during the phase change from liquid to solid;
- the energy it loses as ice as it drops further from 0°C to −10.00°C.

What we know/what we don't. (from Tables 11-1 and 11-2)

$c_{ice} = 2090$ J/(kg · °C) $c_{water} = 4186$ J/(kg · °C) $L_{SL, water} = 2.256 \times 10^6$ J/kg

$m_{ice} = m_{water} = 1.60$ kg $T_{i, water} = 8.00°C$ $T_{f, water} = T_{i, ice} = 0°C$

$T_{f, ice} = −10.00°C$ $\Sigma Q = ?$

The mathematical solution. The terms on the left side of Equation 11-10 must represent the three energy transfers that we identified:

$$m_{water}c_{water}(T_{f, water} − T_{i, water}) − m_{water}L_{SL, water} + m_{ice}c_{ice}(T_{f, ice} − T_{i, ice}) = \Sigma Q$$

The minus sign in the second term denote a loss of energy. Inserting values, we get

$$(1.60 \text{ kg})(4186 \text{ J/[kg · °C]})(0°C − 8.00°C) − (1.60 \text{ kg})(2.256 \times 10^6 \text{ J/kg})$$
$$+ (1.60 \text{ kg})(2090 \text{ J/[kg · °C]})(−10.00°C − 0°C) = \Sigma Q$$
$$= −5.36 \times 10^4 \text{ J} − 3.61 \times 10^6 \text{ J} − 3.34 \times 10^4 \text{ J} = \mathbf{−3.70 \times 10^6 \text{ J}}$$

Making sense of the results. Notice that all three terms in the sum have negative values; the H_2O loses energy during all three stages. The change-of-state term is much larger than the other two. The ice needs to get back that much energy to melt again. That's why in cooler climates ice seems to take forever to melt in the spring.

◆ Related homework: Problems 11-46, 11-47, and 11-48.

11-7 Modes of Heat Transfer

In everyday experience, we find that heat can be transferred to or from a system in three distinct ways.

◆**CONVECTION** In colonial times, people used to warm their beds before going to sleep by filling a long-handled metal pan with hot coals or embers from the fireplace and placing the pan under the covers. If we think of the bed and its covers as one system and the fireplace as another, energy is being transferred from a hotter to a cooler system (which is how we defined heat transfer) because the matter within the pan is literally *carrying* or *conveying* the energy as it moves from one system into the other.

Definition: **Convection** (the noun form of *conveying*) occurs when traveling matter carries the energy along with it.

Convection also occurs (Figure 11-13) when water cooled in the radiator of your car is pumped through the car engine and carries away energy from the hotter engine to reduce the engine temperature. It occurs when a cold air mass from one system (such as southern Canada) moves into another (such as New England), displacing the warmer air that was present. Because the energy carried out by the warm air mass is greater than the energy brought in by the cold air mass, the total heat transfer to New England is negative, and there is cooling. Convection occurs whenever a system warms up or cools down because matter is moving into or out of the system and carrying energy with it.

In your car engine and in many home heating or cooling systems, where the substance that carries the energy (water or air) is forced to move by a pump or fan, the convection that occurs is called **forced convection.** A contrasting

(a)

Hose to engine

Hose from engine

(b)

Figure 11-13 Instances of convection. (*a*) The stream of air over the reclining figure carries energy away from the body. This allows for faster cooling than if the air were stationary and results in a *windchill factor.* (*b*) Water circulating between the car radiator and the engine transfers heat from the engine to the radiator by convection. The radiator in turn transfers heat to the still cooler surroundings, in part by radiation. But air convection caused by the fan also transfers heat away from the radiator. Despite the name *radiator,* more heat is transferred by convection than by radiation.

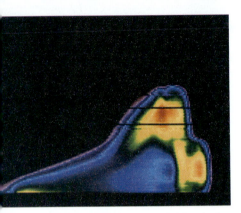

Thermograms can provide vital information about heat transfer. This NASA thermogram shows the heat shielding of the space shuttle during re-entry into the atmosphere.

situation is the energy transfer through the air surrounding baseboard heating or the air surrounding your heated home on a cold day. When the body of air nearest the heater or the home is warmed, it expands, becoming less dense than the adjacent body of cooler air. The buoyant force then causes it to rise, just as an air bubble rises in a denser liquid, and cooler air moves in under it, to be warmed in turn. The convection of energy away from the warming body in these instances is called **natural convection. STOP&Think** Why are heating units installed around baseboards rather than near the ceiling? ◆

◆**CONDUCTION** Here is a contrasting situation. You take a warm bottle of a soft drink and you put it in a chest of ice water. The liquid inside the bottle eventually reaches the same temperature as the liquid outside. But no liquid passes into or out of the bottle, so this is not heating by convection. What is happening here?

On-The-Spot Activity 11-2

Take an all-metal fork or knife, or any strip of metal of comparable dimensions. Wet it and put it in the freezer until a thin layer of frost forms on it. Before removing it from the freezer, prepare a cup of near-boiling water. When you take the frosted metal strip out of the freezer, hold it by one end with two fingers so that you disturb the frost as little as possible, and quickly immerse just the very tip of the other end into the hot water (Figure 11-14a). Carefully observe how the frost melts, paying

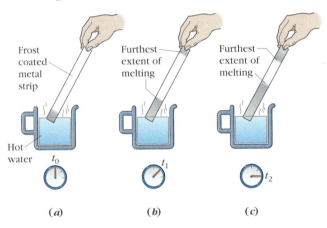

Figure 11-14 Time sequence of observations in On-the-Spot Activity 11-2.

(a)   (b)   (c)

attention to both ends of the metal strip. Does the melting occur everywhere at once, or does it start at one or more specific places and then spread? If it starts in specific places, does it spread at the same speed from each of these places? As you watch the melting spread, what do you think is actually moving from place to place, and what do you think its speed depends on?

In the activity, no metal moves from place to place as the melting spreads from the ends, so we can take our observations as evidence that energy is flowing from the ends. We commonly express this by saying the metal *conducts* heat. Likewise, the glass of the soft drink bottle *conducts* heat from the warm soft drink within to the ice water outside.

Definition: **Conduction** occurs when a material transports energy along its own length or thickness from a warmer region to a cooler one, while the material itself goes nowhere.

This is in contrast to convection, in which the energy is carried along with the material as it moves. We discussed how people in colonial times warmed their beds by convection by carrying coals to the bed in long-handled pans. But once the pan is in the bed, the last stage of heat transfer from the pan contents to the bedding occurs by conduction.

You should also have observed (Figures 11-14*b* and *c*) that the melting spreads more quickly from the hot-water end than from the hand end. The hot water is warmer than your hand, so the evidence suggests that the *rate* of energy transfer from one region to another is greater when you create a larger temperature difference between the regions. **STOP&Think** As heat transfers from one region to another, what happens to the temperature difference between the regions? How does that in turn affect the rate of heat flow between the two regions? Should there be any net heat transfer between two regions that are at the same temperature? ◆

◆**RADIATION** Convection and conduction don't account for all heat transfer. There is a negligible amount of matter in the empty space between Earth and the sun, and only a very few particles on an atomic or subatomic scale actually travel from the sun to Earth. Yet almost all the energy on which life on Earth depends comes or has come from the sun. The sun emits energy in all directions, and we are warmed by the tiny fraction of it that Earth intercepts. Something that travels outward from a source in all directions is said to **radiate** from the source. The word *radiate* applied to energy suggests that the energy travels outward along every possible radius of a sphere centered at the source (Figure 11-15). Usually physicists reserve the term **radiated energy** to refer to energy that spreads out in this way even when no medium is available to conduct it. Visible light, infrared, ultraviolet, radio waves, microwaves, X-rays, and gamma rays all satisfy this definition. For now we will just recognize *radiation* as another mode of heat transfer; in later chapters we will have accumulated enough evidence to inquire into the nature of these various kinds of radiation and what they have in common.

(*a*) The radii of a circle radiate from the center C.

(*b*) The spokes of a wheel radiate from the hub.

(*c*) Light radiates from a bulb.

(*d*) Heat radiates from a hot stove.

(*e*) Sound radiates from a crying child.

Figure 11-15 Things that radiate from a central point or source.

11-8 Some Factors Affecting Heat Transfer

We can define the **rate of heat transfer** (or of energy flow) straightforwardly when the transfer or flow is all in one direction, as in Figure 11-16. We mentally position ourselves at any point P along the flow path and ask, "What additional amount of energy ΔQ passes us during each additional time interval Δt that our clock ticks off?" The rate will then be $\frac{\Delta Q}{\Delta t}$, and its SI units will be joules per second (J/s), or watts (W).

The amount of energy that can pass point P each second depends on how big a cross-sectional area A it has to pass through. If we mentally divide a thick wall into some number n of these rods (Figure 11-16), the same energy will flow through each, and the total rate of flow will be n times as great because the total cross-sectional area is nA. In other words,

the total flow or transfer rate is proportional to the cross-sectional area.

This is true no matter what is flowing, whether energy, water, electric charge, or sewage. But for heat transfer, we have also observed that

the flow rate is *proportional to the temperature difference.*

In addition, the rate of flow will depend on how much thickness d of material it must pass through to get from where the temperature is T_1 to where it is T_2. The greater the thickness, the slower the flow rate. The more layers of insulation you put in your attic, the more slowly heat flows out of your house to melt the snow on your roof. In an analogous situation, if you use two coffee filters instead

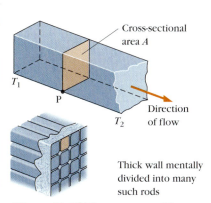

Thick wall mentally divided into many such rods

Figure 11-16 The amount of heat that can flow between the ends of a rod depends on its cross-sectional area.

of one, the rate at which water flows through into your cup will be halved. The flow rate is inversely proportional to the thickness:

The flow rate is proportional to $\frac{1}{d}$.

We now construct an equation that says the heat transfer rate is proportional to all of the things we've highlighted:

$$\frac{\Delta Q}{\Delta t} = KA\left(\frac{1}{d}\right)(T_2 - T_1) \tag{11-12a}$$

| heat flow rate | is proportional to (= a constant times ...) | area | inverse of thickness | temperature difference |

Simplifying the algebra, we can rewrite this as

$$\frac{\Delta Q}{\Delta t} = KA\frac{(T_2 - T_1)}{d} \tag{11-12b}$$

This equation involves joint proportionality. To explore fully what this means, work through WebLink 11-1.

The proportionality constant K depends on what material the heat is flowing through. A metal cup permits energy to flow out of your coffee more quickly than a glass mug, and much more quickly than an equal thickness of Styrofoam. Of these materials, K is greatest for metals, the best conductors of heat, and lowest for Styrofoam, the best insulator. In fact, Styrofoam consists mostly of air, trapped in a little bit of plastic, so its K value is close to that of air (compare polyurethane foam and air in Table 11-3). K is called the **thermal conductivity** of the material. Table 11-3 gives some typical values. The units shown are necessary to make the transfer rate come out in J/s = W when distances are in meters and temperatures in °C.

For **WebLink 11-1:**
Heat Transfer Equation,
go to
www.wiley.com/college/touger

Table 11-3
Thermal Conductivities of Common Materials at Room Temperature

Material	$K\left(\dfrac{W}{m \cdot °C}\right)$
Copper	400
Aluminum	236
Iron	80
Water	0.60
Air (dry)	0.025
Human skin	0.37
Muscle or bone	0.042
Fat	0.021
Glass (typical window)	0.8–1.0
Brick (typical)	0.6
Wood (typical pine)	0.12
Fiberglass	0.040–0.048
Polyurethane foam	0.024

Many values here are approximate, because they come from various sources listing values at slightly different "room" temperatures, and because properties of manufactured materials have considerable variability.

Example 11-12 *A Cool View*

The Joneses have bought a house in which the outer wall of one room faces a pretty lake, but it has no window. The Joneses decide to install a picture window. This wall of the room is 4.5 m wide by 2.5 m high, and contains 0.080-m-thickness polyurethane foam insulation (assume the insulating value of the surface materials is negligible). They can get a bargain price on a window made of 0.0050-m-thick glass, 1.5 m wide by 1.0 m high.

a. What is the rate at which heat will be lost through the window on a winter day when the outside temperature is −10°C and room temperature is kept at 20°C? For these temperatures, the temperature difference between the interior and exterior glass surfaces will be about 0.35°C and the temperature difference between the interior and exterior polyurethane surfaces will be about 29.5°C.

b. Is the window really a bargain? Estimate the additional heating cost of maintaining room temperature under these conditions once the window is in place if you get approximately 10^8 J of energy from burning a gallon of heating oil and heating oil costs roughly $2.00/gallon.

Solution

Choice of approach. For part **a,** we need to look up a typical value for the thermal conductivity of glass so we can apply Equation 11-12 (a or b). For part **b,** we need the thermal conductivity of polyurethane foam so we can find the rate at which heat was being lost before the window was installed. We do not need the dimensions of the whole wall! The window is only replacing a section of the wall the same size as the window. The difference between the loss rate through the window and the loss rate through the section of wall is the additional rate at which heat is being lost. Once we find this additional rate, we multiply by the number of seconds in a day to find the total additional energy loss that must be replaced by burning heating oil. We then calculate the number of gallons required to provide that many joules of energy, and then the cost at $2.00/gallon.

What we know/what we don't.

a. $A = 1.0 \text{ m} \times 1.5 \text{ m} = 1.5 \text{ m}^2$ $d_{glass} = 0.0050 \text{ m}$ $\Delta T_{glass} = -0.35°C$ (a drop)

$K_{glass} = 0.8 \text{ W}/(\text{m} \cdot °C)$ $\left(\dfrac{\Delta Q}{\Delta t}\right)_{glass} = ?$

b. $A = 1.0 \text{ m} \times 1.5 \text{ m} = 1.5 \text{ m}^2$ $d_{foam} = 0.080 \text{ m}$ $\Delta T_{foam} = -29.5°C$ (a drop)

$K_{foam} = 0.024 \text{ W}/(\text{m} \cdot °C)$ $\left(\dfrac{\Delta Q}{\Delta t}\right)_{foam} = ?$ $\left(\dfrac{\Delta Q}{\Delta t}\right)_{foam} - \left(\dfrac{\Delta Q}{\Delta t}\right)_{glass} = ?$

Joules in a day $= ?$ Number of gallons $= \dfrac{? \text{ J}}{10^8 \text{ J/gal}}$ Cost $= ?$

The mathematical solution.

a. $\left(\dfrac{\Delta Q}{\Delta t}\right)_{glass} = K_{glass} A \dfrac{\Delta T_{glass}}{d_{glass}}$

$= (0.8 \text{ W}/[\text{m} \cdot °C])(1.5 \text{ m}^2) \dfrac{(-0.35°C)}{0.0050 \text{ m}} = \mathbf{-84 \text{ W}}$

(negative because energy is being lost)

b. $\left(\dfrac{\Delta Q}{\Delta t}\right)_{foam} = K_{foam} A \dfrac{\Delta T_{foam}}{d_{foam}}$

$= (0.024 \text{ W}/[\text{m} \cdot °C])(1.5 \text{ m}^2) \dfrac{(-29.5°C)}{0.080 \text{ m}} = -13.3 \text{ W}$

$\left(\dfrac{\Delta Q}{\Delta t}\right)_{glass} - \left(\dfrac{\Delta Q}{\Delta t}\right)_{foam} = -84 \text{ W} - (-13.3 \text{ W}) \approx -71 \text{ J/s}$

to two significant figures.
At this rate, the number of additional joules used in a day is

$$71 \text{ J/s} \times \dfrac{3600 \text{ s}}{1 \text{ h}} \times \dfrac{24 \text{ h}}{1 \text{ day}} = 6.1 \times 10^6 \text{ J}$$

If the Joneses get 10^8 J from a gallon of heating oil, this requires

$$\dfrac{6.1 \times 10^6 \text{ J}}{10^8 \text{ J/gal}} = 0.061 \text{ gal}$$

which at a $2.00 a gallon costs 0.061 gal $\times$ $2.00/gal = **$0.122**

Making sense of the results. This is a little under 2 gallons a month, or a cost of a little less than $4 a month. For many families, this is an affordable self-indulgence. But for a home or building with many such windows, the fuel consumption and costs might become appreciable.

◆ Related homework: Problems 11-54, 11-56, 11-58, and 11-59.

Heat transfer and calorimetric considerations also bear on the ways living things adapt to their environment. For an example of this, work through WebLink 11-2.

What role do an elephant's ears play in heat transfer? (See WebLink 11-2)

The little hyrax is a distant relative of the elephant. The two animals have very different external features to compensate for their very different surface-area-to-volume ratios. (See WebLink 11-2)

For **WebLink 11-2:**
The Elephant and the Hyrax, go to
www.wiley.com/college/touger

Some questions remain. If a heat transfer to a system is equivalent to an input of mechanical energy, how does this mechanical energy get transferred, and why does such a transfer occur between bodies of differing temperature? What connection exists between temperature and mechanical energy? In the next chapter we will develop an explanatory model that addresses these questions.

◆ SUMMARY ◆

When objects are at different temperatures, there is ordinarily a **transfer of energy** from the hotter to the colder. The energy being transferred *due to the temperature difference* is what we call **heat,** or equivalently, a **heat transfer.** (Work is also an energy transfer but is due to the exertion of a force, not a temperature difference.)

For the time being, we define **temperature** as what we read on a liquid thermometer, and we establish **temperature scales,** such as the Celsius, Fahrenheit, and Kelvin (absolute) scales.

For conversion of *temperature differences,* we use

$$\frac{\Delta T_F}{\Delta T_C} = \frac{\Delta T_F}{\Delta T_K} = \frac{9}{5} \tag{11-1}$$

but to convert single thermometer readings we use

from °C to °F $\quad T_F = \frac{9}{5}T_C + 32° \tag{11-2}$

from °F to °C $\quad T_C = \frac{5}{9}(T_F - 32°) \tag{11-3}$

from °C to K $\quad T_K = T_C + 273.15 \tag{11-4}$

We consider the heat transfers between objects, initially at different temperatures, that reach **thermal equilibrium.** *Heat and temperature are not the same thing.* The amount of heat that must be transferred depends not only on the temperature increase to be produced but on how much *mass* is being heated. It also depends on the *material* to which it is transferred.

1 **calorie** is the amount of heat that raises 1 g of H_2O by 1°C (or 1 K).

1 **BTU** is the amount of heat that raises 1 lb of H_2O by 1°F.

The food Calorie (C) is a kilocalorie (kcal).

The work of Rumford, Joule, and others provides experimental evidence of the "**mechanical equivalent of heat.**" We therefore take heat transfers to be transfers of mechanical energy, with the equivalence to SI units given by

$$1 \text{ calorie} = 4.1868 \text{ joules} \tag{11-5}$$

$$(1 \text{ BTU} = 252 \text{ cal} = 1060 \text{ J})$$

How we calculate the heat transfer to an object (from an object if negative) of mass m depends on the change the added energy produces:

For a *temperature change:* $\quad Q = cm\Delta T = cm(T_f - T_i) \tag{11-7}$

(object made of material with **specific heat** c)

For a *change of state:* $\qquad Q = (\pm)Lm \tag{11-9}$

(object made of material with **heat of transformation** L_{LG} between liquid and gas, L_{SL} between solid and liquid)

To a system that can gain energy from or lose it to its environment, the total heat transfer ΣQ is

$$\Sigma cm(T_f - T_i) + \Sigma(\pm)Lm = \Sigma Q \tag{11-11}$$

In a perfectly insulated environment (e.g., a **calorimeter**), this becomes

$$\Sigma cm(T_f - T_i) + \Sigma(\pm)Lm = 0 \tag{11-10}$$

terms associated with temperature changes
positive when $T_f > T_i$ (warming)
negative when $T_f < T_i$ (cooling)

terms associated with phase changes
+ for melting, evaporating
− for freezing, condensing

Heat can transfer from a higher to a lower temperature region in three ways:

• **Convection:** traveling matter carries (or *conveys*) the energy with it.

• **Conduction:** material transports energy along its own length but the material itself goes nowhere.

• **Radiation:** energy spreads outward in all directions from the source even when no matter is available to carry or transport it.

The following equation summarizes factors affecting the **rate** of heat transfer by conduction:

$$\frac{\Delta Q}{\Delta t} = KA\left(\frac{1}{d}\right)(T_2 - T_1) \tag{11-12a}$$

heat flow rate | is prop. to (= a constant times …) | area | inverse of thickness | temperature difference

or

$$\frac{\Delta Q}{\Delta t} = KA\frac{(T_2 - T_1)}{d} \tag{11-12b}$$

(through object made of material with **thermal conductivity** K)

✦QUALITATIVE AND QUANTITATIVE PROBLEMS✦
Hands-On Activities and Discussion Questions

The questions and activities in this group are particularly suitable for in-class use.

11-1. *Hands-On Activity.* Take two pennies at room temperature. Touch them with your upper lip to make sure the pennies feel like they are at the same temperature (your upper lip is more sensitive to temperature differences than your fingertips). Secure one of the two pennies to a surface that can take a pounding (you are going to hammer the penny) by first putting a loop of tape down on the surface and then pressing the penny down on the loop of tape. Now take a good, heavy hammer and pound on the secured penny as hard and as rapidly as you can for a full minute. As soon as you are done, without delay, use your lip to compare the temperatures of the two pennies.

a. How do the temperatures of the two pennies compare?

b. What kind of energy was transferred to the hammered penny? What happened to this energy? Does this explain what you observed? If so, how?

11-2. *Discussion Question.* You run the furnace of your home heating system until the inside temperature is 20°C and then shut the system off.

a. Compare what happens on a cool fall evening and on the coldest night of winter. What physics principles account for the difference?

b. Compare what happens when you have insulation blown into your walls and when you do not. What physics principles account for the difference?

c. Compare how long it takes you to heat your home to 20°C before and after you have put on a two-room addition. What physics principles account for the difference?

11-3. *Discussion Question.* When someone spills hot coffee on you, would you say that there is heat transfer by conduction, convection, radiation, or by more than one of these modes? Does the question need to be clarified? Explain.

Review and Practice

Section 11-1 Temperature

11-4. Can you have a liquid thermometer that will give you readings for an unlimited range of temperatures? Explain.

11-5.

a. Suppose we change the graph in Figure 11-3 so that the Celsius scale is on the vertical axis and the Fahrenheit scale is on the horizontal axis. Use the coordinates of the two known points to find the equation of the resulting straight line.

b. Use this new equation to find T_C when $T_F = 104$°F.

c. Use Equation 11-2 to find T_F when $T_C = 40$°C. (Are your results in *b* and *c* consistent with each other?)

11-6. Professor Smart is a superb linguist and a hot tub enthusiast but has never had much interest in mathematics. Touring Iceland, he comes across a sign at a thermal hot spring, which he translates from the Icelandic as "Water temperature 80°." Back in the United States, this would represent a pleasant soak. By a calculation, determine how pleasant or unpleasant soaking in this hot spring would be.

11-7. What is your normal body temperature on the Celsius scale?

11-8. What is your normal body temperature on the Kelvin scale?

11-9. SSM An international resort hotel tells its Fahrenheit-using guests that the water temperature in the swimming pool is 12° above the air temperature. In reporting this to its Celsius-using guests, how much above the air temperature should the hotel say the water temperature is?

11-10. A BBC weather report from London, where the Celsius scale is used, says the temperature there is 15° above normal. How high above normal would this be on the Fahrenheit scale?

11-11. At normal atmospheric pressure, liquid nitrogen boils at 77 K.

a. What is its boiling point on the Celsius scale?

b. What is its boiling point on the Fahrenheit scale?

11-12. In the desert it is not uncommon for the temperature to drop by 25°C between noon and midnight. How big a drop is this on the Kelvin scale? How big a drop is it in °F?

11-13. A column of air trapped under an oil drop in a glass capillary tube (see Figure 11-4) remains at constant pressure as its temperature changes.

a. If it has a height of 2.4 cm at a temperature of 150 K, what will the height be when the temperature is raised to 200 K?

b. If it has a height of 2.4 cm at a temperature of 150°C, what will the height be when the temperature is raised to 200°C?

c. Are your answers to *a* and *b* the same or different? Briefly explain why.

Section 11-2 Temperature Differences and Heat Transfer (Qualitative)

11-14. Describe a situation that demonstrates a significant difference between heat and temperature.

11-15. Would an aluminum can make a good calorimeter? What criteria for a good calorimeter does (or doesn't) it meet?

SSM Solution is in the Student Solutions Manual   **WWW** Solution is at http://www.wiley.com/college/touger

Section 11-3 Heat Transfer and Energy (Qualitative)

11-16.

a. When you leave a cup of hot tea on a kitchen counter, does the air temperature in the kitchen go up by a measurable amount?

b. Describe how you would design a set-up to show that as a hot cup of tea cools down, it transfers heat to the air around it.

11-17.

a. Why did Count Rumford conclude that we cannot think of heat as a substance (caloric) that is contained within an object and can flow from one object to another? What evidence did he cite? How did he reason on the basis of this evidence?

b. Cold object A and hot object B are placed in good thermal contact. Based on Rumford's conclusion, does it make sense to say that after they have remained in contact for a while, object A has more heat than before and object B has less? If not, what might be a better way to describe what happens?

11-18. A deep pot and a shallow pot, both filled with room temperature water, are placed on identical electrical stove burners at identical settings. (The bottoms of the pots are the same size, shape, and material.)

a. If the burners are at the same settings, is energy delivered at a greater rate to one pot than the other? If so, which?

b. Does one pot take longer than the other to reach 80°C? If so, which?

c. Which pot, if either, must have a greater total amount of energy transferred to it to reach 80°C? Briefly explain how this follows from your answers to *a* and *b*.

d. Which pot, if either, has more total internal (mechanical) energy at 80°C?

e. Part *c* of this question refers to an amount of energy in transit from the burner to the pot of water. Part *d* refers to an amount of energy contained within the pot of water. In which of these two circumstances (or both, or neither) can the term *heat* be applied to the amount of energy?

11-19. Distinguish between *heat* and *internal energy*.

Section 11-4 Units

11-20.

a. There are 130 Calories in a cup of 2% low-fat milk. If your body were 100% efficient in directing this energy into lifting, how much weight would it enable you to transfer from the floor to a tabletop 0.80 m above it?

b. At 30% efficiency (which is more realistic), how much weight would it enable you to transfer from the floor to the tabletop?

11-21. SSM WWW

a. If your body were 100% efficient in converting the energy content of food into mechanical energy, approximately how many Calories would you require to stand up from a sitting position? Give a brief justification for the values that you use in arriving at your estimate.

b. Your body is actually about 25–35% efficient at making this conversion. Repeat *a* using a more realistic value for the efficiency.

11-22. Mario is 1.7 m tall and his mass is 60 kg. If Mario's body is 30% efficient in converting the energy content of food to mechanical energy, approximately how many calories does he use when he stands up from a sitting position?

11-23. An American air conditioner manufacturer advertises that one of its air conditioners has a cooling capacity of 7000 BTU/h. Assuming it draws from only one room, how many joules of energy does it remove from the room each second?

11-24. A consumer shopping online finds a company in the United Kingdom advertising an air conditioning unit with a cooling capacity of 2.94 kW (kilowatts). An American unit having the same cooling capacity would be rated at how many BTU/h?

Section 11-5 Temperature Differences and Heat Transfer (Quantitative)

11-25. Calculate the total amount of energy that must be released in one day by the chemical reactions in all the cells of a person weighing 700 N to maintain normal body temperature in a 20°C room. Assume that under the given conditions, if the person died, it would take 1.6 h for his body to cool down to room temperature. Approximate the specific heat of his body to be the same as for water.

11-26. As Molly exercises strenuously for a short time interval, her body temperature increases from 37.0°C to 37.5°C. Assume her body's specific heat is 3470 J/kg · °C. Molly's coach estimates that 10% of the energy released in the chemical reactions in her cells has gone into raising her body temperature. Based on this estimate, what was the total energy output of her cells during this interval?

11-27. SSM A 300 W heating coil is immersed in a well-insulated cup containing 0.25 kg of water. How long will it take the coil to raise the water temperature from 20°C to 50°C?

11-28. Suppose a modified Joule's apparatus contains 2.0 kg of water. Figure 11-9 shows the original apparatus. In the modified apparatus (Figure 11-17), instead of the ropes having weights on the end, the apparatus is turned by exerting a constant horizontal force of 40 N on a heavy cord wrapped around the axle. How far would the cord have to be pulled for the rotating paddles to raise the water temperature by 1.0°C?

Figure 11-17
Problem 11-28

11-29. When a handful of hot glass beads are dropped into an equal mass of unknown liquid in a calorimeter, the temperature of the liquid rises less than the temperature of the glass drops. Assuming the calorimeter is ideal and it contains no other materials, is c_{glass} less than, equal to, or greater than c_{liquid}? If not enough information is given to compare the two values, tell what else is needed. Otherwise, briefly explain your choice.

11-30. When two liquids, A and B, are mixed together in a calorimeter, the temperature of A increases more than the temperature of B decreases. Assuming the calorimeter is ideal and it contains no other materials, is c_A less than, equal to, or greater than c_B? If not enough information is given to compare the two values, tell what else is needed. Otherwise, briefly explain your choice.

11-31.

a. If a 0.50-kg block of aluminum loses 4500 J of energy by heat transfer, by how much does its temperature drop?

b. If the block is submerged in 1.00 kg of water in a well-insulated container, by how much does the water temperature increase?

11-32. When a loosely wound coil of heated copper wire with a mass of 0.100 kg is dropped into a calorimeter containing 0.200 kg of water, the copper temperature drops by 30.00 K and the water temperature goes up by 1.32 K.

a. According to these measurements, what would be the specific heat of copper?

b. How does this value compare with the standard value given in Table 11-1?

11-33. SSM WWW 2.5 kg of water in a car radiator are on the verge of freezing. If antifreeze has a specific heat of 2600 J/(kg · K), how much room temperature (20°C) antifreeze must be added to the water to raise its temperature to 1.5°C?

11-34. 0.15 kg of copper wire is boiled in water for a while, then quickly removed and immersed in 0.50 kg of water in an ideal calorimeter. The water is initially at 15°C. What temperature is eventually reached by each of the materials?

11-35. 0.10 kg of copper at 150°C and 0.20 kg of aluminum at 120°C are added to 0.50 kg of water in a well-insulated container. If the water is initially at 20°C, what final temperature is reached by the materials in the container?

11-36. A 40-g block of iron (A), a 20-g block of iron (B), and a 20-g block of aluminum (C) are boiled in water for a while and then quickly removed and placed in 100 g of water in an ideal calorimeter.

a. Which blocks have the same initial temperature at the instant when materials are added to the calorimeter? (For each part of this question, the choices are: none, A and B, A and C, B and C, all three.)

b. Which blocks undergo the same change in temperature?

c. Which blocks have approximately the same amount of heat transferred to or from them?

d. Which blocks arrive at the same final temperature?

11-37. Usually, when measurements are done in a calorimeter, there is some air in the calorimeter. When we have done calorimeter calculations, such as those in Examples 11-7 and 11-8, we have not considered the heat lost or gained by the air as it changes temperature. Should we be concerned about this? Briefly explain.

Section 11-6 Changes of State and Heat Transfer

11-38. As you read the graph in Figure 11-10c left to right, it tells a story of what is happening to the mothballs. What are the main events in that story?

11-39. SSM The graph in Figure 11-18 shows the temperature plotted against time for a material that is being heated at a constant rate, so that a longer time interval represents more

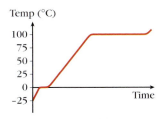

Figure 11-18 Problem 11-39

heat being added. The first plateau occurs during melting, the second during evaporation.

a. Is the heat of transformation for the solid-to-liquid phase change greater than, equal to, or less than for the liquid-to-gas phase change? Briefly explain how you can tell.

b. Is the specific heat of the material in its solid state greater than, equal to, or less than the specific heat in its liquid state? Briefly explain how you can tell.

11-40. In the set-up in Figure 11-10, suppose that the flame is used to keep the water at 80°C as a test tube containing finely crushed mothballs is immersed in it. Sketch a graph plotting the Celsius temperature T_C of the mothballs against time from the instant of immersion until the test tube and its contents have reached the same temperature as the water.

11-41. Equal masses of two materials at different temperatures are brought together in a perfect calorimeter. Is it ever possible that the temperature of only one of the two materials will change? Briefly explain.

11-42. Why are the units for heat of transformation different than those for specific heat capacity?

11-43. When a scoop of very hot metal chips are dropped into an equal mass of unknown liquid in a calorimeter, most of the liquid evaporates. The amount that the temperature of the liquid rises to reach the evaporation point is much less than the temperature of the metal drops. Assuming the calorimeter is ideal and it contains no other materials, is c_{metal} less than, equal to, or greater than c_{liquid}? If not enough information is given to compare the two values, tell what else is needed. Otherwise, briefly explain your choice.

11-44. A 2-kg block of ice is on the verge of melting. By how much 50°C water must the block be surrounded in a perfectly insulated container to melt the block completely?

11-45. Is the amount of energy required to evaporate a liter of water greater than, equal to, or less than the amount required to evaporate a liter of liquid nitrogen if neither substance changes temperature?

11-46. Calculate the energy output when 0.25 kg of water vapor, initially heated to 120°C, condenses to water at 90°C?

11-47. What minimum energy is required to turn a 5-kg block of ice entirely into steam if it is initially at −10°C?

11-48. A 0.030-kg ice cube is removed from a freezer set at −8.0°C. Meanwhile, 0.050 kg of iron filings are removed from a 150°C oven. The ice and the iron are added to a well-insulated calorimeter containing 0.20 kg of water at 18°C. What final temperature is reached by the materials in the calorimeter?

Section 11-7 Modes of Heat Transfer

11-49. Give a specific example of how heat transfer by convection plays a role in regulating your body temperature.

11-50. Give examples of how different modes of heat transfer—conduction, radiation, and convection—are used in cooking.

11-51. SSM Give examples of how different modes of heat transfer—conduction, radiation, and convection—are used in home heating.

11-52. Of conduction, convection, and radiation, which type(s) of heat transfer (one or more) is/are involved in each of the following situations? Briefly explain your answers.

a. The application of an ice pack to a sore shoulder.

b. The warm updrafts or *thermals* on which hawks and vultures soar.

c. The change of temperature of a region of the moon's surface that is dark when there is a crescent moon but lit when there is a full moon.

d. Circulation of blood through the human body.

e. The increase in temperature under the boardwalk at Coney Island from before dawn to noon on a clear day. .

f. The working of a microwave oven.

11-53. In a typical room in a heated home, the temperature will be measurably higher near the ceiling than near the floor. (Try measuring this with an ordinary household thermometer.) Explain why this is so.

Section 11-8 Some Factors Affecting Heat Transfer

11-54. In Figure 11-19, a round object is maintained at temperature T_1 and three surrounding block-shaped objects are maintained at a different temperature T_2. Three solid copper cylinders (I, II, and III) connect the round object to each of the others. The diameters of cylinders I and II are equal; the diameter of cylinder III is twice as great. Compare the rates at which heat is transferred past various points in the diagram by finding the following ratios:

a. $\dfrac{\text{rate past point A}}{\text{rate past point C}}$

c. $\dfrac{\text{rate past point A}}{\text{rate past point B}}$

b. $\dfrac{\text{rate past point C}}{\text{rate past point D}}$

d. $\dfrac{\text{rate past point B}}{\text{rate past point C}}$

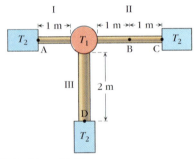

Figure 11-19 Problem 11-54

11-55. A solid metal rod is connected between two large containers of water, one at 20°C and one at 80°C (Figure 11-20). Heating and cooling mechanisms maintain the temperatures of the two bodies of water despite the heat transfer between them.

a. After the rod has been connected between the two containers for a while, would you expect the heat transfer rate past a given point to be the same or different at different points along the length of the rod? Briefly explain your reasoning.

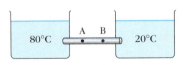

Figure 11-20 Problem 11-55

b. Consider two points A and B along the rod. If the rate of heat transfer were always greater at A than at B, would the part of the rod between points A and B get warmer, get cooler, or remain at the same temperature?

c. Suppose the temperature of the rod is 20°C before being connected. When (or a nanosecond after) the rod is first connected between the two containers, would you expect the heat transfer rate past a given point to be the same or different at different points along the length of the rod? Briefly explain your reasoning.

d. Again suppose the temperature of the rod is 20°C before being connected. After a while, what temperature would you expect its left end to reach? What temperature would you expect its right end to reach? What temperature would you expect its midpoint to reach?

e. Once the midpoint reaches this temperature, could it remain at this temperature if the rate of heat transfer toward it on one side was different than the rate of heat flow away from it on the other side?

f. Compare the rates of heat transfer that Equation 11-12a or b would give you **i.** between the 80°C end and the 20°C end, and **ii.** between the midpoint at 50°C and the 20°C end. Check to make sure your answer to part *f* is consistent with your answers to parts *a–e*.

11-56. A 0.30-m-long copper wire with a diameter of 0.002 cm is connected between a body of boiling water and a block of ice. Find the rate of heat transfer along the wire.

11-57. Two metal blocks, one 50°C hotter than the other, are separated by a thick insulating wall. Two copper wires pass through the wall as they run from one block to the other. Wire A has double the radius of wire B and is four times as long. Compare the amount of energy that the two wires transfer from one block to the other each second. Express the comparison as a ratio $\frac{\text{amount transferred by wire A}}{\text{amount transferred by wire B}}$.

11-58. Figure 11-21 shows four copper blocks. The left end of each block is kept at 80°C and the right end of each block is kept at 20°C. Rank the four blocks according to the rate at which energy flows from the left to the right end. Start with the lowest rate and indicate any equalities.

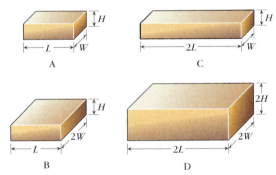

Figure 11-21 Problem 11-58

11-59. An architectural engineer is planning to insulate a wall 3.5 m wide by 2.1 m high using a polyurethane foam with a thermal conductivity of 0.024 W/(m · °C). His goal is to reduce the total energy flow through the wall to 10 J/s when the temperature inside is 20°C and the outside temperature is −17°C.

a. Assuming that the inner and outer wall surfaces contribute negligibly to preventing heat flow, how thick will the wall have to be to accommodate enough foam to achieve the architect's goal?

b. Is this a practical solution?

11-60. Two identical baked potatoes A and B have the same temperature when taken out of the oven. Potato A is placed in a snug Styrofoam container (of negligible mass), and potato B is left uncovered. Both potatoes are left out on the kitchen counter for 48 h (hours).

a. For which potato, if either, would the rate of heat transfer to the surrounding air be greater a few minutes after being left on the kitchen counter? Explain.

b. How would the rates of heat transfer from the two potatoes to the surrounding air compare 48 h after being left on the kitchen counter? Explain.

c. How would the total amount of energy transferred from the two potatoes to the surrounding air compare a few minutes after being left on the kitchen counter? Explain.

d. How would the total amount of energy transferred from the two potatoes to the surrounding air compare 48 h after being left on the kitchen counter? Explain.

11-61. SSM Living creatures make use of thermal properties of materials in their everyday behaviors. Make use of appropriate data from the tables in this chapter in justifying your answers to each of the following. (What property of what material is relevant?)

a. Why do small birds puff out their feathers on cold days?

b. Why is a duck likely to need (and have) more fat than a chicken?

11-62. Objects like the bricks surrounding a fireplace are sometimes said to have a large thermal mass.

a. What do you suppose "thermal mass" means, and why is this analogous to how inertial mass affects an object's motion?

b. What quantity or quantities would this thermal mass depend on? Justify your answer.

Going Further

The questions and problems in this group are not organized by section heading, so you must determine for yourself which ideas apply. Some of them will be more challenging than the Review and Practice questions and problems (especially those marked with a • or ••).

11-63. In our homes, we take advantage of the thermal properties of water and of insulating materials, using the water in our heating pipes to carry energy from the basement furnace to an upstairs bedroom and slowing down its transfer out of our homes by packing our walls with insulation.

a. Our bodies also make use of the thermal properties of water. When we get "overheated," we become flushed and perspire. Describe what is happening in these bodily functions in terms of transfer of energy and relevant properties of water. In doing so, identify any property or properties that are involved. (Keep in mind that your blood is predominantly water.) Compare and contrast these roles of water in your body with the role it plays in a home heating system.

b. Identify some insulating materials that are found in or on living species. Describe the energy transfers that are slowed down by these materials.

11-64. Liquid thermometers can be made using any liquid. Of water, mercury, and ethyl alcohol, which would be the best material for a liquid thermometer used to measure

a. very low temperatures?

b. very high temperatures? Obtain information from relevant tables in this chapter to justify your choices.

11-65. A liquid thermometer has a large bulb at the bottom. From here, the liquid rises in a very narrow glass tube. An outer glass wall of substantially greater diameter protects the fragile inner tube.

a. How does the volume of liquid in the bulb compare to the volume of liquid in the narrow glass tube?

b. What happens to the liquid in the bulb when the temperature increases? How does this affect the liquid in the tube?

c. Liquid thermometers A, B, and C in Figure 11-22 are all shown at −5°C. Rank these thermometers in order from least to greatest of how high the liquid rises in the thermometer when the surrounding temperature increases to 30°C. Indicate any equalities.

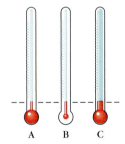

Figure 11-22
Problem 11-65

11-66.

a. Redraw the graph in Figure 11-3 with the equivalent Kelvin temperatures plotted horizontally, then use the coordinates of the known points to find the equation of the resulting straight line, that is, an equation expressing Fahrenheit temperature in terms of Kelvin temperature.

b. Use the equation obtained in **a** to find the value of absolute zero on the Fahrenheit scale.

11-67. SSM WWW Suppose you actually performed the experiment described in Example 11-7, using a calorimeter that was not sufficiently insulated. Will the value that you obtain for the specific heat of aluminum be correct, too high, or too low? Briefly explain your reasoning.

11-68. Equal masses of finely divided silver (a dust or fine powder) and finely divided iron are mixed together in an evacuated and insulated chamber. The two materials start out at different temperatures. Will their common final temperature be closer to the initial temperature of the silver, closer to the initial temperature of the iron, or midway between the two initial temperatures? Briefly explain your reasoning.

11-69. Based on the specific heat values in Table 11-1, explain why the temperature changes from day to night tend to be much more extreme in the desert than they are at the seashore.

•11-70. Estimate the mass of the block of ice that when placed in your room will cause the temperature in your room to drop by 1°C. Assume your room is perfectly insulated.

11-71. The curator of the State Museum of Exquisite Artifacts has acquired a statuette that appears to be solid gold, but she fears it might turn out to be gold plate over a body of base metal, such as iron. How can she use a calorimeter (admittedly not the best way of doing this) to figure out what the statuette is made of without scratching the finish on it?

11-72. The erect plates along the back of a stegosaurus (Figure 11-23) are often cited as evidence that these dinosaurs were warm-blooded. Judging by their form, what thermal purpose might these plates have served?

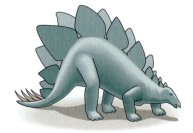

Figure 11-23 Problem 11-72

11-73. Two neighboring homes are identically constructed, and their owners keep them heated to the same temperature. Both houses have the same exposure to the sun, but the snow on one house's roof has melted, whereas the snow on the other remains. How would you explain the difference?

11-74.

a. The air pressure inside a certain sealed container is 5.0×10^4 N/m^2 at a temperature of 200 K. What will the pressure be when the temperature is raised to 250 K?

b. The air pressure in another sealed container is 5.0×10^4 N/m^2 at a temperature of 200°C. What will the pressure be when the temperature is raised to 250°C?

c. Are your answers to **a** and **b** the same or different? Briefly explain why.

11-75. A student is told that the volume of a certain body of air doubled when its Kelvin temperature doubled. The student replies, "The volume is proportional to the Kelvin temperature, but the pressure is also proportional to the Kelvin temperature. So that must mean that the air pressure exerted by the body of air has also doubled." Critique this student's reasoning.

11-76. How much room temperature water could be boiled using the energy output from the complete combustion of a piece of chocolate cake having 600 Calories?

11-77. 0.150 kg of iron filings are added to 0.280 kg of water in a calorimeter. The iron filings lose 1350 J of energy. If the mixture ends up at a temperature of 26°C, at what temperature **a.** did the iron filings start out? **b.** did the water start out?

11-78. The calorimetry experiment in Example 11-8 is repeated. Without the experimenter knowing it, someone has substituted antifreeze (with a specific heat of 2600 J/(kg · K)) for an equal portion of the water. If the experimenter makes no other mathematical errors, how will this affect the value that the experimenter calculates for the mass of the aluminum beads? Will it be too high, too low, or correct? Briefly explain.

11-79. 0.100 kg of copper pennies heated to a temperature of 130°C are added to 0.150 kg of water in a calorimeter. If the addition is just enough to double the Celsius temperature of the water, what is the final temperature of the water?

11-80. A laboratory assistant adds 0.60 kg of crushed ice to 1.20 kg of water at 44.0°C and ends up with 1.80 kg of water at 2.0°C. Find the initial temperature of the ice.

11-81. An Antarctic explorer spills a cup of hot tea into a hole drilled down to the permafrost. Sketch a labeled graph plotting the Celsius temperature T_C of the tea against time from the instant it is spilled until the temperature has stopped changing.

11-82.

a. Consider the kinetic energy associated with the random motion of the molecules making up the mothballs in Case 11-1. If you were to sketch a graph plotting the total kinetic energy of the molecules against time as the melted mothballs cooled to room temperature, would it look more like the graph in Figure 11-10b or the graph in Figure 11-10c? Briefly explain.

b. Now consider the total distributed energy (thermal + chemical) of the molecules. If you were to sketch a graph plotting this total distributed energy against time, would it look more like the graph in Figure 11-10b or the graph in Figure 11-10c? Briefly explain.

11-83. To cool the engine of a car, water is circulated through the water jacket of the engine and from there through the radiator. Where in this system is there

a. heat transfer by convection?

b. heat transfer by conduction?

c. heat transfer by radiation?

d. For the situations you've identified in **a** and **b**, describe what the molecules are doing to accomplish the heat transfer.

11-84. The two arrangements of small cubes in Figure 11-24 have been raised to the same temperature, which is considerably higher than the surrounding air temperature.

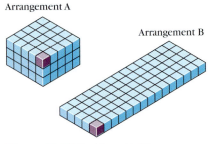

Figure 11-24 Problem 11-84

a. In which of the two arrangements is a small cube closer to the surface on the average?

b. Based on **a**, which arrangement transfers heat to the surrounding air at a faster rate?

c. Which of the two arrangements in Figure 11-24 has the larger surface-area-to-volume ratio?

d. Based on **c**, which arrangement transfers heat to the surrounding air at a faster rate?

e. Do you reach the same conclusion by both lines of reasoning? Comment.

11-85. List the following samples of H_2O in rank order of how much energy they transfer to their surroundings. Order them from least to greatest, indicating any equalities.

Sample A: a kg of H_2O that drops from 55°C to 45°C in an hour

Sample B: a kg of H_2O that drops from 105°C to 95°C in an hour

Sample C: a kg of H_2O that drops from 55°C to 45°C in two hours

Sample D: 2 kg of H_2O that drop from 55°C to 50°C in two hours

Sample E: a kg of H_2O that drops from 5°C to −5°C in an hour

••11-86. *Heat conduction through multiple layers:* An opening in the exterior wall of a house is closed off with two layers of insulation, labeled I and II in Figure 11-25. The two layers are made of materials I and II, which have different insulating properties. Equation 11-12 (a or b) involves the following quantities:

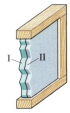

Figure 11-25
Problem 11-86

$$\frac{\Delta Q}{\Delta t}, K, A, d, \Delta T$$

(we've written $T_2 - T_1$ as ΔT).

a. Consider these quantities one by one. Tell whether each is the same or different for the two insulating layers and give a reason for each part of your answer.

b. Suppose that

- ΔT is the temperature difference between the inside of the house and the outside;
- $(\Delta T)_{\mathrm{I}}$ is the temperature difference across insulating layer I;
- $(\Delta T)_{\mathrm{II}}$ is the temperature difference across insulating layer II.

How does ΔT relate to $(\Delta T)_{\mathrm{I}}$ and $(\Delta T)_{\mathrm{II}}$? Write this relationship as an equation, and label it Equation 1.

c. Rewrite Equation 11-12a twice, once for layer I and once for layer II. In the two equations, put appropriate subscripts (I or II) on all the quantities that are different for the two layers. Do *not* put subscripts on any quantities that are the same for both layers.

d. Solve the two equations you wrote in part **c** to obtain expressions for $(\Delta T)_{\mathrm{I}}$ and $(\Delta T)_{\mathrm{II}}$. Use these expressions to substitute for $(\Delta T)_{\mathrm{I}}$ and $(\Delta T)_{\mathrm{II}}$ in Equation 1, which you obtained in part **b**. Show that the equation that results from this substitution can be rewritten as

$$\frac{\Delta Q}{\Delta t} = \frac{A}{d_{\mathrm{I}}/k_{\mathrm{I}} + d_{\mathrm{II}}/k_{\mathrm{II}}} \Delta T$$

11-87. Convection occurs when water cooled in the radiator of your car is pumped through the car engine and carries away energy from the hotter engine. Does conduction also occur in this situation? (Would there be any energy transfer if the pump stopped operating so that the water in the circulation system remained stationary?) Briefly explain.

11-88. Based on the data in Table 11-3, why do some pots have copper bottoms?

11-89.

a. Consider the amount of energy lost by the body over an hour and a half in Example 11-5. How many 100-watt bulbs (to the nearest whole bulb) would be needed to emit the same amount of energy over the same time interval?

•b. If the body emits as much energy as this many bulbs during the same time interval, why doesn't the body feel as warm as the bulbs? (*Hint:* Think about irradiance at the surface.)

11-90. Equal amounts of hot water are poured from the same pot into two identical Styrofoam cup calorimeters. Using identical thermometers that have both been at room temperature, students A and B measure the temperature of the water in the two cups. Student A dips the bulb of her thermometer just below the water surface in one cup, while student B extends the bulb of her thermometer almost to the bottom of the other cup. Will the temperature that student A reads will be higher than, equal to, lower than the temperature that student B reads? Briefly explain. (Based on an idea in J. Gash, *Physics Teacher*, **40**, 74 [2002].)

11-91. In room A the room temperature is 22°C, and in room B it is 32°C. To maintain a normal body temperature of 37°C when you are in room B, how many times as much energy must be released by the chemical reactions in all the cells of your body as when you are in room A?

11-92. In Example 11-8, suppose that the calorimeter was not sufficiently insulated. Would the mass of aluminum beads that you calculated based on the temperature reading be too high, too low, or correct? Briefly explain.

11-93. A bowl of warm water is put in a freezer until it becomes completely frozen. If the water lost exactly as much energy in turning from liquid to solid as it did in dropping from its initial temperature to the freezing point, at what temperature did it start out?

11-94. A well-insulated water tank is divided into two equal spaces by a thin copper wall. The wall has an area of 0.060 m^2 and a thickness of 0.0020 m. One space is filled with water at 40°C, the other with water at 10°C.

a. At what rate does energy transfer from the warmer to the cooler water when the tank is first filled?

b. When the temperature of the warmer water has dropped to 30°C, at what rate does energy transfer from it to the cooler water?

c. Will more energy transfer through the copper wall during the first minute or the fifth minute after the two spaces are filled? Briefly explain.

11-95. Down, or fluffy feathers, is often used in winter clothing. An inventive student, impressed by how warm he is kept by his down jacket on a cold day, decides to design an improved down jacket. He comes up with the two ideas below. Determine whether each of these is a good idea or not. State your reasoning, and use relevant data from an appropriate table in the chapter to support your reasoning.

a. Flatten the layer of feathers with which the jacket is filled so that the jacket will not be so thick and bulky.

b. Keep the thickness the same, but pack in lots more down by squeezing out the space between the feathers. The student's idea is that if feathers provide good insulation, a higher density of feather material will provide better insulation.

11-96. Molly owns a fake fur winter coat. One day, she finds it keeps her quite warm. The next day she lends her coat to a friend who gets caught in a downpour. When Molly puts the coat on, the rain has stopped, and the weather is now identical to the day before, but the fur is still quite wet and matted down. You wish to determine whether the coat will keep her as warm, assuming that putting it on doesn't make her wet. Which two of the following pieces of information would it be most relevant to compare in making this judgment: *the specific heat of fur; the thermal conductivity of fur; the specific heat of air; the thermal conductivity of air; the specific heat of water; the thermal conductivity of water?*

11-97.

a. Table 11-3 shows that the thermal conductivity of human skin is much higher than the thermal conductivity of fat or bone. What practical value does this have for humans?

b. For what sort of measuring instrument might you want a key part of it to have high thermal conductivity? Do you see any connection between this question and part ***a?***

11-98. A visitor from planet Ork informs you that according to his Orkian thermometer, water freezes at 50° Orkian and boils at 80° Orkian. What will his thermometer read in a physics classroom where the temperature is 22°C?

11-99. Given the data in Problem 11-98, what temperature on the Orkian scale corresponds to absolute zero?

•**11-100.** *Metabolic rate of a spherical duck:* Many species of duck are perfectly at home in water near the freezing point. Their fat and feathers enable them to maintain a body temperature of 41°C, higher than that of humans. The duck's *metabolic rate* is the rate at which the chemical reactions in its cells must output energy. The minimum rate is that required to replace the energy lost through the duck's body insulation; the rate must in fact be greater than this to sustain other activity. To estimate the minimum rate, approximate a duck by a sphere with a radius of 0.10 m covered by a 0.008-m-thick layer of fat. Because feathers trap air, which has a thermal conductivity close to that of fat, we will not need to distinguish between fat and feathers. Based on this approximation, what must be the duck's minimum metabolic rate in J/s when its environment is at 0°C?

Problems on WebLinks

11-101. Suppose that in WebLink 11-1, the temperature T_H is maintained at 50°C. Then for the energy flow rate $\frac{\Delta Q}{\Delta t}$ to remain constant, T_C must do which of the following: *increase at a constant rate; decrease at a constant rate; either increase or decrease at a constant rate; remain unchanged?*

11-102. (See WebLink 11-2.) Find the surface-area to volume ratio for each of the following shapes to determine whether it is proportional to its linear dimension (its scale):
a. a sphere of radius *r*;
b. a cylinder whose height is equal to its radius;
•***c.*** a rectangular solid whose sides are in the ratio 1:2:3.

11-103. *Of Mice and Men:* (See WebLink 11-2.) Biologist J.B.S. Haldane points out that the mass of an adult human being is roughly equal to that of 5000 mice.

a. Which will lose a greater percentage of its body's internal energy each second, a human or a mouse? Briefly explain.

b. Of mice and humans, which would you expect to have to eat a greater fraction of their body weight each day? Briefly explain why.

c. Are larger or smaller animals generally better suited to living in the Arctic? Briefly explain.

11-104. By the reasoning in WebLink 11-2, in which animal—a dog, a cow, or a mouse—would body temperature drop most quickly on a cold day and in which would it drop most slowly?

CHAPTER TWELVE

The Kinetic Theory of Gases, Entropy, and Thermodynamics

In Chapter 11, we dealt with some thermal properties of matter. In this chapter we ask, "What underlying mechanisms enable matter to exhibit the properties that we observe?" For example, why should the pressure or volume of a gas change with temperature? When the liquid or gas in a fluid thermometer rises, what is actually happening on a microscopic scale as the thermometer reading changes, or in other words, what *is* temperature, beyond being "what the thermometer reads"? In fact, what is a gas like that it should exert pressure at all?

To address these questions, we try to picture what is happening on a scale too small to observe directly; that is, we construct a mental model. We then test the explanatory power of this microscopic model by seeing how well it explains other patterns of observed behavior that we have not previously considered, such as those having to do with order and disorder.

"Why should the pressure or volume of a gas change with temperature?"

12-1   The Kinetic Theory of Gases (Qualitative)

In Figures 11-4 and 11-5, we saw experimental data showing that at constant pressure, the volume of a gas is proportional to its temperature in kelvins, and that when its volume is held constant, the pressure a gas exerts is proportional to its temperature in kelvins. What characteristics does a gas have that cause it to behave in these ways? In fact, why should a gas exert pressure at all?

What else do we know about how gases behave? You can squeeze a balloon flat when its neck is open but not when it is tied shut; that tells you there is something inside (we call it air). You can walk through air or sweep your hand through it, even in a sealed room. Now suppose a gas is made up of parts, or building blocks. It is usual to call these **molecules,** but by naming them we are making no assumptions about what they are like. In fact we are asking, "What must these building blocks be like for a gas to have the properties it has?"

When you move your hand through the air, the air must be yielding it enough space to pass through. You can argue that this happens because your hand compresses the individual molecules in its path or because there is plenty of space between the molecules (Figure 12-1*a*). Based on this one piece of evidence, either explanation could be correct, so we need to examine more evidence before deciding.

Figure 12-1 tries explaining two other pieces of evidence with each of these models. Neither model fully succeeds in explaining both. Some people say that warming increases the air pressure in a fixed volume because the molecules "want to expand." But what does that mean? Obviously, we don't think molecules sit around saying, "Gee, it's hot. I feel like expanding." They have no capacity to want. The question is, what do they actually *do* to exert increased pressure? Model II in Figure 12-1 is also inadequate so far. If there is space between them, how can they raise the piston in Figure 12-1*b* to increase the volume? And when the container volume is fixed, as in Figure 12-1*c*, the molecules are confined to the same space and cannot spread out.

Two more pieces of evidence offer some valuable clues.

1. A student lets a tiny drop of red ink fall from a dropper into a beaker of still water. The student is careful not to disturb the water. The red color spreads,

Observation	Model I: Molecules of air fill the space but can expand or be compressed.	Model II: Molecules are tiny and have a lot of space between them.
(*a*) Hand can pass through air.	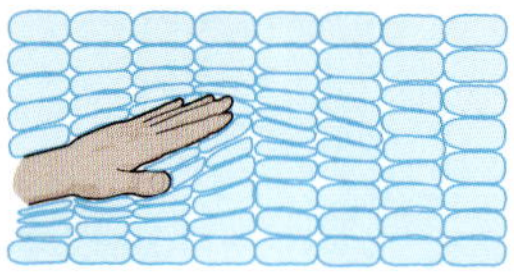	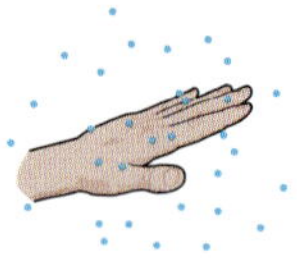
(*b*) Volume of gas increases when warmed at constant pressure.	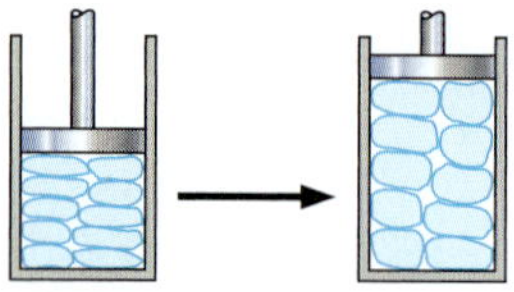 at lower temperature at higher temperature	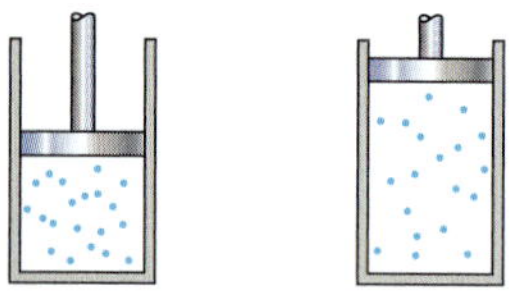 at lower temperature at higher temperature
(*c*) Pressure exerted by a gas increases when warmed at constant volume.	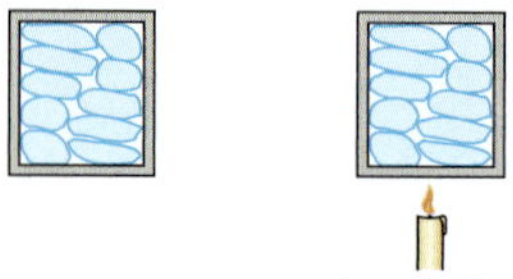 molecules "want to expand?"	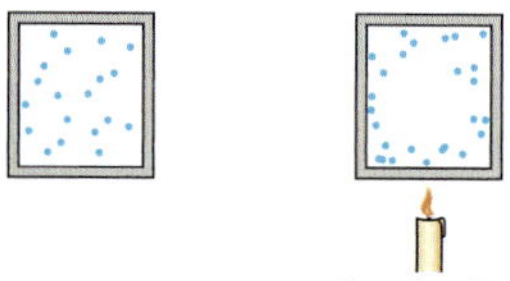 molecules "want to spread out?"

Figure 12-1 Which model explains the observations?

or **diffuses,** very slowly from its point of entry. It may take hours before the color is more or less uniformly distributed throughout the beaker.

2. A professor opens a large jar of ammonia in the front of a lecture room and stirs it a bit. The students in the front rows smell it after a few seconds. Those in the back rows smell it a number of seconds later.

Observation 2 suggests that the molecules of evaporated ammonia are either traveling or expanding until they reach the back rows. But if air molecules fill the room, how do the ammonia molecules get through? We may also ask, do the molecules of a liquid expand (grow larger) when the liquid evaporates, or do they remain the same size but get farther apart? If the molecules grew larger on evaporation, it would be more difficult for molecules of one gas to spread through another than for molecules of one liquid to spread through another, as red ink does through water. The spread of one gas through another would be slower than the spread of one liquid through another. That is the opposite of what we observe. The evidence compels us to choose the model with lots of empty space between the molecules.

This model was incomplete before. To explain the spread of the ammonia smell, we have built an additional assumption into the model: *the molecules travel.* Let's refine this assumption. At any given distance, people on the left and right sides of the lecture room start smelling the ammonia at the same time. So we have to assume the molecules are moving in all directions. This can occur if their motion is *random.*

STOP&Think How can a model that views a gas as widely separated molecules in random motion explain why gases exert pressure? In thinking about this question, you need to ask yourself, pressure exerted on what? How did we define pressure? What is involved in exerting pressure? ◆

What happens when the molecules reach the walls of their container? They collide with the walls and bounce back. Each colliding molecule exerts an impulse on the wall. When very large numbers of molecules move randomly, the wall receives very nearly the same number and distribution of impulses (some harder, some gentler) during each tiny time interval Δt, and the pressure on it—the cumulative effect of all these impulses—remains the same.

Humans are acutely sensitive to the defensive secretions sprayed by skunks. The human nose can detect the thiols that are the offending molecules in skunk spray at concentrations as low as about 10 parts per billion.

Example 12-1 *Increasing Volume at Constant Temperature and Increasing Temperature at Constant Volume*

A tight-fitting piston traps a body of air within a metal cylinder open at the top (Figure 12-2*a*).

a. Suppose that as the piston is raised (Figure 12-2*b*), the air within the cylinder is kept at constant temperature by close thermal contact with a large bath of water. What happens to the air pressure within the cylinder?

b. Next suppose that the piston is kept at the same height, but the water is cooled, cooling the air in the cylinder along with it. Now what happens to the air pressure within the cylinder?

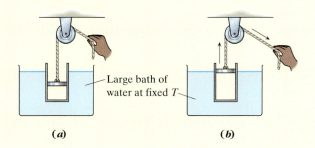

(*a*) (*b*)

Figure 12-2 Increasing volume at constant temperature. Here the volume is increased isothermally.

EXAMPLE 12-1 *continued*

Solution

Choice of approach. We use the model of a gas that we have developed so far to reason about what would happen. We follow the behavior of a typical molecule and consider its contribution to the pressure.

The qualitative solution.

a. For simplicity, let's follow a molecule that is approaching the underside of the piston straight on. When it bounces back from the piston, it moves toward the bottom of the cylinder. On the average, it will keep bouncing back and forth between the piston and the bottom. We neglect collisions with other molecules, even though they are frequent, because the other molecules are moving randomly in all directions. Therefore the average total effect of multiple collisions with other molecules is zero.

   When the piston is raised higher, each molecule travels further between collisions. Then if its speed doesn't change, it can make fewer trips and hit the piston fewer times each second. With fewer collisions, each molecule contributes less to the total pressure on the piston, and the pressure is reduced.

b. Now the temperature is lowered. The distance between piston and cylinder bottom remains unchanged, so nothing will happen to the pressure *if* our explanatory model is correct and complete. We know (Figure 11-5) that at constant volume the pressure decreases as the temperature drops. So predicting on the basis of our model gives us incorrect results. We must either replace or refine our model.

   ◆ Related homework: Problems 12-4 and 12-5.

How often our typical molecule collides at top or bottom depends not just on how far apart the piston and the cylinder bottom are but on how fast the molecule is moving. Our model *will* explain the drop in temperature if we include a further assumption: that on average, the molecules slow down when the temperature drops and speed up when the temperature rises. This does not mean that at a given temperature all the molecules have the same speed, only that their speeds are distributed about some average value that rises and falls with temperature.

The model we have been developing is known as the **kinetic theory of gases.** We now summarize its assumptions and some of its consequences:

The Kinetic Theory of Gases

Assumptions

1. Gases are made of large numbers of tiny (negligible volume) building blocks called molecules. (The theory makes no assumptions about the internal structure of atoms or molecules. It treats atoms and molecules as equivalent.)

2. The distances between molecules are very large compared to the size of the molecules; gases are mostly empty space.

3. The molecules are in constant random motion.

4. The molecules of a gas collide elastically and practically instantaneously with each other and with the surfaces in contact with the gas. They do not interact in other ways.

5. The molecules obey Newton's laws of motion.

Consequences

1. Because gas molecules keep colliding with all surfaces in contact with the gas, the gas exerts pressure on those surfaces.

2. Their average speed increases as the temperature of the gas increases.

We must assume the collisions are elastic because otherwise the molecules would gradually be slowed down by repeated collisions with the surfaces that they hit. We still need to explain why the second consequence follows from the assumptions; we will do this shortly.

◆**LIMITATIONS OF THE THEORY** We will shortly see that this model can explain the behaviors of gases graphed in Figures 11-4 through 11-6. A gas that exhibits those behaviors is called an **ideal gas.** Remember that the low-temperature parts of those graphs did not represent actual data but were obtained by extending the line segments actually determined by the data points. Real gases do not drop to zero temperature or pressure at absolute zero; under suitable conditions at sufficiently low temperatures, they condense and become liquids. That means the molecules bond; their interactions are no longer elastic collisions. But until the temperature drops very close to the boiling temperature of the liquid, bonding forces generally have negligible effect, so our model provides a very good approximation of what is happening.

In Chapter 10, we stated that *even if* the tiny volume to which the material could be reduced were exactly zero and *even if* the gases never condensed to form liquids, we could still get no lower than the temperature we call absolute zero. We can now state this in terms of kinetic theory: *Even if* the molecules themselves occupied zero volume and *even if* there were no bonding interactions, the temperature could not go below absolute zero. In the next section, we will see why.

Example 12-2 *Why Volume Increases with Temperature at Constant Pressure*

If the piston on top of the column of air in Figure 12-3 can move freely, the enclosed air is kept at constant pressure as the air temperature increases, because neither downward force changes as the piston moves. Explain why the volume of the air increases under these conditions.

Solution

Choice of approach. We use the kinetic theory of gases as our explanatory model. **STOP&Think** Why is the pressure the same before and after the temperature change? ◆

The qualitative solution. The piston is in translational equilibrium both before and after the temperature increase. Then at each time (Figures 12-3a and c), the sum of the vertical forces in the free-body diagram is zero: $\Sigma F_y = P_{eq}A - mg - P_{atm}A = 0$, and the upward force $F = P_{eq}A$ on the bottom of the piston must be equal and opposite to the weight of the piston plus $P_{atm}A$. So the equilibrium pressure P_{eq} must be the same before and after. But the piston is not in translational equilibrium *while* it is moving up. **STOP&Think** Why not? How is the molecular activity affected by the temperature change? How does that alter any of the forces? ◆

Between the two equilibrium positions (Figure 12-3b), there is a net upward force, and the pressure P on the piston is greater than P_{eq}. Kinetic theory provides the rest of the explanation. As the air is warmed, the molecules speed up. At greater speed, they collide with the underside of the piston more frequently and with a greater average impulse per collision, so the pressure increases. The resulting net upward force moves the piston upward. **STOP&Think** Why does the total upward force eventually become zero again? ◆

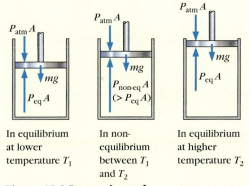

In equilibrium at lower temperature T_1

In non-equilibrium between T_1 and T_2

In equilibrium at higher temperature T_2

Figure 12-3 Increasing volume at constant pressure. The enclosed air pressure is the same in each equilibrium position (P_{eq} is its equilibrium value) because neither of the downward forces change.

➥**A note on language:** We are careful to speak of translational equilibrium here, meaning $\Sigma\vec{F} = 0$, to distinguish it from *thermal equilibrium,* the state reached when the average speed and distribution of the molecules stops changing and quantities such as pressure and volume have reached their final values.

EXAMPLE 12-2 *continued*

As the piston moves upward, its distance from the bottom of the cylinder increases. Though the molecules are traveling faster, they have to go further between collisions, so the frequency of collisions drops. So although the impulse per collision is greater, the number of collisions during any Δt gets smaller, until the total impulse in each Δt is again the same as before the temperature increase, and equilibrium is restored.

◆ Related homework: Problems 12-6 and 12-7.

12-2   The Kinetic Theory of Gases (Quantitative)

The molecules in the kinetic theory of gases are subject to the same laws of physics as any other objects. In this section, we use what we already know about Newton's second law and energy to analyze what the kinetic theory predicts about the behavior of gases. Based on the theory, what should we expect to observe about the behavior of gases? Does that agree with the patterns of observation that we have actually found?

To incorporate the assumptions of the kinetic theory into our analysis, we will use our basic mechanics principles to think about what happens in a cubic container containing a gas of monatomic (single-atom) molecules moving at an average speed v. Ultimately, we will be interested in the average translational kinetic energy of the molecules, which depends on the average value of v^2. Then for our average speed, we want the square root of the average value of v^2. This kind of average is called the **root-mean-square value** (or *rms* value). For example, if two molecules have speeds $v_1 = 5$ and $v_2 = 12$, we square the values ($5^2 = 25$, $12^2 = 144$), add the squares ($25 + 144 = 169$), divide by the number of values ($169/2 = 84.5$), and then take the square root ($\sqrt{84.5} = 9.2$). Note that this gives a different value than the arithmetic mean (the most common kind of average), which gives us $\bar{v} = \frac{v_1 + v_2}{2} = \frac{5 + 12}{2} = 8.5$.

To be guided step by step through the analysis, go to WebLink 12-1. A more concise version follows. We assume the cubic container has sides of length a, and there are N molecules in the enclosed gas. To simplify the analysis, we assume all the molecules are moving at the rms speed v. We then consider what happens to the right wall (Figure 12-4) over a time interval Δt long enough for each molecule to have made a number of back-and-forth trips between the left and right walls. Over this interval, the total impulse $F_{av}\Delta t$ on the right wall will be

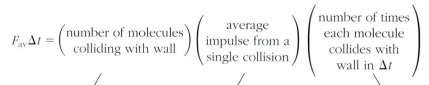

$$F_{av}\Delta t = \left(\begin{array}{c} \text{number of molecules} \\ \text{colliding with wall} \end{array} \right) \left(\begin{array}{c} \text{average} \\ \text{impulse from a} \\ \text{single collision} \end{array} \right) \left(\begin{array}{c} \text{number of times} \\ \text{each molecule} \\ \text{collides with} \\ \text{wall in } \Delta t \end{array} \right)$$

For **WebLink 12-1: Quantitative Reasoning about an Ideal Gas,** go to www.wiley.com/college/touger

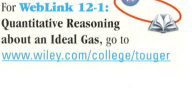

Figure 12-4 The path of a typical gas molecule in a closed cubic container of side *a*.

The number of molecules that can collide with the right wall is N, the number of molecules in the box.	As Figure 12-4 shows, the momentum of each molecule colliding with the right wall has a component $p_x = mv_x$ before the collision and $p_x = -mv_x$ after. The impulse on the *molecule* is $\Delta p_x = (-mv_x) - (mv_x) = -2mv_x$, so the equal and opposite impulse on the *wall* is $+2mv_x$.	A molecule collides with the right wall after each two-way trip. The total horizontal distance for each two-way trip is $2a$. Over the interval Δt, it travels a total horizontal distance $v_x\Delta t$. When we divide this total distance by the distance for each trip, we obtain the number of two-way trips: $v_x\Delta t/2a$.

Using these partial results in our equation for the total impulse, we get

$$F_{av}\Delta t = N(2mv_x)\left(\frac{v_x\Delta t}{2a}\right)$$

By some algebraic manipulation, we can rewrite this as

$$F_{av} = \left(\frac{2N}{a}\right)(\tfrac{1}{2}mv_x^2)$$

If v is the rms value, the average translational kinetic energy of a single molecule is $\tfrac{1}{2}mv^2 = \tfrac{1}{2}mv_x^2 + \tfrac{1}{2}mv_y^2 + \tfrac{1}{2}mv_z^2$. But because the molecular motion is random, v_y^2 and v_z^2 must each have the same average value as v_x^2. Then $\tfrac{1}{2}mv^2 = 3(\tfrac{1}{2}mv_x^2)$, or $\tfrac{1}{2}mv_x^2 = \tfrac{1}{3}(\tfrac{1}{2}mv^2)$, and

$$F_{av} = \left(\frac{2N}{3a}\right)(\tfrac{1}{2}mv^2)$$

Because the area A of the right wall is a^2 and the volume V of the cube is a^3, the pressure P on the right wall is

$$P = \frac{F_{av}}{A} = \frac{F_{av}}{a^2} = \left(\frac{2N}{3a^3}\right)(\tfrac{1}{2}mv^2) = \left(\frac{2N}{3V}\right)(\tfrac{1}{2}mv^2)$$

so that

$$PV = \frac{2N}{3}(\tfrac{1}{2}mv^2)_{av} \tag{12-1}$$

where $(\tfrac{1}{2}mv^2)_{av}$ means $\tfrac{1}{2}mv_{rms}^2$

Equation 12-1 is a relationship between the pressure and volume of an ideal gas and the average translational kinetic energy of its molecules. To see what it actually tells us, it will be useful to rewrite it two different ways:

$$\textbf{a. } P = \frac{2N}{3V}(\tfrac{1}{2}mv^2)_{av} \qquad \textbf{b. } V = \frac{2N}{3P}(\tfrac{1}{2}mv^2)_{av}$$

N is constant, so when the volume V is constant, version **a** says

$$P = (\text{a constant})(\tfrac{1}{2}mv^2)_{av}$$

The pressure is proportional to the average translational kinetic energy of the molecules at constant volume.

When the pressure P is constant, version **b** says

$$V = (\text{another constant})(\tfrac{1}{2}mv^2)_{av}$$

The volume is proportional to the average translational kinetic energy of the molecules at constant pressure.

Compare these two statements to the patterns, based in part on actual observations (measurements), that are displayed in the graphs in Figure 11-6. In words, these patterns are **a.** *the pressure is proportional to the absolute temperature (in kelvins) at constant volume,* and **b.** *the volume is proportional to the absolute temperature (in kelvins) at constant pressure.*

Both pairs of statements can be true only if the absolute temperature of a gas is proportional to the average translational kinetic energy of its molecules. In that case, we are led to the following:

The absolute temperature is a measure of the average translational kinetic energy of the molecules.

We have now justified listing this as a consequence in our summary of kinetic theory. In developing a self-consistent explanatory model, we have arrived at a meaning of temperature that goes deeper than simply "what the thermometer reads."

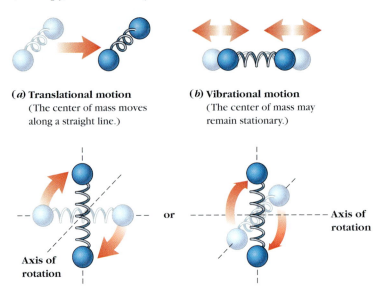

(*a*) **Translational motion**
(The center of mass moves along a straight line.)

(*b*) **Vibrational motion**
(The center of mass may remain stationary.)

Axis of rotation   or   Axis of rotation

(*c*) **Rotational motion** around either of two perpendicular axes or around an axis with components directed along each of these axes
(The center of mass may remain stationary.)

Figure 12-5 Types of motion for a diatomic (two-atom) molecule. There is kinetic energy associated with each of these kinds of motion, but pressure depends only on the translational kinetic energy of the center of mass.

The proportionality between the average translational kinetic energy of the molecules and the absolute temperature is commonly written

$$\left(\tfrac{1}{2}mv^2\right)_{av} = \tfrac{3}{2}kT \tag{12-2}$$

(Without a subscript, T always means T_K.) The constant k (sometimes written k_B) is called **Boltzmann's constant.** In SI units, its numerical value is $k = 1.38 \times 10^{-23}$ J/K, so that the product kT has units of energy.

So far, we have addressed only the average *translational* kinetic energy of the molecules. In general, the thermal energy, the microscopically distributed kinetic energy of the molecules, has contributions associated with three different kinds of motion, as shown in Figure 12-5. When the molecules are monatomic, however, they are very nearly point objects, so we can ignore rotational and vibrational motion. In that case, the total thermal energy of a gas of N monatomic molecules is simply $N\left(\tfrac{1}{2}mv^2\right)_{av} = \tfrac{3}{2}NkT$, the total translational kinetic energy of the molecules. It turns out that for diatomic and larger molecules, the total internal energy is also proportional to the Kelvin temperature T, but the proportionality constant changes.

The proportionality constant in Equation 12-2 is written as $\tfrac{3}{2}k$ so that when we substitute $\tfrac{3}{2}kT$ for $\tfrac{1}{2}mv^2$ in Equation 12-1, the resulting relationship, called *the ideal gas law,* has no fractions:

Ideal gas law: $PV = NkT$ (12-3)

The ideal gas law incorporates both of the relationships that we noted for ideal gases:

• At constant volume, the pressure $P = \tfrac{Nk}{V}T$, and so is proportional to T.
• At constant pressure, the volume $V = \tfrac{Nk}{P}T$, and so is proportional to T.

Like those relationships, the ideal gas law tells us how the measured values of pressure, volume, and temperature must be related. These are *macroscopic* properties of the gas—properties we can measure directly. Because they describe the condition or state of a gas (or of a solid or liquid), they are called **state variables,** and an equation of relationship between state variables is called an **equation of state.** The ideal gas law is an example of an equation of state.

➡ German physicist Ludwig Boltzmann (1844–1906) made major contributions to our understanding of thermal physics.

Great trumpet players, like the late Louis Armstrong, have an intuitive sense of pressure–volume relationships.

The study of the interrelationships among state variables is called **thermodynamics.** Thermodynamics focuses on macroscopic, directly measurable quantities and the connections between them. In contrast, we cannot directly measure the velocities of molecules, but we arrived at the ideal gas law by applying our knowledge of mechanics to the molecules in the gas. In doing so, we dealt with *averages,* and to that extent we were applying *statistics*. Statistics deals with obtaining numbers that tell us something about a population as a whole when there is great variability in the population. For example, the molecules that "populate" the gas may have widely varying individual velocities, but the average velocity is a characteristic of the population as a whole. **Statistical mechanics** is an approach that treats macroscopic behavior as the statistical outcome of applying mechanics principles to huge numbers of microscopic events. The discovery that these two approaches, thermodynamics and statistical mechanics, lead us to the same conclusions about the laws by which nature behaves was one of the great achievements of nineteenth-century physics.

In many applications of the ideal gas law, like Example 12-3, proportional reasoning eliminates the need to know the values of N and k.

Example 12-3 *The Pressure Cooker*

The wall thermometer in the kitchen of a seaside restaurant reads 23°C. An absent-minded chef clamps the top shut on an empty pressure cooker and turns the burner on under it. What is the air pressure within the pressure cooker when the air in it reaches 90°C?

Solution
Choice of approach.

1. The ideal gas law relates pressure to temperature.

2. The ideal gas law deals with absolute temperature, so the temperatures must first be converted from °C to kelvins (from T_C to T).

3. In applying the ideal gas law, it is always important to recognize what remains constant and what changes. The cooker's volume can be treated as constant—its expansion with temperature is negligible.

4. Because the restaurant is at sea level, the initial pressure is normal sea level pressure (1.01×10^5 Pa = 1.01×10^5 N/m^2).

What we know/what we don't. $(T_C)_1 = 23°C \quad T_1 = ? \quad P_1 = 1.01 \times 10^5$ N/m^2

$$(T_C)_2 = 90°C \quad T_2 = ? \quad P_2 = ?$$

The mathematical solution. Because T with no subscript means T_K, we apply Equation 11-4:

$$T_1 = (T_C)_1 + 273.15 = 23°C + 273.15 = 296 \text{ K}$$

$$T_2 = (T_C)_2 + 273.15 = 90°C + 273.15 = 363 \text{ K}$$

We now rearrange the ideal gas law (Equation 12-3) so that all the constants are on right side and all the variables on the left:

$$\frac{P}{T} = \frac{Nk}{V} \qquad \text{(check by cross-multiplying)}$$

Because the right side remains constant, the ratio on the left side must also remain constant as the values of P and T vary (that's what it means to say P and T are proportional), so

$$\frac{P_1}{T_1} = \frac{P_2}{T_2} \left(= \frac{P_3}{T_3} = \frac{P_4}{T_4} = \cdots \text{ etc.} \right) \qquad (12\text{-}4)$$

EXAMPLE 12-3 *continued*

Solving for P_2 gives us

$$P_2 = \left(\frac{P_1}{T_1}\right)T_2 = \frac{1.01 \times 10^5 \text{ N/m}^2}{296 \text{ K}}(363 \text{ K}) = \mathbf{1.24 \times 10^5 \text{ N/m}^2}$$

◆ Related homework: Problems 12-15 and 12-16.

Equation 12-4 is sometimes called **Charles' law,** but it is just a rewriting of the ideal gas law for the specific case when the volume is constant. Example 12-3 suggests a general approach to applying the ideal gas law, which we can use rather than memorizing what it looks like in various specific cases.

PROCEDURE 12-1

Applying the Ideal Gas Law

1. From the conditions in the specific problem, determine which quantities in Equation 12-3 are variable and which are constant. For example,
   a. If the container is rigid, the volume V remains constant.
   b. If the gas is trapped under a freely moving weight, the pressure P remains constant.
   c. If the gas in good thermal contact with a temperature reservoir (defined below) and the process proceeds slowly, the temperature T remains constant.
   d. If no gas enters or leaves the container, the number of molecules N remains constant.
2. Rearrange Equation 12-3 so that all the variables are on the left of the equal sign and all the constants are on the right. The right side is then constant. Because the two sides of an equation are equal, the expression on the left side must be constant too (i.e., it is the same in one state as in another).
3. Write an equation (such as $P_1 V_1 = P_2 V_2$) that says that this expression is the same in state 1 as in state 2.

A **temperature reservoir** is a body that can exchange any reasonable amount of heat with an ordinary-sized object without undergoing a significant change in its own temperature. Often this is because the reservoir has a very large mass. Suppose the reservoir gains or loses an amount of heat Q. Because $Q = mc\Delta T$, a large enough mass m will make ΔT negligibly small. The higher body temperature of a lone swimmer will have no measurable effect on the temperature of a chilly mountain lake, though the swimmer, or many such swimmers, will certainly be chilled by the lake.

In other cases, the reservoir's temperature remains constant because the thermal energy transferred from it is replenished: For example, the thermal energy that an electric stove burner transfers to a pot is replenished by an electrical energy input. Good thermal contact with a temperature reservoir keeps an object at the temperature of the reservoir. Example 12-4 involves an application of Procedure 12-1 in which the surrounding air acts as a temperature reservoir.

Example 12-4 *Squeeze Play*

**For a guided interactive solution, go to Web Example 12-4 at
www.wiley.com/college/touger**

Unbeknownst to Uncle Lucius, a prankster has sealed all the openings in his accordion. When it is fully spread out, the air pressure within is 1.04×10^5 N/m². If the seams of the accordion can withstand a maximum pressure of 1.80×10^5 N/m² before bursting, to what fraction of its fully spread volume can Uncle Lucius compress his accordion without doing damage? Assume that because he suspects a prank, he compresses the accordion *very* slowly.

Brief Solution

Choice of approach. We again apply the ideal gas law, following the steps of Procedure 12-1. Step 1: We treat the temperature as constant, because compressing the accordion slowly allows enough time for heat transfer to the surrounding air to prevent a temperature increase. The variables are P and V. Steps 2 and 3 are part of the mathematical solution below.

What we know/what we don't. $P_1 = 1.04 \times 10^5$ N/m^2 $\qquad P_2 = 1.80 \times 10^5$ N/m^2

$$V_1 = ? \qquad\qquad V_2 = \underline{?}\ V_1$$

The mathematical solution. This time $PV = NkT$ already satisfies the goal of Step 2: The left side has only variables, and the right has only constants. Following Step 3, we write that the left side, because it equals a constant, is the same in State 1 as it is in State 2:

$$P_1 V_1 = P_2 V_2 \tag{12-5}$$

We now solve for V_2.

$$V_2 = \frac{P_1 V_1}{P_2} = \frac{(1.04 \times 10^5 \text{ N/m}^2)V_1}{1.80 \times 10^5 \text{ N/m}^2} = \mathbf{0.578\ V_1}$$

Making sense of the results. Note the inverse relationship between pressure and volume; as one decreases, the other increases.

◆ Related homework: Problems 12-17 and 12-18.

Equation 12-5 is often called **Boyle's law** and is another form that the ideal gas law takes under specific circumstances. But if you learn to *reason* using the ideal gas law, there is no point in memorizing all the forms it can take.

◆**EXPRESSING THE IDEAL GAS LAW IN TERMS OF MOLES AND AVOGADRO'S NUMBER** Another common way of expressing the ideal gas law makes use of some results from atomic physics. A molecule of water (H_2O) has roughly the mass of 18 protons, a hydrogen molecule (H_2) roughly the mass of 2. We say their **molecular masses** are about 18 and 2, respectively. Because the two molecules have a mass ratio of 9:1, If we take larger samples of the two substances that have the same mass ratio of 9:1, the two samples will have equal numbers of molecules. In particular, 18 g of water and 2 g of hydrogen will contain the same number of molecules. There is nothing special about water and hydrogen: If we take a number of grams of any substance that is numerically equal to its molecular mass, it will also contain the same number of molecules. That number of grams is called the substance's *gram-molecular mass,* which is usually shortened to a **mole** of the substance.

By its definition, a mole of any substance contains the same number of molecules. A mole represents a particular amount, just as a dozen does. By a number of independent experimental methods, atomic physicists have been able to determine that number. It is 6.02×10^{23}, a number so huge as to be beyond anything in direct human experience, and it has been named **Avogadro's number** (denoted N_A).

If n is the number of moles of a substance, the total number of molecules is the number of moles times the number of molecules in each mole:

$$N = n N_A$$

With this we can rewrite the ideal gas law as

$$PV = n N_A k T$$

➡**Molecular masses:** Exact values of molecular masses are established by assigning a carbon atom with 6 protons, 6 neutrons, and 6 electrons a mass of exactly 12.00000 and assigning a value to each other substance based on the ratio of its mass to carbon's. So the relative molecular mass of water is 18.02 and that of molecular hydrogen is 2.02.

➡Amadeo Avogadro (1776–1856), a professor at the University of Turin in Italy, was the first to conjecture that molecules of hydrogen, nitrogen, and oxygen had two atoms.

We can simplify this by combining the two constants into one and calling it R, that is, we let $N_A k \equiv R$. We then get

$$PV = nRT \tag{12-6}$$

$R = 8.314$ J/(K · mole) and is called the **universal gas constant.** Example 12-5 uses the ideal gas law in this form.

Example 12-5 *A Lot of Hot Air*

For a guided interactive solution, go to Web Example 12-5 at www.wiley.com/college/touger

A family on a winter vacation arrives at their one room cabin and finds that the inside wall thermometer reads $-11.0°C$. They start a fire in the fireplace, and after a while the temperature in the cabin reaches $19.0°C$. The cabin is 5.80 m by 4.80 m and has an average ceiling height of 2.40 m. If the cabin is otherwise well sealed, what mass of air goes out the chimney as the cabin warms up? Assume a mole of air has a mass of 29 g and that the cabin is at sea level. Also, neglect the molecules (oxygen) that are removed from the air in combustion reactions and therefore do not go out the chimney.

Brief Solution

Choice of approach. We can reason with the ideal gas law to understand *why* air goes out the chimney: Because the chimney allows air flow between interior and exterior, the interior remains at normal atmospheric pressure as it warms. The volume doesn't change either, so the left side of Equation 12-6 remains constant. This establishes a fixed value for the right side. To maintain this value, the number of moles of air in the cabin must drop as the temperature rises. We can use Equation 12-6 to calculate the number of moles in the cabin at each temperature. Knowing how many grams are in each mole, we can determine how much mass is lost.

What we know/what we don't. At sea level, $P = 1.01 \times 10^5$ N/m^2

$V = 5.80$ m $\times$ 4.80 m $\times$ 2.40 m	$R = 8.314$ J/K · mole	1 mole air = 29 g
$(T_C)_1 = -11.0°C \qquad T_1 = ?$	$(T_C)_2 = 19.0°C$	$T_2 = ?$
$n_1 = ? \qquad n_2 = ?$	$n_1 - n_2 = ?$	$m_1 - m_2 = ?$

The mathematical solution. Converting temperatures to kelvins, we get

$$T_1 = -11.0°C + 273.15 = 262 \text{ K} \qquad T_2 = 19.0°C + 273.15 = 292 \text{ K}$$

The volume $\qquad V = 5.80$ m $\times$ 4.80 m $\times$ 2.40 m $= 66.8$ m^3

Now we use Equation 12-6 to solve for the number of moles:

$$n_1 = \frac{PV}{RT_1} \quad \text{and} \quad n_2 = \frac{PV}{RT_2}$$

so $\quad n_1 - n_2 = \dfrac{PV}{R}\left(\dfrac{1}{T_1} - \dfrac{1}{T_2}\right)$

$$= \frac{(1.01 \times 10^5 \text{ N/m}^2)(66.8 \text{ m}^3)}{8.314 \text{ J/K} \cdot \text{mole}}\left(\frac{1}{262 \text{ K}} - \frac{1}{292 \text{ K}}\right) = 318 \text{ moles}$$

This equals an air mass of 318 moles $\times \frac{29 \text{ g}}{1 \text{ mole}} = 9222$ g or **9.2 kg**

◆ Related homework: Problems 12-11, 12-22, and 12-23.

12-3 Extending Kinetic Theory to Liquids and Solids

If the temperature of a gas is a measure of the average kinetic energy of the molecules in a gas, what does it tell us about a liquid or a solid? To address this, we can ask: What else does our model (Figure 12-6a) need to account for heat transfer from one body of gas to another through a solid barrier?

We already know that heat is transferred along or through a metal by conduction. But how? A solid's molecules cannot be traveling all over the place like those of a gas; the molecules must be sufficiently bound to one another for the solid to keep its shape. A mechanism presents itself if we picture the bonds to be spring-like (Figure 12-6b). We are not assuming they are actually connected by tiny springs but rather that, like springs, the interatomic forces between a molecule and all its neighbors tend to push it back into position when it starts to move away. This means that each atom or molecule in the solid will jiggle about some equilibrium position if it is disturbed. The more it jiggles, the more kinetic energy it has.

This is intended as a first glimpse of what is happening on a microscopic scale; the complete picture is substantially more complicated. In our simplified picture, we can think of a metal as a crystal lattice of ions with spring-like links between them, plus a gas of conduction electrons that can move from one ion site to another. When the temperature of a metal increases, there is more kinetic energy associated with the allowable vibrations in the lattice (the conduction electrons also gain energy). When faster gas molecules hit the metal wall in Figure 12-6a from above, they collide with the wall's outermost ions and affect their vibrational motion. Because of the spring-like forces between neighbors, the increased jiggling is passed on until the innermost molecules of the wall jiggle with greater average kinetic energy as well. This means they can transfer more energy to the gas molecules below that collide with them, somewhat as the bumpers in a pinball game transfer energy to the ball and speed it up.

In general, a change in the temperature of a material means a change in its total microscopically distributed internal energy. For an ideal monatomic gas, this is purely translational kinetic energy. For an ideal gas of diatomic molecules it may involve rotational and vibrational kinetic energy as well (recall Figure 12-5). For a liquid or solid, or even for a very dense gas, energies associated with interactions between molecules, atoms, ions, and so on, are also involved. But at temperatures approaching room temperature and higher, what increases with temperature in a solid is predominantly the mechanical energy associated with the vibrations transmitted from atom to atom. For atoms linked together by spring-like forces, each oscillating atom's energy converts back and forth between kinetic energy and elastic potential energy, so on average each kind of energy makes up half of the total. For solids, the temperature is an indicator of the level of this vibrational energy.

Now we can provide a rough qualitative explanation of the plateaus in the cooling curves in Figures 11-10c and 11-12. During a plateau, the cooling material continues to raise the temperature of its surroundings, even while its own temperature is not dropping. This shows it is continuing to transfer energy to their surroundings—it is losing energy. But because the temperature isn't dropping, we know it is not losing kinetic energy.

However, during each plateau the material is either condensing or solidifying. Stronger bonds are forming. When you break bonds, you pull things apart against the attractive forces they exert on each other, and you therefore increase their potential energy, just as you do when you pull a chair and Earth apart by lifting the chair (recall Figures 6-6 and 6-7 or WebLink 6-2). As the objects come back together, bonds are strengthened or new bonds are formed, and there is a decrease in potential energy. So during the plateaus, the energy of the substances continues to drop, but it is potential rather than kinetic energy. (*Note:* We are not talking here about the elastic potential energy associated with the oscillations.)

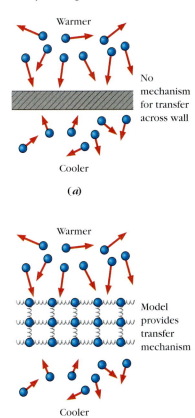

Figure 12-6 Modeling a solid capable of transferring heat.

Although this depiction of chemical bonding is not completely accurate, our explanation does give a reasonable qualitative picture of what happens during the plateau. The main idea is that a heat transfer to or from a system can change its thermal energy (evidenced by a change in temperature), its bonding situation (evidenced by a change of state or a chemical reaction), or both.

12-4 Work and Other Energy Aspects of Thermodynamics

We now focus on situations where the only changes in the energy of a system are changes in the distributed energies—thermal and chemical—rather than PE or KE changes associated with the overall motion of objects. For these situations, we can rewrite our conservation of energy equation (Equation 6-12) in a form called *the first law of thermodynamics*:

➡**Notation:** In other textbooks you may see this written as $Q - W = \Delta E_{\text{system}}$ or $Q - W = \Delta U$. U is another common symbol for the distributed energy of the system. With a $-$ instead of a $+$ sign, it means that W in the equation represents work done *by* the system rather than work done *on* the system.

First law of thermodynamics

$$Q + W = \Delta E_{\text{system}} = \Delta E_{\text{thermal}} + \Delta E_{\text{chemical}} \qquad (12\text{-}7)$$

Because heat and work both represent inputs to (or if negative, outputs from) the system, it says the total change in a system's energy is zero if there is no energy transfer to or from its surroundings; in other words, energy cannot be created or destroyed within the system.

When the only changes within the system are changes of temperature and changes of state—that is, when there are no chemical reactions involving a net input or output of energy—the right side of Equation 12-7 is identical to the left side of Equation 11-11 ($\Sigma cm\Delta T + \Sigma(\pm)Lm = \Sigma Q$). The left side of Equation 11-11 is really the sum of all the changes in the distributed energy of the system. Equation 12-7 is more general than Equation 11-11 because it allows for work as well as heat as an energy input, and it allows for other changes in chemical energy than those associated with changes of state.

To apply the first law of thermodynamics, we first have to identify a system. The heat and work on the left side of the equation are the energy inputs to the system; the right side of the equation tells us what becomes of the total energy input once it enters the system.

Case 12-1 ◆ *Revisiting Rumford*

Count Rumford (Section 11-3), struck by the high temperatures reached by the brass of the guns during the cannon-boring process, had contrived a way of submerging the boring tool and the part of the cannon being bored in a body of water. In doing so, he established a *system* on which he could make temperature measurements. In Figure 12-7, the system's boundaries are indicated by a dashed line. The work done by the horses to drive the boring mechanism is a positive energy input into the system. As the water temperature rises, some heat flows out of the system, so Q is negative. The *rate* of heat transfer $\frac{\Delta Q}{\Delta t}$ varies with conditions. We can imagine two extreme cases.

One extreme occurs when the rate of heat transfer $\frac{\Delta Q}{\Delta t}$ is much slower than the rate $\frac{\Delta W}{\Delta t}$ at which work is being done on the system. Then over a given time interval, the combined energy input $Q + W \approx W$. This was what happened in Rumford's experiment. As work

was done on the system, the heat transfer from the water bath to the surroundings was small, so almost all of the energy added to the system by doing work became distributed energy of the system, both thermal energy (as the system's temperature rose) and chemical energy (as the water boiled).

Processes in which the rate of heat transfer is negligibly small, so that $W \approx \Delta E_{\text{system}}$, are called **adiabatic** processes.

Figure 12-7b shows a more completely adiabatic version of the Rumford experiment.

The other extreme occurs when the rate of heat transfer is great enough to keep the system's temperature from changing. To keep the temperature of the cannon bore from rising, Rumford might have set things up as shown in Figure 12-7c. If the rapid pumping of

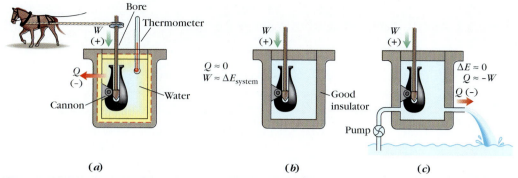

Figure 12-7 **Rumford's experiment viewed as a system.** (*a*) We can identify the region bounded by the dashed line as a system. (*b*) The usual version of the experiment is adiabatic: the heat transfer *Q* out of the system is negligible. (*c*) In this isothermal version of the experiment, the flow of water makes possible an energy output equal in magnitude to the work input.

water removes energy quickly enough, the temperature of the system will not rise at all.

> A process in which a system remains at constant temperature is called **isothermal.**

For isothermal processes in ideal gases, $\Delta E_{\text{system}} = 0$, so $Q = -W$.

For a step-by-step guided approach to the content of Case 12-1, go to WebLink 12-2.

Recognizing if a situation is adiabatic or isothermal can simplify our calculations. Think about whether either condition applies as you work through Example 12-6.

For **WebLink 12-2:** **Revisiting Rumford,** go to www.wiley.com/college/touger

Example 12-6 *The Beater Heater*

For a guided interactive solution, go to Web Example 12-6 at www.wiley.com/college/touger

A physics student, having discovered that a work input can increase the temperature of a system, decides to use an electric eggbeater to heat 0.30 kg of water for her coffee. The eggbeater is rated at 115 W. If the eggbeater were able to work with perfect efficiency, and the student were able to keep the water perfectly insulated, how long would it take the student to raise the water temperature from room temperature (20°C) to nearly boiling (95°C)?

Brief Solution

Choice of approach. The perfect insulation tells us this is an adiabatic process, so the work input equals the change in distributed energy. Knowing the mass and specific heat of water, we can find the energy input needed to raise its temperature by a known ΔT. Then because the power rating (115 watts) is the *rate* $\frac{\Delta W}{\Delta t}$ at which electrical energy is put in (or work is done), we can figure out how much time is required for the *total* energy input.

What we know/what we don't.

For water $m = 0.30 \text{ kg}$ $c = 4186 \text{ J/(kg} \cdot \text{°C)}$ $T_i = 20\text{°C}$ $T_f = 95\text{°C}$

         $W \approx \Delta E_{\text{system}} = ?$ $P = 115 \text{ W}$ $\Delta t = ?$

The mathematical solution. The only change in distributed energy is the one associated with a temperature change:

$$W \approx \Delta E_{\text{system}} = mc(T_f - T_i) = (0.30 \text{ kg})(4186 \text{ J/[kg} \cdot \text{°C]})(95\text{°C} - 20\text{°C})$$

$$= 94\ 300 \text{ J}$$

EXAMPLE 12-6 *continued*

Because power is defined as $P = \frac{\Delta W}{\Delta t}$,

$$\Delta t = \frac{\Delta W}{P} = \frac{94\ 300\ \text{J}}{115\ \text{J/s}} = \textbf{820 s} \text{ or } 13.7 \text{ min.}$$

◆ Related homework: Problems 12-26 and 12-28.

The first law of thermodynamics is expressed in terms of the work done on (not by) a system. This distinction is important when a body of gas at some particular pressure P changes volume. For example, suppose our system is the gas expanding with temperature at constant pressure in Figure 12-1*b*. As the piston moves upward a distance Δs, the work done *on* the gas is

$$W = F_{\text{by piston on gas}}\Delta s$$

The gas exerts an upward force $F_{\text{by gas on piston}} = +PA$ on the piston, so by Newton's third law, the piston exerts a downward force $F_{\text{by piston on gas}} = -PA$ on the gas. Then the work done on the gas is $W = -PA\Delta s$. Because $A\Delta s$ is the change ΔV in the volume,

$$W = -P\Delta V \tag{12-8}$$

ΔV is positive for an expanding gas. Although the work done *by* an expanding gas on its surroundings is positive, the work done *on* the gas is negative. In contrast, if the gas is being compressed, ΔV will be negative, making W positive. As you would expect, you have to do work to compress a gas.

Note: Equation 12-8 is only completely correct when the volume changes at constant pressure, as in Figure 12-3. Otherwise, it remains approximately correct only to the extent that ΔV is small enough so that the pressure changes negligibly.

Example 12-7 *The Bicycle Pump*

If you pump a bicycle pump rapidly, you can feel the pump heat up (try it). Why?

Solution
Interpreting the question. You need to recognize when words are being used with their everyday meanings rather than with the more limited definitions that physicists apply because of their need to make the language precise. The phrase "heat up" is everyday language and means that the temperature is increasing. It does *not* refer to the transfer of energy from a warmer to a cooler region that physicists call heat. In fact, as we will explain in greater detail, the pump "heats up" precisely because there is too little time for a significant heat flow out of the pump.

Choice of approach. With each stroke of the pump, you compress the air within (Figure 12-8). Work must be done on a system to change its volume. Because the only energy changes to consider here are changes in distributed energy, we can apply the first law of thermodynamics to relate the work done to other changes/inputs/outputs. The reference to *rapid* pumping prompts us to think about the rate at which these occur.

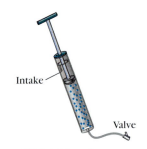

Intake

Valve

Figure 12-8 A bicycle pump.

The details. Each compression means a negative change in the volume of the air. If ΔV is negative, $W = -P\Delta V$ tells us the work done on the air is positive; although P is not constant here, the sign relationships still hold. If heat flows outward from the system, Q is negative. Then the total change in the system's energy ($\Delta E_{\text{system}} = W + Q$) will be positive if heat cannot flow outward as quickly as work is done. This will happen if the work is done too quickly because the rate of heat transfer outward ($\Delta Q/\Delta t = KA\Delta T/d$) is limited by the dimensions and thermal conductivity of the metal walls. If we just consider what happens during one very rapid stroke, the outward heat transfer is negligible. To the extent that it is negligible, the process is adiabatic. The change in the energy of the system is thermal ($\Delta E_{system} = cm\Delta T$) because there are no changes in chemical energy, so the temperature goes up.

♦ Related homework: Problem 12-27.

12-5 Irreversible Processes and the Tendency Toward Disorder

The first law of thermodynamics is a kind of balance sheet for keeping track of energy. The ability to keep track of different kinds of energy allows us to say a great deal about various sequences of events that occur in nature:

- *Sequence A:* A flower pot loses potential energy and gains kinetic energy as it falls (Figure 12-9a). When the flowerpot hits the ground, its kinetic energy is dissipated and transferred to its surroundings. Some of that energy becomes the activation energy needed to break the bonds that held the flowerpot together. Ultimately, the molecular store of energy that the flower requires for its life functions is also dissipated.

- *Sequence B:* Humpty Dumpty has his great fall (Figure 12-9b). The energy description is much the same as for the flower pot.

- *Sequence C:* A warm six-pack is tossed into an ice chest. Gradually the beverage cools off and the ice melts. Eventually everything reaches the same temperature, as energy gets transferred from the beverage to the ice.

In each sequence, we can account for much of what happens by keeping track of energy, and we are able to keep track because we know energy is conserved. If energy of one type is lost, energy of another type is gained: If one thing loses energy, something else must gain energy. These sequences display normal, real-world behavior.

If energy is not created from start to finish, neither is it created or destroyed from finish to start. So the same energy principles hold true for the sequences of events shown in Figure 12-10. But experience compels us to reject these as examples of real-world behavior. All the king's

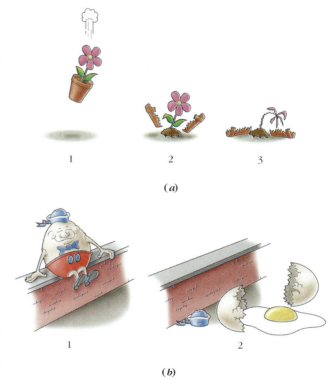

Figure 12-9 Two sequences that obey the law of energy conservation.

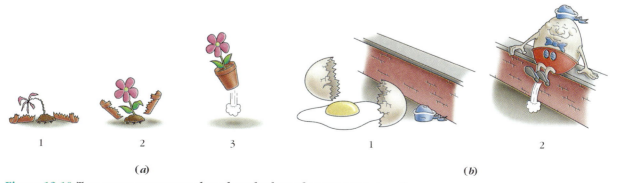

Figure 12-10 Two more sequences that obey the law of energy conservation.

horses and all the king's men could *not* put Humpty Dumpty together again. The question is, why not? Why can the sequence happen in one direction and not the other? Why can't we start out with things at the same temperature and end up with one thing hotter and the other colder? The energy gained by one thing (the six-pack) would still equal the energy lost by the other (the water cooling and turning to ice), so conservation of energy would still be obeyed.

But the evidence is that we can't. The sequences are not equally probable forward and backward. We call these **irreversible processes.** One way of stating the apparent prohibition of the "backward" sequence is that we can't reverse time. However we say it, we need to examine more closely what we mean.

Again we're looking for the pattern, the underlying rule by which nature is playing in all these situations. What distinguishes the sequences that can occur in the real world from those that can't? All of the irreversible situations that we have discussed so far involve changes on an atomic or molecular level. In trying to answer the question, we might find a clue from a sequence involving macroscopic objects. When we look at macroscopic objects, however, we must keep in mind that we are only looking for useful analogies to help shape our thinking. We must be careful not to assume that the same thermodynamic laws are involved. Remember that thermodynamics involves properties of systems that derive from the behavior of the molecules making up the system.

With that caution, consider the following sequence. When we unwrap a new deck of cards, the cards are all in order. When we shuffle the cards repeatedly, the cards become more and more disordered. Repeated shuffling seems never to bring us back to the original ace-through-king arrangement of each suit. **STOP&Think** If you have just one card, is it meaningful to ask whether it's in order or not? How about two cards? ◆

➡How close to impossible? Suppose you could shuffle the cards thoroughly every five seconds. If it took 8×10^{67} shuffles for the original order to come up again, it would take you more than 10^{61} years, which is more than 10^{50} times the age of the universe. Although this isn't *exactly* never, you'd be foolish to wait for it to happen.

The idea of order and disorder does not apply if you have only a single object; it applies only if you have enough objects so that being "in order" or "out of order" is meaningful. With two cards there are two possible arrangements, both equally likely. There is still no point in saying one is more ordered than the other. But for a deck of 52 cards, an incredibly huge number of arrangements (about 8×10^{67}) are possible, and all are equally likely. Your original, new-deck arrangement is only one of this number of equally likely possibilities. This number is so vast the odds are overwhelmingly against a particular arrangement or occurring again even over trillions of years of shuffling.

The card example shows us that when there are many objects in a system, and many different arrangements or states of the system can be achieved by a random process, the system tends spontaneously toward greater disorder. This is because the arrangements or states that we consider ordered are such a tiny fraction of the total number of possible states that their likelihood of occurring is negligibly small. The word *spontaneously* is important here. Certainly we could do work to put the deck back in order, but it does not spontaneously happen.

Now think about the falling objects in Figure 12-9. While they are falling, all like molecules of the object have roughly the same translational kinetic energy. None of the molecules in the object's surroundings have any of this kinetic energy. After the collision, this kinetic energy is much more randomly distributed as vibrational kinetic energy of molecules in both the object and its extended surroundings—the ground, the air, the eardrums of nearby observers, and so forth. For much the same reason, processes where motion is reduced by friction are irreversible.

What about the six-pack in the ice chest? To understand what happened when objects changed temperature, we had to think about what happens to the molecules. So the question now is, what happens when you take a collection of faster-moving molecules (the warm six-pack) and plunk it down in the middle of a collection of slower-moving molecules (the mixture of ice and cold water)? We can think about what happens in general when you begin with separate collections of faster and slower molecules and then bring them in contact by looking at a simpler example.

In Figure 12-11 a warm gas and a cool gas are separated by an insulating partition. The molecules in each gas are moving randomly in all directions, some bumping against the partition, but the molecules in the warm gas are on average moving more rapidly. When the partition is removed, molecules that would have collided with it pass into the other gas, and a mixing of faster and slower molecules occurs. The result is an in-between average speed for the mixture and therefore an in-between temperature reflecting the new average kinetic energy of the molecules.

At the beginning, the faster and slower molecules were sorted out by the barrier, like having all black cards on the top of the deck and all red ones on the bottom. In effect, the faster and slower molecules become "shuffled." We have gone from a more ordered to a less ordered (or more random) arrangement or state of the molecules. And because there are typically 10^{23} or more molecules in a sample of material, the number of possible states is vastly greater than for the cards, and the ordered states are vastly more improbable.

For heat to flow from a cooler to a warmer object, the hotter object would have to continue to increase its already excessive share of the faster molecules. This never occurs spontaneously. It can happen, but not without assistance. The mechanism of a refrigerator or an air conditioner, for example, creates a heat flow from an already cooler interior to a warmer exterior (put your hand on the coils on the back of your refrigerator to verify this), but in these devices an input of energy is required. It is possible, with an input of energy, to increase order locally, but this cannot happen without an increase in disorder elsewhere, such as a breakdown of fuel molecules at the power plant that provides the electricity for your refrigerator.

We have seen that systems with a large number of objects tend spontaneously toward arrangements or states that are vastly more probable and in general more disordered. When applied to a system of huge numbers of molecules, this underlies an observed fact that is so fundamental that it is called the *second law of thermodynamics:*

Second law of thermodynamics

Heat can never flow unassisted from a cooler object to a warmer object.

This will turn out to be only one of several equivalent ways of stating this law.

12-6 Entropy

The tendency toward disorder or randomness, in which systems go from low-probability highly ordered states to high-probability disordered states, was called "time's arrow" by the physicist Arthur Eddington. Here we must distinguish between **macroscopic states** and **microscopic states.** In gases, for example, a macroscopic state has to do with the gas as a whole; a microscopic state has to do with how the individual molecules contribute to the gas's overall condition. Macroscopic states are observable. For instance, when the partition in Figure 12-11 is removed, the gas reaches a uniform temperature that can be measured with a thermometer. But this average could result from different microscopic states: It could occur if all the molecules had the same speed or if some were faster and some were slower.

As with the macroscopic arrangements of cards, there are many more disordered microstates than ordered microstates, so systems tend spontaneously toward disorder. Statistical mechanics defines a quantity called the **entropy** of a system, which serves as an indicator of this tendency. We will not give the mathematical definition here, but its value for a given macroscopic state depends on the number of different microscopic states that give rise to the macroscopic state. Thus, when a system goes spontaneously from lesser to greater disorder, its entropy increases.

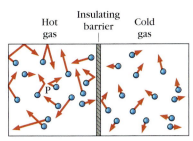

Figure 12-11 Temperature as a measure of molecular motion in gases. This is a schematic "time exposure" of gas molecules over a short time interval. The arrows show the motions of the molecules during the interval. The faster molecules go further during the interval, so their arrows are longer. The bent arrows indicate molecules rebounding off the walls. Within each gas, the speeds and directions of the molecules vary. But *on average,* as the arrow lengths show, the molecules in the hot gas travel faster; on average, they have more kinetic energy.

➡**Ordered versus uniform:** If you mix black sand and white sand, at a distance it will look uniformly gray. Some of you might intuitively consider this more ordered, but this is not what we mean by ordered here. As we are using the term, *ordered* means sorted out in some way. So if you sort out the grains of sand so you can see separate areas of white sand and black sand, we call that more ordered.

➥German physicist Rudolph Clausius (1822–1888) coined the word *entropy,* which he derived from the Greek word for transformation. He consciously chose a word that resembles *energy* to suggest it had comparable importance: As the first law of thermodynamics deals with energy, so the second law deals with entropy.

We have seen that those spontaneous processes in which entropy increases are irreversible. Prior to the development of statistical mechanics, Rudolph Clausius introduced the concept of entropy as a thermodynamic quantity associated with the macroscopic state of a system. His definition did not consider microstates or probabilities and was motivated by the related notion that spontaneous heat flow from hot to cold is irreversible. Actually, Clausius provided a definition not for entropy (denoted by S) but for ΔS. As is true of potential energy, the zero level is arbitrary according to his definition, and only changes in value have physical significance. He defined the change in the entropy of a system in terms of the heat Q flowing into the system and the system temperature T in kelvins:

Clausius's definition:

Change in entropy of a system $\Delta S \equiv \dfrac{Q}{T}$ (12-9)

Strictly speaking, this definition is valid only when the system changes reversibly. There can be no friction to dissipate energy; otherwise, heat transfers could not be reversed. However, we will not deal with situations in which this is an important consideration. Note also that Equation 12-9 involves a single temperature T, so strictly speaking it only applies when T is unchanging, as in phase changes. However, when the change ΔT is small compared to the initial and final temperatures, we can use the temperature's average value T_{av} during the process:

$$\Delta S \approx \frac{Q}{T_{av}} \qquad (\Delta T_{small})$$ (12-9a)

Clausius's definition may appear a bit strange and arbitrary at first. To get a feeling for its meaning, let's look at how its value changes from one situation to another.

Example 12-8   *Changes of Entropy during Melting and Freezing*

a. Find the change of entropy of a 2.0-kg block of ice that melts completely while remaining at the melting temperature of 0°C.
b. Find the change of entropy of a 2.0-kg puddle of water that freezes completely while remaining at 0°C.

Solution
Choice of approach. Knowing the mass and heat of transformation of the H$_2$O, we can find Q, keeping in mind that it will be positive when energy is added to the system (melting) and negative when energy is lost (freezing). We can then apply the definition of ΔS. Remember that T in the definition must be in kelvins.

What we know/what we don't.

$$m = 2.0 \text{ kg} \quad L_{SL} = 3.335 \times 10^5 \text{ J/kg}$$

$$T = 0°C = 273 \text{ K} \qquad Q = ? \qquad \Delta S = ?$$

The mathematical solution.
a. For melting,

$$Q = mL_{SL} = (2.0 \text{ kg})(3.335 \times 10^5 \text{ J/kg}) = 6.67 \times 10^5 \text{ J}$$

Then $\Delta S = \dfrac{\Delta Q}{T} = \dfrac{6.67 \times 10^5 \text{ J}}{273 \text{ K}} = \mathbf{2.44 \times 10^3 \text{ J/K}}$

b. For freezing, $Q = -mL_{SL}$, so $\Delta S = \mathbf{-2.44 \times 10^3 \text{ J/K}}$

◆ Related homework: Problems 12-35 and 12-36.

We see that ΔS is positive when the material changes from the more structured solid state to the more disorderly liquid state and is negative when the reverse happens. But that's also true for Q. What does ΔS tell us that Q by itself does not? To address this question, let's now find the total entropy change of a system in which a warmer body and a cooler body exchange energy.

Example 12-9 *Allowable and Forbidden Heat Exchanges*

Your well-insulated home is kept at a comfortable 21°C while the air temperature outside is a nippy −9°C. Suppose that during one second, 10.0 J of heat flow through the walls, windows, and roof of your home, far too little to significantly change either the inside or outside temperature. **STOP&Think** Does this indicate that your insulation is very good, very bad, or somewhere in between? How can you tell? ◆ Find the total change in the entropy of the system consisting of your home and the surrounding air if

a. as expected, the heat flowed from the warmer interior to the cooler air outside.

b. in violation of the second law of thermodynamics, the heat flowed from the cooler air outside to the warmer interior.

Solution

Choice of approach. The amount of heat that flows out of one body flows into the other; it is positive for the body it enters and negative for the body it leaves, so the total Q for the combined system is zero. The equal and opposite Q's, however, are divided by different temperatures, so the two changes of entropy that make up the total are not equal and opposite.

What we know/what we don't. (subscripts: H = home, E = exterior air)

a. $Q_H = -10.0$ J $\quad T_H = 21°C = (273 + 21)K = 294$ K $\quad \Delta S_H = ?$

$\quad Q_E = +10.0$ J $\quad T_E = -9°C = (273 - 9)K = 264$ K $\quad \underline{\Delta S_E = ?}$

$\qquad\qquad\qquad\qquad\qquad\qquad\qquad\qquad\qquad\qquad$ total $\Delta S = ?$

b. Same as above, except that $Q_H = +10.0$ J and $Q_E = -10.0$ J.

The mathematical solution.

a.
$$\Delta S_H = \frac{Q_H}{T_H} = \frac{-10.0 \text{ J}}{294 \text{ K}} = -0.0340 \text{ J/K}$$

and
$$\Delta S_E = \frac{Q_E}{T_E} = \frac{+10.0 \text{ J}}{264 \text{ K}} = \mathbf{+0.0379 \text{ J/K}}$$

so the total system change is $\Delta S_H + \Delta S_E = \mathbf{+0.0039 \text{ J/K}}$

b. Similarly, when the signs on Q_H and Q_E are reversed, we will get

$$\Delta S_H = +0.0340 \text{ J/K} \qquad \text{and} \qquad \Delta S_E = -0.0379 \text{ J/K}$$

so that the total system change $\Delta S_H + \Delta S_E = \mathbf{-0.0039 \text{ J/K}}$

◆ Related homework: Problems 12-37 and 12-64.

In Example 12-9, entropy increases (ΔS is positive) when the allowable heat transfer takes place, and it would decrease only if the forbidden heat transfer from colder to warmer were to take place (it never does). There is nothing special about the objects or the numbers in this example. The result we obtained is totally general, depending only on the fact that the two objects are at different temperatures and that the heat input to one is equal and opposite in sign to the heat output from the other. Therefore the entropy of a system—any system, including the universe as a whole—always increases when there are heat exchanges between the objects making up the system. A system's entropy can decrease only if it exchanges heat with matter outside the system. But for the universe, there is no "outside the system." The statement that heat can never flow unassisted from a cooler object to a warmer object is thus equivalent to this alternate statement of the second law of thermodynamics:

The entropy of the universe increases (or remains the same) in all physically possible processes.

The generality of the result in the last example is also not limited by the fact that because we considered the amount of heat transferred in just one second, we could consider the temperatures of the two bodies as remaining essentially constant. If the heat transfer continued over a long time, the temperature within the house would drop. But the temperature change during each second of that longer time period would still be negligible. We could let the temperature drop from any one-second interval to the next, ignoring the change in temperature during each second, so that for each one-second interval we could calculate $\Delta S = \frac{Q}{T}$, and then add up all the individual ΔS's to obtain a total ΔS for the entire time period. The signs would be the same for each second as they were for the second we considered in Example 12-9, so the entropy would increase for the allowable heat exchange and decrease for the forbidden one over the longer time period as well.

Analogies have their limitations. If we try to apply this thinking to the macroscopic situations involving cards that we considered as analogies, we must recognize that those sequences involved neither heat transfers nor changes of microscopic states. Even though there were macroscopic changes in order, $\Delta S = 0$ for these situations by Clausius's definition.

When systems move toward greater entropy, it lessens the usefulness of the energy in the system for doing work. In Figures 12-12a and b, the red block is released from the same height and approaches the green one. In a, it will collide with the green one and do work on it. In b, let us suppose that the friction force brings the red block to a stop just before it reaches the green one. As a result of the friction, the energy is dissipated. Rather than all the molecules having the same macroscopic displacement or the same velocity, the molecules are set to jiggling more as temperature increases, resulting in a random distribution of microscopic displacements and velocities. Distributed in this way, the energy is no longer useful for doing work on the green block. This reduction in the usefulness of energy for doing work that accompanies an increase in disorder is sometimes called the **degradation of energy.**

The Newcomen atmospheric steam engine. Perhaps the first steam engine of the Industrial Revolution was designed by Thomas Newcomen in England in the early 1700s. The boiler was connected to a cylinder, in which a piston was lifted by the pressure of steam. At the top of the stroke, cold water was sprayed into the cylinder. This condensed the steam, causing a vacuum, so that the atmospheric pressure at the top of the open cylinder forced the piston down again, completing the cycle. The device was called an atmospheric engine because the downward stroke, driven by atmospheric pressure, delivered the work output.

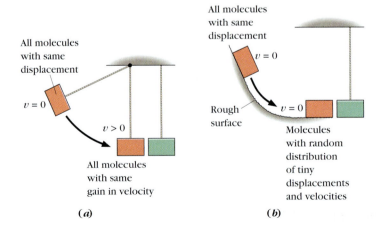

All molecules
with same
displacement

$v = 0$

All molecules
with same
gain in velocity

$v > 0$

(a)

All molecules
with same
displacement

$v = 0$

Rough
surface

$v = 0$

Molecules
with random
distribution
of tiny
displacements
and velocities

(b)

Figure 12-12 Whether energy remains useful for doing work (*a*) or not (*b*) depends on the extent to which disorder is increased.

12-7 Heat Engines and Refrigerators

The first law of thermodynamics, $Q + W = \Delta E_{system}$, is of special interest when applied to devices that receive a continuous heat input and produce a continuous work or mechanical energy output (so that Q is positive and W is negative). The steam engine and the internal combustion engine of an automobile are devices of this type. Most electric power plants are also steam engines at heart. Just as in an old-fashioned steam locomotive, a fuel (coal, oil, or a nuclear fuel) is used to boil water and produce steam that drives the moving parts. The principle is the same, whether the moving parts include the wheels of the locomotive or the rotating turbine of an electric generator.

Any device that converts a heat input into a work output in this way is called a **heat engine.** How can we tell if something is a heat engine? To explore this, let's consider the primitive set-up in Figure 12-13a. The burner provides a heat input, boiling the water in the kettle to produce the steam that turns the paddle wheel, so that work is done on the bicycle wheel, which is external to the system. The work output ultimately becomes the kinetic energy of the bicycle wheel. If there were nothing more to the set-up, the water would quickly be totally evaporated and there would be no further work output. A device that operates on a one-shot basis like this is not a heat engine.

We can improve on the set-up by adding the parts depicted in green in Figure 12-13b, turning it into a closed system in which the steam is condensed and the water cycled back to the kettle to be boiled again. A cyclical aspect that allows the work output to be ongoing is a required feature of all heat engines. In this cycle, the burner serves as a **high-temperature reservoir,** and the air cooling the parts where condensation takes place serves as a **low-temperature reservoir.** If these were not reservoirs, they would eventually reach a common in-between temperature, and repetition of the cycle would cease. Another feature of the cycle is this: Because heat is transferred to the cooler reservoir, not all of the energy put into the system as heat emerges as useful work. The heat engine is not 100% efficient. We can now summarize:

A **heat engine** is a device in which a working substance cyclically

1. absorbs heat from a reservoir at hotter temperature T_h,
2. does work on something external to the engine, and
3. transfers heat to a reservoir at cooler temperature T_c.

The cycle is often summarized by a schematic **flow diagram** like the one in Figure 12-14a, which shows the energy flowing into and out of the system. The

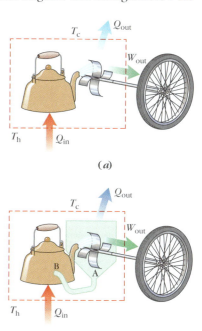

(a)

(b)

Figure 12-13 Turning a one-shot device (a) into a cyclical heat engine (b).

The Corliss steam engine at the Centennial Exposition in 1876. This giant steam engine was a central attraction at the exposition celebrating 100 years of American independence; it was viewed as symbolic of the nation's growing industrial might. The two cylinders drove a 56-ton flywheel 30 feet in diameter.

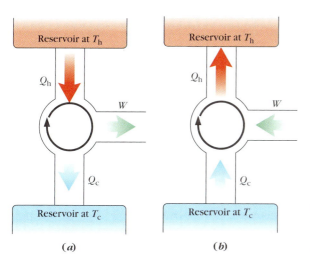

(a)   *(b)*

Figure 12-14 Schematic flow diagrams showing energy flow for (a) a heat engine and (b) a refrigerator.

Under the hood. Engines like this power state-of-the art motor vehicles today.

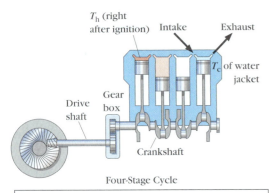

T_h (right after ignition) Intake Exhaust

T_c of water jacket

Drive shaft

Gear box

Crankshaft

Four-Stage Cycle

1) *Intake*: Air and gasoline admitted to cylinder as piston goes down

2) *Compression*: Piston moves up, compressing air-gasoline mixture

3) *Ignition*: Spark plug ignites mixture, creating pressure that drives piston downward

4) *Exhaust*: Products of combustion are forced out as piston moves back up

Figure 12-15 Main features of of a gasoline (internal combustion) engine. Combustion of fuel within each cylinder raises air pressure and forces the piston downward. The pistons move through a four-stage cycle (see box) in staggered sequence, thus turning the crankshaft.

For **WebLink 12-3:**
Flow Diagrams for Heat Engines and Pumps, go to www.wiley.com/college/touger

➥Freon and other chemicals of a group called *cholorofluorocarbons* (CFCs) were used as coolants in refrigerators and air conditioning systems for many years, until it was realized that their release into the upper atmosphere was causing the breakdown of the ozone layer.

working substance is the material that is cyclically warmed and cooled in the engine. For example, the water in a steam engine or the combustible mixture of air and vaporized fuel that exerts pressure on the pistons in an automobile engine (Figure 12-15).

A device that reverses the directions of the inputs and outputs in Figures 12-13 and 12-14*a* is the **refrigerator** (Figure 12-16). We show its flow diagram (Figure 12-14*b*) next to that of the heat engine for comparison. To develop this comparison further, work through WebLink 12-3. We will use the term **heat pump** to describe any device that, like the refrigerator, transfers heat from a cooler to a warmer place. As the second law of thermodynamics tells us, this cannot happen spontaneously; it requires an input of work.

Figure 12-16 shows the principles of operation of a common refrigerator. Its working substance is a liquid (the *coolant*) that boils at a fairly low temperature. The boiling temperature varies with the pressure. Work is done to compress the evaporated coolant so that it condenses, transferring heat to the air outside the refrigerator as it does so. It is then allowed to expand through an expansion valve into a larger volume, so that the pressure drops and the coolant evaporates again, drawing heat from the air within the refrigerator as it does so.

Another heat pump, the *air conditioner,* works on the same principle, but provides for regulated air flow into and out of the space being cooled. **STOP&Think** What would be the effect of running this kind of heat pump in reverse? Could that be useful? ◆

Operating in reverse, the same heat pump draws heat from the outdoor space and delivers it indoors. Thus a reversible heat pump can serve as both an air conditioner in warm weather and a heater in cool weather. In Texas, schools and other public buildings are retrofitting air conditioning systems with such heat pumps, using ground water as a temperature reservoir.

Because of the cyclical operation of engines and refrigerators, the change in their internal energy over a complete cycle is zero. Therefore the heat input is always just equal to the heat output plus the work done:

$$Q_h = Q_c + W$$

The **efficiency** of any device is the fraction of the total energy input that is converted to useful work: $eff = \frac{W_{out}}{E_{in}}$. Because energy cannot be created in the device, it is always a fraction ≤ 1 (or $\leq 100\%$). We define the **efficiency of a heat engine** to be the fraction of the heat input that is converted to useful work during each cycle:

$$eff = \frac{|W|}{|Q_h|} = \frac{|Q_h| - |Q_c|}{|Q_h|} = 1 - \frac{|Q_c|}{|Q_h|} \qquad (12\text{-}12)$$

We include the absolute value signs so that we are always working with positive numbers, simplifying calculation and interpretation.

Because the heat engine in Figure 12-14*a* had a nonzero heat output Q_c, its efficiency was less than one (or less than 100%). No heat engine ever created has been 100% efficient, and the laws of thermodynamics tell us that none ever will be. **STOP&Think** If your car had a perfectly efficient engine, would it also need a radiator? ◆ The heat engine could be 100% efficient only if the heat transferred to the cold reservoir returned to the hot reservoir of its own accord. But that would violate the second law of thermodynamics, which prohibits the unaided flow of heat from a colder to a warmer

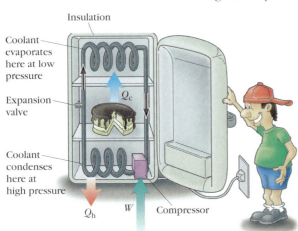

Insulation

Coolant evaporates here at low pressure

Expansion valve

Q_c

Coolant condenses here at high pressure

Q_h W Compressor

Figure 12-16 The refrigerator: a heat engine in reverse.

object. Therefore, we can state the second law of thermodynamics in yet other equivalent ways:

No heat engine is 100% efficient.

or

It is impossible to construct a cyclical heat engine that receives heat from a reservoir as its sole energy input and turns all of it into work.

or

There is no such thing as a perfect engine.

Although 100% efficiency is not possible, zero efficiency is. If a copper rod, insulated except at its ends, is connected between a hot and a cold reservoir, all the heat that enters the hot end is conducted through the rod and leaves the cold end, so $|Q_h| = |Q_c|$ and *eff* = 0. No work is done in this case.

Different heat engines operating between the same hot and cold temperature reservoirs may have different efficiencies, depending on the details of their cycles. In 1824, French engineer N. L. Sadi Carnot determined that to maximize the efficiency of the engine's cycle, both the system and the heat reservoirs had to return to exactly the states in which they had begun the cycle. In this sense, the engine is *reversible*. Such an engine is now called a **Carnot engine** and its cycle a **Carnot cycle.** If the engine is restored to the same state at the end of each complete cycle, the change in the entropy of the engine over a cycle must be zero. For each reservoir, $\Delta S = \frac{Q}{T}$. Because there is a heat output (−) from the hot reservoir and a heat input (+) into the cold reservoir, $Q_c = |Q_c|$, but $Q_h = -|Q_h|$. Then the total change is

$$\Delta S_{total} = \frac{Q_c}{T_c} + \frac{Q_h}{T_h} = \frac{|Q_c|}{T_c} - \frac{|Q_h|}{T_h} = 0$$

which after a couple of algebra steps becomes

$$\frac{|Q_c|}{|Q_h|} = \frac{T_c}{T_h}$$

This says that the minimum value of $\frac{|Q_c|}{|Q_h|}$ is $\frac{T_c}{T_h}$. Because the efficiency *eff* = $1 - \frac{|Q_c|}{|Q_h|}$ is a maximum when $\frac{|Q_c|}{|Q_h|}$ is a minimum, it follows that the maximum efficiency is

$$eff_{max} = 1 - \frac{T_c}{T_h} \tag{12-13}$$

(temperatures in K)

for any engine operating between reservoirs at the specific temperatures T_c and T_h.

The maximum efficiency is an upper limit on what nature will allow us to achieve, no matter what we do. Because T_c is never absolute zero, no real engine is ever 100% efficient, and most real-world engines operate at substantially less than this theoretical maximum. However, the Carnot engine provides a standard against which the efficiency of real engines can be judged.

Example 12-10 *Efficiency of a Diesel Engine*

Diesel engines compress air to a far greater pressure than ordinary internal combustion gasoline engines, and therefore the air and fuel are raised to a greater temperature, typically 700–900°C. What is the maximum efficiency of a diesel engine in which the air is raised to a temperature of 800°C?

EXAMPLE 12-10 *continued*

Solution

Choice of approach. Equation 12-13 tells us that we can calculate the upper limit on the efficiency of an engine if we know the temperatures of both the hot and cold reservoirs in kelvins. We therefore have to assume a reasonable temperature T_c for the cold reservoir and then convert both temperatures to kelvins. The cold reservoir is likely to be the temperature of the coolant or of adjacent air warmed by the engine, rather than the surrounding air temperature. A typical coolant temperature might be 100°C, because antifreeze and the increased pressure inside the cooling system would raise the boiling temperature of water used as a coolant to above 100°C.

What we know/what we don't.

$$T_h = 800°C = (800 + 273)\,K = 1073\,K$$

$$\text{Assume } T_c = 100°C = (100 + 273)\,K = 373\,K \qquad eff_{max} = ?$$

The mathematical solution.

$$eff_{max} = 1 - \frac{T_c}{T_h} = 1 - \frac{373\,K}{1073\,K} = \mathbf{0.652} \quad (\text{or } 65.2\%)$$

Making sense of the result. If we had used an ambient temperature of 20°C for T_c, we would have gotten an efficiency value of 0.727. Because we are estimating an upper limit on the efficiency, and these numbers are in the same general ballpark, either one of them will serve, because in any event the efficiencies of real diesel engines will be substantially lower. Nevertheless, diesel engines do offer greater efficiency than other types of internal combustion engines.

◆ Related homework: Problems 12-38 and 12-39.

The cooling towers of a nuclear power plant. Water used as a coolant in power plants becomes very hot. Because hot water released directly can have a negative environmental impact, cooling towers like these contain elaborate mechanisms for cooling the water before it is returned to the environment.

In general, in closed systems, as Example 12-9 showed, entropy is increased as objects within the system reach the same temperature. Ultimately, this means that all hot and cold reservoirs within the closed system move toward the same temperature, and the efficiency $1 - \frac{T_c}{T_h}$ of heat engines within the system approaches $1 - 1 = 0$. This is degradation of energy in the extreme; the thermal energy is not gone, but it is useless. This end result is sometimes called the "heat death" of the system. To avoid this, the system must be open rather than closed to allow energy inputs from its surroundings. As we noted earlier, living organisms accomplish this by eating.

The efficiency $\frac{|W|}{|Q_h|}$ of a heat engine compares the energy transfer (work) that we want to get out of it to the energy input Q_h. Heat pumps such as refrigerators and air conditioners are also rated by a ratio that compares the energy transfer we wish them to produce to an energy input that they require. In this case, work is an input and the transfer we want is Q_c, the heat transfer out of the cold reservoir.

	Input	Desired Transfer	Ratio Comparing the Two
Heat engine	Q_h	Work W external object	$\dfrac{|W|}{|Q_h|} = eff$
Air conditioner or refrigerator	W	Energy Q_c removed from cold reservoir	$\dfrac{|Q_c|}{|W|}$

The ratio used to rate air conditioners and refrigerators is called the *coefficient of performance*.

For a heat pump used to cool an interior space:
$$\text{Coefficient of performance} = \frac{|Q_c|}{|W|} \qquad (12\text{-}14a)$$

STOP&**Think** Can the coefficient of performance ever be greater than 1? ◆

Although the efficiency can never be greater than 1 (because you cannot get more total output than total input), the same constraint does not apply to the coefficient of performance, because it compares two inputs. A value of 3 or more might be reasonable for a home air conditioner: That means that for every joule of electrical energy input, 3 or more joules of thermal energy are transferred from cold to hot.

We noted earlier that heat pumps are sometimes used in reverse to warm interior spaces in winter. In that case, the coefficient of performance compares Q_h, the heat delivered to the interior space, to the work done:

For a heat pump used to warm an interior space:
$$\text{Coefficient of performance} = \frac{|Q_h|}{|W|} \qquad (12\text{-}14b)$$

Figure 12-17 Flow diagrams of two heat engines in series.

Flow diagrams like those in Figure 12-14 can be used as a basis for further reasoning, because when systems are combined we can just total their inputs and outputs. This can be done by combining flow diagrams, as in Figure 12-17. To see this approach in greater step-by-step detail, work through WebLink 12-4. The approach is applied in the following example.

For **WebLink 12-4: Combining Flow Diagrams**, go to www.wiley.com/college/touger

Example 12-11 *No Perfect Refrigerators*

For a guided interactive solution, go to Web Example 12-11 at www.wiley.com/college/touger

The less work input a refrigerator requires to make heat flow from a cooler to a warmer place, the better the refrigerator is. A perfect refrigerator is one that requires zero work input to produce this cooler-to-warmer heat flow. The second law of thermodynamics tells us that there are no perfect engines. Use flow diagrams to show that if a perfect engine existed, it could be combined with a real refrigerator to produce a perfect refrigerator.

Brief Solution
Choice of approach. We select a perfect engine that has a work output just equal in absolute value to the work input required by our refrigerator. We then combine these two devices into a single system and look at the features of the resulting flow diagram.

What we know/what we don't.
$$|W_{\text{out (from engine)}}| = |W_{\text{in (to refrig.)}}|$$

Flow diagram for combined system looks like ??

The diagrammatic solution. The combining of diagrams is shown in Figure 12-18. For the engine, since $Q_c = 0$, $|Q_{h1}| = |W_{\text{out}}|$; therefore $|Q_{h1}| = |W_{\text{in (to refrig.)}}|$. But the output of the ordinary refrigerator is made up of two inputs, one of which is W_{in}, so that $|Q_{h2}| > |W_{\text{in}}|$. Then $|Q_{h2}| > |Q_{h1}|$, and because the output is greater than the input they combine to give us a net *output* $Q_{h2} - Q_{h1}$. The resulting diagram has only a heat input from a cooler reservoir and a heat output to a warmer refrigerator; in other words, it is a *perfect refrigerator*.

EXAMPLE 12-11 *continued*

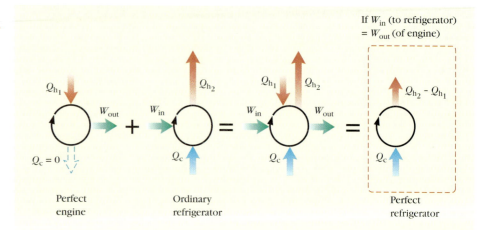

Figure 12-18 Flow diagrams for Example 12-11.

Perfect engine   Ordinary refrigerator   Perfect refrigerator

Making sense of the results. Our result tells us that if there were perfect engines (but there are not), there would be perfect refrigerators. If we read Figure 12-18 right to left, it tells us that if there were a perfect refrigerator, it could be decoupled into an ordinary refrigerator and a perfect engine. Because we cannot have a perfect engine, we cannot have a perfect refrigerator.

◆ Related homework: Problems 12-46 and 12-47.

✦ SUMMARY ✦

The Kinetic Theory of Gases

Assumptions

1. Gases are made of large numbers of tiny (negligible volume) building blocks called molecules (with no internal structure assumed).

2. The distances between molecules are very large compared to the size of the molecules, and also compared to the distance between molecules when the gas liquefies; gases are mostly empty space.

3. The molecules are in constant random motion.

4. The molecules of a gas collide elastically with each other and with the surfaces the gas is in contact with. They do not interact in other ways.

5. The molecules obey Newton's laws of motion.

Consequences

1. Because gas molecules keep colliding with the surfaces the gas is in contact with, the gas exerts pressure.

2. The molecules' average speed increases as the temperature of the gas increases.

The **kinetic theory of gases** is developed as a microscopic model on the basis of which we can explain the observed macroscopic patterns of behavior in gases (including *P proportional to T at constant V, V proportional to T at constant P*). A gas that behaves according to this model at all temperatures (never condensing to liquids as real gases do) is called an **ideal gas.** For the model to predict quantitatively how gases behave above their condensation temperatures, a further connection must hold true:

The absolute temperature is a measure of the average translational kinetic energy $\frac{1}{2}mv_{rms}^2 = (\frac{1}{2}mv^2)_{av}$ of the molecules;

that is,

$$(\tfrac{1}{2}mv^2)_{av} = \tfrac{3}{2}kT \qquad (12\text{-}2)$$

where **Boltzmann's constant** k_B or $k = 1.38 \times 10^{-23}$ J/K. Without a subscript, T always means T_K. With this assumption, we get the *ideal gas law.*

Ideal gas law	$PV = NkT$	(12-3)
or	$PV = nRT$	(12-6)

(N = total number of molecules, n = total number of moles, **Avogadro's number** $N_A = 6.02 \times 10^{23}$ is the number of molecules in a mole, and the **universal gas constant** $R \equiv N_A k = 8.314$ J/(K · mole))

See **Procedure 12-1** for guidelines on applying the ideal gas law.

The ideal gas law is an *equation of state,* relating state variables P, V, and T. **Thermodynamics** is the study of relationships among macroscopic state variables.

Temperature is a measure of the average translational kinetic energy associated with the random motion of atoms and molecules in liquids and solids as well as gases. This energy is **thermal energy.** During changes of state (solid–liquid or liquid–gas), there is a change in **chemical energy** (somewhat like potential energy) rather than thermal (kinetic) energy, so no temperature change accompanies the energy input or output.

When only the microscopically distributed energies—thermal and chemical—change in a system, the law of conservation of energy takes a form called the *first law of thermodynamics.*

First law of thermodynamics

$$Q + W = \Delta E_{system} = \Delta E_{thermal} + \Delta E_{chemical} \quad (12\text{-}7)$$

inputs	outputs
Q positive: heat transfer *into* the system	Q negative: heat transfer *out of* the system
W positive: work done *on* the system	W negative: work done *by* the system

In **adiabatic processes,** the rate of heat transfer to or from the system is negligibly small, so for short time periods $Q \approx 0$. In **isothermal processes,** the temperature of the system remains constant, so $\Delta E_{thermal} = 0$. When a gas expands against a resisting surface, the work it does is

$$W = -P\Delta V \quad (12\text{-}8)$$

Although the amount of energy in the universe always stays the same, the amount of disorder or randomness in the universe is always increasing. Faster and slower molecules become more randomly mixed when things come to the same temperature. If the opposite happened, it would violate the tendency toward disorder, which leads us to the first of the many ways of stating the second law of thermodynamics:

The second law of thermodynamics
(equivalent statements)

1. *Heat can never flow unassisted from a cooler object to a warmer object.*

   or

   There are no perfect refrigerators.

   or

   The entropy of the universe increases (or remains the same) in all physically possible processes.

2. *No heat engine is 100% efficient.*

   or

   It is impossible to construct a cyclical heat engine that receives heat from a reservoir as its sole energy input and turns all of it into work.

   or

   There are no perfect heat engines.

Processes that can only proceed in one direction and are prohibited by the second law from going in reverse are called **irreversible processes.**

Entropy can be defined in a way that treats it as a measure of disorder: Its value for a given macroscopic state depends on the number of different microscopic states that give rise to the macroscopic state. This definition is equivalent to *Clausius's thermodynamic definition.*

Change in entropy of a system
$$\Delta S \equiv \frac{Q}{T} \quad (12\text{-}9)$$

By either definition, ΔS is positive for irreversible processes and zero for reversible processes. The reduction in the usefulness of energy for doing work that accompanies an increase in entropy is sometimes called the **degradation of energy.**

A **heat engine** is a device in which a working substance cyclically

1. absorbs heat from a reservoir at hotter temperature T_h,
2. does work on something external to the engine, and
3. transfers heat to a reservoir at cooler temperature T_c.

A **temperature reservoir** is an object that can exchange any reasonable amount of heat with an ordinary-sized object with a negligible change ΔT in its own temperature.

The **efficiency** of any device is defined as

$$eff = \frac{W_{out}}{E_{in}}.$$

The **efficiency of a heat engine** is the fraction of the heat input that is converted to useful work during each cycle:

$$eff = \frac{|W|}{|Q_h|} = \frac{|Q_h| - |Q_c|}{|Q_h|} = 1 - \frac{|Q_c|}{|Q_h|} \quad (12\text{-}12)$$

The *maximum or Carnot efficiency* is

$$eff_{max} = 1 - \frac{T_c}{T_h} \quad (12\text{-}13)$$

(temperatures in K)

for any engine operating between reservoirs at the specific temperatures T_c and T_h.

A **heat pump** is any device that, like the **refrigerator,** transfers heat from a cooler to a warmer place. This always requires an input of work. The *coefficient of performance* is used to rate heat pumps:

For a heat pump used to cool an interior space:
$$\text{Coefficient of performance} = \frac{|Q_c|}{|W|} \quad (12\text{-}14a)$$

For a heat pump used to warm an interior space:
$$\text{Coefficient of performance} = \frac{|Q_h|}{|W|} \quad (12\text{-}14b)$$

The cycles of heat engines and refrigerators are often summarized by schematic **flow diagrams** (Figure 12-14), showing the system's energy inputs and outputs. These diagrams can be used for reasoning about composite systems (Example 12-11).

◆QUALITATIVE AND QUANTITATIVE PROBLEMS◆
Hands-On Activities and Discussion Questions

The questions and activities in this group are particularly suitable for in-class use.

12-1. *Discussion Question.* A student observes that the pressure in the tires of an automobile is greater when they are very hot. The student explains this by saying, "When the temperature goes up, the molecules want to spread out. But the tire walls are stopping them from doing that, so the pressure increases." Critique this explanation. As you do so, you might want to consider the following questions.

a. When the tires get hotter, do you think there are more air molecules closer to the walls of tires than further in, or do you think the molecules are uniformly distributed? In each case, how would the pressure vary from one location within the tire to another? What would happen to the air molecules in a region of higher pressure if there were a region of lower pressure next to it?

b. What, if anything, is the student suggesting that the molecules of a gas must do to exert pressure?

12-2. *Discussion Question.* *The effect of concentration on the net movement of molecules.* In this chapter, we introduced the idea that the molecules of a fluid are in random motion. Suppose a gas-filled chamber is separated into two equal compartments by a wall. Each compartment contains the same kind of gas at the same temperature, but compartment A contains a higher concentration of gas than compartment B. (The **concentration** is the amount of mass of a substance in each unit of volume [in kg/m^3], but we don't speak of it as density when the substance is dispersed through the volume with space or other matter in between.)

a. Suppose an opening is formed in the dividing wall between the two chambers. Will some molecules travel from each chamber into the other? Why?

b. Suppose that during a time interval Δt, a certain fraction of the molecules from chamber A make it into chamber B, and a certain fraction of the molecules from B make it into A. Will these two fractions be equal, or will one be greater than the other, and if so, which? Briefly explain your reasoning.

c. During the interval described in *a,* will the total number of molecules that go from B to A be less than, equal to, or greater than the total number that go from A to B? In other words, will there be a net movement of molecules in one direction?

d. A very long time after the opening is formed, would you expect that the concentration of molecules would still be greater in chamber A? Make sure that your answers to the previous parts of this problem are consistent with what you expect to happen.

Review and Practice

Section 12-1 The Kinetic Theory of Gases (Qualitative)

12-3. When air within a solid rigid container is warmed, which of the following remains constant? *the average distance between molecules; the average speed of the molecules; the air pressure within the container; the average impulse that a molecule delivers to the container walls.*

12-4. Assume a model of a gas in which the molecules are spaced far apart and are in random motion. Suppose that, contrary to the kinetic theory of gases, the speed of the molecules did not change with temperature.

a. Could this model explain why the air pressure within a closed space increases when the volume is reduced at constant temperature? Briefly explain.

b. Could this model explain why the air pressure within an enclosure having fixed volume decreases when the temperature is lowered? Briefly explain.

12-5. SSM Suppose the piston in Figure 12-2 is lowered while the water bath keeps the air temperature within the cylinder constant. Tell whether each of the following increases, decreases, or remains the same as this happens.

a. the average speed of the molecules striking the underside of the piston;

b. the average impulse due to the molecules striking the underside of the piston;

c. the number of molecules striking the underside of the piston each second;

d. the total pressure exerted on the underside of the piston.

12-6. A body of air is trapped below a piston that is free to move frictionlessly within a cylinder, as in Figure 12-1*b*. The cylinder starts out at height h_i. If the cylinder and the air within it are cooled, the cylinder moves until it reaches a final height h_f. Is the pressure that the air within the cylinder exerts on the piston when it is at h_f less than, equal to, or greater than when it is at h_i? Briefly explain.

12-7. A body of air is trapped below a piston that is free to move within a cylinder, as in Figure 12-1*b*. Suppose the piston starts out at height h_1. When the air temperature is raised by some fixed amount, the piston rises until it reaches height h_2. Then

a. is the air pressure against the bottom of the cylinder when it is at height h_2 greater than, equal to, or less than the pressure when it is at height h_1?

b. is the number of molecules striking the bottom of the cylinder each second when it is at height h_2 greater than, equal to, or less than the number when it is at height h_1?

c. is the average speed of the molecules striking the bottom of the cylinder each second when it is at height h_2 is greater than, equal to, or less than the average speed when it is at height h_1?

12-8. Tell in your own words why volume must increase to restore equilibrium in Example 12-2.

Section 12-2 The Kinetic Theory of Gases (Quantitative)

12-9. Containers A and B each hold the same number of gas molecules. The average speed of the molecules is the same in both containers. The containers themselves are identical except that container A is twice as long as container B. What effect (if any) does this difference in length have on the pressure that the two gases exert on the interior walls of their containers? Express your comparison as a ratio P_A/P_B of the pressures, and explain your answer by considering the relevant aspects of what molecules do to exert pressure.

12-10. In the kinetic theory of gases, a gas's temperature is a measure of the average random kinetic energy of its molecules, which are colliding with great frequency with one another and with the surfaces they come in contact with. We might expect the molecules to be slowed down by all these collisions. That would result in a lower average kinetic energy and therefore a drop in temperature. But gases do *not* spontaneously keep getting cooler. According to the theory, why doesn't this happen?

12-11.
a. How much volume is occupied by a mole of air at a pressure of 1 atm $(1.01 \times 10^5 \text{ m/s}^2)$ and a temperature of 293 K? If $1 \text{ L} = 10^{-3} \text{ m}^3$, how many liters is this?
b. Helium balloons rise in air. If you redid the calculation in *a* for helium at the same temperature and pressure, would the volume that you would get be greater than, equal to, or less than the volume you found for air? Briefly explain your answer.

12-12.
a. Use the result of Problem 12-11a and the fact that the density of air at this temperature and pressure is 1.21 kg/m^3 to find the mass of a mole of air.
b. What is the average mass of any one molecule in the air?
c. What is the average speed of an air molecule at this temperature? How does this compare with other speeds that you are familiar with?
d. How far would an air molecule at this temperature go in one second if it didn't bump into anything?
e. Would you expect an air molecule to actually travel that far each second? Briefly explain.

12-13. At a normal room temperature of 294 K, which (if either) would be moving faster on average, the oxygen molecules or the carbon dioxide molecules? Briefly explain.

12-14. Does a single molecule have a temperature? Briefly explain.

12-15. **SSM** A gas is contained in a sealed container whose volume changes negligibly on heating. At what temperature will the gas pressure be double what it is at 20°C?

12-16. A column of air is trapped under an oil drop in a capillary tube open at the top. The column has a height of 4.2 cm when the temperature is 300 K. The oil drop is 0.2 cm top to bottom. If the capillary tube has a total length of 7.0 cm, at what temperature will oil begin to come out of the top of the capillary tube? Assume the expansion of the oil drop itself is negligible, as is expansion of the capillary tube.

12-17. Suppose you seal off the outlet of a bicycle pump and slowly push the handle in so that the space occupied by the air inside is reduced from 400 cm^3 to 350 cm^3. If the air inside is at normal atmospheric pressure when the handle is all the way out, what is the air pressure inside after you've pushed the handle in?

12-18. Two sealed cylindrical containers A and B are filled with air. Compare the values of the air pressures P_A and P_B in the two containers (express the comparison by the ratio P_A/P_B) in each of the following cases, in which the only difference between the two situations is the one stated.
a. The molecules are moving twice as fast in container A.
b. The radius of container A is $\frac{2}{3}$ of the radius of container B.
c. The air in container A is 20% denser than the air in container B.
d. The air temperature is 20°C in container A and 60°C in container B.

12-19. **SSM WWW** A sample of gas is enclosed in a compressible container. If its pressure triples as its volume is halved,
a. by what numerical factor does its temperature in kelvins get multiplied? $(T_2 = \underline{\ ?\ } T_1)$
b. Does its temperature in °C get multiplied by the same number? Briefly explain.

12-20. The ideal gas law can tell us how air pressure changes with volume in a body of air under which of these conditions?
a. the air is trapped under a rising piston in a thin-walled metal container immersed in ice water.
b. the air is trapped under a descending piston in a well-insulated container.
c. the air is contained within a rapidly pumped bicycle pump.
d. the air is contained within a sealed aluminum can that is warmed very slowly.

12-21. The air pressure in a certain airtight container increases by 75% when the temperature in kelvins doubles. What else changes, and how many times its original value is it after the change?

12-22. In Example 12-5, how many molecules of air leave the chimney?

12-23. An open, empty one-liter bottle sits in a room with an air temperature of 22°C. If the bottle is then immersed in ice water with the opening of the bottle barely above the level of the water,
a. does a net amount of air enter or leave the bottle?
b. how many moles of air enter or leave?

Section 12-3 Extending Kinetic Theory to Liquids and Solids

12-24. A fire can be thought of as a body of gas that is so hot that it glows. When an iron pot filled with water is placed over a fire, the water warms up. Describe the mechanism by which the fast-moving molecules of the gas cause the water molecules to have increased kinetic energy.

12-25. The mass of an iron atom is 9.277×10^{-26} kg. At what temperature will the atoms in a block of iron be jiggling at an average (rms) speed of 600 mi/h? (A useful conversion relationship is 1.000 mi/h = 0.447 m/s.)

Section 12-4 Work and Other Energy Aspects of Thermodynamics

12-26. When an electric light bulb is first switched on, it gets hot. After a brief initial period, its temperature levels off to a constant value. Suppose we make the following crude approximation: For the first second after a 60-W bulb has been switched on, we have an adiabatic process, but after it has been on for a while the energy transfer process becomes isothermal. Based on this approximation, find ΔQ and $\Delta E_{thermal}$ for this system

a. during the first 0.1 s the bulb is on.

b. during a 0.1-s time interval after the bulb has remained on for several minutes.

12-27. SSM WWW When the plunger of a certain bicycle pump is all the way out, the column of air contained in the pump has a mass of 7.6×10^{-4} kg and a length of 0.50 m. When the plunger is pushed all the way in, the length of the column of air is halved. Suppose that when the plunger is all the way out, the air outlet of the pump is blocked off to prevent any air from escaping. Then to push the plunger all the way in, you find you must exert an average force of 88 N. If you do this extremely rapidly, calculate the approximate amount the temperature of the column of air would increase, and briefly explain why your calculation is only an approximation.

12-28. A simplified version of Joule's apparatus (Figure 11-9) is set up, with only a single descending weight and not very good insulation. In this simplified set-up, there are 0.80 kg of water in the container, and the weight is 15 N. The set-up is mounted on an upper-story window ledge of a tall building, so that the weight can fall a distance of 40 m in a single descent at essentially constant speed. Over one such descent, the water temperature increases by 0.10°C. How much heat is transferred from the water to its surroundings during this descent?

12-29. A *winch* is an iron crank-and-pulley mechanism for raising heavy weights. A particular winch, known to be 75% efficient, is used to lift a 300-N crate a distance of 12 m.

a. The raising is done slowly enough so that the temperature of the winch itself does not increase. How much heat is transferred to the surroundings as the winch lifts the crate?

b. When the crate is raised a bit more quickly, the temperature of the winch, which has a mass of 2.0 kg, increases by 0.25°C. How much heat is transferred to the surroundings during this lifting?

12-30. Suppose the cylinder in Figure 12-1*b* is filled with 0.045 kg of air and is very well insulated. The piston is then pushed down until the air temperature increases by 0.30°C.

a. Is this process adiabatic, isothermal, both, or neither?

b. How much work has been done on the air?

12-31. Suppose the enclosed gas in Figure 12-3 is kept at a constant pressure of 1.1×10^5 Pa. If the enclosed gas initially occupies a volume of 9.3×10^{-3} m^3 and the piston does 88 J of work on the gas as it is cooled, what volume does the gas occupy after cooling?

Section 12-5 Irreversible Processes and the Tendency Toward Disorder

12-32. A crate is tossed onto the back of a moving pick-up truck. The crate slides until it reaches the same velocity as the truck; that is, until it has zero velocity relative to the truck. Briefly explain why this is not a reversible process by describing

a. the macroscopic features that make it irreversible.

b. what happens to the molecules of the involved objects and their surroundings.

12-33. Your bath water has cooled off. To warm it up, you run the hot water (which is very hot) for a few seconds, and then shut it off. You do not stir the water, so when you first close the faucet the water is much warmer near the faucet. Eventually the temperature of the water becomes more uniform. Is this a shift toward greater order or greater disorder? Briefly explain your answer in terms of what happens to the molecules.

Section 12-6 Entropy

12-34. When a system undergoes an adiabatic process, is the change in entropy always positive, always zero, always negative, or can it vary in sign?

12-35. SSM A jewelry maker pours 0.040 kg of molten silver into a mold.

a. Calculate the change in entropy that takes place as the silver solidifies. Refer to appropriate tables as needed.

b. Is the final state of the silver more or less disordered than the initial state? Briefly explain how the value you calculated tells you that.

12-36. Liquid nitrogen is commonly used as a coolant by physicists who study the properties of materials at very low temperatures. Dr. Myriam Sarachik, a well-known condensed matter physicist, has immersed her sample container in 2.5 kg of liquid nitrogen.

a. Calculate the change in entropy that will take place as all the nitrogen evaporates.

b. Is the final state of the nitrogen more or less disordered than the initial state? Briefly explain how the value you calculated tells you that.

12-37. A tank for storing water is separated into two compartments by a good but not perfect insulating wall. Each compartment contains 1.0 kg of water. The water in one compartment is at room temperature (293 K). A heating element keeps the water in the other compartment at 373 K.

a. Suppose there is a heat transfer of 100 J through the wall between the two compartments. Will a heat transfer of this size significantly affect the water temperature in either compartment? How do you know?

b. Consider the water in both tanks as a total system. What would be the change in entropy in the system if the 100 J are transferred from the warmer to the cooler water?

c. What would be the change in entropy in the system if the 100 J are transferred from the cooler to the warmer water?

d. Can the heat transfer in *c* occur spontaneously? What aspect of the entropy change that you calculated tells you whether it is or is not?

Section 12-7 Heat Engines and Refrigerators

For the problems below, recall that 1 horsepower = 1 hp = 746 W.

12-38. Compare the maximum possible efficiency that can be achieved by a steam engine on a hot day and on a cold day. Identify the factors you considered in your reasoning.

12-39. If the boiler of a steamship's engine maintains a temperature of 523 K on a day when the outside air temperature is 273 K, what is the maximum possible efficiency the engine can achieve on that day?

12-40. An electric jigsaw draws a maximum of 300 W of electric power and its maximum mechanical power output is given as Max. hp: $\frac{1}{7}$. Operating at maximum output, how much energy does the saw draw each second, how much work does it do each second, and what is its efficiency?

12-41. **SSM** The power output of an automobile is given as 100 hp.
a. How much work does the engine do in 1 second? Answer in SI units.
b. If the engine is 28% efficient, how much heat is provided by the ignition and combustion of fuel in 1 second?
c. How much heat is transferred out of the engine in 1 second?

12-42. The engine of a Ford Model A car was rated by a 1931 handbook at "well over 24.03 horsepower." According to this same handbook, "The loss through the water jacket of the average automobile power plant [engine] is over 50% of the total fuel efficiency. This means that more than half of the heat units that should be available for power are absorbed and dissipated by the cooling water. Another 16% is lost through the exhaust valve, and but $33\frac{1}{3}$% of heat units do useful work." (V. Pagé, *Ford Model "A" Car and "AA" Truck*, N. W. Henley, New York, 1931).
a. How much useful work did this engine do in 1 second?
b. How much heat was provided by the ignition and combustion of fuel in 1 second?
c. How much heat was transferred out of the engine in 1 second?

12-43. Referring to suitable energy flow diagrams, compare and contrast the operation of a heat engine and a refrigerator.

12-44.
a. In Figure 12-16, is the amount of heat that is transferred to the air outside a refrigerator when the coolant condenses less than, equal to, or greater than the amount of heat that is removed from the refrigerator's interior when the coolant evaporates? Briefly explain.
b. If you left the refrigerator door open, would the overall effect of the refrigerator on the air in the room be to cool it, warm it, or leave its temperature unchanged? Briefly explain.

12-45. An air conditioner has a coefficient of performance of 3.2 and a power rating of 620 W. How long would it take this air conditioner to remove 1 000 000 J of thermal energy from the room that it was cooling?

12-46. Heat engine B operates on the waste heat from heat engine A. Of each 10 J that engine A draws from a hot reservoir, it eliminates 6 J as waste heat. For each 6 J that engine B draws from engine A, it eliminates 4 J as waste heat. Draw the energy flow diagrams for the two engines separately and for the combined system.

12-47. In problem 12-46,
a. what is the efficiency of each individual engine?
b. what is the overall efficiency of the two-engine system?

12-48. For each 4 J of work done by an air conditioner in a bedroom, 15 J of thermal energy are transferred to the air outside the house. Does this violate conservation of energy? Briefly explain.

12-49. For each 100 J of work done by a wall freezer unit in a grocery store, 350 J of heat are transferred to the air outside the store. Find the freezer unit's coefficient of performance.

Going Further

The questions and problems in this group are not organized by section heading, so you must determine for yourself which ideas apply. Some of them will be more challenging than the Review and Practice questions and problems (especially those marked with a • or ••).

12-50. A piston is pushed down on a trapped body of air within a well-insulated cylindrical container. (The basic arrangement is like Figure 12-1b.) Which of the following statements correctly tells what happens and also gives the correct reason?
a. The temperature of the gas increases because the ideal gas law $(PV = NkT)$ tells us that temperature increases as pressure increases.
b. The temperature of the gas decreases because the ideal gas law $(PV = NkT)$ tells us that temperature decreases as volume decreases.
c. The temperature of the gas increases because the work input exceeds the heat output.
d. The temperature of the gas decreases because negative work is done on the gas.

12-51. When the two ends of a long gas-filled tube are maintained at very different temperatures (assume the left end is warmer), the gas within the tube will be less dense near one end than near the other. At which end will the gas be denser? Explain why this is so in terms of the detailed behavior of the molecules, assuming they behave in accordance with the kinetic theory of gases.

12-52. The density of air is typically about 1.2 kg/m^3 at 1 atmosphere (about 1×10^5 N/m^2) of pressure at 20°C. Calculate the speed of a typical molecule of this air. How does this compare with the speeds of any everyday objects that you are familiar with?

12-53.
a. Suppose you have a mixture of hydrogen and oxygen gas at 300 K. According to Equation 12-2, which molecules, if either, will tend to be moving faster on average?
b. In the kinetic theory of gases, molecules collide elastically. Suppose two molecules A and B with unequal masses $(m_A > m_B)$ travel in opposite directions at the same speed and have a head-on collision. After the collision, does one molecule have greater speed than the other, and if so, which one?
c. Your result in *b* tells what happens in a typical event in a mixture of gases. Is this what you would have expected based on your answer to *a*? Briefly explain.

•**12-54.**

a. Suppose that in Problem 12-53**b**, $m_A = 2m_B$. In considering how the speeds of the two molecules compare after the collision, express the comparison by finding their after-collision ratio $\frac{\text{speed of A}}{\text{speed of B}}$.

b. Does either molecule have the same kinetic energy after the collision as it had before?

c. Is the average (mean) of the two kinetic energies different after the two collisions than it was before?

12-55. Argon atoms have about twice the mass of neon atoms. A laboratory apparatus has two chambers separated by a thin copper wall. One of the chambers is filled with argon, the other with neon, and the entire apparatus is kept at room temperature. How does the average speed of the neon atoms compare to the average speed of the argon atoms? Express the comparison by finding their ratio $\frac{\text{average speed of neon atoms}}{\text{average speed of argon atoms}}$.

12-56. Three identical rigid-walled tanks are filled with equal masses of oxygen and sealed. A year later, all the tanks are still sealed. Tank A is at room temperature in a sea-level research lab, tank B is in the open basket of a weather balloon at an altitude of 10 000 m above Earth's surface, and tank C has accompanied astronauts to the moon, where it remains in a compartment of their rocket that is kept at room temperature. Rank the three tanks A, B, and C in order of the pressure exerted by the oxygen within each tank. Order them from least to greatest, noting any equalities. Briefly state reasons for your ranking.

12-57. When a liquid is placed in the bottom of a closed container, liquid will evaporate until the pressure of the vapor is high enough to prevent further net evaporation. At this pressure an equilibrium condition is reached: The rates of evaporation and condensation are the same, so there is no net change in the amount of vapor. This equilibrium pressure is called the **saturated vapor pressure.** The saturated vapor pressure of water at a room temperature of 293 K is about 2.3×10^3 N/m^2 or 2.3×10^3 Pa.

a. Suppose that a sealed chamber is partially filled with room temperature water, so that 0.50 m^3 of unfilled space remain above the water's surface. How many moles of the water will evaporate?

b. If the ceiling of the chamber is now lowered at constant (room) temperature until only 0.35 m^3 of unfilled space remain above the water's surface, how many moles of the water vapor will condense?

12-58. A column of air is trapped under a drop of water in a narrow, thick-walled glass tube that is open at the top. Its temperature undergoes three successive temperature drops:

i. from 15°C to 5°C

ii. from 5°C to −5°C

iii. from −5°C to −15°C

Rank the three temperature drops i, ii, and iii in order of each of the following. In each part, order them from least to greatest, noting any equalities.

a. The volume drops that they bring about in the column of air. (Briefly explain your reasoning.)

b. The pressure drops that they bring about in the column of air. (Briefly explain your reasoning.) (*Hint:* Did you take the freezing point of water into account?)

12-59. A body of gas is contained in a thin-walled metal cylinder closed off by a piston. A sudden rapid motion of the piston reduces the volume the gas occupies by 20%. At the instant the motion is completed, the pressure is observed to have increased by 30%. If the gas was at room temperature (293 K) before the motion, what is its temperature immediately after?

12-60.

a. If there is no further motion of the piston in Problem 12-59, which of the following happens to the temperature of the gas over the next several minutes? *It continues to have the value found in **a**; it returns to 293 K; it increases; it drops slightly.* Briefly explain why.

b. Expressed as a multiple of the gas pressure before the piston motion, what would be the pressure several minutes after the motion is completed? ($P_{\text{after}} = \underline{\ ?\ }\, P_{\text{before}}$)

12-61.

a. Because the lid in Figure 12-19 is held in place by a spring, it serves as a release valve to prevent the air pressure in the container from exceeding a certain maximum

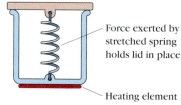

Figure 12-19 Problem 12-61

value. Find this value if the interior of the container has a radius of 0.060 m and depth of 0.150 m, and the spring exerts a force of 400 N when the lid is in place. (Assume the weight of the lid is negligibly small compared to this force.)

b. The air within the container has normal atmospheric pressure at a room temperature of 293 K. Apply the ideal gas law to this air. If the temperature increases to 350 K, what other quantity in the ideal gas law changes? Find its value at 350 K.

c. The heating element now continues to raise the air temperature in the container. At what temperature is the maximum permitted pressure first reached?

d. After the air temperature reaches the value obtained in **c**, it increases by an additional 20%. During this interval, what other quantity in the ideal gas law changes? Find the value of this quantity at the final temperature.

12-62. A piston slowly advancing into an air-filled metal cylinder reduces the volume occupied by the air from 4.00×10^{-3} m^3 to 2.50×10^{-3} m^3 (*Note:* 1 liter = 1.00×10^{-3} m^3).

a. If the air pressure in the cylinder is 1.00×10^5 Pa at the instant the cylinder begins its motion, find the pressure at the instant the motion is completed.

b. If the cylinder and piston were made of a perfectly insulating material rather than metal, would the pressure you calculated in **a** have been higher, lower, or the same? Use the ideal gas law to explain your answer.

12-63. During a very small time interval on a hot summer day, an air conditioner transfers 100 J of energy from the interior of a movie theater to the air outside. Consider a system consisting of both the theater and the air outside.

a. If the air within the theater remains constant at 20°C and the air temperature outside is 31°C, find the overall change in the entropy of this system.

b. Does your answer to **a** tell you that disorder is increased or decreased by this energy transfer?

c. Does this energy transfer violate the second law of thermodynamics? Briefly justify your response.

12-64. In Example 12-9, you are told that during one second, 10.0 J of heat flow through the walls, windows, and roof of your home, far too little to significantly change either the inside or outside temperature. Does this indicate that your insulation is very good, very bad, or somewhere in between? How can you tell?

•12-65. In one reference work, we find that the amount of energy released by the combustion of a kilogram of coal is 29×10^6 J. In another, we find that it was common for coal-fueled steam engines to require "5 pounds of coal per horsepower per hour." Find the efficiency of such a steam engine.

12-66. What is wrong with the flow diagram in Figure 12-20?

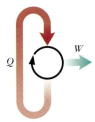

Figure 12-20
Problem 12-66

12-67.

a. Describe a situation in which your body could be considered a temperature reservoir.

b. Describe a situation involving your body in which it does not meet the criteria for a temperature reservoir.

12-68. The door of an oven with a volume of 0.14 m³ is left open a crack while the oven is raised from a room temperature of 70°F to a roasting temperature of 350°F.

a. How many moles of air leave the oven as it warms up?

b. How many molecules is this?

c. Assuming there are 0.029 kg in a mole of air, what is the mass of the air that has left the oven?

12-69. Suppose that while the volume in an auto engine cylinder is reduced by half, the pressure increases by a factor of eight because of the combustion going on in the cylinder. Under these conditions, how does the rms speed of the molecules change? ($v_{final} = \underline{?}\ v_{initial}$)

12-70. A container is made up of rigid walls and bottom and a lid that can be raised or lowered within it like the piston in an auto engine cylinder.

a. How must the average distance between air molecules in the container be changed to increase the air pressure in the container from 1 atm to 8 atm while the air remains at room temperature? (Final average distance = $\underline{?}$ initial average distance.)

b. Suppose the lid is fixed in a particular position when the air within is at normal atmospheric pressure and the temperature within is raised from 300 K to 400 K. How does the average distance between air molecules in the container change? (Final average distance = $\underline{?}$ initial average distance.)

12-71. Find the mean distance between air molecules at a normal atmospheric pressure of 1.01×10^5 Pa and a normal room temperature of 293 K. (*Hint:* It may help to assume that each molecule occupies a cube of space and then find the length of an edge of the cube. If you do so, draw a sketch that shows that if the molecules were equally spaced, the distance between them would be equal to the edge of such a cube.)

12-72. The kinetic theory of gases assumes that the collisions between molecules are elastic. If the collisions were not

perfectly elastic, would the temperature of a gas increase, decrease, or remain the same? Briefly explain.

12-73. *Diffusion.* In Section 12-1, we discussed the diffusion of ammonia vapor through air and red ink through water. Use the reasoning from Problem 12-2 to think about what happens when one fluid diffuses through another.

a. If the concentration of ink is greater in one region of water than another, will there be a net movement of ink molecules between the two regions? If so, in which direction? Explain your answer in terms of the random movement of molecules.

b. Suppose you increase the concentration of ink in the first region, where it was already greater. How will that affect the movement of ink molecules between the two regions? Briefly explain.

c. If one substance is diffusing through another, one can define its rate of diffusion, in kg/s, as $\Delta m/\Delta t$. Would you expect the rate $\Delta m/\Delta t$ at which ammonia diffuses from point A to point B in the air to depend only on the concentration of ammonia at point A, only on the concentration of ammonia at point B, or on the difference between the concentrations at the two points? Briefly explain.

d. In Equation 11-12a or 11-12b, we can think of the temperature as a measure of the concentration of internal energy. There the flow of energy (heat) depends on a change from one point to another in the concentration of internal energy. Now we are talking about how the flow of mass of a substance (diffusion) depends on how the concentration of mass changes along the path of flow. Using C to represent concentration, try to write an equation for the diffusion rate $\frac{\Delta m}{\Delta t}$ that is analogous to Equation 11-12b for the heat flow rate $\frac{\Delta Q}{\Delta t}$. Use D rather than K as your symbol for the proportionality constant. In usual terminology, D is called the **diffusion constant,** and the equation that you obtain, if correct, is called **Fick's law of diffusion.**

12-74. A 40-W compact refrigerator in a student's room transfers 160 J of heat energy each second to the air in the room. Calculate the refrigerator's coefficient of performance.

12-75. Two sealed tanks of gas having identical volumes are kept at the same temperature. An individual molecule of gas A has three times the mass of an individual molecule of gas B. But the total mass of gas A is three-quarters of the total mass of gas B. How does the pressure exerted by gas A compare to that exerted by B? ($P_A = \underline{?}\ P_B$)

12-76.

a. Suppose the high-temperature reservoir of a Carnot engine is kept at 420°C and its low-temperature reservoir engine is kept at 140°C. How much work is done by this engine for each joule of energy that is transferred from the high-temperature reservoir to the engine?

b. Suppose that valves have accidentally been left open along a pipe connecting the two reservoirs, allowing heat to be transferred by convection directly from one reservoir to the other, bypassing the engine. By the time the engine operators realize it, the high-temperature reservoir has dropped to 350°C and the low-temperature reservoir has risen to 210°C. For each joule of energy that could originally be transferred from the high-temperature reservoir to the engine before the mishap, how much work has become unavailable?

12-77. SSM Two sealed containers hold equal volumes of argon gas. In container A, the temperature is 300 K and the pressure is 2.0×10^5 Pa. In container B, the temperature is 600 K but the pressure is only 1.0×10^5 Pa. If an argon atom is selected at random (but not removed) from each container, is one of these atoms moving at greater speed (if so, which one?) or do they have the same speed, or is it impossible to know for sure whether one of them is traveling faster? Briefly explain.

Problems on WebLinks

12-78. In WebLink 12-1, we analyze what happens to the N gas molecules moving about within a cube of side a. Figure 12-4 is a reminder of what we assume happens to a typical molecule in the gas.

a. Why do you suppose we can neglect the effect of gravitational forces in this analysis? Explain your reasoning.

b. Is our neglecting of gravitational forces exactly correct, or is it simply a good approximation? Can you think of any real-world observations you can offer as evidence to support your answer to this question?

12-79. In WebLink 12-1, consider a molecule in the box that is traveling in a purely vertical direction. Suppose the molecule's speed v is fast enough to go up and down between the top and bottom of the box many times in one minute, and that over that one minute interval, the molecule delivers a total impulse I to the top of the box. If the molecule's speed is then doubled to $2v$, what total impulse will it deliver to the top of the box in a minute? Express your answer as a multiple of I (new total impulse = _?_).

12-80. In Problem 12-79, if the speed remains at v and the linear dimension a of the cube is doubled, what total impulse will the molecule deliver to the top of the box in a minute? Express your answer as a multiple of I.

12-81. In Problem 12-79, if the speed remains at v and the volume of the cube is doubled, what total impulse will the molecule deliver to the top of the box in a minute? Express your answer as a multiple of I.

12-82. Consider the two extreme cases presented in WebLink 12-2. Of the three situations below, which is analogous to extreme case 1, which is analogous to extreme case 2, and which is analogous to neither?

a. A well-insulated calorimeter contains a mixture of water and small metal beads. The calorimeter is placed on a high-power speaker at a rock concert, making it and its contents vibrate intensely.

b. When work is done electrically on the speaker in *a* to make it vibrate, the speaker in turn does a smaller amount of work on the surrounding air. The remaining energy input causes the speaker elements to heat up.

c. Fans and heat sinks are required to prevent high-power speakers from heating up excessively. A speaker equipped in this way has been on for a while and has stopped getting hotter.

12-83. Below are several statements about the heat engine cycle in WebLink 12-3. Select one of the following to go in each of the blanks in these statements: *high-temperature reservoir; low-temperature reservoir; coolant; wheel assembly.*

a. Q_C represents energy that is transferred from the _____.

b. Q_H represents energy that is transferred from the _____.

c. W represents energy that is transferred from the _____.

12-84. Below are several statements about the refrigerator cycle in WebLink 12-3. Select one of the following to go in each of the blanks in these statements: *the coolant; the air inside the refrigerator; the air outside the refrigerator.*

a. Q_C represents an energy transfer from _____ to _____.

b. Q_H represents an energy transfer from _____ to _____.

c. W represents work done on _____.

12-85. When heat engines 1 and 2 are connected as in WebLink 12-4, which two of the following are always equal: Q_{H1}, Q_{C1}, W_1, Q_{H2}, Q_{C2}, W_2?

12-86. Suppose heat engines 1 and 2 are connected as in WebLink 12-4. Suppose a high-temperature reservoir transfers 1500 J of energy to engine 1. Engine 2 does 500 J of work and transfers 200 J of energy to a low-temperature reservoir. How much work is done by engine 1?

12-87. In Problem 12-86, think of the two engines together as a combined system.

a. Find the work done by the combined system.

b. Find the heat input to the combined system.

c. Find the heat output from the combined system.

CHAPTER THIRTEEN

Periodic Motion and Simple Harmonic Oscillators

Uniform circular motion is **periodic**—the same sequence of stages or **phases** of the motion repeats over and over, and each repetition takes the same time interval or **period.** Many other physical situations exhibit cyclical or periodic behavior: A pendulum swings back and forth; your heart pumps rhythmically; your vocal cords vibrate rapidly when you hum a note.

In this chapter, we will focus on some of the simplest periodic behaviors, like that of a block bobbing up and down on a spring. We will identify their important characteristics and see what they have in common with one another and with uniform circular motion. These connections will lead us to a mathematical description that fits all such situations. By treating simple situations first, we will develop ways of thinking that in later chapters can extend to more complex phenomena, such as sound and electromagnetic waves.

You should pay close attention to the links between the mathematical description that we develop and the correct application of physical principles such as Newton's laws and energy conservation.

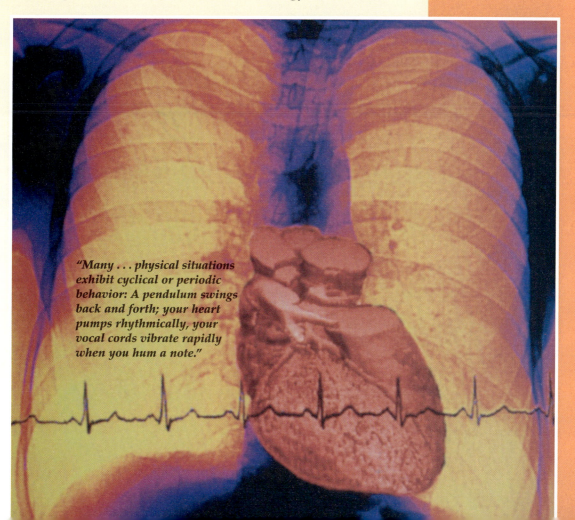

"Many . . . physical situations exhibit cyclical or periodic behavior: A pendulum swings back and forth; your heart pumps rhythmically, your vocal cords vibrate rapidly when you hum a note."

Figure 13-1 The archetypical oscillator: a block bobbing up and down on a spring.

For **WebLink 13-1:**
Your Basic Oscillator,
go to
www.wiley.com/college/touger

13-1 Oscillators and Their Importance

Consider a block bobbing up and down on a spring (Figure 13-1). If it is set in motion with an initial upward push, it will move up and down as shown in the sequence of stop-action pictures in Figure 13-2a. The sequence of pictures is in effect a graph showing how its position varies with time. In the absence of friction and air resistance, this up-and-down motion, or **oscillation,** will go on indefinitely (Figure 13-2b). To see step by step how the picture develops over time, go to WebLink 13-1.

Why devote so much attention to a block bobbing on a spring? It is not particularly interesting for its own sake, but it is the simplest and most easily grasped example of the pattern of behavior graphed in Figure 13-2b. That pattern is pervasive in the natural world. If we understand why it occurs in the block-on-a-spring situation, we can then think about any situation in which the underlying mechanism is not so easy to picture and ask ourselves, "How is it like the block-on-a-spring? What in the new situation corresponds to so-and-so in the block-on-a-spring?" Drawing on our familiarity with the block-on-a-spring, we can use the power of analogy to understand an array of phenomena for which our intuitions are less well-developed.

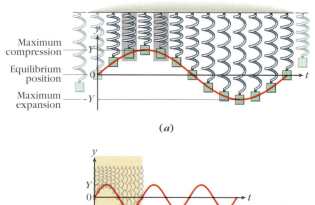

(a)

(b)

Figure 13-2 The behavior of an oscillator (a) over a single cycle (b) over repeated cycles.

◆**A FORWARD LOOK** What *are* those other situations? To motivate spending time and effort on the block-on-a-spring, we will take a brief forward peek.

There are other motions that are cyclical in the same way that the block's motion is cyclical:

- If the simple frictionless pendulum (Figure 13-3a) is given a gentle push to the right, it swings back and forth between the extreme positions B and C. The angle ϕ made by the pendulum rod with the vertical is then seen to vary between two extreme values, which we denote by $\phi = +\Phi$ and $\phi = -\Phi$. It varies with time t (Figure 13-3b) the way y does in Figure 13-2b.

- If the disk in Figure 13-4 is given a small angular velocity ω_o in the direction shown, the string will twist until the disk is brought to a stop, and then will untwist, setting the disk to rotating the other way. Because the disk is still rotating when the string is fully untwisted, it keeps going until the string is sufficiently twisted the other way to bring the disk to a stop, and the untwisting starts again. Figure 13-3b shows how the angle of rotation ϕ varies in this situation as well. A device like this is called a **torsion pendulum.** Examples of torsion pendulums include the oscillating parts (called **balances**) of wind-up watches and the suspended portion of the Cavendish balance (Figure 8-16).

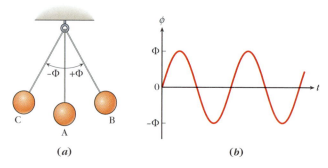

Figure 13-3 The behavior of a simple pendulum.

(a)

(b)

• Our "basic picture" (Figure 13-2*b*) can also describe the motion of a boat riding at anchor (Figure 13-5*b*) as a wave passes under it. In this case, *y* would represent the displacement of the boat (or of the water directly under it) above (+) or below (−) its undisturbed level. As Figure 13-5*a* shows, our "basic picture" may also describe the behavior of the bar producing the wave—the wave's *source*. As the bar goes up and down, it produces a rising and falling disturbance that travels away from the bar. The traveling disturbance or wave is shown at three successive instants in Figure 13-5*b*. The boat rises and falls as the wave passes under it; it is at position P_1 at instant t_1, at P_2 at instant t_2, and at P_3 at instant t_3. Thus, as the bar goes up and down, the boat replicates that up-and-down motion, but it does so after a time lag, because it takes time for the disturbance to travel from the source to the boat. Working through WebLink 13-2 will help you visualize in detail how the motion of the boat depends on the motion of the source as the wave travels.

Note: Disturbances are not always periodic. If the bar's displacement follows a less regular pattern than in Figure 13-2*a,* the traveling disturbance in Figure 13-5 will show a corresponding lack of regularity. The vibrations of your sound system speakers must be anything but repetitive to reproduce the richness of music and human speech.

In Chapter 14, we will view waves as disturbances that travel from a source to distant points where detectors may in turn be "disturbed." A *detector* here means any object that, by responding in some visible way, indicates the presence of the disturbance. If the disturbance at the source is oscillatory, the detectors are likewise set to oscillating. Figure 13-2*b* shows how the size of an oscillatory disturbance varies over time. The size of the disturbance is not always a linear displacement or distance. In general, it tells how much a particular quantity deviates from its undisturbed level: the position of the block-on-a-spring, the height of the water, or the twist angle of the torsion pendulum. We will later see that it can also refer to deviations from normal air pressure or density in the case of sound waves, and deviations from background electric and magnetic field intensity in the case of electromagnetic waves (which, it turns out, include light, radio waves, and X rays, among others). Understanding oscillators is a necessary first step toward understanding wave phenomena, and understanding oscillators will be easier if we first develop a detailed understanding of one typical oscillator whose behavior is easy to picture: the block-on-a-spring.

Recall also that in extending kinetic theory to solids (Section 12-3), we thought about the kinetic energy associated with the oscillations of the molecules about fixed positions. Thus, understanding oscillators is a starting point for understanding the thermal properties of solid matter on a microscopic scale.

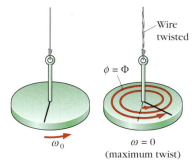

Figure 13-4 A torsion pendulum.

For **WebLink 13-2: Making Waves,** go to www.wiley.com/college/touger

These examples of oscillatory motion have become popular with both youngsters and collectors.

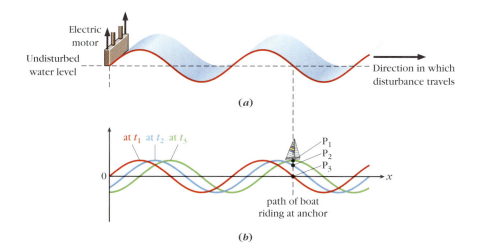

Figure 13-5 **An oscillator setting up a wave.** (*a*) The set-up in stop-action at the instant t_1. (*b*) The wave travels.

Figure 13-6 A horizontal oscillator.

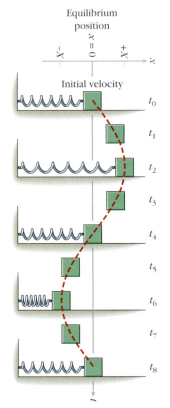

Figure 13-7A Phases of an oscillating block-on-a-spring I.

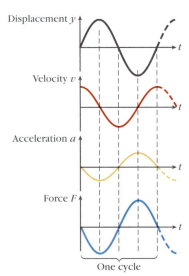

Figure 13-8 Simulated range finder and force probe data for a block on a spring.

13-2 How a Typical Oscillator Behaves (Qualitative Discussion)

We now examine the behavior of the block-on-a-spring in greater detail. For simplicity, consider a block that oscillates horizontally on a frictionless surface, so that the only force on it in the direction of motion is the spring force. This situation is reasonably well approximated by the glider connected by a compressible spring to one end of the air track in Figure 13-6. (We must specify *compressible* because many real springs are manufactured so that they can be stretched but not compressed.)

We assume that for the range of stretches and compressions we are considering, the spring obeys Hooke's law:

$$F_x = -kx \qquad (6\text{-}17a)$$

a relationship between (+ or −) *components*

STOP&Think First let's see how well you can apply the concepts you have already learned correctly and consistently to such an oscillator. Figure 13-7A shows a block-on-a-spring set-up at nine instants (t_0 to t_8) during a complete oscillation. We choose the undisturbed or equilibrium position as the origin. You should write down answers to the following questions, along with a brief explanation of your reasoning, before continuing your reading. Then, as you proceed, if you encounter anything that disagrees with your present thinking, you should try to resolve the discrepancies. In this way you can work toward a more consistent understanding of the oscillator's behavior.

a. At which instants is the velocity positive, and at which instants is it negative? Also, at which instants, if any, is the velocity zero?

b. During which intervals is the velocity increasing, and during which intervals is it decreasing?

c. During which intervals is the acceleration positive, and during which intervals is it negative? Also, at which instants, if any, is the acceleration zero? (In answering **c**, did you think about Newton's second law? Did you think about the direction of the force the spring is exerting during the various intervals?)

d. During which intervals is the total force on the block in Figure 13-7A positive, and during which intervals is it negative? Also, at which instants, if any, is this total force zero? ◆

Since

$$\Sigma F_x = ma_x \qquad (13\text{-}1)$$

it follows that

$$ma_x = -kx \qquad (13\text{-}2)$$

Then if x is positive, what must be the sign of a_x? How must the sign of ΣF_x relate to the sign of a_x? Review your answers to **STOP&Think** questions **a–d**. Were they consistent with these sign relationships? ◆

Notice that the force on the block is always opposite in direction to the displacement. Because it tends to restore the block to its equilibrium position, it is called a **restoring force.** We will see that in general, a restoring force is a necessary condition for oscillatory motion.

◆**WHAT DO WE OBSERVE?** The range finder we described in Chapter 2 (Figure 2-16) can be used to obtain plots of the oscillator's position (x), velocity, and acceleration against time. By connecting an electronic force probe between the block and the spring to measure the elastic force, the software can also plot this force against t. The resulting graphs are displayed in Figure 13-8.

Are these observed results what you expected? Let's verify that they make sense in terms of the physics. For instance, Newton's second law requires that the acceleration must at every instant be in the same direction as the force. We see that this is so. This can also be seen in Figure 13-7B, which shows the velocity, acceleration, and total force vectors at each instant. As usual, each of these

quantities is positive when the vector is directed to the right and negative when it is directed to the left. **STOP&Think** Fill in the missing information at t_7 and t_8 in Figure 13-7B for yourself. ◆ Doing Problem 13-10 will help you review some of the important ideas in these figures.

Notice that over each interval and at each instant, the sign of the acceleration accurately describes how the velocity is changing. For example, the acceleration must be negative whenever the velocity is decreasing. Remember—we describe the velocity as decreasing both when it is becoming less positive (between t_0 and t_2) and when it is becoming more negative (between t_2 and t_4). **STOP&Think** Can you give a similar argument connecting the sign of the acceleration with the way the velocity is changing between t_4 and t_6? Between t_6 and t_8? ◆ At t_2, the velocity is instantaneously zero as it goes from positive to negative. It is therefore *decreasing* at this instant, so its rate of change—the acceleration—must be negative. **STOP&Think** Give a similar argument for why the acceleration is positive at t_6. ◆

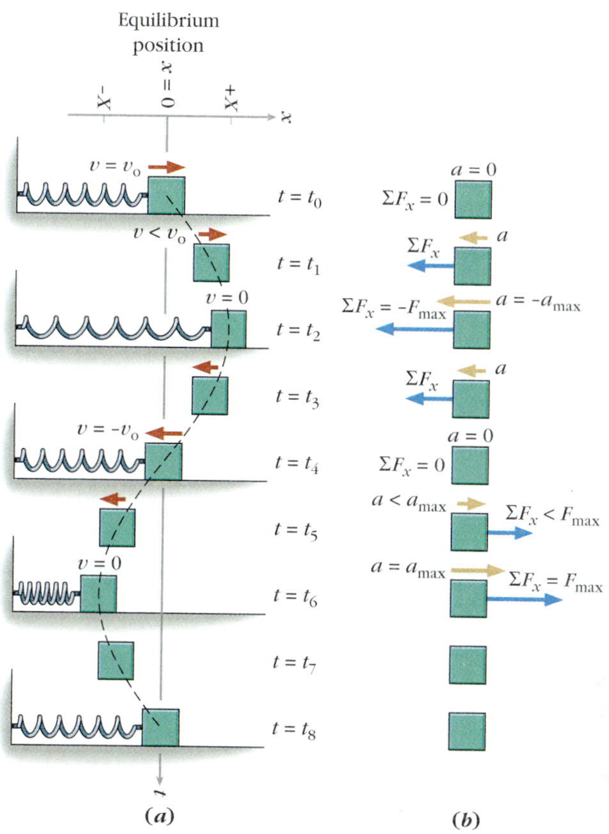

◆**THE VERTICAL BLOCK-ON-A-SPRING** Now let's consider a block suspended from a spring so that it oscillates vertically. If you support the block so that the spring is unstretched (Figure 13-9a) and then very slowly lower your hand, the spring will stretch until it exerts enough upward force to support the weight of the block. The spring is therefore stretched an amount that we call y_0 at equilibrium—the unstretched position and the equilibrium position are not the same. In equilibrium, $\Sigma F_y = 0$. Then $ky_0 - mg = 0$, and therefore $ky_0 = mg$ (see free-body diagram in Figure 13-9b). If the block is displaced upward a distance y from its equilibrium position, it is stretched by an amount $y - y_0$ (Figure 13-9c) and the magnitude of the elastic force is $k(y - y_0)$. The free-body diagram then shows that the *total* force on the block in this position is

$$\Sigma F_y = -k(y - y_0) - mg = -ky + ky_0 - mg$$

But because $ky_0 = mg$, it follows that

$$\Sigma F_y = -ky$$

where y is the displacement from equilibrium. The *total force* still has a Hooke's law–like dependence on displacement, but keep in mind that in the vertical case ΣF_y is not the elastic force alone. With only slight modifications, you can produce figures like 13-7A and 13-7B to describe the vertically oscillating block-on-a-spring. (Problem 13-11 summarizes some of the key features for the vertical situation as Problem 13-10 does for the horizontal situation.)

Figure 13-7B Phases of an oscillating block-on-a-spring II. This revisits Figure 13-7A. Part (*a*) shows the positions plotted against time and the velocity in each phase. Part (*b*) shows the total force on the block and the acceleration. The maximum values of position, velocity, acceleration, and ΣF_x are denoted by X, v_{max}, a_{max}, and F_{max}.

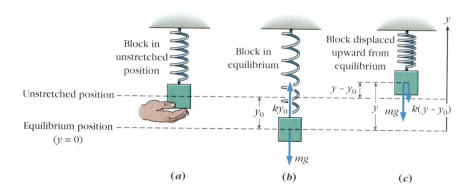

Figure 13-9 Forces on a block suspended from a spring.

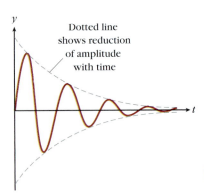

Figure 13-10 A damped oscillation.

Dotted line shows reduction of amplitude with time

Oscillators that behave like Figure 13-2b are usually called **simple harmonic oscillators.** They are called *harmonic* because oscillators of this type, including the prongs of tuning forks and the vibrating strings of guitars, were first studied by the ancient Greeks in connection with the production of musical sounds. Their most basic oscillations, which can occur in combinations, are called *simple*. These and other periodic oscillations conserve mechanical energy. This is in contrast with **damped** oscillations, which gradually diminish over time as their energy is dissipated (Figure 13-10).

13-3 A Model for Describing Oscillators Mathematically

In this section we develop a way of describing oscillatory motion mathematically that is consistent with the physics. From observation, we already have the *graphic* description of y, the oscillator's displacement from equilibrium, as a function of t (Figures 13-2 and 13-8). We need to fit a mathematical expression to this picture.

Figure 13-2a shows a simple harmonic oscillator at various stages, or **phases,** of its cycle. The term *phases* has been used traditionally to describe the successive stages of other cycles, such as the cyclical appearance of the moon as it circles Earth (Figure 13-11). In fact, motion in a circle is a kind of cycle—and it is one we have already described mathematically. We will see how it can guide our attempt to describe the cycles of oscillators.

We have observed (Figure 13-2b) that the value of the displacement y varies back and forth between a maximum value Y and an extreme opposite value $-Y$. We call this maximum displacement the **amplitude.** The displacement y is always some fraction, positive or negative, of the amplitude Y; that is, it is always Y multiplied by some number between 1 and -1:

$$y = Y \times \left(\begin{array}{l} \text{some number between 1 and } -1\text{, the value of which depends} \\ \text{on the phase, or on the instant } t \text{ at which that phase occurs} \end{array} \right) \quad (13\text{-}3)$$

Our problem is to determine what fraction between 1 and -1 we have at any given time t, or equivalently, at any given phase of the cycle. We can think of our task as trying to develop a table that looks like this:

Phase	Fraction between 1 and -1

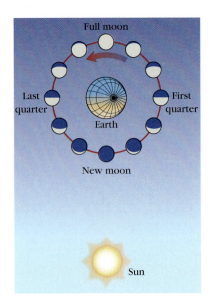

Figure 13-11
Phases of the moon.

Full moon

Last quarter

First quarter

Earth

New moon

Sun

When completed, the table would give the fraction between 1 and −1 as a function of phase, because mathematically, a function is anything that permits us to obtain a unique output value for any input value. The problem is, how do we get the numbers that go in the table?

We will assume (and demonstrate later) that this table is the same for all simple harmonic oscillators (all oscillators behaving like the one in Figure 13-2*b*). What varies from one instance to another in Equation 13-3 are the amplitude *Y* (and the physical quantity that it represents) and the amount of time required to go through the phases of a cycle. Thus, once we complete the table for one oscillator that displays this behavior, we can use it for *all* such oscillators. The oscillator we choose, which we'll call our reference oscillator, is produced by the set-up in Figure 13-12. The transparent circular disk has a radius of length *r* = 1. As the radius rotates, the projectors cause it to cast horizontal and vertical shadows. These shadows are our reference oscillators. As Figure 13-13 shows, they oscillate in size as the radius rotates and can match an oscillating block-on-a-spring phase by phase. The circle that generates this shadow behavior, abstracted to a simple geometric diagram in Figure 13-14, is called the **unit reference circle** (*unit* means one, because *r* = 1). WebLink 13-3 will guide you in greater detail through the ideas summarized in Figures 13-12 to 13-14 and will show in detail how this set-up gives us the values that go in the table. We will shortly see here as well how it generates these values.

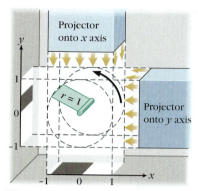

Figure 13-12 A set-up to generate "standard" oscillations between 1 and −1.

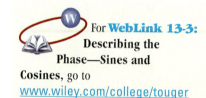

For **WebLink 13-3: Describing the Phase—Sines and Cosines,** go to www.wiley.com/college/touger

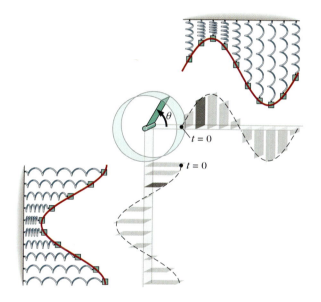

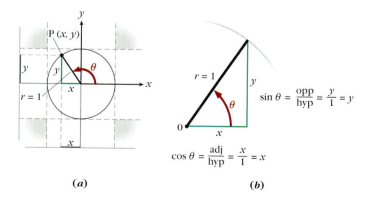

Figure 13-13 Variation over time of shadows produced by the set-up in Figure 13-12. Notice how the variation of the shadows matches the motion of the oscillators.

When the rotating radius of the unit reference circle is in a particular phase (positioned at an angle *θ*), its endpoint P has coordinates *x* and *y* (Figure 13-14). These coordinates *x* and *y* give the sizes of the horizontal and vertical shadows cast by the radius. Because *r* = 1, these are fractions between 1 and −1; they

$$\sin \theta = \frac{\text{opp}}{\text{hyp}} = \frac{y}{1} = y$$

$$\cos \theta = \frac{\text{adj}}{\text{hyp}} = \frac{x}{1} = x$$

(*a*)   (*b*)

Figure 13-14 An abstraction of Figure 13-12. This is called the unit reference circle because the radius has "unit size" *r* = 1.

If the sun were directly overhead, each car of the ferris wheel would cast a shadow on the ground directly below it. What would the motion of this shadow be like as the ferris wheel went around?

are the fractions we were looking for. As θ varies, x and y also vary—these fractions are *functions of θ.*

These functions are commonly called the **cosine** and the **sine** of the angle. Hence, because we define them that way,

$$\text{(when } r = 1) \qquad \cos \theta \equiv x \qquad \text{and} \qquad \sin \theta \equiv y \qquad \text{(13-4c, 13-4s)}$$

Compare these definitions with the definitions based on the sides of a right triangle: $\cos \theta = \frac{\text{adj}}{\text{hyp}}$ and $\sin \theta = \frac{\text{opp}}{\text{hyp}}$. For the triangle in Figure 13-14b, the latter definitions also give us $\cos \theta = x$ and $\sin \theta = y$. The advantage of equations 13-4c and 13-4s is that they don't limit us to the angles between 0 and 90° that occur in right triangles. Since point P in Figure 13-14a can go clockwise or counterclockwise around the unit reference circle any number of times, we can get values of its coordinates x and y for any value of θ from $-\infty$ to $+\infty$.

We treated r as unitless, so x and y are likewise unitless: A value of $\cos \theta$ or $\sin \theta$ is a pure number between -1 and 1. Thus, we can relabel the table

as

Phase	Fraction between 1 and −1
θ	$\cos \theta$

or

Phase	Fraction between 1 and −1
θ	$\sin \theta$

Example 13-1 shows how you can use the unit reference circle to find the values in these tables.

Example 13-1 *Finding Sines and Cosines by Measurement*

For a guided interactive solution, go to Web Example 13-1 at www.wiley.com/college/touger

Using a cm ruler and protractor, find $\sin 125°$ and $\cos 125°$ to two decimal places.

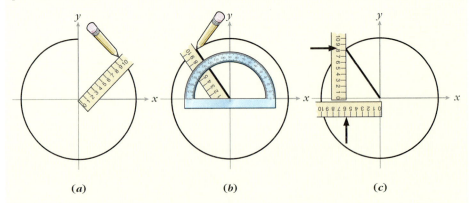

(a)   (b)   (c)

Brief Solution

Choice of approach. To use the definitions of the sine and cosine that we developed above (Equations 13-4c and 13-4s), we first draw a circle of radius 1 dm (part *a* of figure) centered on an xy coordinate frame. Note that

1 decimeter = 10 cm, so 1 cm = 0.1 dm. We then use the protractor to draw the radius 125° clockwise from the +x axis (part *b* of figure), and measure the *x* and *y* coordinates of the tip of the radius (part *c* of the figure). The *x* coordinate is negative because it is measured to the *left* of the origin.

The mathematical solution. We measure *x* and *y* in dm because *r* = 1 in dm. For example, *y* = 8.2 cm = 0.82 dm. But the sine and cosine are unitless. Then from the measurements in part *c* of the figure, we see that

$$\cos \theta = x = \mathbf{-0.57} \quad \text{and} \quad \sin \theta = y = \mathbf{0.82}$$

◆ Related homework: Problems 13-13 and 13-14.

By repeating Example 13-1 for other angles, we can complete our sine and cosine tables. We can then refer to the tables (or our calculators) whenever we need a value.

With $\cos \theta \equiv x$ and $\sin \theta \equiv y$, the sequences of shadows in Figure 13-13 are simply graphs of $\sin \theta$ versus θ (across the page) and $\cos \theta$ versus θ (down the page). The size of the rotating radius can be multiplied by $Y(r = 1 \times Y = Y)$ in any units to make it the amplitude of any oscillation. Then the size of each shadow is also multiplied by Y. As Figure 13-15 shows, the graphs then become graphs of $Y \sin \theta$ versus θ and $Y \cos \theta$ versus θ. Either $\cos \theta$ or $\sin \theta$ can be the "number between 1 and -1" in Equation 13-3, giving us the following:

$$y = Y \sin \theta \tag{13-5s}$$

if the oscillator starts out in equilibrium with an initial upward velocity (and so behaves like the vertical shadow)

$$y = Y \cos \theta \tag{13-5c}$$

if the oscillator starts out with its maximum positive displacement (and so behaves like the horizontal shadow).

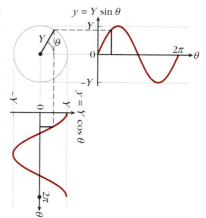

Figure 13-15 Oscillator descriptions generated by a reference circle of radius Y.

Note: Because the sine and cosine are unitless ratios, Equations 13-5s and 13-5c remind us that the displacement *y* and the amplitude *Y* must always have the same units.

An oscillator can also start out in between these two phases, but we can simply choose to start our stopwatch when the oscillator is in one of these two phases (Figure 13-16). Both the sine and the cosine are **sinusoidal** (sine-like) **functions.** The only difference between them is the horizontal placement of the origin.

➡**Initial phase:** Because $\cos \theta = \sin (\theta + \pi/2)$, using the cosine is equivalent to using the sine with an initial phase $\theta_o = \pi/2$ added to θ. In general, we could add whatever initial phase matches the initial conditions, and write $y = Y \sin (\theta + \theta_o)$.

◆**WHAT DOES θ MEAN FOR THE BLOCK-ON-A-SPRING?** The angle θ tells you where the rotating radius is in its cycle. But what does it tell you about a block oscillating along a horizontal or vertical line where there is no angle to consider? We must return to the idea that a given angle identifies a particular *phase* in the cycle of the rotating radius. Just as we can divide a cycle of this kind into either 360 or 2π equal divisions, we can arbitrarily divide any cycle

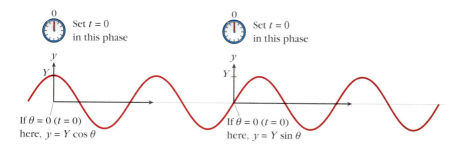

Figure 13-16 The sine and cosine differ only in the choice of initial phase.

into 360 or 2π divisions, and θ will tell you how far along you are in that cycle—that is, it will tell you the phase. That means it will tell you which of the stop-action pictures in Figure 13-2a represents the oscillator at the instant in question. In a complete cycle the block goes through each possible phase once, moving continuously from one phase to the next. (The two phases darkened in Figure 13-2a are different, even though the block's position is the same in both, because the block is moving upward in one phase and downward in the other.)

> In general, θ represents the *phase* of an oscillator, and $\sin \theta$ or $\cos \theta$ tells what fraction of the amplitude ($+$ or $-$) gives you the actual displacement in that phase.

Although θ is sometimes called the oscillator's **phase angle**, the "angle" exists physically only for the rotating radius whose shadow mimics the oscillation.

Example 13-2 *How Sinusoidal Functions Describe Oscillators I*

A block-on-a-spring set-up is stretched to a maximum displacement of 0.30 m and then released. What is its displacement when it is $\frac{7}{8}$ of the way through its first cycle?

Solution

Restating the question. If a complete cycle is divided into 360 or 2π equal divisions, then $\frac{7}{8}$ of a cycle is $\frac{7}{8}(360°) = 315°$ or $\frac{7}{8}(2\pi) = \frac{7}{4}\pi$. Because it will prove more convenient later on, we will choose to work in radian measure. The question is then, *What is y when* $\theta = \frac{7}{4}\pi$?

Choice of approach. We need to address three questions:
1. In what phase is the oscillator? (Find θ.)
2. By what fraction ($+$ or $-$) is the amplitude multiplied when it is in this phase? (Find $\sin \theta$ or $\cos \theta$.)
3. What is the displacement? (Multiply the amplitude by the fraction found in 2.)

What we know/what we don't.

$$Y = 0.30 \text{ m} \qquad \theta = \tfrac{7}{8}(2\pi) = \tfrac{7}{4}\pi \qquad y = ?$$

The mathematical solution. We have already answered question **1.** To answer question **2,** we must decide whether to apply Equation 13-5s or 13-5c. Because the oscillator starts off at its maximum displacement, it behaves like the cosine, so we choose Equation 13-5c. Using tables or a calculator, we find that $\cos \frac{7}{4}\pi = -0.707$. Now we can multiply the amplitude by this fraction to get the displacement:

$$y = Y\cos \theta = (0.30 \text{ m})(-0.707) = \textbf{−0.21 m}$$

◆ Related homework: Problems 13-19, 13-20, and 13-21.

◆**THE DEPENDENCE OF PHASE ON TIME** The time T that it takes for an oscillator to go through one complete cycle is its period of oscillation. The fraction $\frac{t}{T}$ gives the number of cycles that have gone by between $t = 0$ and the instant t. For example, if the period $T = 6$ s, then at $t = 3$ s, $\frac{t}{T} = \frac{1}{2}$, and half a cycle has elapsed. Because there are 2π radians in a cycle, multiplying 2π by the number of cycles gives the phase in radians:

$$\theta = 2\pi \frac{t}{T} \qquad (13\text{-}6)$$

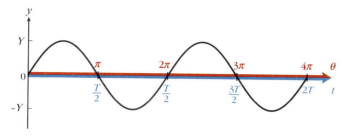

Figure 13-17 **Changing a graph of** *y* **versus** *θ* **to a graph of** *y* **versus** *t* **is simply a change of scale.**

so it is equivalent to plot *y* against *θ* or *t* (Figure 13-17), and Equations 13-5s and 13-5c become

$$y = Y\sin\left(2\pi\frac{t}{T}\right) \quad \text{and} \quad y = Y\cos\left(2\pi\frac{t}{T}\right) \qquad \text{(13-7s, c)}$$

The following procedure outlines how these equations are used.

PROCEDURE 13-1

Calculating the Displacement of an Oscillator

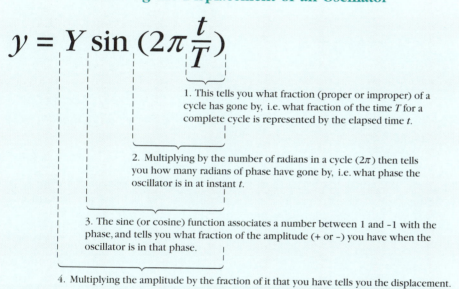

$$y = Y\sin\left(2\pi\frac{t}{T}\right)$$

1. This tells you what fraction (proper or improper) of a cycle has gone by, i.e. what fraction of the time *T* for a complete cycle is represented by the elapsed time *t*.

2. Multiplying by the number of radians in a cycle (2π) then tells you how many radians of phase have gone by, i.e. what phase the oscillator is in at instant *t*.

3. The sine (or cosine) function associates a number between 1 and –1 with the phase, and tells you what fraction of the amplitude (+ or –) you have when the oscillator is in that phase.

4. Multiplying the amplitude by the fraction of it that you have tells you the displacement.

Example 13-3 *How Sinusoidal Functions Describe Oscillators II*

For a guided interactive solution, go to Web Example 13-3 at www.wiley.com/college/touger

A block suspended from a spring, initially at rest, is set to oscillating by giving it an upward velocity at *t* = 0. If it has a maximum displacement of 0.28 m and takes 0.40 s to complete a full cycle, what is its displacement after 0.15 s?

Brief Solution

Restating the question. You need to remember definitions. The "maximum displacement" is the amplitude *Y*. You must distinguish between *t*, the actual time elapsed, and the period *T*, the time it takes "to complete a full cycle." Then the question is, *y* = ? *when t* = 0.15 s.

Choice of approach. We must address the same three questions as in Example 13-2. The phase $\theta = 2\pi\frac{t}{T}$; to answer question **1**, we apply Procedure 13-1 to calculate the value of this expression. We choose Equation 13-7s rather than

EXAMPLE 13-3 *continued*

13-7c to fit the initial conditions (see notes on Equations 13-5s and 13-5c). Finding $\sin\theta$ answers question **2**.

What we know/what we don't.

$$y = ? \qquad Y = 0.28 \text{ m} \qquad t = 0.15 \text{ s} \qquad T = 0.40 \text{ s}$$

The mathematical solution. The phase

$$\theta = 2\pi\frac{t}{T} = 2\pi\frac{0.15 \text{ s}}{0.40 \text{ s}} = \tfrac{3}{4}\pi \text{ in radian measure (or } 135°\text{)}$$

This answers question **1**. Using your calculator or tables,

$$\sin\left(\tfrac{3}{4}\pi\right) = 0.707$$

This answers question **2**. Now we can apply Equation 13-7s to answer question **3**:

$$Y = Y\sin\left(\tfrac{3}{4}\pi\right) = (0.28 \text{ m})(0.707) = \mathbf{0.20 \text{ m}}$$

◆ Related homework: Problems 13-23 and 13-25.

Over a single cycle, $\Delta\theta = 2\pi$ and $\Delta t = T$. Then the rate ω at which the phase changes is

$$\omega = \frac{\Delta\theta}{\Delta t} = \frac{2\pi}{T} \qquad (13\text{-}8)$$

We will call ω the **phase frequency,** though it is often called the *angular velocity* or *angular frequency.* Equations 13-7 (s and c) then become

$$y = Y\sin\omega t \qquad \text{and} \qquad y = Y\cos\omega t \qquad (13\text{-}9\text{s, c})$$

The phase frequency measures how rapidly the oscillator oscillates in rad/s, just as the frequency measures it in cycles or complete oscillations per second (Hz). Equation 9-8 ($\omega = 2\pi f$) is essentially a unit conversion and lets us write

$$y = Y\sin 2\pi ft \qquad \text{and} \qquad y = Y\cos 2\pi ft \qquad (13\text{-}10\text{s, c})$$

If we write sin ___ and cos ___, whatever goes in the blank tells us how to calculate the phase θ, and we must do that before we can use tables or calculators to find the sine or cosine, the number between 1 and -1 that goes with that phase.

Example 13-4 *Reading Information about Oscillators from Sinusoidal Functions*

For a guided interactive solution, go to Web Example 13-4 at www.wiley.com/college/touger

The function $\phi = 12\sin 4t$ (with ϕ in degrees) describes the oscillation of a certain simple pendulum (Figure 13-3). Find the maximum angle to which the pendulum swings, the period of oscillation, and the phase at which the maximum angle first occurs. Assume that the phase is in radians.

Brief Solution

Choice of approach. In Equations 13-7s, 13-9s, and 13-10s, what is in the Y position is always the amplitude—in this case we may call it Φ—and the multiplier of t is always $\omega = 2\pi f = \frac{2\pi}{T}$. Remember that the displacement and amplitude need not be linear distances; they may be *any* quantity that varies, as shown in Figure 13-4b. Be careful to distinguish between ϕ and θ. The displacement ϕ is a physical angle—an amount of rotation, in contrast to the phase θ, which in this case is $\theta = 4t$. Nothing in this situation rotates by an amount θ.

What we know/what we don't.

Number in the Y position $= 12° =$ amplitude Φ Multiplier of $t = 4 = \dfrac{2\pi}{T}$

The behavior of $\sin\theta$ versus θ (Figure 13-17)

The mathematical solution. We have already determined that the amplitude $\Phi = \mathbf{12°}$.

As Figure 13-17 shows, the sine always reaches its first maximum ($\sin\theta = 1$) when the phase is $\theta = \frac{\pi}{2}$ (or **90°**).

Since $4 = \frac{2\pi}{T}$, $T = \frac{2\pi}{4} = 1.57$. If we assume SI units, $T = \mathbf{1.57 \ s}$.

Making sense of the results. Be aware that there are two very distinct quantities involved here that can be measured in degrees or radian measure. When the amplitude Φ is 12°, the phase $\theta = 4t$ is $\frac{\pi}{2}$ rad. The 12° is the actual spatial angle that the pendulum makes with the vertical. The phase $\frac{\pi}{2}$ (or 90°) is an angle only in the metaphorical sense and tells you that the pendulum has gone through a quarter of a cycle.

◆ Related homework: Problems 13-24, 13-26, and 13-27.

Example 13-5 *Determining Frequency*

A block on a spring is raised 0.200 m from its equilibrium position and released. It first reaches a position 0.040 m below its equilibrium after 0.10 s. Find its frequency of oscillation.

Solution

Choice of approach. The block starts at maximum displacement, so we choose a cosine description of the displacement. Because we are interested in f, the most suitable cosine description is Equation 13-10c. We must use this equation to

1. find the value of the sine, then

2. use our tables or calculator in reverse to find the phase θ that has this sine, and finally,

3. because $\theta = 2\pi ft$, solve for f.

What we know/what we don't.

$$y = -0.040 \ m \qquad Y = 0.200 \ m \qquad t = 0.10 \ s \qquad f = ?$$

The mathematical solution. From Equation 13-10c,

$$\sin 2\pi ft = \frac{y}{Y} = \frac{-0.040 \ m}{0.200 \ m} = -0.20$$

The angle in the second quadrant (why this quadrant?) that has this sine is $1.77 = \frac{1.77}{\pi}\pi = 0.56\pi$ in radian measure. The 2π in $\theta = 2\pi ft$ tells us we are working in radian measure. Solving this equation for f gives us

$$f = \frac{\theta}{2\pi t} = \frac{0.56\pi}{2\pi(0.10 \ s)} = \mathbf{2.8 \ s^{-1}} \qquad \text{(or 2.8 cycles/s or 2.8 Hz)}$$

◆ Related homework: Problems 13-28 and 13-29.

13-4 Checking the Mathematical Model Against the Physics

In the previous section, we introduced a mathematical model—a reference circle—that appeared to mimic the behavior of a simple harmonic oscillator. In this section we raise further questions to see how well our model matches what oscillators actually do. What does it imply about the velocity and acceleration, and is it consistent with the observations? How does the period (or the frequency or the phase frequency) depend on the physical properties of the oscillator?

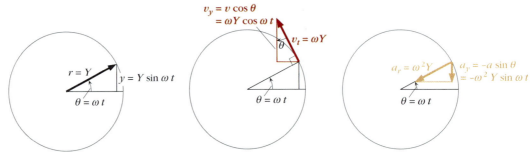

Figure 13-18 Vertical components of y, v, and a for the tip of a rotating radius.

Figure 13-18 addresses the first of these questions. The tip of the rotating radius has a velocity v that is entirely tangential and an acceleration a that is entirely radial. The figure shows that when the radius has rotated an amount $\theta = \omega t$, the quantities r, v, and a have vertical components

$$y = Y \sin \omega t \tag{13-9s}$$

$$v_y = v \cos \omega t = \omega Y \cos \omega t \tag{13-12}$$

and

$$a_y = -a \sin \theta = -\omega^2 Y \sin \omega t \tag{13-13}$$

Note: Just as Equation 13-5s has alternate forms (Equations 13-7s, 13-9s, and 13-10s), we can write Equations 13-12 and 13-13 with θ expressed as ωt, $2\pi ft$, or $2\pi \frac{t}{T}$.

Important: Because $\cos \omega t$ and $-\sin \omega t$ both have a maximum value of 1, v_y and a_y have maximum values

$$(v_y)_{\text{max}} = \omega Y \tag{13-14}$$

and

$$(a_y)_{\text{max}} = \omega^2 Y \tag{13-15}$$

To see in greater detail how we arrive at these expressions for the velocity and acceleration, go to WebLink 13-4.

In Figure 13-19 we plot the above expressions for y, v_y, and a_y against t. The maximum values are indicated on the vertical axes. Notice that the three quantities do not peak at the same time. From one graph to the next, the values of θ at which the peaks occur shift $\frac{\pi}{2}$ (or 90°) to the left—they are "out of phase" by $\frac{\pi}{2}$. If we compare these graphs to the range finder plots of y, v, and a for a block-on-a-spring (Figure 13-8), we see the same offset in the peaks (Figure 13-20). For example, in Figure 13-7B we note v is zero whenever y (like x) and a are at their extreme values, and vice versa. In both sets of graphs, we see that the acceleration and displacement are always opposite in sign (they are out of phase by π or 180°), but apart from sign the acceleration is a maximum when y is. This fits with our understanding of oscillators because for oscillators, the restoring force—and hence the acceleration—is greatest when the displacement from equilibrium is greatest. Thus it boosts confidence in our model. In Equations 13-12 through 13-15 for the oscillator, we can just write v and a instead of v_y and a_y. Example 13-6 provides practice working with the equations for velocity and acceleration.

For **WebLink 13-4:**
Velocity and Acceleration from the Reference Circle, go to
www.wiley.com/college/touger

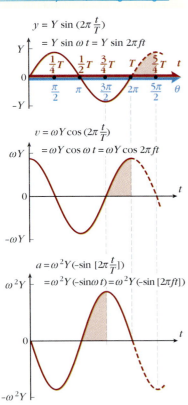

Figure 13-19 Phase relationships among quantities describing the motion of simple harmonic oscillators. The position y, the velocity v, and the acceleration a are plotted against t.

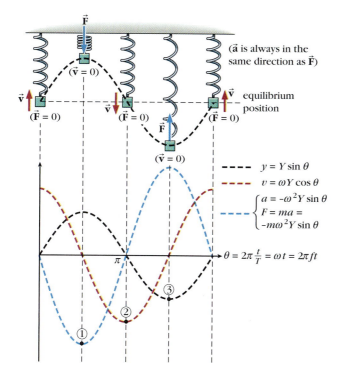

$$y = Y \sin \theta$$
$$v = \omega Y \cos \theta$$
$$\begin{cases} a = -\omega^2 Y \sin \theta \\ F = ma = \\ -m\omega^2 Y \sin \theta \end{cases}$$
$$\theta = 2\pi \frac{t}{T} = \omega t = 2\pi f t$$

($\vec{a}$ is always in the same direction as $\vec{F}$)

equilibrium position

Figure 13-20 Connecting the phase differences with the physics. The acceleration $\vec{a}$ is always opposite in direction to the force $\vec{F}$. At ①, the spring is most compressed, so F, which varies as $-\sin \theta$, has its extreme downward value. The *acceleration* also has its extreme downward value here, since a varies as F. At ②, $\frac{\pi}{2}$ later, the *velocity* has its extreme downward value. At ③, another $\frac{\pi}{2}$ later, the *displacement* has its extreme downward value.

Example 13-6 *Using Sinusoidal Functions to Reason about Oscillators III*

For a guided interactive solution, go to Web Example 13-6 at www.wiley.com/college/touger

Find the velocity and acceleration of the oscillator in Example 13-3 0.15 s after it is set in motion.

Brief Solution

Choice of approach. We now apply Equations 13-12 and 13-13, with θ expressed as $\omega t = 2\pi \frac{t}{T}$, which also tells us that $\omega = \frac{2\pi}{T}$. Again we must answer the three questions in Example 13-2, except now we substitute "maximum value" for "amplitude" and "velocity" or "acceleration" for "displacement." Otherwise we proceed as in Example 13-3.

What we know/what we don't.

$$y = ? \qquad Y = 0.28 \text{ m} \qquad t = 0.15 \text{ s} \qquad T = 0.40 \text{ s}$$

The mathematical solution. As before, the phase is

$$\theta = 2\pi \frac{t}{T} = 2\pi \frac{0.15 \text{ s}}{0.40 \text{ s}} = \tfrac{3}{4}\pi \qquad \text{(or 135°)}$$

This answers question **1.** Using your calculator or tables,

$$\cos\left(\tfrac{3}{4}\pi\right) = -0.707 \qquad \text{and} \qquad \sin\left(\tfrac{3}{4}\pi\right) = 0.707$$

This answers question **2** for the velocity and the acceleration, respectively. The maximum value of the velocity is

$$v_{\max} = \omega Y = \frac{2\pi}{T} Y = \frac{2(3.14)}{0.40 \text{ s}} (0.28 \text{ m}) = 4.4 \text{ m/s}$$

and the maximum value of the acceleration is

$$a_{\max} = \omega^2 Y = \left(\frac{2\pi}{T}\right)^2 Y = \left(\frac{2[3.14]}{0.40 \text{ s}}\right)^2 (0.28 \text{ m}) = 69 \text{ m/s}^2$$

EXAMPLE 13-6 *continued*

We now use Equations 13-12 and 13-13 to answer question **3** for the velocity and the acceleration:

$$v = v_{max} \cos \theta = \omega Y \cos\left(2\pi \frac{t}{T}\right) = (4.4 \text{ m/s})(-0.707) = \mathbf{-3.1 \text{ m/s}}$$

and

$$a = a_{max}(-\sin \theta) = \omega^2 Y\left(-\sin\left[2\pi \frac{t}{T}\right]\right) = (69 \text{ m/s}^2)(-0.707) = \mathbf{-49 \text{ m/s}^2}$$

Making sense of the results. Verify that these results agree with the situation at t_3 in Figures 13-7A and 13-7B.

◆ Related homework: Problems 13-36, 13-37, 13-30, 13-31, and 13-32.

On-The-Spot Activity 13-1

As you do this activity, ask yourself: What is the physical meaning of the slope of a graph of y versus t? What is the physical meaning of the slope of a graph of v versus t? We addressed this in Chapter 2.

a. On the graph of y versus t in Figure 13-19, find the points where the slope is zero. Mark each of these points with an A.

b. Now, on the graph of v versus t, identify the points that occur at the same instants t as the points marked A on the graph of y versus t. Mark each of these with a B. What is the value of v at each of these points? Compare the information about the velocity that points A and points B provide.

c. On the graph of y versus t, find those points where the slope is steepest upward. Mark them with a C.

d. On the graph of v versus t, find the points where the velocity reaches its extreme positive value. Mark these with a D. Do the points marked C and the points marked D occur at the same instants or different instants? Why?

e. On the graph of y versus t, mark with an E each point where the slope is steepest downward.

f. On the graph of v versus t, mark the points where v reaches an extreme negative value with an F. Do the points marked E and the points marked F occur at the same instants or different instants? Why?

◆**THE MATHEMATICAL MODEL AND NEWTON'S SECOND LAW** Now let's see if our model satisfies the requirements of Newton's second law. The second law applied to the vertical block-on-a-spring tells us that $\Sigma F = -ky = ma_y$, so that

$$\frac{a_y}{y} = -\frac{k}{m} \tag{13-16}$$

Because k and m are both positive constants, the ratio of a_y to y must be a negative constant. Equations 13-13 and 13-9s tell us that in our model,

$$\frac{a_y}{y} = \frac{-\omega^2 Y \sin \theta}{Y \sin \theta} = -\omega^2$$

Here also the ratio of a_y to y is a negative constant because ω is constant. The model matches the physics. But what's more, the match is complete if we let the radius of the reference circle rotate with an angular velocity $\omega = \sqrt{\frac{k}{m}}$, to make the two constants equal. But then $\sqrt{\frac{k}{m}}$ must also be the phase frequency of the block-on-a-spring: We have found how this phase frequency depends on the oscillator's physical properties (the mass of the block and the elastic constant).

For this oscillator,

$$\omega = \sqrt{\frac{k}{m}} \tag{13-16a}$$

and so

$$f = \frac{\omega}{2\pi} = \frac{1}{2\pi}\sqrt{\frac{k}{m}} \tag{13-16b}$$

and

$$T = \frac{1}{f} = 2\pi\sqrt{\frac{m}{k}} \tag{13-16c}$$

Example 13-7 *Determining the Spring Constant of a Spring*

A block suspended from a spring is pulled down a distance of 0.05 m and released. The block goes up and down seven times in the next 5 s. If the block has a mass of 0.12 kg, find the spring constant.

Solution

Choice of approach. We can apply either Equation 13-16b or 13-16c because we can use the same information to find either f or T. The frequency is the number of cycles in each second and so is $\frac{7}{5\,s} = 1.4\ s^{-1}$. Then T, the number of seconds per cycle, is $\frac{1}{f} = \frac{5\,s}{7} = 0.71$ s. (We wrote the definition of frequency as $f = \frac{\Delta n}{\Delta t}$ in Chapter 9, so that $T = \frac{1}{f} = \frac{\Delta t}{\Delta n} = \frac{5\,s}{7}$. But if you understand what frequency and period mean, you do not have to remember formulas to find the values of f and T.) We will arbitrarily pick Equation 13-16b to use below.

What we know/what we don't. Seven times in 5 s means $\frac{7}{5} = 1.4$ times each second, so $f = 1.4\ s^{-1}$. Restated, 5 s for seven times means $\frac{5\,s}{7} = 0.71$ s for one time, so $T = 0.71$ s.

$$f = 1.4\ s^{-1} \qquad T = 0.71\ s \qquad m = 0.12\ kg \qquad Y = 0.05\ m \text{ (not relevant!)} \qquad k = ?$$

The initial displacement is Y, but we have no need for it in the calculation. In most real-world situations, there is all sorts of information around (although not necessarily the information you need), and *you* have to decide which of it is useful for finding out what *you* want to know.

The mathematical solution. Solving Equation 13-16b for k, we get

$$k = (2\pi f)^2 m = (2 \times 3.14 \times 1.4\ s^{-1})^2(0.12\ kg) = 9.3\ kg/s^2 = 9.3\frac{(kg \cdot m)/s^2}{m}$$

$$= \mathbf{9.3\ N/m}$$

◆ Related homework: Problems 13-41, 13-42, and 13-43.

13-5 The Mathematical Model and Conservation of Energy

The elastic force is a conservative force, so that for a simple harmonic oscillator, the total mechanical energy must be conserved. It is easiest to consider a block-on-a-spring moving horizontally without friction. Because the elastic potential energy is $\frac{1}{2}kx^2$ (Equation 6-18), it follows that the system's total energy is

$$\text{Total Energy} = KE + PE = \tfrac{1}{2}mv^2 + \tfrac{1}{2}kx^2 \tag{13-17}$$

Because it is conserved, $(\text{Total Energy})_1 = (\text{Total Energy})_2$ for any two states during the oscillation. When the restoring force returns the block to its equilibrium position (where $PE = 0$), it also restores the block to its maximum KE.

Case 13-1 • *Total Energy of a Vertical Block-on-a-Spring*

The vertical block-on-a-spring is more complicated because its gravitational PE also varies as it bobs up and down, but we will show that we can write its total energy in much the same form as in Equation 13-17. At rest (Figure 13-9*b*), the spring is stretched an amount y_o *below* its unstretched position. If we take the equilibrium position as the origin, $PE_{grav} = 0$ at this position and $PE_{total} = PE_{elas} = \frac{1}{2}ky_o^2$. If the block is now displaced upward a distance y from equilibrium (Figure 13-9*c*), it is a distance $y - y_o$ above its unstretched position. PE_{grav} depends on its distance from the origin (the *equilibrium* position) but PE_{elas} depends on its distance from the *unstretched* position. Thus its total PE is

$$PE_{grav} + PE_{elas} = mgy + \frac{1}{2}k(y - y_o)^2$$
$$= mgy + \frac{1}{2}k(y^2 - 2yy_o + y_o^2)$$
$$= mgy + \frac{1}{2}ky^2 - kyy_o + \frac{1}{2}ky_o^2$$
$$= y(mg - ky_o) + \frac{1}{2}ky^2 + \frac{1}{2}ky_o^2$$

But (see Figure 13-9*b*) $mg = ky_o$, so

$$PE_{total} = PE_{grav} + PE_{elas} = \frac{1}{2}ky^2 + \frac{1}{2}ky_o^2$$

The total PE differs from its equilibrium value by an amount $\frac{1}{2}ky^2$. Because we only care about changes in PE as a system goes from one state to another (recall Figure 6-15), we can effectively change our zero point for PE by subtracting $\frac{1}{2}ky_o^2$ from the total PE in each state. With this change, $PE_{total} = \frac{1}{2}ky^2$, and

$$\text{Total Energy} = KE + PE = \frac{1}{2}mv^2 + \frac{1}{2}ky^2 \qquad (13\text{-}18)$$

which looks like Equation 13-17. Both x in Equation 13-17 and y in Equation 13-18 represent the displacement from equilibrium, but whereas x represents the actual distance the spring is stretched or compressed, y does not. However, once we adopt this zero point for PE, the vertical block-on-a-spring can be described by the same equations as the horizontal one—that is, the energy conservation equation and the equations for x, v, and a as functions of time. You can use these equations with either x or y.

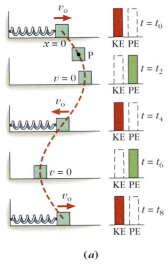

Figure 13-21 Keeping track of a simple harmonic oscillator's energy. The graphs show the back-and-forth interchange between potential and kinetic energy during each cycle of the oscillator's motion. Here v_{max} is the initial velocity v_o.

(a)

When the block is at point P

(b)

Let's consider the energy of the horizontal block-on-a-spring in Figures 13-7A and 13-7B (Figure 13-21):

- At t_2 and t_6: displacement is an extreme $(x = \pm X)$
  velocity is zero
  $$\text{Total Energy} = KE + PE = 0 + \frac{1}{2}kX^2 = \frac{1}{2}kX^2$$

- At t_0, t_4, and t_8: displacement is zero
  velocity is an extreme $(v = \pm v_{max} = \pm \omega X$ by Equation 13-14)
  $$\text{Total Energy} = KE + PE = \frac{1}{2}mv_{max}^2 + 0$$
  $$= \frac{1}{2}mv_{max}^2 = \frac{1}{2}m\omega^2 X^2$$

- At any in-between state: both the KE and PE are nonzero
  $$\text{Total Energy} = \frac{1}{2}mv^2 + \frac{1}{2}kx^2$$

Because energy is conserved,

$$(\text{Total Energy})_{\text{in between}} = (\text{Total Energy})_{\text{2 or 6}} = (\text{Total Energy})_{\text{0 or 4 or 8}}$$

that is,

$$\frac{1}{2}mv^2 + \frac{1}{2}kx^2 = \frac{1}{2}kX^2 = \frac{1}{2}mv_{max}^2 \qquad (13\text{-}19x)$$

The last two terms can be equal only if $v_{max} = \sqrt{\frac{k}{m}}X = \omega X$, again consistent with our model. Example 13-8 verifies that the total energy $\frac{1}{2}mv^2 + \frac{1}{2}kx^2$ must remain constant as θ varies.

Example 13-8 *Energy Conservation by a Block-on-a-Spring*

For a guided interactive solution, go to Web Example 13-8 at
www.wiley.com/college/touger

For a block suspended from a spring, the phase frequency ω is given by $\omega^2 = \frac{k}{m}$ (k = spring constant, m = mass of block). Show that if the block's displacement $x = X \sin \theta$ and its velocity $v = \omega X \cos \theta$, then the total energy of the oscillator is constant—that is, it is the same for any value of θ.

Brief Solution

Choice of approach. The total energy $KE + PE$ for this oscillator is $\frac{1}{2}mv^2 + \frac{1}{2}kx^2$. We substitute our expressions for x and v into this result and then use algebra and trigonometry to simplify the result to see if it is something we recognize as a constant.

The mathematical solution. Substituting $x = X \sin \theta$ and $v = \omega X \cos \theta$, we get

$$\text{total energy} = \tfrac{1}{2}m\omega^2 X^2 \cos^2 \theta + \tfrac{1}{2}kX^2 \sin^2 \theta$$

$$= \tfrac{1}{2}kX^2 \cos^2 \theta + \tfrac{1}{2}kX^2 \sin^2 \theta \qquad \left(\text{since } \omega^2 = \frac{k}{m}\right)$$

$$= \tfrac{1}{2}kX^2 (\cos^2 \theta + \sin^2 \theta) = \tfrac{1}{2}kX^2$$

because by the Pythagorean theorem for a triangle of hypoteneuse 1 (Figure 13-14*b*), $\cos^2 \theta + \sin^2 \theta = 1$. The amplitude X and the spring constant k are both constant, so this total is constant.

◆ Related homework: Problem 13-44.

Example 13-9 *Energy Conservation on a Bungee Cord*

For a guided interactive solution, go to Web Example 13-9 at www.wiley.com/college/touger

A bungee jumper of mass 50.0 kg (part *a* of figure) jumps from a bridge to which he is secured by a bungee cord. (This is *not* an On-the-Spot Activity!) At the point where the cord is stretched just enough to support his weight, the jumper has a downward speed of 24.0 m/s. If the cord has an elastic constant of 300 N/m, how much further will the jumper have descended when his speed has been reduced to 15.0 m/s?

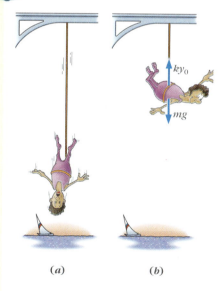

Brief Solution

Choice of approach. We can get a good approximate solution by neglecting air resistance and (as stating a spring constant implies) assuming the bungee cord obeys Hooke's law as it stretches. So although a bungee cord doesn't compress, it behaves like the block-on-a-spring while the spring is stretching. We thus can apply the same conservation-of-energy reasoning. The "point where the string is stretched just enough to support his weight" is by definition the equilibrium position (part *b* of figure), which we take as $y = 0$.

What we know/what we don't.

State 1		State 2		Constants
$y_1 = 0$	$v_1 = -24.0$ m/s	$y_2 = ?$	$v_2 = -15.0$ m/s	$m = 50.0$ kg
$(PE_{\text{total}})_1 = 0$		$(PE_{\text{total}})_2 = \tfrac{1}{2}ky_2^2$		$k = 300$ N/m
$KE_1 = \tfrac{1}{2}mv_1^2$		$KE_2 = \tfrac{1}{2}mv_2^2$		

(a)   (b)

The mathematical solution. Applying conservation of energy (using Equation 13-18 or 13-19), we write

$$(\text{Total Energy})_1 = (\text{Total Energy})_2$$

$$\tfrac{1}{2}mv_1^2 + 0 = \tfrac{1}{2}mv_2^2 + \tfrac{1}{2}ky_2^2$$

Solving for y_2^2 and then taking the negative root because the jumper is below the equilibrium point, we get

$$y_2 = -\sqrt{\frac{m}{k}(v_1^2 - v_2^2)} = -\sqrt{\frac{50.0 \text{ kg}}{300 \text{ N/m}}\left([-24.0 \text{ m/s}]^2 - [-15.0 \text{ m/s}]^2\right)} = \mathbf{-7.65 \text{ m}}$$

◆ Related homework: Problems 13-45 and 13-46.

Saying an oscillation is undamped is equivalent to saying the total mechanical energy remains constant, because each time $v = 0$, the total energy is $\frac{1}{2}kX^2$, which cannot remain constant if the amplitude X is diminishing (Figure 13-10). Damping occurs whenever there are mechanisms, such as frictional or resistance forces, that convert mechanical energy to internal energy or result in a transfer of energy from the oscillator to its surroundings.

13-6 What All Simple Harmonic Oscillators Have in Common

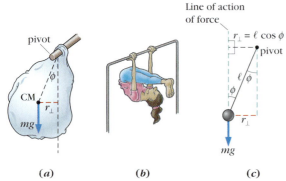

Line of action of force

$r_\perp = \ell \cos \phi$
pivot

Figure 13-22 Pendulums. (*a*) and (*b*) show physical pendulums; (*c*) is a simple pendulum.

We have treated the block-on-a-spring in detail so that it can help us understand other oscillators that are somehow like it. We must now ask: In what ways are other oscillators like this one, and how are they different? We will address this for one other oscillator, the simple pendulum, and then generalize.

A *pendulum* is any body that can swing freely when hung from a pivot point. If it is rotated so that its center of mass is no longer directly below the pivot point (Figure 13-22*a*), the gravitational force on it produces a torque opposite to the angular displacement ϕ, which tends to restore the body to hanging "straight down."

If the body's mass is not all at one distance from the pivot, we call the body a **physical pendulum.** A gymnast hanging from a bar (Figure 13-22*b*) is a physical pendulum. If the mass *is* all at one distance from the pivot, like a point object suspended from an essentially massless string, we call the arrangement a **simple pendulum.** The pendulums of many old clocks are fair approximations of simple pendulums.

If we displace a simple pendulum of length l by an angle ϕ (Figure 13-22*c*), the moment arm of the gravitational force mg is $r_\perp = l \sin \phi$. The torque is therefore

$$\tau = -mgl \sin \phi \qquad (13\text{-}20)$$

The minus sign indicates that the torque is opposite in rotational direction to ϕ; it is a restoring torque rather than a restoring force. We will show that this results in motion that is approximately simple harmonic when the angle ϕ remains small.

A useful approximation. For small angles, if ϕ is in radians,

$$\sin \phi \approx \phi \qquad (13\text{-}21)$$

Few physics exams are as scary as the pendulum in this Arthur Rackham illustration for Edgar Allen Poe's classic horror tale, "The Pit and the Pendulum." (From *Poe's Tales of Mystery and Imagination*, Lippincott, Philadelphia, c. 1920.)

➡**How good is the approximation in Equation 13-21?** By what percent does $\sin \phi$ differ from ϕ when $\phi = 0.25$ rad? When $\phi = 0.79$ rad? What is the value of each of these angles in degrees? Using your calculator to answer these questions will give you a sense of how good the approximation is for different size angles.

We show this using the unit reference circle in Figure 13-23. As the angle ϕ becomes smaller, the arc of length s (in red) becomes less distinguishable from the straight line of length y (in blue), just as over small enough arcs on Earth's round surface, the Earth appears flat. Because $y = \sin \phi$ and $s = \phi$, the approximation $\sin \phi \approx \phi$ becomes more and more accurate as ϕ gets smaller. Equation 13-21 can be verified by using a calculator to compare small values of $\sin \phi$ and ϕ in radian measure (see Problem 13-74). It is not valid if ϕ is in degrees.

With this approximation, Equation 13-20 becomes

$$\tau \approx -mgl\phi \qquad (\phi \text{ small and in radians}) \qquad (13\text{-}22)$$

By Equation 9-14, $\tau = I\alpha$. The moment of inertia for a point object at a distance l from a pivot is $I = ml^2$, so

$$\tau = -ml^2\alpha \qquad (13\text{-}23)$$

If we now equate these two expressions for τ, we get $ml^2\alpha \approx -mgl\phi$, or

$$\frac{\alpha}{\phi} \approx -\frac{mgl}{ml^2} = -\frac{g}{l} \qquad (\text{a negative constant}) \qquad (13\text{-}24)$$

Because angular displacement and acceleration are analogous to linear displacement and acceleration, $\frac{\alpha}{\phi}$ is analogous to $\frac{a}{y}$ in Equation 13-16. The fact that the ratio $\frac{\alpha}{\phi}$ is a negative constant means that, like a and y, α and ϕ are π (or 180°)

out of phase. In general, this phase relationship is satisfied if the displacement varies like a sine or cosine:

If the displacement (or its analog) is given by . . .	. . then the velocity (or its analog) is given by . . .	. . and the acceleration (or its analog) is . . .	. . so that the ratio of acceleration to displacement (or of their analogs) is . . .
$y = Y \sin \omega t$	$\omega Y \cos \omega t$	$\omega^2 Y(-\sin \omega t)$	$-\omega^2$ (a negative constant)
$y = Y \cos \omega t$	$\omega Y(-\sin \omega t)$	$\omega^2 Y(-\cos \omega t)$	$-\omega^2$ (a negative constant)

By analogy, then, Equation 13-24 is satisfied by either

$$\phi = \phi_{max} \sin \omega t \qquad \text{or} \qquad \phi = \phi_{max} \cos \omega t$$

Recall the distinction we made in Example 13-4 between the physical angle ϕ and the phase θ. We also need to distinguish between the rates of change of these two quantities. If we use $\omega = \frac{\Delta \theta}{\Delta t}$ to denote the *phase frequency,* we can use $\omega_\phi = \frac{\Delta \phi}{\Delta t}$ to denote the *angular velocity.* The angular velocity ω_ϕ varies sinusoidally with t, but $\omega \approx \sqrt{\frac{g}{l}}$ is a constant. In this notation, the above table tells us that if a pendulum's displacement is $\phi \approx \Phi \cos \omega t$, then $\omega_\phi \approx -(\omega_\phi)_{max} \sin \omega t$.

The last column in the table tells us that the ratio of *angular* acceleration to *angular* displacement is $\frac{\alpha}{\phi} = -\omega^2$. Comparing this with Equation 13-24, we get $\omega^2 \approx \frac{g}{l}$, or

$$\omega \approx \sqrt{\frac{g}{l}} \tag{13-25}$$

which in turn means that

$$f \approx \frac{1}{2\pi}\sqrt{\frac{g}{l}} \qquad \text{and} \qquad T \approx 2\pi\sqrt{\frac{l}{g}}$$

A pendulum's period will be greater if the pendulum is longer (Figure 13-24).

Reminder: The pendulum is *approximately* a simple harmonic oscillator—the approximation is only good when ϕ is small enough that $\sin \phi \approx \phi$.

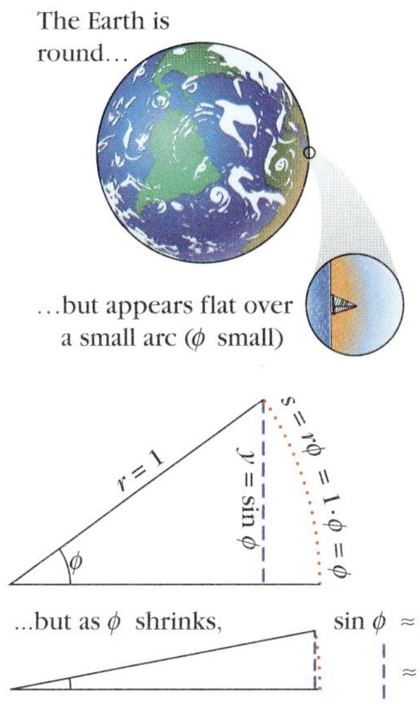

The Earth is round...

...but appears flat over a small arc (ϕ small)

...but as ϕ shrinks, $\sin \phi \approx \phi$

Figure 13-23 The "flat Earth" approximation: $\sin \phi \approx \phi$.

Figure 13-24 The ticking of a pendulum clock depends on its length. A pendulum clock ticks twice during each period (when $\phi = +\Phi$ and when $\phi = -\Phi$). Because the period increases with length, there is more time between ticks for the stately grandfather clock than for the more workaday shelf clock; it ticks more slowly.

On-The-Spot Activity 13-2

Make a simple pendulum by tying a small weight to the end of a string. Hold the string by one end with your right hand, and let the weight swing freely (but with not too large an amplitude). Note its approximate period. Now, with your left hand, grab the string about halfway between your right hand and the suspended weight, and let the weight continue to swing. This shortens the length of your pendulum. How does the period of oscillation compare to before? How does the frequency of oscillation compare to before? Do these observations agree qualitatively with the above equations for T and f?

Example 13-10 *Using the Pendulum to Determine g*

A pendulum can be used to do accurate measurements of the gravitational acceleration g. Such measurements are of great geological interest, because variations in the value of g from one locality to another can indicate variations in the density of underground matter. This could, for example, indicate the presence of an oil deposit, which would have a different density than solid rock.

A carefully calibrated 1.0000-m pendulum in an evacuated chamber is set in motion with a very small amplitude at the top of Pike's Peak. The pendulum is watched for exactly 10 000 s, during which it swings back and forth $4979\frac{1}{2}$ times. What is the value of g on Pike's Peak?

Solution

Choice of approach. You must consider
1. the relationship between ω and g,

2. the relationship between ω and f or T, and

3. how to determine f or T from the given data.

What we know/what we don't.

$$\text{number of cycles } \Delta n = 4979.5 \qquad l = 1.0000 \text{ m}$$

$$\Delta t = 1.0000 \times 10^4 \text{ s} \qquad \omega = ? \qquad f = ? \qquad g = ?$$

The mathematical solution. The frequency is the number of cycles each second:

$$f = \frac{\Delta n}{\Delta t} = \frac{4.9795 \times 10^3}{1.0000 \times 10^4 \text{ s}} = 4.9795 \times 10^{-1} \text{ s}^{-1}$$

Because $\omega = 2\pi f$ and $\omega = \sqrt{\frac{g}{l}}$ as well, we can equate the two expressions and solve for g:

$$g = 4\pi^2 f^2 l = 4(3.14159)^2 (4.9795 \times 10^{-1} \text{ s}^{-1})^2 (1.0000 \text{ m})$$

$$= \mathbf{9.7888 \ m \cdot s^{-2}} \quad \text{or} \quad \mathbf{9.7888 \ m/s^2}$$

Notes: (1) To maintain accuracy, we must keep π to at least as many significant figures as the other quantities involved in our calculation. (2) The very small amplitude is necessary to ensure that the pendulum's motion is sinusoidal to this degree of approximation.

◆ Related homework: Problems 13-49, 13-50, and 13-53.

Velocity is the rate of change of displacement and acceleration is the rate of change of velocity. If a quantity (like ϕ) is analogous to displacement, its rate of change (like ω_ϕ) is the analog to velocity. The rate of change of this rate of change (as a is the rate of change of ω_ϕ) is the analog to acceleration. We can summarize the procedure we followed for the pendulum in general terms.

PROCEDURE 13-2

Identifying Sinusoidal Oscillatory Behavior of a Physical Quantity

1. See if the ratio

$$\frac{\text{analog to acceleration}}{\text{analog to displacement}} = -c \quad (c \text{ a positive constant}) \qquad (13\text{-}26)$$

This is always the result of a restoring force (or force-like quantity, such as torque).

2. If so, the quantity's behavior over time will be described by a sinusoidal function, such as those in Equations 13-9s and 13-9c.

3. By looking at the quantities that make up c for the particular oscillator (such as k and m for the block-on-a-spring or g and l for the pendulum), and setting $c = \omega^2$, you can determine how the phase frequency ω depends on the physical properties of that oscillator.

4. Using $\omega = 2\pi f = \frac{2\pi}{T}$, you can then determine how the frequency f and period T depend on those properties.

◆**POTENTIAL ENERGY OF A SIMPLE PENDULUM** Because $\omega^2 = k/m$, the PE of the block-on-a-spring (Equation 13-18) is

$$PE = \tfrac{1}{2}ky^2 = \tfrac{1}{2}m\omega^2 y^2$$

Purely by means of our linear → rotational analogies ($m \rightarrow I$, $y \rightarrow \phi$), we can anticipate that the corresponding equation for a rotational oscillator such as a pendulum might be $PE = \tfrac{1}{2}I\omega^2\phi^2$. (Contrast this with its kinetic energy, $\tfrac{1}{2}I\omega_\phi^2$.) Since $I = ml^2$ and $\omega^2 \approx g/l$ for the simple pendulum, its total energy would be

$$E_{\text{total}} = KE + PE \approx \tfrac{1}{2}ml^2\omega_\phi^2 + \tfrac{1}{2}mgl\phi^2 \qquad (13\text{-}27)$$

Example 13-11 *A Frictionless Pendulum Conserves Energy*

If a simple pendulum of length 1.2 m is displaced 2.0° from the vertical and released, what angular velocity does it reach at the bottom of its swing?

Solution

Choice of approach. We apply conservation of energy in much the same way that we did in Chapter 6; that is, we begin with the conservation of energy equation $KE_1 + PE_1 = KE_2 + PE_2$. For the simple pendulum swinging with small amplitude, $KE + PE \approx \tfrac{1}{2}ml^2\omega_\phi^2 + \tfrac{1}{2}mgl\phi^2$. Remember that ϕ must be in radians. If state 1 is when the angular displacement is a maximum, $KE_1 = 0$. At the bottom of the swing, $\phi = 0$, so $PE_2 = 0$.

EXAMPLE 13-11 *continued*

What we know/what we don't.

State 1	State 2	Constants
$\phi_1 = 2.0° \times \dfrac{2\pi}{360°} = \dfrac{\pi}{90}$	$\phi_2 = 0$	$l = 1.2 \text{ m} \qquad m = ?$
$\omega_{\phi 1} = 0$	$\omega_{\phi 2} = ?$	$g = 9.8 \text{ m/s}^2$

The mathematical solution. Because $KE_1 = 0$ and $PE_2 = 0$, $KE_1 + PE_1 = KE_2 + PE_2$

becomes

$$\tfrac{1}{2} mgl\phi_1^2 = \tfrac{1}{2} ml^2\omega_{\phi 2}^2$$

or

$$g\phi_1^2 = l\omega_{\phi 2}^2$$

Note that the mass has dropped out of the calculation. The result does not depend on the mass, so we don't need to know it. Solving for $\omega_{\phi 2}$ gives us

$$\omega_{\phi 2} = \sqrt{\frac{g}{l}}\,\phi_1 = \sqrt{\frac{9.8 \text{ m/s}^2}{1.2 \text{ m}}}\left(\frac{\pi}{90}\right) = \mathbf{0.032\pi/s} \text{ or } \mathbf{0.10 \text{ rad/s}}$$

◆ Related homework: Problems 13-56 and 13-57.

13-7   Forced Oscillations and Frequency Matching

Half a cycle later

Figure 13-25 A fanciful arrangement for exerting a roughly sinusoidal torque on a swing.

The simple harmonic oscillator, like the body with no external forces exerted on it, is an idealization; that is, a circumstance which for macroscopic systems may be achievable to a high degree of approximation but never exactly. To keep a real-world oscillator going requires a recurrent energy input to replace the energy that is lost. Typically, doing work on the oscillator provides this input.

A child on a swing is an oscillator—a physical pendulum. Here we reduce it to a simple pendulum for simplicity. The parent pushing the child does work with each push. The pushes are not random. If they were, the force $\vec{\mathbf{F}}$ would be exerted parallel to the displacement $\Delta\vec{\mathbf{s}}$ (yielding positive work) as often as antiparallel (yielding negative work). The net energy gained from each such pair of inputs would be $F\Delta s - F\Delta s = 0$, and the damping of the oscillation would not be prevented.

Instead, even a minimally competent parent will time each push to occur as the child begins to swing forward. This pushing has the same frequency as the swing's **natural frequency** of oscillation ($\frac{1}{2\pi}\sqrt{\frac{g}{l}}$ for the simple pendulum). In a **forced oscillation,** one in which the exertion of a force compensates for damping, the amplitude gets the biggest boost if we *match the frequency of the driving force to the natural frequency of the oscillator.* This matching is called **resonance.** An even better match would be a torque that varied sinusoidally in sync with the swing's angular displacement, as in the fanciful arrangement in Figure 13-25.

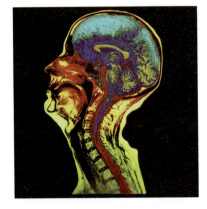

Magnetic Resonance Imaging (MRI). In a magnetic field, the hydrogen nuclei (protons) in the water molecules of your body can oscillate between two energy states. A radio wave input at the resonance frequency can be absorbed by the protons, driving them into the upper state. The protons then reemit radio waves, which can be detected and computer-analyzed to produce an image showing the distribution of the emitting protons. Images like this one of the human brain have become a valuable diagnostic tool for modern medicine. The lighter blue area shows *atrophy*, the shrinking and wasting away of tissue. This can be a result of stroke or senile dementia.

Case **13-2** ◆ *The Tunable Swing*

This being an era of robotics, let's suppose the parent is replaced by a robot pusher (Figure 13-26*a*), who pushes the three children on the variable swing set in Figure 13-26*b*. If the pusher is set so that the frequency of the force it exerts matches the natural frequency of pendulum (swing + child) B, it will not do a very good job of pushing A or C, because the lengths, on which their natural frequencies depend, are different.

We can think of the push as a kind of *signal,* and the amplitude of the swing (the receiver) as a measure of how strongly the signal is detected or received. Child B gets good reception; A and C get lousy reception. One way to deal with this is to design the robot so that the push frequency can vary. But this means the robot needs to know which detector is going to receive its signal. This would be like a TV station having to send out a signal of different frequency to every

receiver (TV set). The more practical solution would be for the TV set owner to be able to adjust the frequency *it* can receive and then be able to select the frequency of one or another station's signal.

The natural frequency of the swing in Figure 13-26*c* can be varied continuously by varying the length on which it depends. It is *tunable;* that is, it can be adjusted to match the frequency of any input signal. The tuning circuits of radio and TV sets are essentially electrical oscillators. Their oscillations are driven by the radio waves transmitted by different stations (TV also uses radio waves), each at a different basic frequency. When you turn your channel selector, you are varying the property on which the natural frequency of your tuning circuit depends to match the frequency of the signal you wish to receive.

Height of CM of child (treated as point object) ℓ_1 ℓ_2 ℓ_3

A   B   C

Figure 13-26 Matching the frequency of the driving force to the natural frequency of the oscillator.

(*a*) The robot pusher is set to push at a fixed frequency.

(*b*) Only one of the oscillators of various frequencies is well suited to "receive the signal".

(*c*) The solution – the TUNABLE SWING (varies ℓ to vary natural frequency of oscillator to match push frequency)

◆ SUMMARY ◆

In this summary, we start with the ideas we reached toward the end of the chapter to provide a framework for seeing how the ideas that preceded it fit together. We have looked at situations in which an object's displacement varies cyclically with time in a **sinusoidal** or sine-like pattern (Figure 13-2). In each instance, the condition that gives rise to this behavior is (or is analogous to)

$$\frac{\text{acceleration}}{\text{displacement}} = -c \quad \text{(c a positive constant)} \quad (13\text{-}26)$$

where $\qquad\qquad c = \omega^2$

then gives us the natural phase frequency of the oscillation. This condition occurs when there is a **restoring force** (or a restoring torque). Objects satisfying this condition are called **simple harmonic oscillators** and include

1. the *block-on-a-spring* governed by a Hooke's law–like total force $\Sigma F_y = -ky$ (Equation 13-1), so that by Newton's

second law, $-ky = ma_y$, and

$$\frac{a_y}{y} = -\frac{k}{m} \quad \text{(which is like 13-26)} \quad (13\text{-}16)$$

Figure 13-7B describes the behavior of this system in detail.

2. the *simple pendulum with small amplitude,* on which the restoring torque

$$\tau = -mgl \sin \phi \approx -mgl\phi \quad (13\text{-}20)$$

because $\qquad \sin \phi \approx \phi \quad$ (ϕ small and in radians) $\quad (13\text{-}21)$

It follows from $\qquad \tau = I\alpha = ml^2\alpha \quad (13\text{-}23)$

that $\qquad \dfrac{\alpha}{\phi} \approx -\dfrac{g}{l} \quad$ (which is also like 13-26) $\quad (13\text{-}24)$

At any instant *t*, when the oscillator is in a particular **phase** of its cycle, the displacement

$$y = Y \times \begin{pmatrix} \text{some number between 1 and } -1, \text{ the value} \\ \text{of which depends on the phase, or on the} \\ \text{instant } t \text{ at which that phase occurs} \end{pmatrix} \quad (13\text{-}3)$$

Y is the *maximum* displacement, or **amplitude.** To model this behavior mathematically, we use the **unit reference circle** and *define*

$$\cos \theta \equiv x \quad \text{and} \quad \sin \theta \equiv y \quad (13\text{-}4c, 13\text{-}4s)$$
$$\text{(assuming } r = 1)$$

The model implies that in general, the displacement has this form:

$$y = Y \sin \theta \qquad (13\text{-}5s)$$

if the oscillator starts out in equilibrium with an initial

upward velocity (and so behaves like the vertical shadow).

$$y = Y \cos \theta \qquad (13\text{-}5c)$$

if the oscillator starts out with its maximum positive displacement (and so behaves like the horizontal shadow).

The *phase* θ may be written as $2\pi(t/T)$ or ωt or $2\pi f t$

(T = period, f = frequency, $\omega = 2\pi f$ = phase frequency)

From the reference circle model (Figures 13-12 to 13-14 and WebLink 13-4) we find that

If the displacement (or its analog) is given by . . .	. . then the velocity (or its analog) is given by . . .	. . and the acceleration (or its analog) is . . .	. . so that the ratio of acceleration to displacement (or of their analogs) is . . .
$y = Y \sin \omega t$	$\omega Y \cos \omega t$	$\omega^2 Y(-\sin \omega t)$	$-\omega^2$ (a negative constant)
$y = Y \cos \omega t$	$\omega Y(-\sin \omega t)$	$\omega^2 Y(-\cos \omega t)$	$-\omega^2$ (a negative constant)

Procedure 13-1 outlines what is involved in using these sinusoidal functions to reason about oscillators. Comparing Equations 13-16 and 13-24 with the last column above, we get the **phase frequencies:**

$$\omega = \sqrt{\frac{k}{m}} \text{ for a block-on-a-spring}$$

$$\omega \approx \sqrt{\frac{g}{l}} \text{ for a simple pendulum}$$

(You can then find the period T and frequency f using

$$\omega = \frac{2\pi}{T} \text{ and } \omega = 2\pi f.)$$

Each of the successive sinusoidal functions (sin, cos, −sin, −cos) differs in phase with the next ("lags behind") by $\frac{\pi}{2}$ in radian measure, so that x (or y), v, and a are correspondingly *out of phase* in the ways shown in Figure 13-7B (and summarized in Problems 13-10 and 13-11).

Simple harmonic oscillators are those that behave sinusoidally; they have constant amplitude and therefore *conserve mechanical energy:*

For the block-on-a-spring:

$$E_{\text{total}} = KE + PE = \tfrac{1}{2}mv^2 + \tfrac{1}{2}ky^2 = \tfrac{1}{2}mv^2 + \tfrac{1}{2}m\omega^2 y^2$$
$$(y = \text{displacement from equilibrium})$$

For the simple pendulum:

$$E_{\text{total}} = KE + PE \approx \tfrac{1}{2}ml^2\omega_\phi^2 + \tfrac{1}{2}mgl\phi^2$$

A **damped** oscillator (Figure 13-10), in contrast, loses or dissipates mechanical energy. To maintain its amplitude, and hence its total mechanical energy, work must be done on it by the exertion of an external driving force. In such a **forced oscillation,** the amplitude gets the biggest boost if we *match the frequency of the driving force to the natural frequency of the oscillator,* a condition called **resonance.**

◆QUALITATIVE AND QUANTITATIVE PROBLEMS◆
Hands-On Activities and Discussion Questions

The questions and activities in this group are particularly suitable for in-class use.

13-1. *Discussion Question.* Discuss the similarities and differences between a perfectly elastic bouncing ball and a block bobbing up and down when suspended from a vertical spring. In your discussion, be sure to address specific quantities that describe or affect motion.

13-2. *Hands-On Activity.* Take a light wooden or plastic ruler and press one end of it down firmly against a desk or table-

top. Let the rest of the ruler stick out unsupported. Near the free end, make a small but conspicuous mark. Then twang the free end to set it vibrating, and watch the mark. You will see two images of the mark that gradually come together as the oscillation damps out. But as it oscillates, the mark passes through all the positions between the two extremes where the images appear. Why do you see the mark only faintly, if at all, at the in-between positions but very plainly at the extremes?

Review and Practice

Section 13-1 Oscillators and Their Importance
13-3.
a. List all the similarities you can between a block bobbing up and down on a vertical spring and a person rocking in a rocking chair. Your list should include references to specific

physical quantities, and you should take note of when a quantity used to describe one of these two situations is analogous to a quantity used to describe the other.

b. List any other situations you can think of in the real world that have similar characteristics.

Section 13-2 How a Typical Oscillator Behaves (Qualitative Discussion)

13-4. When a block is suspended from a spring, is the block in equilibrium when the spring is at its unstretched length? Briefly explain.

13-5. Complete Figure 13-7B by providing the missing information about the velocity, the acceleration, and ΣF_x at the instants t_7 and t_8.

13-6. Assume that the equilibrium position of the air track glider in Figure 13-6 is at $x = 0$. When the glider is oscillating back and forth through this position, its velocity is (*always*, *sometimes*, or *never*) negative when $x = 0$. Explain your choice.

13-7. A block of mass 0.25 kg hanging on a spring with elastic constant 40.0 N/m is set to oscillating vertically.
a. Find the total force on the block when the spring is neither stretched nor compressed.
b. Find the total force on the block as it is passing the equilibrium position.
c. How far from the equilibrium position is the block in part **a**?

13-8. In Figure 13-7A,
a. how does the velocity of the block at instants t_5 and t_7 compare? Address both the magnitude and the sign. Briefly explain.
b. how does the acceleration of the block at instants t_5 and t_7 compare? Address both the magnitude and the sign. Briefly explain.

13-9.
a. When a block suspended from a spring oscillates vertically without damping, the acceleration is (*always, sometimes,* or *never*) zero when the velocity is zero. Briefly explain.
b. When the same block oscillates vertically *with* damping, the acceleration is (*always, sometimes,* or *never*) zero when the velocity is zero. Briefly explain.

13-10. The table below lists four quantities—the position x, the velocity v_x, the total horizontal force ΣF_x, and the acceleration a_x—that describe the horizontal oscillation depicted in Figures 13-7A and 13-7B. To complete the table, indicate whether each of these is positive ($+$), negative ($-$), zero (0), has its most positive value ($+$max), or has its most negative value ($-$max) at each of the nine instants shown (t_0 to t_8).

	t_0	t_1	t_2	t_3	t_4	t_5	t_6	t_7	t_8
x									
v_x									
ΣF_x									
a_x									

13-11. Repeat Problem 13-10 for the vertical oscillation depicted in Figure 13-27, substituting y for x in each of the quantities in the left-hand column.

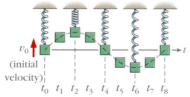

Figure 13-27 Phases of a vertically oscillating block-on-a-spring. Problem 13-11

Section 13-3 A Model for Describing Oscillators Mathematically

13-12.
a. When the rotating radius in Figure 13-13 first makes an angle of $\frac{\pi}{3}$ with the $+x$ axis, what is the phase of either of the block-on-a-spring oscillators in the figure? Why?
b. Explain in your own words how the angular velocity of the rotating radius is related to the phase frequency of either block-on-a-spring in the figure.

13-13. SSM Sketch a unit reference circle and by reading coordinates from it, find the sines and cosines of 0, 90°, 180°, 270°, and 360°.

13-14.
a. Repeat Example 13-1 to find the sines and cosines of the following angles: 30°, 60°, 120°, 150°, 210°, 240°, 300°, 330°, and 351°.
b. Use these values and the values obtained in Problem 13-13 to complete the following tables:

θ	$\cos \theta$		θ	$\sin \theta$

c. Use the values in your tables to plot graphs of $\cos \theta$ versus θ and $\sin \theta$ versus θ.

13-15. A block with a pencil through its center is suspended from a vertical spring. To one side is a vertically mounted turntable, as in Figure 13-28. A peg is mounted on the rim of the turntable. When the block hangs in equilibrium, the pencil points along a line through the center of the turntable. Suppose that at $t = 0$, the peg on the rotating turntable passes through position A. At that instant the block is released from rest from a position where the pencil is pointing directly at the peg, as shown in the figure. The turntable's angular velocity is set so that the pencil always points directly at the peg as they both move.

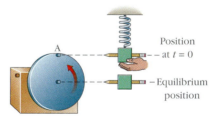

Figure 13-28 Problem 13-15

a. To accomplish this, how should the angular velocity of the turntable compare with the phase frequency of the block-on-a-spring? Is it equal? If not, how many times as great should it be? Explain your answer.
b. If the block and pencil have a combined mass of 0.30 kg and the spring constant is 120 N/m, what is the angular velocity of the turntable?
c. Under these conditions, how long does it take the peg to make one complete circle?
d. How long does it take the block to make one complete cycle?

13-16. A harmonic oscillator has a frequency of $0.40\ s^{-1}$ (0.40 cycles/s). Find its phase frequency and its period.

13-17. The positions of four oscillators A, B, C, and D are described as follows: $y_A = 5\sin(3t)$, $y_B = 4\sin(5t)$, $y_C = 4\cos t$, $y_D = \cos(3t)$. Rank the four oscillators in order of their periods of oscillation. Rank them from least to greatest, noting any equalities.

13-18. Suppose the air track glider in Figure 13-6 has a mass of 0.25 kg and the spring connected to it has a spring constant of 4.0 N/m. If the glider were pulled along the track until the spring was stretched a short distance and then released, what would be the frequency and the period of the glider's oscillation?

13-19. SSM Suppose the block-on-a-spring in Figure 13-7A is given an initial velocity in the $+x$ direction at $t = 0$. If it reaches a maximum displacement of 0.18 m, what is its displacement after $\frac{7}{12}$ of a cycle?

13-20. Suppose the block-on-a-spring is instead given an initial velocity in the $-x$ direction at $t = 0$. Again it reaches a maximum displacement of $+0.18$ m.
a. What is its displacement after $\frac{7}{12}$ of a cycle?
b. What is its phase the first time it reaches its maximum displacement?
c. What is its phase the second time it reaches its maximum displacement?

13-21. The phase of a certain oscillator is 0 when $t = 0$. If the oscillator has a period of 20.0 s, what is the phase of the oscillator after 3 s?

13-22. The phase of a certain simple harmonic oscillator is 0 when $t = 0$.
a. If the oscillator has a frequency of $5.00\ s^{-1}$, when is its phase $\frac{\pi}{6}$?
b. At what instant does it next have the same position and velocity as in *a*?

13-23. A certain block-on-a-spring oscillates horizontally on a frictionless surface. Its displacement from equilibrium is given by $x = (0.12\ \text{m})\sin(2\pi\frac{t}{T})$. Its period of oscillation is 0.40 s.
a. State what phase θ this oscillator is in at each of the following instants: $t = 0$, $t = 0.05$ s, $t = 0.10$ s.
b. The maximum displacement of this oscillator is 0.12 m. What fraction of the maximum displacement is the block's actual (instantaneous) displacement at each of the instants in *a*?
c. Find the block's displacement at each of the instants in *a*.

13-24. The displacement of a certain simple harmonic oscillator is $y = 15\sin 8t$.
a. Find the amplitude, period, frequency of oscillation, and phase frequency of this oscillator.
b. How long does it take the phase to change by $\frac{\pi}{4}$?
c. If the oscillator began its oscillations at $t = 0$, what are the first two instants at which $y = 5$? The first two instants at which $y = -5$?
d. In what phase will the oscillator's acceleration reach its extreme positive value for the first time?

13-25. A block suspended from a spring, initially at rest, is set to oscillating by giving it an upward velocity at $t = 0$. If it takes the block 0.10 s to reach its maximum displacement of 0.12 m,

a. what is its displacement after 0.075 s? After 0.35 s? After 0.45 s?
b. Can you answer *a* by setting up a ratio? Briefly explain.
c. If your answer to *b* is yes, do you get any displacement values greater than 0.12 m? (If so, consider your results in light of the given information that the maximum displacement is 0.12 m.)

13-26. For the pendulum oscillation described in Example 13-4, find
a. the frequency of the oscillation;
b. the pendulum's phase when it returns to its equilibrium position for the first time after $t = 0$; and
c. the instant t at which the phase in *b* occurs.

13-27. The function $\phi = -30\cos 6\pi t$ (with ϕ in degrees) describes the oscillation of a certain torsion pendulum (Figure 13-4). Assume the counterclockwise direction to be positive. Find
a. the maximum angle to which the torsion pendulum swings;
b. the frequency of oscillation; and
c. the phase at which the torsion pendulum first reaches its greatest counterclockwise rotation.

13-28. Suppose the block-on-a-spring in Figure 13-7A is first set in motion as shown at $t_0 = 0$, and its oscillation has an amplitude of 0.15 m. If it first has a displacement of -0.050 m at $t = 0.90$ s, what is its frequency of oscillation?

•13-29. A simple pendulum is displaced 8.0° from the vertical and released at $t = 0$. At $t = 0.40$ s it has a displacement of 3.0° from the vertical for the first time.
a. Find the pendulum's period.
b. Find the pendulum's period if at $t = 0.40$ s it has a displacement of 3.0° from the vertical for the *second* time.

Section 13-4 Checking the Mathematical Model Against the Physics

13-30. A block hangs stationary from a vertical spring until $t = 0$. At that instant it is given a small displacement downward and then released. During the resulting oscillation, when the block's displacement is most negative, its velocity is ____ (*positive; negative; zero; more than one answer is possible*). Explain.

13-31. A block hangs stationary from a vertical spring until $t = 0$. At that instant it is given a small displacement downward and then released. During the resulting oscillation, when the block's displacement is most negative, its acceleration is ____ (*positive; negative; zero; more than one answer is possible*). Explain.

13-32. A block hangs stationary from a vertical spring until $t = 0$. At that instant it is given a small displacement downward and then released. During the resulting oscillation, when the block's acceleration is most positive, its velocity is ____ (*positive; negative; zero; more than one answer is possible*). Explain.

13-33. Complete each of the sentences below by filling in the blank with one of these four choices:
  i. a positive maximum only
  ii. a negative maximum only
  iii. zero
  iv. either a positive maximum or a negative maximum
A block hangs stationary from a vertical spring until $t = 0$. At that instant it is given a small displacement downward and then released. During the resulting oscillation,

a. when the block's acceleration is most negative, its velocity is ____.

b. when the block's acceleration is zero, its velocity is ____.

c. when the block is moving upward through its equilibrium position, its velocity is ____.

d. when the spring is at maximum compression, its acceleration is ____.

e. when the force on the block is a positive maximum, its displacement is ____.

13-34. The glider in Figure 13-6 is set to oscillating back and forth. Assume the positive direction is toward the right. At a certain instant, the glider has its greatest positive acceleration. A quarter of a period later, the glider will ____ (*have zero velocity; be in equilibrium; have its greatest negative acceleration; have zero velocity and be in equilibrium*). Explain.

13-35. SSM Student X argues as follows: "I know that the maximum velocity of a simple harmonic oscillator is ωY and its maximum acceleration is $\omega^2 Y$. If I know that the velocity is a maximum at a certain instant and I know its value, then all I have to do is multiply it by ω to find the acceleration at the same instant." Is there anything wrong with this reasoning? Explain.

13-36. Find the maximum velocity and the maximum acceleration of a block-on-a-spring whose displacement in cm is $y = 3.6 \cos\left(\frac{\pi t}{3}\right)$.

13-37. Find the maximum angular velocity and the maximum angular acceleration of the torsion pendulum in Problem 13-27.

13-38. A glider of mass 0.220 kg is connected to one end of a horizontal air track by a spring. The glider is released from rest when the spring is stretched a distance of 0.180 m and is observed to have a speed of 0.806 m/s when the spring is compressed 0.080 m. Find the spring constant.

13-39. Consider the graphs of v versus t and a versus t in Figure 13-19.

a. On the graph of v versus t, find those points where the slope is steepest upward. Mark them with a G.

b. On the graph of a versus t, find the points where the acceleration reaches its extreme positive value. Mark these with an H. Do the points marked G and the points marked H occur at the same instants or different instants? Why?

13-40. Consider the graphs of v versus t and a versus t in Figure 13-19.

a. On the graph of a versus t find the points where a has its most extreme values, either positive or negative. Mark those points with a J.

b. Now, on the graph of v versus t, identify the points that occur at the same instants t as the points marked J on the graph of a versus t. Mark each of these with a K. What is the value of v at these points?

c. Give a physical reason why v has the value it does at the points marked K. (*Hint:* Assume the graphs are describing the motion of a block-on-a-spring, and think about what is happening to the block at the points marked K.)

13-41. Identical blocks are suspended from springs A and B. The block on spring A is pulled down a distance of 0.06 m, and the block on spring B is pulled down a distance of 0.12 m. If both blocks are then released at the same instant, each takes 4.5 s to go through six complete oscillations. Is the spring constant of spring B less than, equal to, or greater than the spring constant of spring A?

13-42. Identical blocks are suspended from springs C and D. The block on spring C is pulled down a distance of 0.06 m and the block on spring D is pulled down a distance of 0.12 m. If both blocks are then released at the same instant, it takes the block on spring C twice as long as the block on spring D to complete five oscillations. Compare the two spring constants by finding the numerical multiplier in $k_C = \underline{\ ?\ } k_D$.

13-43. When a block of mass 0.24 kg is hung on a certain spring and set to oscillating, it takes the block 0.30 s to complete each oscillation. Find the spring constant of the spring.

Section 13-5 The Mathematical Model and Conservation of Energy

13-44. For a certain block on a spring, the block's displacement $x = X \sin\theta$ and its velocity $v = \omega X \cos\theta$. If the block has a mass of 0.20 kg and the spring constant is 45 N/m, calculate the total energy when *a.* $\theta = 0°$. *b.* $\theta = 30°$. *c.* $\theta = 90°$. *d.* Determine the total energy when $\theta = 115°$ without doing any further calculations.

13-45. A carnival performer of mass 50.0 kg hangs in equilibrium at the end of a giant bungee cord that obeys Hooke's law with an elastic constant of 450 N/m for stretches of up to a few meters.

a. If she is given an initial downward velocity of 2.4 m/s, how far below her equilibrium position does she go?

b. Where is she when she has an *upward* velocity of 1.5 m/s?

c. What is her acceleration at the position described in *a*?

d. What is her acceleration at the position described in *b*?

13-46. In Example 13-9, if the water is 11.0 m below the equilibrium point, determine by a calculation whether the bungee jumper will get wet.

Section 13-6 What All Simple Harmonic Oscillators Have in Common

13-47. Explain the difference(s) between the angular velocity and the phase frequency of a simple pendulum.

13-48. The table below lists four quantities—the angular displacement ϕ, the angular velocity ω_ϕ, the total torque $\Sigma\tau$, and the angular acceleration α—that describe the oscillation depicted in Figure 13-29. The extreme positions occur at t_2 and t_6 in the figure. The instant t_1 is understood to be between t_0 and t_2, t_3 is between t_2 and t_4, and so on. To complete the table, indicate whether each of these is positive (+), negative (−), zero (0), has its most positive value (+max), or has its most negative value (−max) at each of the nine instants shown (t_0 to t_8).

Figure 13-29 Phases of a simple pendulum. Problem 13-48

	t_0	t_1	t_2	t_3	t_4	t_5	t_6	t_7	t_8
ϕ									
ω_ϕ									
$\Sigma\tau$									
α									

13-49. Suppose a simple pendulum consists of a small weight (called a *pendulum bob*) suspended from a 0.40-m-long string. If the pendulum bob were displaced to one side by a small angle and then released,
a. what would be the frequency and the period of the pendulum's oscillation?
b. what would be its phase frequency?

13-50. A small child and her heavyset father swing on identical swings at the neighborhood playground. Both are content to swing through fairly small arcs. At a certain instant while they are swinging, both of them stop pumping. Assume that the centers of mass of father and daughter are both at the same distance above the seat of the swing. Compare the amount of time it takes each of them to complete three more back-and-forth swings. Briefly explain.

13-51. Find the maximum angular velocity and maximum angular acceleration of a simple pendulum oscillating with an amplitude of 0.7° and a period of 0.80 s.

13-52. An antique collector has bought a grandfather clock in which the pendulum consists of a small metal disk with a mass of 0.20 kg at the end of a stiff metal wire 1.5 m long. Assuming the radius of the disk is negligible compared to the length of the wire, how long does it take for the pendulum to swing from its left-most position to its right-most position?

13-53. SSM Suppose you are designing a pendulum clock using a simple pendulum. You would like the clock to tick twice each second (the clock ticks when the pendulum is furthest from the vertical in either direction). How long should the pendulum be?

13-54.
a. A small ball is placed in the bottom of a hemispherical bowl. It is then displaced a small distance along the surface of the bowl and released. If dissipative forces are negligible, does the ball behave like a simple harmonic oscillator? Explain.
b. Draw a free-body diagram of the displaced ball in *a* and compare it to the free-body diagram of the bob of a simple pendulum when it has a small displacement. Reconsider your answer to *a* if necessary.

13-55. A simple torsion pendulum can be made by taking a one-foot length of cellophane tape and wrapping the last two inches of one end around a ruler (Figure 13-30). (You may find that actually making such a pendulum and seeing how it behaves will help you answer the following questions.) The top view in the figure shows the ruler in its equilibrium position (A) and at its most extreme displacements (B and C). Assume that the positive direction of rotation in the top view is counterclockwise. As the ruler oscillates, in which (one or more) of the positions shown is the ruler's angular acceleration
a. most positive?   *b.* most negative?   *c.* zero?

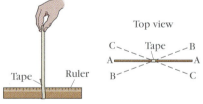
Figure 13-30 A simple torsion pendulum. Problems 13-55 to 13-60

13-56. As the ruler in Problem 13-55 oscillates, in which (one or more) of the positions shown is the ruler's angular velocity
a. most positive?   *b.* most negative?   *c.* zero?

13-57.
a. As the ruler in Problem 13-55 oscillates, in which (one or more) of the positions shown is the torque exerted on the ruler by the twisted tape most negative?
b. To which (one or more) of your answers to Problem 13-55 and/or Problem 13-56 should this answer correspond? Briefly tell why.

13-58.
a. As the ruler in Problem 13-55 oscillates, is the angular displacement positive, negative, or zero when the angular acceleration is most negative?
b. Is this what you expect of a simple harmonic oscillator? Briefly explain.

13-59.
a. As the ruler in Problem 13-55 oscillates, is the angular displacement positive, negative, or zero when the angular velocity is most negative?
b. Is this what you expect of a simple harmonic oscillator? Briefly explain.

13-60.
a. As the ruler in Problem 13-55 oscillates, is the angular acceleration positive, negative, or zero when the angular velocity is most negative?
b. Is this what you expect of a simple harmonic oscillator? Briefly explain.

13-61. SSM WWW At an instant when the angular displacement of a simple pendulum is half its maximum value, the angular speed of the pendulum must be what fraction of its maximum angular speed?

13-62. A simple pendulum of total length 2.0 m consists of a small ball with a mass of 0.10 kg suspended from a string with negligible mass. The pendulum is given an initial angular displacement of 0.08π and released.
a. What is the angular displacement when the pendulum reaches its maximum angular speed?
b. What is the value of this maximum angular speed?
c. What is the value of the angular speed when the pendulum's angular displacement is 0.04π?

Section 13-7 Forced Oscillations and Frequency Matching

13-63. SSM A block of mass 2.00 kg on a spring of elastic constant 50.0 N/m undergoes damped oscillations. The lost energy can be restored by giving the block equal taps in the same direction at uniform intervals. To optimize the effectiveness of this procedure, how long should the interval between taps be?

13-64. *Coupled oscillators:* Two simple pendulums consist of identical bobs suspended on stiff metal wires. The wires are both soldered at their top ends to the same taut horizontal wire, which can experience a small amount of twist when either pendulum swings. Suppose the first pendulum is set to swinging. Will it affect the second pendulum more during the first few swings if it is slightly longer than the second one or if both pendulums are the same length? Briefly explain.

Going Further

The questions and problems in this group are not organized by section heading, so you must determine for yourself which ideas apply. Some of them will be more challenging than the Review and Practice questions and problems (especially those marked with a • or ••).

13-65. In Chapter 3 we wrote the equations $v_y = v_{oy} - gt$ and $y = y_o + v_{oy}t - \frac{1}{2}gt^2$ to describe vertical motion. Are these equations applicable to the vertical block-on-a-spring? Briefly explain.

13-66. Which of the following remain constant for the block-on-a-spring while it is oscillating vertically? Choose *all* correct responses: *the position; the velocity; the spring constant; the acceleration of the block; the gravitational acceleration; the elastic potential energy; the gravitational potential energy; the gravitational force on the block; the total force on the block; the total energy; the momentum of the block.*

13-67. If the vibrating prongs of a tuning fork were perfect simple harmonic oscillators, could you hear any sound from the tuning fork? Briefly explain.

13-68. Three identical blocks-on-springs are set to oscillating vertically with different amplitudes. At a certain instant, they have the positions and velocities shown in Figure 13-31. Are the forces on these blocks equal at the instant shown? Briefly explain. (Adapted from L. Viennot, *European Journal of Science Education,* 11, 205 [1979].)

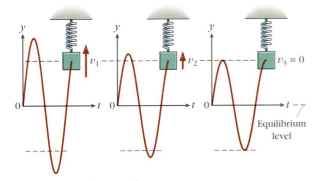

Figure 13-31 Problem 13-68

13-69. How would the appearance of the sequences of horizontal and vertical shadows in Figure 13-13 be changed if the rotating radius casting the shadows gradually sped up instead of rotating at a constant rate?

13-70. If the block in Problem 13-23 begins oscillating at $t = 0$ (and still has a period of 0.40 s),

a. at what instant of time does the block first have a displacement of −0.12 m?

b. what is the next instant of time at which the block again has a displacement of −0.12 m?

c. at what instant of time after the block is set in motion does it first have an instantaneous velocity of zero?

13-71. **SSM WWW** The displacement of a certain simple harmonic oscillator is $y = 6 \cos 7\pi t$ (in m).

a. If the oscillations began at $t = 0$, in what phase is the oscillator when the displacement is +3 m for the first time?

b. In what phase is the oscillator when the displacement is +3 m for the second time?

c. Does it have the same velocity at these two instants? Briefly explain.

d. In what phase will the oscillator's velocity reach its extreme positive value for the first time?

13-72.

a. If $y = 15 \sin 8t$ (from Problem 13-24) is the displacement in meters of a block on a spring with an elastic constant of 60 N/m, find the block's mass.

b. A different block oscillating on a spring has displacement $y = 6 \cos 7\pi t$ (in m), as in Problem 13-71. Find this block's mass.

13-73. Springs A, B, and C are suspended from a horizontal bar. Figure 13-32 shows that they have the same unstretched length. (This assumes the springs have negligible mass, so we can ignore stretching due to their own weight.) When the blocks in Figure 13-32 are hung from these springs and allowed to reach equilibrium, the blocks hang stationary in the positions shown. (The block suspended from spring B has twice the mass of either of the other two blocks.) Suppose that each of these blocks is then given a small vertical displacement from its equilibrium position and then released. Rank the three blocks-on-a-spring in order of their periods of oscillation. Rank them from least to greatest, noting any equalities.

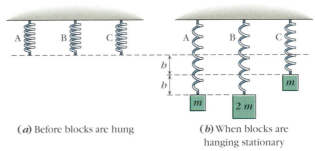

(*a*) Before blocks are hung   (*b*) When blocks are hanging stationary

Figure 13-32 Problem 13-73

13-74. Calculate the percent error caused by using Equation 13-21 to substitute ϕ for $\sin \phi$ **a.** when $\phi = \frac{1}{360}\pi$. **b.** when $\phi = \frac{1}{36}\pi$. **c.** when $\phi = \frac{1}{3}\pi$.

13-75.

a. If a pendulum has an amplitude of 0.70° and a period of 0.80 s when it is near Earth's surface, find its period when it is near the surface of the moon.

b. If the pendulum is given the same maximum angular velocity as on earth, what will its amplitude be on the moon?

c. When it reaches its Earth amplitude under these conditions, how much angular velocity will it still have?

13-76.

a. The Jungfrau, a mountain in the Swiss Alps, rises more than 2 km above the city of Interlaken. A pendulum clock is set to keep perfect time in a shop in town. It is then transported up to the railroad station near the top of the mountain for use by the stationmaster, who complains that it does *not* keep perfect time. Does he find it running fast or slow? Why?

b. The pendulum clock can be finely adjusted by rotating a turnscrew at the bottom of the pendulum bob (Figure 13-33) so that it moves up or down the screw post. Should the stationmaster raise or lower the turnscrew? Briefly explain.

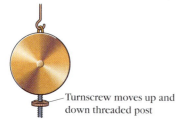

Turnscrew moves up and down threaded post

Figure 13-33 Problem 13-76

13-77. Is a child jumping on a spring-loaded pogo stick a simple harmonic oscillator? (Assume that the dissipative forces are negligible.) Briefly explain.

•**13-78.** *Another simple harmonic oscillator:* Imagine a tunnel drilled the length of a diameter of Earth (Figure 13-34). Assume also that the Earth is a sphere of uniform density. To the extent that this is so, it is possible to show that if you are at point P, the shell of matter that is further out than you (in yellow) exerts a total gravitational force of zero on you because various parts of it pull you equally in different directions. The gravitational force on you is therefore due only to the sphere of matter (in brown) that is closer to the center than you.

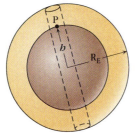

Figure 13-34
Problem 13-78

a. Show that the mass of this sphere is $M_E r^3/R_E^3$, and that the gravitational force on a body of mass m at point P is $F = -GM_E mr/R_E^3$ (the minus sign denotes downward).

b. This expression for F satisfies the criterion for a simple harmonic oscillator (Step 1 of Procedure 13-2). Describe the oscillation that would occur. In doing so, be sure to consider Steps 3 and 4 of Procedure 13-2.

c. You are walking with your friend when your friend falls into a tunnel of this type. How long must you wait until your friend returns?

••**d.** Suppose that a second tunnel is drilled along any chord through Earth that is *not* a diameter. (A chord is a straight line between two points on the surface of a sphere.) If the walls of this tunnel are frictionless, and if we could continue to assume Earth is a sphere of uniform density, show that someone falling into this tunnel would experience simple harmonic motion with the same period as in *a–c*.

13-79.

a. Use the fact that $\sin\theta \approx \theta$ for very small angles or phases to find the approximate slope $\frac{\Delta(\sin\theta)}{\Delta\theta}$ of a graph of $\sin\theta$ versus θ near $\theta = 0$.

b. Use this result to find the slope $\frac{\Delta y}{\Delta\theta}$ of a graph of $y = Y\sin\theta$ versus θ near $\theta = 0$.

c. Note that because $\theta = \omega t$, $\Delta\theta = \Delta(\omega t) = \omega\Delta t$ if ω is constant. From the result of **b**, find $\frac{\Delta y}{\Delta t}$ near $\theta = 0$ (and thus near $t = 0$).

d. What is the physical meaning of $\frac{\Delta y}{\Delta t}$?

e. Find the approximate velocity near $\theta = 0$ from $v = \omega Y\cos\theta$.

f. Compare the results of **c** and **e**.

13-80. *The oscilloscope revisited:* Section 3-1 discussed how the horizontal and vertical positions of a point of light on an oscilloscope screen are separately controlled. Fully describe what you would see on the screen in each of the following cases. In all cases, assume that the center of the screen is the origin (0,0). (*Hint:* If you are having difficulty visualizing, you may be able to use the materials of Activity 1 in Section 3-1 to try things out on paper.)

a. The horizontal position of the dot remains constant at $x = 0$. The variation of the dot's vertical position is described by $y = Y\sin 2\pi ft$, where $Y = 3.0$ cm and $f = 1.0$ s^{-1} (1.0 cycles/s).

b. The same as **a**, except that $f = 120$ s^{-1}.

c. The vertical position of the dot remains constant at $x = 0$. The variation of the dot's *horizontal* position is described by the graph in Figure 13-35.

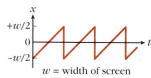

w = width of screen

Figure 13-35 Problem 13-80

d. The variation of the dot's vertical position is described by $y = Y\sin\left(2\pi\frac{t}{T}\right)$ ($Y = 3.0$ cm, $T = 1.0$ s). The variation of the dot's *horizontal* position is described by the graph in Figure 13-35.

e. The variation of the dot's vertical position is described by $y = Y\sin 2\pi ft$ and the variation of the dot's horizontal position is described by $x = X\cos 2\pi ft$ ($X = Y = 3.0$ cm, $f = 120$ s^{-1}). (*Hint:* Recall Figures 13-14 and 13-15.)

f. The variation of the dot's vertical position is described by $y = Y\sin 2\pi ft$ and the variation of the dot's horizontal position is described by $x = X\sin 2\pi ft$ ($X = Y = 3.0$ cm, $f = 1.0$ s^{-1}). Both are now sines.

g. The same as **f**, except $X = 3.0$ cm and $Y = 4.0$ cm.

•**h.** The same as **e**, except $X = 3.0$ cm and $Y = 4.0$ cm.

•**13-81.** A horizontal turntable with a radius of 0.080 m rotates at a constant angular velocity of 5.0 rad/s. A peg is mounted on the outer rim of the turntable, and a simple pendulum is suspended from a point directly over the turntable's center (Figure 13-36). The pendulum consists of a small heavy bob suspended from a light string. As the pendulum swings, we would like the pendulum bob to be directly above the peg each time the peg is at position A (as shown in the figure), but not at any other time.

Figure 13-36
Problem 13-81

a. To accomplish this, how long should the string be and what should be the maximum angular displacement of the pendulum bob?

b. Suppose we did not specify "but not at any other time." Find a different solution to **a** that satisfies the rest of the conditions.

13-82. As you do this activity, think about the physical meaning of the slope of a graph of v versus t. (We addressed this in Chapter 2.)

a. On the graph of v versus t in Figure 13-19, find the points where the slope is zero. Mark each of these points with a P.

b. Now, on the graph of a versus t in the figure, identify the points that occur at the same instants t as the points marked P. Mark each of these with a Q. What is the value of a at each of these points? Compare the information about the acceleration that points P and points Q provide.

c. On the graph of v versus t, mark with an R each point where the slope is steepest downward.

d. On the graph of a versus t, mark the points where a reaches an extreme negative value with an S. Do the points marked R and the points marked S occur at the same instants or different instants? Why?

13-83. Blocks A and B are suspended vertically from separate springs (the positive direction is upward). Their oscillations are 90° (or $\frac{\pi}{2}$) out of phase. When block A has its greatest upward displacement, block B has (*zero velocity; zero acceleration; its greatest negative velocity; or its greatest negative acceleration*). Explain.

13-84. Blocks C and D are suspended vertically from separate springs (the positive direction is upward). Their oscillations are 180° (or π) out of phase. When block C is moving upward and slowing down, block D is (*moving upward and slowing down; moving downward and slowing down; moving upward and speeding up; moving downward and speeding up*). Explain.

13-85. Blocks A and B are suspended vertically from separate springs. If they bob up and down with the same frequency, (*the spring constants of the two springs must be equal; the masses of the two blocks must be equal; their oscillations must have equal amplitudes; all of the first three answers are true; none of the first three answers are true*). Explain.

13-86. The displacements of two pendulums A and B are given by $\phi_A = 0.20 \cos 3t$ and $\phi_B = -0.15 \cos 3t$. Are the motions of the two oscillators in phase? If so, how do you know? If not, by how much are they out of phase?

13-87. A circus performer with a mass of 40 kg sits on a trapeze suspended from ropes 5.0 m long. The trapeze is given an initial angular displacement of 0.08π and released.

a. What is the phase frequency of the trapeze with this performer on it?

b. What is the average (rms) angular speed of the trapeze with this performer on it?

13-88. [*Choose the pair of terms that correctly fills in the two blanks.*]

The _____ of a simple harmonic oscillator does not depend on the value of its _____.

a. maximum velocity . . . . period

b. period . . . frequency

c. period . . . amplitude

d. amplitude . . . total energy

13-89. A block oscillates on a spring with a frequency of 0.80 s^{-1}. If the block is moving at maximum speed at $t = 3.0$ s, at what instant will it next be moving at maximum speed?

13-90. A block oscillates on a spring. The graph in Figure 13-37 shows how y, its displacement from equilibrium, varies with time. Seven points A–G are labeled. Rank these points in order of the block's velocity at each point. Rank them from most negative to most positive, noting any equalities.

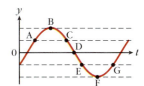

Figure 13-37 Problems 13-90, 13-91, and 13-92

13-91. SSM In Problem 13-90, rank points A–G in order of the block's acceleration at each point. Rank them from most negative to most positive, noting any equalities.

13-92. In Problem 13-90, rank points A–G in order of the net force on the block at each point. Rank them from most negative to most positive, noting any equalities.

13-93. Suppose that in Figure 13-26, $l_1 = 2.1$ m. At what frequency should the robot push child A to maximize the amplitude of the child's swings?

Problems on WebLinks

13-94. The first part of WebLink 13-1 shows one complete oscillation of a block-on-a-spring. Assume this oscillation starts at $t = 0$ and is completed at $t = 1.00$ s, so that, for example, a quarter of a cycle is completed at $t = 0.25$ s.

a. During which time intervals is the block speeding up? Express each interval as "from $t =$ __ s to $t =$ __ s."

b. During which time intervals is the block slowing down?

13-95. Make the same assumptions as in Problem 13-94 about the oscillation in WebLink 13-1.

a. During which time intervals is the acceleration positive (directed upward)? Express each interval as "from $t =$ __ s to $t =$ __ s."

b. During which time intervals is the acceleration negative?

c. Are your answers to **a** all the same as your answers to 13-94**a**?

d. Are your answers to **b** all the same as your answers to 13-94**b**?

13-96. Make the same assumptions as in Problem 13-94 about the oscillation in WebLink 13-1.

a. During which time intervals is the force on the block upward? Express each interval as "from $t =$ __ s to $t =$ __ s."

b. During which time intervals is the force on the block upward?

c. Are your answers to **a** all the same as your answers to 13-95**a**?

d. Are your answers to **b** all the same as your answers to 13-95**b**?

13-97. Make the same assumptions as in Problem 13-94 about the oscillation in WebLink 13-1.

a. During which time intervals is the force on the block increasing in magnitude? Express each interval as "from $t =$ __ s to $t =$ __ s."

b. During which time intervals is the force on the block decreasing in magnitude?

13-98. Make the same assumptions as in Problem 13-94 about the oscillation in WebLink 13-1.

a. At which instants is the acceleration zero?

b. At which instants is the block in equilibrium?

c. Are your answers to *b* all the same as your answers to *a*?

13-99. At the end of WebLink 13-2, the bar is at its lowest position and point D on the traveling disturbance has reached the boat. When the bottom of the bar is next at the undisturbed water level,

a. the boat will be _____ (*at its highest position; at its lowest position; at the undisturbed water level and moving upward; at the undisturbed water level and moving downward*).

b. the boat will be _____ (*at the same horizontal position; a quarter wavelength further to the right; a half wavelength further to the right; a wavelength further to the right*).

13-100. Suppose that in WebLink 13-3, the block-on-a-spring oscillates with an amplitude of 5.0 cm. When the block is 2.0 cm above its equilibrium position, what will be the size of the corresponding shadow cast by the opaque unit radius?

13-101. When the tip of the radius in WebLink 13-4 is moving horizontally to the left, in what direction is its acceleration?

13-102. Rework the line of reasoning in WebLink 13-4 (which treated the vertical components) to show that the *horizontal components* of the position, velocity, and acceleration of the tip of the radius are $y = Y \cos \omega t$, $v = \omega Y(-\sin \omega t)$, and $a = \omega^2 Y(-\cos \omega t)$.

CHAPTER FOURTEEN

Waves and Sound

In trying to understand how the game of nature is played, physicists often must think about how things behave in domains not directly accessible to our senses: over time intervals or distances too small for human perception, on astronomical scales too vast for human experience to encompass, or at speeds too fast to observe directly. We move into these domains guided by indirect evidence, asking, "What kinds of behavior in those inaccessible domains could account for what we actually *see?*"

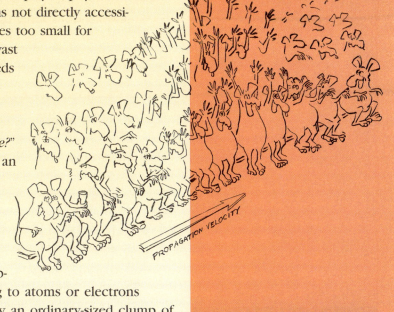

Constructing models or theories is largely an activity of envisioning. It is a creative and imaginative process but necessarily draws on experience. If, for example, we say that atoms jiggle or electrons spin, we draw on our acquaintance with things jiggling or spinning in our everyday world. We can see how ascribing appropriate aspects of jiggling or spinning to atoms or electrons might allow us to draw conclusions about how an ordinary-sized clump of

"To decide whether something is wavelike, we must know precisely what we mean by wave behavior in observable situations."

matter containing some 10^{23} or 10^{24} atoms or electrons might behave. We can make these connections rigorous and logical only if we adequately understand the mechanics of spinning and jiggling on an ordinary scale.

You undoubtedly have a passing acquaintance with things in nature that people speak of as waves—microwaves, for example, or radio waves. We don't see these waves. How do we know they are there, and what does it mean to say they are wavelike? To decide whether something is wavelike, we must know precisely what we mean by wave behavior in observable situations. Only then can we ask what wavelike behavior we can discern when not all aspects of a situation are accessible to our senses.

In this chapter, we will draw on careful observation of waves in every-day situations to build a quantitative description of waves. We will then use that description to help us understand sound. In the next chapter, looking at some less familiar aspects of wave behavior will give us a basis for investigating questions like, "Do we find evidence that light behaves like a wave?"

14-1 Traveling Disturbances: Some Basic Observations

Suppose you and another student hold a long spring stretched between you. The person on the right plucks the spring with her free hand, momentarily deforming it, then releasing it. The resulting deformation or disturbance would travel from right to left as shown in Figure 14-1. If several ribbons were tied to the spring at intervals, one and then another and then another would be disturbed. Although the disturbance travels from right to left, no matter moves from right to left. The horizontal positions of the ribbons never change. Neither do the horizontal positions of any coil of the spring. But as each coil goes up and down, it exerts a vertical force on the next coil and sets it in motion. This passing on of the motion is called the **propagation** of the disturbance.

There are two distinct motions here, and you must be careful to distinguish between them (Figure 14-2):

1. The disturbance travels away from the hand (the *source* of the disturbance) along the spring. At any particular instant, the disturbance has a **path position** x—the distance it has traveled from the source (at $x = 0$).

2. As the disturbance passes the ribbon, the ribbon moves away from and back to its undisturbed position. At any particular instant, this motion results in a **displacement** (denoted by y) from its undisturbed position. In Figure 14-2, it will be useful to describe the displacement as positive if the ribbon is above its undisturbed position ($y = 0$) and negative if below.

Ribbon

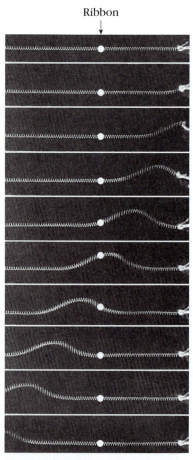

Figure 14-1 The motion of a pulse from right to left along a spring with a ribbon tied around it.

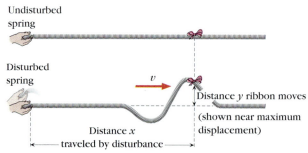

Figure 14-2 Distinguishing between y, the displacement, and x, the distance that the disturbance travels from its source (the hand).

Example 14-1 *Path Position and Displacement at Successive Instants*

Figure 14-3 shows a disturbance traveling along a spring in a sequence of stop-action pictures taken at four different instants.

a. Find the path position and displacement of the pink ribbon at each of the four instants shown.

b. Find the path position of the maximum disturbance at each of the four instants shown.

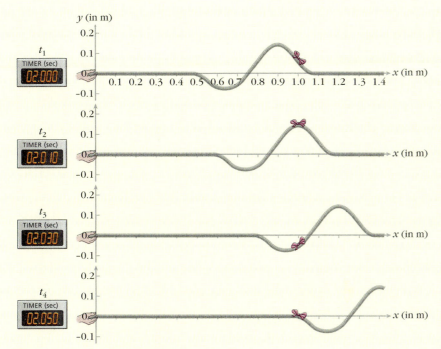

Figure 14-3 A traveling pulse.

Solution

Choice of approach. In **a** you are simply being asked to measure the coordinates of the ribbon in each picture. The values of t_1, t_2, t_3, and t_4 are not relevant here. In **b**, you must find the horizontal coordinate of the peak in each picture.

The measurements. You should confirm for yourself that

a. at t_1, the ribbon is at $x = $ **1.0 m**, $y = $ **0.07 m**.   **b.** at t_1, the peak is at $x = $ **0.9 m**.

at t_2, the ribbon is at $x = $ **1.0 m**, $y = $ **0.15 m**.   at t_2, the peak is at $x = $ **1.0 m**.

at t_3, the ribbon is at $x = $ **1.0 m**, $y = $ **−0.04 m**.   at t_3, the peak is at $x = $ **1.2 m**.

at t_4, the ribbon is at $x = $ **1.0 m**, $y = $ **0**.   at t_4, the peak is at $x = $ **1.4 m**.

(Note that the four pictures are not separated by equal time intervals.)

Making sense of the results. In **a** the ribbon's path position x is fixed, and the variation of y with t describes the up-and-down motion of the ribbon. In **b**, x varies with t as the disturbance moves along the spring. (In this case, the size of the peak, $y = 0.15$ m, stays the same as the peak travels.)

◆ Related homework: Problems 14-6 through 14-10.

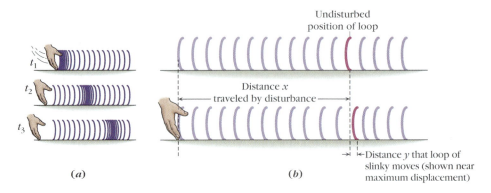

Figure 14-4 A longitudinal disturbance. (*a*) When the Slinky is struck by the hand at one end, a compression travels along it. (*b*) For a longitudinal disturbance, we must also distinguish between *y*, the displacement, and *x*, the distance that the disturbance travels from the source (compare with Figure 14-2).

◆**TRANSVERSE AND LONGITUDINAL WAVES** Figure 14-4*a* shows a somewhat different kind of disturbance. When you tap one end of a Slinky toy sharply with the palm of your hand, the loops are pressed together, and this compression travels the length of the Slinky. If we focus on one loop (shown in pink in Figure 14-4*b*) of the Slinky, we see that as the disturbance reaches it, it is displaced from its undisturbed position. In this case, the displacement is *along* the direction in which the disturbance propagates. But like the ribbon in Figure 14-2, the loop returns to its undisturbed position as the disturbance keeps traveling to the right. Even though they are along the same axis in this case, it avoids confusion if we keep using *x* for path position and *y* for displacement, the size of the disturbance at a particular path position.

When the displacement is perpendicular to the direction of travel or propagation (as in Figure 14-2), we say that we have a **transverse** disturbance. When the displacement is along the direction of propagation (as in Figure 14-4), we call it a **longitudinal** disturbance. Graphs of *y* versus *x* can be very similar for transverse (Figure 14-3) and longitudinal (Figure 14-5) disturbances.

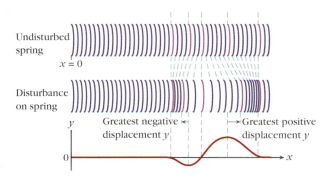

Figure 14-5 What the graph of a longitudinal disturbance shows. If *y* denotes the size of the disturbance (no matter what its direction), then graphs of *y* versus *x* may look alike for transverse and longitudinal disturbances (compare with Figure 14-3).

14-2 Energy in Traveling Disturbances

STOP&Think Before going on, look at the situation depicted in Figures 14-1 and 14-2. Try to describe as completely as you can the energy transfers and energy conversions that occur in this situation.◆

◆**TRANSFER** In Figure 14-6*a*, adjacent segments of the spring exert equal and opposite forces on each other. During a tiny interval of upward motion, the energy input to each segment is the work ΔW done on it. The energy input to the right-hand segment (ΔW_R) is positive, but the energy input to the left-hand segment ($\Delta W_L = -|F_y|\,\Delta y$) is negative. The energy lost by the left-hand segment is gained by the right, so the energy is transferred left to right along the spring.

◆**CONVERSION** Figure 14-6*b* concentrates on the tiny segment of the spring that has the ribbon attached. Once it is given a transverse velocity v_y, its kinetic

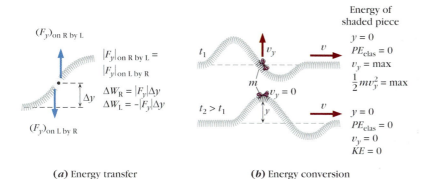

Figure 14-6 Energy transfer and conversion along a spring.
(*a*) Energy transfer (*b*) Energy conversion.

(*a*) Energy transfer

(*b*) Energy conversion

energy is converted into elastic potential energy, which in turn is converted to kinetic energy as the segment comes back down. But unlike an undamped simple harmonic oscillator, the sum of the segment's PE and KE does not remain constant; it decreases as the segment does work on the next segment of the spring.

◆**PROPAGATION SPEED** The **propagation speed** is the speed with which a disturbance propagates or travels along a medium such as a spring. If a one-time disturbance, or **pulse,** retains its shape, the propagation speed is the speed of any particular point (such as the crest) on the "profile" of the disturbance; that is, on the shape's outline. This speed depends on the ability of the medium to pass on the disturbance and is constant (average = instantaneous) for a uniform medium. Then we can write

$$\text{Propagation speed} \qquad v = \frac{\Delta x}{\Delta t} \qquad (14\text{-}1)$$

Example 14-2 compares the propagation speed in two related situations.

Example 14-2 *Speed of a Pulse on a Spring*

For a guided interactive solution, go to Web Example 14-2 at
www.wiley.com/college/touger

Two students stand 8.0 m apart with a long spring like the one in Figure 14-1 stretched between them. When one student plucks the spring to create a pulse, the pulse makes five back-and-forth trips in 8.0 s. They then stretch the spring until they are standing 12.0 m apart and repeat the experiment. Again they find that the pulse makes five back-and-forth trips in 8.0 s. Find the propagation speed when they are 8.0 m apart and when they are 12.0 m apart.

Brief Solution
Anticipating results. Before going on, do you expect to get the same propagation speed in both instances? Explain, based on your current thinking.

Choice of approach. To apply Equation 14-1, we first find the total distance Δx in each case: If there are five back-and-forth trips, there are $n = 10$ *one-way* trips along the length l of the spring.

What we know/what we don't.

	First instance		Second instance	
	$n = 5 \times 2 = 10$	$l = 8.0$ m	$n = 5 \times 2 = 10$	$l = 12.0$ m
		$\Delta t = 8.0$ s		$\Delta t = 8.0$ s

EXAMPLE 14-2 *continued*

The mathematical solution. In the first instance,

$$\Delta x = nl = 10(8.0 \text{ m}) = 80 \text{ m} \qquad \text{and} \qquad \frac{\Delta x}{\Delta t} = \frac{80 \text{ m}}{8.0 \text{ s}} = \textbf{10 m/s}$$

In the second instance,

$$\Delta x = nl = 10(12.0 \text{ m}) = 120 \text{ m} \qquad \text{and} \qquad \frac{\Delta x}{\Delta t} = \frac{120 \text{ m}}{8.0 \text{ s}} = \textbf{15 m/s}$$

Making sense of the results. The propagation speeds in the two cases are *not* the same. The propagation speed depends on the medium, and by stretching the spring more tautly you have altered the medium.

◆ Related homework: Problems 14-14, 14-15, and 14-16.

◆PROPAGATION SPEED VERSUS DISPLACEMENT SPEED STOP&Think

If a pulse travels at a speed of 15 m/s on a long spring having a mass of 0.40 kg, can you calculate the amount of energy carried by this disturbance? ◆ It would seem a simple matter to write down $KE = \frac{1}{2}mv^2$ and plug in 0.40 kg for m and 15 m/s for v. But think again! Does any of the mass of the spring travel at 15 m/s? In Figure 14-3, the segment with the ribbon has a positive (upward) velocity v_y during the interval from t_1 to t_2, zero velocity at t_2, and a negative velocity between t_2 and t_3. Moreover, not all segments move at once. The kinetic energy of the disturbance is due to the movement of mass toward or away from its undisturbed position; it involves the **displacement speed** $|v_y|$, *not* the **propagation speed** v along the spring. The propagation speed tells us how quickly the energy of the disturbance is transferred along the length of the spring, but it tells us nothing about how much energy is being transferred. The propagation speed and the displacement speed are totally unrelated.

◆DISTURBANCES TRAVELING IN MORE THAN ONE DIMENSION

Of all waves, water waves are the most familiar. Everyone has seen ripples traveling outward from a pebble dropped in the water (Figure 14-7*a*). What do these waves have to do with a disturbance moving along a string or a spring?

A pulse on a spring only travels back and forth in one dimension—along the spring. But a pulse in water caused by a dropped pebble travels outward in all directions. (They also travel back inward, if they reach a not-too-distant outer boundary. You can see this reflection back inward if you drop a small wad of wet paper dead center in a toilet bowl.) Figure 14-7*b* sets up a somewhat analogous situation: a large number of springs in a horizontal plane, all radiating outward from a central ring. If we give the ring a quick vertical shake, it sends a pulse out along each of the springs. If the springs are identical, the pulses all travel at the same speed, so at any instant all will have gone the same distance. If the springs are packed closely enough together, the pulses appear to form a circular ridge that is expanding outward. The water extends in all directions, as

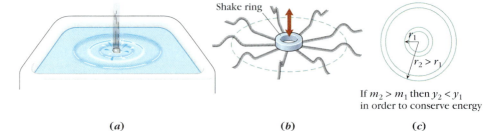

Figure 14-7 Analogy between a pulse on a water surface (two-dimensional) and pulses traveling along many radial (one-dimensional) springs.

(*a*)　　　(*b*)　　　(*c*)

If $m_2 > m_1$ then $y_2 < y_1$ in order to conserve energy

the multiple springs do, and the disturbance can travel along each possible path provided by the medium.

There is an important difference, however. If we think of the water surface as made up of rings of equal width (Figure 14-7c), the rings that are further out have greater circumference and therefore greater mass. Suppose we think just of potential energy transferred outward as each ring in turn is maximally displaced. If no energy is dissipated, $m_1 g\, y_1 = m_2 g\, y_2$. So if m_2 is greater than m_1, y_2 must be smaller than y_1. Thus, conservation of energy tells us that the height of a ripple must decrease as it travels outward.

14-3 From Pulses to Periodic Waves

A pulse is a one-time disturbance. But it is also possible to have a continuous disturbance. If you hold one end of the spring in Figure 14-1 and keep shaking your wrist up and down, pulses will travel away from your hand one after the other. Likewise, in a two-dimensional situation, if a succession of pebbles fall into the water in Figure 14-7a, a succession of ridges or crests will travel outward from the point of impact. A continuous disturbance may be very irregular: You can slosh your hand around in the water any old way and the disturbance will continue to travel outward from your hand. But many of the most interesting situations in physics involve disturbances that are regularly recurring, or **periodic.** We will concentrate on situations in which a source produces a periodic disturbance and the disturbance travels in all available directions along an **isotropic** medium. By *isotropic,* we mean that the properties of the medium are the same in all directions, so that the disturbance propagates at the same speed in all directions.

If you shake your hand up and down in a regularly repeated way (Figure 14-8a), the regular succession of pulses traveling away from it is called a **wave.** We assume for the moment that the other end of the spring is so far away that in the short term we do not have to worry about pulses reflecting back from the far end. Think of Figure 14-8a as a stop-action or freeze-frame picture of this wave—the picture will be changed a fraction of a second later when each of the pulses has traveled a short distance to the right.

We can think of what we see as a superposition of simpler motions, as we did in Section 3-1. In Figure 14-8b, a pencil is embedded in a block bobbing up and down on a spring. Much as in an *xy*-recorder (Figure 3-1), a scrolling sheet of paper carries the pencil's "disturbance" to the right. Here the medium (the scroll) itself travels, whereas in Figure 14-8a the disturbance gets passed along a stationary spring. But the resulting picture is the same, and changes over time in the same way if we match the scroll speed to the propagation speed. The height of the disturbance under the fixed wire in Figure 14-8b corresponds to the displacement of the ribbon in Figure 14-8a.

In Figure 14-8c, a vibrating bar suspended by rubber bands produces a similar wave in water. Like the ribbon on the spring, the little raft does not get swept along in the direction of propagation, but bobs up and down as the succession of crests passes under it. Strictly speaking, water particles move in elliptical paths as the wave travels (Figure 14-9). The ellipse is the superposition of two motions, an up-and-down (transverse) motion and a back-and-forth (longitudinal) motion. We will only consider the up-and-down motion.

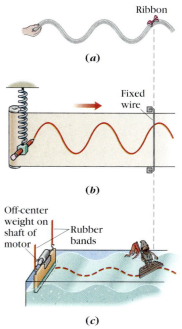

(a)

(b)

(c)

Figure 14-8 Production of continuous one-dimensional waves. In (c), the long straight bar oscillates up and down as a small electric motor with an off-center weight on its shaft rotates. This is a standard way of producing waves in ripple tanks for demonstration purposes.

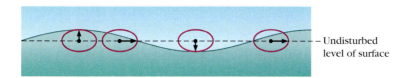

Figure 14-9 Motion of water particles subjected to a continuous traveling wave. The arrows show the displacement of a particle at different positions along a stop-action picture of the wave. As the wave travels, the particle goes through each displacement in sequence, resulting in the elliptical path shown.

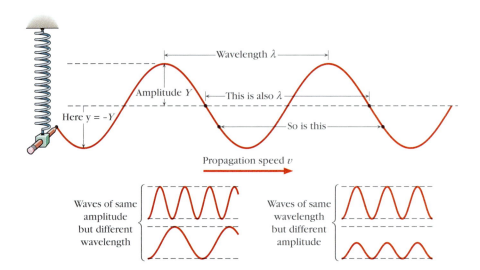

Figure 14-10 Defining constants describing overall properties of a continuous wave.

The sources of all of the waves in Figure 14-8 are simple harmonic oscillators or very nearly so. The oscillation of the source has a fixed period T, and the **crests** (high points) and **troughs** (low points) are therefore evenly spaced. Moreover, when there is just one direction of travel, there is no "spreading out" of the energy as there was in Figure 14-7c, so the crest size remains constant (assuming no damping). Under these conditions, we can picture a standard idealized "waveform" or "wavetrain" (Figure 14-10)—a pattern of regularly repeating crests and troughs—and we define the following quantities to describe it:

Wavelength (λ) is the distance between successive repetitions of equivalent points in the pattern (e.g., between one crest and the next, or between any point and the next point having the same phase). The symbol λ is the Greek letter *lambda* (lower case).

Amplitude (Y) is the maximum displacement (whether above or below, forward or back) from the undisturbed level. The displacement y thus varies between $+Y$ at each crest and $-Y$ at each trough; Y itself is always positive.

Frequency (f) and **period** (T) mean the same here as for simple harmonic oscillators. They can describe either the oscillation of the source or the oscillation of the distant object (the ribbon or the boat) as the wave passes under it.

Propagation speed (v) was defined by $v = \frac{\Delta x}{\Delta t}$ (Equation 14-1).

An important relationship among λ, f, and v follows directly from the definitions. In Figure 14-11, the number of down-and-up cycles the wheel experiences each second—that is, the frequency—is the same as the number of wavelengths

Figure 14-11 A model for visualizing how the relationship $v = f\lambda$ follows from the definitions. (This model is animated in WebLink 14-1.) If two wavelengths, each 3 m long, are pulled past the wheel each second, the wheel goes up and down twice each second ($f = 2 \text{ s}^{-1}$) and the waveform's speed is $v = 6$ m/s.

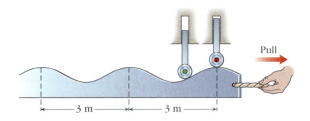

that pass under it each second. So the latter is also the frequency. As the figure shows,

$$\text{the propagation speed} = \text{the number of wavelengths passing each second (this is the frequency)} \times \text{the wavelength}$$

Working through WebLink 14-1 will help you visualize this relationship. We can write this word equation in symbols as

$$v = f\lambda \tag{14-2}$$

Because $f = 1/T$, we can also rewrite this as

$$v = \frac{\lambda}{T} \tag{14-3}$$

You should satisfy yourself that this makes sense because the wavelength is the distance traveled in one period ($\Delta x = \lambda$ when $\Delta t = T$).

For **WebLink 14-1**: **Wavelength, Frequency, and Propagation Speed**, go to www.wiley.com/college/touger

Example 14-3 *How Fast Does the Disturbance Travel?*

a. Suppose that when the hand in Figure 14-8a goes up and down 28 times in 10 s, the crests on the spring are 6.0 m apart. How fast does the wavetrain travel?

b. How fast would a single pulse travel on the same spring stretched by the same amount? What could you say about the wavelength and frequency in this case?

Solution

Restating the question. To answer **a**, you need a working familiarity with relevant definitions. The distance between crests is the wavelength λ. The number of up-and-downs (or cycles) divided by the time interval in seconds gives the frequency, so the question is: $v = ?$ *when* $\lambda = 6.0$ m *and* $f = \frac{28}{10\ \text{s}}$.

Choice of approach. Now we can apply Equation 14-2.

What we know/what we don't.

$$f = \frac{\Delta n}{\Delta t}, \text{ where } \Delta n = 28 \text{ and } \Delta t = 10 \text{ s}$$

$$\lambda = 6.0 \text{ m} \qquad v = ?$$

The mathematical solution. **a.** By Equation 14-2,

$$v = f\lambda = \left(\frac{28}{10\ \text{s}}\right)(6.0 \text{ m}) = \textbf{17 m/s}$$

b. Again, $v = $ **17 m/s**. The medium determines the propagation speed, so pulses have the same speed whether they travel alone or one after the other in a wavetrain. For a single pulse, however, wavelength and frequency have no ordinary meaning. We cannot measure the distance *between* crests when there is only one crest. Equivalently, we can say the wavelength is infinite—you never get to the next crest. Either way, if the pulse doesn't recur, we cannot speak of the number of cycles each second.

◆ Related homework: Problems 14-20, 14-21, 14-22, and 14-23.

STOP&Think Could you have used Equation 14-2 to do Example 14-2? Could you have used Equation 14-1 to do Example 14-3? ◆

14-4 Fully Describing Waves Mathematically

In this section we will develop a full mathematical description of an idealized (undamped) periodic wavetrain with a single direction of propagation (a "one-dimensional") wave. Figure 14-12 shows stop-action or freeze-frame pictures of such a wave at two different instants. The quantities that describe each stop-action picture—amplitude Y and wavelength λ—do not change from picture to picture. The propagation speed v, frequency f, and period T also remain constant.

But the displacements of the boats do *not* remain constant from picture to picture. The displacements of the boats at $x = 1$ and $x = 3$ decrease in magnitude from $t = 0$ to $t = t_1$, whereas the displacement of the boat at $x = 2$ increases during this interval. The displacement y not only is different at each position x in a single picture (i.e., at a single instant) but also differs from instant to instant at the same position. A complete mathematical description will give the displacement y at any position x and at any instant t. Then our task is to develop an equation that looks like

$$y = \left(\begin{array}{c} \text{a bunch of stuff in which the only } \textit{variables} \text{ are } x \text{ and } t, \text{ so that} \\ \text{for any position } x \text{ and instant } t, \text{ we can find a numerical value} \\ \text{for this side of the equation} \end{array} \right) \qquad (14\text{-}4)$$

Because y will be a *function* of x and t that describes the wave, it is called a **wave function.**

In developing the desired equation, we will assume the following:

- The source of a simple periodic wave oscillates up and down (or back and forth if the disturbance is longitudinal) at a fixed frequency and disturbs some medium.

- The medium is uniform and carries the disturbance at a constant propagation speed v.

- Therefore the disturbance at any position x along the medium replicates or copies the disturbance at the source. With no damping, the disturbance will have the same form (the same up-and-down or back-and-forth pattern) and the same amplitude.

- Whatever is happening at the source (at $x = 0$) will happen later at a distant position $x > 0$, because it takes time for the disturbance to travel to that position from the source. Thus there is a *time lag* between what the medium is doing at a distant point and what the source is doing. The time lag depends on how far away (x) the point is and how fast (v) the medium can carry the disturbance there.

In working out the details of Equation 14-4, we will (1) write an equation to describe the cyclical behavior of the source (we'll assume it to be at $x = 0$) and then, because the same thing is happening at any position x except for the time lag, we will (2) try to build in the time lag. The time lag will depend on how far (x) the position is from the source, so we will end up with an equation that lets us calculate y if we know x and t; that is, with y as a function of x and t.

Figure 14-12 Describing an undamped traveling wavetrain. To change x while keeping t constant, go from boat to boat in the same picture. To change t while keeping x constant, look at the same boat in different stop-action pictures.

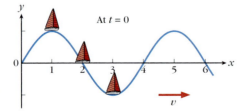

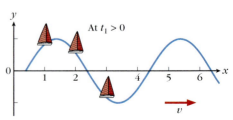

We have already done step (1) for the simple harmonic oscillator. In Section 13-3, we found that if we choose the right phase to be at $t = 0$,

$$y = Y \sin\left(2\pi \frac{t}{T}\right) \quad \text{or} \quad y = Y \sin \omega t \quad \text{or} \quad y = Y \sin 2\pi f t$$

Now we build in the time lag t_{lag}. If it takes t_{lag} for the disturbance to travel from the source to a position x, then at instant t, the disturbance at x will be doing what the source was doing t_{lag} ago; that is, at instant $t - t_{\text{lag}}$. To describe this disturbance, we just substitute $t - t_{\text{lag}}$ for t and get (for example)

$$y = Y \sin\left(2\pi \frac{t - t_{\text{lag}}}{T}\right)$$

We still need to see how t_{lag} depends on position x. But t_{lag} is the time interval the disturbance takes to travel from the source to x. So the propagation speed $v = \frac{\Delta x}{\Delta t} = \frac{x - 0}{t_{\text{lag}}} = \frac{x}{t_{\text{lag}}}$. Solving for t_{lag} gives us

$$t_{\text{lag}} = \frac{x}{v} \tag{14-5}$$

and the equation for y becomes

$$y = Y \sin 2\pi \left(\frac{t - \dfrac{x}{v}}{T}\right) \tag{14-6}$$

We can rewrite Equation 14-6 as

$$y = Y \sin 2\pi \left(\frac{t}{T} - \frac{x}{vT}\right)$$

But from Equation 14-3, $\lambda = vT$, so

$$y = Y \sin 2\pi \left(\frac{t}{T} - \frac{x}{\lambda}\right) \tag{14-7}$$

We have now provided the right side of Equation 14-4. In this wave function, the argument of the sine, $\theta = 2\pi(\frac{t}{T} - \frac{x}{\lambda})$, is the phase at a given instant t and position x. Note that since $f = 1/T$, Equation 14-6 can also be written as

$$y = Y \sin 2\pi f \left(t - \frac{x}{v}\right) \tag{14-8}$$

Example 14-4 *Reading Information from a Wave Function*

For a guided interactive solution, go to Web Example 14-4 at www.wiley.com/college/touger

In SI units, the wave function for a particular periodic wave on a very long wire is $y = 8.00 \sin\left(5t - \frac{x}{4}\right)$. Find the amplitude, wavelength, frequency, period, and propagation speed of this wave.

Brief Solution

Choice of approach. The key here is to recognize that a number in a particular position always means the same thing. For example, the number multiplying the entire sine function is always the amplitude Y. The number multiplying t is always $\frac{2\pi}{T}$ (as it is in Equation 14-7), and knowing T you can find f.

The mathematical solution. We compare the given equation to Equation 14-7; that is, $y = 8.00 \sin\left(5t - \frac{x}{4}\right)$ to $y = Y \sin 2\pi\left(\frac{t}{T} - \frac{x}{\lambda}\right) = Y \sin\left(\frac{2\pi t}{T} - \frac{2\pi x}{\lambda}\right)$. Matching terms and putting in suitable SI units,

the amplitude $Y =$ **8.00 m**

EXAMPLE 14-4 *continued*

The multiplier of t must be equal in both: $5 = \frac{2\pi}{T}$, so the period $T = \frac{2\pi}{5} = $ **1.26 s**. The multiplier of x must be equal in both: $\frac{1}{4} = \frac{2\pi}{\lambda}$, so the wavelength

$$\lambda = \frac{2\pi}{\frac{1}{4}}$$

that is, $\qquad\qquad\qquad\qquad \lambda = 8\pi = \textbf{25.1 m}$

Now we rely on other basic relationships.

Because $f = 1/T$, the frequency $f = \dfrac{1}{1.26 \text{ s}} = \textbf{0.794 s}^{-1}$

and since $v = f\lambda$, the propagation speed $v = (0.794 \text{ s}^{-1})(25.1 \text{ m}) = \textbf{19.9 m/s}$

◆ Related homework: Problems 14-27 and 14-28.

PROCEDURE 14-1

Calculating the Displacement from a Sinusoidal Wave Function

$$y = Y \sin 2\pi \left(\frac{t}{T} - \frac{x}{\lambda} \right)$$

1. This tells you what fraction (proper or improper) of a cycle has gone by at a position x along the wave's path after an elapsed time t.

2. Multiplying by the number of units in a cycle (2π) then tells you how many units of phase have gone by, i.e. what phase the oscillation is in at instant t at a position x along the path of the wave.

3. The sine (or cosine) function associates a number between 1 and –1 with the phase, and tells you what fraction of the amplitude (+ or –) you have when the oscillation is in that phase.

4. Multiplying the amplitude by the fraction of it that you have tells you the displacement.

Procedure 14-1 tells you how to use Equation 14-7 to calculate the displacement at a given point along the wave path at a given instant (compare with Procedure 13-1). To try the procedure out, work through Example 14-5.

Example 14-5 *Finding Displacement at a Distance from a Source*

For a guided interactive solution, go to Web Example 14-5 at www.wiley.com/college/touger

A periodic wave on a long cable is described by Equation 14-7. Its wavelength is 8.0 m and its frequency is 2.0 s^{-1}. If a ribbon tied to the cable 9.0 m to the right of the origin has a maximum displacement is 0.14 m, find its displacement at the instant $t = 1.0$ s.

Brief Solution

Choice of approach. Because T is the inverse of the frequency, and the amplitude Y means the maximum displacement, we can put in values for all the constants in Equation 14-7. Then we can follow the steps of Procedure 14-1.

What we know/what we don't.

$$\lambda = 8.0 \text{ m} \qquad f = 2.0 \text{ s}^{-1} \qquad T = ?$$

$$Y = 0.14 \text{ m} \qquad x = 9.0 \text{ m} \qquad t = 1.0 \text{ s} \qquad y = ?$$

The mathematical solution. The period $T = 1/f = 1/(2.0 \text{ s}^{-1}) = 0.50$ s. Substituting the values of all constants into Equation 14-7 gives us

$$y = (0.14 \text{ m}) \sin 2\pi \left(\frac{t}{0.50 \text{ s}} - \frac{x}{8.0 \text{ m}} \right)$$

We now follow the steps of Procedure 14-1:

1. At the given x and t,

$$\frac{t}{T} - \frac{x}{\lambda} = \frac{1.0 \text{ s}}{0.50 \text{ s}} - \frac{9.0 \text{ m}}{8.0 \text{ m}} = 2.0 - \frac{9}{8} = \frac{7}{8}$$

(The disturbance has gone through $\frac{7}{8}$ of a cycle at the given position and instant.)

2. Then $2\pi(\frac{t}{T} - \frac{x}{\lambda}) = 2\pi(\frac{7}{8}) = \frac{7}{4}\pi$. This is the phase at this position and instant.

3. $\sin 2\pi(\frac{t}{T} - \frac{x}{\lambda}) = \sin \frac{7}{4}\pi = -0.71$. In this phase, the displacement is this (negative) fraction of the amplitude.

4. Thus $y = Y(-0.71) = (0.14 \text{ m})(-0.71) = \mathbf{-0.099 \text{ m}}$.

◆ Related homework: Problem 14-29.

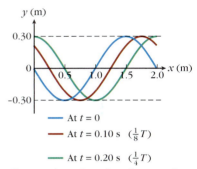

Figure 14-13 Graphs of a traveling wave at successive instants. Here $Y = 0.30$ m, $T = 0.80$ s, and $\lambda = 2.0$ m. Using $y = Y \sin 2\pi(\frac{t}{T} - \frac{x}{\lambda})$ to obtain graphs of y versus x at three different instants, we see that the wave advances to the right over time.

If we have values for Y, λ, and T, then at a given instant t we can use Equation 14-7 to find the value of y at each position x, just as we did at a single position x in Example 14-5. This gives us a profile of the wave. The green graph in Figure 14-13 shows the profile of one such wave at $t = 0$. If we then do the same thing at later instants $t = 0.10$ s and $t = 0.20$ s, we see that the wave profile shifts to the right as t increases. In other words, the wave travels. To see the details of how these graphs are obtained and to see how Equation 14-7 gives us a frame-by-frame "video" of a traveling wave, go to WebLink 14-2.

At $x = 0$, Equation 14-7 reduces to $y = Y \sin 2\pi \frac{t}{T}$ and thus gives the displacement of the oscillating source at that position. At $t = 0$, Equation 14-7 becomes

$$y = Y \sin 2\pi \left(-\frac{x}{\lambda} \right) = -Y \sin \left(2\pi \frac{x}{\lambda} \right) \tag{14-9}$$

The argument of the sine is now $\theta = 2\pi\frac{x}{\lambda}$. The disturbances at two positions where the values of θ differ by 2π are identical and are said to be **in phase.** As Figure 14-14 shows, these positions are λ apart (when $x = \lambda$, $\theta = 2\pi$; when $x = 2\lambda$, $\theta = 4\pi$; etc.) It follows that the wavelength is the distance between any two successive positions where the disturbances are in phase (as is true of two crests). In Figure 14-14, boats A and B are a wavelength apart; they bob up and down in phase. **STOP&Think** Boat C is *not* a wavelength away from boat B, but it has the same displacement. Are boats B and C moving in phase? Are they always doing the same thing at the same time? ◆

For **WebLink 14-2:** **Generating Wave Pictures from the Equation,** go to www.wiley.com/college/touger

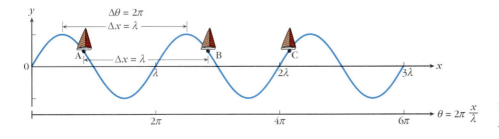

Figure 14-14 Boats A and B are *in phase.* Are boats B and C in phase?

14-5   Standing Waves and Superposition

Case 14-1 ◆ *Another Kind of Disturbance on a Spring*

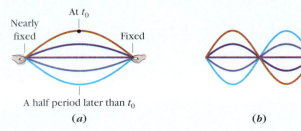

Nearly fixed   At t_0 Fixed

A half period later than t_0

(a)   (b)   (c)

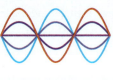

Figure 14-15 Multiple exposures of standing waves on a spring.

The two students in Example 14-2 again hold the ends of the spring 8.0 m apart. One of them tries moving her wrist up and down with a regular rhythm. She discovers that if she does this 10 times in a 16-s interval, she produces the disturbance shown in multiple exposure in Figure 14-15a. If she speeds up to 20 up-and-downs in 16 s, she gets the disturbance in Figure 14-15b, and at 30 up-and-downs in 16 s, she gets the disturbance shown in Figure 14-15c. At any one instant, each of these disturbances looks like a segment of a periodic wavetrain. But instead of traveling along the spring, the crests stay at the same position and vary in size. This kind of disturbance is called a **standing wave.**

Because the wave doesn't travel, there are positions, called **nodes,** where the displacement is always zero. Any position where the spring is held fixed must be a node. The end shaken by the student's wrist moves only slightly compared to the amplitude of the standing wave, so that end is approximately a node. In contrast, the positions where the displacement reaches a maximum or minimum are called **antinodes.**

If you start, say, at an antinode (any A in Figure 14-16), you have to advance by *two* antinodes to go a whole wavelength. Successive antinodes are only a *half wavelength apart*. So are successive nodes (N). If we think of Figures 14-15a, b, and c as showing one, two, and three "loops," each loop is $\frac{1}{2}\lambda$ in length.

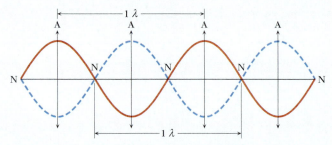

Figure 14-16 Nodes (N) and antinodes (A).

The students realize that from the number of up-and-downs in the 16-s time interval, they can find the frequency of each disturbance. From the number of $\frac{1}{2}\lambda$ loops along the 8.0 m of spring, they can find the wavelength in each case. From λ and f, they can compute v. Their calculations are as follows:

Disturbance in Fig. 14-15a	Disturbance in Fig. 14-15b	Disturbance in Fig. 14-15c
$f = \frac{10}{16\,s} = \frac{5}{8}\,s^{-1}$	$f = \frac{20}{16\,s} = \frac{10}{8}\,s^{-1}$	$f = \frac{30}{16\,s} = \frac{15}{8}\,s^{-1}$
$\frac{1}{2}\lambda = 8.0$ m	$\frac{2}{2}\lambda = 8.0$ m	$\frac{3}{2}\lambda = 8.0$ m
$\lambda = 2\,(8.0\text{ m}) = 16$ m	$\lambda = 8.0$ m	$\lambda = \frac{2}{3}(8.0\text{ m}) = \frac{16}{3}$ m
$v = f\lambda = (\frac{5}{8}\,s^{-1})(16\text{ m})$	$v = f\lambda = (\frac{10}{8}\,s^{-1})(8.0\text{ m})$	$v = f\lambda = (\frac{15}{8}\,s^{-1})(\frac{16}{3}\text{ m})$
$= 10$ m/s	$= 10$ m/s	$= 10$ m/s

"Aha!" says student A. "We get the same speed as in Example 14-2 because the spring is stretched the same amount, and it's the medium—the spring—that determines the speed."

"There's only one problem", says student B, "v is the propagation speed, not the displacement speed."

"So?"

"So the propagation speed tells you how fast the crests travel *along* the spring. But these crests just get bigger and smaller—they *aren't going anywhere*. What can propagation speed possibly mean for a wave that isn't traveling?"

"Then we've also got another problem."

"What's that?"

"The frequencies and wavelengths we got were for the *standing* waves. Yet when we used them to calculate the propagation speed, we got the correct value for a *traveling* wave on the same spring. How can that be?"

"Hmmm," says student B, "there must be some kind of connection between standing waves and traveling waves."

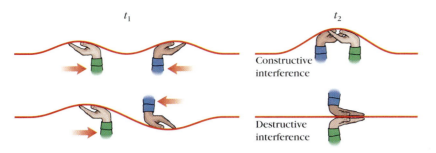

Constructive
interference

Destructive
interference

**Figure 14-17 Superposition of
disturbances.**

Following up on student B's conjecture, we will see if perhaps some combination of traveling waves can result in a standing wave. To do that, we must first consider how waves combine.

◆**SUPERPOSITION OF WAVES** Just as motions in general can be superimposed to produce more complex motions, wave motions can also be superimposed to produce a single resulting pattern of motion. Imagine traveling disturbances created by two hands pushing upward while sliding along a stretched tarp (Figure 14-17). When the hands are far apart, there are two separate bulges or displacements. When the hands are together, they produce a single displacement. Twice as much force is exerted, and the displacement is doubled. If one hand pushes up while the other pushes down, the displacements are opposite in direction. When the hands meet, they push against each other and the resulting displacement is reduced—to zero if the forces are *equal* and opposite.

In general, displacements arriving at the same position at the same instant must be added together, taking sign into account, to obtain the actual displacement observed at that position. For two waves traveling along the same path, you must add the disturbances separately at each position. For example, in Figure 14-18*a*, waves 1 and 2 individually would cause different displacements at position x_1. The disturbance we actually observe at x_1 is the sum of these two displacements. Likewise, adding the individual displacements at x_2 gives you the observed displacement at x_2. If waves 1 and 2 are traveling in opposite directions (Figure 14-18*b*), the disturbances at a later instant t_2 are changed at each position, and the *new* displacements must be added. This point-by-point addition of displacements to obtain the resulting waveform (a graph of y versus x) is called **superimposing** or **superposition of waves.**

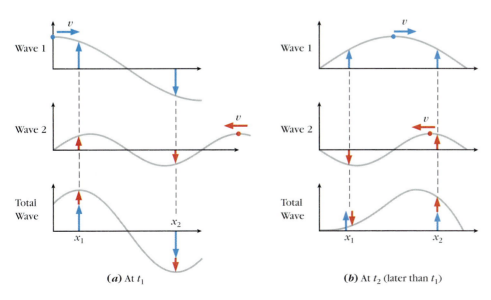

Wave 1

Wave 2

Total
Wave

x_1

x_2

(*a*) At t_1

Wave 1

Wave 2

Total
Wave

x_1

x_2

(*b*) At t_2 (later than t_1)

**Figure 14-18 Superposition of
waves.** (*a*) The point-by-point addition of disturbances at instant t_1. (*b*) The point-by-point addition of the same disturbances at a later instant t_2. Looking at how the red and blue dots have shifted position from (*a*) to (*b*) will help you see how wave 1 and wave 2 are traveling.

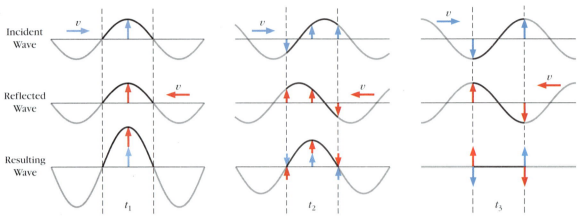

Incident Wave

Reflected Wave

Resulting Wave

t_1 t_2 t_3

Figure 14-19 Superposition of an undamped wave and its reflection to produce a standing wave.

For **WebLink 14-3:**
Connecting Traveling Waves and Standing Waves,
go to
www.wiley.com/college/touger

What happens if we superimpose a continuous undamped wave on a string and its reflection from a distant endpoint—that is, two identical waves traveling in opposite directions? Figure 14-19 shows that the resulting superposition is a standing wave that has the same wavelength λ and frequency f as the two traveling waves. (To see what happens in step-by-step detail, work through WebLink 14-3.) For this reason, using standing wave values in the product $f\lambda$ gives us the propagation speed v of the traveling wave. We have resolved the dilemma in Case 14-1.

Example 14-6 *Propagation Speed from Standing Wave Data*

The figure below shows a stop-action picture of a standing wave on a 12.0-m spring. The ribbon tied to the spring is observed to go up and down 2.5 times per second.

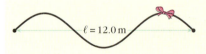

$\ell = 12.0$ m

a. With what speed does the crest of a standing wave travel along this spring?
b. With what speed does the crest of a traveling wave travel along the spring?

Solution

Choice of approach. Again we must be attentive to definitions. The speed at which the crest travels along the spring is the propagation speed v of a *traveling* wave. We can take the product $v = f\lambda$ in either case, but crests of a standing wave don't travel along the spring. The answer to **a** is **zero**. To apply $v = f\lambda$ to **b,** we must first recognize that the number of up-and-downs per second is the frequency and that the wavelength is not 12.0 m. **STOP&Think** Why not? Remember that the loops between successive nodes are $\frac{1}{2}\lambda$ long. ◆

What we know/what we don't.

$$f = 2.5 \text{ s}^{-1} \qquad l = 12.0 \text{ m}$$

The figure shows three $\frac{1}{2}\lambda$ loops along l, so $l = 3(\frac{1}{2}\lambda)$ $\qquad v = ?$

The mathematical solution. Because $l = 3(\frac{1}{2}\lambda)$,

$$\lambda = \frac{2}{3}l = \frac{2}{3}(12.0 \text{ m}) = 8.0 \text{ m}$$

Then $\qquad\qquad v = f\lambda = (2.5 \text{ s}^{-1})(8.0 \text{ m}) = \textbf{20 m/s}$

◆ Related homework: Problems 14-35, 14-36, and 14-37.

14-6 Introduction to Sound

Hum while your fingers are on your throat and feel the vibration. Pluck a guitar string or twang a tightly stretched rubber band and see the vibration die down as the sound does. Feel the vibration of the speaker diaphragm on any home sound system. All sound sources are vibrating bodies. They create disturbances that travel through one or more media (air, walls, etc.) to your ear. In turn, there are parts of your ear that are easily set to vibrating.

Consider the speaker diaphragm's vibration in more detail. Each time it bulges outward, it compresses the air next to it. Each time it bows inward, the adjacent air is given more room and becomes rarefied (regions where the density has dropped below normal are called **rarefactions**). These disturbances in the density of the air, and therefore in the local air pressure, are propagated through the medium—the air—just as shaking your hand (Figure 14-20) can set up a series of compressions and rarefactions (of the coils) that travel along the Slinky. When the increases and decreases in air pressure reach your eardrum—another flexible diaphragm—they cause it to bow inward or bulge outward. Once again, the vibration of the detector mimics the source, but with a time lag. Sound, too, has a propagation speed that depends on the medium. (We will discuss the specific aspects of the medium that affect it a little later.)

The approximate speed of sound is evident to our senses when we watch fireworks going off in the distance and hear the bang after we see the flare (Figure 14-21). There is a delay of about 1 s for each 340 m of distance, so

$$v_{\text{sound in air}} \approx 340 \text{ m/s}$$

This value is fairly exact for dry air at sea level at a temperature slightly over 14°C. Changes in altitude, moisture, and temperature would alter the properties of the medium on which the propagation speed depends (Table 14-1).

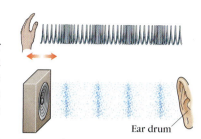

Figure 14-20 Compressions and rarefactions along a longitudinal wave. Just as the hand produces successive compressions and rarefactions of the Slinky coils, the speaker diaphragm produces successive compressions and rarefactions of air molecules, experienced as increases and decreases in pressure, that arrive at your eardrum.

Figure 14-21 We hear the bang after we see fireworks explode because sound travels at only a tiny fraction of the speed of light.

Example 14-7 *Now You See It . . . Now You Hear It*

At a particular instant you observe a bolt of lightning. If the lightning strikes a tree 2 km from where you are standing, how much later do you hear the thunderclap?

Solution

Choice of approach. Because Equation 14-1 applies to any kind of traveling disturbance, we can use it to find Δt for sound. We assume that light travels fast enough to reach your eye virtually instantaneously.

What we know/what we don't. $v_{\text{sound in air}} \approx 340$ m/s.

$\Delta x = 2$ km $= 2000$ m (All values must be in SI units) $\Delta t = ?$

The mathematical solution. From $v = \frac{\Delta x}{\Delta t}$, it follows that

$$\Delta t = \frac{\Delta x}{v} \approx \frac{2000 \text{ m}}{340 \text{ m/s}} = \textcolor{red}{5.9 \text{ s}}$$

◆ Related homework: Problems 14-38 and 14-39.

The rarefactions and compressions along the Slinky are due to the longitudinal displacement of the coils. Likewise, the variations in density, and therefore in pressure, are due to the longitudinal displacement of the molecules of the medium. For a periodic sound wave, the longitudinal displacements and the

Table 14-1 Speed of Sound in Some Representative Media

Medium	Speed (m/s)
Gases (at 1 atm)	
Carbon dioxide (0°C)	259
Oxygen (0°C)	317
Air (0°C)	331
Air (20°C)	343
Methane (0°C)	430
Helium (0°C)	965
Liquids (25°C)	
Methyl alcohol	1140
Water (distilled)	1497
Sea water	1531
Solids (at room temperature)	
Wood (oak, along grain)	3850
Aluminum (along thin rods or wires)	5000
Iron/steel (along thin rods or wires)	5200

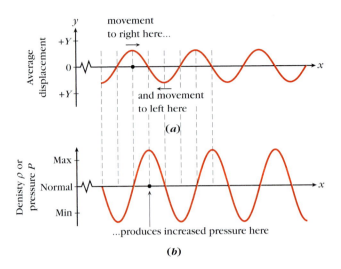

Figure 14-22 Variation of (*a*) longitudinal displacement of molecules and (*b*) density and pressure (proportional to density) along a continuous sound wave. The longitudinal displacement at a position x is really the average displacement of molecules in a tiny region around x, because in any region the speeds and directions of the molecules vary.

density or pressure both vary sinusoidally (Figure 14-22), and we can use Equation 14-7 to describe either. But if it is describing the "pressure wave" (Figure 14-22*b*), the displacement and amplitude are no longer y and Y but ΔP and $(\Delta P)_{max}$ and now have units of pressure. They represent deviations above or below undisturbed pressure. In general, displacement and amplitude need not refer to distances. They can tell us how any quantity deviates from its undisturbed value as the disturbance travels.

Example 14-8 *The Physics Underlying the Mathematics*

Figure 14-22 shows the instantaneous graphs of longitudinal displacement and pressure to be 90° out of phase. For example, the pressure maxima occur where the average longitudinal displacement of molecules is zero. Use basic physics principles to explain why this is so.

Solution

Choice of approach. Pressure is proportional to density, so we need to think about how the particles near a certain position must be displaced if the density at that position is to be a maximum.

The explanation. If the density at a certain point is a maximum, the molecules to the left and right of the point must be closer to the point than usual. To be closer, those molecules to the left of the point must be displaced toward the right and those to the right of the point must be displaced toward the left (draw a sketch to picture this). If molecules to the left of the point have positive displacements, and molecules to the right have negative displacements, the displacement must pass through zero *at* the point in question.

◆ Related homework: Problems 14-41 and 14-42.

We can hear an interesting consequence of the superposition of sound waves if we hook up speakers to two audio (sound) signal generators and vary the frequency of one output so that it gets further from or closer to the frequency of the other. When we do this, the sound pulses or wavers: It repeatedly gets louder and lower with a fixed frequency, which depends on how far off the two

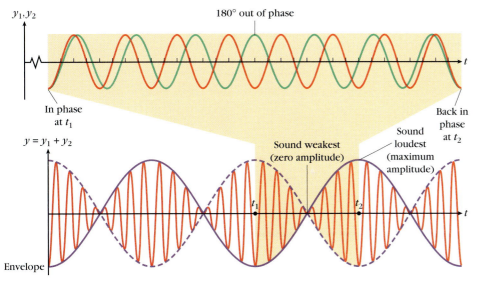

Figure 14-23 The production of beats. (*a*) If two signals are in phase at t_1, and their frequencies differ only slightly, it takes many cycles before they come back in phase at a later instant t_2. (*b*) Looking at the superposition of the two signals over a much longer time interval, we see that the two signals go in and out of phase repeatedly. Both here and in (*a*), the time interval from to t_1 to t_2 is highlighted in yellow. There is a beat—the two signals come back in phase—after each like time interval. The total signal is loudest—has maximum amplitude—when the two contributing signals are phase and is weakest when they are 180° out of phase or nearly so.

contributing frequencies are from each other. This wavering phenomenon is called **beats.** The reason for this is outlined briefly in Figure 14-23. For a fuller understanding of why beats occur, work through WebLink 14-4. In Figure 14-23*b*, each occurrence of a maximum denotes one beat. It can be shown that the beat frequency is

$$f_{\text{beat}} = f_1 - f_2 \qquad (14\text{-}10)$$

which typically is very small compared to the two contributing frequencies. The "envelope" in Figure 14-23 shows the slow overall variation that we hear. Our ears cannot detect the more rapid variation that occurs during each beat, just as we hear only a steady sound, not the variation in displacement, when the signal has constant amplitude and frequency.

For **WebLink 14-4:** **The Key Goes Off, the Beat Goes On,** go to www.wiley.com/college/touger

14-7 Resonance and Sources of Musical Sound

On-The-Spot Activity 14-1

On a long rope or flexible garden hose, or on a giant spring if one is available, see if you can set up standing waves that look like each of the ones in Figure 14-15. Use the *same length* of rope or hose for all three. What changes about the motion of your hand as you go from producing the standing wave with one loop to the standing wave with two? From a standing wave with two loops to one with three?

Do you get a well-formed standing wave for any frequency of your hand motion? Using a watch that records seconds, find the number of up-and-downs your hand must make in a 20-s interval to produce each of the standing waves in the figure. Then calculate the frequency of your hand motion in each case.

Now try moving your wrist at a frequency about halfway between the two lowest values. Repeat at a frequency about halfway between the two highest values. What do you get on the rope? Can you get a well-formed standing wave at any in-between frequencies? What about at frequencies higher than those you've tried?

Standing waves on a given rope can occur only at certain frequencies and not those in between. Why? Because the ends are fixed or (where the hand moves)

nearly so, there must be a node at either end of the rope. The distance from each node to the next is half a wavelength, so the distance between the nodes at the ends must always be some whole number (n) of half wavelengths: if L is the length of the rope,

$$L = n\left(\tfrac{1}{2}\lambda\right) \qquad (n = 1, 2, 3, \ldots) \qquad \text{(14-11)}$$

Solving for λ, we find that the only permissible values of wavelengths are

$$\lambda = \frac{2L}{n} \qquad (n = 1, 2, 3, \ldots) \qquad \text{(14-12)}$$

If we substitute this into $v = f\lambda$ and solve for f, we find the frequencies are also limited to certain values:

$$f = \frac{nv}{2L} \qquad (n = 1, 2, 3, \ldots) \qquad \text{(14-13)}$$

A standing wave on an extended body such as a rope is like an oscillation, except that the parts of the extended body do not all oscillate in phase. Just as a swing should be pushed at its natural frequency (recall Section 13-7 on resonance), you must match the driving frequency of your wrist to one of the allowable frequencies of the rope. The allowable frequencies are called **resonant frequencies,** or sometimes simply **resonances.**

If you are not pumping your wrist at a resonant frequency, it is much like the parent pushing out of sync with the child's swing. Sometimes you are pushing with the natural motion and sometimes you are pushing against it. The work you are doing is sometimes positive and sometimes negative, so there is no significant build-up in the total energy (analogous to $\tfrac{1}{2}kY^2$) of the oscillation or its the amplitude. You can get a sense of this if you first get your rope oscillating at a resonant frequency and then try to slightly slow down or speed up the frequency of your wrist motion to make it out of sync with the rope's motion.

Example 14-9 *Allowable Frequencies*

You and a friend are setting up standing waves on a long spring stretched between you to a length of 9.0 m. You have found that a single pulse travels along it at a speed of 30 m/s. Find the three lowest frequencies at which you can shake your wrist to set up a well-formed standing wave on this spring.

Solution

Choice of approach. You must shake your wrist at a resonance frequency. Although we can use Equation 14-13 directly, the underlying concepts are Equations 14-11 and $v = f\lambda$. (It's better to know fewer equations but be able to deduce the others by reasoning.) The frequencies are lowest when the wavelengths are longest (by $v = f\lambda$). The wavelengths are longest when there are fewest of them along the string. (How does Equation 14-11 tell you that?) Then we want to find the wavelengths when there are $n = 1, 2$, and 3 half-wavelengths along the string, and then find the corresponding frequencies.

What we know/what we don't.

$$L = 9.0 \text{ m} \qquad v = 30 \text{ m/s.}$$

When $n = 1$, $\lambda = ?$ and $f = ?$ When $n = 2$, $\lambda = ?$ and $f = ?$

When $n = 3$, $\lambda = ?$ and $f = ?$

The mathematical solution. Solving $L = n\left(\tfrac{1}{2}\lambda\right)$ for λ

$$\text{when } n = 1 \text{ gives } \lambda = \frac{2L}{1} = 2(9.0 \text{ m}) = 18 \text{ m}$$

$$\text{when } n = 2 \text{ gives } \lambda = \frac{2L}{2} = 9.0 \text{ m}$$

$$\text{when } n = 3 \text{ gives } \lambda = \frac{2L}{3} = \frac{2}{3}(9.0 \text{ m}) = 6.0 \text{ m}$$

Then

$$\text{when } n = 1, f = \frac{v}{\lambda} = \frac{30 \text{ m/s}}{18 \text{ m}} = \mathbf{1.7 \text{ s}^{-1}}$$

$$\text{when } n = 2, f = \frac{v}{\lambda} = \frac{30 \text{ m/s}}{9.0 \text{ m}} = \mathbf{3.3 \text{ s}^{-1}}$$

$$\text{when } n = 3, f = \frac{v}{\lambda} = \frac{30 \text{ m/s}}{6.0 \text{ m}} = \mathbf{5.0 \text{ s}^{-1}}$$

◆ Related homework: Problems 14-46, 14-47, and 14-48.

Example 14-10 *How Frequency Depends on Length*

For a guided interactive solution, go to Web Example 14-10 at www.wiley.com/college/touger

Suppose you and your friend continue to set up standing waves on the stretched spring of Example 14-9. But now, while you continue to grasp one end, your friend firmly holds the middle of the spring instead of the other end. Assuming you have not altered the degree of stretch, how does this alter the lowest frequencies possible for standing waves on the length of spring stretched between you?

Brief Solution

Choice of approach. The length of spring between the two fixed points is now 10 m rather than 20 m. Although you could simply redo Example 14-9 with this value of L, it would be better to determine *in general* how the frequency changes with L. The steps from Equation 14-11 to Equation 14-13 bring us to a relationship between f and L.

The mathematical solution. Equation 14-13 says that $f = nv/2L$. Then $fL = nv/2$. For any given n, the right-hand side involves only constants. Therefore the left-hand side must be constant: f and L may individually change, but the product cannot. Then $f_1L_1 = f_2L_2$. Our question is then how does f_2 compare to f_1 if $L_2 = \frac{1}{2}L_1$? Solving for f_2, we get $f_2 = f_1L_1/L_2$. Then when $L_2 = \frac{1}{2}L_1$,

$$f_2 = \frac{f_1L_1}{L_2} = \frac{f_1L_1}{\frac{1}{2}L_1} = \mathbf{2f_1}$$

Each allowable frequency doubles when the length is halved.

◆ Related homework: Problem 14-49.

In general, as in Example 14-10, the smaller L is, the higher the frequency will be. Human ears are capable of detecting vibrations at frequencies from about 20 s^{-1} to $20\,000 \text{ s}^{-1}$. This **audible range** is more commonly written as 20 Hz to 20 kHz, where

$$1 \text{ hertz (Hz)} = 1 \text{ oscillation or cycle per second} = 1 \text{ s}^{-1}$$

(1 kHz is 1 kilohertz). The sensitivity of the human ear to different frequencies depends on the loudness of the sound. At sound levels of ordinary human speech, the range of audible frequencies is less, roughly from 50 Hz to 15 kHz. This range diminishes with age and with excessive exposure to loud noise.

Figure 14-24 **Characteristic producers of high- and low-pitched sounds.**

Ultrasound imaging is used to monitor the progress of a fetus in the womb of an expectant mother.

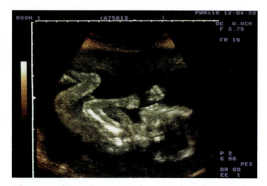

Figure 14-25 **Blue whales produce subsonic (or infrasonic) signals that carry for thousands of kilometers.** These signals are now being detected by U.S. Navy hydrophones formerly used for submarine detection.

The higher the frequency, the higher the pitch of the sound we hear. Therefore, shorter sources tend to produce higher-pitched sounds, as examination of the various sources in Figure 14-24 will bear out.

The hearing of other creatures is not limited to the same range of frequencies as that of humans. Dogs respond to dog whistles that produce **ultrasonic** (above 20 kHz) vibrations inaudible to humans. Bats' sensory organs are capable of detecting signals at up to 150 kHz, including the echoes or reflections of signals that they themselves emit. Detection of these reflected signals enables bats to locate objects around them (a process called *echolocation*) in a manner much the way reflected radio waves are used in radar detection.

Because very-high-frequency sounds have very small wavelength, they can be used for detection of smaller objects. In a process akin to echolocation, **ultrasound,** typically having frequencies of 1–15 megahertz, is used medically to image anatomical features within the body. Perhaps the most familiar use of this is monitoring the progress of the fetus in the womb of an expectant mother.

Some biologists believe elephants communicate extensively at **subsonic** (below 20 Hz) frequencies. Underwater microphones originally developed to detect Soviet submarines during the Cold War have already revealed that whale sounds, mostly too low-pitched for the human ear to detect (Figure 14-25), fill the oceans every bit as much as the higher-pitched sounds of land-bound wildlife fill tropical rain forests. Although higher frequency sound waves "damp out" over shorter distances (recall Figure 13-10), sounds produced by the blue whale are of sufficient amplitude and low enough frequency to carry in undisturbed ocean depths for thousands of miles: "A blue whale in the Bahamas can easily be heard by another on the Grand Banks off Newfoundland."[1]

◆**LOUDNESS** Different aspects of our experience of sound depend on different physical properties. What we call pitch, whether the sound is high like a soprano's high C or low like a bass fiddle, depends on the frequency. But the human experience of pitch depends not only on the physical properties of the signal but on the nature of the receiver—the ear—and the highly complex system that interprets the signal—the human brain. The same is true of loudness. Our experience of loudness depends principally on the intensity (or energy flux density, see Section 6-5) of the signal incident on our eardrums. The traveling disturbance we call sound consists of particle oscillations. We know that the total energy ($\frac{1}{2}kY^2$) of a simple harmonic oscillator is proportional to the square of its amplitude. It follows that the sound intensity will also be proportional to the amplitude squared: Intensity = cY^2. If Y is the maximum

[1]Whale researcher C. Clark, quoted in the *Boston Sunday Globe,* July 18, 1993, p. 13.

displacement of each little bit of the medium, c depends on the propagation speed, the density for a medium such as air, and the square of the frequency. (Recall that for the simple block-on-a-spring oscillator, the spring constant $k = \frac{1}{2}m\omega^2 = \frac{1}{2}m(2\pi f)^2$ also depends on the square of the frequency.)

✦**SOUND LEVEL** "This is the way the world ends, / Not with a bang but with a whimper." So wrote the poet T. S. Eliot. The sound intensity reaching your ear from an explosion at the threshold of pain will be 10^{10} times as great as from a whimpering child a meter away from you. To humans, though, the explosion does not sound 10^{10} times as loud as the whimpering child, but only about 10 (the value in the exponent) times as loud, although different individuals' subjective responses may vary somewhat. The exponent is a logarithm in base 10. Our subjective **sound level** scale depends on the logarithm of the actual wave intensity. A sound level scale that uses this fact is the **bel** (named after Alexander Graham Bell) scale. This scale compares sounds to the **threshold of hearing,** the least intense sound (with $\text{Intensity}_o \approx 1 \times 10^{-12}$ J/(s · m²) or 1×10^{-12} W/m²) audible to the typical human ear. For a given intensity,

$$\text{sound level} = \log_{10}\left(\frac{\text{Intensity}}{\text{Intensity}_o}\right) \quad \text{(bel scale)} \qquad (14\text{-}14)$$

where $\quad \text{Intensity}_o \approx 1 \times 10^{-12}$ J/(s · m²) or 1×10^{-12} W/m² $\quad$ (threshold of hearing)

For finer gradation, physicists and engineers commonly measure by tenths of a bel, or **decibels** (1 bel = 10 decibels or 10 dB). In decibels,

$$\text{sound level} = (10\text{ dB})\log_{10}\left(\frac{\text{Intensity}}{\text{Intensity}_o}\right) \quad \text{(decibel scale)} \qquad (14\text{-}15)$$

➥**Reminder:** We can write any positive number as 10 to some exponent. The necessary exponent is called $\log_{10}$ of that number. For example, $20 = 10^{1.3010}$, so the logarithm of 20 is 1.3010.

Normal conversations are typically in the 50–70 dB range. Chainsaws and jackhammers typically produce sound in the 100–120 dB range, where regular exposure for more than a minute can do permanent damage to your hearing. Rock concerts commonly produce sound in this range. The threshold of pain is 125 dB. Jet take-offs and some rock concerts exceed this threshold.

✦**MUSICAL SOUNDS** The wavelength of a standing wave on a string can only have certain values because the displacement must be zero where the string's ends are fixed. In general we call such a requirement a **boundary condition.**

The two most common boundary conditions for one-dimensional waves are shown in Figure 14-26. In $a1$ and $b1$, the boundary condition at the right end prevents any particle displacement: In $a1$, the string is fixed; in $b1$, as in some organ pipes, air molecules cannot move past the closed end of the column. In $a2$ and $b2$, displacement is unrestrained at the right end: The ring can slide up and down, and air molecules can move in and out of the open end of the column. In $a1$ and $b1$, there must be a node at the right end. In $a2$ and $b2$, there must be an antinode at

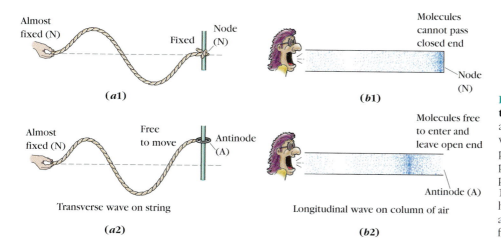

(a1)

(b1)

(a2)

(b2)

Figure 14-26 Boundary conditions. Whether there is a node or antinode at a boundary depends on whether the boundary is fixed in place or free to move. Because pressure and density are 90° out of phase with displacement (Figure 14-22 and Example 14-8), we can have maximum or minimum density at a displacement node, as this figure shows.

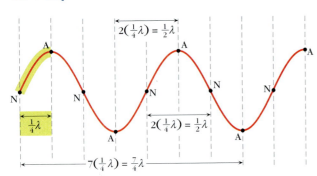

Figure 14-27 You can picture standing waves as being made up of a whole number of quarter-wavelengths.

or very near the right end. In the case of a column of air, if the width of the column is not negligible compared to the length, the antinode actually occurs slightly beyond the open end. This is observed for organ pipes. For musical instruments like trumpets and trombones, where the opening is more funnel shaped, the situation is further complicated.

To see how these boundary conditions affect frequencies, it is useful to break up the waveform mentally into quarter-wavelengths (Figure 14-27). A quarter-wavelength $(\frac{1}{4}\lambda)$ is the distance between a node (N) and its nearest neighbor antinode (A). To go from one A to another or one N to another always takes an *even* number of quarter-wavelengths. Between an A and an N, there are always an *odd* number of quarter-wavelengths.

Suppose you have N's or A's as boundary conditions at the ends of a one-dimensional vibrating body. If both ends are the same (N or A), the body's length L must be an even number of quarter-wavelengths:

$$L = n(\tfrac{1}{4}\lambda), \; n = 2, 4, 6, \ldots (n \text{ even}) \tag{14-16a}$$

Solving for λ then gives the wavelengths that satisfy this condition:

$$\lambda = \frac{4L}{n} \quad (n \text{ even}) \tag{14-17a}$$

Substituting this into $v = f\lambda$ and solving for f gives us the permissible frequencies:

$$f = \frac{v}{\lambda} = \frac{nv}{4L} \quad (n \text{ even}) \tag{14-18a}$$

If the boundary conditions at the two ends of the vibrating body are opposite (an N and an A), the reasoning is exactly as before except that n must be odd:

$$L = n(\tfrac{1}{4}\lambda), \; n = 1, 3, 5, \ldots (n \text{ odd}) \tag{14-16b}$$

$$\lambda = \frac{4L}{n} \quad (n \text{ odd}) \tag{14-17b}$$

$$f = \frac{v}{\lambda} = \frac{nv}{4L} \quad (n \text{ odd}) \tag{14-18b}$$

We can apply the reasoning entailed in these equations systematically:

PROCEDURE 14-2

Finding Allowable Frequencies					
1	2	3	4	5*	6
Starting with physical situation,	Identify boundary conditions.		For each permissible n, solve for λ_n.		Then use $v = f\lambda$ to find f_n.
string fixed at both ends	both N: $L = n(\frac{1}{4}\lambda_n)$ (n even)	Sketch waveform for each n.	$\lambda_n = \frac{4L}{n}$ (n even)	If v not known, determine from properties of medium.	$f_n = \frac{v}{\lambda_n} = \frac{nv}{4L}$ (n even)
organ pipe open at both ends	both A: $L = n(\frac{1}{4}\lambda_n)$ (n even)		$\lambda_n = \frac{4L}{n}$ (n even)		$f_n = \frac{v}{\lambda_n} = \frac{nv}{4L}$ (n even)
organ pipe open at one end	A and N: $L = n(\frac{1}{4}\lambda_n)$ (n odd)		$\lambda_n = \frac{4L}{n}$ (n odd)		$f_n = \frac{v}{\lambda_n} = \frac{nv}{4L}$ (n odd)

*We will discuss this step shortly.

(λ_n and f_n are the wavelength and frequency corresponding to a particular number n of quarter-wavelengths between the endpoints.)

Example 14-11 *A Pair of Pipes*

**For a guided interactive solution, go to Web Example 14-11 at
www.wiley.com/college/touger**

On a warm day in the old Oriental Theater, when the speed of sound in air
is 350 m/s, the organist is playing the great Wurlitzer organ. Two of its pipes,
one closed at one end and the other open at both, are otherwise identical.
Both are 2.75 m long. What are the three lowest single frequencies produced
by each pipe?

Brief Solution

Choice of approach. We follow Procedure 14-2. Although it is possible to jump
from step 1 to step 6, it is wiser to follow all the steps, thereby reaching step 6
by reasoning from the basic principles rather than relying on memorization.
We'll call the pipe open at one end Pipe A and the other Pipe B.

What we know/what we don't.

$$v_{\text{in air}} = 350 \text{ m/s}$$

Pipe A (open one end; odd values of n): $L = 2.75$ m $f_1 = ?$ $f_3 = ?$ $f_5 = ?$

Pipe B (open both ends; even values of n): $L = 2.75$ m $f_2 = ?$ $f_4 = ?$ $f_6 = ?$

**Sketching standing
wave diagrams for
the organ pipes.**
The N's and A's
(nodes and antinodes)
must alternate.
Start with the fewest N's
and A's consistent with
boundary conditions.
Put two more quarter-
wavelengths (one more
N and one more A)
along the length for
each higher frequency.
More wavelengths
means each wavelength
is smaller. As l
decreases, f increases.

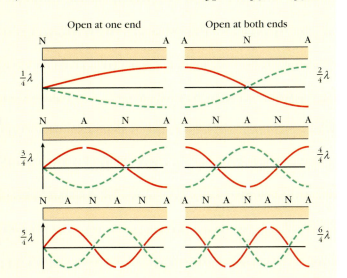

The mathematical solution. To develop a sense of how the boundary conditions affect allowable wavelengths, it is valuable to sketch wave diagrams as
shown here. To fit more quarter-wavelengths between the ends of the pipe,
that is, to increase n, each wavelength must be smaller. Then for a fixed
v, $v = f\lambda$ tells us we should expect f to increase with n.

The three lowest frequencies occur for Pipe A when $n = 1, 3, 5$, and for
Pipe B when $n = 2, 4, 6$. We do the remaining calculations systematically as
follows:

	Step 2	Step 4	Step 6
Pipe A:			

$n = 1$ $L = 1(\frac{1}{4}\lambda)$ $\lambda = \dfrac{4L}{1} = 4(2.75 \text{ m}) = 11.0 \text{ m}$ $f_1 = \dfrac{v}{\lambda_1} = \dfrac{350 \text{ m/s}}{11.0 \text{ m}} =$ **31.8 Hz**

$n = 3$ $L = 3(\frac{1}{4}\lambda)$ $\lambda = \dfrac{4L}{3} = \dfrac{4(2.75 \text{ m})}{3} = 3.67 \text{ m}$ $f_3 = \dfrac{v}{\lambda_3} = \dfrac{350 \text{ m/s}}{3.67 \text{ m}} =$ **95.4 Hz**

$n = 5$ $L = 5(\frac{1}{4}\lambda)$ $\lambda = \dfrac{4L}{5} = \dfrac{4(2.75 \text{ m})}{5} = 2.20 \text{ m}$ $f_5 = \dfrac{v}{\lambda_5} = \dfrac{350 \text{ m/s}}{2.20 \text{ m}} =$ **159 Hz**

EXAMPLE 14-11 *continued*

Pipe B:

$$n = 2 \quad L = 2\left(\tfrac{1}{4}\lambda\right) \quad \lambda = \frac{4L}{2} = \frac{4(2.75 \text{ m})}{2} = 5.50 \text{ m} \quad f_2 = \frac{v}{\lambda_2} = \frac{350 \text{ m/s}}{5.50 \text{ m}} = \textbf{63.6 Hz}$$

$$n = 4 \quad L = 4\left(\tfrac{1}{4}\lambda\right) \quad \lambda = \frac{4L}{4} = \frac{4(2.75 \text{ m})}{4} = 2.75 \text{ m} \quad f_4 = \frac{v}{\lambda_4} = \frac{350 \text{ m/s}}{2.75 \text{ m}} = \textbf{127 Hz}$$

$$n = 6 \quad L = 6\left(\tfrac{1}{4}\lambda\right) \quad \lambda = \frac{4L}{6} = \frac{4(2.75 \text{ m})}{6} = 1.83 \text{ m} \quad f_6 = \frac{v}{\lambda_6} = \frac{350 \text{ m/s}}{1.83 \text{ m}} = \textbf{191 Hz}$$

Notice that for the pipe with the same condition at both ends (B), the two higher frequencies are two and three times the lowest (apart from rounding error). All integer multiples of the lowest frequency occur. For the pipe with contrasting conditions at the two ends (A), the two higher frequencies are three and five times the lowest; only odd integer multiples of the lowest frequency occur.

◆ Related homework: Problems 14-51, 14-52, and 14-53.

✦**HARMONICS** In music, the standing waves at allowable frequencies are called **harmonics.** The lowest allowable frequency (that of the first harmonic) is called the **fundamental frequency.** The traditional descriptive language can be cumbersome. Although doing so requires different equations for different boundary conditions, harmonics are traditionally numbered using the values of n from either Equation 14-13 or 14-18b. For example, for a string fixed at both ends, we get the same frequency whether we use $n = 3$ in Equation 14-13 or $n = 6$ (the third *even* integer) in Equation 14-18a. But the language of harmonics calls this the *third* harmonic. In this usage, all numbered harmonics are possible for an organ pipe open at both ends, but by Equation 14-18b only odd-numbered harmonics are possible for an organ pipe closed at one end. The fundamental frequency for the two-ends-open case, which we labeled f_2 in Example 14-11, would traditionally be labeled f_1. Also, harmonics above the fundamental frequency are sometimes called **overtones** (e.g., the second harmonic on a string is the first overtone).

In musical instruments, the sounds produced are complex. Even a simple one-dimensional vibrating body—a plucked string or a long, thin column of air—has many allowable harmonics, and the actual vibration will typically be a superposition of these harmonics—a much more complicated waveform. The specific combination of fundamental frequency and higher harmonics varies with the instrument producing it, so a particular note played on one kind of instrument will sound different if played on another.

Furthermore, real instruments have other vibrating parts. For instance, guitar strings pass over a hole in a soundbox roughly shaped like a figure eight. When you pluck a string, the vibration is carried by the air in the soundbox. These vibrations can be large if one of the resonant frequencies of the enclosed body of air is matched. Because the soundbox is irregularly shaped, various distances between walls of the soundbox are possible, each with an associated resonant frequency. A well-designed soundbox with a continuous range of wall-to-wall distances can resonate at all the frequencies that the strings can produce.

✦**DETERMINATION OF PROPAGATION SPEED FROM PROPERTIES OF MEDIUM** In Procedure 14-2, we indicated that the propagation speed in a medium can be found from the properties of the medium. Which properties? How do we do this? We will address this in detail for the case of a stretched string.

Case 14-2 ◆ *Factors Affecting Propagation Speed on a Stretched String: A Qualitative Exploration*

It will be valuable to set this up and do the observations with some fellow students if possible.

The two set-ups in Figure 14-28 are identical except that one string is thicker—described by some students as "heavier"—than the other. When the buzzers are turned on, the vibrating hammers set up standing waves on the strings. We can then see what kinds of changes in the set-up change the number of loops, and we can reason as follows. Because the frequency is set by the vibrating hammer of the buzzer, it is the same in both set-ups. The fewer loops there are on a fixed length of string, the longer the wavelength. An increased λ with a fixed frequency means (by $v = f\lambda$) a greater propagation speed. Thus, the number of loops decreases as v increases, and vice versa. Figure 14-29 shows what we observe.

Observation 1: We see more loops on the "heavier" string. Then the propagation speed on that string is less. **STOP&Think** Should "heavier" refer to the total weight of the string? ◆

Observation 2: We move the buzzer further from the pulley until we double the length of string between buzzer and pulley. The number of loops then doubles. But in this case, because there is twice the length of string, the length of each loop is unchanged, so neither λ nor v is changed. Then the propagation speed does *not* depend on the length of the string.

But haven't we put more string between the pulley and the buzzer? So isn't it "heavier?" What hasn't

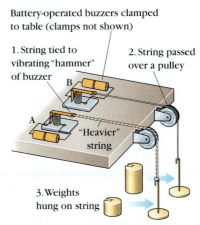

Battery-operated buzzers clamped to table (clamps not shown)

1. String tied to vibrating "hammer" of buzzer

2. String passed over a pulley

3. Weights hung on string

"Heavier" string

Figure 14-28 Set-up for Case 14-2.

changed is the mass *of each meter of length* of the string. But the mass per unit length *is* different for the strings in A and B, so the number of loops along the same length of string differs in the two set-ups. We conclude from Observation 1 that when the mass per unit of length (also called the **linear mass density** μ_{lin}) is increased, the propagation speed decreases.

Observation 3: When we hang more weight on the hanger of either set-up, each loop gets longer and the number of loops is reduced. The amount of weight suspended determines the tension in the string. Therefore, when the tension is increased, the propagation speed increases.

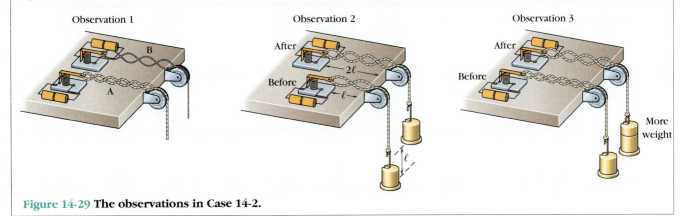

Observation 1

Observation 2

Observation 3

Figure 14-29 The observations in Case 14-2.

By applying kinematics, Newton's second law, and geometry, it is possible to show that

$$v = \sqrt{\frac{T}{\mu_{\text{lin}}}} \qquad (14\text{-}19)$$

Consistent with our observations, we see that (1) because the tension is in the numerator, v increases with T; and (2) because the linear mass density is in the denominator, v decreases as μ_{lin} increases.

Example 14-12   *The Medium Is the Message*

Find the wavelengths and frequencies of the two lowest harmonics on a 0.010 kg string if the string has a length of 1.25 m and is subjected to a force of 80.0 N. If the sounds produced on the string are no louder than ordinary human speech, what is the lowest audible frequency of the string?

Solution

Choice of approach. We again follow Procedure 14-2, including step 5. We determine the propagation speed from the string's properties using Equation 14-19. Because μ_{lin} is the mass per unit of length, we obtain it by dividing the string's total mass by its total length. The lowest even values of n are 2 and 4.

What we know/what we don't.

$$m = 0.010 \text{ kg} \qquad L = 1.25 \text{ m} \qquad \mu_{lin} = \frac{m}{L}$$

$$T = \text{applied force} = 20.0 \text{ N} \qquad \lambda_2 = ? \quad f_2 = ? \quad \lambda_4 = ? \quad f_4 = ?$$

The mathematical solution. In step 2 of Procedure 14-2, the boundary conditions require that

$$\text{when } n = 2, L = 2(\tfrac{1}{4}\lambda) \qquad \text{when } n = 4, L = 4(\tfrac{1}{4}\lambda)$$

The sketches of the waveforms—step 3 of the procedure—should look like Figure 14-15a and b. Solving for the wavelengths (step 4) gives us

$$\lambda_2 = \frac{4L}{2} = \frac{4(1.25 \text{ m})}{2} = 2.50 \text{ m} \qquad \lambda_4 = \frac{4L}{4} = \frac{4(1.25 \text{ m})}{4} = 1.25 \text{ m}$$

Before finding the frequencies, we must find the propagation speed. Because $\mu_{lin} = \frac{m}{L} = \frac{0.010 \text{ kg}}{1.25 \text{ m}} = 0.0080 \text{ kg/m}$,

$$v = \sqrt{\frac{T}{\mu_{lin}}} = \sqrt{\frac{80.0 \text{ N}}{0.0080 \text{ kg/m}}} = 100 \text{ m/s}$$

The two lowest allowable frequencies are then

$$f_2 = \frac{v}{\lambda_2} = \frac{100 \text{ m/s}}{2.50 \text{ m}} = 40 \text{ Hz} \qquad f_4 = \frac{v}{\lambda_4} = \frac{100 \text{ m/s}}{1.25 \text{ m}} = 80 \text{ Hz}$$

Because most people at human speech sound levels cannot hear sounds below about 50 Hz, the **80 Hz** harmonic will be the lowest audible one.

◆ Related homework: Problems 14-60, 14-61, 14-62, and 14-63.

14-8   The Doppler Effect

As you wait on the platform of a train station, a train roars by without slowing down. You notice that once the train is past you, it sounds less high-pitched than when it was approaching. Why should the sound depend on whether its source is moving toward or away from you? This effect, called the **Doppler effect,** gives us an opportunity to explain a surprising phenomenon using basic wave concepts. To understand qualitatively why it happens, let's consider the analogy depicted in Figure 14-30. (WebLink 14-5 presents a more detailed and animated presentation of the ideas in this figure.)

In Figure 14-30a, the conveyor belt (the medium) sets the speed. If donuts drop more frequently, each will have gone less far when the next one drops. They will be closer together. The crests of the waves in Figures 14-30b and c

For **WebLink 14-5:**
Doppler Donuts, go to
www.wiley.com/college/touger

will be closer together for the same reason: The wavelength is smaller because the frequency has increased.

If the machine is moving away, the next donut drops further from the previous one (Figure 14-31). In the analogous situations, the wavelength is greater. But because the conveyor belt is still advancing at the same speed, if the donuts or crests are further apart they reach the detector less often. (At constant v, if λ increases, f must decrease, and vice versa.)

If the *source* is stationary and the detector travels toward the source, the spacing between donuts (or the wavelength) is unaffected. But just as a cyclist pedaling against traffic comes up on an oncoming car too quickly, the baker's helper experiences the donuts coming at him more quickly. Their speed *relative to him* is greater than the speed v of the conveyor belt. Now λ remains fixed, so f must increase.

Now let's develop these ideas quantitatively.

◆**STATIONARY DETECTOR, MOVING SOURCE** The time interval between successive donuts dropping (or successive crests forming) is just the period of the source (s): $\Delta t = T_s = 1/f_s$. During this interval, the first donut (crest) travels at the conveyor belt speed (propagation speed) v to the right and goes a distance $\Delta x_1 = v\Delta t$ (Figure 14-31). Meanwhile the source moves away (at speed v_s) from where it formed the first crest and produces a second one at a distance $\Delta x_2 = v_s\Delta t$ from this point. Then as the figure shows, the total distance between successive crests—the wavelength—is $\lambda = \Delta x_1 \pm \Delta x_2$ (+ when source moves away from detector, − when toward). Then

$$\lambda = v\Delta t \pm v_s\Delta t = (v \pm v_s)\Delta t = \frac{v \pm v_s}{f_s}$$

Crests with this spacing reach the detector at a speed v, so the frequency f_d with which they reach the detector is

$$f_d = \frac{v}{\lambda} = \frac{v}{\dfrac{v \pm v_s}{f_s}}$$

or
$$f_d = f_s \frac{v}{v \pm v_s} \qquad (14\text{-}20)$$

(+ when source moves *away* from detector, − when *toward*)

If your ear is the detector, f_d is the frequency of the sound you hear. Frequencies are always positive, so Equation 14-20 is valid only when $v/(v \pm v_s)$ is positive. Because the speeds are positive, this will be true except when $v - v_s$ is negative; that is, except when $v_s > v$, in which case the source coming toward you would reach you before the signal it emitted.

◆**STATIONARY SOURCE, MOVING DETECTOR** If the source is stationary, the spacing between crests remains fixed at $\lambda = v/f_s$. But if the detector is moving, it changes the speed at which the crests approach the detector. As Figure 14-32 shows, if the detector moves a distance $\Delta x_2 = v_d\Delta t$ while the donut (crest) travels a distance $\Delta x_1 = v\Delta t$ (v_d = detector speed, v = propagation speed), the total change in the separation between crest and detector is $\Delta x = \Delta x_1 \pm \Delta x_2 = (v \pm v_d)\Delta t$ (+ if detector going toward source, − if away). Then the *relative* speed, the rate at which this separation changes, is $\Delta x/\Delta t = v \pm v_d$. The frequency f_d experienced by the detector is

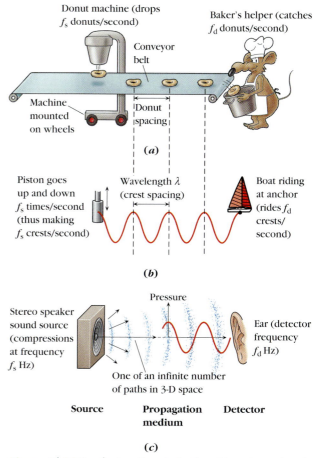

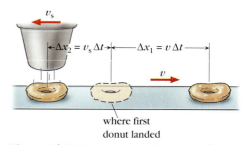

Figure 14-30 Exploring the analogies. (An animated and interactive version appears in WebLink 14-5.)

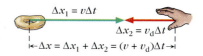

Figure 14-31 A source moving away from a stationary detector.

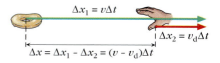

Figure 14-32 Stationary source; moving detector.

the speed at which the crest reaches the detector—the relative speed—divided by the wavelength:

$$f_d = \frac{v \pm v_d}{\lambda} = \frac{v \pm v_d}{\dfrac{v}{f_s}}$$

or

$$f_d = \frac{v \mp v_d}{v} f_s \tag{14-21}$$

(+ if detector going *toward* source, − if *away*)

As in Equation 14-20, the multiplier of f_s in Equation 14-21 is assumed positive. It does not apply when $v - v_d$ is negative. **STOP&Think** What would be happening if $v - v_d$ were negative? ◆

◆SOURCE AND DETECTOR BOTH MOVING RELATIVE TO THE MEDIUM

If the source and the detector of a sound are both in motion relative to the medium through which the sound propagates, then in the previous calculation we must use $\lambda = (v \pm v_s)/f_s$, the wavelength when the source is in motion, instead of $\lambda = v/f_s$. This gives us

$$f_d = \frac{v \mp v_d}{v \pm v_s} f_s \tag{14-22}$$

(v_d + if detector going *toward* source, − if *away*.
v_s + if source going *away* from detector, − if *toward*.
In each case, the upper sign corresponds to *away from,* the lower to *toward*.)

Like Equations 14-20 and 14-21, Equation 14-22 assumes that $v - v_d$ and $v - v_s$ are positive.

Example 14-13 *Doppler-Shifted Sounds of the City*

For a guided interactive solution, go to Web Example 14-13 at
www.wiley.com/college/touger

An ambulance approaches the scene of an accident from the east and a police rescue unit approaches from the west. The ambulance siren produces a 3500-Hz blare.

a. What frequency is heard by the police officers in the rescue unit when they momentarily come to a stop behind a moving van while the ambulance is racing toward the accident site at 24.0 m/s?

b. A minute later the ambulance has arrived at the accident site, but its siren is still going. The police car is still heading toward the site at 24.0 m/s. What frequency do the police officers hear?

c. When rescue operations are complete, ambulance and rescue unit race off in opposite directions, each at 24.0 m/s. What frequency do the police officers hear?

Brief Solution
Choice of approach. Equation 14-22 can be applied to all parts. Identify the source and the detector. In part **a,** because $v_d = 0$, Equation 14-22 becomes Equation 14-20. In part **b,** where $v_s = 0$, Equation 14-22 becomes Equation 14-21. In all parts, you must pay careful attention to signs. We'll assume conditions for which the speed of sound in air is 340 m/s.

What we know/what we don't.

$$f_s = 3500 \text{ Hz} \qquad v = 340 \text{ m/s}$$

a. $v_s = 24.0$ m/s   (− because toward) $v_d = 0$
b. $v_s = 0$ $v_d = 24.0$ m/s   (+ because toward)
c. $v_s = 24.0$ m/s (−) $v_d = 24.0$ m/s (+)

The mathematical solution.

a. $f_d = \dfrac{v}{v - v_s} f_s = \dfrac{340 \text{ m/s}}{340 \text{ m/s} - 24.0 \text{ m/s}} (3500 \text{ Hz}) =$ **3770 Hz**

The sound frequency is higher from an approaching source.

b. $f_d = \dfrac{v + v_d}{v} f_s = \dfrac{340 \text{ m/s} + 24.0 \text{ m/s}}{340 \text{ m/s}} (3500 \text{ Hz}) =$ **3750 Hz**

The sound is also higher to a detector approaching the source, but the value is not the same as in **a.**

c. Because the detector and the source are moving away from each other,

$$f_d = \dfrac{v - v_d}{v + v_s} f_s = \dfrac{340 \text{ m/s} - 24.0 \text{ m/s}}{340 \text{ m/s} + 24.0 \text{ m/s}} (3500 \text{ Hz}) = \textbf{3040 Hz}$$

◆ Related homework: Problems 14-64 through 14-68.

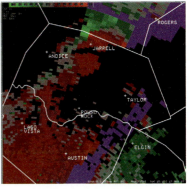

Monitoring tornado formation with Doppler radar. Although any radar can detect storm systems because the waves reflect off the water droplets, ice pellets, or other particles, using the Doppler effect enables forecasters to monitor wind speeds and direction relative to the overall motion of the storm. On May 27, 1997, a violent tornado struck Jarrell, Texas. This Doppler radar image shows the mesocyclone, the circulation of air from which the tornado formed. The brightest red and green indicate wind speeds of 50 knots (57.5 mi/h) in opposite directions. The opposite colors just north of Jarrell indicate counterclockwise rotation.

The Doppler shift occurs with any kind of wave, including water waves, light waves, and all other waves that move at the speed of visible light, from X-rays and gamma rays to radio waves. (The equations are different, however, for waves that travel at the speed of light.) The Doppler shift for radio waves is used in police radar units (RADAR stands for RAdio Detection And Ranging) and in the Doppler radar used in weather forecasting. Because the wave reflected from a moving vehicle or from a moving storm front differs in frequency from the wave transmitted by the radar unit, the two waves together produce beats. The speed of the moving object can be determined from the beat frequency (see Problems 14-85 and 14-102).

✦ SUMMARY ✦

Waves are regularly repeating traveling disturbances. The passing on of the disturbance is called **propagation** and takes place at a **propagation speed** $v = \Delta x / \Delta t$ that depends on the medium. **Transverse** disturbances are perpendicular to and **longitudinal** disturbances are along the propagation direction. As the disturbance travels, energy is transferred in the direction of propagation.

Quantities useful in describing waves also include the following:

- **Wavelength** (λ) is the distance between successive repetitions of equivalent points in the pattern (e.g., between one crest and the next).
- **Displacement** (y) is the size of the disturbance (not necessarily a distance) at a particular **path position** x along the path of the wave at a particular instant t. (This changes at a displacement velocity $v_y = \Delta y / \Delta t$ as the disturbance passes path position x.)
- **Amplitude** (Y) is the maximum displacement (whether above or below, forward or back) from the undisturbed level. The displacement y varies between $+Y$ and $-Y$; Y itself is always positive.
- **Frequency** (f) and **period** (T), defined as for simple harmonic oscillators, describe either the source oscillation producing the wave or the oscillation of a distant object disturbed by (detecting) the passing wave. They are inverses of each other: $f = 1/T$.

A fundamental relationship follows from the above definitions:

$$v = f\lambda \quad (14\text{-}2) \qquad \text{or} \qquad v = \frac{\lambda}{T} \quad (14\text{-}3)$$

A complete mathematical description of an undamped one-dimensional wave produced by a simple harmonic oscillator source can be written in alternate ways:

$$y = Y \sin 2\pi \left(\frac{t}{T} - \frac{x}{\lambda} \right) \quad (14\text{-}7)$$

or

$$y = Y \sin 2\pi f \left(t - \frac{x}{v} \right) \quad (14\text{-}8)$$

Procedure 14-1 describes the use of these equations to find the displacement y at any instant t at any path position x. Example 14-4 shows you how to read information from these **wave functions.**

The position-by-position addition of displacements from individual waves to obtain the resulting waveform that we actually observe is called **superposition of waves.** The allowable stationary vibrations of a bounded, extended body (such as a string fixed at its ends) are called **standing waves.** These may be understood as the superposition of a traveling wave and its reflection, and thus have the same wavelength and frequency as traveling waves. For stationary vibrations, there are **nodes** (where the displacement is always zero) and **antinodes** (where the oscillation is greatest) in fixed positions. The boundary conditions determine their locations, and therefore fix the *allowable wavelengths* and *allowable frequencies*: These are the **resonant frequencies** of the vibrating body. We can produce standing waves of substantial amplitude by matching the driving frequency to a standing wave frequency. **Procedure 14-2** outlines the reasoning involved in finding allowable frequencies of standing waves.

Sound is produced by vibrating sources that produce longitudinal particle displacements in the surrounding medium, resulting in variations in pressure and density. These variations propagate through the medium ($v_{\text{sound in air}} \approx 340$ m/s at 15°C) and cause vibrations on a detector (such as your eardrum) that mimic the source vibrations. The range of **audio frequencies** (those detectable to the human ear) is 20 Hz to 20 kHz. We hear higher frequency (shorter wavelength) sounds as higher pitched. Allowable or resonant frequencies for sound sources are called **harmonics.** The lowest harmonic is the **fundamental** frequency.

The propagation speed of sound depends on the medium:

On a string
$$v = \sqrt{\frac{T}{\mu_{\text{lin}}}} \qquad (14\text{-}20)$$

(T = tension, μ_{lin} = mass per unit of length)

Beats occur with **beat frequency** $f_{\text{beat}} = f_1 - f_2$ as two sustained notes of slightly different frequencies drift into and out of phase over time.

The **intensity** of sound (on which our perception of loudness depends) varies as the square of the amplitude Y ($I = cY^2$) while the perception of loudness (**sound level**) varies as the logarithm (power of 10) of the intensity:

$$\text{sound level} = (10 \text{ dB})\log_{10}\left(\frac{\text{Intensity}}{\text{Intensity}_o}\right) \quad \text{(decibel scale)} \quad (14\text{-}15)$$

A **Doppler shift** in the detected frequency f_d occurs if the source and/or the detector is in motion along the path between them. If f_s is the source frequency, and v, v_s, and v_d are the speeds of the wave, the source, and the detector relative to the medium, then

$$f_d = \frac{v \mp v_d}{v \pm v_s}f_s \qquad (14\text{-}22)$$

(v_d + if detector going *toward* source, − if *away*.
v_s + if source going *away* from detector, − if *toward*.
In each case, the upper sign corresponds to *away from*, the lower to *toward*.)

Equations 14-22 is the same as Equation 14-20 when the detector is stationary ($v_d = 0$) and it is the same as Equation 14-21 when the source is stationary ($v_s = 0$).

◆ QUALITATIVE AND QUANTITATIVE PROBLEMS ◆
Hands-On Activities and Discussion Questions

The questions and activities in this group are particularly suitable for in-class use.

14-1. Discussion Question. When a pebble is dropped in a pool, a circular ripple spreads outward from the pebble.
a. Why does the ripple spread outward? (What is the mechanism by which this happens?)
b. Suppose a large plastic hoop was floating on the surface of the water, and you dropped the pebble at the center of the hoop. What would happen to the ripple that was produced? Why?

14-2. Hands-On Activity. The following activity will work well with paper straws and with some plastic straws. You will produce a primitive wind instrument by doing the following (as shown in Figure 14-33).
a. Take a straw, flatten it at the tip, and cut out two side pieces. The result is that one end of the straw separates into two flaps (see figure). The flaps will vibrate when you put the cut end in your mouth and blow into the straw, setting up a vibration of like frequency on the column of air in the straw.
b. Try blowing into the straw at the cut end until you can reliably produce a sound (with some straws, you may have to vary the degree of flattening or the depth of the cuts).
c. What will happen to the sound if you cut off a couple of cm from the far end of the straw? (Answer the question before you try it.) Why?
d. Now try it. While you are blowing and producing a sound, continue to cut off pieces of the straw 1 to 2 cm long from the far end. What happens to the sound as you do so? Explain your observations.
e. Suppose you begin with a longer instrument, produced by attaching a second straw to the far end of the first to

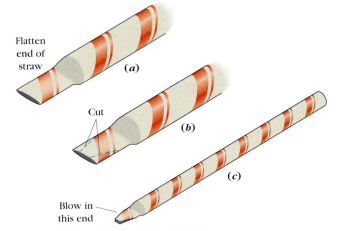

Flatten end of straw

Cut

Blow in this end

Figure 14-33 Problems 14-2 and 14-52

extend it (wrap tape around the joint). How would the sound of this longer instrument compare to that of the single straw? Try it. Progressively snip pieces from your longer instrument as you blow into it, and explain what you hear.

14-3. Hands-On Activity. Why is y in Equation 14-9 a *negative* sine function? To get a feel for this, try the following.
a. Cut a vertical slit about 5 cm high in a piece of lightweight cardboard (cardboard from a cereal or cracker box will do nicely) and mark it as shown in Figure 14-34a.
b. When you move a pencil point from 0 up to $+Y$, then down to $-Y$, and back to O again, it acts roughly like an oscillator described by $y = Y \sin 2\pi\frac{t}{T}$ going through one complete oscillation.

The set-up

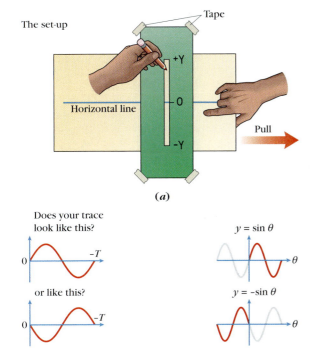

(a)

Does your trace look like this?

or like this?

(b)

Figure 14-34 Problem 14-3

$y = \sin \theta$

$y = -\sin \theta$

(c)

c. Draw a horizontal line on a strip of paper. Tape the cardboard in place on a flat surface, so that the paper strip can be pulled freely underneath it (Figure 14-34a) with the horizontal line always showing through the slit at the O position. The traveling paper will "carry the disturbance." (Remember that with real waves, only the disturbance travels, not the medium.)

d. Now place your pencil point at the O position and make a heavy dot on the horizontal line. Label this dot $-T$.

e. Beginning with your pencil point on this heavy dot, perform the pencil motion described in step 2 while pulling the paper strip to the right at uniform speed. Make a heavy dot on the horizontal line where your motion finishes and label it 0. Imagine $t = 0$ to be the instant you start your stopwatch, although the disturbance has been carried to the right since $t = -T$.

f. Undo the bottom pieces of tape on the cardboard and look at the trace drawn by your pencil. Which of the graphs in Figure 14-34b does your trace more nearly resemble? You have just stopped the action at $t = 0$. Your pencil trace is the stop-action picture at this instant. Is this picture fairly well described by Equation 14-9? (Check the graphs of $\sin \theta$ and $-\sin \theta$ in Figure 14-34c.)

g. Which point on the trace that you drew was made last? Was your trace drawn right-to-left or left-to-right? Notice that the two graphs in Figure 14-34c are left-right reversals of each other.

Review and Practice

Section 14-1 Traveling Disturbances: Some Basic Observations

14-4. Two people holding opposite ends of a long spring stretch it to a length of several meters. How does the disturbance get passed on from one part of the spring to the next? (What is the mechanism by which this happens?)

14-5. A pulse is sent along a spring, as in Figure 14-1. If a third person firmly gripped the center of the spring so that it couldn't move, would the pulse still travel from one end of the spring to the other? Briefly explain.

14-6. A traveling disturbance is shown at a particular instant in Figure 14-35. Find the displacement at each of the following positions: 10 cm, 18 cm, 26 cm.

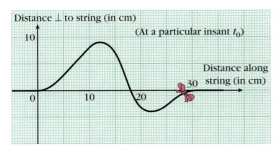

Figure 14-35 Problems 14-6 through 14-9 and 14-75

14-7. Suppose the disturbance in Figure 14-35 is shown at a particular instant t_o as it travels along a string to the right. A ribbon is tied to a particular point on the string. If the disturbance continues to the right, describe the motion of the ribbon from t_o forward.

14-8. Suppose the traveling disturbance in Figure 14-35 is shown at a particular instant t_o. Find the displacement at $x = 18$ cm after the wave has traveled another 6 cm
a. to the right.   **b.** to the left.

14-9. Suppose Figure 14-35 shows a stop-action picture of a disturbance at a particular instant t_o as it travels along a string to the right. A ribbon is tied to a particular point on the string. Which of the graphs in Figure 14-36 correctly plots the ribbon's displacement against time?

Distance ⊥ to string

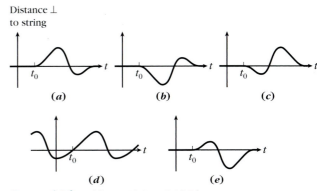

(a)   (b)   (c)

(d)   (e)

Figure 14-36 Problems 14-9 and 14-74

14-10. Suppose the pictures of the spring in Figure 14-5 are actual size. Use a cm ruler to find (to the nearest tenth of a centimeter)

a. the maximum displacement for the disturbance shown, and

b. the path position at which this maximum displacement occurs at the instant depicted.

Section 14-2 Energy in Traveling Disturbances

14-11. A pulse travels along a uniform stretched spring. At a particular instant,

a. can the propagation velocity be different at different positions along the spring? Briefly explain.

b. can the displacement velocity be different at different positions along the spring? Briefly explain.

14-12.

a. In Figure 14-6b, what factor(s) or condition(s) or action(s) are responsible for the value of the propagation speed v?

b. What factor(s) or condition(s) or action(s) are responsible for the maximum value of the displacement velocity v_y?

14-13. Find the propagation speed of the pulse in Figure 14-3.

14-14. A spring is stretched between two posts. When a pulse is produced at one end of the spring, it makes three back-and-forth trips along the spring in 2.6 s. If the pulse has a propagation speed of 15 m/s, how far apart are the posts?

14-15. SSM In Figure 14-37, a pulse traveling along a long spring without damping is shown first at the instant $t_1 = 0.030$ s and then at the instant $t_2 = 0.050$ s. A small ribbon with a mass of 1.6 g is tied around the spring at the position $x = 0.50$ m.

a. What is the propagation speed of the pulse?

b. Over the time interval $\Delta t = t_2 - t_1$, what is the average displacement speed of the ribbon?

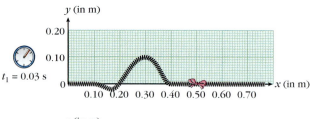

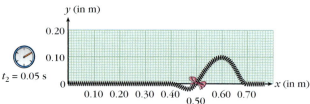

Figure 14-37 Problem 14-15

14-16. Suppose that in Problem 14-15, the ribbon had been tied around the spring at the position $x = 0.60$ m instead of $x = 0.50$ m. How would this have affected your answer to each part of that problem?

a. Would the answer to 14-15***a*** increase, decrease, or stay the same?

b. Would the answer to 14-15***b*** increase, decrease, or stay the same?

14-17. If you increase the vertical height of the pulse in Figure 14-3, you will also increase ____ (*the speed at which the pulse travels horizontally; the maximum speed at which the ribbon travels vertically; both of these speeds; neither of these speeds*). Briefly explain.

Section 14-3 Pulses and Periodic Waves

14-18. The stretched length of a spring is 6.0 m. A ribbon is tied around the spring at a point near its center. A pulse created at one end of the spring makes 12 back-and-forth trips along the spring in 9.00 s.

a. With what frequency does the ribbon go up and down?

b. With what speed does the pulse travel along the spring?

14-19. Figure 14-38 shows stop-action pictures of a boat riding at anchor as three different waves move under it toward the right. Rank the three pictures in order of each of the following. In each case, order them from least (or most negative) to greatest, noting any equalities.

a. The displacement of the boat for each wave.

b. The amplitudes of the waves.

c. The frequencies of the waves.

d. The displacement velocity (or transverse velocity) of the boat.

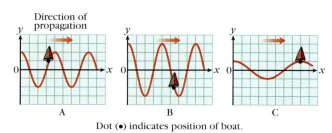

Dot (●) indicates position of boat.

Figure 14-38 Problem 14-19

14-20. Picture a wave with crests 2.4 m apart. If one crest passes you every 3 s, what is the propagation speed of the wave?

14-21. A periodic wave of wavelength 4.0 m travels along a very long string at a propagation speed of 32 m/s. Find the period of oscillation of a speck of dust sticking to the string.

14-22. Suppose a single pulse travels along the string in Problem 14-21. Which (one or more) of the following would be the same for the pulse as for the periodic wave: the wavelength, the propagation speed, the period, the frequency?

14-23. SSM A cork floats on the surface of a tank of water. A periodic wave passes the cork at a speed of 0.30 m/s. As the cork bobs up and down, it takes it 0.05 s to drop from maximum displacement to its undisturbed level.

a. Find the frequency of the wave, and

b. find its wavelength.

14-24. A very long rope is tied to a very long string (Figure 14-39). Both are kept as taut as possible. A vibrating source sets up a periodic wave train that travels left to right from the rope to the string. (Assume the string is so long that we don't have to consider the reflected wave.)

a. Is the propagation speed along the string less than, equal to, or greater than the propagation speed along the rope?

Figure 14-39 Problem 14-24

b. If the propagation speed changes as the wave train passes from the rope to the string, what else changes: the wavelength, the frequency, or both? Briefly justify your answer by telling what is happening physically. (Then see if your answer is consistent with your answers to the remaining parts.)

c. After a crest is set up by the source, what determines how long it takes until the next crest is set up? Is it the source, the rope, the string, or something else? Briefly, how do you know?

d. Based on **c,** what determines the frequency of the wave train? Is it the source, the rope, the string, or something else? Briefly, how do you know?

e. What determines how fast a crest travels along the rope? Is it the source, the rope itself, the string, or something else? Briefly explain.

f. Based on **e,** what determines how far a crest will travel from the source before the next one is set up? Is it the source, the rope itself, the string, or something else?

g. Based on **c** through **f,** what determines the wavelength on the rope? Is it the source, the rope itself, the string, or something else?

h. Is the frequency on the rope the same or different than the frequency on the string?

i. Could the frequency of the knot be any different than that of the rope just to the left of the knot? Than that of the string just to the right of the knot?

j. Could the rope and the string remain attached if they did not go up and down the same number of times each second? (After answering **j,** see if you need to revise your answers to **b** and **i.**)

k. Now let's summarize. In the equation $v = f\lambda$, what changes and what stays the same as the wave train travels from the rope to the string?

Section 14-4 **Fully Describing Waves Mathematically**

14-25. Consider Equation 14-7.
a. Show that at a fixed position $x = x_0$ the argument of the sine (that is, the phase $\theta = 2\pi[\frac{t}{T} - \frac{x}{\lambda}]$) changes by 2π when the clock reading t changes by one period (from t to $t \pm T$).

b. Show that in a stop-action picture at a fixed instant $t = t_0$, the phase changes by 2π when the position x changes by one wavelength (from x to $x \pm \lambda$).

14-26. A certain periodic wave, described by Equation 14-7, has an amplitude of 0.25 m, a wavelength of 1.6 m, and a period of 0.64 s. On the same axes, plot or sketch the wave's profile—a graph of y versus x—at each of the following instants: $t = 0$, $t = 0.08$ s, and $t = 0.16$ s.

14-27. For the wave described (in SI units) by $y = 10 \sin 2\pi(\frac{t}{8} + \frac{x}{4.5})$, find the amplitude, the wavelength, the period, the frequency, and the propagation speed.

14-28. For the wave described (in SI units) by $y = 0.6 \sin(5t + \frac{2x}{3})$, find the amplitude, the wavelength, the period, the frequency, and the propagation speed.

14-29. A periodic water wave like the one in Figure 14-8c has a propagation speed of 0.32 m/s and a wavelength of 0.08 m. A cork floating on the water is lifted 0.005 m from its equilibrium position each time a crest passes under it. If the cork is 0.03 m to the left of the origin, finds its displacement at $t = 0.20$ s.

14-30. Find the speed at which the wave in Figure 14-13 travels to the right **a.** using Equation 14-1. **b.** using Equation 14-2. **c.** Do you get the same value both ways?

14-31. **SSM WWW** A certain wave is described by $y = Y \sin 2\pi(\frac{t}{T} + \frac{x}{\lambda})$. (*Hint:* How does this differ from Equation 14-7?) Like the wave in Problem 14-26, this wave has an amplitude of 0.25 m, a wavelength of 1.6 m, and a period of 0.64 s. On the same axes, plot or sketch the wave's profile—a graph of y versus x—at each of the following instants: $t = 0$, $t = 0.08$ s, and $t = 0.16$ s. How does its motion differ from that of the wave in Problem 14-26?

Section 14-5 **Standing Waves and Superposition**

14-32. Figure 14-40 shows how two disturbances A and B would appear if they occurred on a length of string at different instants. Suppose, however, that the two disturbances occur at the same instant, so that the resulting observed disturbance R on this string is the superposition of A and B. y_A and y_B are the displacements due to the two individual disturbances. Measure y_A and y_B at various representative positions, and then find the total displacement y_R at each of these positions. Use the resulting values to sketch the observed disturbance y_R; that is, sketch a graph of y_R versus x.

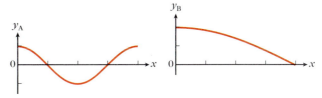

Figure 14-40 Problem 14-32

14-33. In Figure 14-19, the waves are shown at three instants a quarter of a period ($\frac{1}{4}T$) apart. Suppose now that t_4 is a quarter of a period later than t_3, and t_5 is a quarter of a period later than t_4. Continue the figure by sketching the incident and reflected waves and the resulting standing wave at t_4 and t_5.

14-34. Suppose the peaks of the incident and reflected disturbances in Figure 14-19 each travel along the wave path with a propagation speed of 15 m/s. With what speed does a peak of the resulting disturbance (the superposition of these two disturbances) travel along the wave path?

14-35. Suppose the ends of the spring in Figure 14-15 are 7.2 m apart. Suppose also that the hand making the spring go up and down goes through 24 up-and-down motions in 20 s.
a. If the resulting standing wave looks like Figure 14-15b, what would be the propagation speed of a traveling wave on the spring when it is stretched to this length?

b. Repeat **a** for the case in which the resulting standing wave looks like Figure 14-15a.

14-36. A standing wave is set up on a stretched spring. Figure 14-41 shows the spring at three different instants. The spring appears completely horizontal at $t = 0.07$ s but not at any other instant between $t = 0$ and $t = 0.14$ s. (It is completely horizontal at those two instants.)

a. With what speed would the crest of a traveling wave travel along this spring?

b. With what speed does the crest of the standing wave travel along the spring?

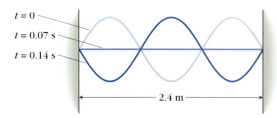

$t = 0$
$t = 0.07$ s
$t = 0.14$ s

2.4 m

Figure 14-41 Problem 14-36

14-37. Suppose that in Problem 14-36, the spring appeared completely horizontal not only at $t = 0.07$ s but also at two other instants between $t = 0$ and $t = 0.14$ s. With what speed would the crest of a traveling wave travel along the spring under these conditions?

Section 14-6 Introduction to Sound

14-38.

a. In Example 14-7, how long after the lightning bolt strikes the tree do you see it if light travels at 3.0×10^8 m/s?

b. Does this satisfy the assumption we made that we see it virtually instantaneously?

14-39. Standing on a roof on July 4, you observe a burst of distant fireworks. You hear the explosion 3.5 s later. Roughly how far are you from the point of the explosion?

14-40. In what SI units can the amplitude of a wave function for a sound wave be expressed?

14-41. Consider a sound wave traveling in air.

a. At points along the wave where the longitudinal displacement of the air molecules is a maximum, is the density of molecules normal, above normal, or below normal?

b. At points along the wave where the air pressure is a maximum, is the density of molecules normal, above normal, or below normal?

14-42. In Figure 14-22, find the positions where the longitudinal displacement y of the molecules is either at a positive or a negative maximum. Notice that these are the positions where the pressure nodes occur (i.e., where there is no deviation from normal pressure). Explain physically why there should be no change from normal pressure at these points. *Hint:* Focus on one such point. How are the molecules in the region to the left of this point displaced? How are the molecules in the region to the right of this point displaced? Is there any difference? How do the displacements of the molecules in these two regions combine to affect the density of molecules in the immediate vicinity of the point you are considering?

14-43. The undisturbed air pressure in a certain long (more than a wavelength) hollow tube is 1.0132×10^5 Pa (1 Pa = 1 N/m²). A particular sound wave produced in this tube is described by the wave function $\Delta P = 20 \sin 2\pi(670t - \frac{x}{0.50})$, with all quantities in appropriate SI units.

a. At any instant of time while this sound is being produced, what is the maximum air pressure in the tube, and what is the minimum pressure in the tube?

b. What is the speed of sound in the air in this tube?

14-44. The undisturbed air in a certain long hollow tube exerts a pressure of 1.0132×10^5 Pa. On a cold day, sound waves travel down this tube at a speed of 330 m/s. Suppose a sound wave with a frequency of 500 s^{-1} (500 Hz) causes the pressure in the tube to vary between 1.0120×10^5 Pa and 1.0144×10^5 Pa. Write a possible equation or wave function describing this wave.

14-45. The background hums of two speakers in a home sound system have the same amplitude, but one has a frequency of 278 Hz while the other has a frequency of 284 Hz. How many beats per second would you hear when you switch on this system?

Section 14-7 Resonance and Sources of Musical Sound

14-46. The fixed ends of a particular string on a musical instrument are 0.60 m apart. When the tension in the string is set so that the propagation speed is 360 m/s, what are **a.** the wavelengths and **b.** the frequencies of the first three harmonics?

14-47. A string on a child's violin has a length of 0.30 m. The child has "tuned" the violin so that the propagation speed along the string is 1500 m/s. Find the frequency of the eighth harmonic on this string and comment on what this harmonic would sound like.

14-48. You and a classmate are gripping opposite ends of a spring stretched to a length of 8.0 m. Under these conditions, a pulse would travel along the spring at a speed of 12.0 m/s. You now wish to set up standing waves rather than a traveling disturbance. How many times per second must you shake your wrist up and down to set up each of the standing wave patterns shown in Figure 14-15?

14-49. Naomi and Molly stand 8 m apart holding the ends of a spring stretched between them. As they do so, Jerry firmly holds the spring at a point 2 m away from Molly. Let part A be the part of the spring between Molly and Jerry, and let part B be the part of the spring between Jerry and Naomi.

a. If λ_A and λ_B are the longest possible wavelengths that standing waves can have on the two parts of the spring, how do the two wavelengths compare? Answer by giving the numerical multiplier in $\lambda_A = \underline{\ ?\ } \lambda_B$.

b. If f_A and f_B are the frequencies of the standing waves on the two parts of the spring when the wavelengths are λ_A and λ_B, how do these two frequencies compare? Answer by giving the numerical multiplier in $f_A = \underline{\ ?\ } f_B$.

14-50. The propagation speed along a wire stretched tautly between two endpoints 3.50 m apart is 22.0 m/s. When the string is vibrating at a frequency of 30.8 Hz,

a. how many nodes are there between the two endpoints?

b. how many antinodes are there between the two endpoints?

14-51. Of two long, narrow organ pipes A and B, A is open at one end and B at both ends. Both pipes have the same fundamental frequency.

a. If A is 2.4 m long, how long is B?

b. The next lowest frequency (first overtone) of A will then be (*half of; $\frac{2}{3}$ of; $\frac{3}{4}$ of; equal to; $\frac{3}{2}$ of; $\frac{4}{3}$ of; twice*) the next lowest frequency of B.

14-52. A simple wind instrument made from a paper straw (as in Figure 14-33) is 0.14 m long. Assuming the speed of sound in air to be 340 m/s, what is the frequency of the sound you would produce by blowing on this straw?

14-53. Your hearing is most sensitive to frequencies in the vicinity of the fundamental frequency of the auditory canal. The auditory canal, a part of your ear, is basically a pipe closed at one end by the eardrum. The mean length of the human auditory canal is 2.77 cm. What is the fundamental frequency of sound waves in the auditory canal?

14-54. The author's daughter was playing her flute in her high school music room, where other instruments were stored. Whenever she hit a certain note, the cymbals on a nearby set of drums started vibrating. Explain why this happened.

14-55. A steel wire is kept taut by passing it over a pulley and suspending weights on a hanger from its free end. Unlike the strings in Figure 14-28, the fixed end is attached to a rigid object, not a buzzer. Plucking the wire produces a sound. To produce a higher-pitched sound, should you add weights to the hanger or remove weights? Briefly explain.

14-56. In this and the next three problems, suppose that the buzzers in Figure 14-28 are adjustable—they can be made to vibrate more quickly or slowly. Assume that any changes that are made are sufficient to result in another standing wave. How would speeding up the buzzer affect the number of loops in the standing wave?

14-57. How would speeding up the buzzer affect the wavelength of the standing wave in Problem 14-56?

14-58. How would increasing the length of string between the buzzer and the pulley affect the number of loops in the standing wave in Problem 14-56?

14-59. How would increasing the length of string between the buzzer and the pulley affect the wavelength of the standing wave in Problem 14-56?

14-60. Guitar strings A and B are identical except that the diameter of A is twice the diameter of B. How does the frequency of the sound from A compare to the frequency of the sound from B? ($f_A = \underline{\ ?\ } f_B$)

14-61. An aggressive piano tuner increases the tension on a certain piano wire by 25%. How does the frequency of the sound that the wire will produce after this is done compare to the frequency before? ($f_{after} = \underline{\ ?\ } f_{before}$)

14-62. Each meter of a spool of steel wire (see Table 14-1) has a mass of 0.020 kg. A 0.40-m length of the wire is cut from the spool and stretched taut. What tension force must be maintained on this length of wire if it is to have a fundamental frequency of 320 Hz?

14-63. A popular trick at children's parties where there are helium balloons is to draw a breath of helium and then speak. When you do this, you find yourself sounding like a high-voiced cartoon character. Try to explain why by thinking about what affects the speed of sound on a string, and then drawing an analogy for the speed of sound in a gas. Then consider how this would affect the frequency.

14-64. A person in a typical conversation speaks at a sound level of 60 dB. The music at a rock concert often reaches a sound level of 120 dB. How many times as great is the intensity of the music?

14-65. SSM Speakers A and B each produce sound for one minute, A at a sound level of 50 dB, B at a sound level of 70 dB.

a. Compare the total energy outputs of the two speakers for the minute in question. How many times as great is speaker B's energy output?

b. Assuming the two speakers are equally efficient, how many times as great is speaker B's wattage requirement during this minute?

Section 14-8 The Doppler Effect

14-66. Equation 14-21 is not valid when $v_d > v$ and the detector is moving away from the source, because that would result in a negative f_d and frequencies are never negative. What would actually be happening if the detector is moving away from the source at a speed $v_d > v$; that is, what would the detector actually detect under these conditions, and why?

14-67. An up escalator has a stairway on either side (Figure 14-42). Three painters are dripping paint on the steps of the escalator, each at a rate of one drop per second. The painter dripping blue paint is going up one staircase, while the painter dripping red paint is coming down the other. The painter dripping yellow paint is standing still. Three supervisors (one for each painter—obviously political appointees) stand on the top landing counting the drops. Each clicks his clicker each time a drop of his painter's color reaches him. Consider the situation during a time interval when you can ignore end effect questions (like "What happens when the painter reaches the end of the escalator?" or "What happens when the same escalator step starts over again at the bottom?"). If the distances are much longer than shown, this can be a reasonable time interval.

a. Rank the colors in order of how far apart the drops are, from closest together to furthest apart.

b. Rank the supervisors (indicate them by color) in order of how frequently they click, starting with the lowest frequency.

c. Three mechanical toys A, B, and C produce identical humming sounds when stationary. Two are set in motion, A toward you and B away from you, while the third one (C) stands still. Rank the frequencies of the sounds that you hear from the three toys. Briefly explain how this situation is like the situation in **b**.

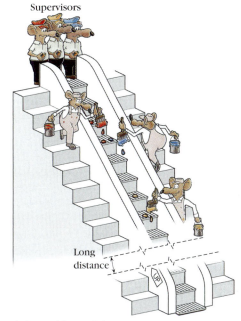

Supervisors

Long distance

Figure 14-42 Problem 14-67

14-68. In the nineteenth century, Dutch meteorologist C.H.D. Buys-Ballot set up an experiment in which trumpeters, with their instruments previously tuned to be in unison, were placed on a train and on several stations along the railway line. If each musician played the note of A_2 (440 Hz) while the train sped past a station at a speed of 18 m/s, would observers on the station platform hear the trumpet on the train as being higher or lower than the one on the platform, and by how much would the trumpet on the train sound off-key

a. when the train was approaching the station?

b. when the train was going away from the station?

•**c.** How fast would the train have to be going for observers on the platform to hear a beat frequency of 1.2 Hz?

14-69. Problem 14-68 asks what observers on the station platform would hear. Now determine whether observers on the train hear the trumpet on the train as being higher or lower than the one on the platform, and by how much the trumpet on the platform would sound off-key

a. when the train was approaching the station.

b. when the train was going away from the station.

14-70. Suppose that in the experiment described in Problem 14-69, an additional trumpeter was stationed on a second train going in the opposite direction on the adjacent track. Train A is eastbound at 18 m/s. Train B is westbound at 12 m/s. Would the passengers on train A hear the trumpeter on train B producing a higher or lower note than the trumpeter on their own train, and by how much would the trumpeter on train B sound off-key

a. when the two trains were approaching each other?

b. just after the two trains had passed each other?

14-71. A pedestrian standing at the curb hears a musical note coming from the radio of a car approaching her at 20 m/s. For her, the note has a frequency of 3000 Hz.

a. At what frequency does the driver hear this note?

b. After the car passes her, the same note is repeated. At what frequency does she hear the note this time?

Going Further

The questions and problems in this group are not organized by section heading, so you must determine for yourself which ideas apply. Some of them will be more challenging than the Review and Practice questions and problems (especially those marked with a • or ••).

14-72.

a. Can the displacement at a point along the path of a wave ever have different units than the wave's amplitude? Briefly explain.

b. Can the units of displacement for a particular wave ever be different than the units of path position? Briefly explain.

14-73. SSM WWW Think of a single stop-action picture, taken at some particular instant, of a periodic traveling wave on a very long horizontal spring.

a. Sketch the picture and put a dark dot at each point on the spring where the displacement is instantaneously zero.

b. Are the disturbances at all of these positions in phase? If so, why? If not, which of them *are* in phase?

c. Suppose the wave is traveling to the right. A tiny fraction of a second later than the instant depicted, which of the points that you darkened will have moved upward? Which will have moved downward? Are the disturbances at the positions where there is upward motion in phase with the disturbances at the positions where there is downward motion? If there is any contradiction between your answers to **b** and **c,** try to resolve it.

•**14-74.** A ribbon tied to a spring is disturbed by a pulse that travels along the spring. Graph *b* in Figure 14-36 shows how the displacement of the ribbon varies with time. A particular instant t_o is marked on the *t* axis for reference. A graph of *y* versus *x* at any instant shows us what the spring with the pulse on it looks like at that instant. Sketch the graph of *y* versus *x* at the instant t_o.

•**14-75.** At the instant at which the disturbance in Figure 14-35 is depicted,

a. are there any positions between $x = 0$ and $x = 25$ cm where the *displacement velocity* is zero? If so, what are they?

b. at what position is the displacement speed greatest?

c. between what pair(s) of positions is the displacement velocity negative?

••**14-76.** Energy is transferred through a medium as a pulse travels. The transfer is ongoing when there is a continuous periodic wave. In that case, the energy transferred each second (the power P) to each unit of mass of the medium is proportional to the square of the amplitude A (see end of Section 14-2): $P/m \sim A^2$. Then $P \sim mA^2$.

a. Suppose the medium is a two-dimensional surface, such as a pond surface carrying water waves. We can picture the surface as made up of concentric rings of equal width, with the source at the center (Figure 14-43a). A ring passes on all its energy to its outer neighbor as it receives all the energy from its inner neighbor. Because the energy is spreading at uniform speed in all directions, the energy per second (power) each ring transfers outward must be the same, or else energy would not be conserved. But $P \sim mA^2$, and the rings of greater circumference have greater mass ($m \sim 2\pi r$). Combining these facts, show that for energy to be conserved, the amplitude must decrease with r and that this relationship is expressed by $A \sim 1/\sqrt{r}$.

b. When energy spreads through a uniform *three*-dimensional medium, we can use the same reasoning as in a if we picture the medium to be made up of concentric spherical shells of equal thickness (Figure 14-43b). In that case the mass of each shell is proportional to its surface area ($m \sim 4\pi r^2$). Using similar reasoning to **a,** show that in the three-dimensional case the variation of the amplitude with r is given by $A \sim 1/r$.

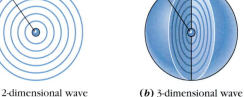

(a) 2-dimensional wave   (b) 3-dimensional wave

Figure 14-43 Problem 14-76

c. Recalling that $P/m \sim A^2$, compare the results of *b* with the relationship developed in Problem 6-72. What reasons can you give for any similarities that you notice?

14-77. The radio antenna on your car is an old-fashioned long, straight type. It is attached to the car at its lower end and is extended to a length of 0.9 m. A rude neighborhood kid twangs the antenna as he goes by. You observe that the antenna oscillates exactly 12 times in 5.0 s. Assuming the antenna to be uniform along its length, how fast can such disturbances travel along it?

14-78. The prairie warbler is a small North American bird whose song is a series of notes going right up the scale from low to high. As it runs through the notes, its head tilts further and further back. Speculate about what might be happening anatomically to produce the change in pitch.

14-79. Two space explorers find that the atmospheric pressure on a newly discovered planet is the same as on Earth. One of them has just measured the temperature to be 0°C and fires a flare to catch his colleague's attention. His colleague, standing about a kilometer away, hears the shot $2\frac{1}{3}$ s after she sees the flare. Of what gas could the atmosphere of this planet be composed?

14-80. When a 33 rpm phonograph record is played at 78 rpm, the music comes out not only faster but higher pitched. Explain why this is so with careful reference to what changes in the equation $v = f\lambda$.

14-81. **SSM**

a. A standing wave on a string that can carry a disturbance at a speed of 18.0 m/s has a frequency of 20 Hz. What is the distance between any antinode along this wave and its nearest-neighbor node?

b. What can you determine about the length of the string from the given information: the actual numerical value of the length, the different possible values the length can have, or nothing at all? Briefly explain.

14-82. A tall, very narrow, empty bottle is 0.20 m tall. When you blow across the opening, you produce a sound.
a. Find the wavelength and frequency of this sound.
b. What assumptions did you make in answering *a*?
c. How could you get the bottle to produce a higher frequency sound? Explain.

14-83. Longer-wavelength sounds travel further; that is, they are not damped as much over a given distance. Using the equation $v = f\lambda$ as a basis for your arguments, give some of the reasons why the sounds of blue whales carry through the ocean depths for such great distances (up to thousands of kilometers). Can you suggest any reasons that this equation doesn't address?

14-84. The key of C in the lower octave of the treble clef corresponds to a frequency of 264 Hz. In an organ, if a pipe open at both ends is used to produce this tone, another pipe open at both ends must be $\frac{2}{3}$ as long to produce the key of G in this octave. To what frequency does the key of G correspond?

•14-85. A student walks toward the chalkboard in front of the classroom carrying a vibrating 440-Hz tuning fork. The chalkboard reflects sound. The rest of the class, who are in their seats, hear four beats per second. How fast is the student walking? (Based on a note by Tom Greenslade in *Physics Teacher,* **31**, 443 [1993])

14-86. A wave with a wavelength of 2.0 m travels along a one-dimensional path with a propagation speed of 4.2 m/s. Two observers stand beside the path at different points. Observer A reports being passed by two crests during a certain one-second interval. Observer B reports being passed by three crests during the same interval. The best evaluation of the two reports is that (*both observers' reports; only observer A's report; only observer B's report; neither observer's report*) can be correct.

•14-87. A sinusoidal traveling wave follows a one-dimensional path with a propagation speed of 3.6 m/s. During a particular 2.0-s time interval, an observer standing beside the path sees five crests go by. What are the minimum and maximum possible wavelengths of this wave?

14-88. By shaking your wrist, you can produce a pulse that travels along a long spring, as in Figures 14-1 and 14-2*a*. To make the pulse travel faster, you can (*stretch the spring more; shake your wrist harder; do either of these because both will work; not do either of these because neither will work*).

14-89. You can produce a longitudinal pulse that travels along a Slinky by tapping it at one end, as in Figure 14-2*b*. Without replacing the Slinky, what change can you make to make the pulse travel faster?

14-90. A periodic sound wave is traveling down a long tube of air. At $t = 0$, the air pressure has its normal value of 101 000 Pa at $x = 0$ and next has that value at $x = 0.40$ m. Halfway between, the pressure at this instant is 101 015 Pa. Write the equation describing this wave.

14-91. The sound intensity reaching your ear from an explosion at the threshold of pain will be 10^{10} times as great as from a child whispering a meter away from you. How much larger are the molecular oscillations caused by the explosion than those caused by the whisper?

14-92. A piano tuner uses a tuning fork with a frequency of 440 Hz while playing the note "concert A" on Professor Touger's old upright piano. The tuner hears three beats per second, indicating that the piano is off-key. What is the fundamental frequency of the note played on the piano?

14-93. Where they reach your ear (taken to be $x = 0$), two sound signals are described by $y = 20 \sin 480\pi t$ and $y = 20 \sin 486\pi t$. How many beats per second do you hear?

14-94. Answer the following by counting crests and/or troughs in Figure 14-44. Assume that the time interval from t_1 to t_2 is one second.
a. How many cycles does y_1 go through in a second? How many cycles does y_2 go through in a second?
b. How many cycles does the amplitude of y (the superposition of y_1 and y_2) go through in a second? (How does this number relate to the two numbers you obtained in *a*?)
c. How many beats occur during the second between t_1 and t_2, that is, what is the beat frequency?

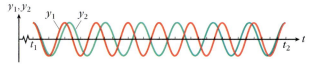

Figure 14-44 Problem 14-94

14-95. A buoy is floating in a harbor. A periodic wave with a wavelength of 1.25 m passes it at a speed of 0.3 m/s. How long does it take the buoy to rise from its undisturbed position to its maximum displacement?

•**14-96.** A sinusoidal wave with a period of 2.4 s is traveling past point P. At $t = 0.60$ s, the displacement at P equals half the wave's amplitude. At what instant will the displacement at P next be 0.866 times the amplitude?

14-97. SSM Alex and Maritza stand 8.0 m apart holding the ends of a long spring that is stretched between them. When Alex plucks the spring at his end, it takes the pulse 2.4 s to travel to Maritza and come back to him. If they remain standing the same distance apart, what is the lowest frequency at which they can produce a standing wave on the spring?

14-98. In Figure 14-45, the two gas-filled chambers have equal lengths L, and are separated by a thin metal wall. Both gases are kept at standard temperature and pressure. A sound signal takes 0.014 s to travel from point A to point B. Find the length L of each chamber. (Assume the time it takes for sound to pass through the wall between the chambers is ≈ 0.)

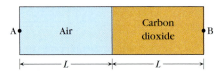

Figure 14-45 Problem 14-98

14-99. The intensity of sound at the threshold of hearing is normally considered to be 1.0×10^{-12} W/m^2. Suppose the sound level you hear when your friend is standing a meter away from you and speaking in a normal voice is 60 dB. What is the intensity of this sound when it reaches you?

14-100. Figure 14-46 shows the same traveling wave at two different instants. Find the frequency of this wave.

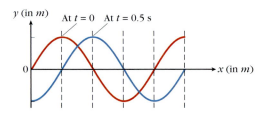

Figure 14-46 Problem 14-100

14-101. The human *tympanum* or eardrum vibrates with an amplitude of about a micron (10^{-6} m) at the threshold of hearing.

a. If the response of your eardrum were linear—simply proportional to the intensity rather than logarithmic—then compared to the threshold of hearing, how many times as great would the vibrational energy of your eardrum be when you are hearing amplified music at a level of 110 dB?

b. The vibrational energy of your eardrum is proportional to the square of the amplitude of its vibrations. Compared to the threshold of hearing, how many times as great would the amplitude of your eardrum's vibrations be when you are hearing sound at this level?

c. If the response of your eardrum were linear, how large (in meters) would the amplitude of your eardrum's vibrations have to be to be able to detect sound at this level?

d. Could you hear such a large range of intensities if your eardrum's response was linear rather than logarithmic?

14-102. A police radar unit emits a radio wave that reflects off a moving target. The moving target acts as the source of the reflected wave. Suppose the radar unit is in a stationary patrol car and emits a radio wave at a frequency of $1.060\,000\,000\,0 \times 10^{10}$ Hz. Radio waves, like light waves, travel at a speed of 3.0×10^8 m/s.

a. If a truck is coming toward the patrol car at a speed of 26.8 m/s (about 60 mi/h), what frequency would the radar unit detect for the reflected wave? (Do not round off.)

b. Because the reflected wave differs in frequency from the wave transmitted by the radar unit, the two waves together produce beats. What is the beat frequency? (The radar units actually detect the beat frequency and calculate the speed of the vehicle from that.)

Problems on WebLinks

14-103. In WebLink 14-1, if the red wheel starts and ends at its lowest position, the number of times it returns to its lowest position is (*always; sometimes; never*) equal to the number of wavelengths that pass under it.

14-104. If the animation that opens WebLink 14-1 had begun when the timer read 0.00 s and ended when it read 5.00 s, and if the crests of the waveform were 1.4 m apart, at what speed would the waveform have been moving?

14-105. Graph the stop-action pictures of the wave in WebLink 14-2 at the instants $t = 0.50$ s and $t = 0.70$ s.

14-106. A certain wave is described by $y = Y \sin 2\pi(\frac{t}{T} + \frac{x}{\lambda})$. (*Hint:* How does this differ from Equation 14-7?) Like the wave in WebLink 14-2, this wave has an amplitude of 0.20 m, a wavelength of 10 m, and a period of 0.80 s. Obtain stop-action pictures of the wave at $t = 0$, $t = 0.10$ s, and $t = 0.20$ s and sketch them on the same set of axes. How does its motion differ from that of the wave in WebLink 14-2?

14-107. As the red and blue waves in WebLink 14-3 travel in opposite directions,

a. the total displacement at points A and B will (*always; sometimes; never*) be zero.

b. the total displacement at the point midway between points A and B will (*always; sometimes; never*) be zero.

14-108. Suppose that in WebLink 14-3, point A is at $x = 0$ and point B is at $x = 1.00$ m. Suppose also that the red and blue waves each have an amplitude of 0.15 m. Then as the waves travel, the maximum displacement at the position $x = 0.40$ m will be ___ (*zero; between 0 and 0.15 m; 0.15 m; between 0.15 m and 0.30 m; 0.30 m*).

14-109. Which (one or more) of the following are the same for the red and blue waves in WebLink 14-3 as they are for the resulting standing wave: *the frequency; the amplitude; the wavelength; the speed at which the crests travel in the horizontal direction?*

14-110. In the graphs of y_1 versus t and y_2 versus t in WebLink 14-4, suppose that the number of crests between the starting and ending times remained the same, but that the ending time was $t = 0.20$ s instead of $t = 1.0$ s. In other words, the only change in the graphs is the numerical value shown on the horizontal axis.

a. How many times does signal y_1 get lower and louder in a second?

b. How many times does signal y_2 get lower and louder in a second?

c. How many times does the combined signal $y_1 + y_2$ (what you hear when the sounds are produced together) get lower and louder in a second?

d. What is the frequency of signal y_1?

e. What is the frequency of signal y_2?

f. What beat frequency would you hear?

14-111. Suppose that the donut machine in WebLink 14-5 moves to the left from position A to position B, and then stops.

a. The frequency with which the donuts reach the baker's helper is (*less; the same; greater*) when the donut machine is stationary at B than when it is stationary at A.

b. The frequency with which the donuts reach the baker's helper is (*less; the same; greater*) when the donut machine is moving from A to B than when it is stationary at B.

CHAPTER FIFTEEN

Wave Optics

In Chapter 14 we saw that the displacement resulting from the superposition of two waves is sometimes smaller than from either wave individually; where that happens, the combined intensity is also smaller. In this way, disturbances (including waves) behave differently from conventional particles of matter, such as grains of sand or droplets of spray paint. A flow of such particles arriving over a small area from two different sources always results in a greater rate of input to the area than from either one alone, never less.

Based on this difference, we can decide whether certain observations about light indicate particlelike or wavelike behavior. In this chapter we will explore these observations

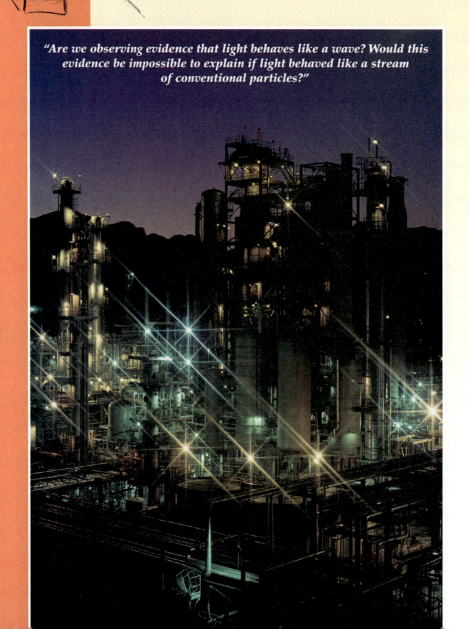

"Are we observing evidence that light behaves like a wave? Would this evidence be impossible to explain if light behaved like a stream of conventional particles?"

in detail: what do we see, and how can we explain what we see? Are we observing evidence that light behaves like a wave? Would this evidence be impossible to explain if light behaved like a stream of conventional particles? In other words, which rules of nature does light appear to obey, the rules that govern particle behavior or the rules that govern wave behavior?

15-1   Does Light Travel? If So, How? And How Fast?

To most young children, and to some people of all ages, the notion that light travels is not obvious. Why not suppose that something goes out from your eye to an object when you see it (don't we "direct our gaze" at an object?), or that light somehow instantaneously fills the space between a lamp and your eye when you turn it on? **STOP**&**Think** When you see the sun, how do you know that light is traveling from the sun to your eye? When you see a chair, does light travel from the chair to your eye? *What is the evidence?* If you cannot answer these questions with confidence, you should discuss them with classmates or with your instructor. ◆ Physicists assume that light travels; you should be able to justify this assumption on the basis of commonsense arguments about ordinary experience.

But what does it mean to say that light travels? What actually goes from the source to the observer? A substance? A stream of particles? Or does the source set up some kind of disturbance or wave that gets carried to our eyes?

Our visual awareness of events seems instantaneous, as though light were traveling infinitely fast from the observed objects to our eyes. But to speak of light traveling *like particles* or *like waves* implies that like either of those, it must travel at finite speed. (In our reference frame, an ordinary particle traveling at infinite speed would take zero time to go anywhere. It would be at its departure point and at its arrival point at the same instant, so there would be no need to say it travels!) **STOP**&**Think** What would be the wavelength of a wave traveling at infinite speed from a vibrating source with a fixed frequency? Would it even be meaningful to speak of this as a wave? ◆

◆**MEASURING THE SPEED OF LIGHT** Experience tells us that if the speed of light is not infinite, it must at least be very great. A lag between an event happening and being seen would become evident (i.e., long enough to be measurable) only if light traveled extremely far. Thus the first measurements of light speed involved light traveling over great distances. We will treat one historic measurement here; for another, see Problems 15-6 and 15-7.

Not surprisingly, the first lag to be recognized between an event and its observation involved interplanetary distances. The events in this case were the eclipses of Io, the innermost of Jupiter's four largest moons. The moons are in eclipse when Jupiter block's the suns light from reaching them (Figure 15-1). Because each moon orbits with a constant period (Table 8-2), one should be able to predict when the moon will next be seen emerging from Jupiter's shadow. Suppose we determine Io's period using observations made when Earth is closest to Jupiter (at or near point A in the figure) and use that period to predict the subsequent times when Io should emerge from eclipse. It turns out that when Earth is at B or C, each of these times occurs about 1000 s later than predicted. (At B and C, Earth is nearly at its maximum distance from Jupiter, but its view of Jupiter is not yet blocked by the sun.) In 1675, Olaf Roemer, a Dane, deduced that a lag due to light's finite speed could explain this. (Some historians attribute this work to his successors.)

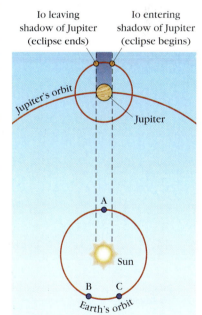

Figure 15-1 The basis for Roemer's calculation of the speed of light. When Earth has traveled from A to B, Io's emergence from the shadow of Jupiter appears late to an Earth observer because light from Io has had to travel the additional distance from A to B.

(a)

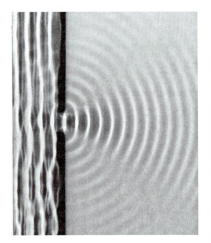

(b)

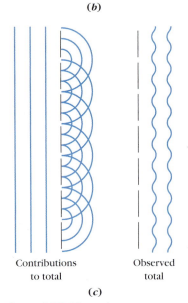

Contributions   Observed
to total   total

(c)

Figure 15-2 Observing water waves illuminated from below. Similar patterns of water waves spread from (*a*) a point source and (*b*) a narrow opening in a barrier. When (*c*) there are enough openings in the barrier, the waves to the right of the barrier begin to look as they would with no barrier present.

Roemer reasoned as follows. The instants at which Io is seen emerging from Jupiter's shadow should be one period apart (Io's period is about 42.5 hours). Suppose one such instant occurs when Earth is at A. If a certain number of periods occur as Earth travels from A to B, then at B we should be able to predict when Io should next be seen emerging from Jupiter's shadow. It emerges about 1000 s late because the light from Io had to travel the additional distance from A to B before we saw it. This distance is approximately the diameter of Earth's orbit. Because the value of this diameter that Roemer and followers had available was somewhat inaccurate, the value they got for the speed of light differed from the currently accepted value. If we apply Roemer's reasoning using the currently accepted value of 2.99×10^{11} m, we would find the speed of light c to be $c = \frac{2.99 \times 10^{11} \text{ m}}{1000 \text{ s}} = 2.99 \times 10^8$ m/s. We have been considering light traveling through space, so this is the speed of light *in a vacuum;* its speed in other media is less. Based on more precise measurements, the meter was redefined in 1983 so that the speed of light in a vacuum is exactly 299 792 458 m/s. We will use this value rounded to three significant figures:

Speed of light in a vacuum $c = 3.00 \times 10^8$ m/s

Example 15-1 *Light-Years Away*

A **light-year** is the distance traveled by light in one year. The nearest bright star to our own sun is α Centauri ("alpha Centauri"), 4.3 light-years away; that is, the light reaching us now from α Centauri left the star 4.3 years ago. What is the distance to α Centauri in meters?

Solution

Choice of approach. From the definition of a light-year, it takes light 4.3 years to reach us from α Centauri, traveling at speed c. Once we convert 4.3 years into SI units (s), we can use the definition of average speed to find the distance traveled by anything at speed c, whether particle or wave.

What we know/what we don't.

$$v = c = 3.00 \times 10^8 \text{ m/s}$$

$$\Delta t = 4.3 \text{ years} = ? \text{ s} \qquad \Delta x = ?$$

The mathematical solution. Converting from light-years to seconds,

$$4.3 \text{ years} \times \frac{365 \text{ days}}{1 \text{ year}} \times \frac{24 \text{ h}}{1 \text{ day}} \times \frac{3600 \text{ s}}{1 \text{ h}} = 1.4 \times 10^8 \text{ s}$$

Then $\Delta x = v\Delta t = (3.00 \times 10^8 \text{ m/s})(1.4 \times 10^8 \text{ s}) = \mathbf{4.2 \times 10^{16} \text{ m}}$

or about 42 trillion km.

◆ Related homework: Problems 15-10 and 15-8.

15-2   Waves in Two Dimensions

Before asking whether light is wave-like, we must know more about how waves behave in more than one dimension. To build this understanding, we will look at the behavior of water waves because they are so readily observable.

When a point object bobs up and down periodically at the surface of an otherwise still body of water, it produces a series of expanding circular crests and troughs (Figure 15-2*a*) because the medium carries the disturbance away from the source at the same speed in all directions (compare with Figure 14-7). In Figure 15-2*b*, a straight bar (unseen at left) bobs up and down in the water. The disturbance now travels from each point along the straight bar, to produce the straight crests and troughs. The disturbance is carried beyond the barrier only where there is a small opening. Because the water beyond the barrier is also uniform, the disturbance travels in all directions only from the water rising and falling in the opening. The water in the opening is essentially a point source, so beyond the barrier, the disturbance in Figure 15-2*b* looks like the disturbance from a point source in Figure 15-2*a*. If the opening were at a different point along the barrier, the circular waves would spread outward with *that* point as their center. If there were several openings, each one would act as a point source, and the observed disturbance would be the total of these individual disturbances. If there are enough of these openings (Figure 15-2*c*), the total disturbance begins to look as it would if there were no barrier at all. This should not be surprising, because before the barrier the disturbance that we see comes from the infinitely many point sources that make up the length of the oscillating bar.

To see how point sources contribute to a total disturbance, let's look at the water wave pattern produced by two identical small balls bobbing up and down in unison. The two sources oscillate in phase, with the same amplitude and frequency. Figure 15-3*a* shows a stop-action photo of such a pattern. The crests (bright) and troughs (dark) are actually moving outward from the sources. Like a boat riding at anchor, a fleck of sawdust at point A would bob up and down repeatedly as successive crests and troughs passed it.

But there is another feature in this pattern that we didn't see in Figure 15-2. Notice the gray lines fanning outward from the midpoint between the sources. There are no crests and troughs moving outward along these lines. The water along them is virtually still. A fleck of sawdust at point B would not bob up and down. We will refer to these lines as lines of minimal disturbance. They are also called *nodal lines*. It is useful to think of them as radiating outward like the ribs of a folding fan (Figures 15-3*c* and *d*), with the angle between ribs describing the spread; that is, how open or closed the fan is.

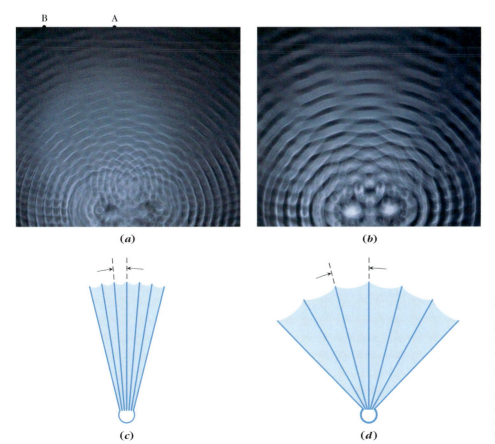

(*a*) (*b*)

(*c*) (*d*)

Figure 15-3 Water wave patterns produced by a pair of point sources. Note the difference between the patterns for (*a*) a shorter wavelength and (*b*) a longer wavelength. In comparing the lines of minimal disturbance in the two situations, picture the lines of minimal disturbance as (*c* and *d*) the ribs of a folding fan as you think about their "spread."

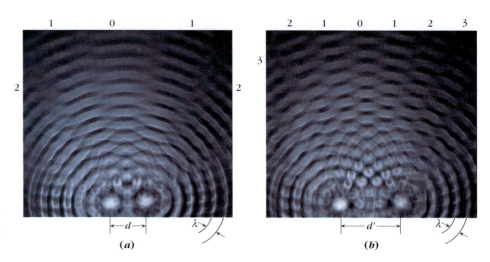

Figure 15-4 The spread of lines of minimal disturbance decreases as source separation increases.

(a)   (b)

By experimentation we find that there are two ways that we can alter the spread of the lines of minimal disturbance:

1. We can reduce the frequency of the two balls bobbing up and down. **STOP&Think** How does this affect the wavelength? ◆ When we do this, the lines of minimal disturbance fan out more.

2. When we move the two sources further apart (from a separation d to a separation d' in Figure 15-4), the lines of minimal disturbance become less spread out.

How can we explain these observations? First recall how superimposing two traveling waves of amplitude Y along a one-dimensional wave could result in a pattern—a standing wave—with nodes and antinodes:

	Contribution y from each wave at any instant	Total displacement at any instant	Total amplitude (maximum displacement)	When there is a crest from one wave . . .
At the nodes	Equal and opposite	0	0	. . . there is a trough from the other.
At the antinodes	Equal and in same direction	$2y$	$2Y$	. . . there is a crest from the other.

• At certain points, there is a crest from one wave whenever there is a trough from the other. At these points the resulting oscillation has minimum amplitude, and we say that there is **destructive interference.** At a node, the crest and trough are of equal amplitude, and there is *total* destructive interference.

• Where crest always meets crest and trough always meets trough, the oscillations have maximum amplitude—the sum of the individual wave amplitudes—and we say that there is **constructive interference.**

This suggests a similar model for the two-dimensional situation, namely, that the pattern in Figures 15-3 and 15-4 is a superposition of the waves (now circular) that the two sources would produce individually. Here, too, there are places of destructive interference that always remain relatively undisturbed, and places of constructive interference where the up-and-down oscillation of water reaches maximum amplitude.

To be valid, our model must agree with the observations. We assume that the waves spread out uniformly in all directions—via an infinite number of outward paths—from identical sources S_1 and S_2. To explain what is happening at a particular point, we must ask what the waves look like along the paths reaching that particular point from the two sources. To do so, we imagine cuts in the waves (shown in red in Figure 15-5) that reveal the contour or profile of the wave along each path. We can then see how each contributes at the point of intersection.

In the following On-the-Spot Activity we will test our model, to see if the superposition of two identical circular traveling waves really gives us the patterns that have been photographed in Figures 15-3 and 15-4.

On-The-Spot Activity 15-1

In this activity, you will investigate what should happen at various representative points on the water's surface if our model is correct.

Step 1: Near the bottom of a large blank sheet of paper, draw two points 10.0 cm apart to represent the two sources. Label them S_1 (left) and S_2 (right), as in Figure 15-6a.

Step 2: Make photocopy enlargements of the two wave profiles in Figure 15-7 (so the numbers are really 5 cm apart) and cut them out. Fold the tabs over so that they stick out perpendicular to the wave profiles (Figure 15-6b). It would be best to copy the profiles onto a transparent plastic sheet, like those used with overhead projectors, so that the pattern can be seen from both sides.

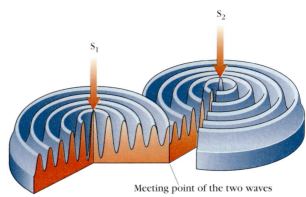

Figure 15-5 To think about what happens at a typical meeting point of two waves, we picture what happens along the paths that pass through that point. Note that the crests and troughs along each path decrease in amplitude as the circular waves spread out from the source, because the energy is distributed over an increasingly large circumference. Therefore, the crest and trough at the meeting point are opposite but not of equal magnitude, and the resulting displacement, though small, is not exactly zero. However, the change in amplitude becomes very gradual beyond the first few crests from each source. If we are careful not to consider regions too close to the two sources, the variation in crest and trough size is negligible.

 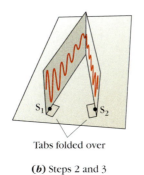

Figure 15-6 Steps of activity.

(**a**) Step 1

(**b**) Steps 2 and 3

(**c**) Designating a point where the two waves meet

Step 3: Pin the profiles in place at "sources" S_1 and S_2. The profiles should stand upright with the printed sides facing each other (Figure 15-6b).

- Notice that when your wave profiles are standing up, they look like the shaded cuts in Figure 15-5. The point where they meet can be represented by the paper clip in Figure 15-6c. The numbered edges of the profiles (the edges with the cm scales along them) tell you how far this "meeting point" is from each source.

- Notice that your wave profiles have a wavelength of 3.0 cm. Each time the distance changes by 3.0 cm, it changes by one wavelength.

Step 4: Your wave profiles can pivot on the pins to allow them to meet at any point on the large sheet of paper. Rotate them until the numbered edges meet at a point that is the same distance from S_1 as from S_2. Place a zero (0) on the sheet of paper at this point. Compare the vertical displacements due to the two individual waves at this point.

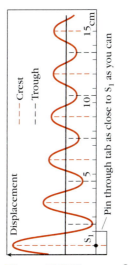 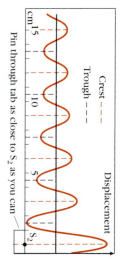

Figure 15-7 Wave profiles.

Step 5: Now see how many other places you can have the numbered edges meet that are equally distant from S_1 and S_2. Put a zero at each such point, then connect the zeroes dot-to-dot with a solid line. Are the two displacements equal at each of these points?

Step 6a: Rotate your wave profiles until the numbered edges meet at a point on the large sheet of paper that is 3.0 cm (one wavelength) further from S_2 than from S_1. Mark this point with a "1." Find several other points that are 3.0 cm further from S_2 than from S_1. Mark each of these with a "1" as well. Connect the 1's dot-to-dot with a solid line.

Step 6b: Repeat Step 6a, but this time for points that are 3.0 cm further from S_1 than from S_2.

Step 7: Repeat Steps 6a and b, but this time for points that are 6.0 cm (*two* wavelengths) further from one source than the other. This time you should be marking each point with a "2" and connecting the 2's dot-to-dot with a pair of solid lines.

Step 8: Do the same steps for points that are 9.0 cm (*three* wavelengths) further from one source than the other. **STOP&Think** All your solid lines should show you places where constructive interference occurs. Why? ◆

Step 9: Repeat Steps 6a and b for points that are 1.5 cm (a half-wavelength) further from one source than the other. Mark each such point with "$\frac{1}{2}$" and connect the points with *dotted* lines. Then do the same for points that are 4.5 cm (three half-wavelengths) further from one source than the other, marking each such point with "$\frac{3}{2}$" and connect the points with dotted lines. How do the displacements due to the two waves compare at each of the points you located in this step? **STOP&Think** What kind of interference is going on at every point along the dotted lines? Why? ◆

Step 10: Look at the pattern of solid and dotted lines appearing on your paper. Compare this pattern to the photographed patterns in Figure 15-3a. Do your lines fan out the way the directions of constructive and destructive interference fan out in the photograph? Do your lines represent directions in which there is either constructive or destructive interference? (*Additional activities using this set-up appear in Problems 15-2 and 15-3.*)

Point by point, you have located positions where the two displacements are either the same (giving constructive interference everywhere along the solid lines) or opposite (giving destructive interference everywhere along the dotted lines), depending on how much further the point is from one source than from the other. Figure 15-8 summarizes some of the relationships you have observed. If disturbances start out at two sources in phase and travel the same distance from

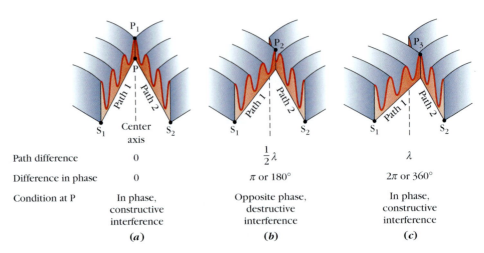

Figure 15-8 Superposition of circular waves: different points (P_1, P_2, P_3), same instant ($t = 0$).

	(a)	(b)	(c)
Path difference	0	$\frac{1}{2}\lambda$	λ
Difference in phase	0	π or 180°	2π or 360°
Condition at P	In phase, constructive interference	Opposite phase, destructive interference	In phase, constructive interference

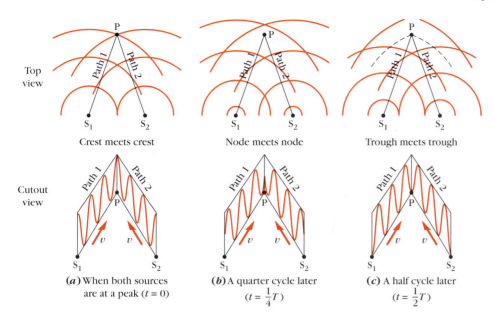

**Top
view**

Crest meets crest   Node meets node   Trough meets trough

**Cutout
view**

(a) When both sources   *(b)* A quarter cycle later   *(c)* A half cycle later
are at a peak (t = 0) $(t = \frac{1}{4}T)$ $(t = \frac{1}{2}T)$

Figure 15-9 Superposition of circular waves: different instants,
same point P.

each, they will be in phase. This is the case at any point along the center axis
(Figure 15-8*a*). The further you swing to the right (or left) of the central axis to
choose a point, the more paths 1 and 2 will differ in length. What happens at
the point depends on this **path difference** (p.d. = |length$_1$ − length$_2$|). For
example, if the wave has traveled $\frac{1}{2}\lambda$ further from one source than from the other,
the phase of the disturbance arriving from the more distant source will have
changed by an additional 180°, and the two disturbances will reach that point
180° out of phase, resulting in *destructive* interference. Further to the right (or
left), where the path difference is a whole wavelength, the phase from the more
distant source will have changed by an additional 360°, so the disturbances will
be back in phase, giving us *constructive* interference.

Figure 15-9 shows that these phase relationships do not change over time if
the sources have the same frequency. As the waves travel outward from the two
sources, the phases of the two individual disturbances at point P change by the
same amount, so the phase *difference* remains unchanged. If both disturbances
are at a peak (no phase difference), then a half period later they will be at a
trough (still no phase difference). For a further look at some of the visual reasoning we've been using, work through WebLink 15-1.

For **WebLink 15-1:**
**Understanding
Wave Interference**
through Pictures I, go to
www.wiley.com/college/touger

Example 15-2 *The Importance of Path Difference*

**For a guided interactive solution, go to Web Example 15-2 at
www.wiley.com/college/touger**

Two sources, identical and in phase, produce water waves with a wavelength
of 0.08 m. Describe the type of interference that is observed at a point on the
water's surface that is
a. 0.96 m from one source and 1.20 m from the other.
b. 0.96 m from one source 1.00 m from the other.
c. 1.00 m from one source 0.98 m from the other.

Brief Solution

Choosing an approach. The two extreme conditions are constructive interference,
when the disturbances at a point are in phase, and destructive interference, when
they are 180° out of phase. To figure out the phase difference, we must (1) find
the path difference, (2) find *n*, the number of wavelengths the path difference
is equal to, and then (3) determine by how much the two disturbances are out
of phase when one has traveled this many wavelengths further than the other.

EXAMPLE 15-2 *continued*

What we know/what we don't. $\lambda = 0.08$ m
a. path 1 = 0.96 m, path 2 = 1.20 m. $n = ?$
b. path 1 = 0.96 m, path 2 = 1.00 m. $n = ?$
c. path 1 = 1.00 m, path 2 = 0.98 m. $n = ?$

The mathematical solution. Because only the absolute value of the difference between the two path lengths matters, we can simply subtract the shorter path from the longer.
a. p.d. = path 2 − path 1 = 1.20 m − 0.96 m = 0.24 m

$$n = \frac{p.d.}{\lambda} = \frac{0.24 \text{ m}}{0.08 \text{ m}} = 3$$

Because the path difference is a whole number (3) of wavelengths, the phases of the two contributing disturbances differ by a whole number times 360°—that is, they are *in phase,* so there is **constructive interference**.
b. p.d. = path 2 − path 1 = 1.00 m − 0.96 m = 0.04 m

$$n = \frac{p.d.}{\lambda} = \frac{0.04 \text{ m}}{0.08 \text{ m}} = \tfrac{1}{2}$$

Because the path difference is a half wavelength, the two contributing disturbances are 180° *out of phase,* so there is **destructive interference**.
c. Now p.d. = path 1 − path 2 = 1.00 m − 0.98 m = 0.02 m

$$n = \frac{p.d.}{\lambda} = \frac{0.02 \text{ m}}{0.08 \text{ m}} = \tfrac{1}{4}$$

Now the two contributing disturbances are $\tfrac{1}{4}$ of a cycle or 90° out of phase, so the amplitude of the resulting disturbance is not zero: It is larger than at points where there is destructive interference, but smaller than where there is constructive interference.

◆ Related homework: Problems 15-13, 15-14, 15-15, and 15-16.

STOP&Think Would you change any of the solutions in Example 15-2 if you found out that the two sources were 0.22 m apart? In that case, how much further (at most) could you be from one source than from the other? (Sketch it if you need to.) Would the path difference found in **a** be possible? ◆

The reasoning we applied in Example 15-2 is summarized in Procedure 15-1.

PROCEDURE 15-1

Finding the Interference Condition at a Point in an Interference Pattern due to Two Identical Sources

1. Determine the length of the path from each source to the point.
2. Subtract the shorter path length from the longer to find the path difference (p.d.).
3. Divide the p.d. by λ to find how many more wavelengths one disturbance has traveled than the other to reach the point.
4. If the sources are in phase, then
   a. there is constructive interference at the point if

$$\frac{p.d.}{\lambda} = 0, 1, 2, 3, \ldots \qquad (15\text{-}1ci)$$

(or $\dfrac{p.d.}{\lambda} = n$, where n represents "zero or any positive integer")

b. there is destructive interference at the point if

$$\frac{\text{p.d.}}{\lambda} = \tfrac{1}{2}, \tfrac{3}{2}, \tfrac{5}{2}, \tfrac{7}{2} \dots \tag{15-1di}$$

(or $\dfrac{\text{p.d.}}{\lambda} = n + \tfrac{1}{2}$. Since $n = 0, 1, 2, 3, \dots, n + \tfrac{1}{2}$ represents

any odd number of halves)

c. For other values of p.d./λ, the amplitude is neither a maximum nor a minimum.

Note: The main points of step 4 may also be stated as in Table 15-2 in the chapter summary.

Example 15-3 *Detecting the Disturbance at Different Points in an Interference Pattern*

For a guided interactive solution, go to Web Example 15-3 at
www.wiley.com/college/touger

Suppose that in the vicinity of one meter from the sources, the amplitudes of the individual waves produced by the sources in Example 15-2 are both fairly constant at 1.20 cm. Let's also suppose that a crest occurs at each of the sources at $t = 0$. Away from the sources, a fleck of sawdust floating at any point on the water's surface "detects" the displacement at that point.

a. Find the displacement of a fleck of sawdust at $t = 0$ if it is located at the position given in Example 15-2**a.**

b. Find the displacement of a fleck of sawdust at this same position after a quarter of a cycle ($t = \tfrac{1}{4}T$) and after half a cycle ($t = \tfrac{1}{2}T$).

Brief Solution

Choice of approach. The rules of superposition of waves apply. At each point we add the *displacements* y_1 and y_2 (*not* amplitudes) from the two waves to get the total displacement y detected by the fleck of sawdust. In each case we must determine how the individual displacements relate to the amplitude Y. This depends on how many wavelengths each point is from the source and how many cycles or periods have gone by.

What we know/what we don't. $Y = 1.2$ cm for each wave

Displacement $= Y$ at $t = 0$ at each source. $\lambda = 0.08$ m

a. path 1 = 0.96 m   path 2 = 1.20 m $t = 0$

$n_1 = ?$ $n_2 = ?$ $y_1 = ?$ $y_2 = ?$ $y = ?$

b. path 1 = 0.96 m   path 2 = 1.20 m $t = \tfrac{1}{4}T, \tfrac{1}{2}T$

$n_1 = ?$ $n_2 = ?$ $y_1 = ?$ $y_2 = ?$ $y = ?$

The mathematical solution.

a. The number of wavelengths the disturbance has traveled by path 1 is

$$n_1 = \frac{\text{path 1}}{\lambda} = \frac{0.96 \text{ m}}{0.08 \text{ m}} = 12$$

and by path 2 is $n_2 = \dfrac{\text{path 1}}{\lambda} = \dfrac{1.20 \text{ m}}{0.08 \text{ m}} = 15$

Because the sawdust fleck is a whole number of wavelengths from each source, the disturbances reaching it are in phase with the disturbances at

EXAMPLE 15-3 *continued*

the two sources, so $y_1 = y_2 = Y$, just like the displacements at the two sources. The total displacement of the sawdust fleck is then

$$y_1 + y_2 = 2Y = 2(1.2 \text{ cm}) = 2.4 \text{ cm}$$

b. Think about a traveling waveform. If a crest arrives at the fleck from each source at $t = 0$, then after a quarter of a period, when each wave has advanced by a quarter of a wavelength, a zero point will arrive from each source (Figure 15-9). After a half period, when each wave has advanced by half a wavelength, a trough will arrive from each source. Thus,

$$\text{at } t = \tfrac{1}{4}T\text{: } y_1 + y_2 = 0 + 0 = \mathbf{0}$$

$$\text{at } t = \tfrac{1}{2}T\text{: } y_1 + y_2 = (-Y) + (-Y) = -2Y = \mathbf{-2.4 \text{ cm}}$$

Making sense of the results. Notice in **b** that the displacement at a point of constructive interference may instantaneously be zero.

◆ Related homework: Problems 15-17 and 15-18.

We have been looking at a wave pattern produced on the water's surface by two identical oscillating sources. The pattern is made up of places of constructive interference, places of destructive interference, and in-between places. Such a pattern is called an **interference pattern.** In it, the lines along which constructive interference and destructive interference occur spread out from the midpoint between the two sources like the ribs of a folding fan. By experimenting, we observe that these lines are

1. more spread out when the sources are closer together, and

2. more spread out when the wavelength is greater.

We need to examine how our model, the superposition of two circular wave patterns, accounts for these two observations. Figure 15-10 shows pictorially how it

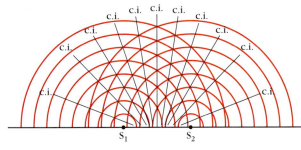

(*a*) Superposition of two circular waves giving rise to lines of constructive and destructive interference. Only the lines of constructive interference (c.i.) are drawn.

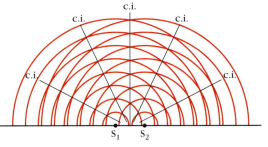

(*b*) When the sources are closer together, the lines of constructive interference are more spread out.

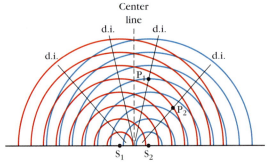

(*c*) Lines of *destructive* interference (d.i.). Here the two circular wave patterns in (*b*) are shown in contrasting colors for easier viewing.

Figure 15-10 Lines of constructive and destructive interference.

accounts for observation 1. Doing the following activity will help you see how it accounts for observation 2.

On-The-Spot **Activity 15-2**

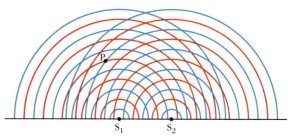

View Figure 15-11 through a red or blue filter. How do the wavelengths compare when viewed with and without the filter? How does the spread of the lines of constructive interference compare in the two cases? *Note:* If you do not have access to a filter, just trace Figure 15-11, but trace only the red crests, not the blue ones.

The filter doubles the wavelength by removing every other crest from view, so that the remaining crests are twice as far apart. With half the crests eliminated, there are fewer points where crest meets crest and fewer directions along which these points occur. So with increased wavelength, the lines of constructive interference are more spread out. Again, our model leads to conclusions that agree with what we observed.

Figure 15-11 The "spread" of lines of constructive interference. See On-the-Spot Activity 15-2 for instructions. (Point P is considered in Problem 15-22.)

To see more completely how changing the wavelength or the distance between the sources affects the interference pattern, work through WebLink 15-2. (We will later develop a mathematical description consistent with the pictures.)

For **WebLink 15-2: Varying an Interference Pattern**, go to
www.wiley.com/college/touger

Example 15-4 *Lines of Destructive Interference*

For a guided interactive solution, go to Web Example 15-4 at www.wiley.com/college/touger

a. In Figure 15-10*b,* sketch the first two lines of *destructive* interference to the right of the center line.
b. Select a point on the first of these two lines and express the path difference for this point as a multiple of the wavelength. Repeat for a point on the second of the two lines.

Brief Solution

Choice of approach. Destructive interference occurs wherever the two signals arriving at a point are 180° out of phase. Of all such points, the easiest to identify are the points where a crest from one source meets a trough from the other. Troughs lie midway between each two crests. To find the path difference for such a point, count crests to find the length in wavelengths of the path from source S_1 to the point, do the same for the path from S_2, then subtract the shorter from the longer path.

The pictorial and mathematical solution.
a. At point P_1 in Figure 15-10*c,* a blue crest is arriving from source S_2, but the point is midway between two red crests, so a trough is arriving from S_1. Because a crest and a trough meet, this is a point of destructive interference. Examine the straight line passing through this point. Wherever it crosses a crest of one color, it is midway between two crests of the other color, so there is destructive interference all along this line.

Point P_2 lies on a *red* crest, but is midway between two *blue* crests. It is likewise a point of destructive interference, at which a crest meets a trough. **STOP**&**Think** Verify that there is destructive interference everywhere along the straight line passing through P_2. ◆
b. Point P_1 is midway between the fifth and sixth red crests away from S_1, so it is $5\frac{1}{2}\lambda$ from S_1. It is on the fifth blue crest away from S_2, so it is 5λ from S_2. The path difference here is therefore

$$\text{p.d.} = 5\tfrac{1}{2}\lambda - 5\lambda = \tfrac{1}{2}\lambda \qquad (\text{at } P_1)$$

EXAMPLE 15-4 *continued*

Point P_2 is midway between the third and fourth blue crests away from S_2, so it is $3\frac{1}{2}\lambda$ from S_2. It is on the fifth (red) crest away from S_1, so it is 3λ from S_1. The path difference here is therefore

$$\text{p.d.} = 5\lambda - 3\tfrac{1}{2}\lambda = \tfrac{3}{2}\lambda \qquad \text{(at } P_2\text{)}$$

Making sense of the results. The paths from the two sources to any point on the center line are of equal length. As you move clockwise or counterclockwise from the center line, one path becomes increasingly longer than the other: The path difference increases. Each time it becomes an odd number of half wavelengths ($\frac{1}{2}\lambda$ then $\frac{3}{2}\lambda$), there is destructive interference.

◆ Related homework: Problems 15-19, 15-21, and 15-22.

15-3 Mathematical Description of the Two-Source Interference Pattern

We will now develop a mathematical description of the interference pattern produced by two identical point sources oscillating in phase. Recall that because it takes a half-wavelength to go from a crest to a trough, there is constructive interference at points where the path difference satisfies p.d./$\lambda = n = 0, 1, 2, 3, \ldots$ and destructive interference where the path difference satisfies p.d./$\lambda = n + \frac{1}{2} = \frac{1}{2}, \frac{3}{2}, \frac{5}{2}, \frac{7}{2}, \ldots$ We now want to ask, "If I pick *any* point, how much further is it from one source than from the other, and what does this path difference depend on?" This is a question about geometry; we can answer it whether the sources are producing waves or not. To do so, we follow the steps in Figures 15-12a–d.

If a point P lies in a direction described by an angle θ (Figure 15-12b), then in the triangle sketched in Figure 15-12d, the source separation d is the hypotenuse, and the path difference is the side opposite the angle β, which is equal to θ. From this triangle, we get the following relationship.

path difference $\qquad \text{p.d.} = d \sin \theta \qquad$ (15-2)

where d is the distance between the two sources

(To see in greater detail how we obtain this relationship, work through WebLink 15-3.) Again, it is useful to think of a folding fan. The angle θ tells us how far from the center line a given rib of the fan is rotated. Because the situation is the same to either side of the center line, we need only think about angles between 0 and 90°. Note that for points far enough from the two sources, the path difference does not depend on how far the point is from the sources, but only on the direction in which the point lies.

STOP&Think The path difference depends on the source separation d. Why should this be? To get a feel for this, pick a point in Figure 15-12a directly to the right of S_2. Label it A. How much further is point A from S_1 than from S_2? Does Equation 15-2

For **WebLink 15-3: Finding the Path Difference**, go to www.wiley.com/college/touger

Figure 15-12 Steps in obtaining Equation 15-2. (*a*) To describe the direction in which a point P is located, we use the angle θ that the direction makes with the center line. (*b*) We draw the paths from the two sources to P. (*c*) If we mark the shorter path off onto the longer one, the path difference is what is left over. (*d*) For small values of θ, the actual path difference is approximately equal to one side of the right triangle marked off here.

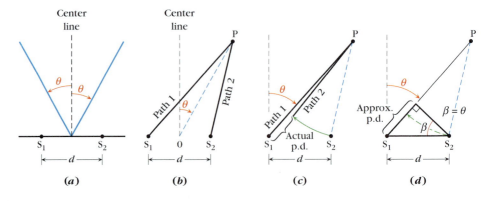

give you this value for the path difference? (What is the value of θ at point A? of $\sin \theta$?) For these two sources, is there any point where the path difference can have a larger value than it does at point A? ◆

Example 15-5 *What's Going On Here?*

Suppose the two sources in Figure 15-12a are 0.042 m apart.
a. If a point lies far from the sources in a direction 30° counterclockwise from the center line, how much further is this point from source S_2 than from source S_1?
b. What kind of interference will take place at this point if the sources are oscillating so that they produce waves with a wavelength of 0.0070 m?
c. Repeat **b** for a wavelength of 0.0140 m.

Solution

Choice of approach. You should understand the definition of path difference well enough to realize that this is what **a** is asking for (use Equation 15-2). Once you have found the path difference, you can answer **b** and **c** by determining whether the path difference equals a whole number of wavelengths (indicating constructive interference), an odd number of half-wavelengths (destructive interference), or something in between at this point.

What we know/what we don't.
a. $d = 0.042$ m $\theta = 30°$ p.d. = ?
b. p.d. = value from **a.** If $\lambda = 0.0070$ m, is p.d. a whole number of wavelengths ... ?
c. p.d. = value from **a.** If $\lambda = 0.0140$ m, is p.d. a whole number of wavelengths ... ?

The mathematical solution.
a. p.d. = $d \sin \theta$ = (0.042 m) $\sin 30°$ = (0.042 m)(0.500) = **0.021 m**.
b. We divide this by λ to find out how many wavelengths are in the path difference.

$$\frac{\text{p.d.}}{\lambda} = \frac{0.021 \text{ m}}{0.0070 \text{ m}} = 3 \quad \text{(a whole number)}$$

Because the path difference is a whole number of wavelengths, there is **constructive interference** at this point.

c.
$$\frac{\text{p.d.}}{\lambda} = \frac{0.021 \text{ m}}{0.0140 \text{ m}} = 1.5 \text{ or } \tfrac{3}{2}$$

Because the path difference is an odd number of half wavelengths, there is **destructive interference** for this wavelength at the given point.

Making sense of the results. The path difference at the given point is the same in all parts because the point and the sources don't change position. But what's happening at the point changes if the wavelength changes.

◆ Related homework: Problems 15-24 and 15-25.

Because the path difference is always $d \sin \theta$, the condition for constructive interference (Equation 15-1ci) can be written as

$$\frac{\text{p.d.}}{\lambda} = \frac{d \sin \theta}{\lambda} = n \quad (n = 0, 1, 2, 3, \ldots)$$

that is, for constructive interference,

$$\sin \theta_n = \frac{n\lambda}{d} \quad (n = 0, 1, 2, 3, \ldots) \tag{15-3}$$

The subscript n indicates that there is a different value of θ for each value of n.

Example 15-6 *Finding the Directions in Which Constructive Interference Occurs*

For a guided interactive solution, go to Web Example 15-6 at
www.wiley.com/college/touger

Suppose that the sources in Figure 15-12a are 0.150 m apart, and produce waves of wavelength 0.0620 m. Find the three smallest angles away from the center line at which constructive interference occurs.

Brief Solution

Choice of approach. This is a seemingly straightforward application of Equation 15-3, but we must be careful in interpreting the results.

What we know/what we don't.

$$n = 1, 2, 3 \qquad \lambda = 0.0620 \text{ m} \qquad d = 0.150 \text{ m}$$

$$\sin \theta_1 = ? \qquad \sin \theta_2 = ? \qquad \sin \theta_3 = ? \qquad \theta_1 = ? \qquad \theta_2 = ? \qquad \theta_3 = ?$$

The mathematical solution. $\sin \theta_n = n\lambda/d$, so

$$\sin \theta_1 = \frac{1\lambda}{d} = \frac{0.0620 \text{ m}}{0.150 \text{ m}} = 0.413 \qquad \theta_1 = \textbf{24.2°}$$

$$\sin \theta_2 = \frac{2\lambda}{d} = \frac{2(0.0620 \text{ m})}{0.150 \text{ m}} = 0.827 \qquad \theta_2 = \textbf{55.8°}$$

$$\sin \theta_3 = \frac{3\lambda}{d} = \frac{3(0.0620 \text{ m})}{0.150 \text{ m}} = 1.24 \qquad \theta_3 = \textbf{???}$$

The sine of an angle can never be greater than one, so there is no possible value of θ_3. **There is no third angle**—constructive interference only occurs in two directions on each side the center line.

Making sense of the results. The path difference increases from zero at points along the center line to a maximum when the point is along the same line as the two sources, that is, when $\theta = 90°$. There the difference between the two path lengths is just the distance d between the sources. Because d is the largest possible path difference, constructive interference can only occur when the path difference is both a whole number of wavelengths ($n\lambda$) and equal to or less than d (in other words, only when $n\lambda \leq d$). Equation 15-3 says the same thing mathematically: There is no angle at which constructive interference occurs when $n\lambda > d$ because no angle has a sine greater than one.

◆ Related homework: Problem 15-26.

Example 15-7 *Determining the Wavelength of Unseen Waves*

Figure 15-13 represents an aerial surveillance photo of a harbor at night. Only the boats lined up at anchor (at right) and two oil pumping stations (S_1 and S_2) are illuminated. The two stations pump in unison, producing waves, but they only pump at night, so in the darkness the waves themselves cannot be seen. While they are pumping, the boats marked with two-headed arrows (↕) bob up and down most vigorously. The boats where there are zeroes (0) report negligible bobbing. Boats anchoring at in-between positions would report experiencing motion that is in between these extremes.

In the aerial view in the figure, 1 cm represents 20 m. After taking the needed measurements directly from the surveillance "photo," calculate the wavelength of the waves produced by the pumping stations. (To aid you in

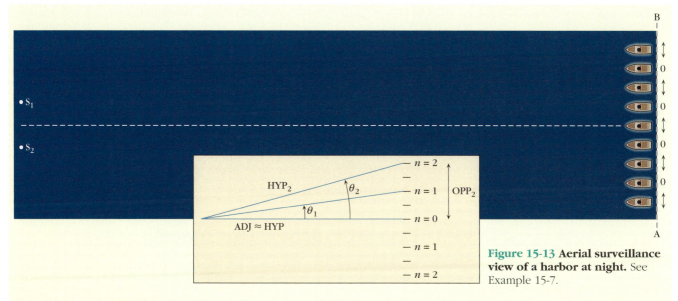

Figure 15-13 **Aerial surveillance view of a harbor at night.** See Example 15-7.

your measurements, dots have been placed on each object to indicate their positions, and a dashed center line has been drawn.)

Solution

Choice of approach. The two pumping stations are point sources of waves, so we expect an interference pattern like those in Figure 15-3. Although the pattern cannot be seen in darkness, it results in positions of constructive and destructive interference, which we can identify by the behavior of the boats. The boats with maximum up-and-down motion are at positions of constructive interference. The inset in Figure 15-13 shows the angular positions of the first two boats away from the center line where constructive interference occurs. We can use either one as the basis for the calculation, but because it is easier to make measurements accurately for the larger angle, we will measure the sides of the triangle necessary to determine the sine of the angle θ_2. We can then use Equation 15-3 with $n = 2$ to find λ. *Note:* Because the adjacent side is very nearly equal to the hypotenuse for small angles, it will be simpler to measure ADJ and use that measurement as the approximate value of HYP_2.

What we know/what we don't.

$$n = 2 \qquad \sin \theta_2 = ? \qquad \lambda = ?$$

From measurements (verify these values by measuring them for yourself):

$$d = \text{distance from } S_1 \text{ to } S_2 = 1.2 \text{ cm} \; \frac{20 \text{ m}}{1 \text{ cm}} = 24 \text{ m}$$

$$\text{OPP}_2 = 2.0 \text{ cm} \times \frac{20 \text{ m}}{1 \text{ cm}} = 40 \text{ m} \qquad \text{HYP} \approx \text{ADJ} = 18 \text{ cm} \times \frac{20 \text{ m}}{1 \text{ cm}} = 360 \text{ m}$$

The mathematical solution.

$$\sin \theta_2 = \frac{\text{OPP}_2}{\text{HYP}} \approx \frac{40 \text{ m}}{360 \text{ m}} = 0.11$$

We can now apply Equation 15-3 (which only requires the value of the sine, not the value of the angle itself). With $n = 2$, the equation becomes $\sin \theta_2 = 2\lambda/d$. Solving for λ then gives us

$$\lambda = \frac{d \sin \theta_2}{2} = \frac{(24 \text{ m})(0.11)}{2} = \textbf{1.3 m}$$

◆ Related homework: Problems 15-27 and 15-28.

In Example 15-7, the waves themselves are not visible, but the row of boats serves as a detector. The boats' motions, implying that they are at positions of constructive and destructive interference, are the only evidence we have that there are waves originating at the two pumping stations. Just as we don't see the waves traveling from the pumping stations to the boats, we do not see the light traveling from a bulb to our eyes (the detectors) if there is nothing in between to "catch the light." (Dust specks and droplets of moisture in the air may reflect light, causing a glow along a light path when the air is damp or dusty, but when the air is clean and dry this does not happen. Think about the sight of electrically lit houses on a distant hillside on a clear, dark night.) But if we could find evidence of constructive and destructive interference, it would imply that there is something wavelike about the way light travels to our eyes.

15-4 Does Light Behave Like Waves?

Suppose we wished to look for evidence of constructive and destructive interference to determine whether light is wavelike. How would we recognize constructive and destructive interference for light if we saw it? And what could we do to produce such an interference pattern?

If light travels like a wave, we must think of it as somehow being a traveling disturbance, even if we don't know what kind of disturbance. Then where there is a disturbance, we see light; no disturbance, no light. We should then expect an interference pattern to appear as a pattern of light and darkness.

We have seen with water waves that two sources can produce an interference pattern. But not *any* two sources: The two sources must be identical. If they are in phase, they must remain in phase. If they begin with a difference in phase, the difference cannot vary. (Disturbances that maintain a constant phase difference are said to be **coherent.**) And even that isn't enough. Equation 15-3 tells us that if $\lambda > d$, the sine will be greater than 1 even if $n = 1$, so there will be no place except on the center line where constructive interference occurs.

On the other hand, what if λ is much smaller than d? Equation 15-3 then tells us that $\sin \theta$ (and therefore θ itself) will be very small for small values of n. If the angles are small enough, the places of constructive interference will be too close together (like a collapsed folding fan) for us to distinguish one from another. We wouldn't know whether constructive and destructive interference were occurring because we couldn't see the evidence. Therefore, although λ must be less than d, it cannot be too much smaller.

To summarize, our criteria for a set-up where we could observe light interference if it happened are

- two identical, coherent sources, and
- a source separation d that is greater than λ, but not too much greater.

The second criterion is guesswork, of course, because we cannot know about wavelengths of light before we have even shown that light is wavelike.

In devising a set-up, we have to think about a further complication. Light doesn't just travel on a two-dimensional surface, as water waves do; it spreads out from a source in all directions in three-dimensional space. But we can produce a simplified three-dimensional picture using a long straight-line wave source. Waves spread uniformly in all directions from the line, so successive crests (Figure 15-14a) will be cylindrical. Looking at these cylinders from on top, the cylinders look like circles and the source looks a point, so that a picture much like the picture of the water wave situation can be used to represent this three-dimensional case.

If we have two line sources, and they are identical, they produce the pattern in Figure 15-14b. The top view (in plane ABEF) is the now-familiar two-source interference pattern in two dimensions. The view through "screen" ABCD is a

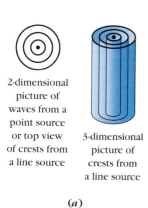

2-dimensional
picture of
waves from a
point source
or top view
of crests from
a line source

3-dimensional
picture of
crests from
a line source

(a)

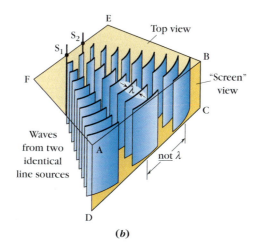

Top view

"Screen"
view

Waves
from two
identical
line sources

not λ

(b)

Figure 15-14 Waves from line sources.

series of stripes (physicists have come to call them **fringes**) where there is alternately a disturbance or no disturbance. If light were a wave-like disturbance, this would mean alternate places of light and no light (darkness). To see the ideas in Figure 15-14 developed in greater detail, go to WebLink 15-4.

The screen aligned along line AB is the detector, where such a pattern might be viewed if it existed, just as the row of boats placed along line AB in Figure 15-13 serves as a detector for the water-wave interference pattern. The screen here could be literally a screen, or it could be the retina of your eye, the screen at the back of your eye where the eye's lens projects the image that you see. What happens on the screen could provide indirect evidence of unseen waves reaching it from the two sources, just as the bobbing of illuminated boats provided evidence of water waves traveling unseen through the darkness in Example 15-7.

For **WebLink 15-4:**
**Understanding
Wave Interference**
through Pictures II, go to
www.wiley.com/college/touger

STOP & Think Many light fixtures in schools and workplaces use two or more long fluorescent tubes. Do we see any patterns of light and darkness produced by these fixtures? Would they be very good light fixtures if we could? Why might these arrangements of light sources be inadequate for observing evidence of interference? How might we improve on them? ◆

When we view a pair of long fluorescent tubes in an overhead light fixture, we see no pattern of light and dark fringes. But we can't rule out the possibility that light is wave-like from this observation. There will only be an observable interference pattern if the two sources are coherent (they maintain a zero or constant phase difference), and their separation d is greater than the wavelength (which we don't know yet) but not too much greater. Maybe the separation d between the two tubes is totally in the wrong ballpark. Worse, there is nothing to assure us that the phase(s) of waves (if they exist) from one tube bears any relationship to the phase(s) of waves from the other. In fact, it turns out that each tube gives off many randomly timed bursts of waves rather than one continuous wave.

15-5 Young's Double-Slit Experiment

Englishman Thomas Young (1773–1829) was the first to grapple successfully with these difficulties. Young surmised that wavelengths of visible light might be extremely small. He was also aware that when a wave reaches a barrier, each tiny opening in the barrier essentially becomes a point source for waves beyond the barrier (recall Figure 15-2). If two points on the original wave are in phase when they reach the two openings, they automatically become in-phase sources for the waves beyond the barrier. Thus, reasoned Young, if you use a *single* line

Figure 15-15 Preparing a double slit with a known slit separation. Painting one side of a glass microscope slide with a graphite-in-alcohol solution leaves a black coating when the alcohol evaporates. Double-edged razor blades are then used to make a pair of parallel slits one razor blade thickness (about 10^{-4} m or a tenth of a millimeter) apart.

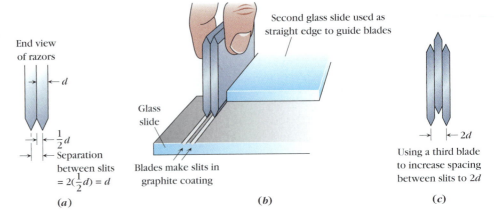

End view of razors

Separation between slits $= 2(\frac{1}{2}d) = d$

Glass slide

Second glass slide used as straight edge to guide blades

Blades make slits in graphite coating

Using a third blade to increase spacing between slits to $2d$

(a) (b) (c)

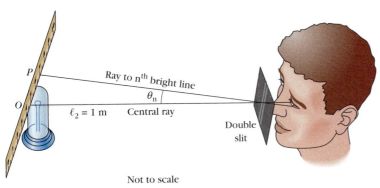

Figure 15-16 **Making a double-slit observation.**

P

Ray to n^{th} bright line

θ_n

O $\ell_2 = 1$ m Central ray

Double slit

Not to scale

Figure 15-17 **Double-slit observations for white, red, and blue light.**

source and allow the light to pass through two parallel slits in an opaque barrier, the two slits will of necessity be identical, coherent sources. Young's double-slit experiment in 1801 was a landmark in the history of physics.

Double slits can be prepared in various ways. Figure 15-15 shows one simple method of producing homemade slits with a known separation. You can then observe a straight filament bulb—a line source of white light—through the double slit (Figure 15-16). By wrapping a red or a blue filter around the bulb, you can turn the bulb into a source of red or blue light as well.

The observations for white, red, and blue light are shown in Figure 15-17. We do indeed see alternating light and dark fringes. This is the "screen" view that Figure 15-14b tells us to expect if light is wave-like. If light were a stream of particles passing through the two slits, there would be the greatest accumulation of particles (the most light) at just two locations centering around the two slits, and it would get progressively darker away from those locations. There would not be multiple areas of light and darkness. These observations can only be explained by a wave model; that is, by assuming that light is a disturbance that varies in space and time as water, sound, and other waves do. But what kind of disturbance? Young's experiment gives us no clue! This remained an unanswered question until the work of Maxwell two-thirds of a century later. The answer emerged from a very different line of inquiry and will be addressed in Chapter 24.

To verify that the light pattern we are observing through the double slit is really a two-source interference pattern, we must see if it has other features that we observed for the two-source pattern. For example, does the pattern become less spread out if the sources—the slits—are farther apart? (When the slits are prepared as in Figure 15-15, part c shows how to double the separation between slits.) When we increase the slit separation, the light and dark fringes do in fact become more closely packed, as we would expect for wave interference.

When we considered two-source water-wave interference patterns, we also found that those patterns spread out when we increased the wavelength. How can we do this for light, if we don't know what governs the wavelength for light? But wait! The red light pattern (Figure 15-17) packs fewer light and dark fringes than the blue into the same space. It is more spread out. This observation makes sense if we assume that red light has a greater wavelength than blue. As we experience sound waves of different wavelength or frequency as having different pitches,

we experience light of different wavelengths as being different colors.

Important: The fringes in Figure 15-17 are the sites of constructive and destructive interference; they are *not* crests and troughs. In Figure 15-14*b* the separation between crests *in the top view* is a wavelength. The separation between fringes of constructive interference in the "screen" view is *not* a wavelength. What is depicted in the top view can never be directly observed for light. We can only determine light wavelengths *indirectly* from measurements we make in the screen view.

◆**DIFFRACTION** In our wave picture of light, we have assumed that at each slit, the light behaves like the water waves in Figure 15-2*b*. It does not just go straight ahead but spreads out to the sides beyond the slit. This spreading means that the edges of the barrier on either side of the slit do not cast sharp shadows: We can trace paths of light that bend around the edges. This bending is called **diffraction.** Because diffraction occurs at each slit, an arrangement of slits for producing a light interference pattern is called a **diffraction grating.**

When we see the light interference pattern, it looks like it is at the same distance as its source, the glowing bulb filament in Figure 15-18. The angle θ_n is between the center line and the *n*th bright fringe away from the center. The figure shows that this angle is the same whether we measure it in front of the slits (where we actually see this fringe) or behind (where the image forms on the retina of your eye). So we can make the measurements needed for calculating $\sin \theta_n$ in the region in front of the slits.

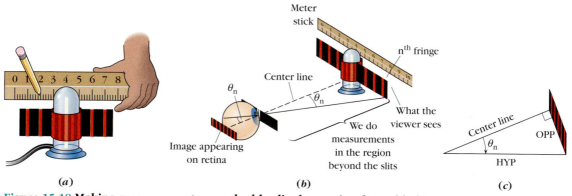

(a)   *(b)*   *(c)*

Figure 15-18 Making measurements on a double-slit observation for red light.

Example 15-8 *"Measuring" the Wavelength of Red Light*

For a guided interactive solution, go to Web Example 15-8 at
www.wiley.com/college/touger

The figure shows two views of a double-slit set-up for red light. The slits are separated by a single-razor-blade thickness, which we will take to be 1.0×10^{-4} m. Using the cm rulers in the figure as your scale, do whatever measurements are necessary on these views to find the wavelength of the light source and then calculate the wavelength.

Brief Solution

Choice of approach. Because the picture is analogous to the two-source water wave situation, the same equation (Equation 15-3) describes it. We pick a bright fringe far enough from the center so that we can easily measure its distance from the center. Then we find its number *n* by counting fringes from the center (0th) bright fringe located at the filament. θ_n gives the angular position of

Enlarged detail

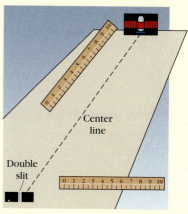

Center line

Double slit

EXAMPLE 15-8 *continued*

this fringe. If we measure two sides of the right triangle containing the angle θ_n (see Figure 15-18c), we can calculate $\sin \theta_n$ as a ratio of sides. We can then use Equation 15-3 to calculate λ.

What we know (or can measure)/what we don't.

We will arbitrarily pick the eighth fringe from center line: $n = 8$.

Distance from double slit to filament (measured along center line) = 0.20 m.

Distance from 0th to 8th bright fringe = 0.010 m (see enlarged detail).

(Verify the above two distances by measuring them yourself in the figure.)

$$d = 1.0 \times 10^{-4} \text{ m} \qquad \sin \theta_n = \sin \theta_8 = ? \qquad \lambda = ?$$

The mathematical solution. In the right triangle containing θ_8, we measure the side opposite θ_8 to be 0.010 m and the adjacent side to be 0.20 m. Then by the Pythagorean theorem, the hypotenuse is $\sqrt{(\text{OPP})^2 + (\text{ADJ})^2} = \sqrt{(0.010 \text{ m})^2 + (0.20 \text{ m})^2} \approx 0.20$ m to two places. In other words, for small θ_n, HYP $\approx$ ADJ, and $\sin \theta_n \approx \tan \theta_n$. So $\sin \theta_8 \approx \frac{0.010 \text{ m}}{0.20 \text{ m}} = 0.050$.

With $n = 8$, Equation 15-3 becomes $\sin \theta_8 = 8\lambda/d$. Solving for the wavelength, we get

$$\lambda = \frac{d \sin \theta_8}{8} = \frac{(1.0 \times 10^{-4} \text{ m})(0.050)}{8} = \textbf{6.3} \times \textbf{10}^{-7} \textbf{ m}$$

Making sense of the results. There are actually a range of wavelengths that our eyes and brain respond to by seeing red. This is a value within this range. It is less than a thousandth of a millimeter, far too small to be seen with the naked eye. This reinforces the point that you are not seeing crests and troughs when you look at a fringe pattern, because you clearly see the separation between successive bright fringes.

◆ Related homework: Problems 15-32 and 15-34.

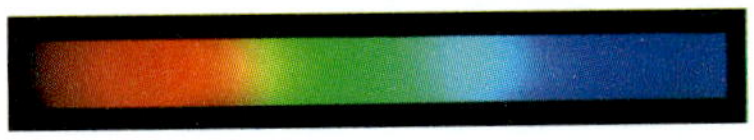

Figure 15-19 The visible spectrum.

◆THE VISIBLE SPECTRUM Measurements like the one in Example 15-8 show that visible light comes in a continuous range, or **spectrum,** of wavelengths (Figure 15-19). The range extends from about 4.0×10^{-7} m, the shortest wavelength of violet light, to 7.5×10^{-7} m, the longest wavelength of red light. Table 15-1 summarizes the colors we see at different wavelengths. This range varies somewhat with the individual and changes with age.

Again compare the red and blue light patterns in Figure 15-17 or Figure 15-20a. The centers of the $n = 1, 2, \ldots$ etc. red and blue fringes are offset from each other, but because the fringes have width, they overlap. Where they overlap, you cannot see the separate colors. So if both red and blue light were passing through the slits at the same time, you couldn't distinguish the separate patterns very well. It would be even worse if we also had in-between wavelengths (colors). Where all the colors overlapped, you would see white.

If we used a double slit with a smaller slit separation (Figure 15-20b), the centers of the fringes would be further apart. But so would the left and right "edges" of any single fringe, so the fringes are widened and nothing is gained. (The "edges" are not really sharp edges; the transition from light to dark occurs gradually.) To see what wavelengths are emitted by a light source, we would like to further offset the centers of the bright fringes from one another, and at the same time narrow each bright fringe, so that they don't overlap (Figure 15-20c). This is accomplished by using more than two slits.

Table 15-1 Wavelengths of Light in a Vacuum or in Air*

Violet	4.0–4.5 $\times 10^{-7}$ m
Blue	4.5–5.0 "
Green	5.0–5.7 "
Yellow	5.7–5.9 "
Orange	5.9–6.1 "
Red	6.1–7.5 "

*Although the wavelengths are very slightly different in air than in a vacuum, the difference does not appear until the fifth significant figure.

◆MULTIPLE SLITS If we have multiple slits in a barrier, each the same tiny distance d from the next, the paths from all the slits to a point on a distant ($\gg d$)

screen will be very nearly parallel (Figure 15-21) and their directions described by the same angle θ. Any two nearest-neighbor paths are like paths from a double slit: The path difference is $d \sin \theta$, and there is constructive interference between the two where this path difference equals $n\lambda$. Constructive interference at a point means that light from both slits reaches the point *in phase*. But if light from the first and second slits arrives in phase, then light from the second and third slits arrives in phase, and so on. In short, light from all the slits will be in phase. So the criterion for constructive interference from multiple, equally spaced slits is Equation 15-3, just as it is for two slits:

$$\sin \theta_n = \frac{n\lambda}{d} \qquad (n = 0, 1, 2, 3, \dots)$$

There is an important difference, however. In the pattern produced by a double slit, the bright fringes are wide because for some distance away from the center of each bright fringe, where constructive interference is total, the phases of the waves arriving by the two paths remain close enough so that the interference between them is still substantially constructive. But with multiple slits, as soon as you move away from *total* constructive interference, slits that are further apart are more out of phase with one another. Then the combined amplitude from all the slits drops away from its maximum value much more rapidly as you move away from the center of the fringe. The resulting bright fringes therefore appear narrower, as Figure 15-20c shows.

Spectroscopes, devices for viewing the spectra from various light sources, often use multiple slits (also called *diffraction gratings*) with such a small slit separation d that θ_n is already quite large when $n = 1$. A typical inexpensive student spectrometer uses a multiple slit with 13 200 lines per inch (Figure 15-22). This is approximately 5.2×10^5 lines/m, so $d \approx \frac{1}{5.2 \times 10^5}$ m = 1.9×10^{-6} m. When we view white light through such a spectroscope, different colors interfere constructively at different values of θ_1 (shown to the left and right of the entry slit in the figure), indicating that *white light is made up of all these colors with a continuous range of wavelengths*. If you use a strip of photographic film as the detector, rather than your eye, we should expect the film to be exposed where we see light and remain unexposed where we see darkness. However, many types of film show exposure well beyond both ends of what we see as the light area. The film detects light, or something lightlike, where our eyes don't. What exposes the film immediately beyond the red is called **infrared;** what exposes the film immediately beyond the violet is called **ultraviolet.** Because we know the angular positions where the film is exposed, we can use Equation 15-3 to find the wavelengths of these waves. Infrared wavelengths are longer than those of red (so the corresponding frequencies are lower: *infra-,* meaning "below," refers to the frequencies) and ultraviolet wavelengths are shorter than those of violet (so the frequencies are higher: *ultra-* means "above"). The ranges of wavelengths that other living organisms can detect may differ somewhat from the range humans can see; for some it may extend into the near parts of the infrared

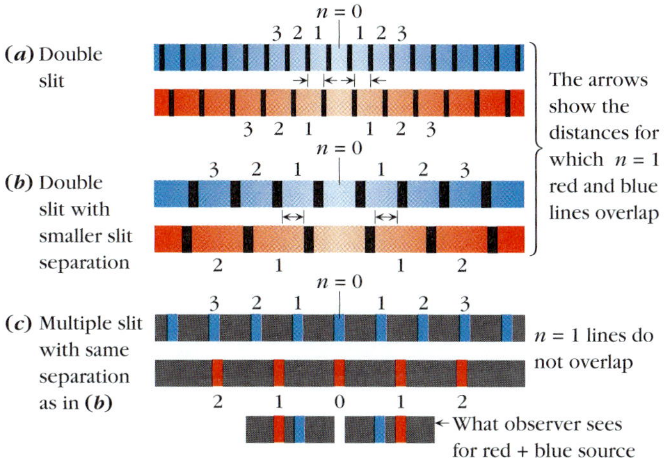

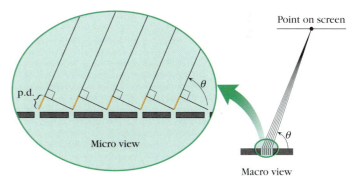

Figure 15-20 Separating fringes of different colors. To eliminate the overlap among fringes of different colors (a), the separation between the centers of bright fringes must be increased by reducing the slit separation (b), and the number of slits must be increased so that the brightness of each fringe will drop off more sharply away from its center (c).

Figure 15-21 Path differences between paths from a multiple-slit diffraction grid to a point on a distant screen.

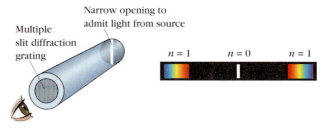

(*a*) A simple spectroscope   (*b*) What you see for white light through the spectroscope

Figure 15-22 A simple spectroscope. (*a*) The spectroscope. (*b*) What you see when viewing white light through the spectroscope.

or ultraviolet regions. However, with slight variation, light-detecting living things are in general most sensitive to the range of wavelengths that are emitted with greatest intensity by the sun.

Example 15-9 *Viewing Laser Light*

A helium-neon laser emits red light at a single wavelength of 6.328×10^{-7} m. Suppose light from such a laser shines through the multiple-slit diffraction grating of a spectroscope onto a screen. (Looking directly through the grating at laser light will damage your eyes.)

a. At what angular displacement from the center line will the first bright fringe from the center appear on the screen?

b. At what linear distance from the center fringe will this first bright fringe appear if the screen is 2.00 m from the multiple slit?

Solution

Choice of approach. In **a**, we apply the criterion for constructive interference (Equation 15-3) to find θ_1. We can then use a trigonometric function of θ_1 (e.g., sin or tan) to relate the known side of a right triangle (recall Figure 15-18), the distance to the screen, to an unknown side, the distance asked for in **b**.

What we know/what we don't.

a. $\lambda = 6.328 \times 10^{-7}$ m $d = 1.9 \times 10^{-6}$ m $n = 1$

$$\sin \theta_n = \sin \theta_1 = ?   \theta_1 = ?$$

b. $\sin \theta_1$ known from solution to **a**   HYP ≈ ADJ = 2.00 m   OPP = ?

The mathematical solution. **a.** By Equation 15-3,

$$\sin \theta_1 = \frac{1\lambda}{d} = \frac{6.328 \times 10^{-7}\ \text{m}}{1.9 \times 10^{-6}\ \text{m}} = 0.33,   \text{so}   \theta_1 = \mathbf{19°}$$

b. $\sin \theta_1 = \frac{\text{OPP}}{\text{HYP}}$, so

$$\text{OPP} = (\text{HYP}) \sin \theta_1 = (2.00\ \text{m})(0.33) = \mathbf{0.66\ m}$$

◆ Related homework: Problems 15-35, 15-37, and 15-38.

Example 15-10 *Reviewing the Double-Slit Pattern for White Light*

In the double-slit pattern for white light (Figure 15-17), we begin to see colors at the edges of the white fringes further from the center. Why?

Solution

Choice of approach. We must use the fact that white light is made up of all visible wavelengths and think about what we know about the individual colors and where they would appear.

The analysis. Figure 15-23 shows where the colors at the two extremes of the visible spectrum (red and violet) would appear if they occurred alone. When they are components of white light, they retain these same wavelengths, but your eye–brain circuitry is designed to interpret the combination of all the colors as white. When we view white light, the red and violet patterns (as well as the patterns for all the in-between wavelengths) occur at once. We see that there are places where the colors overlap and places where they don't. Where they overlap (the in-between colors will also overlap where the two extremes do), our eyes will see white. Where they don't overlap, our eyes see only the

This...

...superimposed on this...

...gives this.

Figure 15-23 Explaining the double-slit pattern observed for white light. See Example 15-10 for explanation. Only the colors at the two ends of the visible spectrum are shown here, but the in-between colors are assumed to be present contributing to the resulting white light.

colors that are present. As we move away from the center of the pattern, the colors become more offset from each other. The regions where there is no overlap grow, and it is there that we see the colors.

◆ Related homework: Problem 15-40.

15-6 Other Instances of Diffraction

If we have an extended light source, light, like all waves, spreads out from each point on it as though it were a point source (recall Figure 15-2b). The resulting wave is a superposition of the waves from all these point sources. The method of obtaining the resulting wave geometrically in this way is called **Huygens' construction.** As we have seen for the diffraction grating, this superposition can produce an interference pattern.

The diffraction grating is a peculiar case of an extended source because it is interrupted by the barriers between the slits. If we remove the barriers, we end up with a single wider slit. But there are still phase differences between the light arriving at a point from different locations across the width of this slit. Although the mathematical details of superposition are more difficult in this case, we still get an interference pattern (Figures 15-24a and b).

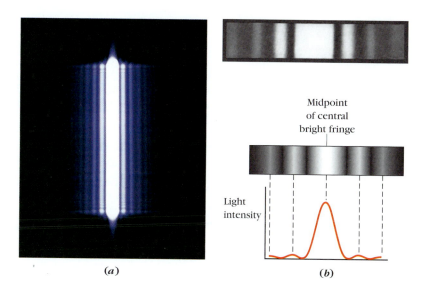

Figure 15-24 A single-slit diffraction pattern.

Figure 15-25 shows how this pattern changes as the slit widens: The interference effects are observable in an increasingly narrow region, and it looks more and more like the borders of the slit are casting a clear-cut shadow. Now imagine extending the right edge of the slit out to infinity, so that the left edge is just

Figure 15-25 How the single-slit diffraction pattern for light of wavelength λ varies with the width of the slit. The graphs show how the relative intensity (the fraction of maximum intensity) varies with θ for three different values of the slit width a: the wider the slit, the narrower the central maximum.

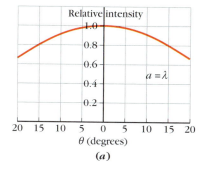

(a)

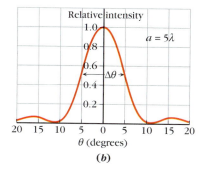

(b)

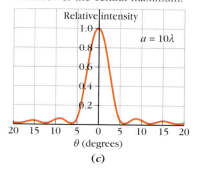

(c)

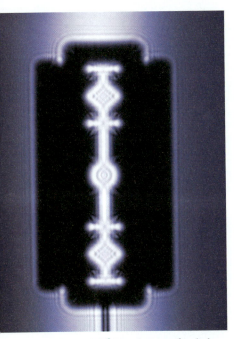

Figure 15-26 Maxima and minima produced by diffraction of monochromatic light at the edges of a razor blade.

the straight edge of an opaque object, and light passes freely to its right. But near this edge, the wave behavior is still similar to the wave behavior near the edges of the slit, so we see interference at the edges of a backlit opaque object, as Figure 15-26 shows.

On-The-Spot Activity 15-3

Hold up a stiff piece of paper or card with writing on it so that it is backlit by a very bright light. You should see one or two thin lines close to but beyond the backlit edge. In case you think this is because your hand is unsteady, notice that you do not see any doubling of the writing on the page. The edge effect is an interference effect due to diffraction.

Diffraction effects can result from reflection as well as transmission. A row of equally spaced posts in water will be point sources of reflected waves (Figure 15-27), just as slits act as point sources of transmitted waves. The interference patterns produced in Figures 15-27a and b will look the same.

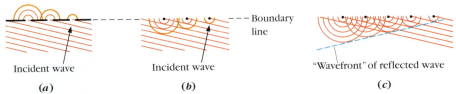

(a)   (b)   (c)

Figure 15-27 Two analogous situations. Just as (a) narrow slits act as point sources for transmitted waves, (b) posts act as point sources for reflected waves. If the reflecting points are more closely packed (c), the lead crests make up a "wavefront" of the resultant reflected wave. Note that the angle of the reflected wavefront equals the angle of the incident wavefront.

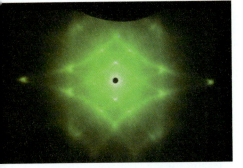

Figure 15-28 Photo of the X-ray diffraction pattern produced by a crystal.

Like for the slits, the spacing between the point reflectors must be smaller, but not too much smaller, than the wavelength to produce an observable interference pattern. This principle was used by Max von Laue in 1912 to establish the wave nature of X rays. The reflectors used were the regularly spaced atoms in a crystal. (Both the wavelength and the spacing were in the ballpark of 10^{-10} m, much tinier than the wavelengths of visible light.) A photographic plate exposed to the reflected X rays revealed a distinctive interference pattern (Figure 15-28).

In Figure 15-27, if the wavelength is known, the spacing between posts or even between rows of posts can be determined. Similarly, with an X ray of known wavelength, the resulting diffraction pattern makes it possible to calculate spacings between atoms or planes of atoms. This type of analysis is called **X-ray crystallography** and is used by chemists to determine the atomic structure of crystals and even of complex molecules with repeating structures, such as DNA.

15-7   Interference due to Reflection: Thin Films

In the double- and multiple-slit interference patterns, light from a single source travels by more than one path to any given point in the pattern. The *phase difference* between the waves reaching a point by different paths results from the difference in path length—the path difference—and dictates whether the interference at that point will be constructive, destructive, or something in between. Therefore any method of splitting a beam of light so that it travels to a point by more than one path can produce an interference condition at the point.

Light from a single source is identical at all points on a wavefront. In the multiple-slit arrangements, the slits "pick out" different points on the wavefront as secondary sources from which the light travels to each point of observation. This picking out can be accomplished by reflecting surfaces as well as by slits (Figure 15-29).

Splitting can also be accomplished by a surface that reflects part of the light and permits the rest to pass through. Such splitting occurs, for example, at every pane of ordinary window glass. Stand in front of a shop window. You can see your own reflection, telling you some of the light is reflected, but the people in the shop can also see you, telling you that some of the light is transmitted, or passes through. Devices that split light beams in this way are called **amplitude-splitting interferometers.** In general, interferometers are so-called because we can make use of the interference patterns they produce to measure wavelengths, as we did in Example 15-8.

Two arrangements that produce interference patterns by means of reflection are shown in Figure 15-30. In each, the half-silvered mirrors (sometimes used in one-way mirrors—see Problem 15-65) transmit about half the light and reflect the rest. Figure 15-30a shows that light from two identical points P_1 and P_2 on a wavefront can reach a single observation point O from parallel reflecting surfaces. The light from P_1 reaches O via a longer path. Whether or not the path difference is a whole number of wavelengths determines the interference condition at O.

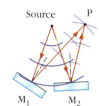

Figure 15-29 Two-source interference produced by two mirrors (M_1 and M_2) rather than two slits.

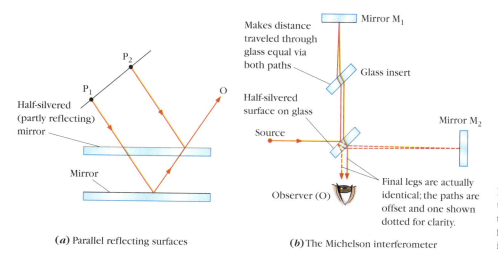

(*a*) Parallel reflecting surfaces

(*b*) The Michelson interferometer

Figure 15-30 Two arrangements that produce interference patterns by reflection. (*a*) Parallel reflecting surfaces. (*b*) The Michelson interferometer.

A more sophisticated arrangement (Figure 15-30b) allows both the reflected and transmitted light from a half-silvered surface to reach the same observation point O. This device, called the **Michelson interferometer,** is named for its designer, American physicist A. A. Michelson (1852–1931).

Interference produced by parallel or nearly parallel reflecting surfaces is responsible for a variety of interesting effects that we observe in nature (Figure 15-31). The reflecting surfaces may be those of an oil slick and the underlying pavement

(*a*)

(*b*)

Figure 15-31 Examples of interference patterns due to reflection. (*a*) Light reflects from a soapy water film across an upright loop. The topmost part of the film is dark because it is too thin to produce constructive interference. The bands of color across the lower part are interference fringes. The irregularities indicate liquid is moving within the film (predominantly in what direction?). (*b*) The iridescent feathers of this broad-billed hummingbird have no pigment; the colors result from interference of reflected light from the layered structure of the feathers.

or water, the inner and outer surfaces of a soap bubble, the tiny scales in a butterfly's wing, or the microstructure of the feathers in a variety of birds. In each case, we can consider the interference to be between light reflecting from the nearer and further surfaces of a thin transparent layer or film (or of several such layers), whether of soap, oil, or simply air.

In the more controlled environment of optical technology, **thin films** are used to coat lenses, mirrors, or other optical elements when there is a need to regulate which wavelengths are reflected and which are transmitted. For example, we don't want reflections of visible light from "nonreflecting" glass in picture frames, and we don't want infrared wavelengths to reflect back from movie or slide projector lenses and damage the film.

In the most visually dramatic instances, we see a pattern of iridescent color. Understanding why the pattern occurs requires some analysis. The kind of interference that occurs is the result of (1) a path difference occurring when the light travels by more than one path and (2) that path difference being equal to some number of wavelengths.

We must therefore ask:

• On what does the path difference depend in the case of thin films?

• How is wavelength affected by the medium making up the thin film?

We will address the second of these questions first. In media other than a vacuum, the speed of light varies somewhat with its frequency (we treat this in Chapter 16), but for now we will assume this variation is small enough for us to neglect. Remember that if a medium carries the wave more slowly, the wave goes less far before another crest is produced, and therefore the wavelength is smaller. If the speed of light is $c = f\lambda$ in a vacuum and is $v = f\lambda'$ in a "slower" medium, then dividing the second equation by the first gives us

$$\frac{v}{c} = \frac{\lambda'}{\lambda} \qquad \text{or} \qquad \lambda' = \lambda\left(\frac{v}{c}\right) \qquad (15\text{-}4)$$

If the medium is not a vacuum or air, we must use this smaller wavelength λ'.

To address path difference, we observe that the path to the lower surface and back is longer by the amount highlighted in yellow in Figure 15-32. When light that arrives nearly normal (perpendicular) to both surfaces is reflected almost straight back (Figure 15-32a), the path difference is well approximated by

$$\text{p.d.} = 2d \qquad (15\text{-}5)$$

where d is the distance between the two surfaces. If the light arrives at a substantial angle to the normal direction, its path bends as it passes from one medium to another (Figure 15-32b) for reasons we will discuss in the next chapter. *The path difference varies with the incident angle of the light.* For simplicity, we will restrict our quantitative treatment to the nearly normal case.

As in the double-slit situation, each whole wavelength of path difference results in an additional 360° of phase difference, which returns us to the same phase. But when reflection is involved, the conditions of reflection can result in a further phase difference. Figure 15-33a shows how this comes about. When a pulse travels from one medium to another, part of the disturbance gets reflected at the boundary. Suppose the media are a "light" string and a "heavy" rope (strictly speaking, we are referring to linear mass density here). Then when an upward pulse traveling in the "lighter" medium reaches the "heavier" medium, it is snapped back downward, as though the upward-moving string segment at the

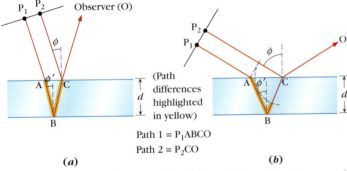

Path 1 = P₁ABCO
Path 2 = P₂CO

(a)

(b)

Figure 15-32 Path differences for light reflected from two parallel surfaces. The light is in phase at P_1 and P_2, two points on the same wavefront. In (a) the light is almost normal to the surfaces; the angles ϕ and ϕ' are tiny ($\phi, \phi' \approx 0$) but are exaggerated in the diagram for readability. In (b) the light strikes the surfaces at a substantial angle to the normal.

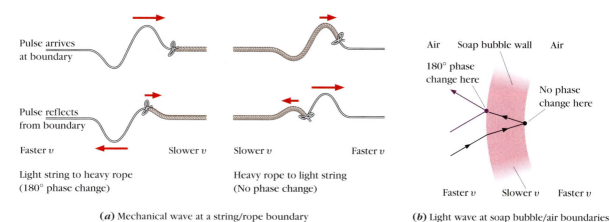

Figure 15-33 Phase change at a reflecting surface. (*a*) Mechanical waves at a string/rope boundary. (*b*) A light wave at soap bubble/air boundaries.

boundary had run into a hard-to-move obstacle. This results in a 180° phase change (a peak to a trough) at the boundary. This phase change does *not* occur when an upward pulse traveling in the "heavier" medium reaches the "lighter" medium. In this case the "heavier" rope segment at the boundary tends to carry the adjacent "lighter" string segment upward with it. We can think of the moving boundary as a point source from which disturbances travel to the right (the part of the pulse transmitted along the string) and to the left (the part reflected back along the rope).

Waves in one dimension travel faster along media having smaller linear mass densities. Put in terms of wave speed, the pulse undergoes a 180° phase change when reflected off a medium in which it would travel more slowly, but undergoes no phase change when reflected off a "faster" medium. This holds true for all waves, whether in one, two, or three dimensions, including light. Light travels faster in a vacuum or in air than in other media, a point we treat more fully in the next chapter. Thus, the light that reflects off the outer surface of a soap bubble wall (Figure 15-33*b*) undergoes a 180° phase change, but the part of the light that continues through the wall and then reflects off the *inner* surface does not change phase.

Because of the additional 180° phase change, a path difference of $n\lambda'$, which would ordinarily have resulted in the two reflected waves arriving in phase, now results in their arriving 180° out of phase, and we get destructive instead of constructive interference. (The additional path length is within the medium, so we have to use the wavelength λ' in that medium.) On the other hand, a path difference of $(n + \frac{1}{2})\lambda'$, which would have resulted in the waves arriving 180° out of phase, now results in their arriving in phase, and we get constructive instead of destructive interference. In short, *with a reflection off a "slower" medium at one of the two surfaces, the path difference requirements for constructive and destructive interference are reversed*. However, if light reflects off a "slower" medium at *both* surfaces, as when light traveling in air reflects off a wet pane of glass $(v_{glass} < v_{water} < v_{air})$, there are two 180° phase changes, and we are back to the usual path difference requirements for constructive and destructive interference.

Example 15-11 *A Slick Calculation*

For a guided interactive solution, go to Web Example 15-11 at www.wiley.com/college/touger

Light travels more slowly in water than in air, and slower still in natural oils, such as corn oil or olive oil. In water, $v = 0.75c$. In these oils, $v_{oil} = 0.68c$. A careless picnicker has spilled some oil on the surface of a pond. Where the

EXAMPLE 15-11 *continued*

oil slick is 9.8×10^{-8} m thick, what color will be seen when the sun is almost directly overhead? (Problems 15-48 and 15-49 treat the more usual petroleum-based oil slick.)

Brief Solution

Choice of approach. Color is determined by wavelength (Table 15-1). We will see the color corresponding to the wavelength for which there is constructive interference. In this example, light is going from a "faster" to a "slower" medium at the air-to-oil surface of the oil slick, but from a "slower" to a "faster" medium at the oil-to-water surface. Then because there is a 180° change at one surface but not the other, the path difference must be $(n + \frac{1}{2})\lambda'$ for *constructive* interference. (We must use λ', the wavelength in oil, because that is where the extra distance is traveled.) We can set this equal to the path difference $2d$, and solve for the wavelength λ'. We must find the equivalent wavelength in air to use Table 15-1. To do so, we must recall that $v/c = \lambda'/\lambda$.

What we know/what we don't.

$$v_{oil} = 0.68c$$

$$n = 0, 1, 2, \ldots \qquad d = 9.9 \times 10^{-8}\,\text{m} \qquad \lambda' = ? \qquad \lambda = ?$$

The mathematical solution. Equating the two expressions for p.d. again gives us

$$2d = (n + \tfrac{1}{2})\lambda'$$

Considered only from the point of view of the mathematics, this leads to a different value of λ' for each value of n. We'll start with $n = 0$ and see what happens. Solving for λ' when $n = 0$ gives us

$$\lambda' = \frac{2d}{n + \frac{1}{2}} = \frac{2(9.9 \times 10^{-8}\,\text{m})}{0 + \frac{1}{2}} = 3.96 \times 10^{-7}\,\text{m}$$

This is the wavelength in oil. But $\frac{v}{c} = \frac{\lambda'}{\lambda}$, the equivalent wavelength in air is

$$\lambda = \frac{\lambda'}{\dfrac{v}{c}} = \frac{3.96 \times 10^{-7}\,\text{m}}{0.68} = \mathbf{5.8 \times 10^{-7}\,m}$$

Table 15-1 tells us this is a wavelength in the **yellow** range. When $n = 1$, the denominator is $1 + \frac{1}{2} = \frac{3}{2}$ rather than $\frac{1}{2}$, giving us a value of λ one-third as large as when $n = 0$. But as Table 15-1 shows, a wavelength this small is not in the visible range, and values of λ for $n > 1$ will be smaller still, so no other colors will be *seen*.

Making sense of the results. **STOP&Think** Why do you usually see a number of colors in an oil slick rather than just one? If the sun directly overhead were the sole point source of light, would you see any reflection when you stood off to one side of the oil slick? (Think of the slick as being like a mirror.) ◆ In fact, because Earth has an atmosphere, light from the sun reaches us indirectly from all parts of the sky and reflects at all angles. The light reaching your eye from different points on the oil slick surface has traveled at different angles, and thus for different distances, through the oil, so the path differences are different for light reaching you from these different points.

◆ Related homework: Problem 15-46, 15-47, 15-48, and 15-49.

It is also important to realize that when light reaches a thin film on the surface of a transparent medium, what is not reflected is transmitted. If there is total constructive interference for a particular wavelength of reflected light, none of that wavelength is transmitted. But if there is total destructive interference for that reflected light, all of it is transmitted. The wavelengths that are not reflected by the oil slick continue through it into the water; the wavelengths that are not reflected by a lens coating continue through the lens.

✦ SUMMARY ✦

The experiments of Roemer and others established that **light travels at a finite speed**

$$c = 3.00 \times 10^8 \text{ m/s} \qquad (15\text{-}1)$$

(or more precisely $c = 2.997925 \times 10^8$ m/s) **in a vacuum** (the speed of light in other media is less, but in air is only very slightly less).

Before investigating whether light travel is wavelike or particlelike, we needed a fuller understanding of how waves behave, so we looked at water wave **interference patterns** produced by two point sources (Figure 15-3). In these patterns, there is **constructive interference** at some points on the water and **destructive interference** in others, depending on whether waves from the two sources reach the point **in phase**, 180° **out of phase**, or in between. If the sources are identical and in phase, the phase difference between waves arriving at a particular point depends on the path difference

path difference   p.d. = length of longer path
$\qquad\qquad\qquad\qquad$ − length of shorter path

because a wave returns to the same phase after each whole wavelength λ. **Procedure 15-1** summarizes how you use these ideas to find the interference conditions at a given point. Part of the procedure is restated in Table 15-2.

Table 15-2 Interference Conditions between Waves from Two Identical (in phase) Point Sources

At points where the path difference (p.d.) is . . .	the phase difference due to the p.d. is . . .	and the type of interference that occurs at this point is . . .
$n\lambda$ ($n = 0, 1, 2, \ldots$)	$n(360°)$ (equiv. to 0)	constructive
$(n + \frac{1}{2})\lambda$	$(n + \frac{1}{2})(360°)$	destructive
in between	in between	in between

The path difference at any point an angle θ away from the center line of the pattern is

$$\text{p.d.} = d \sin \theta \qquad (15\text{-}2)$$

(d = separation distance between the two sources)

We observe that the interference pattern spreads out (like the ribs of a folding fan) when either (1) the wavelength λ is increased or (2) d is decreased. This is because the path difference $d \sin \theta$ must equal a whole number of wavelengths ($n\lambda$) to produce constructive interference along directions described by angles θ_n (between 0 and 90°) such that

(for constructive interference) $\quad \sin \theta_n = \dfrac{n\lambda}{d} \quad (n = 0, 1, 2, 3, \ldots)$ (15-3)

Thus, for each value of n, θ_n increases if either λ increases or d decreases. No interference pattern is observed if $\lambda > d$ (no

angle has a sine >1) or if $\lambda \ll d$ (the directions of constructive interference cannot be seen as separate if they are too close together).

If a row of boats were spread across a water interference pattern, individual boats would bob up and down or not, depending on whether they were in a direction of constructive or destructive interference. Similarly, if light were wavelike, individual points on a screen spread out across a light interference pattern would be lit (if constructive interference) or not (if destructive interference). **Young's double-slit experiment** met the necessary conditions to observe light interference if it existed, and the observed pattern of light and dark stripes, or **fringes,** provides convincing *evidence for the wavelike nature of light.*

Because the spread of the pattern depends on wavelength, the fact that the patterns for different colors have unequal spreads is evidence that

the color we see depends on the wavelength(s) of the light.

Equation 15-3 describes the positions of constructive interference both for double slits and for **multiple slits (diffraction gratings)**, but the bright fringes become narrower as the number of slits increases. **Spectroscopes** make use of diffraction gratings with very small slit separations to produce large θ_1's for the various colors (wavelengths) from a light source and thus separate them for identification. We observe that white light is made up of a continuous range of all visible wavelengths. Wavelengths just below (**ultraviolet**) or just above (**infrared**) those in the visible spectrum will expose photographic film, providing evidence of their existence.

Interference patterns also occur when light reaches points by more than one path because of reflection. **Amplitude-splitting interferometers,** such as the **Michelson interferometer** (Figure 15-30b), use a half-silvered mirror to split light into reflected and transmitted beams traveling by different paths. Many interference phenomena in nature (Figure 15-31) come about when light is reflected off two or more parallel surfaces:

For a **thin film** (with incident light nearly normal to the film), the path difference is just

$$\text{p.d.} = 2d \qquad (15\text{-}5)$$

d = the distance separating the two surfaces of the film

Because further 180° phase changes occur at surfaces where light traveling in a "faster" medium reflects off a "slower" medium, the condition for constructive interference will be p.d. = $(n + \frac{1}{2})\lambda'$ instead of p.d. = $(n\lambda)$ if light reflects off a "slower" medium at one but not both of the two surfaces. Here λ' is the wavelength in the medium making up the thin film, and

$$\frac{v}{c} = \frac{\lambda'}{\lambda} \qquad \text{or} \qquad \lambda' = \lambda\left(\frac{v}{c}\right) \qquad (15\text{-}4)$$

◆QUALITATIVE AND QUANTITATIVE PROBLEMS◆
Hands-On Activities and Discussion Questions

The questions and activities in this group are particularly suitable for in-class use.

15-1. *Discussion Question.* What evidence can you offer from everyday experience to support the assumption that light travels?

15-2. *Hands-On Activity.* (Use the materials in the set-up for On-the-Spot Activity 15-1. See Figures 15-6 and 15-7.)

a. On a sheet of paper, draw sources S_1 and S_2 18.0 cm apart instead of 10 cm apart. Now repeat Steps 4–9 of On-the-Spot Activity 15-1 to obtain the first several lines of constructive interference.

b. How does the spread of these lines compare with the spread you obtained when the separation between sources was 10.0 cm?

c. How does the angle θ that describes the spread change as the source separation increases?

15-3. *Hands-On Activity.* (Use the materials in the set-up for On-the-Spot Activity 15-1. See Figures 15-6 and 15-7.)

a. On a sheet of paper, draw sources S_1 and S_2 10.0 cm apart, as in the original activity. Now redraw the two cut-out wave profiles in Figure 15-7 so that the crests are only half as far apart (keep a crest at each source). Repeat Steps 4–9

of On-the-Spot Activity 15-1 using the redrawn wave profiles to obtain the first several lines of constructive interference.

b. How does the spread of these lines compare with the spread you obtained when the separation between sources was 10.0 cm but the crests were twice as far apart?

c. How does the angle θ that describes the spread vary with wavelength?

15-4. *Hands-On Activity.* Look at a distant street lamp through a window screen, or through the tight weave of a sheer curtain or of an umbrella (one made of fabric, not a plastic one). Describe the pattern that you observe, and explain why it occurs.

15-5. *Discussion Question.*

a. We can think of the wall of a soap bubble as a thin film between two regions of air. Figure 15-31*a* shows the reflection from a flat bubble stretched on a wire rim (like the kind children blow bubbles with) when white light strikes it head on. Notice that the colors change from top to bottom. What does this tell you about the bubble wall? What is different about it at the bottom than at the top?

b. If the photo had been taken several seconds later, the unstriped region would extend further down. What would cause this?

Review and Practice

Section 15-1 Does Light Travel? If So, How? And How Fast?

15-6. In 1849, a French physicist named Armand H. L. Fizeau performed an experiment in which a short pulse of light was directed to a distant mirror. The pulse was aimed between evenly spaced teeth on the rim of a wheel and reflected back between the teeth to the observer. If the toothed wheel was rotated, the reflected pulse might or might not be seen by the observer, depending on the angular speed at which it rotated. Why does this fact provide evidence that the speed of light is finite?

•**15-7.** The first Earthbound measurement of the speed of light was performed outside of Paris by Fizeau in 1849. The key elements in his apparatus were a toothed wheel, on which the 720 teeth and the spaces between were of equal width, and a mirror 8663 m (about 5 miles) away. A pulse of light could pass between two teeth of the wheel, reach the mirror, and be reflected back between the two teeth to an observer (lenses were used to keep the rays directed). Fizeau found that the minimum angular speed at which the toothed wheel prevented him from seeing the reflected pulse was 12.6 rev/s. Find the speed of light indicated by this finding and the information about Fizeau's apparatus. (How does it compare with the currently accepted value?)

15-8.

a. Earth's mean distance from the sun is about 1.5×10^{11} m. How long does it take light to reach Earth from the sun?

b. Based on the data available in his own time, Roemer determined that it took light "about 22 minutes" to travel the length of a diameter of Earth's orbit around the sun. By what percent does the result from calculating the speed of light using this figure differ from the currently accepted value of the speed of light?

15-9. In Section 15-1, we found that when Earth is near its furthest point from Jupiter, each eclipse of Io, which orbits Jupiter with a period of 1.769 Earth days, is seen by Earth observers to occur about 1000 s later than predicted. Ganymede, another of Jupiter's large moons, has a period of 7.155 Earth days. If we observed the eclipses of Ganymede instead, the lag between predicted and observed times would be (*less than, equal to, greater than*) 1000 s.

15-10. Rigel, an easily seen blue-white star in the Orion constellation, is about 5.1×10^{18} m from Earth. If you were to look at Rigel tonight, in what year would the light reaching your eye have left Rigel? How many light-years from us is Rigel?

Section 15-2 Waves in Two Dimensions

15-11. **SSM** Explain in your own words why when a wave reaches a barrier with a very tiny opening, the wave beyond the opening looks like a wave from a point source. (To answer this non-trivially, you must talk about the mechanism by which a disturbance gets passed or carried along from point to point.)

15-12. Consider the crests approaching the barrier from the left in Figure 15-2*b*. Suppose that instead of being straight,

---

SSM Solution is in the Student Solutions Manual   **WWW** Solution is at http://www.wiley.com/college/touger

these crests were circular like those in Figure 15-2*a*. What would the wave pattern beyond the slit look like? Why?

15-13. Suppose there are two identical point sources of waves, S_1 and S_2. In the following steps, you will need to think about the path difference at various locations for waves arriving from these two sources.

a. What is meant by path difference, and why is path difference important?

b. On a piece of paper, draw two points separated by some distance and label them S_1 and S_2 to represent the two sources. In your drawing, show the direction in which the path difference is least. What is the value of the path difference in this direction?

c. Now show the direction(s) in which the path difference is greatest. Making any necessary measurement on your drawing, find the value of this path difference.

d. Now draw a point P at a distance from the two sources in a direction somewhere between the directions considered in *b* and *c*. Making whatever measurements are necessary, find the path difference for waves arriving at P.

15-14. As a horizontal rod suspended by rubber bands bobs up and down, two identical balls projecting below the rod disturb the surface of a tank of water and create waves of wavelength 3.0 cm (Figure 15-34). What kind of interference (total constructive, total destructive, or something in between) occurs at a point on the water's surface that is

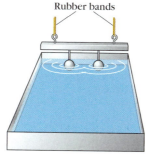
Rubber bands

Figure 15-34 Problem 15-14

a. 48.0 cm from one ball and 54.0 cm from the other?

b. 52.4 cm from one ball and 47.9 cm from the other?

c. equidistant from both balls?

15-15. Two identical sources produce waves of wavelength 12 cm on the surface of a swimming pool. A small rubber duck floating in the pool will be undisturbed by the waves (or very nearly so) if it is ____ from one source and ____ from the other.

a. 36 cm . . . 36 cm;   *c.* 45 cm . . . 63 cm;

b. 42 cm . . . 42 cm;   *d.* 48 cm . . . 72 cm

15-16.

a. In Figure 15-10*a,* find the furthest point to the left at which a crest from source S_1 intersects with a crest from S_2. Mark it with an A. How many wavelengths away from S_1 is point A? How many wavelengths away from S_2 is it? How many wavelengths is the path difference for point A?

b. From point A, go one crest closer to S_1 and find the furthest point to the left where *this* crest is intersected by a crest from S_2. Mark this point with a B. How many wavelengths away from S_1 is point B? How many wavelengths away from S_2 is it? How many wavelengths is the path difference for point B?

c. Identify any other points where two crests intersect and where the path difference is the same as it is for point A. What can you say about the location of all of these points? What can you say about interference at these points?

d. Consider the point midway between A and B. What kind of interference is occurring here? Briefly explain how you know.

15-17. SSM WWW Suppose that in the vicinity of one meter from the sources, the amplitudes of the individual waves produced by the sources in Example 15-2 are both fairly constant at 1.20 cm, and, as in that example, the wavelength of the waves is 0.08 m. Let's also suppose that at each source, there is a crest at $t = 0$. Away from the sources, a fleck of sawdust floating at any point on the water's surface "detects" the displacement at that point. Find the displacement of the fleck of sawdust at $t = 0$ if it is

a. 1.14 m from one source and 0.98 m from the other.

b. 1.10 m from one source and 0.98 m from the other.

c. 1.00 m from one source and 0.98 m from the other (as in Example 15-2**c**).

15-18. Find the displacement of the fleck of sawdust in Problem 15-17 after a quarter of a cycle ($t = \frac{1}{4}T$) and after half a cycle ($t = \frac{1}{2}T$) if it is at the position given in

a. Problem 15-17**b**   *b.* Problem 15-17**c**.

15-19. SSM

a. Trace or photocopy Figure 15-10*a* and clearly mark several points on it that are a half-wavelength ($\frac{1}{2}\lambda$) further from one source than from the other.

b. Now, using a different color pen or pencil, mark several points that are $\frac{3}{2}\lambda$ further from one source than from the other.

c. Sketch the first two lines of destructive interference on each side of the center line.

15-20. A small motor makes two balls bob up and down in unison to produce a two-source interference pattern in a tank of water. How (and why) will the spread of the pattern change if

a. you speed up the motor?

b. you replace the water with a liquid in which waves travel more slowly?

c. the balls break the surface of the water more forcefully?

d. a third ball is made to bob up and down midway between the original two?

15-21.

a. In Figure 15-10*a,* find a point on the third line of constructive interference to the right of the center line. Express the path difference for this point as a multiple of the wavelength (p.d. = __?__ λ).

b. Now do the same for a point on the second line of *destructive* interference to the right of the center line.

15-22.

a. When both the red crests and the blue crests are present in Figure 15-11, find the path difference at point P and express it as a multiple of the wavelength (p.d. = __?__ λ).

b. Is there constructive or destructive interference at P when all the crests are present?

c. In On-the-Spot Activity 15-2, we eliminated all the blue crests to double the wavelength. Find the path difference at point P when only the red crests are present, and express it as a multiple of the wavelength (p.d. = __?__ λ).

d. Is there constructive or destructive interference at P when only the red crests are present?

Section 15-3 Mathematical Description of the Two-Source Interference Pattern

15-23. In the equation $\sin \theta_n = n\lambda/d$, explain the significance of the
a. product $n\lambda$.   **b.** subscript n on θ_n.

15-24. Suppose the two sources in Figure 15-12a are speakers placed 4.0 m apart, and a listener stands in a direction 36.87° from the center line. Determine by a calculation whether the listener is at a point of constructive interference, destructive interference, or something in between for sound with a wavelength of **a.** 1.6 m. **b.** 1.2 m.

15-25.
a. Two point sources of water waves are 0.050 m apart. If a point lies far from the sources in a direction making an angle of 24° with the center line, how much further is this point from one source than from the other?
b. Assuming the two point sources oscillate in phase, what is the largest wavelength for which there can be constructive interference at this point?
c. What is the largest wavelength for which there can be destructive interference at this point?
d. What is the second largest wavelength for which there can be constructive interference at this point?

15-26. Two identical point sources produce water waves with a wavelength of 0.018 m. The sources are 0.060 m apart.
a. Find the two smallest angles away from the center line at which constructive interference occurs.
b. Find the second smallest angle away from the center line at which destructive interference occurs.
c. In all, how many lines of constructive interference will there be on each side of the center line?

15-27. A long floodgate protects the opening to a harbor (Figure 15-35). The floodgate is being buffeted by a series of long, straight crests (parallel to the gate) arriving 8 m apart. Someone has forgotten to close two narrow vertical openings in the floodgate. The openings are 35 m apart.

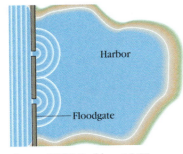

Figure 15-35 Problem 15-27

Find a position in the harbor where a small boat would be least disturbed by waves; that is, describe the position by giving its directional angle relative to the center line.

15-28. Redo the calculation in Example 15-7 using
a. $n = 1$.
b. $n = 3$ (assume there are more boats in the row, with the same spacing between them).

Section 15-4 Does Light Behave Like Waves?

Section 15-5 Young's Double-Slit Experiment

15-29. Why do we take the observations of light and dark fringes in the double-slit experiment to be evidence that light is wavelike?

15-30. What evidence is there that different colors of light correspond to different wavelengths?

15-31. When we look at a ceiling light fixture with two long parallel fluorescent tubes, we do not observe any evidence of interference. Why not?

15-32. In Figure 15-36, depicting a Young's double-slit set-up,
a. which (one or more) of the numbered points are three wavelengths further from one slit than from the other?
b. at which (one or more) of the numbered points is the path difference an odd number of half-wavelengths?

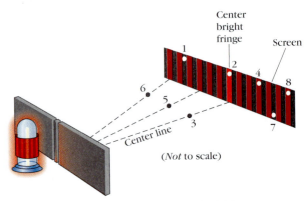

Figure 15-36 Problems 15-32, 15-33, and 15-35

15-33. In Figure 15-13, some of the boats are bobbing up and down because of waves produced by pumping stations S_1 and S_2. Others are not. Compare the features in this figure to the features in the double-slit experiment (Figure 15-36) by selecting the correct term from the list of choices following the table to fill in each blank in the second column below. (*Choices may be used more than once.*)

Feature of double-slit experiment	Corresponding feature in Figure 15-13
Bulb	
Dark fringes	
Bright fringes	
Double slit	
Wavelength of the light	

Choices:
   Pumping stations
   Boats bobbing up and down
   Boats not bobbing up and down
   Distance between nearest neighbor boats
   Distance between one bobbing boat and the next
   No corresponding feature shown in the figure

15-34. A double-slit arrangement is used to determine the wavelength of light from a monochromatic (single-wavelength) source. The pattern of fringes is viewed on a screen 2.50 m from the slits, which are 3.0×10^{-5} m apart. The distance from the center fringe to the seventh bright fringe away from the center is 3.75×10^{-3} m.
a. At what angle away from the center line is the seventh bright fringe located?
b. Find the wavelength emitted by the source.

15-35. In Figure 15-36 (not drawn to scale), suppose point number 8 is 10 cm to the right of the center line. The screen is positioned 200 cm from the slits. The source emits red light with wavelengths in the vicinity of 6.2×10^{-7} m.

a. Calculate the separation between the slits.

b. What would happen to the angular spacing of the bright fringes on the screen if the screen were moved closer to the double slit?

c. What would happen to the linear distance between bright fringes on the screen if the screen were moved closer to the double slit?

d. Consider the fringe on which point 8 appears. How far from the center line would this fringe be if the screen were moved up to a distance of 150 cm from the slits?

15-36. Suppose you wish to use a double-slit set-up to measure the wavelength of light emitted by a nearly monochromatic yellow light source. You know the double slit has been scratched out on a blackened slide using two double-edged razor blades pressed together. You have only measuring instruments that measure distances. What distances must you measure, and how would you calculate the wavelength from these distances?

15-37. SSM WWW When blue light from a certain filtered light source is passed through a diffraction grid with 1000 slits per centimeter, it produces a pattern of fringes on a screen. Figure 15-37 shows the actual size of the pattern when the screen is placed 0.25 m from the grid. Find the wavelength of this blue light.

Figure 15-37 Problems 15-37 and 15-38

15-38. If we replaced the blue light source in Example 15-37 with a helium-neon laser, at what angular displacement from the center line would the first bright fringe from the center appear on the screen? (A helium-neon laser emits red light with a wavelength of 6.328×10^{-7} m.)

15-39.

a. When light from a hydrogen lamp is directed through a multiple-slit diffraction grid onto a screen, four bright lines or fringes of different colors appear on the screen, with black areas in between. What does this tell you about the composition of the light from this source, and how does it differ from white light?

b. When a length of photographic film is stretched across the screen, the film is darkened or exposed where the colored lines are seen. In addition, several lines of darkening occur on the film to the left and right of these lines, where no light was visible. Explain why these lines occur.

15-40. Suppose that the line spectra in Figure 15-23 are shown as they appear on a screen when the screen is a certain distance from the light sources. If we doubled the distance between the screen and the sources, would that reduce the extent to which the red and violet fringes overlap? Explain.

Section 15-6 Other Instances of Diffraction

15-41. A small rectangular panel is set up on a darkened stage. Light from a spotlight is directed perpendicular to the panel,

so that it strikes the panel head on. Ignoring reflections from other surfaces, does any light from the spotlight get behind the panel? Briefly explain.

15-42. When we treat the straight filament of a bulb as a line source of light, are we assuming that the distribution of light from the filament looks most like part **a, b**, or **c** of Figure 15-25? Briefly explain.

15-43. SSM Light passes through a small, round pinhole. Think of this pinhole not as a point source but as a circle having a tiny (but nonzero) diameter. Based on what you know about single-slit diffraction patterns, what would you expect to see?

15-44. Suppose we draw a perpendicular to the row of posts in Figure 15-27. When the wavelength is 0.30 m, constructive interference occurs along the perpendicular, and next occurs in a direction at an angle of 20° to the perpendicular. Find the distance between nearest-neighbor posts.

Section 15-7 Interference due to Reflection; Thin Films

15-45. Explain the similarities and differences between the causes of the following two observations.

Observation 1: Areas of light appear to either side of (and also above and below) a distant street light viewed through a fine mesh curtain at night.

Observation 2: A rainbow effect is seen in an oil slick on a paved road surface.

15-46. Light from a single point source travels to a distant point P via two different paths. When it travels by path A, it reflects off surface A before reaching P. When it travels by path B, it reflects off surface B before reaching P.

a. Draw a labeled sketch of a set-up that fits this description and results in constructive interference where the path difference is $n\lambda$.

b. Draw a different labeled sketch of a set-up that also fits the given description but results in constructive interference where the path difference is $(n + \frac{1}{2})\lambda$.

15-47. SSM In Figure 15-38, light is partly reflected both at the interface between medium 1 and medium 2 and at the interface between medium 2 and medium 3. If light travels at speeds $v_1 = 0.7c$, $v_2 = 0.9c$, and $v_3 = 0.8c$ in the three media, there will be constructive interference at point P for light arriving in the direction shown. Will there be constructive interference, destructive interference, or something in between at point P if instead the speeds of light in the three media are

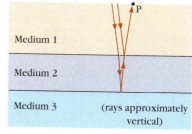

Figure 15-38 Problem 15-47

a. $v_1 = 0.9c$, $v_2 = 0.8c$, and $v_3 = 0.7c$?

b. $v_1 = 0.8c$, $v_2 = 0.6c$, and $v_3 = 0.7c$?

c. $v_1 = 0.6c$, $v_2 = 0.7c$, and $v_3 = 0.8c$?

15-48. The speed of light in ordinary petroleum-based oil is $0.80c$. If the oil making up the oil slick in Example 15-11 were this type of oil rather than vegetable-based oil, what would be the wavelength in air of the light that would be seen when the sun is almost directly overhead?

15-49. The speed of light in ordinary petroleum-based oil is $0.80c$. Suppose sunlight strikes a particular oil spill on the surface of a still body of water. If the sun is directly overhead, the light reflected straight back upward from a particular location on the spill has a wavelength of 5.2×10^{-7} m.

a. What is the smallest possible thickness of oil at this location?

b. What is the next smallest thickness that is possible at this location?

15-50. In the Michelson interferometer (Figure 15-30*b*), an air-to-glass reflection produces a phase change of 180° at each of the two mirrors. In the resulting interference pattern, will you observe constructive interference where the path difference is $n\lambda$ or where it is $(n + \frac{1}{2})\lambda$? Explain.

15-51. Compare Equations 15-2 and 15-5. How do the meanings of d in the two equations differ?

15-52. The speed of light is different in different types of glass. A coating material in which the speed of light is $0.60c$ is used to coat two flat glass surfaces. Surface A is made of glass in which light travels at a speed of $0.66c$; surface B is made of glass in which light travels at a speed of $0.56c$. A beam of red light ($\lambda = 6.328 \times 10^{-7}$ m) from a helium-neon laser is directed normal to each surface.

a. What minimum thickness of coating material must be applied to each surface to produce destructive interference for the reflected light?

b. What is the next smallest thickness of coating material that will also produce destructive interference for light reflected from each surface?

15-53. Overhead sunlight illuminates a petroleum-based oil slick (in which $v_{oil} = 0.80c$) on a still puddle. If the oil slick has a thickness of 2.32×10^{-7} m, for what color of normally reflected light in the visible part of the spectrum will there be constructive interference?

15-54. A thickness of material on a camera lens must be a minimum of 1.05×10^{-7} m thick to produce destructive interference for normally reflected green light that has a wavelength of 5.5×10^{-7} m in air. Assuming that light travels faster in the coating material than in the lens glass, find the speed of light in the coating material.

Going Further

The questions and problems in this group are not organized by section heading, so you must determine for yourself which ideas apply. Some of them will be more challenging than the Review and Practice questions and problems (especially those marked with a • or ••).

15-55. Suppose that in Figure 15-4*a*, we placed a long, straight barrier in the water so that it ran along the top of the picture. Assume there is more water beyond the top of the picture. Suppose there was a single narrow slit in the barrier.

a. Describe or sketch the wave pattern that would appear beyond the slit if the slit were placed at one of the positions labeled 1.

b. Now describe or sketch the wave pattern that would appear beyond the slit if the slit were placed midway between positions 0 and 1.

15-56. Suppose that the two oscillating sources in Problem 15-26 are not in phase. They still have the same frequency and amplitude, but source 1 is always a quarter of a cycle ahead of source 2. In the resulting interference pattern, what is the smallest path difference for which constructive interference can occur?

15-57. Explain why a phase change of $n(360°)$ is equivalent to no phase change at all.

15-58. In Problem 15-17, assume there is sawdust sprinkled on the water's surface in the region around the positions given in *a* and *b*. This region is far enough from the sources so that we can assume that in this region the amplitudes of the waves from the individual sources have a roughly constant value of 0.80 cm. Suppose there is a crest at each source at $t = 0$, and the sources take 0.12 s for each complete oscillation.

a. Find the displacement of a fleck of sawdust at $t = 0$ if it is located at the position given in Problem 15-17*a*.

b. Find the displacement of a fleck of sawdust at this same position at $t = 0.03$ s. Repeat for $t = 0.06$ s.

c. Repeat *a* and *b* for a fleck of sawdust located at the position described in Problem 15-17*b*.

15-59. **SSM** A two-source interference pattern is sketched in Figure 15-39. Points P_1, P_2, and P_3 show the positions of three detectors. Rank these points from least to greatest, indicating any equalities, in order of

a. the path difference at each point. Briefly explain your reasoning.

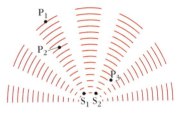

Figure 15-39 Problem 15-59

b. the amplitude of the disturbance detected at each point. Briefly explain your reasoning.

15-60. Two identical sources produce waves along the surface of a tank of water. At all points along the center line between these two sources, there is constructive interference. Let's pick one point on this center line and call it P. The nearest points to P at which *destructive* interference occurs are 10.7 cm from one source and 11.9 cm from the other. What is the wavelength of the waves from the two sources?

15-61. At a particular point of constructive interference in a two-source interference pattern, is the displacement a maximum at every instant? Briefly explain.

15-62.

a. By reasoning about the picture or diagram (without referring to any equations), locate the points in Figure 15-13 at which the path difference is a maximum. Then, by doing measurements on the diagram with a centimeter ruler, find the value of the path difference at one such point. (Drawing the two paths to the point may help you answer this.)

b. Now refer to Equation 15-3. Does this equation lead you to the same conclusion that you reached in *a*? Briefly explain.

15-63. Two identical sources are set 0.24 m apart on the surface of a large tank of water. When they oscillate in phase, they produce waves with a wavelength of 0.053 m.

a. Find the greatest angle from the center line (see Figure 15-12*a*) at which constructive interference will occur.

b. Find the greatest angle from the center line at which destructive interference will occur.

15-64. Light from a helium-neon laser (wavelength 6.328×10^{-7} m) passes through a double slit, producing an alternating pattern of red and unlit fringes. The separation between slits is 2.5×10^{-5} m. How far away from the double slit must you place the screen if you want the first bright fringe on either side of the center fringe to appear 1.0 cm from the center fringe? (Assume that distances between fringes are measured center to center.)

15-65. The education departments of some colleges and universities run lab schools for young children. In some of these, education majors and their professors use one-way mirrors to observe the children without the children seeing them. Discuss the transmission and reflection of light by such a one-way mirror. Is it really one way under all conditions? Might it even be one way the other way under some conditions? Explain.

15-66.

a. Is it possible to obtain destructive interference from a double slit when the wavelength is slightly greater than the separation between the slits? When the wavelength is slightly more than twice the separation between the slits? Briefly explain.

b. Is it possible to obtain destructive interference due to reflection from two parallel surfaces when the wavelength is slightly greater than the separation between the surfaces? When the wavelength is slightly more than twice the separation between the surfaces? Briefly explain.

15-67. An observer using a Michelson interferometer set up as in Figure 15-30*b* sees an interference pattern. The observer notices that there is constructive interference at a particular point P in the pattern.

a. What will the observer see happening at point P if the mirror M_2 is gradually moved toward the source (without changing its orientation)?

b. What will the observer see at point P if the glass insert shown in the figure is removed?

15-68. In Problem 15-52, what minimum thickness of coating material on each glass surface would produce destructive interference not for the reflected light but for transmitted light, that is, for the light continuing through the glass?

15-69. A projector used as an illumination source shines a beam of light through a multiple-slit diffraction grating onto a distant screen. When a red filter is placed in front of the projector and diffraction grating A is used, it is observed that the bright red lines in the interference pattern on the screen are a certain distance apart. When the red filter is replaced by a blue filter, and diffraction grating A is replaced by diffraction grating B, the bright lines in the observed interference pattern are the same distance apart as before. How do the slit separations in the two diffraction gratings compare? Briefly explain.

15-70. A certain commercial source produces microwaves with a wavelength of 5 cm. Two of these sources emitting microwaves in phase can be used to produce a two-source interference pattern. To observe such a pattern, it would be best to set up the sources with a distance of (*30 cm; 2 cm; 1000 m; one-razor-blade thickness*) between them. Give a reason for your answer.

15-71. How is a two-source interference pattern for waves of wavelength λ affected if

a. λ is greater than the source separation d?

b. λ is many orders of magnitude smaller than the source separation d?

15-72. Two identical, in-phase sources are producing water waves in a ripple tank. Point P is equidistant from the two sources. Point Q is a quarter-wavelength further from one source than from the other. (Assume that the distance between P and Q is very small compared to either point's distance from the two sources, so that differences in the amplitudes of the individual contributions at the two points are negligible.) Then the amplitude of the combined disturbance at Q is approximately (*double; 1.4 times; equal to; 0.7 times; half*) the amplitude of the combined disturbance at P.

•15-73. Two sources of water waves, identical and in phase, are 0.32 m apart. The wavelength of the waves they produce is 0.10 m.

a. Find the smallest path difference that will result in destructive interference.

b. Find the largest path difference that will result in destructive interference.

15-74. Two balls bobbing up and down in unison in a tank of water produce an interference pattern. The balls are 4λ apart. The path differences to four points on the water's surface are as follows:

Point	Path Difference
P_1	0.75λ
P_2	1.50λ
P_3	2.25λ
P_4	3.00λ

Rank these points from least to greatest in order of the amplitude of the disturbance at each point. Indicate any equalities.

15-75. In ripple tank A, two balls bobbing up and down produce an interference pattern in water. In ripple tank B, the water is replaced by a liquid in which the propagation speed is half of what it is in water. To produce an identical interference pattern in tank B, the balls must _____ (*be half as far apart; bob up and down with double the frequency; produce waves with half the wavelength; bob up and down with double the amplitude*). Give a reason for your answer.

15-76. Diffraction grid A has 500 equally spaced slits over a total width of 1 cm. Diffraction grid B has 1000 equally spaced slits over a total width of 2 cm. A line source of red light is 1 m away from you. Suppose you view it first through grid A, then through grid B. Then when you view it through grid B, _____ (*the central bright fringe will appear brighter; the pattern will appear more spread out; the pattern will appear less spread out; the central bright fringe will appear dimmer*). Give a reason for your answer.

15-77. A single microphone feeds simultaneously into two speakers, placed a meter apart, facing an outdoor audience (there are no reflecting walls). Assuming the input to the mike is human speech with a frequency of 500 Hz, at what angle to the forward direction would the speaker output be most difficult to hear? Give your answer in degrees.

15-78. In an architect's design for an aquarium exhibit, a pool is enclosed in a courtyard. Viewers look down on the pool from a raised bridge. What should be the minimum depth of the water in the pool to minimize the amount of reflected sound that viewers hear from below? (*Hints:* Use Table 14-1. Assume that the speed of sound in other solids is in the same ballpark as those listed.)

15-79. SSM WWW Two identical balls are producing waves in a ripple tank. The first direction of total constructive interference to the right of the center line is observed to occur at an angle of 18°. In all, how many directions of total constructive interference will there be to the right of the center line?

15-80. At each of the reflecting surfaces in Figure 15-30*a*, a fraction r of the light intensity that reaches the surface is reflected and the remaining fraction t is transmitted. Suppose we want the intensities of the two beams arriving at O to be equal. If we neglect the effects of multiple reflections, which of the following correctly expresses the relationship between the two fractions r and t?

$$r = t \qquad r = t^2 \qquad r = \sqrt{t} \qquad r = 2t$$

15-81. Suppose point P is a point where there is constructive interference of waves from two identical sources. If the two contributing disturbances both have nonzero amplitudes, can the displacement ever be zero at P? Briefly explain.

15-82.
a. If an oil slick on water is too thin, and the sun is directly overhead, it will not produce constructive interference for any color of visible light. Why not?
b. Will this oil slick necessarily produce no constructive interference for visible light coming indirectly from other regions of the sky? Explain your answer. A sketch may help.

15-83. Suppose that in Figure 15-13, the two sources are 40 m apart and produce waves with wavelength 10 m, and the row of boats is 400 m from the sources. At approximately what minimum linear spacing between the boats will the three middle boats all experience constructive interference?

15-84. Suppose that the two wave sources in Figure 15-13 are 180° out of phase, and the row of boats is 400 m from the sources. At what minimum linear distance from the center line will a boat experience destructive interference?

•**15-85.** Suppose that the two wave sources in Figure 15-13 are 90° out of phase. Both produce waves of wavelength 8.0 m.
a. What are the four smallest path differences for which constructive interference will occur?
b. Do the lines of interference to one side of the center line form the same angles with the center line as the lines of interference to the other side? Briefly explain.

15-86. Light from a monochromatic source passes through a double slit. In the resulting interference pattern, point P is located on the first line of destructive interference away from the center line. Here are two pieces of information about the set-up:

A. The slits are 3.0×10^{-4} m apart.

B. The light has a wavelength of 6.3×10^{-7} m.

Which of these pieces of information (*A only, B only, both,* or *neither*) are necessary to answer each of the following questions?
a. How much further is point P from one slit than from the other?
b. What is the phase difference between light reaching point P from one slit and light reaching it from the other?
c. What angle does the line of interference that passes through point P make with the center line?

15-87. Two identical sources of water waves are positioned 0.40 m apart. Point P is a point on the water's surface at a distance from the two sources. If the frequency of the two sources is increased, will the path difference to P increase, decrease, or remain the same? Briefly explain.

Problems on WebLinks

15-88. In WebLink 15-1, the path difference increases as point P is moved to the right. If we continued moving P to the right beyond where we left it at the end of WebLink 15-1,
a. constructive interference would next occur when the path difference to P is equal to how many wavelengths?
b. destructive interference would next occur when the path difference to P is equal to how many wavelengths?

15-89. In WebLink 15-2, the two sources are moved closer together until they are located at the same point. If we were to continue moving the two sources in the directions they were being moved previously, the lines of constructive interference would (*become more spread out; become less spread out; not change their spread; not reappear*).

15-90. WebLink 15-3 shows that the path difference is $d \sin \theta$ only when _____ (*θ is small; one path length is greater than the other; both path lengths are much greater than d; θ is large; point P is to the right of the center line*).

15-91. In WebLink 15-4, each white fringe occurs where the crests of two (*circular; cylindrical; spherical; straight-line*) waves intersect.

CHAPTER SIXTEEN

The Geometry of Wave Paths and Image Formation: Geometric Optics

Images are a central concern in the study of light—both naturally occurring images, such as pond reflections and mirages, and images formed by low- and high-tech devices ranging from your bathroom mirror to microscopes, projectors, and the imaging systems on the Hubble Space Telescope (Figures 16-1 and 16-2).

The woman in the photo on this page is looking in a shop window in a Portuguese square on a sunny day. She has her back to the square, which is seen in the reflected image in the shop window, so that the sign letters appear reversed. But the window transmits as well as reflects light, so that lights within the shop are also visible,

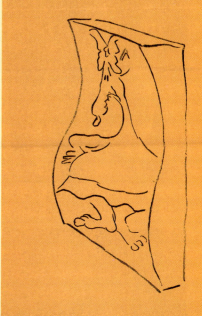

"So many images! But . . . what are images?"

Figure 16-1 Images do not always look like the objects themselves.

Figure 16-2 An image obtained by the NASA Hubble Space Telescope. This image reveals a pair of one-half-light-year-long interstellar "twisters," eerie funnels and twisted-rope structures in the heart of the Lagoon Nebula, 5000 light-years from Earth.

seemingly superimposed against the sky. The camera formed an image of this on film. Other images were formed in the process of reproducing the photo for this book. Images of the photo and these words are being formed right now on the retina of your eye. So many images!

STOP&Think What *are* images? Why do they form in the various instances we've mentioned? ◆ You may find yourself unable to come up with a satisfactory answer at this point, but to design optical devices such as microscopes, cameras, or projectors, we need to know precisely what an image is and what conditions yield what kinds of images: sharp or fuzzy, enlarged or reduced, upright or inverted, and so on. That is the aim of this chapter, and geometric optics is a means to this end.

16-1   Why Geometry? Looking at Shadows

Although shadows have small-scale edge effects due to diffraction (recall Figure 15-26), a typical shadow has fairly distinct boundaries that mimic the outline of the object casting it. You know how these shadows behave from everyday experience, but it is worth taking the time to do some review observations in order to clarify and systematize this knowledge.

On-The-Spot Activity 16-1

1. Take a single-bulb light source (the smaller the size of the bulb, the better), and place an object with a sharp, well-defined outline between the bulb and the wall so that the object casts a distinct shadow on the wall. Make sure that the distance from the bulb to the object is always much greater than the diameter of the bulb, so that you can treat the bulb approximately as a point source.

2. Mentally draw a straight line from the bulb through the object to the wall. Move the object back and forth along this line. What happens to the shadow when you move the object further from the wall? When you move it closer to the wall?

3. Now consider the situation quantitatively. With the object in a particular position, measure the distances in the numerator and denominator of each of the following ratios:

$$\frac{\text{distance from bulb to object}}{\text{distance from bulb to wall}} \qquad \frac{\text{height of object}}{\text{height of shadow on wall}}$$

4. Calculate the two ratios. How do they compare?

5. Move the object either closer to or further from the wall and repeat Steps 3 and 4. Has the value of each ratio changed? Are the two ratios in the new situation equal to each other?

6. Can you diagram your set-up in a way that could explain why these two ratios should always remain equal?

Figure 16-3*a* shows a sample set-up for the above activity. Figure 16-3*b* suggests a simple model to explain why the ratios remain equal. The model, called the **ray model,** assumes that

1. *Light travels in straight-line paths, or rays, in all directions from the source.*

2. *The object intercepts a cluster of rays shaped like an irregular cone spreading out from the point source.*

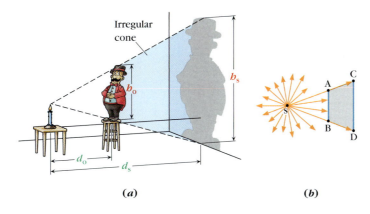

Rays that are blocked by the object do not reach the wall. The resulting diagram contains two similar triangles, △SAB and △SCD, framed by the boundary rays of the intercepted cone of light. The heights of the object and shadow, h_o and h_s, and their distances from the source, d_o and d_s, are sides of these triangles. Because the ratios of corresponding sides of similar triangles are equal, $\frac{d_o}{d_s} = \frac{h_o}{h_s}$. These are just the two ratios that you calculated in the activity. The equality should agree with your results.

Example 16-1 *Applying the Geometry of Light Rays*

Suppose the man on the stool in Figure 16-3 is 2.0 m from the wall and 3.0 m from the candle. If he is 1.5 m tall, how tall is his shadow?

Solution
Choice of approach. Use the fact that the ratios of corresponding sides of similar triangles are equal to find h_s, one side of the larger triangle.

What we know/what we don't.

$$h_o = 1.5 \text{ m} \qquad h_s = ?$$
$$d_o = 3.0 \text{ m} \qquad d_s = ?$$
$$\text{distance from man to wall} = 2.0 \text{ m}$$

(This is not d_s; d_s is the distance from the candle to the wall)

The mathematical solution. It is easier to use $\frac{d_o}{d_s} = \frac{h_o}{h_s}$ in inverted form, $\frac{d_s}{d_o} = \frac{h_s}{h_o}$, so that we just have to multiply both sides by h_o to solve for h_s:

$$h_s = h_o\left(\frac{d_s}{d_o}\right)$$

Note that if the man is 2.0 m from the wall and 3.0 m from the candle, the wall is 5.0 m from the candle: $d_s = 5.0$ m. Then

$$h_s = 1.5 \text{ m}\left(\frac{5.0 \text{ m}}{3.0 \text{ m}}\right) = \textbf{2.5 m}$$

◆ Related homework: Problems 16-2 through 16-6.

The ray model led us to a conclusion $\left(\frac{d_o}{d_s} = \frac{h_o}{h_s}\right)$ that matches observation. It is useful because it enables us to do calculations based on the geometric diagrams we can draw. Approaching light and image formation in this way is called

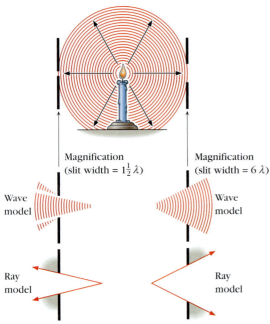

Figure 16-4 **Connections between wave and ray models of light from a point source.**

For **WebLink 16-1:**
Waves and Rays, go to
www.wiley.com/college/touger

geometric optics. But how valid is this model in general, and how does it fit with the wave model of light we explored in Chapter 15?

If we consider light waves traveling uniformly in all directions from a point source (Figure 16-4), then

the rays show the wave paths—the directions in which the waves are traveling—and are always perpendicular to the wavefronts.

The magnified insets in Figure 16-4 remind us what the diffraction patterns from single slits look like. When the slit width is not much greater than the wavelength of light, the spreading out of the light after it passes the slit contradicts what we expect from our ray model. But when the slit width is at least several times the wavelength of light, the illuminated region looks very much like what the ray model would give us. This assumes that the parts that block the light—the opaque regions between the slits—are of at least comparable size. For a visual summary of the ray model and how it connects with our wave model of light, work through WebLink 16-1.

In general, the larger the opaque regions and apertures (openings) of an object are in comparison to the wavelength(s) of the light striking it, the more nearly the resulting pattern of illumination and shadow predicted by the wave model looks like the all-or-nothing blocking out of light that we infer from ray diagrams. For objects whose dimensions are large compared to visible light wavelengths, we can confidently apply the ray diagram reasoning of geometric optics. We do this in the following case.

Case 16-1 ◆ *Shadows Due to Extended Light Sources*

(If possible, do this as though it were an On-the-Spot Activity.)

1. Select an extended light source: either a window admitting *indirect* daylight or (if at night) a doorway admitting indirect light from an adjacent room. You must be able to eliminate the light by either pulling down the shade or closing the door.

2. Stand by a wall opposite the light source and hold your hand close enough to the wall so that it casts a well-defined shadow. Carefully observe how the shadow changes as you gradually move your hand further from the wall (i.e., toward the light source). Attempt to draw a ray diagram to explain your observations (you may find it difficult at this stage).

3. Now pull down the shade (or close the door). Then place a point source of light such as a lamp with a single bare bulb at each of the two outer edges of the window or doorway (Figure 16-5), and repeat Step 2. Carefully draw a sequence of sketches to record how the shadow of your hand changes as you move it away from the wall. How

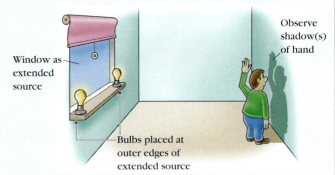

Figure 16-5 **Set-up for Step 3 of Case 16-1.**

is this like what you saw in Step 2? How is it different?

4. Figure 16-6a outlines a ray diagram with your hand positioned near the wall. To see how this accounts for the shadows you see, complete the diagram as follows. Identify the cone of rays your hand intercepts from source S_1. Lightly shade the region between hand and wall that is blocked from receiving the rays in this cone. Now identify the cone of rays your hand intercepts from source S_2. Lightly

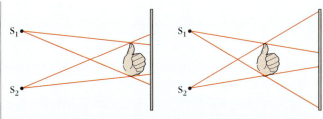

(**a**) Hand closer to wall   (**b**) Hand further from wall

Figure 16-6 Ray diagrams for Case 16-1. To show the regions of total and partial shadow cast by a hand in the presence of two point light sources, complete these diagrams as directed in Steps 4 and 6 of Case 16-1.

shade the region that is blocked from these rays. If this overlaps a region that you previously shaded, darken the shading where it overlaps.

5. Your diagram should now show lighter shadow where light arrives from only one of the two sources, and darker shadow where light is not arriving from either source. Is this like what you observed?

6. Now apply the directions of Step 4 to Figure 16-6*b*, with the hand further from the wall. How does the pattern of shadow change?

7. What would happen to the pattern of shadow in Step 4 if you placed a third source of light between the other two? If you placed an entire string of point sources between the outer two? Increasing the number of sources brings us closer to the continuous extended light source we used in Steps 1 and 2, and take us from separate clearly defined areas of lighter and darker shadow to more continuous gradations of shadow (Figure 16-7). We can now summarize this approach:

An extended light source must be treated as though it were made up of an infinite number of point sources.

In principle you must consider separately the cone of light intercepted from each of these points, but you can generally do this by considering the cones of light from just the outermost points.

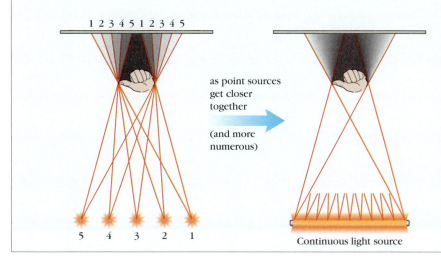

as point sources get closer together

(and more numerous)

Continuous light source

Figure 16-7 The shadows that are cast when there are multiple point sources start to look like the shadows that are cast when there is a single extended light source.

The region of shadow where an object, like the hand in Steps 5 and 6 of Case 16-1, has blocked off all light directed toward the region from an extended source is called the **umbra.** The region where there are various gradations of lighter shadow, because light reaches that region from some but not all points on the extended source, is called the **penumbra.** Figure 16-8 shows what happens when the sun is the extended light source and the moon or Earth casts the shadow.

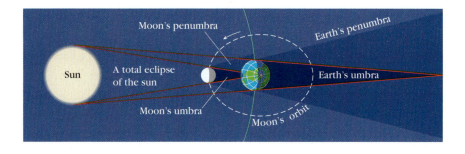

Figure 16-8 An eclipse of the sun. Just the small portion of Earth's surface that lies in the umbra of the moon experiences a total eclipse. The much larger region that lies in the moon's penumbra experiences a patial eclipse.

16-2   Reflection and Mirror Images

Case 16-2 ◆ *Images of a Thumbtack*

It is best to do the steps described here yourself as though it were an On-the-Spot Activity. (To do so, you will need a small unframed rectangular mirror, a sheet of lined paper, a sheet of tracing paper, tape, a thumbtack, a rigid ruler, colored pencils, a protractor, and a flat working surface—a newspaper would do—that you can stick the tack into without damaging any furniture.)

1. Set up the paper, tracing paper, thumbtack, and mirror as shown in Figure 16-9a. The tracing paper goes *over* the lined paper. The lines of the paper should make an acute angle with the edge of the mirror. The lined paper should be able to rotate freely about the thumbtack under the tracing paper. The edges of the tracing paper should be taped to your working surface to keep it still. Trace the bottom mirror edge with a black line on the tracing paper so that when you lift the mirror up to rotate the lined paper, you can put it back down in the same position.

2. Now place your eye so that *in the mirror* you are sighting the image of the tack along one of the paper lines (Figure 16-9b). Place your ruler on edge and line it up with one end against the mirror, as

shown: The ruler and the paper line should form a continuous straight line ⓐ. This is your line of sight to the image of the tack.

3. Record this arrangement as follows. Run a red pencil along the ruler to record the line of sight ⓑ on the tracing paper; this is the *reflected ray*. Also, identify the line ⓒ on the paper that goes from the tack to where the ruler touches the mirror, and trace it in red; this is the *incident ray*. Then from the point where the ruler touches the mirror, draw a dashed perpendicular ⓓ to the mirror in blue. This is the *normal* to the mirror.

4. Measure the angles ⓔ and ⓕ that the two red lines make with the dashed blue perpendicular. How do they compare?

5. Now rotate the sheet of lined paper so that its lines make a different angle with the mirror (Figure 16-9c) and repeat Steps 2–4.

   At this stage, if you remove the sheet of tracing paper, it should look like Figure 16-9d. Mark the position of the object (the hole left by the tack) with an O. Light rays traveled in all directions from the

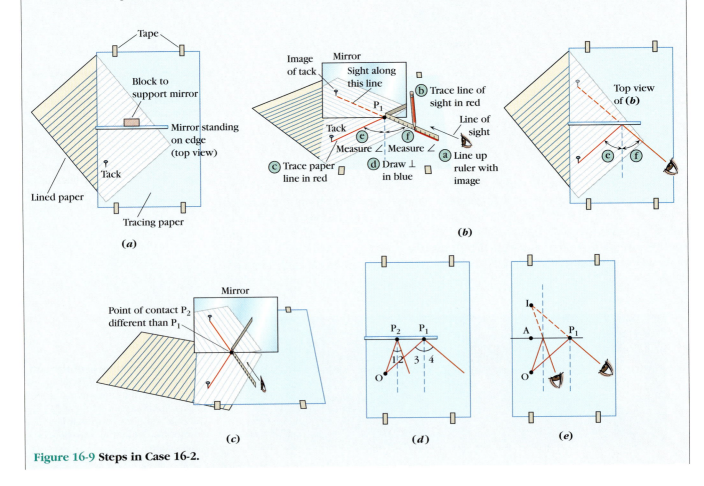

(a)

(b)

(c)

(d)

(e)

Figure 16-9 Steps in Case 16-2.

object. (Although this object—the tack—is not itself a source of light, we see it by the light reflected from it, which travels outward from it in all directions as though from a source.) The two red lines, each bent at the mirror, describe the light paths that arrive at the two positions where you have placed your eye. As you probably found in steps 4 and 5, Figure 16-9d shows that ∠1 = ∠2 and that ∠3 = ∠4. In each case the incident ray and the reflected ray make the same angle with the perpendicular. (The rays hitting a surface are called **incident rays.**) Thus the rays obey the *law of reflection*.

Law of Reflection
∠of incidence = ∠of reflection   (16-1)

In three dimensions, the reflected ray must be in the plane determined by the incident ray and the normal.

(This agrees with Figure 15-27c if you sketch in the directions in which the crests are traveling.)

6. Now extend the two sight lines behind the mirror. Draw the extensions as *dashed* red lines (Figure 16-9e) and mark the point where they meet with an I. Both lines of sight are directed at this point; you see the image here *no matter where you place your eye*. I is thus the **image point,** the point where the two rays reaching the different positions of your eye *appear to be coming from.*

Figure 16-9e shows that the image point I is as far behind the mirror as O is in front of it. (We could prove this is so by doing geometry on the ray diagram.) This is where ordinary experience tells us an object seen in a plane mirror "appears" to be. There is no point in saying, "That's where the image appears to be, but where is it really?" Images *are* appearances. They are not things and have no other reality. With images, what you see is what you get.

(This activity is derived from activities developed for an AAPT workshop by Dewey Dykstra. It is informed by research on students' understanding of optics by Fred Goldberg and Lillian C. McDermott.)

In Case 16-2, you could easily have positioned your eye to view additional reflected rays, with the same result. The pupil of your eye always subtends a slender cone of rays from any given point. Our reasoning applies not just to the two rays that we have considered in detail but to the entire cone of rays reaching your eye. Figure 16-10 shows that, assuming perfect reflection, the cone of light reaching your eye from the image point is identical to the cone of light that would reach your eye from a real tack placed at that point (with no mirror). This is the identifying property of the image point.

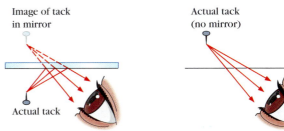

Figure 16-10 Rays reach the eye from an image as though from an object at the same position.

The **image point** is the point from which the light rays appear to spread out, or diverge, as they approach your eye.

Because the image point is where these rays meet, you always need to consider at least two rays to identify an image point.

The procedure for tracing rays from an object to locate its image(s) works equally well whether the object is itself a source of light or we see it because it reflects light from some external source. When the surface of an object is sufficiently irregular (not smooth and flat like a mirror), the object doesn't look shiny, and light reflects from it in all directions. This is called **diffuse reflection.** If the reflection from a point object is diffuse, light travels in all directions from that point just as though it were a point source.

➤Note: In Figures 16-9e and 16-10, no light from the object actually travels where the line is dashed. We will always use dashed lines in our diagram to indicate apparent paths of light where no light is actually traveling.

◆**REAL AND VIRTUAL IMAGES** When an image is formed by a plane mirror (Figure 16-11a), no light actually passes through the image point, so it is called a **virtual image point.** In contrast, when we consider curved mirrors and lenses we will see that rays sometimes diverge to your eye from a point that light actually passes through (Figure 16-11b). Such a point is called a **real image point.**

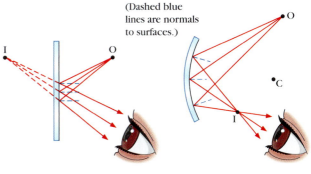

(*a*) Virtual   (*b*) Real

Figure 16-11 Real and virtual image points.

Type of image point	Light reaching your eye from the point . . .	So all the rays drawn through the point are . . .
Real	all passes through the point	drawn as solid lines
Virtual	doesn't pass through point	drawn as dashed lines

For **WebLink 16-2:**
Image Points, go to
www.wiley.com/college/touger

To develop these ideas more fully, work through WebLink 16-2.

◆**IMAGE OF AN EXTENDED OBJECT** In dealing with light coming from an extended source when we considered shadows, we had to consider the cone of rays diverging from each point source making up the extended source, or from a judicious selection of those point sources. Likewise, if light is arriving at a plane mirror from an extended object, we must consider the cones of rays from different points on that object. Typically, as in Figure 16-12, it is sufficient to consider the two endpoints of the object and assume that the other points will line up appropriately in between.

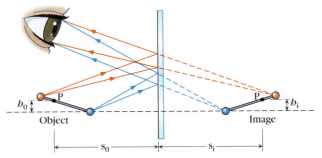

Figure 16-12 The eye receives a cone of rays from an image point for each object point. The two endpoints of the rod are shown in different colors. The blue rays are drawn to locate the image of the blue end; the red rays are drawn to locate the image of the red end.

We see from the diagram that the image formed by reflection from a plane mirror is the same size as the object. Each point P on the image is at the same distance s from the mirror plane and at the same height h as the corresponding point on the object:

> For a **plane mirror** (o = object, i = image)
>
> $$s_o = s_i \tag{16-2}$$
>
> and
>
> $$h_o = h_i \tag{16-3}$$

For this reason, we will often find it useful to speak of *object points* and their corresponding *image points*.

When the image in one mirror is in turn reflected in a second mirror, we can treat the image in the first mirror as the object for the second mirror, because the rays arrive at the second mirror as though they were coming from the first image. Multiple images can be treated by successive repetitions of this reasoning (Figure 16-13).

Figure 16-13 Multiple mirror images in a barber shop. The front and back walls of the shop are in different colors to help you distinguish.

Example 16-2 *Multiple Images Formed by Perpendicular Mirrors*

For a guided interactive solution, go to Web Example 16-2 at
www.wiley.com/college/touger

Find all the images of the point object O formed by the two mirrors M_1 and M_2 in part *a* of the figure on the next page.

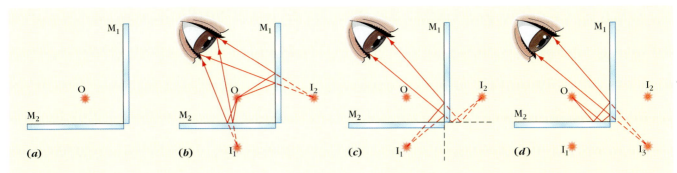

(a)   (b)   (c)   (d)

Brief Solution

Choice of approach.

1. Place an eye at a convenient position in the figure. (Think back to Case 16-2. Will the image positions depend on where you place your eye?)

2. Draw the two rays reflected by M_1 to the two outer edges of the pupil of the eye (part *b* of figure). Make sure the reflected angle is equal to the incident angle for each ray. Using dashed lines, extend these two rays to the point where they intersect. This is the position of I_1, the image seen in M_1.

3. Repeat Step 2 for M_2 to obtain the position of I_2.

4. Now treat I_1 and I_2 as objects, and by the same procedure as Step 2, find the image of each in the other's mirror (part *c* of figure). Based on the geometry of the situation and the fact that $s_o = s_i$, you should satisfy yourself that you get the same position for each of these images; that is, they are in fact a single image, I_3. Part *d* of the figure shows that the two rays that are double-reflected from O to the outer edges of the eye do indeed appear to diverge from I_3.

◆ Related homework: Problems 16-11, 16-12, and 16-13.

We may see objects because they emit light themselves or because they reflect light from some external source. The procedure for tracing rays from an object to locate its image(s) works equally well in both cases.

16-3 Reflections from Curved Mirror Surfaces

When rays are incident on a curved reflecting surface, the direction of the normal or perpendicular to the surface is different at each point where a ray strikes (Figure 16-14*a*).

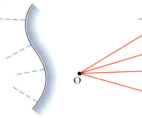

(*a*) Construct normals at each surface point...

(*b*) ...then apply law of reflection at each point

Figure 16-14 **Reflection from curved surfaces.**

PROCEDURE 16-1 _____

Drawing Ray Diagrams to Determine Images Resulting from Reflection

For each ray striking a point on the surface, identify
1. the direction of the normal at that point and
2. the angle that the particular incident ray makes with the normal,
3. so that you can apply the law of reflection separately at that point (Figure 16-14*b*).

For a spherical mirror, one for which the reflecting surface is a segment of a sphere, this procedure is particularly straightforward because *the radii of a sphere are perpendicular to its surface* and therefore *are* the normals. A spherical

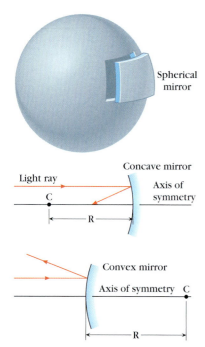

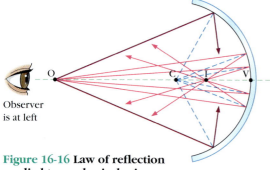

Figure 16-15 Concave and convex mirrors. A spherical mirror is one shaped like a segment cut from a sphere. For a concave mirror, the reflecting surface is part of the inside surface of the sphere; for a convex mirror, it is part of the outside surface.

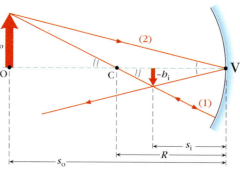

Figure 16-16 Law of reflection applied to a spherical mirror surface.

mirror is usually classed as **concave** or **convex,** depending on whether its reflecting surface was a part of the inside or outside surface of a sphere (Figure 16-15). If you look at a concave mirror surface such as a make-up or shaving mirror, it bows away from you; a convex mirror surface bulges toward you. For either kind of mirror, its **center of curvature** C is the center of the sphere from which it came, and its **radius of curvature** R is the radius of that sphere.

Figure 16-16 follows Procedure 16-2 for the case where light from a point object O is reflected from a concave spherical mirror. The point V is the central point, or **vertex,** of the mirror surface. The dashed green line is the axis of symmetry of the mirror, also called the **principal axis.** For convenience, we will refer to it simply as the mirror's axis. In addition to the rays shown, a ray from O traveling to V along the axis would be reflected back along the axis. For that ray, the incident and reflected angles are both zero.

The reflections of the rays close to the axis all converge at very nearly the same point I and subsequently diverge from that point to the viewer at the left of the figure. Point I is thus an *image point*. Because the reflected light reaching the eye actually passes through this point, this is a *real* image point, in contrast to the virtual image produced by the plane mirror. By looking at an actual concave mirror with sufficient curvature, you should verify that in contrast to "ordinary" (plane) mirrors, where the image appears behind the mirror, you can get an image that appears in front of a concave spherical mirror; that is, between the mirror surface and your eye. A simple way to see this is to take a pencil with its point toward the mirror and move the pencil point along the axis toward the mirror until the pencil point's image appears to come out to meet it.

The rays do not all meet exactly at I. The **paraxial rays** (those nearly parallel to the axis) all converge at very nearly the same point, but when incident rays (shown darker in Figure 16-16) strike the mirror substantially further from V, the reflected rays do not converge to the same point. (The only exception is when the object is at the center C of the sphere, in which case all reflected rays will converge to C.) This deviation from perfect convergence is called **spherical aberration.** The more spherical aberration you have, the blurrier the image. We state without proof that a *parabolic* mirror does provide perfect convergence for parallel rays from objects "at infinity"—that is, at a distance much greater than any of the other distances or lengths involved—and more nearly perfect convergence for paraxial rays from nearer objects. Parabolic mirrors are used in telescopes and as the reflectors behind the bulbs in many flashlights and automobile headlights.

We will use the paraxial rays (which is all of them if the mirror subtends a small solid angle) to determine the characteristics of the image of an extended body: the arrow in Figure 16-17a. We already

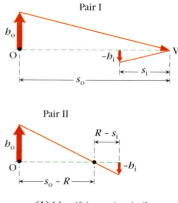

Figure 16-17 Ray diagram geometry for a spherical concave mirror.

(a) The ray diagram (numbers refer to principal rays)

(b) Identifying pairs similar triangles in ray diagram

know that the image of the tail, like that of the point object in Figure 16-16, will be on the axis. We therefore draw a ray diagram (Figure 16-17a) to determine where the image of the arrow tip will be positioned. Because we need the intersection of two rays to determine the image point, we could in principle choose any two of the infinite number of rays from the arrow tip. In practice, we choose the two, called the **principal rays,** for which it is easiest to determine the direction of the reflected rays.

PROCEDURE 16-2

Identifying Principal Rays for Spherical Mirror Diagrams

1. **The ray from the object point through the center of curvature C.** Because this travels along a radius of the sphere, which is perpendicular to the surface, the reflected ray returns along the same radius.

2. **The ray from the object point to the vertex V.** The axis is perpendicular to the surface at V, so the reflected ray makes the same angle with the axis as the incident ray.

We can now do geometry on the resulting diagram to find the vertical position h_i (measured from the axis) and the horizontal position s_i (measured from V) of the image point if the object point has a height h_o and is a distance s_o from V. If we take the upward direction as positive, then h_i is negative.

We now isolate two pairs of similar triangles in the ray diagram (Figure 16-17b). In each pair, the ratios of corresponding sides are equal:

$$\textbf{Pair I: } \frac{-h_i}{h_o} = \frac{s_i}{s_o} \qquad \textbf{Pair II: } \frac{-h_i}{h_o} = \frac{R - s_i}{s_o - R} \qquad \text{(16-4,I and 16-4,II)}$$

(Note that the ratios involve the *lengths* of sides, so that if the position of the image tip is h_i and h_i is negative, the length of the image is $-h_i$.) Equating the right sides of these two equations gives us

$$\frac{s_i}{s_o} = \frac{R - s_i}{s_o - R}$$

On-The-Spot Activity 16-2

Try to do the necessary algebra to reduce this equation to the following:

$$\textbf{Mirror Equation} \qquad \frac{1}{s_o} + \frac{1}{s_i} = \frac{2}{R} \qquad \text{(16-5)}$$

To use this mirror equation to find the image position s_i, we need to know R as well as s_o. Where we see the image depends on how curved the mirror is.

It is also useful to define the ratio of "heights" (actually vertical positions measured from the axis, so sign must be considered) to be the **magnification:**

$$\textbf{Magnification} \qquad M \equiv \frac{h_i}{h_o} \qquad \text{(16-6)}$$

M will be positive or negative depending on whether h_o and h_i have the same or opposite signs:

M is + if the image is upright.	$\lvert M \rvert > 1$ if the image is enlarged.
M is − if the image is inverted.	$\lvert M \rvert < 1$ if the image is reduced in size.

It follows from Equations 16-6 and 16-4,1 that

$$M = -\frac{s_i}{s_o} \qquad (16\text{-}7)$$

In Example 16-3 you will see how we can set up a sign convention for s_o and s_i that is consistent with our sign conventions for b_o, b_i, and M.

In applying these equations, the following steps are important:

PROCEDURE 16-3

Doing Geometric Optics Calculations

1. Carefully sketch a ray diagram showing the *principal rays* to establish an expectation of what the image should look like.
2. Do the calculation *with careful attention to sign conventions* and make sure the results agree with your expectations: Is the image

      real or virtual?   *upright or inverted?*   *enlarged or reduced?*

Example 16-3 *A Close-Up Image in a Concave Spherical Mirror*

For a guided interactive solution, go to Web Example 16-5 at www.wiley.com/college/touger

Find the position and height of the image formed by a concave spherical mirror with a radius of curvature of 0.40 m when a 0.10 m tall bottle stands 0.15 m in front of the mirror.

Brief Solution

Choice of approach. Follow Procedure 16-3. In Step 2 of the procedure, apply Equation 16-5 to find s_i, then apply Equation 16-7 to find M, and finally apply Equation 16-6 to find b_i.

What we know/what we don't.

$$s_o = 0.15 \text{ m} \qquad b_o = 0.10 \text{ m} \qquad R = 0.40 \text{ m}$$
$$s_i = ? \qquad M = ? \qquad b_i = ?$$

The ray diagram. From the top of the bottle we draw the principal rays through V and C (Figure 16-18). Because the bottle is closer to the mirror than C is, the ray through C must reflect off the mirror *before* passing through C. We then take the rays that diverge toward the viewer and extend them (dashed) back to the point where they intersect to locate the image's bottle top. In doing so, we get a virtual image that is erect and enlarged.

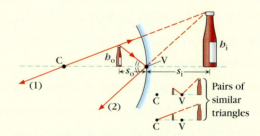

Figure 16-18 Ray diagram for an object close to a concave mirror.

The mathematical solution. By Equation 16-5,

$$\frac{1}{s_i} = \frac{2}{R} - \frac{1}{s_o} = \frac{2}{0.40 \text{ m}} - \frac{1}{0.15 \text{ m}} = -1.67 \text{ m}^{-1}$$

so that
$$s_i = \frac{1}{-1.67\ m^{-1}} = \textbf{-0.60 m}$$

This agrees with the ray diagram if we interpret the minus sign to mean that the image position is *behind* the mirror, so that the image is *virtual*. Moreover, we see by Equation 16-7 that a negative s_i will give us a positive M, indicating an *upright* image, in agreement with the ray diagram:

$$M = -\frac{s_i}{s_o} = -\frac{-0.60\ m}{0.15\ m} = 4.0$$

Then from Equation 16-6, $h_i = Mb_o = 4.0(0.10\ m) = \textbf{0.40 m}$.

Making sense of the results. Figure 16-18 shows that an object close to a concave mirror has an upright virtual image. For s_i to be negative, $\frac{2}{R} - \frac{1}{s_o}$ must be negative in our calculation. This occurs when $\frac{1}{s_o} > \frac{2}{R}$, or when $s_o < \frac{R}{2}$. Figure 16-17 showed that an object at $s_o > \frac{R}{2}$ has a real inverted image. You should experiment with an actual concave mirror to verify that this is what you really see.

◆ Related homework: Problems 16-20 and 16-21.

STOP&Think What does Equation 16-5 tell you the image position s_i will be when $s_o = R$? What do you expect to see in this case? Try to test out your expectations with an actual concave mirror. ◆

◆**FOCAL POINT AND FOCAL LENGTH** Consider Equation 16-5 again. The further the object is from the mirror, the closer $\frac{1}{s_o}$ is to zero. When $\frac{1}{s_o} = 0$, $\frac{1}{s_i} = \frac{2}{R}$, so that $s_i = \frac{R}{2}$. This is the position of the image when the object is "at infinity." Recall that "at infinity" really means its distance from the mirror is much greater than any of the other distances or lengths involved. In a practical situation where other distances are on the order of several cm, an object across the room is effectively at infinity.

Figure 16-19a shows that the further the object is from the mirror, the more nearly the rays in the cone intercepted by the mirror are parallel to one another. The point where rays reaching the mirror parallel to one another meet after being reflected (Figure 16-19b) is called the **focal point** F. The distance from the vertex V to the focal point is called the **focal length** (f) of the mirror. Because rays reaching the mirror from a point object at infinity are effectively parallel to one another, the focal point is at the image position (shown above to be $\frac{R}{2}$) of an object at infinity.

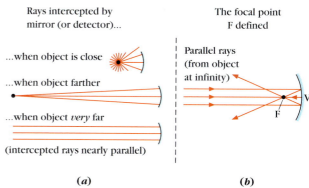

Rays intercepted by mirror (or detector)...

...when object is close

...when object farther

...when object *very* far

(intercepted rays nearly parallel)

The focal point F defined

Parallel rays (from object at infinity)

(a)   *(b)*

Figure 16-19 Parallel rays and the focal point.

Focal length of a mirror:	$f = \dfrac{R}{2}$	(16-8)

so that
$$\frac{1}{s_o} + \frac{1}{s_i} = \frac{1}{f} \qquad (16\text{-}9)$$

Because the focal length is a particular image position, the sign convention for image positions also applies to focal lengths. Equation 16-8 requires that R have the same sign as f, so this sign convention also applies to R. Figure 16-20 summarizes our sign conventions for mirrors.

According to the sign convention for R, the position R of the center of curvature of the convex mirror in the following example is negative. As the example

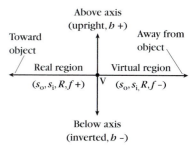

Above axis (upright, b +)

Toward object

Away from object

Real region $(s_o, s_i, R, f +)$

Virtual region $(s_o, s_i, R, f -)$

Below axis (inverted, b –)

Figure 16-20 Sign conventions for mirrors.

shows, this produces calculated results that agree with what a ray diagram leads us to expect.

Example 16-4 *Reflection in a Convex Mirror*

Find the position and height of the image formed by a convex mirror, given the data of Example 16-3. Is the image formed in this case real or virtual, upright or inverted, enlarged or reduced?

Solution

Choice of approach and What we know/what we don't. Same as Example 16-3, except that here we must take R to be negative.

$$s_o = 0.15 \text{ m} \qquad h_o = 0.10 \text{ m} \qquad R = -0.40 \text{ m} \qquad f = \frac{R}{2} = -0.20 \text{ m}$$

$$s_i = ? \qquad\qquad M = ? \qquad\qquad h_i = ?$$

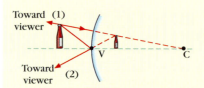

Toward (1) viewer

Toward (2) viewer

The ray diagram. From the top of the bottle we again draw the principal rays through V and C (see figure). The ray "through C" reflects directly back from the mirror without reaching C, but is along the line (shown dashed) that passes through C. Again, to locate the image of the bottle top, we extend the rays that diverge toward the viewer back to the point where they intersect. We now get a virtual image that is upright but reduced.

The mathematical solution. By Equation 16-5,

$$\frac{1}{s_i} = \frac{2}{R} - \frac{1}{s_o} = \frac{2}{-0.40 \text{ m}} - \frac{1}{0.15 \text{ m}} = -11.67 \text{ m}^{-1},$$

so that

$$s_i = \frac{1}{-11.67 \text{ m}^{-1}} = \mathbf{-0.086 \text{ m}}$$

The minus sign tells us the image is virtual. By Equation 16-7,

$$M = -\frac{s_i}{s_o} = -\frac{-0.086 \text{ m}}{0.15 \text{ m}} = 0.57$$

Because M is positive and <1, the image is upright but reduced, as we expect from the ray diagram. Finally, from Equation 16-6,

$$h_i = Mh_o = 0.57(0.10 \text{ m}) = \mathbf{0.057 \text{ m}}$$

◆ Related homework: Problems 16-22 through 16-29 and 16-31.

◆**APPLYING MIRROR EQUATIONS TO PLANE MIRRORS** The greater a mirror's radius of curvature $|R|$, the more gradually it curves. Carried to the extreme, this says that when $R = \pm\infty$ the mirror doesn't curve at all: it's a plane mirror. If $R = \infty$, Equation 16-8 tells us that $f = \infty$ as well. Does this make sense? Remember that the focal point is where rays reaching the mirror parallel to one another meet after being reflected. The law of reflection tells us that when parallel rays strike a plane mirror, the reflected rays are also parallel. They never meet, or equivalently, they meet at infinity.

If $R = f = \infty$, $\frac{2}{R} = \frac{1}{f} = 0$. In that case, Equations 16-5 and 16-9 both become $\frac{1}{s_o} + \frac{1}{s_i} = 0$, and it follows that $s_i = -s_o$. If s_o is positive, s_i is negative. The image is as far behind the mirror plane as the object is in front, as we expect for a plane mirror.

16-4 Refraction

Rays are affected by what they encounter in their paths. When they reach the surface of an object, they may reflect off the surface, as they do in the case of mirrors, or they may pass through the object, as they do in the case of a lens or a glass of water. (Actually, they do both to some extent in all of these cases, but we are concentrating on the dominant behavior.) In this section, we will concentrate on what happens when light passes through a material or medium and what happens when it passes from one medium into another.

Figure 16-21 shows the paths of several rays of light from a laser as they pass from air through a block of glass and into air again. We see that:

1. The rays are straight within each medium. Light appears to travel in straight line paths within a medium (this is true only if the medium is uniform).

2. The rays bend at the points where they pass from one medium into another. The surface where two media (plural of *medium*) meet is called the **interface** between the two media. The bending at this surface is called **refraction.** We say light is refracted at the interface between two media.

Remember that reflection also takes place at interfaces, such as the surface of a still pond—an air-to-water interface. The geometry of ray diagrams is therefore shaped by the events—reflections and refractions—that take place at interfaces. Each time a ray reaches an interface, we must examine what happens to it. The law of reflection tells us which way the ray travels if it is bounced back into the medium in which it has been traveling. To say which way it travels if it passes into the next medium, we need some rule relating the **angle of refraction** (the angle the refracted ray makes with the normal) to the angle of incidence. If we make measurements on Figure 16-21, we can look at the numbers to see if such a rule or pattern exists.

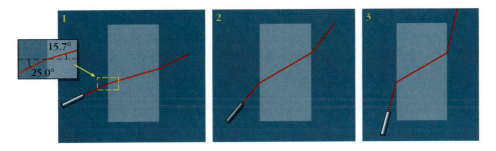

Figure 16-21 **The bending or refraction of rays passing from one medium into another.** Light is refracted on passing from air into a block of glass and is refracted again on leaving the glass.

On-The-Spot Activity 16-3

At each point in Figure 16-21 where a ray crosses the air-to-glass interface, draw a dashed normal. Then measure the angles θ_i and θ_r that the incident and refracted rays make with the normal. We have done this for one ray in the blow-up view in the figure. To make sure you are measuring angles correctly, do the measurements for this ray first and see if your values are within a degree of the given values. Record your data for all three rays in the first two blank columns of the table below.

Ray #	θ_i	θ_r	$\dfrac{\theta_r}{\theta_i}$	$\sin \theta_i$	$\sin \theta_r$	$\dfrac{\sin \theta_r}{\sin \theta_i}$
1	25.0°	15.7°				
2						
3						

Clearly $\theta_i \neq \theta_r$ for refraction. Finding a pattern may be a trial-and-error process. The remaining columns set up some things to try. Might θ_i and θ_r be proportional?

Calculate the ratio θ_r/θ_i for each ray and enter the values in the next column. Is the ratio the same for all rays? In the next two columns, find and enter the sines of θ_i and θ_r for each ray. In the last, calculate and enter the ratio of each pair of sines. Does *this* ratio have the same value for all rays?

With care, you should have found that the ratios in the last column all have the same value. For small incident angles, the ratios θ_r/θ_i also have roughly this value. Let us provisionally label this constant value $k_{1 \text{ to } 2}$, because it is the value for light passing between two particular media 1 and 2. If we did the same table for rays passing between two different media (say, from water to clear plastic), all the values in the last column would again equal one another but would not equal the value that we got for air-to-glass. For refraction, then,

$$\frac{\sin \theta_r}{\sin \theta_i} = k_{1 \text{ to } 2} \quad \left(\text{for small angles, } \frac{\theta_r}{\theta_i} \approx k_{1 \text{ to } 2}\right) \quad (16\text{-}10)$$

STOP&Think Look at what is happening at the glass-to-air interface in Figure 16-21. How does this compare with what is happening at the air-to-glass interface? What should be the numerical value of $k_{1 \text{ to } 2}$ for rays traveling from glass to air? What would be the numerical value of $k_{1 \text{ to } 2}$ if the media on both sides of the interface were identical? ◆

◆**A WAVE MODEL EXPLANATION FOR THE OBSERVED PATTERN** The velocity of a wave depends on the medium in which it is propagating and therefore changes as it travels across an interface between two media. We can get an idea of the effect of the change by considering what happens to the front axle of a vehicle that rolls off the edge of a paved parking lot at the beach onto the surrounding sand (Figure 16-22a). If the car approaches at an angle to the pavement–sand interface, there is an interval during which one wheel moves on pavement while the other moves on sand. Because the wheel still on pavement is not slowed down by the sand, it travels further during that interval. The effect is to alter the direction of the car. The wheels are like two points on any one of the wavefronts in Figure 16-22b, so the direction of travel of the wavefronts (indicated by the red arrow) likewise changes at the interface. A ray indicating the direction of travel is therefore bent toward the normal to the interface (Figure 16-22c) when traveling from a "faster" to a "slower" medium. Traveling the other way across the interface (Figure 16-23), the first and second media are interchanged, and the ray bends away from the normal.

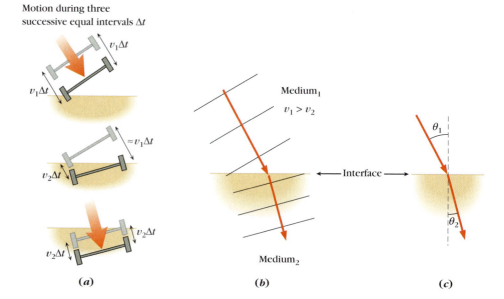

Motion during three
successive equal intervals Δt

$v_1\Delta t$

$v_1\Delta t$

$\approx v_1\Delta t$

$v_2\Delta t$

$v_2\Delta t$

$v_2\Delta t$

Medium$_1$

$v_1 > v_2$

Interface

Medium$_2$

θ_1

θ_2

(a)   (b)   (c)

Figure 16-22 A mechanical analogy to the bending of rays at an interface.

A ray traveling from medium 1 into medium 2 . . .

. . . bends toward the normal if $v_2 < v_1$

. . . bends away from the normal if $v_2 > v_1$

The angle that the wavefront or axle makes with the interface is the same angle that the ray makes with the normal (Figure 16-24). From the geometry in that figure (abstracted from Figure 16-22a),

$$\sin \theta_1 = \frac{v_1 \Delta t}{d} \quad \text{and} \quad \sin \theta_2 = \frac{v_2 \Delta t}{d}$$

Then

$$\frac{\sin \theta_1}{v_1} = \frac{\sin \theta_2}{v_2} \quad (16\text{-}11)$$

because both sides are equal to $\Delta t/d$.

Equation 16-11 can be rewritten as $\frac{\sin \theta_2}{\sin \theta_1} = \frac{v_2}{v_1}$. Because both velocities are constant, this is the same pattern expressed in Equation 16-10 except for a trivial change in the subscript notation. We thus have an underlying picture or model for why that pattern occurs.

◆**INDEX OF REFRACTION AND SNELL'S LAW** It is usual to multiply both sides of Equation 16-11 by c, the speed of light in a vacuum, and to define the ratio c/v that results for each medium as the *index of refraction n* of that medium:

Index of refraction of a medium $\qquad n \equiv \dfrac{c}{v} \qquad (16\text{-}12)$

Equation 16-11 then becomes

Snell's law $\qquad n_1 \sin \theta_1 = n_2 \sin \theta_2 \qquad (16\text{-}13)$

This is also called the **law of refraction.** In a vacuum (and approximately, in air), $v = c$ so $n = c/c = 1$. Because the speed of light in a medium never exceeds its value in a vacuum, n is always ≥ 1 (see Table 16-1). In Equation 16-13, if one

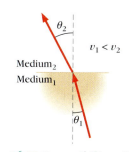

Figure 16-23 Reversibility of light paths. Compare with Figures 16-22b and c.

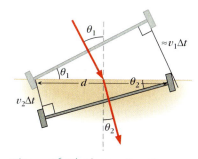

Figure 16-24 Abstracting the geometry from Figure 16-22a.

Table 16-1 Speed and Index of Refraction of Light in Selected Media*

Medium	Speed ($\times 10^8$ m/s)	Index of Refraction $n = c/v$
Vacuum (exact)	2.997925	1.00000
Air (STP)	2.99706	1.00029
Carbon dioxide	2.9966	1.00045
Water (20°C)	2.25	1.33
Aqueous humor and vitreous humor of human eye	2.244	1.336
Ethyl alcohol	2.20	1.36
Sugar solution (25%)	2.18	1.37
Cornea of human eye	2.179	1.376
Crystalline lens of human eye (nonuniform)	2.163 to 2.132	1.386 (inner core) to 1.406 (cortex)
Sugar solution (75%)	2.03	1.48
Glass, crown (typical)	1.98	1.52
Sodium chloride	1.95	1.54
Polystyrene	1.94	1.55
Glass, flint (heavy)	1.82	1.65
Sapphire	1.69	1.77
Glass, flint (heaviest)	1.59	1.89
Diamond	1.24	2.42

*For yellow light ($\lambda = 5.89 \times 10^{-7}$ m).

medium is air, then the bigger the index of refraction of the other medium is, the more the two angles will differ; that is, the more the path of light will be bent. Thus, a medium's index of refraction is an indicator (or "index") of how much bending occurs at an interface between air and that medium.

Because the frequency of a wave is not changed by crossing an interface, $v_1 = f\lambda_1$ and $v_2 = f\lambda_2$, so that Equation 16-11 can also be rewritten as

$$\frac{\sin \theta_1}{\lambda_1} = \frac{\sin \theta_2}{\lambda_2} \tag{16-14}$$

Drawing ray diagrams with refracted rays is identical to the procedure (16-1a) we followed for reflected rays, except that in Step 3 we apply Snell's law instead of the law of reflection.

PROCEDURE 16-1 REVISITED

Drawing Ray Diagrams to Determine Images Resulting from Refraction

For each ray striking a point on the surface, identify
1. the direction of the normal at that point and
2. the angle that the particular incident ray makes with the normal,
3. so that you can apply Snell's law (the law of refraction) separately at that point.

Because rays are refracted when they cross an interface from one medium to another, they appear to diverge from a point other than the source. When you view something through an interface, the image you see will not be located where the object is. For example, when viewed from above, a coin at the bottom of a swimming pool will look closer if the pool is filled with water than if it is drained (Figure 16-25).

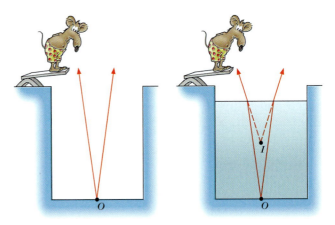

Figure 16-25 Apparent depth due to refraction. The difference between the actual and apparent depths is exaggerated here.

Example 16-5 *The Sparkle of a Jewel*

For a guided interactive solution, go to Web Example 16-5 at www.wiley.com/college/touger

Many jewels are cut with flat surfaces or *facets* to "catch the light" and give them sparkle. To see how this comes about, consider a ray of light that has entered a sapphire and is approaching a flat surface from the inside. Find the angle of refraction (measured from the normal) in air if the incident angle is **a.** 30°, **b.** 60°.

Brief Solution

Choice of approach. We can apply Snell's law, using values of n from Table 16-1. Because the ray is crossing the interface *from* sapphire *into* air, medium 1 is sapphire and medium 2 is air.

What we know/what we don't.

$$n_1 = 1.77 \qquad n_2 = 1.00 \qquad \theta_1 = 30° \textbf{ (a)} \text{ or } 60° \textbf{ (b)} \qquad \theta_2 = ?$$

The mathematical solution.

a. From Equation 16-11,

$$\sin \theta_2 = \frac{n_1 \sin \theta_1}{n_2} = \frac{1.77 \sin 30°}{1} = (1.77)(0.500) = 0.885$$

The angle with this sine is $\qquad \theta_2 = \textbf{62.3°}$

As expected, when light travels into a medium where its speed is greater, it bends away from the normal.

b. By the same reasoning,

$$\sin \theta_2 = \frac{1.77 \sin 60°}{1} = (1.77)(0.866) = 1.53$$

But there is no angle that has a sine greater than 1. Thus there is no refracted beam. Figure 16-26 shows what does happen. Instead of being partially reflected and partially transmitted (and refracted), if the incident ray makes too great an angle with the normal, the ray is entirely reflected back into medium 1. These internal reflections are what give gemstones their sparkle. The facets are cut so that when light enters the gemstone, as much of it as possible leaves through the surface facing the viewer. Light that might exit elsewhere is redirected by internal reflection.

◆ Related homework: Problems 16-39 and 16-40.

STOP&Think In Example 16-5 the ray can be entirely reflected back into medium 1 only if the speed of light is greater in medium 2. Why? ◆

In Example 16-5, we find that when $\sin \theta_2 > 1$, which is impossible for any actual angle θ_2, there is no refracted ray. So what does happen? We consider this next.

◆**CRITICAL ANGLE AND TOTAL INTERNAL REFLECTION** If the speed of light increases on going from one medium to another, a ray crossing the interface in that direction bends away from the normal: The angle of refraction is greater than the angle of incidence. As the incident angle gets larger, so does the refracted angle (Figure 16-26a). Then for some incident angle θ_1 less than 90°, the refracted ray will make an angle of 90° with the normal, and the refracted ray will skim the interface. This value of θ_1 is called the **critical angle** θ_c (Figure 16-26b). By Snell's law (with $\theta_2 = 90°$),

$$\sin \theta_c = \frac{n_2 \sin 90°}{n_1} = \frac{n_2}{n_1} \qquad (16\text{-}15)$$

For incident angles greater than θ_c, there is **total internal reflection** (as in Example 16-5, the would-be refracted angle has a sine greater than 1, but there is no such angle). This is illustrated in Figure 16-27. For a step-by-step visual presentation of what happens as the critical angle is approached and exceeded, work through WebLink 16-3.

Because $n = c/v$, if v increases from medium 1 to medium 2 then $n_2 < n_1$. Only when this is so can we have a critical angle and total internal reflection. (What does Equation 16-15 tell you about the critical angle if $n_2 > n_1$?) As Table 16-1 indicates, gemstone materials typically have high indices of refraction, giving rise to total internal reflections at relatively smaller incident angles and increasing the likelihood of a ray being internally reflected at one surface and then another and then another. Diamonds and other gemstones are purposely cut or faceted to maximize this effect and enhance their sparkle.

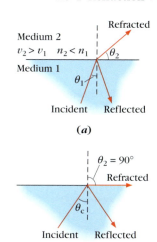

(a)

(b)

Figure 16-26 The onset of total internal reflection. The reflected part of the light always makes the same angle with the normal as the incident ray. When $\theta_1 > \theta_c$, there is only a reflected ray.

Figure 16-27 A total internal reflection. Viewed from below, the surface of the water reflects the portion of the spoon that is below water. One object point and its image point are marked with an ×. (From A. P. Deschanel, *Natural Philosophy,* 13th ed., New York, 1894).

For **WebLink 16-3:**
Critical Angle,
go to
www.wiley.com/college/touger

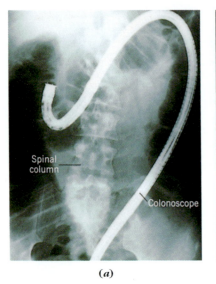

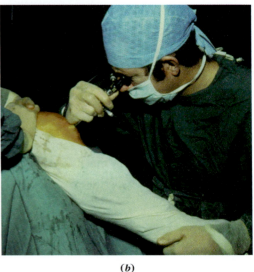

Figure 16-28 Examples of endo-scopes used in medicine. (*a*) This image was obtained using a colonoscope within a human colon. Colonoscopies are performed to look for cancerous abnormalities. (*b*) Here arthroscopic knee surgery is being performed. A fiber-optic arthroscope connected to a miniature television camera can be inserted through a very small incision, permitting the surgeon to see into the joint without making the kind of large incision traditional surgery required.

(*a*)

(*b*)

◆**LASERS AND FIBER OPTICS** Modern technology makes use of total internal reflection in a variety of ways. The production of the light that emerges from a laser requires that the light be reflected back and forth many times between the ends of the laser before it emerges. Each passage stimulates the emission of further light (the word *laser* is in fact an acronym for *l*ight *a*mplification by *s*timulated *e*mission of *r*adiation), but this cannot be effective if light is lost or absorbed through the sides. In ruby lasers, in which the light travels back and forth along the length of a synthetic ruby crystal, the sides of the crystal are highly polished. Light hitting along the sides does so at glancing angles, that is, at large angles with the normal. This, coupled with the fact that gemstone materials such as ruby have high indices of refraction, results in total internal reflection, so that none of the energy of the light beam is lost.

The same principle permits light to travel along great lengths of glass fiber without significant energy loss in a range of applications known collectively as **fiber optics.** Because the critical angle for glass is less than 45°, light initially directed roughly parallel to the axis of a long, thin, flexible glass fiber will always strike the surface at less than the critical angle and be totally internally reflected, so that light signals can be transmitted along bundles of fibers without energy loss. In many applications, fiber optics is replacing electric wiring for communications links, as in interconnected computer systems. In medicine, a variety of fiber optic **endoscopes** (from the Latin "to look within") are used to observe internal areas of the body, such as passages of the digestive system, that could not be accessed by less flexible viewing devices (Figure 16-28).

Water-filled hollow prism in air   Air-filled hollow prism in water

(*a*)   (*b*)

Figure 16-29 Typical and atypical prisms. Rays bend toward the normal at an air-to-water interface and away from the normal at a water-to-air interface.

◆**PRISMS** Prisms are transparent objects with triangular cross-sections that refract light at each surface. In typical prisms, the index of refraction of the material filling the volume of the prism is substantially greater than that of air, and the ray bends *toward the base of the triangle* (Figure 16-29*a*). However, if the inside and outside media are reversed (Figure 16-29*b*), the ray may be bent away from the base.

45°-45°-90° triangular glass prisms are often used in optical instruments. Because the critical angle of glass is less than 45°, light striking the surface from within is totally internally reflected. These prisms can therefore be used for the range of purposes shown in Figure 16-30 with virtually no loss of light. Although mirrors could be used for some of these purposes as well, they do not reflect as great a percent of the light, and their reflecting coatings deteriorate with age. Binoculars (Figure 16-30*d*) often use prisms to redirect rays and turn inverted images upright. **STOP & Think** Where does the image form in Figures 16-30*c* and *d*? ◆

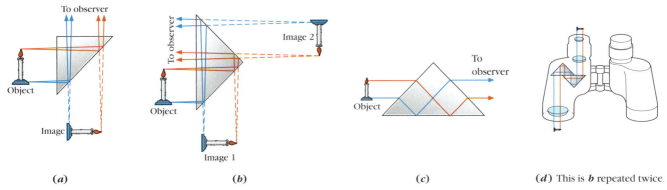

(a)   **(b)**   **(c)**   **(d)** This is **b** repeated twice.

Figure 16-30 **45°-45°-90° prisms are used to change the position or orientation of an image.** The prisms in *a* and *b* allow the observer to see the image in a different direction than the object. The prism in *c* inverts the image. In *d*, two such prisms redirect rays. Because the light-gathering lenses of the binoculars must be large, their centers are widely spaced. The redirection allows the rays to reach your less widely spaced eyes. The object distances, reduced here to fit into the diagram, are generally large, so that the rays reaching the prism are paraxial. In *c* and *d*, the path of only one paraxial ray is shown.

◆**DISPERSION** Although all frequencies of light travel at the same speed in a vacuum, measurements show that these frequencies are not equally slowed down on entering a medium. Because the speeds v differ for different colors in the medium, the indices of refraction $n = c/v$ are also different for different colors.

The values in Figure 16-31 show that in a typical glass, red light is slowed down least and violet most. It follows from Equation 16-11 that at a glass–air interface, red light is bent least and violet light most, whether toward the normal when entering the glass or away from the normal when leaving it. When rays of all wavelengths travel in the same direction, our eyes perceive them collectively as white light. When they enter a prism, they emerge in different directions (Figure 16-31) so that we see the colors spread out as a **spectrum,** an ordered range or display of colors or wavelengths. This spreading out of the colors is called **dispersion.** Dispersion of internally reflected light enhances the sparkle of gemstones, enabling us to see glints of color in diamonds, for example.

Dispersion by a prism enables us to see which colors or wavelengths are present in light from a particular source. In this way, it has been found that light from a hot gas of a single element, such as neon or mercury, contains only certain wavelengths or colors and not the in-between ones (Figure 16-32). Which ones are present is different for each element. In other words, each element has

red ($n = 1.613$)
yellow ($n = 1.621$)
green ($n = 1.628$)
blue ($n = 1.636$)
violet ($n = 1.661$)

Figure 16-31 Dispersion in silicate flint glass.

Neon (Ne)

Mercury (Hg)

Solar spectrum

Figure 16-32 Line emission spectra for different elements contrasted with the spectrum observed for sunlight.

its own unique spectrum, and this provides us with a way of identifying an element. By dispersing light from the sun or a distant star, we are therefore able to determine what elements are present in those bodies.

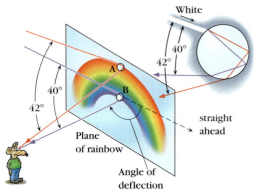

Figure 16-33 Dispersion and partial internal reflection: the ingredients of a rainbow.

◆**RAINBOWS** Rainbows occur when drops of water in the atmosphere refract and disperse rays of light from the sun (Figure 16-33). As the enlargement shows, light reaching a spherical drop is refracted toward the normal on entering, then undergoes a partial internal reflection—like light striking a shop window, part of the sunlight is reflected and part transmitted. Finally, it is refracted away from the normal on leaving. Because of dispersion, the angle between the incoming and outgoing rays differs for the different colors. Only rays leaving drop A at a particular angle with the incoming ray will reach the eye of our observer. From drop B, lower in the sky, only rays leaving at a smaller angle with the incoming ray reach the observer's eye (other rays pass above or below the eye). For the angles shown, our observer will detect red light coming from A, violet light from B, and intermediate colors from the in-between points. When parallel rays of the same wavelength strike different parts of the spherical surface of droplets, they get deflected back toward the viewer at different angles. The detailed explanation of what happens to light in a rainbow by René Descartes (1596–1650) was an early triumph of geometric optics. By drawing careful ray diagrams for many droplets, it is possible, as Descartes did, to show that each wavelength (color) gets deflected most strongly within a small range of angles clustering around the angle of maximum deflection (that is, most deflected from the forward direction) for that wavelength. For example, violet is strongest at a deflection angle of $180° − 40° = 140°$ (Figure 16-33). The region of sky from which light bouncing back at this angle reaches our eyes is shaped like an arc, so that arc of the sky appears purple to us. (Light of other wavelengths reaches our eyes from this arc too, but not as strongly.)

◆ SUMMARY ◆

To explain image formation, we have introduced a model in which

• Light travels in straight-line **rays** in all directions from the source.
• The rays show all the wave paths—all the directions in which the waves travel from the source—and are perpendicular to the wavefronts.
• An object will intercept *a cone of rays* from each point on the source.

The directions of rays are changed at **interfaces** by reflection and/or refraction. The directions of rays reaching your eye are a result of those changes. An **image point** is a point from which the light rays appear to spread out or diverge as they approach your eye.

You need to consider at least two rays to identify an image point.

Rays sometimes appear to diverge from a point that no light actually passes through. The *apparent* paths through

regions where no light actually travels are represented by *dashed lines* in ray diagrams.

Type of image point	The light reaching your eye from the point . . .	So all the rays drawn through the point are . . .
Real	passes through the point	drawn as solid lines
Virtual	doesn't pass through point	drawn as dashed lines

Extended sources must be envisioned as a composite of point sources. We must in principle determine an image point for each object point. (In practice we usually can consider just the outermost points.)

In **geometric optics,** we draw **ray diagrams** to determine images, then we do geometry on these diagrams to determine relationships among linear distances giving the sizes of objects and images and their distances from specified points. Some rays' paths are governed by the **law of reflection** (as in Figures 16-9 to 16-14), others are governed by **Snell's law** (see below) for refraction (Figures 16-21 to 16-24). We must apply these principles in constructing ray diagrams as in **Procedure 16-1** and **Procedure 16-1 Revisited** to determine images resulting from reflection and refraction: *For each ray* striking a point on a surface or interface, identify

1. the direction of the normal at that point and

2. the angle that the particular incident ray makes with the normal,

3. so that, after determining which is appropriate, you can apply the law of reflection or Snell's law for refraction separately at that point.

For mirrors, we apply the **law of reflection:**

$$\angle \text{of incidence} = \angle \text{of reflection} \qquad (16\text{-}1)$$

Doing geometry on ray diagrams for reflected rays, we find these relationships:

For a **plane mirror** (o = object, i = image, s = distance from mirror, b = height)

$$s_o = s_i \qquad (16\text{-}2)$$

and

$$b_o = b_i \qquad (16\text{-}3)$$

For spherical mirrors, the radii of the sphere are the normals to the surface. For these mirrors, paraxial rays (those nearly parallel to the axis) very nearly converge to the same point on reflection, but those further out do not. The deviation from perfect convergence is called **spherical aberration.**

In drawing a ray diagram, we need only draw the **principal rays** (the *incident* rays for which it is easiest to determine the direction of the reflected or refracted ray).

Procedure 16.2 identifies principal rays for spherical mirror diagrams:

1. The ray from the object point through the center of curvature C.

2. The ray from the object point to the vertex V.

Doing geometry on ray diagrams gives us the following for a curved mirror with *radius of curvature R*:

$$\frac{1}{s_o} + \frac{1}{s_i} = \frac{2}{R} \qquad (16\text{-}5)$$

We define magnification:

Magnification	$M \equiv \dfrac{b_i}{b_o} = -\dfrac{s_i}{s_o}$	(16-6 and 16-7)

M is + if the image is upright.

M is − if the image is inverted.

$|M| > 1$ if the image is enlarged.

$|M| < 1$ if the image is reduced in size.

We define the mirror's **focal point** F as the point where rays reaching the mirror parallel to one another meet after being reflected. The distance from the vertex V to the focal point is called the **focal length** (f) of the mirror.

Focal length of a mirror:	$f = \dfrac{R}{2}$	(16-8)

so that

$$\frac{1}{s_o} + \frac{1}{s_i} = \frac{1}{f} \qquad (16\text{-}9)$$

Procedure 16-3 stresses that in doing geometric optics calculations, you should

1. carefully sketch a ray diagram showing the *principal rays* to establish an expectation of what the image should look like, and

2. do the calculation *with careful attention to sign conventions* and make sure the results agree with your expectations: Is the image *real or virtual? upright or inverted? enlarged or reduced?*

Figure 16-20 summarizes the sign conventions for mirrors.

Refraction is the bending of light that takes place at an interface between media because the speed of light is different in the two media.

> A ray traveling from medium 1 into medium 2 . . .
> . . . bends toward the normal if $v_2 < v_1$
> . . . bends away from the normal if $v_2 > v_1$

The **index of refraction** of a medium

$$n \equiv \frac{c}{v} \qquad (16\text{-}12)$$

($c = 3.00 \times 10^8$ m/s is the speed of light in a vacuum)

Refraction is then governed by Snell's law:

Snell's law:	$n_1 \sin \theta_1 = n_2 \sin \theta_2$	(16-13)
or	$\dfrac{\sin \theta_1}{v_1} = \dfrac{\sin \theta_2}{v_2}$	(16-11)
or	$\dfrac{\sin \theta_1}{\lambda_1} = \dfrac{\sin \theta_2}{\lambda_2}$	(16-14)

When the speed of a particular color light is changed, its wavelength $\lambda = v/f$ is correspondingly changed.

When light reaches an interface with a medium having a lower index of refraction, there is **total internal reflection,** and consequently no refracted ray, if the incident angle exceeds a **critical angle** θ_c such that

$$\sin \theta_c = \frac{n_2 \sin 90°}{n_1} = \frac{n_2}{n_1} \qquad (16\text{-}15)$$

In media other than a vacuum, light waves of different colors (different frequencies) travel at different speeds v, so that their indices of refraction $n = c/v$ also differ. As a consequence, the colors making up white light are bent by different amounts on crossing an interface (Figure 16-31). The spreading out that results is called **dispersion.**

◆QUALITATIVE AND QUANTITATIVE PROBLEMS◆
Hands-On Activities and Discussion Questions

The questions and activities in this group are particularly suitable for in-class use.

16-1. Hands-On Activity. In this activity, you will explore images formed by plane mirrors. You should do this activity with a partner. You will need a large wall mirror, a sheet of thick paper or cardboard, a sheet of clear tracing paper or transparent acetate, and a marker. It is important that when you write with the marker, it doesn't show through the paper or cardboard but shows clearly through the tracing paper or acetate.

a. Does a plane mirror reverse left and right? Write down your answer.

b. Write your name in letters large enough to cover most of your sheet of paper and cardboard. Then hold it up so you see your name reflected in the mirror. Are left and right reversed? Explain.

c. Have your partner stand to your right with both of you facing the mirror. Is your partner's image to the left or right of your image? Are left and right reversed in the reflected image? Explain.

d. Put the tracing paper or acetate over your name and trace your name onto it. Now hold the tracing paper up between you and the mirror so that your name reads left to right on the tracing paper as it is facing you. Does your name read left to right or right to left on the mirror image? Are left and right reversed in the reflected image? Explain.

e. Now hold up the paper between you and the mirror and have your partner do the same with the tracing paper/acetate, so that each of you can read your name left to right on the sheet you are holding. Then rotate your paper or cardboard 180° (about a vertical axis) so that it faces the mirror. Have your partner perform the identical motion with the tracing paper. Does your name read right to left or left to right on each of the two mirror images? Has the mirror reversed left and right, or have you?

f. Is your partner's left hand on the left or right side of his or her mirror image? Have your partner turn so that his or her back is to the mirror. *Now* is your partner's left hand on the left or right side of his or her mirror image? Has the mirror reversed left and right, or has your partner?

g. Draw coordinate axes on the tracing paper/acetate, so that the origin is at the center near the bottom, and the $+x$ and $+y$ axes go to the right and upward, respectively. Now have your partner face away from the mirror, holding the tracing paper over his or her head so that the mirror image reads left to right. Your partner should be holding the tracing paper with one hand gripping it at each end. Is your partner's left hand in the $+x$ or $-x$ direction? What about your partner's right hand?

h. Now, moving only his or her feet, your partner should turn to face the mirror. Has the direction of the $+x$ axis as seen in the mirror changed? *Now* is your partner's left hand in the $+x$ or $-x$ direction? What about your partner's right hand? How does this compare with your answers to *g*?

i. Would your answers to *g* and *h* have been the same if your coordinate frame were drawn on the mirror instead of on the tracing paper/acetate? What is your partner reversing by turning around? Is the mirror reversing anything?

j. Why doesn't this activity work very well if your name is OTTO or AVA?

Review and Practice

Section 16-1 Why Geometry? Looking at Shadows

16-2. A tree casts a shadow 40 m long on level ground when the sun is at an angle of elevation 30° above the horizon. How tall is the tree?

16-3. SSM WWW

a. Stella Starlite, a 1.80-m-tall actress, stands 5.00 m from the edge of the stage. A single footlight (assumed to be a point source) at the edge of the stage directly in front of her projects a shadow on the backdrop 7.00 m behind her. How tall is her shadow?

b. Dustin Case, her 1.60-m-tall costar, also has a footlight directly in front of him. How far back from the footlight should he stand if he wants his shadow to be as tall as Ms. Starlite's?

16-4. Suppose that in Problem 16-3*a,* the backdrop is raised while Ms. Starlite is standing in the position described. She then casts a shadow 5.40 m high on the rear wall of the stage. How far behind Ms. Starlite is this wall?

16-5. SSM A room is lit by a single hanging bulb centered over a square table, so that the table casts a shadow on the floor. If the table is then pushed a short distance to one side, will the size of the shadow increase, decrease, or remain the same? Briefly explain.

16-6. A single bulb is hung over one corner of a rectangular table 1.2 m long by 0.8 m wide. The table is 0.6 m high, and the bulb hangs 1.8 m above the tabletop. Find the dimensions of the shadow that the table casts on the floor.

16-7. The wall in Figure 16-34 is lit by two point sources S_1 and S_2. An opaque screen stands in front of the wall. Rank the four labeled points on the wall, from least to greatest, in order of how brightly they are lit. Be sure to indicate any equalities.

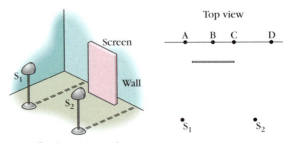

Figure 16-34 Problem 16-7

16-8. A shadow puppet is held at a point between an extended light source and a screen, such that the shadow on the screen consists of both an umbra and a penumbra. Use ray diagrams to answer each of the following:

a. If the screen is moved toward the puppet, what happens to the size of the umbra? What happens to the size of the penumbra?

b. Repeat ***a*** for the case where the screen is moved away from the puppet.

c. Suppose that the screen and puppet both remain fixed, but the light source is moved toward both of them. How do the umbra and penumbra change in size?

16-9.

a. When a humble creature stands in front of an extended light source as shown in Figure 16-35, at what minimum distance behind our humble creature could a screen be placed so that the creature's shadow includes penumbra but no umbra?

b. How tall would the penumbra be at this distance?

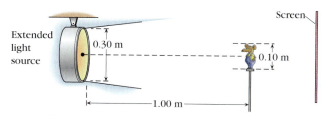

Figure 16-35 Problem 16-9

Section 16-2 Reflection and Mirror Images

16-10. Explain in your own words what is meant by an image.

16-11. A lit candle is placed 1 m in front of a plane mirror.

a. When you "see the flame in the mirror," where does the light reaching your eye actually come from (for example, from the mirror surface, from in front of the mirror surface, from between the mirror surface and 1 m behind it, from 1 m behind the mirror surface, or some other choice)?

b. Where does it *appear* to come from?

c. Which of questions ***a*** and ***b*** is asking, "Where is the image of the flame?"

d. Does light actually travel from the image of the flame to your eye?

16-12. In Figure 16-36, for which (one or more) of objects A, B, C, D, and E can an observer at O see an image in the plane mirror?

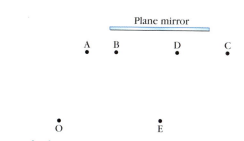

Figure 16-36 Problem 16-12

16-13. SSM In Figure 16-13, suppose one mirror is 1.0 m in front of the barber, and the other mirror is 2.0 m behind him.

Draw ray diagrams to determine the positions of the first four images of the barber.

16-14. Suppose the barber in Figure 16-13 is 0.4 m tall. If one mirror 1.0 m in front of him, and the other is 2.0 m behind him, how tall is each of the first three images?

16-15. SSM WWW A 0.12-m-tall candle stands upright between two vertical plane mirrors A and B. The mirrors are parallel to each other. The candle is 0.20 m from mirror A and 0.30 m from mirror B. (*Hint:* As a first step, sketch the situation.)

a. Find the positions of the two nearest images of the candle in each mirror.

b. Find the height of each of these images.

16-16. You take a photograph of a friend standing a short distance in front of a full-length mirror. In the photograph, your friend appears 12 cm high. The picture of her mirror image is only 8 cm high. Does this contradict the claim that a plane mirror produces an image size equal to the size of the object? Briefly explain.

Section 16-3 Reflections from Curved Mirror Surfaces

16-17. Concave mirrors are used in certain telescopes (called reflecting telescopes) to form images of distant objects. Which of the spherical mirrors in Figure 16-37 would be better for this purpose? Briefly explain why.

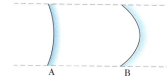

Figure 16-37 Problem 16-17

16-18. When a tiny object is placed 2.0 m in front of the nearest point on a reflecting sphere, its image appears 0.40 m behind that point. What is the radius of the sphere?

16-19. How are the ray diagrams for Examples 16-3 and 16-4 similar? How are they different?

16-20. A concave spherical mirror has a radius of curvature of 0.70 m. Find the position of the image formed for an object placed at each of the following distances in front of the vertex of the mirror:

a. 0.20 m.   ***b.*** 0.35 m.   ***c.*** 0.70 m.   ***d.*** ∞.

16-21. SSM WWW Suppose the object in Problem 16-20 is 0.10 m tall. Find the height of the image formed when the object is at each of the positions in Problem 16-20. Then, in each of the four parts, indicate whether the image is real or virtual, upright or inverted, enlarged or reduced.

16-22. Repeat Problem 16-20 for the same given values if the mirror is convex rather than concave.

16-23. Suppose the object in Problem 16-22 is 0.10 m tall. Find the height of the image formed when the object is at each of the positions in Problem 16-21. Then, in each of the four parts, tell whether the image is real or virtual, upright or inverted, enlarged or reduced.

16-24. In Problem 16-23, where should the object be positioned if the image height is to be a quarter of the object height?

16-25. Sketch a ray diagram to determine the image of a small upright object placed at the center of curvature of a concave spherical mirror.

16-26. An object stands upright in front of a spherical mirror with a radius of curvature of 1.00 m ("in front" implies you should consider only real object positions).

a. If the mirror is concave, for what position or range of positions will the object have a real image and for what position or range of positions will it have a virtual image?

b. If the mirror is convex, for what position or range of positions will the object have a real image and for what position or range of positions will it have a virtual image?

16-27. Suppose the situation is as stated at the start of Problem 16-26.

a. If the mirror is concave, for what position or range of positions will the object have an upright image and for what position or range of positions will it have an inverted image?

b. If the mirror is convex, for what position or range of positions will the object have an upright image and for what position or range of positions will it have an inverted image?

16-28. Suppose the situation is as stated at the start of Problem 16-26.

a. If the mirror is concave, for what position or range of positions will the object have an enlarged image and for what position or range of positions will it have an image reduced in size?

b. If the mirror is convex, for what position or range of positions will the object have an enlarged image and for what position or range of positions will it have an image reduced in size?

16-29.

a. Is there anyplace outside of a reflecting sphere where you can place an object so that the image produced by the sphere is larger than the object? Briefly explain.

b. Suppose you place an object somewhere within a sphere that has a mirrored interior surface. Is there anyplace inside this sphere where you can position the object so that its image will be larger than the object itself? Briefly explain.

16-30. A billiard ball and a basketball are painted with a reflecting coating. Which has the greater focal length?

16-31. An actor is applying make-up using a concave make-up mirror with a focal length of 0.20 m.

a. If the actor sees his eye to be twice its actual size, how far is his eye from the mirror if the image is virtual?

b. How far from the mirror should the actor's eye be to produce a real image that is twice the size of the eye itself?

c. Could the actor actually see such an image? Explain.

16-32. Sketch a ray diagram to determine the image (if any) of a small upright object placed at the focal point of a concave spherical mirror.

16-33. A useful rule of thumb is that an object can be considered to be "at infinity" when it is more than 100 focal lengths from the mirror. When the object is 100 focal lengths from the mirror, by what percent does the image position differ from the focal length f?

16-34. A mirror that has an infinite focal length must be a (*convex; concave; plane*) mirror.

Section 16-4 Refraction

16-35. Suppose light passes from air into blocks A and B. Light travels more slowly in B than in A.

a. Which block has the higher index of refraction?

b. In which block is the path of the light bent more on entering?

16-36. Medium A and medium B both have an index of refraction of 1.40. As a light ray passes from medium A into medium B, it (*bends toward the normal; doesn't change its direction; bends away from the normal*).

16-37. Make a general statement relating the index of refraction of a medium to the degree of bending that a ray experiences when passing from air into the medium.

16-38. Most analogies are limited; the two situations being compared are alike in specific ways, but they may be different in others. In what ways is the behavior of the axle in Figure 16-22 different than the wavefronts with which it is being compared?

16-39. **SSM** A narrow beam of light passes through a thick pane of glass with an index of refraction of 1.52. The surfaces of the pane are parallel, and the medium on both sides of the glass is air.

a. If the beam of light makes an angle of 30° with the normal to the pane of glass, what angle does it make with the normal within the glass?

b. What angle does it make with the normal when it reemerges into the air on the other side of the glass?

16-40.

a. Suppose that in Problem 16-39, the medium on both sides of the glass is water rather than air. Without doing a calculation, decide whether the angle that the beam makes with the normal within the glass will be greater than, the same as, or less than it was before. Briefly explain your reasoning.

b. Now repeat the calculation in Problem 16-39*a* with water rather than air as the surrounding medium, and see whether the calculated result meets the expectations you stated in Problem 16-40*a*.

16-41. A thin ray of light enters a container of room-temperature liquid from the air. Its angle of incidence is 45°. The ray in the water makes an angle of 30° with the normal. Can the liquid be pure water? On what information or numerical value(s) do you base your answer?

16-42. Figure 16-38 shows four rays of light from the same monochromatic laser source. Each ray passes from one medium to another. Water is always one of the two media; the other media (A, B, C, and D) are all different. Rank these media according to each of the following, listing them in order from least to greatest and indicating any equalities.

a. The index of refraction of the medium.

b. The wavelength of the light in the medium.

c. The frequency of the light in the medium.

Figure 16-38 Problem 16-42

16-43. Figure 16-39 shows a ray of light traveling from air into a block of glass. The ray is partly reflected (at an angle ϕ_r to the normal) and partly transmitted (at an angle ϕ_t to the normal) at the surface of the glass. If the glass is then replaced by a different type of glass that has a greater index of refraction, how is each of the following affected

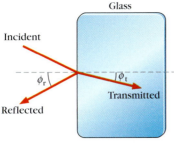

Figure 16-39 Problem 16-43

a. The transmitted angle ϕ_t.

b. The angle of reflection ϕ_r.

c. The speed of the transmitted light.

d. The wavelength of the transmitted light.

e. The frequency of the transmitted light.

16-44. An open beaker is filled with ethyl alcohol. A beam of light from a sodium lamp is directed at the liquid surface from above. At what angle to the normal did the ray enter the alcohol if

a. it makes an angle of 24° with the normal in the alcohol?

b. it is directed along the normal in the alcohol?

c. If light from a sodium lamp has a wavelength of 5.89×10^{-7} m in air, what is its wavelength in ethyl alcohol?

16-45. Calculate the critical angle for total internal reflection in a diamond.

16-46. Light passes through a certain medium. When it encounters an interface between this medium and air, it is totally reflected back into the medium if it strikes the interface at an angle with the normal exceeding 62°. What is the index of refraction of this medium?

16-47.

a. Does a fiber optic strand lose its energy-carrying efficiency if it is bent?

b. Does it matter how sharply it is bent? Briefly explain.

16-48.

a. Find the velocity in the glass of each of the colors of light in Figure 16-31.

b. Of the colors of light shown, which travels slowest in glass, and which travels fastest?

16-49. You are given a water-filled prism and decide to do measurements to determine the index of refraction of water using a helium-neon laser, which produces monochromatic red light of wavelength 6.328×10^{-7} m. Will the value of n that you obtain from your measurements be less than, equal to, or greater than the value given in Table 16-1?

16-50. A 45°-45°-90° prism can be used to reflect parallel rays so that outgoing rays make an angle of 90° with the incoming rays. The same thing can be accomplished with a plane mirror.

a. What angle should the surface of the mirror make with the incoming rays to produce this result?

b. What is the advantage of using the prism rather than a mirror?

Going Further

The questions and problems in this group are not organized by section heading, so you must determine for yourself which ideas apply. Some of them will be more challenging than the Review and Practice questions and problems (especially those marked with a • or ••).

16-51. **SSM** Bright indirect sunlight enters the window of the art gallery in Figure 16-40. A panel for displaying art (shown in end and front views) extends horizontally across the room. By sketching appropriate rays and doing measurements,

a. use the scale given in the figure to determine the height of the umbra, the region of the wall that is in total shadow because of the panel (neglecting reflections from surfaces within the room).

b. determine the minimum distance from the wall at which the panel will produce no umbra on the wall.

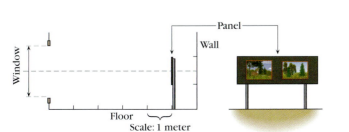

Figure 16-40 Problem 16-51

16-52. Circular disks A and B in Figure 16-41 are coaxial; that is, they have the same center axis. A is a light source; B is opaque and can cast a shadow on the screen. At what distance should B be placed from the screen for

a. the umbra (the region of the screen which is totally in shadow) to have a diameter equal to half the diameter of B?

b. the penumbra (the region of the screen which is in partial shadow) to have an outer diameter equal to half the diameter of B?

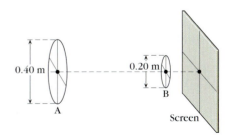

Figure 16-41 Problems 16-52, 16-53, and 16-54

16-53. Again consider the situation in Figure 16-41. Suppose that in Problem 16-52*a* disk B is kept at the same distance from the screen and remains parallel to it but is moved up or down, so that it is no longer centered on the common axis. Will the size of the umbra increase, decrease, or remain unchanged? Explain your reasoning.

•16-54. Suppose that no numerical values are given in Figure 16-41. Instead assume that the coaxial source disk A and opaque disk B have radii r_A and r_B, respectively, and are at distances d_A and d_B from the screen (where $d_A > d_B$). By means of ray geometry, show that the radius r_u of the umbra formed on the screen is

$$r_u = \frac{d_A r_B - d_B r_A}{d_A - d_B}$$

when the numerator of this expression is positive or zero, and that there is no umbra when the numerator of this expression is negative.

16-55. Is it possible to move a pencil so that the shadow it casts has a greater speed than the pencil itself? If yes, briefly explain how. If no, briefly explain why not.

•16-56. Plane mirrors 1 and 2 in Figure 16-42 form a 60° angle. Observer O and object A are positioned on the bisector of this angle, as shown.

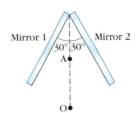

Figure 16-42
Problem 16-56

a. By sketching a ray diagram, find the position of the largest image of object A that observer O sees in mirror 1.

b. Now find the position of the second largest image of object A that observer O sees in mirror 1.

•16-57. A beam of light travels horizontally from a source to a vertical screen. When a pane of transparent material ($n = 1.5$) is placed between the source and the screen as shown in Figure 16-43, the ray strikes the screen 1.2 cm below where it struck when the pane wasn't there. Find the thickness of the pane. (*Note:* The pane's thickness is measured perpendicular to its faces.)

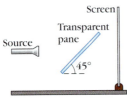

Figure 16-43
Problem 16-57

16-58. You have a large flexible sheet of reflecting material that you are going to use for a carnival mirror. The sheet can be bent about its vertical axis (Figure 16-44a) with its right and left edges either toward or away from you. It can be bent about its horizontal axis (Figure 16-44b) with its top and bottom edges either toward or away from you. Finally, combinations of these bendings, such as the one shown in Figure 16-44c, are possible. How should you bend the sheet if you wish your image to appear

a. tall and thin?   c. short and thin?

b. tall and fat?   d. short and fat?

In each case, tell whether it matters how far from the mirror you stand.

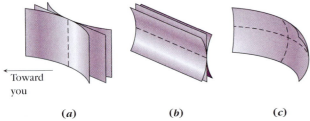

Figure 16-44 Problem 16-58

16-59. **SSM** By doing the necessary algebra, show that if $\frac{s_i}{s_o} = \frac{R - s_i}{s_o - R}$ (the equation immediately preceding Equation 16-5), then $\frac{1}{s_o} + \frac{1}{s_i} = \frac{2}{R}$ (Equation 16-5).

16-60. Equations 16-4,I and 16-4,II were obtained by doing geometric reasoning on the two pairs of similar triangles identified in Figure 16-17.

a. Starting with the pairs of similar triangles in Figure 16-18, see if you can derive the same two equations. (Don't forget about sign conventions.)

b. Identify and sketch the pairs of similar triangles from the ray diagram in Example 16-4 that you could use to derive these equations.

16-61. The inside and outside surfaces of a large hollow sphere are painted with a reflecting coating.

a. Is there any place outside the sphere that you can stand if you wish your image to appear at the center of the sphere? If so, where? If not, why not?

b. Is there any place inside the sphere that you can stand if you wish your image to appear at the center of the sphere? If so, where? If not, why not?

16-62. The blocks of glass (shaded) in Figure 16-45 are surrounded by air. Each has an index of refraction of 1.60. A ray of yellow light enters each block. The entry points and directions are indicated by the arrows. In each of cases *a–f*, (*i*) find the coordinates of the point where the ray leaves the block, and (*ii*) give the direction of the exiting ray by telling what angle it makes with the positive *x* axis.

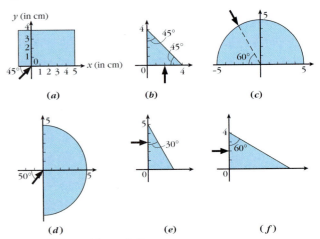

Figure 16-45 Problem 16-62

16-63. The effect of a prism is often represented by diagrams like Figure 16-46. Why is this not a proper ray diagram? How does it differ from Figure 16-30a?

16-64. The atmosphere is densest closest to Earth's surface.

a. Does light from a distant star travel at exactly the same speed all the way from the source to a telescope in a sea-level observatory on Earth? Briefly explain.

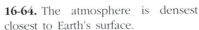

Figure 16-46
Problem 16-63

b. An astronomer in this observatory must direct a telescope at an angle of 45° above the horizontal to observe a certain star. Is the star's actual angle of elevation greater than, equal to, or less than 45°? Briefly explain.

16-65. Two students attempt to find the index of refraction of clear corn syrup by filling an open-topped rectangular glass container with the syrup and shining a narrow beam of light on one of the surfaces, then measuring the angles that the beam makes with the normal in air and in the syrup and applying Snell's law. Student A aims her light through one of the glass sides of the container. Student B aims his light from above. Student B says he is doing this because "otherwise the light will bend as it passes from air into glass, and that will alter the refracted angle that we observe in the syrup." Student A says, "It won't make any difference. If the beam makes a given angle with the normal in air, it will have the same refracted angle in the syrup whether it passes through glass first or not." Which of the two students is correct? Briefly explain why.

16-66. The point object O in Figure 16-47 is surrounded by four mirrors. The reflecting surfaces all face the object. The square grid lets you compare distances. Think only about the primary images—the images each mirror would form if the others weren't there. Rank the mirrors in order, from nearest to furthest, of how far the images are from the object. Indicate any equalities.

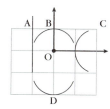

Figure 16-47
Problem 16-66

•• **16-67.** Figure 16-48 shows the shadows cast on a wall by a standing opaque screen 0.8 m wide. Determine how to arrange the screen and two point light sources to produce these shadows. Calculate any distances you need to know. Draw a diagram showing a top view of your arrangement (similar to the top view in Figure 16-34) and indicating all relevant distances.

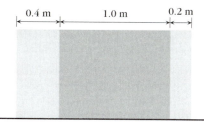

Figure 16-48 Problem 16-67

16-68. How far would the moon have to be from Earth to never experience a total eclipse? (Consult tables to obtain any relevant distances and sizes.)

16-69. When a glass tank filled with concentrated sugar solution is left to stand, the concentration of sugar becomes gradually greater toward the bottom of the tank. If a laser beam is directed into the tank from one side at an angle slightly above the horizontal, which way will the beam bend? Briefly explain. Refer to the values of index of refraction for sugar solutions in Table 16-1.

16-70. On very hot days, the ground and paved surfaces absorb a great deal of heat and transfer that heat to the adjacent air. The air closest to the ground becomes hotter, and therefore less dense, than the air above it.

a. Based on the information in Table 16-1, will the index of refraction of the hotter air be greater than, equal to, or less than the cooler air above it?

b. If a ray of light starts out traveling in a direction 20° below the horizontal, will it bend toward the horizontal, bend away from the horizontal, or continue in the same direction? (This behavior contributes to the formation of mirages and to the illusion of wetness that you sometimes see on highways on very hot days.)

16-71. In Figure 16-49, a 5-m-long boom is mounted at a point 8 m from the building. A floodlight is directed at the building from a point at street level 8 m further from the building. Find the height of the shadow that the boom casts on the building

a. when it is in position A.

b. when it is in position B.

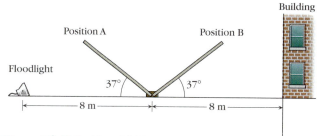

Figure 16-49 Problem 16-71

16-72. You are standing midway between two parallel plane mirrors A and B. The nearest image that you see in A appears 1 m behind the surface of A. How far behind the surface of A will the next nearest image that you see in A appear?

16-73. A single plane mirror reverses _____

a. left and right only.

b. up and down only.

c. both left and right and up and down.

d. the forward and backward directions only.

• **16-74.** A group of students planning for a homecoming event decide that it would be fun to have a set-up where, when you look at yourself in a mirror, you see yourself upside down. Devise a way of accomplishing this using plane mirrors, and sketch a ray diagram in enough detail to show that it indeed inverts your image. What is the minimum number of plane mirrors you could use to accomplish this?

• **16-75. SSM WWW** Suppose the water in the pool in Figure 16-25 is 1.6 m deep. How deep will it appear to be to the individual looking down at it? (Determine the apparent depth by doing geometry on the ray diagram, and using the index of refraction of water.)

• **16-76.** A child at a day camp, a non-swimmer, stands at the edge of the swimming pool. The child's nose is at a height of 1.0 m. From above, the water appears to be 0.80 m deep. Is it safe for the child to jump in? Provide a calculation or other quantitative reasoning to support your answer.

16-77. A tank of exotic fish in Sea World has a glass front and a painted backdrop. The actual distance from the glass to the backdrop is d_1, but the distance appears to be d_2 when seen through the water. If the actual distance d_1 is 0.30 m, what is the apparent distance d_2?

16-78. A baby sits on the floor between two plane mirrors A and B (Figure 16-50) and rolls a ball toward mirror A with a speed of 0.8 m/s. Assuming that the positive direction is toward mirror B, what is the velocity *relative to the ball* of the nearest image of the ball

a. in mirror A?　　*b.* in mirror B?

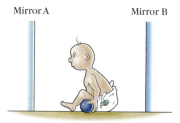

Figure 16-50 Problem 16-78

16-79. The double-slit set-up in Figure 16-51 has a mono-chromatic (single-color) light source S. The set-up is ordinarily used with air filling the region between the double slit and the screen.

a. If a block of glass is then inserted to fill the entire region between the double slit and the screen (Figure 16-51*a*), will

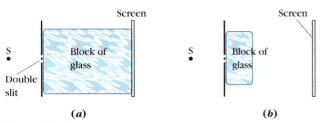

Figure 16-51 Problems 16-79 and 16-80

the spread of the interference pattern increase, decrease, or remain the same? Briefly explain.

b. If the block of glass fills only part of the region between the double slit and the screen (Figure 16-51*b*), will the spread of the interference pattern be greater than, less than, or the same as in ***a***? Briefly explain.

16-80. Suppose the double-slit arrangement in Figure 16-51 is first used without the block of glass present. Under these conditions, the wavelength of light from the source is 5.9×10^{-7} m, and the first line of constructive interference away from the center line occurs in a direction 4.0° from the center line. When a block of glass is then inserted to fill the entire region between the double slit and the screen (Figure 16-51*a*), at what angle from the center line will this first line of constructive interference occur? (Check to see if this agrees with your answer to Problem 16-79***a***).

Problems on WebLinks

16-81. The fact that the ratios of corresponding parts of two similar triangles are equal can be written $\frac{\text{altitude 1}}{\text{base 1}} = \frac{\text{altitude 2}}{\text{base 2}}$. In WebLink 16-1, this is applied to two triangles framed by the rays coming from a candle flame. As the spherical crests of the light waves travel outward from the candle flame, do these two ratios increase, decrease, or remain the same?

16-82. In WebLink 16-2, a candle flame serves as a point source, and images are produced by two different mirrors.

a. The image of the candle that the eye sees is located in front of the mirror, that is, between the mirror surface and the observer _____ (*for neither one of the mirrors; only for one of the two mirrors that was used; no matter which of the two mirrors is used*).

b. The image of the candle that the eye sees is located right at the mirror surface _____ (*for neither one of the mirrors;*

only for one of the two mirrors that was used; no matter which of the two mirrors is used).

16-83. In WebLink 16-3, the transmitted ray disappears when the laser is rotated by a certain amount. Suppose the glass is replaced with a different kind of glass that has a higher index of refraction. Will the laser have to be rotated more, less, or the same amount as before to make the transmitted ray disappear? Briefly explain.

16-84. In WebLink 16-3, light reaching the flat surface of the glass is reflected back into the glass _____ (*only when it strikes the surface at an angle less than the critical angle; only when it strikes the surface at an angle greater than the critical angle; no matter which of these two conditions applies*).

CHAPTER SEVENTEEN

Lenses and Optical Instruments

From microscopes, telescopes, and binoculars to cameras and projectors to endoscopes and other viewing instruments used in medicine, optical instruments produce images for a multitude of purposes. They vary in complexity from a simple magnifying glass to the human eye. The images are formed as light entering the instruments undergoes a series of reflections and refractions. In most of these instruments, refraction occurs as the light passes through one or more lenses. In this chapter, we explore the role of lenses in image formation.

"From microscopes . . . to endoscopes and other viewing instruments used in medicine, optical instruments produce images for a multitude of purposes."

17-1   A Qualitative Picture of What Lenses Do

Before we begin, it will be worthwhile to examine your own ideas about how lenses produce images and to make sure we are all working with the same self-consistent set of ideas.

Case 17-1 ◆ *An Image from a Lens*

(Treat this as an On-the-Spot Activity.) For this activity you will need:

- a convex lens (a simple magnifying glass will work well)
- a light source with an ordinary bare bulb
- a piece of opaque cardboard or similar material
- a second piece of white paper or cardboard that you can set upright as a screen
- any convenient object, such as a bottle, that can serve as a holder to keep the magnifying glass upright
- a partner to help with Step 6

Position bulb, glass, and screen as in Figure 17-1 so that you get an inverted image of the bulb on your screen. The best spacing depends on the magnifying glass you use. Try placing the bulb and screen about $\frac{1}{2}$ m apart for each inch of magnifying glass diameter. You will then have to move the magnifying glass back and forth between them until the image on the screen is sharply focused.

Figure 17-1 Set-up for Case 17-1.

Step 1. What do you suppose will happen to the image on the screen if (*don't do it yet!*) you take the lens away? First write down what you think will happen, along with the *reasons for your prediction* (a sketch should help). Only then should you try it to *observe* what happens. Finally, try to *explain* any differences between what you saw and what you predicted.

In your reasoning, did you consider whether the image is upright or inverted? Is it also flipped left to right? (You can check this with the lens back in place by putting your finger on one side of the bulb and seeing where it appears in the image.)

In the remaining steps of this activity you will follow the same POE sequence: first *predict,* then *observe,* and finally *explain* differences between prediction and observation.

Step 2. Put the lens back in place. What do you *predict* will happen to the image on the screen if you cover the top half of the lens with the opaque cardboard (Figure 17-2)? First write down your reasoning, with a supporting sketch, and only then *observe* what actually happens. Finally, try to *explain* any difference between what you predicted and what you saw.

Figure 17-2
Step 2 of Case 17-1.

Step 3. Make a hole in the cardboard just large enough so that the cardboard will cover the entire lens except for a small region in the center. What do you *predict* will happen to the image on the screen when you place this over the lens with the hole at the center? Again, *explain* your prediction, and then *observe* what happens. Once again, try to *explain* any discrepancy between what you expected and what you saw.

Step 4. What will happen to the image if you cover only the center of the lens? You can do this by stretching a string across the lens like a clothesline, and hanging a small piece of opaque material at the center. Now repeat the same *predict-observe-explain* sequence.

Step 5. What will happen to the image on the screen if you move the screen closer to the bulb? Will it change in size or brightness, or disappear, or something else? *Predict; observe; explain.*

Step 6. Suppose you remove the screen (don't move anything else). Is there any place you can position your eye so that you can still see the image? Predict, and write down your reasons.

Now try standing about $\frac{1}{2}$ m to 1 m beyond the lens so that you are looking back through the lens at the bulb. What do you see (ignore small reflections)? Is this "bulb" an image or the object itself? How do you know?

While you are still at this viewing position, have a partner reposition the screen so that a sharp image of the bulb appears on it. Then have your partner slide the screen sideways until it is only half blocking the "bulb" that you see, so that you see part of the "bulb" while your partner sees part of an image on the screen. Now have your partner move a finger until it looks to you to be positioned at the same distance as the "bulb" that you see when looking back through the lens. Compare the position of the finger and the screen. Is one closer to the lens than the other, or are they at the same distance? In short, do you see an image without the screen, and if so where is it, compared to the image on the screen? Explain.

You will want to review your explanations in light of Section 17-1.

This activity is adapted from workshop materials developed by D. Dykstra for the AAPT and a Woodrow Wilson Summer Institute, which in turn drew on research on student understanding of image formation by F. Goldberg and L. C. McDermott (*American Journal of Physics,* **55,** 108, 1987).

Two concave surfaces (biconcave lens) | Two convex surfaces (biconvex lens) | One plane and one convex surface (plano-convex lens) | One plane and one concave surface (plano-concave lens) | Lenses with one concave and one convex lens (meniscus convex) (meniscus concave)

Figure 17-3 Lenses with one or two spherical surfaces.

Lenses typically have two surfaces. A given lens surface can be a plane surface, a **convex** surface, which bulges outward, or a **concave** surface, which "caves" inward (Figure 17-3). As the figure shows, lenses are named for the types of surfaces they have. (In common practice, biconvex and biconcave lenses are sometimes just called convex and concave lenses.) Each surface of a lens is an interface, most commonly between glass and air, at which refraction occurs.

Convex and concave lenses can be approximated by a pair of prisms. Figure 17-4 traces rays from a point object through these "lenses." Assuming the prism has a greater index of refraction than its surroundings, each ray bends toward the base of the prism it passes through. The rays *converge* on leaving the convex lens; they *diverge* on leaving the concave lens. It is therefore common to speak of these as *converging* and *diverging lenses,* respectively, although the bending of rays is opposite when the prism or lens has a lower index of refraction than its surroundings (recall Figure 16-29).

But in both cases, the rays ultimately diverge toward the viewer. Just as with mirrors, the point from which they appear to diverge toward the viewer is the image point I. Because the rays shown reaching the viewer from the concave lens do not actually pass through I, I is a *virtual* image point and the extensions of the actual rays that meet at I are drawn as dotted lines. In contrast, the image point I for the convex "lens" is a real image point.

Figure 17-5 shows what happens when a convex lens produces an image of an extended object. Rays from each point on the object strike all points on the lens surface. The rays from each object point converge to a corresponding image point. This picture of what happens is developed more fully with step-by-step graphics in WebLink 17-1.

We can use this model of what the lens does to the light to explain your observations in Case 17-1. For example, in Step 2, when you cover the top half of the lens with the cardboard, both the blue and the red rays passing through the bottom half of the lens in Figure 17-5 still converge to the same image points. You still get an image, but it is dimmer because only half as much light reaches

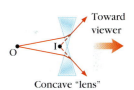

Figure 17-4 Prism approximations of convex and concave lenses.

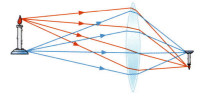

Figure 17-5 Image of an extended object produced by a convex lens. By placing a blue filter over this figure, you can view only the red diagram. Using a red filter, you can view just the blue diagram.

For **WebLink 17-1: The Image of an Extended Object,** go to www.wiley.com/college/touger

For **WebLink 17-2:**
Explaining Properties
of Images by the Ray Model,
go to
www.wiley.com/college/touger

each point. To see step by step how this model explains all your observations in Case 17-1 (and to check your own explanations), carefully work through WebLink 17.2.

17-2 Quantitatively Analyzing What Thin Lenses Do

When you place an object at a certain distance from a lens, the size and position of the image depend on the material of which the lens is made and the curvature of each of its surfaces. By doing geometry on the rays refracted at each of the two surfaces (Figure 17-6), it is possible to find a relationship between the object and image distances s_o and s_i. If, in contrast to Figure 17-6, the lens's thickness is negligibly small compared to s_o and s_i, and the lens is surrounded by air, the relationship takes the following form:

The **lensmaker's formula** (for thin lenses in air)

$$\frac{1}{s_O} + \frac{1}{s_i} = (n - 1)\left(\frac{1}{R_A} - \frac{1}{R_B}\right) \tag{17-1}.$$

R_A = radius of curvature of near surface (nearer to original object)
R_B = radius of curvature of far surface
n = index of refraction of lens

Figure 17-6 Image resulting from refraction at two lens surfaces. The image point viewed by a would-be observer within the lens material (compare with Figure 16-25) becomes the object point for light reaching surface 2.

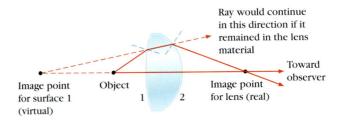

This equation holds for lenses with any combination of convex and concave surfaces if we follow the sign convention in Figure 17-7, in which s_i, R_A, and R_B are negative in the region on the near side of the surface, so that a negative s_i means a virtual image. Example 17-1 shows how these sign conventions work.

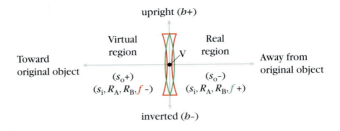

Figure 17-7 Sign conventions for thin lenses.

Example 17-1 *Can the Same Lens Produce Both Real and Virtual Images?*

For a guided interactive solution, go to Web Example 17-1 at www.wiley.com/college/touger

The radius of curvature of each surface of a biconvex glass lens ($n = 1.6$) is 0.30 m. Find the position of the image if a point object is placed along the lens's principal axis
a. 0.40 m to the left of the lens.
b. 0.20 m to the left of the lens.

Brief Solution

Choice of approach. Use the lensmaker's formula (17-4), paying careful attention to sign conventions (Figure 17-11). The center of curvature of the second surface is to the left of the lens so you must take R_B to be negative.

What you know/what you don't.

$$R_A = +0.30 \text{ m} \qquad R_B = -0.30 \text{ m} \qquad n = 1.6$$

In **a**, $s_o = 0.40$ m, $s_i = ?$ In **b**, $s_o = 0.20$ m, $s_i = ?$

The mathematical solution. From Equation 17-1,

$$\frac{1}{s_i} = (n - 1)\left(\frac{1}{R_A} - \frac{1}{R_B}\right) - \frac{1}{s_o}$$

Let's first insert values common to both parts:

$$\frac{1}{s_i} = (1.6 - 1)\left(\frac{1}{0.30 \text{ m}} - \frac{1}{-0.30 \text{ m}}\right) - \frac{1}{s_o} = 4 \text{ m}^{-1} - \frac{1}{s_o} \qquad (7\text{-}2)$$

Then for **a**, $\frac{1}{s_i} = 4 \text{ m}^{-1} - \frac{1}{0.40 \text{ m}} = 1.5 \text{ m}^{-1}$, which we must invert to obtain

$$s_i = \frac{1}{1.5 \text{ m}^{-1}} = \textbf{\textcolor{red}{0.67 m}}$$

The positive value tells us this is a *real* image appearing to the *right* of the lens. But for **b**, $\frac{1}{s_i} = 4 \text{ m}^{-1} - \frac{1}{0.20 \text{ m}} = -1.0 \text{ m}^{-1}$, so that inverting gives us

$$s_i = \frac{1}{1.5 \text{ m}^{-1}} = \textbf{-1.0 m}$$

Now the image appears to the *left* of the lens and is *virtual*.

(*a*)

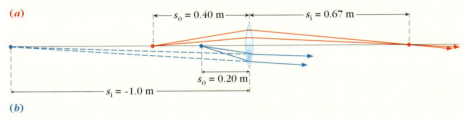

(*b*)

Making sense of the results. Our results correspond to the two ray diagrams in the above figure: for part **a** in red above the axis and for part **b** in blue below the axis. The rays from both object positions are bent toward the axis, but the rays from the object in **b** strike the lens at a greater angle to the normal, so that bending them toward the axis is not sufficient to make them converge on the side away from the object.

◆ Related homework: Problems 17-15, 17-16, and 17-17.

We found that for the lens in Example 17-1, $\frac{1}{s_i} = 4 \text{ m}^{-1} - \frac{1}{s_o}$. It follows that whenever $\frac{1}{s_o} > 4 \text{ m}^{-1}$, s_i will be negative and the image will be virtual. This will hold true whenever $s_o < 0.25$ m. In the paragraphs that follow, we will consider the significance of this value.

◆**FOCAL LENGTH AND FOCAL POINT OF A LENS** When the object is so far from the lens that its distance is effectively infinite, so that $\frac{1}{s_o} = \frac{1}{\infty} = 0$, it follows from Equation 17-1 that

$$\frac{1}{s_i} = (n - 1)\left(\frac{1}{R_A} - \frac{1}{R_B}\right)$$

But as we discussed for mirrors, rays arriving from infinitely far away are effectively parallel, and the point where such rays converge is called the **focal point** F. The distance from the vertex to this point is the **focal length** f. In short, when $s_o = \infty$, $s_i = f$, and the above equation tells us the following:

The **focal length** f of a lens surrounded by air or a vacuum is given by

$$\frac{1}{f} = (n - 1)\left(\frac{1}{R_A} - \frac{1}{R_B}\right) \qquad (17\text{-}3)$$

Because the focal point is a special case of an image position, the same sign convention applies for f as for s_i (Figure 17-7).

Equation 17-3 tells us that we can rewrite Equation 17-1 as

$$\frac{1}{s_o} + \frac{1}{s_i} = \frac{1}{f} \qquad (17\text{-}4)$$

the same equation we had for mirrors. In both cases it tells us that the focal point is the image position when the object is at infinity. But for a lens, you calculate the focal length very differently than for a mirror. The focal length of an object always depends on its light-affecting properties. A mirror is just a single reflecting surface. The focal length $\frac{R}{2}$ depends only on the curvature of that surface. The medium of which the mirror is made doesn't matter because the rays do not enter the medium. But the incoming rays do pass through the medium of which a lens is made and are refracted at both surfaces, so the focal point depends on the index of refraction of the medium as well as the curvature of each surface.

Example 17-2 *The Focal Length of a Biconvex Lens*

The radius of curvature of each surface of a biconvex glass lens ($n = 1.6$) is 0.30 m. Find its focal length.

Solution
Choice of approach. This is the same lens as in Example 17-1. Apply Equation 17-3, remembering that R_B must be negative.

What we know/what we don't.

$$R_A = +0.30 \text{ m} \qquad R_B = -0.30 \text{ m} \qquad n = 1.6 \qquad f = ?$$

The mathematical solution. Much as in Example 17-2, substituting these values into Equation 17-3 gives us $\frac{1}{f} = 4 \text{ m}^{-1}$ (note that this is Equation 17-2 with s_o taken as ∞ for parallel rays). Then

$$f = \frac{1}{4 \text{ m}^{-1}} = \textbf{0.25 m}$$

Making sense of the results. In Example 17-1, we found that when the object was closer than this distance from the lens, the image was virtual. When it was further away, the image was real. You can verify this by looking at a small bulb through a magnifying glass held at arm's length. When the glass is close to the bulb, the image of the bulb is clearly on the same side of the glass as the bulb. If you then back away from the bulb, keeping the magnifying glass at arm's length, you pass a point where the image seems to blur and disappear completely. Backing away still further, you see an image which may seem to hover on your side of the glass (recall what you saw in Case 17-1 when you stood beyond the lens and took away the screen).

Mathematically, because $\frac{1}{s_o} + \frac{1}{s_i} = \frac{1}{f}$, if $\frac{1}{s_o} > \frac{1}{f}$ then $\frac{1}{s_i}$ must be negative, and so must s_i. But $\frac{1}{s_o} > \frac{1}{f}$ only when $s_o < f$.

◆ Related homework: Problems 17-18 and 17-19.

◆**CONVERGING AND DIVERGING LENSES** If rays that arrive parallel at a lens converge on leaving it, the lens is called a *converging lens*. If the same incident rays diverge on leaving the lens, it is called a *diverging lens*.

Example 17-3 *Is the Lens Converging or Diverging?*

For a guided interactive solution, go to Web Example 17-3 at
www.wiley.com/college/touger

Determine whether the lens shown in part *a* of the figure is converging or diverging. Then repeat for the case when the lens is turned around as in part *b* of the figure.

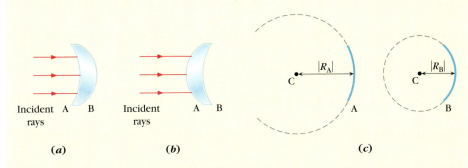

(*a*)   (*b*)   (*c*)

Brief Solution

Choice of approach. We consider Equation 17-3 and use the sign conventions in Figure 17-7. If the rays converge, the focal point will be to the right of the lens and f will be positive. If they diverge, as from a point to the left of the lens, then f will be negative.

The mathematical solution. We have no numbers to work with, but we can apply geometric and algebraic reasoning. Part *c* of the figure shows that the radius of curvature of surface A in *a* must have a greater absolute value than that of surface B. But because the center of curvature of each surface is to the left of each surface, the sign convention requires us to treat both R_A and R_B as negative: $R_A = -|R_A|$ and $R_B = -|R_B|$. Equation 17-3 tells us the sign of f will be the same as the sign of $\frac{1}{R_A} - \frac{1}{R_B}$. Now

$$\frac{1}{R_A} - \frac{1}{R_B} = \frac{1}{-|R_A|} - \frac{1}{-|R_B|} = \frac{1}{|R_B|} - \frac{1}{|R_A|}$$

But because $|R_A| > |R_B|$, $\frac{1}{|R_A|} < \frac{1}{|R_B|}$ (just as $4 > 2$, so $\frac{1}{4} < \frac{1}{2}$). Therefore $\frac{1}{R_A} - \frac{1}{R_B}$ is positive. So f is positive and the **lens is converging**.

When the lens is positioned as in *b*, the surfaces are reversed, so that $|R_B| > |R_A|$. But now the centers of curvature are to the *right* of the surfaces, so R_A and R_B are both positive. Then $\frac{1}{R_B} < \frac{1}{R_A}$, and again $\frac{1}{R_A} - \frac{1}{R_B}$ is positive. Thus the lens is converging no matter which way it is placed.

◆ Related homework: Problem 17-20.

For **WebLink 17-3:**
From Prisms to Lenses,
go to
www.wiley.com/college/touger

For more on how we can tell which kind of lens we have, work through WebLink 17-3.

In general, we could show the following for thin lenses with spherical surfaces:

If the lens is . . .	parallel rays from the left . . .	and the focal length f is . . .
thickest at the center (as a biconvex lens is),	converge toward the right (lens is a converging lens)	+
thinnest at the center (as a biconcave lens is),	diverge toward the right (lens is a diverging lens)	−

These properties are reversed if the lens has a lower index of refraction than its surroundings (not a possibility when the surrounding medium is air or a vacuum).

Example 17-4 *Locating Image Position for a Biconcave Lens*

Find the position of the image when an object is placed 0.10 m to the left of a biconcave lens with a focal length of magnitude 0.25 m.

Solution

Choice of approach. Apply Equation 17-4 with careful attention to sign conventions. A biconcave lens is a diverging lens, and therefore has a negative focal length.

What we know/what we don't. $s_o = 0.10$ m, $f = -0.25$ m, $s_i = ?$

The mathematical solution. From Equation 17-4, we obtain

$$\frac{1}{s_i} = \frac{1}{f} - \frac{1}{s_o} = \frac{1}{-0.25\,\text{m}} - \frac{1}{0.10\,\text{m}} = -4.0\,\text{m}^{-1} - 10\,\text{m}^{-1} = -14\,\text{m}^{-1}$$

Then (don't forget to invert!)

$$s_i = \frac{1}{-14\,\text{m}^{-1}} = \mathbf{-0.071\ m}$$

The minus sign tells you the image is to the left of the lens and is virtual.

◆ Related homework: Problems 17-21 and 17-22.

The lenses of early cameras such as this one were commonly mounted on bellows that could be extended. What was the purpose of this?

Example 17-5 considers an object that is not just a point but extends along the central axis.

Example 17-5 *That's about the Size of It*

For a guided interactive solution, go to Web Example 17-5 at www.wiley.com/college/touger

A biconvex lens has a focal length of 0.50 m. Suppose you lay a 0.20-m-long pencil along the principal axis of the lens, so that the pencil point is 0.70 m from the lens and points toward the lens (part *a* of the figure). How long is the image of the pencil?

Brief Solution

Choice of approach. For an extended object, you must in principle find an image point for each object point. In practice, you need only apply Equation 17-4 to the two ends of the pencil. The eraser is 0.70 m + 0.20 m = 0.90 m from the lens. Note that although we are asking about the size of the image, we are *not* asking about h, because h gives the linear dimension perpendicular to the axis rather than along it.

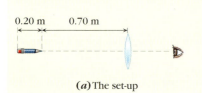

0.20 m 0.70 m

(a) The set-up

(b) The ray diagram

What we know/what we don't.

(Instead of o = object, we here use p = point, e = eraser,
i of p = image of point, etc.)

$$s_p = 0.70 \text{ m} \qquad s_e = 0.90 \text{ m} \qquad f = 0.50 \text{ m} \qquad s_{i \text{ of } p} = ? \qquad s_{i \text{ of } e} = ?$$

$$\text{length} = |s_{i \text{ of } e} - s_{i \text{ of } p}| = ?$$

The mathematical solution. From Equation 17-4, $\frac{1}{s_i} = \frac{1}{f} - \frac{1}{s_o}$. Then

$$\frac{1}{s_{i \text{ of } p}} = \frac{1}{f} - \frac{1}{s_p} = \frac{1}{0.50 \text{ m}} - \frac{1}{0.70 \text{ m}} = 0.57 \text{ m}^{-1}$$

and

$$\frac{1}{s_{i \text{ of } e}} = \frac{1}{f} - \frac{1}{s_e} = \frac{1}{0.50 \text{ m}} - \frac{1}{0.90 \text{ m}} = 0.89 \text{ m}^{-1}$$

Inverting these results gives us

$$s_{i \text{ of } p} = \frac{1}{0.57 \text{ m}^{-1}} = 1.75 \text{ m} \qquad s_{i \text{ of } e} = \frac{1}{0.89 \text{ m}^{-1}} = 1.12 \text{ m}$$

and so $\quad \text{length} = |s_{i \text{ of } e} - s_{i \text{ of } p}| = |1.12 \text{ m} - 1.75 \text{ m}| = \textbf{0.73 m}$

Making sense of the results. Our calculations show that although the pencil point is nearer than the eraser to the lens, the image pencil point is further than the image eraser from the lens. Does this make sense? Try looking at distant objects with a magnifying glass to see that this is what we should expect. It also agrees with the ray diagram (part *b* of the figure).

◆ Related homework: Problems 17-23 and 17-24.

In the next section, we consider the images of extended objects that have dimensions perpendicular to the axis as well as along it.

17-3 Lens Images of Extended Objects

As we did with mirrors, we will draw ray diagrams to see what kinds of images lenses form of extended objects. It will again prove useful to pick a conspicuous object point and identify the rays from that point whose paths are easiest to determine.

1. A ray through the center of a thin lens experiences the center as though it were a flat pane of glass (Figure 17-8). The ray is bent equally but oppositely at the two surfaces, so the emerging ray is in the same direction as the incident ray: *A ray through the center of a thin lens does not change direction.* (Actually, the ray is offset by the amount labeled *d* in Figure 17-8, but if the thickness of the lens is negligible, so is *d*.)

2. Incident rays that are parallel to the axis converge to the focal point F.

3. We have defined the point where parallel rays converge as the focal point. Parallel rays arriving from the far side of the lens (away from the object) would pass through a focal point on the opposite side of the lens from F. We will label it F'. So a parallel ray from P' would pass through P in Figure 17-9. But all optical paths are reversible—I can't look you in the eye without you seeing my eye in return. Therefore a ray (3) traveling from P through F' would emerge from the lens parallel to the axis and pass through P'.

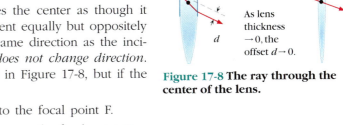

Thin lens approximation
Approximately flat near center
(1)
As lens thickness → 0, the offset $d → 0$.

Figure 17-8 The ray through the center of the lens.

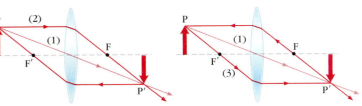

Figure 17-9 Optical paths are reversible.

Rays (1), (2), and (3) in Figures 17-8 and 17-9 are the principal rays for lens diagrams.

PROCEDURE 17-1

Identifying Principal Rays for Lens Diagrams

1. **The ray from the object point through the center of the lens.** This ray does not change direction.

2. **The incident ray that leaves the object point parallel to the axis.** The refracted ray (or its extension) then passes through focal point F. F is on the far side (further from the object) for converging lenses and on the near side for diverging lenses.

Drawing rays (1) and (2) is sufficient; (3) can serve as a check.

3. **The incident ray that goes from the point object toward or through F′.** The refracted ray is then parallel to the axis.

For **WebLink 17-4:**
Drawing a Ray Diagram for a Lens,
go to
www.wiley.com/college/touger

Procedure 17-1 has been followed in producing each of the ray diagrams in Figure 17-10. The rays in the diagram are numbered as above. Ray (3) is drawn fainter because it is serving as a check on the other two. Make sure you understand these diagrams and how they are drawn. Work through WebLink 17-4 to produce a diagram like Figure 17-10b step by step.

In all three ray diagrams in Figure 17-10, triangles AOP and AIP′ are similar, so that the ratios of corresponding sides are equal. Because AO = $|s_o|$ and AI = $|s_i|$, it follows that $\frac{|b_i|}{|b_o|} = \frac{|s_i|}{|s_o|}$. But b_i is negative for the inverted image in Figure 17-10a. To agree with this, we must write that the magnification

$$M = \frac{b_i}{b_o} = -\frac{s_i}{s_o} \qquad (17\text{-}5)$$

just as for mirrors. Then we can interpret M as we do for mirrors:

If M is +, the image is upright.   If $|M| > 1$, the image is enlarged.
If M is −, the image is inverted.   If $|M| < 1$, the image is reduced in size.

STOP&**Think** What are the signs on b_o, b_i, s_o, and s_i in Figures 17-10b and c? Check to see if $\frac{b_i}{b_o}$ has the same sign as $-\frac{s_i}{s_o}$ for each of these situations as well. ◆

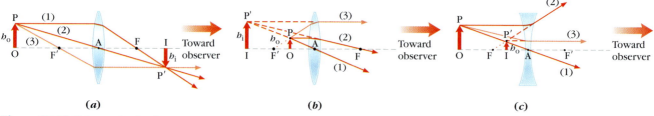

Figure 17-10 Using principal rays to construct ray diagrams for lenses.

Example 17-6 *Same Size Object and Image*

A converging lens has a focal length of magnitude 0.30 m. Where must an object be placed along its principal axis if the image is to be the same size as the object, but inverted?

Solution

Restating the question. This says that the object size $|b_o|$ and the image size $|b_i|$ are the same, but because the image is inverted they are opposite in sign. The magnification is therefore $M = \dfrac{b_i}{b_o} = -1$, and the question can be restated as

$$s_o = ? \ when \ M = -1.$$

Choice of approach. Equation 17-4 relates object and image distances to focal length. Equation 17-5 relates the same two distances to the magnification. We must use these two equations together.

What we know/what we don't.

$$M = -1 \qquad f = 0.30 \ m$$

$$s_o = ? \qquad s_i = ?$$

The mathematical solution. Because $M = -1 = -\dfrac{s_i}{s_o}$, it follows that $s_i = s_o$. We can use this to substitute for s_i in Equation 17-4. Then

$$\frac{1}{s_o} = \frac{1}{f} - \frac{1}{s_o} \qquad or \qquad \frac{2}{s_o} = \frac{1}{f}$$

Inverting, we get $\qquad \dfrac{s_o}{2} = f \qquad$ or $\qquad s_o = 2f$

This result is true no matter what the value of f. In this particular case,

$$s_o = 2(0.30 \ m) = \textbf{0.60 m}$$

◆ Related homework: Problem 17-27.

Example 17-7 *An Image Produced by a Biconcave Lens*

**For a guided interactive solution, go to Web Example 17-7 at
www.wiley.com/college/touger**

A 0.20-m-tall rodent stands 0.30 m from a biconcave lens with a focal length of magnitude 0.60 m. Find the position and height of the rodent's image, and determine whether the image is real or virtual, upright or inverted, and enlarged or reduced.

Brief Solution

Choice of approach. Before doing your calculation, you should sketch a ray diagram to give you a qualitative sense of what to expect from your calculation. **STOP**&**Think** How is the ray diagram for this situation similar to Figure 17-10c? How is it different? Does the difference have any significant consequences? ◆

We apply Equations 17-4 and 17-5 to find s_i and b. We can then use the conventions in Figure 17-7 and in the boxes before Example 17-4 and following Equation 17-5 to answer the additional questions about the image. In particular, note that f is negative for a biconcave lens.

What we know/what we don't.

$$M = ? \qquad f = -0.60 \ m$$

$$b_o = 0.20 \ m \qquad b_i = ? \qquad s_o = 0.30 \ m \qquad s_i = ?$$

The mathematical solution. From Equation 17-4,

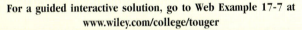

$$\frac{1}{s_i} = \frac{1}{f} - \frac{1}{s_o} = \frac{1}{-0.60 \ m} - \frac{1}{0.30 \ m} = -5.0 \ m^{-1}$$

EXAMPLE 17-7 continued

Inverting, we get

$$s_i = \frac{1}{-5.0 \text{ m}^{-1}} = -0.20 \text{ m}$$

Then from Equation 17-5,

$$b_i = -\frac{s_i}{s_o} b_o = -\frac{-0.20 \text{ m}}{0.30 \text{ m}}(0.20 \text{ m}) = +0.13 \text{ m}$$

and

$$M = \frac{b_i}{b_o} = \frac{0.13 \text{ m}}{0.20 \text{ m}} = 0.65$$

Because . . .	the image is . . .
s_i is negative,	virtual.
M is positive,	upright.
$|M| < 1$,	reduced.

Making sense of the results. Even though the object is closer to the lens than the focal point, the image is smaller and closer to the lens than the object, just like when the object is beyond the focal point (Figure 17-10c). It is possible to show that this will be the case for a biconcave lens for any object position (see Problem 17-50).

◆ Related homework: Problems 17-28, 17-29, 17-30, and 17-31.

17-4 Aberration in Lenses

Lenses do not generally produce the perfectly sharp images that we would like. The focal points that we have found have been based on paraxial and thin-lens *approximations.* All rays arriving parallel at real lenses do not in fact converge precisely to a single point. The ways in which lenses deviate from perfect focus are called **aberrations.**

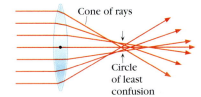

Figure 17-11 Spherical aberration. Rays reaching the lens parallel do not all converge to the same point. The rays come closest to forming a single image point at the circle of least confusion.

◆**SPHERICAL ABERRATION** We have shown parallel rays incident on different parts of a lens to converge at a single point. But just as for mirrors, this is an approximation that is good only near the center of the lens. In fact, rays that are incident further from the center are refracted more and therefore converge closer to the lens (Figure 17-11). This effect is called **spherical aberration.** If a screen is slid along the axis of the lens in Figure 17-11, the circle of light would be smallest (most point-like) where shown. The circle at this point is called the *circle of least confusion.* It can be reduced by eliminating the rays from the outermost parts of the lens. This is accomplished in a camera by reducing the aperture, or opening.

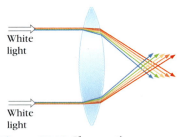

Figure 17-12 Chromatic aberration.

◆**CHROMATIC ABERRATION** Up until now, we have been treating the light as though it were monochromatic, that is, all of the same wavelength. In fact that is rarely the case. The focal point of a lens depends on the extent to which rays are refracted at its surfaces. But the index of refraction of the lens depends on the color or wavelength of the light (Figure 16-31). So, too, must the focal length. This effect is called **chromatic aberration.** The greater the index of refraction, the more the light is bent and the closer the focal point (Figure 17-12). Tables commonly give the index of refraction for a particular wavelength of yellow light (5.89×10^{-7} m, produced by a sodium lamp) near the middle of the visible spectrum, and the given focal lengths of lenses are based on these index of refraction values.

Because not all of the light from a nearly point source of white light is focused at the same distance, a screen placed at the image position where focus is best will show an image with a halo around it. That is because although light of a particular wavelength from each object point is converging to a single image point to form the image, light of other wavelengths has not completely converged and therefore arrives within a small region around each image point. In addition, these regions around the image points overlap. As you move the screen from this position, the image will become more blurred; the outer edges will take on a reddish or bluish cast, depending on whether the screen is moved toward or away from the lens.

Because the chromatic aberration is opposite for converging and diverging lenses, lenses are often combined to reduce aberration. Paired converging and diverging lenses used in this way are called *doublets*.

17-5 Optical Instruments

The general purpose of optical instruments is to enable us to observe something by producing a suitably sized image of it in a desired place. Sometimes the image is produced where we can simply look at it, as in the case of a magnifying glass or a microscope. Other times it is produced where light-sensitive detectors at the image position can record and/or transmit the optical information that they receive. For example, the image light pattern may be recorded when it causes a photochemical reaction on film placed at the image position in a camera, or it may be recorded electronically on magnetic tape in camcorders after being converted to electrical signals by photocells at the image position. Similarly, optical information may be transmitted to your brain via the optic nerve after being received by suitable detectors, called rods and cones, at the image position in your eye. In this section, we will concentrate on how optical elements—lenses, mirrors, prisms, and so on—are combined to produce an image of desired size in the desired image position.

The simplest device for producing an image, loosely speaking, is a pinhole in an otherwise opaque barrier. In the idealized case, an infinitesimally small hole, rays from each object point can only pass through the hole to a screen in a single direction (Figure 17-13a). The reflected rays then reach an observer on the screen side of the barrier as though from a point on the screen (the "image point"[1]). The *camera obscura*, Latin for "dark chamber" or "dark room," was a box that used this principle (Figure 17-13b). It was the ancestor of the

Figure 17-13 Pinhole images. (*a*) An image produced by a pinhole. (*b*) A nineteenth-century depiction of the camera obscura, an enclosed and darkened chamber that produces an image with a pinhole. (From Deschanel's *Natural Philosophy*, 13th ed., New York, 1894)

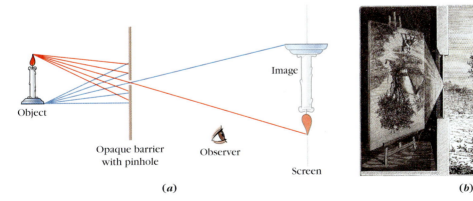

(a)

(b)

---

[1]To the extent that the pinhole is truly a point, there is only a single ray traveling from the object point to the point on the screen, and thence to your eye, so the point on the screen is not strictly speaking an image point in the sense of a point from which multiple rays diverge. Nevertheless, light does travel to your eye from this point as though from the object itself.

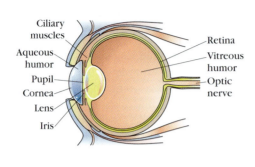
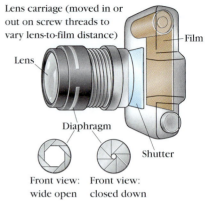
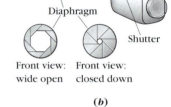

Figure 17-14 The human eye and the camera compared. (*a*) The human eye (*b*) The camera.

(*a*)

(*b*)

The widely dilated pupils of the big-eyed tarsier are necessary to provide enough light-gathering for nocturnal activity

photographic camera. **STOP&Think** Why does the image scene in the figure look dimmer? Why is the easel on which the image appears placed in an enclosing box? ◆

In the human eye and in the camera (Figure 17-14), the opening is not infinitesimally small but is finite in diameter and therefore admits not just a single ray but a cone of rays from each object point. A converging lens is therefore necessary to converge each such cone of rays to a single image point. Table 17-1 shows that in broad outline, the principles governing the operation of the eye and a simple photographic camera are the same.

Table 17-1 Comparison of Eye and Camera

	Eye	Camera
Image is produced by . . .	lens and cornea (including aqueous humor) on retina	lens (often compound) on film
Image position is changed by . . .	ciliary muscles changing concavity of lens to vary focal length (s_i changed by changing f in Equation 17-4)	varying lens position by means of a bellows or a screwing mechanism (s_i changed by changing s_o in Equation 17-4)
Image is detected by a process in which . . .	light energy is absorbed by pigment molecules in millions of rods and cones distributed over the retina (the energy is "captured" and initiates signals along the optic nerve toward the brain)	light provides the activation energy for breaking down silver compounds in the emulsion on the film and freeing the silver, resulting in a blackening (although bulk metallic silver is shiny, finely dispersed silver is black) which increases with light intensity (producing a "negative" image on the film)
Brightness increases (and sharpness decreases) with the increased opening of . . .	the iris, which determines the diameter of the eye's opening, called the pupil	the diaphragm, which governs the diameter of the camera's opening, called the aperture (this is what is indicated by "f-stop")
Entry of light is permitted or prevented by the opening or closing of . . .	the eyelid	the shutter

The elements of the eye have evolved to a high degree of optical usefulness. The cornea and lens both serve a lens-like function. Ray convergence, primarily produced by the cornea, is augmented and fine-tuned by the flexible lens in focusing images on the screen-like retina, in the back of the eye. We can think of the cornea and lens together as functioning like a compound lens in a camera. The aqueous humor, the fluid that fills the region behind the cornea, and the vitreous humor, which fills the region between the lens and the screen, also affect how light is refracted as it passes through the eye.

The retina is covered by millions of light-sensitive rods and cones. Because each rod or cone separately "measures" the intensity of light falling at its location, millions are needed to produce a high-resolution picture, so the screen-like retina must be large enough to accommodate all those rods and cones sufficiently spaced from one another to make non-overlapping readings. For similar reasons, early photographers put their light-sensitive emulsions on large glass plates to achieve high resolution (these glass plate negatives are now prized by collectors), and fine-grain emulsions on glass are widely used today in making holograms. Nature has not evolved more miniaturized detectors, so small creatures like mice have relatively larger eyes for their head size. Further enlargement is not required for bigger creatures, so the ratio of eye to head size becomes much smaller in elephants and whales.

Muscles in the eye control *focusing* and *light entry*. We will now consider how they do this.

◆**FOCUSING** If an object moves toward you, s_o decreases in Equation 17-4. If f were fixed, s_i would correspondingly *increase*. Because the lens-to-retina distance is fixed, the image would be in focus behind the retina, not on it. To compensate, the focal length of the lens must change to produce a focused image at the required s_i. To accomplish this, fibers of the ciliary muscles encircling the flexible lens contract. This relaxes the ligament that holds the lens in position, permitting the lens to bulge outward as an object comes closer. When the surfaces are more rounded, their radii of curvature are smaller, and the focal length decreases (see Equation 17-3). The rapid adjustment of the lens shape to viewing distance is called **accommodation.**

◆**LIGHT ENTRY** The colored iris is essentially a diaphragm of muscle that controls the diameter of the pupil, the eye's aperture. Thus when you come out of a darkened room, your pupils appear large but become much smaller in bright sunlight, when a much narrower light cone needs to converge to a given point on the retina to produce the same intensity. The same reasoning applies to a camera, where the ratio of focal length f to diameter D, called the **f-number** or "f-stop" ($f = \frac{f}{D}$), is commonly used to indicate how open the aperture is. As D increases, the f-number increases. Then to reduce the aperture (or "stop down") when you go from dim indoor light to very bright sunlight, you might change from an f-number of 1.4 to an f-number of 22 (commonly written as f/1.4 and f/22 although they are not fractions).

Applying lens equations to the eye is complicated by three important considerations:

- We are dealing with a compound lens system (cornea and lens), though it is certainly possible to consider an effective focal length for the combination.

- Under ordinary circumstances, the medium is not the same on both sides of this lens system: in front is air; behind is the aqueous humor, which has essentially the same index of refraction as water (though when you open your eyes underwater, you have roughly the same medium—water—on both sides of your lens system). In general, when the media on the two sides of the lens do not both have an index of refraction of 1, we must use a more general form of Equation 17-4:

$$\frac{n_A}{s_o} + \frac{n_B}{s_i} = \frac{1}{f} \tag{17-4a}$$

in which n_A and n_B are the indices of refraction of the media nearer to and further from the object, and f is the focal length with those two media present.

• It is possible to apply this more general form of the thin lens equation (as in the next example) to the eye's compound lens system. But this still gives only a rough approximation of what happens in the eye, because as Figure 17-14 shows, the thickness of the cornea + lens system is by no means negligibly small compared to the lens-to-retina distance. This system is *not* a thin lens.

Example 17-8 *Using Your Eyes*

For a guided interactive solution, go to Web Example 17-8 at
www.wiley.com/college/touger

Using a reasonable estimate of the lens-to-retina distance in your eye, estimate the required effective focal length of the cornea + lens system of your eye when you are viewing an object 10 m away and when you are viewing an object 0.20 m away. In which case must the radii of curvature of the lens surfaces be greater?

Brief Solution

Choice of approach. The focal length in question is the focal length of the cornea + lens system. Because light passes through this lens system from air into the vitreous humor, we must use Equation 17-4a: Medium A is then air, and medium B is the vitreous humor. Because this is a thin lens equation, it cannot give accurate results for such a thick lens system, but it can give us a rough estimate. Two centimeters (0.020 m) seems a fair estimate for the lens-to-retina distance—the desired image distance. We can compare the radii of curvature qualitatively by following Procedure 17-1 to sketch ray diagrams.

What we know/what we don't.

$$n_A = n_{air} = 1.000 \qquad n_B = n_{vitreous\ humor} = 1.336 \text{ (from Table 16-1)}$$

When $s_o = 10$ m, $s_i = 0.020$ m (assumed) and $f = ?$

When $s_o = 0.20$ m, $s_i = 0.020$ m (assumed) and $f = ?$

The mathematical and qualitative solutions: By Equation 17-4a,

$$\frac{n_A}{s_o} + \frac{n_B}{s_i} = \frac{1.000}{10 \text{ m}} + \frac{1.336}{0.020 \text{ m}} = \frac{1}{f}$$

so $0.1 \text{ m}^{-1} + 66.8 \text{ m}^{-1} = 66.9 \text{ m}^{-1} = \dfrac{1}{f}$, and $f = \dfrac{1}{66.9 \text{ m}^{-1}} = $ **0.015 m**

(the result is negligibly less than 0.02 m). In the second case,

$$\frac{n_A}{s_o} + \frac{n_B}{s_i} = \frac{1.000}{0.20 \text{ m}} + \frac{1.336}{0.020 \text{ m}} = \frac{1}{f}$$

so $5.0 \text{ m}^{-1} + 66.8 \text{ m}^{-1} = 71.8 \text{ m}^{-1} = \dfrac{1}{f}$

and $f = \dfrac{1}{71.8 \text{ m}^{-1}} = $ **0.014 m**

The focal length is reduced by a few percent.

We compare the two situations by means of ray diagrams in Figure 17-15. As the object draws nearer the eye (going from *a* to *b* in the figure), the direction of ray 2 changes, so that the intersection of rays 1 and 2 falls behind the retina. It is therefore necessary to "bulge" the lens (Figure 17-15*c*) to shorten

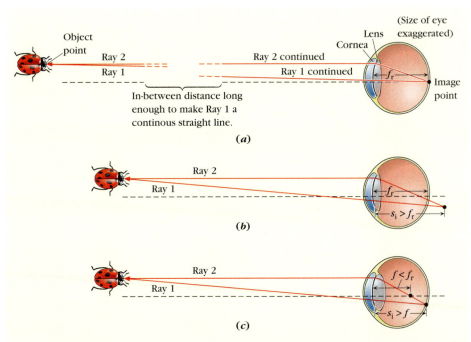

(a)

(b)

(c)

Figure 17-15 How the eye's lens adjusts to object distance. (*a*) When the object is at infinity, the image is at the focal point, a distance from the cornea + relaxed lens. The retina is also at this distance f_r from the cornea + lens, so an image forms on the retina. (*b*) (If lens were rigid) When the object draws nearer, the image distance increases ($s_i > f_r$). But the retina is fixed at a distance f_r from the cornea + lens, so the image falls beyond the retina. (*c*) But when the lens bulges, f is reduced. Now $s_i > f$, but $f < f_r$, so the image of the nearer object can fall on the retina.

the focal length, changing the direction of the ray through the focal point (ray 2) so that it once again intersects with ray 1 at the retina. The closer the object comes to the eye, the closer the focal point must be to the lens.

◆ Related homework: Problems 17-35 and 17-36.

As Example 17-8 points out, the lens's flexibility is limited. If an object is brought closer to the eye than a so-called **near point**, a sharp image cannot be produced on the retina. As flexibility declines with age, this minimum distance at which an object can be seen in sharp focus increases, from about 0.20 m in early adolescence (assuming good eyesight) to perhaps triple this value in one's forties and more than ten times this value by one's sixties. The eyes' lenses then have to be augmented by additional converging lenses, namely, reading glasses. For consideration of how eyeglasses are used to correct for nearsightedness (**myopia**) and farsightedness (**hyperopia**), see Problem 17-37.

◆**THE MAGNIFYING GLASS** The magnification $M = \frac{b_i}{b_o}$ is the *linear* magnification, the ratio of the linear dimensions of object and image. But this doesn't necessarily tell us whether the image *appears* larger to your eye. A hand in front of your face totally blocks your view of a distant skyscraper; it appears larger than the skyscraper. The objects in Figure 17-16*a* all appear the same size to the observer peering along the table surface. The same reasoning would apply to

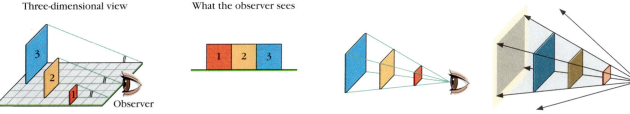

(a) Objects or images of different sizes may *appear* the same size when at different distances.

(b) Likewise, objects at different distances from a light source can cast the same size shadow.

Figure 17-16 When objects subtend the same angle. (*a*) Objects or images of different sizes may appear the same size when at different distances. (*b*) Likewise, objects at different distances from a light source can cast the same size shadow.

images at the same positions. As the diagram shows, they appear the same size because they all subtend the same *angle*. The vertex of this angle is at the vertex of the outer surface of the lens of your eye. An optical instrument magnifies an object by producing an image that subtends a larger angle than the object does. The ratio of these two angular sizes is called the **angular magnification** (M_A) of an optical instrument:

$$M_A \equiv \frac{\theta'}{\theta} = \frac{\text{angular size of image seen through instrument}}{\text{angular size of object seen by unaided eye}} \qquad (17\text{-}6)$$

In principle, you could keep making an object appear bigger by continuing to bring it closer to your eye. But closer than the near point, the object cannot be kept in focus by your eye. A **magnifying glass** is the simplest device that produces an image of increased angular size *at a position further than the near point*. Recall that when an object is closer to a convex lens than its focal point (Figure 17-10*b*), the image is virtual, upright, and enlarged. Again, we consider only the rays close to the axis (the paraxial approximation), where angles are small enough so that if θ is in radians, $\theta \approx \sin \theta \approx \tan \theta$.

The **magnifying power** (*MP*) of the lens is the value of M_A that compares the angles (θ and θ') in two particular situations:

- In the *reference situation* (Figure 17-17*a*), there is no magnifying glass, and the object is at the near point. The closer an object is to your eye, the greater its angular size, so it occupies more of your field of view. But if it gets too close, you cannot see a sharp image. So for your unaided eye, the angular size of a sharp image reaches a maximum at the near point.

- In the *test situation* (Figure 17-17*b*), the lens is held close to your eye, so that distances from your eye and the lens are essentially the same.

In the reference situation,

$$\theta \approx \tan \theta = \frac{h_o}{N} \qquad (N = \text{distance to near point})$$

In the test situation,

$$\theta' \approx \tan \theta = \frac{h_i}{s_i} \qquad \text{(from the larger triangle)}$$

and also $\qquad \theta' \approx \tan \theta = \frac{h_o}{s_o} \qquad \text{(from the smaller triangle)}$

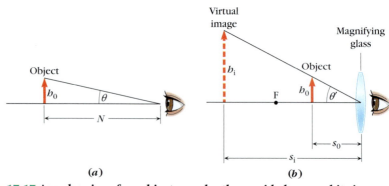

(a) $\qquad\qquad\qquad\qquad\qquad\qquad$ (b)

Figure 17-17 Angular size of an object seen by the unaided eye and its image seen through a magnifying glass. (*a*) Without a magnifying glass, the object's angular size θ is largest when the object is at the near point, a distance N from the eye. Closer than this, the object cannot be seen clearly. (*b*) An object closer than the near point has a larger angular size θ'. If the object is nearer than the focal point F, the magnifying glass produces an enlarged image that has the same larger angular size. Though the object is too close to be seen clearly, the image is not.

Then $MP = \dfrac{\theta' \,(\text{test situation})}{\theta \,(\text{reference situation})} = \dfrac{\dfrac{b_o}{s_o}}{\dfrac{b_o}{N}}$, or simplifying the fraction,

magnifying power: $\qquad MP = \dfrac{N}{s_o} \qquad (s_o \text{ in } test \ situation) \qquad$ (17-7)

which increases as the object is brought closer, reducing s_o. What limits the range of values for s_o? The following example addresses this.

Example 17-9 *How Close Is Too Close?*

For many optical applications, it is common to assume a near point distance of 0.25 m. Suppose a magnifying glass with a focal length of 0.20 m is held to your eye.

a. How close to your eye can an object be and still give you a sharp magnified image? What is the magnifying power of the glass for this object position?

b. What is the glass's magnifying power when the image is as far away as possible?

Solution

Choice of approach. **a.** Your eye doesn't distinguish between an object and an image: either is "too close" when it is closer than the near point. We can use Equation 17-4 to find the object position when the image is at the near point, keeping in mind that a virtual image has a negative image position. We then apply Equation 17-7 to find MP.

b. As we move the object to the focal point, the image moves to $-\infty$. So now we use $s_o = f = 0.20$ m in Equation 17-7.

What we know/what we don't.

$$\text{near point} = 0.25 \text{ m} \qquad f = +0.20 \text{ m}$$

$$\text{In } \textbf{a}: s_i = -0.25 \text{ m} \qquad s_o = ? \qquad MP = ?$$

$$\text{In } \textbf{b}: s_o = +0.20 \text{ m} \qquad MP = ?$$

The mathematical solution. **a.** In Equation 17-4,

$$\frac{1}{s_o} + \frac{1}{-0.25 \text{ m}} = \frac{1}{0.20 \text{ m}}$$

so $\qquad \dfrac{1}{s_o} = \dfrac{1}{0.20 \text{ m}} - \left(\dfrac{1}{-0.25 \text{ m}} \right) = 9.0 \text{ m}^{-1}$

and $\qquad\qquad\qquad\qquad s_o = \textbf{0.11 m}$

When $s_o = 0.11$ m, $\qquad MP = \dfrac{0.25 \text{ m}}{0.11 \text{ m}} = \textbf{2.3}$

b. When $s_o = f = 0.20$ m, $\qquad MP = \dfrac{0.25 \text{ m}}{0.20 \text{ m}} = \textbf{1.3}$

Making sense of the results. The magnification is reduced as you move the object further than its position in **a.** But the muscles regulating the bulge of your eye's lens become fully relaxed as what your eye is viewing, be it object or image, moves toward infinity. So in practice, there is a trade-off between comfort and optimum magnification.

◆ Related homework: Problem 17-38.

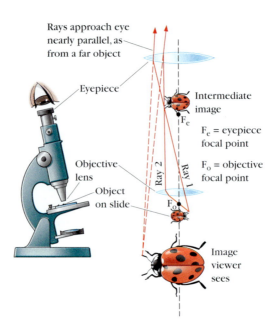

Figure 17-18 The compound microscope.

◆**MICROSCOPES AND TELESCOPES** In an optical system, the image produced by one optical element can be the object "seen" by the next, as when you see the mirror image of a mirror image (Figure 16-13). Likewise, the object viewed by a magnifying glass can in fact be an image produced by one or more other optical elements. A magnifying glass serving in this way is called an **eyepiece** or **ocular.** Like the object in part **b** of the last example, the image it views must be formed at or near its focal point to provide strainless viewing for your eye. Under these conditions, $MP = \frac{N}{s_o} = \frac{0.25 \text{ m}}{f}$.

This is what happens in microscopes and telescopes. In the **compound microscope** (Figure 17-18) and the **refracting telescope** (Figure 17-19), the image viewed by the eyepiece is formed by a second lens, called the **objective lens** (it views the *object*). Table 17-2 compares the principal features of these two instruments. Differences in design follow from the fact that microscopes are used to view small objects that are placed very close to the objective lens, whereas telescopes are used to view very large objects that are very far away.

Because the objects viewed by microscopes are very close, we can illuminate them artificially, so we can keep the lens diameter small to minimize aberration and still see a sufficiently bright image. But we cannot illuminate the distant objects we view through a telescope. Therefore the diameter of the telescope's objective lens is made large to allow it to receive a wider cone of rays from the observed object. (Recall that in Case 17-1, the image was brighter when there was more uncovered lens surface.)

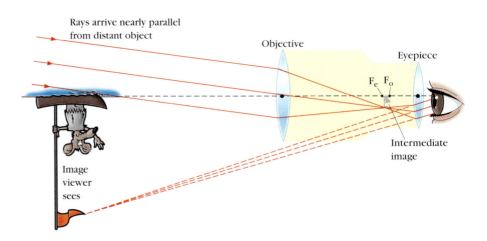

Figure 17-19 The refracting telescope.

Table 17-2 Comparison of Compound Microscope and Refracting Telescope

	Compound Microscope	Refracting Telescope
Position of object	very close (near focal point of objective lens)	very far (at ∞)
Objective lens	small diameter; f_{obj} very short ($<$1 cm)	large diameter; f_{obj} very long (several tens of cm or more)
Eyepiece	$f_{eye} \approx$ a few cm	$f_{eye} \approx$ a few cm
Distance between objective and eyepiece	$= f_{obj} + f_{eye} + L$ (where L, the "tube length" or distance between the two interior focal points, is commonly a standard 16.0 cm)	$\approx f_{obj} + f_{eye}$
First image forms	at a distance f_{eye} or a bit less from eyepiece	at a distance f_{eye} or a bit less from eyepiece
and is . . .	real, inverted, enlarged	real, inverted, reduced
Second image forms	far from viewer (beyond near point)	far from viewer (beyond near point)
and is . . .	virtual, inverted (compared to object but not compared to first image), and further enlarged	virtual, inverted (compared to object but not compared to first image), and further enlarged

The other differences in the design of the two instruments are governed by the relationship $\frac{1}{s_o} + \frac{1}{s_i} = \frac{1}{f}$, as you will see in the next example.

Example 17-10 *Positioning the Eyepieces of a Microscope and a Telescope*

For a guided interactive solution, go to Web Example 17-10 at www.wiley.com/college/touger

The eyepieces of microscopes and telescopes are essentially magnifying glasses, so the images they view must be formed at or near their focal points. If we assume "at," how far from the focal point of the eyepiece must the manufacturer place the focal point of the objective lens

a. in a microscope that has an objective lens focal length of 0.58 cm and is used to view a specimen on a glass slide 0.60 cm below the objective lens?

b. in a telescope used to view a star a few light-years away?

Brief Solution

Choice of approach. Equation 17-4 tells us how the relevant distances are related. Keep in mind that the object and image distances in Equation 17-4 are measured from the vertex of the lens, not from the focal point. It is important to sketch the situation (Figure 17-20) to see how the distance in question relates to the distances in the equation.

Figure 17-20 Ray diagrams showing placement of the object and the image formed by the objective lens. In both the microscope and the telescope, the eyepiece must be placed so that the image formed by the objective lens is at or almost at the focal point F_e of the eyepiece.

For microscope: (Objective has short focal length)

For telescope: (Objective has long focal length)

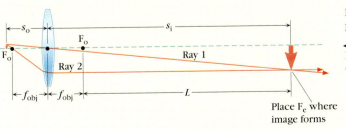

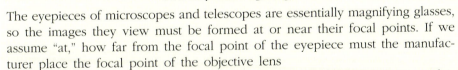

EXAMPLE 17-10 *continued*

What we know/what we don't.

Microscope: $s_o = 0.60$ cm $f_{obj} = 0.58$ cm $s_i = ?$ $L = ?$

Telescope: $s_o \approx \infty$ $f_{obj} = ?$ $s_i = ?$ $L = ?$

The mathematical solution. **a.** From Equation 17-4,

$$\frac{1}{s_i} = \frac{1}{f_{obj}} - \frac{1}{s_o} = \frac{1}{0.58 \text{ cm}} - \frac{1}{0.60 \text{ cm}} = 0.0575 \text{ cm}^{-1}$$

so that

$$s_i = \frac{1}{0.0575 \text{ cm}^{-1}} = 17.5 \text{ cm}$$

and the distance from the focal point (see Figure 17-20) is

$$L = s_i - f_{obj} = 17.5 \text{ cm} - 0.58 \text{ cm} = \textbf{16.9 cm}$$

slightly larger than the usual value of 16.0 cm in commercial microscopes.
b. In the case of the telescope, if $s_o = \infty$, $s_i = f_{obj}$, and

$$L = s_i - f_{obj} = \textbf{0}$$

The focal points of the two lenses would coincide. (Telescopes are usually designed so that the focal point of the objective lens is slightly closer to the eyepiece, so that the image the viewer sees is not at infinity.)

◆ Related homework: Problems 17-40 and 17-41.

For both microscope and telescope, Figure 17-20 summarizes the positions of the object and the image formed by the objective lens. The image position in turn determines where the eyepiece must be placed.

The objective lens of a microscope must provide a high degree of magnification; that is, $M = -\frac{s_i}{s_o}$ must have an absolute value much greater than one. Then s_i must much greater than s_o. As we saw in Example 17-10, this is accomplished by placing the object only slightly further from the objective lens than its focal point (s_o slightly greater than f_{obj}), so that the difference $\frac{1}{f_{obj}} - \frac{1}{s_o} = \frac{1}{s_i}$ is much less than one, and its inverse s_i is consequently very large. **STOP&Think** Why wouldn't you want to position your specimen *nearer* than the focal point to the objective lens? ◆ Because objects viewed by a microscope are placed very close to the objective lens, and the distance s_o must be slightly greater than f_{obj}, it follows that f_{obj} must be very small in microscopes.

The eyepiece of the microscope provides further magnification. The overall magnification of a compound microscope is usually taken to be the product of the linear magnification M of the objective lens and the angular magnification $\frac{0.25 \text{ m}}{f_{eye}}$ of the eyepiece.

In contrast to what happens in the microscope, the intermediate image produced by the objective lens of the telescope is greatly reduced; that is, $M = \frac{b_i}{b_o} \ll 1$. **STOP&Think** Why is this obvious? What types of objects do you usually view with a telescope? Would you expect the image to be larger or smaller than such an object? ◆ Also, the viewed object is typically a huge distance away, so $M \ll 1$ because $M = -\frac{s_i}{s_o}$ and $|s_i| \ll |s_o|$. Though M is necessarily very small, it can still be made many times greater than the magnification produced by the human eye, so that the *angular* magnification can be much greater than one. For very distant objects, $s_i \approx f$, so $M = -\frac{s_i}{s_o} \approx -\frac{f}{s_o}$, so this is accomplished by making f_{obj} many times larger than the focal length of the human eye. The latter is roughly the distance from the lens to the retina, less than 2 cm, so f_{obj} in a telescope will be tens or even hundreds of cm. The eyepiece of the telescope then produces further angular magnification of the intermediate image.

The World's Largest Reflecting Telescope. The Keck telescope at Mauna Kea, Hawaii, has a mirror diameter of 9.8 m, giving it about ten times the diameter or a hundred times the light-gathering area of the largest objective lens for a refracting telescope.

In the **reflecting telescope,** parallel rays from the distant object are intercepted and focused by a reflecting concave mirror rather than a refracting lens. One popular configuration of mirror and eyepiece, called the Newtonian after its designer, is shown in Figure 17-21. For astronomical telescopes, the mirror affords several advantages over the objective lens. A mirror is easier to produce with a larger diameter so that it can intercept rays crossing a larger area and direct them to the eyepiece. The mirror can be made parabolic to reduce spherical aberration. Aberration is further reduced because passage through one layer of glass (the objective lens) is eliminated.

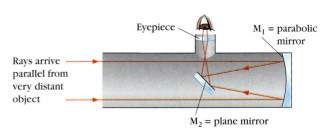

Figure 17-21 The reflecting telescope.

More refined instruments may include additional elements. For example, the telescopes described here produce inverted images, which is fine for astronomical purposes, but not great for looking at an Earthbound object like a distant ship. Therefore, a *terrestrial telescope* includes additional lenses to produce a second, compensating inversion.

◆ SUMMARY ◆

Rays arriving at each point of each surface (A and B) of a lens are **refracted** in accordance with **Snell's law.** Doing geometry on the ray diagrams that result from drawing the incident and refracted rays, we get a key relationship for a thin lens in air:

Lensmaker's formula
$$\frac{1}{s_o} + \frac{1}{s_i} = (n-1)\left(\frac{1}{R_A} - \frac{1}{R_B}\right) \quad (17\text{-}1)$$

(R = radius of curvature, s_o = object distance, s_i = image distance, n = index of refraction, with **sign convention** shown in Figure 17-7)

The **focal length** (the value of s_i when $s_o = \infty$) of a thin lens surrounded by air or a vacuum is given by

$$\frac{1}{f} = (n-1)\left(\frac{1}{R_A} - \frac{1}{R_B}\right) \quad (17\text{-}3)$$

so we can rewrite Equation 17-1 as

$$\frac{1}{s_o} + \frac{1}{s_i} = \frac{1}{f} \quad (17\text{-}4)$$

When the media interfacing with the two lens surfaces have indices of refraction n_A and n_B, this generalizes to

$$\frac{n_A}{s_o} + \frac{n_B}{s_i} = \frac{1}{f} \quad (17\text{-}4a)$$

For thin lenses with spherical surfaces,

if the lens is . . .	parallel rays from the left . . .	and the focal length f is . . .
thickest at the center (as a biconvex lens is),	converge toward the right (lens is a converging lens)	+
thinnest at the center (as a biconcave lens is),	diverge toward the right (lens is a diverging lens)	−

These properties are reversed if the lens has a lower index of refraction than its surroundings (not a possibility when the surrounding medium is air or a vacuum)

Images of extended objects are determined by locating the image points of representative object points. We do this by finding the points where **principal rays** intersect; see **Procedure 17-1: Identifying Principal Rays for Lens Diagrams.**

The **magnification** of images formed by lenses is

$$M = \frac{h_i}{h_o} = -\frac{s_i}{s_o} \quad (17\text{-}5)$$

If M is +, the image is upright.	If $|M| > 1$, the image is enlarged.
If M is −, the image is inverted.	If $|M| < 1$, the image is reduced in size.

The ways in which lenses deviate from perfect focus are called **aberrations:**

• **Spherical aberrations** occur because parallel rays arriving near the center of a spherical lens converge to within a small circle, which is only approximately a single point and is less so as rays that are incident further from the lens's center are included.

• **Chromatic aberrations** occur because the index of refraction (and thus the focal length) of a lens is different for different colors (wavelengths).

A range of **optical instruments** use combinations of *reflecting* and *refracting* elements to produce images. In the human eye, a *lens* focuses an image on the *retina*, producing *photochemical reactions* in detectors called *rods and cones*. In the photographic *camera,* the lens similarly focuses light on the film, where photochemical reactions in the emulsion on the film provide the means of detection. The **near point** is the closest object position for which the lens of your eye can produce a focused image on the retina.

How big something appears to you is determined by how much of your field of vision it takes up, that is, what angle it subtends. The extent to which a magnifying glass or other instrument makes an object appear bigger is the instrument's **angular magnification:**

$$M_A \equiv \frac{\theta'}{\theta} = \frac{\text{angular size of image seen through instrument}}{\text{angular size of object seen by unaided eye}}$$
(17-6)

When an object or intermediate image is closer to a magnifying glass than its focal point, the glass produces *an angularly magnified image beyond the near point of the viewer*. The magnifying glass serves as the **eyepiece** of instruments such as the compound microscope and the reflecting and refracting telescopes, in each case viewing an intermediate image that another optical element has formed of the object.

Instrument	Object viewed	Optical element producing intermediate image
Compound microscope	small, very near	very short focal length objective lens
Refracting telescope	large, far away	very long focal length objective lens
Reflecting telescope	large, far away	concave mirror (parabolic)

◆QUALITATIVE AND QUANTITATIVE PROBLEMS◆
Hands-On Activities and Discussion Questions

The questions and activities in this group are particularly suitable for in-class use.

17-1. Hands-On Activity. If you have a pair of binoculars, you can consider the linear and angular magnification of the binoculars in the following way. Select an object about 30–35 cm (about a foot) high and view it from a distance of about 8 m (about 25 ft). With both eyes open, use one eye to view the object directly, and the other eye to view the object through one side of the binoculars.

a. Does the object subtend a larger angle (does it fill more of your field of vision) when viewed directly or through the binoculars? Based on your answer, is the angular magnification greater or less than one?

b. Continue to use one eye to view the object directly and the other eye to view the object through one side of the binoculars. Hold the binoculars with one hand. Move the index finger of your free hand away from you until it looks to be at the same distance as the image that you see through the binoculars. The distance of your finger from your eye is the image distance. Keeping your hand at this distance, spread your thumb and index finger until they are level with the top and bottom of the object's image. That gives you the image height. How does it compare with the height of the object? Based on your answer, is the linear magnification less than or greater than one?

17-2. Hands-On Activity. Make a small hole with a pin or a needle in the center of the bottom of a plastic film canister (Figure 17-22a). (The lid of the canister is useful for pushing the needle through the bottom without sticking your finger.) Now bring your eyes gradually closer to this page until the

printing on the page appears so blurred you can no longer read it. Call this eye-to-page distance D. With your eyes still at this distance D from the page, place the canister over one eye, as in the figure, and close the other so that you are viewing the page through the hole.

a. How does the printing on the page appear now? Is it clearer than before?

b. Try to explain any differences in what you see when you look through the pinhole.

c. Take the dot on one letter *i* on the page to be a point object (Figure 17-22b). Sketch a ray diagram showing what happens to the light between this object and the retina of your eye when your eye is too close to it to see it clearly. Be sure to include multiple rays from the object.

d. Now add the film canister with the hole in it to your diagram, as in Figure 17-22c. Do all the rays that reached your retina before still do so? What difference does this make? Does this help you to answer part **b**?

e. How does the brightness of the page appear when viewed through the pinhole, in comparison to its appearance when viewed without the canister? How do you explain this difference? Are the ray diagrams helpful in explaining the difference?

17-3. Hands-On Activity.

a. Suppose you enlarged the diameter of the pinhole in Problem 17-2. What change(s) would you expect in the appearance of the printing on the page as seen through the pinhole? Predict what would happen and explain your reasoning before you try it.

b. Now enlarge the pinhole using a pencil point or similar instrument, and again view the printing through the pinhole at the same distance D. Record your observations, and explain any differences between what you observe and what you predicted.

17-4. Discussion Question. A biology class is using hand magnifying glasses to examine small specimens in the field. On reaching a very shallow pool, some of the students hold their magnifying glasses under water to look at specimens at the bottom of the pool. They are puzzled that their magnifying glasses are now giving them getting poorer magnification. What accounts for the change?

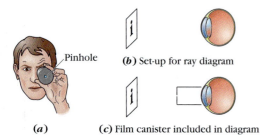

(a)   (b) Set-up for ray diagram   (c) Film canister included in diagram

Figure 17-22 Problem 17-2

Review and Practice

Section 17-1 **A Qualitative Picture of What Lenses Do**

17-5. A biconvex lens is used to project an image of a bulb onto a screen.

a. What happens to the image that you see on the screen when you extend your finger across the center of the lens? Briefly explain.

b. How does the image differ from the image in *a* if you extend your finger across the center of the bulb rather than across the center of the lens? In which case(s) (one? both? neither?) do you see your finger in the screen image?

c. Explain any differences between your answers to *a* and *b*.

17-6. When you use a biconvex lens to project an image of some object, and you then cover part of the lens,

a. why does the image become dimmer?

b. why doesn't any part of the image disappear?

17-7. Why do you ordinarily put slides into a slide projector upside down?

17-8. A biconcave lens produces a virtual image on the same side of the lens as the object.

a. If you are on the other side, do you see the image disappear if you blacken the center of the lens? Briefly explain.

b. Does it matter on which side of the lens you blacken it? Briefly explain.

17-9. A bulb, a biconcave lens, and the image of the bulb are lined up so that their centers all lie on the central axis of the lens.

a. From which part(s) on the bulb does light pass through the center of the lens? Briefly explain.

b. From which part(s) of the bulb does light pass through the lens near the bottom? Briefly explain.

c. From which part(s) of the lens does light arrive at the uppermost point on the image? Briefly explain.

17-10. A slide projector projects the image of a slide onto a small screen a short distance in front of it. The screen is free-standing and is not near a wall. If the screen is then removed, would it still be possible to see the image? If so, where would you have to position your eye to see it? Or if not, why not? In either case, explain. (*Note:* If you wish to test this, you should put the projector on a dimmer switch. Do *not* stare into a projector at full brightness.)

17-11. A slide projector projects the image of a slide onto a freestanding screen set up halfway between the projector and the wall. If the screen is removed, the image would _____ (*disappear; remain where the screen had been; appear on the wall; none of the above*).

17-12. When we speak of a particular shape lens as being converging or diverging, we mean that it is converging or diverging *under usual conditions*. What are these usual conditions? Under what conditions, for example, could a biconvex lens be a diverging rather than a converging lens? (Try to answer this question with no hints. If you are no closer to an answer after 24 hours, look at Figures 17-4 and 16-29.)

17-13. SSM WWW Replace the lens in Figure 17-5 with a concave lens, and draw the resulting ray diagram.

Section 17-2 **Quantitatively Analyzing What Thin Lenses Do**

17-14. A and B are two biconvex lenses with the same diameter. The surfaces of B bulge outward more than the surfaces of A. Which lens's surfaces have a greater radius of curvature? Briefly explain.

17-15. SSM The radius of curvature of each surface of a biconvex glass lens ($n = 1.6$) is 0.40 m. How far to the left of the lens should an object be positioned if its image is to appear 0.50 m to the

a. right of the lens?　　*b.* left of the lens?

17-16. Repeat Example 17-1 for a biconcave lens with the same index of refraction and the same amount of curvature of its surfaces.

17-17. Suppose your slide projector has a biconvex projecting lens with an index of refraction of 1.62. Both surfaces have a radius of curvature of magnitude 0.20 m. Ordinarily, when you focus the projector, the lens moves toward or away from the slide. But your lens is stuck at a distance of 0.17 m from the slide. How far away from the lens do you have to place your screen to view the image?

17-18.

a. Find the focal length of a biconcave lens ($n = 1.65$) if the magnitudes of the radii of curvature are 0.40 m and 0.60 m.

b. In your solution, did it matter which was surface A and which was surface B? Briefly explain.

c. For a biconcave lens, is the focal length f always, sometimes, or never positive?

17-19. A bulb is placed 1.20 m from a screen. A student wishes to place a biconvex lens halfway between the bulb and the screen to produce a sharp image of the bulb on the screen. What focal length lens should the student use?

17-20. Repeat Example 17-3 for the lens shown in Figure 17-23.

a. When the lens is placed as in Figure 17-23a, is it converging or diverging? Briefly explain.

b. When the lens is placed as in Figure 17-23b, is it converging or diverging? Briefly explain.

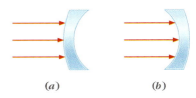

(a)　　　　　　　　*(b)*

Figure 17-23 Problem 17-20

17-21. A biconvex spherical lens has a focal length of 0.35 m. Find the position of the image formed for an object placed at each of the following distances in front of the vertex of the lens

a. 0.20 m.　　*b.* 0.35 m.　　*c.* 0.70 m.　　*d.* ∞.

17-22. Repeat Problem 17-21 for the same given values if the lens is biconcave rather than biconvex.

17-23. Repeat Example 17-5 for the case where the pencil point is 0.20 m from the lens and points toward the lens.

17-24. If the pencil in Example 17-5 is placed so that its point is 0.40 m from the lens and its eraser is 0.60 m from the lens, you can still find the image positions for the eraser and the point. Is the distance between these two positions equal to the length of the image of the pencil in this case? Briefly explain.

Section 17-3 Lens Images of Extended Objects

17-25. In Procedure 17-1, why is ray (3) not needed for determining the image point if you draw rays (1) and (2)?

17-26. Estimate the image position and the image height for each of the situations in Figure 17-24 by sketching the principal rays in each case.

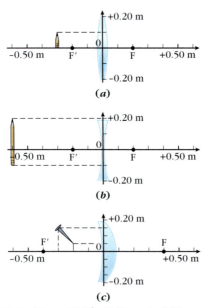

Figure 17-24 Problems 17-26, 17-28, and 17-29

17-27. SSM WWW A converging lens has a focal length of magnitude 0.50 m. Where must an object be placed along its principal axis if the image is to be
a. half the size of the object, but inverted?
b. upright and 20% larger than the object?

17-28.
a. Calculate the position and height of the image formed of the pencil in Figure 17-24a.
b. Calculate the position and height of the image formed of the pencil in Figure 17-24b.

•**17-29.** Calculate the length of the image that is formed of the nail in Figure 17-24c.

17-30. Suppose the object in Problem 17-22 is 0.10 m tall. Find the height of the image formed when the object is at each of the positions in Problem 17-21, and in each case, tell whether the image is real or virtual, upright or inverted, enlarged or reduced.

17-31. Suppose the object in Problem 17-21 is 0.10 m tall. Find the height of the image formed when the object is at each of

positions **a** and **c** in Problem 17-21. In each case, tell whether the image is real or virtual, upright or inverted, magnified or reduced.

Section 17-4 Aberration in Lenses

17-32. People who wear mild reading glasses often find they don't need them to read in bright sunlight. What happens to your pupils in bright sunlight? What effect does this have on the extent to which spherical aberration of light passing through your cornea + lens system affects the apparent sharpness of the printing on a page? Explain.

17-33. Redraw the ray diagram in Figure 17-11 to show spherical aberration for a biconcave lens.

17-34. Redraw the ray diagram in Figure 17-12 to show chromatic aberration for a biconcave lens.

Section 17-5 Optical Instruments

17-35. When completely relaxed, a normal eye forms an image of a very far object on the retina. For the average person, the distance from the cornea to the retina is 1.7 cm. Find the focal length of the cornea + lens system under these conditions.

17-36. Light passes from a candle flame several meters in front of you to the retina of your eye.
a. Based on the data in Table 16-1, at which of the following interfaces does the light bend most: air to cornea, cornea to aqueous humor, aqueous humor to lens, lens to vitreous humor? Briefly explain.
b. Which part of the eye contributes most to the convergence of rays to form an image of the candle flame on your retina? Briefly explain.

17-37. SSM WWW When completely relaxed, a normal eye forms an image of a very far object on the retina. As the object approaches the eye (see Figure 17-25) the ciliary muscles bulge the lens to change the focal length, so that the image position remains at the retina. *Nearsighted* (or **myopic**) people see near objects clearly, but the images of distant objects form in front of the retina. In contrast, *farsighted* (or **hyperopic**) people see far objects clearly, but the images of near objects form behind the retina. Use ray diagrams to figure out which of the lenses depicted in Figure 17-25 corrects which of these conditions.

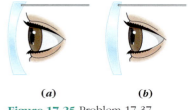

(a)　　　(b)

Figure 17-25 Problem 17-37

17-38. You find a magnifying glass at a yard sale. When you hold the glass a negligible distance from your eye and look at your fingertip, you find that you cannot see your fingertip clearly when you bring it closer than 0.14 m from your eye. What is the focal length of the magnifying glass?

17-39. Why are telescopes much longer than microscopes?

17-40. What possible alteration(s) could you make in the design of the microscope in Example 17-10 if you want the distance between the focal points of the two lenses to be the 16.0 cm conventionally chosen by manufacturers of optical instruments?

17-41. SSM A microbiology student is using a microscope in which the focal points of the objective lens and the eyepiece are 16 cm apart and the objective lens has a focal length of 0.50 cm. How far from the objective lens should a specimen be positioned if its image is to appear at the focal point of the eyepiece?

17-42. When you view a specimen through a microscope, you usually position it so that it is just slightly beyond the focal point of the objective lens. Why would you not want to position it so that it is slightly closer than the focal point?

17-43.
a. In a reflector telescope (Figure 17-21), why doesn't the plane mirror block out all light from part of the field of view and prevent it from reaching the parabolic mirror?
b. What would be the effect on the image you observed in the reflector telescope if the plane mirror were increased in area?

Going Further

The questions and problems in this group are not organized by section heading, so you must determine for yourself which ideas apply. Some of them will be more challenging than the Review and Practice questions and problems (especially those marked with a • or ••).

17-44. The diagram in Figure 17-26 is taken "as is" from an actual 1940s physics textbook. What is wrong with this diagram?

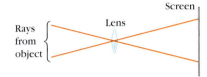

Figure 17-26 Problem 17-44

17-45. (*This problem can also be treated as a **Hands-On Activity.***) The partially coated convex lens in Figure 17-27 is used to project the image of a bulb onto a screen. (Any opaque material can be used instead of the black coating.) When the opaque mask with a slit in it is placed over the lens, light reaches the screen only through the part of the clear center strip left uncovered by the slit. Suppose the screen is positioned so that a focused image is obtained when the center strip is uncovered. Then the slit is gradually pulled left to right across the center strip, as shown. What happens to the image on the screen as the slit moves? For instance, does it move in a particular direction, does it brighten or dim, does it remain unchanged, or what? Briefly explain.

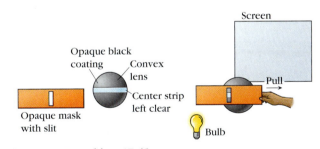

Figure 17-27 Problem 17-45

17-46. The diagrams in Figure 17-28 each show rays from a point object O incident on a biconcave lens. The lens is transparent and made of a material that has an index of refraction of 1.33. Which diagram correctly shows

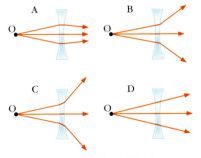

Figure 17-28 Problem 17-46

what happens to the rays if the lens and object are surrounded by
a. air?
b. a concentrated sugar solution?
c. pure water? Briefly explain.

17-47. In Figure 17-29a, a biconvex lens produces an image of a small light bulb (small enough to be considered a point source) on a screen. The screen is positioned where the image is sharpest. In Figure 17-29b, the bulb is replaced by a laser pointer, with its light outlet positioned at the point where the bulb was. The laser pointer essentially produces a single directed ray, as shown. At the place where the ray strikes the screen, there will be a small red dot. If the pointer is then rotated from position 1 to position 2, will the dot move upward on the screen, move downward on the screen, or remain in the same position? Briefly explain.

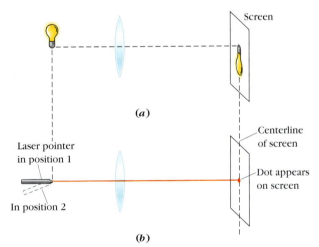

Figure 17-29 Problem 17-47

17-48. Sketch a lens with one concave and one convex surface that functions as a diverging lens in air. Then sketch a ray diagram in sufficient detail to demonstrate that your lens is indeed diverging.

17-49. Sketch a lens with one concave and one convex surface that functions as a converging lens in air. Then sketch a ray diagram in sufficient detail to demonstrate that your lens is indeed converging.

•**17-50.** Use Equation 17-4 to show that the image will always appear on the same side of a biconcave lens as the object and that it will always be closer to the lens than the object.

17-51. SSM Figure 17-30 shows four possible object positions (A–D) to the left of a biconvex lens. The focal point F of the lens is also shown. Rank the four

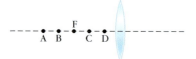

Figure 17-30 Problem 17-51

labeled points on the axis in order, from most negative to most positive, according to the image positions that would occur for these object positions. Indicate any equalities.

17-52. A wedge is placed along the axis of a biconvex lens as shown in Figure 17-31. The lens has a focal length of 0.40 m. Do whatever calculations are necessary to determine completely the appearance (placement and dimensions) of the image of the wedge.

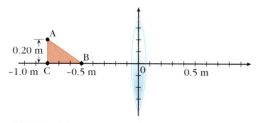

Figure 17-31 Problems 17-52 and 17-53

17-53. Repeat Problem 17-52 for a biconcave lens with the same focal length.

17-54. In each of two identical set-ups, a convex lens is used to produce an image of a light bulb on a screen. Then the lenses are each partially blackened on one surface. On lens A, an opaque black coating is applied over the 50% of the surface area closest to the center. On lens B, the same type of coating is applied over the 50% of the area that is furthest from the center. Compare the images produced by the two lenses with regard to
a. brightness.   **b.** sharpness.

17-55. Suppose you view the moon with an ordinary pair of binoculars (looking in the right end). Is the linear magnification produced by this optical instrument less than, equal to, or greater than one? Is the angular magnification less than, equal to, or greater than one? Briefly explain.

•**17-56.** Show that if the distance between the objective lens and the eyepiece of a refracting telescope is exactly $f_{obj} + f_{eye}$, if the observed object is at infinity then the image the viewer sees will also be at infinity. (A telescope set up like this is called *afocal* because the combination of lenses does not have a finite focal length.)

17-57. Because optical paths are reversible, it is possible to position the pencil in Figure 17-32 so that the rays from the

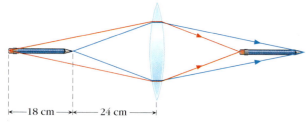

Figure 17-32 Problem 17-57

blue point and red eraser are mirror images of each other ("mirrored" in the plane of the lens), and the distances from the lens at which point and eraser are located are reversed for object and image. For the pencil length and distance given in the figure, what focal length must the biconvex lens in the figure have to produce this symmetrical result?

17-58. A *planoconvex* lens has one plane (flat) and one convex surface. A certain planoconvex lens is made of glass with an index of refraction of 1.48, and its curved surface has a radius of curvature of 0.50 m.
a. What is the radius of curvature of the plane surface?
b. Find the focal length of this lens.

••**17-59.** When the media on the two sides of a lens are not the same, we must use a more general form of the thin lens equation:

$$\frac{n_A}{s_o} + \frac{n_B}{s_i} = \frac{1}{f}, \text{ where } \frac{1}{f} = \left(\frac{n_{lens} - n_A}{R_A} + \frac{n_B - n_{lens}}{R_B} \right)$$

R_A = radius of curvature of surface A (surface nearer to object)
R_B = radius of curvature of surface B (further from object)
n_A = medium at surface A n_B = medium at surface B

Try to make use of the ideas in Problem 17-58 along with these equations, to solve the following: At the bottom of a giant fish tank in the Municipal Aquarium are two round viewing portholes. In each, the "window" is a section of a hollow Plexiglas sphere of radius 1.0 m. But one is convex (it bulges outward toward the visitor), and the other is concave.
a. If a rainbow parrotfish is 1.40 m behind the center of the convex "window," how close to the window does it look to visitors?
b. If it is 1.40 m behind the center of the concave "window," how close to the window does it look to visitors?

17-60. Show that in air, the focal length of a planoconvex lens (see Figure 17-3) is $f = \frac{|R|}{n - 1}$ where n is the index of refraction of the lens material and $|R|$ is the absolute value of the radius of curvature of the convex surface.

17-61. SSM Show that in air, the focal length of a planoconcave lens (see Figure 17-3) is $f = \frac{|R|}{1 - n}$ where n is the index of refraction of the lens material and $|R|$ is the absolute value of the radius of curvature of the concave surface.

17-62. When a solid glass lens is submerged in water, is its focal length less than, equal to, or greater than when it is surrounded by air?

••**17-63.** Suppose light from an object point passes through two thin lenses in sequence with negligible space between them (so that the combined thickness of the two lenses is still negligible compared to the object distance s_o and the image distance s_i from the combination of two lenses).
a. The image formed by the first lens becomes the object for the second lens. Taking sign conventions into account, how does the image distance s_{i1} for the first lens compare to the object distance s_{o2} for the second lens?
b. Show that the thin lens formula for the combination of two lenses becomes $\frac{1}{s_o} + \frac{1}{s_i} = \frac{1}{f_1} + \frac{1}{f_2}$
c. Show that if we rewrite this as $\frac{1}{s_o} + \frac{1}{s_i} = \frac{1}{f_{eff}}$ where f_{eff} is the effective focal length of the combination of two lenses, then $f_{eff} = \frac{f_1 f_2}{f_1 + f_2}$

17-64. Suppose you open your eyes under water (*not* salt water or chlorinated water). Refer to the data in Table 16-1 in answering the following:

a. If the lens of your eye were not flexible, would your eye produce more or less convergence of rays from a distant object than when you view the object through air?

b. Because the lens *is* flexible, would it bulge more or less than when you were viewing the same object through air?

c. Would its radius of curvature therefore be greater or less than when you were viewing the same object through air?

d. Will the object appear closer or further than it would in air?

17-65. Your instructor is standing across the room from you, and an image of your instructor is formed on the retina of your eye.

a. Is the image on your retina real or virtual?

b. Is it upright or inverted?

c. Is it enlarged or reduced?

17-66. Suppose that in the previous problem, your instructor begins to walk toward you.

a. If the lens of your eye was rigid and could not adjust, the image of your instructor would end up _____ (*in front of; on; behind*) the retina.

b. To keep the image on the retina rather than behind it, the muscles connected to the lens must _____ (*increase the radii of curvature of the lens surfaces by making the lens bulge more; decrease the radii of curvature of the lens surfaces by making the lens bulge more; increase the radii of curvature of the lens surfaces by making the lens bulge less; decrease the radii of curvature of the lens surfaces by making the lens bulge more*).

17-67. SSM Nearsighted people cannot see distant objects clearly because the focal lengths of their relaxed lenses are too short, so the image falls in front of the retina. The maximum distance at which a nearsighted person can see clearly is called the person's *far point*. For such a person, the role of corrective glasses is to enable them to see more distant objects (out to infinity) by placing the images of these objects at or closer than the person's far point. Find the focal length of the corrective lens required for a nearsighted person whose far point is 6.0 m. (Neglect the distance between the lens and the person's eye.)

17-68. Consider the following lenses: the lens of a camera, the objective lens of a microscope, the eyepiece of a microscope, the objective lens of a telescope. Which (one or more) of these lenses will produce an image at its focal point for an object placed at infinity? Briefly explain.

17-69. Suppose that the objective lens of a microscope has a focal length of 0.45 cm, and its eyepiece has a focal length of 0.030 m. How far apart should the two lenses be positioned when viewing a specimen on a slide placed 0.47 cm below the objective? Is this practical?

17-70. A collector with a near point of 0.25 m examines a postage stamp using a magnifying glass with a focal length of 0.40 m. The collector holds the glass up against his eye, and holds the stamp so that its image appears at his near point. Under these conditions, what angular magnification is produced by the glass?

17-71. For the observer in Figure 17-13*a* to be able to view the image of the candle, should the screen have a diffuse reflecting surface or a mirror-like surface? Briefly explain your answer.

17-72. A manually operated camera has a lens with a focal length of 52 mm. The user adjusts the f-number from f/2.8 to f/11. To what diameter is the aperture open

a. before the adjustment?   *b.* after the adjustment?

•**17-73.** A small bright bulb and a screen are 150 cm apart. A biconvex lens with a focal length of 24 cm is placed between the bulb and the screen. At what two distances from the bulb could the lens be placed for a focused image of the bulb to appear on the screen?

••**17-74.** Suppose that in Problem 17-73, a biconvex lens with a focal length of 80 cm is used instead. Is there any place between the bulb and the screen that you could place the lens to produce a focused image of the bulb on the screen? Support your answer with a suitable calculation.

17-75. The objects in Figure 17-33 can be moved to various positions along the track. Values of x refer to the position scale along the track's edge. Suppose the bulb is positioned at $x = 10$ cm and the lens, which has a focal length of 30 cm, is positioned at $x = 60$ cm. At what value of x should the screen be placed to show a focused image of the bulb?

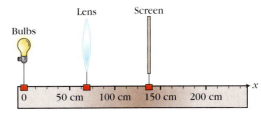

Figure 17-33 Problems 17-75 and 17-76

17-76. If the bulb in Figure 17-33 is positioned at $x = 20$ cm, where should a lens with a focal length of 35 cm be placed if the rays from the bulb are to emerge parallel to the right of the bulb? (See Problem 17-75 for details about the track.)

17-77. A biconvex lens has a focal length of magnitude 25 cm when it is between water and air. Suppose the lens is floating on the surface of the water in a fish tank, and a fish passes 80 cm below it. Where will the image of the fish appear?

17-78. Repeat Problem 17-77 for a biconcave lens with a focal length of the same magnitude.

17-79. Two thin convex lenses, one with a focal length of 0.40 m and one with a focal length of 0.30 m, are positioned at $x = 0$ with negligible space between them. The x axis is the principal axis for both lenses.

a. Using the result of Problem 17-63, find the position of the image for an object placed at $x = -0.25$ m.

b. Is the image real or virtual?

c. If you used either of the two lenses by itself, would the image be real or virtual?

17-80. Repeat Problem 17-79*a* for the case in which the lens with the 0.30 m focal length is concave instead of convex.

Problems on WebLinks

17-81. In WebLink 17-1, the rays arriving at the point where the image of the flame forms come from ____ (*only the top half of the lens; only the bottom half of the lens; both halves of the lens; only the center of the lens*).

17-82. In WebLink 17-2, if the screen were removed, an image of the candle flame would ____ (*form where the screen was; form to the left of where the screen was; form to the right of where the screen was; not form at all*).

17-83. In WebLink 17-2, if you covered all but a region in the center of the lens, ____ (*the center of the image would vanish; only the center of the image would remain; the whole image would remain but be dimmer; the image wouldn't change at all*).

17-84. Which (one or more) of the following are approximated by combinations of triangular prisms in WebLink 17-3: biconcave lenses; biconvex lenses; meniscus convex lenses; meniscus concave lenses? (See Figure 17-3 for pictures of these lenses.)

17-85. WebLink 17-4 traces three rays. Match what each ray does after passing through the lens with what it does before reaching the lens.

Before reaching lens	After passing through lens
a. travels along line through center of lens	**i.** travels parallel to central axis
b. travels along line through near focal point	**ii.** travels along line through center of lens
c. travels parallel to central axis	**iii.** travels along line through far focal point

CHAPTER EIGHTEEN

Electrical Phenomena: Forces, Charges, Currents

In Chapter 4, we introduced the idea of interactions at a distance or equivalently of action-at-a-distance forces: gravitational, electrical or electrostatic, and magnetic. We have treated one of these, the gravitational force, in considerable detail. But although gravitational forces may be far more conspicuous in your direct experience, electrical forces are every bit as pervasive in the universe. This was true even before the electronic age—indeed, even before the emergence of life on Earth—because electrical forces are central not only to the operation of electronic devices but to the fabric of matter itself.

In this chapter we will develop qualitatively an underlying model that will enable us to explain a range of electrical behaviors. In the chapters that follow, we will formulate these ideas quantitatively.

". . . electrical forces are central not only to the operation of electronic devices but to the fabric of matter itself."

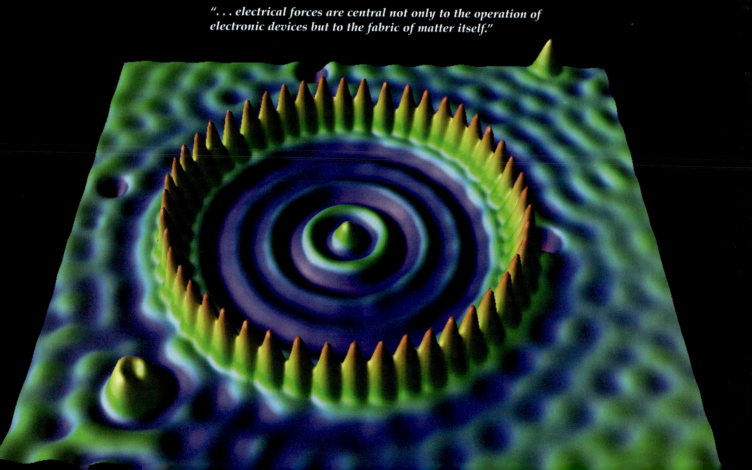

18-1 Developing an Underlying Model to Account for the Observations

The ancient Greeks made the first known observations of electrostatic effects. Just as you may observe that a comb run through your hair attracts bits of tissue paper, the ancient Greeks found that a briskly rubbed piece of amber would attract small bits of various materials. The Greek word for amber is *elektron,* so the forces were called electric or electrical.

Englishman William Gilbert (1540–1603) found that a variety of materials, when rubbed, would display this same property: sulfur, wax, resinous substances such as amber, glass, and precious stones. In the seventeenth and eighteenth centuries, such materials were called *electrics,* because they were amber-like in their behavior. Numerous modern materials display the same behavior, among them cellophane, Styrofoam, the polyurethane of plastic bags, Mylar, audio or video tape, and synthetic fabrics with their well-known static cling.

In the late 1600s, Otto von Guericke of Magdeburg (in what is now Germany) discovered that a body, after being attracted to an *electric,* might later be repelled by it and not be attracted again until it had come in contact with yet another body. The following sequence of events repeats his observations with everyday materials.

Case 18-1 ◆ *Comb and Foil: Some Observations in Need of an Explanation*

Important: Although the following steps are fully described, reading a description cannot substitute for making the observations for yourself. You are strongly urged to do this as an On-the-Spot Activity.

Materials needed: a plastic comb, a human hair (about 5 to 10 cm long) or fine thread, a small piece of aluminum foil about 1 cm $\times \frac{1}{2}$ cm, a little glue or substitute (see below), and an inch of tape.

Apply some glue to one side of the aluminum foil—or make it sticky by wiping it against a freshly licked stamp or hard candy—and fold it in half around one end of the hair, with the sticky side in (Figure 18-1a). Press the halves firmly together. You are now fully equipped.

Step 1: Use the tape to hang the hair by the free end so that the aluminum foil at the other end dangles freely. Then run the comb briskly through your hair.

Step 2: Bring the comb, teeth forward, to within a cm of the foil. You will observe the foil attracted strongly toward the comb. It may remain in

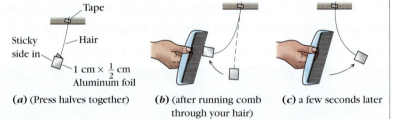

(a) (Press halves together) (b) (after running comb through your hair) (c) a few seconds later

Figure 18-1 Evidence of electrical forces in simple materials.

contact with the comb anywhere from a few seconds to a minute or more (Figure 18-1b).

Step 3: But then, without your doing anything more, the foil will jump sharply away from the comb (Figure 18-1c). If you move the comb slightly toward the foil at this stage, the foil will edge further away from the comb.

Step 4: Touch the aluminum foil firmly with the fingers of your other hand (the one without the comb). Maintain this contact for a couple of seconds, then release it so that the foil once again dangles freely from the hair. Bring the comb close to the foil again. You will observe that the foil is once again attracted.

How can we explain what we observed in Case 18-1? You could readily verify with a refrigerator magnet that aluminum does not respond to magnets, so the forces involved here are not magnetic forces. Your massive physics book doesn't attract the foil either; gravitational forces between objects this size are far too weak to produce these effects. So the electrostatic force is clearly a different kind of force.

To what is it due? Much as we assumed that a property of matter, called gravitational mass, is responsible for the gravitational force, we might assume there is some other property of matter or substance within matter giving rise to the electrical or electrostatic force. In older English usage, a charge meant a load carried by anything, whether the carrier was a cannon or a horse. Because this property or substance was "carried" by matter, it was called the **electric charge.** This doesn't tell us *what* was being carried. We could as well have called it the electric whatsit; calling it *charge* doesn't mean we understand its nature. In fact, for the same reason, we could equally well have referred to gravitational mass as *gravitational charge*. Some modern authors have used this term to stress the conceptual similarity, but its use is not common practice.

Gravitational forces are always attractive, but our observations show evidence of both *attractive* and *repulsive* electrostatic forces. One kind of charge, always behaving the same way, could not account for both. A simple assumption consistent with the evidence is that there are *two* kinds of charge: We call these **positive** and **negative.** We can account for our observation of both attraction and repulsion by supposing that opposite charges attract each other, and like charges repel each other. For further evidence of this, work through Problems 18-1, 18-2, and 18-3.

STOP&Think Suppose there are equal amounts of positively and negatively charged matter in the universe. How would this matter tend to rearrange itself over time if *like* charges attracted and *opposites* repelled? What if opposite charges didn't react to each other at all (as gravitational charge doesn't react to electric charge), but instead charges of one kind (say, positive) were mutually attractive and charges of the other kind were mutually repulsive? ◆

Most objects show no evidence of exerting electrostatic forces on each other. We call such matter **neutral.**

Because the comb doesn't attract the foil until we rub it, it does not seem to be charged until that point. But we didn't do anything to the foil. Why should *it* have become charged? Or did it?

Our model so far has made two assumptions:

- *Matter ordinarily contains equal amounts of positive (+) and negative (−) charge.*
- *Opposite charges attract; like charges repel.*

Suppose we now make two further assumptions:

- *Charge can neither be created nor destroyed.*
- *Charge can move through matter.*

The first of the new assumptions is called the **law of conservation of electric charge.** It fits with the evidence that the appearance of positive charge in one place is always accompanied by the appearance of equal negative charge somewhere else. The last assumption provides a way that the usual neutral situation can be altered if we can't "create" charge. The negatively charged comb attracts the positive charge and repels the negative charge in the neutral aluminum foil. If either kind of charge can move, this causes a charge separation (Figure 18-2*a*), with more of the negative charge further from the comb. (We will see later that in metals, only the electrons, which are negative, can move.) But if there is repulsion as well as attraction, why is the aluminum foil attracted to the comb? We must add yet another assumption to the model:

- *Electrostatic forces decrease with distance.*

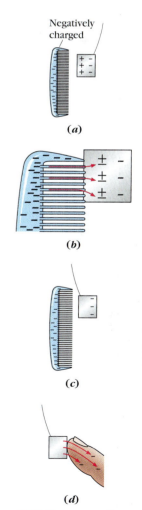

Negatively charged

(*a*)

(*b*)

(*c*)

(*d*)

Figure 18-2 Charge configuration diagrams for Case 18-1.

For **WebLink 18-1:**
A Model to Explain
Electrical Attractions
and Repulsions, go to
www.wiley.com/college/touger

With this assumption, if the negative charges are further from the comb, they will be repelled less strongly than the positive charges are attracted. The total force is then toward the comb, and the foil is attracted to it. But the negative charges on the comb also repel one another. Once the foil is touching the comb, some of these negative charges can move onto the aluminum foil. Now both bodies have net negative charges, and they repel each other. We have now explained Steps 1 to 3 of Case 18-1. For a graphic step-by-step presentation of the reasoning that explains these observations, go to WebLink 18-1.

When you bring your fingers in contact with the foil (*Step 4* and Figure 18-2*d*) after it has gained negative charge from the comb and been repelled, mutually repelling negative charges on the foil can now get farther from one another by traveling into your body and possibly spreading from there to the ground or Earth. So little charge remains on the foil that it is once again effectively neutral. This explanation makes yet another assumption—namely, that charges could travel through your finger and body much more readily than through the hair connected to the foil. We therefore add another assumption to our model:

- *Materials differ as to how readily they permit the passage of charge.*

We will shortly consider other evidence of this.

Figure 18-2 diagrams the arrangements or configurations of charge at different stages of the observations in Case 18-1. After each new action, we try to show the system when there is no longer any net movement of charge. At that stage, we say that the system is in **electrostatic equilibrium.** The changes in configuration from one diagram to the next are governed by our assumptions about how charges behave. Drawing these diagrams is important for visualizing what is going on in the underlying model, which we adopt because it explains the behaviors that we actually observe. We are again in the business of looking for the basic rules by which the game of nature is played.

PROCEDURE 18-1

Charge Configuration Diagrams

1. Diagrams should show the distribution of charges on each body. Plus ($+$) or minus ($-$) signs in a region indicate an excess of positive or negative charge in a region. Regions that are neutral and have a uniform mix of positive and negative charges are unmarked.
2. In each diagram, the distribution of charges will be governed by the nature of electrostatic forces and the extent to which the charges can move in response to those forces:
   a. Opposite charges attract and like charges repel each other.
   b. Like charges will therefore move as far from one another as they can if paths are available.
   c. Charges exert weaker forces on one another when they are further apart.
   d. Charges will move in response to these forces if there is a path by which they can travel.
3. For sequences of events, you should draw a sequence of diagrams in which the progression from one to the next is clearly determined by factors 2a–d.
4. Arrows (in contrasting color) indicate how charge has moved.

The diagrams in Figure 18-2 illustrate point 3. Example 18-1 illustrates the significance of point 2d.

Example 18-1 *Conducting and Insulating Rods*

For a guided interactive solution, go to Web Example 18-1 at
www.wiley.com/college/touger

A horizontal rod is suspended at its center from a metal ceiling (Figure 18-3a). At one end, the rod is in contact with a piece of aluminum foil that hangs on a thread or wire. A negatively charged rubber rod is briefly brought in contact with the other end of the rod and then removed. This procedure is repeated for three different instances:

	Horizontal rod made of . . .	and hung from . . .
Instance I	metal	nylon thread
Instance II	glass	nylon thread
Instance III	metal	copper wire

STOP&Think Before continuing, see if you can figure out what will happen to the piece of foil in each case. ◆

In instance I, the piece of foil is immediately repelled by the horizontal rod and remains repelled when the rubber rod is removed. In instances II and III, no repulsion is observed. Use charge configuration diagrams to explain these behaviors.

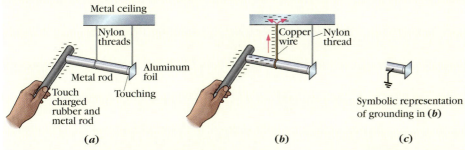

Figure 18-3 Grounding a conductor. If the aluminum foil in this arrangement were suspended by an insulating thread (*a*), negative charge would spread from the rubber rod through the metal rod to the aluminum foil. The negatively charged metal rod and aluminum foil would repel each other. But in (*b*), the copper wire allows the negative charge to spread over a much larger conducting region, preventing enough concentration of charge to produce a visible repulsion. Preventing an accumulation of charge in this way is called **grounding.**

Brief Solution

Choice of approach. The mutual repulsion of like charges and mutual attraction of opposite charges must govern the behavior that we see.

Diagrams and reasoning. If two bodies repel, there must be like charge on them. The only source of excess charge is the rubber rod. If the metal rod permits the passage of charge, the mutually repelling negative charges that start out on the rubber in Figure 18-3a will get as far away from one another as they can and thus distribute themselves over the metal rod and the foil. Once rod and foil are both negatively charged, they will repel each other. We therefore must assume that the metal rod permits the passage of charge very quickly.

The only difference between instance II and instance I is that the horizontal rod is made of glass. If no repulsion takes place, we must infer that excess negative charge has *not* spread to the far end of the rod and to the foil, and therefore glass does not readily permit the passage of charge.

In instance III there is no repulsion even though the rod is metal. Why? In instances I and II, we didn't consider the thread supporting the rod as a possible path for the excess charges. Like glass, nylon does not permit the passage of charge. But copper is a metal. The charges can therefore get farther

EXAMPLE 18-1 *continued*

away from one another by going up the copper wire and spreading out over the metal ceiling (recall point 2b in Procedure 18-1), as in Figure 18-3b. The excess charge quickly becomes so spread out that the excess negative charge in any one location is insufficient to produce visible repulsions.

◆ Related homework: Problems 18-12 through 18-15.

In the example we see that some materials, such as metals, permit the passage of charge very quickly. A material that does this is called a **conductor.** Metals are good conductors. Other materials, such as glass, do not readily permit the passage of charge. A material of this type is called an **insulator.** Like *conductor, insulator* is not an absolute term. Materials may be conducting or insulating to various degrees. Until we make these ideas quantitative, we will make do by speaking of good or poor conductors and insulators.

A wire connected to the ground or Earth would have the same effect as the copper wire in the above example or in the similar situation in Figure 18-3a and *b,* so any large body where excess charge can spread out when connected by a wire is called **ground** (the British call this *earth*). Connecting an object to ground is called **grounding.** When you use jumper cables to start your car, one terminal must always be connected to the engine block or car frame, which serves as ground, to prevent a dangerous build-up of excess charge. The usual symbol for ground in diagrams is ⏚ (Figure 18-3c). It is possible to charge an object connected to ground by bringing an already charged object near it, without transferring any charge from that second object. (For the details of this process, called **charging by induction,** see Problem 18-46.)

In Case 18-1 (Figure 18-2), we looked at situations in which electrostatic forces produce a separation of charge. Figure 18-4 shows that this separation can result either from positive charge moving one way or negative charge moving the opposite way. Because it is a pain in the neck to be worrying about minus signs all the time, physicists often take advantage of this fact to describe situations in terms of the movement of positive charge, even when negative charge is moving the other way. We have not yet established whether one or both kinds of charge are able to move in various conducting materials. So in the following example, for instance, in which positive charge is effectively transferred to a region of a conducting sphere, it may actually be that negative charge is being transferred away from that region or that both are occurring.

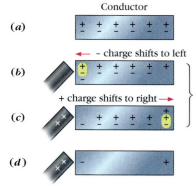

Conductor

(a)

← - charge shifts to left

(b)

+ charge shifts to right →

(c)

(d)

Both movements result in the same *net* negative charge on the left and the same net positive charge on the right. (highlighted areas neutral)

Summary charge configuration for either (*b*) or (*c*)

Figure 18-4 A flow of negative charge in one direction is equivalent to a flow of positive charge in the other.

Example 18-2 *Charge Distributions on Conductors, I*

Figure 18-5a shows an isolated conducting sphere. Some positive charge is effectively transferred to a small region of the sphere at some instant. How is the charge distributed when the sphere reaches electrostatic equilibrium? **STOP&Think** Attempt to do the reasoning yourself before reading the solution. ◆

Solution
Choice of approach. We can apply point 2b of Procedure 18-1.

Diagrams and reasoning. The excess positive charges repel one another, so that the charge spreads out as far as the conducting material permits. It thus ends up distributed uniformly over the surface of the sphere (Figure 18-5b). For the same reason, *any excess charge on a conducting body in equilibrium will always be located on the surface,* although not always distributed uniformly.

◆ Related homework: Problem 18-16.

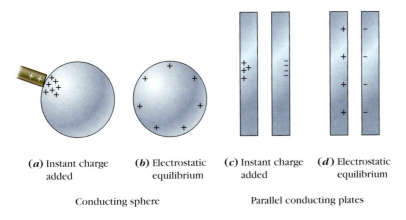

(*a*) Instant charge added

(*b*) Electrostatic equilibrium

Conducting sphere

(*c*) Instant charge added

(*d*) Electrostatic equilibrium

Parallel conducting plates

Figure 18-5 Figures for Examples 18-2 and 18-3.

Example 18-3 *Charge Distributions on Conductors, II*

Figure 18-5*c* shows an edge view of two thin conducting metal plates (the thickness is exaggerated in the diagrams) separated by a nonconducting gap. Equal and opposite amounts of charge are effectively transferred to small regions of the plates at some instant. How is this charge distributed when the system reaches electrostatic equilibrium? **STOP&Think** Once again, try to do the reasoning yourself before reading the solution. ◆

Solution

Choice of approach. We can again apply Procedure 18-1.

Diagrams and reasoning. As in Example 18-2, and for the same reason, the excess charge on each conductor spreads out over its surface, but this time *not* uniformly over the entire surface. Because the opposite charges on the two plates attract each other, they will move as close to each other as they can. Thus nearly all the excess charge will end up on the *inner* faces of the two plates (Figure 18-5*d*). The distribution will be essentially uniform over the inner faces, except near the edges.

◆ Related homework: Problem 18-17.

If the charges in Figure 18-5*d* are truly in equilibrium, there must be constraining forces on the charges that are equal and opposite to the forces that the positive and negative charges exert on each other. The constraining forces are exerted either by the conducting surfaces or the intervening medium; we will not concern ourselves with the details. If the two opposite concentrations of charge become sufficiently large, the electrostatic forces on the charges will exceed the maximum constraining forces, and charges will jump across the gap between the charged objects. This happens when charges build up on clouds, causing an opposite buildup of charge on Earth's surface, until finally there is the discharge that we know as *lightning*.

The forces that individual charges exert on each other are directed toward each other if attractive, away from each other if repulsive. When like charges are on a flat or gently curving surface of a conductor, the repulsive forces they exert are essentially *along* the surface (Figure 18-6*a*), causing the charge to spread out on the surface. On a needle-like projection (Figure 18-6*b*), the forces those same charges exert on each other would be essentially *perpendicular* to the surface and thus would not contribute significantly to spreading. Consequently, on irregularly shaped conducting bodies, charge tends to concentrate at sharp projections. It is for this reason that lightning tends to strike tall trees or lightning rods rather than flat expanses of ground or roof.

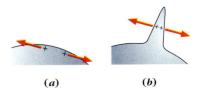

(*a*)　　　　(*b*)

Figure 18-6 Repulsive forces that surface charges on conductors exert on each other. (*a*) Forces mostly parallel to gently rounded surface. (*b*) Forces mostly perpendicular to the surface of a sharp point.

18-2 Charge Carriers

In addition to the terms we've defined so far to describe our underlying model, we will also suppose (and later provide evidence) that there are charge *carriers:*

> **Charge carriers:** Particles that have the property of charge as an inherent and unalterable trait. As the carriers move, their charge goes with them.

➡**A note on language:** Although *electric current* has sometimes been called *electricity,* the word *electricity* in common usage has a variety of related meanings, such as the study of all electrical phenomena. To avoid confusion, we introduce a less ambiguous term.

➡**A note on language:** We use the term *electrostatic forces* even when charge carriers are in motion, because unless their speeds exceed 0.1*c,* the electrical forces on them are essentially the same as when they are stationary or static. Such high speeds are not a concern in electric circuits.

In general, we will refer to the movement or flow of charge from one place to another as an **electric current,** or simply a **current.** This definition is preliminary. Later we will define *current* quantitatively, but we will retain the notion that there is a flow of charge if and only if there is a flow of *charge carriers.*

Strictly speaking, a flow of charge doesn't require us to assume the existence of charge-carrying particles. The 1771 *Encyclopaedia Britannica,* reflecting the prevalent views of the time, described electricity as a "very subtle fluid . . . capable of uniting with almost every body." Whether it be fluid or particles, we assume it gets set in motion because an electrostatic force is exerted on it. If we also assume that the relationship between forces and motion is always governed by Newton's second law of motion, we must attribute inertial mass as well as electric charge to whatever is moving.

Subsequent experimentation validated viewing electrical phenomena in terms of forces and motion. Eighteenth-century investigators found that by rapidly rotating a large glass globe against a suitable rubbing surface using a wheel and belt mechanism (Figure 18-7*a*), the rubbing would result in a very substantial charge on the globe. A modern device that makes more sophisticated use of a continuous belt conveyor to develop charge on a metallic globe is the Van de Graaff generator (Figure 18-7*b*), invented by Robert J. Van de Graaff in 1929. A conductor brought in contact with the globe also becomes charged (Figure 18-8). Investigators using either device have found that when a large enough charge is developed on the globe and a conducting body is brought sufficiently close, a visible and audible spark jumps from the globe to the conducting body (Figure 18-7*b*), much as lightning jumps from clouds to Earth. The charges that remain after the spark are found to be reduced, indicating that charge has jumped the gap, and providing evidence of motion brought about by electrostatic forces.

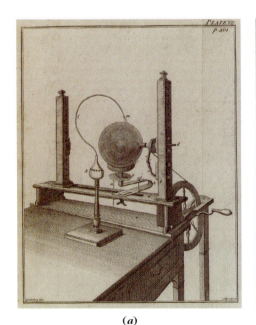

(*a*)

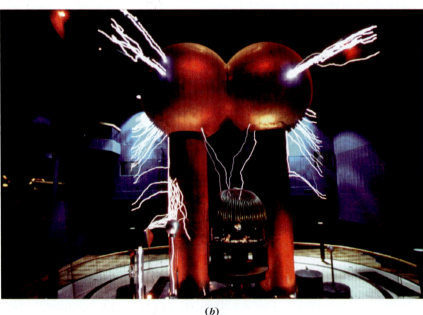

(*b*)

Figure 18-7 Devices for generating large build-ups of electric charge. (*a*) 18th century: Joseph Priestly's electrical machine (plate 73 of vol. 2 of 1771 *Encyclopaedia Brittanica*). (*b*) 20th century: The giant Van de Graaff generator at the Boston Museum of Science.

STOP&Think Why won't the spark jump to an insulator? Draw diagrams showing the configuration of charges on the conductor before and after it is brought near a positively charged sphere. Does your "after" configuration account for a force that could cause charge to jump the gap? Why can't you get the same "after" configuration for the insulator? ◆

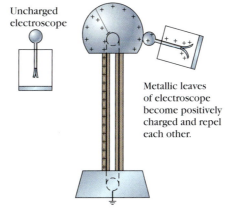

Uncharged electroscope

Metallic leaves of electroscope become positively charged and repel each other.

Figure 18-8 **Transfer of charge from Van de Graaff generator to conducting bodies.**

By the late nineteenth century, investigators were able to obtain not just brief sparks but sustained "rays" or beams between two conducting terminals on which opposite charges were continuously replenished. These beams were produced within sealed glass chambers and caused a glow where they struck the fluorescent material coating the glass surface at one end (Figure 18-9). Because the negatively charged terminal was called a *cathode,* the beams were called **cathode rays,** and the glass chambers **cathode ray tubes.** These were the primitive ancestors of most television, personal computer, and oscilloscope tubes (before widespread use of liquid crystal displays), in which, by rapidly varying the direction of the beam, a pattern of light is produced on the end of the tube that the viewer sees.

English physicist J. J. Thomson (1856–1940) observed that when oppositely charged plates were positioned above and below the beam, the beam deflected toward the positive plate, indicating that the beam itself was negatively charged. Moreover, for any given amount of charge on the plates, the beam deflected a fixed amount; there was no spreading out of the beam. Thomson reasoned as follows: (1) Imagine the beam to be a stream of separate tiny particles. Thomson called these particles "cathode corpuscles." (2) The electrostatic force will be the same on all particles having the same charge. (3) The force will produce an acceleration, and thus a deflection $y = \frac{1}{2}at^2$, perpendicular to the beam direction. (4) But by Newton's second law, the acceleration varies inversely as the mass. (5) Thus, if there were no relationship between charge and mass, that is, if particles having equal charge could have different masses, the particles would undergo different accelerations. They would therefore deflect by different amounts, and there would be a spread. (6) But *there is no spread.* So particles with the same charge must have the same mass (and vice versa), and so the ratio of charge to mass must have a fixed value.

Thomson's cathode corpuscles are today called **electrons,** and for establishing its charge-to-mass ratio (treated in Chapter 24), Thomson is generally credited with being the electron's discoverer. His reasoning established clearly that a given amount of charge moves with a fixed amount of mass. This supports our idea of charge carriers. Thomson further surmised that his particles were fundamental components of all matter, as we now know electrons to be.

But there is nothing in this reasoning that logically requires the carriers to be separate particles rather than a continuous flow of matter. (People knew there

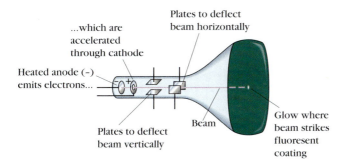

...which are accelerated through cathode

Heated anode (–) emits electrons...

Plates to deflect beam horizontally

Plates to deflect beam vertically

Beam

Glow where beam strikes fluoresent coating

Figure 18-9 **The cathode ray tube.**

was a fixed ratio of hydrogen to oxygen in water long before they knew that what we experience as a continuous fluid was made up of individual molecules, each containing two hydrogen atoms and one oxygen atom.) The particle nature was confirmed, however, by American physicist Robert A. Millikan (1865–1953).

In his famous **oil drop experiment** (1910–1913), Millikan put oil in an atomizer to obtain tiny individual droplets that could acquire electric charge from contact with ions in the air. By allowing the droplets to enter a region between oppositely charged plates, he could measure their terminal velocities. Knowing all the other forces on each droplet, he could determine the electrostatic force on it and from that its charge. He found that the charges on the droplets were always integer multiples of the same value, which he inferred must be the smallest possible "chunk" of charge. This he took to be the charge of the electron. Charge, then, was made up of pieces of fixed size, and if the ratio of mass to charge was constant, then the fixed amounts of charge must be carried by pieces of matter with fixed amounts of mass, in other words, particles.

Experiments such as these began to shape our present-day ideas about the structure of matter. Atoms contain positive nuclei surrounded by negative electrons. The nuclei are in turn made up of positive protons and neutral neutrons. The positive and negative particles exert electrostatic forces on one another. Under some circumstances, an outer electron of an atom may be more attracted to the positive nucleus of a neighbor atom than to its own, and will be pulled to the other nucleus, giving its new home atom a net negative charge and leaving its previous home atom with a net positive charge. Atoms charged in this way are called positive and negative **ions.**

Moving ions are also charge carriers. Electric currents in liquids generally involve the movement of ions. This is the case in many processes that you are likely to encounter in more detail in your chemistry course, such as electrolysis, electroplating, and the operation of certain types of batteries. In your own body and in all living things, which are mostly water, some kinds of ions can pass through particular cell membranes more readily than others (the membrane is said to be *permeable* to these ions). This can result in net positive and negative charges on opposite sides of a membrane. Such charge distributions have great importance for biological systems. In neurons (nerve cells), for example, the properties of a membrane may change in response to a stimulus such as heat or light or a signal from a neighboring neuron. This "information" to the cell membrane may itself take the form of the arrival of charge carriers to the membrane, which in turn can react chemically with components of the membrane. When the membrane changes properties, its permeability for different ions changes, and there is an ion flow across the membrane. This flow, or current, passed on from neuron to neuron, is the signal that carries information in the nervous system.

Chemical reactions are likewise the result of electrostatic forces. In ionic bonding, ions of opposite charge attract each other electrostatically; in covalent bonding, electrostatic forces are exerted on shared electrons by more than one nucleus. Roughly speaking, chemical reactions—changes in the bonding situation—are a change in the configuration of charge carriers (positive nuclei and negative electrons) in response to electrostatic forces. (We will need to refine this description in Chapter 27.) This is true both of the simplest reaction between two atoms and all the multistep reactions among complex molecules that make up the processes of life. (The totality of all these reactions in a living thing is called its *metabolism*.)

In certain very large molecules, shared electrons can move among many atoms that make up the molecule. This is true, for example, in pigment molecules, such as rhodopsin in the rods and cones of the human eye and chlorophyll in green plants. In a sense, over distances smaller than their total size, these molecules are conductors, and this property is critical to their role in vision or in photosynthesis.

Metals carry this property to an extreme. The outermost electrons are shared by *all* the atoms making up a piece of metal and so can travel throughout the

Nerve cells or neurons. An electric current consisting of a flow of ions from neuron to neuron is the signal that carries information in the nervous system. The neurons here are magnified 2500 times by a scanning electron microscope.

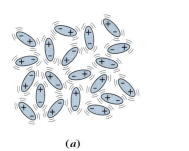
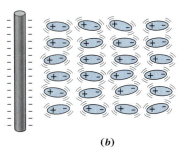

(a)   *(b)*

Figure 18-10 **Polarization: a model to explain the attraction of non-conducting materials to charged bodies.** (Adapted from A. Arons, *Development of Concepts in Physics*, Addison-Wesley, 1965.) (*a*) Thermal motion ordinarily keeps dipole molecules randomly oriented. (*b*) But they partially align in the presence of a charged object. (Thermal agitation prevents complete alignment.)

metal. This is why metals are conductors. The rest of each atom—a positive ion—remains stationary. To fit with this picture, we must refine the model we have developed so far. The charge carriers in metals, and therefore in the wires of ordinary electrical devices, are *electrons*. When electrons move from a region, they leave positive ions behind. Thus, if a negatively charged object is brought near a piece of metal, the nearest part of the metal's surface becomes positive not because positive charge carriers move onto that part but because electrons leave it. For a step-by-step graphic presentation of this reasoning, work through WebLink 18-2.

Because most electrical devices use metal wires in which the carriers are electrons, the currents we deal with most often turn out to be flows of negative charge. But as Figure 18-4 showed, *a flow of negative charge in one direction is equivalent in its effect to a flow of positive charge in the other*. The flow of positive charge is called a **conventional current** (*conventional* means it is something that we agree on) and is usually what is meant by current in quantitative treatments.

For **WebLink 18-2: Movement of Charge in Metals**, go to
www.wiley.com/college/touger

◆**POLARIZATION** We mentioned earlier that after you run a comb through your hair, it attracts small bits of paper. But paper is not a conductor. Placing paper between copper wires prevents current from passing from one to the other. Cardboard cylinders serve as insulators in the bulb sockets of older light fixtures. In explaining why aluminum foil is attracted to the comb, we assumed that a separation of charge could occur in the aluminum (Figure 18-2*b*) because charge can move through a conductor. Then what happens in the paper?

Understanding that molecules have both positive and negative parts enables us to propose a satisfactory model of what goes on in this case. We suppose that although the molecule is neutral overall, it may have more of its positive charge at one end and more of its negative charge at the other. Because it has two oppositely charged "poles," it is called an **electric dipole.**

Like other atoms and molecules, these dipoles will jiggle about in thermal motion, and will ordinarily be randomly oriented (Figure 18-10*a*). But a charged rod brought near them causes them to pivot into alignment with the oppositely charged poles drawn towards the rod and the like charged poles repelled (Figure 18-10*b*). The resulting separation between the average positions of the positive and negative constituents is called **polarization.** Because the positive charges on average are slightly closer to the rod, there is a small net attraction.

➡ Some molecules (called polar molecules) are dipolar even in electrically neutral surroundings; others become polarized only if another charged body exerts opposite electrical forces on their positive and negative components.

18-3 The Electron Gas and the Effect of Uneven Charge Distributions

An image that physicists sometimes find useful in describing metals is that of a negative *electron gas* pervading a connected lattice of positive ions (Figure 18-11). The electrons are pictured as mobile in somewhat the same way that molecules in an ordinary gas are mobile, and, like the molecules in an ordinary gas, the electrons move around randomly when there is no net flow overall.

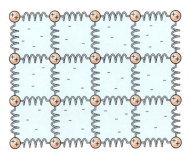

Figure 18-11 **Simple model of a metal: lattice of positive ions + electron gas.**

Like an ordinary gas, an electron gas can also be made to flow in some direction, but the causes of the flow are different. In an ordinary gas, molecules flow from regions where there is a greater density of molecules to regions of lesser density. The gas thus flows from higher to lower pressure (the pressure is a collective effect of molecules colliding with one another and with their surroundings). In an electron gas, electrons flow from regions where there is a greater density of electrons to regions of lesser density. What drives the motion is not a pressure difference but electrostatic forces. Electrons will be repelled by regions where there is a surplus of electrons and attracted toward regions where a deficit of electrons leaves unbalanced positive charges.

Net flow in ordinary gases is affected by differences in density; the excess molecules in regions of higher pressure are distributed throughout the region's volume. Net flow in an electron gas is affected by differences in *charge* density. But excess charges on a conductor are always located on its surface—never in the interior—because mutual repulsion keeps them as far apart as possible (recall Examples 18-2 and 18-3). So the charge density that affects flow is a **surface charge density.** These surface charges exert forces on the electrons in the interior of the conductor and may cause a flow of the interior electrons, but the flow occurs in such a way that the charge density remains zero throughout the interior. The next chapter details how this happens.

◆**PUMPS AND BATTERIES** If two gas-filled chambers are connected (Figure 18-12*a*), the pressure will be uniform throughout. It is possible to increase the pressure in one chamber and decrease the pressure in the other by connecting them through an air pump (Figure 18-12*b*). If we then open a valve connecting the two chambers, there is a rush of gas from the higher-pressure chamber to the lower (Figure 18-12*c*).

If we similarly connect two identical conducting wires (Figure 18-12*d*), charge will distribute uniformly over both. If we then connect a *battery* between the two wires (Figure 18-12*e*) and bring the free ends of the wires together (Figure 18-12*f*), there will be a flow of electrons. If the contact is broken, we may observe a spark across the gap. We have already noted that a spark is evidence of a flow of charge. The two systems in Figure 18-12 are analogous. To explore the analogy more fully, work through WebLink 18-3.

In Figure 18-12*e*, the two wires provide very little surface on which to place excess charge, so the charge density becomes sufficient to oppose the pumping

For **WebLink 18-3:**
Pumps and Batteries,
go to
www.wiley.com/college/touger

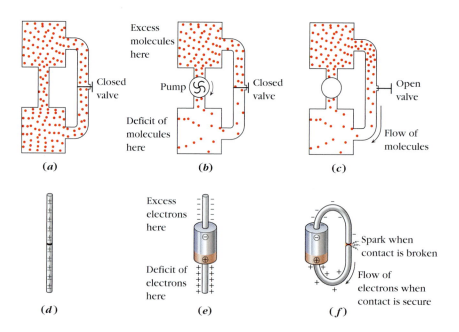

Figure 18-12 Comparing the effect of a pump and a battery.

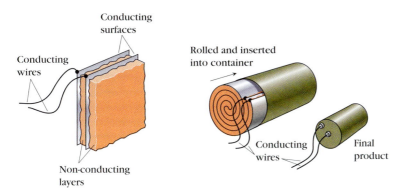

Figure 18-13 **What goes into a typical commercial capacitor.**

of further charge when only a tiny amount of charge has been pumped from one wire to the other. At this stage, the charges already on either wire exert a total electrostatic force on any further charge that is equal and opposite to the force exerted by the battery. Because the latter force results from the battery's chemistry, we call it a **non-electrostatic** force.

Suppose we now take the free ends of the two wires in Figure 18-12*a* and connect them to the plates in Example 18-3. We now have a mechanism, the battery, that can transfer charge from one plate to another to produce the charge configuration in Figure 18-5*d*. The plates, because of their extended surface areas, can take on much more excess charge before the charge density builds up sufficiently to oppose the pumping effect of the battery. Because of this greater capacity to hold charge, any device consisting of two conducting surfaces separated by a small nonconducting gap is called a **capacitor.**

The capacitor in Example 18-3 is called a parallel plate capacitor. In many commercial capacitors (Figure 18-13), the conducting layers or plates and the intervening nonconducting layers are rolled together and put into a cylindrical container to provide large surface areas within a small space. The extent to which a capacitor can store charge is called its **capacitance** (we will provide a quantitative definition later on). Large surface areas are clearly an important factor contributing to large capacitance.

Now let's see what happens when we string together some of the objects we have discussed. An unbroken loop of these objects is called an **electric circuit,** because it is a path around which charge carriers can *circu*late or move in *circ*les. If the loop includes a capacitor, it is not in the strictest sense a circuit because charge cannot pass across the nonconducting layer or gap, but in ordinary usage we call it a circuit anyway for lack of a better term. The first circuit we look at will consist of a battery, a capacitor with large capacitance, connecting wires, and in addition, a small bulb and socket. To understand how charge can move through the bulb, it is important for you to understand the connections that make up the conducting path (Figure 18-14).

Using a defibrillator. A defibrillator is a medical device that applies an electric shock to restore the rhythm of a heart that is not beating properly. The shock is provided by the discharge of a large-capacitance capacitor in the defibrillator.

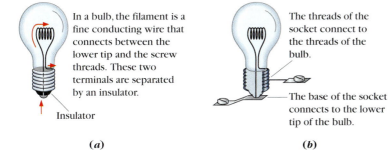
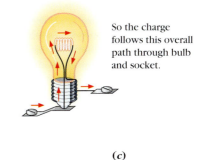

(*a*) (*b*) (*c*)

Figure 18-14 **Conducting path through a light bulb.**

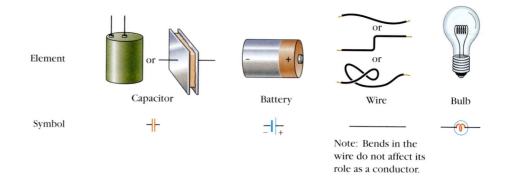

Element	Capacitor	Battery	Wire	Bulb

Note: Bends in the wire do not affect its role as a conductor.

Figure 18-15 Standard symbols for circuit elements.

Physicists usually draw electrical circuits using stylized symbols. Figure 18-15 shows the circuit elements (the building blocks of the circuit) we have discussed so far and their usual symbols. Figure 18-16a shows these elements connected to form a circuit. The accompanying symbolic drawing of the circuit is called a **schematic diagram,** or simply a **schematic.**

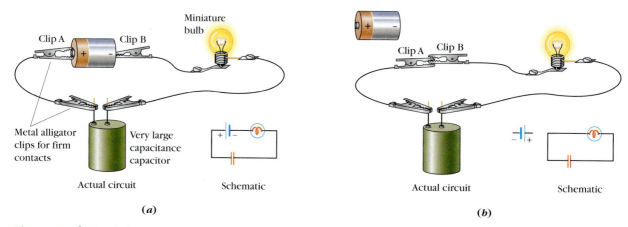

(a)   (b)

Figure 18-16 Circuit for Case 18-2.

Case 18-2 ◆ *Using Batteries to Charge and Discharge Large Capacitance Capacitors*

If you have the materials shown in Figure 18-16a available to you, you should do the following steps for yourself as an On-the-Spot Activity. Then, keeping in mind the ability of the two plates of a capacitor to store opposite charges, you should try to figure out how our underlying model can account for what you observe. You will be more likely to "own" the reasoning if you can work it out for yourself.

Step 1: Assemble the circuit in Figure 18-16a, but leave clip B unconnected. **STOP&Think** What do you expect to happen when you connect clip B to the battery? Will the bulb light up? Will it remain lit? ◆ Now connect clip B to see what happens.

Step 2: Suppose that after all contacts have been in place for a couple of minutes, you disconnect clips A and B from the battery and bring them together as in Figure 18-16b (eliminating the battery from the circuit). **STOP&Think** What do you expect

to happen? Will the bulb light up? Will it remain lit? ◆ Now do it to see what happens.

Step 3: After all contacts have again been in place for a couple of minutes, replace the bulb (call it bulb A) with a different bulb B (e.g., one having a different wattage) and repeat steps 1 and 2.

The Observations (see if these agree with yours)

For Step 1: When clip B is connected, the bulb glows briefly, gradually dying out. The glow may last anywhere from less than a second to over half a minute, depending on the specific capacitor, bulb, and battery used.

For Step 2: Here, too, we observe that the bulb glows briefly and then dies out!

For Step 3: In each of steps 1 and 2, bulb B glows more brightly but dims more quickly. (Or, if we chose a bulb that was initially *less* bright, it would dim more *slowly.*)

How can our model account for the observations? Check your own reasoning against the following.

In step 1: When the battery is connected, it pumps electrons from the positive to the negative capacitor plate. When enough electrons are concentrated on the negative plate to repel the arrival of further charge, and enough of an electron deficit is created on the positive plate so that a net positive charge holds back the remaining electrons there, the flow ceases. The bulb, it appears, glows when there is charge flowing through it.

In step 2: If the glow indicates a flow of charge—a current—how can the bulb glow again after the battery is removed? Doesn't the battery supply the current? The charge imbalance between the two plates can be maintained only as long as the battery is pumping to maintain it. When the pump is removed, the mutual repulsion of electrons on one plate will cause electrons to flow back to the plate where there is an electron deficit. That electrostatic repulsion was also there when the battery was there, but now it is no longer opposed by the non-electrostatic force that the battery exerted.

> A battery "pumps" charge along; it does not supply the charge carriers that make up the current. Current will flow whenever an uneven distribution of charge exerts a net force on available charge carriers, whether caused by a battery or not.

In step 3: We see that when the light is dimmer—less is given off each second—it is emitted for more seconds. In developing our model, we may surmise that the battery "pump" is building up the same total amount of charge on the capacitor plate, but is doing it more slowly through bulb B. Bulb B seems to be more resistant than A to charge passing through it, so the charge passes through it more gradually. This is similar to what happens when you allow water to pass through two paper coffee filters rather than one: The water seeps through more slowly, but ultimately you end up with the same amount of water collected in your pot. In this case, the same amount of charge ultimately passes through the bulb to collect on the capacitor plate.

Pattern of charge behavior

	Charge flowing through bulb each second (current)		Duration of flow		Total accumulation of charge on plates
Bulb A	more	×	less	=	same
Bulb B	less		more		same

A 100 W bulb is brighter than a 15 W bulb. Watts are a unit of power, the *rate* of energy use. The brightness of a bulb is associated with its power rating, the amount of energy it draws and emits *each second*. We begin to see that the pattern of energy emission reflects the pattern of charge behavior:

Corresponding pattern of energy emission

	Bulb brightness (power, or energy emitted each second)		Duration of glow		Total energy use
Bulb A	more	×	less	=	same
Bulb B	less		more		same

Comparing the above patterns, we infer that a bulb's brightness indicates the rate at which charge flows through it, not how much total charge passes through. The rate of charge flow is what we call *current*. A bulb that is more resistant to the passage of charge carriers glows less brightly (we refer to this property as **resistance**). Thus, if bulbs are placed one at a time in the circuit in Figure 18-16, the bulb with higher resistance will permit a smaller current to pass through. (Much of Case 18-2 is adapted from the work of Melvin Steinberg.)

We saw that there is less current through a higher-resistance bulb when each is connected individually to the battery, so that the same difference in surface charge density is maintained between the ends of each bulb filament. But it is *not* true when the two bulbs are placed in the same circuit one after the other, as in Figure 18-17. Now the two bulbs are along the same conducting path. What passes through one must also pass through the other, so the current through both must be the same. But the total change in surface charge density from the beginning of one filament to the end of the other is not distributed equally between the two filaments. The higher-resistance bulb can sustain a greater charge density difference between its two ends, so most of the change is across this bulb. In fact,

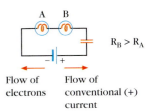

$R_B > R_A$

Flow of electrons   Flow of conventional (+) current

Figure 18-17 Bulbs along same conducting path ("in series").

it takes a greater electrostatic force to move charge carriers through the higher-resistance bulb. The greater charge density difference is necessary to exert this greater electrostatic force. Because charge carriers must be pushed harder through this bulb, the rate of energy expenditure (power) is greater. Thus, *the higher resistance bulb glows more brightly when the same current passes through both*.

In a circuit with no capacitor, and therefore no nonconducting gap, the battery can pump charge continuously around the resulting loop. The simplest such circuit consists of a battery with a wire connected between its terminals (Figure 18-12*f*). (The wire should be made of a fairly high-resistance alloy, such as nichrome, to avoid rapid depletion of the battery.) When the circuit is closed, the surface charge density along the wire, instead of occurring abruptly as in Figure 18-12*e*, tends to be very gradual (Figure 18-12*f*).

In the *steady state,* when the current is not changing, the charge density changes uniformly along the length of the wire. We call this a constant charge density **gradient.** The direction of decrease is the direction of the *negative gradient.* A gas will flow in the direction of a negative pressure or density gradient. Positive charge or conventional current will flow in the direction of a negative surface charge density gradient. Electrons, in contrast, flow in the direction of the positive gradient.

With no resistance in the wire, the configuration of surface charge in Figure 18-18 would exert a nonzero force on electrons in the wire's interior, and they would accelerate in the direction of increasingly positive surface charge. But because there is resistance, they do not accelerate unhindered. Obstacles or conditions that they encounter in their path recurrently slow them down, so they level off to a certain average velocity, called their **drift velocity. STOP&Think** If electrons are drifting through the interior of a current carrying wire, does that mean the wire's interior has a net charge? ◆

Along a closed conducting path, electrons leaving any region are replaced by electrons entering it, so the net charge in the interior of a conductor remains zero, even when there is a current. The predominant movement of charge is through the neutral interior of the circuit, with each tiny region contributing the electrons that drift into the next. The drifting electrons that constitute the current come from sites all along the conducting path, from the bulb filament and the connecting wires as well as the interior of the battery.

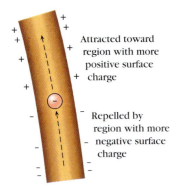

Figure 18-18 The changing charge distribution (charge density gradient) along a wire's surface results in a net electrostatic force on electrons in the wire's interior.

Attracted toward region with more positive surface charge

Repelled by region with more negative surface charge

◆ SUMMARY ◆

In this chapter, we developed an underlying model to explain a range of observed behaviors. The model involves a number of key ideas:

Electrostatic forces are a class of action-at-a-distance forces, both attractive and repulsive, that bodies exert on one another, which are demonstrably not gravitational or magnetic. We call these forces electro*static* because they occur even when there is no motion of the **charge carriers** responsible for the forces. The forces are weaker when the bodies are further apart.

Electric charge is a property we attribute to matter that exerts electrostatic forces (just as *gravitational mass* is responsible for gravitational forces). We postulate that it comes in two varieties, **positive** and **negative.** Oppositely charged bodies attract each other, and bodies of like charge repel each other. **Neutral** bodies contain equal amounts of positive and negative charge.

We can explain certain observations by supposing that charge can move through some matter in response to electrostatic forces. We developed specific vocabulary for describing this motion.

Electric current, or simply **current,** is the rate of movement or flow of charge from one place to another (later

we will make this quantitative). **Electrostatic equilibrium** is the state or condition of a body or system in which there is no net movement of charge (and consequently no net current).

Conductors are materials that permit a current to pass through them. **Insulators** are materials that in contrast to conductors do not permit the passage of a current. *Conductor* and *insulator* are not absolute terms. Materials may be conducting or insulating to various degrees. **Ground** refers to a conducting body (such as the Earth) that is large in comparison to the other bodies in a situation, so that charge may flow onto it or from it without apparent limit.

To put these concepts together in a coherent, connected picture, it is useful to try to explain behaviors by means of **charge configuration diagrams** (see Procedure 18-1).

Charge carriers are particles that have the property of charge as an inherent and unalterable trait. In a current, as the carriers move, their charge goes with them.

Experiments such as **Thomson's cathode ray experiment** and **Millikan's oil drop experiment** led to a model of matter in which atoms are understood to be made up of positive nuclei and negative electrons. In metals, which are conductors, the mobile charge carriers are electrons.

Conventional current is a flow of positive charge equal and opposite to the flow of charge carried by electrons.

A useful model of a metal is one in which a negative **electron gas** is free to move about a lattice of positive nuclei. Excess charge always ends up on the *surfaces* of conductors.

Conventional current will flow from higher to lower **surface charge density.** When positive charges concentrate in one region, their mutual repulsion will produce a net movement away from this region—a flow of charge or current—if a conducting path is available.

In explaining the behavior of circuits, we looked at several common circuit elements:

Batteries set up a surface charge density difference by exerting *non-electrostatic* forces to "pump" charge from one terminal to the other. The resulting charge distributions in turn exert electrostatic forces on carriers in regions where there is a charge density **gradient,** causing them to reach a certain average **drift velocity.**

Current will flow whenever there is a net force on available charge carriers due to an uneven distribution of charge, whether caused by a battery or not.

Capacitors are charge storage devices consisting of two conducting surfaces separated by a nonconducting gap. **Capacitance** is a measure of the extent to which they can hold charge.

Bulbs contain filaments that glow visibly when sufficient current passes through them. The filaments are characterized by their **resistance** to the passage of current. If two bulbs are connected one at a time between the same charge density difference, the higher-resistance bulb will allow charge to pass through at a lesser rate, and will glow less brightly (see Case 18-2). But if the two bulbs are connected so that the same current passes through them sequentially, most of the charge density difference will be across the higher-resistance bulb, so more work will be done and energy will be dissipated at a greater rate in that bulb, and it will glow more brightly.

◆ QUALITATIVE AND QUANTITATIVE PROBLEMS ◆
Hands-On Activities and Discussion Questions

The questions and activities in this group are particularly suitable for in-class use.

In doing the activities in Problems 18-1 to 18-3, try to forget anything you know ahead of time about how like or opposite charges affect each other. The goal of these activities is to deduce what they do from what you observe. To do these activities you will need several strips of ordinary transparent sticky tape (Magic Scotch tape works best) about 8 cm (3 in) in length, with one end of each doubled over (Figure 18-19a) to provide a nonstick handle. These should be used just after being torn off the roll. Use fresh strips for each question.

(These activities are adapted from A. Arons, *A Guide to Introductory Physics Teaching* [Wiley, 1990] and *The Electrostatics Workshop* [AAPT, 1991], with additional helpful suggestions by Robert Morse.)

18-1. *Hands-On Activity.* Attach two strips of tape (described above) in different locations on a smooth, flat, clean, dry surface (Formica and plastic surfaces work especially well) so that the entire length of each strip except for the "handle" is stuck flat to the surface. Run your thumbnail over the strips to ensure good uniform contact.

a. Suppose you were to take both strips by their handles and pull them quickly off the surface in the same way (*don't do it yet*). Should the two strips have like or opposite charges? Explain.

b. Now pull the strips off by their handles and bring the free ends very close (less than a cm) to each other. Do they

attract or repel each other? What does this tell you about the way like charges respond to each other?

18-2. *Hands-On Activity.* See instructions before Problem 18-1. Attach the two strips of tape to each other front to back (sticky side to nonsticky side) for their entire lengths, handles excepted. Run your fingers along them to remove any surplus charge, so that you can start out knowing the pair of strips is neutral.

a. If you were now to pull the two strips apart (*don't do it yet*), could they both end up with the same charge? Could they end up with opposite charges? What does "neutral" tell you about the charges that are initially present?

b. Now quickly separate the two strips of tape, and then bring them close together. Do they attract or repel each other? Does this demonstrate anything about how opposite charges respond to each other? If so, what?

18-3. *Hands-On Activity.* See instructions before Problem 18-1. In this activity you will rub two suitable objects together to charge them. A comb and your hair, a Styrofoam cup and your hair, a plastic (polyethylene) bag and a garment made of synthetic fabric are good possibilities. But first, produce two oppositely charged strips of tape as in Problem 18-2, and suspend both from a table edge about a foot apart.

a. Suppose you brought a charged object near each of the two strips in turn without touching and both responded in the same way (attraction or repulsion). Could the object have the same charge as either strip of tape? What would you have to conclude about whether there are just two kinds of charge?

b. Now let's see if the behavior we hypothesized in *a* actually occurs. Rub various objects to charge them and bring each object near (but never touching) each strip in turn. (If you spend much time on this, you may have to replace the tape strips with a fresh pair produced in the same way.) How do the two strips respond to each object? Do they ever respond in the same way to the same object?

c. Is there any evidence in your observations for a third kind of charge?

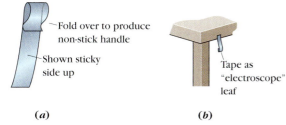

(a)   *(b)*

Figure 18-19 Problems 18-1, 18-2, and 18-3

18-4. *Discussion Question.*
Figure 18-20 shows a loop of conducting wire. The wire is neutral, but nine typical conduction electrons, equally spaced, have been singled out for your attention.

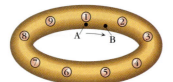

Figure 18-20 Problem 18-4

a. Suppose something (such as a battery) exerts a non-electrostatic force on electron 1 and moves it from point A to point B. How does moving electron 1 affect the total force on electron 2? How does electron 2 move in response?

b. How does moving electron 1 affect the total force on electron 9? How does electron 9 move in response?

c. Are electrons 2 and 9 slow to respond to the motion of electron 1, or is the response more nearly instantaneous? Explain.

d. When electrons 2 and 9 move, how does that in turn affect electrons 3 and 8?

e. Are electrons 4 and 7 ultimately affected? Are 5 and 6? Sketch the resulting arrangement of all nine electrons on the loop.

f. Suppose a battery between A and B continues to pump electrons from A to B. How do the electrons elsewhere in the loop respond?

g. Suppose now that all the electrons are drifting slowly in the direction shown by the arrow. Suppose the resistance between A and B is increased. How would that affect the motion of electron 1?

b. Because electron 1 contributes to the total electrostatic force on electrons 2 and 9, how is the motion of *those* electrons affected? How, in turn, is the motion of the remaining electrons affected?

i. What effect does increasing the resistance between A and B have on the current in the loop? Is the current more affected in one part of the loop than in another? Explain.

j. Would any of your reasoning be changed if instead of a flow of electrons, you considered a conventional current of positive charges moving the other way? Briefly explain.

18-5. *Discussion Question.* Popular descriptions of science are not always accurate and may generate confusion when you wish to use concepts carefully for orderly reasoning. A staff member of a major science museum in a large U.S. city made the following statement when introducing their electricity show to a large audience: "When positive and negative attract, the force generated in that process is electricity."

a. What inaccuracies can you detect in this statement? First try to answer this without going on to the specific hints in part *b.*

b. Is a force something different than an attraction or a repulsion? Does it mean anything to say that an attraction "generates" a force? Is electricity a force, or charge, or a current, or what? Which of these terms have we defined carefully? Did we define *electricity*?

c. How many different meanings can you find for the word *electricity* in ordinary (non-physics) usage?

d. Can you suggest any incorrect conclusions to which the staff member's statement might lead a listener?

Review and Practice

Section 18-1 Developing an Underlying Model to Account for the Observations

18-6. A positively charged rod is brought toward two small conducting spheres suspended by insulating cords. Sphere A is attracted to the rod; sphere B is repelled by it. Indicate whether each of the following statements about the total charges on the spheres is definitely true, possibly true, or false:

a. A is positively charged.   *d.* A is neutral.

b. B is positively charged.   *e.* B is neutral.

c. A is negatively charged.

18-7. SSM Suppose two objects A and B start out neutral, but when they are rubbed together, object A becomes negatively charged.

a. If charge cannot be created or destroyed, where does A's charge come from?

b. What is the effect of this on object B?

c. Does a neutral object contain any positive charge? Does it contain any negative charge? Briefly explain.

18-8. In Figure 18-2c, what happens to the arrangement of charge on the piece of foil if the foil is carried far from the comb without being brought in contact with a conductor?

18-9. In Step 2 of Case 18-1, the comb is replaced by a bar magnet with its north pole forward. How will the piece of aluminum foil be affected when the magnet is brought near it?

18-10. After you run a comb through your hair, the comb will attract tiny bits of both newspaper and aluminum foil. If the bits are of equal mass, which would you expect to be attracted more strongly to the comb? Briefly explain.

18-11. Suppose you have tiny bits of aluminum foil and scraps of newspaper all having the same mass. You charge a plastic comb by running it through your hair. What could you do experimentally to determine which is more strongly attracted to the comb? Be very specific about your procedure and how it permits you to make distinguishable observations for the two situations.

18-12.

a. Describe a situation that will cause electrons to flow onto a neutral object when it is grounded. What will the resulting charge on the object be?

b. Describe a situation that will cause electrons to flow away from a neutral object when it is grounded. What will the resulting charge on the object be?

18-13. SSM A negatively charged rod is brought near a conducting sphere suspended by a copper wire from a metal ceiling. The sphere is attracted to the rod.

a. Was the sphere initially *positively charged*, *negatively charged*, or *neutral*? Briefly explain.

b. The piece of aluminum foil in Case 18-1 was subsequently repelled by the rod. Will the same thing happen to the conducting sphere in this situation? Briefly explain.

18-14. A conducting metal rod is suspended by an insulating nylon wire (Figure 18-21). A small, lightweight metal ball suspended by a nylon thread is very close to one end of the rod but not touching it. If a positively charged rod is briefly touched to the other end of the metal rod, what will happen to the ball if it is initially **a.** has a small positive charge (compared to the charge on the rod)? **b.** has a small negative charge? **c.** is neutral?

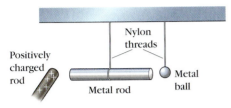

Nylon threads

Positively charged rod

Metal rod

Metal ball

Figure 18-21 Problem 18-14

18-15. Sketch charge configuration diagrams to explain what is happening in each part of Problem 8-14.

18-16. A negatively charged rod is held a small distance to the left of a positively charged solid conducting sphere. Because the sphere is conducting, the excess charge on the sphere will be mostly ____ (*within the left half of the sphere; on the surface of the left half of the sphere; within the right half of the sphere; on the surface of the right half of the sphere*).

18-17. SSM Two flat metal plates are initially very far apart. Plate A is positively charged; plate B is neutral. Plate B is then moved until it is parallel to A and very close to it. Sketch the initial and final charge configurations on both plates.

18-18. Use our underlying model for electrostatic forces to explain in detail why the hair of the person in Figure 18-8*b* appears as it does.

18-19.
a. Why are you at greater risk of being struck by lightning if you are standing in the middle of an open field?
b. If you cannot quickly get to cover, what could you do to reduce the risk?

Section 18-2 Charge Carriers

18-20. When radioactivity was first discovered, three different kinds of radioactive "rays" were identified; they were called *alpha* (α), *beta* (β), and *gamma* (γ) rays. When a beam of each kind of ray was passed between oppositely charged plates, the alpha ray beam bent very slightly toward the negative plate, the beta ray beam bent quite substantially toward the positive plate, and the gamma ray beam passed through without bending at all.
a. Which of these rays might possibly be made up of electrons? Briefly explain.
b. If the alpha and beta rays both turned out to be made up of particles, which particles would you expect to have a greater charge-to-mass ratio? Briefly explain.

18-21. Are the charge carriers always electrons when there is an electric current? Briefly explain.

18-22. In your own words, tell what is meant by a conventional current.

18-23. If a conventional current flows from left to right along a copper wire, which way are the electrons flowing?

18-24. Explain in your own words why a flow of electrons with a certain total charge in one direction along a wire is equivalent to the flow of an equal amount of positive charge in the opposite direction.

18-25.
a. Compare the movement of charge carriers in large molecules such as pigments and in metals.
b. Comment on the validity of the idea that "what goes on in a metal is like one giant covalent bond among all of its atoms."

18-26. Positive and negative electrodes are connected to the two terminals of a battery and dipped into a tank containing a solution of sodium chloride (table salt) in water. This means there are equal numbers of positive sodium ions and negative chloride ions in the water. When the electrodes are dipped into opposite ends of the tank, a current flows between them. Suppose we could chemically remove all the chloride ions, leaving the number of sodium ions the same as before. When the electrodes are dipped into the tank as before, it would produce ____ (*a current greater than the current that occurred with both kinds of ions present; a current equal to that current; a current less than that current but not zero; no current at all*). Briefly explain.

18-27. How does the permeability of a cell membrane to various chemical species affect the electrical current crossing the membrane?

Section 18-3 The Electron Gas and the Effect of Uneven Charge Distributions

18-28. A student argues as follows: "Batteries are the source of current in simple electric circuits. There is a certain amount of current that the battery can provide, and when the battery dies, or if I take the battery away, there is no more current." Comment on the correctness or incorrectness of this argument. In your comments, be sure to address the question of whether there can ever be a current when there is no battery in the circuit.

18-29. SSM WWW Suppose you start out with two identical capacitors and two bulbs, A and B. First you charge the capacitors, one at a time. Then you discharge one capacitor through each bulb. What can you conclude about either the capacitors or the bulbs if **a.** bulb A glows more brightly but for a shorter time interval than bulb B? Briefly explain. **b.** A glows more brightly and also for a longer time interval than B? Briefly explain.

18-30. When you connect the ends of a wire to a battery so that a current flows through the wire, does the total charge on the wire remain zero or is the wire electrically charged? Briefly explain.

18-31. It is common to speak of "charging a battery." Are you transferring any charge to the battery when you do this? Briefly explain.

18-32.
a. When you "charge a capacitor" by connecting it to a battery, where does the charge go, and where does it come from?
b. Does the battery act as a source or supplier of charge? Explain?

18-33. When a battery is connected through a small bulb to a capacitor, the fact that the bulb dims and goes out after a brief interval is evidence that the capacitor plates do not keep gaining charge indefinitely. Use what you know about electrostatic forces on electrons to explain why *a.* the negative plate of the capacitor doesn't keep getting more negative indefinitely. *b.* the positive plate of the capacitor doesn't keep getting more positive indefinitely.

18-34. A team of students finds that in Case 18-2, when they charge and discharge the capacitor through bulb A, the bulb glows very brightly. When they charge and discharge it through bulb B, this bulb glows less brightly. They hypothesize that because bulb A glows more brightly, it is permitting more total charge to pass through and be stored on the capacitor. To test this hypothesis, they decide to charge the capacitor through bulb B and discharge it through bulb A.

a. If their hypothesis is right, how would the brightness of bulb A during discharge compare with its brightness during discharge when the capacitor was also charged through bulb A? Why?

b. When they do the test, they observe that when they discharge the capacitor through bulb A, it glows just as brightly and for just as long as when the capacitor was charged through bulb A. What can you conclude from this observation? (Adapted from M. Steinberg *et al.*, *Electricity Visualized*, 1990.)

18-35. Suppose that in the circuit in Case 18-2, the capacitor is replaced by one with plates that are twice as big. If nothing else is changed, how would this affect *a.* the maximum brightness of the bulb? Use the underlying model to briefly explain your answer. *b.* the time it takes to dim to half its initial brightness? Use the underlying model to briefly explain your answer.

18-36. After the circuit in Figure 18-16*a* has been connected for a few minutes, does the battery produce a surface charge density difference? What effect, if any, does this have on current flowing in the depicted circuit? Justify your answers by sketching the location(s) of any excess charge.

18-37. When the last connection has just been made in the circuit in Figure 18-16*b* and the capacitor is just beginning to

discharge, tell whether each of the following is positively charged, negatively charged, or neutral:

a. the surface of the wire 1 cm from the positive plate of the capacitor

b. the interior of the wire 1 cm from the positive plate of the capacitor

c. the surface of the wire 1 cm from the negative plate of the capacitor

d. the interior of the wire 1 cm from the negative plate of the capacitor

e. the entire capacitor (consider the net overall charge)

f. the entire circuit (consider the net overall charge)

18-38. Figure 18-22 shows a simple closed circuit consisting of a bulb, a battery, and connecting wires. Three points in the circuit are labeled (point B is on the filament of the bulb). Rank the three labeled points according to each of the following, listing them in order from least to greatest and indicating any equalities: *a.* the density of surface charge at each point. *b.* the current at each point.

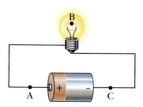

Figure 18-22
Problem 18-38

18-39. Figure 18-23 shows a simple closed circuit consisting of a two bulbs A and B, a battery (labeled C for cell), and connecting wires. Rank the three circuit elements in order, from least to greatest, according to the current that passes through each one. Indicate any equalities.

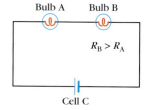

Figure 18-23 Problems 18-39 and 18-40

18-40.

a. In which of the two bulbs in Figure 18-23 is the rate of energy expenditure greater? Briefly explain.

b. Which of the two bulbs will glow more brightly?

Going Further

The questions and problems in this group are not organized by section heading, so you must determine for yourself which ideas apply. Some of them will be more challenging than the Review and Practice questions and problems (especially those marked with a • or ••).

18-41. Suppose there are equal amounts of positively and negatively charged matter in the universe.

a. How would this matter tend to rearrange itself over time if *like* charges attracted and *opposites* repelled? Does this agree with or contradict what we actually observe in the universe?

b. What if opposite charges didn't react to each other at all (as gravitational charge doesn't react to electric charge), but instead charges of one kind (say, positive) were mutually attractive and charges of the other kind were mutually repulsive? How would this matter tend to rearrange itself

over time? Does this agree with or contradict what we actually observe in the universe?

18-42. Two objects A and B are attracted to each other. When a negatively charged body C is brought near them, both are attracted to C. What can you conclude about the charges on A and B?

•**18-43.** A student holds an iron bar magnet in her hand. If a positively charged Styrofoam sphere is initially half a centimeter from the north pole and is free to move, it will be _____ (*attracted to the north pole; repelled by the north pole; attracted then repelled; unaffected by the magnet*).

18-44.

a. In Case 18-1, estimate the gravitational force that the comb and the piece of aluminum foil exert on each other at a separation of 0.01 m.

b. Estimate the time it takes for the piece of aluminum foil to jump the gap to the comb and use this to find an approximate value for the average acceleration of the foil during this interval.

c. Calculate the force necessary to produce this acceleration.

d. Compare the values you've obtained in **a** and **c**. Express the comparison as a ratio $\frac{F \text{ found in } a}{F \text{ found in } c}$. Briefly explain the significance of this comparison (or this ratio).

18-45. SSM WWW A common device for detecting electric charge is the *electroscope* (Figure 18-24a). The metal knob at the top is connected via the conducting metal post below it to two metal foil leaves suspended from the bottom of the post.

a. When a charge is transferred to the knob, what will happen to the foil leaves? Why? How does this compare to the effect shown in Figure 18-8b?

b. Suppose you temporarily ground the electroscope by touching the knob with a finger to neutralize it. You then remove your finger and bring a positively charged rod very near but not touching the knob of the electroscope. What happens to the leaves when the rod is close? Sketch a charge diagram to explain what happens. Is there a net charge on the electroscope? How is it distributed?

c. What happens to the leaves when the rod is moved a little further from the knob? Explain your answer in terms of the forces that charge carriers exert on one another.

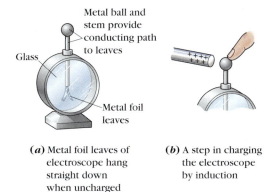

(a) Metal foil leaves of electroscope hang straight down when uncharged

(b) A step in charging the electroscope by induction

Figure 18-24 Problems 18-45, 18-46, and 18-67

18-46. *Charging by induction.* A positively charged rod is again brought close to the knob of a neutral electroscope without touching it, and the leaves respond as in Problem 18-45.

a. With the rod kept in this position, a student places her finger in contact with the knob, as in Figure 18-24b. What happens to the foil leaves? Why? Draw a charge configuration diagram to support your explanation. Is there a net charge on the electroscope? How is it distributed?

b. The student then removes her finger while the rod remains in the same position. Now what happens to the foil leaves? Why? Again, draw a charge configuration diagram to support your explanation. Is there a net charge on the electroscope? How is it distributed?

c. Finally the rod is removed. At this point there is a net charge on the electroscope. Is it positive or negative? How is this charge distributed? What would you see as evidence that the electroscope is charged?

d. This is called charging by **induction.** We say the final charge on the electroscope was **induced** by the charged rod because it was not transferred from the rod, but it was caused by the presence of the rod. Where did the final charge on the electroscope come from? Briefly explain.

18-47. Are fabrics that develop static cling conductors or insulators? Briefly explain how you can tell.

18-48. Golfers are frequent lightning victims. What are some of the factors that contribute to this? (Assume golfers have as much common sense as other people.)

18-49. A lightning rod always has a wire connected to its lower end. To what should the other end of the wire be connected? Why?

18-50. Jumper cables are used to recharge a car battery by connecting it to the battery of another car. It is possible (but not safe) to do this by connecting one cable between the positive terminals of the two batteries and the other between the negative terminals of the two batteries.

a. Why do the safety instructions tell you that instead, you should connect an end of one cable to your engine block or car frame?

b. What would happen if you ran each cable between the positive terminal of one battery and the negative terminal of the other?

18-51. SSM A positively charged rod is brought close to a grounded conducting sphere (Figure 18-25). Unlike the usual charge configuration diagram that shows only excess charge, the figure shows representative charges on the sphere in its initial (neutral) state.

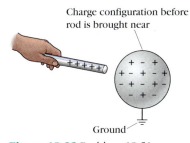

Figure 18-25 Problem 18-51

Show what happens to these charges by drawing the resulting charge configuration on the sphere

a. if only positive charges are free to move in conductors.

b. if only negative charges are free to move in conductors.

c. if both positive *and* negative charges are free to move in conductors.

d. Now draw a usual charge configuration diagram showing only the final distribution of *excess* charge for each of these cases. Does the distribution of excess charge depend on whether you think about a flow of electrons or about a conventional current of positive charge? Explain.

18-52. What advantage is gained by making the upper part of the Van de Graaff generator spherical?

18-53. If the bottom of a small bulb is placed against one terminal of a battery, and no other connections are made, **a.** will there be any flow of charge at any time? Briefly explain. **b.** will the bulb light up? Briefly explain.

18-54. If the circuit in Figure 18-16a were assembled using a capacitor with only a thousandth of the capacitance of the one shown, **a.** would there still be a transient current through the bulb? Briefly explain. **b.** would you still see the bulb light up? Briefly explain.

18-55. In the circuit in Figure 18-26, bulbs A and B are connected between the same points 1 and 2. Bulb B has greater resistance than bulb A.

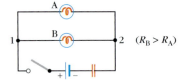

Figure 18-26 Problems 18-55, 18-56, and 18-57

a. Sketch a charge configuration diagram showing the distribution of excess surface charge along this circuit at an instant when the capacitor has just begun to charge (after the switch is closed).

b. Repeat **a** for an instant when the capacitor is fully charged.

18-56. In the circuit in Figure 18-26, bulbs A and B are connected between the same points 1 and 2. Bulb B has greater resistance than bulb A.

a. At an instant when the capacitor has just begun to charge, is the difference in surface charge density between the ends of bulb A less than, equal to, or greater than the difference between the ends of bulb B? Briefly explain your answer.

b. Through which bulb will more current flow? Briefly explain your answer.

c. Which bulb will glow more brightly? Briefly explain your answer.

d. Consider the extreme case in which the resistance of bulb B's filament is so great that the filament is in effect an insulator. In this case, through which bulb will more current flow?

e. In the case described in **d**, which bulb will glow more brightly?

f. Are your answers to **d** and **e** the same as your answers to **b** and **c**? Should they be? Briefly explain.

18-57. Compare and contrast the behavior of the circuits in Figures 18-26 and 18-17.

18-58. Batteries make use of chemical reactions in which one chemical species gives up electrons and another acquires them.

a. Why does a battery wear out when an external conducting path is established between its terminals?

b. Why is it especially unwise to keep a very low-resistance conductor like a short length of copper wire connected between the terminals of the battery?

18-59. **SSM** The circuit in Figure 18-27 is identical to the circuit in Figure 18-16a, except that a second capacitor identical to the first one has been connected between points 1 and 2.

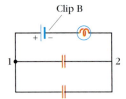

Figure 18-27 Problem 18-59

a. Compare the brightness of the bulbs in the two circuits just after clip B is connected to complete the circuit. Briefly explain.

b. Compare how long the bulbs in the two circuits glow once clip B is connected. Briefly explain.

18-60. The circuit in Figure 18-28 has the same elements as the circuit in Figure 18-16a, but they are arranged differently.

a. Compare the brightness of the bulbs in the two circuits just after clip B is connected to complete the circuit. Briefly explain.

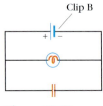

Figure 18-28 Problem 18-60

b. Compare how quickly the capacitors in the two circuits charge once clip B is connected. Briefly explain.

18-61. Draw the simplest schematic circuit diagram you can for the circuit depicted in Figure 18-29.

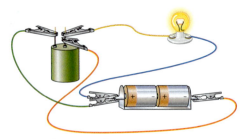

Figure 18-29 Problems 18-61

18-62. In a series of popular lectures in 1934, British Nobel laureate W. L. Bragg proposed the following analogy: "Now suppose I take the spare cash in my pocket and hand it over to you. Ought I to say that a current of riches has passed from me to you, or that a current of poverty has passed from you to me? Unfortunately . . . it was as if everyone agreed to say that the current of poverty passed from you to me, before it was realized that the actual cash . . . went from pocket to pocket in the opposite direction" (W. L. Bragg, *Electricity*, Macmillan, New York, 1936, p. 57). To what situation in physics was Bragg referring? What in the physics situation is like the cash in the situation Bragg described? What in the physics situation is like the "current of poverty" in this situation?

18-63. When a positively charged rod is brought near some tiny shreds of newspaper, the shreds are attracted to the rod. What would happen if the rod were negative instead? Why?

18-64. Suppose you have only the bulb, battery, and single piece of flexible conducting wire that are shown in Figure 18-30. Very carefully sketch a way of arranging these circuit elements that would make the bulb light. Be sure that your sketch shows clearly where the wire is making contact with other objects.

Figure 18-30 Problem 18-64

18-65. A simple closed circuit consists of a battery connected to a bulb by wires. Within the battery, does the conventional current flow from the negative terminal to the positive terminal or from the positive terminal to the negative terminal? Briefly explain.

18-66. In each of two set-ups, a capacitor is charged by connecting its plates to the two terminals of a battery. The set-ups are identical, except that the wires in set-up B have greater resistance than the wires in set-up A. Is the charge acquired by the capacitor in A less than, equal to, or greater than the charge acquired by the capacitor in B? Briefly explain.

18-67. The leaves of an electroscope (Figure 18-24a) spread out when a positively charged rod is brought close to but not touching the ball at the top. In this situation, the ball is _____ and the leaves _____. Briefly explain.

a. *positively charged . . . are both negatively charged*

b. *negatively charged . . . are both positively charged* *c. neutral . . . are both negatively charged*

d. neutral . . . are both positively charged

e. neutral . . . have equal but opposite charges

18-68. In Case 18-1, we assumed that the suspended piece of aluminum foil started out neutral before being attracted to the negatively charged comb. We can best show that this is actually the case by bringing (*a positively charged rod; other negatively charged objects; a second, identically treated piece of aluminum foil*) close to it and seeing what happens. Give a reason for your choice.

18-69. If you think of your body's circulatory system as being analogous to an electric circuit, *a.* what in your circulatory system is analogous to a battery in the electric circuit? *b.* what in the circuit is analogous to blood pressure in your circulatory system?

18-70. Example 18-2 (Figure 18-5) involved an effective transfer of positive charge. Suppose that the conducting sphere is metallic, so that the charge carriers that are free to move are actually negative electrons. How do the electrons move? Sketch diagrams showing the stages of the motion.

18-71. A positively charged rod is held a small distance to the left of a neutral solid conducting sphere. Because the sphere is conducting, there will be excess electrons _____ (*within the left half of the sphere; on the surface of the left half of the sphere; within the right half of the sphere; on the surface of the right half of the sphere*).

Problems on WebLinks

18-72. Suppose you wanted to show that the piece of aluminum foil involved in the observations in WebLink 18-1 started out neutral.

a. Could you accomplish this by putting it next to a charged object to see what happened? If so, tell whether the object should be positively or negatively charged. If not, briefly explain why not.

b. Could you accomplish this by putting it next to an identical piece of aluminum foil to see what happened? Briefly explain.

18-73. Which (one or more) of the following ideas is necessary to explain why the aluminum foil is attracted to the charged comb in WebLink 18-1?

a. Opposite charges attract.

b. Like charges repel.

c. The foil acquires an opposite charge from the comb.

d. Charge can travel in conductors.

e. Charges exert less force on each other when further apart.

18-74. Suppose the aluminum foil in WebLink 18-2 is replaced by a kind of metal foil that has only one conduction electron contributed by each atom. Just before the foil actually touches the comb, is the number of conduction electrons in the part of foil furthest from the comb less than, equal to, or greater than the number of positive ions in that part of the foil?

18-75. In WebLink 18-3, the conduction electrons on the lower wire surface don't all get "pumped" to the upper wire surface because as more electrons accumulate on the upper wire _____ (*the battery runs down; the electrostatic force they exert on other electrons increases; the pressure builds up; the battery runs out of electrons*).

CHAPTER NINETEEN

Electrical Field and Electrical Potential

In Chapter 18, we explained a range of observations by developing a model in which charged carriers exert electrostatic forces on one another. In this chapter, we will formulate the concepts of electrostatic force and electric charge quantitatively. As we did for gravitational forces moving bodies over distances, we will develop related ideas of work and potential energy. But suppose we want to know, say, how a battery affects the charge passing through it. We don't know what the battery will be connected to or for how long. In cases such as this, we must think about what happens to each unit of charge. We will introduce quantities that tell how much force is exerted on each unit of charge and how much potential energy it gains or loses.

"... suppose we want to know, say, how a battery affects the charge passing through it."

19-1 Making Electrostatic Force and Charge Concepts Quantitative

French physicist Charles Augustin de Coulomb (1736–1806) put the concepts of electrostatic force and electric charge on a quantitative basis. (The historical record seems to indicate that Cavendish did this earlier, but Coulomb's results were the first to be widely disseminated in print.) Coulomb reported a series of measurements on charged bodies using a torsion balance apparatus much like the one Cavendish used (Figure 8-16) to determine the universal gravitational constant G. In Coulomb's apparatus (Figure 19-1), if the fiber is twisted, it exerts a restoring torque proportional to the angle of twist. The restoring torque it has to exert to keep the "barbell" in equilibrium is in turn proportional to the electrostatic force that sphere 1 exerts on sphere 2 when both are charged. It had previously been suspected, by analogy with gravitational forces, that the electrostatic force between charged point objects varied inversely as r^2. By observing the angle of twist with the spheres at different separations r, Coulomb showed that this was indeed the case.

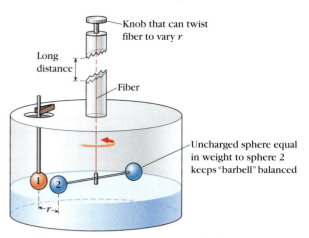

Knob that can twist fiber to vary r

Long distance

Fiber

Uncharged sphere equal in weight to sphere 2 keeps "barbell" balanced

Figure 19-1 Coulomb's torsion balance. Based on drawing in Coulomb's paper in *Histoire et Mémoires de l'Académie Royal des Sciences,* 1785.

Coulomb could also reduce the charge on either sphere by half to see how the force varied with charge. He did this by touching the sphere in question with a conducting sphere identical to it but uncharged. He reasoned that the excess charges on the charged sphere would by mutual repulsion spread out equally over both spheres. Then when the introduced sphere is removed, it should take half the excess charge with it. In this way he found the electrostatic force to be proportional both to the charge on sphere 1 and the charge on sphere 2. These relationships are summarized by Coulomb's law.

Coulomb's law: The magnitude $F = |\vec{F}_{\text{on 1 by 2}}| = |\vec{F}_{\text{on 2 by 1}}|$ of the electrostatic force that each of two point objects with charges q_1 and q_2 exerts on the other at a separation r_{12} is given by

$$F = k\frac{|q_1||q_2|}{r_{12}^2} \qquad (19\text{-}1)$$

Note: The magnitude gives the strength of the interaction.

Direction: The forces on the two bodies are directed along the line connecting them. The force on either body is toward the other body if the two have opposite charges and away if they have like charges.

Note that the signs of the charges are used only in determining the force's direction, not to find the magnitude of the force.

◆**UNITS** In SI, the unit of charge, not surprisingly, is called the *coulomb* (C). (Its definition follows from ideas that we will treat later on.) The value of the proportionality constant k could in principle be found by measuring the force that 1 C charges exert on each other at a separation of 1 m. The resulting force would be 9×10^9 N (verify for yourself that this is about a million tons!). This tells us that one coulomb is an extremely large amount of charge. It then follows from Equation 19-1 that

$$k = F\frac{r_{12}^2}{|q_1||q_2|} = (9 \times 10^9 \text{ N})\frac{(1 \text{ m})^2}{(1 \text{ C})(1 \text{ C})}$$

or

$$k = 9 \times 10^9 \frac{\text{N} \cdot \text{m}^2}{\text{C}^2} \qquad (19\text{-}2)$$

We cannot physically do the measurement using 1 C charges, but because k is a constant, we could get the same value from a much smaller measured force between much smaller charges.

With the humongous coulomb as our unit of charge, the size of nature's basic unit of charge, that of a single electron, is just a tiny fraction of a coulomb:

$$e = 1.602 \times 10^{-19} \text{ C} \approx 1.6 \times 10^{-19} \text{ C} \tag{19-3}$$

(We'll generally use the more rounded-off value.) The symbol e represents the magnitude only: The charge of a proton is $+e$; that of an electron is $-e$. Electrostatic forces now join the inventory of forces that we add as vectors when we consider $\Sigma\vec{F}$ in Newton's second law.

Example 19-1 *Vector Addition of Electrostatic Forces*

For a guided interactive solution, go to Web Example 19-1 at www.wiley.com/college/touger

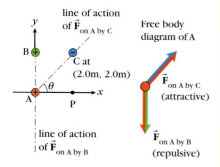

line of action of $\vec{F}_{\text{on A by C}}$

Free body diagram of A

B C at (2.0m, 2.0m)

θ

A P

$\vec{F}_{\text{on A by C}}$ (attractive)

line of action of $\vec{F}_{\text{on A by B}}$

$\vec{F}_{\text{on A by B}}$ (repulsive)

The charges and masses of the three point objects shown below are

$$q_A = 2.4 \times 10^{-6} \text{ C} \qquad q_B = 1.3 \times 10^{-6} \text{ C} \qquad q_C = -2.0 \times 10^{-6} \text{ C}$$

$$m_A = 0.001 \text{ kg} \qquad m_B = 0.002 \text{ kg} \qquad m_C = 0.002 \text{ kg}$$

a. If bodies B and C are fixed in their positions, find the total force on A when it is at the origin.

b. Find the acceleration of A at the instant when it is at this position.

Brief Solution

Choice of approach. **a.** We must (1) identify the individual forces on A and their directions in a *free-body diagram* (see figure), (2) determine the magnitude of each of these forces using Coulomb's law, and then (3) find the x and y components and use them to find the resultant vector. In (1), we use the signs on the charges to determine whether each force is attractive or repulsive and thus establish its direction along its line of action.

b. To find the acceleration, we use Newton's second law.

What we know/what we don't.

To find $F_{\text{on A by B}}$	To find $F_{\text{on A by C}}$	Constants
$|q_A| = 2.4 \times 10^{-6}$ C	$|q_A| = 2.4 \times 10^{-6}$ C	$k = 9 \times 10^9 \dfrac{\text{N} \cdot \text{m}^2}{\text{C}^2}$
$|q_B| = 1.3 \times 10^{-6}$ C	$|q_C| = 2.0 \times 10^{-6}$ C	
$r_{AB} = 2.0$ m	$r_{AC} = \sqrt{(2.0 \text{ m})^2 + (2.0 \text{ m})^2} = 2.8$ m	

Second law applied to A

$$m_A = 0.001 \text{ kg}$$

$$\text{total } F_{\text{on A}} = ? \qquad \theta_F = ? \qquad a = ?$$

The mathematical solution. **a.** We find magnitudes by Equation 19-1:

$$F_{\text{on A by B}} = k\frac{|q_A||q_B|}{r_{AB}^2} = 9 \times 10^9 \frac{\text{N} \cdot \text{m}^2}{\text{C}^2} \frac{(2.4 \times 10^{-6} \text{ C})(1.3 \times 10^{-6} \text{ C})}{(2.0 \text{ m})^2}$$

$$= 7.0 \times 10^{-3} \text{ N}$$

$$F_{\text{on A by C}} = k\frac{|q_A||q_C|}{r_{AC}^2} = 9 \times 10^9 \frac{\text{N} \cdot \text{m}^2}{\text{C}^2} \frac{(2.4 \times 10^{-6} \text{ C})(2.0 \times 10^{-6} \text{ C})}{(2.8 \text{ m})^2}$$

$$= 5.5 \times 10^{-3} \text{ N}$$

In finding the x and y components of these forces, we note that $F_{\text{on A by B}}$ is in the negative y direction, and from the geometry of triangle APC, $\theta = 45°$. Thus,

$$(F_{\text{on A by B}})_x = 0 \qquad (F_{\text{on A by B}})_y = -7.0 \times 10^{-3} \text{ N}$$

$$(F_{\text{on A by C}})_x = (F_{\text{on A by C}}) \cos 45° = (5.5 \times 10^{-3} \text{ N})(0.707) = 3.9 \times 10^{-3} \text{ N}$$

$$(F_{\text{on A by C}})_y = (F_{\text{on A by C}}) \sin 45° = (5.5 \times 10^{-3} \text{ N})(0.707) = 3.9 \times 10^{-3} \text{ N}$$

Now we add components to get the components of the total force on A:

$$(F_{\text{on A}})_x = (F_{\text{on A by B}})_x + (F_{\text{on A by C}})_x = 0 + 3.9 \times 10^{-3} \text{ N} = 3.9 \times 10^{-3} \text{ N}$$

$$(F_{\text{on A}})_y = (F_{\text{on A by B}})_y + (F_{\text{on A by C}})_y = -7.0 \times 10^{-3} \text{ N} + 3.9 \times 10^{-3} \text{ N}$$

$$= -3.1 \times 10^{-3} \text{ N}$$

Finally, we convert to a magnitude and direction description:

$$F_{\text{on A}} = \sqrt{(F_{\text{on A}})_x^2 + (F_{\text{on A}})_y^2} = \sqrt{(3.9 \times 10^{-3} \text{ N})^2 + (-3.1 \times 10^{-3} \text{ N})^2}$$

$$= \mathbf{5.0 \times 10^{-3} \text{ N}}$$

$$\tan \theta_F = \frac{(F_{\text{on A}})_y}{(F_{\text{on A}})_x} = \frac{-3.1 \times 10^{-3} \text{ N}}{3.9 \times 10^{-3} \text{ N}} = -0.79$$

so
$$\theta_F = \mathbf{-38°} \quad (38° \text{ below the } +x \text{ axis}).$$

b. By Newton's second law,

$$a = \frac{F_{\text{on A}}}{m_A} = \frac{5 \times 10^{-3} \text{ N}}{1 \times 10^{-3} \text{ kg}} = \mathbf{5 \text{ m/s}^2}$$

◆ Related homework: Problems 19-7, 19-8, 19-9, and 19-11.

19-2 Electric Fields

For an important new perspective on the calculation we did in Example 19-1, notice that we multiplied by the charge q_A in finding each of the forces $F_{\text{on A by B}}$ and $F_{\text{on A by C}}$. We used these in turn to find the components of these forces and then to find the total components $(F_{\text{on A}})_x$ and $(F_{\text{on A}})_y$ and the magnitude $F_{\text{on A}}$ of the resultant force. So $|q_A|$ was built in as a multiplier in finding all of these. Thus, each of the electrostatic forces on the charge q_A, and each of their components, can be viewed as the product of $|q_A|$ and "everything else" multiplying it. For example we can write the total force on A ($\vec{F}_{\text{on A}}$) very informally as

$$(F_{\text{on A}})_x = |q_A| \left(\frac{\text{everything}}{\text{else}} \right)_x \qquad (F_{\text{on A}})_y = |q_A| \left(\frac{\text{everything}}{\text{else}} \right)_y$$

It then follows that the magnitude $F_{\text{on A}}$ of the total electrostatic force is likewise the product of $|q_A|$ and "everything else." If q_A is at a point P, the "everything else" multiplying it involves the other charges, their distances from P, and because we consider directions when we find components, the directions from P to those charges. In short we can say that the "other stuff" involves the electrical environment of the point P at which q_A was placed. To see this reasoning presented in graphic step-by-step detail, go to WebLink 19-1.

Recall from Chapter 4 that a force is one side of an interaction. A body and its environment exert equal and opposite electrostatic forces on each other. We may consider the magnitude of either force to be a measure of the strength of the interaction. The strength of the interaction is thus the product of a property of the object itself (its charge) and a term involving the electrical properties (charges and their configuration) of its environment. This latter term is what we

For **WebLink 19-1:** **The Electric Field— The Charge Carrier's Environment,** go to www.wiley.com/college/touger

call the **electrical field,** and we denote its magnitude by E:

$$F_{\text{on A}} = |q_A| \left(\begin{array}{c} \text{``everything}\\ \text{else''} \end{array} \right)$$

or

$$F_{\text{on A}} = |q_A| \quad E \qquad (19\text{-}4)$$

strength of   property   properties of A's
electrical   of A   environment
interaction
between A and its
environment

(We use the absolute value of q_A because F and E are magnitudes, so they are always positive.)

You can associate the symbol E with *electrical environment,* or with "*everything else.*"

Notice that if q_A were moved to a different position, its distance and direction from each other charge would change, that is, the "other stuff" contributing to the field E would change. In essence, q_A experiences a different environment—its surroundings appear different—when it is in a different position. The mathematical way of saying this is: *The electric field is a function of position.* Changing position affects *each component* of the force on A, so in two dimensions we can write an equation like 19-4 for each direction

$$F_{x \text{ on A}} = q_A E_x \qquad F_{y \text{ on A}} = q_A E_y \qquad (19\text{-}5x, \ 19\text{-}5y)$$

Here we use the charge rather than its absolute value because the x and y components of a force can be either positive or negative. We summarize the two component equations by the *vector* equation

$$\vec{\mathbf{F}}_{\text{on A}} = q_A \vec{\mathbf{E}} \qquad (19\text{-}5)$$

STOP&Think If after doing Example 19-1, you had to find the force on a different charge (call it q_D) placed at the origin, would you have had to do the entire calculation over again using the value of q_D? ◆

The above question suggests there is an advantage to factoring the force as in Equation 19-5. If we did the calculation in Example 19-1 with the factor q_A left out, we could as easily have multiplied by q_A or q_D at the end. In other words, we could first have calculated the field at a particular point, and then multiplied by the charge to find the force on any charge carrier placed at that point.

From Equation 19-4, the magnitude of the electric field is $E = \frac{F_{\text{on A}}}{|q_A|}$, so it is measured in N/C. This is the number of newtons *on each* coulomb of charge.

The electric field at a point is the force that will be exerted on each coulomb of charge placed at that point.

In Equation 19-5, the force vector is the product of the scalar charge and the vector field. Multiplying a vector by a positive scalar preserves its direction; multiplying it by a negative scalar reverses its direction:

If the charge is positive, the electrostatic force on it is in the same direction as the field; if the charge is negative, the force is opposite in direction to the field.

Another important point follows from this:

The direction of the electric field at any point is the direction of the total electrostatic force that a positive charge placed at that point would experience.

◆**ELECTRIC FIELD DUE TO A POINT CHARGE** For two isolated charges q_1 and q_2, Equation 19-4 becomes

$$F_{\text{on 2 by 1}} = |q_2| E_{\text{due to 1}} \qquad (19\text{-}6)$$

where $E_{\text{due to 1}}$ is the field at the point—let's call it P—where q_2 is placed. The field is due to the environment of charges surrounding P; in this case the only charge in the environment is q_1. Comparing Equation 19-6 with Coulomb's law

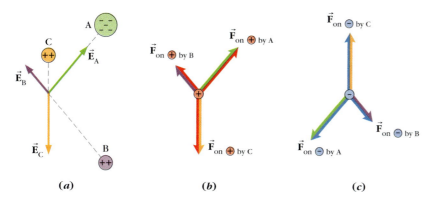

(a)   (b)   (c)

Figure 19-2 Field inventory and free-body diagrams. (*a*) Field inventory diagram showing fields at point P due to surrounding charges. (*b*) Free-body diagram of point object with charge +1 C placed at P. (*c*) Free-body diagram of point object with charge −1 C placed at P.

(Equation 19-1), we see that $E_{\text{due to } 1} = k\frac{|q_1|}{r_{12}^2}$. Because the field is the force on each coulomb (C) placed at P, and because 1 C is a positive amount of charge, the field at P is in the direction of the force that any positive charge placed there would experience. Simplifying our notation, we can now summarize.

Electric field at a position a distance r from a point charge q

$$\text{Magnitude:} \quad E = k\frac{|q|}{r^2} \tag{19-7}$$

Direction: Away from q if q positive,
toward q if q negative.

If the environment of a point P is made up of many point charges, the total field at P is the vector sum of the fields due to the individual charges. An important first step is to draw a diagram showing the individual fields contributing to the sum. In the diagram in Figure 19-2a,

1. $\vec{E}_B$ and $\vec{E}_C$ are directed away from the positive charges on B and C (see rules for direction under Equation 19-7) but $\vec{E}_A$ is toward the negative charge on A;
2. although A and B are the same distance from point P, $E_A > E_B$ because the magnitude of the charge on A is greater; and
3. although B and C are equally charged, $E_B < E_C$ because B is further from P (so r in Equation 19-7 is larger).

The diagram is much like a free-body diagram but inventories fields rather than forces. Another important difference is this: There need not be any object at P. The environment of point P is present whether or not anything is at P to interact with it. In contrast to the two-tone arrows representing forces in Figures 19-2b and c, the field arrows are a single color to remind you that fields represent only one contribution to the interaction. If a charged body is then placed at P, it provides the other contribution, resulting in a force. We then get the free-body diagrams in Figures 19-2b and c, showing the forces on a positively and a negatively charged body at P. Note that for the negatively charged body, the forces are opposite to the fields.

PROCEDURE 19-1

Finding the Electric Field at a Point P
Due to an Array of Point Charges

1. Draw a field inventory diagram (such as Figure 19-2a) showing the fields at P due to each individual charge. The directions are away from positive charges and toward negative charges.
2. Find the distance from P to each charge.
3. Use Equation 19-7 to calculate the magnitude of each field.
4. Find the x and y components of each field and do vector addition to find the resultant field vector (Procedure 3-4).

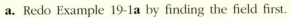

Example 19-2 *A Field Approach to Example 19-1*

For a guided interactive solution, go to Web Example 19-2 at
www.wiley.com/college/touger

a. Redo Example 19-1a by finding the field first.

b. Repeat for the case when point object A is an electron.

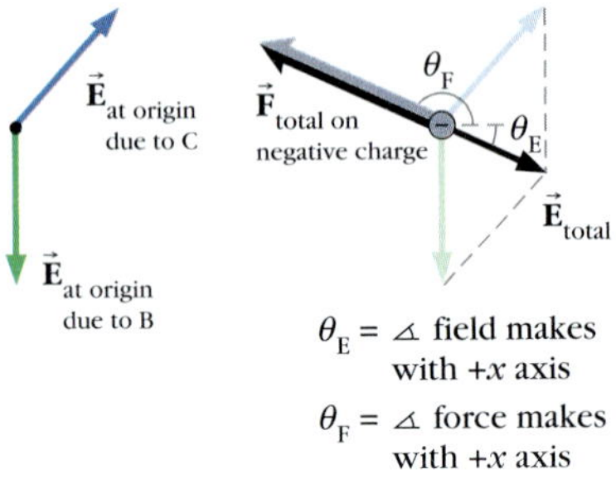

Brief Solution

Choice of approach. In following Procedure 19-1, we repeat the calculation we did for Example 19-1a but leaving out the factor $|q_A|$ until the end. **STOP&Think** How does the above field inventory diagram compare with the free-body diagram in Example 19-1? ◆

What we know/what we don't. See Example 19-1.

The mathematical solution. **a.** (Compare with Example 19-1) Apply Equation 19-7 at the origin:

$$E_{\text{due to B}} = k\frac{|q_B|}{r_{AB}^2} = 9 \times 10^9 \frac{\text{N} \cdot \text{m}^2}{\text{C}^2} \frac{(1.3 \times 10^{-6}\text{C})}{(2.0\text{ m})^2} = 2.9 \times 10^3 \text{ N/C}$$

$$E_{\text{due to C}} = k\frac{|q_C|}{r_{AC}^2} = 9 \times 10^9 \frac{\text{N} \cdot \text{m}^2}{\text{C}^2} \frac{(2.0 \times 10^{-6}\text{C})}{(2.8\text{ m})^2} = 2.3 \times 10^3 \text{ N/C}$$

$$(E_{\text{due to B}})_x = 0 \qquad (E_{\text{due to B}})_y = -2.9 \times 10^3 \text{ N/C}$$

$$(E_{\text{due to C}})_x = (E_{\text{due to C}})\cos 45° = 1.6 \times 10^3 \text{ N/C}$$

$$(E_{\text{due to C}})_y = (E_{\text{due to C}})\sin 45° = 1.6 \times 10^3 \text{ N/C}$$

As before, we add components to get the component description of the total field at the origin (the position where A is or will be placed).

$$E_x = (E_{\text{due to B}})_x + (E_{\text{due to C}})_x = 1.6 \times 10^3 \text{ N/C}$$

$$E_y = (E_{\text{due to B}})_y + (E_{\text{due to C}})_y = -1.3 \times 10^3 \text{ N/C}$$

and convert to a magnitude and direction description:

$$E = \sqrt{E_x^2 + E_y^2} = 2.1 \times 10^3 \text{ N/C} \qquad \tan\theta_E = \frac{E_y}{E_x} = -0.81 \text{ so } \theta = \mathbf{-39°}$$

E is in the same direction as the force in Example 19-1 (except for error due to rounding off). Also agreeing with Example 19-1, the magnitude of the force is

$$F_{\text{on A}} = |q_A|E = (2.4 \times 10^{-6}\text{ C})(2.1 \times 10^3 \text{ N/C}) = \mathbf{5.0 \times 10^{-3} \text{ N}}$$

b. Because we calculated the field first, we need only multiply by the different charge ($|q_A| = e$ for an electron) placed at the origin:

$$F_{\text{on e}^-} = eE = (1.6 \times 10^{-19}\text{ C})(2.1 \times 10^3 \text{ N/C}) = \mathbf{3.4 \times 10^{-16} \text{ N}}$$

The direction of the force on a negative charge is reversed (changed by 180°; see part *b* of figure) from that of the field, so in this case

$$\theta_F = -39° + 180° = \mathbf{141°}$$

◆ Related homework: Problems 19-13, 19-14, 19-15, and 19-19.

Left margin figure

$\vec{E}_{\text{at origin due to C}}$

$\vec{F}_{\text{total on negative charge}}$

θ_F

θ_E

$\vec{E}_{\text{total}}$

$\vec{E}_{\text{at origin due to B}}$

$\theta_E = \angle$ field makes with +x axis

$\theta_F = \angle$ force makes with +x axis

(a) Field inventory diagram

(b) Directions of resultant field and force vectors

19-3 Fields Due to Continuous Charge Distributions

When the source of the field is not a point object, we may have to consider fields due to charges spread out more or less continuously over part or all of the body. We will use qualitative reasoning to draw conclusions about some common distributions of charge.

◆**CONDUCTORS IN ELECTROSTATIC EQUILIBRIUM** In Example 18-2, we deduced that if a body is a conductor in electrostatic equilibrium, all excess charge must be on the body's surface. If there is an electric field in the interior of such a body, positive charge or conventional current will flow in the direction of the field. In electrostatic equilibrium, there is no net motion of charge, so we must conclude:

There is no electric field anywhere in the interior of a conductor in electrostatic equilibrium.

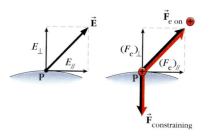

Figure 19-3 Fields and forces at the surface of a conductor.

What about at the surface? Suppose a field at a point P on the surface (Figure 19-3) has components parallel and perpendicular to the surface. By Equation 19-5, the electrostatic force F_e on any excess charge at that point would also have parallel and perpendicular components. But although $(F_e)_\perp$ would be opposed by the constraining forces that keep the charges from leaving the body, $(F_e)_{//}$ would be unopposed, so charge would move along the surface in the direction of $(F_e)_{//}$. But then the conductor would not be in electrostatic equilibrium. We avoid this contradiction only if $(F_e)_{//}$, and therefore $E_{//}$, is zero.

The electric field must be perpendicular to the surface at all points on the surface of a conductor in electrostatic equilibrium.

Applying Procedure 19-1 in detail to a continuous charge distribution would require calculus. But we can apply Procedure 19-1 in part, together with symmetry arguments, to determine the *direction* of the field due to some simple charge configurations. Fortunately, these are the configurations commonly found in electric circuits.

Example 19-3 *The Electric Field along the Axis of a Uniformly Charged Circular Ring*

Find the direction of the electric field along the axis of a circular ring with a uniformly distributed positive charge (Figure 19-4a).

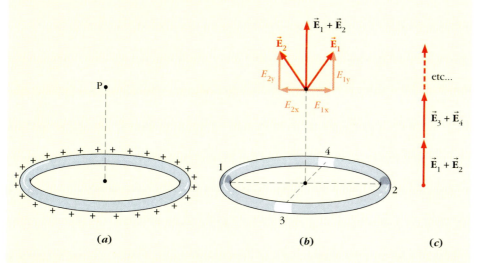

Figure 19-4 The electric field along the axis of a uniformly charged circular ring. See Example 19-3.

Solution

Choice of approach. Mentally we can break up the ring of charge into pieces so tiny that they are nearly point charges. Guided by symmetry considerations, we can select representative pieces. We will draw a field inventory diagram for a point P on the axis (Step 1 of Procedure 19-1), and then sketch the vector addition to determine the direction of the total field.

The reasoning. We can choose pairs of pieces (e.g., 1 and 2, 3 and 4 in Figure 19-4b) that are diametrically across from each other. The field due to each

EXAMPLE 19-3 *continued*

piece is directed away from that piece. Figure 19-4b shows that due to the symmetry in our choice, the horizontal components of the fields $\vec{E}_1$ and $\vec{E}_2$ (due to charges 1 and 2) are equal and opposite. Then the resultant of these two fields is *along the axis*. The same reasoning applies to each such pair of charges, so the vector sum of the contributions of all pairs (Figure 19-4c)—the field due to the entire ring—is likewise along the axis. By considering the surface of a conducting wire to be made up of many such rings, we can build on this result to find the field along the axis of the wire—see Example 9-4.

◆ Related homework: Problem 19-22.

◆**CONDUCTORS NOT IN ELECTROSTATIC EQUILIBRIUM** By maintaining a difference in the surface charge density at its two poles, a battery in a simple circuit prevents the conducting path from reaching electrostatic equilibrium, so there continues to be a net motion of charge, that is, a current. Because electrostatic equilibrium is not reached, the interior field does not become zero. In Example 19-4, we find its direction by extending the approach of Example 19-3.

Example 19-4 *Electric Field in a Wire Connected to a Battery*

For a guided interactive solution, go to Web Example 19-4 at www.wiley.com/college/touger

In Chapter 18 we showed that when a conducting wire is connected between the terminals of a battery, the battery causes a surface charge density gradient along the wire.

a. Find the direction of the electric field in the interior of the wire. In your reasoning, consider a long, straight segment of the wire (Figure 19-5a).

b. Find the direction of the electrostatic force on an electron moving through the wire, and on a positive charge carrier (such as we consider in a conventional current) moving through the wire.

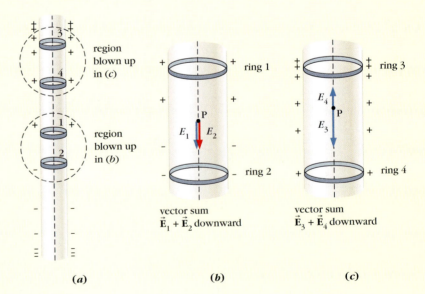

Figure 19-5 Electric field in a straight wire. (Summary of main field ideas in Example 19-4).

Brief Solution

Choice of approach. We can find the direction of the field by the same approach as the previous example. Because a straight wire is cylindrical, we can think of the charge on its surface as being made up of a stack of loops of charge, each like the one in Example 19-3.

The details.

a. First consider the field at a point P on the axis. From Example 19-3, we know the field contribution from each ring is along the axis at point P. Consider pairs of rings at equal distances above and below P. If the ring above is positive and the ring below is negative, as in Figure 19-5*b*, the contribution due to each ring is in the positive-to-negative direction, as is their sum. If both rings are positive, as in Figure 19-5*c*, the contributions are in opposite directions, but because the lower ring is *less* positive, its field contribution is weaker, so the sum of these two contributions is likewise in the positive-to-negative direction. We can similarly show that the field due to each pair of rings chosen in this way, and therefore the total field, is *in the direction of decreasing surface charge density.*

b. Electrostatic forces are in the same direction as the field on positive charges and opposite on negative charges. Thus, conventional current flows in the direction of decreasing surface charge density, but electrons flow oppositely.

◆ Related homework: Problem 19-23.

19-4 Picturing Electric Fields

It is often useful to get an overall picture of the electric field, which we can do as follows. We can sample the effects of the field by placing a point positive charge—a "test charge"—at different positions and observing how the magnitude and direction of the force on it vary from point to point. This is shown in Figure 19-6. The test charge in Figure 19-6*a* is sampling the field due to the point charge in the center; the figure shows the force per unit charge on it (the field) at several different points. Figure 19-6*b* shows a more detailed mapping, but it is still representative; if we made the map complete, the arrows would totally blacken the region and make the map unreadable. So we have to imagine the arrows at the in-between points; in reality, the field is continuous, changing gradually in magnitude and direction from one point to the next except right at the point charge(s) giving rise to the field. The procedure and some of the implications of the mapping that we get are shown in step-by-step detail in WebLink 19-2.

The mappings are sometimes summarized by drawing a continuous set of lines, called **electric field lines** (Figure 19-6*c*). The field vectors in Figure 19-6*b* would be directed along these lines. For fields due to more complicated arrangements of charge, the field lines might be curved. The field vector at each point would then be directed tangent to the field line.

Because the field lines in Figure 19-6*c* don't show the lengths of the field vectors, they must communicate the field's strength in another way. Figure 19-6*d* suggests how this happens. There, because the field is due to a greater point

For **WebLink 19-2:** **Mapping an Electric Field,** go to www.wiley.com/college/touger

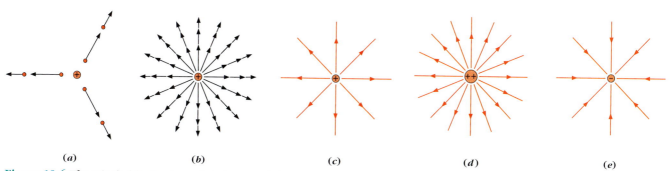

| (*a*) | (*b*) | (*c*) | (*d*) | (*e*) |

Figure 19-6 Electric fields due to point charges. (*a*) Force per unit charge (field) on a test charge at several different positions. (*b*) A more detailed mapping of field vectors at different positions. (*c*) Electric field lines for the mapping in *b*. (*d*) Electric field lines for a greater point charge. (*e*) Electric field lines for a negative point charge.

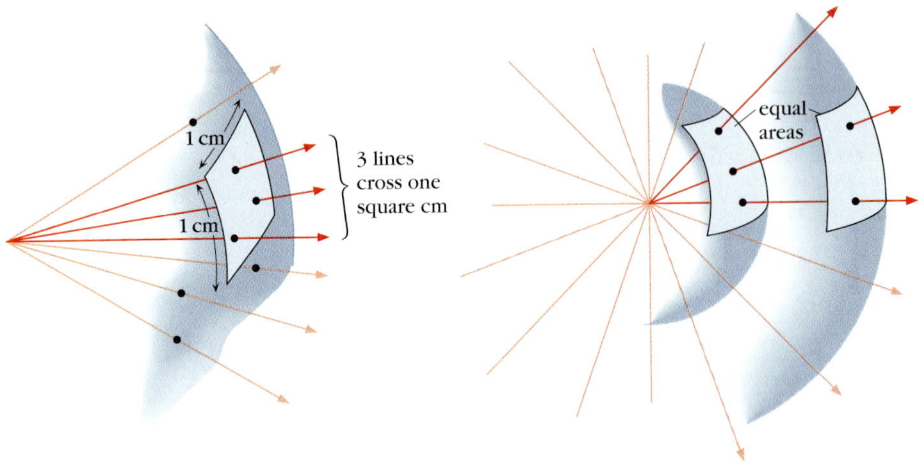

Figure 19-7 The density of field lines indicates the magnitude of the field. (*a*) The density at this surface is 3 lines/cm². (*b*) The same *total* number of lines crosses each sphere, so fewer lines cross each cm² of the larger spherical surface.

(*a*) Density at this surface is 3 lines/cm²

(*b*) Same *total* number of lines crosses each sphere, so fewer lines cross each cm² of larger spherical surface.

charge, the lines are more densely packed. The *density* of field lines visually communicates the strength of the field. Notice, too, that in both Figures 19-6*c* and 19-6*d* the lines fan out (become less dense) as they go further out from the charge. By doing so, they show us that the field gets weaker as the distance r from a point charge increases. Figure 19-6*e* shows that the direction of the field is reversed if the point charge is negative.

In summary, the field line "map" communicates information because they follow a few basic rules:

- *Rule 1:* Field lines start and end at charge carriers (or else go to infinity). They are directed away from positive charge and toward negative charge.
- *Rule 2:* The *direction* of the field line at any point (or of the tangent at that point if the field line is curved) is simply the *direction of the field vector* at that point.
- *Rule 3:* The density of field lines (how closely packed they are) in a small region about a point is drawn to be proportional to the *magnitude of the field vector* at that point.

Quantitatively, the density of field lines is the number of field lines crossing each unit of area perpendicular to the lines in a three-dimensional picture (Figure 19-7*a*). The following line of reasoning demonstrates that with this definition, the depicted field due to a point charge (Figure 19-7*b*) is proportional to $1/r^2$, as Coulomb's law requires it to be. In preparation for our argument, the figure shows imaginary spheres of different radii r centered at the point charge; these provide the surface areas that we picture the field lines crossing.

Argument Statement in Words	Equivalent Mathematical Statement
1. The same number of field lines crosses both spheres in Figure 19-7*b*.	Number of field lines $= c_1$ (a constant)
2. The number of field lines crossing each unit area on a particular sphere of radius r is the total number of lines divided by the sphere's surface area $4\pi r^2$.	Number of lines/unit area $= \dfrac{c_1}{4\pi r^2}$
3. The number of field lines per unit area that we draw is proportional to the field.	$c_2 E = \dfrac{c_1}{4\pi r^2}$ (c_2 is another constant)
4. Then the field due to the point charge is inversely proportional to r^2.	$E = \left(\underbrace{\dfrac{c_1}{4\pi c_2}}_{\text{a constant}}\right)\dfrac{1}{r^2}$ that we call k

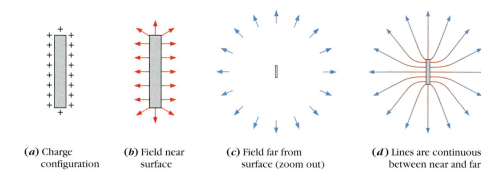

(*a*) Charge configuration | (*b*) Field near surface | (*c*) Field far from surface (zoom out) | (*d*) Lines are continuous between near and far limits

Figure 19-8 Electric field lines due to a charged conducting plate.

For a more intuitive approach to this $\frac{1}{r^2}$ dependence, work through Problem 19-1 at the end of the chapter.

◆**GAUSS'S LAW** Doing Problem 6-55 shows that the irradiance due to a point light source is proportional to $\frac{1}{r^2}$. The irradiance is also called the **energy flux density.** In this terminology there is the same total flow or *flux* of energy per second across each sphere, rather than the same "total flow" of field lines. Then we can speak of the electric field as the **electric flux density.** A total *electric flux,* proportional to the total number of lines, can be obtained for the point charge by multiplying the field at any distance r by the spherical surface to which it is everywhere perpendicular, giving us $\left(\frac{kq}{r^2}\right)(4\pi r^2) = 4\pi kq$ (a constant times q). Thus, the total flux across any of the imaginary spheres enclosing the charge is proportional to the charge enclosed. By deforming the spheres, we could show that the total flux across an enclosing surface of *any* shape is proportional to the total amount of charge enclosed, however the charge is configured. This generalization is called **Gauss's law,** after the great German mathematician Karl Friedrich Gauss (1777–1855). It is useful for finding the electric fields due to various continuous arrangements of charge, but its detailed mathematical treatment is beyond the scope of this book.

A charged conductor is not a point object. To get a picture of the field due to such an object (Figure 19-8*a*), we use the fact that close to a conducting object, the field must be perpendicular to the surface (Figure 19-8*b*). At a great distance, though, the object looks like a point object, so the field at this distance must look like the field due to a point object (Figure 19-8*c*). Because field lines always start and end at charge carriers or else go to infinity, there can be no breaks in the lines between the near and far regions, so we can fill in the connecting parts (Figure 19-8*d*). For a more detailed graphic presentation of this reasoning, go to WebLink 19-3.

In general, we can think of an extended charged object as though it were built up of point charges. Each point charge has a field like the fields in Figure 19-6. We now consider how these fields combine to give us a total field.

◆**SUPERPOSITION OF FIELDS** The electric field lines due to an arrangement of charges communicate information in the same way (Rules 1–3) as the electric field lines due to a point charge. But now the rules tell us how to read the direction and comparative magnitude of the *total* electric field from the field lines. From the field lines for each of two point charges, we can find the two fields at each point in space and then combine them by vector addition (Figure 19-9*a*).

For **WebLink 19-3: Electric Field Due to a Charged Conducting Plate,** go to www.wiley.com/college/touger

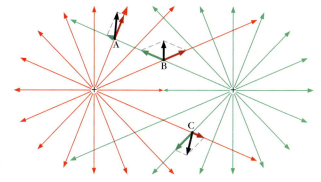

(*a*) The two fields add vectorially at each point.

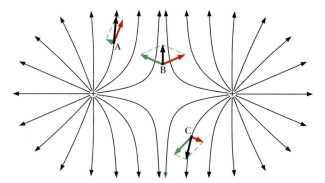

(*b*) The field lines are a map of the *total* field

Figure 19-9 Superposition of fields due to two point charges. (*a*) The two fields add by vector addition at each point. (*b*) The field lines are a map of the *total* field.

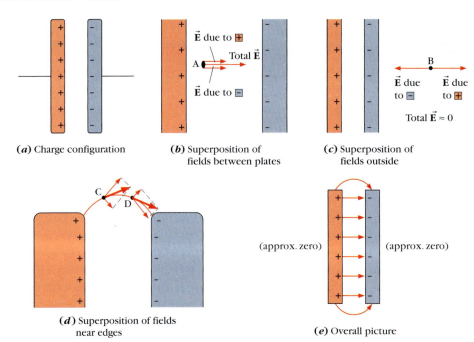

Figure 19-10 Electric field of a charged parallel plate capacitor.

(a) Charge configuration

(b) Superposition of fields between plates

(c) Superposition of fields outside

(d) Superposition of fields near edges

(e) Overall picture

The resulting electric field lines for the combination of charges map information about the resultant of the two field vectors at each point (Figure 19-9b). This point-by point vector addition is called a **superposition** of the two fields. To see how we apply this reasoning to obtain the result in Figure 19-10, work through Example 19-5.

Example 19-5 *Electric Field of a Parallel Plate Capacitor*

For a guided interactive solution, go to Web Example 19-5 at www.wiley.com/college/touger

Draw the electric field lines for a charged parallel plate capacitor, in which the distance between the plates is very small compared to the dimensions of either plate.

Brief Solution

Choice of approach.

1. Draw a charge configuration diagram.

2. Consider the superposition of fields in three separate regions: **a.** between the plates, **b.** to the left and right of both plates, and **c.** around the edges.

The details.

1. As in Example 18-3, the charge distributes over the inner faces of the two plates, forming two opposite planes of charge (Figure 19-10a).

2. A typical point A between the plates (away from the edges) is close enough to each plane of charge so that they both look effectively infinite from A (Figure 19-10b). The uniform fields due to the two planes are both away from positive and toward negative charge, as is the resultant field.

At a typical point B a short distance to either side of both plates, the planes of charge still look more or less infinite (Figure 19-10c). But now their fields, still away from positive and toward negative and still equal in magnitude, are opposite in direction, for a total field of zero. At larger distances from both plates, the fields due to the two planes of charge are very nearly (but not exactly) equal and opposite, so the total field is nearly (but not exactly) zero.

At points near the edges, such as C and D in Figure 19-10*d,* the fields are not equal and opposite, and the total field varies in direction from point to point. This effect is called **fringing.**

Figure 19-10*e* shows the overall field lines resulting from these superpositions.

◆ Related homework: Problems 19-27, 19-28, and 19-29.

Terminal Speed, Electrophoresis, and DNA Testing

An object falling through a fluid (such as air) in a gravitational field is subject to a drag force that increases with speed. When the magnitude of the drag force becomes as great as the gravitational force, the total force on the object becomes zero and the object stops accelerating. The speed it has reached at this stage is called the **terminal speed.** If the magnitude of the drag force is proportional to the speed ($F_d = \zeta v$, where the proportionality constant or **drag coefficient** ζ, the Greek letter zeta, depends on the object's size and shape), then at terminal speed v_t, $\zeta v_t = mg$, and $v_t = \frac{mg}{\zeta}$.

In the same way, when a uniform electric field is applied to ions in solution (also a fluid), the electric force on the ions is opposed by a drag force. As in air, the drag coefficient will be greater for ions with larger cross-sections, and those objects will reach smaller terminal speeds. In this case, the *electrical* force equals the drag force at terminal speed: $\zeta v_t = qE$, and $v_t = \frac{qE}{\zeta}$. Because these terminal speeds are typically reached over very small distances, each type of ion travels through most of the solution at its terminal speed—a different terminal speed for each one. The ions thus separate out. The process of separating out ions in solution by applying an electric field is called **electrophoresis.**

If the ions are permitted to pass through a porous medium, such as a gel, procedures such as staining will show how far the different ions have traveled through the gel. In a given time interval, ions of different sizes, with different terminal speeds, travel different characteristic distances (see Problem 19-72). The resulting separation, and the density of each ion that shows up on staining, makes possible analysis of which ions are present and to what extent. (See photo.) This process is called **gel electrophoresis.**

DNA fragments typically occur in an ionized state in certain kinds of solutions and can therefore be analyzed by gel electrophoresis. Performing gel electrophoresis on DNA, however, is complicated by the fact although the drag coefficient ζ increases with fragment size, so does the total charge q. This is because there is a charge on each of the base pairs making up the rungs of the DNA ladder, so the total charge depends on how many pairs are strung together to make up the total length of the fragment. To the extent that both ζ and q are proportional to length, the ratio $\frac{q}{\zeta}$ doesn't change with length, and the terminal speed $v_t = \frac{qE}{\zeta}$ remains the same for fragments of all lengths in a field E. In this case, a gel with smaller pores is used, so that larger fragments are likely to become trapped over a shorter distance. Gel electrophoresis is a standard procedure for DNA analysis. It is used for such purposes as establishing who is the father of a child, or identifying blood samples in legal cases such as rape or murder trials.

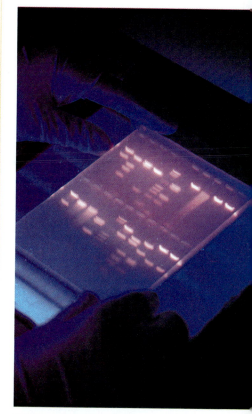

DNA testing. The photo shows the separation of DNA components by gel electrophoresis.

Figure 19-11 Small, irregularly shaped shepherd moons keep the material in Saturn's rings moving in formation.

19-5   Electrical and Gravitational Systems: Similarities and Differences; Electrical Potential Energy

Gravitational and electrostatic interactions are due to different inherent properties of the particles that constitute matter: gravitational mass and electric charge. But mathematically, they obey similar inverse square laws (Coulomb's law and Newton's universal law of gravitation).

Just as the total electrostatic force on a charged body is the vector sum of the forces exerted on it by all other charged bodies, the total gravitational force on a body is the sum of the gravitational forces exerted on it by all other masses. For example, the sun and Earth both exert significant gravitational forces on the moon, and the paths of the chunks of debris that make up the rings of Saturn are affected by several significant gravitational forces. These include the forces exerted by Saturn, by the rest of the debris in the ring, and quite significantly by small moons that rotate with the rings and help keep the debris in ring-shaped configurations (Figure 19-11). Because they "herd" the debris in this way, they are called *shepherd moons*.

Suppose we find the force on one chunk of debris, and moments later a chunk with different mass is at the same point. Rather than recalculate the total force from scratch, we can be guided by what we did for electrostatic forces. We can calculate the contribution to the total force at a point from all the surrounding bodies—the force *per unit of mass*—and then multiply it by the mass of whatever we place at that point. Analogous to Equations 19-4 and 19-7, we can again write equations we used to describe gravitational forces:

$$
\underset{\substack{\text{strength of} \\ \text{gravitational} \\ \text{interaction} \\ \text{between A and its} \\ \text{environment}}}{F_{\text{on A}}} = \underset{\substack{\text{property} \\ \text{of A}}}{m_A} \quad \underset{\substack{\text{property} \\ \text{of A's} \\ \text{environment}}}{g} \qquad \text{(5-1, like 19-4)}
$$

or
$$\vec{\mathbf{F}}_{\text{on A}} = m_A \vec{\mathbf{g}}$$

where $\vec{\mathbf{g}}$ has magnitude
$$g = G\frac{m}{r^2} \qquad \text{(8-8 or 8-9, like 19-7)}$$

Continuing the analogy, we call $\vec{\mathbf{g}}$ the *gravitational field*. Near Earth's surface $\vec{\mathbf{g}}$ has the roughly constant magnitude $g = G\frac{m}{R_E^2} = 9.8$ m/s². It is the gravitational acceleration of a freely falling body. We could treat this value as constant over trajectories near Earth's surface small enough so we can regard Earth as flat. How does this compare with electric fields? Viewed from far away, Earth looks like a point object, and its gravitational field lines (Figure 19-12a) look like the electric field lines due to a *negative* point charge. **STOP & Think** Why not a positive point charge? ♦ Viewed close up, Earth looks flat, and the field lines (Figure 19-12b)—everywhere perpendicular to the surface—look like those near a plane of charge (Figure 19-8b). The lines are parallel, not spreading out or converging, so the density of lines, indicating the magnitude of the field, doesn't change. The field is essentially constant in a flat Earth approximation. (Continuing to treat gravitational field lines as we did electrical, we could deduce Gauss's law for *gravitational* fields: The total *gravitational* flux across an enclosing surface is proportional to the *mass* enclosed.)

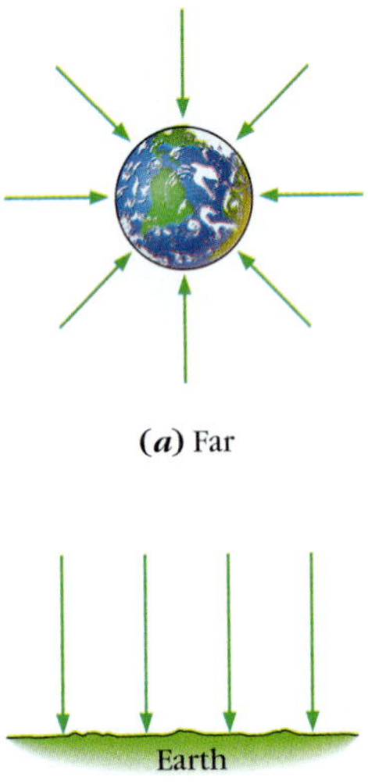

By considering the work done to move bodies in opposition to gravitational forces within a system, we were able to derive expressions for gravitational potential energy when g is constant (e.g., over short distances from a planet's surface)

$$PE_{\text{grav}} = mgy \qquad \text{(6-7)}$$

(a) Far

Earth

(b) Near

Figure 19-12 Gravitational fields far from and near Earth's surface.

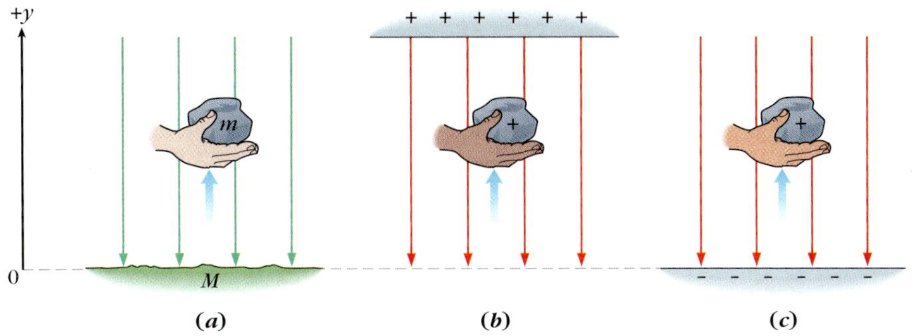

Figure 19-13 **Positive work is done by lifting either a mass or a positive charge against a field directed in the negative y direction.**

and when we must consider g's variation with r

$$PE_{grav} = -G\frac{m_1 m_2}{r} \qquad (8\text{-}10)$$

In either case, we can think of the gravitational PE as the energy stored by moving one body in a system against the gravitational field due to the rest of the system.

Purely by analogy (using the correspondences $m \leftrightarrow q$, $g \leftrightarrow E$, and $G \leftrightarrow k$), we can obtain expressions for the *electrical PE* (energy stored by moving one charged body in a system against the electric field due to the rest of the system):

When $\vec{E}$ is constant: $\qquad PE_{elec} = qEy \qquad (19\text{-}8)$

(*if* we take the direction of the electric field to be the *negative y* direction, as we did for the gravitational field $\vec{g}$)

When we must consider E's variation with r:

$$PE_{elec} = +k\frac{q_1 q_2}{r} \qquad (19\text{-}9)$$

Equations 6-7 and 19-8 have the same sign because we chose our coordinate system so that the field direction is the same in both cases. Thus, in all parts of Figure 19-13, the hand increases the PE of the body by lifting it. Note that Figure 19-13c is really the same as Figure 19-14a or b: In each case the hand is pulling the positive charge away from the plane of negative charge that is attracting it; in each case we pick our coordinates so that the field is in the $-y$ direction.

In the r-dependent situations described by Equations 8-10 and 19-9, however, we are stuck with keeping the object that is the source of the field at the center ($r = 0$). The hand in Figure 19-15 does positive work either by pulling two mutually attractive positive masses (masses are *always* positive) apart or by pushing two mutually repulsive positive charges together. In the gravitational case PE increases as r increases, but in the electrical case PE increases as r *decreases*, so the signs in Equations 8-10 and 19-9 must be opposite.

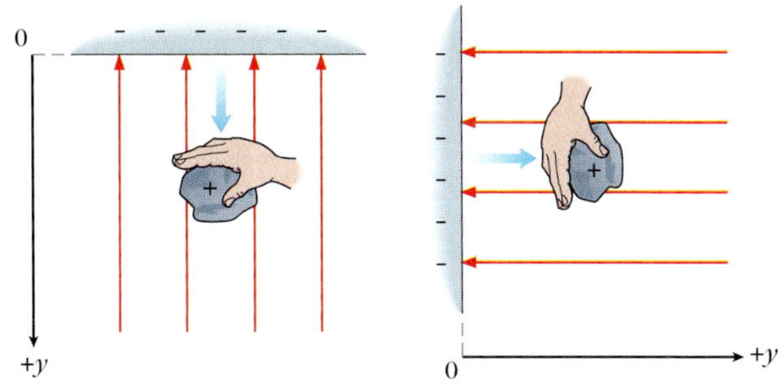

By rotating the page, you can turn Figure 19-13c into either of these pictures

Figure 19-14 **The y direction is chosen so that the field is in the negative y direction.** Compare with Figure 19-13c.

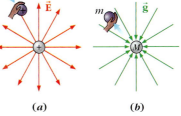

(a)   *(b)*

Figure 19-15 **The hand does positive work (r-dependent PE increases) in both cases.**

Example 19-6 *Energy of a Proton–Electron System*

For a guided interactive solution, go to Web Example 19-6 at
www.wiley.com/college/touger

A proton and an electron are a distance r apart.
a. Compare the electric and gravitational PEs of the proton–electron system.
b. If the distance r is increased, does the electrical PE increase or decrease? Does the gravitational PE increase or decrease?

Brief Solution
Restating the question. The most meaningful way to compare two quantities of very different sizes is by a ratio. In **a.** we ask, $\dfrac{PE_{elec}}{PE_{grav}} = ?$

EXAMPLE 19-6 *continued*

Choice of approach. We can treat the proton and electron as point objects, so the two PEs are $PE_{elec} = +k\frac{q_1 q_2}{r}$ (Equation 19-9) and $PE_{grav} = -G\frac{m_1 m_2}{r}$ (Equation 8-8). The masses and charges are known constants.

What we know/what we don't.

$$r = ? \qquad \frac{PE_{elec}}{PE_{grav}} = ?$$

Known constants:

$$\begin{cases} m_p = 1.67 \times 10^{-27} \text{ kg} \qquad m_e = 9.11 \times 10^{-31} \text{ kg} \qquad G = 6.67 \times 10^{-11} \frac{\text{N} \cdot \text{m}^2}{\text{kg}^2} \\[2mm] q_p = e = 1.6 \times 10^{-19} \text{ C} \qquad q_e = -e = -1.6 \times 10^{-19} \text{ C} \qquad k = 9 \times 10^9 \frac{\text{N} \cdot \text{m}^2}{\text{C}^2} \end{cases}$$

The mathematical solution. **a.** By Equations 19-9 and 8-8,

$$\frac{PE_{elec}}{PE_{grav}} = PE_{grav} = \frac{+k\dfrac{q_1 q_2}{r}}{-G\dfrac{m_1 m_2}{r}} = \frac{+k q_1 q_2}{-G m_1 m_2}$$

$$= \frac{\left(9 \times 10^9 \dfrac{\text{N} \cdot \text{m}^2}{\text{C}^2}\right)(1.6 \times 10^{-19} \text{ C})(-1.6 \times 10^{-19} \text{ C})}{-\left(6.67 \times 10^{-11} \dfrac{\text{N} \cdot \text{m}^2}{\text{kg}^2}\right)(1.67 \times 10^{-27} \text{ kg})(9.11 \times 10^{-31} \text{ kg})}$$

$$= \mathbf{2.30 \times 10^{39}} \text{ (unitless)}$$

PE_{elec} is vastly greater than PE_{grav}; PE_{grav} is negligibly small in comparison. **b.** Both PEs are negative, PE_{elec} because q_e is negative and PE_{grav} because it has a − sign. As r increases, the absolute value of each PE decreases. Then both PEs become less negative, that is, **both increase**.

◆ Related homework: Problems 19-31, 19-32, 19-33, 19-36, and 19-37.

Example 19-6 shows that gravitational interactions are insignificantly weak compared to electric interactions and are only evident between very massive bodies. Thus, on an atomic scale electrostatic forces are the binding forces that hold matter together, but on an astronomical scale it is mutual gravitational attraction that causes great clouds of hot gas to condense into stars and planets.

Like the gravitational force, the electrostatic force is conservative, so Equations 6-10 (conservation of mechanical energy) and 6-16 (the work-energy theorem) remain valid under the same conditions as before when the potential energy is partly or totally electrical.

Example 19-7 *Repelling Klingon Space Rats*

A Klingon spacecraft hovers 18 m from the vast flat surface of a space station. Klingon attack rats, each with a mass of 2.00 kg, can propel themselves across the gap at speeds of up to 12 m/s. But space station engineers have figured out that in the process of propelling themselves, the rats acquire a positive charge of 5.4×10^{-4} C. They therefore devise a way of charging up the plane surface of the station so that it will have a uniform electric field. Find the magnitude and direction of the minimum electric field required to repel all rats.

Solution

Choice of approach. We apply conservation of mechanical energy (Equation 6-10) to the case when the PE is purely electrical and is due to a uniform field. (Remember that for other arrangements of charge, like a point charge, the field is not uniform and we cannot use Equation 19-8.) We desire a final state in which the rat's velocity is reduced to zero. We first sketch the situation (Figure 19-16) to show the necessary direction of the field so that the force on the positively charged rat opposes its motion. We choose our coordinates so that, consistent with Equation 19-8, the positive y direction is opposite to the electric field.

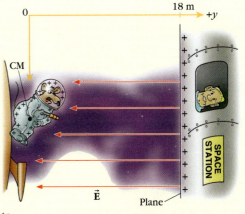

Figure 19-16 The physical situation. With the field as shown, the force opposes the rat's motion.

What we know/what we don't.

State A		State B		Constants
$y_A = 0$	$v_A = 12$ m/s	$y_B = 18$ m	$v_B = 0$	$m = 2.00$ kg
$PE_A = qEy_A$	$KE_A = \frac{1}{2}mv_A^2$	$PE_B = qEy_B$	$KE_B = 0$	$q = 5.4 \times 10^{-4}$ C
$E = ?$		$E = ?$		

The mathematical solution.

$$(\text{Total Energy})_A = (\text{Total Energy})_B$$

or

$$PE_A + KE_A = PE_B + KE_B \qquad (6\text{-}10)$$

becomes

$$qEy_A + \tfrac{1}{2}mv_A^2 = qEy_B + 0$$

Solving for E,

$$E = \frac{mv_A^2}{2q(y_B - y_A)} = \frac{(2.00\ \text{kg})(12\ \text{m/s})^2}{2(5.4 \times 10^{-4}\ \text{C})(18\ \text{m} - 0)} = \mathbf{1.5 \times 10^4\ \text{N/C}}$$

◆ Related homework: Problems 19-34, 19-35, 19-38, and 19-39.

19-6 Potential and Potential Differences

Just as the electrostatic force on any body may be treated as the product of two contributions, one from the body itself (its charge Q) and the other from the electrical environment (the field), the electric PE may likewise be separated into two contributions:

$$PE_{\text{elec}} = Q\left(\frac{PE_{\text{elec}}}{Q}\right) \qquad (19\text{-}10)$$

body's   environment's
contribution   contribution

Suppose the body is small enough for us to consider Q a point charge. As the electric field is a function of position giving the force on each unit of charge placed at that position, so the quantity $\frac{PE_{\text{elec}}}{Q}$, which is called the **potential,** is *a function of position giving the potential energy of each unit of charge at that position.* The units of potential are units of energy (joules) divided by units of charge (coulombs) and are therefore J/C.

It follows from Equations 19-8 and 19-9 that

$$\text{potential} \qquad \frac{PE_{\text{elec}}}{Q} = \frac{QEy}{Q} = Ey \qquad (19\text{-}11)$$

(in a region of constant field in the $-y$ direction, that is, near an effectively infinite plane of charge or close enough to a uniformly charged surface so that it appears flat)

➡ We will shortly introduce another name for J/C, but not yet, because we want the units to serve as a reminder that we are talking about PE per unit charge.

$$\text{and} \qquad \text{potential} \qquad \frac{PE_{\text{elec}}}{Q} = \frac{\dfrac{kQq}{r}}{Q} = \frac{kq}{r} \qquad (19\text{-}12)$$

(at a great enough distance r from a body with charge q so that
the body may be regarded as a point object)

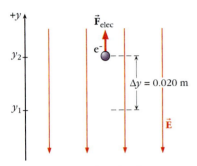

+y
y_2
y_1
$\vec{F}_{\text{elec}}$
e^-
$\Delta y = 0.020\ \text{m}$
$\vec{E}$

Figure 19-17 A sketch of the physical situation for Example 19-8.

Example 19-8   Change of Potential Experienced by an Electron

An electron is released at rest at a point where the potential is 3.0 J/C. The point is in a region of uniform electric field of magnitude 75 N/C. Set in motion by the field, the electron reaches a point 0.020 m from its starting point. What is the potential at this point?

Solution

Choice of approach. We sketch the situation with $\vec{E}$ in the negative y direction (Figure 19-17) so that we can apply Equation 19-11. Because the force on a negative charge is opposite to the field, the electron moves $\Delta y = 0.020$ m in the positive y direction. Because $\frac{PE_{\text{elec}}}{Q} = Ey$,

$$\Delta\left(\frac{PE_{\text{elec}}}{Q}\right) = E\,\Delta y \qquad (E \text{ constant}) \qquad (19\text{-}13)$$

What we know/what we don't.

$$\Delta y = 0.020\ \text{m} \qquad E = 75\ \text{N/C}$$

$$\left(\frac{PE_{\text{elec}}}{Q}\right)_1 = 3.0\ \text{J/C} \qquad \left(\frac{PE_{\text{elec}}}{Q}\right)_2 = ?$$

The mathematical solution.

$$\Delta\left(\frac{PE_{\text{elec}}}{Q}\right) = E\,\Delta y = (75\ \text{N/C})(0.020\ \text{m}) = 1.5\ \text{J/C} \quad (1\ \text{J} = 1\ \text{N}\cdot\text{m})$$

$$\left(\frac{PE_{\text{elec}}}{Q}\right)_2 = \left(\frac{PE_{\text{elec}}}{Q}\right)_1 + \Delta\left(\frac{PE_{\text{elec}}}{Q}\right) = 3.0\ \text{J/C} + 1.5\ \text{J/C} = \textbf{4.5 J/C}$$

Making sense of the results. To lose *potential energy,* negatively charged bodies must move from lower to higher *potential.* Symbolically, $\Delta\left(\frac{PE_{\text{elec}}}{Q}\right)$ must be positive for $\Delta PE_{\text{elec}} = Q\Delta\left(\frac{PE_{\text{elec}}}{Q}\right)$ to be negative when Q is negative.

◆ Related homework: Problem 19-43.

In Chapter 6, we pointed out that it is *changes* in potential energy rather than PE values at a single point that have physical meaning. Likewise, the change in potential energy per unit charge—the change in potential from one point to another—is what has physical significance, not its value at either point. That's why we have not assigned potential its own symbol, but we now choose V_{AB} as our symbol for the **potential difference** between two points A and B:

$$V_{\text{AB}} \equiv \left(\frac{PE_{\text{elec}}}{Q}\right)_{\text{B}} - \left(\frac{PE_{\text{elec}}}{Q}\right)_{\text{A}} = \frac{\Delta PE_{\text{elec}}}{Q} \qquad (19\text{-}14)$$

Units of potential difference, like potential, are J/C. A more common name for these units, especially in the context of electric circuits, is **volts** (V):

$$1\ \text{volt} \equiv 1\ \frac{\text{joule}}{\text{coulomb}} \qquad (1\ \text{V} \equiv 1\ \text{J/C}) \qquad (19\text{-}15)$$

Thus, **voltage** is a common expression for potential difference and motivates our choice of symbol.

Another unit of energy, often used when applying these ideas on an atomic scale, is the **electron volt** (eV). Because an amount of potential difference multiplied by an amount of charge gives us an amount of energy, the electron volt is defined as 1 volt multiplied by the magnitude of the charge of one electron.

$$1 \text{ eV} = (1.6 \times 10^{-19} \text{ C})(1 \text{ V}) = 1.6 \times 10^{-19} \text{ J} \qquad (19\text{-}16)$$

We must emphasize that voltage is a *difference;* it does not tell us the PE per unit charge at one point but how much it changes between *two different points.* For example, the voltage of a battery is the increase (a positive difference) in energy that each unit of charge gets in passing from one terminal to the other. Because the manufacturers don't know what will be connected to a battery or for how long, the useful thing to tell you is how much *nonelectrostatic* work the battery does on each unit of charge (and how much energy it thus transfers to each) passing between its endpoints.

Case 19-1 ◆ *Electric and Gravitational Circuits: Exploring an Analogy*

In Figure 19-18, the escalator does the same amount of nongravitational work on each unit of a population of mice (we may take our unit of mass in this case to be one standard mouse) and so increases the gravitational PE of each unit of mass by the same amount, no matter how many mice ride the escalator in all.

In the gravitational circuit, there is a value of the coefficient of friction between mouse and slide for which the amount of energy dissipated exactly equals the

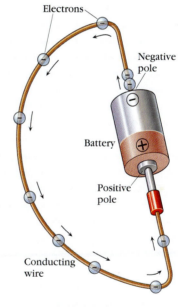

gravitational PE lost, so that the mice move at constant speed rather than gaining or losing KE. Somewhat analogously, because there is resistance in the wire, the electrons travel at an unchanging drift velocity. But like all analogies, this one has its limitations. If the coefficient of friction is increased, the mice will be slowed down, ultimately causing a back-up of mice at the top of the escalator because the escalator keeps carrying the mice at the same speed as before. But in the

(*a*) Gravitational circuit   (*b*) Electrical curcuit

Figure 19-18 Gravitational and electrical circuits compared.

Gravitational Circuit	**Electrical Circuit**
Similarities	
Each mouse gains PE when the electromechanical mechanisms of the escalator do work to move the mice in opposition to gravitational forces.	Each electron gains PE when the chemical mechanisms of the battery do work to move the electrons in opposition to electrostatic forces.
The energy gained is dissipated because of friction along the path.	The energy gained is dissipated because of resistance along the path.
We can characterize the escalator by how much PE in joules it provides to each unit of mass (mouse or kg).	We can characterize the battery by how much PE in joules it provides to each unit of charge (electron or coulomb). (1 J/C = 1 volt)
Differences	
The current or flow of mice consists of a few mice doing complete laps of the circuit.	The net flow of electrons is the result of vast numbers of electrons shifting very small distances at very slow drift velocities.

Case **19-1** ◆ *Electric and Gravitational Circuits: Exploring an Analogy (continued)*

electrical circuit, if electrons are slowed down by greater resistance and begin to back up, they repel the electrons coming up behind them, slowing them down as well. This feedback continues throughout the circuit so the electrons travel the entire circuit at a slower speed (resulting in a smaller current).

There are other differences. Individual mice may board the escalator at the bottom and do repeated laps of the gravitational circuit in a short time. In the electric circuit, in contrast, *all* parts of the circuit contribute electrons to the flow. These electrons actually drift extremely slowly through the circuit, typically traveling on the order of 10 cm in an *hour* (we'll do the calculation in the next chapter). But there are so many of them that, for example, under ordinary conditions 10^{19} of them will pass any point on the filament of a 75 W bulb each second. Remember also that the individual electrons do not travel at constant speed; they are accelerated by the field and slowed down by collisions. The drift velocity is an average.

To further clarify the connection between voltage and energy concepts, work through Example 19-9.

Example 19-9 *A Photoelectron*

For a guided interactive solution, go to Web Example 19-9 at
www.wiley.com/college/touger

Light striking a metal surface may under certain conditions provide some surface electrons with sufficient energy to break away from the surface. This effect is called the *photoelectric effect,* and the resulting emission of electrons is called *photoemission.* Suppose that when light strikes one of two parallel plates, electrons are emitted with a maximum velocity of 4.8×10^5 m/s. What voltage must be maintained between the plates if none of the emitted electrons are to reach the opposite plate?

Brief Solution

Choice of approach. Because voltages are PE differences per unit charge, we focus on energy. We can apply conservation of mechanical energy: The loss in KE as the electron is brought to a stop must equal the gain in PE. The force on a negative charge is opposite the field, so the field is *toward* the opposite plate.

What we know/what we don't.

State A	State B	Constants
$v_A = 4.8 \times 10^5$ m/s	$v_B = 0$	$m_e = 9.11 \times 10^{-31}$ kg
$PE_A = ?$ $KE_A = \frac{1}{2}mv_A^2$	$PE_B = ?$ $KE_B = 0$	$q = -e = -1.6 \times 10^{-19}$ C

$$V_{AB} = \frac{PE_B - PE_A}{q} = ?$$

The mathematical solution.

$$PE_A + KE_A = PE_B + KE_B \qquad (6\text{-}10)$$

becomes

$$\tfrac{1}{2}mv_A^2 = PE_B - PE_A$$

Then

$$V_{AB} = \frac{PE_B - PE_A}{q} = \frac{\tfrac{1}{2}mv_A^2}{q}$$

$$= \frac{(0.5)(9.11 \times 10^{-31}\ \text{kg})(4.8 \times 10^5 \text{m/s})^2}{-1.6 \times 10^{-19}\ \text{C}} = \mathbf{-0.66\ V}$$

Making sense of the result. The decrease in potential, when multiplied by a negative charge, will yield an increase in PE.

◆ Related homework: Problems 19-46 and 19-48.

◆**POTENTIAL DIFFERENCE AND SURFACE CHARGE DENSITY GRADIENTS** In Chapter 18 we established that a current flows in a conductor if the surface charge density changes along the conductor. In Example 19-4 we saw that the surface charge density gradient along the conductor results in an electric field directed from higher to lower surface charge density along the conductor's axis. The net field at any interior point depends on *differences* between the surface charge upstream and downstream from the point. So where the density is changing uniformly (the gradient is constant), the field is likewise constant (except near the ends of the region of uniform change). Where the field is constant, the voltage or potential difference between any two points along the conductor is given by Equation 19-13, which we can rewrite as

$$V_{AB} = E\Delta y \qquad (19\text{-}17)$$
$$(\Delta y = y_B - y_A)$$

If the surface charge densities at points A and B differ, there will be a field along the path between the points. If there is a field, there will be a potential difference or voltage between the points. The voltage, then, is a measure of the surface charge density difference between A and B. We can think of it as a measure of the battery's ability to "pump" charge to maintain this difference (somewhat as an ordinary air pump maintains a difference between the *volume* density of molecules at opposite ends of the pump). Whatever we said of the surface charge density gradient in Chapter 18 also holds true for voltage. For example, if there is a nonzero *voltage* (now replacing the words *surface charge density gradient*) between two points on a conductor, there will be a movement of charge (a current) between the points.

Again, a gravitational analogy is useful. In Figure 19-19 we retain the convention that the electric and gravitational fields are both in the negative y direction. Then the gravitational and electrical PE of a body both decrease in this direction. The change in PE *per unit* of mass is $\frac{\Delta PE_{grav}}{m} = \frac{mg\Delta y}{m} = g\Delta y$; It is a decrease when the object is falling and Δy is negative. We usually choose our coordinates so that $\frac{PE_{grav}}{m} = 0$ when $y = 0$. In the same way in the electrical case, we choose our coordinates so that $\frac{PE_{elec}}{q} = 0$ when $y = 0$. Like ground level in the gravitational case, this is electrical "ground."

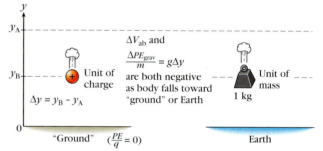

Figure 19-19 Change in PE when going from higher to lower potential in electrical and gravitational systems.

Analogies can also help in thinking about circuits. For example, as is true for gravitational PE, the change in electric PE is the same for any path between the same two points (recall the discussion of Figures 6-17 and 6-18). So it must also hold true for the voltage, the change in electric PE per unit charge. In electric circuits, then, we know that any wires or devices connected between the same two points are subject to the same voltage. We will use this idea extensively in Chapter 21.

19-7 Batteries

How do batteries produce potential differences? Batteries increase electric PE by separating opposite charges that are ordinarily drawn toward each other. Where does this energy come from? **STOP&Think** Do batteries create the energy? ◆

Recall from Section 6-1 (Figures 6-6 and 6-7 or WebLink 6-2) that chemical reactions can release energy that had been stored as chemical energy. What occurs is an energy conversion, not a violation of energy conservation. Common reactions in batteries include the dissolving or ionization to differing degrees of two different metals immersed in a solution, which is often acidic. In such a case the two pieces of metal, such as a piece of copper and a piece of zinc, form the poles of the battery. As the positive ions go into solution, their abandoned outer electrons remain behind on the metals. Because the two metals ionize to different

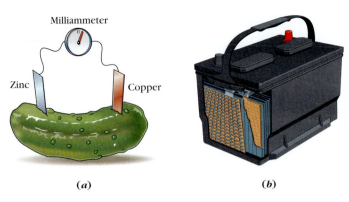

Milliammeter

Zinc   Copper

(a)   *(b)*

Figure 19-20 Two examples of batteries. (*a*) "Pickle power": The pickle's juice is an acid solution. The deflection of the meter indicates a small current between the dissimilar metal poles of this battery. (*b*) A 12.6-V lead-acid storage battery for an automobile.

degrees, the net negative charges left behind on the pieces of metal are different. Typically only one metal (like the copper) dissolves significantly in an acid solution. Random (thermal) motion carries some of the positive ions away from the dissolving pole. Then if a wire is connected between the two poles, mutual repulsion will cause the electrons left behind to flow to the other pole (Figure 19-20). Ions from the first pole that make it to this one will bind to its surplus electrons, thus adhering to and "plating" the second pole. (Copper plating is done in this way.)

➡**A note on language:** Originally a *battery* of cells referred to a number of cells used in combination. Today we commonly speak even of a single cell as a battery.

In some batteries, the regions near the two poles are separated by a mechanism like a porous barrier, which permits larger ions to pass less readily than smaller ions. If the positive and negative ions are of substantially different sizes, the differential flow of the two kinds of ions through the barrier will give rise to a difference in charge density across the barrier. In the same way, the fact that the membranes of living cells are not equally permeable to different ions can give rise to potential differences across cell membranes.

Figure 19-21 shows typical movements of charge carriers and the resulting conventional currents for a type of battery called a galvanic or voltaic cell. Note that although the charge carriers in the external wire are electrons, the carriers within the battery may be ions. Although the current may be equal everywhere in the circuit, the carriers comprising the current can be different in different parts of the circuit.

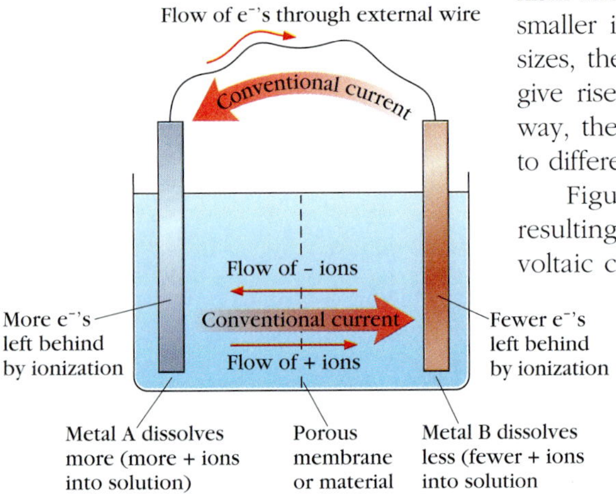

Flow of e⁻'s through external wire

Conventional current

Flow of – ions

More e⁻'s left behind by ionization   Conventional current   Fewer e⁻'s left behind by ionization

Flow of + ions

Metal A dissolves more (more + ions into solution)   Porous membrane or material   Metal B dissolves less (fewer + ions into solution)

Figure 19-21 Principle features in the functioning of a galvanic or voltaic cell.

The increase in PE from one pole to the other can be equated to an amount of work, commonly described as work due to *non-electrostatic forces*. Actually, electrostatic forces *are* indirectly involved because they govern chemical reactions, but no electrostatic forces due to external application of an electric field are involved.

✦SUMMARY✦

The electrostatic force between point charges q_1 and q_2 is described by Coulomb's law.

Coulomb's law: The *magnitude*

$$F = k\frac{|q_1||q_2|}{r_{12}^2} \quad (19\text{-}1)$$

$$(k = 9 \times 10^9 \text{ N} \cdot \text{m}^2/\text{C}^2)$$

The *direction* of the force on either body is toward the other body if the two have opposite charges, and away if they have like charges.

The force *per unit charge*, or **electric field** $\vec{E}$, is defined by $\vec{F}_{\text{on A}} = q_A \vec{E}$ and is a function of position. If a charged body A is placed at that position,

$$\underset{\substack{\text{strength of}\\\text{electrical}\\\text{interaction}\\\text{between A and its}\\\text{environment}}}{F_{\text{on A}}} = \underset{\substack{\text{property}\\\text{of A}}}{|q_A|} \quad \underset{\substack{\text{properties of}\\\text{environment}\\\text{of point where}\\\text{A is placed}}}{E} \quad (19\text{-}4)$$

Electric field at a position a distance r from a point charge q

$$\textit{Magnitude:} \quad E = k\frac{|q|}{r^2} \quad (19\text{-}7)$$

Direction: away from q if q positive, toward q if q negative

Because electric fields are vectors, finding the electric field at a point P due to an array of point charges (**Procedure 19-1**) involves vector addition.

For a *conductor in electrostatic equilibrium*,

- *there can be no electric field at any point in its interior;*
- *at all points on its surface, the electric field must be perpendicular to the surface.*

But if a conducting wire is not in electrostatic equilibrium, the interior field is in the direction of decreasing surface charge density (the direction of the negative *gradient*).

Electric field lines communicate information because they follow a few basic rules:

- *Rule 1:* The *direction* of the field line at any point (or of the tangent at that point if the field line is curved) is simply the *direction of the total field vector* at that point.
- *Rule 2:* The density of field lines (how closely packed they are) in a small region about a point is drawn to be proportional to the *magnitude of the total field vector* at that point.
- *Rule 3:* Field lines start and end at charge carriers (or else go to infinity). They are directed away from positive charge and towards negative charge.

Because Coulomb's law and Newton's universal law of gravitation have similar forms, we can treat many aspects of electrical systems like gravitational systems (and vice versa—we can treat $\vec{g}$ as a gravitational field, map gravitational fields, etc.). Using the correspondences $m \leftrightarrow q$, $g \leftrightarrow E$, and $G \leftrightarrow k$, we obtain

$$PE_{\text{elec}} = qEy \qquad (19\text{-}8)$$

*if $\vec{E}$ constant in the *negative y* direction*

and

$$PE_{\text{elec}} = +k\frac{q_1q_2}{r} \qquad (19\text{-}9)$$

for point charges a distance *r* apart

The electric potential energy per unit charge is called the **electric potential** or simply the **potential:**

$$\text{Potential} \quad \frac{PE_{\text{elec}}}{Q} = Ey \qquad (19\text{-}11)$$

(in a region of constant field in the $-y$ direction, that is, near an effectively infinite plane of charge or close enough to a uniformly charged surface so that it appears flat)

and

$$\text{Potential} \quad \frac{PE_{\text{elec}}}{Q} = \frac{kq}{r} \qquad (19\text{-}12)$$

(at a great enough distance *r* from a body with charge *q* so that the body may be regarded as a point object)

As is true of PE, only *differences* in potential have physical meaning. The **potential difference** or **voltage** between two points A and B is

$$V_{\text{AB}} \equiv \left(\frac{PE_{\text{elec}}}{Q}\right)_{\text{B}} - \left(\frac{PE_{\text{elec}}}{Q}\right)_{\text{A}} = \frac{\Delta PE_{\text{elec}}}{Q} \qquad (19\text{-}14)$$

$$\text{Units: } 1 \text{ \textbf{volt}} \equiv 1\frac{\text{joule}}{\text{coulomb}} \qquad (1 \text{ V} \equiv 1 \text{ J/C}) \quad (19\text{-}15)$$

The **electron volt** (eV), a unit of energy, is then defined as

$$1 \text{ eV} = (1.6 \times 10^{-19} \text{ C})(1 \text{ V}) = 1.6 \times 10^{-19} \text{ J} \quad (19\text{-}16)$$

In a constant field (taken to be in the negative *y* direction)

$$V_{\text{AB}} = E\Delta y \qquad (19\text{-}17)$$

$$(\Delta y = y_{\text{B}} - y_{\text{A}})$$

From this it follows that the electric field is the negative gradient of the potential. The table below summarizes the conditions in a conductor.

	Conditions		
	Total change	Rate of change (in direction of decrease)	Response of positive charge carriers to conditions (negative do the opposite)
Underlying picture	Surface charge density difference	Negative density gradient	Movement of positive charge in direction of negative gradient
Formal concept	Potential difference	Negative potential gradient = field	Conventional current in direction of field

Batteries establish potential differences by a combination of chemical means and mechanisms resulting in unequal flow of positive and negative ions.

The table below contains some important things to remember about the quantities we introduced in this chapter.

	Vector: adds by vector addition	Scalar: adds by scalar addition	
Total amount	Electrostatic force $\vec{F}$	Electric potential energy PE_{elec}	Work done moving charge a distance Δs in one dimension: $\Delta W = F\Delta s$
Amount per unit of charge	Electric field $\vec{E}$	Electric potential $\dfrac{PE_{\text{elec}}}{q}$	Potential difference between endpoints of Δs: $V_{\text{ab}} = \dfrac{\Delta W}{q} = \dfrac{F\Delta s}{q} = E\Delta s$

✦QUALITATIVE AND QUANTITATIVE PROBLEMS✦
Hands-On Activities and Discussion Questions

The questions and activities in this group are particularly suitable for in-class use.

19-1. Discussion Question.
Exploring an analogy: In Figure 19-22a, the paint from the painter's spray can spread out to board A in a cone-shaped region. If board A is removed, the cone will continue to spread to board B, twice as far from the nozzle of the can.

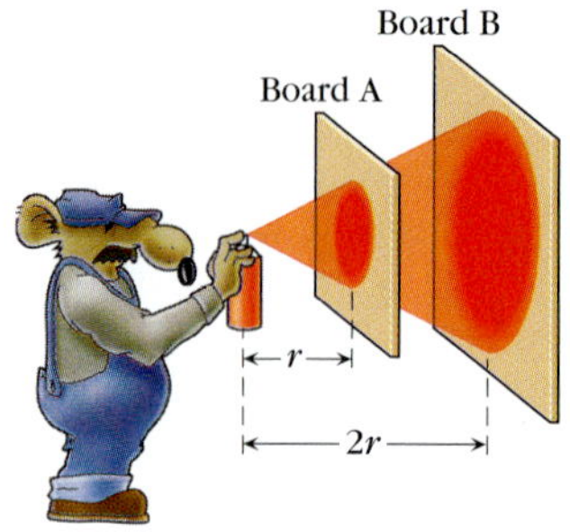

a. When board A is removed, how will the total amount of paint reaching board B each second compare to the total amount of paint that previously reached board A each second?

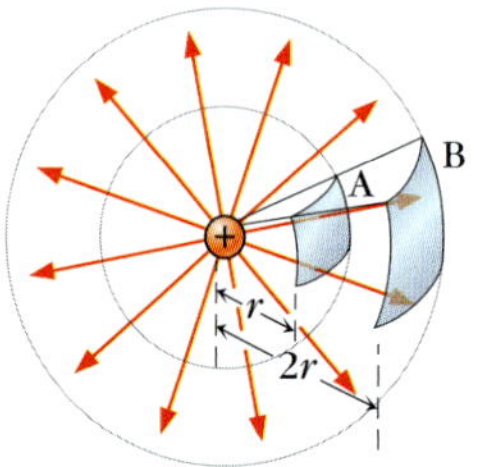

b. How will the area that gets painted on board B compare with the area that gets painted on board A? Express as a ratio $\frac{\text{area B}}{\text{area A}}$.

c. If each board got sprayed for 1 s, how will the thickness of the paint on the two boards (the amount of paint on each square centimeter of board) compare? Express as a ratio $\frac{\text{thickness B}}{\text{thickness A}}$.

d. Now compare the spray paint situation to the situation of electric field lines directed outward from a positive point

Figure 19-22
Problem 19-1

charge in Figure 19-22b. The field lines cross two imaginary spheres, both centered at the point charge. To which quantities in the paint situation can each of the following be compared: the areas of the two spheres? the total number of field lines crossing each sphere? the magnitude of the electric field? Briefly explain each.

e. How does the magnitude of the electric field compare at the surfaces of the two spheres? Express as a ratio $\frac{E \text{ at B}}{E \text{ at A}}$, and briefly explain your reasoning.

19-2. Discussion Question.
a. Other than in superconductors, can there ever be a current between two points A and B if the potential difference V_{AB} is zero?

b. Figure 18-8 shows the anode and cathode in a cathode ray tube as two parallel plates, with a hole in the negative cathode that permits electrons to pass through. Assume the plates are effectively infinite in size compared to the width of the gap between them. The electron is accelerated from the negative cathode to the positive anode. Why isn't the electron slowed down by attraction to the anode once it moves to the right of it?

c. Is there a potential difference between the anode and the screen?

d. Is there an electric field to the right of the anode? What does Equation 19-16 tell you about your answer to **c**?

e. Is there a current between the anode and the screen?

f. Are your answers to **c** and **e** consistent with your answer to **a**? If not, make any necessary corrections.

Review and Practice

Section 19-1 Making Electrostatic Force and Charge Concepts Quantitative

19-3. Two point charges exert an attractive force of 5.0 N on each other. How much force would they exert on each other if the distance between them were

a. ten times as great?

b. one-tenth as great?

c. How would your answers to **a** and **b** be altered, if at all, if the forces the charges exert on each other were repulsive rather than attractive?

19-4. How many electrons must be brought together to constitute a total charge (ignoring sign) of one coulomb?

19-5.
a. Find the electrostatic force that two electrons exert on each other when they are 1 mm apart.

b. Compare this with the gravitational force they exert on each other at the same separation ($m_e = 9.11 \times 10^{-31}$ kg). Express your comparison as a ratio $\frac{F_{\text{elec}}}{F_{\text{grav}}}$.

19-6. What must be the charge on each of two identical bodies if they are to exert a repulsive force of 1 N on each other at a separation of 5 cm?

19-7. **SSM WWW** A point object with a charge of 1.00×10^{-8} C is placed at the origin.

a. Where must a proton be positioned to experience a force of 1.44×10^{-15} N directed to the left?

b. Where must an electron be positioned to experience a force of 1.44×10^{-15} N directed to the left?

19-8. Three charged point objects A, B, and C are all placed along the x axis. A is at the origin, and B is at $x = 3$ cm. A and C have charges of 1.0×10^{-6} C; B's charge is 1.6×10^{-6} C. Find the magnitude and direction of the total electrostatic force on C if C is placed 1 cm **a.** to the left of B. **b.** to the right of B. **c.** to the left of A. **d.** to the right of A.

19-9. Suppose that in Problem 9-8, object C has a mass of 5.0×10^{-3} kg. What is its acceleration when it is positioned 1 cm to the left of A?

19-10. Charged point objects A and B are placed along the x axis at $x = -0.04$ m and $x = +0.04$ m, respectively. The magnitude of the charge on A is double that of B. Where on the x axis will a positively charged point object experience zero total force if **a.** both A and B are positive? **b.** A is positive and B is negative?

19-11. Repeat Problem 19-8 for the case in which A and B remain where they were but C is placed along the y axis at $y = 4$ cm.

19-12. At a certain instant, a proton approaching another proton has an acceleration of 4.0×10^5 m/s². What is its acceleration at a later distance when it is only half as far away?

Section 19-2 Electric Fields

19-13. Two charged point objects A and B are placed along the x axis. A is at the origin and B is at $x = 3$ cm. A's charge is 1.0×10^{-6} C, and B's charge is 1.6×10^{-6} C. Find the magnitude and direction of the total electric field due to these charges at a point 1 cm *a.* to the left of B. *b.* to the right of B. *c.* to the left of A. *d.* to the right of A.

19-14. Use the results of Problem 19-13 to find the force on a point object C placed 1 cm to the right of B if C has a charge of *a.* 1.0×10^{-6} C. *b.* -2.0×10^{-6} C.

19-15. SSM Consider charges A and B in Problem 19-13. *a.* Find the magnitude and direction of the total electric field due to these charges at the point ($x = 0, y = 4$ cm). *b.* Find the force on an electron placed at this point.

19-16. Find the magnitude and direction of the electric field at a certain point if a force of 8.0×10^{-11} N to the right is exerted on *a.* a sodium ion placed at that point. *b.* a chloride ion placed at that point.

19-17. Suppose an oil droplet in the Millikan experiment has two excess electrons on it. If the total non-electrostatic force on the droplet is 4.0×10^{-15} N, what electric field (magnitude and direction) must be applied if the droplet is to descend at constant velocity?

19-18. A 0.0050-kg ball of metal foil is suspended from an insulating thread. When a charged rod is held at a fixed distance directly to the left of the ball, the thread hangs at an angle of 15° to the right of vertical. Calculate the electrostatic force that the rod exerts on the ball.

19-19. SSM WWW For each of the following questions, either answer it or tell why it cannot be answered.
a. A point object has a charge of -2×10^{-18} C. Find the magnitude of the electric field at a position 1 m from this object if there is nothing at that position.
b. Repeat *a* for the case when there is an electron at the position in question.
c. Two point objects, each with a charge of 0.003 C, are positioned 0.40 m apart. Find the magnitude of the electric field.

19-20. Figure 19-23 shows the positions of two point objects with equal positive charges $+Q$. Four points in the vicinity of these objects are labeled A, B, C, and D.
a. Rank these points in order according to the *magnitude* of the total electric field at each point. Rank them from least to greatest, making sure to indicate any equalities.
b. Repeat *a* for the case when the charge at $x = 5$ cm is negative instead of positive.
c. Suppose now that the charge at $x = 1$ cm is negative and the charge at $x = 5$ cm is positive. If you listed the points

in rank order as in *a* and *b,* how would the order compare to the orderings you listed for those two situations?

19-21. A singly charged positive ion is located at the origin and a singly charged negative ion is located on the x axis at $x = 6.00 \times 10^{-3}$ m.
a. Find the electric field (magnitude and direction) at the point $y = 1.44 \times 10^{-2}$ m on the y axis.
b. Find the instantaneous acceleration (magnitude and direction) of an electron as it passes through this point.

Section 19-3 Fields Due to Continuous Charge Distributions

19-22. Consider a square loop of an imaginary material having the peculiar property that any charge placed on it remains where it is put. Suppose that a certain amount of positive charge is uniformly distributed over two adjacent sides of this square and an equal magnitude of negative charge is uniformly distributed over the other two sides. By symmetry arguments, determine the direction of the electric field at the center of the square.

19-23. Figure 19-24 shows part of a closed circuit with the wires connected to the terminals of the battery greatly enlarged. The diagram focuses on the surface charge density on four ring-shaped regions of the wire surfaces, although there is charge distributed over the remainder of the wire surfaces as well. Ring B has a greater density of positive charge than ring A; ring C has a greater density of negative charge than ring D. Point P_1 is midway between rings A and B; point P_2 is midway between rings C and D.
a. What is the direction of the electric field at point P_1 due to ring A alone? (Call this $\vec{E}_A$.)
b. What is the direction of the electric field at point P_1 due to ring B alone? (Call this $\vec{E}_B$.)
c. At point P_1, what is the direction of the total electric field $\vec{E}_A + \vec{E}_B$?
d. The rest of the surface charge distribution on the wire to the battery's left can be divided into rings of charge similar to A and B. What is the direction of the total field at point P_1 due to any one pair of rings? What is the direction of the total field at point P_1 due to all such pairs of rings combined?
e. What is the direction of the electric field at point P_2 due to ring C alone? (Call this $\vec{E}_C$.)
f. What is the direction of the electric field at point P_2 due to ring D alone? (Call this $\vec{E}_D$.)
g. At point P_2, what is the direction of the total electric field $\vec{E}_C + \vec{E}_D$?
h. The rest of the surface charge distribution on the wire to the battery's right can be divided into rings of charge similar to C and D. What is the direction of the total field at point P_2 due to any one pair of rings? What is the direction of the total field at point P_2 due to all such pairs of rings combined?
i. Write a sentence to describe the direction of the electric field along the entire closed loop of the circuit, external to the battery.

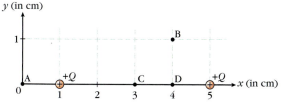

Figure 19-23 Problem 19-20

Figure 19-24 Problems 19-23 and 19-64

Section 19-4 Picturing Electric Fields

19-24. Figure 19-25 shows the electric field lines due to two point objects 1 and 2 having equal positive charges. If points A and B are equidistant from object 1, sketch the electric field *vectors* at these two points. (Adapted from P. Shaffer, G. Francis, and L. C. McDermott, paper abstracted in *AAPT Announcer*, 22, no. 2, p. 51 [1992].)

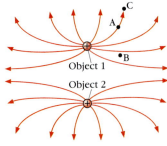

Figure 19-25 Problems 19-24 and 19-25

19-25. Figure 19-25 shows the electric field lines due to two point objects 1 and 2 having equal positive charges. Points A and B are equidistant from object 1.

a. Is the magnitude of the field at A less than, equal to, or greater than the field at B? Briefly explain.

b. If another positively charged point object is placed at point A and released from rest, will the object pass to the left of, to the right of, or directly through point C? Briefly explain.

19-26. Two tiny spheres with equal and opposite charges are separated by a distance much greater than their diameters. Carefully sketch the electric field lines representing the total field due to these charges.

19-27.

a. Carefully sketch the electric field lines due to a thin straight rod of length L with a charge $+Q$ distributed uniformly along it.

b. Repeat **a** for a similar rod with a charge $-2Q$ distributed uniformly along it. What are the significant differences?

19-28. Figure 19-26 shows a metal cylinder in both front and top views. A negative charge is placed on this cylinder and is allowed to reach electrostatic equilibrium.

a. Sketch what the electric field due to this cylinder looks like in the front view.

b. Sketch what the field looks like in the top view.

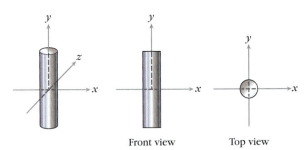

Front view   Top view

Figure 19-26 Problem 19-28

19-29. Figure 19-27 shows two conducting objects A and B. The two objects are charged so that the fields at their surfaces have equal magnitude. Suppose a small test charge A is placed at the center of object A and then moved 1 cm to the right. An identical test charge B is placed at the center of object B and then likewise moved 1 cm to the right. C marks the center

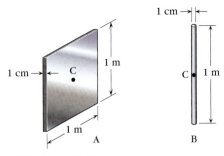

Figure 19-27 Problem 19-29

in each case. On which of the two test charges does the force change by a larger percent? Briefly explain.

Section 19-5 Electrical and Gravitational Systems: Similarities and Differences; Electrical Potential Energy

19-30.

a. In Figure 19-18b, in what part(s) of the electric circuit (if any) does the conventional current of positive charges flow oppositely to the direction of the electric field?

b. In Figure 19-18a, in what part(s) of the gravitational circuit (if any) does mass flow oppositely to the direction of the gravitational field?

c. Is it gravitational forces or other forces that causes mass to flow opposite to the gravitational field in **b**? Briefly explain.

d. Is it electrostatic (Coulomb) forces or other forces that causes conventional current to flow opposite to the electric field in **a**? Briefly explain.

19-31. Two electrons are a small distance apart. When we rearrange this system to increase its electrical potential energy, does the gravitational potential energy of the system increase, decrease, or remain the same? Briefly explain why.

19-32. When the distance between a proton and an electron is doubled, the force on the electron changes in magnitude from F_1 to F_2, and the potential energy of the two-particle system changes from PE_1 to PE_2. The final value of each quantity is a certain fraction its initial value. When we compare the two fractions, do we find that $\frac{PE_2}{PE_1}$ is less than, equal to, or greater than $\frac{F_2}{F_1}$? Briefly explain.

19-33. An atom of the most common isotope of hydrogen consists of a single proton and an electron. Suppose the electron starts out at a distance of 5.29×10^{-11} m from the proton, and after absorbing radiated energy it ends up at four times that distance. By how much (in J) does the potential energy of this atom change?

19-34. The electric field due to the uniform plane of charge in Figure 19-28 has a magnitude of 400 N/C. At a particular instant, a charged particle is in the position shown and has a speed of 800 m/s toward the plate.

a. If the charged particle is a singly charged positive ion and the

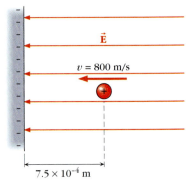

Figure 19-28 Problem 19-34

plane's charge is negative, find the change in potential energy as the particle travels from its present position to the point where it strikes the plane.

b. Find the change in kinetic energy.

c. If the ion has a mass of 2.0×10^{-25} kg, calculate the speed with which it strikes the plane.

19-35. SSM WWW

a. Calculate the speed with which the ion strikes the plane in Problem 19-34 if everything is the same except that the plane's charge is positive.

b. Calculate the speed with which the ion would strike the positively charged plane if the ion were negative.

19-36. Write a problem involving gravitational rather than electrostatic forces and fields that is completely analogous to Problem 19-34.

19-37.

a. Find the change in the electrical potential energy of a system of two electrons during an interval in which the distance between them decreases from 3.0×10^{-10} m to 2.0×10^{-10} m.

b. Find the change in the gravitational potential energy of the system over this same time interval.

19-38. The electric field between the two oppositely charged plates in Figure 19-29 has a magnitude of 1500 N/C.

a. If a proton is released from point P in the figure, which plate would it strike and at what speed?

b. If an electron is released from point P, which plate would it strike and at what speed?

6.0×10^{-4} m

P

2.0×10^{-4} m

Figure 19-29
Problem 19-38

19-39. Two protons are instantaneously at rest at a distance of 5.0×10^{-10} m from each other. If the protons are never affected by any objects except each other, what maximum speed will each of the protons reach? *Hint:* How do the speeds of the two protons compare at any instant? Why?

Section 19-6 Potential and Potential Differences

19-40. In Figure 19-30, the charge Q on the object in the center is 1.00×10^{-6} C. Calculate the change in potential in going

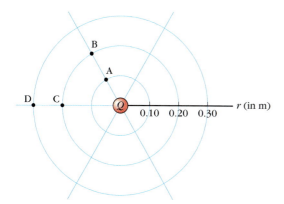

Figure 19-30 Problems 19-40 and 19-41

a. from A to B. **b.** from C to D. **c.** from A to D. Briefly tell what relationship there is between this last answer and your answers to **a** and **b**.

19-41. SSM In Figure 19-30, the charge Q on the object in the center is 1.00×10^{-6} C. In each of the following cases, calculate the change of potential energy and put a sign on each answer to indicate whether the change is an increase (+) or a decrease (−). (*Hint:* Think about whether you can use any of your results from Problem 19-40.)

a. A proton moves outward from A to B.

b. A proton moves outward from C to D.

c. An electron moves outward from C to D.

d. An electron moves inward from D to A.

19-42. The electric field due to a charged point object has a magnitude of 22.5 N/C at a distance of 0.020 m from the object.

a. Find the potential due to this object at the same distance.

b. At what distance would the electric field be doubled? At what distance would the potential be doubled?

19-43. Repeat Example 19-8 for the case in which a *proton* is released at the given point and reaches a point 0.020 m from its starting point.

19-44. When we calculate the electric field by $E = \frac{F}{q}$ in SI units, we obtain E in N/C. When we calculate the field from $V_{ab} = E\Delta y$ in SI units, we obtain E in V/m. Show that N/C and V/m are really the same units.

19-45. An electron in a hydrogen atom gains 10.2 eV of energy in going from its lowest energy level to its first excited state.

a. How much is this in joules?

b. How many electrons in a sample of hydrogen gas could be raised to their first excited states by an input of 1 J of energy?

19-46. How much potential difference must be established between two parallel plates if an electron leaving one of the two plates with negligible speed is to acquire a speed of 8.5×10^{4} m/s on reaching the opposite plate?

19-47. The potential difference between two parallel plates is 12 V. Find the magnitude of the electric field between the plates if the plates are 0.0025 m apart.

19-48. If the electric field in Example 19-8 is produced by two equally but oppositely charged plates separated by a distance of 0.10 m, what potential difference must be applied between the plates?

Section 19-7 Batteries

19-49. A student watches the instructor make a battery using two metal electrodes and an acidic solution. The student later tries to explain it to a friend who missed class. "If I take these two pieces of copper, I can use them as my electrodes. Then, when I insert them into opposite ends of this acid bath, I will be able to measure a potential difference between the ends of the electrodes that stick out of the bath." Will there be a potential difference for the set-up this student is describing? Briefly explain.

19-50. How does the random motion of molecules (thermal motion) contribute to the working of a battery?

19-51. The cell membrane of a living organism is not equally permeable to different kinds of ions. How might this fact give rise to a potential difference across the cell membrane?

Going Further

The questions and problems in this group are not organized by section heading, so you must determine for yourself which ideas apply. Some of them will be more challenging than the Review and Practice questions and problems (especially those marked with a • or ••).

19-52. A positive ion is positioned at the origin. An electron approaches it from the right. As the electron draws closer,
a. does its speed increase, decrease, or remain the same?
b. does its acceleration increase, decrease, or remain the same?

19-53. **SSM** Two spheres of equal size have charges +2 μC and +6 μC. Which of the diagrams in Figure 19-31 most accurately shows the electrical forces that the two spheres exert on each other?

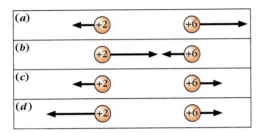

Figure 19-31 Problem 19-53

19-54. Figure 19-32 shows five points in a uniform electric field directed toward the right. Rank these points according to each of the following, listing them in order from least to greatest and indicating any equalities:

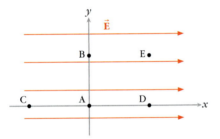

Figure 19-32 Problem 19-54

a. the magnitude of the electric field at each point.
b. the electric potential at each point.

19-55. **SSM** Figure 19-33 shows five points in the vicinity of an infinite plane of uniformly distributed positive charge. Rank these points according to each of the following, listing them in order from least to greatest and indicating any equalities:

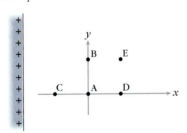

Figure 19-33 Problem 19-55

a. the magnitude of the electric field at each point.
b. the electric potential at each point.

19-56.
a. In Figure 19-34, all three charges have equal magnitude. If object A is instantaneously at rest when it is at the origin, in what direction will it move away from that point?

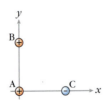

Figure 19-34 Problem 19-56

••b. If instead, object A passes through the origin with a velocity of 5.0 m/s to the left, estimate the direction in which it will be moving 1.0×10^{-3} s later.

19-57. The "leaves" of a certain electroscope are two very small metallic spheres of mass 3.0×10^{-3} kg suspended from 0.20 m lengths of fine conducting wire. When the spheres hang as shown in Figure 19-35, what is the magnitude of the charge on each of the spheres (assuming the two spheres are equally charged)?

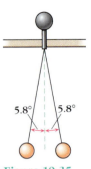

Figure 19-35
Problem 19-57

19-58. A charged sphere (mass 6.0×10^{-4} kg) suspended by a thread hangs at an angle of 3.6° to the left of the vertical when in the presence of a 650 N/C electric field directed to the right. Calculate the charge on the sphere.

19-59. A parrot is on a dry wooden perch in a metal birdcage. The cage is struck by lightning. The parrot is unharmed. Why?

19-60. Redraw Figure 19-5c for the case where both rings are negative but the lower ring is more negative. Sketch the field contribution due to each ring and the resultant field due to the combined effect of the two rings.

••19-61.
a. A uniformly charged circular ring with radius R has a total charge of Q. The ring is positioned perpendicular to the x-axis with its center at the origin (Figure 19-36). Show that at any position x along the positive x-axis, the magnitude of the electric field is

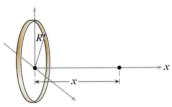

Figure 19-36 Problem 19-61

$$E = \frac{kQx}{(x^2 + R^2)^{3/2}}$$

State any assumptions you make in your reasoning.

b. Find the approximate value of E when x is so large that R is negligibly small in comparison. Can you suggest a reason why you might have expected this value?

•19-62. A point object with a small positive charge is placed between equally and oppositely charged plates. Assume the point object's charge has a negligible effect on the configuration of charge on the plates. Which of the sketches in Figure 19-37 most accurately shows possible electric field lines due to this arrangement of charge? Assume the plate dimensions are large compared to the distance between them, so that the plates extend beyond the top and bottom of the figure. *Note:* Only a few field lines are shown in each sketch.

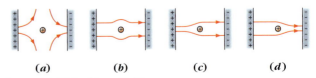

Figure 19-37 Problem 19-62

You might try including some of the others to see whether the sketch makes sense.

19-63. Student A says, "When I connect a bulb by two wires to the terminals of a battery, the electric field that causes charge to flow through the filament of the bulb is due entirely to the concentrations of positive and negative charge at the terminals of the battery. There are no concentrations of charge on the wire itself; they are not needed to cause the current. I picture it like this (Figure 19-38a)."

Student B says, "I don't agree. I think the battery will cause a change in surface charge density along the length of the wire, like this (Figure 19-38b). Most of the change in surface charge density takes place over the length of the filament, because that's where most of the resistance is."

The two circuits in Figure 19-38c provide us with a way of seeing which student's model is better. The circuits are identical, except that in circuit I, the bulb is very near one terminal of the battery, whereas in circuit II the bulb is fairly distant from either terminal.

a. How should the electric field in the bulb filament compare in the two circuits if we assume that student A's model is correct? How should the brightness of the bulb then compare in the two circuits? Briefly explain.

b. How should the electric field in the bulb filament compare in the two circuits if we assume that student B's model is correct? How should the brightness of the bulb then compare in the two circuits? Briefly explain.

c. In real circuits, does the bulb brightness depend on how far it is from the battery? (Check this experimentally if you're not sure.) Which conclusion agrees with what we actually observe, the one based on student A's model or the one based on student B's model? (Problem based on comments received in private communication from Bruce Sherwood.)

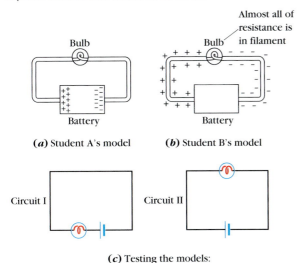

(a) Student A's model   (b) Student B's model

(c) Testing the models:

Figure 19-38 Problem 19-63

19-64.

a. In Figure 19-24, what is the direction of the electric field associated with electrostatic force inside the battery?

b. In what direction does conventional current flow through the battery?

c. Under what circumstances, if any, can the electric field associated with electrostatic forces and the conventional current be in opposite directions?

19-65. A certain wire has a right-angle bend in it. Along each straight segment, away from the bend, the surface charge density gradient is constant, so the field is constant along the axis of the wire. Roughly sketch the configuration of charge that is required in the region of the bend for the field lines to turn the corner in this region.

19-66. An electron is given an initial velocity parallel to a flat plate with a positive charge distributed uniformly over its surface (Figure 19-39). The distance d in the figure is very small compared to the dimensions of the plate.

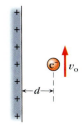

Figure 19-39 Problem 19-66

a. Sketch the trajectory of the electron after it is given its initial velocity. What is the shape of this trajectory? Why?

b. Describe a situation involving only gravitational forces that is analogous to this situation. Tell what feature of the gravitational situation corresponds to each feature of this situation. Does the moving object in the situation you describe follow the same shape trajectory as the electron?

19-67.

a. Make a list of the correspondences between features of the gravitational and electric circuits in Figure 19-18.

b. What aspects of the electric circuit do not have corresponding features in the gravitational circuit?

19-68. How can an electron be made to follow a parabolic trajectory in the presence of a negatively charged plate? Draw a sketch to clarify what you describe. Specify any conditions that you think are required.

19-69. Just as we can speak of non-electrostatic forces exerted on oppositely charged charge carriers in a battery, we can speak of the non-electrostatic force per unit of charge as a non-electrostatic field. Estimate the magnitude of the non-electrostatic field inside a standard 1.5 V D-cell battery and give its direction (toward which pole).

•**19-70.** A simple model of a hydrogen atom consists of an electron of charge $-e$ and mass m_e held in a circular orbit of radius r by a proton of charge $+e$ and mass m_p. Show that the total energy E of this arrangement is $E = -k\frac{e^2}{2r}$.

19-71. The DNA molecule, often described as a double helix, can be pictured as a twisted ladder. The two long strands that wind around each other are connected by bonds between pairs of bases, one base on each strand. These are the "rungs" of the ladder. People studying DNA often look at DNA fragments, which are themselves basically chains of these base pairs. DNA fragments in a neutral pH solution are negatively charged; there will be two excess electrons per base pair. If a 0.1 micron $(1.0 \times 10^{-7}$ m) DNA fragment has a charge of -9.0×10^{-17} C, how many base pairs are there along this fragment?

19-72. Hemoglobin is the protein molecule found in red blood cells. Under suitable conditions, the ratio of charge q to drag coefficient ζ for hemoglobin molecules in solution is $\frac{q}{\zeta} = -2.0 \times 10^{-10} \frac{\text{C·m}}{\text{N·s}}$. An electric field of 1000 N/C is applied to the solution in an electrophoresis apparatus.

a. What terminal speed is attained by the hemoglobin molecules?

b. About how far will the hemoglobin molecules travel in a day?

19-73. An electron passes a point where the electric field is directed toward the right and has a magnitude of 300 N/C. If the electron interacts only with the sources of this field, find the magnitude and direction of the electron's acceleration at the instant it passes through this point.

19-74. An electric field of 150 N/C is maintained between two equally and oppositely charged plates with a distance of 1.2×10^{-2} m between them. The region between the plates is evacuated, and the negative plate is sufficiently heated so that it occasionally emits electrons.
a. If an electron is emitted from the negative plate at negligible speed, with what speed does it reach the positive plate?
b. How long does it take the electron to reach the positive plate?

19-75. SSM A potential difference of 120 V is maintained between two equally and oppositely charged plates. The region between the plates is evacuated, and the negative plate is sufficiently heated so that it occasionally emits electrons. Answer whichever of the following questions it is possible to answer. If a question cannot be answered, explain why not.
a. If an electron is emitted from the negative plate at negligible speed, with what speed does it reach the positive plate?
b. How long does it take the electron to reach the positive plate?

19-76. Two lengths of identical resistance wire are each connected between the terminals of a 1.5 V D-cell battery. (The wires are said to be connected *in parallel.*) Length A is 25 cm long, and length B is 50 cm long.
a. How does the voltage between the ends of length A compare to the voltage between the ends of length B?
b. How does the electric field in length A compare to the electric field in length B?
c. How does the resistance of length A compare to the resistance of length B?
d. How does the resistivity of length A compare to the resistivity of length B?
e. How does the current through length A compare to the current through length B?

19-77. Two lengths of identical resistance wire are each connected between the terminals of a 9.0 V battery. Length A is 0.30 m long, and length B is 0.90 m long. Calculate the electric field in each wire.

19-78. We have spoken of setting up a uniform charge density gradient along the surface of a wire. Each of the choices below shows the surface charge density (in C/m^2) along the surface of a wire. Which one shows a uniform charge density gradient?

a.	2	2	2	2	0	−2	−2	−2	−2
b.	8	4	2	1	$\frac{1}{2}$	$\frac{1}{4}$	$\frac{1}{8}$	$\frac{1}{16}$	
c.	6	5	4	3	2	1	0	−1	−2
d.	2	2	2	2	2	2	2	2	2

19-79. We have spoken of setting up a uniform charge density gradient along the surface of a wire. In each of the graphs in Figure 19-40, surface charge density is plotted against distance along the wire. Which one of these graphs shows a uniform, nonzero charge density gradient?

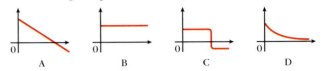

Surface charge density plotted vertically, distance along wire plotted horizontally

Figure 19-40 Problem 19-79

19-80.
a. From point B in Figure 19-9, in what direction would you travel to experience the potential increasing most rapidly with position?
b. The direction you found in *a* is defined to be the direction of the potential gradient at point B. How does the direction of the electric field at B compare with the direction of the potential gradient?

19-81. A proton and an electron are a distance *r* apart. If the distance *r* is increased, does the ratio $\frac{PE_{elec}}{PE_{grav}}$ increase, decrease, or remain the same? Briefly explain.

19-82. Figure 19-41 shows an infinite plane of uniformly distributed negative charge. The arrows in the figure show four possible paths that can be traveled in the vicinity of the plane. Each path starts at the tail of the arrow and ends at the head of the arrow. Rank these paths according to each of the following, listing them in order from least to greatest and indicating any equalities:
a. the change in *potential* from the beginning to the end of the path.
b. the change in *electrical potential energy* that would occur if an electron traveled from the beginning to the end of the path.

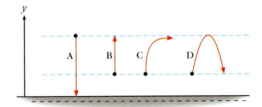

Figure 19-41 Problem 19-82

19-83. *Gravitational and electrical systems compared.* The electrical system in Figure 19-42*a* consists of two wires connected between the terminals of a battery. Wire 2 is twice as long as wire 1. In the gravitational system in Figure 19-42*b*, an object at point P can reach the ground by falling vertically downward

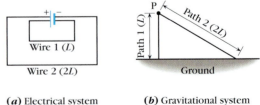

(a) Electrical system   **(b)** Gravitational system

Figure 19-42 Problem 19-83

(path 1) or sliding down a frictionless ramp (path 2). Path 2 is twice as long as path 1.

a. In the electrical system, suppose an electron travels from the negative to the positive pole of the battery. Compare the loss of potential energy when it travels by wire 1 and when it travels by wire 2.

b. In the electrical system, compare the change in potential between the two terminals of the battery along wire 1 and along wire 2.

c. In the electrical system, compare the electrical field along wire 1 to the electrical field along wire 2.

d. In the gravitational system, suppose an object descends from point P to the ground. Compare the loss of potential energy when it descends by path 1 and when it descends by path 2.

e. Compare the accelerations that the object would experience along the two paths.

f. In the gravitational system, compare the gravitational field along path 1 to the gravitational field along path 2.

g. Do the two systems behave analogously?

Problems on WebLinks

19-84. In WebLink 19-1, if we replace charge carrier A with a charge carrier with twice as much charge,

a. will the magnitude of the electric field at the origin increase, decrease, or remain the same?

b. will the magnitude of the force on the charge carrier at the origin increase, decrease, or remain the same?

19-85. In WebLink 19-1, if we replace charge carrier A with a charge carrier with a charge of opposite sign, will the electric field at the origin change direction? Briefly explain.

19-86. Suppose that when a sphere of radius 1 m is centered at the positive point charge in WebLink 19-2, 720 field lines in all cross the surface of the sphere. If this sphere is now replaced by a sphere of radius 2 m centered at the same point

charge, how many field lines in all cross the surface of the new sphere?

19-87. Suppose that when a sphere of radius 1 m is centered at the positive point charge in WebLink 19-2, 60 field lines cross each square unit of the sphere's surface. If this sphere is now replaced by a sphere of radius 2 m centered at the same point charge, how many field lines cross each square unit of the new sphere's surface?

19-88. In WebLink 19-3, does the electric field increase, decrease, or remain the same as you go from *a.* point A to point B? *b.* point B to a point very far to the right of point B? *c.* point A to a point in the interior of the charged plate?

CHAPTER TWENTY

Quantitative Treatment of Current and Circuit Elements

In Chapters 18 and 19, we constructed an underlying model based on charge carriers and the forces they exert on one another to explain the behavior we see. According to this model, wherever there is a surface charge density difference and consequently a potential difference between two points along a conducting path, an electric current will flow.

This is true even if there is not a complete and unbroken closed electrical conducting path. For example, when a capacitor is connected to a battery (Figure 20-1), the circuit is interrupted by the gap between the plates, but current still flows in the conducting parts until the capacitor is fully charged. At that stage, the build-up of charge evens

". . . wherever there is a surface charge density difference—and consequently a potential difference—between two points along a conducting path, an electric current will flow."

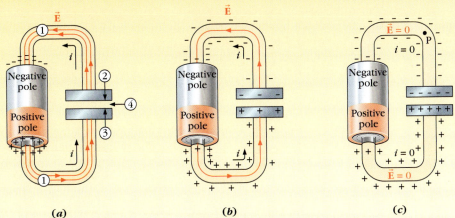

(a)   **(b)**   **(c)**

Figure 20-1 What happens when a capacitor is charged? (a) The circuit at $t = 0$, the instant when the wires are connected to the battery. Within the wires (①), electrons move oppositely to the field direction, which is equivalent to the conventional current i flowing in the field direction. The inner surface of one plate (②) starts to become negative as electrons repelled by the negative pole move toward it. The inner surface of the other plate (③) starts to become positive as electrons from this plate are attracted toward the positive pole. No electrons cross the gap between the plates (④). (b) The circuit slightly later. Why is the field in the wires weaker than at $t = 0$? (c) The circuit many seconds later. The surface charge density has become uniform between each battery pole and the plate connected to it. An electron in the wire, at a point such as P, experiences no net force.

out the charge distribution. With no more surface charge density gradient, the potential difference between each plate and the battery pole connected to it drops to zero, and current ceases to flow. Now positive charge that is pushed toward the positive plate by the positive pole of the battery is as strongly repelled by the concentration of positive charge already on the plate.

In an unbroken conducting path, such as a bulb and battery circuit, there is no charge build-up. The battery continues to maintain a surface charge density gradient, and thus a potential difference, so current continues to flow. In this chapter, you will learn to treat these ideas quantitatively.

20-1   Electric Current

A measure of flow is necessarily a *rate*—it must tell you how much of something is going by in each unit of time. If we envision an observer stationed at a point P along a conducting path (Figure 20-2a), the **electric current,** a measure of

➡**A note on language:** We use the word *current* to refer to both the process of flow and the quantitative measure I of the rate of flow. I might better be called the *current intensity,* as it is in French, but this is not usual usage in English.

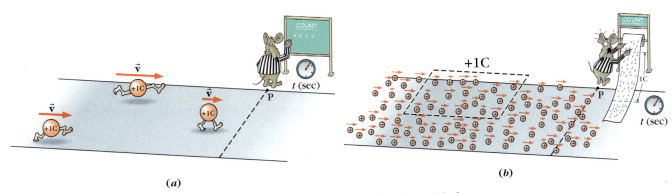

(a)   **(b)**

Figure 20-2 Making the measurements from which current is calculated . . . (a) if each coulomb came as a single lump. (b) when there are huge numbers of carriers, each with a tiny charge.

charge flow, tells you how much charge goes by the observer in each unit of time. If an additional amount of charge Δq goes by in each time interval Δt, then the average rate is $\frac{\Delta q}{\Delta t}$. We define this rate to be the *current,* and denote it by I:

$$\text{Electric current} \qquad I \equiv \frac{\Delta q}{\Delta t} \qquad\qquad (20\text{-}1)$$

In SI, charge is in coulombs and time in seconds, so current is in coulombs per second (C/s). Following the usual physics custom of giving derived units names that obscure their meaning, a coulomb per second is called an **ampere** (A) or "amp" after the French pioneer in the study of electricity André Marie Ampère (1775–1836):

$$1 \text{ ampere} \equiv 1 \text{ coulomb/second} \qquad (1 \text{ A} = 1 \text{ C/s}) \qquad (20\text{-}2)$$

Remember that current is an average. The charge carriers do not all move at the same speed. Nor do typical carriers, such as electrons, carry more than a minuscule fraction of a coulomb. So huge numbers of carriers must pass the observer in Figure 20-2*b* for an entire coulomb of charge to go by. To determine the current, the observer must (a) count carriers during a measured time interval and (b) know how much charge is carried by each carrier.

Example 20-1 *Current across a Cell Membrane*

An average of 50 sodium ions and 20 ferrous ions are transported in the same direction across a certain cell membrane each second. What is the total current across the membrane?

Solution

Choice of approach. Sodium ions are singly charged and ferrous ions are doubly charged. We must therefore find their contributions separately and then total them.

What we know/what we don't.

$$\text{number of sodium ions } \Delta N_{\text{Na}^+} = 50 \qquad q_{\text{Na}^+} = e = 1.60 \times 10^{-19} \text{ C}$$

$$\text{number of ferrous ions } \Delta N_{\text{Fe}^{++}} = 20 \qquad q_{\text{Fe}^{++}} = 2e = 3.20 \times 10^{-19} \text{ C}$$

$$\Delta t = 1 \text{ s}$$

The mathematical solution. The total charge going by in each additional second is

$$\Delta q = (\Delta N_{\text{Na}^+})(q_{\text{Na}^+}) + (\Delta N_{\text{Fe}^{++}})(q_{\text{Fe}^{++}})$$

$$= (50)(1.60 \times 10^{-19} \text{ C}) + (20)(3.20 \times 10^{-19} \text{ C}) = 1.44 \times 10^{-17} \text{ C}$$

The total *rate* is therefore

$$I = \frac{\Delta q}{\Delta t} = \frac{1.44 \times 10^{-17} \text{ C}}{1 \text{ s}} = \mathbf{1.44 \times 10^{-17} \text{ A}}$$

◆ Related homework: Problems 20-3, 20-4, and 20-6.

Case 20-1 ◆ *Case 18-2 Revisited: A Graphic Interpretation*

In Case 18-2, we saw that when a battery is connected through a bulb to a capacitor, as in Figure 18-15*a*, the bulb dims as the capacitor becomes charged. We asso-ciated the brightness of the bulb with the current. Now that we've defined current, we can interpret this graphically. Figure 20-3*a* shows the charge *q* on the

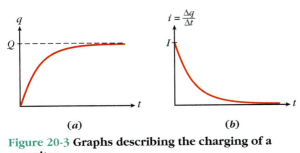

Figure 20-3 Graphs describing the charging of a capacitor.

capacitor building up to a final value Q (see note on notation in margin). Each additional bit of charge Δq that flows through the bulb becomes part of the accumulating total on the capacitor plate, so the current $\frac{\Delta q}{\Delta t}$, the rate at which the charge flows, is the same as the rate at which it accumulates on the capacitor. As the charge levels off, the slope of q versus t in Figure 20-3a is reduced to zero, which means that the rate of change—that is, the current—levels off to zero, as Figure 20-3b shows.

These pictures are still qualitative, because we don't yet know what numbers to put on the axes. How much charge accumulates on the capacitor depends not just on the battery but on the properties of the capacitor that determine its ability to hold charge. How quickly the charge can flow through the bulb depends on the properties of the filament that determine its resistance. We examine these two important characteristics of electric circuit components in the next two sections.

➡**Notation:** When electrical quantities vary with time, we will in general represent their instantaneous values by lowercase letters and their maximum values by uppercase letters.

20-2 Characteristics of Circuit Components I: Capacitance

If, in the circuit in Cases 18-2 and 20-1, we use two batteries end on end, the glow of the bulb appears brighter and remains visible longer (in part because if it is brighter at each instant, it will remain longer at above the minimum brightness that your eye can detect). This indicates that more total charge $\pm Q$ flows to and accumulates on either plate of the capacitor if we double the pumping ability of the batteries—that is, we double the potential difference V that they maintain. In fact the magnitude Q of the charge stored on either plate of the capacitor is proportional to the voltage V across it:

$$Q = \text{constant} \times V$$

This constant is called the **capacitance** (C) of the capacitor, so that $Q = CV$, or

$$C \equiv \frac{Q}{V} \tag{20-3}$$

The capacitance tells you how many coulombs of charge a capacitor can hold for each volt of potential difference between its plates; it is a measure of the capacitor's ability to store charge. The units of coulombs per volt are called **farads** (F) after Michael Faraday (1791–1867), a largely self-educated Englishman who from humble beginnings became a major investigator into electric and magnetic phenomena and one of the giants of nineteenth-century physics.

$$1 \text{ farad} = 1\frac{\text{coulomb}}{\text{volt}} \quad \left(1 \text{ F} = 1\frac{\text{C}}{\text{V}}\right) \tag{20-4}$$

Farads are very large units of capacitance. Commercial capacitors commonly range from microfarads (μF $= 10^{-6}$ F) to millifarads (mF $= 10^{-3}$ F), but in recent years new materials technology has made possible the production of low-cost capacitors of up to 1 F or more. These have made possible the observations we discussed in Cases 18-2 and 20-1; with a capacitor holding only a thousandth as much charge, the flow of charge to the plates would be so reduced that there would never be a visible glow of the bulb filament as evidence of what is happening in the circuit.

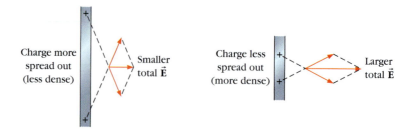
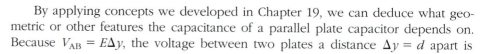

Figure 20-4 How the field depends on the charge density on each plate.

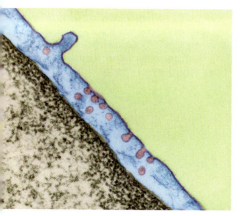

The cell membrane serves as a capacitor in the body's circuitry. The high level of phospholipid molecules in cell membranes makes them relatively impermeable to ions. Therefore, if a charge separation occurs across the cell membrane, it tends to remain separated. We can think of the phospholipid-rich membrane (blue) as the dielectric between the membrane's inner and outer surfaces.

By applying concepts we developed in Chapter 19, we can deduce what geometric or other features the capacitance of a parallel plate capacitor depends on. Because $V_{AB} = E\Delta y$, the voltage between two plates a distance $\Delta y = d$ apart is

$$V = Ed \tag{20-5}$$

The field E between the plates is uniform (recall Example 19-5 and Figure 19-10e). Figure 20-4 shows that the total field—a vector sum—depends on how spread out the charge is on the plates. The field therefore increases as the **charge density** or charge per unit area $\frac{Q}{A}$ on each plate increases. It can be shown by Gauss's law to be directly proportional to $\frac{Q}{A}$: $E = $ (constant)$(\frac{Q}{A})$. The constant of proportionality is commonly written as $\frac{1}{\epsilon}$, and the value of ϵ (a lowercase epsilon) depends on what medium—air, glass, and so on—fills the space between the plates. Thus

$$E = \frac{1}{\epsilon}\frac{Q}{A}, \quad \text{so that} \quad V = Ed = \frac{1}{\epsilon}\frac{Qd}{A}$$

and

$$C = \frac{Q}{V} = \frac{A\epsilon}{d} \tag{20-6}$$

The capacitance thus depends on the geometry of the capacitor (here described by A and d) and the medium in the gap (Figure 20-5). This is true for all capacitors, though the specific equation of relationship may be different for a different geometric arrangement, such as two oppositely charged concentric cylinders. **STOP&Think** How is the capacitance affected by doubling A? By doubling d? ◆ As Equation 20-6 shows, the larger the constant ϵ, the more charge the medium *permits* the capacitor to store; ϵ is therefore called the **permittivity** of the medium.

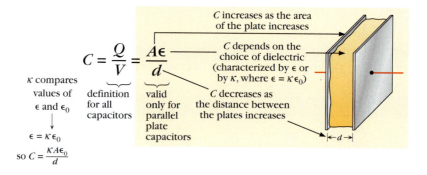

Figure 20-5 Summary of quantities affecting capacitance.

◆**WHAT DETERMINES THE VALUE OF ϵ FOR A MEDIUM?** The nonconducting medium between the plates, commonly called the **dielectric,** may be polarizable (recall Section 18-2 on polarization) to various degrees. As Figure 20-6 shows, the dielectric polarizes in response to the charges on the plates. This results in a thin layer at each surface of the dielectric (highlighted in orange) that has a net charge opposite to (and smaller in magnitude than) that on the plate closest to it. These two layers act like plates within plates. Like the capacitor plates, they produce a uniform field in the region between. This field is weaker than and opposite in direction to the field due to the capacitor plates. The result is a smaller *net* field between the plates, corresponding to a larger ϵ (because $E = \frac{1}{\epsilon}\frac{Q}{A}$).

If the potential difference between the plates is established by an external source, such as a battery, the net field must be the same as when there is no dielectric, because $V = Ed$ and neither V nor d is changed. Mathematically, because $V = \frac{Q}{C}$, and $C = \frac{A\epsilon}{d}$, it follows that $V = \frac{Qd}{A\epsilon}$, so that if ϵ increases Q must correspondingly increase.

The value of ϵ in a vacuum is denoted by ϵ_o and is called the **permittivity of free space**:

$$\epsilon_o = 8.85 \times 10^{-12} \text{ C}^2/\text{N} \cdot \text{m}^2 \qquad (20\text{-}7)$$

The capacitance with a vacuum (or, approximately, with air) between the plates is then $C_o = \frac{A\epsilon_o}{d}$. The ratio $\frac{C}{C_o} = \frac{\epsilon}{\epsilon_o}$ is called the **dielectric constant** and is denoted by κ (lowercase kappa):

$$\kappa \equiv \frac{\epsilon}{\epsilon_o} \qquad (20\text{-}8)$$

so that

$$C = \kappa C_o$$

Note in Table 20-1 that the dielectric constant is exactly 1 for a vacuum (why?) and higher for other materials.

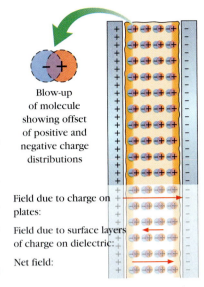

Blow-up of molecule showing offset of positive and negative charge distributions

Field due to charge on plates:

Field due to surface layers of charge on dielectric:

Net field:

Figure 20-6 The polarization of the dielectric contributes to the total electric field between the plates.

Table 20-1: Dielectric Constants κ and Dielectric Strengths of Various Materials

Material	κ	Dielectric Strength (V/m)
Vacuum	1.0	∞
Dry air (at 1 atm)	1.00059	3×10^6
Teflon	2.1	6×10^7
Nylon	3.4	1.4×10^7
Paper (typical)	3.7	1.6×10^7
Glass (various)	5–10	1.3×10^7

The dielectric strength is the value of the electric field above which the dielectric breaks down and begins to conduct. When this happens in air on a small scale, we see a spark; when it happens on a large scale, we see lightning.

Example 20-2 *How Capacitance Must Vary with Voltage to Hold a Fixed Charge*

If a 5.00 mF capacitor holds a certain charge when connected to a 9.00 V battery, what should be the terminal voltage that you would connect to a 120 mF capacitor if you wished it to hold the same amount of charge (assume you are not limited to off-the-shelf batteries)?

Solution

Choice of approach. Because $C = \frac{Q}{V}$, or $V = \frac{Q}{C}$, the capacitance and voltage are inversely proportional to each other if the charge is kept constant. From either equation, we obtain $Q = CV$. Because Q is fixed, $Q = C_1V_1$ for capacitor 1, and $Q = C_2V_2$ for capacitor 2. Therefore $C_1V_1 = C_2V_2$. For the product to remain constant, as either C or V increases the other must decrease. **STOP&Think** Do we need to know the value of Q? ◆

What we know/what we don't.

Capacitor 1	Capacitor 2
$C_1 = 5.00$ mF $V_1 = 9.00$ V	$C_2 = 120$ mF $V_2 = ?$

EXAMPLE 20-2 *continued*

The mathematical solution. From $C_1V_1 = C_2V_2$, we obtain

$$V_2 = \frac{C_1V_1}{C_2} = \frac{(5.00 \text{ mF})(9.00 \text{ V})}{120 \text{ mF}} = \textbf{0.375 V}$$

Note that because millifarads cancel in the numerator and denominator, it was unnecessary to convert to SI units of farads.

◆ Related homework: Problems 20-13, 20-16, and 20-17.

Example 20-3 *A Student-Made Capacitor*

For a guided interactive solution, go to Web Example 20-3 at
www.wiley.com/college/touger

You have made a capacitor by sandwiching a sheet of paper between two sheets of metal foil, and then cutting out a square of "sandwich" 2 cm on a side. The paper comes from a pack of 500 sheets. The pack is 3.20 cm thick. Your meter tells you that the terminal voltage of a fresh "1.5 V" D-cell battery is actually 1.54 V. How much charge will each plate of your capacitor hold when connected to this battery?

Brief Solution
Choice of approach. (1) We first need to determine the capacitance of the capacitors from the materials and geometry. (2) We can then use the definition of capacitance to find the charge it can hold at a given voltage.

What we know/what we don't.

Step 1	Step 2
$A = (2.00 \times 10^{-2} \text{ m})^2 = 4.00 \times 10^{-4} \text{ m}^2$	$C = $ [from step (1)]
$d = \dfrac{3.20 \text{ cm}}{500 \text{ sheets}} \cdot \dfrac{1 \text{ m}}{100 \text{ cm}} = 6.40 \times 10^{-5} \text{ m}$	$V = 1.54 \text{ V} \quad Q = ?$
$\epsilon_{paper} = ? \quad \epsilon_o = 8.85 \times 10^{-12} \text{ C}^2/\text{N} \cdot \text{m}^2$	
$\kappa_{paper} = 3.7$ (from tables)	

The mathematical solution.

Step 1: $C = \dfrac{A\epsilon}{d} = \dfrac{A\kappa\epsilon_o}{d}$

$$= \frac{(4.00 \times 10^{-4} \text{ m}^2)(3.7)(8.85 \times 10^{-12} \text{ C}^2/\text{N} \cdot \text{m}^2)}{6.40 \times 10^{-5} \text{m}} = 2.05 \times 10^{-10} \text{ F}$$

Step 2: $Q = CV = (2.05 \times 10^{-10} \text{ F})(1.54 \text{ V}) = \textbf{3.16} \times \textbf{10}^{\textbf{-10}} \textbf{ C}$

◆ Related homework: Problems 20-19 through 20-22.

20-3 Characteristics of Circuit Components II: Resistance

When a current flows along a conducting path, the charge carriers (e.g., electrons) are typically contributed in great numbers by every piece of the conducting path and, as we shall show shortly, move along the path at very slow drift speeds. To determine the current in a conductor, such as a wire, we must know something about each of the following:

1. the number of carriers n contributed by each unit of volume of conducting material (i.e., the carrier density) along the path,

2. the charge carried by each carrier ($-e$ for electrons and $+e$ for the equivalent charge carriers making up a conventional current in the opposite direction; when the carriers are ions, we would have to allow for charges of magnitude $2e$ or $3e$ as well), and

3. the drift speed v_d (which, as we will see, depends on the voltage).

Suppose an observer at point P in a conducting wire counts the charge carriers that move left to right (Figure 20-7) across the cross-sectional area (perpendicular to the axis) at P. In a time interval Δt, a typical charge carrier can travel a distance $v_d \Delta t$. If we simplify our reasoning by assuming all carriers move at this average speed, then all the carriers in the region extending a distance $v_d \Delta t$ to the left of P will pass P in an interval Δt. The volume of this region (cross-sectional area × length) is $A v_d \Delta t$. Then just as total mass = mass density × volume (Equation 10-2),

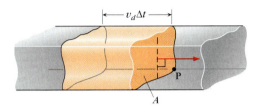

Figure 20-7 Volume of charge moving across area A in interval Δt.

$$\text{total number of carriers} = \text{carrier density} \times \text{volume} = nAv_d\Delta t$$

Multiplying the number of carriers by the charge e of each carrier gives the total amount of charge passing P during an interval Δt:

$$\Delta q = neAv_d\Delta t$$

and thus the current

$$I \equiv \frac{\Delta q}{\Delta t} = neAv_d \tag{20-9}$$

How do we know n? This requires some knowledge from chemistry. Take copper (Cu) as an example. Its density is about 8900 kg/m³. Because the atomic mass of Cu is 63.5, there are 63.5 g in a mole of Cu, and there are always Avogadro's number (6.02×10^{23}) of atoms in a mole. Each Cu atom contributes one outermost or conduction electron—one carrier. Putting all this together just requires unit conversion—repeated multiplication by one in suitable form:

$$n = \frac{8900 \text{ kg}}{\text{m}^3}\left(\frac{1000 \text{ g}}{\text{kg}}\right)\left(\frac{1 \text{ mole}}{63.5 \text{ g}}\right)\left(\frac{6.02 \times 10^{23} \text{ atoms}}{1 \text{ mole}}\right)\left(\frac{1 \text{ carrier}}{\text{atom}}\right)$$

$$= 8.44 \times 10^{28} \text{ carriers/m}^3 \text{ (conduction electrons/m}^3\text{)}$$

This is a typical value for a good conductor.

How do we know v_d? It is common for the #12 copper wire in home wiring to carry a current of 10 amps. Standard tables tell us that #12 wire has a diameter of 2.05 mm = 2.05×10^{-3} m, so its cross-sectional area is $A = \pi r^2 = \pi(\frac{d}{2})^2 = 3.31 \times 10^{-6}$ m². Then using n from the above calculation, Equation 20-9 gives us

$$v_d = \frac{I}{neA} = \frac{10 \text{ A}}{(8.44 \times 10^{28} \text{ electrons/m}^3)(1.60 \times 10^{-19} C/\text{electron})(3.31 \times 10^{-6} \text{ m}^2)}$$

$$= 2.24 \times 10^{-4} \text{ m/s (about } \tfrac{1}{4} \text{ mm per second)}$$

The drift speed is extremely slow. But the movement of charges in one region of wire alters the charge distribution, thus altering the field and causing a like movement of charge further along the wire (see Problem 18-4). Although the electrons themselves move slowly, this *signal* moves almost at the speed of light. Moreover, even at this very low speed, huge numbers of electrons will pass any point along the wire each second, as the following example shows.

Example 20-4   *A Slow-Moving Throng*

For a guided interactive solution, go to Web Example 20-4 at www.wiley.com/college/touger

A circuit breaker allows up to 10 A of current before interrupting the circuit to prevent overheating. What maximum number of electrons pass any given point along the circuit each second?

Brief Solution

Choice of approach. This involves understanding clearly what we mean by a current. Current is defined by $I \equiv \frac{\Delta q}{\Delta t}$. The charge Δq going by in an interval Δt is just the charge of one electron multiplied by the number of electrons ΔN going by in that interval: $\Delta q = e \Delta N$.

What we know/what we don't.

$$I = 10.0 \text{ A} \qquad \Delta t = 1.00 \text{ s}$$
$$e = 1.60 \times 10^{-19} \text{ C} \qquad \Delta q = ? \qquad \Delta N = ?$$

The mathematical solution. From the definition of current, $\Delta q = I\Delta t$. But Δq also equals $e\Delta N$. Equating the two expressions for Δq and solving for ΔN gives us

$$\Delta N = \frac{I\Delta t}{e} = \frac{(10.0 \text{ A})(1.00 \text{ s})}{1.60 \times 10^{-19} \text{ C/electron}} = \textbf{6.25} \times \textbf{10}^{19} \textbf{ electrons}$$

(a huge number).

◆ Related homework: Problems 20-24 and 20-25.

In still air, the molecules move randomly in all directions. If there is a breeze in one direction, the molecules on average have an extra bit of velocity in the direction of the breeze. The magnitude of this extra velocity is their drift speed. Electrons acquire a net drift speed (somewhat like a wind speed for the electron gas) as they are accelerated by the electric field and slowed down by collisions with "objects" they encounter in their paths. These "objects," which are collectively responsible for the resistance of the conducting material, are in fact any irregularities or deviations from perfectly periodic structure in the conductor. If the lattice structure of the conductor were perfectly periodic—the ions always perfectly evenly spaced—the electrons would encounter *no* resistance, although showing this requires understanding quantum aspects of electron behavior that we will not treat in this course.

Two types of irregularities contribute to the total resistance:

1. impurities, that is, occasional atoms or ions of other elements (the potential will be different at a different ion, so the environment experienced by an electron doesn't precisely repeat at an impurity); and

2. any displacement of atoms from perfectly repeated spacing. These include not only permanent structural flaws but also the displacements from their equilibrium positions that the ions have at any instant because of their thermal jiggling

motion. On average, these displacements become larger with increasing temperature. This model of a thermal contribution to the resistance fits with the observation that, with rare exceptions, the resistance of a metal wire increases when the wire is heated.

We can form a very simplified picture of what happens to conduction electrons in wires by thinking about a simple gravitational analog in which marbles roll down a ramp studded with nails (Figure 20-8). Because they experience repeated collisions, the marbles don't accelerate continuously, but reach an average drift speed. On average, the marbles roll down faster when the ramp is steeper and the field $g \sin \phi$ along the ramp is correspondingly greater. Similarly, the drift speed acquired by electrons increases with the electric field applied and is often proportional to it ($v_d = cE$, where c is often constant).

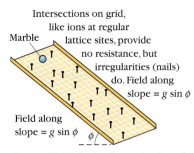

Intersections on grid, like ions at regular lattice sites, provide no resistance, but irregularities (nails) do. Field along slope = $g \sin \phi$

Field along slope = $g \sin \phi$

Figure 20-8 Limited gravitational analog to resistance encountered by electrons in a wire.

◆**RESISTANCE** Between two points A and B a distance $\Delta y = L$ apart, the potential difference $V_{AB} = E\Delta y = EL$. Putting this together with the fact that $I = nev_d A$, where $v_d = cE$, a little algebra (see Problem 20-71) leads to the result

$$V_{AB} = I\left(\frac{1}{nec}\frac{L}{A}\right) \qquad (20\text{-}10)$$

For many materials (but not all—see Figure 20-9), the quantity in parentheses does not change when the voltage V_{AB} is varied. For these materials, Equation 20-10 says that V_{AB} is proportional to I, and the expression in parentheses is the proportionality constant. Equation 20-10 tells us that for a given potential difference V_{AB}, as this constant increases, the current I must decrease. Because this is the behavior that we associate with resistance, we label the constant R and *define* it to be the **resistance.** Then

$$V = IR \qquad \text{or} \qquad \frac{V}{I} = R \qquad (20\text{-}11)$$

where

$$R = \frac{1}{nec}\frac{L}{A}$$

Materials for which R is constant if you keep L and A constant are called *ohmic* materials, and the relationship $V = IR$ is then called **Ohm's law** after German physicist Georg Simon Ohm (1789–1854), an early investigator of conduction in metals. (If the current and voltage are varying, we write $v = iR$. The lowercase letters denote their instantaneous values; the uppercase letters denote maximum values.) Ohm's law is not really a law; it is a mathematical pattern found in the behavior of many but not all materials. Ohm's law holds true only for *ohmic* materials.

When V is in volts and I is in amps, the units of resistance are **ohms** (denoted by Ω, an uppercase omega):

$$1\,\Omega \equiv \frac{1\,\text{V}}{1\,\text{A}} \qquad (20\text{-}12)$$

Because we won't use all the details of $R = \frac{1}{nec}\frac{L}{A}$, we write it in shorthand as

$$R = \rho\frac{L}{A} \qquad (20\text{-}13)$$

(ρ is the Greek letter rho.) Equation 20-13 tells us that the resistance of a sample of material depends on the dimensions of the sample (L and A). The quantities making up ρ depend only on the characteristics of the material itself. They are said to be *intrinsic* to the material; that is, they are the same for all samples of the material, regardless of their size or shape. Therefore we can produce a

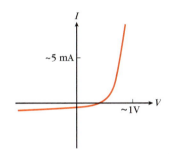

Figure 20-9 _I_ versus _V_ for a diode (a non-ohmic device). The behavior of such devices is said to be *nonlinear*.

➥**Important:** In accordance with usual usage, we have dropped the subscript AB on the potential difference. Without the subscript, you must *remember* that the potential difference is a difference between the potentials at two points, and you must be able to identify the two points (in this case, the endpoints of length L). When potential differences between several pairs of points are being considered, you should reintroduce the subscripts.

standard table of values of ρ for different materials (Table 20-2). To distinguish it from resistance, ρ is called the **resistivity.**

Resistivity is a general property of a material; resistance is a property of a particular piece of material with a particular size and shape.

Table 20-2: Room Temperature Resistivities of Selected Materials

Color code for "charge carrier traffic": Green for good conductors, Amber for semiconductors, and Red for good insulators

Material	ρ ($\Omega \cdot$ m)	at T (°C)
Silver	1.586×10^{-8}	20
Copper	1.678×10^{-8}	20
Gold	2.24×10^{-8}	20
Aluminum (99.996%)	2.655×10^{-8}	20
Tungsten	5.65×10^{-8}	27
Constantan*	48×10^{-8}	0
Nichrome*	150×10^{-8}	20
Carbon (graphite)	1.375×10^{3}	0
Germanium	4.6×10^{-1}	22
Silicon	3×10^{6}	0
Glass (various types)	10^{7}–10^{14}	20
Lucite	$>10^{13}$	20
Amber	5×10^{14}	20
Quartz (fused)	7.5×10^{17}	20

*Alloys. Constantan is a copper-nickel alloy commonly used in thermocouples. Nichrome is a nickel-chromium alloy used extensively in heating elements. Other values for metals and semiconductors are for pure elements unless otherwise indicated. Even a sample that is 99.5% pure can have substantially altered values because impurity sites are important contributors to resistivity.

STOP&Think According to Equation 20-13, how does the resistance of a sample of material change if you double its length? If you double its area? ◆

For **WebLink 20-1:**
Resistance in Wires and Water Drains, go to
www.wiley.com/college/touger

The analogy in Figure 20-10 is intended to give you a feel for what Equation 20-13 tells us about resistance. To develop this qualitative understanding more fully, work through WebLink 20-1 and see Problem 20-1.

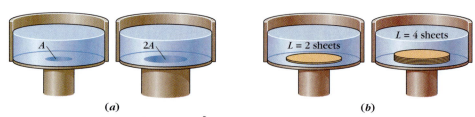

(a)   *(b)*

Figure 20-10 Understanding $R = \rho \frac{L}{A}$ by analogy. Here the rate of water flow (Δ volume/Δt) is like the rate of charge flow (current). In (*a*) the water flows more slowly down the drain with the smaller cross-sectional area. Because A is in the denominator, a smaller A means greater resistance to flow. In (*b*) the water drains more slowly where it must travel a greater distance L through the stack of filter papers; a larger L means greater resistance to flow.

Example 20-5 *Finding the Resistance of a Sample of Wire*

For a guided interactive solution, go to Web Example 20-5 at
www.wiley.com/college/touger

At room temperature (20°C) the resistivity of nichrome, a nickel-chromium alloy, is $1.00 \times 10^{-6} \, \Omega \cdot m$. What is the resistance of a 50-cm length of #18 nichrome wire? (#18 wire has a diameter of 1.024 mm.)

Brief Solution

Choice of approach. We apply Equation 20-13 directly, being careful to express all quantities in SI units.

What we know/what we don't.

$$\rho = 1.00 \times 10^{-6} \, \Omega \cdot m \qquad L = 0.50 \, m \qquad A = \pi r^2 = \pi \left(\frac{d}{2}\right)^2$$

$$d = 1.024 \times 10^{-3} \, m \qquad R = ?$$

The mathematical solution. The area

$$A = \pi r^2 = \pi \left(\frac{d}{2}\right)^2 = (3.14)\left(\frac{1.024 \times 10^{-3} \, m}{2}\right)^2 = 8.23 \times 10^{-7} \, m^2$$

Then $\quad R = \dfrac{\rho L}{A} = \dfrac{(1.00 \times 10^{-6} \, \Omega \cdot m)(0.50 \, m)}{(8.23 \times 10^{-7} \, m^2)} = \textbf{0.61 } \boldsymbol{\Omega}$

Making sense of the results. Although the resistivity is a very small number in SI units, the resistance of an ordinary sized sample of nichrome wire is not, because for a typical wire such as this, the cross-sectional area—the denominator—is also a very small number in SI units.

◆ Related homework: Problems 20-26, 20-27, 20-28, and 20-29.

Example 20-6 *Going to Great Lengths*

Because wires overheat when they carry too much current, you would like to select a length of #18 nichrome wire that will allow only 0.10 A of current to flow when there is a 3.0 V potential difference between its ends. How long a wire do you need?

Solution

Choice of approach. (1) By Ohm's law, current decreases as resistance increases (find R). (2) Resistance in turn depends on the length of the wire (use Equation 20-13 to find L).

What we know/what we don't.

Step 1		Step 2

$V = 3.0 \, V \qquad I = 0.10 \, A \qquad R = ? \qquad\qquad \rho = 1.00 \times 10^{-6} \, \Omega \cdot m$ (from Ex. 20-5)

$$A = 8.23 \times 10^{-7} \, m^2 \text{ (from Ex. 20-5)}$$

$$R = \text{value from step (1)} \qquad L = ?$$

The mathematical solution. Step 1: From Ohm's law $(V = IR)$,

$$R = \frac{V}{I} = \frac{3.0 \, V}{0.10 \, A} = 30 \, \Omega$$

Step 2: From $R = \frac{\rho L}{A}$, it follows that

$$L = \frac{RA}{\rho} = \frac{(30 \, \Omega)(8.23 \times 10^{-7} \, m^2)}{1.00 \times 10^{-6} \, W \cdot m} = \textbf{24.7 m}$$

EXAMPLE 20-6 *continued*

Making sense of the results. Equation 20-13 says a wire's resistance is proportional to its length. You can compare the above result to the result of the previous example to verify that $\frac{R_2}{R_1} = \frac{L_2}{L_1}$.

◆ Related homework: Problems 20-30 and 20-31.

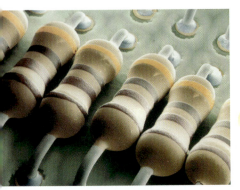

Figure 20-11 Some typical resistors.

➡**Caution:** The peculiar abbreviation EMF occurs because it was traditionally called the electromotive *force,* but that is a misnomer—it is *not* a force but the amount of energy per unit charge that the battery mechanisms provide.

◆**RESISTORS** An object specifically designed to provide resistance along a circuit path and thus control the amount of current is called a **resistor.** Some typical resistors are shown in Figure 20-11. The standard symbol for a resistor in circuit diagrams is —⋀⋀⋀—. Any object may be considered a resistor when its resistance is its only feature of interest in analyzing a particular circuit.

20-4 Characteristics of Circuit Components III: EMF

When the positive charges comprising conventional current travel through a battery from the negative to the positive pole, the resulting PE increase can be thought of as the equivalent of work done by so-called non-electrostatic forces. The battery is characterized by a voltage, or change in potential, equal to this PE gain *per unit charge.* It is a measure of the battery's ability to "pump" or move electric charge along the circuit, sometimes called the battery's *electromotive* ability. The voltage representing this ability is sometimes called the battery's **EMF** and is denoted by $\mathcal{E}$. (Be careful not to confuse this with the permittivity ϵ or the electric field E.)

The EMF is the voltage between a battery's terminals when nothing is connected between them. However, if the battery is connected in a circuit, so that a current I passes through it, there is also some energy per unit charge *lost* in the battery because the battery itself has some resistance. If we denote the battery's **internal resistance** by r, then the voltage between the terminals is reduced:

$$\textbf{terminal voltage } V_t = \mathcal{E} - Ir \qquad (20\text{-}14)$$

In fresh batteries, the internal resistance tends to be small and may be considered negligible in many circuit situations. With use, however, their internal resistance increases and may become quite substantial.

Example 20-7 *EMF and Terminal Voltage*

The potential difference between the terminals of a battery is found to be 1.53 V when the battery is disconnected, and 1.49 V when a 0.80 A current flows through it. Find the internal resistance of the battery.

Solution
Choice of approach. We apply Equation 20-14. The important thing here is distinguishing between the actual terminal voltage when a current is flowing, and the EMF, the terminal voltage when the battery is disconnected.

What we know/what we don't.

$$V_t = 1.49 \text{ V} \qquad \mathcal{E} = 1.53 \text{ V} \qquad I = 0.80 \text{ A} \qquad r = ?$$

The mathematical solution. From Equation 20-14, we find that

$$r = \frac{\mathcal{E} - V_t}{I} = \frac{1.53 \text{ V} - 1.49 \text{ V}}{0.80 \text{ A}} = \textbf{0.050 } \Omega$$

◆ Related homework: Problems 20-36, 20-37, and 20-38.

20-5 Power and Energy in Circuit Components

The potential at a point is the potential energy that each unit of charge placed at that point would have. Electrons move very slowly along a wire, but when an amount of charge Δq enters a region at a point A and an equal amount leaves it at a distant point B, it is equivalent to the same amount of charge going the entire distance from A to B (Figure 20-12). For a fuller development of the reasoning in the figure, work through WebLink 20-2. If there is a potential difference V between A and B, then when an amount of charge Δq is in effect transferred between the two endpoints, the energy gained or lost is $\Delta W = V\Delta q$. The *rate* at which energy is gained or lost, that is, the power, is $P \equiv \frac{\Delta W}{\Delta t} = V\frac{\Delta q}{\Delta t}$. But because $\frac{\Delta q}{\Delta t}$ is our definition of the current I,

$$P = VI \qquad (20\text{-}15)$$

Power, a rate, is here a product of two other rates, as Figure 20-13 illustrates. Remember: Any quantity that states an amount of something per unit of something else is a rate. Rates are not always per unit of time.

If the charge experiences an increase in potential due to the internal chemistry of a battery, the potential difference $V = \mathcal{E}$, the battery's electromotive ability. Then

$$P = \mathcal{E}I \qquad (20\text{-}16)$$

is the rate at which those electrochemical mechanisms provide energy.

If the charge experiences a decrease in potential as it passes through some resistance, the potential difference $V = IR$. Then

$$P = (IR)I = I^2R \qquad (20\text{-}17)$$

is the rate at which the charge carriers lose energy on passing through. This energy is generally dissipated to the surroundings, producing a temperature

For **WebLink 20-2:** Movement of Electrons between Points at Different Potentials, go to www.wiley.com/college/touger

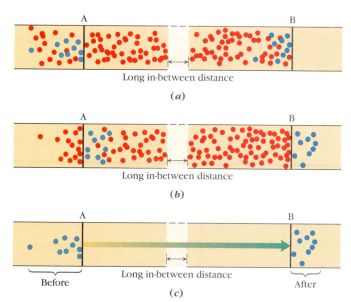

Figure 20-12 Movement of electrons between two points (A and B) with a potential difference V between them. (*a*) The conduction electrons that will pass points A and B during the next brief interval Δt are highlighted in blue for easy identification. (*b*) The conduction electrons travel at different speeds. The average drift velocity is to the right when the electric field is to the left. This is a picture of the same electrons Δt later. (*c*) Electrons are indistinguishable from one another, so although the individual electrons travel very short distances, it is just as if the electrons shown—with a total charge Δq—had gone the entire difference from A to B.

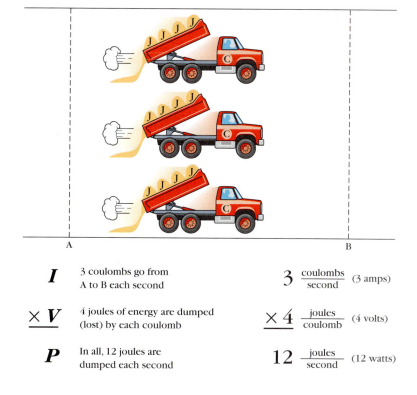

	I	3 coulombs go from A to B each second	$3 \dfrac{\text{coulombs}}{\text{second}}$ (3 amps)
	$\times V$	4 joules of energy are dumped (lost) by each coulomb	$\times 4 \dfrac{\text{joules}}{\text{coulomb}}$ (4 volts)
	P	In all, 12 joules are dumped each second	$12 \dfrac{\text{joules}}{\text{second}}$ (12 watts)

Figure 20-13 A way of picturing how the product of two rates (V in joules per coulomb and I in coulombs per second) results in a further rate P (in joules per second). Each dump truck represents charge carriers with a total charge of 1 C traveling between A and B. Each sandpile represents a joule of energy that the truck loses (dumps) between A and B. Three dump trucks pass A (and therefore travel between A and B) during each second.

➥**Heat and light:** Heat and light do not denote two completely distinct kinds of energy, but refer to the two different ways we experience the energy. One of our body's detection systems, the sense of touch, experiences it as heat; another, our sense of vision, experiences it as light. When both detection systems are able to detect the emitted energy, as is the case with a hot light bulb or the burner of an electric stove when it is set on high, we experience it as both heat and light.

increase and, if it is hot enough, light as well. This power *output* is therefore sometimes called *joule heating*.

In Case 18-2, we concluded that when each of two bulbs is connected in turn to a battery, the bulb that glows brighter is the one that permits more current to pass through it. Equation 20-15 agrees with this conclusion; it says that if V is the same across either bulb, the rate of energy dissipation per second P is proportional to I. **STOP**&**Think** Which will glow brighter when each is connected to the battery in turn, a bulb with lower or higher resistance? ◆

If the battery is unchanged, Ohm's law tells us that $I = \frac{V}{R}$, so that as the denominator R decreases, I increases. The power output is then

$$P = V\left(\frac{V}{R}\right) = \frac{V^2}{R} \tag{20-18}$$

and likewise varies inversely as the resistance.

Bulbs and appliances are generally labeled with a power consumption rating in watts. The rating assumes that the object will be used with a specified voltage. What if you plug it in where the voltage is different? Work through Example 20-8 and see.

Example 20-8 *The "Buck Saver" Bulb*

For a guided interactive solution, go to Web Example 20-8 at www.wiley.com/college/touger

Ms. Vásquez goes to the hardware store to buy a 100 W bulb. She picks up a bulb described as "100 watt 130 volt." Her household voltage is 120 V, which is standard in the United States. By a calculation, determine whether this bulb will function as a 100 W bulb in Ms. Vásquez's home.

Brief Solution
Choice of approach. The power consumption of a bulb depends on the voltage across it. The printed rating means that $P = 100$ W when $V = 130$ V. We use these values in Equation 20-18 to find the resistance. The resistance is a property of the filament, so if we neglect small variations with temperature, it doesn't change. Then we can use this resistance value in Equation 20-18 to find P approximately when $V = 120$ V.

Restating the question. If $P = 100$ W *when* $V = 130$ V, $P = ?$ *when* $V = 120$ V.

What we know/what we don't.

Under conditions supposed on bulb package	Under conditions in Ms. Vásquez's home
$V = 130$ V $R = ? \longrightarrow$	$R = ?$ $V = 120$ V
$P = 100$ W	$P = ?$

The mathematical solution. Solving Equation 20-18 for R, we get

$$R = \frac{V^2}{P} = \frac{(130 \text{ V})^2}{100 \text{ W}} = 169 \text{ } \Omega$$

Therefore, when $V = 120$ V,

$$P = \frac{V^2}{R} = \frac{(120 \text{ V})^2}{169 \text{ } \Omega} = 85.2 \text{ W}$$

In Ms. Vásquez's home, this is *not* a 100 W bulb.

◆ Related homework: Problems 20-43 and 20-44.

◆**ENERGY STORAGE IN CAPACITORS** As always when we separate positive and negative charge, there is a build-up of PE when electrons are "in effect" (as in Figure 20-12) moved from one of two initially neutral capacitor plates to the other. Again, the energy gained or lost by an amount of charge Δq passing between two points is $\Delta W = v\Delta q$. But now, the potential difference between the plates depends on the charge already on the plates (the lowercase v and q indicate that they vary). As the charge q builds up to a maximum value Q, the potential difference between the plates builds up from zero to a maximum value V. The average potential difference experienced by a bit of charge going from one plate to the other is $V_{avg} = \frac{0+V}{2} = \frac{V}{2}$. If we think of moving *all* the charge (Q) through this average potential difference, the total PE built up or stored is

$$W = V_{avg}Q = \frac{QV}{2} \tag{20-19a}$$

Because the capacitance $C = \frac{Q}{V}$, we can substitute for Q or V to obtain

$$W = \frac{Q^2}{2C} \tag{20-19b}$$

or

$$W = \tfrac{1}{2}CV^2 \tag{20-19c}$$

It is also possible to express the energy stored in a capacitor in terms of the electric field E between the plates. We use the facts that $C = \frac{A\epsilon}{d}$ and $V = Ed$ to substitute for C and V in Equation 20-19c:

$$W = \tfrac{1}{2}\left(\frac{A\epsilon}{d}\right)(Ed)^2 = \tfrac{1}{2}Ad\epsilon E^2$$

We can divide by Ad, the volume of the region between the plates, to obtain the energy *per unit of volume*, that is, the *energy density* w_{vol}:

$$w_{vol} = \tfrac{1}{2}\epsilon E^2 \tag{20-20}$$

The energy density does not depend on the dimensions of the capacitor, only on the material between the plates and the magnitude of the field in that material. This suggests that we can generalize this result beyond the limited context of capacitors: In any region where there is an electric field, there is stored energy associated with that field.

An enormous power output is required of loudspeaker systems in large public venues. The total amplification output at SuperBowl XXXVI was 17 000 W. When vibrational energy is produced at such a great rate, so is thermal energy, and its removal to prevent overheating is an important aspect of loudspeaker design.

◆ SUMMARY ◆

The concepts that we need to analyze simple circuits have now been defined quantitatively. We established the following definitions and SI units:

$$\textbf{Electric current} \quad I \equiv \frac{\Delta q}{\Delta t} \tag{20-1}$$

is measured in **amperes:**

$$1 \text{ ampere} \equiv 1 \text{ coulomb/second} \quad (1\text{ A} = 1\text{ C/s}) \tag{20-2}$$

$$\textbf{Capacitance} \quad C \equiv \frac{Q}{V} \tag{20-3}$$

is measured in **farads:**

$$1 \text{ farad} = 1\frac{\text{coulomb}}{\text{volt}} \quad (1\text{ F} = 1\text{C/V}) \tag{20-4}$$

$$\textbf{Resistance} \quad R \equiv \frac{V_{AB}}{I} \tag{20-11}$$

is measured in **ohms:** $1\,\Omega \equiv 1V/A \tag{20-12}$

Materials for which R remains fixed as V and I vary satisfy

$$\textbf{Ohm's law:} \quad V = IR \tag{20-11}$$

EMF ($\mathcal{E}$) is the potential difference between the terminals of a battery due to the electrochemical mechanisms (sometimes loosely characterized as exerting *non-electrostatic forces*) by which a battery separates opposite charges. It is measured in **volts.**

We examined the physical characteristics on which these quantities depend:

• In a parallel plate capacitor of plate area A and plate separation d with a potential difference $V = Ed$ across it, the capacitance

$$C = \frac{Q}{V} = \frac{A\epsilon}{d} \tag{20-6}$$

$\epsilon = \kappa\epsilon_o$ is the **permittivity** of the material or medium between the plates (where κ is the *dielectric constant* of the medium and

$$\epsilon_o = 8.85 \times 10^{-12} \text{ C}^2/\text{N} \cdot \text{m}^2 \tag{20-7}$$

is the permittivity of free space), so that $C = \kappa C_o$.

• A *current* may involve any kind of charge carrier (e.g., electrons in metal wires, ions in certain batteries), and depends on the density of carriers n, the charge of each carrier (which we've assumed to be e), their drift speed v_d, and the cross-sectional area A of the wire or other path:

$$I = neAv_d \qquad (20\text{-}9)$$

- The *resistance* of a circuit element depends on the material (characterized by an intrinsic property called its **resistivity** ρ) and dimensions (L and A):

$$R = \rho \frac{L}{A} \qquad (20\text{-}13)$$

Batteries have **internal resistance** (r), so that the

terminal voltage $V_t = \mathcal{E} - Ir \qquad (20\text{-}14)$

The **power** $P \equiv \frac{\Delta W}{\Delta t}$ gained or lost in a circuit element is in general

$$P = VI \qquad (20\text{-}15)$$

If gained from a battery of EMF $\mathcal{E}$, then

$$P = \mathcal{E}I \qquad (20\text{-}16)$$

If dissipated through a resistance R, then

$$P = I^2 R \qquad \text{or} \qquad P = \frac{V^2}{R} \qquad (20\text{-}17,\ 20\text{-}18)$$

The *energy stored in a capacitor* is

$$W = \frac{QV}{2} = \frac{Q^2}{2C} = \tfrac{1}{2}CV^2 \qquad (20\text{-}19)$$

The energy density (energy per unit of volume) due to the field in *any* region where there is an electric field E is

$$w_{vol} = \tfrac{1}{2}\epsilon E^2 \qquad (20\text{-}20)$$

✦ QUALITATIVE AND QUANTITATIVE PROBLEMS ✦
Hands-On Activities and Discussion Questions

The questions and activities in this group are particularly suitable for in-class use.

20-1. Hands-On Activity.

For this activity, you will need a hollow cylinder open at both ends and some material you can pack into it tightly enough so that water will seep through it but not too quickly (Figure 20-14). For a short while, the cardboard tube from a roll of toilet paper, aluminum foil, and so on, will serve as a sufficiently water-tight cylinder. The material packed inside it can be paper towels or a tightly rolled kitchen sponge. You will also need to be able to measure the height b to which the material is packed (see figure) and the total mass of the water that collects at the bottom (or you can measure the volume of this water with a measuring cup and then use $m = \rho_{water}V$ to determine its mass).

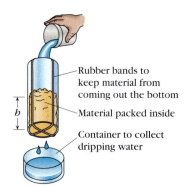

Figure 20-14 Problem 20-1

a. Let the water drain through the material for 60 s (or for a shorter measured Δt if the water is draining too quickly). Measure the mass of water that has accumulated at the bottom, and find the average rate $I = \frac{\Delta m}{\Delta t}$ (in kg/s) at which the water has passed through the material.

b. Measure b in m. Then find the amount of PE that each kilogram of water loses in going from the top to the bottom of the packed material. Write your result in J/kg. To what in an electrical circuit is this result analogous?

c. Assume there is an Ohm's law–like relationship between the values you found in *a* and *b*. Divide the result of *b* by the result of *a*. To what in an electrical circuit is this new result analogous? What does it tell you about the material you packed into the cylinder?

20-2. Discussion Question.

A simple water capacitor (Figure 20-15) can be constructed of two plastic one-liter bottles connected by a plastic soda straw near their bottoms. When you blow into bottle A, the water level in bottle B rises. The levels in the two bottles equalize when you stop blowing. Any flow between the two bottles is prevented if you pinch the straw, and can resume when you let go. (If you wish to try this out, you can easily construct the water capacitor using hot glue, epoxy, or any other reasonably tight sealant where the straw enters the bottles.) Discuss the analogy between this situation and a situation in which you charge a parallel plate electric capacitor with a DC battery, then discharge it through a resistor. What features or quantities are comparable in the two situations, and what aspects of the two situations are different?

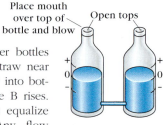

Figure 20-15
Problems 20-2 and 20-50

Review and Practice

Section 20-1 Electric Current

20-3. When ordinary table salt dissolves, it dissociates into positive sodium ions and negative chloride ions, both singly charged.

a. If 100 sodium ions pass through a membrane in 2.5 s, what is the average current across the membrane during this interval?

b. If 50 chloride ions also pass through in the same direction during the interval, what will the average current be during this interval?

c. Repeat *b* for the case where the chloride ions pass through the membrane in the *opposite* direction.

20-4. Alpha particles are helium atoms that have lost both of their electrons. A beam of alpha particles is directed through a tube at a target (there are no misses). If 3.0×10^{20} alpha particles strike the target during a 1-minute interval, what is the average current passing through the tube during this minute?

20-5. SSM *An analogy to electric current.* Suppose that at a certain traffic density (situation A), cars are able to maintain a speed of 55 mph on the freeway. When the number of cars on the road, and thus the density, is doubled (situation B), traffic slows to 25 mph. In situation B, is the flow of traffic (in cars per second passing a stationary observer) less than, equal to, or greater than the flow in situation A? Briefly explain.

20-6. In one full day, 4.0×10^{23} more conduction electrons pass a certain point in a wire in one direction than in the opposite direction. What is the current in this wire?

20-7. A certain capacitor is discharged through a bulb. It takes the capacitor 20 s to lose all but a negligible fraction of its charge. During this time interval the average current through the bulb is 0.075 A. How much charge was on the positive plate of the capacitor before discharging?

20-8. A metal plate has an insulated coating penetrated by two lead wires A and B (i.e., the wires lead to or from the plate). During a 5.0 s interval, an average current of 16 μA flows toward the plate through wire A, and an average current of 20 μA flows away from the plate through wire B. If the plate was neutral at the beginning of this interval, what is the charge on the plate at the end of the interval?

20-9. You spend five minutes trimming your hedges with an electric hedge trimmer connected by a long extension cord to an outlet on the side of the house. During this time, is it likely that any of the electrons flowing through the motor of the trimmer also passed through the outlet? Briefly explain.

20-10. In the circuit in Figure 18-16a, the most likely source of an electron arriving at the negative plate of the capacitor is ____ (*the battery; the wire connected to the negative plate; the positive plate of the capacitor; the wire connected to the positive plate; the bulb*).

Section 20-2 Characteristics of Circuit Components I: Capacitance

20-11. The terminals of a 1.5 V battery with negligible internal resistance are connected to the plates of a capacitor. What is the total potential difference between the terminals of the battery *a.* when the capacitor is fully charged? *b.* when the capacitor has just begun to be charged?

20-12. The terminals of a 1.5 V battery with negligible internal resistance are connected to the plates of a capacitor. Sketch charge configuration diagrams showing the arrangement of charge on the connecting wires and capacitor plates *a.* when the capacitor has just begun to be charged. *b.* when the capacitor is fully charged.

20-13.
a. The terminals of a 1.5 V battery with negligible internal resistance are connected to the plates of a 0.20 F capacitor. What is the charge on the capacitor when current has stopped flowing through the connecting wires?

b. Repeat *a* for the case in which a 9.0 V battery is used instead of the 1.5 V battery.

c. Does the amount of charge on a capacitor when it is fully charged depend on the properties of the capacitor alone? Explain.

20-14. When a certain capacitor has been connected to a particular battery for a long while, the capacitor is said to have a charge of 4×10^{-3} C.
a. How much charge is there on the negative plate of the capacitor at that time?

b. What is the total charge on both plates of the capacitor at that time?

c. When we speak of the charge on a capacitor, do we mean the total charge on both plates? Briefly explain.

20-15.
a. Sketch a graph of the charge on the capacitor plotted against time reading t if $t = 0$ is the instant when the circuit in Case 20-1 is reconnected as in Figure 18-16b.

b. By considering what happens to the slope of the graph you drew in *a*, sketch a graph of the *current* versus t for the same time period.

20-16. How much charge can be stored on either plate of a 20 mF capacitor when it is connected to a 1.5 V battery?

20-17. SSM A variable voltage DC power supply is connected in turn across different capacitors. If it is set at 4.8 V when it is connected across a 3.5 mF capacitor, to what voltage should it be set when it is connected across an 8.0 mF capacitor if you wish the second capacitor to hold the same amount of charge that the first did?

20-18. Find the electric field between the plates of a 3.0 μF capacitor with a 5.0×10^{-4} m plate separation when it is storing 1.8×10^{-5} C of charge.

20-19. A certain parallel plate capacitor has square plates separated by air. How does the capacitance of a second capacitor compare to this one (express the comparison as a ratio $\frac{C_2}{C_1}$) if the second capacitor is identical to the first except that *a.* the sides of the squares are doubled? *b.* the separation between the plates is reduced by half?

20-20. The six capacitors in Figure 20-16 all have the same dielectric material filling the region between the plates. Rank these capacitors in order of their capacitance, from least to greatest, making sure to indicate any equalities.

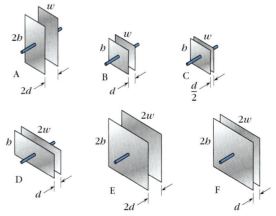

Figure 20-16 Problem 20-20

20-21. A simple parallel plate capacitor is produced by pressing a 2.0×10^{-4} m thickness of paper between two smooth squares of metal foil 0.015 m on a side. **a.** Find its capacitance. **b.** Could you find the capacitance as accurately if, instead of the pieces of foil, two new dimes were used for the capacitor plates? Briefly explain.

20-22. A parallel-plate capacitor consists of two rectangular conducting plates, each 3.0 cm by 2.0 cm. A coating of Teflon 4.0×10^{-5} m thick is applied to the inner surface of one plate to provide a dielectric layer, and then the plates are pressed together. How much voltage must be applied between the plates of this capacitor to establish a charge of 5.0×10^{-10} C on each plate?

Section 20-3 Characteristics of Circuit Components II: Resistance

20-23. Marbles accelerated down the ramp in Figure 20-8 by a gravitational field quickly reach an average drift speed because of their multiple encounters with nails acting as "collision centers".
a. What irregularities do electrons encounter as collision centers when traveling through a conducting wire with an electric field directed along it?
b. If those collision centers weren't present, what effect would the electric field have on the electrons?

20-24. If 1.00×10^{19} electrons pass each point along the filament of a bulb in 1 minute, what is the current through the bulb?

20-25. A strand of copper wire has a cross-sectional area of 1.5×10^{-6} m². The density of conduction electrons in the copper used for the wire is 8.4×10^{28} electrons/m³. When the electrons are typically taking 40 s to advance 1 cm along the wire, what is the current in the wire?

20-26. Two lengths of wire are drawn from the same metal.
a. Compare the resistance of the second length of wire to the first (express the comparison as a ratio $\frac{R_2}{R_1}$) if it is identical to the first except that it is three times as long.
b. Compare the resistance of the second length of wire to the first if it is identical to the first except that its diameter is reduced by half.
c. If you connected the two wires in **b** between the terminals of the same battery, how would the currents through the two wires compare? Express the comparison as a ratio $\frac{I_2}{I_1}$.

20-27. The five resistors in Figure 20-17 are all made of the same material. Rank these resistors according to each of the following, listing them in order from least to greatest and indicating any equalities: **a.** their resistance. **b.** their resistivity.

20-28.
a. Find the resistance at room temperature of a sample of copper wire that has the same dimensions as the sample of nichrome wire in Example 20-5.
b. What length of #18 copper wire would be required to have the same resistance as the sample of nichrome wire had?

20-29. For a 1-m length of copper wire to have a resistance of 1.0 Ω, what would its diameter have to be?

20-30. What voltage must be applied across a 200 Ω resistor to drive a 0.030 A current through it?

20-31. **SSM WWW** A particular 2.0-m length of nichrome heating wire has a diameter of 0.70 mm. What voltage must be applied across to drive a 0.030 A current through it?

20-32. Is Ohm's law a universally valid law? Briefly explain.

20-33. In certain materials, the number of available charge carriers in a sample of the material increases when the voltage across it is increased. Would such a sample obey Ohm's law? Briefly explain how Equation 20-10 bears on your answer.

20-34.
a. How much current will flow through a 40 Ω resistor when it is connected between the terminals of a 1.5 V battery with negligible internal resistance?
b. You could draw twice as much current with a suitable replacement for either the given resistor or the given battery. Describe each of these suitable replacements.

20-35. A 5 A fuse will melt and break a circuit when a current exceeding 5 A passes through it. An appliance is designed for use with a 120 V outlet. If the fuse is connected between the appliance and one terminal of the outlet, what minimum resistance must the appliance have to keep from "blowing" the fuse?

Section 20-4 Characteristics of Circuit Components III: EMF

20-36. In the circuit in Figure 20-18, the potential difference between the terminals of the battery is 1.55 V when the switch is open. When the switch is closed, a 0.12 A current flows through the bulb and the potential difference between the terminals drops to 1.52 V. **a.** Find the internal resistance of the battery. **b.** Find the resistance of the bulb.

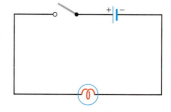

Figure 20-18 Problem 20-36

20-37. A variable resistor with settings from 10 Ω to 50 Ω may be connected between the poles of a battery with a small (but not negligible) internal resistance. You have a multimeter that will read the terminal voltage of the battery. If you wish to know the battery's EMF ($\mathcal{E}$), should you take a reading when the variable resistor is not connected, connected at its lowest setting, or connected at its highest setting? Briefly explain.

20-38. When a battery with an internal resistance of 0.30 Ω is connected to a 40 Ω bulb, a current of 0.224 A passes through the bulb. **a.** What is the terminal voltage of the battery? **b.** What is the EMF of the battery?

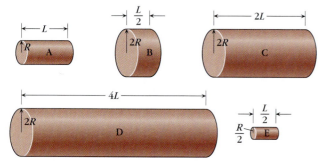

Figure 20-17 Problem 20-27

20-39. SSM WWW A coil of #18 nichrome wire is produced by wrapping insulated nichrome wire around a nonconducting cylindrical core. (See data on #18 nichrome wire in Example 20-5.) The wire makes 100 complete turns around the core, which is 3 cm in diameter.

a. Find the resistance of the coil at room temperature.

b. Find the current that flows through this wire when it is connected to a battery that has a terminal voltage of 1.48 V while the current is flowing.

c. If the internal resistance of the battery is 0.40 Ω, what voltage would a voltmeter read between the terminals of the battery if it were disconnected from the nichrome wire?

Section 20-5 Power and Energy in Circuit Components

20-40. How much energy is dissipated each second when a 0.30 A current flows through a 50 Ω resistor?

20-41. SSM A battery has an EMF of 1.54 V and a 0.25 Ω internal resistance.

a. What power does it deliver to an external load if a current of 0.16 A flows through it?

b. How much energy does it deliver to the external load each minute?

20-42. A particular 1.5 V battery has negligible internal resistance. How much work does it do each second if a bulb connected between its terminals has a resistance of *a.* 15 Ω? *b.* 40 Ω?

20-43. What is the resistance of the filament in a 100 W, 120 V bulb?

20-44. A guest from abroad has brought a traveling iron marked 1200 W, 220 V. Wishing to iron the rumpled clothing in his suitcase, he plugs it into the 120 V outlet in your home. What power does it draw? Neglect variations in resistance with temperature.

20-45.

a. What is the capacitance of a capacitor that stores exactly 1 J of energy when connected to a 12.6 V storage battery?

b. What is the charge on this capacitor? (Try to calculate it two different ways.)

20-46.

a. How much energy is stored in a 100 mF capacitor if each plate holds 3×10^{-2} C of charge?

b. If the capacitor is discharged through a 50 Ω resistor, we may estimate that it is fully discharged after 20 s. What is the average value of i^2 (the square of the current) during this interval?

Going Further

The questions and problems in this group are not organized by section heading, so you must determine for yourself which ideas apply. Some of them will be more challenging than the Review and Practice questions and problems (especially those marked with a • or ••).

•**20-47. SSM WWW** In a simple model of the hydrogen atom developed by Niels Bohr, an electron in a circular orbit of radius 5.29×10^{-11} m about the nucleus has a KE of 13.6 eV. Find the average current in this orbit.

20-48. To determine the dielectric constant of a material, a physicist cuts a rectangular slab of the material 0.4 cm by 0.5 cm and 1.0 mm thick. She places the slab between two sheets of metal foil having the same area dimensions. She is able to determine that the resulting capacitor stores 2.1×10^{-6} C of charge when a 12 V battery is connected between the sheets of foil. What value does she then calculate for the dielectric constant of the material?

20-49. We have calculated that there are 8.4×10^{28} conduction electrons per m^3 of copper.

a. What is the drift speed of the conduction electrons in a length of copper wire 1.2×10^{-3} m in diameter if there is a 20 mA current flowing through it?

b. Under these conditions, how many electrons will pass any given point along the wire each second?

20-50. A simple water capacitor (Figure 20-15) is constructed of two plastic one-liter bottles connected by a plastic soda straw near their bottoms. When you blow into bottle A, the water level in bottle B rises.

a. Consider the water that appears above the zero (0) level in bottle B when you blow into A. Did this water come from bottle A? Briefly explain.

b. When a parallel plate capacitor is charged, do the excess electrons that appear on the negative plate come from the positive plate? Briefly explain.

20-51. The collision centers that electrons encounter in a conductor include any irregularities or deviations from the perfectly periodic crystal lattice structure of the conductor.

a. Would you therefore expect the resistance of a metal wire to increase, decrease, or remain the same when the wire becomes hot?

b. Is the amount of current that passes through the filament of a bulb when it is first turned on more than, less than, or the same as after the bulb has been lit for a while? Briefly explain.

c. Is the power dissipated in the filament of a bulb when it is first turned on more than, less than, or the same as after the bulb has been lit for a while? Briefly explain.

20-52. A piece of silver wire is melted down. The silver is then drawn into a new piece of wire twice as long as the original. No silver is lost or added. How does the resistance of the new piece of wire compare to the original? Express as a ratio $\frac{R_{new}}{R_{old}}$.

20-53. Ionization of air molecules occurs because of energy inputs from the sun and because of collisions with protons in regions of Earth's magnetic field called the Van Allen Radiation Belts. As a result of this, there is a static electric field directed downward toward Earth that has an average value of about 120 N/C near ground level. Assuming Earth to be a conductor, find the average charge on a square meter of Earth's surface. (Data from W. R. Bennett Jr., "Cancer and Power Lines," *Physics Today,* April 1994, p. 20.)

20-54. Equations 20-6 and 20-13 tell how capacitance and resistance, respectively, depend on materials and geometry. Compare and contrast what these two equations tell us about the circuit elements they describe.

20-55.

a. What is the resistance of an ideal capacitor?

b. What is the capacitance of a length of ideal conducting wire?

20-56. A current of 0.20 A passes through a length of resistance wire connected between the terminals of a battery. Another piece of the same wire, twice as long as the first, is connected to a battery with twice the terminal voltage of the original. How much current does it draw? (Neglect internal resistance.)

20-57. A capacitor has circular plates with a radius of 0.065 m and a thin layer of glass with a dielectric constant of 8.0 sandwiched between them. If the electric field between the plates is gradually increased, what maximum charge will build up on the capacitor just before the dielectric breaks down and a current surges across it?

20-58. A multimeter is used to measure the voltage between the terminals of an unconnected battery. When a capacitor is then connected between the terminals of the battery, the multimeter reading drops slightly, then gradually goes back up to its original value. Explain why this happens.

20-59. Assuming its internal resistance is negligible, how much energy will a 9.0 V battery deliver each second to a 0.0045 W electronic device designed to be operated on that battery?

20-60. An electric curling iron is stamped "150 W 120 VAC" (VAC = volts alternating current).
a. Why would it have been inadequate simply to stamp it "150 W"?
b. The owner of the curling iron has brought it along on a trip to Europe, where wall outlets typically provide 220 V. Explain in detail what would happen if the curling iron were plugged into such an outlet.

20-61. Student X gets a brainstorm. "The power ratings printed on all my appliances really depend on their resistance. The more resistance I have, the less current I draw and the less power I use. Therefore if I leave all my appliances on, the resistances will be additive. With a higher total resistance I'll draw less current. Then since $P = VI$, I'll use less power and save on my electricity bills!" Identify the flaw(s) in Student X's reasoning. (*Note:* This problem will be repeated after the next chapter, when there may be additional arguments that you can make.)

20-62. The plates of an air capacitor (one with air between the plates) are squares 1.0 cm on a side and are separated by a distance of 0.30 mm. How much stored energy is released in the discharge that occurs when the electric field between the plates is raised to a value infinitesimally exceeding the dielectric strength of dry air?

20-63. A coffee maker designed for use with a 120 V outlet has a heating element with a resistance of 12 Ω. If the coffee maker is 100% efficient and perfectly insulated, how long will it take to raise the temperature of 0.25 kg of water from 20°C to 80°C?

20-64. In terms of energy loss, if there is a potential difference V between two points A and B, does it matter whether one electron goes the entire distance from A to B or each little piece of the total distance is traveled by a different electron? Briefly explain.

20-65. You are given a length of wire made of yellow shiny metal. Design an experiment in which you would use current readings to determine whether the metal was gold or not.

Describe all measurements you would make, the calculations you would do, and any outside information you would require.

20-66. A student connects two batteries with a total voltage of 3.00 V through a bulb to a 0.22 F "giant capacitance" capacitor. The student observes the bulb to glow for 10 s and estimates that by the end of 15 s the current to the capacitor is negligible. To the extent that the student's estimate is accurate, find the average power delivered to the capacitor in charging it.

20-67. SSM The four parallel plate capacitors described below all have the same dimensions but are made of different materials.

    A: plates made of copper, paper between the plates
    B: plates made of copper, air between the plates
    C: plates made of aluminum, air between the plates
    D: plates made of copper, aluminum between the plates

Rank these capacitors in order of their capacitance. Order them from least to greatest, making sure to indicate any equalities.

20-68. A capacitor and bulb are connected as in Figure 18-16*a* between the poles of a battery with a small (but not negligible) internal resistance. You have a multimeter that will read the terminal voltage of the battery. If you wish to know the battery's EMF ($\mathcal{E}$), you should take a reading ____ (*when the bulb is glowing brightly; when the bulb is glowing dimly; the instant the bulb stops glowing; quite a few seconds after the bulb stops glowing*). (Assume the battery's EMF is not affected by charging the capacitor.)

20-69. A bulb can be used in lamps plugged into outlets anywhere in the world. The quantity that will have the same value (neglecting minor differences due to heating) no matter where the bulb is used is the ____ (*voltage; power; resistance; current*).

20-70. A bulb is connected to a variable voltage supply. If the voltage is doubled, ____ (*the current will double but the power will remain the same; the power will double but the current will remain the same; the current and the power will both double; none of these will be true*).

20-71. Do the algebra that leads to Equation 20-10; that is, if $V_{AB} = EL$, $I = nev_d A$, and $v_d = cE$, show that $V_{AB} = I(\frac{1}{nec}\frac{L}{A})$.

20-72. In which of the two bulbs in Figure 20-19 is the power output greater? Briefly explain.

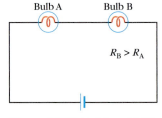

Figure 20-19 Problem 20-72

20-73. SSM Figure 20-20 shows two wires connected between the terminals of a battery. Both are the same diameter of nichrome heating wire, but wire 2 is twice as long as wire 1.
a. Consider one unit of charge traveling from the positive to the negative pole of the battery. Compare the amount of potential energy it loses when it travels by wire 1 and when it travels by wire 2.
b. Compare the power output from wire 1 and from wire 2.

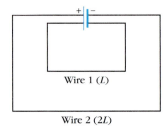

Figure 20-20 Problem 20-73

20-74. A capacitor consists of two metal plates 1.2 mm apart with air between them. When they are connected to a constant voltage source, the charge on the capacitor is 6.0×10^{-13} C. If the plates are moved 0.4 mm further apart while they remain connected to the voltage source,

a. by how much does the charge on the positive plate change?

b. Is this change an increase or a decrease?

Problems on WebLinks

20-75. In WebLink 20-1, the resistance to electric current depends on the length of the wire in the same way that the resistance to water flow (as it drains out of the tank) depends on _____ (*the length of the drainpipe; the area of the drain opening; the total thickness of the stack of filters*).

20-76. In WebLink 20-1, which (one or more) of the following could you increase to increase the flow rate of water down the drain: the length of the drainpipe; the area of the drain opening; the area of each sheet of filter paper; the total thickness of the stack of filters?

20-77. In WebLink 20-2, which (one or more) of the following statements are true of the electrons passing point B?

(i) *They are not the same electrons that passed point A.*

(ii) *They passed A just a tiny fraction of a second earlier.*

(iii) *They are recognizably different in some way than the electrons passing A.*

20-78. Based on the reasoning in WebLink 20-2, as the capacitor is charging in the circuit in Figure 20-21, electrons arriving at the negative plate of the capacitor are most likely to have come from _____ (*the battery; the bulb filament; the positive plate; near point A in the wire; near point B in the wire*).

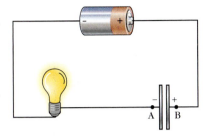

Figure 20-21 Problem 20-78

CHAPTER TWENTY-ONE

Quantitative Circuit Reasoning

In Chapter 20, we used the qualitative models of Chapter 18 and the quantitative definitions of Chapter 19 to develop a more detailed understanding of the behavior of individual circuit elements. The qualitative pictures and the mathematically defined concepts should connect in logically consistent ways when you think about electric phenomena. In this chapter, we use both to analyze the overall behavior of circuits made up of various combinations of these elements.

". . . we [will] analyze the overall behavior of circuits made up of various combinations of these elements."

21-1 Types of Circuit Connections

To begin with, we need to consider the ways individual elements can be connected to one another. Because different types of connections behave differently, we must be absolutely clear about what type of connection we have in any given case. To this end, we now define two important kinds of connections. The definitions that follow hold true in general for circuit elements that have two terminals. These include batteries and other simple voltage supplies, bulbs, capacitors, resistors, meters, and the usual run of electric appliances, but not transistors—which have three terminals—or a range of other complex solid state devices.

◆**IN SERIES** Circuit elements E_1 and E_2 are connected **in series** if a terminal of E_1 is connected to a terminal of E_2 and nothing else connects at this junction (Figure 21-1a). If each of several elements is in series with the next, we speak of them all as being in series.

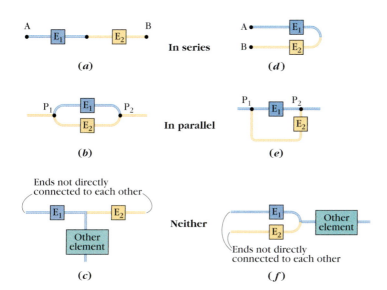

Figure 21-1 Connecting circuit elements in series and in parallel.

◆**IN PARALLEL** Two or more circuit elements are connected **in parallel** if they are connected between the same two points: One terminal of each connects at a single common point P_1 and the other terminal of each connects at a single common point P_2 (Figure 21-1b).

The terminals may be connected either by direct contact (touching) or by conducting wires (Figure 21-2). Wires with negligible resistance can be considered simply as extensions of the terminals. Without resistance, there can be no difference in surface charge density between the ends of the wire. There is then no potential difference—no voltage gain or drop—along the wire. You can treat

➥**Caution:** *In series* and *in parallel* describe the *connections* between elements, not their geometric layout. The circuit segments in Figure 21-1 are shown in pairs exhibiting the same connections. The segments in each pair are electrically equivalent, although they are laid out differently. In segments *b*, *d*, and *f*, elements E_1 and E_2 lie along paths that are *geometrically* parallel, but only in *b* are they *connected in parallel*. In *e*, E_1 and E_2 are *connected in parallel*, even though they do not lie along *geometrically* parallel paths.

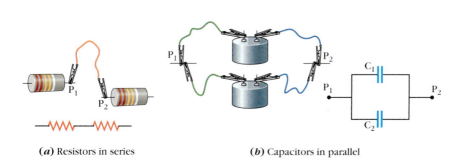

(**a**) Resistors in series   (**b**) Capacitors in parallel

Figure 21-2 Series and parallel connections using wires of negligible resistance. (*a*) Resistors in series. (*b*) Capacitors in parallel.

its two endpoints as equivalent when you apply the definitions of *in series* and *in parallel*:

> *Two points can be treated as a common point if they are connected by a wire with negligible resistance.*

We will assume that connecting wires have negligible resistance unless there is explicit information to the contrary. For practice reasoning with these ideas, work through Example 21-1.

Example 21-1 *Identifying Series and Parallel Connections*

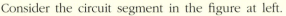

For a guided interactive solution, go to Web Example 21-1 at www.wiley.com/college/touger

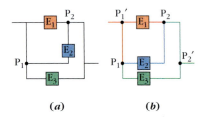

(a)　　　　(b)

Consider the circuit segment in the figure at left.
a. Are elements E_1 and E_2 in series?
b. Are elements E_1, E_2, and E_3 in parallel?

Brief Solution
Choice of approach. In each case we must determine whether the *definitions* of *in series* and *in parallel* are satisfied. Keep in mind that a wire of negligible resistance connected to a terminal can be taken as an extension of that terminal.
a. A terminal of E_1 and a terminal of E_2 connect at point P_2, as the definition of *in series* requires. But the requirement that nothing else connect at this junction is violated, because another branch of the circuit goes off from this point to E_3. E_1 and E_2 are *not* in series.
b. If we draw each element, along with its terminals and the wires that we take to be extensions of its terminals, in a different color (part *b* of the figure), we can see that all three elements are effectively between the same two points P_1 and P_2, and therefore satisfy the definition of *in parallel*.

Making sense of the results. Points P_1 and P_1' are equivalent because there is only a wire with negligible resistance between them. Therefore, nothing changes electrically if we reduce the length of this wire to zero, so that P_1 and P_1' become the same point. The same reasoning applies to points P_2 and P_2'.

◆ Related homework: Problems 21-4, 21-5, and 21-6.

21-2 Measuring Current and Voltage

From the definition of *in series*, it follows that if two conducting elements are connected in series, any charge that flows through one also flows through the other (but if a bulb and a capacitor are in series, current that flows *through* the bulb flows *to* the capacitor plate, but not across the gap).

> *The current must be the same through each conducting element in series.*

$$I_1 = I_2 = \ldots \text{ through elements } E_1, E_2, \ldots \text{ in series} \qquad (21\text{-}1)$$

From the definition of *in parallel*, because the potential difference between the same two points is the same for any path connecting them, we conclude:

> *The potential differences across elements in parallel are equal.*

$$V_1 = V_2 = \ldots \text{ across elements } E_1, E_2, \ldots \text{ in parallel} \qquad (21\text{-}2)$$

"Across an element" is an abbreviated way of saying "between the terminals of the element."

By recognizing when elements are connected in series or in parallel, we can decide if Equation 21-1 or 21-2 (or neither) applies. These equations will be fundamental rules in our analysis of circuits. For instance, they tell us how we must connect instruments to the circuit to measure current and voltage. An instrument called an **ammeter** is commonly used to measure current. A **voltmeter** is commonly used to measure potential difference. The most commonly available meters today for home use are **multimeters** that you can set by a dial to function as either ammeters or voltmeters (Figure 21-3). Although we are not yet ready to explain how these instruments work, we can do measurements with them if we understand the principles governing their proper use.

Figure 21-3 A digital multimeter.

Case **21-1** ◆ *Proper Connection of Ammeters and Voltmeters*

Because current is the rate at which charge flows *through* a circuit element, an ammeter can measure that current only if it flows through the ammeter as well. For the same current to flow through both the element and the ammeter (Equation 21-1), *the ammeter must be placed in series with the element.*

Potential difference is always a difference between the potentials at two different points. In the case of potential difference "across" an element, the two points are the element's end points or terminals. To measure that potential difference (or voltage), the voltmeter must "experience" the same potential difference. It must therefore be connected between the same two points as the element in question, that is, *the voltmeter must be connected in parallel with the element.*

In circuit diagrams, we represent an ammeter by the symbol –Ⓐ– and a voltmeter by the symbol –Ⓥ–. In the circuit in Figure 21-4a, the voltmeter is connected in parallel with bulb 1, the ammeter is in series with bulb 2. **STOP&Think** Is the ammeter also in series with bulb 1? ◆

Strictly speaking, the connection of the ammeter to bulb 1 at point P_2 does *not* satisfy our definition of a series connection because another circuit element, the voltmeter, also connects at that point. But the point of

a series connection is that all the current flowing through one of two elements in series flows through the other. This would still happen if the voltmeter had infinite resistance, so that the current flowing from point P_1 to P_2 would flow entirely through bulb 1. We want to come as close to that as possible, so that connecting the voltmeter doesn't change what's happening in the circuit. Measuring instruments aren't useful if they change what they're measuring. But for the voltmeter to provide a reading, at least a little current must pass through it. (*Something* must happen within the meter to cause it to "read.") Then the voltmeter's resistance cannot be infinite, but it must still be very high, so that *almost* all the current from P_1 to P_2 flows via the bulb, and only a minuscule fraction flows via the voltmeter. By making the resistance of the voltmeter high enough, we can keep the current through the two bulbs not exactly equal but equal to as many significant figures as we care to measure it. Thus, *the voltmeter is always an instrument with very high resistance,* and when connected *in parallel* with a circuit element, negligibly changes the behavior of the circuit.

STOP&Think Suppose we reconnected the voltmeter as in Figure 21-4b. Would the voltmeter and bulb 1 still be in parallel? ◆

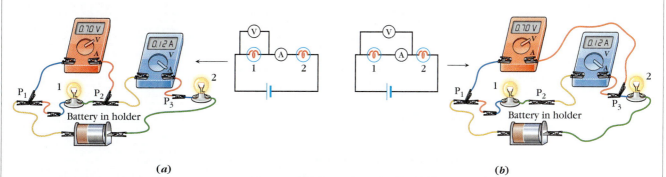

(a)   *(b)*

Figure 21-4 Two permissible ways of connecting an ammeter and a voltmeter. We will always depict the multimeter in red (to remind you of the *energy* connotations of potential difference) when it is serving as a voltmeter and in blue (to remind you of the analogy between current and a flow of *water*) when it is serving as an ammeter.

Case **21-1** ◆ *Proper Connection of Ammeters and Voltmeters (continued)*

The answer would be no (they are no longer connected between the same two points) unless P_2 and P_3 are effectively the same point. But if the ammeter has negligible resistance, we can ignore its presence just as we can ignore the presence of a length of connecting wire, and then P_2 and P_3 *are* effectively the same point.

To meet this requirement, *ammeters always have very low resistance.*

We have shown that when ammeters and voltmeters are properly connected in a circuit, we can ignore their presence in determining whether circuit elements are in series or in parallel.

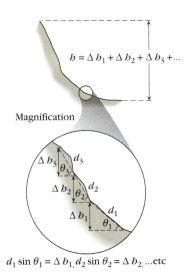

Magnification

$$b = \Delta b_1 + \Delta b_2 + \Delta b_3 + \ldots$$

$$d_1 \sin \theta_1 = \Delta b_1, d_2 \sin \theta_2 = \Delta b_2, \ldots \text{etc}$$

21-3 Series Circuits

In a simple series circuit, the elements are connected end on end to form a single closed path or loop. When a body travels along a continuous path, the change in PE is the sum of the changes that occur over the individual segments that make up the path (review Figure 6-18 at left); that is, if the path goes from point A to point B to point C, and so on, then the

$$\text{total } \Delta PE = \Delta PE_{\text{A to B}} + \Delta PE_{\text{B to C}} + \cdots.$$

The same reasoning applies for potential energy per unit charge—that is, *potential*. With V denoting a change in potential (a potential difference), the above equation translates to

$$V_{\text{path}} = V_{\text{AB}} + V_{\text{BC}} + \cdots$$

If the path is made up of circuit elements in series, the equation states the following:

> *The potential differences across elements in series are additive.*

$$V_{\text{path}} = V_1 + V_2 + \ldots \quad \text{across elements } E_1, E_2, \ldots \text{ in series} \quad (21\text{-}3)$$

We can verify this in any particular situation by using voltmeters to measure the potential differences (Figure 21-5).

Because electrostatic and gravitational forces are both conservative forces, the change in potential energy associated with either force is the same by any path (see Figure 6-18) between the same two points. If one of two paths is followed in the opposite direction, the change in potential energy reverses sign. The total potential energy change around a closed path must therefore be zero. The same holds true for the change in potential:

$$V_{\text{path}} = 0 \quad \text{for any closed path} \quad (21\text{-}4)$$

Some of the potential differences will be increases and others will be decreases; it follows that the sum of the increases must equal the sum of the decreases, so that the total difference or change is zero.

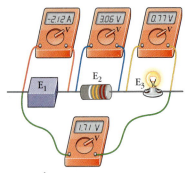

$$V_{\text{path}} \overset{?}{=} V_1 + V_2 + V_3$$
$$1.71 \overset{?}{=} (-2.12 \text{ V}) + 3.06 \text{ V} + 0.77 \text{ V}$$
$$1.71 \text{ V} \overset{?}{=} 1.71 \text{ V}$$

Figure 21-5 Checking additivity of voltages in a series circuit. Each voltmeter is measuring the voltage across one path segment and is connected in parallel with that path segment.

PROCEDURE **21-1**

Identifying Potential Differences *V* along a Path

- The potential increases by $\mathcal{E}$ in going from the negative to the positive terminal of a battery of EMF $\mathcal{E}$ due to work per unit charge done by the so-called non-electrostatic forces.
- By Ohm's law, the potential decreases by IR going in the direction of conventional current across a resistor.

- Therefore there is a total potential increase $V = \mathcal{E} - Ir$ in going from the negative to the positive terminal of a battery of EMF $\mathcal{E}$ and internal resistance r. ($V = \mathcal{E}$ when r is negligible.)
- There is a potential decrease in going from the positive to the negative plate of a capacitor. Because $C = \frac{Q}{V}$, the potential decreases by $V = \frac{Q}{C}$.

    The sign of each of these is reversed if you follow the path the other way.

Example 21-2 *A Simple Series Circuit Calculation*

An external load of 4.06 Ω is connected between the terminals of a battery with a 0.18 Ω internal resistance. Before the circuit is complete, a voltmeter connected across the battery reads 1.46 V. Find the current in the circuit.

Solution
Choice of approach.
1. By Equation 21-4, the sum of the potential differences around the circuit (increases +, decreases −) must be zero.
2. We sketch a diagram of the circuit (Figure 21-6a). The "external load" (whatever is connected outside the battery between its terminals) is characterized by its resistance. As the diagram is drawn, conventional current flows clockwise.
3. Guided by Procedure 2-1, we go clockwise around the circuit identifying all the potential increases and decreases (Figure 21-6b) and set the sum equal to zero.

What we know/what we don't. When the circuit is not complete, the voltmeter simply reads the EMF, because no current can flow ($\mathcal{E} - Ir = \mathcal{E}$ when $I = 0$).

$$\mathcal{E} = 1.46 \text{ V} \qquad r = 0.18 \text{ Ω} \qquad R = 4.06 \text{ Ω} \qquad I = ?$$

The mathematical solution. Taking the potential decreases as negative, Step 3 becomes

$$\mathcal{E} - Ir - IR = 0$$

Solving for I gives us

$$I = \frac{\mathcal{E}}{r + R} = \frac{1.46 \text{ V}}{0.18 \text{ Ω} + 4.06 \text{ Ω}} = \textbf{0.34 A}$$

◆ Related homework: Problems 21-11 and 21-12.

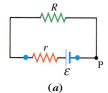

(a)

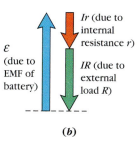

$\mathcal{E}$ (due to EMF of battery)

Ir (due to internal resistance r)

IR (due to external load R)

(b)

Figure 21-6 Diagrams for Example 21-2. (*a*) Circuit diagram. (*b*) Potential increases and decreases (going clockwise around circuit from point P).

◆**SOURCES OF EMF IN SERIES** If batteries or other sources of EMF are connected in series (Figure 21-7a), negative terminal to positive terminal, there is a potential increase from each in the negative-to-positive direction. Then the total EMF provided is simply the sum of the two individual EMFs:

$$\mathcal{E} = \mathcal{E}_1 + \mathcal{E}_2$$

But if terminals of the same sign are connected (say, positive to positive, as in Figure 21-7b), then as we trace the path in the direction of the current, if there is a potential increase across one source of EMF, there is a potential decrease across the other, as the figure shows.

21-4 Resistive Circuits

◆**RESISTORS IN SERIES** If resistors are connected in series, so that the same current I passes through all of them, the potential difference across the combination is again the sum of the individual potential differences:

$$V_{\text{path}} = IR_1 + IR_2 + IR_3 + \cdots = I(R_1 + R_2 + R_3 + \cdots)$$

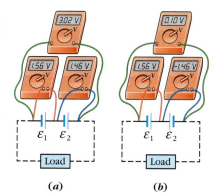

(a)   (b)

Figure 21-7 Sources of EMF connected in series.

Figure 21-8 Equivalent resistance. (*a*) If we measure the voltage *V* across a resistor with a voltmeter and the current *I* through it with an ammeter, we can use $\frac{V}{I} = R$ to find the resistance. (*b*) If we replace the resistance by a box, $\frac{V}{I}$ gives us the resistance *R* of the box. (*c*) If the resistance of the box is due to a group of connected resistors inside it, $\frac{V}{I}$ gives us the resistance *R* of this arrangement of resistors. Replacing the group by a single resistor of resistance *R* would change nothing "outside the box": *R* is the equivalent resistance of the group, whether the box is there or not.

The resulting potential difference is the same as that across a single resistor of resistance $R = R_1 + R_2 + R_3 + \ldots$ We call *R* the *equivalent resistance*.

> The *equivalent resistance* of resistors $R_1, R_2, R_3, \ldots$ in series is
> $$R = R_1 + R_2 + R_3 + \cdots \qquad (21\text{-}5)$$

For **WebLink 21-1:**
Equivalent Resistance,
go to
www.wiley.com/college/touger

Figure 21-8 provides a picture of what an equivalent resistance means. To see this idea developed in greater step-by-step detail, work through WebLink 21-1. In Example 21-3 we use the idea of equivalent resistance in a circuit calculation.

Example 21-3 *Bulbs in Series*

Two miniature bulbs having resistances $R_1 = 14 \ \Omega$ and $R_2 = 42 \ \Omega$ are connected in series to a pair of batteries with a combined EMF of 3.06 V (and negligible internal resistance).
a. Find the current through each resistor.
b. Is the current less than, the same as, or greater than it would be through either bulb alone?

Solution
Choice of approach.
1. The current is the same through each resistance in series (Equation 21-1), so there is only one current to find.

2. The equivalent resistance for bulbs in series is $R = R_1 + R_2$.

3. In setting the sum of potential increases and decreases equal to zero, we use the potential drop across the equivalent resistance instead of the drops across the individual bulbs.

What we know/what we don't.

$$\mathcal{E} = 3.06 \ \text{V} \qquad R_1 = 14 \ \Omega \qquad R_2 = 42 \ \Omega \qquad I = ?$$

The mathematical solution. **a.** $R = R_1 + R_2 = 14 \ \Omega + 42 \ \Omega = 56 \ \Omega$. The sum of potential increases and decreases is $\mathcal{E} - IR = 0$, so

$$I = \frac{\mathcal{E}}{R} = \frac{3.06 \ \text{V}}{56 \ \Omega} = 0.055 \ \text{A}$$

b. The equivalent resistance R is larger than either individual resistance, so in $I = \frac{\mathcal{E}}{R}$, the denominator would be largest if the equivalent resistance is used, and therefore I would be less than through either bulb alone.

◆ Related homework: Problems 21-13, 21-14, 21-15, and 21-16.

◆**RESISTORS IN PARALLEL** A key idea about parallel paths in a circuit is that the potential difference is always the same between the same two points. If resistors are connected in parallel, it means each of them is individually connected between the same two points, and so is the combination of resistors. It follows that

> *Across resistors* $R_1, R_2, \ldots$ *in parallel,* $V = V_1 = V_2 = \cdots$

(V is the potential difference across the combination of resistors).

What happens to the current when a circuit branches into two parallel paths? In Figure 21-9a the charges (Δq) that pass the observer at P_1 during a certain time interval Δt pass the observer at P_2 during a later Δt and pass the observer at P_3 during an even later Δt. So each observer calculates the same total current $\frac{\Delta q}{\Delta t}$. The observer at P_2, however, sees this total moving partly by one parallel path and partly by the other. There is a current along each path, and the total current is the sum of these. The same reasoning applies to the far more numerous charges flowing in Figure 21-9b, showing the current flowing through two parallel resistors. (Think of a marathon viewed from a helicopter, with all the runners in red uniforms following one path and all those in blue following the other.)

In general, the *total current* is

$$I = I_1 + I_2 + I_3 + \cdots \tag{21-6}$$

between two points connected by resistors $R_1, R_2, R_3, \ldots$ in parallel.

From this and Ohm's law, it follows that $\frac{V}{R} = \frac{V_1}{R_1} + \frac{V_2}{R_2} + \frac{V_3}{R_3} + \ldots$, where R is the equivalent resistance of the combination of resistors. But because

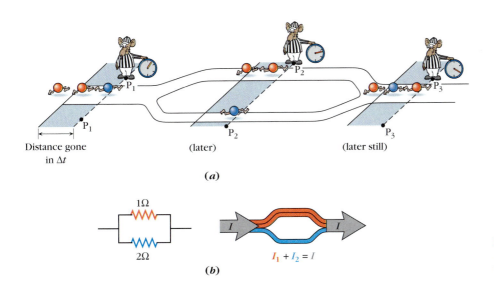

Distance gone
in Δt

(later)   (later still)

(a)

1Ω

2Ω

$I_1 + I_2 = I$

(b)

Figure 21-9 Measuring current along parallel paths. Compare with Figure 20-1.

$V = V_1 = V_2 = \ldots$, we can divide all terms on both sides by the common numerator, giving us this result:

$$\frac{1}{R} = \frac{1}{R_1} + \frac{1}{R_2} + \frac{1}{R_3} + \cdots \qquad (21\text{-}7)$$

for resistors $R_1, R_2, R_3, \ldots$ in parallel.

Example 21-4 *Reasoning about Resistors in Parallel*

For a guided interactive solution, go to Web Example 21-4 at
www.wiley.com/college/touger

The figure and diagram below show part of a complete circuit. If ammeter A_1 reads 0.15 A, ammeter A_2 reads 0.40 A, and the voltmeter V reads 2.4 V, calculate the resistances of the two resistors.

Brief Solution

Choice of approach. The voltmeter is connected between the same endpoints as either resistor or the combination and therefore reads the voltage across any of these. Ammeter A_1 is in series with R_1 and gives the current I_1 through that resistor. Ammeter A_2's position is like that of the observer at P_3 in Figure 21-9 and reads the total current I through the combination. With this information, we can apply Ohm's law to R_1 and the combination to find their resistances, and then use Equation 21-7 to find R_2.

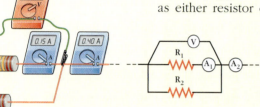

What we know/what we don't.

About R_1		About R_2		About the Combination	
$V_1 = 2.4$ V		$V_2 = 2.4$ V		$V = 2.4$ V	
$I_1 = 0.15$ A	$R_1 = ?$	$I_2 = ?$	$R_2 = ?$	$I = 0.40$ A	$R = ?$

The mathematical solution. By Ohm's law,

$$R_1 = \frac{V_1}{I_1} = \frac{2.4\ \text{V}}{0.15\ \text{A}} = \mathbf{16\ \Omega} \qquad \text{and} \qquad R = \frac{V}{I} = \frac{2.4\ \text{V}}{0.40\ \text{A}} = 6\ \Omega$$

Then from Equation 21-7, we find that

$$\frac{1}{R_2} = \frac{1}{R} - \frac{1}{R_1} = \frac{1}{6\ \Omega} - \frac{1}{16\ \Omega} = 0.167\ \Omega^{-1} - 0.063\ \Omega^{-1} = 0.104\ \Omega^{-1}$$

Inverting gives us $\quad R_2 = \dfrac{1}{0.104\ \Omega^{-1}} = \mathbf{9.6\ \Omega}$

Making sense of the results. Once we know R_2, we can find that

$$I_2 = \frac{V_2}{R_2} = \frac{2.4\ \text{V}}{9.6\ \Omega} = 0.25\ \text{A}$$

which satisfies the requirement (Equation 21-6) that $I = I_1 + I_2$. We also could have begun by using Equation 21-6 to find I_2, and then applied Ohm's law to resistor R_2 to find its resistance. Either method provides a check on our results.

◆ Related homework: Problems 21-17, 21-18, 21-19, 21-20, and 21-22.

As Example 21-4 shows, Equation 21-7 gives an equivalent resistance that is smaller than any of the individual resistances in parallel. Figure 21-10 helps explain why. We can mentally break up a resistor into several parallel pieces, each with the same length L and resistivity ρ. Putting more pieces in parallel increases the total cross-sectional area, and because $R = \frac{\rho L}{A}$, this decreases the overall resistance.

A grouping of connected circuit elements is called a **network.** Some networks (but not all; see Figure 21-11) can be put together by making a succession of series and parallel connections. We need to apply both series and parallel circuit reasoning to understand the behavior of such a network. To see how we do this both qualitatively and quantitatively for a network of resistors, work through Examples 21-5 and 21-6.

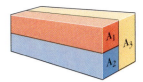

Figure 21-10 Cross-sectional areas of resistances in parallel. The effective cross-sectional area for the combination of resistors is $A = A_1 + A_2 + A_3$.

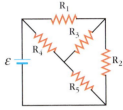

Figure 21-11 A network of resistors that *cannot* be constructed by making a succession of series and parallel connections. Are any two of the resistors in this circuit in series? Are any two of them in parallel?

Example 21-5 *A Network of Christmas Elves (Qualitative)*

For a guided interactive solution, go to Web Example 21-5 at www.wiley.com/college/touger

The O'Flibberty family has purchased four light-up Christmas elves to use as lawn decorations. Each has a resistance of 720 Ω. The manufacturer advises connecting them in parallel (why?), but Uncle Seamus, ever one to take matters into his own hands, has connected the elves as shown. Will the elves glow less or more brightly than if Uncle Seamus had followed the manufacturer's instructions? Explain.

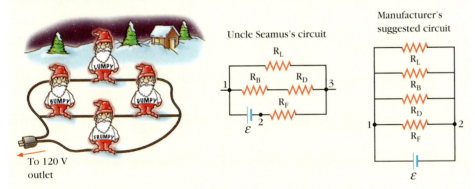

Brief Solution

Choice of approach. Here we will use qualitative reasoning about series and parallel circuits. In the next example we will apply the calculational procedures developed in this chapter, but they should still be guided by your qualitative understanding. For both, a useful first step is to draw clear circuit diagrams (see above) of the contrasting situations. For the present purpose, we can treat the outlet as though it were a 120 V battery.

The qualitative reasoning. In the manufacturer's suggested circuit, the elves are in parallel: The potential difference between points 1 and 2 is the same across any path. No matter how many paths of equal resistance we connect between these two points, the same potential difference drives the same flow of charge through the same resistance along each path. Each elf draws the same current as if it were the only elf connected between 1 and 2. **STOP&Think** Can you keep connecting elves in parallel indefinitely? What happens to the *total* current drawn if you do? ◆

In Uncle Seamus's circuit, the change in potential as we go from 1 to 2 is the change going from 1 to 3 plus the further change going from 3 to 2. Then the differences between 1 and 3 and between 3 and 2 are both less than the potential difference between 1 and 2, which is established by the "pumping"

In the early 1900s in Coney Island, Luna Park was illuminated at night with over 250 000 electric lights.

EXAMPLE 21-5 *continued*

of the wall outlet. Frumpy is connected between 3 and 2: Because there is less of a potential difference across him than in the suggested circuit, charge will flow through him at a lesser rate; the current will be less and he will glow less brightly. Lumpy is connected between 1 and 3; he will glow less brightly for the same reason. Bumpy and Dumpy are connected in series between 1 and 3; the same current flows through each of them. But by this path there is a *greater* total resistance connected between the same potential difference, so the current that flows along this path will be even less than the current through Lumpy. Therefore, Bumpy and Dumpy glow even less brightly than Lumpy.

◆ Related homework: Problems 21-21, 21-23, and 21-24.

Example 21-6 *A Network of Christmas Elves (Quantitative)*

For a guided interactive solution, go to Web Example 21-6 at www.wiley.com/college/touger

Calculate the total current that is drawn in Example 21-5 when Uncle Seamus's network of elves is connected to a 120 V outlet.

Brief Solution

Choice of approach. We need to consider when there are series and when there are parallel connections. We use Ohm's law with a total equivalent resistance obtained by building the resistor network piecemeal: (1) connect R_B and R_D in series and obtain their equivalent resistance R_{BD}; (2) now treat Bumpy and Dumpy as a single resistor of resistance R_{BD}, connect it in parallel with R_L, and find the equivalent resistance R_{BDL} of this combination; (3) connect Frumpy in series with R_{BDL} to obtain the equivalent resistance R_{BDLF} of the total load connected to the outlet. To help keep track, we subscript each equivalent resistance with the resistors contributing to it.

➡ **Notation:** Labeling an equivalent resistance by putting all the resistors that contribute to it in the subscript will help you keep track of which resistances you have combined.

What we know/what we don't.

$$R_B = R_D = R_L = R_F = 720\ \Omega \qquad V = \mathcal{E} = 120\ \text{V} \qquad I = ?$$

The mathematical solution. By Equation 21-5 for resistors in series,

$$R_{BD} = R_B + R_D = 720\ \Omega + 720\ \Omega = 1440\ \Omega$$

By Equation 21-7 for resistors in parallel,

$$\frac{1}{R_{BDL}} = \frac{1}{R_{BD}} + \frac{1}{R_L} = \frac{1}{1440\ \Omega} + \frac{1}{720\ \Omega} = 0.00208\ \Omega^{-1}$$

so

$$R_{BDL} = \frac{1}{0.00208\ \Omega^{-1}} = 480\ \Omega$$

Now connecting R_F in series with this equivalent resistance,

$$R_{BDLF} = R_{BDL} + R_F = 480\ \Omega + 720\ \Omega = 1200\ \Omega$$

Finally, by Ohm's law,

$$I = \frac{V}{R_{BDLF}} = \frac{120\ \text{V}}{1200\ \Omega} = \textbf{0.100 A}$$

Making sense of the results. If all the elves were connected in parallel, we would have $\frac{1}{R_{BDLF}} = \frac{1}{R_B} + \frac{1}{R_D} + \frac{1}{R_L} + \frac{1}{R_F} = \frac{1}{720\ \Omega} + \frac{1}{720\ \Omega} + \frac{1}{720\ \Omega} + \frac{1}{720\ \Omega} = \frac{4}{720\ \Omega} = \frac{1}{180\ \Omega}$.
Then in the manufacturer's recommended circuit, $R_{BDLF} = 180\ \Omega$, and the total

current $I = \frac{120 \text{ V}}{180 \text{ }\Omega} = 0.67$ A. When the four elves are in parallel, one-fourth of the total current $\left(\frac{0.67 \text{ A}}{4} = 0.17 \text{ A}\right)$ flows through each. We showed in Example 12-5 that in Uncle Seamus's network all the current passes through Frumpy and less current passes through each of the elves. Then because the total current in his circuit is 0.10 A, none of the elves draw as much current as they do in the recommended circuit, so none glow as brightly.

♦ Related homework: Problems 21-25 through 21-30.

21-5 Circuits with Capacitors

As we begin to consider circuits with combinations of capacitors, you should be thinking about how they are similar and how they are different from circuits with resistors. Most conspicuously, no current flows through capacitors. Precisely because they have a nonconducting gap, they store charge. So instead of reasoning with $V = IR$, we must reason with $V = \frac{Q}{C}$.

◆**CAPACITORS IN PARALLEL** When capacitors are connected in parallel (Figure 21-12), everything to the left of the dielectrics is a single connected conductor, just as if there were a single positive plate holding the same total charge. In the same way, we could mentally merge all the plates on the right into a single negative plate. We could think of the two combined conductors, separated by the nonconducting region of dielectrics (the gap highlighted in yellow), as though it were one big capacitor with an **equivalent capacitance** C. As the figure shows, there is total charge Q stored on either side of the gap between the plates:

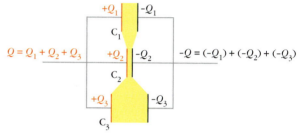

$$Q = Q_1 + Q_2 + Q_3 + \cdots \qquad (21\text{-}8)$$

for capacitors $C_1, C_2, C_3, \ldots$ in parallel

Figure 21-12 Additivity of charge on capacitors in parallel.

In the case where the capacitors have the same ϵ and d, the charges on the individual capacitor plates are proportional to their areas, and Equation 21-8 then follows from the additivity of the areas (Figure 21-13).

Because $Q = CV$ for the equivalent capacitor, it follows that

$$CV = C_1V_1 + C_2V_2 + C_3V_3 + \cdots.$$

But $V = V_1 = V_2 = V_3 = \ldots$ for elements in parallel, so by dividing both sides by this common factor we obtain the equivalent capacitance.

Equivalent capacitance $\qquad C = C_1 + C_2 + C_3 + \cdots \qquad (21\text{-}9)$

for capacitors $C_1, C_2, C_3, \ldots$ in parallel

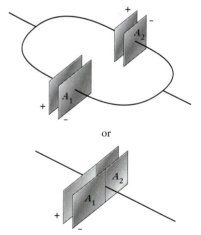

Figure 21-13 Plate areas of capacitors in parallel. The total area available for charge storage on each side of the gap is $A = A_1 + A_2$.

◆**CAPACITORS IN SERIES** As is true for *any* circuit elements in series, the potential differences across capacitors in series are additive. No current flows through capacitors unless they are overloaded. Rather, it follows from the definition of capacitance that $V = \frac{Q}{C}$, that is, the potential difference across each capacitor depends on the *charge* stored on it. The total potential difference across a path consisting of several capacitors in series is then

$$V_{\text{path}} = \frac{Q_1}{C_1} + \frac{Q_2}{C_2} + \frac{Q_3}{C_3} + \cdots \qquad (21\text{-}10)$$

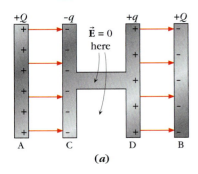

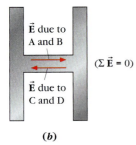

Figure 21-14 Charge configurations and fields for capacitors in series.

For **WebLink 21-2:**
Capacitors in Series,
go to
www.wiley.com/college/touger

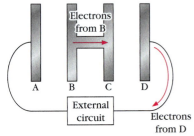

Figure 21-15 Discharging capacitors in series. Only the charge on the outermost plates flows through the external circuit. Before discharging, plates A and C are positive, and B and D are negative.

What can we say about the charges on capacitors in series? To address this question, we turn a single parallel-plate capacitor into two by inserting an **H**-shaped conducting piece between two outer plates (Figure 21-14). Because the battery "pumps" electrons from plate A toward plate B, plates A and B must have equal and opposite charges $\pm Q$. This causes a charge separation on **H**, so plates C and D end up with equal and opposite charges $\pm q$. Because the **H** is conducting, the field inside it must be zero. That means that the electric field due to plates C and D must be equal and opposite to the electric field due to A and B. That can only be true if $q = Q$. If we now think of plates A and C as one capacitor and C and D as the other, the charges on the two capacitors in series are the same. To see this reasoning developed in step-by-step detail, work through WebLink 21-2. Our conclusion holds true for any number of capacitors in series:

$$Q_1 = Q_2 = Q_3 = \cdots = Q \quad \text{for capacitors in series} \quad (21\text{-}11)$$

where Q is the magnitude of the charge on any one plate

When a series combination of capacitors discharges, only the charge on the outermost plates (Q) flows through the external circuit (Figure 21-15), so from the point of view of the external circuit, that is how much charge the combination of plates stores. Equation 21-11 then says:

For capacitors in series, the charge stored by each capacitor individually and by the combination is the same.

From this conclusion and the definition of capacitance, it follows that the capacitance of the combination is $C = \frac{Q}{V_{\text{path}}}$, so that the voltage V_{path} across the combination is $V_{\text{path}} = \frac{Q}{C}$.
Equation 21-10 then becomes

$$\frac{Q}{C} = \frac{Q}{C_1} + \frac{Q}{C_2} + \frac{Q}{C_3} + \cdots$$

We've deleted the subscripts on the charges because they are equal. If we now divide both sides by Q, we get this:

$$\frac{1}{C} = \frac{1}{C_1} + \frac{1}{C_2} + \frac{1}{C_3} + \cdots \quad (21\text{-}12)$$

to calculate the *equivalent capacitance C* of capacitors in series

Example 21-7 *Equivalent Capacitance for Capacitors in Series*

**For a guided interactive solution, go to Web Example 21-7 at
www.wiley.com/college/touger**

If three capacitors having capacitances 20 μF, 30 μF, and 40 μF are connected in series, find the equivalent capacitance of the combination.

Brief Solution
Choice of approach. This is a straightforward application of Equation 21-12.

What we know/what we don't. $C_1 = 20 \ \mu$F $\quad C_2 = 30 \ \mu$F $\quad C_3 = 40 \ \mu$F $\quad C = ?$

The mathematical solution.

$$\frac{1}{C} = \frac{1}{C_1} + \frac{1}{C_2} + \frac{1}{C_3} = \frac{1}{20\ \mu F} + \frac{1}{30\ \mu F} + \frac{1}{40\ \mu F} = \frac{6}{120\ \mu F} + \frac{4}{120\ \mu F} + \frac{3}{120\ \mu F}$$

$$= \frac{13}{120\ \mu F} = 0.108\ \mu F^{-1}$$

Now you must remember to invert to get *C*:

$$C = \frac{1}{0.108\ \mu F^{-1}} = \textbf{9.2\ }\boldsymbol{\mu}\textbf{F}$$

◆ Related homework: Problems 21-31**b**, 21-32**b**, and 21-33.

Example 21-8 *Comparing Combinations of Capacitors*

If you have a 10 mF capacitor and a 15 mF capacitor, how could you combine them to obtain an effective capacitance of **a.** less than 10 mF? **b.** greater than 15 mF?

Solution

Choice of approach. You need to be aware that Equation 21-12 yields an effective capacitance less than any of the individual capacitances of capacitors connected in series, whereas Equation 21-9 yields an effective capacitance greater than any of the individual capacitances of capacitors connected in parallel.

The mathematical solution. We verify this for the given capacitors.
a. When we connect these two capacitors in series, Equation 21-12 tells us that $\frac{1}{C} = \frac{1}{10\ mF} + \frac{1}{15\ mF} = \frac{3}{30}\ mF^{-1} + \frac{2}{30}\ mF^{-1} = \frac{5}{30}\ mF^{-1} = \frac{1}{6}\ mF^{-1}$. Inverting then gives us $C = \textbf{6\ mF}$ (less than the 10 mF, the smaller of the two capacitances).
b. When we connect them in parallel, Equation 21-9 gives us an effective capacitance of $C = C_1 + C_2 = 10\ mF + 15\ mF = \textbf{25\ mF}$ (greater than the 15 mF, the larger of the two capacitances).

◆ Related homework: Problems 21-31**a**, 21-32**a**, 21-36, and 21-37.

21-6 Circuits with Transient Currents

Transient currents are currents of short duration. So far, we have considered constant (nonzero) current situations involving resistors and zero-current situations involving capacitors. But neither of these conditions applies after the switch (___/___) in Figure 21-16*a* is closed. We have considered this situation before. There is a transient current (if the resistor is a bulb, it glows) that drops to zero (and the bulb goes out) as the capacitor becomes fully charged. During the interval when the current is dropping, we can write the potential differences across the resistor and capacitor as iR and $\frac{q}{C}$ (lowercase symbols represent instantaneous values, uppercase represent maximum values). Applying Equation 21-4 and identifying all the potential increases and decreases as we proceed around the circuit, we obtain $\mathcal{E} - iR - \frac{q}{C} = 0$, or

$$\mathcal{E} = iR + \frac{q}{C} \qquad \text{(capacitor charging)} \qquad (21\text{-}13)$$

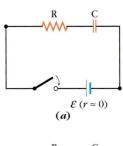
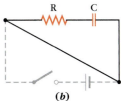

Figure 21-16 Transient current circuits.

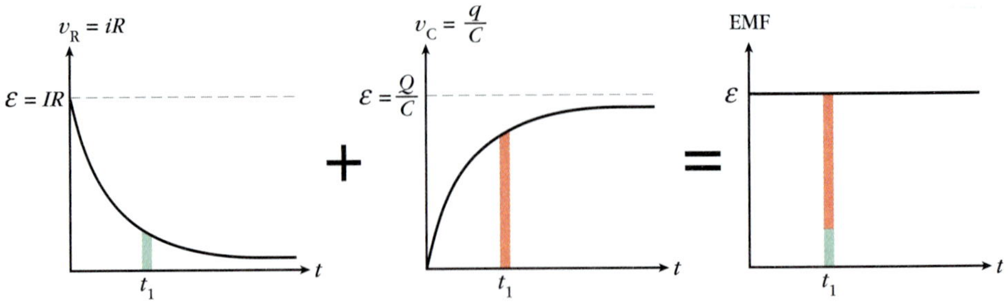

$$v_R = iR \qquad\qquad v_C = \frac{q}{C} \qquad\qquad \text{EMF}$$

Figure 21-17 Additivity of potential differences across elements in a series RC circuit during charging of the capacitor.

As long as the EMF remains constant, the two terms on the right ($v_R = iR$ and $v_C = \frac{q}{C}$) must add up to the same total at every instant (Figure 21-17). This means that i must decrease as q increases. Furthermore, each is at its maximum value when the other is zero:

$$i = I \text{ when } q = 0 \qquad \text{and} \qquad q = Q \text{ when } i = 0$$

For **WebLink 21-3:**
**Potential Differences
in an RC Circuit,** go to
www.wiley.com/college/touger

This mathematical description fits with our observations of the bulb dimming as the capacitor charges up. To reinforce these ideas, work through WebLink 21-3.

Substituting the two pairs of values above into Equation 21-13 gives us

$$I = \frac{\mathcal{E}}{R} \qquad \text{and} \qquad Q = \mathcal{E}C$$

for the maximum values of current and charge. (*Remember:* These maxima do not occur at the same time.)

Example 21-9 *An RC Circuit Calculation*

**For a guided interactive solution, go to Web Example 21-9 at
www.wiley.com/college/touger**

Suppose that in Figure 21-16*a*, $\mathcal{E} = 3.0$ V, $C = 0.10$ F, and $R = 12\ \Omega$. Find the current in the circuit when the charge on the capacitor has reached one-third of its maximum value.

Brief Solution
Restating the question. We can state the question mathematically as

$$i = ? \text{ when } q = \tfrac{1}{3}Q$$

Remember: Lowercase means *at the particular instant in question;* uppercase means *maximum value.*

Choice of approach. To find q, we must first find Q from $Q = \mathcal{E}C$. Once we know q, we can use Equation 21-13 to find i at the same instant.

The mathematical solution. The *maximum* value of the charge is

$$Q = \mathcal{E}C = (3.0\text{ V})(0.10\text{ F}) = 0.30\text{ C}$$

The value of the charge at the instant in question is

$$q = \tfrac{1}{3}Q = \tfrac{1}{3}(0.30\text{ C}) = 0.10\text{ C}$$

Solving Equation 21-13 for i then gives us

$$i = \frac{\mathcal{E} - \dfrac{q}{C}}{R} = \frac{\mathcal{E}}{R} - \frac{q}{RC} = \frac{3.0\text{ V}}{12\ \Omega} - \frac{0.10\text{ C}}{(12\ \Omega)(0.10\text{ F})} = \textbf{0.17 A}$$

Making sense of the results. The maximum current is $I = \frac{\mathcal{E}}{R} = \frac{3.0\text{ V}}{12\ \Omega} = 0.25$ A. Our result tells us that the current has dropped to about two-thirds of its maximum value as the charge has built up to one-third of *its* maximum value. **STOP&Think** What does this mean has happened to the potential difference across the capacitor? Across the resistor? Across the two elements in combination? ◆

   ◆ Related homework: Problems 21-38 and 21-39.

If we substitute a simple conducting wire for the battery and switch in the previous circuit (Figure 21-16b), the capacitor discharges and there is again a transient current. The EMF is now zero, so Equation 21-13 reduces to

$$iR + \frac{q}{C} = 0 \qquad \text{(capacitor discharging)} \qquad (21\text{-}14a)$$

Using the definition of i, this can be rewritten approximately as

$$\frac{\Delta q}{\Delta t} R + \frac{q}{C} \approx 0 \qquad (21\text{-}14b)$$

In Equations 21-14a and b, i is really the instantaneous value of the current, and $\frac{\Delta q}{\Delta t}$ is the average value. The smaller the Δt we choose, the better the average value will approximate the instantaneous value. To add up to zero, one of the two terms on the left must be negative if the other is positive. The math fits the situation, because if q is the charge on the positive plate, it is decreasing, so Δq is negative.

   Equation 21-14b can also be written as

$$\Delta q \approx -\frac{q}{RC} \Delta t \qquad (21\text{-}15)$$

Suppose that the clock by which we keep time keeps advancing by the same fixed interval Δt. Then everything multiplying q on the right side of Equation 21-15 is constant. Then the equation tells us that *the decreases in charge, say, on the positive plate are proportional to the amount of charge remaining on that plate.* **STOP&Think** What does it tell us about the charge on the negative plate? ◆ After each decrease Δq, the charge q that remains is smaller, so the next decrease will be smaller too. Figure 21-18 illustrates this: Over equal intervals Δt the charge drops toward zero by successively smaller amounts. Meanwhile, the charge on the negative plate becomes less and less negative.

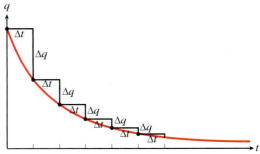

Figure 21-18 Time dependence of charge on a discharging capacitor.

Example 21-10 *Time Required to Discharge a Capacitor*

Suppose that in Figure 21-16, the capacitor is charged by closing the switch in *a* and then discharged by reconnecting the circuit as in *b*. If the resistor is a 15 Ω bulb and the capacitor has a 0.50 F capacitance, roughly how long does it take for the charge on either plate to be reduced by half?

Solution
Choice of approach. The approximation in Equation 21-15 gets better as Δt gets smaller. If we take Δt to be the entire time interval required for the charge to drop by half, the approximation will be very rough indeed, but it can tell us whether it takes a few seconds, a few tenths of a second, or a few hundredths of a second. Mathematicians say the estimate gives us the right order of magnitude; colloquially, it gives us a value "in the right ballpark."

Restating the question. With this choice of Δt, an initial charge $q = q_\mathrm{o}$ will decrease by half, that is, $\Delta q = -\frac{1}{2}q_\mathrm{o}$. Then the question becomes,

$$\text{If } q = q_\mathrm{o} \text{ and } \Delta q = -\tfrac{1}{2}q_\mathrm{o}, \; \Delta t = ?$$

EXAMPLE 21-10 *continued*

What we know/what we don't.

$$q = q_0 \qquad \Delta q = -\tfrac{1}{2}q_0$$
$$R = 15\ \Omega \qquad C = 0.50\ \text{F} \qquad \Delta t = ?$$

The mathematical solution. Solving Equation 21-15 for Δt gives us

$$\Delta t \approx -\frac{RC}{q}\Delta q = -\frac{RC}{q_0}\left(-\frac{1}{2}q_0\right) = \frac{RC}{2} = \frac{(15\ \Omega)(0.50\ \text{F})}{2} = \textbf{3.8 s}$$

Making sense of the results. In this case, because capacitance is so large, the time interval $\Delta t \approx \frac{RC}{2}$ is long enough so that we would see the bulb glow. Before such capacitors became available, commercial capacitors might only have one-thousandth the capacitance of this one. Then the duration of the discharge would be only one-thousandth as long, and we would never see the glow that provides visible evidence of the discharge.

◆ Related homework: Problems 21-41 and 21-92.

In the above example, notice that the product RC has units of time and establishes the "ballpark" of the time interval over which charging or discharging occurs. Thus $\tau = RC$ is called the **time constant** of the circuit.

Exponential Decay

The result we calculated in Example 21-10 doesn't depend on the value of q_0. Therefore, if we reset our clock to zero after the charge has dropped by half, it will require the same time interval for the amount of charge we now have to be reduced by half. Likewise it will require the same time interval for *that* half to be reduced by half, and so on. After n like intervals, the initial amount will be reduced by a factor $\left(\frac{1}{2}\right)^n$, which depends on the exponent n. This pattern of decrease (Figure 21-19) is therefore known as **exponential decay.**

Exponential decay is a pattern typical of many situations in which the decrease in a quantity is proportional to how much remains. In general, we can write

$$\Delta\chi \approx -k\chi\Delta t \qquad (\Delta t\ \text{small}) \tag{21-16}$$

where χ (the Greek letter chi) stands for *quantity*. If, for example, χ stands for the number of radioactive nuclei remaining in a sample of a radioisotope, then the time interval required for each successive reduction of χ by half is called the **half-life** of the radioisotope. (In Equation 21-15, $k = \frac{1}{RC} = \frac{1}{\tau}$. It is in general true that $k = \frac{1}{\tau}$. Using calculus, it is possible show that the half-life $T_{1/2} = 0.6931\,\tau$—not too far off from the rough approximation of $0.5\,\tau = \frac{RC}{2}$ that we obtained in Example 21-10.)

It can be shown that as a capacitor is being charged, the current decays exponentially (as does $v_R = iR$; see Figure 21-17), which accounts for the bulb dimming and going out.

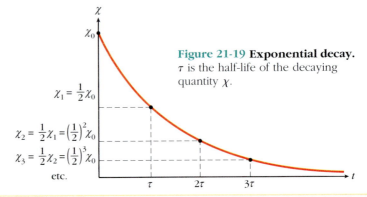

Figure 21-19 Exponential decay. τ is the half-life of the decaying quantity χ.

The current is not transient in all circuits involving capacitors and resistors. In the following case, such a circuit provides a model for a nonelectrical system where we emphatically do not want the flow to be transient.

Case 21-2 ◆ *An Electric Circuit Model for the Human Circulatory System*

Understanding electric circuits can give us insights into systems that behave analogously to electric circuits. The human circulatory system (see Figure 10-25) is also a closed circuit—unless you are bleeding, of course—in which blood flows instead of charge. Figure 21-20 shows how we can think of the human circulatory system analogously to an electric circuit. The blood vessels, like wires, offer resistance to flow. Parts of the circulatory system, specifically the heart and some blood vessels, are elastic; they expand under pressure. These parts are therefore

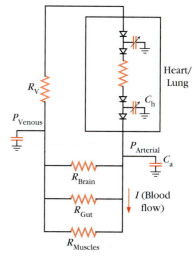

Figure 21-20 Electrical model of the human circulatory system. (Courtesy of S. Massaquoi.)

locations where an additional volume of blood can build up, just as capacitors are locations where additional amounts of charge can build up. When the heart contracts, it sends a flow of blood through the blood vessels, much as a discharging capacitor sends a flow of charge through a circuit. In the diagram, arrows on capacitors indicate that the capacitances are variable. The symbol ⟶▷⟶ indicates a *diode,* a device that permits a flow of charge in only one direction. For a fluid, a *valve* serves the same purpose. Because blood pressure is much greater in the arteries than in the veins and fluids flow from higher to lower pressure, the difference in pressure $P_{arterial} - P_{venous}$ functions like a potential difference.

The analogous quantities in electric circuits and the human circulatory system (note the common root *circu-*) are:

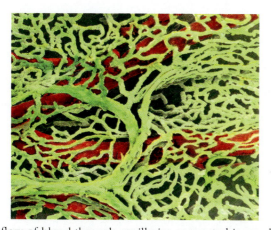

The flow of blood through capillaries connected in parallel can be analyzed analogously to the flow of electrons through wires connected in parallel. This scanning electron micrograph of capillaries in the gall bladder is artificially colored.

In Electric Circuit	In Circulatory System
Electrical charge q	Volume of blood V
$I = \frac{\Delta q}{\Delta t}$ = amount of charge per second that passes a point in a conductor	$I = \frac{\Delta V}{\Delta t}$ = volume of blood per second that passes a point in a blood vessel
$V = \Delta$(potential) = potential difference between one point and another	$P_1 - P_2$ = pressure difference between one point and another
R = electrical resistance due to dimensions and resistivity of conductor ($V = IR$)	R = resistance due to dimensions of blood vessel and viscosity of fluid ($P_1 - P_2 = IR$)
$C = \frac{Q}{V}$ = additional charge held by a capacitor for each volt of potential difference	$C = \frac{V}{P_1 - P_2}$ = additional volume held by a flexible part of the circulatory system for each unit of pressure difference

(*Caution:* Note that the symbol V has different meanings in the two systems. Also, in Chapter 10 we used Q rather than I as our symbol for the volume flow rate $\frac{\Delta V}{\Delta t}$.)

Because the relationships among analogous quantities are the same in the two systems, all of our reasoning about series and parallel circuits is equally valid for the human circulatory system (see Problems 21-69 and 21-70).

21-7   Kirchhoff's Rules for Direct Current Circuits

As Figure 21-11 shows, some circuits cannot be built up of series and parallel combinations. For those, we need to develop some more general rules. Two principles that we have used before can be generalized beyond simple series or parallel circuits.

Equation 21-6 ($I = I_1 + I_2 + I_3 + \cdots$) tells us (1) that at a junction where a circuit splits into several paths, the total current I arriving at the junction equals the sum of the currents leaving it by all possible paths, and (2) that at the junction where the parallel paths come back together, the sum of all currents arriving at the junction equals the total current I leaving it. This is necessary for charge conservation because no charge accumulates at either junction. German physicist Gustav Robert Kirchhoff (roughly pronounced *Keerk-hoff;* 1824–1887) stated this as a simple rule:

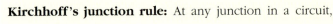

Kirchhoff's junction rule: At any junction in a circuit,

$$\Sigma I_{\text{arriving}} = \Sigma I_{\text{leaving}} \tag{21-17}$$

At the junction in Figure 21-21, for example, this means that $I_1 + I_2 + I_3 = I_4 + I_5 + I_6 + I_7$. In using the junction rule, there is no problem if you assume the wrong direction for one of the currents. You'll find this out if you get a negative value for that current because, for example, -5 A arriving really means $+5$ A leaving.

Equation 21-4 ($\Sigma V = 0$ around any closed path) is true not just for the single closed loop of a simple series circuit but for *each* closed loop in a more complicated circuit like those in Figures 21-22 and 21-23. Kirchhoff stated this as a rule as well:

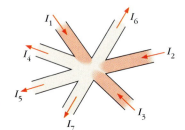

Figure 21-21 All flow of charge through this junction is subject to Kirchhoff's junction rule. $I_1 + I_2 + I_3 = I_4 + I_5 + I_6 + I_7$

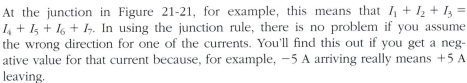

Kirchhoff's loop rule: Around each closed loop in a circuit,

$$\Sigma V = 0 \tag{21-18}$$

(Follow Procedure 21-1 to identify the V's in the sum)

Figure 21-22 shows Kirchhoff's junction rule applied to each junction in a circuit and Kirchhoff's loop rule applied to each complete loop in the circuit. We have begun by drawing an arrow representing the current in each *branch* of the circuit. Junctions are points where different circuit paths meet; *branches* are segments of the circuit that are uninterrupted by junctions. All the circuit elements along a branch are therefore in series, so the current is the same along the entire branch.

We have tried to draw each arrow in the direction in which we think the current is flowing. However, as we noted before, if a choice is wrong, we'll find out by getting a negative value for the current flowing in that direction.

For each loop, we have used Procedure 21-1 to identify the potential differences encountered in going completely around the loop in the direction shown (a ★ indicates the point where we start and end). Notice that our direction through R_3 in the last loop considered is opposite to the assumed current direction, so the + sign on I_3R_3 denotes a potential *increase* in that direction.

If the three branch currents are the only unknowns, we may appear to have more equations than unknowns. But Equations A and B are really the same, and Equation E could be obtained by subtracting Equation D from Equation C, so all the information is really contained just in Equations A, C, and D. In the following summary of our procedure, we suggest a way of keeping down to the minimum needed for solution the number of equations arising from Kirchhoff's rules.

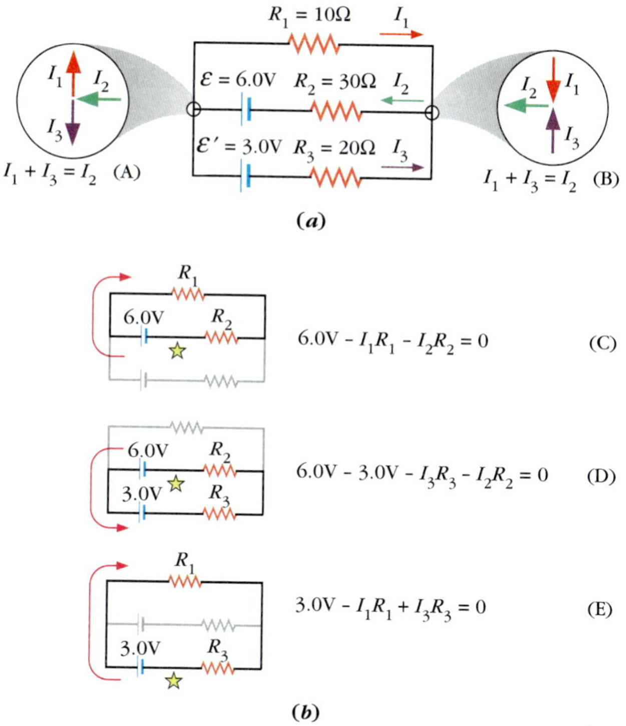

$R_1 = 10\Omega$
I_1
$\mathcal{E} = 6.0\text{V}$
$R_2 = 30\Omega$
I_2
$\mathcal{E}' = 3.0\text{V}$
$R_3 = 20\Omega$
I_3
$I_1 + I_3 = I_2$ (A)
$I_1 + I_3 = I_2$ (B)
(a)
$6.0\text{V} - I_1R_1 - I_2R_2 = 0$ (C)
$6.0\text{V} - 3.0\text{V} - I_3R_3 - I_2R_2 = 0$ (D)
$3.0\text{V} - I_1R_1 + I_3R_3 = 0$ (E)
(b)

Figure 21-22 Applying Kirchhoff's rules to a circuit in which no two resistors are in series or in parallel. (*a*) Kirchhoff's junction rule applied to each junction in the circuit. (*b*) Kirchhoff's loop rule applied to each loop in the circuit, going in the direction indicated by the red arrow.

PROCEDURE 21-2

Solving Problems Using Kirchhoff's Rules[1]

1. Label each branch current (I_1, I_2, etc.) and draw an arrow indicating its assumed direction in the circuit diagram.
2. Apply Kirchhoff's junction rule to enough junctions so that each branch current appears in at least one equation.
3. Apply Kirchhoff's loop rule to enough loops so that each branch current appears in at least one equation if possible.
4. In all, you must obtain as many equations as you have unknowns.
5. Solve the simultaneous equations obtained by steps 2 and 3.

[1]Adapted from A. Van Heuvelen, *Physics,* 2d ed. (Boston: Little, Brown, 1986).

Example 21-11 *Applying Kirchhoff's Rules*

Find the current in each branch of the circuit in Figure 21-22.

Solution

Choice of approach. We apply Procedure 21-2.

What we know/what we don't. See Figure 21-22 for what we know.

$$I_1 = ? \qquad I_2 = ? \qquad I_3 = ?$$

The mathematical solution. In Figure 21-22, Equation A alone satisfies Step 2 of Procedure 21-2, and Equations C and D together satisfy Step 3. In solving these equations simultaneously, we first use C and D to express I_1 and I_3 in terms of I_2.

$$I_1 = \frac{6.0\text{ V} - I_2R_2}{R_1} \quad \text{(from C)} \qquad \text{and} \qquad I_3 = \frac{3.0\text{ V} - I_2R_2}{R_3} \quad \text{(from D)}$$

EXAMPLE 21-11 *continued*

Substituting these into A gives us

$$I_2 = \frac{6.0 \text{ V}}{R_1} - \frac{I_2 R_2}{R_1} + \frac{3.0 \text{ V}}{R_3} - \frac{I_2 R_2}{R_3}$$

so

$$I_2\left(1 + \frac{R_2}{R_1} + \frac{R_2}{R_3}\right) = \frac{6.0 \text{ V}}{R_1} + \frac{3.0 \text{ V}}{R_3}$$

and

$$I_2 = \frac{\dfrac{6.0 \text{ V}}{R_1} + \dfrac{3.0 \text{ V}}{R_3}}{1 + \dfrac{R_2}{R_1} + \dfrac{R_2}{R_3}} = \frac{\dfrac{6.0 \text{ V}}{10 \text{ }\Omega} + \dfrac{3.0 \text{ V}}{20 \text{ }\Omega}}{1 + \dfrac{30 \text{ }\Omega}{10 \text{ }\Omega} + \dfrac{30 \text{ }\Omega}{20 \text{ }\Omega}} = \mathbf{0.14 \text{ A}}$$

Substituting this value back into the expressions we obtained for I_1 and I_3, we readily obtain

$$I_1 = \frac{6.0 \text{ V} - (0.136 \text{ A})(30 \text{ }\Omega)}{10 \text{ }\Omega} = \mathbf{0.19 \text{ A}}$$

and

$$I_3 = \frac{3.0 \text{ V} - (0.136 \text{ A})(30 \text{ }\Omega)}{20 \text{ }\Omega} = \mathbf{-0.054 \text{ A}}$$

Making sense of the results. The minus sign tells us we indeed chose the wrong direction to draw the arrow for I_3. Because the 6.0 V battery pumps harder than the 3.0 V battery, the flow would indeed be as we had first guessed if only the lower loop were present. But with the upper loop present, much of the charge pumped by the 6.0 V battery flows via that loop instead. We can check our results by seeing that they satisfy Equation E in Figure 21-22:

$$3.0 \text{ V} - (0.192 \text{ A})(10 \text{ }\Omega) + (-0.054 \text{ A})(20 \text{ }\Omega) \overset{\checkmark}{=} 0$$

◆ Related homework: Problems 21-42 through 21-46.

Figure 21-23 shows Kirchhoff's rules being applied to every possible junction and every possible loop of a circuit in which no two resistors are in series or in parallel. Following Procedure 21-2, we need only write down those equations shown in bold in the figure. I_6 doesn't appear in any loop equation because there are no resistors along its branch (the reason for "if possible" in Step 3 of the procedure), but as step 4 requires, we have six equations in bold, so we could solve for the six branch currents.

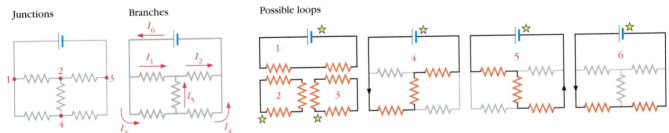

Figure 21-23 Applying Kirchhoff's rules to another circuit in which no two resistors are in series or in parallel.

Junction rule
At junction 1: $I_6 = I_1 + I_3$
At junction 2: $I_1 + I_5 = I_2$
At junction 3: $I_2 + I_4 = I_6$
At junction 4: $I_3 = I_4 + I_5$

Loop rule
Loop 1: $\mathcal{E} - I_1 R_1 - I_2 R_2 = 0$
Loop 2: $-I_3 R_3 - I_5 R_5 + I_1 R_1 = 0$
Loop 3: $-I_4 R_4 + I_2 R_2 + I_5 R_5 = 0$
Loop 4: $\mathcal{E} - I_3 R_3 - I_5 R_5 - I_2 R_2 = 0$
Loop 5: $\mathcal{E} - I_1 R_1 + I_5 R_5 - I_4 R_4 = 0$
Loop 6: $\mathcal{E} - I_3 R_3 - I_4 R_4 = 0$

◆ SUMMARY ◆

To analyze circuits quantitatively, we must identify how elements are connected:

- Circuit elements E_1 and E_2 are connected **in series** if a terminal of E_1 is connected to a terminal of E_2 and nothing else connects at this junction.
- Two or more circuit elements are connected **in parallel** if they are connected between the same two points: One terminal of each connects at a single common point P_1 and the other terminal of each connects at a single common point P_2.

The current must be the same through each conducting element in series.

$$I_1 = I_2 = \ldots \text{ through elements } E_1, E_2, \ldots \text{ in series} \quad (21\text{-}1)$$

The potential difference is the same across all elements in parallel.

$$V_1 = V_2 = \ldots \text{ across elements } E_1, E_2, \ldots \text{ in parallel} \quad (21\text{-}2)$$

An **ammeter** must be connected *in series* with an element to measure the *current through* it. A **voltmeter** must be connected *in parallel* with an element to measure the *potential difference across* it (between its ends or terminals).

Procedure 21-1 summarizes the potential differences across various circuit elements:

- the EMF $\mathcal{E}$ of a battery
- $V = IR$ across a resistor (or a bulb)
- $V = \mathcal{E} - Ir$ across a battery with internal resistance
- $V = \frac{Q}{C}$ across a capacitor

To analyze series circuits, you must know the following in addition to Equation 21-1:

The potential differences across elements in series are additive:

$$V_{\text{path}} = V_1 + V_2 + \cdots \text{ across elements } E_1, E_2, \ldots \text{ in series} \quad (21\text{-}3)$$

$$V_{\text{path}} = 0 \quad \text{for any closed path} \quad (21\text{-}4)$$

The *equivalent resistance* of resistors $R_1, R_2, R_3, \ldots$ in series is

$$R = R_1 + R_2 + R_3 + \cdots \quad (21\text{-}5)$$

The charge is the same on each of capacitors $C_1, C_2, C_3, \ldots$ in series and on the combination:

$$Q_1 = Q_2 = Q_3 = \cdots = Q \quad (21\text{-}11)$$

$$\text{Equivalent capacitance} \quad \frac{1}{C} = \frac{1}{C_1} + \frac{1}{C_2} + \frac{1}{C_3} + \cdots \quad (21\text{-}12)$$

for capacitors $C_1, C_2, C_3, \ldots$ in series

To analyze parallel circuits, you must know the following in addition to Equation 21-2:

$$\text{Total current} \quad I = I_1 + I_2 + I_3 + \cdots \quad (21\text{-}6)$$

between two points connected by resistors $R_1, R_2, R_3, \ldots$ in parallel

$$\frac{1}{R} = \frac{1}{R_1} + \frac{1}{R_2} + \frac{1}{R_3} + \cdots \quad (21\text{-}7)$$

to calculate the *equivalent resistance R* of resistors $R_1, R_2, R_3, \ldots$ in parallel

$$\text{Total charge} \quad Q = Q_1 + Q_2 + Q_3 + \cdots \quad (21\text{-}8)$$

on capacitors $C_1, C_2, C_3, \ldots$ in parallel

$$\text{Equivalent capacitance} \quad C = C_1 + C_2 + C_3 + \cdots \quad (21\text{-}9)$$

for capacitors $C_1, C_2, C_3, \ldots$ in parallel

Many networks can be built up of series and parallel combinations. When this is not the case, we must apply Kirchhoff's rules (see **Procedure 21-2**):

Kirchhoff's junction rule: At any junction in a circuit,

$$\Sigma I_{\text{arriving}} = \Sigma I_{\text{leaving}} \quad (21\text{-}17)$$

Kirchhoff's loop rule: Around each closed loop in a circuit,

$$\Sigma V = 0 \quad (21\text{-}18)$$

(Follow Procedure 21-1 to identify the *V*'s in the sum)

In transient **RC circuits** in which a battery is switched in or out of the circuit (Figure 21-16),

$$\mathcal{E} = iR + \frac{q}{C} \quad \text{(capacitor charging)} \quad (21\text{-}13)$$

$$iR + \frac{q}{C} = 0 \quad \text{(capacitor discharging)} \quad (21\text{-}14a)$$

from which it follows that

$$\Delta q \approx -\frac{q}{RC}\Delta t \quad \text{(capacitor discharging)} \quad (21\text{-}15)$$

This is an instance of the general pattern for **exponential decay**

$$\Delta \chi \approx -k\chi \Delta t \quad (21\text{-}16)$$

The **time constant** $\quad \tau = \frac{1}{k} = RC$

describes the "ballpark" of the time interval over which decay occurs.

◆QUALITATIVE AND QUANTITATIVE PROBLEMS◆
Hands-On Activities and Discussion Questions

The questions and activities in this group are particularly suitable for in-class use.

21-1. Discussion Question. How is the insertion of the **H**-shaped conducting piece between the plates of a capacitor in Figure 21-14 different than the insertion of a slab of dielectric between the plates?

21-2. Discussion Question. *Feedback in flow patterns.*
a. The exits from a traffic circle have been temporarily closed off, leaving the cars going around it bumper to bumper (Figure 21-24*a*). The driver at position P decides she may as well slow down.
b. What effect does that have on the cars immediately behind hers? Why? What effect does it have on the cars immediately in front of hers? Why?
c. In the circuit in Figure 21-24*b*, R_2 is a variable resistor. When the resistance of R_2 is increased, what effect does it have

on the current through resistor R_1? Why? What effect does it have on the current through resistor R_3? Why?

d. The effect that a change in traffic flow at one point has on the traffic behind it is called *feedback*. How does the motion of charge carriers in one part of a circuit affect the motion of charge carriers elsewhere in the circuit? Since the charge carriers are not bumper to bumper (that is, they are at a distance from one another), by what mechanism or means do they produce this effect?

21-3. Discussion Question.
The circuit in Figure 21-25 will become a simple series circuit when the free ends of wires A and B are brought together.

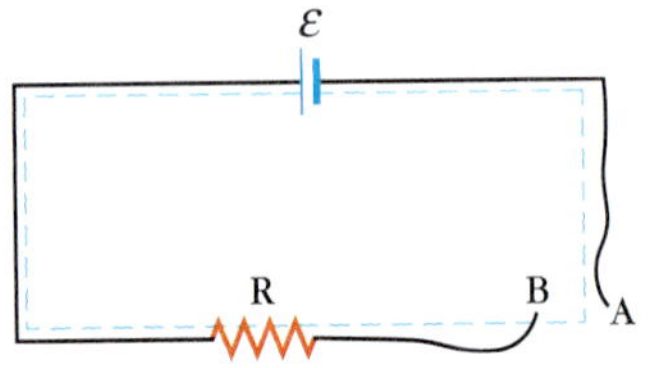

a. Before the two free ends are brought together, is the sum of the potential gains and drops equal to zero around the dotted path? Briefly explain.

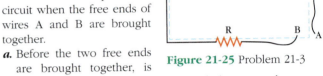

Figure 21-25 Problem 21-3

b. As the free ends of the wire are brought closer together, what happens to the electric field between these ends? Does it increase, decrease, remain zero, or remain constant but not at zero?

c. Tell in a few words what will happen when the two free ends are brought as close to each other as possible without touching. If possible, ask your instructor to set this up as a demonstration to see what actually happens.

d. Did answering *c* make you reconsider your answers to *a* and *b*? Briefly explain.

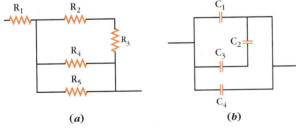

(*a*)   (*b*)

Figure 21-24 Problem 21-2

Review and Practice

Section 21-1 Types of Circuit Connections

21-4. In the circuit segment shown in Figure 21-26*a*, tell whether each of the following pairs of resistors is in parallel, in series, or neither. **a.** R_1 and R_2.   **b.** R_2 and R_3.   **c.** R_2 and R_4.   **d.** R_4 and R_5.

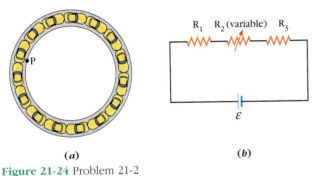

(*a*)   (*b*)

Figure 21-26 Problems 21-4 and 21-5

21-5. SSM Repeat Problem 21-4 for each of the following pairs of capacitors in Figure 21-26*b*. **a.** C_1 and C_2.   **b.** C_2 and C_3.   **c.** C_1 and C_4.   **d.** C_3 and C_4.

21-6. Which of the circuit diagrams in Figure 21-27 correctly represents the circuit shown in the top half of the figure?

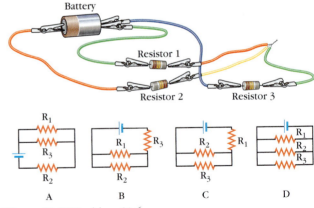

Figure 21-27 Problem 21-6

Section 21-2 **Measuring Current and Voltage**

21-7. In the circuit in Figure 21-28*a,* which meter or meters, if any, correctly read each of the following?
a. The rate at which charge flows through R_2.
b. The current through R_1.
c. The current through the battery.
d. The potential difference across R_1.
e. The terminal voltage of the battery.
f. The EMF of the battery.

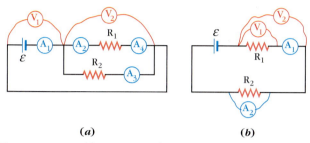

(a) **(b)**

Figure 21-28 Problems 21-7 and 21-8

21-8. In the circuit in Figure 21-28*b,* which meter or meters, if any, correctly read each of the following?
a. The current through R_1.
b. The potential difference across R_1.
c. The current through R_2.

Section 21-3 **Series Circuits**

21-9. In the circuit in Figure 21-29, the capacitor charges when the switch is closed. Consider the following four elements in the circuit: the resistor R_1, the capacitor C, the resistor R_2, and the battery. Suppose you go around the circuit clockwise (the directional of conventional current). Across which (one or more) of these elements is there a
a. potential increase when the switch is first closed?
b. potential decrease when the switch is first closed?
c. potential increase long after the switch is closed?
d. potential decrease long after the switch is closed?

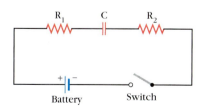

Figure 21-29 Problems 21-9 and 21-10

21-10. At a certain instant after the switch is closed in the circuit in Figure 21-29, the voltage across each resistor is 0.65 V and the terminal voltage of the battery is 1.48 V. At this instant, what is the potential difference across the capacitor?

21-11. SSM A voltmeter connected between the terminals of an otherwise unconnected battery reads 1.54 V. When a small bulb is connected between the terminals as well, it draws a current of 0.105 A and the voltmeter reading drops to 1.47 V.
a. Find the internal resistance of the battery.
b. Find the resistance of the bulb.

21-12. Repeat Example 21-2 if the voltmeter across the battery reads 1.46 V after the circuit is complete rather than before.

Section 21-4 **Resistive Circuits**

21-13. A 60 Ω resistor and a 120 Ω resistor are connected in series to a 9.0 V battery with negligible internal resistance. How much current flows through each resistor?

21-14. Resistor A has twice as much resistance as resistor B. Suppose these two resistors are connected in series between the terminals of a battery. *a.* Compare the current through the two resistors. *b.* Compare the potential difference across resistor A with the potential difference across resistor B. *c.* Compare the potential difference across resistor A with the potential difference across the battery. Express your answer as a ratio $\frac{\text{potential difference across resistor A}}{\text{potential difference across the battery}}$.

21-15. Two resistors A and B are connected in series between the terminals of a battery. The battery's terminal voltage when all connections are secure is 9.06 V, and the current through the circuit is 0.200 A. If the resistance of resistor A is 16.4 Ω, what is the resistance of resistor B?

21-16. A student connects an arrangement of three 20 Ω resistors between the terminals of a 1.50 V battery (with negligible internal resistance). If there is a current of 0.050 A through the battery when all connections are secure, are all three resistors connected in series? Briefly explain.

21-17. A student has a bulb that glows only very dimly when properly connected to a battery. The student reasons, "If I place a second identical battery in parallel with the first, my globe will glow more brightly." Will the student's approach be effective? Briefly explain your reasoning.

21-18. Find the equivalent resistance that is obtained when three resistors whose resistances are 5.0 Ω, 8.0 Ω, and 10.0 Ω, are connected *a.* in series. *b.* in parallel.

21-19. Suppose that the combination of resistors in Problem 21-18 is connected to a 1.5 V battery. How much current will flow through each resistor if the resistors are connected *a.* in series? *b.* in parallel?

21-20. Suppose that in Figure 21-30, the voltmeter reads 3.0 V and ammeter A_2 reads 0.20 A. Find the resistance of resistor R_2 if *a.* $R_1 = 10$ Ω *b.* ammeter A_1 reads 0.12 A. *c.* ammeter A_1 reads 0.

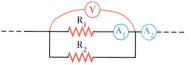

Figure 21-30 Problems 21-20 and 21-47

21-21. Should Christmas tree bulbs be connected in series or in parallel? Why?

21-22. *Reasoning qualitatively and quantitatively about a circuit.* Three identical bulbs of equal resistance R are connected in parallel across a battery of voltage V. If one bulb is removed from its socket, how is the brightness of each of the others affected; does it increase, decrease, or remain the same? First write down what you think, then work through the following lines of reasoning to see if they support your answer. As you do so, try to make connections between the qualitative and quantitative reasoning.

Qualitative Reasoning

a1. If we remove one bulb, does that alter the potential difference between the two terminals of the battery?

b1. Does the potential difference across the other two bulbs change?

c1. How does the potential difference across each remaining bulb compare to the potential difference of the battery?

d1. Does the resistance of either remaining bulb change?

e1. Based on your answers to **c1** and **d1**, does the current through either remaining bulb change when the third one is removed?

f1. What conclusion can you draw about the brightness of the remaining bulbs?

Quantitative Reasoning

(For this part, assume that for each bulb $R = 15\ \Omega$, and for the battery $V = 1.50$ V)

a2. Find the equivalent resistance of the combination of bulbs when all three are present.

b2. Find the *total* current that flows through this combination.

c2. How much of this current flows through each bulb? Give a separate answer for each bulb.

d2. Now find the equivalent resistance of the combination after one bulb is removed.

e2. Find the total current I_t through *this* combination.

f2. How much of this current flows through each remaining bulb? Compare with **c2**. What do you conclude from this comparison about the brightness of the remaining bulbs?

g. Does your answer to **f2** agree with your answer to **f1**?

21-23. In which of the circuits in Figure 21-31 will bulb A glow most brightly?

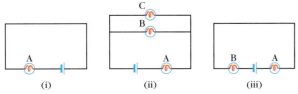

Figure 21-31 Problem 21-23

21-24. In which of the circuits in Figure 21-32 will bulb A glow most brightly?

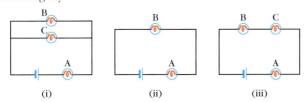

Figure 21-32 Problem 21-24

21-25. Find the equivalent resistance of the arrangement of in the circuit shown in Figure 21-33a.

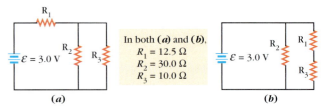

In both (**a**) and (**b**),
$R_1 = 12.5\ \Omega$
$R_2 = 30.0\ \Omega$
$R_3 = 10.0\ \Omega$

Figure 21-33 Problems 21-25, 21-26, 21-27, and 21-28

21-26. For the circuit shown in Figure 21-33a, find **a.** the potential difference across each resistor. **b.** the current through each resistor.

21-27. **SSM WWW** Find the equivalent resistance of the arrangement of in the circuit shown in Figure 21-33b.

21-28. For the circuit shown in Figure 21-33b, find **a.** the potential difference across each resistor. **b.** the current through each resistor.

21-29. In Example 21-6 in Section 21-4,
a. find the current through each elf when the manufacturer's suggested circuit is connected to a 120 V outlet.
b. find the current through each elf when Uncle Seamus's suggested circuit is connected to a 120 V outlet.
c. Do your results agree with the conclusions of Example 21-5 about how the brightness of the elves compares in the two circuits?

21-30. The brightness of the elves in Examples 21-5 and 21-6 depends on the power output of each elf. Using the results of Problem 21-29, find the power dissipated by each elf when **a.** the manufacturer's suggested circuit is connected to a 120 V outlet. **b.** Uncle Seamus's suggested circuit is connected to a 120 V outlet.

Section 21-5 Circuits with Capacitors

21-31. Find the equivalent capacitance that is obtained when three capacitors whose capacitances are 10 mF, 15 mF, and 30 mF, are connected **a.** in parallel. **b.** in series.

21-32. Suppose that the combination of capacitors in Problem 21-31 is connected to a 1.5 V battery. How much charge will be stored on each capacitor if the capacitors are connected **a.** in parallel? **b.** in series?

21-33.
a. Calculate the equivalent capacitance obtained when three capacitors having capacitances 2.0 mF, 3.0 mF, and 4.0 mF are connected in series. How does the result compare with the individual capacitances?
b. How much charge is stored on this combination of capacitors if it is connected between the terminals of a 1.5 V battery?
c. With the connection described in **b**, how much charge is stored on each individual capacitor?

21-34. When two capacitors are connected in series, the equivalent capacitance of the combination is less than either individual capacitance. What happens to the equivalent capacitance of the combination if the capacitance of each individual capacitor is doubled? Briefly explain.

21-35.

a. When two otherwise identical resistors having cross-sectional areas A_1 and A_2 are connected in parallel, what is the effective cross-sectional area of the combination?

b. When two otherwise identical parallel plate capacitors having plate areas A_1 and A_2 are connected in parallel, what is the effective plate area of the combination?

c. Are the plate areas of capacitors directly or inversely proportional to their capacitances?

d. Are the cross-sectional areas of resistors directly or inversely proportional to their resistances?

e. From your answers to **b** and **c**, what can you conclude about the way the capacitances of individual capacitors must be combined to obtain the equivalent capacitance when they are connected in parallel?

f. From your answers to **a** and **d**, what can you conclude about the way the resistances of individual resistors must be combined to obtain the equivalent resistance when they are connected in parallel?

21-36. For the circuit shown in Figure 21-34a, find **a.** the potential difference across each capacitor. **b.** the charge on each capacitor.

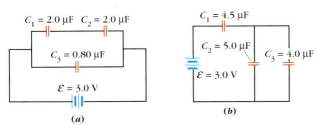

Figure 21-34 Problems 21-36 and 21-37

21-37. SSM For the circuit shown in Figure 21-34b, find **a.** the potential difference across each capacitor. **b.** the charge on each capacitor.

Section 21-6 **Circuits with Transient Currents**

21-38. As an intermediate step in the mathematical solution to Example 21-9, we found that

$$i = \frac{\mathcal{E}}{R} - \frac{q}{RC}$$

a. What does this equation tell you happens to the current as the capacitor becomes charged? **b.** What is the maximum value of the charge, and what does this equation tell you about the current when the charge reaches that maximum value?

21-39. A 0.220 F capacitor is charged by connecting it in series with a 1.54 V battery and a 13.0 Ω resistor. How much current is there through the resistor when the capacitor

a. has acquired $\frac{1}{4}$ of the charge it will have when fully charged?

b. has acquired $\frac{3}{4}$ of the charge it will have when fully charged?

c. is fully charged?

21-40. In Example 21-10, we saw that when we use a giant capacitance capacitor, the time constant RC is long enough so

that if the "resistor" is a small bulb, we can see it glow for a few seconds as the capacitor charges. But we can get the same time constant with a small capacitance if we choose a very large resistance. Why shouldn't we see the same glow under those conditions?

21-41. SSM A 2.0×10^{-4} F capacitor has 6.0×10^{-4} C of charge stored on it. If you then discharge the capacitor through a bulb filament with a resistance of 15 Ω, **a.** what is the current through the filament at the instant when discharge begins? **b.** what is the time constant for this discharge?

Section 21-7 **Kirchhoff's Rules**

21-42. When the switch is closed in the circuit in Figure 21-35, the ammeter reads 0.60 A, the current through resistor R_3 is 0.40 A, and the current through resistor R_4 is 0.32 A. Find the current through each of the remaining resistors.

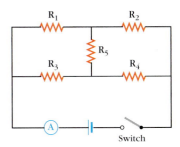

Figure 21-35 Problem 21-42

21-43. Redo Problem 21-26 using Kirchhoff's rules.

21-44. Redo Problem 21-28 using Kirchhoff's rules.

21-45. SSM WWW Calculate the current through each of the resistors in the circuit in Figure 21-36.

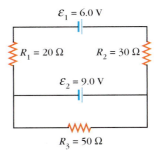

Figure 21-36 Problem 21-45

21-46. Calculate the current through each of the resistors in the circuit in Figure 21-37.

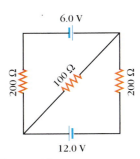

Figure 21-37 Problem 21-46

Going Further

The questions and problems in this group are not organized by section heading, so you must determine for yourself which ideas apply. Some of them will be more challenging than the Review and Practice questions and problems (especially those marked with a • or ••).

•21-47.

a. Suppose that in Figure 21-30, the voltmeter reads 2.7 V and ammeter A_2 reads 0.50 A. What will ammeter A_1 read if $R_2 = 8.1 \Omega$?

b. Can you solve the problem in the same way if A_2 reads 0.16 A? Briefly explain.

21-48. Two storage batteries are connected in series. The EMF of the first is 6.30 V and its internal resistance is 0.60 Ω. The EMF of the second is 12.6 V and its internal resistance is 0.80 Ω. The batteries are connected to a 48.6 Ω external load. What current flows through the resulting circuit if the batteries are connected **a.** negative pole to positive pole? **b.** negative pole to negative pole?

•21-49.

a. Show that because $\frac{1}{C} = \frac{1}{C_1} + \frac{1}{C_2}$ for two capacitors in series, the equivalent capacitance of the combination is less than that of either individual capacitor.

b. Does it follow from this that if several capacitors are connected in series, the equivalent capacitance of the combination is less than that of any one of the individual capacitors? Explain.

21-50. A series circuit contains a switch, two properly connected 1.5 V batteries, a 20 Ω resistor, and a capacitor. If the time constant for the circuit is 5.0 s find **a.** the capacitance of the capacitor. **b.** the maximum current in the circuit. **c.** the charge on the capacitor when the current has dropped to half its maximum value.

21-51. **SSM WWW** A 0.220 F capacitor has a charge of 0.572 C after being connected to a power supply for a long time. The power supply is then disconnected, and a 13.0 Ω resistor is connected between the terminals of the capacitor. How much current is there through the resistor when the capacitor **a.** has $\frac{3}{4}$ of the charge it had when first disconnected from the power supply? **b.** has $\frac{1}{4}$ of the charge it had when first disconnected from the power supply? **c.** is totally discharged?

21-52. Two bulb sockets are connected in parallel between points A and B. A 14 Ω bulb is screwed into one; the other remains empty.

a. Use the formula for resistances in parallel to find the equivalent resistance between A and B. To answer this, what must you assume about the resistance of the empty socket?

b. Is the resistance of this parallel combination less than, equal to, or greater than the resistance of the bulb alone? Is this what you would expect from your qualitative understanding of how resistors work? Briefly explain.

21-53. **SSM** In the circuit in Figure 21-38, a sliding contact P can be moved along a resistance wire stretched between

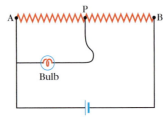

Figure 21-38 Problem 21-53

points A and B. As the contact slides from A to B, does the brightness of the bulb increase, decrease, or remain the same? Briefly explain your reasoning.

21-54. The three bulbs in the circuit in Figure 21-39 are identical. Rank the bulbs in order of brightness, from least to greatest and indicating any equalities, **a.** when the switch is open. **b.** when the switch is closed.

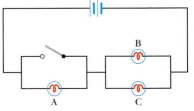

Figure 21-39 Problem 21-54

21-55. The three bulbs in the circuit in Figure 21-40 are identical. If the switch S is closed, **a.** how is the brightness of bulb A affected? Briefly explain why. **b.** how is the brightness of bulb B affected? Briefly explain why.

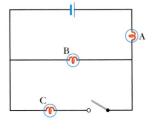

Figure 21-40
Problem 21-55

21-56. What is the effect on the resistance of a combination of resistors if one of the resistors has an effectively infinite resistance if the resistors are connected in **a.** series? **b.** parallel? **c.** What does it mean *physically* to say that a resistor has an effectively infinite resistance?

21-57. What is the effect on the capacitance of a combination of capacitors if one of the capacitors has zero capacitance if the capacitors are connected in **a.** series? **b.** parallel? **c.** What does it mean *physically* to say that a capacitor has zero capacitance?

•21-58. You are given a battery, two resistors, two simple switches S_1 and S_2, and a supply of connecting wire. Design and sketch a single circuit that satisfies both of the following requirements:

i. When S_1 is open and S_2 is closed, the resistors are connected across the battery in series.

ii. When S_1 is closed and S_2 is open, the resistors are connected across the battery in parallel.

21-59. When the switch S is closed in the circuit in Figure 21-41, will the capacitor become charged or not? Briefly explain your reasoning.

•21-60. N identical bulbs of resistance R are connected in parallel between two points having a potential difference V between them. Show by a symbolic calculation that the current through each bulb is the same as it would be if that were the only bulb between the two points.

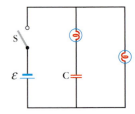

Figure 21-41
Problem 21-59

21-61.

a. The conductance of a circuit element is the inverse of its resistance (conductance $= \frac{1}{R}$). If elements are connected in parallel, write an equation expressing how their conductances combine to give the equivalent conductance of the combination.

b. Repeat for the case in which the elements are connected in series.

c. How do conductances compare to resistances in the way they combine? How do they compare to capacitances?

21-62. A 4 Ω resistor and a 2 Ω resistor are connected in parallel. If you want an equivalent resistance of 3 Ω, how should you connect a third resistor to these (show by a sketch) and what should be its resistance?

21-63. How can you arrange three 40 Ω resistors between the terminals of a 6.0 V dry cell (battery) so that the maximum current that flows through any of the resistors is 0.10 A? Sketch your arrangement.

21-64.

a. A fuse melts (or "blows") if it draws more than the current for which it is rated, leaving an open circuit and preventing a continued flow of excessive current, which would cause overheating. Verify by a calculation that when a 30 Ω hair dryer is connected to a 120 V outlet, it does not draw enough current to blow a 5 A fuse. Give a value for the quantity you've calculated.

b. If two of these hair dryers are connected to the outlet in parallel, what total current will they draw?

c. To prevent the fuse from blowing, a resistor can be connected in series with the pair of hair dryers. What must its minimum resistance be?

21-65. Two bulbs are each rated 40 W, 120 V. Compare the power P_{series} that each bulb dissipates when they are connected in series with the power $P_{parallel}$ that each bulb dissipates when they are connected in parallel. Express your answer as a ratio $\frac{P_{series}}{P_{parallel}}$ of these two quantities.

21-66. You have at your disposal three capacitors. Their capacitances are 20 μF, 30 μF, and 60 μF. How would you connect some or all of these to obtain each of the following equivalent capacitances? In each case, you should sketch a diagram of the capacitor network, and calculate its equivalent capacitance. **a.** 110 μF. **b.** 90 μF. **c.** 10 μF. **d.** 12 μF. **e.** 40 μF. •**f.** 27.27 μF.

21-67. SSM *The potentiometer.* In Figure 21-42, the arrow represents a movable contact with a long resistance wire stretched between a and c.

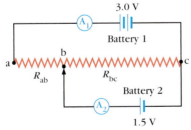

Figure 21-42 Problem 21-67

a. When the contact is positioned so that $R_{ab} = 40$ Ω and $R_{bc} = 80$ Ω, find the current read by each ammeter.

b. To what position should the contact be moved (i.e., what should be the values of R_{ab} and R_{bc}) for ammeter A_2 to read zero? *Note:* When this is the case, then if I is the value read by A_1, $\mathcal{E}_2 = IR_{bc}$. This allows us to calculate $\mathcal{E}_2$ if it is

unknown. Under these conditions, the terminal voltage and the EMF $\mathcal{E}_2$ of the second battery are the same because there is no current through its internal resistance. A device using this principle to measure the EMF of a source without drawing current from it is called a **potentiometer.**

21-68. *The Wheatstone bridge.* In the circuit in Figure 21-43, the resistance wire AB is 100.0 cm long and has a resistance of 100 Ω, so that each centimeter of its length has a resistance of 1.0 Ω. An ammeter is connected between point C and a sliding contact P. Resistor R_1 has a resistance of 20 Ω.

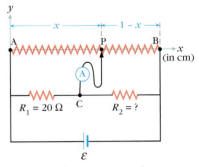

Figure 21-43 Problem 21-68

a. Suppose the contact is slid along the resistance wire until the ammeter reads zero. What is the potential difference between C and P at this position P?

b. How do the potential differences V_{AP} (between points A and P) and V_{AC} (between points A and C) compare when the ammeter reads zero?

c. How do the potential differences V_{PB} and V_{CB} compare when the ammeter reads zero?

d. Suppose the ammeter reads zero when the contact point is at $x = 80$ cm. What is the resistance of the resistor R_2?

e. If you replaced resistor R_2 with a different resistor, how could you use this set-up to measure its resistance? When used to measure unknown resistances, this type of circuit is called a **Wheatstone bridge.**

21-69. Consider the analogies developed in Case 21-2 as you answer these questions.

a. Compare the current in the wires leading away from a capacitor when the capacitor first begins discharging and when the capacitor is almost fully discharged.

b. In a similar way, compare the blood flow rate in the blood vessels leading away from the heart when the heart first starts contracting and when it is almost fully contracted.

c. Compare the potential difference across a capacitor when it first begins discharging and when it is almost fully discharged.

d. There is always a difference between the blood pressure of blood leaving the heart and blood returning to the heart. Compare this pressure difference when the heart first starts contracting to the pressure difference between the same two points when the heart is almost fully contracted. Briefly explain your reasoning.

e. Tell how Ohm's law provides a connection between your answers to **a** and **c.**

f. Is there an analogous connection between your answers to **b** and **d**? Briefly explain.

21-70. *Use the analogies developed in Case 21-2 to answer this question.* Figure 21-44 shows a small cluster of blood vessels. Suppose that blood flows toward the right past point A at a flow rate of Q. If the resistance of the length of blood vessel between points A and B can be expressed as $R_1 = R$, then by comparison the resistances of the upper and lower lengths of blood vessel between points B and C can be expressed as $R_2 = 3R$ and $R_3 = 6R$. Find each of the following, expressed in terms of Q and/or R:

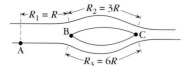

Figure 21-44 Problem 21-70

a. The flow rate through the upper vessel between points B and C.

b. The flow rate through the lower vessel between points B and C.

c. The difference in blood pressure between points B and C.

d. The difference in blood pressure between points A and C.

21-71. Resistor A is connected between the terminals of a battery that has negligible internal resistance.

a. If resistor B, which is identical to resistor A, is taken from a drawer and connected in parallel with resistor A, how will it affect the potential difference across resistor A? Explain your reasoning.

b. If the battery's internal resistance were *not* negligible, how (if at all) would this change your answer to *a*? Explain your reasoning.

21-72. Student X gets a brainstorm. "The power ratings printed on all my appliances really depend on their resistance. The more resistance I have, the less current I draw and the less power I use. Therefore if I leave all my appliances on, the resistances will be additive. With a higher total resistance I'll draw less current. Then since $P = VI$, I'll use less power and save on my electricity bills!" Identify the flaw(s) in Student X's reasoning. (*Note:* This problem is repeated from the previous chapter, but there may be additional arguments that you can make now.)

21-73. What does Equation 21-3 tell us about the changes (differences) in surface charge density that are responsible for the potential differences?

21-74. The bulbs and the batteries in the three circuits in Figure 21-45 are all identical. Rank these circuits in order of how brightly bulb 3 glows when connected in that circuit. Order them from least to greatest, making sure to indicate any equalities.

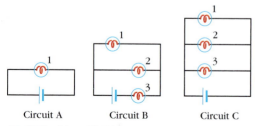

Figure 21-45 Problem 21-74

21-75. In Example 21-5, the correct way to connect the light-up Christmas elves (or any bulbs) is in parallel, because that way each elf draws the same current as if it were the only elf in the circuit. Is there any limit on the number of elves you can connect in parallel? Briefly explain why or why not.

21-76. A 120 V voltage source is connected to a circuit breaker and to a group of objects in a household that are wired in parallel. The circuit breaker is tripped when the current through it exceeds 10 A. Suppose all the objects connected in parallel are identical bulbs, each with a resistance of 240 Ω. **a.** What is the power rating of each bulb? **b.** What maximum number of these bulbs can be connected in parallel without tripping the circuit breaker?

21-77. In Example 21-6, how much energy is used by Uncle Seamus's circuit in one hour?

21-78. A capacitor is charged by connecting it in series with a battery and bulb. At a certain instant, the current has dropped to one quarter of its initial value. At this instant,

a. the charge on the capacitor is _____ (*one-quarter of its initial value; three-quarters of its initial value; one-quarter of its final value; three-quarters of its final value; none of these*).

b. the voltage across the bulb is _____ (*one-quarter of its initial value; three-quarters of its initial value; one-quarter of its final value; three-quarters of its final value; none of these*).

c. the total voltage across the bulb and capacitor in combination is _____ (*one-quarter of its initial value; three-quarters of its initial value; one-quarter of its final value; three-quarters of its final value; none of these*).

21-79. The equation $\Delta q \approx -\frac{q}{RC}\,\Delta t$ describes a capacitor being discharged. When this equation is applied to the charge on the negative plate of the capacitor, the quantity Δq (*is positive and its absolute value gets smaller; is positive and its absolute value gets larger; is negative and its absolute value gets smaller; is negative and its absolute value gets larger*) as the clock keeps advancing by the same fixed interval Δt.

21-80. A voltmeter is connected between the terminals of an otherwise unconnected battery. When a small bulb is connected between the terminals as well, the voltage reading on the voltmeter drops because _____ (*the bulb filament is heating up; the battery has internal resistance; the current is going through the bulb instead of the voltmeter; the voltmeter reads just the EMF of the battery once the bulb is connected*).

21-81. Show by a sketch how three 40 Ω resistors can be connected so that the effective resistance of the three in combination is **a.** 27 Ω. **b.** 60 Ω.

• **21-82.** A student connects an arrangement of three 40 Ω resistors between the terminals of a 1.50 V battery (with negligible internal resistance). There is a current of 0.025 A through the battery when all connections are secure. Sketch the circuit, carefully showing how the resistors can be connected to produce this current.

21-83. A student is given a battery, a bulb, a large-capacitance capacitor, and some connecting wires and is asked to charge the capacitor through the bulb. But instead of connecting the bulb and capacitor in series, the student connects them in parallel. As a result, the bulb _____ (*dims exactly as if the bulb and capacitor were in series; dims but more slowly than if the bulb and capacitor were in series; dims but more quickly than if the bulb and capacitor were in series; doesn't dim at all; doesn't light up at all*).

21-84. A student charges a large-capacitance capacitor by connecting it in series with a battery, a 42 Ω bulb, and an ammeter. About a second after the bulb has stopped glowing, the student notices that the ammeter still shows a small nonzero reading.

a. This indicates that (*the current through the ammeter is not passing through the bulb; there is still a current through the filament but not enough to make it glow; the ammeter readings are delayed; all the current has passed the bulb but some of it hasn't yet reached the capacitor*).

b. If the ammeter is connected in parallel with the bulb instead of in series, then as the capacitor charges ____ (*the bulb won't glow; the bulb will glow but the ammeter will read zero; the bulb will dim more quickly than with the ammeter in series; the bulb will dim more slowly than with the ammeter in series*).

21-85. A 40 W bulb and a 60 W bulb are connected to a 120 V voltage supply. What is the power output from each bulb if the two bulbs are **a.** connected in series to the voltage supply? **b.** connected in parallel to the voltage supply?

21-86. As the number of appliances connected in series to a 120 V voltage supply increases, does the total amount of power drawn increase, decrease, or remain the same? Briefly explain.

21-87. As the number of appliances connected in parallel to a 120 V voltage supply increases, does the total amount of power drawn increase, decrease, or remain the same? Briefly explain.

21-88. How can you arrange the three resistors in Problem 21-16 to get a current of 0.050 A through the battery?

21-89. **SSM** Four resistors each have a resistance of 100 Ω. Sketch an arrangement of these four resistors that has an equivalent resistance of 100 Ω for the entire arrangement.

21-90. Four capacitors each have a capacitance of 0.01 F. Sketch an arrangement of these four capacitors that has an equivalent capacitance of 0.01 F for the entire arrangement.

21-91. A parallel-plate capacitor (Figure 21-46a) is turned into two parallel-plate capacitors by moving the two original plates further apart and inserting an **H**-shaped conducting piece between them (Figure 21-46b). The background grid lets you compare the before-and-after dimensions. Is the equivalent capacitance of the resulting pair of capacitors less than, equal to, or greater than the capacitance of the original capacitor?

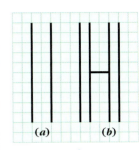

Figure 21-46
Problem 21-91

•**21-92.** In Example 21-10, what is the average power output while the charge on either plate of the capacitor is dropping to half of its initial value? (Base your calculation on the approximation made in the example.)

•**21-93.** In Figure 21-11, suppose that the battery has an EMF of 9.0 V and negligible internal resistance. Suppose also that resistors R_1, R_3, and R_5 each have a resistance of 20 Ω, and resistors R_2 and R_4 each have a resistance of 30 Ω.

a. Find the current through the battery.

b. Find the power output from resistor R_3.

Problems on WebLinks

21-94. Suppose that in WebLink 21-1, when the single resistor is replaced by the connected group of four resistors, the ammeter reading increases to 0.15 A. What would be the equivalent resistance of this group of resistors?

21-95. In WebLink 21-1, the ammeter and voltmeter readings are used to determine the equivalent resistance $R = \frac{V}{I}$ of the connected group of four resistors. The voltmeter in Figure 21-47 provides the value of V. A single ammeter is needed to provide the value of I. Which (one or more) of the four ammeters in the figure could be used to provide this value of I?

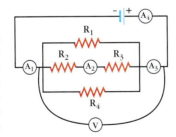

Figure 21-47 Problem 21-95

21-96. Suppose that before the **H**-shaped conductor is inserted between the two original plates in WebLink 21-2, it is given a small positive charge. At a point P within the crossbar of the

H, would the electric fields due to the pair of inner plates and the pair of outer plates still be equal and opposite? Briefly explain.

21-97. In Problem 21-96, will the magnitude of the charge be the same on each of the four plates after the **H**-shaped conductor is inserted between the two original plates? Briefly explain.

21-98. In WebLink 21-3, voltmeters are connected across the battery, the bulb, and the capacitor. As the reading on the voltmeter across the capacitor becomes more negative, the reading on the voltmeter across the bulb ____ (*becomes more negative; becomes more positive; becomes less negative; becomes less positive; remains the same*).

21-99. In Problem 21-98, as the reading on the voltmeter across the capacitor becomes more negative, the reading on the voltmeter across the battery ____ (*becomes more negative; becomes more positive; becomes less negative; becomes less positive; remains the same*).

For a cumulative review, see Review Problem Set II on page 830.

CHAPTER TWENTY-TWO

Magnetism and Magnetic Fields

Probably the most familiar action-at-a-distance force in everyday life is the force that a magnet exerts on an iron nail or a refrigerator door. In ancient times, both the Chinese and the Greeks were familiar with magnets. A magnetic mineral called magnetite or lodestone occurs naturally in many locations. The word *magnet* comes from one of these, a part of Greece called Magnesia. By about AD 1000, the Chinese, having realized that a rotating magnet could show direction, were using magnetic compasses for navigational purposes.

Around AD 1269 in France, Petrus de Maricourt, known as Peter the Pilgrim, mapped the directions in which a small piece of iron pointed when placed at different positions in

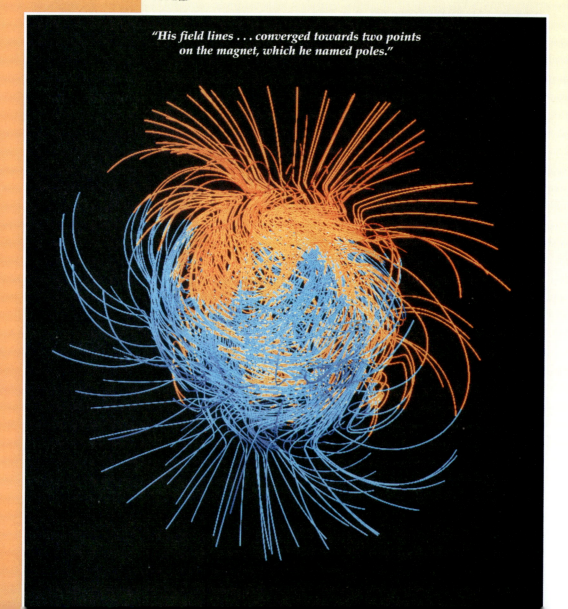

"His field lines . . . converged towards two points on the magnet, which he named poles."

the presence of a magnet. We would today call Peter's mapping a magnetic field. It was perhaps the first recorded mapping of a field of any kind. His field lines, as we shall soon see, converged toward two points on the magnet, which he named *poles*. The pole that ordinarily pointed north came to be known as a *north-seeking* pole, or simply a north pole, and the opposite pole a *south-seeking* or south pole. (See caution about language that follows.) He found that opposite poles attract and like poles repel.

Peter made no connection, however, between electrical and magnetic forces. Remember that magnets have no effect on the little scraps of paper that a comb run through your hair can easily attract. Magnetic and electrical forces are not the same thing. But are they at all connected? And if there is a connection, how might understanding the connection be useful? These are some of the important questions that we will address in this chapter.

Caution about language: Because human laziness has simplified *north-seeking pole* and *south-seeking pole* to the *north pole* and *south pole* of a magnet, you must distinguish carefully between geographic poles and magnetic poles. Ordinarily when we speak of Earth's North Pole, we mean geographic north. But because opposite magnetic poles attract, the north-seeking pole (north magnetic pole) of a compass needle must point to a south magnetic pole. To the extent that Earth's core is a giant magnet, its south magnetic pole is located very near (but not at) the north geographic pole.

22-1 A Qualitative Introduction to the Magnetic Field

Following the lead of Peter the Pilgrim, we can systematically explore the effect of a bar magnet on a compass needle. If you have a bar magnet and compass available, you should try doing the steps for yourself. Figures 22-1*a* and *b* summarize the steps by which the overall picture (Figure 22-1*c*) of the bar magnet's magnetic field emerges. For a more complete step-by-step presentation of the mapping

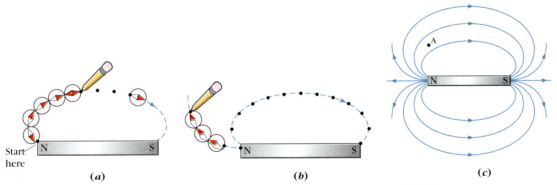

(a) (b) (c)

Figure 22-1 Mapping the magnetic field of a bar magnet. (*a*) Mapping a field line: Starting at the north pole, you can place the compass needles end on end, or using just one compass, you can make a dot at the tip of the compass needle, move the compass needle to the next position, and repeat. (*b*) Mapping a second field line. (*c*) The resulting field.

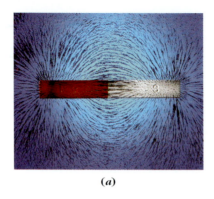 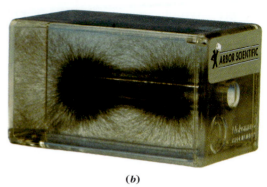

(a)   (b)

Figure 22-2 Some tiny bodies can serve as compass needles to map magnetic fields. (*a*) Iron filings map the field due to a bar magnet. (*b*) The filings are suspended in a silicone oil solution to provide a three-dimensional mapping of the field.

For **WebLink 22-1:**

Mapping the Magnetic Field of a Bar Magnet, go to
www.wiley.com/college/touger

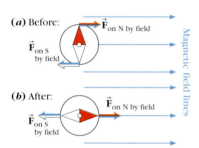

Figure 22-3 The opposite forces on the poles of a compass needle cause it to align with the field.

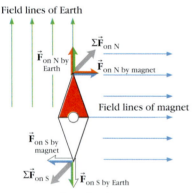

Figure 22-4 Both a bar magnet and Earth's magnetic field affect a compass needle. Vector addition is necessary to find the total force on either pole.

process, work through WebLink 22-1. Figure 22-2 shows that iron filings behave much like compass needles to produce the same magnetic field picture.

The compass needles line up with the magnetic field lines because the forces on the needle's two poles are roughly equal and opposite (Figure 22-3*a*), assuming the needle is much shorter than the bar magnet. The resulting torque causes the needle to rotate until it is in rotational equilibrium; that is, until the torque drops to zero. This happens when the needle and the forces on it are aligned along the field line (Figure 22-3*b*):

The direction of the magnetic field line at a given point represents the direction of the magnetic force on a north magnetic pole placed at that point.

A compass needle will line up with Earth's magnetic field when no other magnet is present. So when the compass is brought near a bar magnet, the force "felt" by either pole of the needle is actually the vector sum of two forces, one exerted by Earth and one by the bar magnet (Figure 22-4). The compass needle will be deflected further from its alignment along Earth's field line where the force exerted by the bar magnet is stronger. This happens when we bring the compass needle closer to either pole of the bar magnet, showing that the field is stronger nearer the poles. Figure 22-1*c* shows that the field lines draw closer together as we approach the poles. The density of lines is an indicator of field strength, just as it is for electric fields:

Magnetic field lines are denser where the field is stronger.

The two highlighted conclusions tell us that qualitatively we read magnetic field mappings much as we do electric field mappings. But we test the fields using different probes: For electric fields we look at the force on a test charge; for magnetic fields, we look at the forces on magnetic poles. Just as opposite charges experience opposite forces in an electric field, so do opposite poles in a magnetic field. But a major difference is this: We can use a positive charge as a probe without an accompanying negative charge, but we can never use a north magnetic pole without a south magnetic pole. Magnetic poles *always* come in pairs. Although physicists have searched for isolated poles (called **magnetic monopoles**), none have ever been detected.

22-2   Connections between Magnetism and Electricity

We have noted some analogies between magnetic and electric fields. Coulomb pursued this in 1785 and showed that the forces that magnetic poles exert on one another obey an inverse square law, just as electrostatic forces do. But finding analogies is not the same as establishing a causal connection. Gravitational forces also behave analogously to electrical forces in many ways, but we do not assume that one is responsible for the other. So electricity and magnetism remained essentially separate areas of inquiry until the nineteenth century, when certain critical observations led to connecting them. In the following case, we will examine some of these observations.

Case 22-1 ◆ *The Effect of a Magnetic Field on a Current-Carrying Wire*

Magnets do not attract loose pieces of copper wire. Does this change if you connect the wire and put a current through it? The set-up in Figure 22-5 allows us to answer this systematically.

The set-up: (If the equipment is available, do this as an On-the-Spot Activity.) On the right tray of the balance scale is a cardboard box. Neodymium magnets, about the size of refrigerator magnets but much stronger, have been pasted to two opposite sides, so that opposite poles are facing each other (see enlargement in Figure 22-5a). The magnetic field points from north to south. Standard masses are placed on the left tray to balance the scale. A copper wire, insulated except at its ends, is suspended from a shelf so that a straight length of the wire passes between the two magnets along a line perpendicular to the line connecting the magnets. (If you are setting this up yourself, taping the length of wire to a pencil-stub splint will keep it straight.) One end of the wire is connected to the negative terminal of a 9-V battery. The other end is positioned so that you can briefly touch it to the positive terminal to complete the circuit. The balance itself should not support any part of this circuit.

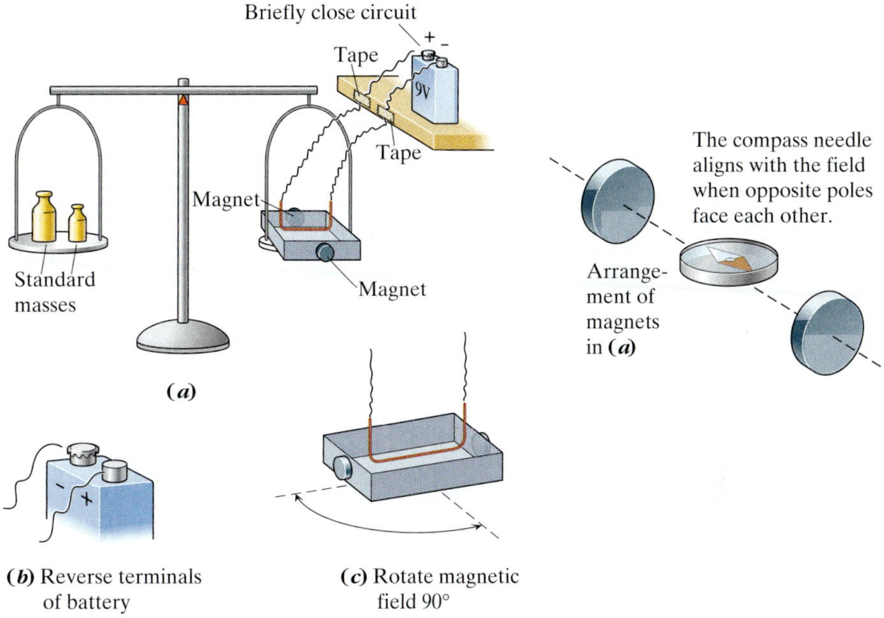

Step 1: Momentarily bring the free end of the wire in contact with the positive terminal of the battery. (If you keep the circuit closed for any length of time, you will have to replace the battery.) When you do this, you should observe that the right tray of the balance scale drops. This indicates that there is now an interaction between the magnets and the wire. If there is a downward force on the magnets, then Newton's third law tells us that the other side of this interaction is an equal and opposite upward force exerted *by* the magnets *on the wire.* Figure 22-6a summarizes this situation. $\vec{B}$ is the customary symbol for the magnetic field.

Figure 22-5 Set-up for Case 22-1.
(b) Reverse terminals of battery
(c) Rotate magnetic field 90°

Step 2: **STOP&Think** What effect would it have if you reversed the terminals of the battery (Figure 22-5b), so that completing the circuit produces a current in the opposite direction? ◆ Answer first, then complete the circuit momentarily as before. You should now see the right tray move up rather than down. This time there is an *upward* force on the magnets and an equal and opposite *downward* force on the wire (Figure 22-6b).

Step 3: Rotate the box by 90° (Figure 22-6c) so that the straight length of wire is now lined up *along* the magnetic field, and repeat Step 1 or Step 2. When you complete the circuit this time, you will see no motion; the scale will remain

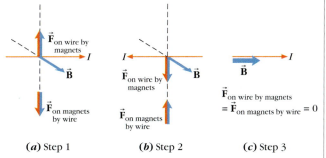

Figure 22-6 Summary of conclusions from Case 22-1.

Case 22-1 ◆ *The Effect of a Magnetic Field on a Current-Carrying Wire (continued)*

balanced. To convince yourself that this is not because your battery died, you can rotate the box back to its original position and repeat Step 1. Figure 22-6c summarizes the conclusions of Step 3.

STOP&Think Is the wire electrically charged when a current is passing through it? ◆

Remember that as electrons drift through a wire, just as many electrons drift into any region as drift out of it, so the total charge remains zero. But the electrons have a nonzero average velocity, whereas the positive lattice ions do not. So a current-carrying wire has a nonzero amount of *moving* charge. The results

summarized in Figure 22-6 therefore tell us that

1. *moving charge traveling perpendicular to a magnetic field experiences a force; the direction of the force is perpendicular to both the velocity of the charge and the direction of the field;*

2. *moving charge traveling along or parallel to a magnetic field does not experience a force.*

Although electric and magnetic forces are not the same thing, we have found a connection between electric charges and magnetic forces. We speak of the former as electro*static* forces to distinguish them from forces that exist only on *moving* charges.

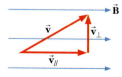

(a) Vector components

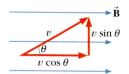

(b) Scalar components (θ is the angle between $\vec{v}$ and $\vec{B}$)

Figure 22-7 Components of $\vec{v}$ parallel and perpendicular to the magnetic field.

If we consider the velocity $\vec{v}$ of a moving charge carrier to be the vector sum of its vector components parallel and perpendicular to the field ($\vec{v} = \vec{v}_{//} + \vec{v}_{\perp}$; see Figure 22-7a), the conclusions of Case 22-1 tell us that only $\vec{v}_{\perp}$ contributes to the magnetic force.

When we dealt with electrical forces, we split the procedure for finding the total electrical force on a charged body into two conceptual stages. In the first, we say that the other bodies create an electrical environment for each point in space, which we call the electric field. In the second, we say that the charged body placed at a particular point is affected by that point's environment and experiences a certain force. We can conceptualize the magnetic force on a moving charged body in the same way: (1) the magnets create a magnetic environment called the magnetic field, and (2) a moving charged body at a particular point experiences a force due to the magnetic field at that point. This has the following advantages. First, we can map the direction of the field, as we did in Figure 22-1. Second, once we know the field direction we can establish clear relationships (Figure 22-6) among

• the direction of the magnetic field at a point;

• the direction of the velocity of a charged body passing through that point; and

• the direction of the resulting force on the charged body.

In the next section, we will put the relationship among these quantities on a quantitative basis.

22-3 Quantitative Treatment of Magnetic Forces

We treated the magnitude of the electric force as the strength of the interaction between a charged object A and its environment

$$F_{\text{on A}} = |q_A| \qquad E \qquad (19\text{-}4)$$

$$\underset{\substack{\text{strength of} \\ \text{electrical} \\ \text{interaction} \\ \text{between A and its} \\ \text{environment}}}{} \qquad \underset{\substack{\text{property} \\ \text{of A}}}{} \qquad \underset{\substack{\text{properties of A's} \\ \text{electrical environment}}}{}$$

where the field $\vec{E}$ involves the electrical properties (charges and their configuration) of the environment. For magnetic forces, the object experiencing the force is a moving charge. The force it experiences depends not just on its charge but on its velocity—more specifically, on $v_{\perp}$, which involves not just the magnitude of the velocity but its orientation relative to the magnetic field. Measurements would show that the magnitude of the magnetic force on the object is proportional to both q_A and $v_{\perp}$. So when we write down the magnetic force as an interaction between the object and its magnetic environment, we must include both of these as properties of object A or quantities describing A:

$$F_{\text{on A}} \quad = \quad |q_A| \, v_{\perp} \qquad \qquad B \qquad \qquad (22\text{-}1)$$

strength of magnetic interaction between A and its environment	quantities describing A	properties of A's magnetic environment

With $v_{\perp} = v \sin \theta$ (see Figure 22-7b), the magnitude of the magnetic force becomes

$$F_{\text{on A}} = q_A v_A B \sin \theta \qquad \text{(ignore signs)} \qquad (22\text{-}2)$$

We can ignore all signs in determining the magnitude of the force vector, because magnitudes are always positive. We will determine direction separately.

To make the direction of the force agree with the conclusions summarized in Figure 22-6, we can use the following rule.

PROCEDURE 22-1

A Right-Hand Rule for Direction of Magnetic Forces

1. Let your right thumb point in the direction of the velocity (Figure 22-8).
2. Line up the remaining fingers of your right hand in the direction of the magnetic field.
3. The palm of your hand then faces in the direction of the force on a positive charge.
4. If the charge is negative, the force is opposite to this direction.

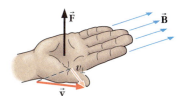

Figure 22-8 **A memory aid: the right-hand rule for magnetic forces.**

Notice in Figure 22-8 that the $\vec{v}$ and $\vec{B}$ vectors determine a plane—the plane of your hand—and the force is always perpendicular to this plane. To try out this right-hand rule in a range of situations, work through WebLink 22-2.

For **WebLink 22-2: The Rule Tool,** go to www.wiley.com/college/touger

◆**UNITS:** If charge, force, and velocity are all expressed in SI units, then the resulting unit of magnetic field is called the **tesla** (T), after Nicola Tesla (1856–1943), a Croatian-American inventor whose work was rooted in an understanding of electromagnetism:

$$1 \text{ tesla} = 1 \text{ T} = 1 \frac{N}{C \cdot m/s}$$

➥**Caution:** If you attempt Procedure 22-1 with your left hand, your left palm will face in the opposite direction. (Try it with both hands at once to convince yourself.)

Table 22-1 gives the values of some typical magnetic fields in teslas.

Because many ordinary magnetic fields are much smaller than a tesla, the **gauss,** another (non-SI) unit of magnetic field, remains in common use:

$$1 \text{ gauss} = 10^{-4} \text{ teslas} \qquad \text{or} \qquad 1 \text{ tesla} = 10^4 \text{ gauss}$$

Table 22-1 Some Approximate Magnetic Field Strengths

At the surface of a neutron star (theoretical)	10^8 T
Near a self-destructing magnet[a]	>200 T
Near a superconducting magnet[b] (used in accelerators and in magnetic resonance imaging [MRI])	5–20+ T
Near a pole of a rare-earth permanent magnet[c]	5×10^{-1} T
Near a pole of a common bar magnet	10^{-2} T
At Earth's surface	$3\text{--}5 \times 10^{-5}$ T

[a]In this "exploding technology," producing fields lasting only a few microseconds, a bank of capacitors with a potential difference of 40–60 kV is discharged across a single copper loop (G. Boebinger, *Physics Today,* June 1996, p. 41).

[b]The upper values have been achieved using NbTi wire and Nb₃Sn wire (ibid.).

[c]Iron-neodymium-boron magnets are available commercially from scientific supply houses.

◆**SHOWING DIRECTION PERPENDICULAR TO THE PAGE** The symbols we will use to show the direction of field lines perpendicular to the page (× and ⊙) are like the crossed tail feathers and the tip of an arrow when viewed end-on:

× × ×   into the page   ⊙ ⊙ ⊙   out of the page
× × ×   (away from you)   ⊙ ⊙ ⊙   (toward you)

Example 22-1 *Magnetic Forces on an Electron–Positron Pair*

For a guided interactive solution, go to Web Example 22-1 at www.wiley.com/college/touger

When a gamma ray passes a heavy nucleus, the energy of the gamma ray is sometimes converted to matter, resulting in the production of an electron and a positron. The positron is a subatomic particle having the same mass as the electron and a positive charge equal in magnitude to the negative charge of the electron. An electron–positron pair is produced in a magnetic field of 1.5 T in the positive y direction. At the instant of production, both particles are traveling at 8.0×10^5 m/s in a direction 20° clockwise from the positive x direction. Find the magnitude and direction of the force on each.

Brief Solution

Choice of approach. The magnitude of each force is given by Equation 22-2. The direction of each is given by the right-hand rule (Procedure 22-1). In determining the magnitude of the force on each particle, the sign of the charge does *not* matter; in determining direction, it *does*. Sketch a diagram to satisfy yourself that you've got the picture. Assume the x and y directions are in the plane of the page.

What we know/what we don't.

$$q_{el} = -e = -1.6 \times 10^{-19} \text{ C} \qquad q_{pos} = +e = +1.6 \times 10^{-19} \text{ C}$$

$$v_{el} = v_{pos} = 8.0 \times 10^5 \text{ m/s} \quad B = 1.5 \text{ T} \quad \theta = -20°, \text{ so } \sin \theta = -0.34$$

The mathematical and hand solution. By Equation 22-2 (remembering to ignore signs), the *magnitude* of the force on each is

$$F_{el} = F_{pos} = qvB \sin \theta = (1.6 \times 10^{-19} \text{ C})(8.0 \times 10^5 \text{ m/s})(1.5 \text{ T})(0.34)$$

$$= \mathbf{6.5 \times 10^{-14} \, N}$$

To determine the *direction* of the force on the positron, align your right hand with $\vec{v}$ and $\vec{B}$ as in Figure 22-8. Your palm then faces **out of the page**. Because the electron is negative, the force on the electron is **into the page**.

Making sense of the results. Your fingers ($\vec{B}$) should point toward the top of the page, so whether $\vec{v}$ (your thumb) makes an angle of $+20°$ or $-20°$ with the $+x$ direction, $\vec{v}_\perp$ is in the same direction (to the right), and that is what the force depends on.

◆ Related homework: Problems 22-1, 22-12, and 22-13.

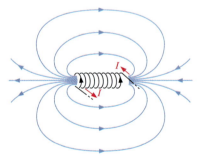

Figure 22-9 Magnetic field due to a current-carrying wire coil. Compare with Figure 22-1*c*.

When we observed that magnets exert forces on current-carrying wires, we noted that the wires exerted equal and opposite forces on the magnets. Does this mean that current-carrying wires also *exert* magnetic forces? Do they give rise to a magnetic field that the neodymium magnets "feel?"

We can answer this by repeating our field-mapping (Figure 22-1) with the bar magnet replaced by a tightly wound coil of wire, as in Figure 22-9. When the switch is closed, a current passes through the coil, and compass needles at various positions around the coil are deflected. The directions of the needles again let us map the field lines. Compare the fields in Figures 22-9 and 22-1. They are essentially the same.

We see that a current-carrying coil acts as an **electromagnet** with poles at either end. If we reverse the direction of the current through the coil, all the needles flip, that is, the field reverses direction. A convenient rule of thumb (literally!) for determining the direction of the field is shown in Figure 22-10. If you wrap the four fingers of your *right* hand around the coil in the direction that conventional (+) current flows, your thumb will point in the direction of the field along the axis of the coil.

This observation that an electric current could produce a magnetic field was first made by Hans Christian Oersted (1771–1851) in 1820 at the University of Copenhagen. Until then, studies of electricity and magnetism were seen as separate and unrelated. In 1820 the two paths of inquiry came together for the first time. We shall see that from that point forward, they become inseparable.

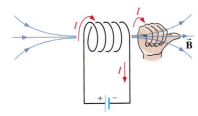

Figure 22-10 Right-hand rule for directions of fields due to coils.

22-4 Magnetic Forces and Circular Motion

Case 22-2 ◆ *The Mass Spectrometer*

A **mass spectrometer** is an instrument that can be used to find the mass of an ion. Chemists use it to determine how much of each isotope of one or more elements is present in a sample. Because isotopes differ by mass, we could think of this as finding the mass spectrum. An environmental lab might use a mass spectrometer to check for harmful lead where old paint has peeled from residential buildings. The lead concentration can be dangerously high where houses are close together; this poses a danger to children playing between the houses.

Understanding how a mass spectrometer works involves applying Newton's second law when magnetic forces are present. Ions from a source are first accelerated by an applied voltage (in the blue region in Figure 22-11).

They then pass into a region (shown in yellow) where crossed electric and magnetic fields determine the velocity at which ions will continue straight ahead. This will occur only if the net force on the ions is zero; that is, only if the electric force qE and the magnetic force qvB are equal and opposite. But $qvB = qE$ for only one velocity; solving for v gives us $v = \frac{E}{B}$. Setting the fields thus selects the velocity of the ions that will continue straight ahead into the next region.

Those that go straight ahead enter into a region (shown in pink) where another magnetic field is applied perpendicular to the page. The magnetic force on the ions is thus perpendicular to their motion and remains so as their path is bent, causing them to follow a semicircular path to a detector, with the magnetic force as the radial force. As Newton's second law tells us (see Example 22-2), when a radial force keeps an object moving in a circular path, the path's radius depends on the mass. Thus, ions of different masses can be separately counted because they strike the detector in different places.

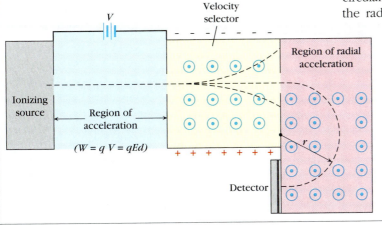

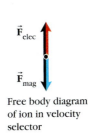

$\vec{F}_{elec}$

$\vec{F}_{mag}$

Free body diagram of ion in velocity selector

Figure 22-11 Main elements of a mass spectrometer.

For a detailed step-by-step development of the main principles involved in the mass spectrometer, work through WebLink 22-3.

The magnetic force on a moving object is always perpendicular to its velocity. How does this affect the object's motion? Recall how the string force changes the direction of the toy car's velocity in Figures 4-11 and 4-12. If the force remains perpendicular to the velocity as the velocity changes direction, the object will move in a circle. To picture how the force remains perpendicular to the velocity,

For **WebLink 22-3:** **The Mass Spectrometer,** go to www.wiley.com/college/touger

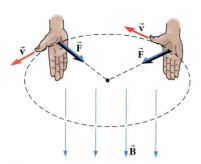

Figure 22-12 Circular motion of a charged particle (+) in a magnetic field. The right-hand rule shows that as the particle circles, the force on it remains perpendicular to the velocity.

run your right hand around any ring-shaped object as in Figure 22-12. If you keep your thumb in the direction of the velocity, your fingers remain in the direction of $\vec{B}$. Then the palm of your hand, indicating the direction of the force, always faces inward toward the center of the circle. The magnetic force is a *radial* force. If $\vec{v} \perp \vec{B}$, Newton's second law with $a_r = \frac{v^2}{r}$ becomes

$$qvB = \frac{mv^2}{r} \qquad (22\text{-}3)$$

Example 22-2 F = ma *in the Mass Spectrometer*

For a guided interactive solution, go to Web Example 22-2 at www.wiley.com/college/touger

Positive ions from the source in a mass spectrometer (Figure 22-11) enter a 'velocity selector' region with crossed electric and magnetic fields. Those that go straight ahead continue into a region where another magnetic field causes them to follow a semicircular path to a detector. Suppose the ion source emits singly charged positive sodium ions of mass 3.82×10^{-26} kg.

a. If the crossed fields in the velocity selector are an electric field of 1000 N/C and a magnetic field of 0.14 T, at what speed do ions continue straight ahead through the velocity selector?

b. How far from their point of entry into the region of radial acceleration do these ions strike the detector if the magnetic field in this region has a magnitude of 0.16 T?

Brief Solution

Choice of approach. We can apply Newton's second law to both parts. The ions that go straight ahead in the velocity selector do not change velocity; they are in translational equilibrium ($\Sigma F = 0$). So in this region, the electric and magnetic forces must be equal and opposite. In the region of radial acceleration, the second law ($\Sigma F_r = ma_r = m\frac{v^2}{r}$) becomes Equation 22-3, which we can solve for r. The distance in question is then $2r$.

What we know/what we don't.

$$m = 3.82 \times 10^{-26} \text{ kg} \qquad\qquad q = +e = 1.6 \times 10^{-19} \text{ C}$$

Fields in **a:** $E = 1000$ N/C, $B_a = 0.14$ T   Field in **b:** $B_b = 0.16$ T

The mathematical solution.

a. $\Sigma F = 0$ becomes

$$qvB_a - qE = 0$$

Solving for v gives us

$$v = \frac{qE}{qB_a} = \frac{E}{B_a} = \frac{1000 \text{ N/C}}{0.14 \text{ T}} = \textbf{7.2} \times \textbf{10}^3 \textbf{ m/s}$$

b. We solve $qvB_b = \frac{mv^2}{r}$ for r to obtain

$$r = \frac{mv^2}{qvB_b} = \frac{mv}{qB_b} = \frac{(3.82 \times 10^{-26} \text{ kg})(7.2 \times 10^3 \text{ m/s})}{(1.6 \times 10^{-19} \text{ C})(0.16 \text{ T})} = 1.1 \times 10^{-2} \text{ m}$$

Then   distance $= 2r = \textbf{2.2} \times \textbf{10}^{-2} \textbf{ m}$

◆ Related homework: Problems 22-14, 22-15, 22-16, and 22-17.

An early cyclotron. This cyclotron, now in Chicago's Museum of Science and Industry, was one of the earliest elementary particle accelerators. An applied magnetic field constrained the charged particles to follow semi-circular paths within each half. For further details, see Problems 22-19 and 22-52.)

If an object has a velocity perpendicular to a magnetic field, it will move in a circle. If it has a velocity parallel to the field, there will be no magnetic force on the field and it will continue in a straight line. If the velocity has components both parallel and perpendicular to the field (Figure 22-13), the circular and linear motions are superimposed, giving us a spiral or helix, just as in Figure 3-3.

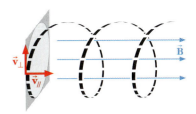

Figure 22-13 Helical motion in a magnetic field.

Magnetic Bottles and Mirrors

"Pinching" the field in Figure 22-13 so that lines come together, as in Figure 22-14, introduces a radial component of $\vec{B}$. Applying the right-hand rule at a point such as A, we see that there is now a force component F_{back} that opposes $v_{//}$ and gradually reverses the particle's movement along the axis of the helix. The point of reversal, where $v_{//} = 0$, is called a **magnetic mirror.** The particle moves back to a second magnetic mirror at the left end, where the radial component of $\vec{B}$, and therefore the force, is opposite in direction. This sends the particle back to the right. The moving charged particles are thus confined within a region bounded by the magnetic field lines as though the region were a bottle with two necks. A magnetic field set up in this way is called a **magnetic bottle.**

For many years, physicists have sought to generate energy by nuclear fusion in gases in which the atoms have sufficient kinetic energy to keep the fusion reaction going. To accomplish this, the gases must be raised to temperatures of about 10^9 K. At such high temperatures the atoms of the gas are completely ionized, and many of these attempts have involved confining this ionized gas, called a **plasma,** within magnetic bottles. A problem occurs because right along the axis there is no radial component of $\vec{B}$ and thus no reversing force, so magnetic bottles leak.

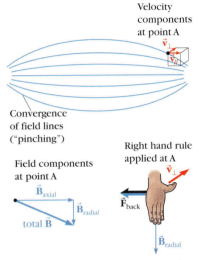

Figure 22-14 "Bottling" effect of pinching the field.

Van Allen Belts and Auroras

On a much larger scale, Earth's magnetic field is also pinched at the poles. Two regions of the field (Figure 22-15), called the **Van Allen belts,** are particularly effective at bottling ions escaping from the sun. Such ions constantly bombard Earth but are particularly numerous during solar flares. Where these bottles leak, ions are released into Earth's atmosphere. Their interaction with atoms in the atmosphere can result in the emission of considerable light. Because this leakage occurs near the magnetic poles, which are situated near the geographic poles, this dramatic emission of light is observed primarily in or near Earth's polar regions, where they are known as the northern lights (or the *aurora borealis*) and the southern lights (the *aurora australis*—see Figure 22-16).

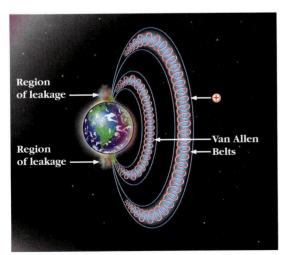

Figure 22-15 Ions from the sun bottled by the converging lines of Earth's magnetic field.

Figure 22-16 The Aurora Australis, or Southern Lights, viewed from orbit.

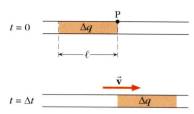

Figure 22-17 Conceptual picture for the derivation of Equation 22-4.

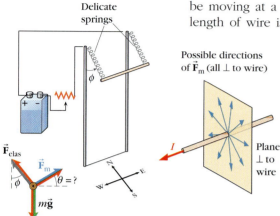

Felas **F**m ϕ $\theta = ?$ $m\vec{g}$

Free body diagram of wire (end view)

Figure 22-18 A magnetic field detector. The lengths of the springs are greatly exaggerated for visibility.

22-5 Magnetic Forces on Current-Carrying Wires

By Equation 22-2, the magnetic force on the amount of moving charge Δq in a small straight length l of current-carrying wire (Figure 22-17) is

$$F = (\Delta q)vB \sin \theta$$

We will assume l is short enough so that the field is effectively constant over the entire length. If it requires a time interval Δt for all this charge to pass the end point P of this length, then the current is $I = \frac{\Delta q}{\Delta t}$. But to do so, the charge must be moving at a speed $v = \frac{l}{\Delta t}$. Then the magnitude of the magnetic force on the length of wire is

$$F = (\Delta q)\frac{l}{\Delta t}B \sin \theta = \frac{\Delta q}{\Delta t}lB \sin \theta$$

that is, $\qquad F = IlB \sin \theta = IlB_\perp \qquad$ (ignore signs) $\qquad$ (22-4)

Because the velocity of positive charge is in the direction of the conventional current, the right-hand rule in Figure 22-8 applies here as well. **STOP&Think** How might you use this result with the set-up in Figure 22-18 to find the strength of a magnetic field? Would this method work if the field is parallel to the wire? ◆

Possible directions of $\vec{F}_m$ (all ⊥ to wire)

Plane ⊥ to wire

22-6 Making Use of Torques on Current Loops

Think about what happens when a current-carrying loop of wire (abcd in Figure 22-19*a*) is exposed to a uniform magnetic field. **STOP&Think** In what direction is the force on side ab? In what direction is the force on side cd? What can you say about the forces on the other two sides? What will happen to the loop if it is free to move? ◆

Before going on, you should satisfy yourself that the loop will rotate toward the position shown in Figure 22-19*b*. If you applied the right-hand rule correctly to each side of abcd in Figure 22-19*a* (which way does the current go around the loop?), you should have found an upward force on side ab, a downward force on cd, and no force on each of the two sides parallel to the field. There is therefore a torque on the loop, which causes it to rotate.

Human ingenuity was quick to take advantage of the fact that a current-carrying loop or coil (many loops) will rotate in a uniform magnetic field. Applications range from electric motors to analog ammeters and voltmeters, where a needle moves with the loop. How far the needle rotates depends on the current. We will examine this dependence in more detail.

Because the field remains perpendicular to sides ab and cd as the loop rotates, the force on each of these sides remains constant, with a magnitude $F = IlB$.

Figure 22-19 A current loop rotating in a magnetic field. Contacts (not shown) maintain a current through the loop as it rotates.

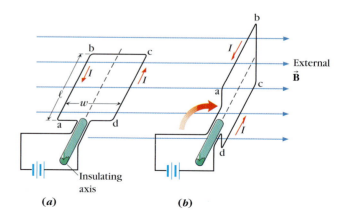

(*a*)   (*b*)

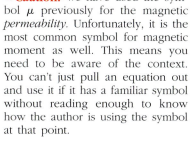

Side cd
(current in)

$\frac{w}{2}$

$\vec{F}_{on\ cd}$

$r_\perp = \frac{w}{2} \sin \theta$

External
$\vec{B}$

$\vec{F}_{on\ ab}$

θ

Side ab
(current out)

Normal
to loop

Figure 22-20 Forces and moment arms resulting in a torque on the loop.

Figure 22-20 shows that when the normal to the loop face makes an angle θ with the field direction, the moment arm $r_\perp = \frac{w}{2} \sin \theta$. The total torque due to the forces on these two sides is

$$\tau = 2r_\perp F = 2(IlB)\left(\frac{w}{2} \sin \theta\right) = I(lw)B \sin \theta$$

or

$$\tau = IAB \sin \theta \qquad (22\text{-}5)$$

where A is the area lw enclosed by the loop.

Although we obtained Equation 22-5 for a rectangular loop, it is valid for a loop of *any* shape enclosing an area A. If N loops are wound around the perimeter to form a coil, there is the same torque on each, so the total torque is

$$\tau = NIAB \sin \theta \qquad (22\text{-}6)$$

The quantity IA, the loop's contribution to the torque, is called the **magnetic moment** or **magnetic dipole moment** μ of the loop. So we can think of the torque as the result of an interaction between a thing's magnetic moment and the field in which the thing is placed: $\tau = \mu B \sin \theta$ on each loop.

A device for measuring current can be constructed as in Figure 22-21. A somewhat flat coil is set between two magnetic poles (we'll assume the field is uniform). A needle mounted on the frame of the coil shows the amount θ that the coil has rotated. A wound spring exerts a torque on the coil opposite in direction to the torque due to the current–field interaction. The torque exerted by the spring is proportional to the angle of rotation. Because it opposes the rotation, it is called a **restoring torque.** The needle will come to rest where the two torques are equal and opposite. Because this depends on the current, we can mark the scale with the current values necessary for equilibrium at the various needle positions. A device that does this is called a **galvanometer.** With minor adaptation (see Problems 22-54 and 22-55), a galvanometer can become either an ammeter or a voltmeter. Modern ammeters and voltmeters commonly use digital readouts rather than needles. In contrast, devices that have needle displacements proportional to the quantities they are measuring are called *analog* devices.

➡**Caution:** We have used the symbol μ previously for the magnetic *permeability.* Unfortunately, it is the most common symbol for magnetic moment as well. This means you need to be aware of the context. You can't just pull an equation out and use it if it has a familiar symbol without reading enough to know how the author is using the symbol at that point.

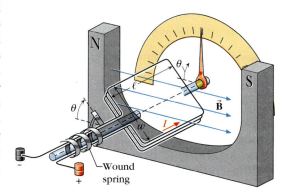

Figure 22-21 A simple galvanometer.

Example 22-3 *Calibrating an Analog Galvanometer*

Consider a coil of 50 turns around a 0.020-m by 0.030-m rectangular frame, set between poles that produce a magnetic field of 0.20 T. Suppose we know that the spring exerts a restoring torque of 6.0×10^{-3} m·N when it is twisted 30°. What should be the current value that appears on the scale where the needle has rotated 30°? What will the full scale reading be if the scale is linear (that is, if θ is proportional to I)?

EXAMPLE 22-3 *continued*

Solution

Choice of approach. The needle comes to rest where the coil is in rotational equilibrium. The torque due to the current–field interaction is then equal in magnitude to the torque exerted by the spring, so we can apply Equation 22-6 (*not* Equation 22-5, because $N \neq 1$).

What we know/what we don't.

$$N = 50 \qquad l = 0.020 \text{ m} \qquad w = 0.030 \text{ m}$$

$$B = 0.20 \text{ T} \qquad \tau = 6.0 \times 10^{-3} \text{ m} \cdot \text{N} \qquad \theta = 30° \qquad I = ?$$

The mathematical solution. The area $A = lw = (0.020 \text{ m})(0.030 \text{ m}) = 6.0 \times 10^{-4}$ m. Therefore we know everything necessary to solve $\tau = NIAB \cos \theta$ for I:

$$I = \frac{NAB \cos \theta}{\tau} = \frac{(50)(6.0 \times 10^{-4} \text{ m})(0.20 \text{ T}) \sin 30°}{6.0 \times 10^{-3} \text{ m} \cdot \text{N}} = \textbf{0.50 A}$$

At full scale, the needle will be rotated 180°—six times 30°—so the current reading at full scale should be six times as great, or **3.0 A**.

◆ Related homework: Problems 22-27, 22-29, and 22-30.

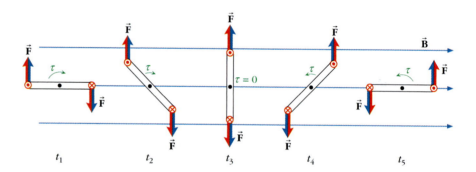

Figure 22-22 Change in torque as the loop in Figure 22-21 rotates.

t_1 t_2 t_3 t_4 t_5

When a rotating loop or coil reaches the position in Figure 22-19b, there is no torque on it. **STOP&Think** Does the coil stop when it reaches this position? ◆ Remember that zero torque results in zero angular acceleration, not zero velocity. Over the time interval from t_1 to t_3 in Figure 22-22, a torque increases its angular velocity. At t_3 there is no torque to change its angular velocity, so it keeps going. Once it passes this position, the torque becomes nonzero again, but in the opposite direction, slowing the coil down.

If we could get it to keep going instead of slowing down, the rotating coil could produce the motion of an electric motor. This would happen if we could reverse the forces as the coil passed through its zero torque position. In turn, the forces would reverse direction if we could reverse the current. A simple way of doing this is shown in Figure 22-23a. The shaft of the loop is split into two

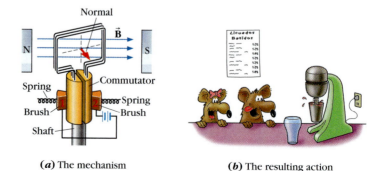

Figure 22-23 The motor of a milkshake machine. (*a*) The mechanism. (*b*) The resulting action.

(*a*) The mechanism   (*b*) The resulting action

copper half-cylinders separated by an insulating layer. As they rotate, the half-cylinders brush against two pieces ("brushes" 1 and 2) that provide their connection to the power supply. But with every half-turn, the half-cylinders reverse the brushes with which they are in contact, so the current flows through the loop in the opposite direction. A device of this type is called a **commutator.** Motors of milkshake machines have commutators much like the one shown (Figure 22-23b).

22-7 How the Magnetic Field Depends on Its Source

A current-carrying coil produces a magnetic field. A moving charge experiences a force in the presence of this field. Taken together, these two facts suggest that ultimately, magnetic forces are exerted *by* moving charges *on* moving charges. We should therefore be able to provide the same type of two-step description for magnetic forces as for electric forces:

1. $F_{\text{on object 1 by object 2}}$ = (property of 1)(field due to 2)

2. How field depends on object 2

Electric force

$$F_{\text{on 1 by 2}} = q_1 E_{\text{due to 2}}$$

where $E_{\text{due to 2}} = k\dfrac{q_2}{r^2}$

Magnetic force

$$F_{\text{on 1 by 2}} = (q_1 v_{1\perp}) B_{\text{due to 2}}$$

or $F_{\text{on 1 by 2}} = (I_1 l_1) B_{\text{due to 2}}$

where $B_{\text{due to 2}} = \mathbf{?}$

To complete this description, we must determine what goes in place of the question mark. How does $B_{\text{due to 2}}$ depend on the relevant properties (charge and velocity of a particle, or current and wire length) of object 2, its "source"? It is mathematically simpler to answer this question for a length of current-carrying wire, so we will limit ourselves to that situation.

Oersted, as we have said, was the first to observe that a length of current-carrying wire produces a magnetic field. He did so by observing that compass needles were deflected when he connected the wire to a voltage source (Figure 22-24). The compass needles flip the other way if the current is reversed. The direction of the field can be found by applying the **right-hand rule for magnetic fields,** illustrated in Figure 22-25a. (Our previous right-hand rule was for magnetic *forces.*)

We can mentally break up any curved wire into many tiny segments, each one small enough so we can

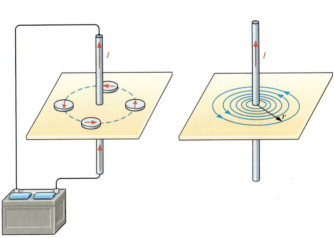

(a) (b)

Figure 22-24 The magnetic field due to a long, straight current-carrying wire. (a) The field lines due to the wire are concentric circles about the wire. One such circular line is indicated. (b) The magnetic field becomes stronger as the radial distance decreases, so the field lines are closer together near the wire.

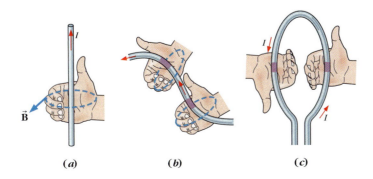

(a) (b) (c)

Figure 22-25 Right-hand rule for the magnetic field due to a straight length of current-carrying wire. (a) Anywhere along a long, straight wire. (b) For each tiny, nearly straight segment of a curved wire. (c) For nearly straight segments of a circular loop. Each segment's contribution to the total field is in the same direction at the center of the loop.

think of it as straight. The right-hand rule for magnetic fields can be applied to each segment (Figure 22-25b). The total field due to the wire is then the vector sum of the fields due to the pieces. In general, doing this in detail requires calculus, but sometimes each piece gives us a contribution in the same direction. This occurs, for example, at the center of the loop of wire in Figure 22-25c. The field due to each segment is directed along the center axis of the loop, so this must be the direction of the total field at the center of the loop. Because this would be true for each loop in a long straight coil of wire, or **solenoid,** the magnetic field within the solenoid is also along the axis, in agreement with Figure 22-9.

By careful measurements, one can show that the magnetic field due to a long straight wire is directly proportional to the current through it and inversely proportional to the distance from the wire: $B = (\text{a constant}) \times \frac{I}{r}$. The constant is commonly written as either $2k'$ or $\frac{\mu}{2\pi}$, so that $k' = \frac{\mu}{4\pi}$. The prime (′) distinguishes k' from the constant k in Coulomb's law. We can thus write

$$B = \frac{2k'I}{r} = \frac{\mu I}{2\pi r} \tag{22-7}$$

The quantity μ depends on the medium in which the magnetic field occurs and is called the *permeability* of the medium. Its value in a vacuum is called the *permeability of free space* and is denoted by μ_o. The value of k' in free space is exactly

$$k' = \frac{\mu_o}{4\pi} = 10^{-7}\frac{\text{T} \cdot \text{m}}{\text{A}}$$

The value in air is very nearly the same; this value is correct for air to several significant figures.

We can now establish a procedure to find the magnetic force that any one object exerts on another. The objects may be individual moving charge carriers or arrangements of current-carrying wires.

PROCEDURE 22-2

Finding the Magnetic Force that Object A Exerts on Object B

1. Identify the position of B.
2. Find the magnetic field due to A at this position.

   Direction: Use the right-hand rule for fields (Figure 22-25).

   Magnitude: (So far, we have limited ourselves to the field due to a long straight wire—Equation 22-7. Table 22-2 will add to the possibilities.)

3. Find the force on B when it is placed at this position.

   Direction: Apply the right-hand rule for forces.

   Magnitude: Use Equation 22-1, 22-2, or 22-4.

➥**Important:** In determining the magnetic field due to object A (step 2 of Procedure 22-2), you need to know its configuration. Table 22-2 gives expressions for some common configurations. You should also make sure that you understand what distances the variables in the particular expression represent.

Note: Step 2 ignores the existence of object B. It simply is finding the magnetic field at a point in space. Step 3 then tells how B, if placed at this point, would interact with the point's environment.

Although finding the magnetic fields due to many configurations of wire requires calculus, it is useful to have at hand some of the results that can be obtained by that procedure. These are listed in Table 22-2.

Table 22-2 Magnetic Fields due to Common Configurations of Wire. (each carrying a current *I*)

Configuration	Diagram showing field lines and relevant distances	Magnitude of field at a point P
Long straight wire		P is at a distance r from the wire: $$B = \frac{2k'I}{r} = \frac{\mu I}{2\pi r}$$
Flat loop of wire of radius R	Thickness assumed negligible	P is at the center of the loop: $$B = \frac{\mu I}{2\pi R}$$ (one turn) $$B = \frac{\mu N I}{2R}$$ (N turns)
Solenoid (long straight coil)		P is inside the solenoid (away from the ends): $$B = \mu \frac{N}{L} I$$

Example 22-4 *Forces That Parallel Straight Wires Exert on Each Other—The Ampere Defined*

Two long straight wires run parallel to each other, carrying equal currents in the same direction.

a. How much current must flow in each wire if, at a separation of 1.00 m, each wire exerts a force of 2.00×10^{-7} N on each 1.00-m length of the other?

b. Describe the direction of the force on each wire.

Solution

Choice of approach. We apply Procedure 22-2. In Step 3, only Equation 22-4 is applicable to the force on a straight length of wire.

What we know/what we don't.

For Field due to B	For Force on a Length of A
$k' = 10^{-7}\dfrac{\text{T} \cdot \text{m}}{\text{A}}$	$F = 2.00 \times 10^{-7}$ N
$r_{\text{A to B}} = 1.00$ m	$l = 1.00$ m
$B_B = ?$ $I_B = ?$	$I_A = I_B = I = ?$

EXAMPLE 22-4 continued

The mathematical solution. The field due to wire B is

$$B_B = \frac{2k'I}{r}$$

The force on a length l of wire A is then

$$F = IlB_B = Il\left(\frac{2k'I}{r}\right) = \frac{2k'I^2l}{r}$$

Solving for I, we get

$$I = \sqrt{\frac{rF}{2k'l}} = \sqrt{\frac{(1.00\ \text{m})(2 \times 10^{-7}\ \text{N})}{2\left(10^{-7}\dfrac{\text{T}\cdot\text{m}}{\text{A}}\right)(1.00\ \text{m})}} = \textbf{1.00 A}$$

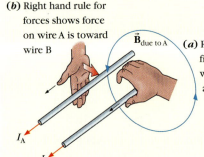

(**b**) Right hand rule for forces shows force on wire A is toward wire B

(**a**) Right hand rule for fields shows field due wire B is downward at wire A

Figure 22-26 Applying appropriate right-hand rules to Example 22-4.

The right-hand rule for fields (Figure 22-26*a*) indicates that the field is downward at the position of wire A and is perpendicular to the wire. Once we know this, applying the right-hand rule for forces (Figure 22-26*b*) shows us that

the force on wire A is toward wire B

STOP&Think Use the right-hand rules to determine the direction of the force that A exerts on B. You should find that the forces the wires exert on each other are equal and opposite. ◆

Making sense of the results. We found that $I = 1.00$ A—*one unit* of current. The **ampere** is actually defined by the set-up for this problem. It is the current in each wire that will cause each to exert a force of 2.00×10^{-7} N on each 1.00-m length of the other at a separation of 1.00 m. The **coulomb** is then defined by $1\ \text{C} \equiv 1\ \text{A} \times 1\ \text{s}$. (From this definition of the ampere, it follows that k' is exactly $10^{-7}\dfrac{\text{T}\cdot\text{m}}{\text{A}}$.)

◆ Related homework: Problems 22-31, 22-32, and 22-34.

Example 22-5   *The Magnetic Force Exerted on an Electron by a Solenoid*

An electron travels within an air-filled solenoid at a speed of 400 m/s, as a current of 0.15 A flows through the solenoid. There are 20 turns per cm along the length of the solenoid. Find the magnetic force on the electron if it is moving
a. perpendicular to the axis of the solenoid.
b. parallel to the axis of the solenoid.

Solution

Choice of approach. Following Procedure 22-2, we find the field due to the solenoid and then find the force on the electron due to this field. Table 22-2 tells us the field within a solenoid is $B = \mu_o \frac{N}{L}I$. Although we don't know N or L individually, $\frac{N}{L}$ is the number of turns per unit of length, which we are told is 20 per cm $\left(\frac{20}{\text{cm}} \times \frac{100\ \text{cm}}{\text{m}} = \frac{2000}{\text{m}}\right)$. We also must consider the angle the direction of travel makes with the field.

What we know/what we don't.

For Field due to Solenoid	For Force on the Electron
$\mu_o = 4\pi \times 10^{-7}\dfrac{\text{T}\cdot\text{m}}{\text{A}}$	$e = 1.6 \times 10^{-19}$ C
$I = 0.15$ A $\dfrac{N}{L} = \dfrac{2000}{\text{m}}$	$v = 400$ m/s
$B = ?$	**a.** $\theta = 90°$ **b.** $\theta = 0$
	$F = ?$

The mathematical solution. The magnetic field within the solenoid is

$$B = \mu_o \frac{N}{L} I = \left(4\pi \times 10^{-7}\frac{\text{T}\cdot\text{m}}{\text{A}}\right)\left(\frac{2000}{\text{m}}\right)(0.15\text{ A}) = 3.77 \times 10^{-4}\text{ T}$$

The force on the electron is a maximum when it is traveling perpendicular to the field ($\theta = 90°$, $\sin\theta = 1$) and zero when it is traveling parallel to the field ($\theta = 0$).

a. The magnitude of the force is

$$F = qvB\sin\theta = (1.6 \times 10^{-19}\text{ C})(400\text{ m/s})(3.77 \times 10^{-4}\text{ T})(1)$$

$$= \mathbf{2.41 \times 10^{-20}\text{ N}}$$

By the right-hand rule, the force will be perpendicular to both the field and the electron's velocity.

b. When $\theta = 0$, $\sin\theta = 0$, so $F = \mathbf{0}$.

◆ Related homework: Problems 22-33, 22-35, and 22-40.

Ampere's Circuital Law

In the equation for the magnetic field due to a long straight wire ($B = \frac{\mu I}{2\pi r}$), we can think of the denominator $2\pi r$ as the circumference or length of a circular path maintaining a distance r from the wire (Figure 22-27a). Then

$$B \times (\text{path length}) = \mu I$$

B decreases as r and the path length increase, so the product is the same no matter which circle we choose, its value determined by the current I.

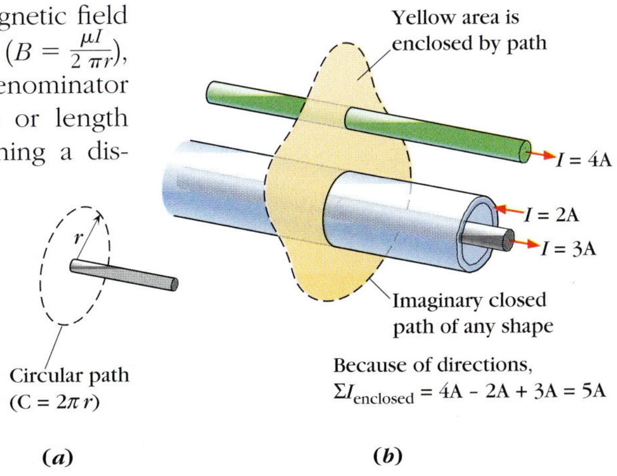

This turns out to be true as long as the path closes back on itself, whether it is circular or not. In fact, if multiple currents cross the area enclosed by the path (Figure 22-27b),

$$B \times (\text{length of closed path}) = \mu\Sigma I_{enclosed} \qquad (22\text{-}8)$$

Currents in opposite directions must have opposite signs.

Figure 22-27 Reasoning with Ampere's circuital law.

This is one form of a relationship called **Ampere's law** or **Ampere's circuital law.** It is a fundamental relationship between the magnetic field and its sources, just as Gauss's law (Section 19-4) is a fundamental relationship between the electric field and its sources.

22-8 Magnetic Materials

We have seen evidence that moving charges exert forces on other moving charges, and that a current-carrying coil of wire gives rise to a magnetic field that is very much like that of a bar magnet. But if a magnetic field is due to moving charge, what is responsible for the magnetic field of a stationary bar magnet? More puzzling

Before:

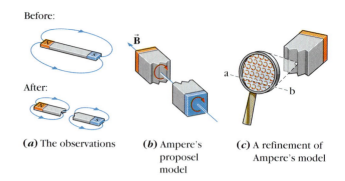

After:

(a) The observations

(b) Ampere's proposel model

(c) A refinement of Ampere's model

Figure 22-28 Fields due to pieces of a bar magnet: observations and an explanatory model.

still, we find that if we break a bar magnet in two (Figure 22-28*a*), each half has the same polarity as the original magnet.

Ampere was the first to propose that somehow, within materials that produce magnetic fields, there must be currents circulating about the N-to-S axis. If this were so, he reasoned, we would observe the same field as if those currents were contained in loops of a coil. Moreover, if the magnet were broken in two (Figure 22-28*b*), the currents in each half would continue circulating in the same direction, producing a field in the same direction as the original magnet's. A refinement of Ampere's model is shown in Figure 22-28*c*. Now the currents within the material are visualized as being very local. Across any line such as ab there is zero total current (which is what we would read with an ammeter) because there is as much current flowing across the line one way as the other. But each loop produces a magnetic field along its own axis in the same direction, just as in Figure 22-10, so there is a nonzero total magnetic field.

Further development of this model had to await a more detailed picture of the structure of atoms and molecules, which was in place by 1913. We will discuss the development of this picture more fully in Chapter 26, but a few of its features are relevant here. The electrons in an atom were now envisioned as orbiting around a positive nucleus. A charged body, such as an electron, making a closed orbit is in effect a little current loop. It has a magnetic moment and gives rise to a magnetic field. Although this picture of the atom later proved to be too simplistic, the essential notion of the electron's *orbital* (meaning orbitlike) behavior remained, as did its magnetic consequences. Later experimental observations by Stern and Gerlach (1921–1922) could be explained by supposing that the electron rotates or "spins" on its own axis as well as orbiting the nucleus (or more precisely, that it had a magnetic moment *as though* it were doing these things). There is a magnetic moment associated with the spin as well. The observations of the Stern-Gerlach experiment implied that in an external magnetic field the spin could have only the two orientations shown in Figure 22-29. In one orientation, called "spin up," the field due to the spinning electron (determined by the right-hand rule in Figure 22-10) is in the same direction as the external field (and so is the magnetic moment). In the other orientation, it is opposite to the external field.

Each electron thus has a magnetic field arising from the combined effects of its orbital motion and its spin. For the most part, the magnetic fields of the electrons in an atom or molecule will cancel one another out. In some materials, the cancellation is not complete and the atoms or molecules have fields of their own. Ordinarily the atoms are in random motion, so the orientations of these fields are random. Then the total field due to a sample of the material containing large numbers of atoms is zero. However, if the sample is placed in an *external* magnetic field, the atomic fields (and also the magnetic moments of the atoms) will tend to shift from being directed randomly toward alignment

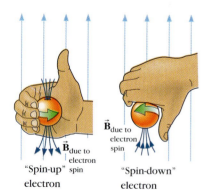

"Spin-up" spin electron

"Spin-down" electron

Figure 22-29 Permissible spin orientations of an electron in a magnetic field. The hand is applying the right-hand rule for magnetic fields due to coils or single loops (Figure 22-10). The direction of the field due to the spinning electron is opposite to the direction of the thumb because the rotating charge is negative.

with the external field, much like little compass needles. Materials that do this are called **paramagnetic** materials. The orientations are still somewhat random because of thermal motion and are increasingly so at higher temperatures. But there is some net alignment, resulting in a small additional field that augments the external field.

In some transition metals, notably iron, cobalt, and nickel, the closeness in the energy levels of electrons in certain orbitals (such as the so-called 3d and 4s orbitals) results in a high degree of interaction between the spins of electrons in these levels. Thus, when these metals are placed in an external magnetic field, the alignment of some atoms causes the alignment of others, and the magnetic field due to the aligned atoms may be 100 to 1000 times as great as the external field causing their alignment. Because iron (*ferrum* in Latin) is the best-known example, these materials are called **ferromagnetic.** For this reason, if a coil is to serve as an electromagnet, it is usual to wind it around an iron core to multiply the strength of its field.

✦SUMMARY✦

Magnetic forces are forces exerted *by* moving charge carriers *on* moving charge carriers. Current-carrying wires or loops do likewise because they contain moving charge carriers. In describing their interactions, we say they give rise to magnetic fields and in turn experience forces and torques in the presence of magnetic fields. On an atomic scale, the orbiting and spinning of the electrons in an object are like current loops. When these are not randomly oriented, as in compass needles or other bits of ferromagnetic metal, the object can exert or experience a nonzero total force or torque.

The direction of the magnetic field line at a given point represents the direction of the magnetic force on a north magnetic pole placed at that point.

In this way, we can map the field due to a bar magnet or a current-carrying loop or coil. The field direction along the axis of a loop or coil is given by the "rule of thumb" (**right-hand rule for coils**) in Figure 22-10.

Force on a moving charge carrier A in a given magnetic field:

Magnitude: $\quad F_{\text{on A}} = q_A v_A B \sin \theta \quad$ (ignore signs) (22-2)

Direction: given by right-hand rule for direction of magnetic forces (**Procedure 22-1** and Figure 22-8).

If $\vec{v} \perp \vec{B}$, the charge carrier will move in a circle with the magnetic force as the radial force, and $\Sigma F_r = ma_r$ becomes

$$qvB = \frac{mv^2}{r} \qquad (22\text{-}3)$$

(applicable, among other places, in mass spectrometers and in magnetic bottles). For charge carriers in a straight current-carrying wire of length l, Equation 22-2 becomes

$$F = IlB \sin \theta = IlB_\perp \qquad \text{(ignore signs)} \quad (22\text{-}4)$$

Such forces give rise to a **torque on a current loop** ($N = 1$) or on a coil of N turns:

$$\tau = NIAB \sin \theta \qquad (22\text{-}6)$$

(θ is the angle the normal to the loop face makes with the field)

(Applications include **electric motors** and the movement of a **galvanometer** needle.)

With $N = 1$, the loop's contribution to the torque is called its **magnetic moment** or **magnetic dipole moment:**

Magnitude: $\qquad \mu = IA$

Direction: that of field due to loop ("rule of thumb," Figure 22-10)

The torque is the result of an interaction between the loop's magnetic moment and the field in which it is placed: $\tau = \mu B \sin \theta$.

A magnetic field depends on its source. If the source is a **long straight current-carrying wire,**

Magnitude: $\qquad B = \dfrac{2k'I}{r} = \dfrac{\mu I}{2\pi r} \qquad (22\text{-}7)$

Direction: given by right-hand rule for magnetic fields (Figure 22-25)

In free space,

$$k' = \frac{\mu_o}{4\pi} = 10^{-7} \frac{\text{T} \cdot \text{m}}{\text{A}}$$

If we can find the field due to one object, and we can find the force on a second object in a given field, we can find the force that the first object exerts on the second—see **Procedure 22-2.**

Ampere's law expresses a fundamental relationship between the magnetic field and the currents that are its sources. If currents cross the area framed by a closed path (Figure 22-27*b*),

$$B \times (\text{length of closed path}) = \mu \Sigma I_{\text{enclosed}} \qquad (22\text{-}8)$$

◆QUALITATIVE AND QUANTITATIVE PROBLEMS◆
Hands-On Activities and Discussion Questions

The questions and activities in this group are particularly suitable for in-class use.

22-1. *Hands-On Activity.* A magnetic field of 0.94 T is directed into the page. A proton traveling in the plane of the page is moving at a speed of 1000 m/s in a direction 50° counterclockwise from the +*x* axis.

a. Find the magnitude and direction of the magnetic force on the proton.

b. Take two pencils and, keeping their eraser ends touching, let one point in the direction of the proton's velocity and let the other point in the direction of the field. What is the angle between these two pencils?

c. How does the angle between the two pencils compare with a right angle—is it larger? Equal? Smaller? Check this by holding the square corner of a page against the angle made

by the two pencils. Is the angle between the two pencils the angle that you used in your calculation in ***a***? If not, redo the calculation.

d. Lay a sheet of paper flat on your desk and draw a point P on the sheet of paper. Then draw a bunch of straight lines passing through P in different directions. Finally, hold your pencil pointing straight up with its eraser end resting on point P. What angle (in degrees) does the pencil make with each of the lines you drew through P? What angle (in degrees) does the field in ***a*** make with any velocity vector in the plane of the paper?

e. When you hold the two pencils as in ***b***, they determine a plane. (You can envision the plane as a stiff sheet of paper pressing flat against both pencils.) Is the direction that you found in ***a*** perpendicular to this plane? If not, use the right-hand rule to reconsider your answer to ***a***.

Review and Practice

Section 22-1 A Qualitative Introduction to the Magnetic Field

22-2. Compare the information given by a magnetic field line mapping and an electric field line mapping. In what ways are they similar? In what ways are they different?

22-3. In Figure 22-1*c,* suppose we take a compass needle that is as long as the bar magnet and place it so that its center is at point A. Will the compass needle line up with a field line? Will the magnetic forces on the two poles of the compass needle be either equal in magnitude or oppositely directed? Explain, and comment on how the situation for this compass needle differs from that of the compass needles shown in Figure 22-1*a.*

22-4. Suppose that in Figure 22-1, you place your compass near the north pole of the bar magnet and center it on the straight line that passes through the two poles. Based on the data in Table 22-1, what angle will the compass needle make with this line if the bar magnet has been aligned in an east–west direction?

22-5. A compass is placed at point A in the presence of two identical bar magnets. (Treat Earth's magnetic field as approximately zero for this problem.) In which direction does the compass needle point when the magnets are arranged ***a.*** as in Figure 22-30*a*? ***b.*** as in Figure 22-30*b*?

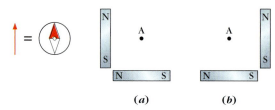

Figure 22-30 Problem 22-5

Section 22-2 Connections between Magnetism and Electricity

22-6. When a straight length of current-carrying wire is subjected to a magnetic field, is there always, sometimes, or never a magnetic force exerted on the wire? Briefly explain.

22-7. Does a length of wire have to be electrically charged to have a magnetic force on it? Briefly explain.

22-8. If a length of wire that is electrically charged is positioned perpendicular to a uniform magnetic field, will there necessarily be a magnetic force on the wire? Briefly explain.

22-9. SSM WWW The length of wire AB in Figure 22-31 is placed between two opposite magnetic poles. Its ends are connected to a variable power supply (not shown). How does the magnetic force on this length of wire change if

a. the voltage produced by the power supply is doubled?

b. the length AB is rotated 90° so that it ends up parallel to the side of the page?

c. the length AB is rotated 90° so that it ends up pointing out of the page?

Figure 22-31
Problems 22-9, 22-10

22-10. Suppose now that the length of wire AB in Figure 22-31 is not connected to any power supply. Instead, it has a positive charge uniformly distributed over its surface. What can you say about the magnitude or direction of the magnetic force on this length of wire when it is placed between the two poles of the magnet as shown in the figure?

Section 22-3 **Quantitative Treatment of Magnetic Forces**

22-11. In Equation 22-1, what exerts the force on A?

22-12. Find the direction of the force on the charged particle in each of the situations depicted in Figure 22-32 (particle velocities are in red, magnetic field lines in blue).

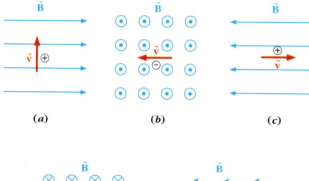

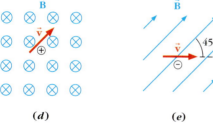

Figure 22-32 Problem 22-12

22-13. The lines of a uniform magnetic field of 0.15 T point west. Find the magnitude and direction of the force on

a. a proton traveling straight down at a velocity of 4000 m/s.

b. an electron traveling due east at a velocity of 2.0×10^5 m/s.

c. a singly charged negative ion heading 30° north of west at a speed of 3.0×10^4 m/s.

d. a neutron traveling straight up at 6500 m/s.

Section 22-4 **Magnetic Forces and Circular Motion**

22-14. In Figure 22-33, a positive ion enters at velocity $\vec{v}$ into a region of uniform magnetic field.

a. In what direction is the force on the ion when the ion first enters this region?

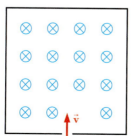

Figure 22-33
Problem 22-14

b. After entering this region, does the ion speed up, slow down, or continue at the same speed?

c. When the direction of the ion has changed by 10°, will the angle that the ion's velocity makes with the field be less than, equal to, or greater than 90°?

d. When the direction of the ion has changed by 10°, by how much will the direction of the magnetic force on the ion have changed?

22-15. SSM A 1.7×10^{-26} kg particle carries a charge of 3.2×10^{-19} C.

a. What field (magnitude and direction) must be applied if the particle is to travel in the xy plane in a clockwise circle of radius 0.25 m at a speed of 4000 m/s?

b. Calculate the particle's acceleration in this field, and describe its direction.

22-16. A singly charged ion traveling in the region of radial acceleration of a mass spectrometer (Figure 22-11) follows a path of radius 0.15 m at a speed of 4.0×10^4 m/s. If there is a 0.20 T magnetic field in this region, what is the mass of the ion?

22-17. A charged particle is traveling due east at a constant speed of 3200 m/s through a chamber in which there is a uniform magnetic field of 0.60 T toward the south. What *electric* field (magnitude and direction) must be applied in this region to keep the particle traveling in a straight line?

22-18. *Bubble Chamber Photographs.* Physicists interested in elementary particles use *bubble chambers* in conjunction with particle accelerators to get charged particles to leave visible tracks. The chambers contain a superheated liquid, which is brought to boiling in the vicinity of an ion. An ion passing through it therefore leaves a trail of bubbles. The bubble chamber photograph sketched in Figure 22-34 was taken in the presence of a magnetic field directed out of the paper. (Assume each particle starts out traveling from the left.)

a. Which tracks were left by positively charged particles?

b. Which tracks were left by negatively charged particles?

c. If tracks 2 and 3 were left by particles of equal charge and mass, which particle was moving faster? Briefly explain.

d. If tracks 1 and 3 were left by singly charged particles traveling at the same speed, which particle was more massive? Briefly explain.

e. What does the fact that tracks 4 and 5 spiral inward tell you is happening to these particles? Briefly explain.

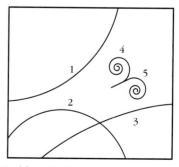

Figure 22-34 Problem 22-18

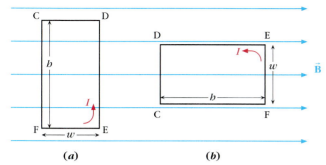

22-19. ***The Cyclotron.*** The cyclotron, developed by Ernest O. Lawrence and M. S. Livingston at the University of California at Berkeley in 1931, was one of the first particle accelerators. Its principal parts (Figure 22-35*a*) were (1) two hollow chambers called "dees" because of their shape, (2) an ion source S that produced the particles to be accelerated, (3) magnets to produce a magnetic field perpendicular to the plane of the dees, and (4) a power supply that could set up a voltage V across the gap between the dees and could reverse the polarity of the voltage at regular intervals. A particle emitted at the source is accelerated across the gap gaining an amount of KE equal to qV. It travels in a semicircular path in the uniform magnetic field within the dee, and so is brought back to the gap. When it arrives back, the voltage is reversed, accelerating the particle the other way across the gap. In regions A and B in Figure 22-35*b,* the uniform magnetic field is uniform and directed into the page. A charged particle follows the path shown in the figure. It travels at uniform speed in region A.

Figure 22-35 The cyclotron Problems 22-19, 22-52

a. Is this particle positively or negatively charged? Briefly explain.

b. What is the radius of curvature of the path the particle follows in region B if while it travels from point 1 to point 2 it is subjected to an electric field that doubles its kinetic energy?

c. What is the radius of curvature of the path the particle follows in region B if it passes through a resistive medium between points 1 and 2 that reduces its kinetic energy by half?

Section 22-5 Magnetic Forces on Current-Carrying Wires

22-20. In the vicinity of the city of Metropolis, Earth's magnetic field is 4.0×10^{-5} T. Find the force on a 20-m length of power cable carrying a current of 1.5 A if the current is flowing ***a.*** due east. ***b.*** due south. ***c.*** in a direction 37° south of east.

22-21. A current of 0.25 A passes through a long straight wire in the presence of a 1.6×10^{-3} T magnetic field. If the force on each meter of the wire is 2.5×10^{-4} N, what angle does the wire make with the field?

22-22. A current of 0.15 A is maintained in a rectangular loop of wire lying flat on a table top. When there is a magnetic field of 1.6×10^{-3} T perpendicular to the tabletop, what is the magnitude of the total magnetic force on the wire loop?

22-23. In Figure 22-36, two oppositely charged plates mounted on insulating posts are separated by a distance of 0.040 m in a uniform magnetic field of 0.60 T (directed upward). The plates are then discharged by connecting a straight wire between them. Find the force (magnitude and direction) on the wire when the discharge current is 0.20 A.

Figure 22-36 Problems 22-23, 22-24

22-24. Suppose that in the previous problem, the charge on each plate initially has an absolute value of 2.0×10^{-4} C. When the wire is connected, it takes 0.0010 s for each plate to lose 90% of its charge. What is the average force on the wire during this interval?

22-25. **SSM WWW** The wire in Figure 22-37 wraps around a cube 4.0 m on a side. A current passes through the wire in the direction indicated by the arrows. A uniform magnetic field is applied in the $+y$ direction, as shown. Rank the segments of the wire (AB, BC, CD, DE, EF) in order of the magnetic force (magnitude only) on each. Order them from least to greatest, making sure to indicate any equalities.

Figure 22-37 Problems 22-25, 22-26

22-26. The wire in Figure 22-37 wraps around a cube 4.0 m on a side. A 0.30 A current passes through the wire in the direction indicated by the arrows. A uniform magnetic field of 0.25 T is applied in the $+y$ direction, as shown. Find the magnitude and direction of the magnetic force on each of the following segments of the wire ***a.*** CD; ***b.*** EF; ***c.*** AB; ***d.*** BC; ***e.*** DE.

Section 22-6 Making Use of Torques on Current Loops

22-27. Suppose that a rectangular loop of wire runs around the edge of this page, and that a current passes through it counterclockwise.

a. In what direction(s) can a magnetic field be applied without resulting in any torque on this loop?

b. In what direction(s) must a magnetic field be applied to produce the maximum torque on this loop?

22-28. A rectangular loop of wire CDEF is placed in a magnetic field as shown in Figure 22-38*a*. There is a counterclockwise current I in the wire.

a. Find the magnitude and direction of the magnetic force on each side of the wire. Express the magnitudes in terms of

Figure 22-38 Problem 22-28

the current I, the field magnitude B, and the loop dimensions b and w.

b. Now choose a vertical (top to bottom of page) axis of rotation and use the forces and moment arms to find the torque about this vertical axis.

c. Next, choose a different vertical axis of rotation and repeat **b**. How does the torque compare to the torque you found in **b**? How do you expect your results to compare for other vertical axes of rotation?

d. Repeat parts **a** through **c** for the case when the loop is positioned as shown in Figure 22-38b. Although the loop is now turned over on its side, you should still be considering vertical axes of rotation. How do the torques that you find now compare to those you found in **b** and **c**?

22-29. SSM A circular loop of radius 4×10^{-2} m carries a current of 0.50 A. Calculate the torque on the loop in a 2.2 T uniform magnetic field if the field **a.** is perpendicular to the plane of the loop. **b.** is parallel to the plane of the loop. **c.** makes an angle of 15° with the plane of the loop.

22-30. A coil of 10 turns of wire is wound around a hollow core with a rectangular cross-section 0.06 m by 0.08 m. Find the maximum current that can flow through it if the torque on the loop is never to exceed 9.6×10^{-4} m·N in a magnetic field of 0.4 T.

Section 22-7 **How the Magnetic Field Depends on Its Source**

22-31. Suppose two long straight wires run horizontally across this page. The upper one carries a current to the left. The lower one carries an equal current to the right.

a. In what direction is the field that the lower wire experiences due to the upper wire?

b. In what direction is the force on the lower wire?

c. In what direction is the total magnetic field at a point midway between the two wires?

d. If an electron passed through this midway point traveling toward the lower wire, what would happen to the motion of the electron at this point? In other words, what would happen to the magnitude and/or direction of its velocity over the next tiny time interval?

e. Repeat **d** for the case in which the wires carry equal currents in the same direction.

22-32. Suppose two long straight wires run horizontally across this page. The upper wire carries a 0.16 A current to the right.

The lower wire, a distance of 10 cm below the upper one, carries a 0.20 A current to the left.

a. Find the magnitude and direction of the magnetic field that the lower wire experiences due to the upper wire.

b. Find the magnitude and direction of the force on a 20-cm length of the lower wire.

c. Find the magnitude and direction of the magnetic field that the upper wire experiences due to the lower wire.

d. Find the magnitude and direction of the force on a 20-cm length of the upper wire.

e. Find the magnitude and direction of the total magnetic field at a point midway between the two wires.

22-33. SSM Find the direction of the total magnetic force exerted on object 2 by object 1 in each of the situations depicted in Figure 22-39.

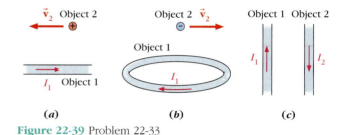

Figure 22-39 Problem 22-33

22-34. Find the direction of the total magnetic force exerted on object 2 by object 1 in each of the situations depicted in Figure 22-40.

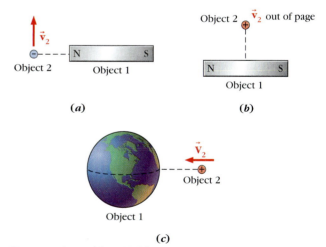

Figure 22-40 Problem 22-34

22-35. Find the direction of the total magnetic force exerted on object 2 by object 1 in each of the situations depicted in Figure 22-41.

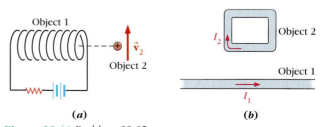

Figure 22-41 Problem 22-35

22-36. A wire running along the x-axis carries a current of 0.050 A toward the left. An electron crosses the y-axis at $y = -0.002$ m traveling to the right at a speed of 10 000 m/s. Find the magnitude and direction of the force on the electron at this instant.

22-37. Figure 22-42a shows the cross-sections of two current-carrying wires that are perpendicular to the plane of a computer screen. (To help you picture this, Figure 22-42b provides a three-dimensional view of the same arrangement.) The wires carry equal currents I, but $\otimes$ indicates a current flowing into the screen, and $\odot$ indicates a current flowing out of the screen.

a. Suppose you draw your field vectors so that at a distance of 3 cm from a single straight wire carrying a current I, the magnetic field is represented by an arrow 1.0 cm long. How long would the arrow representing the magnetic field be at a distance of 6 cm from this single wire? In the following parts, you should draw all your field vectors to this scale.

b. By means of suitable sketches, find the total magnetic field vector at the origin in Figure 22-42a. Roughly how long is the arrow representing this vector, and what is its direction?

c. Repeat **b** for the total magnetic field vector at point A.

d. Repeat **b** for the total magnetic field vector at point B.

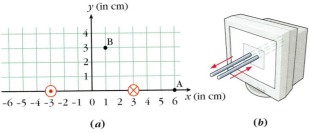

(a)   *(b)*

Figure 22-42 Problems 22-37, 22-38

22-38. Suppose that Figure 22-42a is now changed so that both currents are out of the page. Everything else about the set-up remains the same. Repeat parts **b**–**d** of Problem 22-37 for the changed set-up.

22-39. In the coaxial cable in Figure 22-43, the outer and inner conductors are separated by a layer of insulation. The radius r_1 of the inner conductor is 1.0×10^{-3} m. The inner and outer radii of the outer conductor are $r_2 = 2.8 \times 10^{-3}$ m and $r_3 = 4.1 \times 10^{-3}$ m. If a current of 0.60 A flows through the inner conductor in the $-x$ direction and a current of 0.40 A flows through the outer conductor in the $+x$ direction, find the magnitude of the magnetic field at a distance of **a.** 2.5×10^{-3} m from the

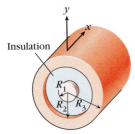

Figure 22-43
Problem 22-39

central axis of the cable. **b.** 4.5×10^{-3} m from the central axis of the cable.

22-40. Use Table 22-2 to find the force (in air) on an electron traveling at a speed of 1000 m/s due to a wire carrying a current of 0.50 A in each of the following circumstances.

a. The wire is long and straight; the electron is traveling parallel to the wire at a distance of 2.0×10^{-3} m from it.

b. The wire is wound as a 0.10-m-long hollow solenoid with 500 turns; the electron is crossing perpendicular to the axis of the solenoid somewhere in its interior.

c. The wire forms a circular loop with a 1.0 cm diameter; the electron is moving in the plane of the loop and passing through its center.

d. The wire forms a circular loop with a 1.0 cm diameter; the electron is passing through the loop's center in a direction making an angle of 37° with the axis.

Section 22-8 Magnetic Materials

22-41. A bar magnet is broken into two pieces as shown in Figure 22-44.

a. What kind of pole (north or south) occurs at the broken end of the piece that has the original north pole?

Figure 22-44
Problem 22-41

b. What kind of pole occurs at the broken end of the piece that has the original south pole?

c. If the two broken ends are kept at a tiny distance from each other, what is the direction of the magnetic field between the two broken ends?

d. If the two broken ends are brought back together, so that the original bar magnet appears whole again, what is the direction of the field at the location of the break?

e. What is the direction of the field at other points along the length of the bar magnet? Briefly explain your reasoning.

22-42. A certain electromagnet consists of a coil of wire wrapped around a hollow cardboard cylinder. A constant current is maintained in the coil. How is the magnetic field due to this electromagnet affected when a solid nickel cylinder is placed within and nearly fills the cardboard cylinder? Briefly explain.

22-43. When a loop of wire carrying an electric current is subjected to an external magnetic field $\vec{B}_{ext}$, a torque will generally be exerted on the loop and the loop will reorient itself until the torque on it is zero. The current in the loop also gives rise to a magnetic field, which we may call $\vec{B}_{loop}$.

a. When the loop is oriented so that the torque on it is zero, how do the directions of $\vec{B}_{loop}$ and $\vec{B}_{ext}$ compare?

b. Is the orientation of the loop like the alignment of a spin-up electron or a spin-down electron (see Figure 22-29)? Briefly explain.

Going Further

The questions and problems in this group are not organized by section heading, so you must determine for yourself which ideas apply. Some of them will be more challenging than the Review and Practice questions and problems (especially those marked with a • or ••).

22-44. How are electrostatic and magnetic forces different?

22-45.

a. Can objects that exert no electrostatic forces on each other ever exert magnetic forces on each other? If so, give an example, or if not, tell briefly why not.

b. Can objects that exert no magnetic forces on each other exert electrostatic forces on each other? If so, give an example, or if not, tell briefly why not.

c. Can subatomic particles that exert no electrostatic forces on each other exert magnetic forces on each other? If so, give an example, or if not, tell briefly why not.

22-46. Two electrons are in motion. The directions in which they are traveling affect the magnitude _____ (*only of the electrostatic forces they exert on each other; only of the magnetic forces they exert on each other; of both kinds of forces; of neither*).

22-47.

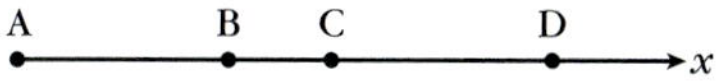

a. Suppose that in Figure 22-45 a positive point charge is placed at point B and a negative point charge of equal magnitude is placed at point C. No other charges are present. Draw arrows to indicate the direction of the *electric* field along line segment AB, line segment BC, and line segment CD.

A ● — B ● — C ● — D ● →x

Figure 22-45 Problem 22-47

b. Suppose now that the point charges are removed and a bar magnet is placed with its north pole at point B and its south pole at point C. Draw arrows to indicate the direction of the *magnetic* field along line segment AB, line segment BC, and line segment CD.

c. Are there any differences between the patterns of arrows you've drawn in **a** and **b**? Comment on the reason(s) for any difference(s).

22-48. A proton traveling at uniform velocity goes a distance of 0.10 m in 2.0×10^{-5} s.

a. What is the maximum amount of work that can be done on the proton during this interval by an electric field with a magnitude of 250 N/C?

b. What is the maximum amount of work that can be done on the proton during this interval by a magnetic field with a magnitude of 0.25 T?

22-49. In dealing with the motion of a charged point particle in a magnetic field, why haven't we found it useful to talk about the particle's magnetic potential energy? *Hint:* What is the relationship between work and potential energy?

22-50. At a particular instant t_0, a proton and an alpha particle (a helium nucleus, made up of two protons and two neutrons) enter a region of uniform magnetic field. At this instant, both are traveling at the same speed in the same direction, perpendicular to the field. If the two particles interact only with the magnetic field, is it possible for their velocities to be perpendicular to each other at some later instant? Briefly explain.

22-51. **SSM** Figure 22-46 shows the path followed by the alpha particle in Problem 22-50 from the instant t_0, when it first enters the region of uniform magnetic field, until a later instant t_1, when it has traveled through a quarter of a circle. Compare this path with the path followed by the proton in Problem 22-50 by answering the following.

a. How does the radius of curvature of the proton's path compare with that of the alpha particle's path? Express your answer as a ratio $\frac{r_{proton}}{r_{alpha}}$.

b. Sketch the path followed by the proton between t_0 and t_1. Draw it to the same scale as Figure 22-46. Between t_0 and

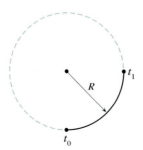

Figure 22-46 Problem 22-51

t_1, what fraction of the distance around its circular path will the proton travel? Why?

c. Compare the directions in which the two particles are traveling at instant t_1. Does this agree with your answer to Problem 22-50?

••22-52. A particle is being accelerated in a cyclotron (see Problem 22-19).

a. How does the path followed by the particle in the left-hand dee compare with the path it traveled previously in the right-hand dee?

b. Describe the subsequent motion of the particle, assuming the voltage reverses to speed it up during each crossing of the gap. *Each time* the particle enters the left-hand dee, how does the path that it follows compare with the path it traveled previously in the right-hand dee?

c. Write an equation for the velocity v with which the particle emerges at point P. The equation should be written in terms of quantities such as the charge q and mass m of the particle, the field B, and the dees' radius R. Does this final velocity depend on the voltage V? Why or why not?

d. Modify the equation in **c** so it gives the velocity v at any distance r from the center. From this equation, find an expression for the frequency f (in revolutions per second) of the particle. Does this result vary with r? What implication does this have for the frequency with which the voltage source must reverse polarity?

22-53.

a. Do utility companies need to be concerned about forces on overhead electric power lines due to Earth's magnetic field?

b. What minimum current would a power line have to carry for there to be a force of 1 N on each 100 m of the line? (After answering this, see if you wish to change your answer to part **a**.)

22-54. ***The Galvanometer as Ammeter.*** To measure the current through a circuit element, you must place an ammeter in series with that element so that the same current passes through the ammeter. If you use a galvanometer as an ammeter, its coil must be in series with the element because the needle deflection depends on the current through the coil.

a. Should the ammeter have high resistance or low? Briefly explain why.

b. The coil has a certain resistance. You can alter the resistance of the galvanometer by placing another resistor either in series or in parallel with the coil. Which should you do in an ammeter? Should this resistor have high or low resistance? Briefly explain your reasoning.

22-55. **SSM WWW** ***The Galvanometer as Voltmeter.*** To measure the voltage across a circuit element, you must place a voltmeter in parallel with the element in question so that it has the same voltage across it. A galvanometer can be adapted for use as a voltmeter, because the voltage across the coil is proportional to the current through it. In that case, the coil must be connected in parallel with the element in question.

a. Should the voltmeter have high resistance or low? Briefly explain why.

b. The coil has a certain resistance. You can alter the resistance of the galvanometer by placing another resistor either in series or in parallel with the coil. Which should you do in a voltmeter? Should this resistor have high or low resistance? Briefly explain your reasoning.

22-56. Devices that use electromagnets to control switches are called **relays.** Figure 22-47 shows a relay that serves as a circuit breaker to protect a device that can be damaged if the current through it becomes too great. Explain in detail how it works.

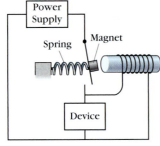

Figure 22-47 Problem 22-56

• **22-57.** A straight piece of wire 0.50 m long is stretched along the x-axis (left to right across the page). A current of 0.30 A flows through it in the +x direction. The wire is in a region of space where there is a uniform magnetic field of magnitude 2.0 T. How many different direction(s) of the magnetic field are possible if the magnetic force on the piece of wire has a magnitude of **a.** 0.30 N and is directed into the page? **b.** 0.18 N and is directed out of the page? **c.** 0.40 N and is directed into the page?

22-58. A straight piece of wire 0.50 m long is stretched along the x-axis (left to right across the page). A current of 0.30 A flows through it in the +x direction. Describe the direction or directions of the magnetic field that would result in a magnetic force on the piece of wire that has a magnitude of 0.24 N and is directed out of the page.

22-59. In each of the five situations in Figure 22-48, a singly charged positive ion is moving in the presence of a current-carrying wire. The current in the wire and the speed and direction of the ion vary from one situation to another. The imaginary cube is placed in each situation to help you visualize the directions. (In situation D, the ion moves along the diagonal of the top face; in E the wire is along a diagonal.) Rank these five situations in order of the magnetic force (magnitude only) that the wire exerts on the ion. Order them from least to greatest, making sure to indicate any equalities.

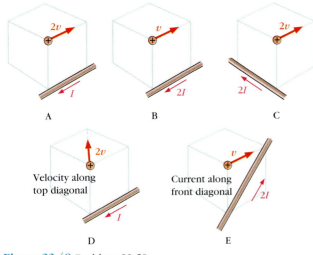

Figure 22-48 Problem 22-59

22-60. In Figure 22-49, what is the direction of the magnetic field at point P if **a.** the ring is stationary, and a current flows through it in the direction shown (Figure 22-49a)?

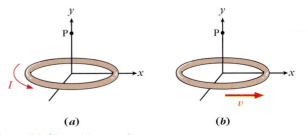

Figure 22-49 Problem 22-60

b. no current flows through the ring, but the ring is negatively charged and is traveling to the right with speed v (Figure 22-49b)?

•• **22-61. The Hall Effect.** Figure 22-50 shows a thin slab of conducting material with top-to-bottom thickness d. It is subjected to a magnetic field of magnitude B directed out of the paper.

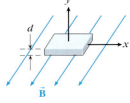

Figure 22-50 Problem 22-61

a. If a conventional current travels through the slab in the +x direction, what will happen to the traveling positive charge carriers? Why? How will this affect the charge distribution on the top and bottom surfaces of the slab?

b. As the distribution of charge changes, what happens to additional positive charge carriers traveling through the slab? Will they be affected as much? Explain.

c. Is there a stage at which additional positive charge carriers continue undeflected? Describe all the forces on a charge carrier that passes through the slab in the +x direction with no deflection. How do the magnitudes and directions of these forces compare?

d. Show that if the positive charge carriers have a drift velocity v_x, the altered distribution of charge results in an electric field in the y direction, with magnitude $E_y = v_x B$, and that there will then be a potential difference $V = v_x B d$ between the top and bottom of the slab. The production of a potential difference in this way is called the **Hall effect,** and V is called the **Hall voltage.** (Because this voltage would be reversed if the charge carriers were negative, it provides a way of determining the sign of their charge.)

22-62. Figure 22-51 shows an alternative to the right-hand rule given in Figure 22-8. Try applying this alternative version to each part of Problem 22-12 and verify that you get the same results by both versions.

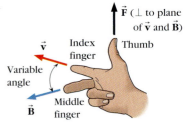

Figure 22-51 Alternative right-hand rule. Problem 22-62

22-63. In Figure 22-52, a bar magnet is suspended from a string. If the positively charged rod is placed in the position shown, the bar magnet will _____ (*rotate clockwise in the plane of the page; rotate counterclockwise in the plane of the page; rotate in a horizontal plane with the string as its axis* *of rotation; not rotate at all*). (Based on an example in a paper presented by L. C. McDermott at the International Conference on Undergraduate Physics Education, University of Maryland, 1996.)

22-64. Explain how each of the following work. Make sure to discuss the role of magnetic forces and torque in your explanation.
a. The galvanometer in Figure 22-21.
b. The electric motor in Figure 22-23*a*.

22-65. In the set-up in Figure 22-18, a straight length of heavy copper wire is suspended by two delicate springs made of fine conducting wire. The springs, which

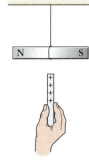

Figure 22-52
Problem 22-63

are very short compared to the straight wire, are magnified in the figure for visibility. (Because the force on a current-carrying wire is proportional to the wire's length, we can therefore assume that any complicating effects of the magnetic forces on the conducting springs are negligibly small.) The other ends of the springs are connected to a 9.0 V battery and a variable resistor. Describe in detail how you could use this set-up to find the strength of a magnetic field. What would you measure, what concepts would you apply, and how would you calculate the field based on these measurements?

22-66. Suppose that in Problem 22-65, the springs are made longer and the straight wire is shortened, so that the magnetic forces on the current-carrying springs cannot be considered negligible.
a. Would both springs always make the same angle with the vertical when an external magnetic field was applied? Briefly explain.
b. Would the springs necessarily remain perpendicular to the straight wire at the points where they connect to the straight wire? Briefly explain.

22-67. Suppose that in the set-up in Figure 22-18, the magnetic field is vertically downward. When the resistor is set so that the current through the 0.20-m-long wire is 0.45 A, the springs make an angle of 35° with the vertical and their stretch tells us that they exert a combined elastic force of 0.48 N. Find the magnitude of the magnetic field.

22-68. Find the force (in air) on a singly charged positive ion traveling at a speed of 300 m/s due to a wire carrying a current of 0.80 A in each of the following circumstances.
a. The wire is long and straight; the ion is moving along a path perpendicular to the wire and intersecting it, and passing a point 0.25 m from the wire.
b. The wire is wound as a 0.10-m-long hollow solenoid with 500 turns; the ion is within the solenoid and is moving perpendicular to the axis of the solenoid.
c. The wire forms two turns of a circular loop with a 5.0-cm diameter; the ion is passing through the center of the loop as it travels along the loop's diameter.

22-69. **SSM** In Figure 22-53, a conducting crossbar is free to slide vertically on a U-shaped conducting wire. A magnetic field is directed perpendicular to the page. The crossbar is 0.20 m long and has a mass of 0.030 kg and a resistance of 24 Ω. The resistance of the U-shaped wire is negligible. A conducting lubricant at the contact points keeps friction negligible while maintaining good electrical contact.

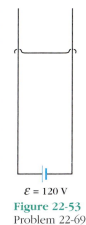

$\mathcal{E} = 120$ V
Figure 22-53
Problem 22-69

a. The crossbar will remain stationary when the magnetic field is large enough if it is directed ____ (*into the page only; out of the page only; either into or out of the page*). Briefly explain.
b. If the magnetic field is properly directed, what must its magnitude be to keep the crossbar stationary?

22-70. Between the instant when an ion leaves the pink region (Figure 22-11) of the mass spectrometer and the instant when it reaches the detector, does its kinetic energy increase, decrease, or remain the same?

22-71. Between the instant when an ion leaves the ionizing source (Figure 22-11) of the mass spectrometer and the instant when it reaches the detector, does its total mechanical energy increase, decrease, or remain the same?

Problems on WebLinks

22-72. If a small bar magnet is placed at point P and is oriented like the bar magnet in WebLink 22-1, which of the pictures in Figure 22-54 correctly shows how the surrounding compass needles will line up?

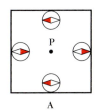
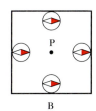
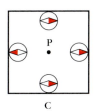
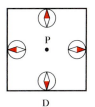
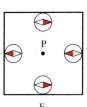

A   B   C   D   E

Figure 22-54 Problem 22-72

22-73. In WebLink 22-2, if the fingers of the Rule Tool remain in the same direction but the angle between the fingers and the thumb increases from 30° to 40°, it indicates that ____ (*the direction of the field has changed by 10°; the direction of the force on the charged particle has changed by 10°; the magnitude of the force on the charged particle has increased; the magnitude of the force on the charged particle has decreased*).

22-74. In WebLink 22-2, if the fingers of the Rule Tool remain in the same direction but the angle between the fingers and the thumb increases from 70° to 160°, it indicates that ____ (*the direction of the field has changed by 90°; the direction of the force on the charged particle has changed by 90°; the magnitude of the force on the charged particle has increased; the magnitude of the force on the charged particle has decreased*).

22-75. In the velocity selector in WebLink 22-3, if the magnetic field is directed toward the top of the page, in what direction must the electric field be?

22-76. In the mass spectrometer in WebLink 22-3, the ions that reach the detector have traveled through regions 1, 2, and 3. In which of the three regions were these ions accelerated? (List all correct answers, or write "none" if they were not accelerated at all.)

CHAPTER TWENTY-THREE

Electromagnetic Induction

Electrically charged objects exert electrostatic forces on one another. We can also state this in terms of electric fields: A charged object experiences a force when placed where there is an electric field due to other charged objects. These fields are responsible for the movement of charge carriers in direct current circuits.

In this chapter we will consider ways charge carriers can experience electric fields that are *not* simply due to other charged objects. We will speak of these as **induced** electric fields. Electric fields produced by induction are responsible for the movement of charge carriers in the *alternating* current circuits typical of power transmission lines, household wiring, home appliances, and most modern electronics and telecommunications (Figure 23-1).

Figure 23-1 Transmitting towers for radio and television. Fields produced by induction play a central role in modern telecommunications.

"Electric fields produced by induction are responsible for the movement of charge carriers in . . . power transmission lines [and elsewhere]."

23-1   Basic Observations Showing the Occurrence of Electromagnetic Induction

Case 23-1 ◆ *Exploring Induction*

(If the equipment is available, you should treat this case as an On-the-Spot Activity.)

Let's consider a test circuit consisting of a coil of insulated wire wound around a soft hollow core and connected at its ends to a galvanometer (Figure 23-2a). It is best to use a fine wire for the coil so you can fit as many turns as possible along each centimeter of the core. (The core should be something soft enough—like the cardboard tube from a roll of toilet

paper—so that you can crush it flat.) The connecting wires are long enough so that the coil can be moved sideways or rotated without disconnecting it. The galvanometer reads zero because there is no voltage supply in the circuit. In the following activity, you will observe how various actions affect the galvanometer reading. In none of these actions will anything actually touch the test circuit (except your hand when moving the coil).

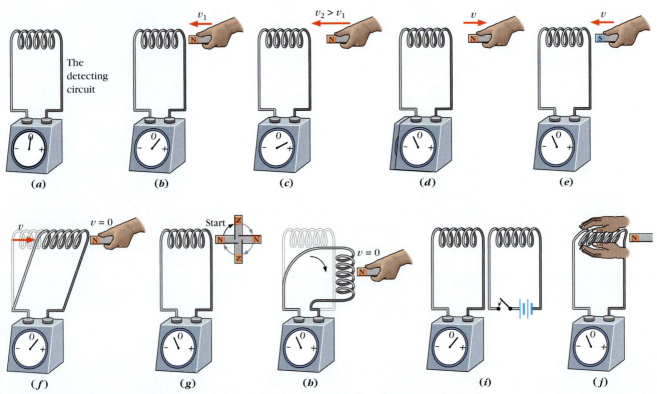

Figure 23-2 Ways of inducing a current. A + deflection of the needle indicates current is flowing left to right through the galvanometer.

1. Line up a bar magnet along the coil's axis with its north pole pointing toward the coil, then move it toward the coil at medium speed (Figure 23-2b). What happens to the needle of the galvanometer while the magnet is in motion? What happens to the needle of the galvanometer when the motion stops?

2. Repeat step 1 at slower and greater speeds (Figure 23-2c). How does the deflection of the needle vary with the speed of the magnet?

3. Repeat step 1 with the magnet moving *away* from the coil (Figure 23-2d). As you do so, note the deflection of the galvanometer needle.

4. Repeat step 1 with the *south* pole of the magnet pointing toward the coil (Figure 23-2e). Again note the deflection of the galvanometer needle.

5. Repeat steps 1 to 4, except that where previously you moved the magnet toward (or away from) the coil, this time move the *coil* toward (or away from) the magnet (Figure 23-2f).

6. Line up the magnet along the coil's axis with its north pole near one end of the coil, then rotate the magnet as shown in Figure 23-2g, stopping after each quarter turn. Repeat with faster quarter turns, then repeat with slower quarter turns.

7. Return the magnet to its starting position for step 6, and this time rotate the *coil* a quarter turn (Figure 23-2*h*).

8. Next to your detection circuit, set up a second circuit containing a coil, batteries, and a switch in series (Figure 23-2*i*). The axes of the two coils should coincide as shown, but no part of one circuit should touch any part of the other. Close the switch and keep it closed for a few seconds, all the while observing the galvanometer needle. Then open the switch.

9. Replace the switch in the second circuit with a variable resistor. Watch what happens to the galvanometer needle when you decrease the resistance, and also when you increase the resistance. **STOP & Think** Based on the observations for step 8, what do you expect to see in this case? ◆

10. Again place the magnet in its starting position for step 6. Now flatten the coil (Figure 23-2*j*). Then try to puff it out again to its original circular cross section. Perform each of these actions as quickly as you can.

The various parts of Figure 23-2 show how the galvanometer needle is deflected when each of these actions is performed. Note when the deflection is in one direction and when it is in the other. Note also that the deflection is greater in Figure 23-2*c* than in 23-2*b*. It is important to recognize that in each of these situations, the needle returns to zero as soon as the action producing the deflection is completed. For any observations you were unable to make yourself, see WebLink 23-1.

What generalizations can we make about our observations? With each of these actions, the galvanometer needle is deflected, indicating a current in the detection circuit. We speak of this as an **induced current** because the cause is external to the circuit. Charge will flow only if there is a potential difference. Then if there is an induced current, there must be an **induced EMF 𝓔.** By connecting a voltmeter or potentiometer across the coil, we would find that this induced EMF occurs even when the coil is not part of a complete circuit, so that no current flows.

In each step in Case 23-1, the induced EMF and the resulting induced current only occur while the situation is changing. When the action stops, the galvanometer needle returns to zero. But *what* is changing that gives rise to the induced EMF? Is it something different in each case? It would be preferable if we could say that the same thing is changing in all cases; that is, if we could find a single basic rule that nature follows in all these different circumstances. The discovery of just such a rule was arguably the greatest of the many achievements of Michael Faraday (1791–1867), a largely self-educated English bookbinder's apprentice who became one of the giants of nineteenth-century physics.

For **WebLink 23-1:** **Observing Electromagnetic Induction,** go to www.wiley.com/college/touger

23-2 Faraday's Law of Electromagnetic Induction

◆**FLUX AND FLUX DENSITY** Faraday made extensive use of fields and lines of force in visualizing electric and magnetic effects. In Faraday's words, "not only are they useful in rendering the vague idea more clear for the time, giving it something like a definite shape, that it may be submitted to experiment and calculation; but they lead on, by deduction and correction, to the discovery of new phenomena."[1] In particular, Faraday made use of the idea that field lines are less dense, or less closely packed, where the field is weaker, and thus diverge from a source such as a point charge or a magnetic pole. Faraday pictured the lines flowing across cross-sections of space. In thinking about flow, he was guided by the resemblance between fluid flow patterns, for example, and patterns of electric lines of force (Figure 23-3).

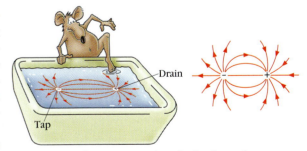

Figure 23-3 Flow patterns in a bathtub analogous to the pattern of lines of force due to an electric dipole. When the water flows from a point source (a tap emerging from the bottom of the tub) to a point sink (the drain), the flow pattern is much like the field pattern due to a pair of equal and opposite point charges.

[1]Quoted in A. Arons, *Development of Concepts of Physics* (Addison-Wesley, Reading, MA, 1965), p. 537.

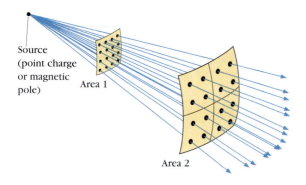

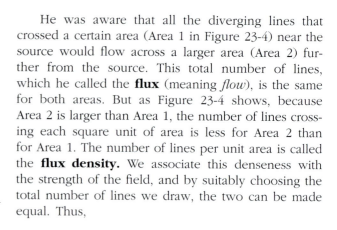

Flux across Area 1 = 16 lines   Flux density at Area 1 = $\frac{16 \text{ lines}}{1 \text{ sq. unit}} = 16 \frac{\text{lines}}{\text{sq. unit}}$

Flux across Area 2 = 16 lines   Flux density at Area 2 = $\frac{16 \text{ lines}}{4 \text{ sq. units}} = 4 \frac{\text{lines}}{\text{sq. unit}}$

Figure 23-4 Flux ("flow" of field lines) and density of field lines.

He was aware that all the diverging lines that crossed a certain area (Area 1 in Figure 23-4) near the source would flow across a larger area (Area 2) further from the source. This total number of lines, which he called the **flux** (meaning *flow*), is the same for both areas. But as Figure 23-4 shows, because Area 2 is larger than Area 1, the number of lines crossing each square unit of area is less for Area 2 than for Area 1. The number of lines per unit area is called the **flux density.** We associate this denseness with the strength of the field, and by suitably choosing the total number of lines we draw, the two can be made equal. Thus,

electric field = electric flux density

magnetic field = magnetic flux density

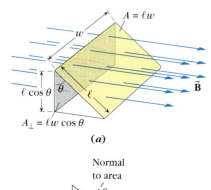

(a)

(b)

Figure 23-5 The cross-sectional area $A_\perp$ is encountered "head-on" by the flow of field lines.

To obtain the total flux, we can multiply the number of lines crossing each square unit of area by the number of square units crossed, or more precisely, by the cross-sectional area perpendicular to the flow. Figure 23-5a shows why. The same lines cross both A and $A_\perp$ (the perpendicular "shadow" of A), even though $A_\perp < A$. We can think of $A_\perp$ as the face that A presents to the flow; that is, its *cross-sectional area*. As the figure shows,

$$A_\perp = A \cos \theta \qquad (23\text{-}1)$$

where θ (in Figure 23-5b) is the angle the field lines make with the *normal* to the area. We now multiply the number of lines per unit area (the field) by this cross-sectional area to obtain the total flux (denoted by Φ):

Electric flux $\Phi_{\text{elec}} = EA \cos \theta$ Magnetic flux $\Phi = BA \cos \theta$ (23-2e,m)

(We omit a subscript for the magnetic flux, because that is what we will be discussing)

Figure 23-6 shows how Faraday pictured the magnetic flux across the area enclosed by each loop of the coil (or in briefer language, "across each loop" or "through each loop"). In reviewing the actions that could induce an EMF in a coil (summarized in Figure 23-2), he realized that an EMF could be induced by changing any one of the three quantities on which this flux depends (see Equation 23-2m): the field strength B, the area A, and the angle θ that the field lines make with the normal to the area. **STOP&Think** Review Case 23-1 and identify which of the three quantities changes in each step. ◆

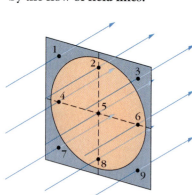

Figure 23-6 The magnetic flux viewed as the field lines crossing an area. An imaginary plane is crossed by magnetic fields running perpendicular to it. The shaded area in the plane is bounded by a loop. This is what it means for a current loop to bound an area. Five of the nine lines shown here cross the area bounded by the loop (1, 3, 7, and 9 do not).

• Changing B: When the bar magnet and coil are brought closer together (Figures 23-2b, c, e, f), the field crossing each loop of the coil is stronger. Consequently the number of field lines (the flux) crossing each loop increases (Figure 23-7b).

• Changing A: When the coil is crushed or puffed out, the area A enclosed by each loop is reduced (to zero if the coil is completely flattened) or increased, and so is the number of lines that can cross it (Figure 23-7a).

• Changing θ: When either the coil or the magnet is rotated, θ changes, and the number of field lines that can be intercepted by the enclosed area likewise changes (Figure 23-7c).

In each case the induced EMF depends on a change in the flux. We also saw that the needle deflection is greater—the induced EMF is larger—when the change is more rapid. Thus, Faraday concluded, the induced EMF is proportional to the

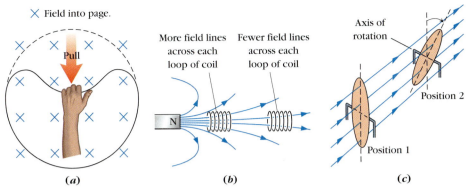

Figure 23-7 Changing the magnetic flux Φ = BA cos θ crossing a loop by changing B across the loop, A, or θ. (a) Changing the area A: The area after deformation is missed by field lines crossing the original area, which extended to the dashed line. (b) Changing the field B: More lines cross the area bounded by each turn or loop of the coil as the coil moves closer to the pole of the magnet. (It doesn't matter whether the magnet moves toward the coil or the coil moves toward the magnet.) (c) Changing the angle θ. A loop is mounted on an axis so that it can rotate. The loops is shown in two positions. In position 1, all four of the field lines cross the loop. After it is rotated to position 2, only two of the field lines cross the loop.

rate $\frac{\Delta \Phi}{\Delta t}$ at which the magnetic flux changes. With units chosen to give the proportionality constant a value of one, we can write this in equation form:

Faraday's law of electromagnetic induction for a single loop

$$\text{Induced EMF } \mathcal{E} = -\frac{\Delta \Phi}{\Delta t} = -\frac{\Delta (BA \cos \theta)}{\Delta t} \qquad (23\text{-}3)$$

If there are N turns in the coil, the EMFs induced in the individual loops or turns are in series and are therefore additive. Hence,

for a coil of N turns: $\quad$ induced EMF $\mathcal{E} = -N\frac{\Delta \Phi}{\Delta t} \qquad (23\text{-}3N)$

(We have not yet addressed the reason for the minus sign. We will do so after considering some examples.) To explore the effects of changing B, A, and θ individually in Faraday's law, work through the next three examples. In doing these examples, the procedure below may help you with the math.

PROCEDURE 23-1

A Useful Mathematical Rule When Using Faraday's Law

In general, $\Delta(BA \cos \theta) = B_2 A_2 \cos \theta_2 - B_1 A_1 \cos \theta_1$. But typically B, A, and θ don't all change. Suppose that in the change of a product, say $\Delta(XY)$, one of the factors remains constant; that is, $X_1 = X_2 =$ the constant value X. Then $\Delta(XY) = X_2 Y_2 - X_1 Y_1 = XY_2 - XY_1 = X(Y_2 - Y_1) = X\Delta Y$. For example, if the constant value is X = 8, this says that $\Delta(8Y) = 8Y_2 - 8Y_1 = 8(Y_2 - Y_1) = 8\Delta Y$. In short, *you can always bring the constant value outside the difference as a common factor.*

Example 23-1 *Turning on an Electromagnet*

For a guided interactive solution, go to Web Example 23-1 at www.wiley.com/college/touger

A small, flat coil consisting of three circular turns of high-resistance wire is inserted into the opening between the poles of a toroidal (doughnut-shaped)

EXAMPLE 23-1 *continued*

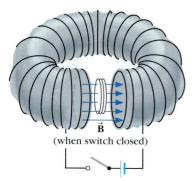

B
(when switch closed)

Figure 23-8 In this figure, Φ varies with B.

electromagnet with 500 turns (Figure 23-8). The coil has a radius of 2.0 cm and a total resistance of 2.3 Ω. When the electromagnet is turned on, the magnetic field reaches its steady state value of 1.4 T in a hundredth of a second.
a. Find the average induced EMF during this interval.
b. Find the average induced current during the same interval.

Brief Solution

Choice of approach.

a. We apply Faraday's law (Equation 23-3N), noting that the field crossing the area enclosed by the loop goes from zero to its final value over the 0.01-s interval. A and θ, the other quantities on which the flux depends, do not change. The field is perpendicular to the loop, so it makes an angle of zero with the normal ($\cos 0 = 1$). **STOP&Think** In Equation 23-3N, which value should we use for N? ◆

b. Once we know the EMF $\mathcal{E}$, we use the fact that the sum of the potential gains and losses around the coil must be zero: $\mathcal{E} - IR = 0$.

What we know/what we don't.

$$r = 2.0 \text{ cm} = 0.020 \text{ m} \qquad \Delta t = 0.01s \quad R = 2.3 \text{ Ω}$$

$N = 3$ (the number of turns in the coil in which the EMF is being induced!)

At beginning of Δt	at end of Δt
$B_1 = 0$	$B_2 = 1.4$ T
$A_1 = A = \pi r^2 = \pi (0.020 \text{ m})^2$	$A_2 = A = \pi r^2 = \pi (0.020 \text{ m})^2$
$\theta_1 = \theta = 0$	$\theta_2 = \theta = 0$

(Drop subscripts on quantities that don't change.)

For **a:** $\mathcal{E} = ?$ For **b:** $I = ?$

The mathematical solution.

a. We apply Procedure 23-1, noting that B is the only factor in the product $BA \cos \theta$ that is not constant:

$$\mathcal{E} = -N\frac{\Delta \Phi}{\Delta t} = -N\frac{\Delta(BA \cos \theta)}{\Delta t} = -N\frac{A \cos \theta \, \Delta B}{\Delta t}$$

$$= -3\frac{\pi(0.020 \text{ m})^2 \cos 0 (1.4 \text{ T} - 0)}{0.01 \text{ s}} = \mathbf{-0.53 \ V}$$

b. From $\mathcal{E} - IR = 0$, we get

$$I = \frac{\mathcal{E}}{R} = \frac{-0.53 \text{ V}}{2.3 \text{ Ω}} = \mathbf{-0.23 \ A}$$

◆ Related homework: Problems 23-7 and 23-14.

Example 23-2 *EMF from a Sliding Contact*

A wire bent into the shape of a rectangle open at one end is placed on a 37° incline in the presence of a uniform 0.90 T magnetic field directed vertically upward (Figure 23-9a). A metal rod laid horizontally across the two parallel sides completes a closed conducting path. The rod is permitted to slide downhill. If the rod begins in the position shown and slides 0.10 m downhill in 0.40 s, calculate the average EMF induced during this interval.

Solution

Choice of approach. In this case, A is the only factor in the product $BA \cos \theta$ that is not constant. The area enclosed by the circuit decreases from 0.20 m by 0.50 m to 0.20 m by 0.40 m as the rod slides 0.10 m downhill.

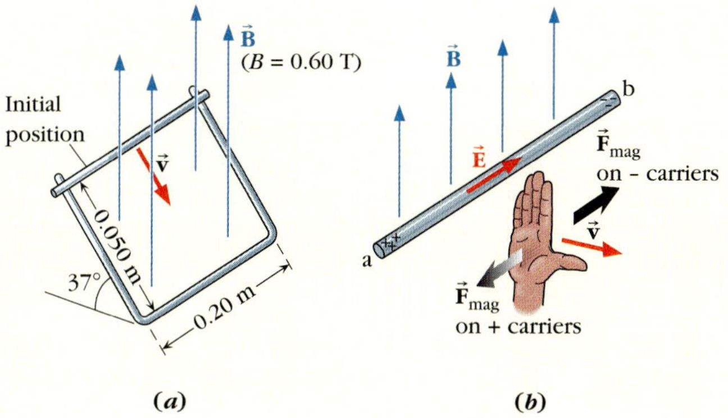

Figure 23-9 In this figure, Φ varies with A.

What we know/what we don't.

$$A_1 = 0.20 \text{ m} \times 0.50 \text{ m} \qquad A_2 = 0.20 \text{ m} \times 0.40 \text{ m}$$

$$\theta_1 = \theta_2 = \theta = 37° \qquad \text{(The normal makes the same angle with the vertical field that the incline makes with the horizontal.)}$$

$$B_1 = B_2 = B = 0.90 \text{ T}$$

$$\Delta t = 0.40 \text{ s} \qquad N = 1 \qquad \mathcal{E} = ?$$

The mathematical solution. In this case, applying Procedure 23-1 gives us

$$\mathcal{E} = -N\frac{\Delta \Phi}{\Delta t} = -N\frac{\Delta(BA \cos \theta)}{\Delta t} = -N\frac{B \cos \theta \, \Delta A}{\Delta t}$$

$$= -(1)\frac{(0.90 \text{ T})(\cos 37°)(0.20 \text{ m} \times 0.40 \text{ m} - 0.20 \text{ m} \times 0.50 \text{ m})}{0.40 \text{ s}}$$

$$= 0.036 \text{ V}$$

Making sense of the results. In this particular case we can also understand why an EMF is induced by picturing what happens to the charge carriers in the conducting rod. Remember that there is a force $qvB \sin \theta$ on a charge moving in a magnetic field. Because the charge carriers in the rod move downhill with the rod, they experience magnetic forces. As Figure 23-9b shows, the forces on oppositely charged carriers are toward opposite ends of the rod. This separation of charge gives rise to an electric field directed along the rod and therefore to a potential difference Ed (d = length of rod). The electrostatic force on any charge carrier in this field is opposite in direction to the magnetic force on it. As the electric field builds up, the electrostatic force will become equal in magnitude to the magnetic force. The charge carriers in the rod will then be in equilibrium, so there will be no further separation of charge. The role of the magnetic force in separating charge is like that of the non-electrostatic force in a battery.

◆ Related homework: Problems 23-10 and 23-11.

Example 23-3 *Detecting Stinky*

For a guided interactive solution, go to Web Example 23-3 at www.wiley.com/college/touger

Your neighbor's rambunctious child, Stinky, comes into your yard to use the swing when he thinks nobody is home. You decide to make use of the fact that the swing faces north and that Earth's magnetic field in your yard is horizontal and has a magnitude of 4.0×10^{-5} T. To detect Stinky's presence, you

EXAMPLE 23-3 *continued*

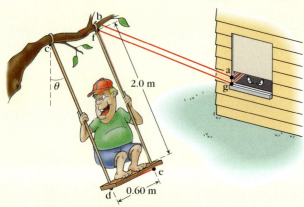

2.0 m

θ

0.60 m

Figure 23-10 In this figure, Φ varies with θ.

run a wire (abcdefg in Figure 23-10) around the swing and connect a potentiometer (a sensitive device for measuring EMF) between a and g. If your potentiometer cannot read voltages of less than 10^{-4} V,

a. through how large an angle from the vertical can Stinky swing in a tenth of a second without being detected?

b. How might you improve the wiring to make your detector sensitive to smaller movements?

Brief Solution

Choice of approach. In this case θ is the only factor in the product $BA \cos \theta$ that is not constant. If we assume that the swing hangs vertically ($\theta_1 = 0$) at $t_1 = 0$, then question **a** becomes: $\theta_1 = ?$ at $t_2 = 0.10$ s.

STOP&Think Try to answer **b** yourself before reading the solution. ◆

What we know/what we don't.

$$A_1 = A_2 = 2.00 \text{ m} \times 0.60 \text{ m}$$

$$B_1 = B_2 = B = 4.0 \times 10^{-5} \text{ T}$$

$$\theta_1 = 0 \qquad \theta_2 = ?$$

$$\Delta t = 0.10 \text{ s} \qquad N = 1 \qquad \mathcal{E} = 10^{-4} \text{ V}$$

The mathematical solution. With B and A both constant,

$$\mathcal{E} = -N\frac{\Delta \Phi}{\Delta t} = -N\frac{\Delta (BA \cos \theta)}{\Delta t} = -N\frac{BA\Delta(\cos \theta)}{\Delta t}$$

$$= -N\frac{BA\Delta(\cos \theta_2 - \cos \theta_1)}{\Delta t}$$

Then

$$\frac{-\mathcal{E}\Delta t}{NBA} = \cos \theta_2 - \cos \theta_1$$

so

$$\cos \theta_2 = \cos \theta_1 - \frac{\mathcal{E}\Delta t}{NBA}$$

$$= \cos 0 - \frac{(10^{-4} \text{ V})(0.10 \text{ s})}{(1)(4.0 \times 10^{-5} \text{ T})(2.00 \text{ m} \times 0.60 \text{ m})} = 0.79$$

and

$$\theta_2 = \mathbf{38°}$$

b. No matter how wild a kid Stinky is, he is unlikely to be swinging fast enough to swing through an arc of 38° in a tenth of a second, so he is pretty safe from detection. To improve detection, you want a smaller motion to produce a larger EMF. You can multiply the EMF by any factor N simply by wrapping the wire N times around the rectangle bcde.

◆ Related homework: Problems 23-12, 23-13, 23-15, and 23-16.

23-3 Determining the Direction of Induced Current Flow

Case 23-2 ◆ *Which Way Does the Current Flow?*

Consider the situation in Figure 23-2b. Can the induced current flow as shown in Figure 23-11a? What would be the consequences? By the right-hand rule for fields, the magnetic field resulting from the current would be in the same direction as the field due to the bar magnet. The coil would act as an electromagnet with its north pole pointed in the same direction as the bar magnet's. Then its south pole would be toward the bar magnet's north pole, and the two magnets would attract. If the bar magnet acquired its initial velocity from a slight shove, as in Figure 23-11a, the attraction would speed it up. The flux across the loops of the coil would change more quickly. A greater current would therefore be induced. The electromagnet would be stronger and would exert a greater attractive force on the bar magnet, causing it to speed up even more. This logical feedback loop (Figure 23-11b) would keep getting repeated, and the speed of the bar magnet would build up without limit. Its kinetic energy would keep increasing without any compensating loss of energy. This can't happen; it violates conservation of energy. So our initial supposition must be wrong. The induced current must be the other way, so that the force exerted by the electro-

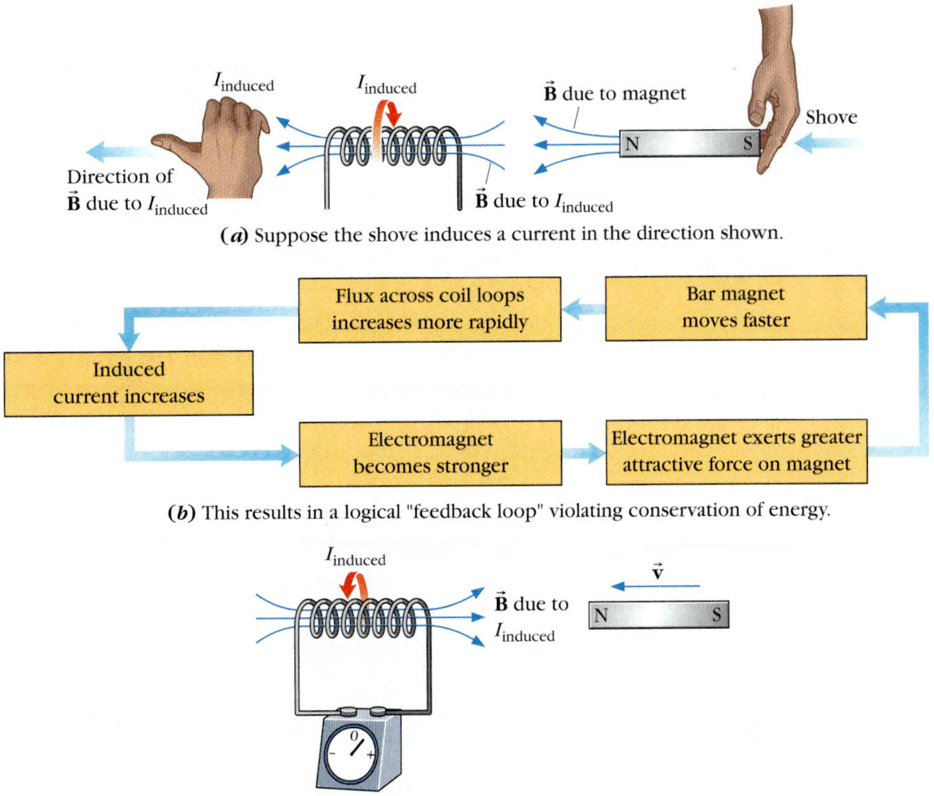

(a) Suppose the shove induces a current in the direction shown.

(b) This results in a logical "feedback loop" violating conservation of energy.

(c) The current must therefore be in the opposite direction.

Figure 23-11 Rejecting a supposition that violates conservation of energy.

magnet *opposes* the bar magnet's motion (Figure 23-11c). There must be negative rather than positive feedback.

As Case 23-2 shows,

We must always choose the current direction that is permitted by the law of conservation of energy.

German scientist H. F. E. Lenz (1804–1864) came up with a useful rule to ensure this, known as Lenz's law.

Lenz's law: When a certain cause produces an induced current, the current in turn will have some effect. The current must flow in a direction such that the effect opposes the cause.

What does this mean in specific instances? The cause of an induced current in a loop is always a change of flux across the area bounded by the loop. The induced current always gives rise to a magnetic field—its effect. Then one interpretation of Lenz's law is

The induced current will flow so as to produce a magnetic field opposing the <u>change</u> in flux.

In Case 23-2, moving the north pole toward the coil increases the number of field lines—the flux—crossing the loops right to left. In opposition to this, the current in the coil must give rise to a field directed left to right. The right-hand rule for coils then gives us the current direction shown in Figure 23-11c (which agrees with the needle deflection in Figure 23-2b).

The number of field lines flowing right to left across the loops of the coil is also increased by the actions depicted in Figures 23-2c, f, and i. Therefore the directions of the induced current and the resulting galvanometer needle deflection are the same in those situations as in the case we have considered. If, on the other hand, the current is induced by decreasing the number of field lines flowing right to left (as in Figure 23-2d) or by increasing the number of field lines crossing the loops left to right (toward the south pole in Figure 23-2e), the change must be opposed by a current giving rise to a field directed toward the left. Such a current would be in the opposite direction, as the galvanometer needles in those figures indicate.

In many of the cases we have investigated, the cause of the induced current involves the motion of some object. In those cases, a useful interpretation of Lenz's law is:

> *If the motion of an object causes an induced current, the current must flow in such a direction that its effect is a force that opposes the object's motion.*

For instance, in Figure 23-11c, we could reason as follows: (1) The induced current is caused by the bar magnet moving toward the left. (2) Then the current must result in a force on the bar magnet toward the right, opposing its motion. (3) To exert a force in this direction, the electromagnet's north pole must point toward the bar magnet's north pole. (4) Then by the right-hand rule, the current must be in the direction shown in the figure.

Example 23-4 *A Sliding Contact, Revisited*

In Figure 23-9, does the current flow from a to b or from b to a across the sliding rod?

Solution
Choice of approach. We will try using each interpretation of Lenz's law.
Reasoning by first interpretation.

1. The component of the field that contributes to the flux across the enclosed area is normal to the area. Its direction is shown in Figure 23-12a.

2. The flux—the total number of field lines crossing the area in this direction—decreases as the area itself decreases.

3. The field due to the current must oppose this decrease and therefore must be in the same direction as the original field.

4. With your thumb in this direction, the right-hand rule for a loop then shows the current flowing across the rod **from b to a** (Figure 23-12a).

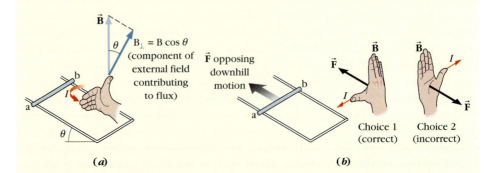

Figure 23-12 Using right-hand rules with Lenz's law.

(a)   (b)

Reasoning by second interpretation.

1. The current must result in an uphill force on the rod (or a force with an uphill component) to oppose its downhill motion.

2. Apply the right-hand rule for magnetic forces with your fingers in the direction of the field and your thumb in one of the two possible directions for the current (Figure 23-12*b*). Choice 1 (the correct choice) gives us an uphill force component, whereas Choice 2 results in a downhill component. In Choice 1, the current direction is **from b to a,** agreeing with our previous conclusion.

Note: Every application of Lenz's law requires the use of at least one of the right-hand rules.

◆ Related homework: Problems 23-17, 23-18, 23-21, and 23-22.

As we noted earlier, if there is not a complete circuit there may be an induced EMF but no induced current. But we can still apply Lenz's law:

The polarity of the induced EMF is always such that if the circuit were completed, the current would flow in the direction determined by Lenz's law.

(Remember that conventional current flows from higher to lower potential.)

The minus sign in Faraday's law of electromagnetic induction serves as a reminder that if the induced EMF results in a current, the induced current will flow so as to produce a magnetic field opposing the change in flux.

23-4 The Electric Generator

The extent of our dependence on electric generators is dramatized whenever there is a power outage and we are cut off from the electric power generated at distant electric power plants (Figure 22-13). If the outage is for a few hours and fairly local, it may just be a source of great inconvenience. If it is more extensive, like the outages that affected many parts of California between November 2000 and May 2001, it can be life threatening. If it is as extensive as the blackout that hit most of the northeastern United States in August 2003, it could raise serious questions of national security.

Electric generators make simple and direct use of Faraday's law. In Figure 23-7*c*, an EMF is induced when the loop is rotated. This also happens in the

Figure 23-13 The dependence of modern society on electric generators. (*a*) Huge generators such as these at Hoover Dam in Nevada provide electric power to very large regions. (*b*) When delivery of power from these generators fails, life is seriously disrupted, and the impact can be life-threatening. The photo shows a New York City street during the blackout of 2003, the glow of crawling traffic contrasting with the darkness of the buildings.

(*a*)

(*b*)

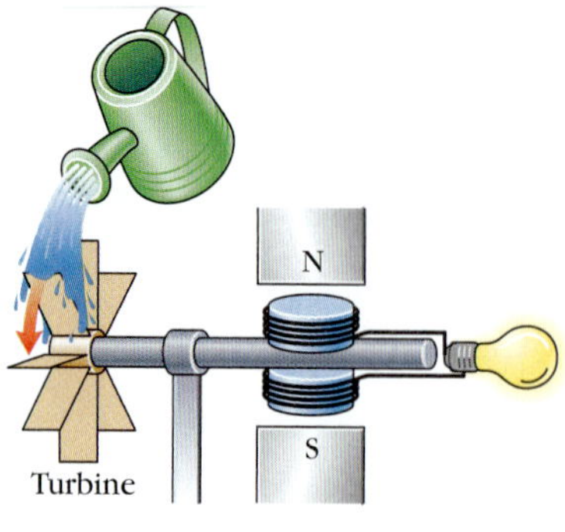

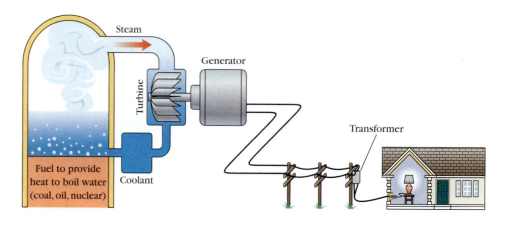

Figure 23-14 Induction in hydro-electric plants. (*a*) A homemade hydroelectric plant. (*b*) Hoover Dam harnesses the power of the Colorado River.

simple homemade hydroelectric plant in Figure 23-14*a*. If it is mounted on a **turbine,** a device that can be turned by external means, the loop can be kept rotating, and thus becomes an electric generator. The principal difference between one kind of electric power plant and another is this: What turns the turbine that turns the generator? Where there are swift-moving rivers, waterfalls, or tides, water can be harnessed to turn the turbines (*hydro-* means water). This is the principle employed, for example, at massive public works projects like Hoover Dam in Nevada, the great Tennessee Valley Authority dams in the southeastern United States (Figure 23-14*b*), or the Aswan High Dam on the Nile in Egypt, all built to bring electrification to their surrounding regions. A power plant on the Rance estuary in France harnesses the giant tides of the Brittany coast.

More commonly, turbines are turned by steam (Figure 23-15). The difference between fossil fuel (coal and oil) fired plants and nuclear power plants is simply the choice of fuel for providing heat to boil the water.

If the loop in Figure 23-16*a* is rotating at a constant angular velocity $\omega = 2\pi f$ (recall Section 9-2), $\theta = \omega t$ and the magnetic flux crossing the loop is

$$\Phi = BA \cos \theta = BA \cos(\omega t)$$

In a typical generator, B and A remain constant, so the induced EMF

$$\mathcal{E} = -N \frac{\Delta \Phi}{\Delta t} = -NBA \frac{\Delta \cos(\omega t)}{\Delta t}$$

Because a graph of $\cos(\omega t)$ versus t is not a straight line, the slope of any tiny segment of it—the rate of change $\frac{\Delta \cos(\omega t)}{\Delta t}$—is not constant. To see how it varies with time, we reason as follows. The slope of a graph of $\sin \theta$ versus θ is zero at 0, π, and 2π (Figure 23-16*b*). Careful measurement would show that it equals 1 for the tiny segment at $\frac{3\pi}{2}$, so by symmetry it is -1 for the segment at $\frac{\pi}{2}$. If we then plot the slope $\frac{\Delta \cos \theta}{\Delta \theta}$ against θ (Figure 23-16*c*), we get a graph that we can recognize as an upside-down sine curve; that is, a graph of $-\sin \theta$ versus θ. This reasoning is developed in step-by-step detail in WebLink 23-2. In fact, as

For **WebLink 23-2:** **Sinusoids and How They Change,** go to www.wiley.com/college/touger

Figure 23-15 A typical commercial power plant arrangement.

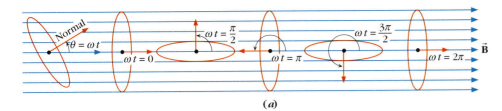

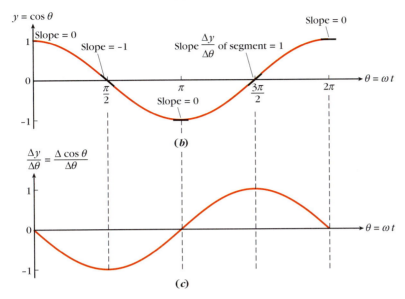

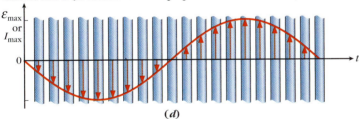

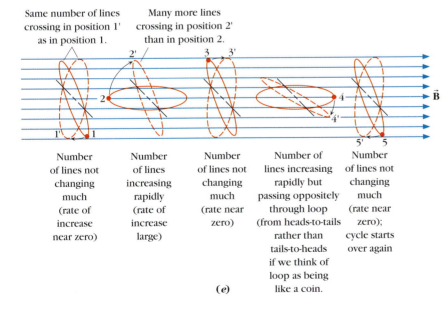

Figure 23-16 Generation of an alternating current by induction.
(*a*) A coil rotates in a magnetic field. (*b*) The flux across the coil varies as $\cos\theta = \cos(\omega t)$. (*c*) Plotting the slope of y versus θ at different values of θ gives us a graph we recognize as a plot of $-\sin\theta$ versus θ. (*d*) A time sequence of pictures showing the magnitude of the induced current or voltage (indicated by the size of the arrow) and the direction of flow at various stages or phases of the cycle. (*e*) The magnetic flux changes at different rates as the loop rotates from position 1 to 1′, from 2 to 2′, and so on.

WebLink 23-2 shows, the rate of change of any **sinusoidal** function ($\pm$ cosine, $\pm$ sine) is always another sinusoidal function. In particular, we have shown that $\frac{\Delta \cos \theta}{\Delta \theta} \approx -\sin \theta$. Now we can substitute ωt for θ. When ω is constant, $\Delta(\omega t) = \omega \Delta t$ (recall Procedure 23-1). Multiplying both sides by ω, we get

$$\frac{\Delta \cos(\omega t)}{\Delta t} \approx -\omega \sin(\omega t) \qquad (23\text{-}5)$$

It follows that the induced EMF is

$$\mathcal{E} = \underbrace{-\omega NBA}_{\substack{\text{max.} \\ \text{EMF}}} \underbrace{\sin(\omega t)}_{\substack{\text{varies} \\ \text{between } \pm 1}} \qquad (\omega = 2\pi f) \qquad (23\text{-}6)$$

The EMF consists of an amplitude or maximum value multiplied by a number—the sine (or the negative sine)—which oscillates between ± 1 over time.

In a closed circuit with a resistance R, the induced current is (as in Example 23-1**b**)

$$i = \frac{\mathcal{E}}{R} = -\underbrace{\frac{\omega NBA}{R}}_{\substack{\text{max.} \\ \text{current } I}} \sin(\omega t)$$

As Figure 23-16*d* shows, a current that varies sinusoidally flows first in one direction then the other. It is therefore called an **alternating current.** In the same way, the polarity of the EMF keeps reversing, and is called an **alternating EMF** or **alternating voltage.** Equation 23-6 also tells us that the induced EMF is proportional to the angular velocity ω. This is consistent with all our observations showing that the induced EMF or current was greatest when the motion or change was most rapid. Figure 23-16*e* shows how the above equations fits with the idea that the induced EMF depends on the rate at which the flux through the loop is changing. To see this connection developed more fully, view WebLink 23-3.

For **WebLink 23-3:**
As the Coil Turns,
go to
www.wiley.com/college/touger

Example 23-5 *The Hand-Cranked Generator*

In a hand-cranked generator, a coil of 50 turns of wire, each enclosing an area of 10 cm^2, is rotated in a 0.60 T magnetic field. With effort, you are able to turn the handle five full revolutions each second. At this rate,
a. what is the maximum EMF you can generate?
b. what is the instantaneous value of the EMF at $t = 0$, $t = \frac{1}{20}$ s, $t = \frac{2}{20}$ s, $t = \frac{3}{20}$ s, $t = \frac{4}{20}$ s? (Assume $t = 0$ at an instant when the EMF is going from $+$ to $-$.)

Solution
Choice of approach. In Equation 23-6, $\mathcal{E}$ is the instantaneous value of the EMF (it depends on t); ωNBA is its maximum value. 5 revolutions/s (or 5 s^{-1}) is the frequency f of rotation, and $\omega = 2\pi f$.

What we know/what we don't.

$$f = 5 \text{ s}^{-1} \qquad \omega = ? \qquad N = 50$$

$$B = 0.60 \text{ T} \qquad A = 10 \text{ cm}^2 = 10(0.01 \text{ m})^2 = 0.0010 \text{ m}^2 \qquad \mathcal{E}_{\text{max}} = ? \qquad \mathcal{E} = ?$$

The mathematical solution.
a. The maximum EMF

$$\mathcal{E}_{\text{max}} = \omega NBA = 2\pi f NBA = 2\pi(5 \text{ s}^{-1})(50)(0.60 \text{ T})(0.0010 \text{ m}^2) = \mathbf{0.94 \ V}$$

b. By Equation 23-6,

$$\mathcal{E} = -\mathcal{E}_{\text{max}} \sin(\omega t) = -\mathcal{E}_{\text{max}} \sin(2\pi f t) = -(0.94 \text{ V})\sin([10 \ \pi \text{s}^{-1}]t)$$

We can now substitute specific values of *t* to find the EMF at different instants:

t	*ε*
0	$-(0.94\text{ V})\sin 0 = \mathbf{0}$
$\frac{1}{20}$ s	$-(0.94\text{ V})\sin\left(10\pi\text{s}^{-1}\times\frac{1}{20}\text{s}\right) = -(0.94\text{ V})\sin\frac{\pi}{2} = -(0.94\text{ V})(1) = \mathbf{-0.94\text{ V}}$
$\frac{2}{20}$ s	$-(0.94\text{ V})\sin\left(10\pi\text{s}^{-1}\times\frac{2}{20}\text{s}\right) = -(0.94\text{ V})\sin\pi = -(0.94\text{ V})(0) = \mathbf{0}$
$\frac{3}{20}$ s	$-(0.94\text{ V})\sin\left(10\pi\text{s}^{-1}\times\frac{3}{20}\text{s}\right) = -(0.94\text{ V})\sin\frac{3\pi}{2} = -(0.94\text{ V})(-1) = \mathbf{+0.94\text{ V}}$
$\frac{4}{20}$ s	$-(0.94\text{ V})\sin\left(10\pi\text{s}^{-1}\times\frac{4}{20}\text{s}\right) = -(0.94\text{ V})\sin 2\pi = -(0.94\text{ V})(0) = \mathbf{0}$

Making sense of the results. If there are five revolutions in 1 s, there is one revolution in $\frac{1}{5}$ s $= \frac{4}{20}$ s. The table shows how the values of the EMF alternate over one complete revolution or cycle.

◆ Related homework: Problems 23-25, 23-26, 23-27, and 23-28.

23-5 Inductance

When the current through a coil changes, the magnetic field that it sets up also changes. Wherever the changing field lines cross the area enclosed by a loop of wire, they constitute a changing flux across the loop and therefore induce an EMF in the loop. This is true for any loop; it makes no difference whether the loop is in another coil or is part of the original coil (Figure 23-17).

Thus, a changing current in a coil induces an EMF in that same coil. This is in addition to any other EMFs in the coil circuit, including the EMF giving rise to the original current. Because the magnetic field is everywhere proportional to the current producing it (as in Equation 22-7 or any of the equations in Table 22-2), so is the flux. If we write the proportionality as $\Phi = \text{constant}_1 \times I$,

$$\varepsilon = -N\frac{\Delta\Phi}{\Delta t} = -N\frac{\Delta(\text{constant}_1 \times I)}{\Delta t} = -N \times \text{constant}_1\frac{\Delta I}{\Delta t}$$

This says that the induced EMF is proportional to the change in current, because $N \times \text{constant}_1$ is likewise a constant. This constant is called the **inductance** of the coil and is denoted by L (you should remember that it depends on the number of turns N). The above equation is then written as

$$\varepsilon = -L\frac{\Delta i}{\Delta t} \tag{23-7}$$

(We have switched to a lowercase *i* to indicate a current that varies with *t*. In such cases, as previously, we will reserve *I* to denote the maximum value of *i*)

In SI units (*ε* in volts, *i* in amperes, and *t* in seconds), the unit of inductance L is the **henry** (H), named for Joseph Henry, an American contemporary of Faraday who also investigated induction.

The minus sign is a reaffirmation of Lenz's law that the effect opposes the cause. If a current toward the right is increasing, this EMF would tend to drive a current toward the left, to oppose the change in current. If a current toward the right is *decreasing,* this EMF would tend to drive a current toward the right, to oppose the decrease. This **back EMF,** as it is sometimes called, is large in a coil that has a great many turns so that its inductance L is large.

A coil of wire wound around a cylinder and capable of carrying a current is called a **solenoid.** Solenoids and other coils with significant inductance are called **inductance coils.** They act as a kind of ballast in a circuit, slowing down changes

As the magnetic field produced by the lower coil changes…

…the flux across this turn changes, and…

…the flux across this turn also changes.

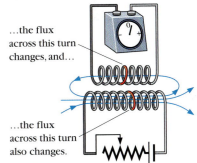

Figure 23-17 Turns of wire affected by changing magnetic flux.

in the current. Any circuit element that, because of its inductance, functions in this way is called an **inductor.** To the list of potential differences that we can identify along a circuit path (Procedure 21-1), we must add:

PROCEDURE **23-2**_____

Continuation of Procedure 21-1

- If a current flowing in a particular direction through an inductor (e.g., a coil) is increasing, there is a potential *drop* of magnitude $V = L\frac{\Delta i}{\Delta t}$ along the coil in the direction of the current.
- If a current flowing in a particular direction through an inductor is decreasing, there is a potential *increase* of magnitude $V = L\frac{\Delta i}{\Delta t}$ along the inductor in the direction of the current.

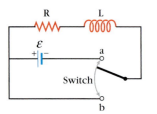

Figure 23-18 A circuit to produce transient effects in a series RL circuit.

Let's look at how the inclusion of an inductance coil affects a series circuit.

◆**A SERIES RL CIRCUIT** In Figure 23-18, the battery is included in the circuit when the switch is in position a, but is replaced by a plain wire when the switch is in position b. The resistor is in series with a coil serving as an *inductor* (represented as ‿〇〇〇〇〇‿ in circuit diagrams). We take the resistance of the coil wire and the internal resistance of the battery to be negligible. When the switch is moved to position a, a current will begin to flow (will increase from zero) clockwise. We apply Procedure 21-1, remembering that all the potential increases and decreases around a closed loop must add to zero at any instant. Although $V = L\frac{\Delta i}{\Delta t}$ is an average value over a time interval Δt, we can use this V to approximate an instantaneous value if Δt is small enough. Then proceeding clockwise from the negative terminal of the battery we get $\mathcal{E} - L\frac{\Delta i}{\Delta t} - iR \approx 0$, or

$$\mathcal{E} \approx iR + L\frac{\Delta i}{\Delta t} \qquad (\Delta t \text{ small}) \qquad (23\text{-}8)$$

Because the EMF $\mathcal{E}$ of the battery remains constant, the two terms on the right add up to the same total at every instant. Thus, as Figure 23-19 shows, as the current builds up, $\frac{\Delta i}{\Delta t}$ must drop. Note that as the graph of i versus t levels off, its slope approaches zero. Because $\frac{\Delta i}{\Delta t}$ is the slope of the graph, the second graph reflects this fact. **STOP&Think** Compare the graphs in Figure 23-19 with the first two graphs in Figure 21-17. What are the similarities? How are the similarities reflected in the behavior of the circuits described by the two pairs of graphs? ◆

STOP&Think What happens to Equation 23-8 when the switch in Figure 23-18 is moved to position b? If this is done at $t = 0$, what will a graph of I versus t look like? What will a graph of $\frac{\Delta i}{\Delta t}$ versus t look like? ◆

Calculations involving "RL" and RC circuits are similar, as you will see by working through Example 23-6.

Figure 23-19 Behavior of i and $\frac{\Delta i}{\Delta t}$ when the switch in Figure 22-18 is moved to position a at $t = 0$.

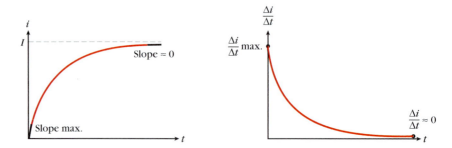

Example 23-6 *An RL Circuit Calculation*

For a guided interactive solution, go to Web Example 23-6 at
www.wiley.com/college/touger

Suppose that in Figure 23-18, $\mathcal{E} = 3.0$ V, $L = 1.0$ mH $= 1.0 \times 10^{-3}$ H, and
$R = 12 \ \Omega$.

a. Find the current in the circuit at the instant when it is increasing at a rate
   of 1000 A per second.
b. Find the potential differences across the resistor and across the coil at this
   instant.

Brief Solution

Restating the question. Because $\frac{\Delta i}{\Delta t}$ is the rate at which the current increases or
decreases, the question is: $i = ?$ *when* $\frac{\Delta i}{\Delta t} = 1000$ A/s.

The individual potential differences in part **b** are the terms iR and $L\frac{\Delta i}{\Delta t}$ in
Equation 23-8. We can label these v_R and v_L.

Choice of approach. We use the fact that the sum of the potential changes
around a closed loop is zero. For the given circuit, this fact is equivalent to
Equation 23-8, which says the gain equals the total drop.

The mathematical solution.
a. Solving Equation 23-8 for i gives us

$$i = \frac{1}{R}\left(\mathcal{E} - L\frac{\Delta i}{\Delta t}\right) = \frac{1}{12 \ \Omega}(3.0 \text{ V} - [1.0 \times 10^{-3} \text{ H}][1000 \text{ A/s}]) = \textbf{0.17 A}$$

b. $v_R = iR \approx (0.17 \text{ A})(12 \ \Omega) = \textbf{2.0 V}$

$v_L \approx (1.0 \times 10^{-3} \text{ H})(1000 \text{ A/s}) = \textbf{1.0 V}$

Making sense of the results.
1. Note that the potential differences across the resistor and the coil add up
   to the potential difference across the battery.
2. The maximum current is $I = \frac{\mathcal{E}}{R} = \frac{3.0 \text{ V}}{12 \ \Omega} = 0.25$ A. **STOP&Think** How can the
   current increase at 1000 A/s when the maximum current is so small? ◆

◆ Related homework: Problems 23-33, 23-34, 23-35, and 23-36.

By doing a calculation analogous to Example 21-10 (see Problem 23-65), we
find that the amount of time it takes the current to build up to most of its final
value is roughly $\frac{L}{R}$. The constant $\tau = \frac{L}{R}$ is called the **time constant** of an RL circuit. It is analogous to the time constant $\tau = RC$ that we found for a resistor and
capacitor in series in Section 21-6.

23-6 Transformers and Other Applications of Induction

We have seen that generators produce alternating currents that are sinusoidal; that
is, they vary as $\pm\sin(\omega t)$ or $\pm\cos(\omega t)$. A sinusoidally varying current gives rise to
a sinusoidally varying field. As Figure 23-17 illustrates, a varying field induces an
EMF not only in each turn of the coil producing the field but in each turn of
any other coil placed in its vicinity. We will refer to these two coils as the **primary coil** and the **secondary coil.** The EMF induced in the secondary coil is
proportional to $\frac{\Delta\Phi}{\Delta t}$. Because $\Phi = BA_\perp$, and $A_\perp$ doesn't change, $\frac{\Delta\Phi}{\Delta t} = A_\perp\frac{\Delta B}{\Delta t}$. Then
the EMF is proportional to the field's rate of change.

Now we make use of the point illustrated in Figure 23-16*b* (and treated in
WebLink 23-2), namely, that *the rate of change of a sinusoidal function is another*

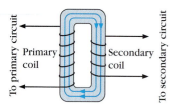

Figure 23-20 A transformer.

Transformers such as this are commonly used to "step down" from the high voltages of the power lines.

sinusoidal function. The EMF in the secondary coil is proportional to the rate of change of the sinusoidally varying field produced by the primary. So that EMF must also vary sinusoidally. It will drive a sinusoidal current if the secondary coil is part of a complete circuit. But then that second current produces a sinusoidally varying field that crosses the areas enclosed by each turn in the primary coil. This induces an EMF in the first coil. In short, the sinusoidally varying current in each coil induces a sinusoidal EMF in the other (this is sometimes called **mutual induction**).

In a **transformer,** the two coils are wound around a common core made of a highly magnetizable material, such as iron (Figure 23-20). The wire is insulated, so the circuits are separate. But in a well-designed transformer, the flux is the same all along the core at any instant and therefore varies in the same way across the loops of each coil:

$$\text{If in the primary} \qquad \mathcal{E}_1 = -N_1 \frac{\Delta \Phi}{\Delta t}$$

$$\text{then in the secondary} \qquad \mathcal{E}_2 = -N_2 \frac{\Delta \Phi}{\Delta t}$$

Dividing the first equation by the second, we get

$$\frac{\mathcal{E}_1}{\mathcal{E}_2} = \frac{N_1}{N_2} \tag{23-10}$$

The transformer is useful because it enables us to obtain a maximum EMF in the secondary circuit that is higher (if the secondary coil has more turns) or lower (if the secondary coil has fewer turns) than in the primary circuit. A transformer used to increase the voltage is called a **step-up transformer;** one used to decrease the voltage is called a **step-down transformer.**

Example 23-7 *From High-Voltage Transmission Lines to Household Voltage*

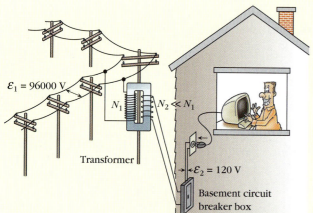

Figure 23-21 "Stepping down" from high voltage to household voltage. The stepping down ordinarily occurs in several stages.

Suppose that high-voltage power lines from an electrical generation plant, with a potential difference of 96 000 V, came directly to your home. If you plug your personal computer into a 96 000 V outlet, you had better have a fire extinguisher at hand. You would need a transformer (Figure 23-21) to step the voltage down to the 120 V at which your computer is designed to operate. If the primary coil has 12 000 turns, how many turns must there be in the secondary coil of the transformer?

Solution

Choice of approach. This is a straightforward application of Equation 23-10.
What we know/what we don't.

$$\mathcal{E}_1 = 96\ 000 \text{ V} \qquad \mathcal{E}_2 = 120 \text{ V}$$
$$N_1 = 12\ 000 \qquad N_2 = ?$$

The mathematical solution. Solving for N_2 is easier if we first invert both sides of Equation 23-10 to get the unknown into the numerator, giving us

$$\frac{\mathcal{E}_2}{\mathcal{E}_1} = \frac{N_2}{N_1}$$

Then $\qquad N_2 = N_1 \frac{\mathcal{E}_2}{\mathcal{E}_1} = 12\ 000 \left(\frac{120 \text{ V}}{96\ 000 \text{ V}} \right) = \textbf{15}$

◆ Related homework: Problems 23-41 and 23-42.

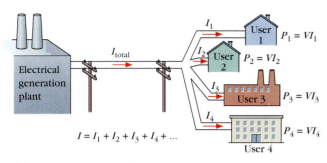

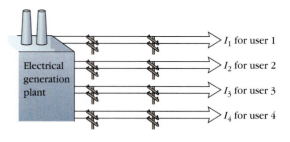

(*a*) No good. Wire would burn out because of high total current.

(*b*) No good. Multiple lines use too much wire (too expensive).

Figure 23-22 What electric power distribution would be like without transformers.

Why should utility companies use 96 000 V transmission lines if your bulbs and appliances are designed to operate at 120 V? The transmission lines must carry enough power to provide not only for your electricity needs but the needs of everyone else in your community as well. Remember that $P = VI = \mathcal{E}I$. With a smaller potential difference V, the transmission lines would have to carry a greater current I to provide the same power P. To provide power at reasonable cost, the utility company can afford to have only a small percent of the power it generates dissipated before it gets to you. The rate at which energy is dissipated as heat in a wire is $P = VI = (IR)I = I^2R$. If the current were increased tenfold, the rate of power loss would increase a hundredfold ($10^2 = 100$). Furthermore, the wire can only carry so much current without overheating.

Without transformers, the current carried by the transmission lines would have to be the total of all the current drawn by all the users in parallel (Figure 23-22*a*). Even if the transmission line didn't burn out, most of the power provided at the generation plant would be dissipated as heat. We could reduce the current in the transmission wires by a factor of, say, 1000, if we ran 1000 times as many wires (Figure 23-22*b*), but then the cost of the wires would be prohibitively high.

Instead, transformers are used. In a typical real transformer, the percent of power dissipated is quite small. To a good approximation, it behaves like an ideal transformer, which doesn't dissipate any power. To the extent that it does, the power in equals the power out:

$$V_1I_1 = V_2I_2 \qquad (23\text{-}11)$$
(maximum values)

Therefore if the power line voltage across the primary coil is 1000 times greater, the current in the power lines can be a thousand times less. At a generating station, the voltage may be stepped up to several hundred thousand volts for transmission, and then stepped down in two stages, first to perhaps a few thousand volts at a substation servicing a local geographic area—say, your town or neighborhood—and then to 120 V at the transformer on the utility pole near your home.

Example 23-8 *Home Electrical Service*

For a guided interactive solution, go to Web Example 23-8 at www.wiley.com/college/touger

The Wang family's home has "100 amp" service, as do their neighbors' homes. The 100 A is a maximum; however, at a certain time of day the Wangs and several neighbors may *together* be drawing a total of 100 A. Assume these homes have only 120 V outlets. Suppose there is a 4800 V potential difference between the wires of the transmission line that runs from the nearest substation to the transformer on the utility pole on the Wangs' street. How much current must the transmission line carry?

➡Utility companies determine need at various times of day based on averages over large numbers of customers. They must also accommodate the fact that average use is much higher at peak hours (e.g., dinner time and the hours following) than at noon or 3 AM.

EXAMPLE 23-8 *continued*

Brief Solution

Choice of approach. An ideal transformer conserves energy. This means that power out = power in: $V_1 I_1 = V_2 I_2$.

What we know/what we don't.

$$V_1 = 4800 \text{ V} \qquad V_2 = 120 \text{ V}$$

$$I_1 = ? \qquad I_2 = 100 \text{ A}$$

The mathematical solution. Solving for I_1 in Equation 23-11 gives us

$$I_1 = \frac{V_2 I_2}{V_1} = \frac{(120 \text{ V})(100 \text{ A})}{(4800 \text{ V})} = \textbf{2.5 A}$$

This results in much less power dissipated as heat.

◆ Related homework: Problems 23-43 and 23-62.

A variation on the transformer is the ignition system of an automobile. Unlike the usual transformer, it operates with direct current. It is an important use of electromagnetic induction.

Case 23-3 ◆ *A Vintage Automobile's Ignition System*

We will first examine a simpler situation in which a change in the current in one circuit loop induces a current in a second circuit loop. We will then show that older automobile ignition systems were a refinement of this basic circuit. The basic circuit looks like this.

a. The circuit is set up so that at the instant switch S_1 opens, switch S_2 closes. The current in loop 1 rapidly drops to zero, resulting in a rapidly changing magnetic field in the iron core. This induces an EMF

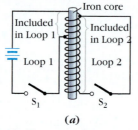

(a)

in each of the turns in loop 2. Because there are many more turns included in loop 2, the total induced EMF is briefly very large.

b. When there is a tiny gap in loop 2, the very large EMF will cause a spark across the gap. (The capacitor now included in loop 1 prevents sparking when switch S_1 is opened.)

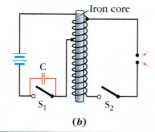

(b)

c. In the automobile, switch S_1 is open or closed depending on the position of a rotating piece that has projecting corners called **points.** Switch 2 is replaced by a **distributor,** which rotates as the points do. The distributor is called this because it connects one at a time to any of several gaps. Each gap is at one of the **spark plugs,** and a spark across it ignites the fuel in one of the cylinders of your engine (recall Figure 12-15). Switch S_1 is timed to open as each of the spark plug loops is closed in succession, producing a spark that "fires" the corresponding cylinder.

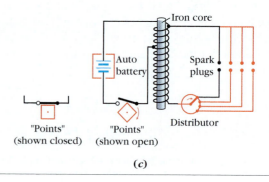

(c)

A variety of devices make use of the fact that in the presence of a magnet, a current may be induced in a wire coil if either the magnet or the coil is in motion. A **microphone** is one such device. If either the coil or the magnet is mounted on a membrane that can be set to vibrating by an incident sound wave, the vibrations induce a current that varies with the relative motion, and is therefore responsive to both the amplitude and frequency of the sound wave. An electrical signal is thus produced that reproduces the form of the sound signal.

When the vibrations that cause the relative motion of the coil and the magnet are produced by earthquakes or explosions rather than musical sound or human speech, the device that puts out an electrical signal in response to this vibrational input is called a **seismograph** (or a *geophone*). Seismographs are used by geologists and meteorologists to detect earth tremors and by military surveillance to monitor bomb tests, whether above or underground.

A **metal detector** also makes use of electromagnetic induction. When a current-carrying coil functioning as a strong electromagnet is brought near a metal object, it will cause some magnetization in the metal. This gives rise to a magnetic field due to the metal object. The onset of this field induces a momentary EMF in the coil of the metal detector, which produces a momentary but detectable change in current. The **pickup** of an electric guitar works on the same principle. But here the metal objects are the vibrating steel guitar strings, and the induced current varies in synch with their vibrations.

◆ SUMMARY ◆

Induced EMFs and **induced currents** occur in circuits as a result of causes external to those circuits. An EMF is induced in a circuit whenever the number of magnetic field lines (the **magnetic flux**) crossing the area enclosed by the circuit is changing.

$$\text{Magnetic flux} \quad \Phi = BA \cos \theta \qquad (23\text{-}2\text{m})$$

($A \cos \theta = A_\perp$, the cross-sectional area "seen" by the field lines)

Faraday's law of electromagnetic induction

$$\text{Induced EMF } \mathcal{E} = -N \frac{\Delta \Phi}{\Delta t} = -N \frac{\Delta (BA \cos \theta)}{\Delta t} \qquad (23\text{-}3\text{N})$$

We can therefore induce an EMF by changing either B (as in Example 23-1), A (as in Example 23-2), or θ (as in Example 23-3).

The direction of an induced current is the one that does not violate the law of conservation of energy. You can determine this direction by Lenz's law:

> **Lenz's law:** When a certain cause produces an induced current, the current in turn will have some effect. The current must flow in a direction such that the effect opposes the cause.

Some interpretations of Lenz's law are useful in specific circumstances:

• *The induced current will flow so as to produce a magnetic field opposing the <u>change</u> in flux.*

• *If the motion of an object causes an induced current, the current must flow in such a direction that its effect is a force that opposes the object's motion.*

A device in which EMF is continuously induced because rotating coils in the presence of a magnetic field changes θ sinusoidally (see Figure 23-16) in Equation 23-4 is called an **electric generator.** *The rate of change of a sinusoidal function*

is another sinusoidal function (for example, $\frac{\Delta \cos \theta}{\Delta \theta} \approx -\sin \theta$). Then if a turbine rotates at a constant ω, $\theta = \omega t$ and

$$\mathcal{E} = \underbrace{-\omega NBA}_{\substack{\text{max.} \\ \text{EMF}}} \underbrace{\sin (\omega t)}_{\substack{\text{varies} \\ \text{between } \pm 1}} \qquad (\omega = 2\pi f) \qquad (23\text{-}6)$$

The current $I = \frac{\mathcal{E}}{R}$ likewise varies between ± 1 times its maximum value. It is an **alternating current,** flowing alternately one way then the other, and the EMF that drives it is called an **alternating EMF.**

The changing flux is often due to a changing current. An EMF is induced in a coil or other **inductor** when the current in the coil itself changes. It is proportional to the rate of current change:

$$\mathcal{E} = -L \frac{\Delta i}{\Delta t} \qquad (23\text{-}7)$$

(i indicates a current that varies with t; I denotes the maximum value of i.)

We add this to the list of potential gains and drops in Procedure 21-1 (see **Procedure 23-2**).

In a *series RL circuit*, using Procedure 21-1 gives us

$$\mathcal{E} \approx iR + L \frac{\Delta i}{\Delta t} \qquad (\Delta t \text{ small}) \qquad (23\text{-}8)$$

($i = I$ when $\frac{\Delta i}{\Delta t} = 0$ and $\frac{\Delta i}{\Delta t} = \left[\frac{\Delta i}{\Delta t}\right]_{\max}$ when $i = 0$).

In **transformers,** a **primary coil** (with N_1 turns) and **secondary coil** (with N_2 turns) are wound about a common core of highly magnetizable material, so that a common flux along the axis of the core crosses the loops of both coils. Then

$$\frac{\mathcal{E}_1}{\mathcal{E}_2} = \frac{N_1}{N_2} \qquad (23\text{-}10)$$

A transformer used to increase voltage is called a **step-up transformer;** one used to decrease voltage is called a **step-down transformer.** Power into an ideal transformer equals power out:

$$V_1 I_1 = V_2 I_2 \qquad (23\text{-}11)$$

(maximum values)

◆QUALITATIVE AND QUANTITATIVE PROBLEMS◆
Hands-On Activities and Discussion Questions

The questions and activities in this group are particularly suitable for in-class use.

23-1. *Discussion Question.* Is it easier to induce an alternating current or a direct current for an extended period of time? Explain your reasoning.

23-2. *Discussion Question.* In Figure 23-23, the wire coil contains hundreds of turns wrapped around an iron core, and the voltage supply provides a large voltage. A metal ring is placed on the core to the right of the coil. What would you see happen when the switch was closed? Explain your reasoning, and explain why conservation of energy would be violated if the opposite happened.

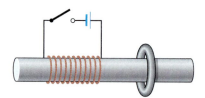

Figure 23-23 Problem 23-2

Review and Practice

Section 23-1 Basic Observations Showing the Occurrence of Electromagnetic Induction

23-3. Can a current be made to flow in a closed circuit consisting only of resistors and connecting wires? Briefly explain.

23-4. If two electric circuits have no conducting path connecting them, is it possible to affect the current in one of the circuits by changing the current in the other? Briefly explain.

Section 23-2 Faraday's Law of Electromagnetic Induction

23-5. **SSM** Suppose that the magnet being brought toward the coil in Figure 23-2*b* starts at position 1 (furthest from the coil) and stops at position 2 (closest to the coil).
a. Will the magnetic flux through the coil be greatest when the magnet is at position 1, between position 1 and position 2, or at position 2?
b. Will the galvanometer reading be greatest when the magnet is at position 1, between position 1 and position 2, or at position 2?

23-6. In each of the following cases, determine whether a current is induced in the square loop of wire (in red in Figure 23-24), and explain your reasoning. Assume the loop remains parallel to the end(s) of the magnet in all parts except *c*.

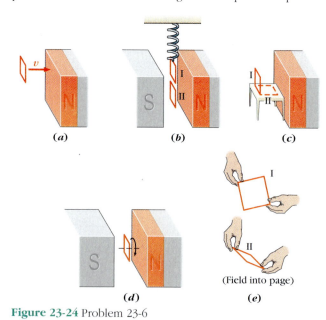

Figure 23-24 Problem 23-6

a. The loop moves at speed v toward the north pole of the magnet.
b. The loop oscillates up and down between positions I and II.
c. The loop falls over from position I to position II.
d. The loop is rotated about an axis perpendicular to the ends of the magnet.
e. The corners of the loop are pulled until the loop assumes shape II in Figure 23-24*e*.

23-7. For each part of Problem 23-6, which (if any) variable on the right side of Equation 23-2 changes during the described motion?

23-8. The wire loops in Figure 23-25 are shown in both front and edge views. If the flux across loop A is Φ_A, express the flux across each of the other loops as a multiple of Φ_A; that is, **a.** $\Phi_B =$ __?__ Φ_A. **b.** $\Phi_C =$ __?__ Φ_A. **c.** $\Phi_D =$ __?__ Φ_A. (Answers may be approximate.)

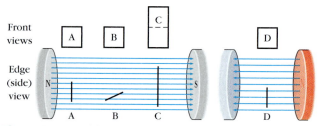

Figure 23-25 Problem 23-8

23-9. Two loops A and B are placed in the same uniform magnetic field. Loop A has an area of 20 cm^2, and its normal makes an angle of 37° with the field. Loop B is perpendicular to the field. If the magnetic flux is the same through both loops, find the area of loop B.

23-10. The flexible loop of wire in Figure 23-26 is held so that a magnetic field of 0.16 T is perpendicular to the plane of the

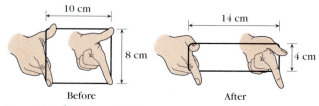

Figure 23-26 Problem 23-10

loop. If, starting with the before arrangement, it takes 0.080 s to pull the loop into the after arrangement, what average current is induced while the loop is changing shape?

23-11. SSM The loop in Figure 23-27 is made of four stiff pieces of wire, two straight and two semicircular. It has a total resistance of 2.0 Ω. Suppose that a 0.25 T magnetic field makes an angle of 30° with the normal to this page. If it takes 0.04 s to flip the two semicircular pieces from their before position to their after position, how much average current is induced while they are being flipped?

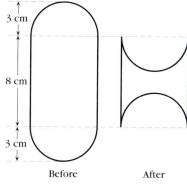

Figure 23-27 Problem 23-11

23-12. A circular loop of wire enclosing an area of 50 cm² (5.0×10^{-3} m²) is placed in a uniform magnetic field (Figure 23-28).

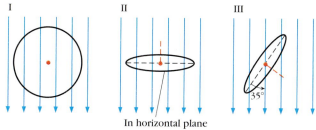

Figure 23-28 Problem 23-12

a. What cross-sectional area does the loop present to the field in each of the positions shown?

b. Calculate the magnetic flux across the loop in each of the positions shown if the field has a magnitude of 0.60 T.

23-13. In Problem 23-12, calculate the average induced EMF over a 2.0-s time interval if **a.** the loop is rotated from position I to position II over this interval. **b.** the loop is rotated from position II to position III over this interval.

23-14. A coil of 20 turns of resistance wire is wrapped around a core with a 0.08 m by 0.15 m rectangular cross-section. Each turn has a resistance of 0.15 Ω. A magnetic field is directed along the core.

a. Calculate the average EMF generated between $t = 0.40$ s and $t = 0.60$ s if the magnetic field drops from 1.20 T to 0.70 T over this interval.

b. If the ends of the coil are connected to an ammeter, what is the average reading of the ammeter over this interval?

23-15. In each of situations A through E in Figure 23-29, the same wire loop is rotated about one of its sides (ab) in the presence of a uniform magnetic field. In all five situations, the field has the same magnitude and is in the +x direction, and the loop takes the same time interval Δt to rotate from position 1 to position 2. Rank the five situations in order of the average current induced in the loop during the interval Δt.

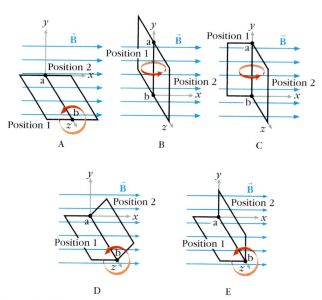

Figure 23-29 Problems 23-15, 23-19, and 23-20

Order them from least to greatest, making sure to indicate any equalities.

•**23-16.** During the 0.10 s interval in Example 23-3a, what would the minimum average radial acceleration of the seat of the swing have to be for Stinky to be detected? Express your answer as a multiple of g.

Section 23-3 Determining the Direction of Induced Current Flow

23-17. The coils in the two circuits in Figure 23-30 are wound around a common core, but the circuits are electrically insulated from each other. Use Lenz's law to determine whether the induced current flows from a to b or from b to a in Circuit 1 when **a.** the switch is closed in Circuit 2. Briefly explain. **b.** with the switch kept closed, the sliding contact of the variable resistor in Circuit 2 is moved from point c to point d. Briefly explain.

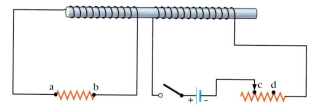

Figure 23-30 Problem 23-17

23-18.

a. In Problem 23-17**a,** you used Lenz's law to find the direction of the induced current. Explain what would happen if the induced current flowed opposite to the direction required by Lenz's law. How would this behavior violate conservation of energy? Do the kind of step-by-step reasoning we did in Example 23-4.

b. Repeat **a** for Problem 23-17**b.**

23-19. SSM The rectangular wire loop in Figure 23-29c is subjected to a uniform magnetic field in the +x direction. As the loop rotates from position 1 to position 2, does the induced current flow through side ab from a to b or from b to a? Briefly explain.

23-20. The rectangular wire loop in each situation in Figure 23-29 is subjected to a uniform magnetic field in the $+x$ direction. During which (one or more) of the depicted rotations does the current induced in the loop reverse direction?

23-21. As the rectangular wire loop in Figure 23-31 is moved from position I to position II, does an induced current flow through side ab of the loop from a to b or from b to a? Briefly explain.

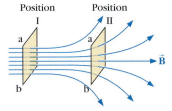

Figure 23-31 Problem 23-21

23-22. For each of Figures 23-2g (first quarter-rotation only) and 23-2h, use Lenz's law to explain why the galvanometer needle is deflected in the direction shown. In each case, do the kind of step-by-step reasoning we did in Example 23-4.

23-23. As the box in Figure 23-32 gets crushed by the safe, does the induced current flow *clockwise* or *counterclockwise* through the wire coil wrapped around it? Briefly explain.

23-24. In Problem 23-23, what would be the effect if the induced current flowed the opposite way, and how would that violate conservation of energy?

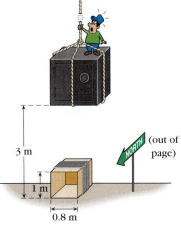

Figure 23-32 Problems 23-23, 23-24, and 23-49

Section 23-4 The Electric Generator

23-25. In Figure 13-13, suppose that there is a uniform magnetic field directed from the top of the page to the bottom, and suppose also that a closed loop of wire runs around the perimeter of the rotating rectangular radius.
a. Describe the relevance of Equation 23-1 to this set-up. In particular, what happens to $A_\perp$ as the radius rotates and how is this shown in the figure?
b. Why does this rotation induce an EMF in the loop of wire?
c. Sketch the position of the radius at which the EMF is instantaneously at a maximum.
d. Sketch the position of the radius at which the EMF is instantaneously zero.

23-26.
a. Find the maximum EMF produced by an electric generator making 5 revolutions each second if the generator consists of 100 turns of wire wrapped around a cylindrical core with a radius of 0.20 m rotating in a uniform magnetic field of 1.5 T.
b. If the EMF is at this maximum value at a certain instant, what is its value a quarter of a revolution later?
c. A half revolution later?
d. A whole revolution later?

23-27. **SSM** The rectangular coil in Figure 23-33 rotates with an angular velocity of 50 radians/s in the presence of a uniform magnetic field of 0.20 T directed into the page. The coil has three turns with a total resistance of 1.0 Ω and is connected in series with an external resistance of 19 Ω. What is the maximum value of the current through the external resistance?

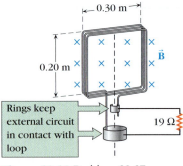

Figure 23-33 Problem 23-27

23-28. A generator that takes 0.0500 s to complete each revolution produces a maximum EMF of 12.0 V. If $t = 0$ is an instant when the EMF is going from positive to negative, what is the EMF at *a.* $t = 0.0100$ s? *b.* $t = 0.0125$ s?

23-29. Repeat Problem 23-28 for the case when the EMF is at its maximum value at $t = 0$.

23-30. An inline skating enthusiast conceives the idea of having a tiny generator turned by the wheels of her skates to light flashlight bulbs mounted on the toe and heel for night skating. She wants a peak EMF of 3.0 V when she is skating at a speed that keeps the wheels turning at 80 revolutions/s. She wants to wrap wire around a core with a cross-sectional area of 4 cm² $(4 \times 10^{-4} \text{ m}^2)$, and she has compact magnets that will provide a field of 0.10 T. How many turns of wire must she wrap around the core to generate the desired EMF?

23-31. A certain electric generator has 10 turns of wire, each enclosing an area of 0.050 m², rotating in a uniform magnetic field perpendicular to the axis of rotation. It produces an EMF in volts described by the equation $\mathcal{E} = -50 \sin(60\pi t)$.
a. Find the angular velocity and the frequency of rotation of this generator.
b. Find the maximum value of the EMF.
c. If the generator starts up at $t = 0$, at what instant is this maximum value first reached?
d. What is the value of the EMF at $t = \frac{1}{300}$ s?
e. Find the magnitude of the magnetic field.

23-32. Dmitri the Electrical Dancer (Figure 23-34) whirls rapidly in a uniform magnetic field. The field is perpendicular to

Figure 23-34 Problem 23-32

the plane of the page. On Dmitri's hat is a coil with its ends connected to a bulb. An identical arrangement is mounted on the toe of his boot. Which bulb, if either, glows more brightly? Briefly explain why.

Section 23-5 Inductance

23-33. Suppose that the circuit in Figure 23-18 contains a 9.0 V battery with negligible resistance, a 10.0 Ω resistor, and a coil that has an inductance of 5.0×10^{-3} H and a resistance of 2.0 Ω. Suppose that at some instant you flip the switch from position b to position a.

a. What is the value of the current at the instant the switch makes contact at a?

b. At what rate is the current changing at this instant?

c. At the instant described in **a,** find the potential difference across the resistor and the potential difference across the coil.

d. What is the sum of these two potential differences?

23-34.

a. In Problem 23-33, what is the value of the current several seconds after the switch is flipped to position a?

b. At what rate is the current changing at this later instant?

c. Find the potential difference across the resistor and the potential difference across the coil at this instant.

d. What is the sum of these two potential differences?

23-35. SSM WWW

a. When the current has reached 60% of its final value in Problem 23-33, at what rate is it changing?

b. At the instant described in **a,** find the potential difference across the resistor and the potential difference across the coil.

c. What is the sum of these two potential differences?

23-36. Consider the set-up in Problem 23-33 once more. Problems 23-33**d,** 23-34**d,** and 23-35**c** asked for the sum of the potential difference across the resistor and coil at three different instants. Compare the values of the sum at these three instants; that is, rank them in order from least to greatest, making sure to indicate any equalities.

23-37. Find the time constant for the RL circuit described in Problem 23-33.

23-38. Suppose the switch in Figure 23-18 is moved to position a at $t = 0$. **a.** Draw graphs in which the voltages (v_R and v_L) across the resistor and the coil are plotted against time. Clearly indicate the maximum values of these voltages on the graphs by expressing them in terms of $\mathcal{E}$, R, and L. **b.** At any instant t, what is the sum of these two voltages? How does this situation compare with the situation of the RC series circuit represented by the graphs in Figure 21-17?

Section 23-6 Transformers and Other Applications of Induction

23-39. Explain the difference between the wiring of a step-up transformer and the wiring of a step-down transformer.

23-40. You are traveling to Europe and have brought along your electric hair dryer. Your dryer is designed to operate at 120 V, but European outlets are 220 V. If you want to use this hair dryer, what additional piece of equipment should you bring along and why?

23-41. SSM A student optics kit uses a bulb designed to operate at 12 V and 3.0 A. The lab has standard 120 V outlets and a variable transformer in which the primary coil has 500 turns. A sliding contact determines how many turns are included in the secondary coil, up to a maximum of 500.

a. To use the lamp, should the primary coil be connected to the bulb or the outlet? Should the secondary coil be connected to the bulb or the outlet?

b. How many turns should be included in the secondary coil in order to meet the lamp's specifications?

23-42. If the transformer and bulb in Problem 23-41 are properly connected and set, find the current that flows from the outlet.

23-43. A substation of a generating plant receives a power input of 100 MW (megawatts) from the power generation facility. The substation in turn transfers this power to the power lines of the transmission grid.

a. If the transmission lines are to carry a current of 400 A, what must be the transmission line voltage?

b. If the input voltage to the substation is 30 kV (kilovolts), what is the input current?

23-44. Two coils A and B are wound around the same core, which has a radius of 0.020 m. Coil A has 400 turns; coil B has 20 turns. Both coils carry 60 cycle/s (60 Hz) alternating current. The magnetic field in the core, which has a maximum amplitude of 0.15 T, reverses with the same frequency as the current.

a. What is the maximum magnetic flux in the core?

b. How many times each second does the magnetic field reverse direction?

c. Find the average EMF in each coil during a single reversal of the magnetic field direction.

23-45. Compare the induction coil in an automobile ignition system to the typical electric company transformer found on a utility pole near someone's home. In what ways are they alike? In what ways are they different?

Going Further

The questions and problems in this group are not organized by section heading, so you must determine for yourself which ideas apply. Some of them will be more challenging than the Review and Practice questions and problems (especially those marked with a • or ••).

23-46. The wire arrangement in Figure 23-35 is in a uniform magnetic field of 0.50 T. The slide wire has a resistance of 2.0 Ω. The resistance of the ⊐-shaped wire is negligible. The

slide wire is moved from position I to position II over a 0.25-s interval. Find the average current induced over this interval if the magnetic field **a.** is directed out of the page. **b.** is directed left to right across the page. **c.** makes an angle of 37° with its direction in **a.**

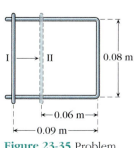

Figure 23-35 Problem 23-46 and 23-47

• **23-47.** The wire arrangement in Figure 23-35 is in a uniform magnetic field of 0.50 T pointing out of the page. The slide wire has a resistance of 2.0 Ω and is moved from position I to position II over a 0.25-s interval. Now the resistance of the ⊐-shaped wire is no longer negligible. What data would you need for the ⊐-shaped wire to find the average current through the slide wire while it is being moved?

23-48. The wire arrangement in Figure 23-36 is in a uniform magnetic field directed upward. Only the straight slide wire has a nonzero resistance, and it slides at constant speed along the track formed by the rest of the wire. How do the readings of the ammeter compare when the slide wire is at position I and when it is at position II? Briefly explain.

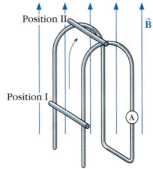

Figure 23-36 Problem 23-48

•• **23-49.** A coil of 100 turns of wire is wrapped around a very flimsy empty cardboard box (Figure 23-32). The box sits on the ground facing north at a location where Earth's magnetic field points due north and has a magnitude of 4.0×10^{-5} T. The rope supporting the 2-ton safe suspended over the box is about to break. When the rope breaks, the box will be crushed flat by the falling safe. Calculate the average EMF induced during the tiny time interval when the box is being crushed.

23-50. The axis of a coil with 500 turns of wire is aligned with a uniform magnetic field of 0.90 T. Each turn encloses an area of 8.0×10^{-3} m. Through what angle must you rotate the coil over a 0.10-s interval to induce an average EMF of 18 V over this interval?

23-51. **SSM** The laboratory rat in Figure 23-37 is running in a treadmill connected to a coil consisting of 10 square turns 0.25 m on a side. A uniform magnetic field of 0.80 T is directed upward. If the circumference of the treadmill cage is 0.90 m, how fast must the rat run (without slipping) to generate an alternating voltage with a maximum value of 30 V?

Figure 23-37 Problem 23-51

•• **23-52.** Write an equation for the EMF generated by the generator in Problem 23-25, assuming $t = 0$ at an instant when the EMF is going from positive to negative. Then use the equation to answer the following question: If the radius is rotating at an angular velocity of $4\pi s^{-1}$, what fraction of its maximum value will the EMF have at $t = 1.25 \times 10^{-3}$ s?

23-53. Devise and describe a set-up in which a constant induced EMF (rather than an alternating EMF) is generated over a short interval. Then comment on whether your set-up is practical for continuing to generate a constant EMF over a very long period of time.

23-54. A battery has been connected between the terminals of a capacitor for a long time. The battery is then disconnected, and a long length of resistance wire is connected between the capacitor's terminals (Figure 23-38*a*). When this connection is completed, the needle of the galvanometer G remains visibly deflected, indicating the presence of a current, for 5 s. The same sequence is then repeated, but this time part of the resistance wire has been wound around an iron bar (Figure 23-38*b*). Will the galvanometer needle remain visibly deflected for a time interval longer than, equal to, or shorter than 5 s? Briefly explain your reasoning.

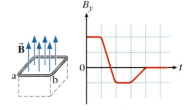

(*a*)   (*b*)

Figure 23-38 Problem 23-54

23-55. **SSM** The magnetic field crossing the loop in Figure 23-39 is always parallel to the *y*-axis (assume upward is positive) but varies with time in a manner shown by the graph in the figure. On the same axes, sketch a graph showing how the induced current (plotted vertically) varies with time. Take the current to be positive when it is traveling left to right through side ab of the loop.

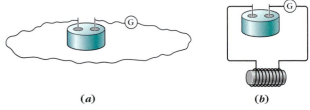

Figure 23-39 Problem 23-55

23-56. In Figure 23-40, one of the two ⊔-shaped wires is anchored to a block suspended from a spring. As it oscillates up and down between positions I and II, its ends maintain sliding contact with the other ⊔-shaped wire to form a complete circuit. If this motion occurs in a uniform magnetic field directed toward the right, describe the current that is generated in as much detail as you can.

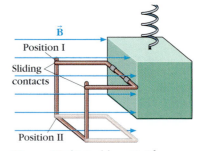

Figure 23-40 Problem 23-56

• **23-57.** Starting with the defining equations for inductance and resistance ($\mathcal{E} = -L\frac{\Delta i}{\Delta t}$ and $V = IR$), show that in SI (international system of metric units) the units of $\tau = \frac{L}{R}$ are seconds.

23-58. A certain transformer has 100 turns in the primary coil and 1000 turns in the secondary coil. The primary coil is connected to a 120 V outlet. The secondary coil is connected to a lamp with a 60 W, 120 V bulb.

a. What will happen to the bulb when the lamp is switched on? Why?

b. Does your answer in any way depend on the resistance of either coil? Explain. After doing this problem, do Problem 23-59 and see if your solution to that problem is consistent with your answer to this one.

23-59. A certain transformer has 100 turns in the primary coil and 1000 turns in the secondary coil. The primary coil is connected to a 120 V outlet. The secondary coil is connected to a lamp with a 60 W, 120 V bulb.

a. What current does the bulb ordinarily draw when it is connected directly to the outlet?

b. Suppose the primary coil has a resistance of 60 Ω and the resistance of the secondary coil is negligible. What current flows through the primary coil?

c. Under these conditions, what current flows through the secondary coil? What current flows through the bulb? Compare this to the value you obtained in **a.** Will the bulb burn out, glow more brightly than before, glow just as brightly, or glow less brightly? Briefly explain.

d. If the resistance in the secondary coil is negligible, what must be the resistance of the primary coil for the bulb to glow at its customary brightness?

• **23-60.** The field along the axis of a long straight current carrying coil—or solenoid—is given by $B = \mu \frac{N}{l} I$ (N is the number of turns and l is the length of the coil).

a. Using Faraday's law of electromagnetic induction ($\mathcal{E} = -N\frac{\Delta \Phi}{\Delta t}$) and Equation 23-7 ($\mathcal{E} = -L\frac{\Delta i}{\Delta t}$), which implicitly defines induction, find an equation with which you can calculate the inductance of a coil of wire wound around a hollow cylinder from its properties.

b. In a 1920s set of directions for constructing a crystal radio set, you are told to use a "coil 4″ diam. × 5″ long [with] 109 turns of #22 [copper] wire." Assuming this coil is hollow, find its inductance.

23-61. SSM WWW An electric power substation for a small college campus is a receiver point for long distance power lines with a potential of 96 000 V between them. In the substation transformer, there are 72 times as many windings in the primary coil as in the secondary coil. At a certain time of day, when students are away, the campus requires a total current of 120 A. How much current must there be in the long-distance power lines to meet the campus's needs?

23-62. *Why transformers are needed.* Suppose that the values in Example 23-8 reflect the fact that 5.0% of the power provided by the substation is dissipated in the transmission lines.

a. If 4800 V is the potential difference between the wires of the transmission line where it reaches the transformer near the Wangs' home, what must be the potential difference between the wires of this line where the line leaves the substation?

b. What is the total resistance of the transmission line wires running between the substation and this transformer?

c. If the current in the transmission lines was increased, would the potential drop along the transmission line wires running

between the substation and this transformer increase, decrease, or remain the same? Would that increase, decrease, or not change the potential difference between the wires of the transmission line where it reaches the transformer near the Wangs' home?

d. If one kept increasing the current in the transmission lines, at what level of current would the power dissipated in the lines reach essentially 100% (leaving a negligible amount for consumption by the Wangs and their neighbors)?

e. How does the value you found in **d** compare with the total current that the Wangs and their neighbors require? Is it much less than, about the same as, or much more than they require? Can the utility company meet their customers' needs without the use of transformers?

23-63. Two students trying to design a microphone are arguing about the approach. Both of them agree to use a membrane that can be made to vibrate when struck by sound waves. But student A says, "We need to mount a small permanent magnet on the back of the membrane and let it move within a stationary coil," whereas student B says, "I think we should mount the wire coil on the back of the membrane, so *it* can move while the *magnet* remains stationary." Is one student's approach more likely to produce an induced current than the other? Are there any practical differences between the two students' approaches? Explain.

23-64. The wire coil in Figure 23-41 is wrapped around a cylindrical bar magnet. The handle can be used to rotate the magnet. The wire is bare, so there must be a tiny space between each turn of the coil and a thin insulating wrapper between the bar magnet and the wire. End P of the coil makes continuous contact with a copper ring, which in turn is connected to a conducting rod RS. The other end of the wire has a weight W suspended from it. The hanging portion of the wire makes contact with the conducting rod near end R to complete the circuit. Tell what you would have to do to make this set-up produce a constant direct current through the ammeter A, and explain why that current is produced.

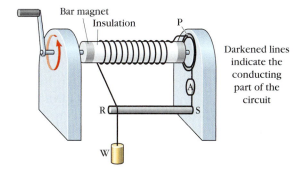

Figure 23-41 Problem 23-64

23-65. In Example 23-6, estimate how long the current takes to reach half its maximum value if the switch is moved to position a at $t = 0$. (Review Example 21-10 for RC circuits and try to proceed by analogy with that example.)

23-66. If the switch in Figure 23-18 is moved from position a to position b at $t = 0$, what will a graph of I versus t look like? What will a graph of $\frac{\Delta i}{\Delta t}$ versus t look like? Sketch both graphs.

23-67. Two circular loops A and B are placed in the same uniform magnetic field. Loop A has a radius of 4.0 cm, and its normal makes an angle of 30° with the field. Loop B is perpendicular to the field. If the magnetic flux is the same through both loops, find the radius of loop B.

23-68. The loop in Figure 23-42 rotates clockwise around edge ab in the presence of a uniform magnetic field directed toward the left. Its position is shown at three instants (t_1, t_2, t_3).

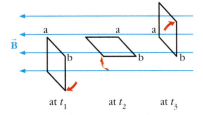

a. At t_1, tell whether the current is flowing from a to b, from b to a, or not at all.

Figure 23-42 Problem 23-68

b. Repeat for instant t_2.

c. Repeat for instant t_3.

23-69. Suppose that the circuit in Figure 23-18 contains a 1.5 V battery with negligible resistance, a 10 Ω resistor, and a coil with an inductance of 5.0×10^{-3} H (and negligible resistance). After the switch has been in position a for a long time, you flip it to position b.

a. What is the current in the circuit at the instant the switch has broken contact at a?

b. At what rate is the current changing at this instant?

c. Find the rate at which the current is changing when the current has dropped to one third of its initial value.

23-70. Repeat Problem 23-38 for the case in which the switch in Figure 23-18 is moved from position a to position b at $t = 0$.

23-71. A certain nonideal transformer—85% efficient—steps down voltage from the 6000 V of the transmission lines along Parkside Avenue to the 120 V used in the apartment building at number 540. For the apartment building to draw a total of 500 A, how much current must the transmission lines carry?

23-72. A folk musician decides to adapt the pickup of an electric guitar for use on a traditional washtub bass. A washtub bass consists of an inverted metal washtub with a broomstick braced against the lip of the tub. A rope—the "string" of the bass—is stretched taut between the upper end of the broomstick and the center of the washtub bottom. To pick up the sound electronically, the pickup should be ____ (*attached to the bottom of the tub; placed near the bottom of the tub but not attached to it; placed near the rope; attached to the broomstick*). Briefly explain why.

23-73. For each graph of magnetic flux versus time in Figure 23-43a, identify the corresponding graph of EMF versus time from Figure 23-43b (choices can be reused).

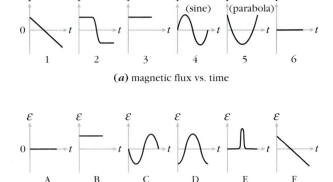

(*a*) magnetic flux vs. time

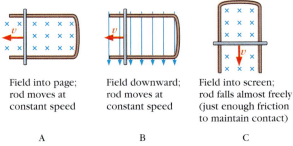

(*b*) EMF vs. time

Figure 23-43 Problems 23-73 and 23-74

23-74. Match each situation in Figure 23-44 to the corresponding graph of magnetic flux versus time in Figure 23-43a.

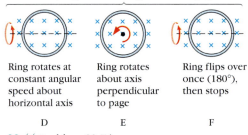

In A-C, conducting rod slides along U-shaped conducting wire

Field into page; rod moves at constant speed   A

Field downward; rod moves at constant speed   B

Field into screen; rod falls almost freely (just enough friction to maintain contact)   C

In D-F, ring is conducting, flux is into page

Ring rotates at constant angular speed about horizontal axis   D

Ring rotates about axis perpendicular to page   E

Ring flips over once (180°), then stops   F

Figure 23-44 Problem 23-74

23-75. A coil of 20 turns of high-resistance wire is positioned so that a uniform magnetic field is directed along its axis. A conducting wire of negligible resistance is connected between its two ends to complete the circuit. When the magnetic field increases at a uniform rate, a current is induced in the circuit. Suppose the situation is repeated with a coil of 40 turns, all other aspects remaining the same. Will the induced current be less than, equal to, or greater than before? Briefly explain.

Problems on WebLinks

23-76.

a. [*Choose all correct answers.*] In WebLink 23-1, the galvanometer needle deflects when _____ (*a switch is first closed; the switch remains closed; a switch is first opened; the switch remains open*).

b. The switch in part ***a*** is _____ (*in series with; in parallel with; not connected to; a part of*) the galvanometer.

23-77. In WebLink 23-1, a number of different changes are made, each of which causes a galvanometer needle to deflect. When any one of these changes is reversed—that is, when you go back to the original situation—the galvanometer needle _____ (*deflects in the opposite direction; deflects in the same direction; goes back to zero*).

23-78. In WebLink 23-2, at points where $y = \sin \theta$ has a value of zero, the slope of the graph _____ (*always has a value of zero; always has a value of $+1$; always has a value of -1; always has a value of either $+1$ or -1*).

23-79. In WebLink 23-2, if the sinusoid is $y = \sin \theta$, its rate of change $\frac{\Delta y}{\Delta \theta}$ is _____ ($\sin \theta$; $-\sin \theta$; $\cos \theta$; $-\cos \theta$)

23-80. As the loop in WebLink 23-3 rotates at constant angular speed, the rate $\frac{\Delta \Phi}{\Delta t}$ at which the number of lines crossing the loop is changing is _____ (*always; sometimes; never*) zero.

For a cumulative review, see Review Problem Set II on page 830.

CHAPTER TWENTY-FOUR

As the Twentieth Century Opens: The Unanswered Questions

The nineteenth century witnessed extraordinary advances in human understanding of the physical world. But by the close of that century, many questions remained unanswered. This chapter will first summarize two major outcomes of nineteenth-century physics:

1. Maxwell's remarkable synthesis of electricity, magnetism, and optics to establish the existence of electromagnetic waves, and

2. a growing appreciation that many macroscopic phenomena could only be explained by a detailed model of how the microscopic constituents of matter behaved.

We will then frame some of the questions that were still challenging physicists as the twentieth century opened.

". . . many macroscopic phenomena could only be explained by a detailed model of how the microscopic constituents of matter behaved."

24-1 The Triumph of Electromagnetism

The nineteenth century had been ushered in by Young's double-slit experiment, demonstrating the wave nature of light at a time when the development of the voltaic cell was fueling the intensifying inquiry into electrical phenomena. By mid-century, the work of Oersted, Ampere, and Faraday had clearly established connections between electricity and magnetism. The idea of the field had come to play a central role in this understanding.

In particular, Faraday's law of electromagnetic induction could be stated as a relationship between the two fields. Faraday's law tells us that a changing magnetic field induces an EMF. But because an EMF along a path indicates that there is an electric field along it, we can also state Faraday's law this way:

A changing magnetic field gives rise to an electric field.

The various laws involving electric and magnetic fields were ultimately brought together into a coherent and more powerful pattern of reasoning by James Clerk Maxwell, a Scottish physicist. He focused on four of these laws: (1) Gauss's law (Section 19-4), (2) a similar law for magnetic fields, (3) Faraday's law of electromagnetic induction, and (4) Ampere's circuital law (Section 22-7). The second of these laws states that the total magnetic flux across an enclosing surface is always zero because magnetic field lines always close on themselves. It is sometimes called *Gauss's law for magnetic fields.*

Details of Maxwell's approach use vector calculus and are well beyond the level of this course, but we can outline the main ideas. Having expressed these laws as equations, Maxwell was able to (1) discern and extend patterns in the mathematical relationships, and (2) draw further conclusions using the if-then reasoning power that the mathematics provides. His equations, when extended as described in this section, are widely known as **Maxwell's equations.**

Maxwell noticed that to some extent, these equations looked the same if he interchanged $\vec{E}$ and $\vec{B}$: There was a partial symmetry in the behavior of the two fields. He realized he could make this symmetry more complete by putting an additional term into Ampere's circuital law. But that changes what the equation says about how the world behaves. Would the new statement still be correct?

In fact, there had been a problem with Ampere's law. Although the two situations in Figures 24-1*a* and *b* produce the same magnetic field, Ampere's law did not always give the same result in the second case (Figure 24-1*c*).

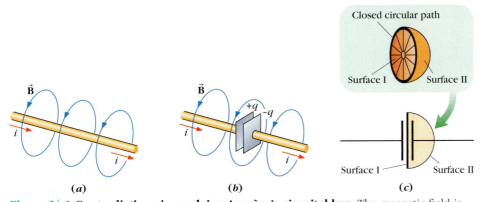

Figure 24-1 Contradictions in applying Ampère's circuital law. The magnetic field is the same—and continuous—around a long straight wire carrying a current *i* whether the current path is (*a*) continuous or (*b*) broken by the gap of a capacitor that is being charged. In (*c*), surfaces I and II, like the surfaces of a grapefruit half, are both bounded by the same closed path but give contradictory results. Applying Ampere's law to surface II, which is crossed by the current, gives the correct magnetic field. But applying it to surface I, which passes through the capacitor gap so no current crosses it, gives zero field.

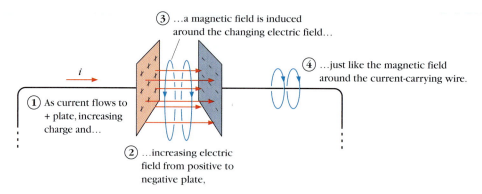

③ ...a magnetic field is induced around the changing electric field...

④ ...just like the magnetic field around the current-carrying wire.

① As current flows to + plate, increasing charge and...

② ...increasing electric field from positive to negative plate,

Figure 24-2 Magnetic fields caused by both a current and a changing electric field.

Maxwell could approach this dilemma as follows: "If there is no break in the cause of the magnetic field as I follow the circuit, Ampere's law gives me the same magnetic field no matter which of these surfaces I use for the calculation. That can still happen if another cause replaces the current where there is a gap." But *what* other cause?

Maxwell knew that as the charge on the plates increases, the electric field between the plates correspondingly increases. Moreover, because the electric field between the plates is proportional to the charge q on the plates, its rate of change is proportional to the rate of change of the charge, that is, the *current* ($\frac{\Delta E}{\Delta t}$ = constant $\times \frac{\Delta q}{\Delta t}$ = constant $\times I$) to or from the plates. The changing electric field across the gap could be the cause of the magnetic field circling the gap (Figure 24-2). Maxwell recognized that this proposed cause would also flesh out the pattern of interconnections between the electric and magnetic fields. Faraday's law said that a changing magnetic field gives rise to an electric field. With Maxwell's added term, Ampere's law becomes the **Ampere-Maxwell law.** The added part tells us,

A changing electric field also gives rise to a magnetic field.

With this inclusion, Maxwell reasoned that the induction of an electric field by a changing magnetic field and the induction of a magnetic field by a changing electric field might occur even in regions of free space away from any charges or conduction currents. In that way, the induction of fields could be passed on from region to region, as in Figure 24-3. In other words, they could propagate through space as waves do—hence, **electromagnetic waves.** His detailed mathematical analysis showed that both fields varied in time and space as

$$y = Y \sin 2\pi f \left(t - \frac{x}{v} \right) \tag{14-8}$$

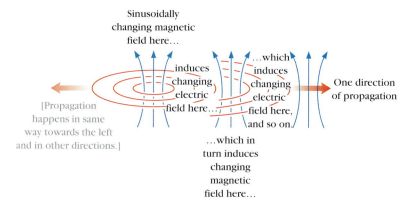

Sinusoidally changing magnetic field here...

...which induces

induces

changing electric field here...

changing electric field here, and so on.

One direction of propagation

[Propagation happens in same way towards the left and in other directions.]

...which in turn induces changing magnetic field here...

Figure 24-3 Induction of fields passed on from region to region.

They are *wave functions,* where now y represents either B or E, and $v = \frac{1}{\sqrt{\mu_o \epsilon_o}}$. Because the values of μ_o and ϵ_o are known, we can calculate the propagation speed at which such a disturbance could travel. The result is $v = 3.0 \times 10^8$ m/s. This is a speed that we should recognize (and Maxwell certainly did)—it is the speed of light!

Recall from Chapter 15 that when light was found to be wavelike, no one knew what kind of wave it was. The existence of interference patterns showed that it was some kind of disturbance that varied sinusoidally, but they provided no clue to what kind of disturbance. Maxwell now offered an answer: *if light travels at the speed predicted for electromagnetic waves, then a light wave is a type of electromagnetic wave.* The 1864 publication of Maxwell's *Dynamical Theory of the Electromagnetic Field* represented another great linking of two fields previously thought to be unconnected—electromagnetism and optics. In fact, since Maxwell, *optics* has come to refer to the study not only of light but of all electromagnetic waves.

Maxwell's conclusions were not experimentally confirmed until 1887, eight years after his death, when German physicist Heinrich Hertz (1857–1894) succeeded in setting up devices for producing electromagnetic waves and detecting them elsewhere in his laboratory. The components of Hertz's experimental set-up were in a sense the first transmitting and receiving antennas for electromagnetic waves. Barely ten years after that, Guglielmo Marconi was transmitting and receiving radio waves over great distances. Modern telecommunications had begun.

◆**THE ELECTROMAGNETIC SPECTRUM** Hertz's electromagnetic waves were not light waves but radio waves. They differ from light waves in that they have much longer wavelengths and correspondingly lower frequencies. In a vacuum all electromagnetic waves travel at the same speed ($c = 3.00 \times 10^8$ m/s). In principle they can occur at all wavelengths and frequencies, subject to the constraint that $\lambda f = c$. The entire range of electromagnetic waves, in order of wavelength or frequency, is called the electromagnetic spectrum.

Electromagnetic waves in different wavelength or frequency ranges typically have different kinds of emitters and different types of detectors. The human eye, for example, is capable of detecting a very narrow band of wavelengths, called the **visible light spectrum,** from about 4.0×10^{-7} m to about 7.5×10^{-7} m (often written as 400 to 750 nm or 4000 to 7500 Å). **Nanometers** (1 nm = 10^{-9} m) are common units for wavelengths of visible light. The **angstrom** (1 Å = 10^{-10} m) —roughly the diameter of the smallest orbit of an electron around a hydrogen atom—is commonly used for wavelengths on an atomic scale. Very hot objects are common emitters of light—the filament of a light bulb, for example, or the hot gases that make up a flame or the surface of the sun. It is those emissions, whether received directly from the emitting objects or reflected off the surfaces of other objects, that our eyes have primarily evolved to detect. By detecting the emissions—by seeing—we gather information about the objects.

One of the principal technological achievements of the nineteenth and twentieth centuries was learning to use the emission and detection of electromagnetic waves. In 1864, when Maxwell published his ideas, the use of chemical emulsions containing materials such as silver nitrate or silver bromide, which blackened on exposure to visible light, had become widespread as a means of capturing images. In that same year, photographer Matthew Brady was using them to record the horrors of America's Civil War (Figure 24-4).

German physicist Wilhelm C. Roentgen first detected X rays in 1895 when photographic plates in his lab were unexpectedly exposed. X rays typically have wavelengths of a fraction of an angstrom to a few angstroms, less than a thousandth of visible light wavelengths. The fact that many materials (such as human flesh) that are opaque to visible light are transparent to X rays made X-ray photography the first major method of medical imaging.

Figure 24-4 An example of how chemical emulsions were used to detect visible light in Maxwell's time. This Civil War photograph by American photographer Matthew Brady was taken c. 1863, the year before Maxwell published his *Dynamical Theory of the Electromagnetic Field.*

Telecommunications, such as radio and television, has generally made use of electromagnetic waves at the long wavelength end of the spectrum, which we call radio waves. The shortest wavelength radio waves are called **microwaves,** because even though their wavelengths are much longer than those of visible light, they are orders of magnitude smaller than the wavelengths used for radio broadcasting. Typical radio waves might be on the order of hundreds of meters or several kilometers.

Radar

When an electromagnetic wave strikes a conducting object, it is scattered or reradiated in all directions. The scattered amount of radiation depends on how the object's size compares to the wavelength. Only a tiny fraction of very large wavelength radiation would be reflected back to the transmitter from an object the size of a ship or plane, so no reflected signal would be detected. But a transmitted microwave signal with wavelength substantially smaller than the target object (Figure 24-5a) *would* be reflected back sufficiently to permit the determination of the direction and distance of the object. One of the first such observations of a "radio echo" was made by Dr. Albert H. Taylor of the U.S. Naval Research Laboratory in 1922. The military value of such detection was quickly recognized, and the technology of *radio detection and ranging* (radar) reached maturity during World War II. In civilian use, radar information has become indispensable for pilots and air traffic controllers. In 1990–91, radar imaging was used by NASA's *Magellan* spacecraft mission to map the surface of Venus (Figure 24-5b), which, because of that planet's thick cloud cover, could not be mapped by visible light photography.

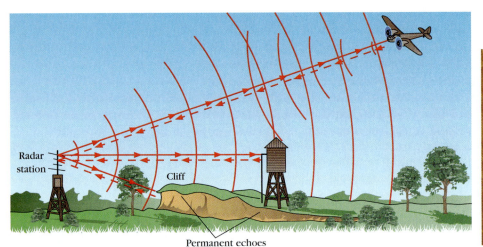

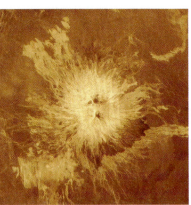

(a)   (b)

Figure 24-5 RADAR transmission and reflection of microwaves. (*a*) Early use (adapted from *Radar Electronic Fundamentals,* published as a restricted document by the U.S. War Department in 1943). (*b*) This view, produced by imaging radar, looks directly down on the volcano *Sapas Mons* on Venus.

Some Modern Uses of Electromagnetic Waves

Over very small distances (a few meters), remote controls for televisions and VCRs operate by the emission and detection of infrared radiation at wavelengths just above the visible range. Over very large distances (many light-years), the detection of emissions from distant objects in space (stars, nebulae,

(a)   **(b)**

Figure 24-6 Some devices for making astronomical observations across the electromagnetic spectrum. (*a*) The Arecibo Observatory in Puerto Rico, the world's largest radiotelescope, is 1000 feet in diameter, enabling it to detect very weak signals from space. (*b*) NASA's Extreme Ultraviolet (EUV) Explorer was launched in 1992.

supernovas, quasars, etc.) over all parts of the electromagnetic spectrum (Figure 24-6) has provided astronomers and astrophysicists with a vastly expanded observational base on which to build their understanding of the universe and its origins. Some of the detection devices, in analogy to light telescopes, bear names like radiotelescopes or X-ray telescopes. Though they hardly fit our conventional image of a telescope, their large reflecting surfaces—like the concave mirror in a reflecting light telescope—gather the received emissions and direct them to a suitable detection element at or near a focal point.

Figure 24-7 summarizes the different parts of the electromagnetic spectrum.

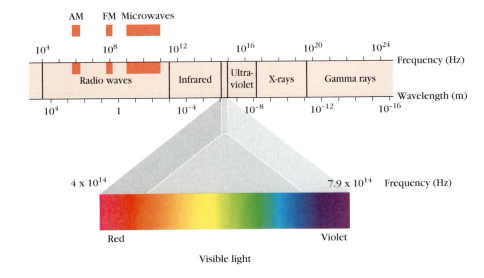

Figure 24-7 An overview of the electromagnetic spectrum.

24-2   Probing the Atomic World

John Dalton (1766–1844) was a mathematics and chemistry teacher in Manchester, England. In 1808, he published *A New System of Chemical Philosophy,* in which he proposed an atomic theory of matter. Dalton's atomic theory gave widespread credibility to the notion that matter was not continuous, that it could not be infinitely subdivided, but that each element was instead made up of basic building blocks called atoms that could not be broken down further. Dalton himself did not envision that the atom might itself have constituent parts, such as protons and electrons. He proposed atoms of fixed weight for each element to explain the observed ratios in which elements combined to form compounds.

By late in the century, John Tyndall (1820–1893), an Irish-born British physicist who wrote about science for popular audiences, was reflecting a prevalent if not yet universally accepted point of view when he wrote,

> Science ought to teach us to see the invisible as well as the visible in nature: to picture in our mind's eye those operations that entirely elude the eye of the body; to look at the very atoms of matter, in motion and in rest, and to follow them forth into the world of senses.

Tyndall's idea of "the invisible in nature" might well have embraced electric and magnetic fields and their oscillations as well as "the very atoms of matter." **STOP&Think** By what means do you suppose Tyndall intended us to "follow [the atoms] forth into the world of senses?" ◆

Looking at atoms in motion was refined into the kinetic theory of gases (recall Chapter 12) by nineteenth-century physicists, notably Joule and Maxwell, enabling them to explain a broad range of thermal phenomena, from energy transfer between objects at different temperatures to the plateaus in cooling curves. (Bernoulli had proposed a particles-in-motion model of a gas a century earlier to explain how a gas could exert pressure.) None of this work required any specific understanding of the *structure* of atoms or molecules.

Evidence that the atom itself might have component parts came only as the nineteenth century was drawing to a close. J. J. Thomson's cathode ray experiment in 1897 (Section 18-2) established the existence of "cathode corpuscles," or what we now call electrons. However, his idea of how they fit into an overall picture of the structure of the atom, his so-called plum pudding model (Figure 24-8), was disproved a dozen years later. For Americans unfamiliar with the plum puddings beloved in Victorian England, a better image might be that the electrons were like negative chocolate chips embedded in a positive mass of cookie dough. Moreover, the idea that the electron carried an elementary fixed unit of charge and had a fixed mass would not be firmly established until Millikan's oil drop measurements (also in Section 18-2) over the years 1910–1913. But Thomson was able to establish that the ratio $\frac{q}{m}$ of charge to mass was constant for electrons. To see how basic physics principles led to this conclusion, work through Example 24-1.

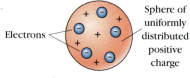

Electrons — Sphere of uniformly distributed positive charge

Figure 24-8 Thomson's plum pudding model of the atom.

Example 24-1 *Finding the Charge Ratio $\frac{q}{m}$ for Cathode Rays*

For a guided interactive solution, go to Web Example 24-1 at www.wiley.com/college/touger

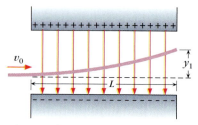

Figure 24-9 The deflection of a cathode ray beam by a uniform electric field.

The cathode ray beam is allowed to pass between two plates (Figure 24-9). When the plates are uncharged, the beam travels to the right undeflected. When the plates are charged as shown, the beam is deflected by some vertical distance y_1 by the time it reaches the right end of the plates. The amount of deflection depends on the electric field strength. If a magnetic field of suitable strength is applied perpendicular to the page, the magnetic force on the electrons will be equal and opposite to the electric force, and the ray will again travel straight ahead. The deflection y_1, the electric field magnitude E, the magnetic field magnitude B, and the horizontal distance L traversed between the plates can all be measured. In terms of these quantities, find

a. the horizontal speed with which the beam of particles enters the region between the charged plates.

b. the ratio $\frac{q}{m}$ of charge to mass for the charged particles making up the beam.

Brief Solution

Choice of approach.

a. When the electric force qE and the magnetic force qvB are equal and opposite, the particles are not deflected; they continue straight ahead at their initial horizontal velocity v_{ox}. Because $\Sigma F = 0$, we can solve for v_{ox}.

b. When the magnetic field is shut off, there is an acceleration a in the y direction due to the electrostatic force qE; then using Newton's second law, we can find a. We can then use kinematics to find the deflection y when the particles have traveled horizontally a distance x. Because $y = y_1$ when $x = L$, we can then solve for the ratio $\frac{q}{m}$ that occurs in the resulting expression for y.

What we know/what we don't. y_1, E, B, and L are taken as known quantities.

$$y = y_1 \text{ when } x = L \qquad \frac{q}{m} = ?$$

The mathematical solution.

a. $\Sigma F = 0$ becomes $qE - qvB = 0$. Setting $v = v_{ox}$ and solving for v_{ox} gives us

$$v_{ox} = \frac{E}{B}$$

b. When there is only an electrostatic force, Newton's second law becomes $qE = ma$, so there is an acceleration $a = \frac{qE}{m}$ in the y direction. We can now apply our basic kinematic relationships for constant acceleration:

$$x = v_{ox}t \qquad \text{and} \qquad y = v_{oy}t + \tfrac{1}{2}at^2 = \tfrac{1}{2}at^2 \qquad \text{since } v_{oy} = 0$$

Because we have found a, we can write

$$y = \tfrac{1}{2}\frac{qE}{m}t^2$$

But from $x = v_{ox}t$ and our previous result that $v_{ox} = \frac{E}{B}$, we find that

$$t = \frac{x}{v_{ox}} = \frac{xB}{E}$$

We can use this to substitute for t in our expression for y:

$$y = \tfrac{1}{2}\frac{qE}{m}\left(\frac{xB}{E}\right)^2 = \tfrac{1}{2}\frac{q}{m}\frac{x^2B^2}{E}$$

If $y = y_1$ and $x = L$,

$$y_1 = \tfrac{1}{2}\frac{q}{m}\frac{L^2B^2}{E}$$

Finally, we solve for $\frac{q}{m}$ to obtain

$$\frac{q}{m} = \frac{2y_1E}{L^2B^2}$$

◆ Related homework: Problems 24-9 through 24-13.

Finding the ratio $\frac{q}{m}$ of these "cathode corpuscles" did not in itself determine either the charge or mass of an electron. Only after Millikan showed that the charges must always be integer multiples of the same value could the particular

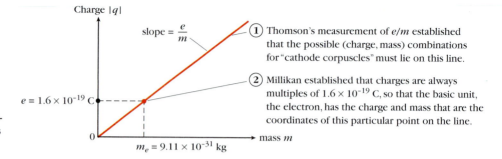

Figure 24-10 The relationship between Thomson's measurements and Millikan's.

charge and mass that had this ratio (out of all possibilities) be convincingly established as the charge and mass of the electron (via the reasoning in Figure 24-10).

Early in the nineteenth century, physicists were illustrating their reports of scientific advances with woodcuts and steel plate engravings; by the end of the century they were able to use photographic reproduction. The new technology created possibilities for new observations. After photographic plates in his lab were unexpectedly exposed, Roentgen established the existence of X rays (1895), so called because their nature remained a mystery for years after their discovery. The next year, exposing photographic plates to crystals of uranium salt led Henri Becquerel of Paris to the discovery of **radioactivity,** which he at first mistook for weak X rays. Subsequently, Marie Sklodowska Curie (1867–1934) and her husband, Pierre, studied a range of radioactive materials and discovered a new element far more radioactive than uranium. They named it *radium.*

Other researchers were quick to explore the properties of the newly discovered radioactivity. At McGill University in Montreal, Ernest Rutherford (1871–1937) investigated the ability of radioactivity to penetrate layers of aluminum foil. He found a quick drop-off for the first few layers, and then a much slower additional drop-off as further layers were added. He concluded that there were two kinds of radiation here, one that could be screened out by the first few layers of foil and another that was far more penetrating. He labeled these **α and β rays** (alpha and beta rays). In 1899, by methods similar to Thomson's, Becquerel showed that β rays had the same charge-to-mass ratio that Thomson had found for cathode rays but greater velocities. If the cathode rays were streams of negative particles (later called electrons), so were β rays. By 1900, Villard in France had found an even more penetrating component of radiation, which he called **γ rays** (gamma rays).

X rays, α, β, and γ rays—what did they tell us about the atoms from which they emerged? The emissions from a microscopic world beyond our direct perception were offering tantalizing clues about the nature of that world, if we could only learn to read them, and a new generation of physicists was quick to take on the challenge. This atmosphere of scientific ferment occupied the stage on which the opening curtain of the twentieth century was to rise.

The questions physicists now had before them could not have been raised a century earlier. In 1800 there were no cathode ray tubes to emit X rays, no photographic plates for detecting radioactivity; there was neither the vacuum pump technology nor the high-voltage coils that cathode ray tubes required. All that technology was spawning a new set of questions. For physicists, the excitement of new technologies continues to be that they raise unanticipated new questions as well as helping answer old ones.

24-3 Atomic Spectra: Patterns and Puzzlements

Not only did these strange new emissions raise questions, there were also unanswered questions raised by that most seemingly ordinary of emissions, visible light. A flame is a region of gas so hot that it becomes luminous; that is, it emits light. By the late nineteenth century, physicists were also able to excite gases to luminosity by applying high voltages to gases contained in cathode ray tubes. Since the middle 1700s, first using prisms and later using diffraction gratings, physicists had observed that the spectra emitted by luminous (and not very dense) gases were different from the continuous spectra emitted by more concentrated hot matter, such as the sun or a piece of metal made "red hot" or "white hot." Rarefied gases of pure elements emitted characteristic **line spectra** (Figure 24-11). What distinguished these line spectra from a **continuous spectrum** (see figure) is that certain colors appeared at the same positions they occupied in the continuous spectrum, but the in-between colors were missing, leaving the individual lines of color isolated or **discrete.** The word *discrete* means distinctly separate, in contrast to *continuous,* and should not be confused with *discreet,* which you might use to describe someone who doesn't tell your secrets. Line spectra are also called **discrete spectra.** Once Young's double-slit experiment had connected color with wavelength, it was recognized that a discrete set of lines meant that only particular wavelengths were emitted, not those in between.

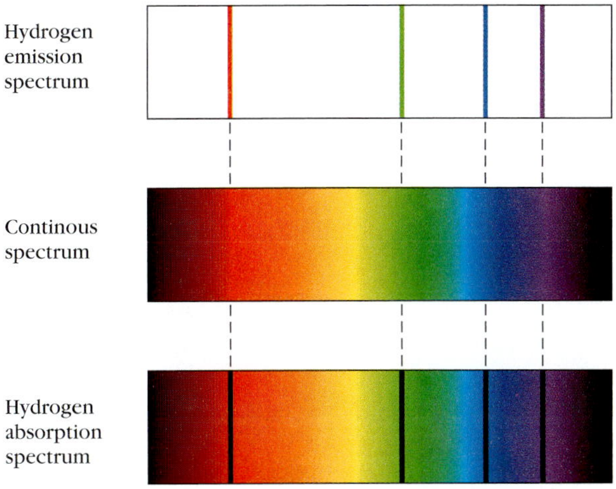

Figure 24-11 **The emission and absorption spectra of a gas (hydrogen) compared to the continuous visible spectrum.**

Line spectra quickly captured the interest of chemists when they realized that each element had its own unique spectrum, so that the presence of a particular set of lines was a sure indicator of the presence of the element that emitted that set of lines. It was the element's identifying signature. The spectroscope (Section 15-5, Figure 15-22) became a tool of chemical analysis. The observation of previously unobserved sets of lines led to the discovery of new elements. In this way in 1855, Kirchhoff and Bunsen (the chemist for whom the Bunsen burner is named) discovered rubidium and cesium.

In addition to discrete lines of colored light, discrete dark lines had been observed in otherwise continuous spectra. Most notably, a set of dark lines in the spectrum emitted by the sun was recorded by German physicist Joseph Fraunhofer over the years 1814–1824. They are still called the **Fraunhofer lines.** In 1859, Kirchhoff reported that when white light from another source passed through the gas of a particular element, the gas absorbed the same wavelengths—the same colors—it emitted when it was made luminous and was itself the source (Figure 24-11). In other words, the wavelengths corresponding to the bright lines of an element's **emission spectrum** are the same as the wavelengths corresponding to the dark lines of its **absorption spectrum.**

Kirchhoff's realization was the key to understanding the Fraunhofer lines. The continuous emission spectrum of light from the sun comes from the hotter, denser layers of the sun. Those emissions then pass through the cooler, less dense gases of the outermost solar atmosphere, which absorb certain of the wavelengths initially present. But if the absorption lines are at the same wavelengths as the emission lines that we use to identify elements, then by observing these absorption lines we can identify what elements are present in the outer layers of the sun! Indeed, subsets of the Fraunhofer lines corresponded to known spectra of particular elements. When one set did not, it was assumed to come from a previously undiscovered element. Only later was this element found on Earth, and it still bears the name *helium,* from the Greek word *helios* meaning sun. Kirchhoff's realization opened up the possibility that we can know more about the sun and stars than simply their motions. *Spectroscopy* (the study of spectra) continues to be one of the most powerful tools that scientists have at their disposal for learning not only about the sun but about more distant stars as well.

Line spectra corresponded to a set of discrete wavelengths that can be measured, just as we used the double slit to measure the wavelength of red light in Chapter 15. Where there is numerical data, curious people will search for patterns in the data. One such searcher was a Swiss schoolteacher named Johann Balmer. He was able to write a general formula

$$\lambda = (3.6456 \times 10^{-7} \text{ m})\frac{n^2}{n^2 - 2^2} \tag{24-1}$$

When the four values 3, 4, 5, and 6 are substituted for n in this formula, it gives the wavelengths of the four lines of the visible hydrogen spectrum to within about one part in 40 000, an astonishing degree of accuracy. Balmer recognized that other wavelengths could be found if 2^2 were changed to the square of a different integer, but no lines with those wavelengths had yet been found when Balmer published his findings in 1885 or for two decades thereafter. They eventually were found in the infrared and ultraviolet parts of the spectrum. Part of the importance of Balmer's pattern was its predictive power.

Where a pattern of behavior is found in the data, physicists have learned to seek mechanisms or structures that could account for that pattern. Newton's universal gravitation varying inversely as r^2 accounted for the pattern found by Kepler that T^2 is proportional to R^3. The kinetic theory of gases was a model that could explain the patterns of proportionality among pressure, volume, and temperature that gases exhibit under suitable conditions. What kind of structure for hydrogen could account for this strange pattern among the wavelengths that it emitted? At the end of the nineteenth century, no answer was yet forthcoming either to this question or to the larger question to which it related.

The larger question was: Why should a particular element emit only particular lines and no others, and why should these lines be different for different elements? If different elements had different constituent atoms, in what way could the atoms of one element differ from those of another that could account for the differing emissions? Surely the differences must be connected with the structure of the atoms themselves. Arriving at a model of the structure of atoms sufficiently detailed to account for the specifics of these emissions was a task that the physicists of the nineteenth century left for the physicists of the twentieth.

Other patterns of behavior relating to wavelengths of light also were perplexing physicists as the nineteenth century came to a close. We will focus on one because of its historical importance. This pattern involves the spectrum of electromagnetic radiation emitted by **condensed matter,** that is, by matter in which the atoms are packed close enough together to interact substantially with one another. Solids, liquids, and dense gases are all condensed matter. The spectrum emitted by such matter changes as its temperature changes. Although the pattern of change is expressible as an equation, we will explore it qualitatively by looking at the kinds of behavior it describes in some simple everyday situations.

On-The-Spot Activity 24-1

If you have an electric stove burner available, turn it on low. It gives off no visible light, but if you bring your nose (which is more sensitive to hotness than your hand) to within an inch or so of the burner, you can readily detect that it is radiating what is commonly called heat—that is, it is giving off infrared radiation. Now raise the burner setting from low to high. Observe that the burner begins to glow red. STOP&Think How do the wavelengths it is emitting now compare to those it emitted when it was cooler? ◆

For the rest of this activity, you will need a bulb connected to a dimmer switch. A bulb in which the filament is plainly visible is best. Set the switch so that the bulb is barely glowing. The emitted light should appear distinctly reddish. Now change the setting gradually until the bulb reaches full brightness. The light should appear quite white at this point. STOP&Think Reflect on how the wavelengths emitted now compare to those the bulb emitted when it was cooler. ◆ White light contains all the wavelengths in the visible spectrum, not just those at the longer-wavelength red end of the spectrum.

In general, as objects become hotter they emit more electromagnetic radiation at ever shorter wavelengths, or equivalently, at ever higher frequencies. What happens when an object becomes as hot as the surface of the sun? Because of the growing concern in recent years about protecting ourselves from the sun's ultraviolet (UV) rays, it has become common knowledge that a substantial part of the sun's emissions are in the still-shorter wavelength UV part of the spectrum. In fact, we can draw inferences about the surface temperatures of other stars from their color. On clear winter nights in the Northern Hemisphere, two stars that you can easily observe and compare are Betelgeuse and Rigel, at opposite ends of the Orion constellation (Figure 24-12). Betelgeuse is a distinctly reddish star; Rigel appears bluish. **STOP**&**Think** Which star is hotter? How do you know? ◆

Objects appear black when none of the light incident on them is reflected back to our eyes; it is all absorbed. Electric stove burners and the outer surfaces of wood stoves are usually black because the best absorbers of radiated energy are also the best emitters. So ideal emitters are often called **black bodies.**

STOP&**Think** If an ideal absorber is also an ideal emitter of light, how can it appear black? ◆ An ideal emitter is an ideal emitter of electromagnetic radiation, not necessarily of *visible* light. Your stove burner radiates in the infrared region (Figure 24-13a) until its temperature increases sufficiently for it to glow red hot.

The observations that we have described qualitatively are summarized quantitatively in Figures 24-13a and b. Graphs for several different temperatures are shown. Notice that the peak, which occurs at the wavelength emitted most strongly, shifts to the left (toward shorter wavelengths) as the temperature increases. A very simple pattern of relationship turns out to exist between the temperature of the emitter and the wavelength λ_{max} at which the graph peaks:

$$\lambda_{max} T = \text{constant} \qquad (24\text{-}2)$$

Again the appearance of a pattern raises the question: Why should this pattern occur? What underlying mechanism or model explains it? Encouraged by the successes of the kinetic theory of gases, physicists sought explanations in terms of behavior at a microscopic level. Because oscillators were known to be the sources of other waves, Maxwell's work had led physicists to think in terms of atomic oscillators as the sources of electromagnetic waves.

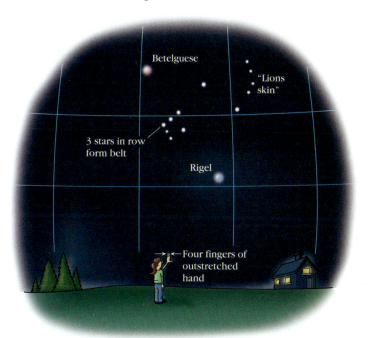

Figure 24-12 Locating Betelgeuse and Rigel. These stars are easily seen in winter in the Northern Hemisphere. First find the three bright stars in a row that make up Orion's belt. Then mentally draw a line segment roughly perpendicular to it and extending about four finger widths to either side of it. You will find Betelgeuse and Rigel at opposite ends of this line segment.

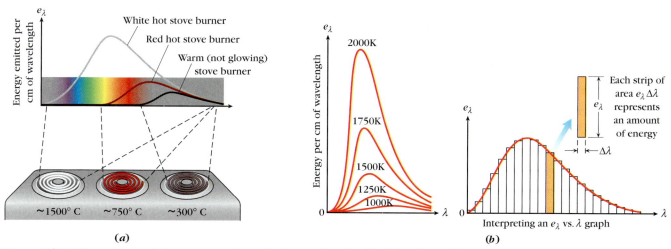

(a)

(b)

Figure 24-13 The pattern of electromagnetic radiation emitted by black bodies at different temperatures.

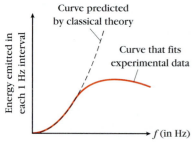

Figure 24-14 Radiated energy distribution from black bodies—classical theory breaks down (the ultraviolet catastrophe).

In Figure 24-14, the typical pattern is plotted against frequency rather than wavelength, so that shorter wavelengths—higher frequencies—are toward the right. Modeling the black body as an array of atomic oscillators gave a good fit with the long-wavelength (low-frequency) part of the pattern. But as we move toward the higher frequencies of the UV range, the amount of emission predicted by the model differs increasingly from the amount actually observed. Where the observed amount of emitted energy drops with increasing frequency, the predicted amount continues to grow toward infinity, which violates conservation of energy. This unacceptable behavior in the UV range and beyond was called the **ultraviolet catastrophe.** It indicated that something was very wrong with the theory that predicted such behavior. The theory that was now being cast in doubt was the very understanding that nineteenth-century physicists had of the microscopic structure and behavior of matter. It could not provide an atomic-scale picture that would accurately account for how matter and electromagnetic radiation interact.

In 1900, German physicist Max Planck came up with a clever mathematical contrivance that could be used to make the predicted pattern of radiation emission fit the observed pattern at higher frequencies as well. Planck considered the energies of the oscillators emitting the radiation and looked at the mathematical consequences of supposing that these energies could only be integer multiples of some basic fixed value E_o ($E = nE_o$, $n = 1, 2, 3, \ldots$), which Planck called a **quantum** of energy. Planck further assumed that the basic value E_o for each oscillator was proportional to the frequency f of its oscillation:

$$E_o = hf \qquad \text{and} \qquad E = nE_o = nhf \qquad (24\text{-}3)$$

The proportionality constant h is called **Planck's constant.** With its value suitably chosen ($h = 6.625 \times 10^{-34}$ J · s), Planck was able to get the collective behavior of his theoretical oscillators to match the experimentally observed results at *all* frequencies, not just low frequencies.

But why should energy only come in integer multiples of these quanta? Recall from Chapter 13 that a block oscillating on a spring has total energy $E = \frac{1}{2}kX^2 = \frac{1}{2}m\omega^2 X^2 = \frac{1}{2}m(2\pi f)^2 X^2$. Nothing in Newtonian physics forbids us from stretching the spring any maximum distance X we wish, so for any frequency f, *all* values of E are permissible. Planck's quantum flies in the face of this reasoning; it suggests that something very different is going on with his atomic oscillators than with any oscillator that is large enough to observe directly. This troubled Planck. It was a puzzle that was left for twentieth-century physics to unravel.

Classical Physics and Modern Physics: A Note on Terminology

Planck's introduction of the quantum marks the beginning of what physicists with increasing inaccuracy call **modern physics,** in contrast to **classical physics.** Classical physics is the physics that had become widely accepted by the end of the nineteenth century. It includes *classical mechanics,* meaning Newton's mechanics and more advanced concepts and methods that build on the Newtonian foundation, and *classical electrodynamics,* which culminates in the work of Maxwell. Physics that builds on these ideas, such as kinetic theory, which applies Newtonian mechanics to the microscopic domain, is considered classical. Modern physics introduces such concepts as the quantum of energy, which have no basis in classical physics and frequently conflict with the ideas of classical physics. Some physicists argue that Einsteinian relativity is an extension of classical physics, not a violation of it. Usually, though, modern physics refers to the new foundation of quantum physics and Einsteinian relativity established in the first decades of the twentieth century (and which we treat in detail in the following chapters). However, the word *modern* becomes increasingly a misnomer as the years advance.

Some seemingly simple questions also remained unanswered. Throughout the preceding chapters, we have listed tables giving numerical quantities of various properties of materials: electrical properties like conductivity, thermal properties like specific heat, optical properties like index of refraction. These values came from careful measurement. But what makes the value different for one material than for another? Why should iron be more attracted to a magnet than gold or silver? Why is copper a better conductor than iron and a far better conductor than silicon? Physicists at the end of the nineteenth century had no model of matter detailed enough to answer any of these questions.

24-4 The Speed of Light: Relative or Absolute?

In addition to these "simple" questions about the properties of matter, there were questions about a more perplexing kind of matter—if matter it was—that nineteenth-century scientists called the ether. The idea of an ether arose as a way of trying to explain action at a distance. Physicists had learned to look for mechanisms to explain the patterns of behavior that they observed. It seemed unacceptable to them that one object could exert a gravitational or electrostatic or magnetic force on another at a distance without some mechanism that communicated the effect across the intervening space. Newton had also wondered how heat might be radiated through a vacuum and speculated about whether it might be "conveyed through the vacuum by the vibration of a much subtler medium than air,"[1] which would remain in a container even after all the air had been pumped out of it. Such a medium, extending throughout the heavens, could carry heat from the sun to Earth. This "subtler medium," which could provide a mechanism for transmitting action-at-a-distance forces, came to be called the **ether.**

It was with this notion of an ether-like medium in mind that Faraday conceived of lines of force (field lines). When he saw iron filings aligning along curved lines in space between the poles of a magnet, he took it as visible evidence of an effect on the medium that occupied the in-between space—the ether. Building on this notion of a field, Maxwell's electromagnetic theory embraced the ether concept. For Maxwell, the ether was the medium through which his electromagnetic waves propagated. In calculating the wave propagation speed $v = \frac{1}{\sqrt{\mu_o \epsilon_o}} = 3.0 \times 10^8$ m/s, he thought of μ_o and ϵ_o as the properties of the medium on which v depended. As the medium that carried the light wave, it was called the *luminiferous* (light-bearing) ether.

We now encounter a dilemma. If a sound wave or a water wave propagates in a medium at a certain speed, the speed detected by different observers will depend on the speed at which they are traveling through the medium. (We showed this was true for the speed of sound in Section 14-8.) But differently moving observers do not obtain different values when they do electrical and magnetic measurements to find the values of μ_o and ϵ_o, so they do *not* obtain different values for the speed of light or other electromagnetic waves. This value is absolute and unique; it does not vary from one observer to another. The fact that sound and water wave speeds, in contrast, *are* relative to the observer follows simply from Newtonian kinematics (Section 14-8). Then the unique value of c violated Newton's mechanics. In the first years of the next century, Einstein rose to the task of reconciling this conflict.

To understand the dilemma more fully—both to see how it was viewed near the end of the nineteenth century and to prepare us for thinking about the implications of Einstein's work in the next chapter—we will briefly review the

[1]From Newton's *Opticks* III, query 18, quoted in A. A. Arons, *Development of Concepts of Physics,* Addison-Wesley, Reading, MA, 1965.

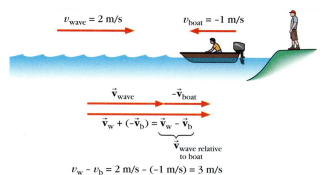

$v_{wave} = 2$ m/s $v_{boat} = -1$ m/s

$\vec{v}_{wave}$ $-\vec{v}_{boat}$

$\vec{v}_w + (-\vec{v}_b) = \vec{v}_w - \vec{v}_b$
$\underbrace{\qquad}$
$\vec{v}_{wave\ relative\ to\ boat}$

$v_w - v_b = 2$ m/s $- (-1$ m/s$) = 3$ m/s
(Boat observer sees crests getting closer by 3 m/s)

Figure 24-15 Velocity of a wave relative to a moving observer.

concept of relative velocity introduced in Section 7-3. Figure 24-15 illustrates the point that the wave velocity that an observer detects (the velocity relative to the observer) is different if the observer is moving relative to the medium (like the person in the boat) than if he is not (like the person standing on shore). If the propagation velocity v_{wave} is 2 m/s to the right, the person on shore sees each crest coming at him at a speed of 2 m/s. But the observer on the boat traveling to the left at 1 m/s sees crests coming toward him at 3 m/s, the velocity of the wave relative to him. (The figure shows this as a vector addition $\vec{v}_{wave} + (-\vec{v}_{boat})$.) If the boat travels to the *right* at 2 m/s, crests never catch up with him because the velocity of the wave relative to him is then $v_{wave} - v_{boat} = 2$ m/s $- 2$ m/s $= 0$. The relative velocity is different still when the boat travels perpendicular to the direction of wave propagation (parallel to the crests).

Example 24-2 *The Relative Velocity of Two Objects Traveling along Perpendicular Paths*

For a guided interactive solution, go to Web Example 24-2 at www.wiley.com/college/touger

A boat travels west at a speed of 1.0 m/s, parallel to a series of long straight crests that are moving north at 2.0 m/s (Figure 24-16a). The boat's motion is perpendicular to the direction in which the crests are moving. Find the magnitude of the wave's velocity relative to the boat.

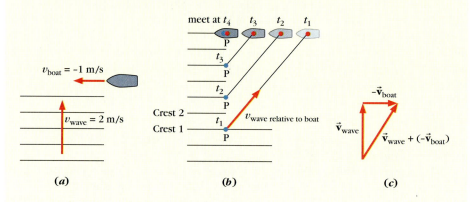

Figure 24-16 Relative velocity in two dimensions.

(a)   *(b)*   *(c)*

Brief Solution

Choice of approach. To help visualize what is happening, we pick out a single point P on one of the crests—a point that will meet the boat—and look at the positions of point P and the boat at successive instants t_1, t_2, t_3 . . ., (Figure 24-16b). The velocity of P relative to the boat tells us how the displacement vector from the boat to the point (see figure) is changing. Because the displacement decreases in magnitude as the boat and the point come together, the velocity vector is in the opposite direction. We end up with a sequence of velocity vectors that are always toward the boat: The observer on the boat sees point P getting closer and closer to the boat at constant velocity until it reaches it.

Formally, the velocity of P relative to the boat is $\vec{v}_{wave} - \vec{v}_{boat}$. We draw the vector diagram, do the vector subtraction separately in each direction to get the components of the relative velocity, and then use these to find its magnitude.

What we know/what we don't.

$$(v_{boat})_x = -1.0 \text{ m/s} \qquad (v_{boat})_y = 0$$

$$(v_{wave})_x = 0 \qquad (v_{wave})_y = 2 \text{ m/s}.$$

$$(v_{wave} - v_{boat})_x = ? \qquad (v_{wave} - v_{boat})_y = ? \qquad |\vec{v}_{wave} - \vec{v}_{boat}| = ?$$

The vectors are sketched in Figure 24-16c.

The mathematical solution. The components are

$$(v_{wave} - v_{boat})_x = (v_{wave})_x - (v_{boat})_x = 0 - (-1.0 \text{ m/s}) = 1.0 \text{ m/s}$$

$$(v_{wave} - v_{boat})_y = (v_{wave})_y - (v_{boat})_y = 2.0 \text{ m/s} - 0 = 2.0 \text{ m/s}$$

Notice that the components are to the east and north, consistent with the direction shown in the figure. The magnitude is

$$|\vec{v}_{wave} - \vec{v}_{boat}| = \sqrt{(v_{wave} - v_{boat})_x^2 + (v_{wave} - v_{boat})_y^2}$$

$$= \sqrt{(1.0 \text{ m/s})^2 + (2.0 \text{ m/s})^2} = \textbf{2.2 m/s}$$

Making sense of the results. The boat in this example has the same speed as in Figure 24-15, and the speed of the wave relative to the water (as seen by an onshore observer who is stationary relative to the water) is the same in both cases. However, in Figure 24-15 the boat is moving perpendicular to the crests and here it is moving parallel to the crests. Because of this, the wave speeds in the two cases differ for an observer on the boat.

◆ Related homework: Problem 24-2 and Problems 24-29 through 24-34.

Near the end of the nineteenth century, most physicists still thought in terms of motion of electromagnetic waves relative to the ether. Suppose that like the boat traveling over the water, Earth travels through the luminiferous ether. Example 24-2 showed that an observer saw a different wave speed when the crests were parallel to the boat's motion than when they were perpendicular to it (Figure 24-16). In the same way, an Earthbound observer will experience two light waves that have the same propagation speed through the ether as having different speeds when the two waves are traveling in mutually perpendicular directions. If indeed Earth is traveling through the ether, we should be able to detect this difference in the speed of light.

The Michelson interferometer (see figure) is a device in which the light is divided into two mutually perpendicular paths by a partially silvered mirror. The phase difference between light reaching the observer via the two paths will depend not only on the difference in the lengths of the two paths but on the difference in speed. This will result in a particular interference pattern of light and dark fringes being observed. If the interferometer is then rotated 90°, the velocities for the longer and shorter paths should reverse, resulting in a measurable shift in the interference pattern; that is, unless there is no difference in velocity in the two directions.

With his colleague E. W. Morley, Michelson performed this experiment in 1887. Their equipment was sensitive enough to detect a shift in the interference pattern one-hundredth as big as the one the ether might be expected to cause. The **Michelson-Morley experiment,** a landmark in the history of physics, found no detectable shift in the interference pattern, indicating there was no difference in the speed of light relative to an Earthbound observer in the two directions. If that was so, there was no point in assuming the existence of a special medium to provide the reference frame in which light had a speed $c = 3.0 \times 10^8$ m/s. The rationale for an ether was gone!

The implication that there was no ether left an even greater dilemma. In Newtonian physics, speed is always relative to some observer. The implication

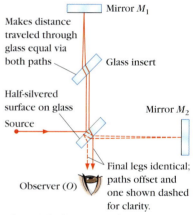

The Michelson interferometer.

➥Although the Michelson-Morley experiment showed it to be unnecessary, the idea of an ether was surrendered only very slowly by the wider scientific community. As late as the 1920s, one can find popular electronics books referring to radio waves as "ether waves."

that light traveled at the same speed for an observer moving in one direction as in another violated what Newton's physics tells us about relative velocity. Here was indeed a problem of major proportions that was left for twentieth-century physics to solve.

The Michelson-Morley experiment did not directly influence Einstein's work addressing this problem; he was motivated rather by the simple fact of $c = \frac{1}{\sqrt{\mu_o \epsilon_o}}$ having a unique value, and the related notion that the laws of electricity and magnetism ought not to depend on the velocity of the reference frame in which they are observed. Nevertheless, the Michelson-Morley experiment did influence the preparedness of other physicists to accept Einstein's work.

24-5 The Past Is Prologue

After the major breakthroughs in electromagnetic theory and thermal physics, many scientists believed that the major part of the task of finding the rules by which nature played had been completed. Michelson, thought to have been quoting Lord Kelvin, said that "An eminent physicist has remarked that the future truths of physical science are to be looked for in the sixth place of decimals." Then came the surge of activity in the last years of the 1800s that showed that we were only beginning to find out how much we did *not* understand. The twentieth century eventually answered many of the questions left unanswered at the end of the nineteenth (as you will see in the chapters that follow), as well as many questions that the physicists of that era had not yet thought to raise, such as cosmological questions about the nature and origins of the universe. In 1900, those cosmological questions seemed rather more the province of religion and philosophy than of hard science. Now, as science addresses them, new and still more perplexing questions are being brought to light. The endeavor to discover the rules by which nature plays is never-ending. It continues to challenge the human intellect and imagination in the twenty-first century.

✦ SUMMARY ✦

This chapter is a prologue to the chapters that follow. In part it summarizes and reorganizes material from earlier chapters. **Maxwell's equations** reorganized and augmented what was known about electric and magnetic fields. In particular, the **Ampere-Maxwell law** could account for situations that Ampere's law could not. Based on these equations, he

• predicted the existence of **electromagnetic waves,** and
• determined that their propagation speed was $\frac{1}{\sqrt{\mu_o \epsilon_o}} =$ 3.0×10^8 m/s, providing evidence that visible light waves are part of a vastly more extensive **electromagnetic spectrum.**

Historical background is presented on

• the beginnings of an understanding of atomic structure;
• the emergence of new evidence—X rays, radioactivity, and the patterns found in emission and absorption spectra—that by 1900 had not yet been synthesized into that understanding.

This material is summarized in Table 26-1 in Section 26-1 as a lead-in to discussing later advances.

A crucial implication of Maxwell's work was that $c = \frac{1}{\sqrt{\mu_o \epsilon_o}}$ had a unique value independent of the motion of the observer. The fact that the speed of other kinds of waves was different relative to observers traveling at different velocities followed directly from Newtonian mechanics. The inability of Newtonian mechanics to explain why it was not true for light sets the stage for Chapter 25 on special relativity.

The **ether** was conceived as a medium in which light waves could travel. It then followed that if Earth moved through the ether, an Earthbound observer should see light as having different speeds in two mutually perpendicular directions. The **Michelson-Morley experiment** showed that this was not the case and helped establish a receptive climate for the work of Einstein that would follow.

✦ QUALITATIVE AND QUANTITATIVE PROBLEMS ✦
Hands-On Activities and Discussion Questions

The questions and activities in this group are particularly suitable for in-class use.

24-1. Discussion Question. a. Identify some properties of matter that can be explained by thinking of matter as being made up of indivisible and indestructible atoms that have no internal structure and cannot be broken down further. Justify your choices. **b.** Identify some properties of matter that cannot be explained by such a simple notion of the microscopic structure of matter, and explain why not.

24-2. *Discussion Question.* A motor boat is traveling at a speed of 3 m/s in a direction parallel to a series of long, straight crests, while the crests travel along the surface of the water at a speed of 2 m/s (Figure 24-17).

a. The boat's driver reasons, "The velocity of these crests is 2 m/s relative to my boat, because each crest gets 2 m closer to me each second." Explain why this reasoning is or is not correct.

b. The boat driver's friend comments, "Well, if that were so then by the same reasoning, if you were traveling parallel to the shore, your velocity would be zero relative to the shore. I don't think that's right." Does the friend's reasoning change your mind? Try to fill in the details of the friend's reasoning. Who is right, the driver or the friend? Why?

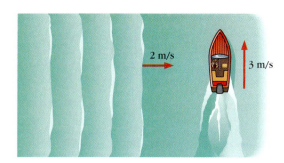

Figure 24-17 Problem 24-2

Review and Practice

Section 24-1 **The Triumph of Electromagnetism**

24-3. At an instant shortly after the switch in Figure 24-18 is closed, the capacitor is still charging. Rank the three points A, B, and C in order of the magnetic field strength at

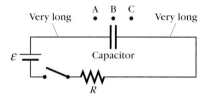

Figure 24-18 Problem 24-3

each point at the instant described. Order them from least to greatest, making sure to indicate any equalities.

24-4. As described by a stationary observer, the crests of a wave are traveling at speed v_{wave} in the direction shown in Figure 24-19. Observers A, B, and C are all traveling at the same slower speed v_{obs}, but in different directions: A is traveling toward the top of the

Figure 24-19 Problem 24-4

page, B toward the bottom, and C toward the left. Rank these three observers from least to greatest, making sure to indicate any equalities, in order of the wave speed that each sees if the wave is ***a.*** a water wave. ***b.*** a light wave.

24-5.

a. Discuss the similarities and differences between Faraday's law of electromagnetic induction and the Ampere-Maxwell law.

b. In particular, under what conditions are these two laws most similar? Under those conditions, what do the two laws tell us about the effect of changing fields (fields that change over time)?

24-6. Maxwell's equations state conditions that electric and magnetic fields must always satisfy. When your statements about the world are mathematical equations, mathematical manipulation serves as the reasoning by which you can reach further equations, that is, new statements or conclusions. By performing such mathematically based reasoning, what important new conclusions was Maxwell able to reach about electric and magnetic fields?

24-7. SSM Is it possible for the electric field to be varying sinusoidally throughout a region in which the magnetic field remains constant? Briefly explain.

24-8. *Propagation of an Electromagnetic Disturbance.* An alternating current flows through wire 1 in Figure 24-20. The rectangles in the figure are not actual objects; they simply mark off regions of two-dimensional space.

a. As a result of the current in wire 1, rectangle A is crossed by lines of what kind of field, and how does this field vary?

b. Because of the varying field across (perpendicular to) rectangle A, what kind of field is induced across rectangle B, and how does it vary?

c. Because of the varying field across (perpendicular to) rectangle B, what kind of field is induced across rectangle C, and how does it vary?

d. If you continue this reasoning from rectangle to rectangle, what kind of field is induced across rectangle H, and how does it vary?

e. If wire 2 is part of a complete circuit, what effect will the varying field across rectangle H produce in wire 2?

f. Would your reasoning have changed at all if we had chosen rectangles of different width? Explain.

g. If wire 2 had crossed the y-axis at point P in the figure, but was still parallel to the z-axis, would the same effect have been produced in it? Explain.

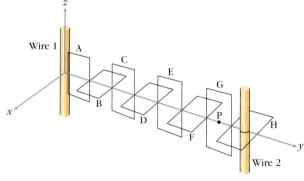

Figure 24-20 Problem 24-8

Section 24-2 **Probing the Atomic World**

24-9. Figure 24-21 shows three possible continuations of the path in Figure 24-9. Which of these diagrams shows the path correctly to the right of point P? Briefly explain.

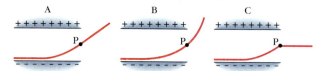

Figure 24-21 Problems 24-30 and 24-31

24-10.

a. In Figure 24-9, if the region of electric field through which the beam travels extends for a distance 2L rather than L, the sideways displacement *y* of the beam (*is not changed; increases slightly; doubles to 2y; increases to 4y*). Briefly explain.

b. If a beam of protons traveled at the same speed, would its sideways deflection be greater, the same, or less? Briefly explain.

24-11. In Example 24-1, *a.* what is the deflection *y* of the beam when the magnetic field *B* is applied? *b.* what is the value of the magnetic field when $y = y_1$?

24-12. The ratio $\frac{q}{m}$ of charge to mass was found by Thomson to have an absolute value of 1.75×10^{11} C/kg for electrons.

a. Which (one or more) of the following particles would satisfy this ratio?

  i. a particle with a charge of -9.11×10^{-31} C and a mass of 1.6×10^{-19} kg

  ii. a particle with a charge of -8.0×10^{-20} C and a mass of 4.56×10^{-31} kg

  iii. a particle with a charge of -3.2×10^{-19} C and a mass of 1.82×10^{-30} kg

  iv. a particle with a charge of -2.4×10^{-19} C and a mass of 1.37×10^{-30} kg

  v. a particle with a charge of -3.2×10^{-19} C and a mass of 1.82×10^{-27} kg

b. If charged objects must in addition satisfy the condition that their charge be an integer multiple of 1.6×10^{-19} C, which of the particles above could consist entirely of one or more electrons?

c. Which of the particles above could *contain* one or more electrons?

24-13. Which of the principles involved in the design of the mass spectroscope did Thomson use in his measurement of the ratio $\frac{q}{m}$ of charge to mass?

24-14. Without knowing what they were made of, how was Rutherford able to determine that β rays were different than α rays?

24-15. At the start of the twentieth century, physicists knew of the existence of *X rays, alpha rays, beta rays,* and *gamma rays.* If you had to sort these "rays" into two groups, which ones would you group together? Briefly explain the basic difference between your two groups.

Section 24-3 **Atomic Spectra: Patterns and Puzzlements**

24-16. How can physicists and astronomers determine what elements are present in the sun or a distant star?

24-17. Explain in your own words how a discrete spectrum differs from a continuous spectrum.

24-18. Explain in your own words the difference between an emission spectrum and an absorption spectrum.

24-19. SSM

a. The visible emission spectrum for sodium consists of just two yellow lines, which are very close together. If white light is passed through sodium gas, what would the absorption spectrum look like?

b. Two lamps, a sodium lamp and an ordinary incandescent white bulb, are connected to dimmer switches. Compare what happens to the spectra emitted from these two lamps as they are dimmed.

24-20.

a. The star Antares appears reddish to the naked eye. What can you infer about how its temperature compares to the temperature of our own sun?

b. Would you expect the temperature of a neon sign, which emits red light, to be comparable to that of Antares? Briefly explain why or why not.

c. Is there a significant difference in the nature of the two light sources? If so, briefly tell what it is.

24-21. If the temperature of a black body in kelvins is doubled, how does this alter the wavelength at which it radiates the most energy?

24-22. As the temperature of an object becomes very high, does it stop emitting radiation at low frequencies? Briefly explain.

24-23. SSM According to the Balmer formula, what happens to the difference between each wavelength and the next as *n* increases? (Does the difference increase, decrease, or remain the same?)

24-24. It occurred to Balmer that if the number 2^2 could occur in the denominator of his formula, the squares of other integers might also occur.

a. If we chose 1^2 instead of 2^2, the formula would read

$$\lambda = (3.6456 \times 10^{-7} \text{ m})\frac{n^2}{n^2 - 1^2}$$

Calculate the wavelength this formula would give you when $n = 2$, and then briefly explain why we don't see a hydrogen line with this wavelength.

b. Now try substituting 3^2 for 2^2 in the original Balmer formula, and calculate the wavelength that this variation of the formula would give you when $n = 4$. Then briefly explain why we don't see a hydrogen line with this wavelength either.

c. Would it make sense to try $n = 1$ with the variation of the Balmer formula that you used in *b*? Briefly explain by telling what would happen to the wavelength.

24-25. Planck looked at the mathematical consequences of supposing that the energies emitted by atomic oscillators were *quantized;* that is, they could only be integer multiples of some basic fixed value E_o ($E = nE_o, n = 1, 2, 3, \ldots$), which Planck called a *quantum* of energy. Tell whether each of the following are quantized.

a. The readings of a clock with a sweep second hand

b. The readings of a digital wristwatch

c. Electric charge

d. Readings on a mercury thermometer

e. The mass of samples of pure oxygen-16

f. The speed of a ball dropped from the roof of an apartment building

g. The kinetic energy of a ball dropped from the roof of an apartment building

h. The potential energy of a block left somewhere on a staircase

i. The energy emitted by one of Planck's atomic oscillators

j. Wavelength values permitted by the Balmer formula (Equation 24-1)

k. The radii at which an artificial satellite could circle Earth

Section 24-4 The Speed of Light: Relative or Absolute?

24-26. A disabled state police cruiser pulled over on the side of the highway keeps its lights flashing and its siren wailing. One observer moves toward the cruiser at high speed; another stands by the roadside. What observations about the emitted sound and light would be different according to the two observers, and what observations would be the same?

24-27. What result or implication of Maxwell's work on electromagnetic waves is in violation of Newtonian mechanics?

24-28. What is the most important conclusion to be drawn from the Michelson-Morley experiment?

24-29. **SSM** Suppose that in Example 24-2 (Figure 24-16*a*), the boat travels as shown, but the crests are traveling toward the bottom of the page.
a. What would be the speed of this wave relative to the boat? Is this the same relative speed that we found for the waves traveling toward shore in Example 24-2?
b. Do water waves traveling in two directions, one perpendicular to the other, reach the boat at the same relative speed?
c. Suppose that a Michelson-Morley set-up and an observer are fixed in place somewhere on Earth as Earth follows its orbit through space. If Earth were traveling through some medium that carried the light—a medium we might call the ether—would the light following the two mutually perpendicular directions in the Michelson-Morley set-up (Figure 15-30*b*) have the same speed relative to the observer?

24-30. What analogies are there between the features of the Michelson-Morley situation and the features of the boat's situation in Example 24-2? (What in one situation is like what in the other?)

24-31. Try redrawing Figure 24-16*b* carefully for a point on crest 2 instead of crest 1. *a.* Will this point meet the boat, pass a short distance in front of the boat, or pass a short distance behind the boat? *b.* Does considering a different point on the crest change the relative velocity? Briefly explain.

24-32. In Figure 24-22, assume that the red car and the coordinate frame fixed to it are yours. The velocities given in the figure are *relative to the road.* The figure shows the car's positions at *t* = 0.
a. Rank the blue cars in order of their velocities relative to your car at the instant *t* = 0. Order them from least to greatest, making sure to indicate any equalities. Briefly justify your ordering.

b. Assume each car continues at the velocity indicated in the figure. Now consider an instant one second after your car has overtaken car C, and repeat *a* for this instant. Briefly justify why your ordering is the same as (if it is) or different than (if it is) in part *a.*

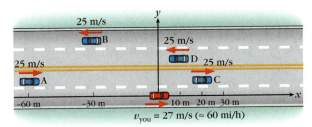

Figure 24-22 Problem 24-32

24-33. **SSM WWW** Whenever we speak of velocity relative to a particular observer, we can picture a coordinate frame anchored to the observer, as in Figure 24-22. If we are calculating the velocity $\vec{v} = \frac{\Delta \vec{x}}{\Delta t}$ relative to that observer, that is the coordinate frame in which we must measure the change in position $\Delta \vec{x}$. In Figure 24-22, assume that the red car and the coordinate frame fixed to it are yours. The velocities given in the figure are *relative to the road.*
a. In your coordinate frame, what is the position of car D as it passes you? What is the position of car C at the same instant? What is the position of your car at this instant?
b. In your coordinate frame, what is the position of car C when you overtake it? What is the position of car A at the same instant? What is the position of your car at this instant?
c. Figure 24-22 shows the positions of all cars at *t* = 0. In your coordinate frame, find the change Δx in each car's position between *t* = 0 and *t* = 1 s.
d. Find the velocity of each of the other cars relative to your car.

24-34. In Figure 24-23, assume that the red car and the coordinate frame fixed to it are yours. The velocities in the figure are *relative to the road* and are constant.
a. Figure 24-23 shows the positions of all cars at *t* = 0. In your coordinate frame, find the components Δx and Δy of the change in each car's position (including your own) between *t* = 0 and *t* = 1 s.
b. Find the velocity (magnitude and direction) of each of the other cars relative to yours.
c. Is there a change in the velocities of cars A and D relative to you after they go through the intersection? Briefly explain.
d. Is there a change in the velocities of cars A and D relative to you after *you* go through the intersection? Briefly explain.

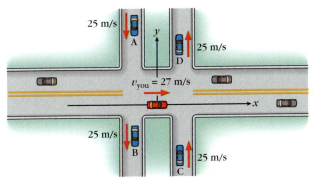

Figure 24-23 Problem 24-33

Going Further

The questions and problems in this group are not organized by section heading, so you must determine for yourself which ideas apply. Some of them will be more challenging than the Review and Practice questions and problems (especially those marked with a • or ••).

24-35. For each of situations *a–e* depicted in Figure 24-24, answer the following two questions.

i. Is there a magnetic field along the dotted line? (If so, tell briefly why it occurs.)

ii. Is there an electric field along the dotted line? (If so, tell briefly why it occurs.)

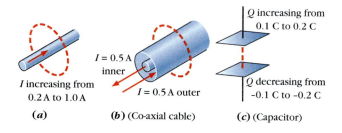

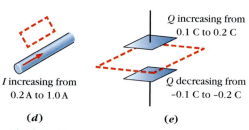

Figure 24-24 Problem 24-35

24-36. An electromagnetic wave traveling toward the right passes wire 1 (which is long and straight) and wire 2 (which forms a rectangular loop) in Figure 24-25. As the wave passes any given point, the magnetic field vector at that point oscillates in a plane perpendicular to the page—that is, it alternates between being into and out of the page.

a. As the wave passes wire 1, what effect does it have on the current in wire 1? Briefly explain.

b. As the wave passes wire 2, what overall effect does its magnetic field component have on the current in wire 2? Does this depend on the size of the square? Briefly explain.

c. As the wave passes wire 2, what overall effect does its electric field component have on the current in wire 2? Briefly explain.

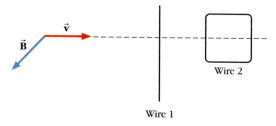

Figure 24-25 Problem 24-36

24-37. Do cathode rays have most in common with α rays, β rays, γ rays, or X rays?

24-38. Which of the following is most different from the others? α rays   yellow light from a sodium flame   γ rays X rays

24-39. Physicists at the start of the twentieth century had learned of several new kinds of "rays": X rays, α rays, β rays, and γ rays. Which (one or more) of these kinds of rays could cause an electron to oscillate? Briefly explain why.

24-40. If the cathode beam in Figure 24-9 had been produced by a higher voltage, how would its path differ from the path shown once it entered the region of constant field shown in the figure?

24-41. If the electric field in Figure 24-9 were replaced by a magnetic field in the same direction, how would the cathode ray beam be affected? Would it be deflected? If so, in which direction? Would its speed change? If so, how?

24-42. Without changing the total electric field, how could you get the cathode ray beam in Figure 24-9 to travel in a straight line? Describe fully the change(s) that you would make in the set-up.

24-43. The light passing through the diffraction grating or prism of a spectroscope can be made to fall on and expose a strip of photographic film, thus producing a record of the spectrum for that light. Suppose this is done twice. The first time, the light passing through the spectroscope is produced by electrically exciting the atoms in a tube containing a very low-density sample of a certain gas. The second time, the light comes from a source of white light but passes through a very dense sample of the same gas before reaching the spectroscope. How would the appearances of the strips of film exposed in these two ways compare? Be as specific as possible.

•24-44. What does the graph in Figure 24-13*b* tell you about the *total* amount of energy given off by a black body at different temperatures? *How* does it tell you that?

24-45. Which of the following is most likely to emit a line spectrum?

a. A red-hot soldering iron (pure iron)

b. A small volume of compressed hydrogen raised to a temperature of 700°C

c. A tiny amount of mercury evaporated in a vacuum tube and subjected to a very high voltage

d. The interior gases of a star in the Big Dipper

24-46. When an ambulance travels away from you, its siren sounds lower-pitched than when the ambulance is stationary. If a star is traveling away from our solar system, how would the hydrogen line spectrum for that star compare with the hydrogen line spectrum observed for light from our own sun?

24-47. Do absorption spectra have to be line spectra? When white light strikes an orange carrot or a green leaf, what part of that light gets returned to your eye? Roughly, what do you expect the absorption spectra of these objects to look like? (A sketch for each object might help.)

24-48. Boat A travels in a direction 45° east of north at a speed of 12 m/s. Boat B travels in a direction 15° west of north at a speed of 16 m/s. Find the velocity (magnitude and direction) of boat B relative to boat A.

24-49.

a. AM radio waves have wavelengths hundreds of meters long. Can AM radio waves be used as radar is used to detect aircraft? Briefly explain.

b. Visible light has wavelengths from about 4.0×10^{-7} m to about 7.5×10^{-7} m. Can visible light be used to detect individual atoms? Briefly explain.

c. Is it possible to use visible light photography to obtain images of individual atoms? Briefly explain.

24-50. When an ordinary light bulb is viewed through a spectroscope as it is dimmed, the blue part of the spectrum will disappear (*before; at the same time as; after*) the red part of the spectrum. Briefly explain.

24-51. Suppose an intergalactic space traveler could travel at a speed of 1.2×10^{8} m/s. If the traveler were heading toward a star at this speed, what would be the speed of light from this star relative to the traveler?

24-52. Oscillator A emits long-wavelength radiation and oscillator B emits short-wavelength radiation. By Planck's reasoning, if E_A and E_B are the lowest total energies that the two oscillators can have, is E_A less than, equal to, or greater than E_B? Briefly explain.

CHAPTER TWENTY-FIVE

Relativity

The idea of *relativity*—namely, that a quantity may be detected as different relative to different observers—exists even in Newton's physics. Newtonian physics tells us that the speed of a wave should be different relative to different observers. Newtonian physics also places no theoretical limit on how fast anything may move. Newtonian physics is thus unable to account for two important observational facts:

- *The speed of light in a vacuum is independent of the observer; its measured value is always $c = 3.0 \times 10^8$ m/s, no matter how the source or the observer may be moving.*

- *Nothing has ever been observed to move at a speed greater than c.*

". . . the speed of light detected by an observer is $c = 3.0 \times 10^8$ m/s, regardless of how the observer is moving."

The first of these facts compelled a reexamination of our understanding of space, time, and the reference frames we use to locate events in space and time. Any new understanding would have to account for the second fact as well. In this chapter, we will see how such a reexamination resulted in Einstein's theory of **special relativity**, a theory that (1) accounted for *both* of these behaviors at speeds at or near *c*, and (2) gave essentially the same results as Newton's mechanics for objects moving at speeds substantially less than the speed of light (the range of speeds for which Newton's mechanics had been extraordinarily successful).

25-1 Reference Frames Revisited

We start by looking more closely at the situations that Newtonian physics failed to explain. In ordinary experience, catching a baseball barehanded (Figure 25-1a) hurts less when your hand is moving away from the ball than when it is moving toward it, because the velocity of the ball *relative* to your hand is less in the first case than in the second. Newtonian physics implies that the speed of a wave should also be different relative to different observers. Maxwell's work, however, implied that the speed of light detected by an observer is $c = 3.0 \times 10^8$ m/s, regardless of how the observer is moving. This was confirmed by the Michelson-Morley experiment, but it violates what classical physics and our everyday experience tell us about relative velocity. The speed of light relative to your eye (Figure 25-1b) does *not* depend on whether your eye is moving toward or away from the source.

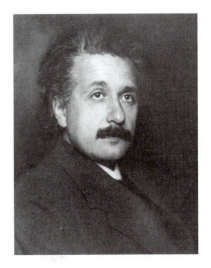

➡Albert Einstein (1879–1955) was Born in Ulm, Germany. While working in obscurity in the Swiss patent office in 1905, he wrote three papers that have become landmarks in the history of physics, one of which introduced his theory of special relativity. Fame followed, as did prestigious positions in German universities. When the Nazis came to power in 1932, he was a visiting scholar at the Institute for Advanced Study in Princeton. Einstein, who was of Jewish background, never returned to Germany. He became an American citizen in 1940.

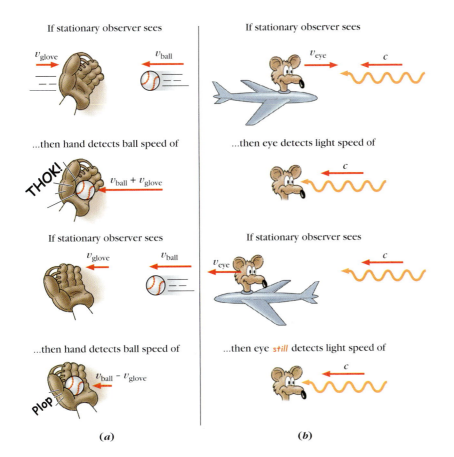

(a)

(b)

Figure 25-1 Newtonian or Galilean relativity is obeyed by everyday objects but not by light.

Newtonian physics also places no theoretical limit on how fast anything may move. But nothing has ever been observed moving faster than 3.0×10^8 m/s. Even when an electron is accelerated to over 99% of the speed of light in a high-energy particle accelerator, any further input energy only brings the electron closer to the speed of light, but its speed never reaches or exceeds c.

We are obliged by this evidence to revisit what we know about velocities. Our determinations of velocities are based on

• position readings on a device like a meter stick, from which we find Δx, and
• clock readings, from which we find Δt.

Up to this point, we have never questioned whether these readings will be the same to all observers. Neither did Newton. In his *Principia*, Newton argued for the existence of **absolute time** and **absolute space**:

> I. Absolute, true and mathematical time, of itself, and from its own nature, flows equably and without relation to anything external, and by another name is called duration. Relative, apparent, and common time is some sensible [i.e., detected by the senses] and external . . . measure of duration by means of motion . . . such as an hour, a day, a month, a year.
>
> II. Absolute space, in its own nature, without relation to anything external, remains similar and immovable. Relative space is some moveable dimension or measure of the absolute spaces, which our senses determine relative to bodies, and which is commonly taken for immovable space.

These commonsense assumptions must come under scrutiny because they lead to Newtonian kinematics and Newtonian relativity (also called Galilean relativity) in which the observed velocity of an object must differ for observers moving relative to one another.

Let us think about things happening in space and time in the following way. We will define an **event** to be anything that happens at a particular point in space at a particular instant. Locating an event in space and time (what we will later call space-time) requires a set of coordinate axes and a clock, which together make up a **reference frame.** It will be useful to distinguish between **inertial reference frames,** in which Newton's first law of motion (the law of inertia) is valid, and **noninertial reference frames,** in which it is not. The following case helps make the distinction.

Case 25-1 ◆ *Inertial and Noninertial Reference Frames, or Who's "Really" Accelerating*

Each of the students in Figure 25-2*a* is the origin of a coordinate system, consisting of two meter sticks held perpendicular to each other, that moves along with the student. Each student also carries a watch. Each student's watch and meter sticks constitute a *reference frame* in which he or she can describe the motion of the other student.

In the sequence shown in Figure 25-2*b,* student B moves to the right relative to student A, and A moves to the left relative to B. The sequence of pictures doesn't tell us whether A is really moving or B is really moving or both. Suppose the students can see only what is shown in Figure 25-2*a.* Everything else is in total darkness so they can see no shifting of position against a backdrop. **STOP&Think** Can the students tell which of them is really moving? In fact, what, if anything, does "really" mean here? ◆

In Figure 25-2*c,* the distance between them grows by a bigger amount each second. Each student sees the other accelerating. Just by looking at each other's motion, neither student can say which one of them is really accelerating; that is, neither one has physically meaningful grounds for saying, "I am accelerating, not you." But they *can* resolve this question if each of them carries a weight on a string. Suppose that A's weight hangs straight down (Figure 25-2*d*), whereas B's hangs at an angle. Then A can reason, "B and her hanging weight are accelerating toward my right. There is a component of the tension toward the right, so there is a force on the weight in the direction of its acceleration. The law of inertia is obeyed in my reference frame, so mine is an **inertial reference frame.** On this basis, I conclude that B is 'really' accelerating."

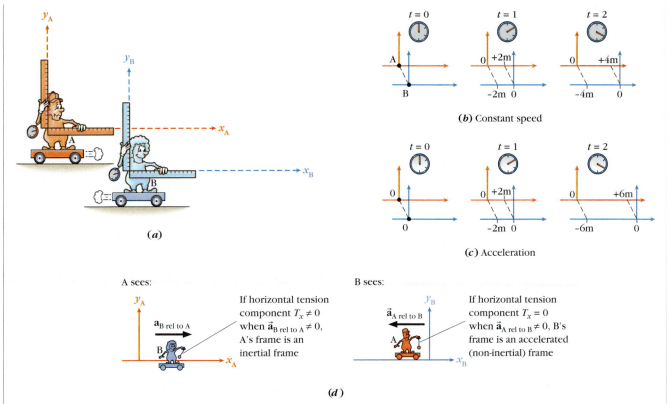

(a)

(b) Constant speed

(c) Acceleration

A sees:

If horizontal tension component $T_x \neq 0$ when $\vec{a}_{\text{B rel to A}} \neq 0$, A's frame is an inertial frame

B sees:

If horizontal tension component $T_x = 0$ when $\vec{a}_{\text{A rel to B}} \neq 0$, B's frame is an accelerated (non-inertial) frame

(d)

Figure 25-2 Inertial and noninertial reference frames.

On the other hand, B sees A's weight hanging straight down. B sees the weight accelerating toward his left, but there is no force component exerted on it toward the left. B concludes, "I'm observing an acceleration without an external force. That violates Newton's first law, so mine is a *noninertial* reference frame. I must be the one who is accelerating, and my reference frame is accelerating with me." *A noninertial frame is an accelerated reference frame.*

Now suppose that the two students move at a *constant* nonzero velocity relative to each other. The first law tells us that no net external force is required to keep an object moving at constant velocity. If one weight hangs straight down, so does the other; they no longer provide a way of distinguishing between the inertial frames. If the first law is obeyed by one of the two frames, it is also obeyed by the other, and *both are inertial frames. Any reference frame moving at constant velocity relative to an inertial frame is also an inertial frame.* Also, because there is no way the students can tell if either frame "really" has zero velocity, we must conclude that *there is no preferred inertial frame.*

In other words, there is no "really" here. It is simply a matter of convenience to say, "I want to treat this object's velocity as zero," even though its velocity is not zero in a different inertial frame, just as it is a matter of convenience to say, "I will let $y = 0$ at ground level." Choosing an inertial frame in which a particular object has zero velocity is as arbitrary as picking the origin of a coordinate system so that the position of a particular object is zero.

To explore the ideas of Case 25-1 further, work through WebLink 25-1.

For **WebLink 25-1:**
Which Elevator Is Accelerating?, go to
www.wiley.com/college/touger

Example 25-1 *Who's Rotating?*

A high school physics student is in the amusement park ride shown here. The center post appears to her to be rotating at a constant rapid angular speed. She wonders whether this is because the center post is really rotating or because the outer cylinder that she is standing against is rotating the other way around the post. How can she tell?

EXAMPLE 25-1 *continued*

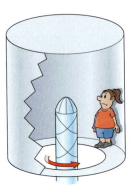

Solution

Choice of approach. We apply Newton's first law, keeping in mind that motion in a circle involves a change in direction and is therefore accelerated motion.

The detailed reasoning. If an object is moving in a circle, there must be a force toward the center of the circle producing the radial acceleration $a = \frac{v^2}{r} = r\omega^2$, which is nonzero even when the angular speed ω is constant. The force toward the center would in this case be the normal force that the cylindrical wall exerts on the student. By Newton's third law, the student exerts an equal and opposite normal force on the wall. When the wall is exerting a normal force, the student describes the experience by saying she feels pressed against the wall. If she does not, it is the center post and not the cylindrical wall that is rotating.

Making sense of the results. A reference frame moving at constant *linear* velocity relative to an inertial frame is also an inertial frame, but a reference frame moving at constant *angular* velocity relative to an inertial frame is not. By "relative to an inertial frame," we mean the quantity is measured using the meter sticks and clock of that inertial frame.

◆ Related homework: Problems 25-4 through 25-8.

Einstein's theory of **special relativity,** which he formulated in 1905 while working in the Swiss patent office in Bern, is limited to what happens in inertial reference frames. In 1916, he formulated a theory of **general relativity,** taking into account accelerated systems. Here we address the special theory.

Einstein assumed that just as Newton's first law must hold true in all inertial frames, so must the other laws of physics, so that different observers will see the same effects as due to the same causes. This was already known to be true in mechanics at ordinary speeds. For example, suppose you are riding a train traveling at constant speed v_t (Figure 25-3a). You see a tennis ball fall vertically from the luggage rack, and it hits you on the head. A viewer standing beside the track would see both you and the ball moving forward with a speed v_t. Her description is different, but Newton's laws of motion and the law of gravitation must

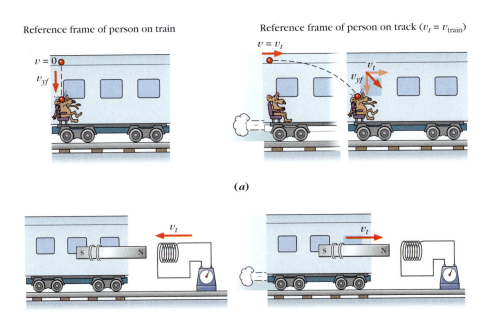

(a)

(b)

Figure 25-3 In different inertial frames, the same events must be due to the same causes.

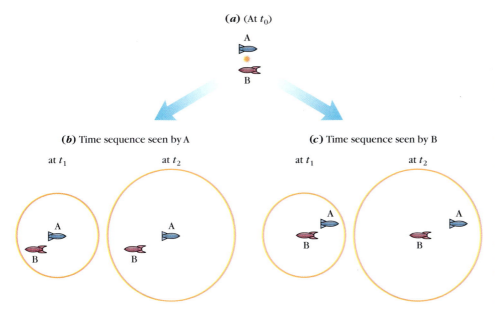

(a) (At t_0)

A

B

(b) Time sequence seen by A

at t_1 at t_2

A
B

A
B

(c) Time sequence seen by B

at t_1 at t_2

A
B

A
B

Figure 25-4 Relative motion if all observers see light traveling at the same speed. In (*a*), rockets A and B pass each other going in opposite directions at great speed. At t_0, the instant they pass each other, they set off a small explosion in the tiny space between them, producing a flash of light. (*b*) Subsequent motion of B and of the circular wavefront spreading from the flash, as seen in A's reference frame. (*c*) Subsequent motion of A and of the wavefront, as seen in B's reference frame. (Adapted from R. Scherr, P. Shaffer, and S. Vokos, *American Journal of Physics,* 70, (2002), 1238–48).

produce the same result for her as for you. If they lead to the conclusion that the ball hits you on the head in your reference frame, they cannot lead to the conclusion that the ball misses you in hers. You cannot escape to a reference frame in which you avoid getting conked.

Importantly, Einstein held that the laws of electricity and magnetism must also produce the same results for all observers. For example, the same deflection is produced on the galvanometer whether the magnet mounted on the front of the train approaches the coil (Figure 25-3*b*) or (as you, the rider, would see it) the coil approaches the magnet. The invariance of the laws of electricity and magnetism in turn made it necessary that the speed of electromagnetic waves be the same for all observers.[1] As the foundation of his special theory, Einstein therefore set down two postulates:

• **Postulate 1.** *The laws of physics are the same in every inertial reference frame.* (Einstein called this the **principle of relativity**)

• **Postulate 2.** *The speed of light in a vacuum is independent of the observer—its measured value is always $c = 3.0 \times 10^8$ m/s, no matter how the source or the observer may be moving.*

Figure 25-4 shows the implications of postulate 2. Although the motions of the two rockets are different in one rocket's reference frame than in the other, the motion of the wave front produced by the flash of light is the same to observers in both frames.

Although Einstein's theory is usually called *relativity,* these are postulates not about what differs relative to different observers but about what remains the same for all observers. At the heart of Einstein's theory was his conviction that the laws of nature must be universal.

The second postulate was implied by a few experimental outcomes, but Einstein assumed it to be universally valid and considered the consequences. In particular, he reasoned, if light travels at the same speed for all observers, we must revise some of our fundamental ideas about time.

---

[1]In the 1905 paper in which he presented this argument, Einstein also asserted that it was superfluous to assume the existence of an ether. He therefore did not consider the outcome of the Michelson-Morley experiment to be surprising—for him it was the expected result.

25-2 It's About Time

We have defined an *event* as something occurring at a particular point in space at a particular instant—every event must be describable by a single set of coordinates (x, y, z) and a single value of t. The following examples qualify as events: a collision of two point objects (Figures 25-5a and c), a spark between two electrodes a negligible distance apart, a particular crest of a wavetrain passing some fixed position. Anything extending over a time *interval* or having non-negligible extent in space is *not* an event (Figures 25-5b and d).

Our conventional way of thinking about things is that events either happen at the same instant (are *simultaneous*) or they don't. Einstein, however, deduced that if light travels at the same speed in all inertial reference frames, then events that are simultaneous for one observer may not be simultaneous for another. The following thought experiment (Case 25-2) demonstrates this point.

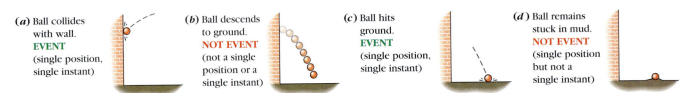

(a) Ball collides with wall.
EVENT
(single position, single instant)

(b) Ball descends to ground.
NOT EVENT
(not a single position or a single instant)

(c) Ball hits ground.
EVENT
(single position, single instant)

(d) Ball remains stuck in mud.
NOT EVENT
(single position but not a single instant)

Figure 25-5 Distinguishing events from nonevents.
This assumes the ball is a point object.

Case 25-2 ◆ *"At the Same Time" Is a Relative Statement*

Figure 25-6a shows two trains T and T' passing each other on parallel tracks. Each train has a velocity of magnitude v relative to an observer standing on the other train. Each train has two electrodes sticking out, and there is a spark whenever two electrode tips pass infinitesimally close together.

Observer O on train T positions himself midway between electrodes A and B. Observer O' on train T' positions herself midway between electrodes A' and B'. Both observe the occurrence of two events:

• **Event AA':** Electrode A passes infinitesimally close to electrode A'.

• **Event BB':** Electrode B passes infinitesimally close to electrode B'.

Each event produces a spark, a light signal of negligible duration.

Suppose the two events are simultaneous for observer O. Consider the sequence of events that follow events AA' and BB'. Figure 25-6 shows these events as observer O would see them. In O's reference frame, the pulses from AA' and BB' reach O simultaneously, but the pulse from BB' reaches observer O' before the pulse from AA' does.

Now suppose that events AA' and BB' are also simultaneous for observer O'. Consider what

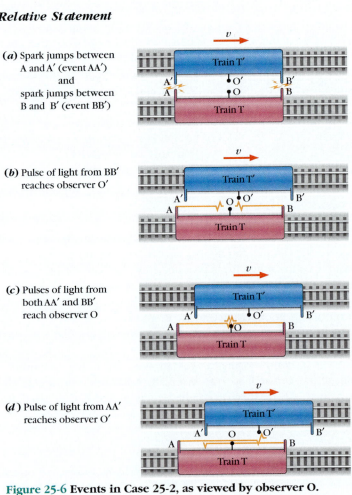

(a) Spark jumps between A and A' (event AA') and spark jumps between B and B' (event BB')

(b) Pulse of light from BB' reaches observer O'

(c) Pulses of light from both AA' and BB' reach observer O

(d) Pulse of light from AA' reaches observer O'

Figure 25-6 Events in Case 25-2, as viewed by observer O.

happens in her reference frame. O' is midway between the electrodes on her train, where the two events occurred. Because light travels at the same speed c from both electrodes, the pulses must reach her at the same time.

Thus, if AA' and BB' are simultaneous for both observers, then (1) the two pulses reach O' simultaneously in her reference frame, but (2) one pulse reaches her before the other in O's reference frame. But (1) and (2) cannot both be true. Both pulses arrive at the same place, and events at the same position must occur in the same order for all observers, otherwise Einstein's first postulate would be violated. Let's see why.

Suppose the pulse from BB' activates a process on the train that produces a fast-acting poisonous gas, and the pulse from AA' activates a vent that disposes of the gas. If the pulse from BB' reaches O' first, the gas will build up before it can be vented and poison O'. But if the pulse from AA' reaches her at the same time or before, there will be adequate venting in time, and O' won't be poisoned. Now, either she is poisoned or she is not; it can't be both ways. This is what Einstein meant in postulating that the laws of physics must be the same in all inertial reference frames.

This compels us to conclude that events AA' and BB' cannot have been simultaneous for O'. O' must receive the pulse from BB' first in her own reference frame as well, but because the two events occurred at the same distance from her, it means that for her BB' must have happened first.

If two events were simultaneous for observer O' instead, we could go through similar reasoning, but treating train T' as stationary so that relative to train T', train T moves to the left. **STOP&Think** If events AA' and BB' were simultaneous according to observer O', which of those two events would occur first according to observer O? ◆

In either case we draw the same conclusion:

If two events at a distance Δx from each other are simultaneous for an observer in one inertial frame, they will <u>not</u> be simultaneous to a second observer whose inertial reference frame is moving relative to the first in the x direction.

The event that the moving observer views as further forward will happen first for the moving observer. For example, if the moving observer is traveling toward the right, she will conclude that the event that takes place further to the right happens first.

Simultaneity is relative to the observer. Just as there is no preferred reference frame for $v = 0$, there is no preferred reference frame for simultaneity.

◆**SYNCHRONIZING CLOCKS** Within each reference frame it is possible to synchronize clocks (Figure 25-7). **STOP&Think** What time does clock A read at the instant the light signal reaches clock B? ◆ A similar procedure can be used to establish the repeated time interval on which a clock is based. We will call such a time interval a *tick,* like the repeated time interval of a traditional clock. Figure 25-8a (next page) sketches the procedure: A pulse of light emitted from the clock at $t = 0$ reflects off a plane mirror and back to the clock. When the reflected pulse is detected by the clock, the clock "ticks" (its hand advances by one division) and it emits a new pulse, repeating the process.

We now compare what happens (1) when the set-up is stationary and (2) when it moves at a horizontal velocity v relative to some observer. The set-up is aligned perpendicular to the motion, so that the vertical distance L_o is the same for both observers. **STOP&Think** Why not align the set-up along the direction of motion? To determine the clock-to-mirror distance, we must find their positions at the same instant. This must be done in each inertial frame. But simultaneity is relative. Suppose the clock is ahead of the mirror in the direction of motion. If an observer moving with them says the clock passes some point P at the same instant that the mirror passes another point Q, would a stationary

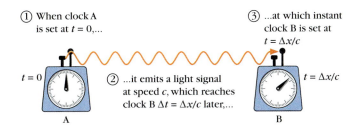

① When clock A is set at $t = 0$,...

② ...it emits a light signal at speed c, which reaches clock B $\Delta t = \Delta x/c$ later,...

③ ...at which instant clock B is set at $t = \Delta x/c$

$t = 0$

$t = \Delta x/c$

A

B

Figure 25-7 Synchronizing two clocks.

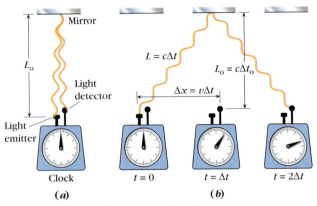

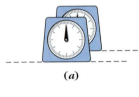

Both clocks set simultaneously at same place at $t = t' = 0$.

(a)

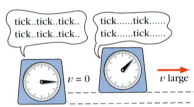

tick..tick..tick..
tick..tick..tick..

tick......tick......
tick......tick......

$v = 0$ v large

Stationary clock   Moving clock

Clock advances | Because each tick
one division | takes longer, moving
with each tick. | clock ticks fewer times.

(b)

Figure 25-9 "Moving clocks run slow."

Figure 25-8 Set-up for comparing time intervals in two inertial frames. In (a) the mirror and clock are stationary. In (b) they are moving to the right at constant speed v.

➦**Note:** So far we have avoided describing either observer as stationary, because both observers can be stationary relative to their own reference frames.

observer agree? Could they then agree about the clock-to-mirror distance? ◆

Suppose now that when the set-up is stationary (that is, the observer moves with it), it takes light Δt_o to travel the distance L_o from the clock to the mirror, and another Δt_o to return, so "one tick" = 2 Δt_o. For an observer who sees the set-up as moving, however, it takes light Δt to travel the distance L (Figure 25-8b), and "one tick" = 2 Δt. Then $L_o = c\Delta t_o$ and $L = c\Delta t$. We can therefore apply the Pythagorean theorem to the right triangle in Figure 25-8b to find an expression that compares Δt_o and Δt:

$$(c\Delta t)^2 = (v\Delta t)^2 + (c\Delta t_o)^2$$

Then
$$(\Delta t)^2(c^2 - v^2) = c^2(\Delta t_o)^2$$

so
$$\Delta t = \sqrt{\frac{c^2}{c^2 - v^2}}\,\Delta t_o$$

or
$$\Delta t = \frac{\Delta t_o}{\sqrt{1 - v^2/c^2}} \tag{25-1}$$

The denominator $\sqrt{1 - v^2/c^2}$ occurs frequently in special relativity. It equals one when $v = 0$, and gets smaller as v increases. So for $v \neq 0$, Δt is always greater than Δt_o. **STOP&Think** What would happen to this denominator if v could be greater than c? ◆

The time interval Δt_o is called the **proper time interval** between two events. It is the time interval measured by an observer who sees the events occur at the same x position, assuming the x direction is the direction of motion. For example, Figure 25-8 shows a sequence of events—the emission, reflection, and detection of the pulse of light—as seen by two different observers. For the observer who sees the sequence as it appears in Figure 25-8a, these three events all occur at the same x position. The time interval that this observer measures between these two events is therefore the proper time interval between them. Equation 25-1 says, *If an observer sees the device recording proper time interval as moving, the interval Δt measured by this observer is always greater than the proper time interval.* This phenomenon is called·**time dilation** (*dilate* means expand).

A consequence of Equation 25-1 is that **"moving clocks run slow"** by a factor $1/\sqrt{1 - v^2/c^2}$. Figure 25-9 shows what this means. Suppose we use the label "stationary" for an observer in a **proper reference frame;** that is, the reference frame in which the proper time interval is measured. If each tick $2\Delta t$ of the moving clock takes longer than if the clock were stationary, then a stationary observer will hear fewer ticks during a given interval. That means the hand of the clock is advancing by fewer divisions—it is "running slow."

Example 25-2 *Time Dilation at "Relativistic" and "Nonrelativistic" Speeds*

For a guided interactive solution, go to Web Example 25-2 at www.wiley.com/college/touger

a. Cosmic rays are high-energy particles that reach Earth from the sun and other stars. In cosmic rays, physicists discovered the existence of elementary particles other than protons, neutrons, and electrons, virtually all of them unstable. One, called a *muon,* has an average lifetime of 2.2×10^{-6} s when at rest. It may come into existence as a product of the spontaneous decay of another short-lived particle called a *pion.* Then the muon itself spontaneously decays into an electron, a neutrino, and an antineutrino. The lifetime is the interval between these two *events*—the two decays. What is the average lifetime for muons in cosmic rays reaching the earth at a speed of 0.85c? What percent increase is this?

b. The pilot of a jet plane traveling at 300 m/s, nearly the speed of sound, finds herself out of radio contact with the ground-based air traffic controller for a 200-s interval. How long does the controller think the pilot is out of contact? What percent increase is this?

Brief Solution

Choice of approach. In both parts, we are being asked to compare the time interval detected by an Earthbound observer to the proper time interval. (For example, "the pilot stops detecting radio signals" is one event; "she resumes detecting them" is another. In her reference frame these two events occur at the same position, so hers is the *proper reference frame*. **STOP&Think** These two events are comparable to the two decays that bracket the lifetime of the muon. In what reference frame do those two decays occur at the same position? ◆ It is easier to answer the second part of each question first, because $\Delta t/\Delta t_o = 1/\sqrt{1 - v^2/c^2}$. If, for example, we were to get $\Delta t/\Delta t_o = 1.25$, that would tell us that the Δt exceeded Δt_o by $0.25\Delta t_o$, or 25% of Δt_o. Once we find the factor $1/\sqrt{1 - v^2/c^2}$, we can multiply Δt_o by this factor to obtain Δt.

What we know/what we don't.

$$c = 3.00 \times 10^8 \text{ m/s}$$

a. proper time interval $\Delta t_o = 2.2 \times 10^{-6}$ s   **b.** proper time interval $\Delta t_o = 200$ s

$v = 0.85c$ $\dfrac{1}{\sqrt{1 - v^2/c^2}} = ?$ $v = 300$ m/s $\dfrac{1}{\sqrt{1 - v^2/c^2}} = ?$

% increase = ? $\Delta t = ?$ % increase = ? $\Delta t = ?$

The mathematical solution.

a. The factor

$$\frac{1}{\sqrt{1 - v^2/c^2}} = \frac{1}{\sqrt{1 - (0.85c)^2/c^2}} = \frac{1}{\sqrt{1 - (0.85)^2}} = 1.90$$

so the lifetime increases by $0.90\Delta t_o$, or **90%** of Δt_o.

Then $\Delta t = \Delta t_o\left(\dfrac{1}{\sqrt{1 - v^2/c^2}}\right) = (2.2 \times 10^{-6} \text{ s})(1.90) = \textbf{4.2} \times \textbf{10}^{-6}$ **s**

b. Now

$$\frac{1}{\sqrt{1 - v^2/c^2}} = \frac{1}{\sqrt{1 - (3.00 \times 10^2 \text{ m/s})^2/(3.00 \times 10^8 \text{ m/s})^2}}$$

$$= \frac{1}{\sqrt{1 - (1.00 \times 10^{-6})^2}} \approx 1.000000000005$$

so the time interval increases by $0.000000000005 \, \Delta t_o = 5 \times 10^{-12} \, \Delta t_o$,

or $\mathbf{5 \times 10^{-10}\%}$ **of** Δt_o

and $\Delta t = \Delta t_o(1.000000000005) = \textbf{200.000000001 s}$

◆ Related homework: Problems 25-15, 25-16, 25-17, and 25-18.

Example 25-2 shows that at speeds that are only a very small fraction of the speed of light, the difference between the time intervals observed by the moving and stationary observers is negligible. Speeds for which the difference is negligible are called *nonrelativistic speeds,* in contrast to *relativistic speeds,* at which the difference becomes significant. At nonrelativistic speeds, the differences between the values given by Newtonian physics and special relativity are too small to be detectable, and Newtonian physics continues to serve us well.

If we want three-place accuracy, a reasonable cutoff is to say that speeds below $0.1c$ are nonrelativistic. A useful approximation at nonrelativistic speeds is

$$\frac{1}{\sqrt{1 - v^2/c^2}} \approx 1 + \frac{1}{2}\frac{v^2}{c^2} \qquad \text{(when } v \ll c\text{)} \qquad (25\text{-}1a)$$

We could have used this to find the numerical value of $1/\sqrt{1 - v^2/c^2}$ in Example 25-2b. (See Problem 25-14.)

25-3 ◆ Length Contraction and Other Spatial Considerations

The measurement of a distance or a length involves *two events*—finding the positions of the endpoints x_1 and x_2—at the same time. Suppose observers in two different inertial frames each keep a record of where the endpoints are at every instant. Each position-time entry is an event. If they disagree about which two of these events are simultaneous, they will also disagree about the resulting length $\Delta x = x_2 - x_1$.

To see what consequences this has, let's return to the trains in Figure 25-6. Each train moves at a speed v relative to the other. First let's compare the way observers O and O′ find the length of train T′ (Figure 25-10).

Suppose O has a clock. O records the instant when electrode B′ passes him, then the instant when electrode A′ passes him, and finds the difference Δt_o between these two clock readings. (For O the two events occur at the same position—the location of his clock—so Δt_o is the *proper time interval*.) He concludes that the length of train T′ is $\Delta x = v\Delta t_o$. He bases his conclusions on two events, electrode A′ passing his clock and electrode B′ passing his clock. For O′, however, the length of train T′ is the **proper length** Δx_o, by which we mean it is the length determined by an observer who sees it as stationary. Because it is stationary for her, she can measure this length whenever she likes with a meter stick.

Figure 25-10 shows the two events that both observers consider in determining the length of train T′. Let's now look at the reasoning that enables O and O′ to *compare* the lengths that they obtain.

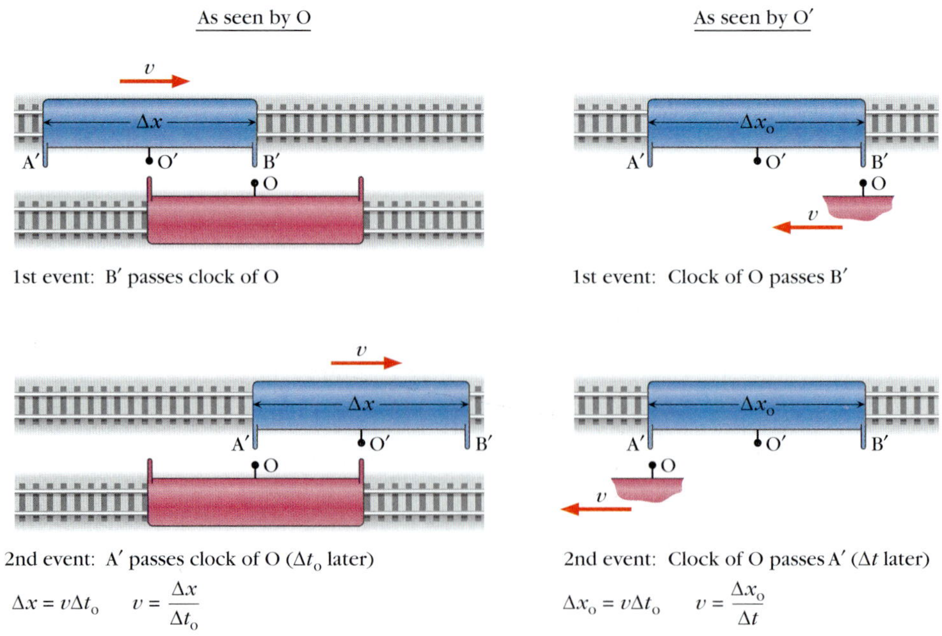

| As seen by O | As seen by O′ |

1st event: B′ passes clock of O

1st event: Clock of O passes B′

2nd event: A′ passes clock of O (Δt_o later)

$$\Delta x = v\Delta t_o \qquad v = \frac{\Delta x}{\Delta t_o}$$

2nd event: Clock of O passes A′ (Δt later)

$$\Delta x_o = v\Delta t_o \qquad v = \frac{\Delta x_o}{\Delta t}$$

Figure 25-10 Observers O and O′ consider the length of the blue train T′.

How Observer O Reasons	How Observer O′ Reasons
I know the speed v at which the train T′ is moving to my right.	I know the speed v at which the train T is moving to my left.
I consider two events: • first electrode B′ passes me (event OB′) • then electrode A′ passes me (event OA′)	I consider two events: • first observer O passes electrode B′ (event OB′) • then observer O passes electrode A′ (event OA′)
My clock tells me these events are Δt_o apart (they occur at the same position in my inertial frame, so this is the proper time interval).	My clock advances by Δt between the two events (they don't occur at the same position in my inertial frame, so this is not the proper time interval).
I calculate Δx between the two events: $\Delta x = v\Delta t_o$.	Because electrodes A′ and B′ are stationary in my inertial frame, I can use my meter stick to measure Δx_o, the proper length of train T′.
I now know v, Δt_o, and Δx.	I now know v, Δt, and Δx_o. Because O travels the distance Δx_o during the time interval Δt, I can use these values to verify that $\Delta x_o = v\Delta t$.

To compare the values Δx and Δx_o, we use the fact that the time interval that O′ measures is not the proper time interval Δt_o but $\Delta t = \Delta t_o(1/\sqrt{1 - v^2/c^2})$. The ratio of the lengths found by the two observers is therefore

$$\frac{\Delta x}{\Delta x_o} = \frac{v\Delta t_o}{v\Delta t} = \frac{\Delta t_o}{\Delta t_o\left(\dfrac{1}{\sqrt{1 - v^2/c^2}}\right)} = \sqrt{1 - v^2/c^2}$$

or
$$\Delta x = \Delta x_o\sqrt{1 - v^2/c^2} \tag{25-2}$$

Because the factor $\sqrt{1 - v^2/c^2}$ is less than 1 whenever $v > 0$, this equation says that the observer for whom the length is moving sees a **length contraction** compared to the proper length Δx_o.

STOP&Think In Case 25-2, were the two trains actually equal in length? Does Figure 25-6 show the proper lengths of the trains? ◆ In fact, each part of the figure shows the trains as viewed by an observer for whom the two "electrodes meeting" events AA′ and BB′ are simultaneous. (Be sure to distinguish between those events and the events OA′ and OB′ that we just considered.) What clues does that provide to the proper lengths of the trains? The following example addresses these questions.

Example 25-3 *The Long and Short of It*

Imagine the trains in Case 25-2 to be capable of relativistic speeds. Suppose that the two sparks (events AA′ and BB′) occur simultaneously for observer O, who sees train T′ going past him at 1.20×10^8 m/s. If train T is 80.0 m long when standing still, what would be the *proper length* of train T′?

Solution

Choice of approach. Because O sees events AA′ and BB′ as simultaneous, the difference between the positions of these two events is both the proper length of his own train and the *contracted* length of train T′ that he sees. We can therefore use Equation 25-2 to find the proper length of train T′.

EXAMPLE 25-3 continued

What we know/what we don't.

$$c = 3.00 \times 10^8 \text{ m/s}$$

$$v = 1.20 \times 10^8 \text{ m/s} \qquad \Delta x = 80.0 \text{ m} \qquad \text{proper length } \Delta x_o = ?$$

The mathematical solution. From Equation 25-2, we get

$$\Delta x_o = \frac{\Delta x}{\sqrt{1 - v^2/c^2}}$$

$$= \frac{80.0 \text{ m}}{\sqrt{1 - (1.20 \times 10^8 \text{ m/s})^2/(3.00 \times 10^8 \text{ m/s})^2}} = \mathbf{87.3 \text{ m}}$$

Making sense of the results. **STOP&Think** Does this result contradict Figure 25-6, in which the lengths of the two trains appear equal? ◆ There is no contradiction. It is the proper lengths of the two trains that are unequal. If the trains are in motion relative to each other, no one view can show the proper lengths of both trains at once. To appear equal in length, the trains must be shown from the point of view of the observer (in this case O) for whom the two pairs of endpoints meet simultaneously (Figure 25-11a). But then the endpoints do not meet simultaneously for observer O' (Figure 25-11b). In fact, O' can only see end B meet end B' before A meets A' if O' observes train T to be shorter than train T'.

O sees

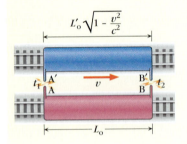

$$L_o'\sqrt{1 - \frac{v^2}{c^2}} = L_o \text{ (so } L_o' > L_o)$$

$t_1 = t_2$ (A meets A' when B meets B')

(a)

Figure 25-11 Figure 25-6 revisited to show the role of length contraction.

O' sees

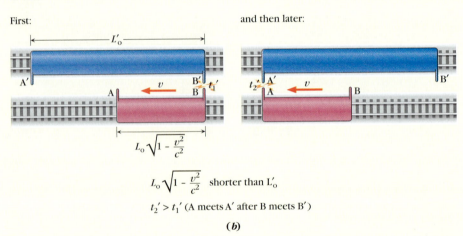

$L_o\sqrt{1 - \dfrac{v^2}{c^2}}$ shorter than L_o'

$t_2' > t_1'$ (A meets A' after B meets B')

(b)

In contrast, if the two proper (rest) lengths were equal, each observer would see the other train as contracted (shorter than his or her own) when they moved past each other, and neither could possibly see events AA' and BB' as simultaneous.

◆ Related homework: Problems 25-23, 25-24, 25-25, 25-28, and 25-29.

◆**WHAT STAYS THE SAME?** Because the speed of light is **invariant**—it remains unchanged from one inertial reference frame to another—time intervals and distances in the direction of relative motion must be altered. What else remains the same? If we apply the Pythagorean theorem to the right triangle in Figure 25-8b, we get

$$(\Delta x)^2 + (L_o)^2 = (c\Delta t)^2 \qquad \text{or} \qquad (\Delta x)^2 - (c\Delta t)^2 = -(L_o)^2$$

Because the proper distance L_o is a constant, the right side of the second equation is invariant, so the left side must also be, even though its terms are not individually invariant. In other words, although $\Delta x \neq \Delta x'$ and $\Delta t \neq \Delta t'$,

$$(\Delta x)^2 - (c\Delta t)^2 = (\Delta x')^2 - (c\Delta t')^2 \qquad (25\text{-}3a)$$

Compare this to what remains invariant in nonrelativistic physics. In Figure 25-12, the length d remains the same in both frames. The Pythagorean theorem tells us that $d^2 = (\Delta x)^2 + (\Delta y)^2$ in one reference frame, and $d^2 = (\Delta x')^2 + (\Delta y')^2$ in the other. Therefore $(\Delta x)^2 + (\Delta y)^2 = (\Delta x')^2 + (\Delta y')^2$. By similar reasoning, $d^2 = (\Delta x)^2 + (\Delta y)^2 + (\Delta z)^2$ is invariant in three dimensions in nonrelativistic physics. In relativistic physics in three dimensions, this must replace the squared length $(\Delta x)^2$ in Equation 25-3a, giving us

$$(\Delta x)^2 + (\Delta y)^2 + (\Delta z)^2 - (c\Delta t)^2 = (\Delta x')^2 + (\Delta y')^2 + (\Delta z')^2 - (c\Delta t')^2 \quad \text{(25-3b)}$$

In contrast to the nonrelativistic three-dimensional invariance relationship, we have had to include the square of a fourth distance $c\Delta t$ or $c\Delta t'$ on either side. This is like considering a *fourth spatial dimension*. The quantity $c\Delta t$, the distance light travels in a given time interval, is as much a measure of time as Δt. It is in this sense that we speak of time as a fourth dimension in relativity theory. Just as $d^2 = (\Delta x)^2 + (\Delta y)^2 + (\Delta z)^2$ is an invariant measure (when we take its square root) of the distance between two points in three dimensional space when $v \ll c$, we may say,

At relativistic speeds $(\Delta x)^2 + (\Delta y)^2 + (\Delta z)^2 - (c\Delta t)^2$ *is an invariant measure of the distance between two points (two events) in* **four-dimensional space-time.**

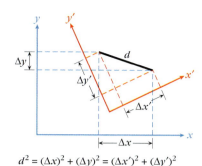

$$d^2 = (\Delta x)^2 + (\Delta y)^2 = (\Delta x')^2 + (\Delta y')^2$$

Figure 25-12 Nonrelativistic invariance in two dimensions.

➥There is nothing mystical or otherworldly about a fourth dimension. A point in ordinary space is fully described by three spatial coordinates, but an *event* requires four coordinates (x, y, z, t). Mathematically, that is what it means to be four-dimensional.

Example 25-4 *Those Trains Again*

Based on the given data and results of Example 25-3, when events AA′ and BB′ occur simultaneously for observer O, how long after event BB′ does event AA′ occur for observer O′?

Solution

Interpreting the question. Because O sees events AA′ and BB′ as simultaneous ($\Delta t = 0$), the difference between the positions of these two events is both the proper length of his own train and the contracted length Δx that he sees for train T′. In Example 25-3, we used length contraction to find the proper length $\Delta x'$ of train T′. We are now asked to find $\Delta t'$ between the two events.

Choice of approach. We make use of the fact that there is a measure of the distance between two events in space-time that is the same in both inertial frames (Equation 25-3a).

What we know/what we don't.

$$c = 3.00 \times 10^8 \text{ m/s} \qquad v = 1.20 \times 10^8 \text{ m/s}$$

$\Delta x = 80.0$ m $\qquad$ proper length $\Delta x' = 87.3$ m (solution to Example 25-3)

$\Delta t = 0 \qquad\qquad \Delta t' = ?$

The mathematical solution. With $\Delta t = 0$, the equation

$$(\Delta x)^2 - (c\Delta t)^2 = (\Delta x')^2 - (c\Delta t')^2$$

becomes $\qquad (\Delta t')^2 = \dfrac{(\Delta x')^2 - (\Delta x)^2}{c^2}$

$$= \frac{(87.3 \text{ m})^2 - (80.0 \text{ m})^2}{(3.00 \times 10^8 \text{ m/s})^2} = 1.36 \times 10^{-14} \text{ s}^2$$

and $\qquad \Delta t' = \mathbf{1.16 \times 10^{-7} \text{ s}}$

◆ Related homework: Problems 25-27 and 25-30.

25-4 Momentum, Mass, and Energy in Relativistic Physics

In this section we ask: If the speed of light c is invariant, what are the consequences for momentum, mass, and energy? We will first look at its implications for an elastic collision between identical objects. Classically, when this happens, momentum is conserved, and if the collision is one-dimensional they simply exchange velocities. The following case shows what happens relativistically.

Case 25-3 ◆ An Elastic Collision at Relativistic Speed

Two ice boats pass each other, each moving at a relativistic speed v relative to the other (Figure 25-13). Each boat's crew launches a puck toward the other boat, and the pucks (A and B) collide elastically midway between the two boats, a distance d from each. To each crew, its own puck is initially moving perpendicular to the motion of the boats at a nonrelativistic speed u. The y component of each puck's velocity after the collision is equal and opposite to what it was before.

Each crew can measure the amount of time it takes puck A to slide the distance d. The two events that frame this time interval, "the puck starts traveling this distance" and "the puck finishes traveling this distance," occur at the same x position in crew A's inertial frame. The time interval crew A measures is therefore the *proper time interval* Δt_o that it takes puck A to go the distance d. Then in crew A's reference frame, puck A's initial velocity $u = \dfrac{d}{\Delta t_o}$.

Likewise, the proper time interval for puck B to travel a distance d is the time interval measured by crew B. For crew A, this time interval is $\Delta t = \Delta t_o / \sqrt{1 - v^2/c^2}$. During this interval, crew A sees puck B undergo a displacement $-d$ in the y direction, so in crew A's reference frame, the y component of puck B's velocity is

$$(u_{B1})_y = \frac{\Delta y}{\Delta t} = \frac{-d\sqrt{1 - v^2/c^2}}{\Delta t_o} = -u\sqrt{1 - v^2/c^2}$$

Because the y components of the two velocities reverse after the collision,

$$(u_{A2})_y = -u \quad \text{and} \quad (u_{B2})_y = +u\sqrt{1 - v^2/c^2}$$

Classically, the momentum in the y direction is $p_y = m\dfrac{\Delta y}{\Delta t}$. The two-puck totals before and after the collision are

• before the collision:
$$\Sigma m\frac{\Delta y}{\Delta t} = mu + m(-u\sqrt{1 - v^2/c^2})$$

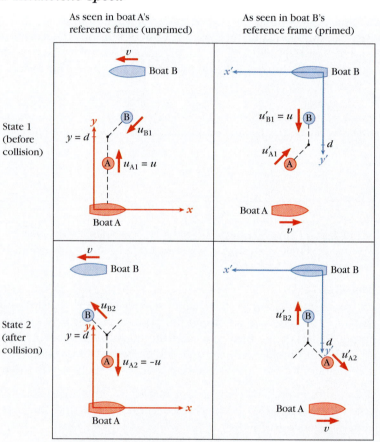

Figure 25-13 An elastic collision viewed in two different reference frames. Ice boats A and B traveling at relativistic speed v relative to each other launch pucks A and B toward each other. Each puck has an initial nonrelativistic speed $u \ll v$ relative to its own boat. For readability, the diagonal paths are shown to be less nearly in the x direction than they would be if $u \ll v$.

• after the collision:
$$\Sigma m\frac{\Delta y}{\Delta t} = m(-u) + m(u\sqrt{1 - v^2/c^2})$$

If we assume the mass of an object is invariant (so that the pucks' masses remain equal regardless of their speeds), the before and after sums are not equal. If we kept $p_y = m\dfrac{\Delta y}{\Delta t}$ as our definition of momentum, momentum would not be conserved.

We will instead adopt a more general definition of the y momentum:

$$p_y \equiv \frac{m\frac{\Delta y}{\Delta t}}{\sqrt{1 - v^2/c^2}}$$

Then before the collision

$$\Sigma p_y = \frac{mu}{\sqrt{1 - 0^2/c^2}} + \frac{m(-u\sqrt{1 - v^2/c^2})}{\sqrt{1 - v^2/c^2}}$$

$$= mu - mu = 0$$

and after the collision

$$\Sigma p_y = \frac{m(-u)}{\sqrt{1 - 0^2/c^2}} + \frac{m(u\sqrt{1 - v^2/c^2})}{\sqrt{1 - v^2/c^2}}$$

$$= -mu + mu = 0$$

With this definition, momentum *is* conserved. Moreover, when $v \ll c$, $\sqrt{1 - v^2/c^2} \approx 1$. Our more general definition then agrees with the classical definition.

For a step-by-step graphic presentation of the reasoning in Case 25-3, work through WebLink 25-2.

The reasoning leading to the adoption of a generalized definition of momentum p_y in Case 25-3 applies to *each* component of momentum and thus to the momentum vector:

For **WebLink 25-2:** **An Elastic Collision at Relativistic Speed,** go to www.wiley.com/college/touger

relativistic momentum $\quad \vec{\mathbf{p}} \equiv \dfrac{m\vec{\mathbf{v}}}{\sqrt{1 - v^2/c^2}}$ (25-6)

(Here $\vec{\mathbf{v}} = \frac{\Delta \vec{\mathbf{r}}}{\Delta t}$. Because $\Delta t = \Delta t_o/\sqrt{1 - v^2/c^2}$, we can also take the definition to be $\vec{\mathbf{p}} \equiv m\frac{\Delta \vec{\mathbf{r}}}{\Delta t_o}$.) This definition agrees with the classical definition $\vec{\mathbf{p}} \equiv m\vec{\mathbf{v}} = m\frac{\Delta \vec{\mathbf{r}}}{\Delta t}$ when $v \ll c$.

◆**RELATIVISTIC AND NONRELATIVISTIC KINETIC ENERGY** Guided by classical physics, we define the kinetic energy of an object to be the amount of work required to give it the momentum $p = mv/\sqrt{1 - v^2/c^2}$ if it starts out at rest. Because calculating this work requires calculus, we simply present the result:

relativistic kinetic energy $\quad KE = \dfrac{mc^2}{\sqrt{1 - v^2/c^2}} - mc^2$ (25-7)

When $v \ll c$, we can use the approximation $1/\sqrt{1 - v/c^2} \approx 1 + \frac{1}{2}\frac{v^2}{c^2}$ (Equation 25-1a). Then at nonrelativistic speeds,

$$KE = mc^2\left(1 + \frac{1}{2}\frac{v^2}{c^2}\right) - mc^2 = mc^2 + \tfrac{1}{2}mv^2 - mc^2 = \tfrac{1}{2}mv^2$$

in agreement with the familiar expression for KE in Newtonian physics.

◆**REST ENERGY OR MASS ENERGY** In Equation 25-7, each term is an amount of energy, but the term mc^2 does not depend on the object's speed v. We can therefore interpret it, as Einstein did, as the amount of energy an object has even when at rest. We can then rewrite Equation 25-7 as

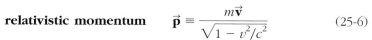

$$E = KE + E_o$$ (25-8)

total   kinetic   rest energy or
energy   energy   mass energy

where $\quad E = \dfrac{mc^2}{\sqrt{1 - v^2/c^2}} \quad$ and $\quad E_o = mc^2$ (25-9a, b)

We began this line of reasoning by *defining* the difference $mc^2/\sqrt{1 - v^2/c^2} - mc^2$ to be the relativistic kinetic energy an object gains as it speeds up from rest to a velocity v. The kinetic energy gained represents an increase in the object's total energy. It is therefore plausible to *define* the quantity $mc^2/\sqrt{1 - v^2/c^2}$ as the total energy E of the object. In the same vein, we refer to $E_o = mc^2$ as the **rest energy** of the particle.

Starting with Equations 25-8, 9a, and 9b and the definition of relativistic momentum (Equation 28-6), a series of algebraic steps (see Problem 25-60) will lead you to another relationship between an object's total energy E and its relativistic momentum p:

$$E^2 = p^2c^2 + E_o^2 \quad \text{or} \quad E^2 = p^2c^2 + (mc^2)^2 \quad (25\text{-}10)$$

Example 25-5 *A High-Energy Electron*

An electron that is accelerated from rest through a potential difference of a million volts is sometimes described as a 1 MeV (million electron volt) electron.
a. By what factor has its energy increased?
b. To what speed is it accelerated?

Solution

Choice of approach. Part **a** is asking, what is E/E_o? As always, when an electron is accelerated through a potential difference, it loses electric potential energy eV and gains an equal amount of kinetic energy. The 1 MeV refers to its KE, which we can find in joules. We can then use the facts that $E - E_o = KE$ and $E = E_o/\sqrt{1 - v^2/c^2}$ to find E/E_o. Once we know E/E_o, we can solve for v. Note that we cannot use $KE = \frac{1}{2}mv^2$ at relativistic speeds.

What we know/what we don't.

$$e = 1.60 \times 10^{-19}\,\text{C} \qquad m_o = 9.11 \times 10^{-31}\,\text{kg}$$

$$c = 3.00 \times 10^8\,\text{m/s} \qquad V = 1.00 \times 10^6\,\text{V} \qquad KE = ? \qquad m = ? \qquad v = ?$$

The mathematical solution.
a. The amount of kinetic energy gained is

$$eV = (1.60 \times 10^{-19}\,\text{C})(1.00 \times 10^6\,\text{V}) = 1.60 \times 10^{-13}\,\text{J}$$

From Equations 25-9a and b, $E/E_o = 1/\sqrt{1 - v^2/c^2}$, so we must find this factor. From Equations 25-8,

$$\frac{mc^2}{\sqrt{1 - v^2/c^2}} - mc^2 = KE$$

Dividing both sides by mc^2 gives us $1/\sqrt{1 - v^2/c^2} - 1 = KE/mc^2$, so

$$\frac{1}{\sqrt{1 - v^2/c^2}} = 1 + \frac{KE}{mc^2} = 1 + \frac{1.60 \times 10^{-13}\,\text{J}}{(9.11 \times 10^{-31}\,\text{kg})(3.00 \times 10^8\,\text{m/s})^2} = \mathbf{2.95}$$

The electron's total energy is nearly tripled.
b. From $E/E_o = 1/\sqrt{1 - v^2/c^2}$, inverting and squaring gives us

$$\frac{1}{(E/E_o)^2} = 1 - \frac{v^2}{c^2}$$

Solving for v gives us

$$v = c\sqrt{1 - \frac{1}{(E/E_o)^2}} = c\sqrt{1 - \frac{1}{(2.95)^2}} = \mathbf{0.941c \text{ or } 2.82 \times 10^8\,m/s}$$

◆ Related homework: Problems 25-31 through 25-35 and 25-37.

Look again at Equation 25-9b. Popularly written without a subscript, this is probably the world's most famous equation. Because c^2 is a very large number, the equation tells us that there is a huge amount of energy associated with even a tiny mass. (For this reason, the rest energy is also called the **mass energy**.)

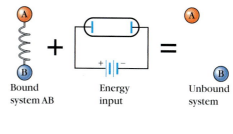

This is significant, though, only if it can be converted to other forms of energy.

For the systems we have considered in previous chapters, it has not mattered whether we included the rest mass energy $E_o = mc^2$ in the total energy of the system, because if it has the same value in the initial and final states we can subtract it from both sides of the conservation equation $E_1 = E_2$. However, in other systems it matters a great deal, as we shall see in the next example. But first, recall from Chapter 6 that reactions typically involve the breaking of old bonds and the forming of new ones. An input of energy is required to break apart a system of particles that is bound together. The same amount of energy is released again when the bonds form anew. This amount of energy is called the **binding energy** E_b. It represents the difference between the total energy of the system when it is bound and the energy of the system when the particles are separated but at rest (Figure 25-14):

$$m_{\text{bound system}} c^2 + E_b = \sum m_{\text{individual particle}} c^2 \qquad (25\text{-}11)$$

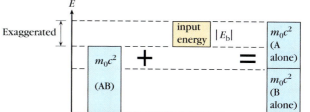

Figure 25-14 Accounting for energy when a bound system of particles is broken apart.

To compare how much mass energy is converted in chemical and nuclear reactions, work through Example 25-6.

Example 25-6 *A Comparison of Two Bound Systems*

For a guided interactive solution, go to Web Example 25-6 at www.wiley.com/college/touger

a. It requires $13.6 \text{ eV} = 2.17 \times 10^{-18} \text{ J}$ of energy to ionize a hydrogen atom. By what fraction is the mass of the atom reduced from the total combined mass of the proton and electron that together form the atom?

b. A helium nucleus, made up of two protons and two neutrons, has a mass of 6.644×10^{-27} kg. How much energy is required to break up this nucleus into its constituent particles, isolated and at rest?

Brief Solution

Choice of approach. In each case we are concerned about the difference between the initial and final masses and its energy equivalent. Equation 25-11, the basic relationship, can be stated in two parts:

$$E_b = c^2 \Delta m, \quad \text{where} \quad \Delta m = \Sigma(m_o)_{\text{individual particle}} - (m_o)_{\text{bound system}}$$

What we know/what we don't.

m of proton $= 1.6724 \times 10^{-27}$ kg $\qquad c = 3.00 \times 10^8$ m/s

m of neutron $= 1.6747 \times 10^{-27}$ kg

m of electron $= 9.1083 \times 10^{-31}$ kg

a. $E_b = 2.17 \times 10^{-18}$ J $\qquad\qquad$ **b.** m of helium $= 6.644 \times 10^{-27}$ kg

$$\Delta m = ? \qquad \text{fraction} = \frac{\Delta m}{m_{\text{bound system}}} = ? \qquad E_b = ?$$

The mathematical solution.
a. Because $E_b = c^2 \Delta m$,

$$\Delta m = \frac{E_b}{c^2} = \frac{2.17 \times 10^{-18} \text{ J}}{(3.00 \times 10^8 \text{ m/s})^2} = 2.41 \times 10^{-35} \text{ kg}$$

EXAMPLE 25-6 *continued*

As a fraction of the mass of the hydrogen atom, which we will approximate by the mass of a proton, this is

$$\frac{\Delta m}{m_{\text{bound system}}} = \frac{2.41 \times 10^{-35}\ \text{kg}}{1.67 \times 10^{-27}\ \text{kg}} = \mathbf{1.44 \times 10^{-8}}$$

b. The total mass of the constituent particles is

$$\Sigma m_{\text{indiv. particle}} = 2(m\ \text{of proton}) + 2(m\ \text{of neutron})$$
$$= 2(1.6724 \times 10^{-27}\ \text{kg}) + 2(1.6747 \times 10^{-27}\ \text{kg})$$
$$= 6.6942 \times 10^{-27}\ \text{kg}$$

$$\Delta m = \Sigma m_{\text{individual particle}} - m_{\text{bound He nucleus}}$$
$$= (6.6942 \times 10^{-27}\ \text{kg}) - (6.644 \times 10^{-27}\ \text{kg}) = 5.0 \times 10^{-29}\ \text{kg},$$

which is nearly 1% of the mass of the helium nucleus. The energy equivalent of this mass change is

$$E_b = c^2 \Delta m = (3.00 \times 10^8\ \text{m/s})^2 (5.0 \times 10^{-29}\ \text{kg}) = \mathbf{4.50 \times 10^{-12}\ J}$$

$$(\text{or } 2.8 \times 10^7\ \text{eV})$$

Making sense of the results. The fractional loss of mass in **a**, which involves an energy of typical size for a chemical reaction, is far too tiny to be detectable in a measurement of the mass of the atom itself. In contrast, in **b**, where the binding forces are nuclear rather than electrostatic forces, the mass difference *is* detectable; it is orders of magnitude greater and so, correspondingly, is the binding energy. This indicates that nuclear forces are much stronger than electrostatic forces.

◆ Related homework: Problems 25-36, 25-38, 25-39, and 25-40.

Because chemical bonds are governed by electrostatic forces, Example 25-6 shows that nuclear bonds are much stronger than chemical bonds. The breaking of old bonds before new ones form can therefore happen much less readily in nuclear than in chemical reactions. This is why chemical reactions are ubiquitous in our everyday life but nuclear reactions long remained unknown, leading to the impression that atoms were indestructible.

The energy required to break up a helium nucleus into individual protons and neutrons is comparable to the energy released when a helium atom is formed in a nuclear reaction. It isn't quite equal because the helium atom will be formed from isotopes of hydrogen, not from isolated protons and neutrons. This reaction is called **nuclear fusion** and typically releases about 2.8×10^{-12} J per helium atom formed. A remarkable amount of energy is involved here. The energy released from the formation of a mole (6.02×10^{23}) of helium atoms (about 4 g) would be

$$(6.02 \times 10^{23}) \times (2.8 \times 10^{-12}\ \text{J}) = 1.7 \times 10^{12}\ \text{J}$$

For comparison, the average American uses less than 4×10^{10} J of electrical energy in a year. Recognize also that this is only the energy equivalent of the *mass difference,* not of the total mass of a mole of helium. Nuclear fusion is the process producing the vast amounts of energy emitted by the sun.

The mass difference

$$\Delta m = \sum_{\substack{\text{protons} \\ \text{and neutrons}}} m\ - m_{\substack{\text{bound} \\ \text{nucleus}}}$$

for a single atom of an element is called the **mass defect,** and the fact the masses of nuclei are less than the sum of the isolated masses of their constituent protons and neutrons is evidence that mass has been converted into an equivalent amount of energy of another kind. We discuss this further in Chapter 28.

* 25-5 General Relativity

After dealing with special relativity (relativity between inertial reference frames), Einstein took on the question of **general relativity;** that is, relativity between *accelerated* reference frames. He assumed that in principle the laws of physics could be stated in a form valid for observers in inertial and accelerated reference frames alike. Also central to his reasoning was the **principle of equivalence,** which says that we cannot distinguish between effects caused by gravitational forces and effects caused by accelerated reference frames. For example, an intergalactic space traveler standing on a bathroom scale in an enclosure reads a weight mg on the scale. The traveler cannot tell whether the enclosure is still on Earth (so that the reading is due to Earth's gravitational pull) or in intergalactic space with an acceleration equal to g, so that the normal force mg exerted by the scale is responsible for the traveler's acceleration. Because of the principle of equivalence, general relativity emerges as a theory about gravitation.

Because gravitational fields vary from point to point in space, it is impossible to choose an accelerated reference frame that eliminates all gravitational effects. To deal with this, Einstein took the radical step of reconsidering fundamental assumptions about the geometry of space-time itself. He replaced Euclidean geometry with Riemannian geometry, which is essentially a geometry of curved space. According to his theory, space-time becomes curved in the vicinity of large concentrations of mass (Figure 25-15). Objects that would follow straight-line paths in Euclidean space are caused to follow curved trajectories. This is not only true of ordinary projectiles. According to Einstein, even the straight-line paths followed by light should be bent by sufficiently massive objects.

A theory will gain acceptance only when it offers a better account of observation than existing theories or explains observations that existing theories cannot. The first such observation was the *precession* of Mercury's orbit (Figure 25-16; also, see Figure 8-2): As Mercury's orbit shifts its orientation about the sun, the point of closest approach changes its angular position by about 500 seconds of arc each century (1 degree of arc = 60 minutes of arc; 1 minute of arc = 60 seconds of arc). There was a 43-second-of-arc discrepancy between this observed value and the predictions of Newtonian theory. General relativity explained the precession in terms of the variation in the gravitational field as Mercury's elliptical path moves closer to and further from the sun and accounted for the 43-s discrepancy.

Einstein also predicted that during a solar eclipse, we should be able to observe a shift in the angular position of stars seen in close proximity to the sun, because the paths of light from those stars to our eye pass close to the sun and

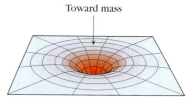

Toward mass

Figure 25-15 Space-time becomes curved in the vicinity of large concentrations of mass.

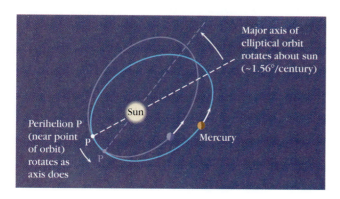

Major axis of elliptical orbit rotates about sun (~1.56°/century)

Sun

Perihelion P (near point of orbit) rotates as axis does

P

P

Mercury

Figure 25-16 The precession of Mercury's orbit.

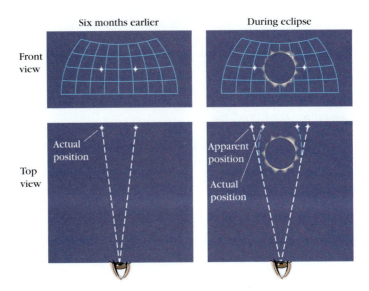

Six months earlier   During eclipse

Front view

Top view

Actual position

Apparent position

Actual position

Figure 25-17 Apparent angular position of stars affected by bending of light from sun.

are bent by it (Figure 25-17). Astronomers observed such a shift (about 1.7 seconds of arc) during the total solar eclipse of 1919. This event so captured the attention of the world press (Figure 25-18) that Einstein quickly gained a celebrity status that persists even today. Nevertheless, his Nobel Prize, which he received two years later, was not for his work on relativity but for his explanation of the photoelectric effect (which we discuss in Chapter 27).

In recent years, the bending of light from distant stars has provided evidence of other objects in space unknown in Einstein's time. When light or other electromagnetic radiation passes through a large accumulation of inter-

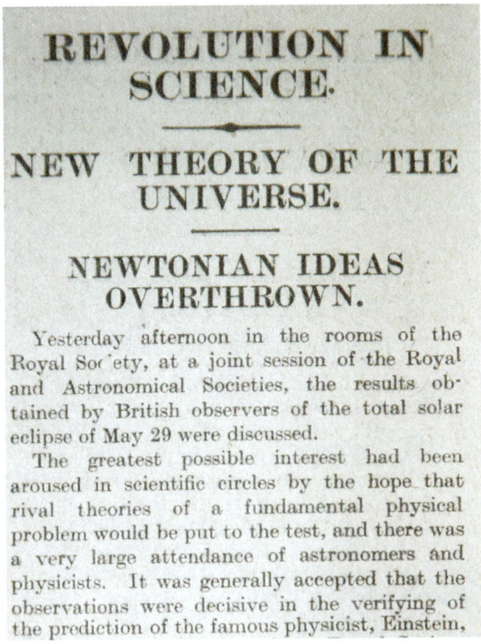

REVOLUTION IN SCIENCE.

NEW THEORY OF THE UNIVERSE.

NEWTONIAN IDEAS OVERTHROWN.

Yesterday afternoon in the rooms of the Royal Society, at a joint session of the Royal and Astronomical Societies, the results obtained by British observers of the total solar eclipse of May 29 were discussed.

The greatest possible interest had been aroused in scientific circles by the hope that rival theories of a fundamental physical problem would be put to the test, and there was a very large attendance of astronomers and physicists. It was generally accepted that the observations were decisive in the verifying of the prediction of the famous physicist, Einstein,

– London Times
7 November 1919

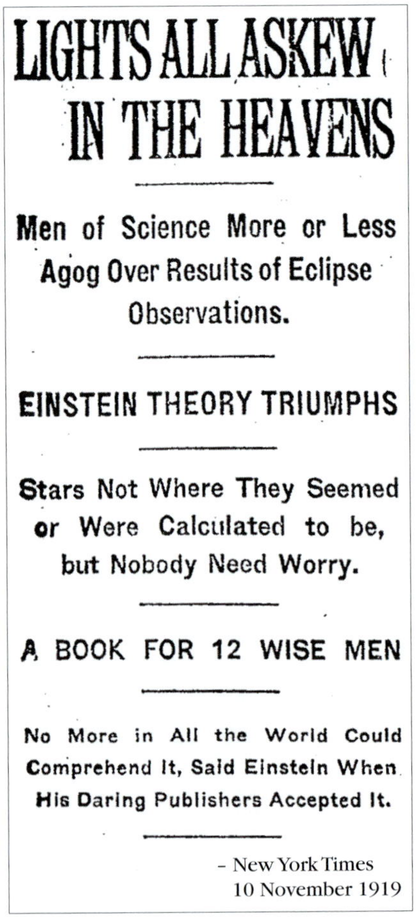

LIGHTS ALL ASKEW IN THE HEAVENS

Men of Science More or Less Agog Over Results of Eclipse Observations.

EINSTEIN THEORY TRIUMPHS

Stars Not Where They Seemed or Were Calculated to be, but Nobody Need Worry.

A BOOK FOR 12 WISE MEN

No More in All the World Could Comprehend It, Said Einstein When His Daring Publishers Accepted It.

– New York Times
10 November 1919

Figure 25-18 Headlines appearing after observations during solar eclipse of 1919 confirmed predictions of general relativity.

stellar matter, it is bent toward the center of mass, much as if the matter were a lens. Such an accumulation of interstellar matter is called a **gravitational lens.** In particular, some extremely massive stars collapse into themselves toward the end of their lifetimes, resulting in a mass density so great that no light can escape. These are the **black holes** that have captured the public imagination. With no electromagnetic radiation of their own to signal their existence, their ability to bend passing light from other sources provides crucial evidence that they are there.

◆ SUMMARY ◆

Newtonian physics fails at speeds approaching the speed of light. It incorrectly allows for the speed of light to be different for different observers and cannot explain why nothing has ever been observed to move at a speed greater than c. Einstein based his special relativity, which accounted for how nature behaves at very high speeds, on two postulates:

- **Postulate 1.** *The laws of physics are the same in every inertial reference frame.*

- **Postulate 2.** *The speed of light in a vacuum is independent of the observer—its measured value is always $c = 3.0 \times 10^8$ m/s, no matter how the source or the observer may be moving.*

In special relativity we (1) carefully identify each **event** as something that happens at a particular point in space at a particular instant, and (2) locate an event in space and time by means of a set of coordinate axes and a clock, which together make up a **reference frame. An inertial reference frame** is one in which Newton's first law, the law of inertia, is valid, in contrast to an **accelerated reference frame.**

Special relativity deals with measurements that vary for observers in different inertial reference frames. *Any reference frame moving at constant velocity relative to an inertial frame is also an inertial frame. There is no preferred inertial reference frame.* **General relativity** deals with accelerated reference frames as well.

There are important consequences of the fact that the speed of light is the same in all inertial reference frames:

1. **Simultaneity is relative:** *If two events at a distance Δx from each other are simultaneous for an observer in one inertial reference frame, they are <u>not</u> simultaneous for an observer in another inertial reference frame moving with respect to the first in the x direction. (The event that the moving observer views as further forward will happen first for the moving observer.)*

2. **Time dilation**

$$\Delta t = \frac{\Delta t_o}{\sqrt{1 - v^2/c^2}} \qquad (25\text{-}1)$$

An inertial reference frame in which events occur at the same position along the direction of motion is called a **proper reference frame** for those events. A clock that is stationary in the proper reference frame reads the **proper time interval** Δt_o between two events. *If an observer sees the device recording the **proper time interval** as moving, the interval Δt measured by this observer is always greater than the proper time interval Δt_o.*

3. **"Moving clocks run slow"** by a factor $1/\sqrt{1 - v^2/c^2}$ (Figure 25-9).

4. **Length contraction**

$$\Delta x = \Delta x_o \sqrt{1 - v^2/c^2} \qquad (25\text{-}2)$$

(The **proper length** Δx_o is the length measured by an observer who sees it as stationary.)

However, as we go from one inertial reference frame to another, the quantity

$$(\Delta x)^2 + (\Delta y)^2 + (\Delta z)^2 - (c\Delta t)^2 =$$
$$(\Delta x')^2 + (\Delta y')^2 + (\Delta z')^2 - (c\Delta t')^2 \quad (25\text{-}3b)$$

(or $(\Delta x)^2 - (c\Delta t)^2 = (\Delta x')^2 - (c\Delta t')^2$ if only one
spatial dimension is involved) (25-3a)

remains an **invariant** measure of the distance between two points in **four-dimensional space-time** (between two events).

In all of these effects, the factor $\sqrt{1 - v^2/c^2} \approx 1$ when $v \ll c$ and approaches zero as v approaches c (so that $1/\sqrt{1 - v^2/c^2}$ becomes infinitely large). It will not have a real value if $v > c$, and so does not permit that as a physical possibility.

To preserve conservation of momentum, we define

relativistic momentum $\qquad \vec{\mathbf{p}} \equiv \dfrac{m\vec{\mathbf{v}}}{\sqrt{1 - v^2/c^2}} \qquad (25\text{-}6)$

To make the work input equal to the KE gained, we define

relativistic kinetic energy $\qquad KE = \dfrac{mc^2}{\sqrt{1 - v^2/c^2}} - mc^2 \quad (25\text{-}7)$

Then

$$E = KE + E_o \qquad (25\text{-}8)$$
$$\text{total} \quad \text{kinetic} \quad \text{rest energy or}$$
$$\text{energy} \quad \text{energy} \quad \text{mass energy}$$

where $\qquad E = \dfrac{mc^2}{\sqrt{1 - v^2/c^2}} \qquad$ and $\qquad E_o = mc^2 \quad$ (25-9a,b)

(At nonrelativistic speeds, $KE = \dfrac{mc^2}{\sqrt{1 - v^2/c^2}} - mc^2 \approx \frac{1}{2}mv^2$.)

It then follows that

$$E^2 = p^2c^2 + E_o^2 \qquad \text{or} \qquad E^2 = p^2c^2 + (mc^2)^2 \quad (25\text{-}10)$$

When particles bind together, say, to form an atom or molecule,

$$m_{\substack{\text{bound} \\ \text{system}}} c^2 + E_b = \Sigma m_{\substack{\text{individual} \\ \text{particle}}} c^2 \qquad (25\text{-}11)$$

or stated in two parts:

the **binding energy** $\qquad E_b = c^2 \Delta m$

where the **mass defect** $\qquad \Delta m = \Sigma m_{\substack{\text{individual} \\ \text{particle}}} - m_{\substack{\text{bound} \\ \text{system}}}$

◆QUALITATIVE AND QUANTITATIVE PROBLEMS◆
Hands-On Activities and Discussion Questions

The questions and activities in this group are particularly suitable for in-class use.

25-1. Discussion Question. A very thin sheet of cardboard slides frictionlessly across a very thin floor with a hole in it. The rest lengths of the cardboard and the hole are equal. The cardboard is moving at a relativistic speed. It must be true that either the cardboard falls into the hole or it does not, and this must be agreed on by all observers, because when all motion stops, the observers can check to see which side of the hole the cardboard is on. Now, according to Observer 1 in the floor's reference frame, the cardboard is contracted by a factor $\sqrt{1 - v^2/c^2}$ and is therefore shorter than the hole. This observer reports that at an instant when the cardboard is entirely over the hole, a downdraft pushes it through. However, Observer 2 in the cardboard's reference frame sees the hole moving in the other direction at the same speed. To Observer 2, the *hole* is shorter by a factor $\sqrt{1 - v^2/c^2}$. Does this mean that for Observer 2, the hole is too short and therefore the cardboard cannot fall through? How can the two observers reach agreement on what they can later verify did or did not happen? Explain.

Review and Practice

Section 25-1 Reference Frames Revisited

25-2. In Figure 25-1a, **a** what is the velocity of the hand relative to the ball once it lodges in the glove? **b** does the hand experience a greater impulse when it is moving toward or away from the ball? Briefly explain. **c** the collision of the ball with the glove occurs over a very small time interval Δt (recall Section 7-1). Is this interval greater when the hand is moving toward or away from the ball? **d** the ball exerts an average force on the hand during the collision. Is this average force greater when the hand is moving toward or *away* from the ball? Briefly explain how your answer follows from your answers to parts **b** and **c**.

25-3. SSM

a A pitcher throws a fast ball at 90 mi/hr as detected by a stationary catcher. If the catcher were instead on a golf cart moving toward the pitcher at 15 mi/hr, what would be the ball speed detected by the catcher?

b Light from α Centauri, the nearest bright star to our own sun, reaches an Earthbound observer at a speed of 3.00×10^8 m/s. What would be the speed of light from α Centauri detected by a spaceship passenger traveling from Earth toward α Centauri at a speed of 1.20×10^8 m/s?

25-4. Can we use reasoning similar to the reasoning in Example 25-1 to figure out whether the sun is going around Earth or Earth is going around the sun? Briefly explain why we can or cannot.

25-5. A person is standing in a totally enclosed elevator that travels the length of an interplanetary spaceship. The person notices that she is gradually feeling lighter. Give two fundamentally different possible reasons why this could be happening.

25-6. Two railroad freight cars A and B are lit by bulbs hanging by flexible electrical cords. A worker on car A sees the bulb on B hanging straight down. During the one second he can see B's bulb, it moves 5.0 m to his right. A worker on car B sees the bulb on A hanging at a small angle to the vertical. During the one second she can see A's bulb, it moves 5.0 m to her left. **a** What can we conclude about how fast each freight car is really moving, and what can we *not* conclude? Briefly explain. **b** What can we conclude about the acceleration of each freight car, and what can we *not* conclude? Briefly explain.

25-7. SSM Tell whether each of the following definitely is, possibly is, or definitely is not an inertial reference frame, and briefly explain your reasoning in each case. Assume there is a clock fixed at the origin of the coordinate axes in each frame described.

a A coordinate frame fixed in an elevator (on Earth) in which you are standing on a correctly calibrated bathroom scale that reads half your actual weight

b A coordinate frame moving at a constant velocity relative to your own

c A coordinate frame maintaining a speed 0.9c faster than a given inertial frame

d A coordinate frame rotating at constant angular speed relative to a given inertial frame

25-8. A physicist is in an isolation booth in which there are two metal spheres connected to each other by a spring balance. The physicist doesn't observe any rotation, but believes that she and the spheres are really all rotating together. Ignoring anything that she herself might feel, on what evidence could she base her belief? What does "really" mean in this case?

25-9. (*Choose all correct answers.*) Observers A and B both witness the motion of the two billiard balls in Figure 25-19. The two observers must agree on whether _____

a *ball A is moving faster than ball B.*

b *the two balls collide.*

c *ball B's velocity has a horizontal component.*

d *ball A is accelerating.*

e *ball A ends up with a velocity opposite in direction to its initial velocity.*

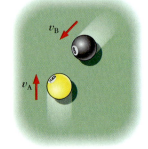

Figure 25-19
Problem 25-9

25-10. Repeat Problem 25-9 for the case when both observers' reference frames are inertial reference frames.

Section 25-2 It's About Time

25-11. A block (assumed to be a point object) slides down a frictionless ramp, at the bottom of which there is a compressible spring (Figure 25-20). In this situation, tell whether each of the following qualifies as an *event* and briefly explain your reasoning in each case.

Figure 25-20 Problem 25-11

a. The block slides down the ramp.
b. The block makes its first contact with the spring.
c. The spring compresses.
d. The spring is at maximum compression.
e. The upper tip of the spring reaches its maximum displacement.

25-12. Figure 25-6 assumes that events AA′ and BB′ are simultaneous for observer O and shows the subsequent events as seen by O. Suppose instead that events AA′ and BB′ were simultaneous for observer O′. Figure 25-6a would then show events AA′ and BB′ as seen in the inertial frame of observer O′. Redraw the remaining parts of the figure to show the subsequent events as they would be seen in the inertial frame of observer O′.

25-13. As a space probe passes the planet Neptune, at a distance of 4.5×10^{12} m from Earth, a timer on the probe is set to $t = 0$. At this instant the timer emits a signal in the form of an electromagnetic pulse. If a timer at NASA headquarters in Houston is to be synchronized with the timer on Earth, to what time should the NASA timer be set when the signal is received?

25-14. By how much does the factor $1/\sqrt{1 - v^2/c^2}$ differ from being exactly 1 when **a.** v is one one-hundredth of the speed of light? **b.** v is one-tenth of the speed of light? **c.** Repeat **a** using the approximation in Equation 25-1a. By what percent do the two results differ? **d.** Repeat **b** using the approximation in Equation 25-1a. By what percent do the two results differ?

25-15. How old are you according to a space traveler who has been traveling away from Earth at a speed of $0.866c$ since before you were born?

25-16. The second as an SI unit of time is defined as the total duration of 9 192 631 770 vibrations of light of a particular wavelength emitted by atoms of the isotope cesium-133.

a. To a stationary observer, how many seconds would it take for this many vibrations to be emitted by a cesium-133 atom moving at a speed of $0.30c$?
b. When 10 s have elapsed on a clock based on stationary cesium-133 atoms, how many seconds will have elapsed on a clock based on the same atoms moving at a speed of $0.30c$?

25-17. SSM WWW You and a mysterious stranger are on trains traveling past each other on parallel tracks at two-thirds the speed of light. As you pass each other, you and the stranger reach out and push the "start" buttons on each other's stopwatches. The stopwatches are identical when both are at rest.

a. If your stopwatch ticks once each second, how long (in your reference frame) will it be between ticks of the mysterious stranger's stopwatch?
b. When your stopwatch reads 40.0 s, what will be the reading (in your reference frame) on the mysterious stranger's stopwatch?
c. According to the stranger's reference frame, what is the reading on your stopwatch when the stranger's stopwatch reads 40.0 s?

25-18. Joe and Josie are born at the same instant, but Joe is born in a hospital on Earth, and Josie is born on an intergalactic spacecraft passing Earth at a speed of 1.2×10^8 s. The spacecraft's speed and direction continue unchanged for the next several decades.

a. When Joe celebrates his tenth birthday, how old is Josie in Joe's inertial reference frame?
b. What speeds or velocities must you assume are negligible to answer part **a**?

Section 25-3 Length Contraction and Other Spatial Considerations

25-19. The measurement of a distance or a length involves two events. What are they?

25-20. Consider the following two events in the situation in Figure 25-6:

• Event BB′: B′ passes B
• Event BA′: A′ passes B

The events are observed by observers O and O′. **a.** In which observer's reference frame can one measure the proper time interval between these two events? Briefly explain. **b.** In which observer's reference frame can one measure the proper distance between the positions where these two events occur? Briefly explain.

25-21.
a. Find the values of $\sqrt{1 - v^2/c^2}$ for $v = 0.01c$, $0.05c$, $0.1c$, $0.5c$, $0.9c$, $0.95c$, and $0.99c$, and use your values to sketch a graph of $\sqrt{1 - v^2/c^2}$ versus v. Comment on the implications that the behavior shown in your graph has for length contraction.
b. Find the values of $1/\sqrt{1 - v^2/c^2}$ for the same values of v, and use your values to sketch a graph of $1/\sqrt{1 - v^2/c^2}$ versus v. Comment on the implications that the behavior shown in this graph has for time dilation.

25-22. In Equations 25-1 and 25-2, what would happen to lengths and time intervals on an object moving at a speed greater than the speed of light?

25-23. A meter stick travels at a speed of $0.80c$ relative to an observer in an inertial frame. According to the observer, how long is the meter stick if it is oriented **a** along the direction of its motion? **b** perpendicular to the direction of its motion?

25-24. According to the observer in Problem 25-23, how long is the meter stick if it is tilted at an angle of 30° to the direction of its motion?

25-25. A stationary observer A sees a rocket traveling to the right at constant speed. Observers B, C, D, E,

Figure 25-21 Problem 25-25

and F are all traveling at the same speed relative to observer A, but in different directions (Figure 25-21). List in rank order the length of the rocket that each of the six observers sees. Order them from least to greatest, indicating any inequalities, by listing the letters of the observers.

25-26. In Case 25-2, how must the velocities of the two trains compare for both observers (O and O′) to conclude that events AA′ and BB′ are simultaneous?

25-27. In Examples 25-3 and 25-4, **a** how far apart are A and A′ (according to O′)? **b** how far apart are B and B′ (according to O′)? **c** Then when B′ meets B, what is the distance from A′ to A (according to O′)? **d** Traveling at the given value of v, how long will it take A′ to reach A? Compare your answer to the result of Example 25-4.

25-28. As his spacecraft overtakes Captain Picard's, Captain Kirk observes that Captain Picard's spacecraft is 10% shorter than his own. He knows that when the two spacecraft were docked together, they were identical.
a How much faster than Captain Picard's spacecraft is Captain Kirk's spacecraft moving?
b What can we say about how fast each spacecraft is really moving? (Answer in a few words.)

25-29. **SSM WWW** While Captain Sisko is in charge of a space station, Captain Janeway's spacecraft goes by at a speed of 1.7×10^8 m/s. The engineers on the space station report to Captain Sisko that the spacecraft has a length of 250 m. If Janeway's spacecraft is to double back and dock at the space station, what minimum length of docking space must Captain Sisko provide?

25-30.
a According to one observer, one of two events happens 2.0 s before another event, which takes place at a point 1.0×10^9 m away. How far apart do the two events occur for an observer who detects them to be simultaneous?
b If the first observer sees events at the same two points happening 6.0 s apart, can the second observer see these events simultaneously? Briefly explain.

Section 25-4 Momentum, Mass, and Energy in Relativistic Physics

25-31. In Example 25-5, what percent error in the electron velocity v would result from mistakenly using the nonrelativistic expression $\frac{1}{2}mv^2$ for the kinetic energy to calculate v?

25-32.
a What is the speed of a 1 MeV proton?
b How does this compare with the speed of a 1 MeV electron? Use the result of Example 25-5 and express your answer as a ratio $\frac{v \text{ of 1 MeV proton}}{v \text{ of 1 MeV electron}}$.

25-33. **SSM** What is the total energy of a 0.75 MeV electron?

25-34. A particle with a rest energy of 1.0×10^{-9} J travels at half the speed of light. **a** What is its relativistic momentum? **b** What is its total energy?

25-35. When a particle is traveling at a speed $v = 0.95c$, is its kinetic energy less than, equal to, or greater than $\frac{1}{2}mv^2$? Briefly explain.

25-36. As fusion reactions occur in the sun's interior, the sun gives off the energy by emitting electromagnetic radiation. Its total power output is 3.92×10^{26} W.
a How much mass is the sun losing each year?
b How does the mass that it loses each year compare to its total mass of 1.99×10^{30} kg? Express your answer as a ratio $\frac{\text{annual mass loss}}{\text{total mass}}$.

25-37. A particle accelerator accelerates an electron to a speed of $0.95c$. **a** What is the electron's total energy when it reaches this speed? **b** What is the electron's mass at this speed?

25-38. In one well-known nuclear fusion reaction, the nuclei of two isotopes of hydrogen—deuterium, with one neutron, and tritium, with two neutrons—fuse to form a helium nucleus (one neutron remains free):

$$^2_1\text{H} + {}^3_1\text{H} \rightarrow {}^4_2\text{He} + {}^1_0n + 17.5 \text{ MeV of energy}$$

(1 MeV $\equiv 10^6$ eV.) How much mass is lost in this reaction?

25-39. **SSM** An atomic mass unit (1 u) is exactly one-twelfth the mass of a $^{12}_6\text{C}$ atom, including its six electrons:

$$1 \text{ u} = 1.660566 \times 10^{-27} \text{ kg}$$

In one of the fusion reactions that takes place in the sun, a common hydrogen nucleus (a proton) fuses with a deuterium nucleus (an isotope of hydrogen that also has a neutron in its nucleus) to form a helium-3 nucleus, consisting of two protons and one neutron: $^1_1\text{H} + {}^2_1\text{H} \rightarrow {}^3_2\text{He}$. In atomic mass units, the masses of the relevant isotopes are:

^1_1H 1.007825 u ^2_1H 2.014012 u ^3_2He 3.016029 u

Calculate the amount of energy released in this fusion reaction. Express your answer in MeV (1 MeV $\equiv 10^6$ eV).

25-40. Helium-3 is an uncommon isotope of helium. A helium-3 nucleus, made up of two protons and one neutron, has a mass of 5.002×10^{-27} kg.
a How large an energy input is required to break up this nucleus into its constituent particles, isolated and at rest?
b What becomes of this energy? (Answer in a few words.)

25-41. Find the momentum of an electron traveling at a speed of $0.6c$.

25-42. Find the momentum of a proton when its total energy is twice its rest mass energy.

*Section 25-5 General Relativity

25-43. A sufficiently large concentration of mass in space will function as a gravitational lens; that is, it bends light in somewhat the same way a convex optical lens would. Is it possible for an arrangement of matter to have the same effect on a cone of light as a *concave* optical lens? Briefly explain.

25-44. During a solar eclipse, a straight line from an Earth-bound telescope to a distant star passes through the sun.
a Could this star be seen through the telescope? Draw a sketch of the situation that explains your reasoning.
b Can you draw any analogy to what you would see if you projected the image of a bulb onto a screen using a convex lens with a blackened center?

25-45. The captain of a spaceship notices that objects aboard the ship weigh 1.2 times their ordinary Earth weights. To determine how well the ship's thrusters are working, he assigns the ship's scientist to figure out what fraction of this apparent weight is due to the thrusters accelerating the spaceship. Assuming the ship's scientist is an expert physicist, she will inform the captain that ____

a. it is not possible to answer his question.

b. the thrusters are almost completely responsible for the apparent weight.

c. the thrusters are responsible for 0.2 times the apparent weight.

d. the thrusters contribute negligibly to the apparent weight.

Going Further

The questions and problems in this group are not organized by section heading, so you must determine for yourself which ideas apply. Some of them will be more challenging than the Review and Practice questions and problems (especially those marked with a • or ••).

25-46. The driver of a car sees a truck in the rearview mirror, and notices that the truck's speed relative to the car is increasing.

a Is there necessarily a nonzero total force on the truck, or can there be another reason for the observed change in the truck's relative speed? (If so, what reason?)

b Another observer stands by the side of the road. Describe as many different scenarios as you can (as seen by *this* observer) that would account for the car driver's observation.

25-47. In his statement about absolute time in Section 25-1, Newton speaks of measuring "relative, apparent, and common time . . . by means of motion." To what motion(s) was he referring?

• 25-48. Suppose you are given three meter sticks and a nail. How can you use these materials to construct a reference frame for two-dimensional space?

25-49. Figure 25-7 shows a method for synchronizing two clocks in which clock B is set after clock A is set. Devise a method for synchronizing these two clocks in which both clocks are set at the same instant. You may introduce any additional equipment that you need.

25-50. If event A happens before event B in your inertial frame, can there be another inertial frame in which event A happens after event B? Briefly explain by referring to situations that illustrate your point.

• 25-51. A flat field appears square to a stationary observer. Observers A and B fly over the field. Both are flying at a speed of 0.5c, but A is traveling parallel to one of the field's sides, whereas B is traveling parallel to the field's diagonal. Compare the shape of the field as observed by A to its shape as observed by B.

• 25-52. A flat field appears square to a stationary observer. Observers A and B fly over the field. Both are flying at a speed of 0.5c, but A is traveling parallel to one of the field's sides, while B is traveling parallel to the field's diagonal. Compare the area of the field as observed by A to its area as observed by B. Express the comparison as a ratio $\frac{\text{area observed by A}}{\text{area observed by B}}$.

25-53. Inertial reference frame A has a speed of 0.5c relative to inertial frame B. A straight rod has the same length to observers in both reference frames. How is this possible?

• 25-54. Can Figures 25-22a and b possibly show the same motions of the same two trains viewed in two different inertial reference frames? Briefly explain.

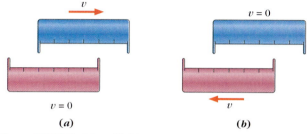

Figure 25-22 Problem 25-54

• 25-55. In Equation 25-1, if Δt is not zero, Δt_o will not be zero, and vice versa. However, in Example 25-4, we obtain a nonzero result for $\Delta t'$ when Δt is zero. Is this in conflict with Equation 25-1? Briefly explain your answer. (*Hint:* Think about the meaning of proper time. Is it possible to establish a proper reference frame for events AA' and BB'?)

25-56. Observer A sees two firecrackers explode. He notices the positions of the two explosions, and observes that the second explosion took place 5.0 s after the first. Observer B agrees with A about the locations of the two explosions. Can she disagree about how much later the second explosion took place? Briefly explain

25-57. The occurrences that we treat as *events* in relativity are commonly only approximately events. For instance, in Case 25-2, the spark passing between electrodes is an event only to the extent that the distance between the electrodes is negligible; otherwise this occurrence takes place over a distance rather than at a single point. Then it would also take place over a time interval rather than at a single instant, because charge carriers travel between the electrodes at finite speed. For each of the following occurrences, tell whether it is an event *exactly*, *approximately*, or *not at all*, and explain your reasoning.

a The tip of a billiard cue collides with a ball

b The origin of one of two inertial frames traveling at different speeds coincides with the origin of the other

c An Olympic sprinter crosses the finish line

d A *scintillation* (a tiny flash of light) occurs when an electron strikes the inside of an oscilloscope screen

e The origin of one of two inertial frames traveling at different speeds moves along the y axis of the other

25-58. Figure 25-23 shows the positions within square ABCD where events 1, 2, and 3 occur. Between which pairs of events (1 and 2, 1 and 3, 2 and 3) could you measure the proper time interval if you were traveling in a straight line at constant speed *a* from point A to point B? *b.* from point A to point C? *c.* from point A to point D? *d.* from point B to point D? *e.* perpendicular to square ABCD?

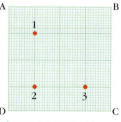
Figure 25-23 Problem 25-58

25-59.

a. The energy released when a mole (6.02×10^{23}) of helium atoms (about 4 g) are formed from hydrogen isotopes is

about 1.7×10^{12} J. Is this *much less than, roughly equal to,* or *much greater than* the mass energy of 4 g of helium?

b. If the average American uses about 4×10^{10} J of electrical energy in a year, how many Americans could be provided with all their electrical energy for a year by the energy output when a mole of helium atoms are formed by fusion? the mass energy of a mole of helium atoms?

25-60. Starting with the magnitude $p = mv/\sqrt{1 - v^2/c^2}$ of the relativistic momentum and Equations 25-8, 9a, and 9b, show that $E^2 = p^2c^2 + E_o^2$.

Problems on WebLinks

25-61. In WebLink 25-1, if the passenger in elevator A observes the motion of elevator B that is shown in the WebLink, which (one or more) of the following are possible explanations of what the passenger is observing?

i. Elevator A is accelerating.

ii. Elevator B is accelerating.

iii. Neither elevator is accelerating.

iv. Both elevators are accelerating.

25-62. To an observer who is stationary relative to the elevator shaft in WebLink 25-1, which (one or more) of the expla-

nations in Problem 25-61 are possible explanations of what the passengers in the two elevators are observing?

25-63. In the collision of the two pucks shown in WebLink 25-2, classical momentum $\vec{p} = m\vec{v}$ is conserved _____ (*only in the x direction; only in the y direction; in neither direction; in both directions*).

25-64. In the collision of the two pucks shown in WebLink 25-2, relativistic momentum $\vec{p} = m\vec{v}/\sqrt{1 - v^2/c^2}$ is conserved _____ (*only in the x direction; only in the y direction; in neither direction; in both directions*).

CHAPTER TWENTY-SIX

Inroads into the Micro-Universe of Atoms

We now return to questions about the structure of matter and of its basic building blocks, atoms. The goal is to deduce how matter must be constructed on an atomic scale and what interactions have to occur at this microscopic level for matter to have the macroscopic properties we can observe. The deduction process must begin with the available evidence. This chapter will review the evidence available at the end of the nineteenth century (recall Chapter 24), and treat additional evidence provided by investigations in the early years of the twentieth century. The following chapter will detail how the evidence led physicists to conclude that the microscopic world was governed by a set of rules—the rules of quantum physics—different from any that governed our everyday macroscopic world.

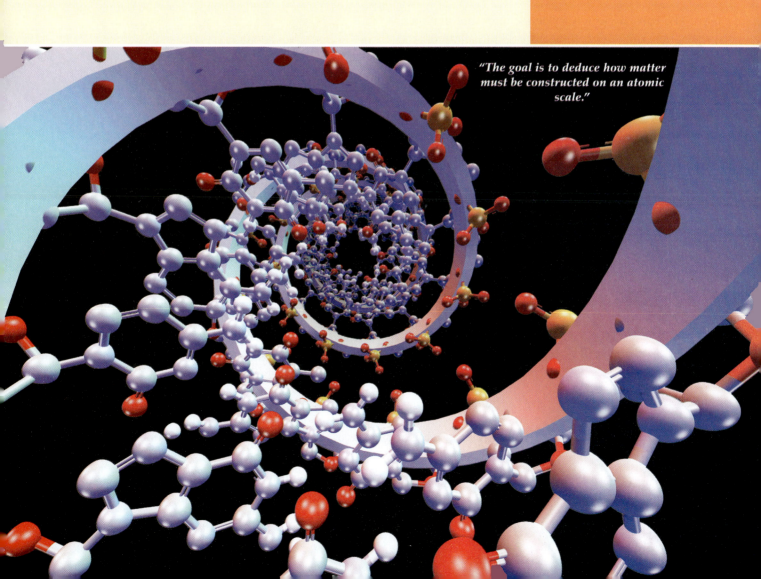

"The goal is to deduce how matter must be constructed on an atomic scale."

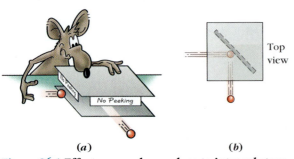

26-1   Probes and Emissions: The Evidence for Atomic Structure

Imagine that you are looking directly down on a box that has an opaque lid and open sides (Figure 26-1). You are not permitted to peek in the sides, so you cannot see directly how the top lid is supported. But by sending in a ball and seeing how it comes out, you can infer how the supports might be arranged. Anything we send into an unseen region in this way to provide information about it is called a **probe.** Sometimes, though, as in the following case, we can draw inferences from what comes out of an unseen region—the *emissions*—even when we have not sent in any probe.

(a)   (b)

Figure 26-1 Effects on probes—clues to internal structure. (*a*) The observation. (*b*) Possible internal structure.

Case 26-1 ◆ *Inferring Structure from Spontaneous Emissions*

In Figure 26-2, a tall enclosure has a hole near the bottom. Every once in a while, a ball comes horizontally out of the hole. The balls—the emissions in this situation—always come out at one of three distinct speeds. A little piece of string is always attached to each ball, and is frayed at its free end. How can we use physics concepts to get at an idea of how things are arranged inside the enclosure?

The balls are stored inside the enclosure until they come out. The frayed string suggests they were held in place by the strings until the strings gave out. When they emerge, each ball has one of three possible speeds, and therefore one of three possible kinetic energies. Conservation of energy tells us that this must previously have been stored energy—potential energy—when the balls were held in place. The balls could have been maintained at three different gravitational potential energies if they were hung at three different heights. This suggests the possible internal structure shown in Figure 26-2b.

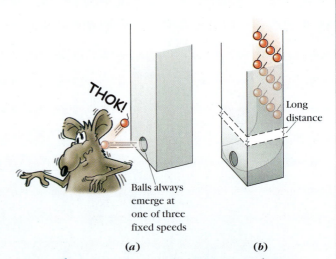

Figure 26-2 Spontaneous emissions—more clues to internal structure. (*a*) The observation. (*b*) Possible internal structure. Assume distances between rows of balls are large.

Because we cannot see directly into an atom, it is basically through the evidence of probes and emissions that a model of its structure has developed. We have already considered some of these emissions: cathode rays, X rays, and α, β, and γ rays, and the discrete wavelengths of visible light to which line spectra attest. The "probes" in some of these cases have been large inputs of electrical energy; in the case of radioactivity, the emissions have been spontaneous. Table 26-1 on the next page summarizes some of the evidence and reasoning we have studied so far that first provided insight into the structure of atoms. We will now discuss some further developments that will extend this table.

Whereas Becquerel had used Thomson's methods to show the ratio q/m to be the same for β rays as for cathode rays, Rutherford used those methods to establish the value of q/m for α rays. He found that even with much greater electric fields, only very small deflections were obtained. Newton's second law (in the form $qE = ma$) made the implications clear: If, like β rays, the α rays were made up of particles, then the α particle had either a much smaller charge than the electron's or a much greater mass. For the α particle, Thomson found the

Table 26-1 Some Early Steps toward an Understanding of Atomic Structure

Line spectra are observed from luminous gases (mid-1700s forward). The **line spectrum** for each element is unique.	Chapter 24
Measurements (late 1700s) show atoms combine to form compounds in characteristic weight ratios. **Dalton's atomic theory** (1808) is proposed to explain those ratios.	Chapter 24
Success of **kinetic theory of gases** (1800s) provides further support for atomicity of matter.	Chapter 12
Maxwell (1864) establishes that light (including emissions of light from matter) is an **electromagnetic wave.**	Section 24-1
Balmer (1885) finds a mathematical pattern of relationship among the wavelengths of the hydrogen line spectrum (the **Balmer series**).	Chapter 24
Roentgen discovers **X rays** emitted by cathode ray tubes at very high voltages (1895).	Chapter 24
Thomson's cathode ray experiment (1897) establishes the existence of negative particles called *electrons* within atoms and shows they have a fixed ratio q/m of charge to mass. Thomson formulates **plum pudding model** of atom.	Section 18-2; Chapter 24
Becquerel discovers **radioactivity** (1896) and Rutherford (1898–1899) and Villard (1900) establish that radioactivity consists of three different kinds of emissions, **α, β, and γ rays.** Becquerel (1899) shows the ratio q/m to be the same for β rays as for cathode rays.	Chapter 24
Electromagnetic radiation from ideal emitters (black bodies) is found empirically to obey the observed relationship $\lambda_{max}T = $ **constant** (1890s).	Chapter 24
Rayleigh's attempt to explain black-body radiation based on atomic oscillators obeying Newtonian physics results in **"ultraviolet catastrophe"** (1900).	Chapter 24
Planck gets theory to fit with experimental data for black-body radiation by introducing the idea of a **quantum** of energy into the atomic oscillator picture (1900).	Chapter 24
Millikan's oil drop experiments (1910–1913) establish the electron as an object of fixed mass with a fixed basic charge.	Section 18-2

q/m ratio to be just a few ten-thousandths of the electron's and opposite in sign. Later experiments eventually showed that when these positive particles captured electrons, the resulting neutral species emitted the line spectrum characteristic of helium. The α particles would prove to be doubly charged helium ions. Example 26-1 shows how basic physics concepts informed Thomson's reasoning.

Example 26-1 *Deflection of α Particles*

**For a guided interactive solution, go to Web Example 26-1 at
www.wiley.com/college/touger**

A typical velocity of α particles is 2×10^7 m/s. Using the present day knowledge that α particles are doubly charged helium ions, determine

a. how strong an electric field is required to deflect a beam of alpha particles 5° from its initial direction if the region between the deflecting plates has a length of 20 cm.
b. how large a voltage must be applied between plates 4 cm apart to produce this field.

EXAMPLE 26-1 *continued*

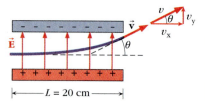

Figure 26-3 Set-up for deflection of a beam of α particles in Example 26-1.

Brief Solution

Choice of approach. A new situation often provides an opportunity to review basic physics. By Newton's second law, a constant electric force qE produces a constant acceleration. Using $\tan \theta = v_y/v_x$ (Figure 26-3), we can relate the angular deflection to the linear quantities in our two-dimensional kinematics equations for constant acceleration and find the acceleration. We can then use the second law to find E. In **b** we use the fact that just as work is force times distance for a constant force, work per unit charge (potential difference or voltage) is force per unit charge (field) times distance for a constant field: $V = Ed$.

What we know/what we don't.

For the kinematics: $v_{ox} = 2 \times 10^7$ m/s $\theta = 5°$ $\tan \theta = 0.087$

$x = 20$ cm $= 0.20$ m $t = ?$ $v_y = ?$ $a = ?$

For the dynamics: $m = 4m_{proton} = 4(1.67 \times 10^{-27}$ kg$) = 6.68 \times 10^{-27}$ kg

$q = 2e = 2(1.6 \times 10^{-19}$ C$) = 3.2 \times 10^{-19}$ C

$a = $ value from kinematics $E = ?$

For **b**: $d = 4$ cm $= 0.04$ m $V = ?$

The mathematical solution. The relevant kinematic equations for constant acceleration are

$$x = v_{ox}t \qquad v_x = v_{ox} \qquad v_y = v_{oy} + at = at \text{ since } v_{oy} = 0$$

Then $$\tan \theta = \frac{v_y}{v_x} = \frac{at}{v_{ox}}$$

But $x = v_{ox}t$ tells us that $t = x/v_{ox}$, so

$$\tan \theta = \frac{a(x/v_{ox})}{v_{ox}} = \frac{ax}{v_{ox}^2}$$

Solving for a, we get

$$a = \frac{v_{ox}^2 \tan \theta}{x} = \frac{(2 \times 10^7 \text{ m/s})^2 (0.087)}{0.20 \text{ m}} = 1.7 \times 10^{14} \text{ m/s}^2$$

Newton's second law tells us that in the y direction $qE = ma$, so

$$E = \frac{ma}{q} = \frac{(6.68 \times 10^{-27} \text{ kg})(1.7 \times 10^{14} \text{ m/s}^2)}{(3.2 \times 10^{-19} \text{ C})} = \mathbf{3.5 \times 10^6 \text{ N/C}}$$

b. This requires a voltage

$$V = Ed = (3.5 \times 10^6 \text{ N/C})(0.04 \text{ m}) = \mathbf{1.4 \times 10^5 \text{ V}}$$

STOP&Think How might you alter the dimensions of the plate arrangement so that less voltage could be applied between the plates to produce the same deflection? ◆

◆ Related homework: Problems 26-4, 26-5, 26-6, and 26-7.

Other work, by William Crookes in 1900 and Rutherford and Frederick Soddy in 1902, focused on uranium and thorium, two elements that emit α and β particles. When isolated, uranium and thorium were expected to lose much of their radioactivity after a few days. But starting out with supposedly pure samples of these elements, they found that the radioactivity was maintained. Rutherford and Soddy surmised that longer-lived radioactive atoms were appearing, indicating "chemical changes . . . occurring within the atom" rather than resulting from chemical interactions between atoms. (Today we would call them nuclear reactions rather than chemical reactions.) The atoms of uranium and thorium were themselves being transformed, decaying into new elements as they emitted α and β particles, and in most cases the resulting new elements then decayed in like manner. This turning of one element into another is called **transmutation of the elements.**

These radioactive transformations were unlike anything known previously. Dalton had viewed the atom as a basic building block of matter that could not be broken down further. Until the discovery of radioactivity, there was no reason to believe otherwise. Dalton's picture required only minor adjustment for the conditions of ordinary chemical reactions, which only involved rearrangements of electrons and left the internal structure of atoms unaltered. But these new circumstances meant that the idea of the atom as indestructible and immutable had to be abandoned. Something was happening to the internal structure of the atoms. We now turn to efforts to probe that internal structure.

In Thomson's plum pudding model of the atom (Section 24-2), the electrons were embedded like raisins in a plum pudding within a sphere of uniformly distributed positive charge. Thomson realized that beams of α and β particles could be used as probes to get a more precise idea of how the electrons were arranged within the positive sphere. Experiments of this type are called **scattering experiments** because probes entering in one direction emerge or are scattered in different directions (Figure 26-1). He predicted that if α and β particles were fired at layers of metal foil just a few atoms thick, the particles would be scattered very little, because a fairly uniform arrangement of charge within atoms would exert only very small deflecting electrostatic forces on the charged probes (Figure 26-4*a*).

The actual experiment was undertaken in 1909 at Rutherford's suggestion by Hans Geiger and Ernest Marsden, two young scientists working in his laboratory, who bombarded a piece of gold foil 4×10^{-7} m thick with a beam of α particles. Rutherford's recollection of the experiment and its outcome is often quoted:

> I remember Geiger coming to me . . . in great excitement and saying, "We have been able to get some of the α-particles coming backwards." It was quite the most incredible event that has happened to me in my life. It was almost as incredible as if you fired a 15-inch shell at a piece of tissue paper and it came back and hit you. On consideration I realized that the scattering backwards must be the result of a single collision, and when I made calculations I saw that it was impossible to get anything of that order of magnitude unless . . . the greater part of the mass of the atom was concentrated in a minute nucleus. It was then that I had the idea of an atom with a minute massive center carrying a charge.[1]

The observations contradicted the predictions of the plum pudding model and led Rutherford to conceive of a kind of solar system model in which the "sun" was a positive nucleus of small enough radius so that the concentrated charge could repel an incoming positive α particle. Only in Rutherford's picture could Coulomb forces that decrease with distance as $\frac{1}{r^2}$ produce the observed pattern of deflections (Figure 26-4*b*). The nucleus had to be massive enough so that it

[1]J. Needham and W. Pagel, eds., *Background to Modern Science* (Macmillan, NY, 1958), quoted in A. Arons, *Development of Concepts of Physics* (Addison-Wesley, Reading, MA, 1965).

Ernest Rutherford (1871–1937). Born on a New Zealand farm, Rutherford worked in England after 1907. He is best remembered for his central contributions to our understanding of the atom. Rutherford won the Nobel Prize for Chemistry in 1908; became director of the Cavendish Laboratory, one of the world's great research centers, in 1919; and was knighted in 1931.

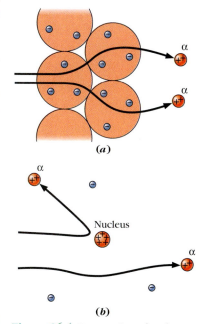

Figure 26-4 Comparing the deflection of α particles that are predicted by two models of the atom. (*a*) What happens according to Thomson's plum pudding model. (*b*) What happens according to Rutherford's solar system model. If drawn to scale, the electrons in Rutherford's model would be thousands of times farther from the nucleus than shown here.

was not significantly displaced by the scattering. Moreover, most of the α particles in the gold foil experiment continued straight through without deflecting, suggesting that most of the atom was empty space. In Rutherford's model the electrons were distributed at distances thousands of times greater than the diameter of the nucleus. The model of the atom that the average person pictures today was beginning to emerge, though Rutherford's model did not say anything further about how the electrons might be arranged.

Example 26-2 *Going for the Gold*

For a guided interactive solution, go to Web Example 26-2 at www.wiley.com/college/touger

Rutherford knew that the charge to mass ratio for α particles was 4.8×10^7 C/kg. From the crossed electric and magnetic fields required to keep the α particle beam undeflected, he was able to calculate a velocity of about 2.1×10^7 m/s for the beam used by Geiger and Marsden. Fired at this speed, how close (measured center of mass to center of mass) can an α particle approach a bare gold nucleus?

Brief Solution

Choice of approach. At large separations $(r \to \infty)$, the electric potential energy $\frac{kq_1q_2}{r}$ is zero. At the point of nearest approach, the kinetic energy of the α particle is zero. The problem is therefore readily solved by conservation of energy.

What we know/what we don't. Atomic number of gold = 79.

Charge of bare gold nucleus $q_1 = 79e = 79(1.6 \times 10^{-19}$ C$) = 1.3 \times 10^{-17}$ C

(Lacking the conclusions of the later Millikan measurements, Rutherford estimated this at about $100e$, because lighter atoms typically had charges [as multiples of e] that were roughly half their atomic masses.)

For α particle, $\dfrac{q_2}{m} = 4.8 \times 10^7$ C/kg $k = 9.0 \times 10^9$ N$\cdot$m^2/C^2

State 1	State 2
$r_1 = 0$ $v_1 = 2.1 \times 10^7$ m/s	$r_2 = ?$ $v_2 = 0$
$PE_1 = 0$ $KE_1 = \frac{1}{2}mv_1^2$	$PE_2 = \dfrac{kq_1q_2}{r_2}$ $KE_2 = 0$

The mathematical solution. The conservation of energy equation

$$PE_1 + KE_1 = PE_2 + KE_2$$

becomes

$$\tfrac{1}{2}mv_1^2 = \frac{kq_1q_2}{r_2}$$

Solving for r_2 gives us

$$r_2 = \frac{2kq_1q_2}{mv_1^2} = \frac{2(9.0 \times 10^9 \text{ N}\cdot\text{m}^2/\text{C}^2)(1.3 \times 10^{-17} \text{ C})(4.8 \times 10^7 \text{ C/kg})}{(2.1 \times 10^7 \text{ m/s})^2}$$

$$= \mathbf{2.5 \times 10^{-14} \text{ m}}$$

Making sense of the results. If the α particle did not penetrate the gold nucleus, then r_2, the distance of closest approach between their centers, had to be greater than the radius of the gold nucleus. This meant that the radius of the gold nucleus could be at most about 2.5×10^{-14} m. The size of an atom was estimated to be on the order of 10^{-10} m (about 10 000 times bigger), hence leading Rutherford to the conclusion that the positively charged portion of the atom (the nucleus, as it could therefore be called) must fill only a tiny part of the space occupied by the atom.

◆ Related homework: Problems 26-8 and 26-10.

26-2 Interactions between Electromagnetic Radiation and Matter: More Evidence in Need of Explanation

In the last part of the nineteenth century, physicists had become aware of a broader range of phenomena involving the interaction of electromagnetic radiation with matter, including

- the selective emission and absorption of visible light by gases of particular elements, resulting in the observation of line spectra.

- the emission of X rays from target materials bombarded by cathode rays (high-speed electrons).

- the ionization of gases (dislodging of electrons from the atoms or molecules of the gas) by X rays.

- the dislodging of electrons from the surfaces of metals by light of sufficiently high frequency (usually ultraviolet) incident on the surface of those metals, an effect called the **photoelectric effect.**

We will consider the photoelectric effect in greater detail here and in the next chapter. In Germany, Philipp Lenard (1862–1947) investigated the effect systematically and published his results in 1902. Figure 26-5 shows the apparatus he used. When electrons are dislodged from the metal plate on the left (Figure 26-5a), the plate is left with a net positive charge. When the emitted electrons reach the opposite plate, an equivalent amount of charge returns to the positive plate. The galvanometer therefore detects a *current* proportional to the number of electrons reaching the opposite plate. In Lenard's full apparatus (Figure 26-5b), a variable voltage could be applied, making the opposite plate increasingly negative *relative to* the emitting plate, and thereby, depending on the applied voltage, stopping some or all of the emitted electrons from reaching the opposite plate (recall Example 19-9). For a fuller description of how the apparatus works, follow the step-by-step visuals of WebLink 26-1.

Photoelectrons are emitted with various kinetic energies up to some maximum. When a voltage V is applied across the gap, electrons moving across it are slowed down: They lose kinetic energy as they gain potential energy up to a maximum amount eV. For the various electrons, the conservation of energy equation becomes:

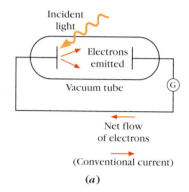

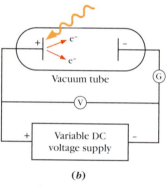

Figure 26-5 Schematic diagram of Lenard's apparatus for investigating the photoelectric effect. (*a*) The main part of the apparatus detects photoemission. (*b*) Application of a large enough voltage can stop electrons from reaching the opposite plate.

For **WebLink 26-1: Investigating the Photoelectric Effect,** go to www.wiley.com/college/touger

for electrons that make it across the gap with KE to spare	$\frac{1}{2}mv_1^2 = \frac{1}{2}mv_2^2 + eV \quad (v_2 < v_1)$	(26-1a)
for electrons that just barely make it across the gap	$\frac{1}{2}mv_1^2 = eV$	(26-1b)
for electrons that don't make it completely across the gap	$\frac{1}{2}mv_1^2 = eV^* \quad (V^* < V)$	(26-1c)

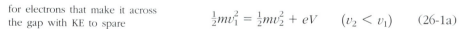

(V^* is the potential difference between the emitting plate and the furthest point reached by electron.)

When the potential difference between the plates is set at a value V_o (called the *stopping potential* or *stopping voltage*) just high enough to stop all electrons from making it across the gap, Equation 26-1b applies to the electrons with maximum kinetic energy:

$$\tfrac{1}{2}mv_{max}^2 = eV_o \tag{26-2}$$

By seeing experimentally what voltage V_o was needed to reduce the galvanometer reading to zero, one could find the maximum kinetic energy of the electrons for a particular set of conditions.

In his experiments, Lenard carefully studied how the effect depended on the frequency of the incident light. He therefore used sources that were as nearly monochromatic (single wavelength) as possible. Some of his observations and those of subsequent experimenters, notably Millikan, were inexplicable in terms of classical physics:

1. He found that for any given metal, electrons could be dislodged by light only at or above a certain minimum frequency or **threshold frequency.** Although even very low-intensity light could produce photoemission when the frequency was above this threshold value, light having frequency below this value would produce no effect, no matter how strong. That means, for example, that if the threshold for a particular metal were somewhere in the yellow part of the visible spectrum, a dim blue bulb would produce an effect, but a powerful red laser would not. The incident light provides the input energy that enables electrons to break away from the metal surface. According to classical physics, energy is provided continuously by the incoming electromagnetic wave and is distributed uniformly to all the surface electrons. By classical physics, the extent to which electrons broke loose should depend on the intensity of the incoming light, not its frequency. The existence of a threshold frequency could not be explained.

2. The maximum kinetic energy reached by the electrons also turns out to depend only on the frequency of the incident light, not its intensity.

3. At frequencies for which the photoelectric effect does occur, it starts up virtually instantaneously (a current is detected within 10^{-9} seconds after the light begins to strike the metal) even at intensities as low as 10^{-10} W/m^2. If (as classical physics assumes) the energy received by the surface electrons is distributed uniformly among them and builds up continuously, a detailed calculation would show that at such a low intensity it would take the electrons hundreds of hours to accumulate enough energy to break away. Here again, the predictions are at odds with the evidence.

To explain these aspects of the photoelectric effect would turn out to require a major break with classical physics. This is the effect's importance to the history of physics; its unexpected features provided clues that a new explanatory model was needed to address the interaction of matter (atoms and electrons) and electromagnetic radiation (light). Other evidence reinforced this need.

◆**LINE SPECTRA** Line spectra are also an important source of evidence of interaction between electromagnetic radiation and the atoms from which it is emitted. Balmer had found that the wavelengths of the visible light line spectrum for hydrogen fit into a mathematical pattern expressible as

$$\lambda = (3.6456 \times 10^{-7}\ \text{m})\frac{n^2}{n^2 - 2^2} \qquad (n = 3, 4, 5, 6) \tag{24-1}$$

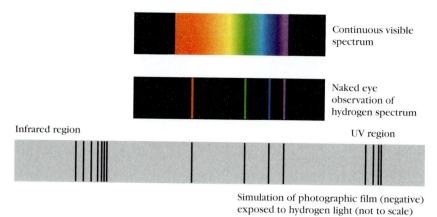

Continuous visible spectrum

Naked eye observation of hydrogen spectrum

Infrared region

UV region

Simulation of photographic film (negative) exposed to hydrogen light (not to scale)

Figure 26-6 Photographic detection of spectral lines beyond the visible range.

In the early 1900s, physicists expanded their knowledge about line spectra beyond the visible range. When we view a continuous spectrum with a spectrometer, we see nothing to the left and right of the visible range because our eyes do not detect ultraviolet or infrared. Likewise, when we view a line spectrum (selected wavelengths from the continuous spectrum), our eyes detect only those lines that are in the visible range. Photographic film, however, responds to ultraviolet and infrared as well as visible light. If we use a suitable strip of photographic film rather than our eyes as the detector, we may expect exposure at whatever wavelengths are present in the infrared and ultraviolet as well as in the visible range (Figure 26-6).

By just such means, Theodore Lyman of Harvard University recorded hydrogen lines in the far ultraviolet range in 1906, and in Germany two years later, Friedrich Paschen (1865–1947) discovered still more hydrogen lines in the infrared range. Later on, additional series of hydrogen lines in the far infrared were found by Brackett (1922) and Pfund (1924). Each series of lines fit a pattern similar to Balmer's, and all of these patterns turned out to be specific instances of a more general pattern.

To get at that general pattern in steps we begin by rewriting Balmer's relationship, inverting and simplifying it:

$$\frac{1}{\lambda} = \frac{1}{3.6456 \times 10^{-7} \text{ m}} \frac{n^2 - 2^2}{n^2} = \frac{2^2}{3.6456 \times 10^{-7} \text{ m}}\left(\frac{1}{2^2} - \frac{1}{n^2}\right), \qquad n = 3, 4, 5, 6$$

We can view this as a specific instance of a more general pattern:

$$\frac{1}{\lambda} = R_H\left(\frac{1}{m^2} - \frac{1}{n^2}\right) \tag{26-3}$$

in which m and n are always unequal integers, and R_H, called the **Rydberg constant for hydrogen,** is simply

$$R_H = \frac{2^2}{3.6456 \times 10^{-7} \text{ m}} = 1.097 \times 10^7 \text{ m}^{-1}$$

Equation 26-3 is called the **Rydberg formula.** For any given m, the allowable values of n are $m + 1$, $m + 2$, $m + 3$, . . . , so that for $m = 2$ (the Balmer case), n can be $2 + 1 = 3$, $2 + 2 = 4$, and so on.

It turns out that when $m = 1$, the Rydberg formula gives the wavelengths of the Lyman lines, and when $m = 3$ it gives the Paschen wavelengths. The Pfund and Brackett lines discovered later fit the pattern for $m = 4$ and $m = 5$. The Rydberg formula even works for spectra of other elements than hydrogen if R_H is changed to an R with a different constant value.

Example 26-3 *Getting a Feel for the Rydberg Formula*

Find the wavelength of the hydrogen line in the Lyman series that has the largest wavelength.

Solution

Choice of approach. To understand the Rydberg formula fully, you need to get a feel for how the wavelength is affected as m and n vary. The wavelengths for the Lyman lines are given when $m = 1$. Because n is always greater than m, the quantity $\frac{1}{m^2} - \frac{1}{n^2}$ is always positive and gets larger as n gets larger (verify this for yourself). Because the wavelength varies inversely as this quantity, the wavelength gets smaller as n gets larger. Then the largest value of λ will correspond to the smallest allowable value of n.

What we know/what we don't.

$$m = 1 \qquad R_H = 1.097 \times 10^7 \text{ m}^{-1}$$

Allowable values of n are $m + 1 = 1 + 1 = 2$,

$$m + 2 = 2 + 1 = 3, \qquad \lambda = ?$$

$$m + 3 = 3 + 1 = 4, \text{ etc.}$$

The mathematical solution. When $m = 1$, the smallest allowable value of n is 2. With these values,

$$\frac{1}{\lambda} = R_H\left(\frac{1}{m^2} - \frac{1}{n^2}\right) = R_H = 1.097 \times 10^7 \text{ m}^{-1}\left(\frac{1}{1^2} - \frac{1}{2^2}\right) = 8.23 \times 10^6 \text{ m}^{-1}$$

Inverting, we get

$$\lambda = \frac{1}{8.23 \times 10^6 \text{ m}^{-1}} = \mathbf{1.22 \times 10^{-7} \text{ m}}$$

Making sense of the results. Recall that the visible range of wavelengths is roughly from 4.0×10^{-7} m to 7.5×10^{-7} m, so this ultraviolet wavelength is shorter than visible wavelengths, as we should expect.

◆ Related homework: Problems 26-17, 26-18, 26-19, and 26-20.

Wavelengths of line spectra were found to fit a broadly applicable mathematical pattern, the Rydberg formula. Classical physics could not even offer an explanation of why particular elements emitted light only at certain frequencies and not those in between, let alone address the question of why the discrete wavelengths should have this mysterious pattern of relationship.

In the next chapter, we will see how a new physics began to emerge to answer questions such as these. It would be a physics much further removed from the commonsense reasoning and intuitions shaped by ordinary experience that we are accustomed to bring to bear when we apply physics to situations in our everyday world. What happens on an atomic or subatomic scale is not part of our ordinary experience. There is no reason to assume that the rules governing macroscopic behavior apply at this scale. For example, pressure and temperature, which we experience directly, are both aggregate quantities describing the state of a gas of some 10^{23} molecules. They have no meaning for individual molecules, and the proportionality of pressure to temperature does not apply at this scale. Other rules might also prove to be less than universal when we look to the atomic scale.

✦SUMMARY✦

The accumulation of evidence shedding light on the microscopic structure of matter and its interaction with electromagnetic radiation (Table 26-1), which had intensified in the last years of the nineteenth century, continued unabated in the early years of the twentieth. Table 26-2 summarizes the additional evidence introduced in this chapter, all of which involves probes and/or emissions.

Table 26-2 More Early Steps toward an Understanding of Atomic Structure

Rutherford used Thomson's methods to establish the value of q/m for α rays, implying their nature as particles (1902) and later shows the α particles to be helium nuclei (1909).	Section 26-1
Rutherford and Soddy show that radioactive emissions are an indication of **radioactive decay** in which nuclei of one element break down into nuclei of another (1902).	Section 26-1
Geiger and Marsden perform **gold foil experiment** (α particle **scattering experiment**); backward scattering of α particles is inconsistent with plum pudding model (1909).	Section 26-1
Rutherford formulates **solar system model** of atom with tiny concentrated **nucleus** and vast amounts of empty space between nucleus and electrons (1911).	Section 26-1
Lenard's investigation of **photoelectric effect** (1902) reveals behavior inexplicable by classical physics:	Section 26-2

- **threshold frequency** must be exceeded for effect to occur.
- maximum kinetic energy of emitted electrons depending only on frequency maximum KE determined by **stopping voltage** V_o needed to reduce photoelectron current to zero

$$\tfrac{1}{2}mv_{max}^2 = eV_o \qquad (26\text{-}2)$$

- nearly instantaneous start-up of effect even at very low intensities.

The Rydberg formula is obtained as a generalization of the Balmer series rule:	Section 26-2

$$\frac{1}{\lambda} = R_H\left(\frac{1}{m^2} - \frac{1}{n^2}\right) \qquad (26\text{-}3)$$

m, n unequal integers; $R_H = \dfrac{2^2}{3.6456 \times 10^{-7}\ \text{m}} = 1.097 \times 10^7\ \text{m}^{-1}$

(Rydberg constant for hydrogen)

The Lyman hydrogen spectrum lines in UV (1906) fit the Rydberg pattern for $m = 1$. Three series of hydrogen lines in infrared found by Paschen (1908), Pfund (1922), and Brackett (1924) fit the Rydberg pattern for $m = 3, 4,$ and 5, respectively.	Section 26-2

✦QUALITATIVE AND QUANTITATIVE PROBLEMS✦
Hands-On Activities and Discussion Questions

The questions and activities in this group are particularly suitable for in-class use.

26-1. *Discussion Question.* Five identical positively charged particles are sent in as probes to obtain information about what is under the rectangle in Figure 26-7. The entry and exit paths of each of the five probes is shown. On the basis of this observed scattering of the probes, what would be a reasonable possibility for what is under the rectangle? Explain your reasoning.

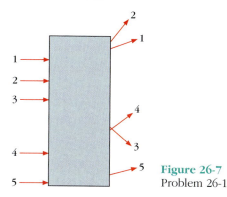

Figure 26-7
Problem 26-1

Review and Practice

Section 26-1 Probes and Emissions: The Evidence for Atomic Structure

26-2. A rectangular region (Figure 26-8) is known to have an overall positive charge Q. A wide beam of positively charged particles is fired in from the left as a probe to determine the distribution of positive charge within the rectangular region. What could you conclude about the distribution of charge

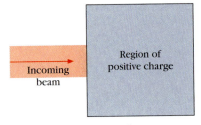

Figure 26-8 Problem 26-2

a. if the beam is bent toward the top of the page when it emerges from the rectangular region? Explain.

b. if most of the particles in the beam continue through undeflected but particles near the center of the beam are deflected by large angles and those right at the center come straight back? Explain.

c. if the particles in the incoming beam come straight back when they enter the rectangular region at low speed but emerge straight ahead at the same speed at which they entered when they enter at high speed? Explain.

26-3. Explain why larger electric fields are required to produce detectable deflections of α rays than detectable deflections of β rays.

26-4. In Example 26-1, if the α particles in the incoming beam are slower, should the electric field be increased, decreased, or left unchanged to maintain the same angular deflection? Briefly explain.

26-5. How might you alter the dimensions of the plate arrangement in Example 26-1 so that less voltage could be applied between the plates to produce the same deflection?

26-6. Radioactive material producing α, β, and γ emissions is encased in a lead block. A long, narrow hole is drilled into the lead, providing the sole direction in which the emissions can escape the lead casing. Suppose the escaping emissions enter a region of uniform electric field perpendicular to the direction of the hole. Rank the three emissions in order of how much they are deflected by this field, starting with the one that is deflected by the smallest angle. (Make sure to indicate any equalities.)

26-7.

a. How strong an electric field is required to deflect beta β particles 8.0° from their initial direction if they have an initial speed of 2.0×10^7 m/s and the region between the deflecting plates has a length of 0.15 m?

b. How large a voltage must be applied between plates 5.0 cm apart to produce this field?

26-8. If a beam of α particles of speed 2.1×10^7 m/s passes between a pair of plates oppositely charged so that a uniform electric field of 6.0×10^5 N/C is maintained between the plates, find the magnitude of the magnetic field that must be

applied if the α particles are not to be deflected from their initial direction.

26-9.

a. Why is Thomson's plum pudding model of the atom not consistent with what was observed in Geiger and Marsden's experiment in which they bombarded thin gold foil with α particles?

b. Why was Rutherford's solar system model more consistent with these observations?

26-10. At what speed would a beam of protons have to be fired at a sheet of silver foil if the protons were able to approach to within 2.5×10^{-14} m of the bare silver nucleus? Silver has atomic number 47.

Section 26-2 Interactions between Electromagnetic Radiation and Matter: More Evidence in Need of Explanation

26-11. **SSM** Electrons are emitted from one of two parallel plates. With what maximum speed could the electrons have been emitted if all are prevented from reaching the opposite plate by a potential difference of 5.0 V maintained between the plates?

26-12. Two plates are 5.0×10^{-2} m apart. Electrons are emitted from the left plate with a speed of 8.0×10^5 m/s. Find the magnitude and direction of the minimum electric field that must be applied to prevent all electrons from reaching the right plate.

26-13. In Problem 26-12, could you apply a magnetic field instead of an electric field and still prevent all electrons from reaching the right plate? Briefly explain.

26-14. A certain metal shows no photoelectric effect when exposed to any visible wavelength of light. Might this metal emit electrons when exposed to sunlight? Briefly explain.

26-15. The photoelectric effect was observed to start up virtually instantaneously for light intensities as low as 10^{-10} W/m^2 = 10^{-10} J/(s·m^2). This problem is intended to give you a feel for how weak light of this intensity is.

a. Suppose we assume a 100 W bulb to be 20% efficient. If this bulb is placed in the center of an imaginary sphere of radius 2 m, what would be the intensity of the light crossing the surface of this sphere?

b. Is the intensity the same at all points on the surface of the sphere?

c. At what distance from the bulb would the intensity of light be 10^{-10} W/m^2?

26-16. Your physics lab has an extremely sensitive photoelectric effect apparatus. An unknown sample of metal is put into position as the emitter plate. One of your classmates bets you that you cannot produce a detectable effect with the first light source that you shine on the emitter bulb. The available sources are an infrared heat lamp, a very intense red laser, and a dim "grow light" that imitates sunlight for growing plants indoors. Which source should you choose? Briefly explain why.

26-17. In the Rydberg formula (Equation 26-3), should m be less than, equal to, or greater than n? Briefly explain why.

26-18. According to the Rydberg formula, what happens to the difference between each wavelength and the next as n increases? Does it increase, decrease, or remain the same?

26-19. **SSM WWW**
a. Find the wavelength of the hydrogen line in the Pfund ($m = 4$) series that has the largest wavelength.
b. Repeat **a** for the Brackett ($m = 5$) series.

26-20. Find the wavelength of each of the following hydrogen lines and tell what part of the electromagnetic spectrum (visible, ultraviolet, beyond ultraviolet, infrared, or beyond infrared) it is in.
a. The hydrogen line in the Paschen series that has the smallest possible wavelength.
b. The hydrogen line in the Paschen series that has the largest wavelength.
c. The hydrogen line in the Lyman series that has the smallest possible wavelength.

Going Further

The questions and problems in this group are not organized by section heading, so you must determine for yourself which ideas apply. Some of them will be more challenging than the Review and Practice questions and problems (especially those marked with a • or ••).

26-21. Based solely on the charge-to-mass ratio, what else could α particles be besides doubly charged helium ions?

•26-22.
a. Estimate the magnetic field that if applied instead of the electric field in Example 26-1, would deflect the α particles by the same angle. (Treat Earth's magnetic field as negligible.)
b. What should be the direction of the applied magnetic field?

26-23. When chunks of radioactive ore are cracked open, small amounts of trapped helium are always released. Why is the helium present?

26-24. If two α particles, each moving at a speed of 2.1×10^7 m/s when far apart, approach each other head on, how closely can their centers of mass approach each other?

26-25. In studying the photoelectric effect, it is important to use light that is as nearly monochromatic as possible to see how the effect varies with frequency. How could you obtain nearly monochromatic light of various frequencies using a multiple slit diffraction grating and an opaque barrier with a single slit? Sketch your proposed set-up.

26-26. Suppose that light of intensity 10^{-10} J/(s·m²) falls uniformly over a 0.040-m² surface of metallic sodium. We may reasonably assume that only conduction electrons are dislodged by the photoelectric effect because the rest are more tightly bound to the individual atomic sites. In metallic sodium, there is one conduction electron per atom, and the atoms are on the order of 10^{-10} m apart.
a. If the classical assumptions that light falls continuously on the surface and is equally shared by all the surface electrons are valid, how much energy is absorbed by each electron in any one second?
b. It is known from chemistry that the binding energy of sodium conduction electrons is -2.3 eV $= -3.7 \times 10^{-19}$ J, meaning a positive energy of this size must be added to bring the total energy to zero so that the electron is no longer bound. This is the amount that must be provided to each electron by the incident light for photoemission to occur. Based on the classical assumptions in **a**, how long would it take for a sodium electron to be dislodged by light of this intensity?

•26-27. **SSM** Suppose that in Problem 26-12, you tried to prevent all the electrons from reaching the right plate by applying a magnetic field instead of an electric field.
a. What minimum magnitude would the magnetic field have to exceed to accomplish this?
b. Could Problem 26-12 be solved using conservation of energy? If not, briefly explain why.
c. Could part **a** of *this* problem be solved using conservation of energy? If not, briefly explain why.

Problems on WebLinks

26-28. When the variable voltage supply in WebLink 26-1 is raised to a higher setting, will the galvanometer reading increase, decrease, or remain the same? Briefly explain.

26-29. When the variable voltage supply in WebLink 26-1 is set at 1.5 V, how much kinetic energy must an electron have when it leaves the plate on the left to reach the plate on the right?

26-30. The reading on the galvanometer in WebLink 26-1 is always proportional to the ____ (*number of electrons leaving the left plate; number of electrons reaching the right plate; voltage to which the variable voltage supply is set; distance between the two plates*).

CHAPTER TWENTY-SEVEN

The Concept of Quantization

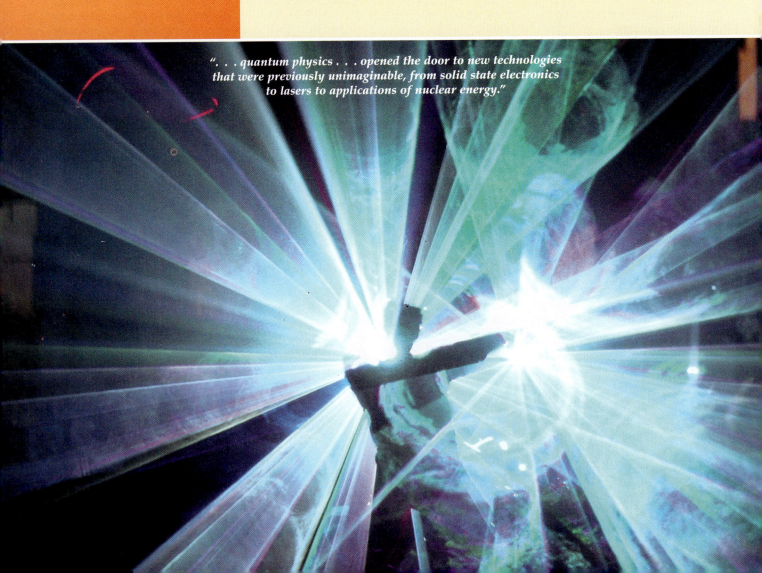

As physicists tried to develop a consistent set of rules for how matter behaved on the microscopic scale of atoms and molecules, the accumulated evidence finally compelled them to conclude that a set of rules different from any governing the macroscopic world had to be in force at the level of atoms and their constituent parts—the rules of quantum physics. The understanding provided by those rules opened the door to new technologies that were previously unimaginable, from solid state electronics to lasers to applications of nuclear energy. This chapter introduces you to the ideas of quantum physics.

"... quantum physics ... opened the door to new technologies that were previously unimaginable, from solid state electronics to lasers to applications of nuclear energy."

27-1 Atomicity and Beyond

In the nineteenth century, physics had embraced the notion of *atomicity* in matter. This means that on a sufficiently microscopic scale, matter is no longer continuous— you cannot take a sample of element X, break it down into ever tinier pieces, and still have something that can be identified by its properties as element X. There is a tiniest possible piece or fundamental building block—a single atom of element X—such that if you break it down further it ceases to have any of the properties that qualify it to be element X.

There was also strong evidence for the atomicity of electric charge. There is a fundamental smallest amount of charge $e = 1.6 \times 10^{-19}$ C, and all other amounts of charge are positive or negative integral multiples of this amount. In dealing with electric circuits, where billions of electrons move along conducting paths, we do not worry about the atomicity of electric charge any more than a plumber needs to worry about the fact that water is made up of individual molecules. In these situations, we treat both charge and matter as continuous.

In everyday experience, radiated energy also seems continuous. Many of our observations of light and other kinds of electromagnetic radiation, and their interactions with matter, like our being warmed by the sun, can indeed be explained by assuming that electromagnetic waves provide a continuous, uniform transfer of energy from the source to the receiver.

By the beginning of the twentieth century, however, physicists were confronted with the accumulation of a considerable body of evidence that could not be explained adequately by this assumption: the existence of line spectra and the mathematical patterns they exhibited, and the inexplicable aspects of the *photoelectric effect*.

Even in developing a model of matter that could correctly account for *continuous* spectra, Planck had to introduce a basic tiniest building block of energy— the *quantum*. The idea of atomicity or discontinuity was creeping into our understanding of radiated energy as well. But though the quantum worked as a mathematical stratagem, nobody at the turn of the century really knew how to interpret it physically. Planck himself recalled his introduction of the quantum as "an act of despair."

Planck's quantum turned out to be the beginning of a conceptual revolution in physics. By the late 1920s, a new **quantum physics** was securely established. It was a physics that described the interaction of electromagnetic radiation and matter on a microscopic scale in terms that made no sense in terms of classical physics. Indeed, it violated the intuitions, shaped by everyday experience, in which classical physics is rooted.

27-2 The Photoelectric Effect and the Idea of the Photon

In 1905, the year he presented the idea of special relativity to the world, Einstein also proposed a revolutionary explanation of the photoelectric effect that successfully accounted for all the experimental observations that were inexplicable by classical physics. Before examining the details of Einstein's proposal, we will need to look more quantitatively at these observations.

Look again at Figure 26-5. Recall that the voltage applied between the plate emitting photoelectrons and the plate opposite can be varied. The *stopping voltage V_o* is the voltage just great enough to prevent even those electrons with maximum kinetic energy from making it across the gap. Because all the kinetic energy is converted to electric potential energy,

$$\tfrac{1}{2}mv_{max}^2 = eV_o \tag{26-2}$$

By measuring V_o, the maximum kinetic energy of the electrons dislodged by any given light source can be determined experimentally. This can be repeated for

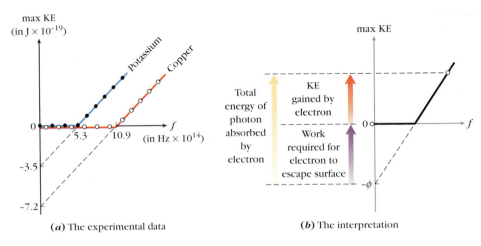

Figure 27-1 Maximum kinetic energy versus frequency for photoelectrons emitted by different materials. (*a*) Experimental data; (*b*) interpretation. Note that in (*a*) there are no data points where the graphs are extended below the axis; kinetic energies can never be negative.

(*a*) The experimental data   (*b*) The interpretation

light of different frequencies. Figure 27-1*a* shows plots of typical data for two different metals. Two features of these graphs are striking:

1. Up to a certain frequency (a different one for each metal), the maximum kinetic energy of the photoelectrons remains zero: The electrons are not dislodged at any velocity. This is the **threshold frequency;** below it, no photoelectric effect occurs.

2. Above this frequency, the graphs for both metals are straight lines with the same slope. We can easily calculate the slope from the data. For potassium, for instance, $\Delta y/\Delta x$ becomes

$$\frac{\Delta(eV_o)}{\Delta f} = \frac{0 - (-3.52 \times 10^{-19}\ \text{J})}{5.3 \times 10^{14}\ \text{s}^{-1} - 0} = 6.6 \times 10^{-34}\ \text{J} \cdot \text{s}$$

You have seen this number before—it is the value of Planck's constant *h*!

The equation of the straight line, written in the form $y = mx + b$, becomes

$$\tfrac{1}{2} mv^2_{\text{max}} = eV_o = hf - \phi \qquad (27\text{-}1)$$

Because the intercept is negative, we have written it as $-\phi$, where ϕ is a positive amount of energy. At the threshold frequency f_o, the maximum kinetic energy is zero, and Equation 27-1 becomes $0 = hf_o - \phi$, or

$$\phi = hf_o \qquad (27\text{-}2)$$

Equation 27-1 then becomes

$$hf = \tfrac{1}{2} mv^2_{\text{max}} + hf_o = \tfrac{1}{2} mv^2_{\text{max}} + \phi \qquad (27\text{-}3)$$

In considering how to interpret this, Einstein was fully aware of Planck's quantum. He saw an inconsistency between the idea of energy flowing continuously and indivisibly into an object as radiated energy and Planck's idea that the emitters and absorbers of radiation were atomic oscillators with energies that changed by discrete amounts. He proposed that

• "the energy of a light . . . [wave] spreading out from a point is not continuously distributed over an increasing space, but consists of a finite number of energy quanta . . .

• which are localized at points in space, . . .

• which move without dividing, and . . .

• which can be produced and absorbed only as complete units."[1]

[1]*Annalen der Physik,* 17 (1905), 132.

In essence, he is describing particles of light energy. These discrete packets of light energy are now called **photons.** To describe their role in the photoelectric effect, Einstein wrote, "The simplest way to imagine this is that a light quantum [a photon] delivers its entire energy to a single electron." In monochromatic light, each photon delivers the same amount of energy to an electron. We can think about this energy in three pieces:

1. the energy expended to get the electron to the surface if it is not already there;

2. the energy required for the electron to break away from the surface once it is there (we can think of this as the *work* that must be done on the electron for it to escape—see Figure 27-1*b*); and

3. the energy the electron retains once it breaks away (this is *kinetic energy*).

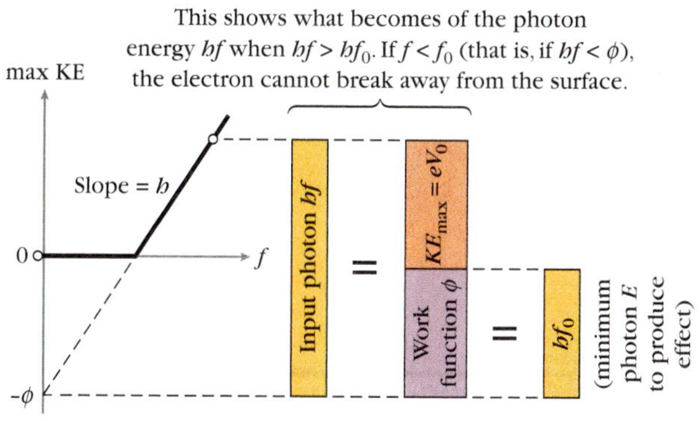

Because the electrons are not all equally close to the surface when they absorb photons, those with the maximum kinetic energy just after breaking away from the surface are the ones that started out at the surface. These are the electrons for which KE_{max} is graphed in Figure 27-1.

Let's see how Einstein's picture fits with the data. Because piece 1 is zero for these electrons,

$$\frac{\text{input energy}}{\text{from photon}} = \frac{\text{work required for electron}}{\text{to escape from surface}} + \frac{\text{maximum KE after breaking}}{\text{away from surface}}$$

we can match the parts of this equality to the parts of Equation 27-3:

$$hf = \phi + \tfrac{1}{2} mv_{max}^2$$

In doing so, we identify hf as the energy of the incident photon. The energy ϕ is commonly called the **work function.** Calculations involving the photoelectric effect generally involve keeping track of energy (Figure 27-2). To see this idea developed more fully, work through WebLink 27-1. If hf is less than the work function $\phi = hf_0$, the electron will not have enough energy to escape from the surface; that is, below the threshold frequency f_0, the effect will not occur.

This shows what becomes of the photon energy hf when $hf > hf_0$. If $f < f_0$ (that is, if $hf < \phi$), the electron cannot break away from the surface.

Figure 27-2 Keeping track of energy in the photoelectric effect.

For **WebLink 27-1: How the Photon's Energy Is Spent,** go to www.wiley.com/college/touger

Reminder on units: The energies of electrons are often expressed in electron volts (eV).

$$1 \text{ eV} = e(1 \text{ V}) = (1.6 \times 10^{-19} \text{ C})(1 \text{ J/C}) = 1.6 \times 10^{-19} \text{ J}$$

EXAMPLE 27-1 *A Basic Photoelectric Calculation*

When light of frequency 9.0×10^{14} Hz is incident on a sodium-surfaced plate, a voltage of 1.4 V is just sufficient to stop electrons emitted by the sodium from reaching an oppositely charged plate. Find **a.** the amount of work that must be done on an electron for it to escape from the sodium surface, and **b.** the threshold frequency for sodium.

Solution

Choice of approach. The conservation of energy equations for the photoelectric effect (Equations 27-1 to 27-3) are summarized graphically in Figure 27-2.

What we know/what we don't.

$$h = 6.6 \times 10^{-34} \text{ J} \cdot \text{s} \qquad e = 1.6 \times 10^{-19} \text{ C}$$

$$f = 9.0 \times 10^{14} \text{ Hz} = 9.0 \times 10^{14} \text{ s}^{-1} \qquad V_0 = 1.4 \text{ V} = 1.4 \text{ J/C} \qquad \phi = ? \qquad f_0 = ?$$

EXAMPLE 27-1 *continued*

The mathematical solution. **a.** Equation 27-1 tells us the maximum kinetic energy is lost as potential energy due to the stopping potential is gained, so

$$\tfrac{1}{2}mv_{max}^2 = eV_o = (1.6 \times 10^{-19}\,\text{C})(1.4\,\text{J/C}) = 2.24 \times 10^{-19}\,\text{J}$$

We can now solve Equation 27-3 to find ϕ:

$$\phi = hf - \tfrac{1}{2}mv_{max}^2 = (6.6 \times 10^{-34}\,\text{J} \cdot \text{s})(9.0 \times 10^{14}\,\text{s}^{-1}) - 2.24 \times 10^{-19}\,\text{J}$$

$$= 5.94 \times 10^{-19}\,\text{J} - 2.24 \times 10^{-19}\,\text{J} = \mathbf{3.7 \times 10^{-19}\,J = 2.3\ eV}$$

b. At threshold, the photon energy is hf_o and $v_{max} = 0$, so Equation 27-3 reduces to $\phi = hf_o$. Then

$$f_o = \frac{\phi}{h} = \frac{3.7 \times 10^{-19}\,\text{J}}{6.6 \times 10^{-34}\,\text{J} \cdot \text{s}} = \mathbf{5.6 \times 10^{14}\,s^{-1}}$$

◆ Related homework: Problems 27-8, 27-9, 27-10, and 27-11.

Einstein's model also explained the other unexpected features of the photoelectric effect. Why did the maximum kinetic energy of the photoelectrons depend only on the frequency and not on the intensity of the incident light? Because the frequency determines the energy of the incoming photon. Why did the photoelectric effect start up virtually instantaneously, even at very low input intensities? If one photon delivers all its energy to one electron, the incoming energy is *not* distributed uniformly and gradually among all surface electrons, and the effect begins as soon as individual electrons absorb photons.

One further aspect of photons is important. Because photons are the constituents of light, they travel at the speed of light and therefore have zero rest mass, because otherwise they would have infinite energy according to relativity theory. Photons are *massless;* they are entirely energy. When they are absorbed, and no longer traveling at speed c, they cease to exist.

Although Einstein's idea of the photon answered some questions, it raised others equally distressing to his fellow physicists. Einstein's words—"[the photons] are localized at points in space . . . [and] move without dividing"—describe *particles,* though they differ from conventional particles in that they have no rest mass. Had not Young's double-slit experiment a century before settled the question of whether light was a stream of particles or a wave? Interference patterns were only explicable by a wave model. But Einstein was asserting that the photoelectric effect was a light phenomenon that could not be explained by a wave model and had to be explained by a particle model.

Einstein's view did not immediately receive universal acceptance. Some physicists felt that photoelectric measurements of greater accuracy would show discrepancies from the Einstein model. But when finally performed, those measurements only served to confirm it. It was for his explanation of the photoelectric effect, not the theory of relativity for which he is most famous, that Einstein received the Nobel Prize in 1921.

But a more subtle question remained. Could light be *both* particlelike and wavelike? This is not the early-nineteenth-century question of whether light *is* a particle or a wave; the question is now whether light has wave*like* properties, particle*like* properties, or both. This might be so if large numbers of photons collectively exhibited wavelike properties. The most current view of photons is that

• They *are* both particlelike and wavelike.

• They are not altogether localized at a single point, as Einstein had proposed.

• Even a single photon can exhibit wave properties in double-slit experiments.

Einstein's model, though an important initial step, was not sufficient to establish this connection. However, it did acknowledge a link between the two models, because the energy hf of the "particle" was proportional to the frequency, a property of the "wave."

27-3 The Bohr Model of the Atom

Danish physicist Niels Bohr (1885–1962) was just twenty years old when Einstein published his explanation of the photoelectric effect. Just a few years later, Bohr worked first with Thomson then with Rutherford, placing himself in the thick of leading-edge inquiries into the nature of atomic structure. In 1913, Bohr brought these two lines of thought together by proposing a model of a photon-emitting atom. The persuasiveness of Bohr's model lay in the fact that it could explain the mathematical pattern

$$\frac{1}{\lambda} = R_H\left(\frac{1}{m^2} - \frac{1}{n^2}\right) \tag{26-3}$$

obeyed by the lines of the hydrogen spectrum.

Bohr's model incorporated Rutherford's solar system model to describe the hydrogen atom and Einstein's photon model to describe the radiation it emits or absorbs. The main steps in his reasoning were these:

1. Using the solar system model, we can calculate the total energy E of the hydrogen atom. (We do this below.)

2. The total energy changes when the atom absorbs or emits a photon of radiated energy. The amount of energy gained or lost is exactly the amount of energy carried by the absorbed or emitted photon:

$$\Delta E = E_f - E_i = \pm hf \tag{27-4}$$
(+ for absorption, − for emission)

3. The angular momentum L of the electron orbiting around the nucleus is quantized; it can only have particular values and not those in between. The allowable values are $L = \frac{h}{2\pi}, 2\frac{h}{2\pi}, 3\frac{h}{2\pi}, \ldots n\frac{h}{2\pi}$ (n = any positive integer). This was a radical new proposal. Although Bohr derived it by subtle reasoning from more basic assumptions, it will be easier for us to treat it as an assumption—one for which we will shortly provide a rationale.

4. When the angular momentum is not changing, the electron is not radiating energy. But an orbiting electron has an acceleration of magnitude v^2/r toward the center of the circle, so this contradicts classical electromagnetic theory, which holds that an electric charge radiates energy whenever it is accelerating. For example, the oscillating charges that we assumed to be the sources of electromagnetic waves have sinusoidal accelerations, just like any other simple harmonic oscillators.

In Bohr's picture, the total energy of the hydrogen atom is the electric potential energy plus the kinetic energy of the electron (if the nucleus is assumed stationary):

$$E = -\frac{ke^2}{r} + \tfrac{1}{2}mv^2$$

We can simplify this expression by applying Newton's second law to the radially accelerated electron. Because the electrostatic force ke^2/r^2 is responsible for the radial acceleration v^2/r, $F = ma$ becomes

$$\frac{ke^2}{r^2} = m\frac{v^2}{r}$$

or (multiplying both sides by $r/2$)

$$\frac{ke^2}{2r} = \tfrac{1}{2}mv^2$$

Substituting for $\tfrac{1}{2}mv^2$ then gives us $E = -\frac{ke^2}{r} + \frac{ke^2}{2r}$ or

$$E = -\frac{ke^2}{2r} \tag{27-5}$$

Let us now explore the implications of making the seemingly strange assumption that the angular momentum $L = n\frac{h}{2\pi}$ (n = any positive integer). Assuming a circular orbit, the angular momentum of the electron (recall Equation 9-19) is just $L = mvr$. Bohr's assumption then says

$$mvr = n\frac{h}{2\pi} \tag{27-6}$$

We can now . . .

. . . write down Newton's second law again: $\quad \dfrac{ke^2}{r^2} = m\dfrac{v^2}{r}$

. . . multiply both sides by mr^3: $\quad mke^2r = m^2v^2r^2 = (mvr)^2$

. . . use $mvr = n\frac{h}{2\pi}$ to substitute for mvr: $\quad mke^2r = \left(n\dfrac{h}{2\pi}\right)^2$

. . . and solve for r: $\quad r = n^2\dfrac{h^2}{4\pi^2kme^2} \quad (n = 1, 2, 3, \ldots)$ (27-7)

This says that

the radius is quantized

because the right side involves only constants and n, which can only have integer values; n is called the **quantum number.** The smallest possible value of the radius, which occurs when $n = 1$, is

$$r_1 = 0.528 \times 10^{-10} \text{ m} \quad (0.528 \text{ Å or } 0.0528 \text{ nm}) \tag{27-8}$$

and is called **the radius of the first Bohr orbit.** Equation 27-7 can then be written as

$$r = n^2r_1 \tag{27-7a}$$

From Equations 27-5 and 27-7, it follows that

the total energy of the atom is also quantized

$$E = -\frac{ke^2}{2r} = -\frac{ke^2}{2}\left(\frac{4\pi^2kme^2}{n^2h^2}\right)$$

or

$$E_n = -\frac{2\pi^2m(ke^2)^2}{n^2h^2} \tag{27-9}$$

or

$$E_n = -\frac{1}{n^2}E_I \tag{27-10}$$

$$\left(\text{where } E_I = \frac{2\pi^2m(ke^2)^2}{h^2}\right)$$

We have used E_I to represent the part of E_n that remains constant. To see how E varies with n, we need only look at how $-1/n^2$ behaves. Figure 27-3 shows this behavior schematically. The energy is most negative when $n = 1$. This

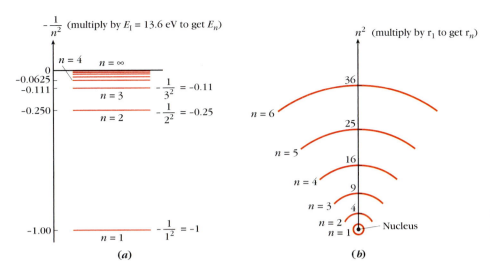

$-\dfrac{1}{n^2}$ (multiply by $E_1 = 13.6$ eV to get E_n)

$n = 4$ $n = \infty$

0
-0.0625
-0.111 $-\dfrac{1}{3^2} = -0.11$ $n = 3$
-0.250 $-\dfrac{1}{2^2} = -0.25$ $n = 2$

-1.00 $-\dfrac{1}{1^2} = -1$ $n = 1$

(a)

n^2 (multiply by r_1 to get r_n)

36
$n = 6$
25
$n = 5$
16
$n = 4$
9
$n = 3$ 4
$n = 2$
$n = 1$ Nucleus

(b)

Figure 27-3 How (a) total energy E and (b) orbital radius r vary with quantum number n in the Bohr model.

lowest energy state is called the **ground state.** Higher energy states are called **excited states,** and one speaks of the atom being excited from a lower to a higher state. This change in the atom's energy occurs as the electron goes from one orbit to another.

We can interpret this variation with n in terms of bound and unbound states (review Case 8-2 and remember that electrostatic and gravitational forces are analogous). The total energy is most negative when the system is in its most tightly bound state and the electron is in its smallest radius orbit. As n increases, the energy also increases, becoming less negative. Because of the additional energy input, the electron is more weakly bound to the nucleus. Finally, at $n = \infty$, the total energy reaches zero. With sufficient energy input, the total energy becomes positive, the system is unbound, and the Bohr model, which describes the bound system, can no longer be used to determine E. Because $E_1 = -E_I$, an energy input $+E_I$ is needed to bring the system energy to zero. It is the amount of energy required to break the electron away from the nucleus from its most tightly bound state—in other words, to *ionize* the hydrogen atom. E_I is therefore called the **ionization energy** (hence the subscript I).

If only certain total energies are permissible, then only certain possible *changes* in energy are possible in going from an initial energy state to a final energy state, as long as the electron remains within the atom (Figure 27-4). In the Bohr model

➡**Notation:** We distinguish between E_I (I for ionization) and E_1, the lowest energy state. $E_1 = -E_I$.

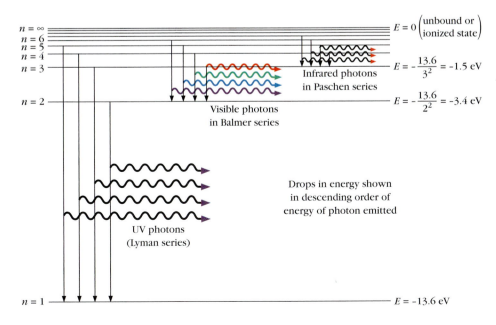

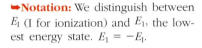

$n = \infty$
$n = 6$
$n = 5$
$n = 4$
$n = 3$

$E = 0$ $\left(\begin{array}{c}\text{unbound or}\\ \text{ionized state}\end{array}\right)$

$E = -\dfrac{13.6}{3^2} = -1.5$ eV

Infrared photons
in Paschen series

$n = 2$

$E = -\dfrac{13.6}{2^2} = -3.4$ eV

Visible photons
in Balmer series

Drops in energy shown
in descending order of
energy of photon emitted

UV photons
(Lyman series)

$n = 1$

$E = -13.6$ eV

Figure 27-4 Differences between energy levels determine the energies of photons that the hydrogen atom can emit (shown) or absorb.

For **WebLink 27-2:**
Bohr Bingo,
go to
www.wiley.com/college/touger

the amount of the energy change is the amount of energy in the emitted or absorbed photon, so this means that photons of only certain energies are possible, and not those in between. Working through WebLink 27-2 should help you visualize this relationship. Because the photon energy $E = hf$, this in turn means that only certain frequencies can be emitted or absorbed, and not those in between. In short, *hydrogen must have a discrete spectrum*. Notice, however, that as n becomes very large, the separations between successive energy levels become very small; the allowable energies come close to having a continuous range of values.

EXAMPLE 27-2   *Photon Emission from the Bohr Hydrogen Atom*

For a guided interactive solution, go to Web Example 27-2 at www.wiley.com/college/touger

Calculate the energy of the photon emitted by the hydrogen atom as it drops from the $n = 3$ state to the $n = 2$ state.

Brief Solution

Choice of approach. The photon carries off all the energy that the atom loses (Equation 27-4). Therefore, we just have to find the energy in the two states (by Equation 27-9 or 27-10), and then calculate the difference. Pay careful attention to sign; the change in energy must be negative because by emitting rather than absorbing a photon, the atom is losing energy.

What we know/what we don't.

$$k = 9.0 \times 10^{9} \ \text{N} \cdot \text{m}^2/\text{C}^2$$

$$m_{\text{electron}} = 9.11 \times 10^{-31} \ \text{kg} \qquad e = 1.6 \times 10^{-19} \ \text{C} \qquad h = 6.6 \times 10^{-34} \ \text{J} \cdot \text{s}$$

$$E_{\text{I}} = ? \qquad E_3 = ? \qquad E_2 = ? \qquad \Delta E = ?$$

The mathematical solution. It is simpler to first calculate E_{I} and then use Equation 27-10 to find E_3 and E_2.

$$E_{\text{I}} = \frac{2\pi^2 m (ke^2)^2}{h^2} = \frac{2\pi^2 m k^2 e^4}{h^2}$$

$$= \frac{2\pi^2 (9.11 \times 10^{-31} \ \text{kg})(9.0 \times 10^{9} \ \text{N} \cdot \text{m}^2/\text{C}^2)^2 (1.6 \times 10^{-19} \ \text{C})^4}{(6.6 \times 10^{-34} \ \text{J} \cdot \text{s})^2}$$

$$= 2.2 \times 10^{-18} \ \text{J}$$

(A calculation using more exact values would give us 2.176×10^{-18} J, or 13.58 eV, the usual accepted value for the ionization energy of hydrogen.) We then get

$$E_3 = -\frac{1}{3^2} E_{\text{I}} = -\frac{1}{9}(2.2 \times 10^{-18} \ \text{J}) = -2.4 \times 10^{-19} \ \text{J}$$

and

$$E_2 = -\frac{1}{2^2} E_{\text{I}} = -\frac{1}{4}(2.2 \times 10^{-18} \ \text{J}) = -5.5 \times 10^{-19} \ \text{J}$$

Because the $n = 2$ state is the final state, Equation 27-4 gives us

$$\Delta E = E_2 - E_3 = -5.5 \times 10^{-19} \ \text{J} - (-2.4 \times 10^{-19} \ \text{J}) = -3.1 \times 10^{-19} \ \text{J}$$

The $-$ sign indicates the loss of energy when a photon is emitted, so the energy hf of the emitted photon is

$$\mathbf{3.1 \times 10^{-19} \ J}$$

◆ Related homework: Problems 27-17, 21-20, 21-22, and 21-25.

Knowing the energies of the photons that can be emitted, we can find their wavelengths in a vacuum. It follows from Equation 27-10 that between an initial state n and a final state m,

$$\Delta E = E_m - E_n = -\frac{1}{m^2}E_I - \left(-\frac{1}{n^2}E_I\right) = -E_I\left(\frac{1}{m^2} - \frac{1}{n^2}\right)$$

The change ΔE is equal to the energy E of the emitted photon. But from $E = hf$ and $c = f\lambda$, it follows that $E = hc/\lambda$. Then from Bohr's model, we conclude that

$$\frac{1}{\lambda} = \frac{E}{hc} = \frac{E_I}{hc}\left(\frac{1}{m^2} - \frac{1}{n^2}\right) \tag{27-11}$$

This conclusion can be checked against the measurements. It has the same form as the Rydberg equation,

$$\frac{1}{\lambda} = R_H\left(\frac{1}{m^2} - \frac{1}{n^2}\right) \tag{26-3}$$

To see if the theory fits the observations, we must check whether $\frac{E_I}{hc}$ and R_H have the same numerical value. Experimental data (see Section 26-2) gave the value $R_H = 1.097 \times 10^7 \text{ m}^{-1}$. The precise value of E_I is 2.176×10^{-18} J (noted in Example 27-2), so

$$\frac{E_I}{hc} = \frac{2.176 \times 10^{-18} \text{ J}}{(6.625 \times 10^{-34} \text{ J}\cdot\text{s})(2.998 \times 10^8 \text{ m/s})} = 1.096 \times 10^7 \text{ m}^{-1}$$

the same value apart from rounding errors. *Bohr's model predicts the wavelengths that are actually found in the hydrogen spectrum.*

The quantization of energy, and now of angular momentum as well, had again succeeded in explaining an observed pattern of data that had been inexplicable by classical physics. But by doing so, it again raised disturbing questions. In Bohr's model the allowable radii of the electron orbit and the corresponding energy values or "energy levels" of the atom were discrete. By the emission or absorption of a suitable amount of energy "packaged" as a single photon, the atom containing the electron could gain or lose energy, going from one allowable energy level to another. It could not take in or lose an amount of energy—a photon—that would leave the electron in an orbit that was not permissible. But if the in-between radii and associated energy levels were forbidden, how could the electron get from one allowable orbit to another? How could the atom get from one allowable energy level to another? The Bohr model had no answer to this. Bohr fully acknowledged the limitations of his theory, which he described as "a preliminary and hypothetical way of representing experimental fact by an expression in the language of the old physics for a rule whose meaning would become clear when the true language of the new physics was discovered."

Bohr could explain the emissions emerging from this subatomic world by making certain assumptions about its structure and the rules by which it behaved, but he could not get those rules to square with our ordinary expectations about how matter should behave. Those expectations, shaped by our everyday experience of the macroscopic world, were not to be satisfied by a world we could experience only indirectly by means of probes and emissions. The notion of electrons in orbits works as a conceptual tool that accounts for the discrete wavelengths of the line spectrum, but we should not take it literally as an accurate picture of the atom. We will see, in fact, that even as a conceptual tool the picture undergoes substantial modification.

◆**THE CORRESPONDENCE PRINCIPLE** Equation 27-7 tells us that the radii of the Bohr orbits increase with n, eventually approaching the macroscopic domain in which classical physics has successfully explained behavior. Bohr reasoned that there must be what he called a *correspondence principle:*

> **The correspondence principle:** At sufficiently large n, the quantum theory must predict the same behavior as classical physics.

For example, using classical theory we would treat the orbiting electron as an oscillator (since the "shadow" of circular motion is simple harmonic motion) and argue that the emitted wave had the frequency of this oscillator. Using the Bohr theory, we would argue that the electrons drop to the nth level principally from the $(n + 1)$th energy level. We would then use the Rydberg formula to determine the frequency. In fact, as n grows very large, the results of these two approaches become very nearly the same. (See Problem 27-56.)

In the Bohr theory, too, the energy levels get closer together as n increases (Figures 27-3 and 27-4), so the allowable energy values become more nearly continuous, as expected by classical physics. Moreover, the quantum numbers refer only to the bound states of the hydrogen atom. Only specific photon energies can excite an atom from a lower to a higher bound state, but if the incoming photon has more than enough energy to ionize the atom, the excess energy carried away by the electron as kinetic energy can have any value at all (Figure 27-3). For that reason, above a certain frequency experimenters find that absorption spectra are continuous.

◆**BEYOND HYDROGEN** The Bohr calculation works well for hydrogen because it is easy to find the electrostatic force and potential energy in a system consisting of a single proton and a single electron. By substituting $2e$, the charge of a helium nucleus, for the charge e of the proton, Bohr was able to calculate correctly the wavelengths of the line spectrum for the singly ionized helium atom as well. Beyond helium, however, we have to deal with multiple electrons and multiple protons, and the calculations become too complicated to do exactly.

His model did, however, offer some insights about larger atoms. Observations of radioactive decay had revealed that the atomic number dropped by two with each emission of an α particle. This had led to the recognition that the atomic number represented the number of positive charges (the number of protons) in the nucleus, and also the number of electrons when the atom was neutral. Bohr realized further that elements in the same column of the periodic table tended to become ionized in the same way, gaining or losing the same number of electrons. He realized that he could account for this by assuming that with each new row of the periodic table, a new "shell" of electrons began to fill in (Figure 27-5), and that electrons in the same shell had similar orbits. The picture of atomic structure was finally reaching the stage where it could begin to account for chemical behavior.

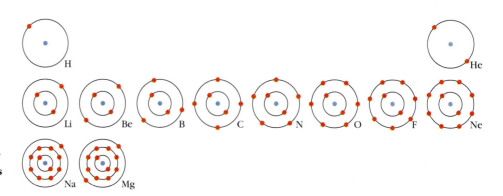

Figure 27-5 Progressive filling of shells in the first dozen elements of the periodic table.

27-4 From Particles *or* Waves to Particle*like* and Wave*like*

Further evidence of the particlelike behavior of radiated energy came from X-ray scattering experiments that American physicist Arthur H. Compton (1892–1962) performed in 1923. If we consider the X ray to be made up of high-energy photons, then scattering means the photons "bounce off" in different directions. The energy of an X-ray photon is orders of magnitude greater than the ionization energy of the atom providing the electron, so virtually all the energy the photon loses to the electron takes the form of kinetic energy. Moreover, because it has lost energy to the electron, the scattered photon has less energy hf than the incident photon, so the frequency of the scattered X rays is reduced, and its wavelength correspondingly increased. The change in wavelength is correctly predicted by applying conservation of momentum and conservation of energy to an elastic collision in which an incident photon collides with and bounces off an initially stationary electron, setting the electron in motion as it does so (Figure 27-6a). The electron's initial momentum is approximately zero because virtually all its kinetic energy comes from the photon. In other words, Compton could explain the observed effect by assuming a particlelike collision in which the photon provides the initial momentum. Because the photon has no rest mass, the equation $E^2 = p^2 c^2 + m^2 c^4$ reduces to $E = pc$. Then

$$p = \frac{E}{c} = \frac{hf}{c} = \frac{hf}{f\lambda}$$

or **photon momentum** $\qquad p = \frac{h}{\lambda}$ (27-12)

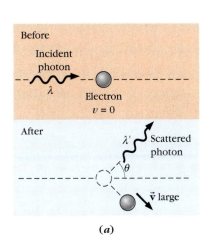

(a)

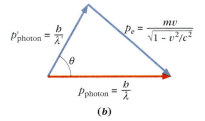

(b)

Figure 27-6 Compton scattering of an X-ray photon by an electron. We can view the scattering as a collision. (a) Before and after states for the collision. (b) Momentum conservation for the collision. The initial momentum (in red) is the vector sum of the two final momenta (in blue).

EXAMPLE 27-3 *Photon and Proton: Comparing Momenta*

At what speed would a proton have to travel to have the same momentum as an X-ray photon of frequency $2.0 \times 10^{18} \text{ s}^{-1}$?

Solution

Choice of approach. We can set the momentum $p = \frac{h}{\lambda}$ of the photon by Equation 27-12 and set it equal to the proton's momentum mv, then solve for v. We can assume our result is correct if the value we get is orders of magnitude less than c, because then the relativistic momentum $mv/\sqrt{1 - v^2/c^2}$ will differ negligibly from mv. Otherwise we must redo the calculation using the relativistic momentum.

EXAMPLE 27-3 *continued*

What we know/what we don't. (P = proton, γ = photon)

$$c = 3.0 \times 10^8 \text{ m/s} \qquad m_P = 1.67 \times 10^{-27} \text{ kg} \qquad h = 6.6 \times 10^{-34} \text{ J} \cdot \text{s}$$

$$f = 2.0 \times 10^{18} \text{ s}^{-1} \qquad \lambda = ? \qquad p_\gamma = ? \qquad p_P = ? \qquad v_P = ?$$

The mathematical solution. The wavelength of the photon is

$$\lambda = \frac{c}{f} = \frac{3.0 \times 10^8 \text{ m/s}}{2.0 \times 10^{18} \text{ s}^{-1}} = 1.5 \times 10^{-10} \text{ m}$$

so its momentum is

$$p_\gamma = \frac{h}{\lambda} = \frac{6.6 \times 10^{-34} \text{ J} \cdot \text{s}}{1.5 \times 10^{-10} \text{ m}} = 4.4 \times 10^{-24} \text{ kg} \cdot \text{m/s}$$

If $p_\gamma = m_P v_P$, then

$$v_P = \frac{p_\gamma}{m_P} = \frac{4.4 \times 10^{-24} \text{ kg} \cdot \text{m/s}}{1.67 \times 10^{-27} \text{ kg}} = \mathbf{2.6 \times 10^3 \text{ m/s}}$$

This is so much less than the speed of light that we need not consider the momentum relativistically.

◆ Related homework: Problems 27-28 and 27-33.

Motivated both by relativity and the idea of a momentum $p = \frac{h}{\lambda}$ for photons, Louis deBroglie, a French doctoral student, proposed in his 1924 dissertation that particles with mass might also have a wavelength associated with them such that their momentum $mv = \frac{h}{\lambda}$ as well:

the deBroglie wavelength $\qquad \lambda = \dfrac{h}{p} = \dfrac{h}{mv}$ (27-13)

(At relativistic speeds mv must be replaced by the relativistic momentum $mv/\sqrt{1 - v^2/c^2}$.)

Just as light exhibits particlelike behavior in its interactions with matter on a microscopic scale but shows wavelike behavior on a large scale, deBroglie speculated that matter might also display both types of behavior, or what physicists call **wave-particle duality**.

What might be the implication of such a suggestion?

Case 27-1 ◆ *A Classical Analogy*

We know that when a string is fixed at both ends, the only allowable frequencies are integer multiples of the fundamental frequency $(f_n = nf_1)$, because only a whole number of half-wavelengths can form on the string. The fact that the string is bounded by two locations where the displacement must be zero gives rise to this *quantization* of the wavelength and frequency. In a similar way (Figure 27-7), if the two ends of the string are fixed to each other (at point P), the string forms a ring. Now the length of the string—that is, the circumference $2\pi r$ of the circle—can only be a whole number of wavelengths $(2\pi r = n\lambda)$ because the "loops" of the wave must be alternately upward and downward. You can't have two

This gives allowable standing wave on ring formed when ends brought together at point P

...but this does not

(a)

Two crests with no trough in between (not possible)

(b)

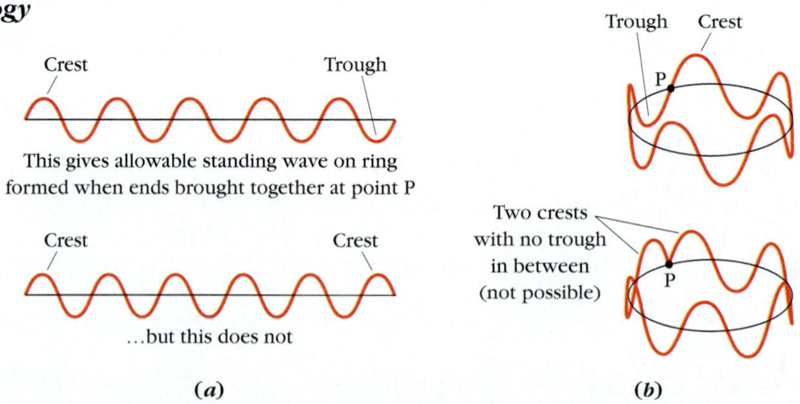

Figure 27-7 How boundary conditions limit allowable wavelengths on a string (*a*) when it is fixed at both ends and (*b*) when the two ends are fixed to each other.

loops of the same kind in a row where the string connects to itself.

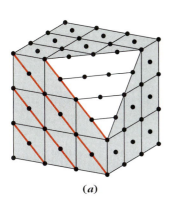

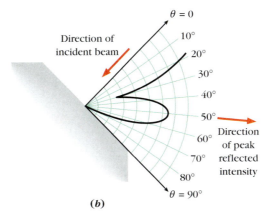

(a)

(b)

Figure 27-8 The Davisson-Germer experiment. (*a*) Layers of atoms of the Ni crystal act as parallel reflecting surfaces for the beam of electrons. (*b*) The observed intensity of the reflected electron beam is plotted against the angle of reflection.

If an electron in an orbit were wavelike, then analogously to waves on the ring in Case 27-1, the electron wavelength must satisfy the condition

$$2\pi r = n\lambda$$

If deBroglie's condition $\lambda = h/mv$ holds true as well, then $2\pi r = nh/mv$. With a little algebraic rearrangement, this becomes $mvr = n\frac{h}{2\pi}$, which is precisely the Bohr model's assertion (Equation 27-6) that the angular momentum of the electron is quantized. Here, too, the quantization arises from the bounded condition (bound state) of the electron. But when it breaks away, does it still have a wavelength? If it does, then beams of electrons should exhibit interference patterns just as beams of electromagnetic radiation do.

In 1927, two separate experiments succeeded in finding evidence of wave interference for a beam of electrons. The first of these, performed by C. H. Davisson and L. H. Germer at Bell Telephone Laboratories, looked at a beam of electrons reflected from the layers of atoms near the cut face of a nickel (Ni) crystal (Figure 27-8*a*). The layers act as a series of equally spaced reflecting surfaces for the incident beam, so if the electron beam is wavelike, the constructive interference condition for parallel reflecting surfaces would apply here as well. The intensity of the reflected beam is found to vary with the reflected angle (Figure 27-8*b*), with a pronounced peak at 50°. This is the angle at which the path difference for reflection from successive layers would be one wavelength for a wave having the expected deBroglie wavelength (see Example 27-4) of the incident electrons, thus resulting in constructive interference.

In the second experiment, G. P. Thomson, inspired by his father J. J. Thomson's gold foil experiment, passed a high-energy beam of electrons through a thin film of material made up of many tiny bits of crystal randomly oriented. He observed interference patterns for electrons much like those observed for X rays incident on polycrystalline powders (Figure 27-9).

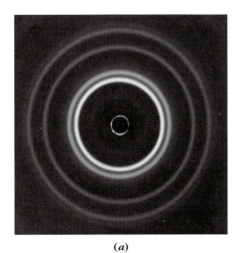

(a)

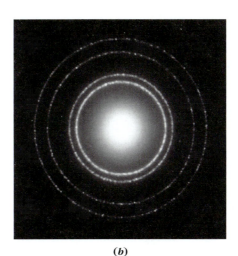

(b)

Figure 27-9 Comparison of (*a*) X-ray and (*b*) electron interference patterns.
(*a*) A diffraction pattern for X rays passing through a thin polycrystalline aluminum foil. (*b*) A diffraction pattern for electrons passing through the same aluminum foil.

The electron, a subatomic particle, had been shown to exhibit wavelike interference properties, just as light, which interference phenomena had shown to be wavelike, was shown to behave like matter. In short, both exhibit a *wave-particle duality*. We are tempted to ask, "Is light *really* a particle or a wave? Is an electron *really* a particle or a wave?" In response to such questions, Arnold Arons has written,

> Our simple modes of speech themselves lead us into a trap. We are tempted to say on one hand that light *is* a wave or on the other that light *is* a stream of particles, and this positive use of the verb "to be", implying a concrete existence of a mode of behavior precisely similar to everyday experiences with macroscopic objects, makes us lose sight of the fact that we are reaching for analogies and comparisons. Our [language] would be clearer if we avoided talking of what light *is* and [just] spoke . . . about how it behaved and what properties it [exhibited] under various circumstances.[2]

Arons is reminding us that we have no direct experience of electrons and photons, and they may not fall neatly into categories such as waves and particles that we have formed to fit the phenomena we experience directly. Observations of emissions and of effects on our probes tell us only that in some circumstances they obey wavelike rules and in others they obey particlelike rules.

Other particles were also ultimately shown to have wavelike properties. For larger particles, they become progressively more difficult to observe because the deBroglie wavelength varies inversely as the mass. As m becomes orders of magnitude greater than the mass of an electron, the deBroglie wavelength becomes orders of magnitudes smaller and is experimentally indistinguishable from zero for macroscopic particles. Example 27-4 reinforces this point.

[2]A. B. Arons, *Development of Concepts of Physics* (Addison-Wesley, Reading, MA, 1965), p. 850.

EXAMPLE 27-4 *Sizes of deBroglie Wavelengths*

For a guided interactive solution, go to Web Example 27-4 at www.wiley.com/college/touger

Find the deBroglie wavelengths of
a. an electron traveling at one one-thousandth of the speed of light.
b. a 0.020-kg marble rolled across the pavement of a playground at a speed of 1.0 m/s.

Brief Solution
Choice of approach. The deBroglie wavelength $\lambda = h/p = h/mv$.

What we know/what we don't.

$$h = 6.6 \times 10^{-34} \, \text{J} \cdot \text{s} \qquad c = 3.0 \times 10^8 \, \text{m/s}$$

a. $m_{\text{electron}} = 9.11 \times 10^{-31} \, \text{kg} \qquad v_{\text{electron}} = 0.001c \qquad \lambda_{\text{electron}} = ?$
b. $m_{\text{marble}} = 0.030 \, \text{kg} \qquad v_{\text{marble}} = 1.0 \, \text{m/s} \qquad \lambda_{\text{marble}} = ?$

The mathematical solution. **a.** The electron's momentum

$$p = mv = (9.11 \times 10^{-31} \, \text{kg})(0.001 \times 3.0 \times 10^8 \, \text{m/s}) = 2.7 \times 10^{-25} \, \text{kg} \cdot \text{m/s}$$

Its deBroglie wavelength is then

$$\lambda = \frac{h}{p} = \frac{6.6 \times 10^{-34} \, \text{J} \cdot \text{s}}{2.7 \times 10^{-25} \, \text{kg} \cdot \text{m/s}} = \textbf{2.4} \times \textbf{10}^{-9} \, \textbf{m}$$

b. The marble's momentum

$$p = mv = (0.030 \, \text{kg})(1.0 \, \text{m/s}) = 0.030 \, \text{kg} \cdot \text{m/s}$$

Its deBroglie wavelength is then

$$\lambda = \frac{h}{p} = \frac{6.6 \times 10^{-34}\,\text{J}\cdot\text{s}}{0.030\,\text{kg}\cdot\text{m/s}} = \mathbf{2.2 \times 10^{-32}\,m}$$

Making sense of the results. For comparison, the size of an atom is on the order of 10^{-10} m and the size of a nucleus is on the order of 10^{-14} m (about 1/10 000 of the size of the atom, as we showed in Example 26-2). So this electron's wavelength is about the size of an atom, but the marble's wavelength is about $\frac{10^{-32}}{10^{-14}} = \frac{1}{10^{18}}$ of the size of the nucleus, which is experimentally indistinguishable from zero.

◆ Related homework: Problems 27-29, 27-30, 27-31, 27-32, and 27-34.

27-5 The Schrödinger Equation

If electrons are waves, they should be describable by wave functions such as $y = Y \sin 2\pi(\frac{t}{T} - \frac{x}{\lambda})$ or $y = Y \sin 2\pi(ft - \frac{x}{\lambda})$. Moreover, because they are particle-like, the frequency and wavelength must be related to the particle properties of total energy and momentum:

Particle Property	Wave Property	Relationship
total energy	frequency	$E = hf$
momentum	wavelength	$\lambda = \dfrac{h}{p}$

Just as Maxwell had arrived at equations for the electric and magnetic fields that were satisfied by wave functions (Section 24-1), Austrian physicist Erwin Schrödinger (1887–1961) deduced an equation incorporating all the conditions the wave function of an electron must meet. These conditions were:

- $E = hf$ and $\lambda = h/p$ for the electron.
- The total energy E of the electron must equal its potential energy (usually labeled V in this context) plus its kinetic energy $\frac{1}{2}mv^2 = \frac{m^2v^2}{2m} = \frac{p^2}{2m}$:

$$E = V + \tfrac{1}{2}mv^2 = V + \frac{p^2}{2m}$$

- The wave function of the electron must be simply sinusoidal in free space, where the potential energy $V = 0$. (This includes the possibility of it being a *complex* function in which the real and imaginary parts are sines or cosines.)

His equation, known as the **Schrödinger equation,** is one of the central equations of modern physics.

Because the Schrödinger equation states the conditions that the wave function must meet, solving it means finding the particular wave function that satisfies those conditions. The solutions are generally complex; they have both real and imaginary parts. For an unbound particle in one dimension, the Schrödinger equation may be satisfied by simple sinusoidal wave behavior such as

$$\text{real part of } \psi = \Psi \sin 2\pi\left(ft - \frac{x}{\lambda}\right) = \Psi \sin \frac{2\pi}{h}(Et - px)$$

(We use the Greek letter *psi*—lowercase [ψ] and uppercase [Ψ]—as we previously used y and Y: ψ gives the instantaneous size of the disturbance and Ψ its maximum value.)

We have seen that when there are boundary conditions (Figure 27-7 and Case 27-1) the wave function is in some way quantized. For instance, the

➥**Complex numbers:** Any complex number can be written as A + iB, where $i \equiv \sqrt{-1}$. A and B are both real numbers, but i is said to be *imaginary,* so in general a complex number has a real part and an imaginary part. When you multiply a complex number $\psi = $ A + iB by its *complex conjugate* $\psi^* = $ A − iB, you get $\psi\psi^* = $ A^2 − i^2B$^2 = $ A^2 − (−1)B$^2 = $ A^2 + B^2, which gives the square of the "size" of ψ. This square is written as $|\psi|^2$, analogously to the square of the absolute value for real numbers.

wavelength may have only certain values and not those in between. In the Schrödinger equation the potential energy sets the boundary condition. It describes the interaction by which the electron is bound. The mathematics by which the equation then gets solved is far beyond the level of this textbook. Some of the solutions, however, are of major importance. We will address one—the solution for the hydrogen atom, where the potential energy is $\frac{ke^2}{r}$.

The Bohr model leads to a single quantum number n. So does fixing the ends of a string, which is a one-dimensional medium. Electrons in the hydrogen atom move in three-dimensional space, so solving the Schrödinger equation leads to three quantum numbers. A fourth quantum number was later included to make the function consistent with observations. Each of these quantum numbers can take on only particular values. Values for all four quantum numbers must be given to fully describe an electron's state. Moreover, in an atom with more than one electron, no two electrons can be in the same state; that is, no two electrons can have all four quantum numbers the same. This principle was first established by Austrian physicist Wolfgang Pauli (1900–1958) even before Schrödinger published his equation and is called the **exclusion principle.** The idea of four quantum numbers, arising first out of attempts to explain observations of multiple spectral lines in detail, had itself preceded Schrödinger's theory. At first it was simply a proposed pattern that seemed to fit the observations, but it was recognized as an idea of some power because, coupled with the exclusion principle, it offered an explanation for the order in the **periodic table.** Specifically, it explained the periodic recurrence of chemical properties as the number of electrons per neutral atom increases. Schrödinger's equation provided a theoretical basis for three of the four quantum numbers.

✦HOW QUANTUM NUMBERS ACCOUNT FOR THE PERIODIC TABLE

When the quantum numbers apply to the electrons of atoms other than hydrogen, they describe not the different possible states of a single electron but the different possible states—one per electron—of all the electrons in a multielectron atom. Unless excited, each electron of the atom is in its lowest possible energy state. Because the Pauli exclusion principle forbids two electrons from having the same two quantum numbers, the electrons in the lowest energy state (ground state) of the atom fill in the allowable electron states from lowest to highest.

Generally speaking, the values of n, as in the Bohr model, designate shells of successively higher energy. But now there are *subshells* (designated by l) within each shell. Each time the number of electrons increases from one atom to the next in the periodic table (Table 27-1), the next state in the outermost subshell is filled. Typically, when a subshell is complete, the next subshell starts to fill in. For $n = 1$ and 2, the number of elements in the nth row of the periodic table is simply the number of allowable states in the nth shell. Beyond $n = 2$, it gets more complicated because, for example, 4s states are less energetic than 3d states and therefore fill in first.

➥**Labeling subshells:** For historical reasons, the values of l are commonly represented by letters, as follows:

value of l	0	1	2	3	4	5	6
letter	s	p	d	f	g	h	i

For example, an electron in the $n = 3$, $l = 1$ subshell is called a 3p electron.

The quantum number l is associated with the electron's orbital angular momentum. We can think of the orbiting electron as a small current loop. Classically, the axis of this loop would line up with a magnetic field so that there is no net torque on it. Work must be done—energy added—to change the loop's orientation. Quantum mechanically, only certain amounts of added energy are possible. An additional quantum number, m_l, is associated with the loop's orientation. To account for various experimental observations, a fourth quantum number, m_s, was also introduced. It can be explained by assuming that the electron has an intrinsic property called **spin.** Strictly speaking, we cannot describe spin in classical terms. A convenient (though not strictly correct) mental picture is that the electron spins on its own axis. In this picture, when a magnetic field is applied, the angular momentum associated with this spin can

Table 27-1 Periodic Table of the Elements

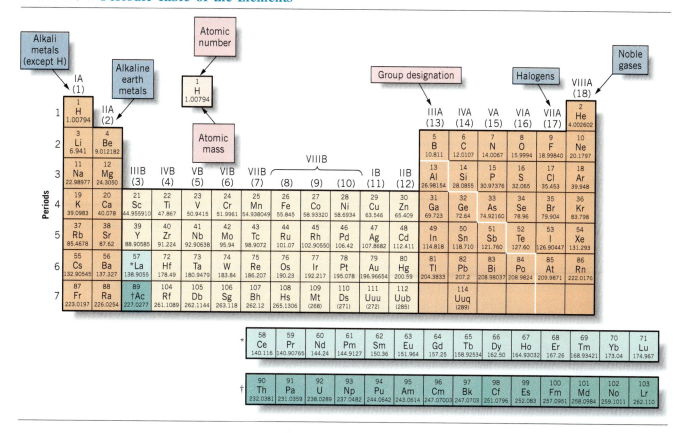

align more nearly parallel (a state called *spin up*) or more nearly antiparallel (*spin down*) to the field (recall Figure 22-9). The quantum number m_s thus has two possible values.

The allowable values for the quantum numbers are found as follows:

PROCEDURE 27-1

Identifying the States of Electrons in an Atom

1. To find n, use the quantization rule $n = 1, 2, 3, 4 \ldots$ (27-14n)
2. For each n, use the rule

$$l = 0, 1, 2, 3, \ldots, n - 1 \qquad (27\text{-}14l)$$

to find the values of l permitted with that value of n.
3. For each l, use the rule

$$m_l = 0, \pm 1, \pm 2, \pm 3, \ldots, \pm l \qquad (27\text{-}14m)$$

to find the values of m_l permitted with that value of l.
4. Each allowable combination of n, l, and m_l can then occur with either

$$m_s = -\tfrac{1}{2} \text{ or } m_s = +\tfrac{1}{2}$$

5. Use the notation $|n\, l\, m_l\, m_s >$ to designate each state by its four quantum numbers. For example, to represent the state in which $n = 2$, $l = 1$, $m_l = -1$, and $m_s = +\tfrac{1}{2}$, write $|2\ 1\ -1\ +\tfrac{1}{2} >$.

This is a brief summary. If you have taken a chemistry course, you have probably seen this in more detail.

Table 27-2 The $n = 3$ States of Electrons in an Atom

n	3								
l	0	1			2				
m_l	0	-1	0	1	-2	-1	0	1	2
m_s	$-\frac{1}{2}$ $+\frac{1}{2}$	$-\frac{1}{2}$ $+\frac{1}{2}$	$-\frac{1}{2}$ $+\frac{1}{2}$	$-\frac{1}{2}$ $+\frac{1}{2}$	$-\frac{1}{2}$ $+\frac{1}{2}$	$-\frac{1}{2}$ $+\frac{1}{2}$	$-\frac{1}{2}$ $+\frac{1}{2}$	$-\frac{1}{2}$ $+\frac{1}{2}$	$-\frac{1}{2}$ $+\frac{1}{2}$

Table 27-2 shows the results of applying the procedure systematically when $n = 3$. Here, $n - 1 = 2$, so the second row of the table shows the values of l going from 0 to $n - 1 = 2$, in accordance with Equation 27-14l. The third row of the table then applies Equation 27-14m for each of these values of l; for example, when $l = 2$, m_l takes on all integer values from $m_l = -2$ to $m_l = 2$. Each of these in turn can occur with a value of $+\frac{1}{2}$ or $-\frac{1}{2}$ for m_s. Each allowable state $|\,n\,l\,m_l\,m_s\,>$ can be identified by combining one of the values of m_s with the values of n, l, and m_l above it. For example, the state that is highlighted is $|\,3\,2\,-1\,+\frac{1}{2}\,>$. Counting across the last row, we see that there are 18 possible states in all. To see how these rules are applied in greater step-by-step detail, work through WebLink 27-3.

For **WebLink 27-3:**
**Building the
Periodic Table,** go to
www.wiley.com/college/touger

27-6 Probability and Uncertainty in the Quantum Universe

The wave-particle duality confronts us with two critical questions that are inextricably linked:

1. How can large numbers of discrete particles collectively display wavelike behavior?

2. If an electron has wavelike properties and can be represented by a wave function ψ, then ψ, like y, tells us the size of something that is varying in space and time, just as y for water waves tells us the height of the water above or below its undisturbed level. But what is it that ψ is the size of?

Figure 27-10 provides a clue. As in Figure 27-9, the bright regions in Figure 27-10c indicate where the film has been exposed. (The exposed regions would be dark on the original negative). In the double-slit pattern for light, the film is most exposed where the light intensity is greatest. But in Figure 27-10a and b, we see something else happening. Here the intensity of exposure builds point by point as more electrons strike the film. When there are relatively few electrons, they appear to be distributed more or less randomly. But when the number of electrons is great enough, it becomes evident that they strike the film more frequently in some regions than others; they have a higher *probability* of striking in those regions. From a particle point of view, then, *the intensity of the beam at any point on the film is proportional to the probability of finding a particle in a tiny region about that point.*

Suppose we now think of the beam as wavelike. As we noted for sound in Section 14-7, the intensity or energy flux density of a classical wave is proportional to the square of its amplitude, just like the energy $\frac{1}{2}kY^2$ of a stationary oscillatory disturbance. Thus, in 1926, German physicist Max Born (1882–1970) proposed an interpretation which, roughly speaking, associated the square of the wave function in a region (actually $|\psi|^2$, because ψ is complex) with the probability of finding a particle in that region. More precisely, $|\psi|^2$ tells us the probability per unit of space (e.g., per m^3 in three dimensions) so it is called the **probability density.** The interference pattern for a very large number of particles shows how the probabilities are distributed.

As Figure 27-9 shows, probability distributions occur for photons as well as for particles with mass. The Schrödinger description only applies to particles with

(a) After 100 electrons

(b) After 3000 electrons

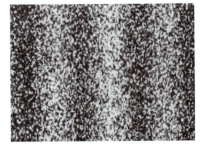
(c) After 70 000 electrons

Figure 27-10 An electron version of Young's double-slit experiment. The familiar fringe pattern gradually becomes apparent as more and more electrons strike the screen.
Source: A. Tonomura, J. Endo, T. Matsuda, and T. Kawasaki, *Am. J. Phys.,* 57 (1989), 117.

mass, but a more advanced description of light (called *quantum field theory*) associates a wave function representing a probability amplitude with the photon, much as the Schrödinger description makes that association for, say, electrons.

◆**PROBABILITY AND UNCERTAINTY** If a particle is bound, it means it is localized within a certain region of space. The probability of finding the particle is high in this region and very small or zero elsewhere.

Suppose a particle has a well-defined momentum. Because $p = h/\lambda$, this particle would also have a well-defined wavelength. What does this mean about how localized the particle is? Simple sinusoidal functions—sines and cosines—meet the criterion of having a well-defined wavelength. They also have a constant amplitude (let's call it A). The real part of a complex wave function may be one such function—say $A \sin kx$—and the imaginary part may be another—say $A \cos kx$. The probability density $|\psi|^2$ combines the squares of the real and imaginary parts:

$$|\psi|^2 = (A \cos kx)^2 + (A \sin kx)^2 = A^2(\cos^2 kx + \sin^2 kx) = A^2$$

$|\psi|^2$ has the same value at all points x in one-dimensional space. A particle with such a probability density could be at any of these points with equal likelihood; it is not localized at all.

We can simplify the remainder of this discussion by considering only the real parts of wave functions. If the amplitude were smaller in some regions than in others, the particle would be less likely to be found in the regions where the amplitude is smaller. This happens as soon as we add two sine waves with slightly different wavelengths as in Figure 27-11*a* and *b* (this is

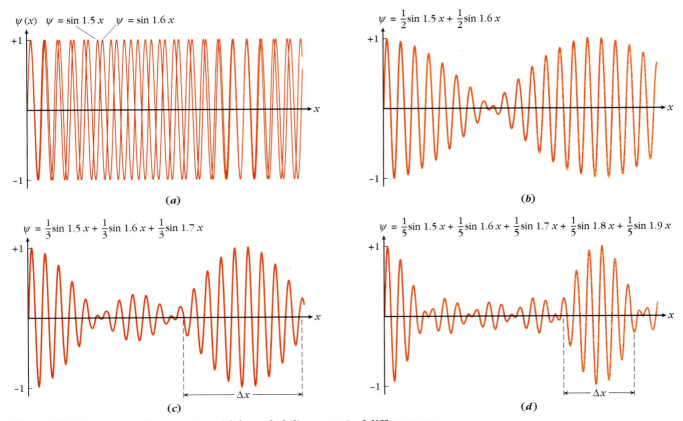

Figure 27-11 As more and more sinusoidal "probability waves" of different wavelengths are superimposed, the probability of finding the object becomes increasingly localized or particlelike. (*a*) Two sine functions of slightly different wavelengths. (*b*) The superposition of the two sine functions. Also shown are the superpositions of (*c*) three sine functions and (*d*) five sine functions of slightly different wavelengths. Note that in (*d*) Δx is smaller than in (*c*). Δx is the region in which ψ has substantial amplitude, so that the probability density $|\psi|^2$ is substantial. A smaller Δx means the probability is more localized.

much like the picture we had for beats in Figure 14-23). (Because there are two contributions, we have multiplied each contribution by $\frac{1}{2}$ to keep the maximum total amplitude the same.) Adding a third sine wave as in Figure 27-11c localizes the particle even more. As waves of more and more wavelengths (and suitable amplitudes) get added to the sum (Figure 27-11d), the region(s) where the wave amplitude of the resulting wave are significantly different than zero become increasingly confined, and so, correspondingly, does the region where there is significant likelihood of finding the particle. This region is called a **wave packet.** The particle is now *localized* to this region. Figure 27-11 suggests that the greater the spread $\Delta\lambda$ of wavelengths contributing to the wave packet, the smaller the length Δx of the wave packet will be, and the more localized the particle will be. One way to express this relationship between Δx and $\Delta\lambda$ is to say that the product of the two must always be at least some minimum value:

$$\Delta x \, \Delta\lambda \geq \text{minimum value} \qquad (27\text{-}15)$$

But because $p = h/\lambda$, the greater the spread of wavelengths is, the greater the spread Δp of momentum values will be (see Problem 27-65), and the less well we can know the particle's momentum. In short, *we can localize the particle—we can be precise about where it is—only by surrendering our ability to be precise about its momentum.*

➥**Young ideas:** Heisenberg was 26 years old when he developed his uncertainty principle. Einstein was 26 when he explained the photoelectric effect. The Bohr model was put forward by Bohr when he was 28. DeBroglie, at 32, was still a doctoral student when he proposed the deBroglie wavelength. The quantum revolution in physics was brought about substantially by young scientists scarcely beyond their student years.

In 1927, using more advanced mathematics and the fact that Δp increases with $\Delta\lambda$, German physicist Werner Heisenberg (1901–1976) was able to specify the "minimum value" and obtain the inequality

$$\Delta x \, \Delta p \geq \frac{h}{2\pi} \qquad (27\text{-}15a)$$

which is called the **Heisenberg uncertainty principle.**

The term *uncertainty*, perhaps better called indeterminacy, refers to the following. A wave packet represents a particle. For the wave packet to be confined within a region of length Δx, it must have a range of wavelengths $\Delta\lambda$ contributing to it, so we cannot specify the particle's momentum $p = h/\lambda$ more precisely than to say that it falls within a range Δp. This Δp is called the uncertainty in the particle's momentum. Conversely, if we know the momentum to within Δp, there is a range of positions Δx where there is a significant likelihood of finding the particle. The uncertainty Δx in its position limits our ability to say where it is. These uncertainties are not experimental errors arising out of the limitations of our measuring equipment. The uncertainty principle tells us that as Δx decreases, Δp increases, and vice versa. The more precisely we know one of the two quantities (position and momentum), the less well we can determine the other. This is a limitation *in principle*, no matter how good our measuring equipment is. We can *never* know the position and momentum exactly *at the same time*.

➥We could infer Equation 27-15b from 27-15a by analogy, because the wave function of a free particle involves sinusoidal functions like $\sin \frac{2\pi}{h}(Et - px)$, in which E relates to the time variation of the wave in the same way that p relates to the variation with position.

For a photon, which by its very nature moves at the speed of light c, $\Delta x = c\Delta t$. Also, because $E = pc$, $\Delta E = c\Delta p$. Then $\Delta p = \frac{\Delta E}{c}$. Therefore, $\Delta p \Delta x = (\frac{\Delta E}{c})(c\Delta t) = \Delta E\Delta t$, so that the uncertainty principle can also take the form

$$\Delta E \Delta t \geq \frac{h}{2\pi} \qquad (27\text{-}15b)$$

This form is valid for particles with mass as well as photons. It tells us that measuring the energy of a particle to within an interval ΔE requires a time interval of at least Δt. Therefore we cannot say precisely at what instant the particle has the measured energy, only that it falls within a certain time interval. Conversely, if we want to say more accurately *when* the particle has this energy, we must give up some accuracy in saying *how much energy* it has.

EXAMPLE 27-5 *Uncertainty*

Find the minimum uncertainty in the momentum of a ground state electron in a hydrogen atom.

Solution

Choice of approach. Strictly speaking, Equation 27-15a should be written as $\Delta x \Delta p_x \approx \frac{h}{2\pi}$. The relationships in the y and z directions are the same, so we will just calculate Δp_x. Roughly, the electron's x coordinate can be anywhere within the diameter $2r_1$ of the first Bohr orbit, so that is the approximate uncertainty in its position. The probability of it being beyond that region is small, but not zero. In finding uncertainties, we typically deal with order of magnitude or ballpark estimates, not with precise values.

What we know/what we don't.

$$r_1 = 0.528 \times 10^{-10} \text{ m (from Equation 27-8)}$$

$$h = 6.63 \times 10^{-34} \text{ J} \cdot \text{s} \qquad \Delta x \approx 2r_1 \qquad \Delta p_x = ?$$

The mathematical solution. From Equation 27-15a,

$$\Delta p_x \approx \frac{h}{2\pi(\Delta x)} \approx \frac{h}{2\pi(2r_1)} \approx \frac{6.63 \times 10^{-34} \text{ J} \cdot \text{s}}{4\pi(0.528 \times 10^{-10} \text{ m})} = \mathbf{1 \times 10^{-24} \text{ kg} \cdot \text{m/s}}$$

Making sense of the results. The lack of additional significant figures in our solution reflects the imprecision in our initial estimate of Δx.

The uncertainty in p_x is close to the average value of $|p_x|$ (see Problem 27-53). The Bohr model tells us nothing about the direction of the momentum. The momentum can be totally in the $+x$ direction, or totally in the $-x$ direction, or it can be partially or not at all in the x direction. For this reason, we should expect the uncertainty in p_x to be in the same ballpark as $|p_x|$.

◆ Related homework: Problems 27-38 through 27-42.

◆**THE EFFECT OF MEASUREMENT** In addition to the uncertainty principle, there is a further limitation to our knowledge at this microscopic scale. On a macroscopic scale, we can watch an object move along a meter stick and in that way determine the object's position at each instant without affecting its motion, or, if we do not wish to be present, we can film the motion. In either case, we might be careful not to nudge the moving object with the meter stick. We don't want our measurement to alter what we are measuring. But to see the object, we must illuminate it, and that means bouncing photons off it. This may have no observable effect on a macroscopic object, but on the microscopic scale even a single photon bouncing off an electron—the minimum requirement for "seeing" the electron—can significantly change the electron's position and momentum. The desire not to have our measurement alter what we are measuring cannot be fulfilled in the microscopic world of quantum theory. What we "see" is unavoidably determined by how we measure it.

◆**PROBABILITY DISTRIBUTIONS FOR SCHRÖDINGER WAVE FUNCTIONS** In the Schrödinger picture, the probability of finding the electron at various positions, even in the hydrogen atom, is more complicated than in the Bohr picture and depends on the quantum numbers of the electron's state. When $l = n - 1$ for a given n, the probability peaks at the Bohr radius for

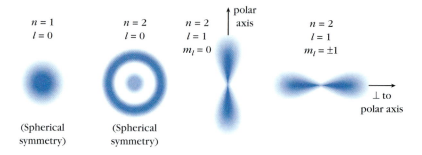

Figure 27-12 Regions of signifi-cant probability of finding the electron for various quantum states of the hydrogen atom. The regions are three-dimensional. The probability is greatest where the regions are darkest.

that n, demonstrating some correspondence with the Bohr picture. But in some states, the electrons are more likely to be found in some directions than in oth-ers; the likelihood is concentrated in dumbbell-shaped regions, as in the $l = 1$ states in Figure 27-12. When no magnetic field is applied, the Schrödinger model gives the same energy levels as the Bohr model, but it is no longer strictly correct to think of the electron as being in well-defined circular orbits, like plan-ets around a sun. Chemists sometimes speak of these more complicated elec-tron probability configurations as **orbitals,** suggesting that in some ways (such as the energies associated with them) they are still orbit*like*. Because of its agree-ment with the Schrödinger picture under certain conditions, the simpler Bohr model remains useful for thinking about what happens to the electrons in many simple situations.

◆QUANTUM STATES OF ELECTRONS IN MOLECULES AND SOLIDS

When atoms come together to form molecules or solids, each electron interacts not just with the nucleus and other electrons of its parent atom but with the nuclei and electrons of the other atoms as well. This leads to a much more complicated set of energy levels, even when the atoms are identical. The probability distribu-tion for the electrons shared in a covalent bond between two atoms, for example, must give them a reasonable probability of being in the vicinity of either nucleus. The probability distribution for the outermost (conduction) electrons in a metal with a repetitive crystal structure must be periodic; that is, it must repeat in the same way at each atomic site to permit the electron to move freely from site to site.

Furthermore, the Pauli exclusion principle prohibits two electrons from being in the same quantum state if they are close enough to interact (i.e., to be "aware" of each other's state). Therefore, the energies even of electrons that were in iden-tical states within two isolated atoms must shift subtly when the atoms are brought together. When many atoms are brought together, each individual energy level splits into a **band** or cluster of energy levels so close together that the allowable energies within the band are practically continuous (Figure 27-13). As the figure shows, the resulting large molecules or solids can then absorb continuous ranges of photon energies.

Figure 27-13 Individual energy levels split into *bands* in large molecules and in solids. (*a*) Allowable energy levels and transitions for a single atom. (*b*) Allowable levels and 2p→1s transitions for n identical atoms. Here, $n = 5$.

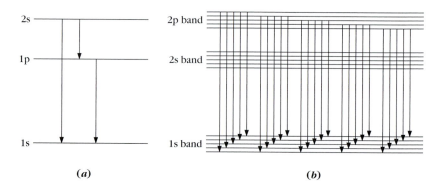

27-7 The Visible World in a New Light

This section presents a brief tour of photon interactions with matter, which ultimately are responsible for the way the visible world looks to us. (Our concern here will be with what happens to the various signals before they reach our eye, and not with the complex way in which the human brain processes those signals to shape our perceptions.)

Some objects that we see, such as the sun and neon signs, are sources of light. We see other objects by virtue of what they do to redirect (by reflection, for example) or otherwise alter light traveling from some source to our eye. The colors that we see depend on the distribution of photons of various energies (or their associated frequencies) that reach our eyes. The line spectrum emitted by neon gas of very low density (Figure 27-14a), in which many red lines predominate, gives rise to the brilliant red light we see from neon signs. (Neon signs are *always* red; other gases are used to produce other colors.) The colors we see from denser sources, in which the energy levels of electrons become so close together as to be essentially continuous, depend on the temperature of the source, because the internal energy of a dense body determines the energy levels to which its electrons are excited. The spectrum of a body that is just warm enough to give off a reddish glow (Figure 27-14b) emits many more photons at the lower-energy (red) end of the visible spectrum. The colors we see from combinations of photons are a consequence of the way our brains process and interpret these signals. For example, we see white when we receive photons in quantity from throughout the visible spectrum or reddish-violet when we receive photons of the four discrete energies giving rise to the predominant lines in the visible hydrogen spectrum.

We see most objects around us not because they themselves give off light but because they are illuminated by the sun or an electric lamp or some other source. The light from the source is reflected from the object to our eyes. We see these objects as various colors. One common reason for the color that we see is the presence of **pigments,** substances that cannot absorb photons to the same extent in all parts of the visible spectrum. Pigments occurring in living things tend to be large, complex molecules characterized by complicated distributions of energy levels like those in Figure 27-13b. The energy levels determine which photons they can absorb. Common pigments include the *chlorophyll* that gives green plants their color and the *melanin,* present in various degrees in people of different geographic origin, that determines the darkness of our skin. The pigments in artists' paints are often semiconducting materials, such as vermilion (HgS), a red pigment, and the cadmium sulfide (CdS) in "cadmium yellow." The energy levels in these substances also result in selective absorption of photons in the visible range.

The photons that do not correspond to differences between energy levels in the pigment molecule are reflected back and received by the eye. For example, the eye receives primarily unabsorbed photons in the green range from the plant in Figure 27-15.

When pigments are in living organisms, the important photons for the organism itself are not the photons that bounce back but the photons that are absorbed by the pigment. This energy is converted and used in essential life processes. In green plants, for example, the energy of the photons absorbed by chlorophyll becomes the input energy (recall Figure 6-8) for the chemical process of *photosynthesis* in which glucose is

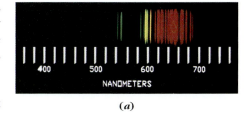

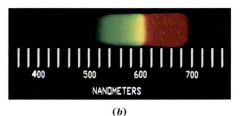

Figure 27-14 Spectra of two red objects. (*a*) Discrete emission spectrum for neon gas. (*b*) Continuous emission spectrum from a red-hot object.

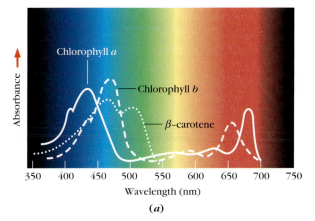

(a)

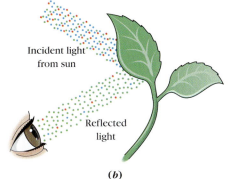

(b)

Figure 27-15 The absorption spectrum of chlorophyll. (*a*) Chlorophyll absorbs primarily in the blue-violet and orange-red parts of the spectrum. (*b*) The green that is *not* absorbed is predominant in the reflected light.

synthesized from carbon dioxide and water. Pigments such as *rhodopsin* in the rods and cones of your eye's retina are capable of absorbing photons across the visible part of the spectrum and converting that energy to forms that can be carried as signals to the brain. These pigments are the *receptors* (receivers) for human vision. In fact, these pigments *define* the visible spectrum, because we can detect only the photons they are capable of absorbing.

Pigments are not the only cause of the various colors that we see in the world around us. We have already taken note (Section 15-7) of the brilliant colors of some birds and butterflies (as well as soap bubbles and oil slicks) resulting from interference patterns due to parallel reflecting surfaces, not from the presence of pigments.

Likewise, there is no pigment in the gases that make up our atmosphere; yet we see vivid blue skies and red sunsets. The colors we see in these instances are due to light being deflected in all directions by molecules and tiny particles in the sky. Just as when beams of particles are deflected in various directions (Section 26-1), this effect is called **scattering.** The same effect is caused by the particles in cigarette smoke, giving it its blue-white appearance. Even before quantum theory, Lord Rayleigh (English physicist John William Strutt, 1842–1919) was able to show that if the particles were small compared to the wavelength of the light, light of shorter wavelengths would be scattered more efficiently (the efficiency is proportional to $1/\lambda^4$). Those are the wavelengths at the blue-violet end of the spectrum. Thus it is predominantly those wavelengths that are redirected toward our eyes by scattering (Figure 27-16a), and we see the regions of the sky from which they reach us as blue. (We don't see those regions as violet because the pigments in our rods and cones absorb a smaller fraction of the violet photons than the blue.) Figure 27-16b shows how the same scattering process is responsible for red sunsets. To see in step-by-step detail how scattering is responsible for what we see in these situations, work through WebLink 27-4.

Although Rayleigh was able to provide a classical explanation for this scattering, we can also think about it in terms of what happens to photons. When Bohr's energy levels are replaced by probability distributions, events that were forbidden by the Bohr model may be merely be improbable, not impossible. There is in fact a small but finite probability that a photon at a frequency well below the resonance frequency (corresponding to an allowable jump from one energy level to another) of an atom or molecule will be absorbed by the atom or molecule and quickly reemitted (a small Δt corresponding to a large ΔE). The reemitted photon will generally have the same energy and frequency as the absorbed photon, but not the same direction; hence it is scattered. Unlike the photons absorbed by pigments, the energy of these photons is not converted after being absorbed.

For **WebLink 27-4:**
ScatterVision,
go to
www.wiley.com/college/touger

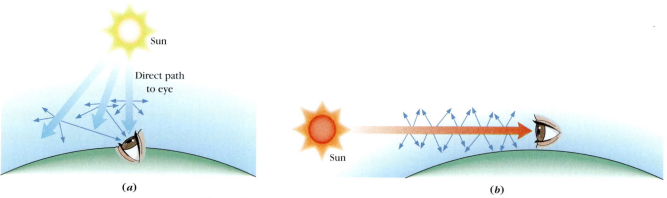

(a)   (b)

Figure 27-16 Blue skies and red sunsets—the effects of scattering of visible light photons by molecules in the atmosphere. (*a*) When light reaching various regions of the sky is redirected toward our eye by scattering, we see those regions as blue. (*b*) When the sun is low in the sky, the light reaching our eye follows a much longer path through the atmosphere, so enough blue photons are scattered out of the light reaching us that the remaining red photons predominate, and we see red sunsets.

✦SUMMARY✦

A quantized picture of electromagnetic radiation and of the energy of electrons and other subatomic particles can explain a range of phenomena involving interactions between matter and radiated energy. In this picture, radiated energy comes in discrete packets of energy (massless particles) called **photons,** with

photon energy $\qquad E = hf$

Atoms, molecules, and so on with bound electrons can have only **discrete energy levels,** and not the energies in between. Atoms go to higher or lower energy levels, as when an electron is **excited** to a higher state or drops back to the **ground state,** by absorbing or emitting individual photons:

$$\Delta E = E_f - E_i = \pm hf \qquad (27\text{-}4)$$
$$(+ \text{ for absorption, } - \text{ for emission})$$

The photon model explains why, in the **photoelectric effect,** (1) no photoelectrons are emitted below a **threshold frequency,** and (2) above this frequency, the maximum kinetic energy of the emitted electrons varies linearly with frequency:

$$\tfrac{1}{2}mv_{\max}^2 = eV_{\mathrm{o}} = hf - \phi \qquad (27\text{-}1)$$

where V_{o} is the **stopping potential.** We can rewrite this as

$$hf \quad = \quad \tfrac{1}{2}mv_{\max}^2 \quad + \quad \phi \qquad (27\text{-}3)$$

input energy = maximum KE after + work required to
from photon = breaking away + escape from surface
$\qquad\quad$ from surface

The **work function** ϕ determines the threshold frequency f_{o}:

$$\phi = hf_{\mathrm{o}} \qquad (27\text{-}2)$$

The equation $E_f - E_i = \pm hf$ predicts correct frequencies for the hydrogen line spectrum when the E values are provided by the **Bohr model of the atom:**

• In the hydrogen atom, the electrostatic attraction between the proton and electron keeps the electron in orbit. The total energy = KE + PE$_{\text{electrical}}$.

• Assuming that the angular momentum of the electrons is quantized

$$mvr = n\frac{h}{2\pi} \qquad (27\text{-}6)$$

results in a picture in which *the radius of the electron orbit is quantized:*

$$r = n^2 \frac{h^2}{4\pi^2 kme^2} = n^2 r_1 \qquad (27\text{-}7,\ 27\text{-}7a)$$

$$(\text{quantum number } n = 1, 2, 3, \ldots)$$

The smallest possible value of the radius is

$$r_1 = 0.528 \times 10^{-10} \text{ m} = 0.0528\ \text{Å} = 0.0528\ \text{nm}$$
(radius of the first Bohr orbit)

The total energy is also quantized:

$$E_n = -\frac{2\pi^2 m (ke^2)^2}{n^2 h^2} = -\frac{1}{n^2} E_I \qquad (27\text{-}9,\ 27\text{-}10)$$

(where E_I = **ionization energy** and $-E_I$ = **ground state** energy)

Electrons above the ground state (with $n > 1$) are said to be in **excited states,** and increase toward zero from more negative values as n increases.

The **correspondence principle** states that *at sufficiently large n, the quantum theory must predict the same behavior as classical physics.*

Compton scattering of X rays from electrons could be explained by assuming that like other particles, photons have momentum:

photon momentum $\qquad p = \dfrac{E}{c} = \dfrac{h}{\lambda} \qquad (27\text{-}12)$

Just as light exhibits particlelike behavior in its interactions with matter on a microscopic scale but shows wavelike behavior on a large scale, deBroglie speculated that matter might also display both types of behavior (**wave-particle duality**). He proposed that for a particle of mass m,

the deBroglie wavelength $\qquad \lambda = \dfrac{h}{p} = \dfrac{h}{mv} \qquad (27\text{-}13)$

(At relativistic speeds mv must be replaced by the relativistic momentum $mv/\sqrt{1 - v^2/c^2}$.)

The **Davisson-Germer** and G. P. Thomson experiments with beams of electrons produced *interference patterns,* showing electrons have wavelike behavior.

The **Schrödinger equation** provided a fuller description of the wavelike aspects of electrons. Its solutions are wave functions in which the wave properties of frequency and wavelength are related to the particle properties of total energy and momentum:

Particle Property	Wave Property	Relationship
total energy	frequency	$E = hf$
momentum	wavelength	$\lambda = \frac{h}{p}$

Solving the Schrödinger equation for the hydrogen atom yields more complicated wave functions characterized by three quantum numbers n, l, and m_l. A fourth quantum number, m_s, is needed to specify the state of an electron completely. The **Pauli exclusion principle** states that no two electrons can have all four quantum numbers the same, and the way in which electrons consequently fill shells and subshells accounts for the periodic repetition of chemical properties (summarized as the **periodic table**) as one goes to elements with increasing numbers of electrons. **Procedure 27-1** tells how to find the allowable values for the quantum numbers.

If ψ is a complex wave function satisfying the Schrödinger equation, $|\psi|^2$ is associated with the **probability** of finding an object at different locations, and is called the **probability density.** To produce a probability that is particlelike—that is, localized within a small region of space Δx—from sinusoidal wave functions distributed uniformly over all space, one must superimpose wave functions with a large spread of wavelengths $\Delta\lambda$, and therefore with a large uncertainty Δp in the momentum. An analogous relationship exists between Δt and $\Delta f = \Delta E/h$. These relationships are summarized by the

Heisenberg uncertainty principle

$$\Delta x \Delta p \geq \frac{h}{2\pi} \qquad \text{and} \qquad \Delta E \Delta t \geq \frac{h}{2\pi} \qquad (27\text{-}15\text{a,b})$$

The photons that reach us from objects determine how the objects look to us. In large molecules and dense matter, because of the exclusion principle, each atomic energy level splits into a **band** of energy levels. The energy levels are packed very close together, becoming effectively continuous over substantial ranges of values; whereas low-density gases emit line spectra, dense sources emit (and absorb) continuous spectra. A class of large molecules called **pigments** absorb selectively in the visible spectrum. The photons they *don't* absorb are reflected to our eyes and determine the color that we see for the pigmented object. Molecules in less dense gases can absorb photons traveling in one direction and reemit identical photons in a different direction. In the atmosphere, this **scattering** of photons results in the sky looking blue and the sun appearing red at sunset.

◆ QUALITATIVE AND QUANTITATIVE PROBLEMS ◆
Hands-On Activities and Discussion Questions

The questions and activities in this group are particularly suitable for in-class use.

27-1. Discussion Question. In considering the energy states of the hydrogen atom, we can represent the energy difference between each state n and the next lowest state $n - 1$ as $E_n - E_{n-1}$. What happens to this difference as n becomes infinitely large? How far apart are successive energies when n is extremely large? When this happens, how meaningful is the distinction that E can only have certain values and not those in between? How do the allowable energy values for large n compare with a continuous range of values?

27-2. Discussion Question. A tiny meteorite crashes into a satellite orbiting Earth and remains embedded in the satellite. Discuss the similarities and differences between this situation and a photon being absorbed by an electron in a Bohr orbit.

27-3. Discussion Question. At a microscopic level, biology becomes biochemistry. The chemical reactions in living things, collectively called *metabolism,* are the basis for growth, reproduction, maintenance, and other processes taking place within each cell. Would life as we know it be possible if electrons did not obey the exclusion principle?

Review and Practice

Reminder on units: The energies of electrons are often expressed in electron volts (eV):

$$1 \text{ eV} = e(1 \text{ V}) = (1.6 \times 10^{-19} \text{ C})(1 \text{ J/C}) = 1.6 \times 10^{-19} \text{ J}$$

Section 27-1 Atomicity and Beyond

Section 27-2 The Photoelectric Effect and the Idea of the Photon

27-4.
a. Why doesn't the existence of a threshold frequency for the photoelectric effect make sense from the point of view of a continuous wave model of light?

b. How does the concept of a photon explain the existence of a threshold frequency?

27-5. When an electron absorbs a photon, the photon ____ (*momentarily loses all its energy; loses some of its energy; changes frequency; ceases to exist*).

27-6. Photons of green light always ____ (*contain more energy; contain less energy; are more numerous; are less numerous*) than photons of orange light.

27-7. In Figure 26-5, a certain voltage V is applied between the plates. A current will then be detected by the ammeter whenever ____
a. the frequency of the incident light exceeds the threshold frequency f_o.
b. the kinetic energy of the photoelectrons exceeds hf_o.
c. the kinetic energy of the photoelectrons exceeds eV.
d. the photon's energy exceeds hf_o.
e. the photon's energy exceeds eV.

27-8. In the photoelectric effect set-up in Figure 26-5, suppose the electromagnetic radiation incident on the emitter plate is 1.0×10^{14} Hz above the threshold frequency. If a voltage of 0.3 V is applied between the plates, will that stop all, some, or none of the photoelectrons from reaching the opposite plate?

27-9. SSM
a. If an electron must expend 4.53 eV of energy to escape the surface of a piece of tungsten, what is the minimum frequency of electromagnetic radiation that will cause tungsten to emit photoelectrons?

b. What frequency of electromagnetic radiation will cause tungsten to emit photoelectrons with a maximum kinetic energy of 6.0 eV?

c. Suppose tungsten is exposed to electromagnetic radiation in the set-up in Figure 26-5. For the situation in *b,* what stopping voltage is required to prevent any photoelectrons from reaching the opposite plate?

27-10. The light striking a particular sheet of metal exceeds that metal's threshold frequency for the photoelectric effect by $2.0 \times 10^{14} \text{ s}^{-1}$.
a. At what maximum speed will photoelectrons be emitted from the surface of the metal?

b. A second sheet of metal is separated from the first by a distance of 3.0×10^{-2} m. What voltage must be applied across the gap between the two sheets to prevent all photoelectrons from the first sheet from reaching the second sheet?

c. What would then be the electric field between the two sheets? Give both the magnitude and the direction.

27-11. Based on the results of Example 27-1, would monochromatic red light ($\lambda = 6.328 \times 10^{-7}$ m) from a helium-neon laser be able to produce a photoelectric effect in sodium? Support your answer with a calculation.

Section 27-3 The Bohr Model of the Atom

27-12. Show by a calculation that when $n = 1$, substituting values for all the constants in Equation 27-6 gives 0.528×10^{-10} m as the value of the first Bohr radius.

27-13. The first Bohr radius has a value of 0.528×10^{-10} m. **a.** Find the energy E_1 of the atom (in joules) when it is in this state. **b.** Find the value of E_1 in electron volts.

27-14. Show that Planck's constant h has the same SI units as the angular momentum mvr of an electron in orbit.

27-15. Which of the following energies is possible for a photon emitted from a hydrogen atom? 3.00 eV; 6.00 eV; 12.00 eV; all of the these; none of the these.

27-16. As the total energy of the electron + nucleus system making up the Bohr hydrogen atom approaches zero from below, the radius of the electron's orbit _____ (*becomes less and less negative; approaches infinity; approaches zero; doesn't change*).

27-17. Consider how Figure 27-3 applies to the following two transitions.

> Transition A: The hydrogen atom goes from the $n = 1$ state to the $n = 2$ state
>
> Transition B: The hydrogen atom goes from the $n = 4$ state to the $n = 5$ state

a. In comparing what happens to the total energy in transitions A and B, we see that it _____ (*increases more in transition A; increases more in transition B; decreases more in transition A; decreases more in transition B*).

b. Does the electron's distance from the nucleus increase more in transition A or in transition B?

c. Does the total energy increase as the electron's distance from the nucleus increases?

d. Is the total energy proportional to the electron's distance from the nucleus?

27-18. Compare the energy a photon must have to raise a hydrogen atom to the next higher state when the atom starts out in its ground state and when it is already excited.

27-19. **SSM** As the energy levels for the Bohr hydrogen atom get closer together, do the corresponding electron orbits also get closer together? Briefly support your answer by reasoning from appropriate equations.

27-20.

a. Calculate the total energy of the hydrogen atom in the $n = 4$ state, then repeat for the $n = 5$ state.

b. In going from the $n = 4$ to the $n = 5$ state, does the energy increase or decrease? Support your answer with a calculation.

c. Calculate the energy of the photon that must be emitted or absorbed (which should it be?) by the atom for it to go from the $n = 4$ to the $n = 5$ state.

d. Find the frequency and the wavelength of this photon.

27-21.

a. Calculate the radii of the second and third Bohr orbits for hydrogen.

b. The value of the first Bohr radius is given in the chapter. Do the orbits get closer together, remain equally spaced, or get further apart as you go outward from the nucleus?

27-22.

a. Calculate the wavelength of electromagnetic radiation emitted by hydrogen atoms dropping from the $n = 4$ to the $n = 1$ state.

b. In what part of the electromagnetic spectrum is this radiation? (Answer in one or two words.)

27-23. **SSM WWW** Find the two lowest frequencies of electromagnetic radiation that can be absorbed by ground state hydrogen.

27-24. It may be useful to refer to Figure 27-3 as you do this problem. Express all your calculated answers in terms of E_1.

a. As n increases from 1 to 2, by how much does the total energy of a hydrogen atom increase? (Increase = _?_ E_1).

b. As n increases from 2 to 3, by how much does the total energy of a hydrogen atom increase? (Increase = _?_ E_1). Is this more than, equal to, or less than the energy increase required to go from the $n = 1$ to the $n = 2$ state?

c. We can represent the energy difference between each state n and the next lowest state $n - 1$ as $E_n - E_{n-1}$. Does this difference become larger, remain the same, or become smaller as n increases?

d. Are successive allowable energies getting further apart, remaining equally spaced, or getting closer together as n increases?

27-25. It may be useful to refer to Figure 27-3 as you do this problem.

a. By how much does the function $-\frac{1}{n^2}$ decrease in value when a hydrogen atom drops from the $n = 2$ state to the $n = 1$ state? What is the energy of the photon emitted when this happens?

b. Find the value of $-\frac{1}{n^2}$ to four significant figures when $n = 1\,000\,000$. Answer the remaining parts of this problem to the same degree of accuracy.

c. By how much does the function $-\frac{1}{n^2}$ decrease in value when the atom drops from the $n = 1\,000\,000$ state to the $n = 2$ state? What is the energy of the photon emitted when this happens?

d. Consider the energies of the photons emitted in parts **a** and **c.** Can a hydrogen atom emit a photon having an energy in between these two values? Explain your answer.

27-26. The ionization energy of a hydrogen atom is 13.58 eV or 2.176×10^{-18} J. **a.** Can an excited hydrogen atom ever lose more than this much energy? Briefly explain. **b.** Can a hydrogen atom in the ground state ever gain more than this much energy? Briefly explain.

27-27. **SSM** A hydrogen atom is initially in the ground state. Determine by a calculation what happens to the atom's electron and to the input energy when the atom is struck by a photon with an energy of
a. 1.632×10^{-18} J. **b.** 1.00×10^{-18} J. **c.** 14.1 eV.

Section 27-4 From Particles *or* Waves to Particle*like* and Wave*like*

27-28. If an electron gains 8.0×10^{-22} J of energy by absorbing a photon, what is the momentum of the photon?

27-29. Estimate your own deBroglie wavelength when you are running to catch a bus. Briefly state the assumptions that go into your calculation.

27-30. Show that if h is in J · s and m and v are also in appropriate SI units, then the deBroglie wavelength (Equation 27-13) will be in meters.

27-31.
a. At what speed would an electron have a wavelength of 10^{-10} m? At this speed, is a nonrelativistic calculation adequate?
b. At what speed would a proton have a wavelength of 10^{-10} m? At this speed, is a nonrelativistic calculation adequate?

27-32. Estimate the deBroglie wavelength of *a.* a compact car traveling at 30 mi/hr. *b.* yourself walking from one class to the next.

27-33. Compare the wavelengths of the photon and the proton in Example 27-3. Express as a ratio $\dfrac{\lambda_{proton}}{\lambda_{photon}}$.

27-34. Find the deBroglie wavelength of a proton that has been accelerated from rest through a potential difference of 100 V.

Section 27-5 The Schrödinger Equation

27-35. A concern that the Bohr model raises is: If the distances from the nucleus at which an electron may be found are discrete, how does the electron get from one distance to the other? How does the Schrödinger model address or modify this concern?

27-36. An electron in an atom has the following four quantum numbers: $n = 3$, $l = 2$, $m = -1$, $m_s = \frac{1}{2}$. Which one of the following can be the set of four quantum numbers for another electron in the same atom?
a. $n = 2$, $l = 0$, $m = 0$, $m_s = 0$
b. $n = 2$, $l = 0$, $m = -1$, $m_s = \frac{1}{2}$
c. $n = 2$, $l = 2$, $m = 1$, $m_s = -\frac{1}{2}$
d. $n = 3$, $l = 1$, $m = -1$, $m_s = -\frac{1}{2}$

27-37. An electron in an atom has the following four quantum numbers: $n = 2$, $l = 1$, $m = -1$, $m_s = \frac{1}{2}$. Here are some other sets of four quantum numbers:

$$|A> \; n = 1, \; l = 0, \; m = 0, \; m_s = \tfrac{1}{2}$$
$$|B> \; n = 2, \; l = 1, \; m = 1, \; m_s = -\tfrac{1}{2}$$
$$|C> \; n = 2, \; l = 1, \; m = -1, \; m_s = \tfrac{1}{2}$$

Which of these sets are possible for another electron in the same atom?

Section 27-6 Probability and Uncertainty in the Quantum Universe

27-38.
a. The angstrom (Å), a unit commonly used for atomic-scale distances, is 1.00×10^{-10} m. What minimum uncertainty must there be in the speed of an electron (assuming negligible relativistic effects) if we are to know its position to within 1 Å? Is this a significant uncertainty (margin of error)?
b. What minimum uncertainty must there be in *your* speed if we are to know *your* position to within 1 Å? Is this a significant margin of error?

27-39. Find the minimum uncertainty in the momentum of the electron in a hydrogen atom when it is in its first excited ($n = 1$) state.

27-40.
a. The energy of a particular electron in an atom is measured to be 6.0×10^{-19} J. If the measurement is made within an interval of 2.0×10^{-15} s, what percent error may we expect in this energy measurement?
b. Repeat *a* for an electron whose energy is measured to be 2.5 eV during an equivalent time interval.

27-41. SSM
a. Find the uncertainty in the momentum of a proton confined to the nucleus of an atom that has a nuclear diameter of about 10^{-14} m.
b. An uncertainty of at least Δp in the momentum of an object means that the magnitude of the actual momentum can be at least Δp above or below the magnitude of the measured momentum. It can only be Δp below the measured momentum if the magnitude of the measured momentum is at least Δp. Therefore, what can be the minimum kinetic energy of the proton in *a*?

27-42. Since the 1930s, hundreds of elementary particles in addition to the proton, neutron, and electron have been discovered. Unlike these three, the additional particles have extremely short lifetimes and therefore cannot be components of the stable structure of atoms. One such particle, the neutral pion, has a lifetime of 8.7×10^{-17} s. Find the uncertainty in the energy of this particle.

Section 27-7 The Visible World in a New Light

27-43. What property of pigments makes it important that they occur in living things?

27-44. Ultramarine is a deep blue pigment traditionally extracted from the mineral lapis lazuli. Estimate the energy of a photon in the visible range that is unlikely to be absorbed by ultramarine.

27-45. What happens to the energy of a photon *a.* if it is scattered by a molecule in the atmosphere? *b.* if it is absorbed by a pigment molecule?

27-46. The water filling a large glass fish tank is made cloudy by the addition of a small amount of milk.
a. When a beam of white light is directed through the water at one end, the light emerging from the sides of the tank (Figure 27-17) has a bluish cast. Why?
b. Will there appear to be any coloration in the light emerging from the opposite end of the tank? Briefly explain.

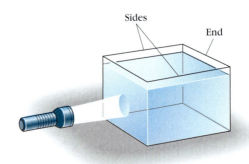

Figure 27-17 Problem 27-46

Going Further

The questions and problems in this group are not organized by section heading, so you must determine for yourself which ideas apply. Some of them will be more challenging than the Review and Practice questions and problems (especially those marked with a • or ••).

27-47. If a particle had half the charge-to-mass ratio of an α particle, what could it be? Give at least two possibilities.

27-48.
a. What is the basic building block involved when we speak of atomicity in matter? in electric charge? in radiated (electromagnetic) energy?
b. How do the basic building blocks of lead and gold differ?
c. How do the basic building blocks of red light and microwaves differ?

27-49. We can generalize the work-energy theorem (Equation 6-14) to read $PE_a + KE_a + E_{ext} = PE_b + KE_b$, where E_{ext} is the sum of all energy inputs from outside the system, including W_{ext}. We use a and b here rather than 1 and 2 to denote the states because if the system goes through several states, the equation holds true between any two of those states.
a. Write what the equation looks like specifically as a photoelectron goes from each of the following states to the next (that is, write one equation for the electron going from state 1 to state 2 and another for state 2 to state 3).

State 1: The electron is at the surface of the metal and has no kinetic energy.

State 2: The electron has absorbed a photon of frequency f and is just leaving the surface. It has had an amount of work ϕ done on it to break it away from the surface, and it has also acquired kinetic energy.

State 3: The electron has traveled across a potential difference V_o and has come to a stop.

b. How do your resulting equations compare with any of the equations in Section 27-2? Comment on the significance of the comparison.

27-50. Is a photon emitted from a hydrogen atom more likely to have an approximate (to within 5%) energy of 10 eV or 5 eV? Briefly justify your answer.

27-51. The electron in the Bohr atom can only occupy discrete energy levels and cannot have in-between energies. Photoelectrons from any given material can have a continuous range of kinetic energies (as shown in Figure 27-1). Are these two situations in contradiction? Briefly explain.

27-52.
a. In the Bohr hydrogen atom, how does the kinetic energy of the electron compare to the total energy of the electron + nucleus system?

b. How do their signs compare?
c. How do their absolute values compare?

27-53.
a. Using the result of Problem 27-52 and the fact that $E_1 = -13.6$ eV, find the magnitude p of the electron's momentum in the ground state.
b. The ground state of the hydrogen atom is spherically symmetrical, so its electron is equally likely to be moving in any direction. From this and the fact that $p^2 = |p_x|^2 + |p_y|^2 + |p_z|^2$, it follows that $|p_x|$ is $\frac{1}{\sqrt{3}}p$. Calculate the average value of $|p_x|$. Compare your result to the uncertainty in p_x, which we found in Example 27-5. See if your results are consistent with the comments at the end of that example.

27-54. Bohr's derivation of the allowable radii and energy levels for the electron in a hydrogen atom can be repeated in a similar way for any chemical species with a single electron, such as a singly ionized helium atom or a doubly ionized lithium atom. The only difference is that if the nucleus has Z protons rather than just one, the electrostatic force between the nucleus is $\frac{k(Ze)e}{r^2} = \frac{kZe^2}{r^2}$ rather than $\frac{ke^2}{r^2}$, so that e^2 must be replaced by Ze^2 in the hydrogen atom expressions for r_n and E_n.
a. Use the Bohr model to find the two innermost allowable radii for an electron in a singly ionized helium atom.
b. Compare their values to the first two radii for the hydrogen atom.

27-55. At atmospheric pressure and room temperature, the average distance between molecules in a gas is about 3.4×10^{-9} m.
a. If the radius of an atom were more than half this distance, the atoms would overlap. At what value of n is the Bohr radius of a hydrogen atom equal to half of this distance? For simplicity, treat hydrogen as though it were a monatomic gas (a gas of single-atom molecules).
b. Within a typical cathode ray type discharge tube (used for producing spectra), the pressure may typically be about $\frac{1}{110}$ of atmospheric pressure at room temperature. Show that under these conditions the average distance between molecules is about 1.6×10^{-8} m. At what value of n is the Bohr radius of a hydrogen atom equal to half of this distance?
c. If a discharge tube filled with hydrogen is used as a light source, estimate the number of spectrum lines in the Balmer series ($m = 2$ in the Rydberg formula) that one can reasonably hope to observe in the laboratory.
d. When Bohr proposed his theory, over 30 lines in the Balmer series had been observed in the spectra of the sun and other stars. About how far apart (at least) must the molecules of a gas be on the average to produce this many lines?
e. How would the brightness of the observed lines be affected by the distance between molecules? What conditions could be provided by a star more readily than by a laboratory setup to compensate for the loss of brightness when the density of the source is reduced?

••27-56.

a. Starting with the Rydberg formula, show that for very large n, the frequency of the photon emitted when an electron drops from the $(n + 1)$th energy level to the nth energy level is approximately $f = 2cR_H/n^3$, where c is the speed of light.

b. Apply Newton's second law to an electron held in a circular orbit of radius $r_n = n^2 r_1$ about a proton (a hydrogen nucleus) by an electrostatic force, which is responsible for the radial acceleration $r\omega^2$. If the electromagnetic wave radiated by the electron has the same frequency f as the orbiting electron, show that the frequency of the wave is

$$f = \frac{\frac{e}{2\pi}\sqrt{k/mr_1^3}}{n^3}$$

where k is the Coulomb force constant and r_1 is the radius of the first Bohr orbit.

c. The values of f given by the classical physics result in **b** will be the same as the values given by the quantum theoretical result in **a** if the numerical values of the numerators of the two expressions for f are the same. Do the necessary calculation to show whether or not this is so.

27-57. Bohr's theoretical formula $\frac{1}{\lambda} = \frac{E}{hc} = \frac{E_1}{hc}(\frac{1}{2} - \frac{1}{2})$ gave the same wavelengths for hydrogen as the Rydberg equation $\frac{1}{\lambda} = R_H(\frac{1}{m^2} - \frac{1}{n^2})$, which was a pattern found from the experimental data, because the numerical value of $\frac{E_1}{hc}$ calculated by Equation 27-9 agreed with the experimentally determined value of R_H. (E_1 is the ionization energy of hydrogen.) For singly ionized helium, the pattern found in the experimental data was $\frac{1}{\lambda} = R_H(\frac{1}{(m/2)^2} - \frac{1}{(n/2)^2})$. If we substitute $2e$ (the charge of a helium nucleus) for the charge e of a proton, the electrostatic interaction with an electron of charge $-e$ involves $2ke^2$ rather than ke^2. Show that if we substitute $2ke^2$ for ke^2 wherever it appears in the expression for E_1, the Bohr formula will agree with the experimental formula for singly ionized helium.

27-58. The beam used by Davisson and Germer was produced by accelerating electrons through a potential difference of 54 V. If electrons are wavelike, what would be their deBroglie wavelength after acceleration?

27-59. As an electron goes to progressively more excited states in the Bohr hydrogen atom, does the uncertainty in its momentum increase, decrease, or remain the same? Briefly explain.

27-60.

a. Estimate the number of photons emitted each second by a 1.0-mW helium-neon laser, which emits monochromatic red light of wavelength 6.328×10^{-7} m.

b. Estimate the number of photons emitted each second by a 40-W bulb. Beyond what you did in **a,** briefly explain what further assumption you had to make to do this estimation simply.

27-61. SSM Radiation from radioactive sources is sometimes spoken of as ionization radiation.

a. Use the Bohr model to explain what this means. In doing so, think about whether this radiation consists of high energy or low energy photons.

b. Ionization radiation can be a serious health hazard. Explain in terms of specific physics or chemistry principles or

mechanisms why exposure to ionization radiation is more hazardous than exposure to visible light.

c. Use your reasoning from **b** to comment on the risks of excessive exposure to sunlight.

27-62. We typically see an object, and so determine where it is, by detecting the light (or other wave) that it reflects. If the wavelength of the incident wave is greater than the size of the object, the incident wave won't be significantly scattered by the object, and our method of detection breaks down. Therefore the uncertainty in its position is given roughly by the wavelength of the wave used for detection.

a. What does this imply about the momentum of the probes required to investigate smaller and smaller subatomic particles? Explain.

b. What does it imply about the energy that must be provided to those probes? Explain.

27-63. On an episode of *Star Trek: Voyager,* a creature "made of photonic energy" moves about the *Enterprise,* endangering its crew. The creature appears rather like a giant octopus made of light and moves its tentacles at octopus speed. Use what you know about photons and light to explain why this is not a realistic possibility.

27-64. In 1967 the General Conference on Weights and Measures established the following internationally accepted definition of the second: "The second is the duration of 9 121 631 770 periods of the radiation corresponding to the transition between the two hyperfine levels of the ground state of the cesium-133 atom." Find the energy difference between those two levels.

27-65. Suppose $\Delta\lambda = \lambda_2 - \lambda_1$ is the difference between two deBroglie wavelengths $\lambda_1 = h/p_1$ and $\lambda_2 = h/p_2$. Assume that λ_1 and λ_2 are both reasonably close to their average value λ_{av}, so that $\lambda_1 \lambda_2 \approx (\lambda_{av})^2$. Show that $\Delta p \approx h\Delta\lambda/\lambda_{av}^2$.

27-66. Find all the allowable $n = 2$ states for an electron in an atom. How many of these states are there? How does this number compare with the number of elements in the second row of the periodic table? Why?

27-67. SSM WWW For historical reasons, the subshells for $l = 0, 1, 2,$ and 3 are often labeled s, p, d, and f, respectively (the letters at first stood for words describing the appearances of different spectral lines). **a.** How many different 4f states are possible in an atom? **b.** How many different 3f states are possible in an atom?

27-68. List all the 4d states (states with $n = 4$ and $l = 2$) that are possible in an atom.

27-69. Find the quantum numbers of all the electrons in the outermost shell of **a.** calcium **b.** neon.

27-70.

a. We can shine light off everyday objects to see where they are. In trying to observe objects on an atomic scale, why is it in principle not possible to measure their positions exactly by directing light (or other electromagnetic radiation) at them?

b. Is it in principle possible to develop any kind of measuring technology that gives exact measurements? Briefly explain.

••27-71. Estimate the deBroglie wavelength of a typical oxygen molecule in a roomful of air at normal room temperature.

Problems on WebLinks

27-72. Consider the metal treated in WebLink 27-1. Call this metal 1. Suppose another metal, metal 2, has a threshold frequency in the green part of the visible light spectrum. If a square meter of each metal were exposed to the same white light, would the number of photoelectrons dislodged from metal 2 be greater than, equal, to, or less than the number dislodged from 1?

27-73. In Problem 27-72, is the work function of metal 2 greater than, equal, to, or less than the work function of metal 1?

27-74. In Problem 27-72, is the maximum kinetic energy of photoelectrons dislodged from metal 2 greater than, equal, to, or less than the maximum kinetic energy of photoelectrons dislodged from metal 1?

27-75. An atom having the fictional energy levels shown in WebLink 27-2 can emit photons having which (one or more) of the following energies: 3.00 eV, 2.00 eV, 1.00 eV, −1.00 eV, −2.00 eV, −3.00 eV?

27-76. An atom having the fictional energy levels shown in WebLink 27-2 can absorb photons having which (one or more) of the following energies: 3.00 eV, 1.85 eV, 0.85 eV, 0.25 eV, −0.40 eV, −3.25 eV?

27-77. WebLink 27-3 shows that the third row of the periodic table, containing the elements from atomic number 11 to atomic number 18, fills in as which (one or more) of the following states become occupied?
a. $n = 2, l = 1$ *b.* $n = 3, l = 0$ *c.* $n = 3, l = 1$ *d.* $n = 3, l = 2$

27-78. WebLink 27-3 shows that the lowest $n = 4$ states become filled ____ (*after all the $n = 3$ states are filled; after most but not all of the $n = 3$ states are filled; before any of the $n = 3$ states are filled*).

27-79. WebLink 27-4 shows that as the sun travels from a point high in the sky at noon to a point low in the sky just before sunset, *a.* the rate at which photons in the red part of the spectrum reach your eye from the sun ____ (*increases; decreases; stays about the same*). *b.* the rate at which photons in the blue part of the spectrum reach your eye from the sun ____ (*increases; decreases; stays about the same*).

CHAPTER TWENTY-EIGHT

The Nucleus and Energy Technologies

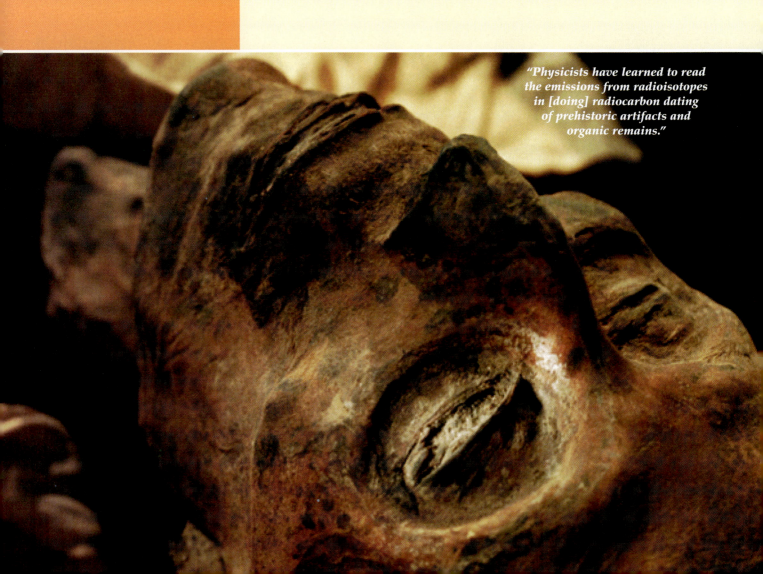

Probably nothing ever made the general public more aware of the work of physicists than the development of nuclear weapons. Nuclear physics, however, goes far beyond weaponry in the scope of its applications. The discovery of radioactivity in 1895 opened the door to the physics of the nucleus. Physicists have learned to read the emissions from radioisotopes in diverse contexts from medical imaging to radiocarbon dating of prehistoric artifacts and organic remains, and to harness nuclear power for peaceful as well as destructive ends. This chapter surveys some of the basic ideas of nuclear physics and their applications.

"Physicists have learned to read the emissions from radioisotopes in [doing] radiocarbon dating of prehistoric artifacts and organic remains."

28-1 Radioactive Decay and Decay Processes

Einstein's recognition of mass-energy equivalence ($E_o = mc^2$) provided one indication of the enormous amounts of energy carried off by radioactive emissions as unstable nuclei decayed. The existence of extremely strong nuclear forces would later be proposed to explain the amounts of energy involved. Years before Einstein, Lord Kelvin had estimated that Earth's lifetime could not exceed 60 000 000 years because thermal physics showed that if Earth were older, its core would have cooled and solidified. His estimate did not take into account the enormous amounts of energy that we now know are released in radioactive decay processes; these decay processes have maintained elevated temperatures in the Earth's core almost 70 times longer than Kelvin thought possible.

The emission of α and β particles and γ rays from radioactive elements at first appeared to violate conservation of energy. Because the first observers of these emissions did not detect their sources diminishing over time, they believed that the radioactive substances might be a perpetual energy source. However, by 1902, Crookes, Rutherford, and Soddy (Section 26-1) had shown that much of the radioactivity from uranium and thorium samples comes from chemically distinct materials within them, which when isolated lose much of their radioactivity after a few days. In the remaining uranium or thorium, however, the radioactivity is maintained. This evidence led them to an underlying picture with the following main features:

- Radioactive emissions are the products of decay processes by which a nucleus of one element is transformed into a nucleus of another (**transmutation of elements**).

- A nucleus of a radioactive element or radioactive isotope (a **radioisotope**) does not keep the same structure forever. It has a finite mean lifetime before it decays into another nucleus or nuclei, which may or may not be radioactive.

- Different radioactive nuclei have different mean lifetimes, some so long that change in the intensity of emissions from a sample may be undetectable over the course of an experimenter's measurements, but others short enough so that a decrease in the intensity of emissions may be detected over a period of days, hours, or even minutes.

➥**Reminder:** Isotopes of an element are atoms having the same number of protons—identifying them as the same element—but different numbers of neutrons.

Suppose we look at a sample of a radioactive substance, and let N represent the number of radioactive nuclei it contains at a particular instant t. After a further time interval Δt, X of these nuclei will decay. If the time interval is small enough so that only a tiny fraction of the existing radioactive nuclei decay, the fraction X/N that decay will be proportional to Δt to a high degree of approximation. We can write this proportionality relationship as $X/N \approx \lambda \Delta t$. The proportionality constant, called the **decay constant,** is generally written as λ (not to be confused with wavelength) in this context. Its value depends on what isotope is decaying. If X nuclei decay, the population N of nuclei remaining radioactive is reduced; that is, $\Delta N = -X$. Therefore, $-\Delta N/N \approx \lambda \Delta t$, or

$$\Delta N \approx -\lambda N \Delta t \quad (\Delta t \text{ very small}) \quad (28\text{-}1)$$

This is a specific instance of the general type of situation

$$\Delta \chi \approx -k \chi \Delta t \quad (21\text{-}16)$$

(see Section 21-6) in which the decrease in a quantity χ is proportional to how much remains. These are *exponential decay* situations. The quantity's decay over time is represented by a characteristically shaped curve (Figure 28-1). Note that because ΔN is proportional to N and N is decreasing, each change ΔN is smaller than the previous ones (the successive ΔNs are shaded in the figure). For a fuller

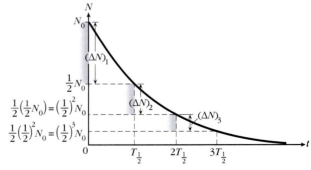

Figure 28-1 Exponential decay of a population N of radioactive nuclei.

For WebLink 28-1:
Exponential Decay,
go to
www.wiley.com/college/touger

Table 28-1 Half-Lives of Selected Radioisotopes

Radioisotope	Half-Life
Element 110*	$\approx 3 \times 10^{-4}$ s
Element 114*	30 s
Oxygen-15	122 s
Carbon-11	20.4 min
Fluorine-18	110 min
Radon-222	3.82 days
Iodine-131	8.04 days
Thorium-234	24 days
Cobalt-60	5.27 years
Strontium-90	28.8 years
Radium-226	1600 years
Carbon-14	5730 years
Technetium-98	4.2×10^{6} years
Uranium-235	7.0×10^{8} years
Potassium-40	1.28×10^{9} years
Uranium-238	4.5×10^{9} years
Thorium-232	1.4×10^{10} years

*Element (still unnamed) does not occur naturally; has been observed only fleetingly under artificially produced conditions. Elements 107–110 were identified by researchers at the Society for Heavy Ion Research (GSI) in Darmstadt, Germany, the last of these as recently as 1994. Element 114 was reported by Russian researchers in 1998.

➥Baryons: Baryons are a class of subatomic particles that include protons and neutrons. They also include some more exotic massive subatomic particles which are *not* ordinarily found in the nucleus.

quantitative treatment of exponential decay that enables us to find N at successive instants, work through WebLink 28-1.

$X/N \approx \lambda\Delta t$ says that the fraction X/N of nuclei that decays is always the same in a given Δt. If one sample starts out with 10 000 radioactive nuclei and another with 20 000 of the same kind, the same fraction of the total (not the same number of nuclei) will have decayed in each sample after Δt. Therefore, it will take the same amount of time for each population of a particular species of radioactive nucleus to decrease by half. If it requires the same time interval for 10 000 radioactive nuclei to decrease by half as for 20 000 nuclei to decrease by half, then when 20 000 is halved to 10 000 it will require the same time interval for *that* 10 000 to decrease by half. Likewise it will require the same time interval for *that* half to be reduced by half, and so on (Figure 28-1). After n like intervals, the fraction N/N_0 of the initial amount that remains will be $(\frac{1}{2})^n$. The time interval required for each halving is called the **half-life** ($T_{1/2}$) of the radioisotope. If $N = N_0$ at $t = 0$, the number of half-lives n that will have gone by at a later instant t is $n = t/T_{1/2}$. For instance, if the half life is 2 days and 6 days have gone by, then $\frac{6}{2} = 3$ half-lives have gone by. Therefore the fraction N/N_0 will be

$$\frac{N}{N_0} = \left(\frac{1}{2}\right)^{t/T_{1/2}} \tag{28-2}$$

after a time interval $t - 0 = t$ has elapsed. Using calculus, it is possible to show that the half-life $T_{1/2} = \frac{0.6931}{\lambda}$ (see Problem 28-57). As Table 28-1 shows, half-lives vary very widely from isotope to isotope.

The half-life is a statistical quantity for large numbers of radioactive nuclei. For one or a few nuclei, it only tells us about the probability of their decaying. Saying that half the nuclei will have decayed after this much time is like saying that if we flip millions of pennies, half of them will come up heads. Not only don't we know *which* individual pennies will come up heads, if we flip only two pennies there is a one in four chance that *both* of them will come up heads.

Notation: In identifying **radioisotopes, we will often use the notation**

$$^A_Z X$$

X is the symbol for the element

Z = atomic number = number of protons

A = nucleon number = atomic mass number

= combined number of protons and neutrons $Z + N$

For example $^{235}_{92}U$, an isotope of uranium (atomic number 92), has 92 protons and 143 neutrons, a total of 235 **nucleons** (a collective name for protons and neutrons). We may write it more concisely as uranium-235 or ^{235}U, omitting the atomic number because it is the same for all isotopes of the same element.

Transmutation of the elements occurs by both α and β emission. A typical process involving α particle emission is the last in a series of decays by which ^{238}U is transformed into lead (^{206}Pb):

$$^{210}Po \rightarrow {}^{206}Pb + \alpha + \Delta E$$

or
$$^{210}_{84}Po \rightarrow {}^{206}_{82}Pb + {}^4_2He + \Delta E$$

This is the general format for writing a **nuclear reaction** (you may have had experience writing chemical reactions in much the same way). It obeys the following rules:

• *The total number of nucleons remains constant* (**baryon conservation**): $\Sigma A_{initial} = \Sigma A_{final}$. In our example, $210 = 206 + 4$.

• *The total charge of all particles involved remains constant* (**charge conservation**): In our example, the charged particles are all protons, so the initial and final proton total must be the same: $84 = 82 + 2$. So far, we have not considered the charge of the electrons in the atoms. The neutral polonium atom has 84 electrons. When the number of protons is reduced by two in the decay process, two of the electrons in the resulting Pb atom are unbalanced; it starts out as an ion. However, it loses its surplus electrons very quickly to surrounding matter. The energy changes involved in this loss are negligibly small compared to the energy changes involved with a change in the nuclear binding situation. We therefore ignore the outer electron situation when we write nuclear reactions.

- *Energy is conserved:* Recall from Section 25-4 that the mass of a nucleus with a certain number of protons and neutrons is less than the total mass of that many *free* protons and neutrons. The resulting **mass defect** Δm changes in a nuclear reaction. The corresponding change in rest-mass energy $(\Delta m)c^2$ must be balanced by other changes in energy, which we have denoted by ΔE above. Typically, rest-mass energy is lost, and ΔE represents the increased kinetic energy of the products. The end product nucleus may initially be produced in an excited state and then drop down to the ground state by emission of a high-energy (γ ray) photon.

The basic reaction resulting in β-particle (high-energy electron) emission is the decay of a neutron into a proton and an electron:

$$\beta \text{ decay (incomplete): } {}_{0}^{1}\text{n} \rightarrow {}_{1}^{1}\text{p} + {}_{-1}^{0}\text{e} \quad (??)$$

(The question marks suggest that we might have to make a later adjustment, because careful measurements showed that as written, this reaction did not conserve energy.)

Unlike the outer Pb electrons in the previous example, the electron here is directly involved in the nuclear reaction and must be accounted for. The -1 on the electron is easiest understood if we mentally reverse the process: The electron combining with a proton to form a neutron would reduce the number of protons remaining, and consequently the atomic number, by 1. Note that the atomic number of the other product is one greater than that of the initial species. This would remain true if the neutron were part of a nucleus, as in the decay reaction

$$ {}_{81}^{206}\text{Tl} \rightarrow {}_{82}^{206}\text{Pb} + {}_{-1}^{0}\text{e}$$

If the β decay were complete as written above, the energy of the products would always be slightly less than the energy of the initial neutron. Energy conservation has been so fundamental a concept in physics that because β decay did not appear to conserve energy, Pauli proposed in 1931 that there must be an additional particle being produced. This additional particle had not been detected because it was chargeless, massless or very nearly so, and so unreactive that there was only a tiny probability of it reacting with matter that it passed through (the reaction would provide the means of detection). Italian physicist Enrico Fermi called this proposed particle a **neutrino,** meaning "little neutral one" in Italian. The proposal of this particle, which at the time was undetectable, seemed far-fetched even to Pauli himself. However, not only did the neutrino preserve conservation of energy, in doing so it also had just the right momentum to satisfy momentum conservation in β decays, and the right angular momentum for angular momentum conservation as well. This was a compelling argument for accepting its existence. It was not detected until the 1950s, when an experiment involving the passage of neutrinos through tons of hydrogen-rich matter produced enough reactions to be detected. The β-decay reaction above can be more fully written as

$$ {}_{0}^{1}\text{n} \rightarrow {}_{1}^{1}\text{p} + {}_{-1}^{0}\text{e} + \bar{\nu}$$

Strictly speaking, $\bar{\nu}$ represents an antineutrino, the antiparticle of the neutrino. We will not be concerned with the distinction at this time.

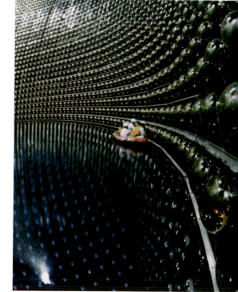

Super-Kamiokande, built into an underground mine in Japan, is the world's largest detector of neutrinos. Neutrinos, nearly massless particles that travel very nearly at the speed of light, reach Earth from the sun and other cosmic sources. The detector's inner and outer volume contain 32 000 tons and 18 000 tons of pure water, respectively. The inner detector has 11 200 photomultiplier tubes.

➡**Antiparticles:** An antiparticle is a fundamental particle that is like another particle in that it has the same mass, but its charge (if charged at all) and certain other properties are opposite to its ordinary counterpart. When a particle and its antiparticle collide, they annihilate each other, resulting in a large release of energy.

EXAMPLE 28-1 *The Decay of a Radioisotope of Bismuth*

For a guided interactive solution, go to Web Example 28-1 at
www.wiley.com/college/touger

When $^{214}_{83}\text{Bi}$ decays, it will undergo an α decay and a β decay, one after the other, in either order. What isotope will remain after these two decays?

Brief Solution

Choice of approach. Apply nucleon number (baryon) conservation and charge conservation to each reaction in turn to find the atomic number Z and nucleon number A of the product. Check the atomic number of the final product on a periodic table to establish its identity.

What we know/what we don't. In the combination of reactions, both an α particle and a β particle are emitted:

$$^{214}_{83}\text{Bi} \rightarrow {}^{A}_{Z}X + {}^{4}_{2}\text{He} + {}^{0}_{-1}\text{e} + \bar{\nu} \qquad Z = ? \quad A = ? \quad X = ?$$

The mathematical solution. Conservation of nucleon number tells us that $214 = A + 4 + 0$, so $A = 210$. Conservation of charge by the nuclei and nuclear products tells us that $83 = Z + 2 + (-1)$, so $Z = 82$. The periodic table tells us that the element with atomic number 82 is lead (Pb), so the end product is $^{210}_{82}\text{Pb}$, a radioisotope of lead.

◆ Related homework: Problems 28-4, 28-5, 28-6, and 28.7.

Recall from Section 25-4 that in a single atom, the mass defect

$$\Delta m = \Sigma(m_{\text{o}}) - (m_{\text{o}})_{\text{bound}} \tag{28-3}$$

$$\underset{\substack{\text{free protons} \\ \text{and neutrons}}}{} \quad \underset{\text{nucleus}}{}$$

and the energy equivalent of this mass is the *binding energy*

$$E_{\text{b}} = c^2 \Delta m \tag{28-4}$$

➡**Reminder:** Bound states of systems have negative energies. E_{b} is the energy input required to break the bonds among bodies whose bound state energy is $-E_{\text{b}}$. The input raises the energy of the system to zero.

Example 25-6 made the point that the binding energy associated with the bond between nucleons in a nucleus is on the order of a million times greater than the binding energy associated with the electrostatic attraction of an electron to the nucleus of its atom. The electrostatic attraction binding energies are typically on the order of a few eV (that is, an energy input of a few eV is required to free an electron from an atom), but the nuclear attraction binding energies are typically on the order of millions of electron volts (MeV).

STOP&Think How can any nucleus be stable? If there are only positive protons and neutral neutrons in a nucleus, shouldn't the protons repel one another? ◆

If only electrostatic forces were involved in the nuclei, all nuclei would blow apart. The fact that most nuclei are stable suggests that another kind of force must also be involved—an *attractive* force between nucleons that we call the **strong nuclear force,** or sometimes simply the *nuclear force.* On the other hand, if nuclear forces were always stronger than electrostatic forces, the nuclei of atoms would attract one another, and they would clump together into a single nucleus. Because this does not happen, we must conclude that the nuclear forces are *short-range:* As the distance r between nucleons increases, the magnitude of the nuclear force of attraction between them must decrease more sharply than the electrostatic or Coulomb forces, which drop off as $\frac{1}{r^2}$. Nuclear forces are strong at the tiny separations that exist within a single nucleus, but at the much larger distances between nuclei they become negligible.

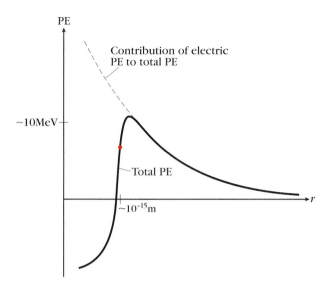

Figure 28-2 Potential energy of an α particle or other positive cluster of nucleons within a nucleus.

The potential energies associated with these forces are additive (Figure 28-2). Electric potential energies associated with the force between like charges are always positive (dashed line in figure). But within a region with a radius on the order of 10^{-14} m, roughly the radius of a nucleus, the total potential energy is negative, because it includes an additional contribution due to the nuclear force. **STOP&Think** Compare this radius to the first Bohr radius of a hydrogen atom. ◆ The resulting graph has a peak. If an α particle had a total energy less than this peak energy, it could never get to where the peak occurs, because at distances where the total energy is less than the potential energy, the kinetic energy would have to be negative. According to classical mechanics, then, the α particle could not cross this *potential energy barrier,* and the nucleus could never decay by α emission. It would remain stable. But quantum mechanically, there is a small probability of the α particle going a small Δr further. If the barrier is narrow enough, the α particle can get through it—an effect called *tunneling*—and the nucleus decays by α emission. The lifetime of the radioactive (stable) nucleus decreases as this probability increases.

If there are A nucleons in the nucleus of the atom, then on the average each nucleon's share of this binding energy will be E_b/A. In general, the nucleus will be most stable when this *binding energy per nucleon* is greatest, so that each nucleon's share of the bound state energy of the nucleus is most negative. The graph in Figure 28-3 displays empirical data showing that the binding energy per

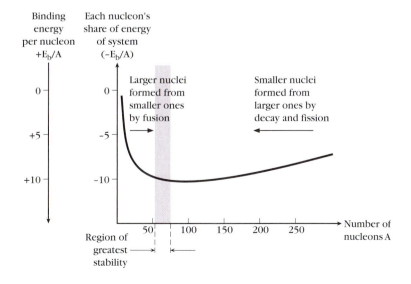

Figure 28-3 The binding energy per nucleon (E_b/A) varies with the number of nucleons A in a nucleus. *Reminder:* E_b is the energy input required to break up a bound system with total energy $-E_b$.

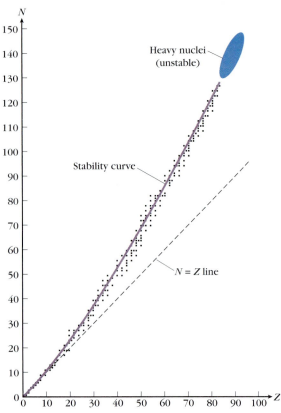

Figure 28-4 The stable nuclei cluster along a well-defined curve (the "stability line") on a plot of neutron number N versus proton number Z.

nucleon varies with the number of nucleons in the nucleus, so that *very small and very large nuclei are less stable than nuclei in the middle range of nucleon numbers.*

Stability results from a balancing between short-range forces of nuclear attraction between all nucleons and electrostatic repulsive forces between protons. As the number of nucleons increases, protons are repelled by all the other protons. However, the nuclear forces are so short-range that they primarily affect nearest-neighbor nucleons, so they are not cumulative to the same extent. This would-be dominance of repulsions over attractions can be reduced if the nucleus contains more neutrons, keeping the protons further apart on the average. This could explain why stable heavy nuclei like $^{206}_{82}Pb$ ($Z = 82$, $N = 206 - 82 = 124$) tend to have more neutrons than protons.

Figure 28-4, based on empirical data, graphs the number of neutrons N against the number of protons Z for nuclei known to be stable. The data points for individual stable nuclei are not randomly scattered but cluster around a curve, which we can call the **stability curve.** For smaller nuclei, the N and Z values are approximately equal, as in the predominant isotopes of carbon and oxygen, $^{12}_{6}C$ and $^{16}_{8}O$. This might seem to suggest that the nuclear forces are strongest when there is a neutron for each proton. But as we have noted, heavier nuclei have more neutrons than protons. Even so, all nuclei beyond $Z = 83$ undergo some kind of decay (bismuth, $^{209}_{83}Bi$, is the heaviest stable nucleus).

Isotopes that do not lie on this curve tend to be unstable and therefore radioactive, because they will decay into end products more nearly on the stability line. For example, the common isotope of carbon, $^{12}_{6}C$, is stable, but $^{11}_{6}C$ and $^{14}_{6}C$ are both radioactive. $^{14}_{6}C$ decays by β emission with a half-life of 5730 years to stable $^{14}_{7}N$. $^{11}_{6}C$ also decays by β emission, but with a difference: The decay process here is

$$^{11}_{6}C \rightarrow {}^{11}_{5}B + {}^{0}_{+1}e + \nu$$

where ${}^{0}_{+1}e$ represents a positron, the antiparticle of the electron. Also, the reaction yields a neutrino, the antiparticle of the antineutrino. Whereas radioisotopes with too high a ratio of neutrons to protons gain protons at the expense of neutrons by the β decay process

$$n \rightarrow p + \beta^{-} + \bar{\nu} \qquad \text{or} \qquad {}^{1}_{0}n \rightarrow {}^{1}_{1}p + {}^{0}_{-1}e + \bar{\nu}$$

radioisotopes with too low a ratio of neutrons to protons gain neutrons at the expense of protons by the β decay process

$$p \rightarrow n + \beta^{+} + \nu \qquad \text{or} \qquad {}^{1}_{1}p \rightarrow {}^{1}_{0}n + {}^{0}_{+1}e + \nu$$

◆**MEASURING RADIOACTIVITY** High-energy radioemissions ionize atoms with which they collide, often dislodging electrons with sufficient energy so that the electrons can then ionize other atoms. This provides a convenient basis for measuring the rate at which radioemission occurs (and the corresponding rate at which the emitter is decaying). An **ionization chamber** consists mainly of a gas-filled cylinder with a straight wire along its axis (Figure 28-5). If a potential difference is applied between the cylinder and the central wire, the gas ordinarily acts as an insulator. However, each time an α or β particle passes through, it frees charge carriers in the gas, giving rise to a current pulse across the gas-filled gap. In the **Geiger counter,** each burst of current registers as an audible click, and the click rate is proportional to the emission rate.

A more efficient and sensitive device for measuring radioemission is the **scintillation counter** (*scintillations* are little bursts of light). Figure 28-6 shows its

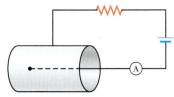

Figure 28-5 The ionization chamber: the basis for the Geiger counter.

main features and operating principles. The *scintillator* is a material in which atoms are excited by the incoming radiation to emit photons. NaI crystals and certain plastics are good scintillators. The incoming radioemission is far more likely to interact with an atom of this solid material than an atom of the much less dense gas in an ionization chamber, hence the greater efficiency. The photons cause photoelectrons to be emitted from a photocathode made of suitable photoelectric material. The **photomultiplier tube** contains a series of electrodes at successively higher potentials, typically increasing by steps ΔV of about 200 V. Each electron accelerated to one of these electrodes strikes it with sufficient kinetic energy to dislodge a few electrons, perhaps two to five, which in turn are accelerated to the next cathode. The number of electrons advancing through the tube thus multiplies at each electrode. **STOP&Think** If there were initially one photoelectron and the number of electrons were multiplied by five at each of six successive electrodes, how many electrons would leave the sixth electrode? ◆

◆**RADIOACTIVE DATING** Because Earth is bombarded by cosmic rays, small numbers of nuclear reactions are always taking place in the atmosphere, producing radioisotopes of the elements commonly found in the atmosphere. The balancing effects of decay of these radioactive nuclei and the production of new ones keep them at a small but roughly constant fraction of the total concentration of that element in the atmosphere. Once the element is removed from the atmosphere—either through metabolism by living things or in geological processes resulting in rock formation—the element is locked into an environment in which the population of radioactive nuclei is no longer maintained by new reactions, and the radioactive fraction of the element undergoes exponential decay.

For example, carbon dioxide is fixed in the food chain of living things by photosynthesis (Figure 6-8) in green plants. When metabolized, the carbon is distributed among the molecules of each living organism. When an organism dies, the carbon remains, no longer renewed by fresh intake from the atmosphere. Thus, after death the fraction of this carbon that is radioactive carbon-14 decays with ^{14}C's characteristic half-life of 5730 years. It is therefore possible to determine the age of formerly living matter thousands of years old (Figure 28-7) by

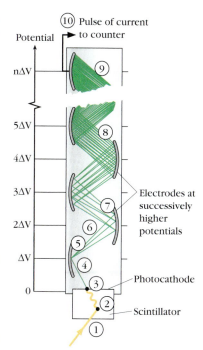

Figure 28-6 A scintillation counter. ① A particle (α or β particle) or a γ-ray photon penetrates scintillator material. ② An atom in the scintillator material is excited by the incoming particle and emits a photon. ③ The photoelectric surface (photocathode) with low work function emits a photoelectron. ④ The photoelectron is accelerated toward an electrode at higher potential ΔV. ⑤ The photoelectron strikes the electrode with enough KE to dislodge *a few* electrons (secondary electrons.) ⑥ The secondary electrons are accelerated toward an electrode at a still higher potential $2\Delta V$. ⑦ Each secondary electron dislodges a few more electrons, multiplying the number that can be accelerated toward the next electrode at an even higher potential. ⑧ The number of secondary electrons continues to multiply at each higher-potential electrode, so that ⑨ perhaps over a million electrons reach the last electrode, from which ⑩ a pulse of current goes to a counter.

↪**A note on language:** Be aware that we have used the word *decay* in two different ways. The process by which a single nucleus breaks down is called a *decay*. In contrast, exponential decay refers to a pattern of decline in a population. Thus, the exponential decay of a population of radionuclei is the cumulative effect of individual nuclear decays.

Figure 28-7 Cave paintings found at several sites in France and Spain, like this one at Lascaux, are thought to be between 10 000 and 15 000 years old. Radiocarbon in formerly living matter could have been used to date the paintings.

comparing the percent of carbon that is ^{14}C in this matter with the percent of carbon that is ^{14}C in atmospheric carbon dioxide. Dating based on ^{14}C is called **carbon dating.**

EXAMPLE 28-2 *Radiocarbon Dating of a Prehistoric Encampment*

For a guided interactive solution, go to Web Example 28-2 at www.wiley.com/college/touger

A newly discovered cave displays evidence of having been used as an encampment, possibly in prehistoric times. Ashes from old wood fires remain. Not long before the fires, the wood had been part of living trees. On the basis of the level of radioactivity measured, the carbon in the ashes is found to contain about 0.70 times the amount of ^{14}C that would be found in an equal mass of carbon taken from living matter. Roughly how long ago were the fires made?

Brief Solution

Restating the problem. We are told that if there were N_o radiocarbon nuclei in the sample from living matter, and N now remaining, the fraction $N/N_o = 0.70$. If we assume the matter was last living at $t = 0$, our problem is to find the present value of t.

Choice of approach. Equation 28-2, $N/N_o = \left(\frac{1}{2}\right)^{t/T_{1/2}}$, relates the remaining number of radioactive nuclei to the number of half-lives $n = t/T_{1/2}$ that have gone by.

What we know/what we don't.

$$\frac{N}{N_o} = 0.70 \qquad T_{1/2} = 5730 \text{ years for } ^{14}C \qquad t = ?$$

The mathematical solution. To solve for an unknown in an exponent, you must take the logarithm of both sides of the equation. You can do this in whatever base you choose; use either the LN or the LOG button on your calculator for logarithms in base e or base 10. The log of x^y, a number raised to a power, is $y \log x$. Therefore, from Equation 28-2,

$$\log\left(\frac{N}{N_o}\right) = \log 0.70 = \frac{t}{T_{1/2}} \log \left(\tfrac{1}{2}\right)$$

If we use the LOG button of the calculator, we get

$$-0.155 = \left(\frac{t}{T_{1/2}}\right)(-0.301)$$

or $t/T_{1/2} = 0.515$. Then

$$t = 0.515 \, T_{1/2} = 0.515(5730 \text{ years}) \approx \textbf{3000 years}$$

(to two significant figures)

◆ Related homework: Problems 28-17, 28-18, 28-19, and 28-21.

28-2 Fission and Fusion

Energy is given off in processes that increase the binding energy per nucleon, so that the total energy is more negative for the resulting nuclei than for those initially present. This happens in three important types of processes: decay of heavy nuclei, nuclear fission, and nuclear fusion.

◆**DECAY OF HEAVY NUCLEI** All nuclei of atomic number greater than 83 are unstable. These heavy nuclei characteristically undergo a chain of α and β decays until a stable end product nucleus (a stable isotope of $_{82}$Pb or $_{83}$Bi) is reached. One such chain is shown in Figure 28-8. With each α decay, the atomic number drops by two, and with β decay it increases by one, so the overall trend is downward.

◆**NUCLEAR FISSION** Fission means splitting. In **nuclear fission,** a heavy nucleus splits into two middle-range nuclei. Because the ratio of neutrons to protons is higher for heavy nuclei than for nuclei in the middle range, excess neutrons are freed in the process. A typical fission reaction is

$$^{236}_{92}\text{U} \rightarrow {}^{90}_{37}\text{Rb} + {}^{143}_{55}\text{Cs} + 3{}^{1}_{0}\text{n} + \Delta E$$

After neutrons were discovered in 1932, physicists tried to bombard heavy nuclei with slow neutrons, with the idea that a slow enough neutron could be taken in or absorbed by a nucleus. They hoped this might yield new elements of higher atomic number, but fission occurred instead (for example, after ^{235}U absorbed a neutron to form ^{236}U). (We will shortly say more about these fission reactions.) Fission can also occur, but with very low probability, when no neutron is absorbed—an occurrence called *spontaneous fission.* In the previous section, we discussed how the possibility of tunneling results in a small probability that a nucleus can emit an α particle spontaneously. This reasoning applies not only to α emission but to the splitting off of larger positive clumps of nucleons; that is, the daughter nuclei produced in fission reactions such as the above.

◆**NUCLEAR FUSION** In **nuclear fusion** reactions, two small nuclei become fused together into one larger nucleus. Because the binding energy per nucleon is greater in the resulting nucleus, rest mass energy is converted into an energy output. An example of a fusion reaction is

$$^{2}_{1}\text{H} + {}^{3}_{1}\text{H} \rightarrow {}^{4}_{2}\text{He} + {}^{1}_{0}\text{n} + 17.5 \text{ MeV}$$

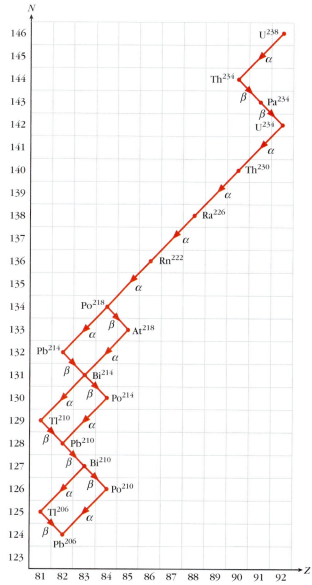

Figure 28-8 The chain of α and β decays from $^{238}_{92}$U to $^{206}_{82}$Pb.

in which deuterium and tritium, two isotopes of hydrogen, combine to form a heavier helium atom and a neutron.

Both fission and fusion reactions have large net energy outputs, but like exothermic chemical reactions (recall Figure 6-6), they do not occur spontaneously but must be initiated by energy inputs (except in the unusual event of spontaneous fission). Here again, the opposing effects of attractive nuclear forces and repulsive electrostatic forces are involved. In *fusion,* the two nuclei start out at a separation distance at which the longer-range electrostatic repulsion between the two positive nuclei predominates; work must be done to bring the nuclei close enough together so that the short-range nuclear attraction will become dominant and the reaction will occur. In *fission,* the attractive nuclear force tends to keep the would-be halves together, and an input of energy is ordinarily needed to move them far enough apart so that the electrostatic repulsion will exceed the nuclear attraction and keep the separation going.

The required inputs, though smaller than the outputs, can be very substantial. The activation energy required to initiate a fusion reaction thermally is not available at naturally occurring Earthbound temperatures. In 1938, Hans Bethe and Carl von Weiszacker independently suggested that fusion was the process responsible for the huge amounts of energy emitted by stars, including our own sun. In stars, the interior temperatures provide adequate input energies, while the strong gravitational forces exerted by the massive stars keep the reactants close enough together for the reaction to occur.

Efforts have been made over the past few decades to generate energy by nuclear fusion in gases in which the atoms have sufficient kinetic energy to keep the fusion reaction going. At the high temperatures (about 10^9 K) necessary for this, ordinary containers cannot be used to confine the gas, but because at these temperatures the atoms of the gas are completely ionized, attempts to confine the ionized gas or *plasma* have made use of the fact that charged particles follow circular or helical paths in magnetic fields (see discussion of magnetic bottles and mirrors in Section 22-4). Many such efforts are based on the Tokamak design originated in the former Soviet Union.

28-3 Bombs and Reactors

Although technology based on fusion is still in the experimental stage, the development of technology based on nuclear fission has been far more extensive. The discovery that unstable nuclei could be split, releasing a large amount of energy, led in turn to the recognition that neutrons freed in the splitting of one such nucleus could in turn cause the splitting of other like nuclei. When this happens repeatedly, it is called a **chain reaction** and results in a prodigious total energy release. Controlled in a nuclear power plant, such a chain reaction can meet the energy needs of a large population, albeit with an element of risk that must be weighed against benefits and against the risks of alternatives. Uncontrolled, it provides the terrifying destructive power of nuclear weapons.

A tremendous energy output can be produced in the following way. We have already noted that when ^{235}U absorbs a neutron to form ^{236}U, the subsequent fission of ^{236}U produces *three* neutrons. These in turn can be absorbed by other ^{235}U atoms, which themselves split and give off more neutrons. In the ensuing chain reaction the number of neutrons builds, more and more atoms split, and the cumulative amount of energy released from multiple splittings rapidly builds up into an explosion (Figure 28-9). For a more detailed picture of how this works, work through WebLink 28-2.

For **WebLink 28-2: Chain Reaction,** go to www.wiley.com/college/touger

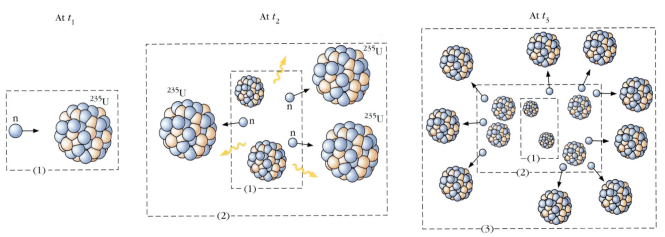

At t_1 At t_2 At t_3

Figure 28-9 A nuclear fission chain reaction. The dashed lines enclose the system we are considering at each instant. The system at t_2 includes all of the system at t_1 plus three additional ^{235}U nuclei. The system at t_3 includes all of the system at t_2 plus nine additional ^{235}U nuclei.

A chain reaction will not occur unless enough ^{235}U is brought together, something that does not occur in nature. It is something like a game in which each person is given three Ping-Pong balls. One person tosses her balls randomly into the air to start the game. Each person who is struck by a ball must then throw his or her Ping-Pong balls into the air. If there are just a few people in a large room, the balls are unlikely to hit anyone, and the chain of events will quickly die off. In a small, densely packed room, on the other hand, most of the balls will find targets and the air will quickly be filled with balls. Likewise, if there are enough ^{235}U atoms close enough together, a condition called **critical mass,** the chain reaction can quickly get out of hand. Recognition of this fact played a pivotal role in the history of World War II.

From Pure Science to the Bomb

In the late 1930s two German chemists, Otto Hahn and Fritz Strassmann, succeeded in splitting the uranium atom and measuring the conversion of lost mass into energy. Lise Meitner, a physicist who had fled Nazi Germany because she was Jewish, collaborated with her nephew Otto Frisch in repeating the experiment more precisely. In Copenhagen, where she found refuge, Meitner worked as a colleague of Niels Bohr. Bohr quickly grasped both the scientific and political implications of these experiments.

Bohr realized that the explosive capacity of chain reactions meant that an atomic bomb was possible, and that because the Germans had split the atom, Hitler's Germany—at least in theory—had the capacity to develop such a bomb. Bohr communicated this to Einstein, who was by that time in the United States. As the most highly regarded physicist of that era—the one physicist, in fact, with celebrity status—Einstein was urged by the physics community in the United States, and especially by the considerable body of leading physicists who had already fled Europe, to communicate these concerns to President Franklin D. Roosevelt. In July 1939, Einstein signed such a letter, urging the President to be watchful and take quick action if necessary.

By October 1939, Roosevelt had approved uranium research in the United States, the first of several steps leading to the establishment of the Manhattan Project to develop an American atomic bomb. That was in September 1942; in December of that year, in a lab under the University of Chicago athletic facilities, Italian emigré physicist Enrico Fermi demonstrated the first sustained nuclear reaction. On August 6, 1945 (the day the author of this book was born), a chain reaction in a critical mass of uranium dropped from an American bomber demolished the Japanese city of Hiroshima. Nearly everything for a radius of over a mile from the center of the explosion was destroyed. Over 70 000 people were killed instantly. The energy stored in the nucleus of the atom was now manifest to the world (Figure 28-10).

Figure 28-10 **A mushroom cloud characteristic of the explosion of a nuclear weapon.**

Missiles capable of delivering nuclear warheads were stockpiled by both the United States and the Soviet Union during the Cold War.

The energy from fusion is easily used for explosives, whether for weaponry or to blast a tunnel or power a spaceship. But using it to provide for sustained energy needs is more difficult, because that requires *containing* a sustained fusion reaction, and that is a requirement still challenging researchers. In 1950, U.S. President Harry Truman authorized the development of the hydrogen or thermonuclear bomb. Because of the high input energy required to activate a fusion reaction, a fission reaction was used to trigger it. In November 1952 the United States produced its first test explosion of a hydrogen bomb at Eniwetok Atoll in the Pacific Ocean. By August 1953, the Soviet Union had exploded its own hydrogen bomb. Today the proliferation of nuclear weapons technology is one of the gravest problems we face.

◆ **NUCLEAR REACTORS** During the work on the bomb in World War II (see box, "From Pure Science to the Bomb"), it became evident to many of the scientists involved that nuclear reactions in a core of uranium at below critical mass might represent a vast and inexpensive source of energy for the future. After the war, work in this direction was begun with great optimism, and with the general public and most politicians insufficiently aware that there might be potential risks as well as benefits.

Nuclear power plants do not fundamentally change the way in which electricity is generated (see Figures 23-14 and 23-15). A fuel is still used to boil the water to produce the steam to turn the turbines. But now the fuel's energy output is released in nuclear rather than combustion reactions.

Figure 28-11 shows the main features of a nuclear power plant. The **reactor core** is where the water is boiled. The core is fueled by many tons of uranium shaped into thousands of long, thin **fuel rods.** A controlled chain reaction in the core continuously releases neutrons. Some of these neutrons sustain the chain reaction. Others collide with protons in the water molecules, thereby increasing their kinetic energy and raising the water temperature. When the water boils, the steam produced turns a turbine generator.

The steam–water circulatory system must be closed; the water that is brought in contact with the uranium fuel cannot be leaked to the outside. To recondense the water, a second flow of water is brought in thermal contact with the first; heat must be exchanged although water cannot be. **STOP&Think** Why not? ◆ As a heat engine, this process is about 30% efficient. The remaining 70% of the energy flow is carried off as waste heat.

Two main mechanisms keep the nuclear reaction in the core controlled. First, high-speed neutrons do not induce fission as readily as low-speed neutrons, which are more easily absorbed by nuclei. The water helps regulate this. If the reaction runs too quickly, the water will bubble and boil off more rapidly. This leaves fewer water molecules to slow neutrons down to optimum speed for inducing fission. Nevertheless, there is the concurrent risk that if the water boils off too

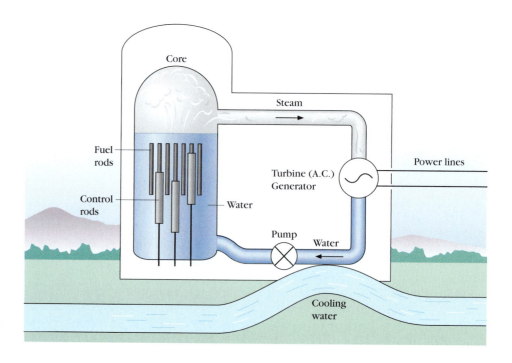

Figure 28-11 Schematic diagram of a nuclear power generator.

rapidly, the loss of coolant can cause the core temperature to rise dangerously. In the worst-case scenario, the fuel rods melt and release the radioactive fission products trapped. Such a *meltdown* took place at the Chernobyl nuclear power plant in the former Soviet Union on April 26, 1986. In that event, the worst nuclear accident in history, the reactor in which the meltdown occurred exploded. Though this was not itself a nuclear explosion, it released 30 to 40 times as much radioactivity as the bombs dropped on Hiroshima and Nagasaki at the end of World War II. Some 200 000 people had to be evacuated, most of them permanently, from a zone of extreme contamination 30 km in radius. In evaluating commercial reactors in the United States, however, it is important to note that they have eliminated design features that contributed to these failures, such as the graphite core and the kind of containment structure used at Chernobyl. **STOP&Think** How does the fact that the uranium fuel is arranged in long, thin rods rather than being lumped into a single solid block help counter the risk of a meltdown? ◆

The risk of excessive energy release necessitates the second controlling mechanism, the use of **control rods** made of good neutron absorbers, such as cadmium or boron. The "absorption" of a neutron is really a nuclear reaction in which the number of neutrons in the absorbing nucleus is increased by one. The control rods can be inserted to various lengths among the fuel rods. When fully inserted, they must be able to absorb enough neutrons to stop the chain reaction completely. This is a necessary design component even when the reactor is operating properly, because it is necessary to shut down the reactor in order to refuel.

Nuclear Power: The First Generation

In early 1954, President Dwight D. Eisenhower initiated legislation to permit electric utility companies in the United States to develop nuclear power plants. On July 17, 1955, for a one-hour demonstration period, Arco, Idaho (pop. 1350) became the first U.S. town to be powered by nuclear energy. The first commercial nuclear power plant began operation in England in 1956; commercial nuclear power began a year later in the United States when the reactor at Shippingport, Pennsylvania, went online. By 1983, nearly 10% of U.S. electrical generating capacity was nuclear, and the percentage was much higher in other industrialized countries that did not have substantial fossil fuel reserves of their own.

28-4 Ionizing Radiation

Ionizing radiation is a general term—it includes radioactive emissions and X rays—for all photons and moving particles that are energetic enough to dislodge electrons from the atoms in the matter that they pass through. Those atoms are ionized, and ions readily react chemically with other ions. Therein lies the danger of ionizing radiation.

Ions play a critical role in the cellular chemistry of all living things. If ions are present that are not normally part of this cellular chemistry, reactions other than the normal ones will occur. These are likely to be life-threatening. If they affect the DNA they can cause genetic damage; when they affect other parts of the cells, the resulting damage is called *somatic* rather than *genetic*. Somatic damage includes the development of cancers, lowered white blood cell count, cataracts of the eyes, lesions and ulceration in body tissue, fibrosis of the lungs, hair loss, and other physical effects.

Because of these health concerns, we need to be able to quantify radioactivity level and its effect on us. A variety of measures exist for this purpose

Table 28-2 Some Measures of Radioactivity

Quantity	Definition	SI Unit	Common Unit
Activity	Number of nuclei that decay per second (measured by detectors such as Geiger counters)	Becquerel (Bq) $\equiv 1$ decay/s	Curie (Ci) Ci $= 3.70 \times 10^{10}$ Bq
Exposure (defined only for X rays and γ rays)	The amount of radiation measured in terms of its effect: how much ionization (measured as the total charge of the ions) it produces in each kilogram of mass	Roentgen (R)[a] $\equiv$ amount of radiation that produces 2.58×10^{-4} C/kg of ionization (C = coulombs)	Roentgen
Absorbed dose or **dose** (D)	The amount of energy absorbed by each kilogram of mass $\left(\frac{\text{total energy absorbed}}{\text{total mass}}\right)$	Gray (Gy) $\equiv 1$ J/kg	rad (1 rad $= 10^{-2}$ Gy) (rad is short for **r**adiation **a**bsorbed **d**ose)
Quality Factor (Q or QF) or **Relative Biological Effectiveness** (RBE)	For a particular kind of emission (e.g. α, β), this tells how many times as much damage will be done by one unit of energy from this type of emission as by one unit of γ- or X-ray photon energy	unitless	unitless
Dose equivalent (H) or **effective dose**	Multiplying the absorbed dose of a particular kind of emission by its RBE gives this measure of the damage to be expected from the absorbed dose	Gy × QF $\equiv 1$ sievert (Sv)	Rad × QF $\equiv 1$ rem (1 rem $= 10^{-2}$ Sv) (rem is short for **r**oentgen **e**quivalent **m**an)

[a]Other definitions of the roentgen may be found in other texts. This is the definition given in *Le Système International d'Unites (SI)* published by the International Bureau of Weights and Measures (the official overseers of SI) in 1991 (and translated in National Institute of Standards and Technology Special Pub. 330, 1991 edition).

(see Table 28-2). Some measures (exposure, absorbed dose) tell us how much *effect* the radioactivity produces in each kilogram of mass. Of these, the absorbed dose is appropriate to all kinds of emissions while the exposure is defined only for X rays and γ rays, so the dose is a more appropriate measure when considering cumulative effects on living things.

But the biological damage done to living matter is not simply proportional to the energy transferred to it. An amount of energy absorbed from one kind of emission might cause more damage than an equal energy input from another kind of emission. For example, α particles will interact with more atoms along each unit of path length than will much faster-moving X-ray or γ-ray photons. The energy from α particles breaks more bonds and produces a greater concentration of inappropriately reactive ions. A range of observations and measurements show the overall damage from α particles to be about 10 times as great as for the same energy input from X-ray or γ-ray photons. For that reason, the dose is multiplied by 10 (a unitless quantity, called the **quality factor** Q or **relative biological effectiveness** RBE) to obtain the **dose equivalent** D, which more directly reflects the amount of damage done:

$$H = QD \tag{28-5}$$

$$\underset{\text{in Sv or rem}}{\text{dose equivalent}} = (\text{quality factor or RBE}) \times \underset{\text{in Gy or rad}}{\text{absorbed dose}}$$

➥**Caution:** Because the quality factor is unitless, H and D appear to have the same units, but the units have different names. This is to remind you that H is not the dose of the emission in question that is absorbed, but the dose of X- or γ-ray photons that would cause equivalent damage.

Strictly speaking, the Système International establishes this relationship to be $H = QND$, where N is the product of any additional corrective multipliers. It is a way of allowing for further corrections. In the absence of further corrections, $N = 1$. Table 28-3 gives quality factor values for some common emissions. They are approximate values based on the experience of accumulated observations.

Public concern about ionizing radiation generally focuses on the exposure to radioactivity resulting from a variety of technologies. But emitters of ionizing radiation occur naturally in our environment as well. Soil typically contains about 3 parts per million (ppm, by mass) of uranium-238 (^{238}U), 10 ppm of thorium-232 (^{232}Th), and 1 ppm potassium-40 (^{40}K). Soil also contains traces of radium, usually in low concentrations. An important decay product of radium is the radioactive gas radon, some of which is trapped in the soil above the radium and some of which occurs in trace quantities in the atmosphere. Where concentrations of radium in rock and soil are higher than average, cracks in basements can act as chimneys for the radon that is produced—chimneys leading into homes rather than out of them.

The top part of Table 28-4 shows that in all, 82% of the average American's exposure to ionizing radiation comes from natural sources, and more than half of it is from radon. Recall also that carbon dioxide fixed in the food chain of living things contains tiny amounts of radiocarbon. Over 10% of the average American's exposure to ionization radiation comes from nuclei within the body. The percentages are very different, of course, where extensive use of radiation for medical purposes is required. The bottom part of Table 28-4 shows dose equivalents for a variety of exposures. In the case of radiation treatment of a cancerous tumor, great care must be taken to restrict exposure to the tumor itself, to protect both the surrounding healthy tissue and the attending medical personnel.

Table 28-3 Quality Factor (Q or QF) or Relative Biological Effectiveness (RBE) of Some Common Emissions

Type of Radiation	Quality Factor*
X rays and γ rays	1.0
Electrons (β particles)	1.0+
Slow neutrons (< 10 KeV)	3–5
Fast neutrons (> 10 KeV)	≤ 10
α Particles and heavy ions	≤ 20

*Values are approximate.

Table 28-4 Exposure to Ionizing Radiation

Sources to which average person in United States is exposed[a]

(100% = an annual effective dose of about 3.6 mSv per person)

From natural sources

Radon	55%
Radioactive elements within body	11%
Rock, soil, and ground water	8%
Cosmic rays	8%

From artificial sources

Medical X rays	11%
Nuclear medicine	4%
Consumer products	3%
Miscellaneous	<1%

Typical dose equivalents for various exposures (in mSv = 10^{-3} Sv)

Total dose in radiation treatment of a cancer patient	≥50 000
CAT scan	50–100
U.S. federal guidelines[b] for upper exposure limits for radiation workers	
for radiation workers, per quarter	30
for radiation workers, per year	50
for general population, per year	5
Annual exposure of a radiation worker in a hospital	2.6–5.4
Total annual exposure for average person in United States	3.6
Chest X ray	0.4–10
Dental X ray	<1

[a]Adapted from *Environmental Encyclopedia*, Gale Research Detroit, 1994.
[b]1986 guidelines common to OSHA, the Nuclear Regulatory Commission (NRC), and the National Council on Radiation Protection and Measurement (NCRP).

EXAMPLE 28-3 *Radiation Exposure in a Mammogram*

For a guided interactive solution, go to Web Example 28-3 at
www.wiley.com/college/touger

The current maximum absorbed dose permitted in the United States for a full mammogram (a set of breast X-ray exposures for early detection of breast cancer) is 300 mrad (1 millirad $= 10^{-3}$ rad), although in practice most mammograms deliver only a fraction of this total. Suppose a woman receives a dose of 100 mrad from a mammogram.

a. How does this compare with the annual dose equivalent to which an average person is exposed (see Table 28-4)?
b. If we estimate that the dose received in the mammogram is distributed over about 5 kg of tissue, what is the total amount of energy (in joules) absorbed?

Brief Solution
Choice of approach.

a. We want to express the comparison as a ratio, which is meaningful only if the quantities are in the same units, so conversion data (Table 28-2) is needed. Because values in Table 28-4 are for *dose equivalents* given in mSv (millisieverts), you need to find the *dose equivalent* in mSv for a given X-ray *dose* in mrad. This requires multiplying by the quality factor QF (Equation 28-5) to get the *dose equivalent* in mrem, then converting from mrem to mSv.
b. The *dose* is the amount of energy absorbed per unit of mass. We need to multiply this by the mass to get the total amount of energy. It will be useful to convert the dose to grays (Gy) first, because 1 Gy = 1 J/kg.

What we know/what we don't.

a. For X rays, $Q = 1.0$ (from Table 28-3) $D = 100$ mrad $H = ?$ (in mrem)

   1 rem $= 10^{-2}$ Sv (from Table 28-2) $H = ?$ (in mSv)

   U.S. average $H_{annual} = 3.6$ mSv (from Table 28-4) $\dfrac{H}{H_{annual}} = ?$

b. $D = 100$ mrad $m = 5$ kg   1 rad $= 10^{-2}$ Gy (from Table 28-2)

   $D = ?$ (in Gy)   Total absorbed $E = ?$

The mathematical solution.

a. Using Equation 28-5 and heeding the cautionary note about units below it, we get

$$H = QD = (1)(100 \text{ mrad}) = 100 \text{ mrem}$$

$$= 100 \times 10^{-3} \text{ rem} \left(\frac{10^{-2} \text{ Sv}}{1 \text{ rem}} \right) = 1 \times 10^{-3} \text{ Sv} = 1 \text{ mSv}$$

$$\frac{H}{H_{annual}} = \frac{1 \text{ mSv}}{3.6 \text{ mSv}} \approx \textbf{0.3 or 30\%}$$

b. Converting the dose to grays gives us

$$100 \text{ mrad} = 100 \times 10^{-3} \text{ rad} = (100 \times 10^{-3} \text{ rad}) \left(\frac{10^{-2} \text{ Gy}}{\text{rad}} \right) = 1 \times 10^{-3} \text{ Gy}$$

Total absorbed $E = (1 \times 10^{-3} \text{ Gy})(5 \text{ kg}) = (1 \times 10^{-3} \text{ J/kg})(5 \text{ kg}) = \textbf{5} \times \textbf{10}^{-3} \textbf{ J}$.

Making sense of the results. The amount of energy is exceedingly small, even compared to the amount of energy you absorb in a few seconds from an ordinary light bulb. The concern is *not* with the total amount of energy but with the fact that the quanta of energy making up the total are large enough to ionize individual atoms.

♦ Related homework: Problems 28-39, 28-40, 28-43, 28-44, and 28-45.

Radioisotopes introduced into the body for medical imaging purposes also emit ionizing radiation. For this reason, they must have very short half-lives and be administered in very low doses. The greatest risk from a particular radioisotope is to the organ(s) where it tends to concentrate or be most reactive. These are called the **critical organs.** For example, the lungs are the critical organ for radon, the bones for radium, the gonads for carbon-14 and potassium-40, the kidney for uranium, and the thyroid gland for iodine-131.

The intake of radioisotopes was also the major risk of "fallout," the dispersal of radioactive matter into the atmosphere, whether from weapons testing or from a mishap at a nuclear power plant that releases radioactive materials into the atmosphere.

Particularly dangerous are those radioactive fission products that can be ingested and metabolized by living organisms. For example, strontium is just below calcium on the periodic table and reacts chemically in much the same way. Strontium-90 (^{90}Sr), one of the radioisotopes produced in fission, was released into the atmosphere during nuclear weapons testing in the 1950s, and consequently ended up along with calcium in cows' milk. Because children's bodies use the calcium in milk to build strong bones and teeth, their bodies used ^{90}Sr, a potential carcinogen, in the same way. Unlike the short-lived radioisotopes used for medical purposes, ^{90}Sr has a half-life of 28 years and therefore poses a long-term health threat.

The leakage of gaseous products of fission is a particular concern associated with the long-term storage of radioactive wastes.

The risks accompanying the use of fission reactions for electric power production ultimately have to be weighed against the risks of *not* using nuclear power to provide for a growing population whose other energy resources (fossil fuels) are diminishing and against the viability of other alternatives. They must also be weighed against the risks associated with using other energy resources, such as the following:

- The burning of fossil fuels (coal and oil) produces potentially cancer-causing (carcinogenic) by-products.
- Potentially fatal diseases, such as silicosis, are occupational hazards for coal miners.
- Naturally occurring traces of uranium in coal get released into the atmosphere or concentrated in the fly ash when coal is burnt.

The relative vulnerability to terrorist actions has also become a major concern.

It is critical that those in decision-making positions be fully informed of both the risks and benefits of all alternatives. The best of analyses, however, can only express the level of risk as a numerical probability. It cannot tell the decision makers whether that level of risk is acceptable. Thus, adequate technical analysis is necessary but not sufficient for making an informed decision; ethics must play an important role as well.

Emissions from a coal-fired power plant in Germany. In assessing the risks and benefits of nuclear power, the risks of competing energy sources, such as emissions from burning coal, must also be considered.

◆ SUMMARY ◆

Radioactive emissions (**α and β particles** and **γ rays**) are the products of decay processes by which an atom of one element is transformed into an atom of another (transmutation of elements). They are evidence that nuclei of **radioisotopes** are not permanently immutable structures but have finite mean lifetimes before decaying into another nucleus or nuclei. The sequence of decays continues until a stable isotope is reached.

For a particular radioisotope, the number of decays in a short time interval is proportional to N, the number of radioactive nuclei present:

$$\Delta N \approx -\lambda N \Delta t \qquad (\Delta t \text{ very small}) \qquad (28\text{-}1)$$

(The **decay constant** λ has a different value for each radioisotope)

so N undergoes *exponential decay*.

The time interval required for each halving of N is the same, and is called the half-life ($T_{1/2}$) of the radioisotope:

$$T_{1/2} = \frac{0.6931}{\lambda}$$

If N has value N_o at $t = 0$, the fraction N/N_o that remains at a later t is

$$\frac{N}{N_o} = \left(\tfrac{1}{2}\right)^{t/T_{1/2}} \qquad (28\text{-}2)$$

In identifying species in **nuclear reactions,** we often use the notation ${}_Z^A X$:

X is the symbol for the element

Z = atomic number = number of protons

A = nucleon number = atomic mass number

  = combined number of protons and neutrons $Z + N$

Nuclear reactions always obey the following rules:

- *The total number of nucleons remains constant* (**baryon conservation**):

$$\Sigma A_{initial} = \Sigma A_{final}$$

- *The total charge of all particles involved remains constant* (**charge conservation**).

- *Energy is conserved* as the **mass defect** Δm changes.

$$\Delta m = \underset{\substack{\text{free protons}\\\text{and neutrons}}}{\Sigma(m_o)} - \underset{\text{nucleus}}{(m_o)_{bound}} \qquad (28\text{-}3)$$

The basic reactions underlying nuclear decay processes are:

α decay:

$${}^A X \to {}^{A-4} Y + \alpha + \Delta E \qquad \text{or} \qquad {}_Z^A X \to {}_{Z-2}^{A-4} Y + {}_2^4 He + \Delta E$$

(*X* and *Y* are different elements)

β^- decay: $n \to p + \beta^- + \bar{\nu}$ or ${}_0^1 n \to {}_1^1 p + {}_{-1}^0 e + \bar{\nu}$

β^+ decay: $p \to n + \beta^+ + \nu$ or ${}_1^1 p \to {}_0^1 n + {}_{+1}^0 e + \nu$

Stability of nuclei involves a balancing between (1) short-range attractive **strong nuclear forces** between all nucleons and (2) longer range repulsive *electrostatic forces* between protons. Resulting empirical rules affecting the comparative stability of different nuclei include:

- The nucleus will generally be most stable when this **binding energy** per nucleon $\frac{E_b}{A}$ is greatest. Empirical data (graphed in Figure 28-3) show that *very small and very large nuclei are less stable than nuclei in the middle range of nucleon numbers.*

- When neutron number N is plotted against proton number Z, the data points for individual stable nuclei cluster around a curve called the **stability curve** (Figure 28-4). For smaller nuclei, N and Z are approximately equal; for larger nuclei the ratio of N to Z increases.

Because radioemissions *ionize* atoms in the material they pass through, the rate of radioemission can be measured using an **ionization chamber** (Figure 28-5) device such as a **Geiger counter,** or using a **scintillation counter** (Figure 28-6).

Radioactive dating (such as **carbon dating**) is possible because the rate of radioemission from elements in the atmosphere remains relatively constant, but decays exponentially with a characteristic half-life once the element has been fixed within solid matter (as when it is metabolized by living things).

Energy is given off in processes which increase the binding energy per nucleon:

1. decay of heavy nuclei

2. **nuclear fission,** in which a heavy nucleus splits into two middle range nuclei, along with excess neutrons.

3. **nuclear fusion** reactions, in which two small nuclei become fused together into one larger nucleus.

Fusion reactions require very high activation energies. They occur naturally in the dense, hot, ionized gases (plasmas) in the interiors of stars and are the source of "solar energy". Efforts are ongoing to confine hot plasmas to obtain energy from controlled fusion reactions.

Fission reactions occur when heavy unstable nuclei *absorb* slow neutrons. Each splitting frees neutrons; these can split more nuclei, resulting in a **chain reaction.** A **critical mass** of radioactive nuclei must be present to sustain the chain reaction. Uncontrolled chain reactions are the basis for atomic bombs. Controlled chain reactions are harnessed in **nuclear reactors.**

Ionizing radiation includes all emissions of sufficient energy to ionize atoms in the matter they pass through. Because ions are highly reactive, such ionization can result in chemical reactions within living things that don't ordinarily occur. Consequences can include genetic damage and cancers. Table 28-2 describes several measures of ionizing radiation and the damage it can cause.

Radioactive nuclei also have a variety of beneficial uses in medicine. In evaluating nuclear technologies, it is important to assess both *benefits and risks.*

✦ QUALITATIVE AND QUANTITATIVE PROBLEMS ✦
Hands-On Activities and Discussion Questions

The questions and activities in this group are particularly suitable for in-class use.

28-1. Discussion Question. When a heat lamp is directed at a sheet of metal, the metal becomes hot, but no photoelectric effect occurs. When a weak ultraviolet light is directed at an identical sheet of metal, it does not become hot but some photoelectrons are emitted. How does this pair of situations compare with the following pair of situations?

 i. A patient is immersed in a hot water bath for 10 minutes.

 ii. The patient undergoes a CAT scan.

In particular, discuss the following:

 a. Which situation in each pair involves the greater transfer of energy to the patient?

 b. Which situation in each pair involves an effect that does not occur in the other situation in the pair?

 c. What is similar about the effects that occur in only one of the situations in each pair?

28-2. Discussion Question. What features are analogous in a chain reaction and a photomultiplier tube? What features are dramatically different?

Review and Practice

Section 28-1 Radioactive Decay and Decay Processes

28-3. SSM Suppose there are nuclei of a particular radioisotope present in a sample. What integer number of half-lives must go by before fewer than a millionth of these radionuclei remain?

28-4. Determine whether the atomic number of a radioisotope increases by 2, increases by 1, remains the same, decreases by 1, or decreases by 2 in each of the following cases. **a.** The radioisotope emits an α particle. **b.** It emits a β^- particle. **c.** It emits a γ ray.

28-5. In each of the types of nuclear reactions below, tell by how much each of the following quantities changes: $Z =$ atomic number, $N =$ number of neutrons, $A =$ number of nucleons (or baryon number or atomic mass number). Answer by filling in the table. Be sure to include signs to indicate whether each change is an increase ($+$) or decrease ($-$).

	Z changes by ____	N changes by ____	A changes by ____
α decay			
β^- decay			
β^+ decay			

28-6. With the help of the periodic table, complete each of the following nuclear reactions. Write the missing species in the same format ($^A_Z X$) as the species that are given; that is, find A and Z and identify the element X.
a. $^{247}_{97}\text{Bk} \rightarrow \underline{\quad} + \alpha$
b. $^3_1\text{H} \rightarrow ^2_1\text{H} + \underline{\quad}$
c. $^1_0\text{n} + ^{235}_{92}\text{U} \rightarrow ^{235}_{92}\text{Ba} + \underline{\quad} + 3^1_0\text{n}$
d. $^2_1\text{H} + ^2_1\text{H} \rightarrow \underline{\quad} + ^1_0\text{n} + 3.3 \text{ MeV}$

28-7. With the help of the periodic table, complete each of the following nuclear reactions. Write the missing species in the same format ($^A_Z X$) as the species that are given.
a. $^3_2\text{He} + \underline{\quad} \rightarrow ^4_2\text{He} + ^1_1\text{H} + ^1_1\text{H}$
b. $\underline{\quad} + ^{232}_{90}\text{Th} \rightarrow ^{233}_{90}\text{Th}$
c. $^{15}_8\text{O} \rightarrow \underline{\quad} + \underline{\quad}$

28-8.
a. If the β decay of a free neutron were fully described as $^1_0\text{n} \rightarrow ^1_1\text{p} + ^{\ 0}_{-1}\text{e}$, which of the following quantities would be conserved in the reaction and which would not? *(baryon or nucleon number; energy; momentum; charge)* Briefly explain your answers.

b. If the reaction as written left out one of the actual products of the decay, what would its charge have to be? What would its baryon number have to be?

28-9. If 2.0×10^{18} nuclei of carbon-14 are present in a sample at a certain instant, how many nuclei of carbon-14 will there be in the sample seven half-lives later?

28-10. A research lab has two samples A and B of iodine. The numbers of radioactive nuclei in the two samples at $t = 0$ are 2.0×10^{16} in sample A and 8.0×10^{16} in sample B. At a later instant, sample A is found to have 1.5×10^{16} radioactive nuclei. At that same instant, we should expect sample B to have (*about* 6.0×10^{16}; *about* 7.5×10^{16}; *exactly* 6.0×10^{16}; *exactly* 7.5×10^{16}) radioactive nuclei.

28-11. Iodine-131 (I-131) decays by β^- decay with a half-life of 8.0 days. Suppose a sample initially has 6.4×10^{22} I-131 nuclei.
a. How many I-131 nuclei will remain after 24 days?
b. How many high-speed electrons will have been emitted after 24 days?
c. What is the decay constant for I-131?
d. How many I-131 nuclei will decay in the first minute?
e. How many high-speed electrons will be emitted during the first minute?
f. Were there any differences in the way that you arrived at your answers to **b** and **e**? Briefly tell why the differences in approach were or were not necessary.

28-12.
a. Find the decay constant (in s^{-1}) of a radioactive substance with a half-life of 1000 s.
b. Find the half-life of a radioactive substance with a decay constant of 1000 s^{-1}.

28-13.
a. As half-life increases, does the decay rate increase, decrease, or stay the same?
b. All other things being equal (which is not always the case), does the risk of short-term radiation exposure increase, decrease, or stay the same as the half-life of the substance to which you are exposed increases?

Note: The data in Table 28-1 may be useful in some of the following problems

28-14. Two radioisotopes of carbon, ^{11}C and ^{14}C, have very different uses. ^{11}C is one of the isotopes commonly introduced into patients in the medical imaging procedure called positron emission tomography (PET). ^{14}C is used for radiocarbon dating of prehistoric objects (and also very old historic objects). Explain why neither of these isotopes can be used the way the other one is used.

28-15. Carbon-11 and fluorine-18 both undergo β^+ decay. A sample of freshly prepared carbon-11 is placed in one of two identical rooms (room A), and a sample of freshly prepared fluorine-18 in the other (room B). If the two samples initially contain equal numbers N_0 of radionuclei,
a. calculate the number of radionuclei of each that will have decayed after 10 s. Express your results in terms of N_0 (number decayed = $\underline{\quad}?\underline{\quad}N_0$).
b. Use your calculations to answer this question: In which room would you be exposed to more radiation over this initial 10-s period?
c. In which room would you be exposed to more radiation over a period of 5 days (assuming you stay at the same average distance from each sample)? Briefly explain your reasoning.

28-16. Radon-222 and uranium-235 (the fissionable substance in atomic bombs) both decay by α emission. Would you rather spend a day exposed to 10 g of radon-222 or 10 g of uranium-235 under equivalent conditions? Briefly justify your answer.

• **28-17.** Roughly 1 out of each 10^{12} atoms in a sample of atmospheric carbon, or of carbon that has recently been metabolized from the atmosphere by a living organism, is radioactive carbon-14.

a. In a sample of living matter containing 6.02×10^{23} carbon atoms, how many radioactive decays will there be in a 1.0-minute interval?

b. The human body is approximately 18% carbon. How many ^{14}C decays will there be in *your* body in a 1.0-minute interval?

28-18. Suppose the radon-222 (^{222}Rn) level is measured in a basement room, and the room is then sealed so that no radon can get in or out. How many days would it take for the (^{222}Rn) level in the room to drop to 10% of the level that was originally measured?

28-19. SSM WWW A squirrel dies in the forest in 2005. Just after its death, a certain fraction of the carbon in its remains is ^{14}C. In what year will this fraction have dropped by 15%?

28-20. Consider a single atom of each of the following radioisotopes: carbon-14, oxygen-15, radon-222. Rank the atoms in order of the probability that each will decay during the next 24 hours. Order them from least to greatest, making sure to indicate any equalities.

28-21.

a. By a measurement it is determined that in human bones found at a burial site, roughly 1 out of each 10^{13} carbon nuclei is carbon-14. How long ago did this person live? Use information from Problem 28-17.

b. Find the decay constant for carbon-14 in SI units.

c. Find the total number of decays to be expected in one minute from one mole of carbon in which one out of each 10^{12} carbon nuclei are carbon-14.

d. Repeat *c* for a mole of carbon in which 1 out of each 10^{13} carbon nuclei are carbon-14.

e. If a detector records an average of 20 decays per minute from a sample of carbon from a recently living creature, how many decays per minute will it record from an equal mass of carbon taken from the bones found at the burial site?

Section 28-2 Fission and Fusion

28-22. What is the basic difference between fission and fusion?

28-23. Based on Figure 28-4, determine which nucleus in each of the following pairs is more stable: *a.* $^{32}_{16}S$ or $^{35}_{16}S$. *b.* $^{65}_{32}Ge$ or $^{72}_{32}Ge$.

28-24. Determine whether each of the following radioisotopes will decay by β^- emission or β^+ emission, and write its decay reaction:
a. $^{13}_{7}N$. *b.* $^{31}_{14}Si$.

28-25. SSM When $^{218}_{84}Po$ decays, it will undergo an α decay and a β decay, one after the other, in either order. Write your answers to each of the following in the form $^A_Z X$.
a. What isotope will remain after these two decays?
b. What will be the intermediate product if α decay occurs first?

c. What will be the intermediate product if β^- decay occurs first?

28-26. What are the important similarities and differences between the ionization chamber in Figure 28-5 and the arrangement for detecting a photocurrent in Figure 26-5?

28-27. Naturally occurring uranium ores contain only a small percent of fissionable uranium. Why is there no danger of a spontaneous chain reaction occurring in these ores?

28-28. Protons and neutrons exert nuclear strong forces on one another. Protons also exert electrostatic (Coulomb) forces on one another. *a.* Which kind of force predominates in a *fission* reaction? Briefly explain. *b.* Which kind of force predominates in a *fusion* reaction? Briefly explain.

28-29. Work, an input of energy, is required to initiate both fission and fusion reactions.

a. In a *fission* reaction, is the work done against the nuclear forces or the Coulomb forces that the nucleons exert on one another? Briefly explain.

b. In a *fusion* reaction, is the work done against the nuclear forces or the Coulomb forces that the nucleons exert on one another? Briefly explain.

Section 28-3 Bombs and Reactors

28-30. What features of a nuclear reactor make it different than a nuclear weapon?

28-31. Why does the uranium in nuclear reactors take the form of many long, thin fuel rods rather than one solid block?

28-32.

a. Which has more time to interact with a nucleus, a low-speed neutron or a high-speed neutron?

b. Which has the greater probability of being absorbed by a nucleus, a low-speed neutron or a high-speed neutron? Briefly explain.

28-33. In what way(s) is a nuclear power plant similar to a coal-fired electric power plant?

28-34. What would be the effect on nuclear weapons if (contrary to fact) the fission of each uranium atom produced exactly one free neutron?

28-35. Both conduction and convection play a role in the cooling of the water that circulates through the core of a nuclear power plant. The cooling cannot be permitted to take place by conduction alone. Explain both of these statements by describing the specifics to which they refer.

28-36. How does the absorption of neutrons in the control rods of a nuclear reactor differ from the absorption of water by a towel or a sponge?

Section 28-4 Ionizing Radiation

28-37. The roentgen is sometimes defined as the amount of ionizing radiation that will produce 2.08×10^9 positive ions (and an equal number of negative ions) in 1 cm^3 of air at standard temperature and pressure. Under these conditions, the density of air is 1.29 kg/m^3. Assuming the positive ions are singly ionized, show that this definition is equivalent to the definition of the roentgen given in Table 28-2.

28-38. Why is it necessary to distinguish between *dose* and *dose equivalent?*

28-39. SSM A person is subjected to a *dose* of 1.2 mGy (milligrays) of X-ray radiation and a dose of 0.7 mGy of α particle radiation. Which dose of radiation is likely to produce more biological damage? Briefly explain.

28-40. A patient is subjected to a *dose equivalent* of 0.5 mSv (millisieverts) of X-ray radiation and a dose equivalent of 0.8 mSv of α particle radiation. Which dose involves the greater transfer of energy to the patient's body? Briefly explain.

28-41. In a reference book, you find the relative biological effectiveness (RBE) of a particular kind of emission defined as the number of rads of γ- or X-ray radiation that would cause the same amount of biological damage as 1 rad of this kind of emission. How does this definition compare to the definition in Table 28-2? Explain.

28-42.
a. Carbon-14 decays by β emission. If the level of activity—that is, radioactivity—due to ^{14}C in the average person's body is 2.7×10^{-10} Bq (becquerels) *per kilogram,* estimate the number of β particles that are emitted in your body in one year.
b. The annual dose equivalent the average human body receives due to carbon-14 within the body is 0.016 mSv.

How much total energy does the carbon-14 transfer to your body in the course of a year?
c. Use the results of *a* and *b* to estimate the average energy of a β particle emitted by a carbon-14 atom.

28-43. Is the dose equivalent greater if you are exposed to a 100 mrad dose of α particles or to a 300 mrad dose of β particles? Briefly explain.

28-44. A worker at a reactor facility is accidentally exposed to a dose of 20 mrad of slow neutrons.
a. What is the dose equivalent of this exposure?
b. What fraction is this of the annual dose equivalent to which an average person is exposed (see Table 28-4)?

28-45.
a. A cancer patient receiving radiation treatment absorbs a total dose equivalent of 50 Sv over the entire course of treatment. Suppose that for this particular treatment, the average quality factor of the emissions is 1.2, and the dose is received by 5.0 kg of tissue. How much total energy is transferred to the patient by means of this radiation?
b. If you were positioned where you absorbed 1% of the energy emitted by an ordinary 100 W light bulb, how long would you have to stay there to absorb the same amount of energy that the patient absorbed over the duration of the radiation treatment?
c. Which energy transfer is riskier? Why?

Going Further

The questions and problems in this group are not organized by section heading, so you must determine for yourself which ideas apply. Some of them will be more challenging than the Review and Practice questions and problems (especially those marked with a • or ••).

28-46. Based solely on the charge-to-mass ratio, what else could α particles be besides doubly charged helium ions?

28-47. The population of radioactive nuclei in a sample is not the only quantity that decays exponentially. The probability that a telephone call will last an amount of time *t* decays exponentially as *t* increases. Why? What is the exponentially decaying population in this situation? What event in the telephone situation corresponds to the decay of a nucleus in the radioactivity situation?

••**28-48.** The activity (radioactivity) due to a particular radioisotope is proportional to how many nuclei of that radioisotope are present. Therefore the activity decays exponentially in the same way. A person is found to have sufficient strontium-90 in his body to produce an activity level of 6.0×10^{-8} Bq (assume the normal amount is negligible compared to this). The intake of the strontium-90 is believed to have occurred because the person happened to be in the vicinity of a nuclear weapon test a week earlier. If the half-life of strontium-90 is 50 days, what was the approximate level of activity due to strontium-90 in this person's body immediately after exposure? (Modify the procedure in WebLink 28-1 as needed to do the calculation, and use $\Delta t = 1.0$ day.)

•**28-49. SSM** At time $t = 0$, there are 6.0×10^8 atoms of substance A and 6.0×10^7 atoms of substance B. Substance A has

a decay constant of 15 minutes, and substance B has a decay constant of 25 minutes. At what time will there be an equal number of atoms of the two substances?

28-50. Thinking of the nuclear reactor as a heat engine, identify *a.* the working substance. *b.* the high-temperature reservoir. *c.* the low-temperature reservoir. *d.* the fuel for maintaining the temperature of the high-temperature reservoir.

28-51.
a. The fission of a nucleus typically results in the release of about 200 MeV of energy (M = mega or million). How many joules of energy would be released in the fission of a kilogram of ^{235}U nuclei?
b. The combustion of a kilogram of crude oil releases 4.3×10^7 J of energy. About how many kilograms of crude oil would be needed to produce the same energy output as a kilogram of nuclear fuel?
c. Based on these results, explain why the United States decided to develop nuclear-powered submarines for military purposes in the 1950s.

28-52. Charcoal drawings of two rhinoceroses and a bison, discovered in the Chauvet cave in southern France in December 1994, were determined to be between 30 000 and 32 000 years old, the oldest such drawings ever found. Why would it have been more difficult to determine the age of these drawings if they had been done entirely in hematite and ocher, the earthy metallic oxides that cave painters used for the red-orange-yellow range of colors?

28-53. When a woman has a mammogram, a padded compression plate is pressed against her breast to facilitate imaging.

a. If two images are taken of each breast in a full mammogram, how much total energy is transferred to the woman's body (see Table 28-4), assuming each dose is absorbed by 1.0 kg of tissue?

b. Using a reasonable guess at the amount of force exerted by the compression plate and the distance the breast is compressed, estimate the amount of work done on the breast by the compression plate.

c. How does this compare with the total energy transferred to the woman's body by the X-ray exposures themselves?

d. If the energy transfers from the X-ray exposures and from the compression plate were both repeated frequently, would the larger of the two energy transfers pose the greater risk to the patient? Briefly explain your reasoning.

28-54.

a. At a certain instant (which we will call $t = 0$), a particular sample of radioactive waste contains equal numbers of cobalt-60 and strontium-90 nuclei. Both are β emitters. At what later instant of time will there be twice as many of one type of nucleus as of the other?

b. Which of the two types of nucleus will be more abundant at that instant?

c. If workers had been exposed to this waste from $t = 0$ until the later instant found in *a,* from which of the two radioisotopes would they have received a greater total dose of β radiation?

28-55. Suppose that each photoelectron in the photomultiplier of a scintillation counter is able to dislodge four other electrons when it reaches the next electrode. If a single photoelectron strikes the first electrode, after how many electrodes will the number of photoelectrons exceed 1 million?

28-56. On average, 2.52 neutrons are emitted in the fission of a U-235 atom.

a. Does the fission of any single U-235 atom emit 2.52 neutrons?

b. If all possible fissions of U-235 atoms emitted either two or three neutrons, would an average of 2.52 neutrons per fission be possible? Briefly explain.

c. The more densely the U-235 atoms are packed, the more likely it will be that a neutron from the fission of one of these atoms will strike another atom. Assuming each such hit causes a fission, what minimum percent of hits must be exceeded to sustain a chain reaction?

28-57. In calculus-based physics texts, you will see Equation 28-2 written as $N/N_0 = e^{-\lambda t}$, where $e = 2.718\ldots$ and λ is a constant.

a. Use your calculator to verify that $e^{-0.6931} = \frac{1}{2}$.

b. Substitute this result for $\frac{1}{2}$ in Equation 28-2.

c. By comparing the equation you got in *b* to $N/N_0 = e^{-\lambda t}$, obtain an equation expressing λ in terms of the half-life $T_{1/2}$.

Problems on WebLinks

28-58. Consider the radionuclei that are present at $t = 1.00$ s in WebLink 28-1. A certain fraction of these decay between $t = 1.00$ s and $t = 2.00$ s. Then a certain fraction of the radionuclei remaining at $t = 1.00$ s decay between $t = 2.00$ s and $t = 3.00$ s. Is the fraction of radionuclei that decay between $t = 2.00$ s and $t = 3.00$ s less than, equal to, or greater than the fraction that decay between $t = 1.00$ s and $t = 2.00$ s?

28-59. In Problem 28-58, is the actual number of radionuclei that decay between $t = 2.00$ s and $t = 3.00$ s less than, equal to, or greater than the number that decay between $t = 1.00$ s and $t = 2.00$ s?

28-60. For a sample that starts out containing 3.0×10^{22} iodine-131 nuclei at $t = 0$, follow the procedure in WebLink 28-1 with $\Delta t = 1.0$ day to find the approximate number of I-131 nuclei remaining in the sample 5.0 days later.

28-61. In WebLink 26-2, after the $^{235}_{92}$U nucleus absorbs a neutron, it momentarily becomes a $^{236}_{92}$U nucleus, then splits into two daughter nuclei. Which of these splittings would make a more vigorous contribution to a chain reaction?

a. $^{236}_{92}$U splits into a $^{92}_{36}$Kr nucleus, a $^{141}_{56}$Ba nucleus, and neutrons.

b. $^{236}_{92}$U splits into a $^{87}_{35}$Br nucleus, a $^{147}_{57}$La nucleus, and neutrons. Briefly explain your choice.

◆ REVIEW PROBLEM SET I ◆
Picking the Principles: Mechanics
(More challenging problems are marked with a • or ••.)

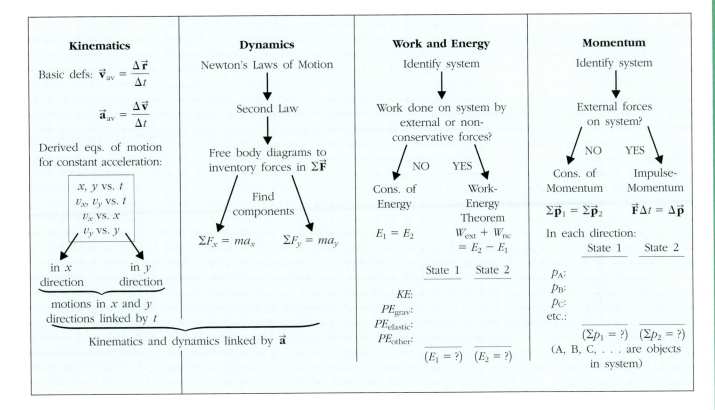

Instructions for Problems I-1 to I-11: The above table outlines some of the fundamental principles of mechanics and some of the main steps necessary for applying them. For each of these problems, first identify the principle(s) and the main steps that you will use to solve the problem. Then use them as a solution strategy to guide your calculations.

I-1. A block slides across a horizontal surface, crossing from a perfectly frictionless region into a long, slightly roughened region. The situation is then repeated with only one change—the second surface is roughened a bit more (the coefficient of friction increased) before repeating. Describe fully how the motion of the block differs in the second situation, and explain your reasoning using appropriate physical principles.

I-2. In a particular situation, a cannon sits on a cliff at a height h above a level plain. The cannon barrel is raised to an angle θ above the horizontal, and a cannonball is fired.
a. How is the velocity with which the cannon strikes the plain altered (by what multiplicative factor is it changed) in a second situation that is like the given one except that the cannon is at a height $h/2$?
b. How is the velocity with which the cannon strikes the plain altered in a situation that is like the given one except that

the barrel's angle with the horizontal is doubled (but still < 90°)?

I-3. A cannon sits on the edge of a cliff overlooking the ocean. The cannon can be raised or lowered so that no matter what angle the cannon fires at, the mouth of the cannon is at a height h above the water's surface. All cannonballs fired from the cannon are identical, and all leave the cannon's mouth at a speed v_o. The first cannonball is fired when the muzzle is at an angle θ above the horizontal, and strikes the water at a certain speed v_1.
a. If the second cannonball is fired at an angle 2θ above the horizontal (but still less than 90°), will it strike the water at a speed less than, equal to, or greater than v_1? Explain your reasoning.
b. If the third cannonball is fired at an angle θ *below* the horizontal, will it strike the water at a speed less than, equal to, or greater than v_1? Explain your reasoning.

I-4. A skateboarder takes a running jump onto a stationary skateboard on level pavement. Just beyond the skateboard is a 15° upgrade. The skate boarder gets up a horizontal speed of 6 m/s just before landing on the skateboard. His weight is 500 N; the weight of the skateboard is 60 N. With no further propulsion (and neglecting friction and air resistance), how far along the upgrade would he go?

I-5. A child pushes a small iceboat in a straight line across a rink. Assume the ice surface is uniform, and that the child exerts a constant force. Three seconds after starting from rest, the boat has gone 4 m and has a speed v_1.

a. Six seconds after starting from rest, how does the speed of the boat compare to v_1?

b. How does the speed of the boat compare to v_1 when the boat has gone a distance of 8 m?

c. Are questions (a) and (b) asking about the same instant or not? Explain.

•I-6. As a 0.15 kg car moves along the semicircular path in Figure I-1, its speed increases uniformly. Its velocities at points A and B are given in the diagram.

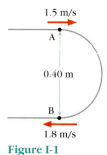
1.5 m/s
A
0.40 m
B
1.8 m/s
Figure I-1

a. Find the magnitude and direction of the car's average velocity over the entire semicircle.

b. Find the magnitude and direction of the average force on the car over the whole semicircular path.

c. Suppose the track is securely mounted on a 0.25 kg platform that sits on a frictionless air table. Tell as precisely as you can what happens to the platform as the car travels from A to B. You should be able to provide a numerical value in your statement of what happens.

•I-7. The spring gun in Figure I-2 has a spring constant of 800 N/m. If it is to fire a 0.20 kg ball into the basket 5.0 m away when it is directed at an angle of 30° to the horizontal, what distance must the spring be compressed?

5.0 m
30°
Figure I-2

•I-8. A nail is propped up vertically so it can be driven into a horizontal board. When a 1.5 kg rock is dropped on the nail from a height of 1.0 m, the nail is driven 1.2 cm into the board. Estimate the average force that the rock exerts on the nail, and state all the assumptions you make in your estimate.

•I-9. When a plane lands on an aircraft carrier, it must hook a cable across the runway, which brings it to a stop. The following data was presented on a public television broadcast about aircraft carriers: The plane, weighing twenty tons, approaches the runway at a speed of 120 mi/hr, and must be brought to a stop over a distance of 200 ft. (Public television has not adopted SI units.)

a. What average force must be exerted by the cable to stop the plane?

b. How long does it take the cable to do this?

••I-10. An acrobatic act makes use of a uniform beam of length $2L$ and mass M balanced on a pivot at its center, high above the circus floor. Suspended from the two ends of the beam are identical rope ladders of length L. Two acrobats step from raised platforms onto the bottom rungs of the ladders simultaneously and begin to climb at constant speed.

a. What happens to the beam if the acrobats have equal masses m but the one on the right climbs twice as quickly as the one on the left?

b. What happens to the beam if the acrobat that is climbing twice as quickly has half as much mass as the other acrobat?

c. In each case, by how much will the potential energy of the system have changed when both acrobats have reached the tops of their respective ladders? (Assume the masses of the rope ladders are negligible.)

••I-11. Two ideal springs with different elastic constants k_1 and k_2 move towards each other, collide end to end, and bounce back from each other. During the collision, when one of the two springs is at maximum compression, will the other spring also be at maximum compression? Use physics principles to justify your answer.

I-12. *What situation can the equations describe?* For each of the equations or sets of equations below, identify a situation (objects, how they are arranged, their relevant characteristics, and what happens to them) that can be represented by the given equation(s.) Fully describe the situation in words and labeled sketches. Do not assume any quantity in the given equations has a value of zero unless no other value is possible.

a. $F_x = f$
$F_y = F_N - mg$

b. $F_x = f$
$F_y = F_N + mg$

c. $F_x = F_N$
$F_y = f + mg$

d. $mgy_1 + \frac{1}{2}mv_1^2 = mgy_2 + \frac{1}{2}k(y_2 - y_1)^2$

e. $F_N \sin\theta = mg$
$F_N \cos\theta = \frac{mv^2}{r}$

f. $0 = m_A v_A + m_B v_B$

g. $m_A v_1 = (m_A + m_B)v_2$
$\frac{1}{2}(m_A + m_B)v_2^2 = (m_A + m_B)gh + \frac{1}{2}(m_A + m_B)v_3^2$

h. $0 = m_A v_A + m_B v_B$
$0 = \frac{1}{2}m_A v_A^2 + \frac{1}{2}m_B v_B^2$

In each of Problems I-13 to I-17, you will be presented with a situation. You will then be asked which principles you would use to solve for different variables in the situation. You are not asked to solve the problems, only to identify the principles you would use to solve them. In each case, the principle or principles should be chosen from the following list:

1a. Newton's second law with $\vec{a} = 0$.

1b. Newton's second law with $\vec{a} \neq 0$.

2. kinematic equations for constant acceleration

3. conservation of energy

4. work-energy theorem

5. conservation of momentum

6. none of the above

I-13. A cannonball with mass m is fired at speed v_o from a cannon mounted on wheels at the edge of a cliff. The cannon makes an angle θ with the horizontal. The mouth of the cannon is at a height h above the plain below. If you have values for all of these quantities, which principle or combination of principles from the above list could you use to solve for each of the following (If there is more than one way of solving, indicate the principle(s) involved in each.):

a. the time the cannonball spends in the air

b. the speed with which the cannonball hits the ground

c. the recoil velocity of the cannon

d. the energy that the cannon provides the cannonball

I-14. A bullet is fired through a block of wood suspended from ropes and comes out the other side. The block swings to a maximum height h_{max} above its initial position. Figure I-3 shows the system at three different instants. Assume you know the masses of the bullet and the block, the length of the ropes, and the length of the block. Which principle or combination of principles from the above list could you use to solve for each of the following? (Just state the principles; do not solve. If there is more than one way of solving, indicate the principle(s) involved in each.):

a. Given the speed of the bullet at t_1 and t_2, find the speed of the block at t_2.

b. Given the speed of the block at t_2, find h_{max}.

c. Given the speed of the bullet at t_1 and t_2, find h_{max}.

d. Given the clock reading t_1 and the speed of the bullet at t_1, find the clock reading t_2.

e. Given the speed of the block at t_2, find how long it takes to reach h_{max}. (Find $\Delta t = t_3 - t_2$).

f. Given the speed of the bullet at t_1 and t_2, find the total tension force exerted by the ropes at time t_2.

g. Given h_{max}, find the total tension force exerted by the ropes at time t_3.

h. Given the speed of the bullet at t_1 and t_2, find the average force exerted by the block on the bullet.

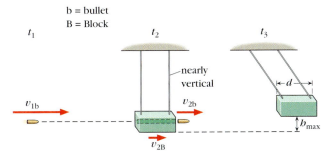

b = bullet
B = Block

t_1 t_2 t_3

nearly
vertical

v_{1b} v_{2b}

v_{2B} h_{max}

Figure I-3

I-15. A child is holding a marble out the window of a train at a height of 2.70 m above the track bed. At an instant when the train has a speed of 20.0 m/s and an instantaneous acceleration of 1.75 m/s² along a straight stretch of track, the child accidentally lets go of the marble. Assuming air resistance is negligible,

a. find the amount of time the marble takes to reach the track bed.

b. find the speed with which the marble hits the track bed.

I-16. The car of an amusement park ride is going around a raised circular track at a constant speed. The track is 6.0 m above the ground and has a diameter of 10.0 m. Figure I-4 shows the position of the car at $t = 0$ and at $t = 1.5$ s. A child is holding his action figure over the edge of the car as he goes around, and accidentally lets go of it at $t = 1.5$ s.

a. Find the speed of the car.

b. Find the total force exerted by the child on the action figure at $t = 0$.

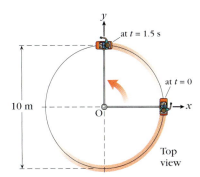

Figure I-4

c. Using the coordinate frame in the figure, find the x and y coordinates of the point where the action figure strikes the ground.

I-17. The runner in Figure I-5 starts from rest at one end of a lightweight, 10 m long platform car. The runner's mass is 50 kg; that of the car is 150 kg. The wheels of the car have very nearly friction-free bearings. When the runner starts from rest, the car is stationary, and point P on the track is directly below him. He increases his speed uniformly until, when he reaches the far end of the car, his speed is 5 m/s. How far is he from point P when he reaches the far end of the car? (Neglect vertical distances.)

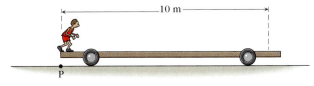

Figure I-5

I-18. *Always, sometimes, or never.* Tell whether the object in each of the following situations is always, sometimes, or never accelerated under the conditions described.

a. The object is slowing down.

b. The object has mass m and its momentum remains constant.

c. The object has mass m and its kinetic energy remains constant.

d. The object exerts a force equal and opposite to a force exerted on it.

e. A person gradually increases the force F that she exerts on the object.

f. The only forces on the object are perpendicular to its direction of motion.

g. As the object slides down a ramp, the potential energy that it loses is all dissipated as heat.

h. The object travels in an elliptical path.

i. The magnitude of the object's momentum remains the same.

j. None of the forces on the object do any work.

I-19. *Always, sometimes, or never.* For each of the situations in the first column below, place an A, S, or N in each of the other columns to indicate whether the column heading is always (A), sometimes (S) or never (N) true for that situation. Assume zero air resistance force in all situations.

Figure I-6

Figure I-7

	Linear momentum is conserved	Total energy is conserved	Kinetic energy is conserved	The velocity remains constant.
a. A roller coaster rolls down a frictionless track from A to B (Figure I-6)				
b. A car goes around a banked circular track at constant speed.				
c. Two identical billiard balls collide. After the collision, they separate at the same relative speed at which they came together.				
d. An object is traveling with only equal and opposite forces exerted on it.				
e. A crate slides down a rough ramp.				
f. A locomotive with a cow-catcher skids along an iced-over (frictionless) track, and continues to skid after catching a cow (Figure I-7).				

Instructions for Problems I-20 to I-28: In each of these problems, you are presented with a set of conditions. In each problem, your task is to describe a single real world situation (objects, how they are arranged, their relevant characteristics, and what happens to them) that satisfies all the conditions for that problem. Where possible, include a sketch of the situation in your description.

I-20. CONDITION 1: An object is accelerating.

CONDITION 2: The object is losing kinetic energy.

I-21. CONDITION 1: An object is accelerating.

CONDITION 2: The object's kinetic energy is remaining constant.

I-22. CONDITION 1: Two objects start out with the same potential energy.

CONDITION 2: Both start out with zero kinetic energy and up with the same non-zero kinetic energy.

CONDITION 3: One object travels by a longer path than the other while gaining its kinetic energy.

I-23. Describe a second situation that meets the conditions of Problem I-22 but involves a different kind of potential energy.

I-24. CONDITION 1: Two objects are dropped from rest.

CONDITION 2: Both situations conserve mechanical energy.

CONDITION 3: After falling equal vertical distances, the two objects are moving at different speeds.

I-25.
a. CONDITION 1: An object's total energy remains constant as it travels but its momentum does not.

CONDITION 2: Its potential energy does not change.

b. CONDITION 1 is the same as for (a) but the second condition changes to

CONDITION 2: Its potential energy changes.

I-26.
a. CONDITION 1: More than one force is exerted on an object.

CONDITION 2: All the forces on the object are perpendicular to the object's motion.

CONDITION 3: The total force on the object is zero.

b. CONDITIONS 1 and 2 are the same as for (a) but the third condition changes to

CONDITION 3: The total force on the object is *not* zero.

•**I-27.** CONDITION 1: Two identical objects start out at the same height.

CONDITION 2: At the start, the same amount of potential energy is provided for each object, and they both start out with zero kinetic energy. The sum of the potential and kinetic energies remains unchanged throughout.

CONDITION 3: They rise to different maximum heights, but they land (at the same elevation) with equal speeds.

•**I-28.** CONDITION 1: Two objects start out with the same gravitational potential energy.

CONDITION 2: Both start out with zero kinetic energy.

CONDITION 3: They end up at $y = 0$ with equal non-zero kinetic energies but with momenta that are *un*equal in magnitude.

I-29. The pendulum bob in Figure I-8 is released from rest at position 1.

a. Carefully sketch the direction of the bob's acceleration when it is in position 2.

•***b.*** Compare the magnitude and direction of the bob's acceleration at positions 1 and 3. Do whatever calculations are necessary to obtain a quantitative comparison ($a_3 =$ ___?___ a_1). *Hint:* It is possible to find the numerical value that goes in the blank even though you are not given any numerical values. (Based on a problem appearing in several research articles of Fred Reif)

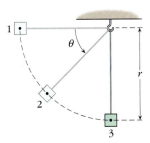

Figure I-8

•**I-30.** The pendulum bob in Figure I-8 is released from rest at position 1. Through what angle θ will the pendulum bob have swung at the instant when the vertical component of its acceleration is zero?

I-31. How realistic are the giant insects in horror movies? Could an insect fly if all its dimensions were increased a hundredfold?

Figure I-9

a. Consider the insect in Figure I-9 and assume it is in equilibrium when airborne in the position shown. Each wing has an area A_{wing}, and the insect's mass is m. Suppose that in this position the average air pressure is P_{upper} at the upper surface of the wing and P_{lower} at the lower surface. Draw a free-body diagram of the insect and express all the forces on the insect in terms of these quantities.

b. Now suppose that the insect is "scaled up" by a factor of 100; that is, each of its linear dimensions is multiplied by 100, but the shape remains the same. Will P_{upper} and P_{lower} change? By what factor does its wing area increase? By what factor do each of the forces due to air pressure increase?

c. By what factor does the insect's volume increase? By what factor does its mass increase?

d. How are each of the forces in your original free-body diagram affected when the insect is "scaled up" by a factor of 100? Is the total upward force on the insect still equal to the total downward force? Can the insect still fly?

✦REVIEW PROBLEM SET II✦

Picking the Principles: Electricity & Magnetism

(More challenging problems are marked with a • or ••.)

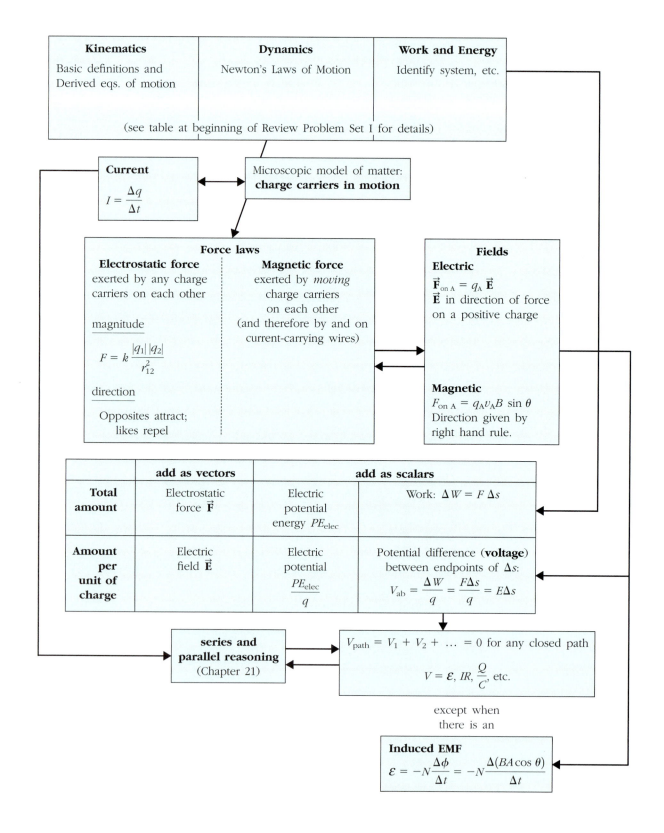

Kinematics	Dynamics	Work and Energy
Basic definitions and Derived eqs. of motion	Newton's Laws of Motion	Identify system, etc.

(see table at beginning of Review Problem Set I for details)

Current

$$I = \frac{\Delta q}{\Delta t}$$

Microscopic model of matter: **charge carriers in motion**

Force laws

Electrostatic force
exerted by any charge
carriers on each other

magnitude

$$F = k\frac{|q_1|\,|q_2|}{r_{12}^2}$$

direction

Opposites attract;
likes repel

Magnetic force
exerted by *moving*
charge carriers
on each other
(and therefore by and on
current-carrying wires)

Fields

Electric
$$\vec{F}_{\text{on A}} = q_A\,\vec{E}$$
$\vec{E}$ in direction of force
on a positive charge

Magnetic
$F_{\text{on A}} = q_A v_A B\sin\theta$
Direction given by
right hand rule.

	add as vectors	**add as scalars**	
Total amount	Electrostatic force $\vec{F}$	Electric potential energy PE_{elec}	Work: $\Delta W = F\,\Delta s$
Amount per unit of charge	Electric field $\vec{E}$	Electric potential $\dfrac{PE_{\text{elec}}}{q}$	Potential difference (**voltage**) between endpoints of Δs: $V_{ab} = \dfrac{\Delta W}{q} = \dfrac{F\Delta s}{q} = E\Delta s$

series and parallel reasoning
(Chapter 21)

$V_{\text{path}} = V_1 + V_2 + \ldots = 0$ for any closed path

$$V = \mathcal{E},\ IR,\ \frac{Q}{C},\ \text{etc.}$$

except when
there is an

Induced EMF
$$\mathcal{E} = -N\frac{\Delta\phi}{\Delta t} = -N\frac{\Delta(BA\cos\theta)}{\Delta t}$$

830

II-1. A certain capacitor has a charge of 2.0×10^{-6}C when it has been fully charged by a 1.5 V D-cell battery. If that battery is then removed and a second, identical battery is connected between the capacitor plates, what is the final charge on the capacitor

a. if the positive pole of the second battery is connected to the same plate as that of the first was?

b. if the poles of the second battery are reversed?

II-2. The wires connecting the circuit elements shown on the board in Figure II-1 are concealed beneath the board. The parts visible on the board are a battery, switches S1 and S2, and identical bulbs A, B, and C. By opening and closing switches and seeing how the bulbs light up

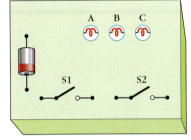

Figure II-1

(brightly, dimly, very dimly, or not at all), it is possible to figure out how the connecting wires might be arranged. Sketch a possible complete circuit that would give you each of the following sets of results.

a.

S1	S2	Bulb A	Bulb B	Bulb C
closed	closed	very dimly	very dimly	very dimly
open	open	not	not	not
open	closed	not	not	not

b.

S1	S2	Bulb A	Bulb B	Bulb C
closed	closed	brightly	brightly	brightly
closed	open	brightly	brightly	not
open	closed	not	not	not

c.

S1	S2	Bulb A	Bulb B	Bulb C
closed	closed	brightly	dimly	dimly
closed	open	brightly	not	not
open	closed	not	dimly	dimly

d.

S1	S2	Bulb A	Bulb B	Bulb C
closed	closed	dimly	dimly	dimly
closed	open	dimly	not	dimly
open	closed	dimly	dimly	not

e.

S1	S2	Bulb A	Bulb B	Bulb C
closed	closed	brightly	not	not
closed	open	brightly	not	not
open	closed	very dimly	very dimly	very dimly

f.

S1	S2	Bulb A	Bulb B	Bulb C
closed	closed	brightly	not	not
closed	open	brightly	not	not
open	closed	dimly	dimly	not
open	open	very dimly	very dimly	very dimly

II-3. Through which of the following is a current least likely to occur under usual conditions? a) a battery b) a capacitor c) a bulb filament d) a resistor.

II-4. Just after the switch is closed in the circuit in Figure II-2, Bulb 1 glows (*more brightly than; less brightly than; equally as brightly as*) bulb 2.

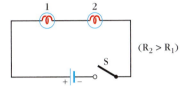

Figure II-2

II-5. Just after the switch is closed in the circuit in Figure II-2, the voltage across Bulb 1 is (*greater than; less than; equal to*) the voltage across bulb 2.

II-6.

a. In each of the situations depicted in Figure II-3, a negative ion is at rest at point P at $t = 0$. At this instant the electric field in the diagram is "switched on" (that is, it builds up almost instantaneously from zero). To the extent possible, describe the path followed by the ion over a tiny time interval starting at $t = 0$ in each situation.

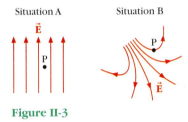

Figure II-3

b. Suppose that instead of being at rest, the negative ion passes through point P at $t = 0$ at a substantial speed v toward the right. Repeat (a) for this new set of conditions.

II-7. A certain parallel plate capacitor with air between the plates has a charge of 4.0×10^{-9} C on it when connected to a 12 V battery. The separation between the plates is increased from 2.0×10^{-4} m to 6.0×10^{-4} m over a 0.20 s interval.

a. Find the average current that flows between the plates during this interval.

b. Starting with the same initial conditions, how would you have to change the plate separation over a like time interval to produce the same average current in the opposite direction?

c. Does the current in parts **a** and **b** pass *through the battery, through the dielectric between the capacitor plates, both*, or *neither*? Briefly explain.

II-8.

a. The circuit shown in Figure II-4a contains four identical DC motors and a battery of suitable voltage so that all the motors turn at observable rotational speeds. If the connection to one of the terminals of motor M_2 is undone, what happens to the rotational speed of M_1? the rotational speed of M_4? the current through the battery? Indicate whether each of these increases, decreases, or stays the same, and explain your reasoning.

b. All connections in Figure II-4a are restored, but motor M_2 is replaced by a motor that is identical except that its coil is made of nichrome instead of copper. What happens to the rotational speed of M_1? the rotational speed of M_4? the current through the battery? Indicate whether each of these increases, decreases, or stays the same, and explain your reasoning.

c. The original circuit is restored, but now the wire shown in Figure II-4b is added. What happens to the rotational speed of M_1? the rotational speed of M_4? the current through the battery? Indicate whether each of these increases, decreases, or stays the same, and explain your reasoning. (Based on a problem in B. Thacker, E. Kim, K. Trefz, and S. Lea, *Amer. J. Phys.* **62**, 627–633 (1994))

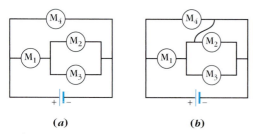

(a)   *(b)*

Figure II-4

II-9. The coil in Figure II-5 is connected to a battery and variable resistor, and placed over a permanent magnet sitting on a table. How is the normal force that the table exerts on the magnet affected when

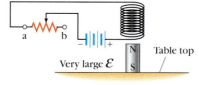

Figure II-5

a. the contact of the variable resistor is slid towards point a?

b. the contact of the variable resistor is slid towards point b?

c. the battery is turned around so that its polarity in the circuit is reversed?

II-10. The antenna circuit shown in Figure II-6 is not a closed loop. Can current flow through it? Explain.

II-11. The simplest electric generators deliver alternating current (AC). Modifications are necessary if you want the generator to deliver direct current. For each of the following uses of the generator, decide whether this modification is necessary, and explain why (or why not):

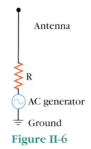

Figure II-6

a. The generator is to be used to light a flashlight bulb.

b. The generator is to be used to charge a capacitor.

•II-12. The two circuits in Figure II-7 are not electrically connected. Circuit 2 is totally insulated except where it is closed by a solder joint. The loop in Circuit 2 wraps snugly around the solenoid in Circuit 1.

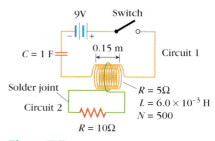

Figure II-7

When the switch in Circuit 1 is closed, it will initially behave like an RL circuit and later behave like an RC circuit because of the very different time constants involved.

a. Find the time constants τ_{RL} and τ_{RC} for these two behaviors.

b. How does the current vary when the switch is first closed? How does it vary when t is much greater than τ_{RL} but still less than τ_{RC}?

c. The current will be approximately (within 1%) at its maximum value at $t = 5\tau_{RL}$. To the same accuracy, what average EMF is induced in Circuit 1 during the time interval from $t_1 = 5\tau_{RL}$ to $t_2 = 5\tau_{RC}$?

d. What average EMF is induced in Circuit 2 during this interval?

e. What average EMF is induced in Circuit 1 during the time interval from $t_1 = 0$ to $t_2 = 5\tau_{RC}$?

II-13. Figure II-8 shows a flexible wire loop. Part of it rests on a slab of insulating material; the rest of it hangs over the edge and swings continuously back and forth. A uniform magnetic field is directed towards the right. What is the direction of the current caused through side ab of the loop at an instant when side cd of the loop

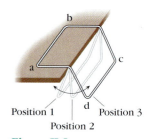

Figure II-8

a. has traveled part way from position 1 to position 2?

b. has traveled part way from position 2 to position 3?

c. is at position 3?

II-14. What are the similarities and what are the differences between

• charging by induction, and

• using electromagnetic induction to induce a voltage or current?

II-15. The two fields in Figure II-9 are uniform; the electric field has a magnitude of 500 N/C and the magnetic field has a magnitude of 0.30 T. As an electron travels from point A to point B in the circular loop of wire, how much work is done on it

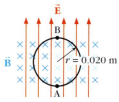

Figure II-9

a. by the electric field?

b. by the magnetic field?

II-16. A point object has a charge of $q = \pm Q$. At a certain instant, it has a velocity $\vec{v}$ in a region where there are a uniform electric field $\vec{E}$ and a uniform magnetic field $\vec{B}$. The three vectors have non-zero magnitudes v, E, and B. Sketch one set of possible directions for the vectors when the total force on the charged object can be described by

a. $F_x = QE$ and $F_y = QvB$ (and $q = -Q$)

b. $F_y = -QE$ and $F_z = -QvB$ (the $+z$ direction is out of the paper, and $q = +Q$)

c. $\vec{F} = 0$ ($q = -Q$. More than one answer is possible.)

d. $F_x = QE - \dfrac{QvB}{\sqrt{2}}$ and $F_y = \dfrac{QvB}{\sqrt{2}}$ (and $q = +Q$)

II-17. The slide wire in Figure II-10 is set in motion at $t = 0$ and moved at constant speed to the left. As it slides, its end points A and B remain in contact with the U-shaped portion of the circuit. Assume all part of the circuit have negligible resistance except the resistor at the right. Sketch graphs of each of the following plotted against t, starting at $t = 0$:

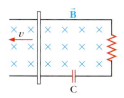

Figure II-10

a. the EMF induced in the slide wire

b. the current through the resistor

c. the charge on the capacitor

d. the potential difference across the resistor

e. the potential difference across the capacitor

•II-18.

a. In the circuit depicted in Figure II-11, the coil has an extremely large number of turns wound around an iron core, but negligible resistance. Describe what happens to the brightness of the bulb from the instant the switch is first closed until a few minutes later, and explain why it happens.

b. Will the bulb light up again if the switch is opened? Briefly explain.

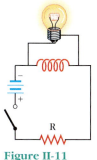

Figure II-11

II-19. The maximum magnetic field that can be produced by an electromagnet is limited by the fact that when the magnetic field is too great, the electromagnet will self-destruct. Why will this happen? Be specific about what physical principles are involved.

II-20. *Transducers:* Often, if we want to measure a physical quantity, we can construct a set-up or device in which a current or voltage is produced that is proportional to the quantity in question. We can then use an ammeter or voltmeter to read the output, but calibrate the reading in units of the quantity of interest. Such a device is called a *transducer.* For example, suppose we want to measure the depth of water in a tank. The ammeter or voltmeter would be calibrated in meters rather than amps or volts.

a. Design a simple device that would enable us to measure the water depth in this way.

b. Suppose you wanted a transducer that produced an electrical output which would be a measure of the rotational speed of a wheel. What physical principle could you take advantage of in designing such a transducer? Explain.

II-21. The capacitance of a capacitor with square plates will double if you ____

a. double each side of the square plate

b. reduce the voltage across it by half

c. reduce the separation between the plates by half

d. let the capacitor finish charging

e. [Two of these answers are correct.]

II-22. When an electron is positioned at a certain point P, the potential at that point is 0.40 V. What is the potential at that point when the electron is removed?

••II-23. A 1.0 m^2 sheet of commercial aluminum foil has a mass of about 40 g. An air capacitor can be made by mounting two such sheets on the facing sides of two non-conducting surfaces separated by a distance of 1.0×10^{-4} m.

a. What net fraction of the *conduction* electrons on one plate are transferred to the other when a pair of 1.5 V D-cell batteries are connected in series between the two plates? (We speak of the *net* fraction transferred because the electrons lost by one plate are not the same electrons that arrive at the other plate.) What fraction of all its electrons is lost by the positive plate? [*Hint:* Avogadro's number tells us how many atoms there are in a mole of atoms of any element. A mole of aluminum has a mass of about 27 g. since Al has an atomic weight of about 27. Assume three conduction electrons per aluminum atom.]

b. The maximum electric field that air can tolerate before it breaks down as a dielectric and sparking occurs (the dielectric strength of air) is 3.0×10^6 V/m. What is the maximum net fraction of the conduction electrons on one plate that can be transferred to the other without such a breakdown occurring? What voltage is required to produce such a transfer?

II-24. Each part below gives you a set of one or more equations, inequalities, and/or other conditions. For each part, identify one real-world situation (objects, how they are arranged, their relevant characteristics, including the sign of any charge, and what happens to them) that can be represented by *all* the given equations, etc. for that part. Fully describe the situation in words and labeled sketches. (Note: In a real world situation, you cannot just say there is a field or a potential difference. You must describe the object or arrangement of objects that is responsible for it.)

a. $F_x = k\dfrac{q_1 Q}{r^2} \cdot \dfrac{x}{r} > 0$

$F_y = k\dfrac{q_2 Q}{r^2} \cdot \dfrac{y}{r} < 0$

b. $F_1 = k\dfrac{q Q_1}{r_1^2}$

$F_2 = k\dfrac{q Q_2}{r_2^2} = 4F_1$

$|Q_1| = |Q_2|$

c. $k\dfrac{q Q}{r^2} = m\dfrac{v^2}{r}$

d. $qEx_1 = qEx_2 + \frac{1}{2}mv_2^2$

e. $qEy_1 + mgy_1 = qEy_2 + mgy_2$

$y_2 > y_1$

f. $x = \frac{1}{2}\left(\dfrac{eE}{m}\right)t^2$

g. $\frac{1}{2}mv_2^2 + qV = \frac{1}{2}mv_2^2$

h. $qvB = m\dfrac{v^2}{r}$

i. $E = vB$

••j. $k_1\dfrac{qQ}{(r + \Delta r)^2} + k_2\Delta r = m\dfrac{v^2}{r + \Delta r}$

II-25. For each of the following sets of conditions, draw a single carefully labeled circuit diagram that satisfies all the conditions in the set. *Note:* When quantities such as current, charge, and potential difference are variable, capital letters denote their maximum values and lower case letters denote their instantaneous values. Different numerical subscripts refer to different individual circuit elements. Quantities for a combination of elements do not have subscripts. For example, $V = V_1 + V_2 + V_3$ is the total voltage across three elements in series, but elements 1 and 2 could be resistors R_1 and R_2, and element 3 could be a capacitor C_3.

a. $I_1 = I_2$
$V = V_1 + V_2$

b. $Q_1 = Q_2$
$V = V_1 + V_2$

c. $I = I_1 + I_2 + I_3$
$V = V_1 = V_2 = V_3$

d. $V_1 = V_2$
$V = V_1 + V_3$
$I_3 = I_1 + I_2$

e. $I_3 = I_1 + I_2$
$V_1 > V_2$

f. $V_1 = V_2$
$Q = Q_1 + Q_2$

g. $V_1 = \frac{\mathcal{E}}{R_1}$
$\frac{\Delta I}{\Delta t} = 0$

b. $\frac{q}{C} + iR = 0$

i. $v_1 = \frac{\mathcal{E}}{R_1}$ and decreasing
v_2 increasing

j. $\frac{\Delta q_1}{\Delta t}$ proportional to q_1
$i_1 = 0$
$i_2 = i_3 \neq 0$

II-26. Each of the following statements describes what happens when a change is made in a particular circuit. Sketch a circuit (after the change) that satisfies the description. If necessary, describe anything else in the circuit's surroundings that affects the circuit's behavior.

a. Adding a second resistor to the circuit increases the current in the circuit.

b. Adding a second resistor to the circuit decreases the current in the circuit.

c. Adding three more resistors identical to the one that was originally in the circuit leaves the current in the circuit unchanged.

d. Adding a second battery to the circuit increases the current in the circuit.

e. Adding a second battery to the circuit decreases the current in the circuit.

f. Adding a second battery to the circuit leaves the current in the circuit unchanged.

g. Adding an ammeter to the circuit leaves the current through the resistor in the circuit very nearly unchanged.

b. Adding an ammeter to the circuit greatly reduces the current through the resistor in the circuit.

i. Adding a second capacitor to the circuit substantially reduces the charge on the capacitor that was there before.

j. Adding a second capacitor to the circuit leaves the charge on the capacitor that was there before unchanged.

k. Closing a switch in the circuit makes a bulb in the circuit go out.

l. Closing a switch that connects a battery to a bulb in the circuit makes the bulb glow for only a few seconds before going out again.

m. Adding a coil to the circuit makes a bulb in the circuit dim more slowly.

n. Adding a capacitor to the circuit has no effect on a lit bulb in the circuit.

o. Without making any change in a series circuit containing a lit bulb, the bulb is made to glow slightly brighter.

p. Increasing a variable resistance in one circuit makes a bulb glow momentarily in a second circuit. There is no conducting path between the two circuits.

q. Discharging a capacitor in one circuit makes a bulb glow momentarily in a second circuit. There is no conducting path between the two circuits.

r. Rotating the circuit by 90° makes a lit bulb in the circuit go out.

Appendix A

Some Basic Math Ideas

Patterns of relationship between variables: With measurement, observations become quantitative, and observed patterns or regularities take the form of mathematical relationships. You should be familiar with the types of relationships we consider here.

Direct proportionality: Consider the numbers in the tables below.

A	B		A	B		A	B		A	A^2	B
1	3		1	−5		1	3		1	1	3
2	6		2	−10		2	12		2	4	12
3	9		3	−15		3	27		3	9	27
4	12		4	−20		4	48		4	16	48
i.			**ii.**			**iii.**			**iii redrawn.**		

Tables ***i*** and ***ii*** exhibit relationships of the general type $B = kA$, or equivalently $\frac{B}{A} = k$, where k remains constant as B and A change. In ***i***, $k = 3$; in ***ii***, $k = -5$. In a relationship like this, we say that B *varies directly as A,* or that *B is directly proportional to A.* The word *proportion* means *ratio.* In direct proportionality, B doubles as A doubles, B triples as A triples, etc., and so forth, so that the ratio of the two values remains the same.

The relationship in ***iii*** is less obvious, but we can find a pattern by adding a column (highlighted) to (iii) listing the value of A^2 for each value of A. We then see that each value in the B column is three times the corresponding value in the A^2 column: $B = 3A^2$, or $\frac{B}{A^2} = 3$. In ***iii*** we can say that B *is proportional to* A^2.

In Figure A-1*a*, B has been plotted against A for table ***i***. The result is a straight line through the origin. A straight line is always equally steep. In Figure A-1*a*, whenever A increases horizontally by 1, B increases vertically by 3. It is much like a staircase (Figure A-2*a*) where if the treads all have the same width, the risers must all have the same height or else the steepness of the stairs will change. Figure A-1*b* is a graph of table ***ii***. The steepness is uniform here too, but now whenever A increases by 1, B *decreases* vertically by 5. If k is

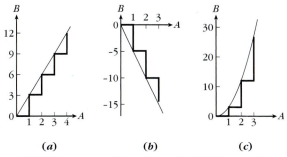

Figure A-2 Staircase analogs of graphs in Figure A-1.

negative, the graph read left to right is like a *descending staircase* (Figure A-2*b*). In contrast, the steepness of Figure A-1*c*, which plots B against A for table ***iii***, is *not* uniform: The graph gets progressively steeper, like the staircase in Figure A-2*c*, in which the treads all have the same width but the risers become progressively higher.

The ratio of the riser to the tread ($\frac{3}{1} = 3$ for Figures A-1*a* and A-2*a*) gives the steepness of the line, and is called the **slope** of the line. If you advance from a point (x_1, y_1) to a point (x_2, y_2) along a graph of y vs. x (see Figure A-3), the x coordinate changes by an amount $\Delta x = x_2 - x_1$ and the y coordinate changes by a corresponding amount $\Delta y = y_2 - y_1$ (Δ denotes "the change in . .", so Δx means "the change in x", and so forth). Then the slope m of this straight line is given by

$$\textbf{Slope} \qquad m = \frac{\Delta y}{\Delta x} = \frac{y_2 - y_1}{x_2 - x_1}. \qquad \text{(A-1)}$$

Any two points (x_1, y_1) and (x_2, y_2) along the straight line will give you the same slope, precisely because a straight line is always equally steep. In contrast, different choices of (x_1, y_1) and (x_2, y_2) in Figure A-1*c* will *not* give you

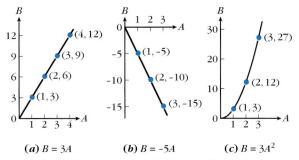

(a) $B = 3A$ **(b)** $B = -5A$ **(c)** $B = 3A^2$

Figure A-1 Graphs of relationships in tables i–iii.
B versus A for (a) $B = 3A$ (b) $B = -5A$ (c) $B = 3A^2$

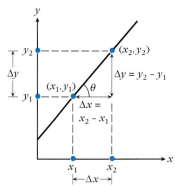

Figure A-3 Graph of a straight line. As you advance from a point (x_1, y_1) to a point (x_2, y_2), your x coordinate changes by an amount $\Delta x = x_2 - x_1$ and your y coordinate correspondingly changes by an amount $\Delta y = y_2 - y_1$.

835

the same value of m; straight lines connecting pairs of points along this graph do not have equal steepness.

If we read left to right, $\Delta x = x_2 - x_1$ is always positive. Then the slope $\frac{\Delta y}{\Delta x}$ is negative when $\Delta y = y_2 - y_1$ is negative. A negative slope tells us the vertically plotted values are going down and the graph looks like a descending staircase. If the graph is horizontal, then the size Δy of the "riser" is zero, and thus the slope $\frac{\Delta y}{\Delta x}$ is also zero.

In the standard x and y notation, x really means "whatever variable you have plotted horizontally" and y really means "whatever variable you have plotted vertically". Thus, for table (1) and Figure A-1a, x becomes A, y becomes B, and the slope becomes $m = \frac{\Delta B}{\Delta A} = \frac{B_2 - B_1}{A_2 - A_1}$. If we let (A_1, B_1) be the origin $(0, 0)$, and let (A_2, B_2) be any other point (A, B) along the line, $m = \frac{\Delta B}{\Delta A} = \frac{B - 0}{A - 0} = \frac{B}{A}$. Any corresponding pair of values from table i gives $\frac{B}{A} = 3$, the proportionality constant we found previously.

> When two variables A and B are directly proportional ($B = kA$), the graph of B (vertically) vs. A (horizontally) is a straight line through the origin, and that the proportionality constant k and the slope m are one and the same thing.

This provides a basis for *interpreting* the slope in different situations. For example, suppose a car's clock and odometer are set to zero at the start of a trip. As the car travels at constant speed, subsequent readings are as follows

Clock Reading (h)	Odometer Reading (km)
0	0
1	80
2	160
3	240
4	320

Its speed is given by $speed = \frac{distance\ traveled}{time\ spent\ traveling}$. Since the speed is constant, we can use any pair of values:

$$speed = \frac{80\ km}{1\ h} = \frac{160\ km}{2\ h} = \frac{240\ km}{3\ h} = \frac{320\ km}{4\ h} = 80\ \frac{km}{h}$$

If we plot the values as in Figure A-4, the slope $m = \frac{\Delta y}{\Delta x}$ here becomes $\frac{\Delta(odometer\ reading)}{\Delta(stopwatch\ reading)}$. Since we have plotted

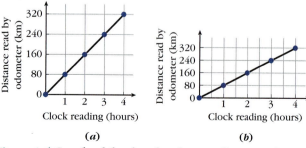

(a)   *(b)*

Figure A-4 Graph of the data for the traveling car. The graph is shown here for two different choices of scale. Both are correct; the choice of scale is arbitrary.

quantities with units, we must retain those units in the calculation of the slope. We can calculate the slope from any two points. Here it is convenient to let the origin $(0,0)$ be one of the points:

$$m = \frac{80\ km - 0}{1\ h - 0} = \frac{160\ km - 0}{2\ h - 0} = \ldots etc. \ldots = 80\ \frac{km}{h}$$

In this case, *calculating the slope gives us the speed.* We can interpret the slope of either graph as the speed.

The slope is the same in both cases, even though one graph looks steeper than the other, *i.e.* the two graphs do not make the same angle with the horizontal. *It is only meaningful to compare the steepness of two graphs that have both the same horizontal scale and the same vertical scale.*

Let's generalize. Speed is a rate of change; it tells you the rate at which the distance changes with respect to time. The word *per* means "in each"; 80 km/h or 80 km *per* hour tells you that the car advances 80 km in each hour of driving time. The slope of a graph is also a rate. It tells you how many units you go up (or down) for each unit that you go across (how many units of riser for each unit of tread):

- The **slope** always tells you the **rate of change** of the vertically plotted variable as the horizontal variable increases.

- The units of the slope are units of the vertical variable per unit of the horizontal variable.

Since the constant of proportionality k in $B = kA$ gives the slope of the graph of B vs. A, it likewise can always be interpreted as a rate.

Consider another example. Suppose you are required to "weigh" a large sand pile. You have at your disposal a scale calibrated in kg and a pail. You use the scale to find the mass of the pail. You then fill the pail with sand, find its mass, and subtract off the mass of the empty pail to find the mass of a "standard pailful" of sand. Now you need only determine the number of times you can fill the pail in the same way, and multiply the mass of the standard pailful by this number to get the total mass of the pile:

$$Total\ mass = (mass\ of\ a\ standard\ pailful)$$
$$\times (number\ of\ pailfuls) \qquad (A\text{-}2)$$

The pail can hold a fixed volume of sand. The "number of pailfuls", like the number of cubic meters or liters in other situations, is the number of units of volume, which ordinarily we just call "the volume". The "mass of a standard pailful" *is* the mass *in each* (or *per*) unit of volume. We can thus rewrite Equation A-2 as

$$Total\ mass = (Mass\ per\ unit\ of\ volume) \times (Volume)$$

The mass per unit of volume is the constant rate in this situation. The total mass is proportional to the total volume of the sand pile, but the mass of each pailful (to within experimental error) is the same, and is a property of the sand itself. This property is called the *density* (ρ) of the sand. We can thus rewrite Equation 1-3 even more concisely as

$$Mass = Density \times Volume \quad \text{or} \quad m = \rho V \quad \text{or} \quad \rho = \frac{m}{V}$$

This is like $B = kA$ or $k = \frac{B}{A}$; density is now the proportionality constant. It is a rate because it tells how much mass of a substance you get *for each* unit of volume.

Proportionality between powers of variables: The graph of B vs. A for table ***iii*** is not a straight line. However, since in ***iii*** B is proportional to A^2, a graph of B vs. A^2 *is* a straight line, as Figure A-5 shows. Note carefully that the numerical values from the A^2 column and not those from the A column, are equally spaced along the horizontal axis. Since x and y are simply stand-ins for whatever variables we plot horizontally and vertically, the slope is in general given by

> $$\text{Slope} = \frac{\Delta(variable\ plotted\ vertically)}{\Delta(variable\ plotted\ horizontally)} \quad \text{(A-3)}$$
> where each "Δ" is a later value of the variable minus an earlier value

In this case, the slope ($m = k = 3$) is $\frac{\Delta B}{\Delta(A^2)}$. We can in general have a direct proportionality between a power of one variable and a power (the same or different) of another.

A	$A^{-1} = \frac{1}{A}$	B	A	$A^{-2} = \frac{1}{A^2}$	B
$\frac{1}{2}$	2	20	1	1	5
1	1	10	2	$\frac{1}{4} = 0.25$	1.25
2	0.5	5	3	$\frac{1}{9} = 0.11$	0.55
	iv			***v***	

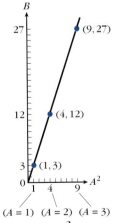

Figure A-5 Graph of B versus A^2 for table iii. For $B = 3 A^2$, the graph of B versus A^2 is a straight line, although the graph of B versus A is not. Note that the evenly spaced values along the horizontal axis are values of A^2, not A.

Inverse and inverse square relationships: These powers can also be negative. In tables ***iv*** and ***v*** A and B are not proportional, but the values in the second and third columns are: In ***iv***, $B = kA^{-1}$ and $k = 10$; in (v), $B = kA^{-2}$ and $k = 5$. We can rewrite these as ***iv*** $B = \frac{k}{A}$ and ***v*** $B = \frac{k}{A^2}$. In (iv), B is said to be *inversely proportional* to A. The relationship between B and A in ***v*** is called an *inverse square relationship*. In both of these relationships, B decreases as A increases. Figure A-6a shows this for ***v***. But in ***v***, B is proportional to $\frac{1}{A^2}$, so a plot of B vs $\frac{1}{A^2}$ is a straight line (Figure A-6b).

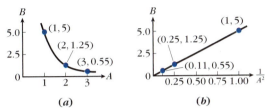

(a)   (b)

Figure A-6 Two graphs of the inverse square relationship $B = 5/A^2$. (a) Plotting B against A shows that B decreases as A increases. (b) Plotting B against $\frac{1}{A^2}$ shows that B is proportional to $\frac{1}{A^2}$. Note that the values of $\frac{1}{A^2}$ must be placed in ascending order left to right along the horizontal axis.

We can rewrite the relationship in ***v*** as $A^2B = k$. For the given values, $k = 5$. If we look at the pairs of values of A and B in ***v*** $1^2 \times 5 = 5$), $2^2 \times 1.25 = 5$, and $3^2 \times 0.55 = 5$. More generally, we can write $A_1^2 B_1 = k$, $A_2^2 B_2 = k$, and so forth.

Linear but not proportional: the initial value problem. Whenever two quantities are directly proportional, a graph of one against the other is a straight line, or *linear*. Does it necessarily follow that when a relationship between two quantities is linear, the quantities are proportional? Consider the following.

The trip graphed in Figure A-4 is repeated at the same constant speed as before. The car clock is set at zero where it previously started the trip, but this time the driver has been visiting with a friend 10 km up the road, and zeroed her odometer when she was there. As the following table shows, her time and trip distance (first two columns) remain proportional, but to each trip distance, we must add 10 km to get her odometer reading.

Clock Reading (h)	Trip Distance (km)	Odometer Reading (in km)
0	0	10
1	80	90
2	160	170
3	240	250
4	320	330

The odometer readings are *not* proportional to the clock readings: if we double the clock reading from 1 h to 2 h, the odometer reading goes from 90 km to 170 km; it

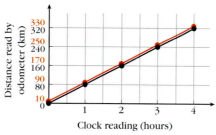

Figure A-7 More data for the traveling car. Distances from the original starting place are in black; distances from the friend's house are in red.

does *not* double. Nevertheless, both graphs in Figure A-7 are straight lines.

Note that the graph of trip distance passes through the origin; the graph of odometer reading does not. At $t = 0$, the odometer reading is 10 km, the *initial distance* from her friend's house when the driver starts her trip. At each subsequent value of t, the odometer reading exceeds the trip distance by this same 10 km. The value at $t = 0$, where the graph crosses the vertical axis, is called the *vertical* or *y intercept*. *The vertical intercept always represents an initial value.* The two graphs are the same, except that one builds from an initial distance value of zero, and the other from an initial distance value of 10 km. However, two linearly related variables are proportional *only* when the vertical intercept—the initial value of the vertically plotted variable—is zero.

The relationship between the values in the first and third columns above, which is graphed in Figure A-7, can be expressed as

$$\frac{odometer}{reading} = \frac{initial\ odometer}{reading} + speed \times \frac{time\ spent}{traveling}$$

This relationship between the vertically and horizontally plotted variables holds true whenever the rate of change (in this instance, the speed) is constant, so that in general

$$\frac{vertically}{plotted} = \frac{initial\ value\ of}{vertically} + \frac{rate\ of}{change} \times \frac{horizontally}{plotted}$$
$$variable \qquad plotted\ variable \qquad \qquad variable$$
$$(A\text{-}4a)$$

But the rate of change is the slope of the corresponding straight line graph, so we can write Equation A-4a symbolically as

$$y = y_0 + mx \qquad (A\text{-}4b)$$

Equation A-4b is a general form of the equation of a straight line, called the *slope-intercept form*. The initial value y_0 is the vertical intercept, the value of y at $x = 0$. It is commonly denoted by b in mathematics texts (so that A-4b appears as $y = mx + b$). More importantly, Equation A-4a is the interpretation of A-4b that gives it meaning when it is used to represent real world situations.

A direct proportionality relationship always has the form

$$one\ variable = constant \times other\ variable$$

Whenever the slope is constant, the relationship $\Delta y = m\,\Delta x$ has this form, so that *changes* in y are proportional to *changes* in x. But unless the vertical intercept y_0 (or b) is zero, the equation for a straight line, $y = y_0 + mx$, does not have this form, and y itself is *not* proportional to x.

A Brief Review of Geometry and Trigonometry

Geometry: In geometry, the quantities that give the "size" of things include perimeter (the circumference is a circle's perimeter—the distance around), area, and volume. Table B-1 summarizes how to calculate these quantities for the most common geometrical shapes.

Table B-1 Useful Information about Geometric Shapes

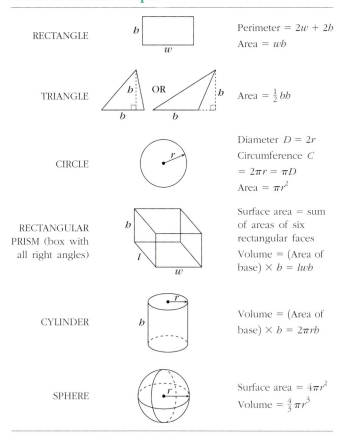

RECTANGLE		Perimeter $= 2w + 2b$ Area $= wb$
TRIANGLE		Area $= \frac{1}{2}bb$
CIRCLE		Diameter $D = 2r$ Circumference C $= 2\pi r = \pi D$ Area $= \pi r^2$
RECTANGULAR PRISM (box with all right angles)		Surface area = sum of areas of six rectangular faces Volume = (Area of base) $\times b = lwb$
CYLINDER		Volume = (Area of base) $\times b = 2\pi rb$
SPHERE		Surface area $= 4\pi r^2$ Volume $= \frac{4}{3}\pi r^3$

Trigonometry: In dealing with the size and arrangements of objects in space, you often need to figure out lengths and/or angles that you don't know from others that you do. The geometric relationships that exist in right triangles enable you to do this.

One useful relationship is that the ratio of corresponding sides is the same for all triangles having the same shape. By nesting the smaller triangles within the larger ones (Figure B-1*a*), we can see that all right triangles with the same acute angle θ have the same shape. For a given θ, we identify the three sides as the *side adjacent to* θ (ADJ in Figure B-1*b*), the *side opposite to* θ (OPP), and the *hypotenuse* (HYP), the longest side, which is opposite the right angle. The most commonly used ratios of sides are named the

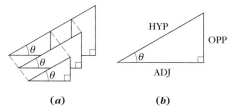

(a)   *(b)*

Figure B-1 Properties of Right Triangles. (*a*) All right triangles with the same acute angle have the same shape. (*b*) Identifying sides of a right triangle.

sine (sin), *cosine* (cos), and *tangent* (tan), and are defined as follows:

$$\sin \theta = \frac{\text{OPP}}{\text{HYP}} \qquad \text{(B-1)}$$

$$\cos \theta = \frac{\text{ADJ}}{\text{HYP}} \qquad \text{(B-2)}$$

$$\tan \theta = \frac{\text{OPP}}{\text{ADJ}} \qquad \text{(B-3)}$$

(The inverses of these three ratios, in the same order, are called the *cosecant* (csc θ), the *secant* (sec θ) and the *cotangent* (cot θ). These are much less commonly used.) Because their values depend on θ, these ratios are *functions* of θ, and are called the **trigonometric functions.** It is generally easiest to use your calculator to find the values of these functions, but you should know the values in Table B-2 and the reasons for them (see Section 13-3).

The angles in a triangle always add up to 180°. The sum of the two acute angles in a right triangle must therefore be 90°. If one of these angles is θ, the other must be $90° - \theta$. A side that is opposite one of these two angles must be adjacent to the other, so

$$\sin \theta = \cos(90° - \theta) \qquad \text{(B-4)}$$

$$\cos \theta = \sin(90° - \theta) \qquad \text{(B-5)}$$

Table B-2 Numerical Values of Trigonometric Functions

Angle θ in degrees	Angle θ in radians	$\sin \theta$	$\cos \theta$	$\tan \theta$
0	0	0	1	0
90	$\pi/2$	1	0	∞
180	π	0	-1	0
270	$3\pi/2$	-1	0	$-\infty$
360	2π	0	1	0

The Pythagorean theorem is a relationship between the sides of a right triangle:

Pythagorean theorem	$OPP^2 + ADJ^2 = HYP^2$ (B-6)

Examples 3-2 (with the marginal note on page 62) and 3-3 in Chapter 3 show how Equations B-1 through B-3 and B-6 are used in practice and provide some suggestions about using your calculator properly for these calculations.

If we divide both sides of Equation B-6 by HYP^2, and then use the definitions in Equations B-1 and B-2, we get another useful form of the Pythagorean theorem:

$$\sin^2 \theta + \cos^2 \theta = 1 \qquad (B-7)$$

If you know either the sine or the cosine of an angle, you can use this equation to find the other trigonometric function.

You can think of an angle as an amount of rotation—a total of 360° for each complete revolution. An angle θ representing an amount of rotation can have any value up to infinity. We can also have negative values of θ, representing rotations in the opposite direction. When we define trigonometric functions in terms of the sides of a right triangle (Equations B-1 to B-3), the definitions are valid only for the angles that can actually occur in right triangles—those between 0 and 90°. In Section 13-3, you can see how we make use of the unit reference circle to develop definitions of the sine and cosine that are valid for all possible values of θ from $-\infty$ to $+\infty$. These definitions are summarized in Figure 13-14. For the sines and cosines of angles between 0 and 90°, they give the same values as Equations B-1 and B-2.

Since you are back to facing in the same direction each time you rotate by 360°, the trigonometric functions have the same value each time you increase or decrease θ by 360°. If an angle is not between 0 and 360°, you can always add an integer multiple of 360° to the angle to find an angle between 0 and 360° that has the same sine and cosine. In addition, for any angle between 0 and 360°, there are simple relationships that will let you find the trigonometric functions of the angle if you have values available for the angles in the first quadrant (those between 0 and 90°). These are summarized in Table B-3.

The following additional formulas are sometimes useful when doing calculations involving the sides and angles of triangles. These formulas are good for any triangles, not just right triangles. In each formula, the side

Table B-3 What To Do When Angles Are between 90° and 360° (between $\frac{\pi}{2}$ and 2π).

Second quadrant: $90° \leq \theta \leq 180°$ $\left(\frac{\pi}{2} \leq \theta \leq \pi\right)$	$\sin \theta = \frac{y}{1}$ $\sin \theta =$ $\sin(180° - \theta)$	$\cos \theta = \frac{-x}{1}$ $\cos \theta =$ $-\cos(180° - \theta)$	$\tan \theta = \frac{y}{-x}$ $\tan \theta =$ $-\tan(180° - \theta)$
Third quadrant: $180° \leq \theta \leq 270°$ $\left(\pi \leq \theta \leq \frac{3\pi}{2}\right)$	$\sin \theta = \frac{-y}{1}$ $\sin \theta =$ $-\sin(\theta - 180°)$	$\cos \theta = \frac{-x}{1}$ $\cos \theta =$ $-\cos(\theta - 180°)$	$\tan \theta = \frac{-y}{-x} = \frac{y}{x}$ $\tan \theta =$ $\tan(180° - \theta)$
Fourth quadrant: $270° \leq \theta \leq 360°$ $\left(\frac{3\pi}{2} \leq \theta \leq 2\pi\right)$	$\sin \theta = \frac{-y}{1}$ $\sin \theta =$ $-\sin(360° - \theta)$	$\cos \theta = \frac{x}{1}$ $\cos \theta =$ $\cos(360° - \theta)$	$\tan \theta = \frac{-y}{x}$ $\tan \theta =$ $-\tan(360° - \theta)$

a (lower case) is opposite angle *A* (upper case) of the triangle, and so forth. Apart from maintaining this consistency, it doesn't matter which letter you use for which angle and side.

Law of sines $\dfrac{\sin A}{a} = \dfrac{\sin B}{b} = \dfrac{\sin C}{c}$ (B-8)

Law of cosines $c^2 = a^2 + b^2 - 2ab \cos C$ (B-9)

In the law of cosines, if the angle $C = 90°$, then the side *c* opposite it is the hypotenuse of a right triangle. Since $\cos 90° = 0$, the law then says

$$\text{HYP}^2 = c^2 = a^2 + b^2$$

in other words, it reduces to the Pythagorean theorem.

Below are some additional formulas which are sometimes of value when working with trigonometric functions.

$$\sin(\theta + \phi) = \sin\theta\cos\phi + \cos\theta\sin\phi \quad \text{(B-10)}$$
$$\sin(\theta - \phi) = \sin\theta\cos\phi - \cos\theta\sin\phi \quad \text{(B-11)}$$
$$\cos(\theta + \phi) = \cos\theta\cos\phi - \cos\theta\cos\phi \quad \text{(B-12)}$$
$$\cos(\theta - \phi) = \cos\theta\cos\phi + \cos\theta\cos\phi \quad \text{(B-13)}$$
$$\sin 2\theta = 2\sin\theta\cos\theta \quad \text{(B-14)}$$
$$\cos 2\theta = \cos^2\theta - \sin^2\theta \quad \text{(B-15)}$$

The last two formulas are simply Equations B-10 and B-12 rewritten for the special case when $\phi = \theta$.

Conversion factors may be read directly from these tables. For example, 1 degree = 2.778×10^{-3} revolutions, so $16.7° = 16.7 \times 2.778 \times 10^{-3}$ rev. The SI units are fully capitalized. Adapted in part from G. Shortley and D. Williams, *Elements of Physics,* 1971, Prentice-Hall, Englewood Cliffs, NJ.

Plane Angle

	degrees (°)	minutes (′)	seconds (″)	RADIAN	rev
1 degree = 1		60	3600	1.745×10^{-2}	2.778×10^{-3}
1 minute = 1.667×10^{-2}		1	60	2.909×10^{-4}	4.630×10^{-5}
1 second = 2.778×10^{-4}		1.667×10^{-2}	1	4.848×10^{-6}	7.716×10^{-7}
1 RADIAN = 57.30		3438	2.063×10^{5}	1	0.1592
1 revolution = 360		2.16×10^{4}	1.296×10^{6}	6.283	1

Solid Angle

1 sphere = 4π steradians = 12.57 steradians

Length

	cm	METER	km	in.	ft	mi
1 centimeter = 1		10^{-2}	10^{-5}	0.3937	3.281×10^{-2}	6.214×10^{-6}
1 METER = 100		1	10^{-3}	39.37	3.281	6.214×10^{-4}
1 kilometer = 10^{5}		1000	1	3.937×10^{4}	3281	0.6214
1 inch = 2.540		2.540×10^{-2}	2.540×10^{-5}	1	8.333×10^{-2}	1.578×10^{-5}
1 foot = 30.48		0.3048	3.048×10^{-4}	12	1	1.894×10^{-4}
1 mile = 1.609×10^{5}		1609	1.609	6.336×10^{4}	5280	1

1 angström = 10^{-10} m
1 nautical mile = 1852 m
= 1.151 miles = 6076 ft

1 fermi = 10^{-15} m
1 light-year = 9.460×10^{12} km
1 parsec = 3.084×10^{13} km

1 fathom = 6 ft
1 Bohr radius = 5.292×10^{-11} m
1 yard = 3 ft

1 rod = 16.5 ft
1 mil = 10^{-3} in.
1 nm = 10^{-9} m

Area

	METER^2	cm^2	ft^2	in.^{2}
1 SQUARE METER = 1		10^{4}	10.76	1550
1 square centimeter = 10^{-4}		1	1.076×10^{-3}	0.1550
1 square foot = 9.290×10^{-2}		929.0	1	144
1 square inch = 6.452×10^{-4}		6.452	6.944×10^{-3}	1

1 square mile = 2.788×10^{7} ft^2 = 640 acres
1 barn = 10^{-28} m^2

1 acre = 43 560 ft^2
1 hectare = 10^{4} m^2 = 2.471 acres

Volume

	METER^3	cm^3	L	ft^3	in.^{3}
1 CUBIC METER = 1		10^{6}	1000	35.31	6.102×10^{4}
1 cubic centimeter = 10^{-6}		1	1.000×10^{-3}	3.531×10^{-5}	6.102×10^{-2}
1 liter = 1.000×10^{-3}		1000	1	3.531×10^{-2}	61.02
1 cubic foot = 2.832×10^{-2}		2.832×10^{4}	28.32	1	1728
1 cubic inch = 1.639×10^{-5}		16.39	1.639×10^{-2}	5.787×10^{-4}	1

1 U.S. fluid gallon = 4 U.S. fluid quarts = 8 U.S. pints = 128 U.S. fluid ounces = 231 in.^{3}
1 British imperial gallon = 277.4 in.^{3} = 1.201 U.S. fluid gallons

Mass

Quantities in the colored areas are not mass units but are often used as such. When we write, for example, 1 kg "=" 2.205 Ib, this means that a kilogram is a *mass* that *weighs* 2.205 pounds at a location where g has the standard value of 9.80665 m/s^2.

	g	KILOGRAM	slug	u	oz	Ib	ton
1 gram = 1	0.001	6.852×10^{-5}	6.022×10^{23}	3.527×10^{-2}	2.205×10^{-3}	1.102×10^{-6}
1 KILOGRAM = 1000	1	6.852×10^{-2}	6.022×10^{26}	35.27	2.205	1.102×10^{-3}
1 slug = 1.459×10^4	14.59	1	8.786×10^{27}	514.8	32.17	1.609×10^{-2}
1 atomic mass unit = 1.661×10^{-24}	1.661×10^{-27}	1.138×10^{-28}	1	5.857×10^{-26}	3.662×10^{-27}	1.830×10^{-30}
1 ounce = 28.35	2.835×10^{-2}	1.943×10^{-3}	1.718×10^{25}	1	6.250×10^{-2}	3.125×10^{-5}
1 pound = 453.6	0.4536	3.108×10^{-2}	2.732×10^{26}	16	1	0.0005
1 ton = 9.072×10^5	907.2	62.16	5.463×10^{29}	3.2×10^4	2000	1

1 metric ton = 1000 kg

Density

Quantities in the colored areas are weight densities and, as such, are dimensionally different from mass densities. See note for mass table.

	slug/ft^3	KILOGRAM/METER^3	g/cm^3	lb/ft^3	lb/in.^{3}
1 slug per foot^3 = 1	515.4	0.5154	32.17	1.862×10^{-2}
1 KILOGRAM per METER^3 = 1.940×10^{-3}	1	0.001	6.243×10^{-2}	3.613×10^{-5}
1 gram per centimeter^3 = 1.940	1000	1	62.43	3.613×10^{-2}
1 pound per foot^3 = 3.108×10^{-2}	16.02	16.02×10^{-2}	1	5.787×10^{-4}
1 pound per inch^3 = 53.71	2.768×10^4	27.68	1728	1

Time

	y	d	h	min	SECOND
1 year = 1	365.25	8.766×10^3	5.259×10^5	3.156×10^7
1 day = 2.738×10^{-3}	1	24	1440	8.640×10^4
1 hour = 1.141×10^{-4}	4.167×10^{-2}	1	60	3600
1 minute = 1.901×10^{-6}	6.944×10^{-4}	1.667×10^{-2}	1	60
1 SECOND = 3.169×10^{-8}	1.157×10^{-5}	2.778×10^{-4}	1.667×10^{-2}	1

Speed

	ft/s	km/h	METER/SECOND	mi/h	cm/s
1 foot per second = 1	1.097	0.3048	0.6818	30.48
1 kilometer per hour = 0.9113	1	0.2778	0.6214	27.78
1 METER per SECOND = 3.281	3.6	1	2.237	100
1 mile per hour = 1.467	1.609	0.4470	1	44.70
1 centimeter per second = 3.281×10^{-2}	3.6×10^{-2}	0.01	2.237×10^{-2}	1

1 knot = 1 nautical mi/h = 1.688 ft/s   1 mi/min = 88.00 ft/s = 60.00 mi/h

Force

	dyne	NEWTON	lb	pdl
1 dyne = 1		10^{-5}	2.248×10^{-6}	7.233×10^{-5}
1 NEWTON = 10^5		1	0.2248	7.233
1 pound = 4.448×10^5	4.448	1		32.17

1 ton = 2000 lb

Pressure

	atm	dyne/cm^2	inch of water	mm Hg	PASCAL	lb/in.^{2}	lb/ft^2
1 atmosphere = 1		1.013×10^6	406.8	760	1.013×10^5	14.70	2116
1 dyne per centimeter^2 = 9.869×10^{-7}		1	4.015×10^{-4}	7.501×10^{-4}	0.1	1.405×10^{-5}	2.089×10^{-3}
1 inch of water^a at 4°C = 2.458×10^{-3}		2491	1	1.868	249.1	3.613×10^{-2}	5.202
1 millimeter of mercury^a at 0°C = 1.316×10^{-3}		1.333×10^3	0.5353	1	133.3	1.934×10^{-2}	2.785
1 PASCAL = 9.869×10^{-6}		10	4.015×10^{-3}	7.501×10^{-3}	1	1.450×10^{-4}	2.089×10^{-2}
1 pound per inch^2 = 6.805×10^{-2}		6.895×10^4	27.68	51.71	6.895×10^3	1	144
1 pound per foot^2 = 4.725×10^{-4}		478.8	0.1922	0.3591	47.88	6.944×10^{-3}	1

^{a}Where the acceleration of gravity has the standard value of 9.80665 m/s^2.
1 bar = 10^6 dyne/cm^2 = 0.1 MPa   1 millibar = 10^3 dyne/cm^2 = 10^2 Pa   1 torr = 1mm Hg

Energy, Work, Heat

	Btu	erg	ft · lb	hp · h	JOULE	cal	kW · h	eV	MeV
1 British thermal unit =	1	1.055×10^{10}	777.9	3.929×10^{-4}	1055	252.0	2.930×10^{-4}	6.585×10^{21}	6.585×10^{15}
1 erg =	9.481×10^{-11}	1	7.376×10^{-8}	3.725×10^{-14}	10^{-7}	2.389×10^{-8}	2.778×10^{-14}	6.242×10^{11}	6.242×10^5
1 foot-pound =	1.285×10^{-3}	1.356×10^7	1	5.051×10^{-7}	1.356	0.3238	3.766×10^{-7}	8.464×10^{18}	8.464×10^{12}
1 horsepower-hour =	2545	2.685×10^{13}	1.980×10^6	1	2.685×10^6	6.413×10^5	0.7457	1.676×10^{25}	1.676×10^{19}
1 JOULE =	9.481×10^{-4}	10^7	0.7376	3.725×10^{-7}	1	0.2389	2.778×10^{-7}	6.242×10^{18}	6.242×10^{12}
1 calorie =	3.969×10^{-3}	4.186×10^7	3.088	1.560×10^{-6}	4.186	1	1.163×10^{-6}	2.613×10^{19}	2.613×10^{13}
1 kilowatt-hour =	3413	3.600×10^{13}	2.655×10^6	1.341	3.600×10^6	8.600×10^5	1	2.247×10^{25}	2.247×10^{19}
1 electron-volt =	1.519×10^{-22}	1.602×10^{-12}	1.182×10^{-19}	5.967×10^{-26}	1.602×10^{-19}	3.827×10^{-20}	4.450×10^{-26}	1	10^{-6}
1 million electron-volts =	1.519×10^{-16}	1.602×10^{-6}	1.182×10^{-13}	5.967×10^{-20}	1.602×10^{-13}	3.827×10^{-14}	4.450×10^{-20}	10^{-6}	1

Power

	Btu/h	ft · lb/s	hp	cal/s	kW	WATT
1 British thermal unit per hour =	1	0.2161	3.929×10^{-4}	6.998×10^{-2}	2.930×10^{-4}	0.2930
1 foot-pound per second =	4.628	1	1.818×10^{-3}	0.3239	1.356×10^{-3}	1.356
1 horsepower =	2545	550	1	178.1	0.7457	745.7
1 calorie per second =	14.29	3.088	5.615×10^{-3}	1	4.186×10^{-3}	4.186
1 kilowatt =	3413	737.6	1.341	238.9	1	1000
1 WATT =	3.413	0.7376	1.341×10^{-3}	0.2389	0.001	1

Magnetic Field

	gauss	TESLA	milligauss
1 gauss =	1	10^{-4}	1000
1 TESLA =	10^4	1	10^7
1 milligauss =	0.001	10^{-7}	1

1 tesla = 1 weber/meter^2

Magnetic Flux

	maxwell	WEBER
1 maxwell =	1	10^{-8}
1 WEBER =	10^8	1

Appendix D

Commonly Used Values of Physical Constants and Physical Properties

Physical constants

Gravitational constant	G	$6.672 \times 10^{-11} \dfrac{\text{m}^3}{\text{kg} \cdot \text{s}^2} \left(\text{or } \dfrac{\text{N} \cdot \text{m}^2}{\text{kg}^2} \right)$
Coulomb constant	$k = \dfrac{1}{4\pi\epsilon_o}$	$8.988 \times 10^9 \dfrac{\text{N} \cdot \text{m}^2}{\text{C}^2}$
Magnetic force constant	$k' = \dfrac{\mu_o}{4\pi}$	$1.000 \times 10^{-7} \dfrac{\text{T} \cdot \text{m}}{\text{A}}$ (exact)
Permittivity of free space	ϵ_o	$8.854 \times 10^{-12} \dfrac{\text{C}^2}{\text{N} \cdot \text{m}^2}$
Permeability of free space	μ_o	$4\pi \times 10^{-7} \text{ N/A}^2$ (exact)
Speed of light in a vacuum	c	$2.998 \times 10^8 \text{ m/s}^2$
Planck's constant	h	$6.626 \times 10^{-34} \text{ J} \cdot \text{s}$
Avogadro's number	N_A	$6.022 \times 10^{23}/\text{mole}$
Boltzmann's constant	k or k_B	$1.381 \times 10^{-23} \text{ J/K}$
Molar gas constant	R	$8.315 \dfrac{\text{J}}{\text{mole} \cdot \text{K}}$
Charge of electron (absolute value)	e	$1.602 \times 10^{-19} \text{ C}$
Mass of electron	m_e	$9.109 \times 10^{-31} \text{ kg}$
Mass of proton	m_p	$1.673 \times 10^{-27} \text{ kg}$
Mass of neutron	m_p	$1.675 \times 10^{-27} \text{ kg}$

Physical properties

Gravitational acceleration at surface of Earth (sea level)	g	$\approx 9.80 \text{ m/s}^2$ (varies slightly with location and altitude)
Density of dry air (STP)		1.293 kg/m^3
Density of water (4 °C)		1000 kg/m^3
Specific heat of water (25 °C) (at constant pressure)		$4186 \text{ J/kg} \cdot \text{K} = 1.00 \text{ cal/g} \cdot \text{°C}$
Atmospheric pressure at sea level		$1 \text{ atm} = 1.01 \times 10^5 \text{ Pa (or N/m}^2)$ (See Appendix C for other units.)
Absolute zero		-273.15 °C
Index of refraction of a vacuum (air the same to four significant figures)	$n_{air} \approx n_{vac}$	1.000
Mass of Earth		$5.98 \times 10^{24} \text{ kg}$
Radius of Earth (mean)		$6.38 \times 10^6 \text{ m}$
Radius of Earth's orbit (mean)		$1.50 \times 10^{11} \text{ m} = 1 \text{ AU}$
Mass of moon		$7.36 \times 10^{22} \text{ kg}$
Radius of moon (mean)		$1.74 \times 10^6 \text{ m}$
Radius of moon's orbit (mean)		$3.84 \times 10^8 \text{ m}$
Mass of sun		$1.99 \times 10^{30} \text{ kg}$
Radius of sun (mean)		$6.96 \times 10^8 \text{ m}$

Mathematical constants

	e	2.718282
	π	3.141593

Directory of Tables of Values and Formulas Found in Chapters

(Tables that are relevant only to the immediate context of the chapters where they appear are not listed here.)

Some physical quantities can be defined in terms of others. For example, speed can be defined in terms of length and time. It is possible to identify a small set of quantities that can be used as the basis for defining all other physical quantities. There is more than one way of doing this. In SI, physicists have agreed on the following set of **basic** or **base quantities:** length [L], mass [M], time [T], electric current [I], thermodynamic temperature, amount of substance, and luminous intensity. (We give symbols in brackets here only for the base quantities we will consider further. All the base quantities are listed in Table 1-1 in Chapter 1.)

Other quantities, called **derived quantities,** can be defined in terms of these. The basic quantities that make up the derived quantity are called its **dimensions.** The brackets on the symbols above indicate that they are dimensions. Because a speed is defined by dividing a length by a time interval, the dimensions of speed are $[\frac{L}{T}]$ or $[LT^{-1}]$. By similar reasoning, the dimensions of area are $[L^2]$, and the dimensions of density are $[\frac{M}{L^3}]$ or $[ML^{-3}]$.

To find the dimensions of a quantity, we can look at the equation that defines the quantity. If we take Newton's second law $\Sigma \vec{F} = m\vec{a}$ as the defining equation for force, then the dimensions of force are dimensions of mass times dimensions of acceleration: $[M][\frac{L}{T^2}] = [\frac{ML}{T^2}]$. Likewise, since $\Delta W = F\Delta s$ is the defining equation for work, work has dimensions of force times dimensions of length: $[\frac{ML}{T^2}][L] = [\frac{ML^2}{T^2}]$. **STOP&Think** What are the dimensions of energy? of power? ◆

Trigonometric functions can be defined as ratios of sides of a right triangle, such as $\frac{OPP}{HYP}$ or $\frac{OPP}{ADJ}$. Trigonometric functions, therefore, have the dimensions $[\frac{L}{L}] = [1]$. Since the dimensions "cancel," we say that the sine, tangent, etc., are **dimensionless.** By the same token, an angle θ in radian measure is defined (Chapter 9) by the relationship $s = r\theta$. It then follows that $\theta = \frac{s}{r}$ also has dimensions $[\frac{L}{L}] = [1]$ and *is likewise dimensionless,* even though it may be said to have units of radians.

Notice also that electric current rather than electric charge is chosen as a base quantity in SI. Although we define current in Chapter 18 as $I = \frac{\Delta q}{\Delta t}$, we could as well begin with current and then write $\Delta q = I\Delta t$, from which it follows that electric charge has dimensions [IT].

Although one equation may define a quantity, the quantity may turn up in many other relationships. Checking dimensions is one way of seeing whether there is an error in a mathematical relationship among physical quantities. This investigation of dimensions is called **dimensional analysis.** For example, suppose you can't remember whether $\frac{v^2}{r^2}$ or $\frac{v^2}{r}$ is the correct expression for a_r, the magnitude of the radial acceleration. If you know that acceleration always has the dimensions $[\frac{L}{T^2}]$, you can check to see which expression has the correct dimensions. The expression $\frac{v^2}{r^2}$ has dimensions

$$\frac{\left[\frac{L}{T}\right]^2}{[L]^2} = \left[\frac{1}{T^2}\right]$$

and does not have L to the first power in the numerator; this tells you it can*not* be an acceleration. But $\frac{v^2}{r}$ has dimensions

$$\frac{\left[\frac{L}{T}\right]^2}{[L]} = \left[\frac{L}{T^2}\right]$$

which *are* dimensions of acceleration. You can rule out the first expression, while the second expression *may* be correct.

Caution: Having the right dimensions *doesn't guarantee* that an expression is correct. If $\frac{v^2}{r}$ has dimensions of acceleration, so do $2\frac{v^2}{r}$, $\frac{v^2}{2r}$ and $\pi\frac{v^2}{r}$. Pure numbers are dimensionless, so multiplying by a pure number does not affect the dimension of a quantity.

You can also use dimensional analysis to check the consistency of an equation. Just as you cannot meaningfully add a length to a time (3 m + 2 s is a nonsense expression), you cannot add or subtract terms unless they have the same dimensions. Therefore, if you write the kinematic equation

$$x = x_o + v_o t + \tfrac{1}{2} at^2$$

since x has dimensions [L], *each* term on the right side of the equation must also have dimensions [L]. (By *terms,* we mean the groupings of symbols separated by plus or minus signs.) For example, $v_o t$ has dimensions $[\frac{L}{T}][T] = L$ and $\frac{1}{2}at^2$ has dimensions $[1][\frac{L}{T^2}][T]^2 = L$.

STOP&Think To make sure that you understand how this may be useful as a check, check the following two equations of motion:

$$v^2 - v_o^2 = 2a(x - x_o) \qquad v^2 - v_o^2 = 2a(x - x_o)^2$$

In which of these do all the terms have the same dimensions? Which of these can*not* be a correct equation of motion? ◆

Answers to Odd-Numbered Problems

(Answers that require a proof, graph, or otherwise lengthy explanation are not included. For additional answers, go to www.wiley.com/college/touger)

CHAPTER 1

1. *a.* No. *b.* No. No finite sample provides proof about all. *c.* No. No guarantee this will keep happening indefinitely. *d.* No. No finite sample can provide proof about *always*.

3. No. It's different at different times of year.

5. 24.6 m/s

7. *a.* 178 cm *b.* 1.8 m *c.* 14 in

9. *a.* 2.0×10^5 m^2 *b.* 2.2×10^6 m^3

11. 42 m/s

13. 647.9 m

15. *a.* 4.9×10^{-9} m/s (at $\frac{1}{2}$ in/month) *b.* 9.5×10^{-10} m/s

17. 3.46

19. 3.9×10^{-3} m^3

21. (in kg/cm^2) $<$ (in g/cm^2) $<$ (in kg/m^2) $<$ (in g/m^2)

CHAPTER 2

3. *a.* change in position of moon and sun *b.* motion of ammonia molecules through lab *c.* motion of molecules *d.* atoms must move when they rearrange to form molecules

5. *a.* -20 km, 20 km *b.* 40 km, 40 km

7. *a.* 500 m *b.* -500 m *c.* -500 m *d.* 500 m

9. *a.* 100 m to right *b.* 600 m to right *c.* 600 m to left *d.* 100 m to right

11. *a.* 160 m *b.* 80 m *c.* 2 m/s *d.* 1 m/s *e.* 0 *f.* 1 m/s (assuming speed constant except for negligible time spent on turns)

13. $t = 6$ s to $t = 8$ s, $t = 2$ s to $t = 4$ s, $t = 4$ s to $t = 6$ s, $t = 0$ s to $t = 2$ s, $t = 8$ s to $t = 10$ s

15. *a.* 10 m/s *b.* -5 m/s *c.* 15 m/s

17. 1200 s

19. *a.* -3.2 m/s *b.* -3.2 m/s *c.* See website.

21. See website.

23. The slope of x vs. t is negative between these two instants.

25. *a.* No. See *c.* *b.* 7.9 (mi/h)/s *c.* 28 000 mi/h^2

27. See website.

29. Approx. 0.6 m/s^2

31. See website.

33. *a.* No. The sign changes because it reverses direction. *b.* Yes. Speed is always positive.

35. 30 m

37. *a.* 6.0 m/s *b.* less *c.* 8.5 m/s

39. *a.* 20 m/s *b.* 5 s

41. See website.

43. $v_o = 25$ m/s, $a = -5$ m/s^2

45. only (*b*)

47. 4

49. *a.* 10.6 m *b.* 2.94

51. See website.

53. *a.* $+0.55$ m *b.* -0.55 m

55. None. Can you tell why not?

57. 29 m/s

59. 6×10^6 s = 69 days

61. A

63. 2.6 m/s

65. See website.

67. See website.

69. *a.* $v_A = 0.83\, v_B$ *b.* $a_A = 0.69\, a_B$ (It may be easier to do *b* first.)

71. *a.* Yes. *b.* No. *c.* Yes. *d.* No. *e.* Yes. *f.* No. *g.* No. (See website for reasons.)

73. *a.* 10 m/s *b.* 3 s *c.* less *d.* 15 m at $t = 3$ s, 60 m at $t = 6$ s (Half of this would be 30 m.)

75. *a.* 246 m *b.* 13.7 m/s^2 *c.* before *d.* $t = 3.0$ s

77. Not satisfactory. Even in free fall, g varies with location.

79. *a.* 78.4 m *b.* 39.2 m/s *c.* 19.6 m/s *d.* 19.6 m. This is $\frac{1}{4}$ (not $\frac{1}{2}$) of the distance in *a.*

81. *a.* 12.0 m/s *b.* 11.9 m/s *c.* 0

83. *a.* B's acceleration is greater. *b.* not enough information to tell

85. A and D (In these, the distance gone during each interval changes uniformly.)

87. Speeding up. The absolute value of its slope is increasing.

89. B. Graph A is of x vs. t so its slope is v, not a. Since v is constant, $a = 0$ in A.

91. C. Convert each to (mi/h)/s and see how long each would take to go from 0 to 60 mi/h.

93. *a.* 1.0 s, 7.0 s *b.* 29.4 m/s at $t = 1.0$ s, -29.4 m/s at $t = 7.0$ s *c.* 98 m

95. No. The writer is speaking of change; acceleration is a *rate* of change.

97. See website.

99. 45.9 m.

101. *a.* positive slope, quantity increasing with time *b.* negative slope, quantity decreasing with time

103. 3.9 m

CHAPTER 3

7. turning cartwheels

9. See website.

11. circular motion around the shaft, linear motion along the shaft

13. Only *b* and *g* are true.

15. 82 cm/s

17. Only displacement, velocity, and acceleration are vectors. v_x is a scalar component of a vector; the others do not have spatial directions.

19. See website.

21. See website.

23. 1835 m, 14°

25. (*c*) $<$ (*d*) $<$ (*a*) $<$ (*b*)

27. See website.

29. *a.* 2450 m, 63° *b.* $x = 1100$ m, $y = 2150$ m

31. *a.* 10 m/s, 53° *b.* 6 m/s, 180° *c.* 10 m/s, 307° or $-53°$ *d.* 10 m/s, 233°

33. *a.* 5, 37° *b.* 5, 37° *c.* 5.7, $-52°$ *d.* 10.0, 121° *e.* 2, 45° *f.* 0

35. 567 km, 3.4° N of E

37. *a.* 5, 143° *b.* 5, $-37°$ *c.* 12.1, 73° *d.* 12.3, $-6.8°$ *e.* 8, 45° *f.* 6, 53°

39. *a.* from 0 to 10 *b.* the angle between $\vec{A}$ and $\vec{B}$ *c.* from 0 to 10 *d.* 8.66

41. *b.* $\Delta x = 400$ m, $\Delta y = 1560$ m *c.* $v_x = 5.0$ m/s, $v_y = 19.5$ m/s *d.* 20.1 m/s, 75.6° *e.* same

43. *a.* See website. *b.* $\Delta x = -500$ m, $\Delta y = 750$ m *c.* $v_x = -6.3$ m/s, $v_y = 9.4$ m/s *d.* 11.3 m/s, 124° *e.* same

45. 34 m/s, 163°

47. *a.* Yes. *b.* same *c.* same *d.* 7.1 m/s^2 (not zero), 225°

49. *a.* negative at all points *b.* positive at all points *c.* C *d.* none *e.* x and v_y

51. See website.

53. See website.

55. See website.

57. 49 m/s, $-24°$

59. *a.* 21.8 m/s   *b.* 20.2 m/s   *c.* 33.7 m

61. See website.

63. 14.4 m/s

65. 265 km, 89.6° N of E

67. *a.* cannot   *b.* might   *c.* cannot

69. No. Show that when $x = 1440$ m, the plane's altitude is 4374 m.

71. *a.* 45.8 m/s   *b.* Yes. The cannonball takes 11.4 s to reach P.

73. 57°

75. *a.* 12°   *b.* 69 s

77. only (*b*)

79. any two vectors representing different quantities, such as a position vector and a velocity vector

81. See website.

83. 0.10 m above and 0.90 m to the right of its position at $t = 4$ s

85. *a.* Yes. If one component is increasing in absolute value and the other decreasing, $V = \sqrt{V_x^2 + V_y^2}$ can remain constant.   *b.* Yes—for example, if V_x increases by becoming less negative while V_x increases by becoming more positive

87. 361 km, 53° N of W ($\theta = 127°$)

89. sometimes (when $\theta_A = \theta_B = 0$)

91. *a.* The directions are the same but the magnitude of $\vec{\overline{v}}$ differs by a scalar multiplier $\frac{1}{\Delta t}$.   *b.* The directions are the same but the magnitude of $\vec{a}_{av}$ differs by a scalar multiplier $\frac{1}{\Delta t}$.

93. *a.* Yes. The $\Delta v'_x$ s and $\Delta v'_y$ s have the same magnitude.   *b.* No. The $\Delta v'_x$ s and $\Delta v'_y$ s change direction.   *c.* Yes. The definition of acceleration is always the same.   *d.* No. They only apply when $\vec{a}$ is constant, not if its direction changes.

95. Student A.

97. A < D < B = C

99. See website.

101. *a.* 10.0 cm, -5.0 cm   *b.* 11.2 cm, $-26.6°$ or 333.4°

103. *a.* 22.5 m/s^2 *b.* 17.7 m/s^2

CHAPTER 4

5. *a.* No. A frictional force opposes it.   *b.* No. Use Newton's first law to explain.

7. Inertia is the tendency of a particle to remain at the same velocity *as if* it remembered what it was doing.

9. *a.* You feel pressed backward. You tend to continue at same speed until seat exerts force on you to speed you up.   *b.* You continue going forward (inertia).   *c.* You lean or slide left relative to car. You are actually tending to continue going forward as car veers right.

d. You do not feel anything pressing on you (no force needed to maintain constant velocity).

11. Yes. Book exerts force on shelf; shelf exerts equal and opposite force on book.

13. It interacts with Earth; Earth and parcel exert equal and opposite gravitational forces on each other.

15. Neither student is right. See website for explanation.

17. No. It requires an *external* force.

19. Equal to. The two sides of an interaction are always equal and opposite.

21. *a.* equal to   *b.* less than

23. It doesn't affect her motion (same force, same acceleration) but it halves the skateboard's acceleration.

25. *a.* B   *b.* 1.4

27. 1.67 kg

29. *a.* 10 m/s^2 *b.* 15 m/s^2

31. *a.* 1.6×10^{-24} m/s^2 *b.* 8.0×10^{-25} m

33. always

35. No. One pound is the *weight* of an object with a *mass* of 454 g.

37. *a.* 15 N, $-37°$ or 323°   *b.* 0.25 m/s^2, $-37°$ or 323°

39. Yes. Any reference frame moving at constant velocity relative to an inertial frame is also an inertial frame.

41. See website.

43. (*c*)

45. *a.* towards bottom of page (perpendicular to side of table)   *b.* perpendicular to edge (along dashed line at each point) at A, C, and D; at B the force is zero.   *c.* Force is perpendicular to barrier (towards center of circle) at each point; velocity is along (tangential to) barrier's edge

47. Infinitely far if there is nothing to exert a force on it to affect its motion (only approximately true in reality)

49. *a.* A's is stretched more. It must exert more force to produce the greater acceleration (slope at t_1) of A's car.   *b.* They are stretched equally. The acceleration is now zero (graph horizontal) so the net force is zero on each car.

51. *a.* (*iii*) by Newton's third law: $F_{\text{on A by B}} = -F_{\text{on B by A}}$ *b.* (*ii*) $m_A a_A = -m_B a_B$ so object with doubled mass has half as much acceleration.

53. See website.

55. *a.* If the end by the thumb points in the direction of the acceleration, the cart's acceleration will be proportional to the force the spring exerts (and thus proportional to its stretch)   *b.* Hold the device so that the spring is perpendicular

to the direction of motion rather than along it.

57. zero

59. *a.* 0.5   *b.* 1

CHAPTER 5

3. *a.* normal force by stool; gravitational force by Earth (your weight)   *b.* the gravitational force you exert on Earth   *c.* the normal force the stool exerts on you   *d.* the normal force you exert on the stool

5. See website.

7. See website.

9. See website.

11. none (scale not in contact with specimen)

13. See website.

15. See pages 118–119.

17. Sometimes

19. *a.* Parts *b–d* are based on a weight of 150b lb   *b.* 667 N   *c.* 68 kg   *d.* 802 N   *e.* 530 N

21. *a.* 0.083 m/s   *b.* 0.37 m

23. *a.* 200 N   *b.* -0.28 m/s^2

25. *a.* right   *b.* right   *c.* no   *d.* left   *e.* yes   *f.* II

27. See website.

29. See website.

31. *a.* 2.0 N   *b.* 0.84 N

33. 3.35 m/s^2

35. 11.2 m/s

37. See website.

39. Neither. Each block's weight equals the tension in the rope.

41. *a.* Yes, by Third Law   *b.* No. They have the same a but not the same m, so ΣF_x must be different on each.   *c.* Yes. They move together.   *e.* Yes. $\Sigma F_y = 0$ on each.

43. *a.* All magnitudes are positive. $F_{\text{on A by B}} = F_{\text{on B by A}} = F_{\text{on B by h}}$ *b.* $F_{\text{on B by h}} < F_{\text{on B by A}} = F_{\text{on A by B}}$

45. *a.* 3.27 m/s^2 *b.* 1307 N   *c.* 5.88 m/s^2 *d.* 1568 N

47. *a.* 133 N   *b.* 892 N

49. *a.* No. Show that child needs to be 2.5 m from pivot and can't be more than 2.2 m from pivot.   *b.* 0.70 m from pivot

51. *a.* It drops to zero.   *b.* It doesn't change.   *c.* It is halved. (Consider the weight's moment arm in each case.)

53. $\tau_1 = 10$ m · N $\tau_2 = -10$ m · N $\tau_3 = 14.1$ m · N $\tau_4 = \tau_5 = 0$ $\tau_6 = -5$ m · N

55. *a.* $F_A = 200$ N $F_B = 300$ N   *b.* The force exerted by the acrobat he approaches increases; the other decreases.

57. *a.* They are equal. *b.* No. There are only two unknown forces in the vertical direction and only one equation ($\Sigma F_y = 0$).

59. *a.* $F_L = 625$ N $F_R = 225$ N

61. *a, b.* There is not a single coefficient of friction. You can find μ_k between the suitcase and the chute and μ_s between the duffle bag and the chute.

63. 0.577

65. 11.3°

67. See website.

69. See website.

71. $|\tau_D| = |\tau_A| < |\tau_C| = |\tau_B|$

73. 2.95 m/s²

75. See website.

77. 1.66

79. 52 N

81. Decreases because the moment arm (radius of remaining coiled hose) decreases

83. $F_R = 250x$ $F_L = 250(3.2 - x)$

85. No. The force exerted by the pivot changes as the distribution of weight changes.

87. *a.* $\frac{1}{2}$ *b.* $\frac{1}{2}$ *c.* Up. The upward force (2 × tension) is greater than in the equilibrium situation (*b*).

89. *a.* 45.2 N *b.* 42.4 N frictional force *c.* your weight

91. No. The knot would untie.

93. *a.* See website. *b.* 36.9°

95. *a.* 3.46 m *b.* If there is no frictional force opposing the normal force the wall exerts on the ladder, $\Sigma F_x \neq 0$ and the ladder slips.

97. See website.

99. *c*

CHAPTER 6

3. See website.

5. See website.

7. See website.

9. See website.

11. *a.* A is right. ($F \cos \theta$) $\Delta s = F (\Delta s \cos \theta)$; $\Delta s \cos \theta$ is the component of $\Delta \vec{s}$ in the direction of $\vec{F}$. *b.* This supports A.

13. *a.* Yes. *b.* No. *c.* Different. The expended energy is not transferred to the wall; no work is done on the wall.

15. No change.

17. 4.98×10^4 J

19. *a.* to the right *b.* at an angle of 60° to the direction of motion *c.* perpendicular to the motion *d.* to the left

21. *a.* $mgy_1 = \frac{1}{2}mv_2^2$. On Earth g is larger so v_2 is larger. *b.* No. We can divide both sides of $mgy_1 = \frac{1}{2}mv_2^2$ by m; the result doesn't depend on m.

23. 11.9 m/s

25. 196 J

27. 12.7 m/s

29. 4.6 m/s

31. 5.3 m

33. 0.15 m

35. *a.* Yes. The energy stored as PE is still returned to KE. *b.* No. It loses energy to the atmosphere if there are resistive forces.

37. *a.* twice as much *b.* the same *c.* half as much

39. 0.06 m

41. 0.87 m/s

43. 60 000 J

45. 0.8 m

47. *a.* 1.4×10^5 N *b.* the same

49. 200 W/m²

51. See website.

53. $W_d < W_b < W_c < W_e < W_a < W_f$

55. *a.* 50 J *b.* 50 J *c.* 50 J *d.* 40 J *e.* 25 J

57. *a.* by changing direction *b.* No. It would have to change the magnitude of its velocity.

59. *a.* $\frac{1}{4}$ *b.* $\frac{1}{2}$ *c.* $\frac{1}{2}$ The forces that accelerate the two books are equal in *a* but not in *b*, where $\Sigma F = mg$.

61. 54 m/s

63. They are equal in absolute value but opposite in sign (assuming g constant and air resistance negligible).

65. 2.9 m/s

67. *a.* always *b.* never

69. 166 N

71. *a.* roughly 1.0–1.5 J *b.* about 6×10^{27} J

73. 3.8×10^{26} W

75. *a.* 0.050 m *b.* -1 m/s² *c.* 0.033 m *d.* 0.017 m expansion

77. *a.* 6.4×10^3 J *b.* 213 N

79. 0.13 m/s

81. *b.* The slope of M versus l^2 is 280. *c.* l^2 *d.* The energy absorbed is proportional to the exposed area.

83. 2.32

85. *a.* 4 *b.* 1.33

87. 5660 J

89. An activation energy input is required to break old bonds before energy can be released as new bonds form.

91. I = II < III

93. *a.* -565 J *b.* 0

CHAPTER 7

3. *a.* No. An impulse requires an interval Δt.

5. *a.* It remains constant unless there is a net outside force on the object.

b. Its rate of change is equal to the net outside force on the object.

7. 2.0×10^{-3} s

9. *a.* equal to *b.* F and Δt

11. *a.* They are equal in magnitude and both downward *b.* the Earth

13. *a.* 4.8×10^{-21} kg · m/s *b.* 7.6×10^{24} kg · m/s *c.* 0.033 kg · m/s *d.* 346 kg · m/s

15. 0

17. 0.0015 s

19. 1.3 m/s

21. 2.73 m/s

23. 0

25. *a.* left *b.* 0.14 m/s

27. It is turned into internal energy and/or dissipated.

29. 2.1 m/s

31. 8.7%

33. 1.3 m

35. *a.* -17 m/s *b.* -52 m/s

37. *a.* -4.1 m/s *b.* 4.1 m/s *c.* $v_A = -1.7$ m/s, $v_B = 2.4$ m/s *d.* 4.1 m/s *e.* equal in magnitude, opposite in sign

39. *a.* -4.1 m/s *b.* 4.1 m/s *c.* $v_A = -3.13$ m/s, $v_B = 1.07$ m/s *d.* 4.1 m/s *e.* equal in magnitude, opposite in sign

41. *a.* Yes *b.* No, because the total KE has increased.

43. The maximum height is unchanged.

45. 0.24 m

47. *a.* The frictional force is exerted on the initially slower one for a longer Δt, so it loses more momentum. *b.* The frictional force does the same work on both, so both lose the same amount of KE, but the same change in KE does not mean the same change in v if the initial v's are different.

49. *a.* 313 m *b.* 76 km/hr

51. $v_{red} = 6.93$ m/s, $v_{green} = 4.00$ m/s

53. *a.* No, because the second cart covers the distance in a smaller Δt. (Why?)

55. The one that bounces higher. It experiences a greater $\Delta \vec{p}$. The bottle's $\Delta \vec{p}$ will be equal and opposite.

57. The cart must roll back and forth, always opposite in direction to the motion of the child to keep the total momentum of child + cart constant.

59. *a.* $v_A = v_B = 1.7$ m/s *b.* $v_A = -2.57$ m/s, $v_B = 7.43$ m/s

61. *a.* 0 *b.* Yes. Each has the same KE as before.

63. See website.

65. *a.* It moves left as the block moves right because the total momentum must remain zero. *b.* Only in the

x direction. There is a net impulse on it in the y direction due to the gravitational force. *c.* Yes. PE is converted to gravitational KE. *d.* The block moves right. *e.* Momentum is not conserved in either direction but mechanical energy is. The stick delivers an impulse but does no work.

67. *a.* 9 N · s *b.* Area under graph

69. Inelastic. It does not bounce back from the belt.

71. 0.16 m/s

73. See website.

75. *a.*

77. I < III < II

79. 3.0 kg

81. Jack's 11 m/s; Naomi's 7 m/s

CHAPTER 8

3. See website.

5. No. Try to apply Newton's first law.

7. *a.* 10.2 m *b.* 5.1 m *c.* the turn in *b*

9. 30 m/s

11. As r decreases, a (and therefore F) must increase.

13. Normal and frictional. Resin increases μ_s.

15. 0.37

17. Force exerted by rope and radial component of $m\vec{g}$ (see website)

19. 11.7 m/s^2

21. 6400 N/m

23. 655 N

25. 13.4 m/s

27. *a.* greater *b.* point B

29. *a.* 17.7 m/s *b.* 25.0 m/s *c.* 80 000 J *d.* 80 000 J *e.* Yes

31. in general "ballpark of" 3×10^{-6} N

33. *a.* 81 *b.* 0.707

35. 6×10^{-4}

37. *a.* 393 N *b.* 0.9999995

39. *a.* 2.8 *b.* 3.0

41. only the mass of Jupiter

43. *a.* -7.7×10^{28} J *b.* less than (because it is bound, PE + KE < 0) *c.* 3.8×10^{28} J

45. *a.* less than (it gains PE as it loses KE) *b.* 5.9×10^4 m/s

47. Conversion of chemical energy adds to the total mechanical energy of the system.

49. *a.* only at A *b.* only at B *c.* at a point between A and B

51. *a.* No. Not an external force. *b.* No. Not an external force.

53. so that a_r and F_r build up gradually from zero

55. It moves higher up the surface of the bowl (see website for explanation).

57. 0.65 m/s and 0.72 m/s

59. Yes

61. See website.

63. 3.4×10^{-6}

65. $\approx 3.6 \times 10^7$ m

67. 1.6×10^{-4}

71. 5.35

73. *a.* 0.44 *b.* 1.5

75. See website.

77. *a.* 3.9×10^{-15} s^2/m^3 (in SI units) *b.* All equal. T^2/R^3 is constant for this system as well.

81. *a.* -5.8×10^8 J *b.* No. At infinity.

83. *a.* ii *b.* ii *c.* i *d.* i

85. 20.6 N

87. It increases.

CHAPTER 9

3. 0.29π, $-\pi/2$, $7\pi/6$

5. 180°, 150°, −30°, 120°, 810°, 165°, 252°

7. 0.5, −0.31, 0, −0.97

9. 3.53 m

11. 2.6π rad/s, 1.3 Hz or 1.3 s^{-1}

13. *a.* 0.10 s^{-1} *b.* 0.20π rad/s *c.* From center out: 0, 0.63 m/s, 1.26 m/s, 1.88 m/s, 2.51 m/s, 3.14 m/s

15. *a.* 2 rad/s *b.* 10 m/s

17. *a.* Yes, when it is traveling at constant speed *b.* Yes, at an instant when it is reversing direction (so tangential v and $a_r = v^2/r$ are zero)

19. 0.53 m/s^2

21. *a.* 0.032 kg · m^2 *b.* 0.064 kg · m^2

23. *a.* $I_A < I_B < I_C$ *b.* $I_A < I_D < I_B < I_C$

25. Less than. If $I_B = MR^2$, $I_A = 2M(\frac{1}{2}R)^2 = \frac{1}{2}MR^2$

27. 1.54 N

29. 0.33 J

31. No. When centered, the axis of symmetry moves at the same speed as the point of contact. An off-center CM may at a given instant lag or lead the point of contact.

33. The same in either position, to the extent that he is like a cylinder in each position.

35. *a.* solid sphere < solid cylinder < hollow cylinder *b.* hollow cylinder < solid cylinder < solid sphere

37. You'll rotate faster. $L = I\omega$ is conserved; reducing I increases ω.

39. *a.* 0.30π rad/s *b.* Yes.

41. *a.* 3.7×10^3 m/s *b.* 3.7×10^3 m/s *c.* Yes.

43. *a.* always away from the center of disk A *b.* 5.0 m/s *c.* 40 rev/s *d.* 0.08 m from center *e.* 0.02 m from center

45. Student A is correct. See website for explanation.

47. It must decrease. (Why?)

49. The torque exerted on the axle about its axis of rotation.

51. Most of the mass (the heaviest atoms) is aligned as though along the rod in Figure 9-9*d*.

53. 16

55. See website.

57. *a.* It rotates counterclockwise (conservation of angular momentum). *b.* If the mouse ran along a cart, the cart would roll in the opposite direction (conservation of linear momentum).

59. Yes. Less force is needed because the total I is reduced and one can show that $f = I\alpha/r$.

61. Yes. It is dissipated during the Δt over which the collision occurs.

63. *a.* ii *b.* v *c.* i *d.* iv *e.* iii

65. None.

67. *a.* $A_2 < A_1 < A_3$ *b.* $A_4 < A_5 < A_6$

69. *a.* 2.8 rad *b.* 1.1 m -1.2π rad/s^2 *c.* 15π

73. Near the rim, because that increases the total I more.

75. $\tau\Delta\theta = \frac{1}{2}I\omega_2^2 - \frac{1}{2}I\omega_1^2$

CHAPTER 10

3. 2117 lb

5. 6.06 N

7. 8.9×10^2 m

9. On the moon. No air pressure outside.

11. 3.23 m

13. *a.* i = ii = iii *b.* iii < i < ii

15. 1.17×10^5 Pa and 1.12×10^5 Pa

17. $F_C < F_A < F_B$

19. (*c*) The water pressure exceeds the surface air pressure by $rg\Delta y$.

21. Water. Ice floats.

23. 0.74

25. See website.

27. *a.* 88 200 N *b.* 44 100 N

29. 9100 N

31. Reread Section 10-3.

33. 1.02 m/s

35. $v_C < v_A < v_B$

37. *a.* Greater for normal blood. In the equation following 10-4b, more energy is converted to internal energy in the more viscous blood. *b.* Greater for anemic blood. Use same equation.

39. 0.160 m

41. *a.* Blood pressure is lower in the vein. *b.* The flow rate depends on ΔP, and

$\Delta P = \rho g \Delta y$ is proportional to the height Δy. *c.* Lower the bag containing the fluid.

43. *a.* 8.33×10^{-7} m³/s *b.* 0.011 m/s *c.* 1.7×10^{-4} m³/s *d.* 0.245 m

45. *a.* 1.6 m/s *b.* 1.33 m

47. See website.

49. $\rho_A = \frac{1}{8}\rho_B$

51. *a.* 257 lb *c.* No. Once an edge is lifted, air is let in eliminating the inside-outside pressure difference

53. See website.

55. 6630 m³

59. 0.46 m/s

61. greater than (apply the continuity equation $v_A A_A = v_B A_B$)

63. *a.* They must be equal. *b.* No.

65. 4

67. Towards. Air moves at greater speed through the narrower space, so pressure is reduced on the truck side of your car.

69. *a.* 1.9 m/s *b.* 0.02 m

71. by less than the mass of the sinker

75. greater than 2 atm but less than 4 atm

77. *a.* 100 *b.* 10 *c.* No. At this level of stress, fracturing is likely to occur.

79. More. It still has to displace its own weight of fluid, and it is no longer displacing any of liquid B.

81. It shouldn't matter. By Bernoulli's equation, the water should have the same velocity (though not the same flow rate).

83. 21 m

CHAPTER 11

5. *a.* $T_C = \frac{5}{9}T_F - 17.8°$ *b.* 40°C

7. 37.0°C

9. 6.7°C

11. *a.* −196°C *b.* −321°F

13. *a.* 3.2 cm *b.* 2.7 cm *c.* Different. The height is only proportional to temperature when the Kelvin scale is used.

15. No. It is not a good insulator.

17. See website.

19. See website.

21. *a.* roughly 0.04 C (0.04 kcal) to 0.07 C, depending on your height and weight *b.* roughly 0.13 C to 0.24 C

23. 2060 J

25. 3.5×10^8 J

27. 105 s

29. less than, in order that $mc_g|\Delta T_g| = mc_{liq}|\Delta T_{liq}|$

31. *a.* 10 K *b.* 1.1 K

33. 0.33 kg

35. 30°C

37. No. Because the total mass of the air is very small, the heat $mc_{air}\Delta T_{air}$ transferred to it will be small.

39. *a.* The first plateau is shorter, so at a constant rate of heat transfer, less total energy is transferred. *b.* Less than. The slope is greater, and the temperature rises more quickly when less energy in needed for each °C of increase.

41. Yes, if the other material undergoes a change of state.

43. We need to know how L_{SL} compares to $c_{liq}\Delta T$

45. Greater than, because L_{LG} is greater.

47. 1.5×10^7 J

49. See website.

51. See website.

53. See website.

55. See website.

57. ratio = 1

59. *a.* 0.65 m *b.* No. The wall would have to be about 2 ft thick.

61. *a.* to trap air, because the thermal conductivity of air is low *b.* Ducks spend time in water, which has a much higher specific heat than air, so more heat would be lost by ducks in coming to the same temperature as their surroundings.

65. *a.* The volume in the bulb is much greater. *b.* It expands and makes the liquid rise in the tube. *c.* C < B < A

67. See website.

69. The ocean (water) has a high specific heat so it changes temperature less readily.

71. See website.

73. The snow melts more quickly on the less well insulated roof.

75. See website.

77. *a.* 46°C *b.* 25°C

79. 7.1°C

81. See website.

83. See website.

85. A = C = D < E < B

87. Yes, but not enough to keep your engine from overheating.

89. *a.* 8 bulbs *b.* See website.

93. 79.7°C

95. See website.

97. See website.

99. −32°

CHAPTER 12

3. the average distance between molecules

5. *a.* remains the same *b.* remains the same *c.* increases *d.* increases

7. *a.* equal to *b.* less than *c.* greater than *d.* equal

9. $\frac{P_A}{P_B} = \frac{1}{2}$, because $P = \left(\frac{N}{V}\right)kT$, and N is halved for A.

11. *a.* 24 l *b.* Equal to. Nothing changes on the right side of $V = \frac{nRT}{P}$

13. The oxygen molecules. To have the same average KE with less mass, they need greater speed.

15. 313°C

17. 1.15×10^5 Pa

19. *a.* 1.5 *b.* No, because V is proportional to T in K, not in °C

21. Volume. $V_2 = 1.14\ V_1$

23. *a.* A net amount enters. *b.* 0.0030 l

25. 161 K or −112°C

27. 29 K

29. *a.* 1200 J *b.* 975 J

31. 8.5×10^{-3} m³

33. Greater disorder. The faster and slower molecules are more randomly arranged.

35. *a.* −3.4 J/K *b.* Less. The minus sign tells you entropy decreases.

37. *a.* No. The specific heat of water tells you 4186 J are needed to change a kg of water by 1 K. *b.* 0.73 J/K *c.* −0.73 J/K *d.* No. A negative ΔS cannot occur spontaneously.

39. 0.478

41. *a.* 7.46×10^4 J *b.* 2.66×10^5 J *c.* 1.92×10^5 J

43. See website.

45. 504 s (or 8.4 min)

47. *a.* eff$_A$ = 0.40, eff$_B$ = 0.33 *b.* 0.60

49. 3.5

51. The warmer end. See website for explanation.

53. *a.* The hydrogen molecules *b.* molecule B *c.* Yes. The collisions would lead to the lighter molecules moving faster on average.

55. 1.4

57. *a.* 0.47 moles *b.* 0.14 moles

59. 305 K

61. *a.* 1.36×10^5 Pa *b.* 1.21×10^5 Pa *c.* 395 K *d.* n drops to 0.059 moles

63. *a.* −0.012 J/K *b.* decreased *c.* No, because there is a work input.

65. 0.041 or 4.1%

67. See website.

69. $v_{final} = 2\ v_{initial}$

71. 3.4×10^{-9} m

73. *a.* Yes. See website. *b.* See website. *c.* See website. *d.* $\frac{\Delta m}{\Delta t} = \frac{DA(C_2 - C_1)}{d}$

75. $\frac{1}{4}$

77. It is impossible to say. We can compare the rms average speeds, but individual atoms can have speeds that differ from the average.

CHAPTER 13

5. See website.

7. *a.* 2.45 N   *b.* 0   *c.* 0.61 m

9. *a.* Never. $v = 0$ only at maximum compression or stretch, so the force and acceleration are at ± maximum. *b.* Sometimes. When the block stops oscillating completely, both v and a are zero. (Strictly speaking, this is at $t = \infty$, but the oscillation will *effectively* have stopped after a finite time.)

11. See website.

13. See website.

15. *a.* It is equal. Both are equal to $2\pi f$. *b.* 20 rad/s   *c.* 0.314 s   *d.* 0.314 s

17. B < A = D < C

19. 0.115 m

21. 0.30π or 0.94 rad

23. *a.* 0, $\pi/4$, $\pi/2$ *b.* 0, 0.707, 1 *c.* 0, 0.085, 0.12

25. *a.* 0.11 m, −0.085 m, +0.085 m   *b.* No. Displacement isn't proportional to t.

27. *a.* 30°   *b.* 3 s^{-1} *c.* π rad

29. *a.* 2.1 s   *b.* 0.49 s

31. Positive. Acceleration is always opposite in sign to the displacement.

33. *a.* iii   *b.* iv   *c.* i   *d.* ii   *e.* ii

35. See website.

37. $\omega_{max} = 565°/s$, $\alpha_{max} = (1.07 \times 10^4)°/s^2$,

39. See website.

41. equal to

43. 105 N/m

45. *a.* 7.2 m   *b.* ±0.62 m   *c.* 65 m/s^2 *d.* +5.6 m/s^2 when $y = -0.62$ m, −5.6 m/s^2 when $y = +0.62$ m

49. *a.* $f = 0.79$ Hz or 0.79 s^{-1}, $T = 12.7$ s *b.* 0.49 rad/s

51. max $\omega_\phi = 5.5°/s$, max $\alpha = 43°/s^2$

53. 0.25 m

55. *a.* C   *b.* B   *c.* A

57. *a.* B   *b.* 13-55*b* because $\Sigma\tau = I\alpha$

59. *a.* zero   *b.* Yes.

61. 0.866

63. 1.26 s

65. No. They are only valid when the acceleration is constant.

67. No. If there were no damping, no energy would be lost. Then none could be transferred to your ear.

69. See website.

71. *a.* $\pi/3$ *b.* $5\pi/3$ *c.* No. The velocities have opposite sign, because the oscillator passes the position going up and going down

73. $T_C < T_A = T_B$

75. *a.* 1.96 s   *b.* 1.71°   *c.* 11.2°/s

77. No. During intervals when contact with the ground is lost, child + stick are a projectile.

79. *a.* 1   *b.* Y   *c.* ωY *d.* velocity   *e.* They are equal.

81. *a.* 0.39 m, 11.8°   *b.* 0.098 m, 54.7°

83. zero acceleration

85. None of the first three answers are true. The values of k and m can be different as long at the ratio k/m is the same for both.

87. *a.* 1.4 rad/s   *b.* 0.35 rad/s

89. 3.63 s

91. $a_B < a_A = a_C < a_D < a_E = a_G < a_F$

93. 0.34 s^{-1}

CHAPTER 14

5. No. The center has to be able to move to propagate the wave beyond that point.

7. It goes down to $y = 4$, then up to $y = 9$, then back down to $y = 4$

9. (c)

11. *a.* No. It is a property of the medium, which is uniform.   *b.* Yes. The different points along the spring are like oscillators in different phases.

13. 10 m/s

15. *a.* 15 m/s   *b.* zero

17. the maximum speed at which the ribbon travels vertically

19. *a.* $y_B < y_A = y_C$ *b.* $Y_C < Y_A < Y_B$ *c.* $f_C < f_A = f_B$ *d.* $v_{yA} < v_{yC} < v_{yB}$

21. 0.125 s

23. *a.* 5 Hz   *b.* 0.06 m

25. See website.

27. $Y = 10$ m, $\lambda = 4.5$ m, $T = 8$ s, $f = 0.125$ Hz, $v = 0.56$ m/s

29. 0.0023 m

31. See website.

33. See website.

35. *a.* 8.6 m/s   *b.* 17.3 m/s

37. 17 m/s

39. roughly 1200 m

41. *a.* normal   *b.* above normal

43. *a.* $P_{max} = 1.0134 \times 10^5$ Pa $P_{min} = 1.0130 \times 10^5$ Pa   *b.* $v = 335$ m/s

45. 6

47. 20 000 Hz

49. *a.* $? = \frac{1}{3}$ *b.* $? = 3$

51. *a.* 1.2 m   *b.* $\frac{2}{3}$ of

53. 30.7 Hz

55. Add weights. See website.

57. It would decrease λ.

59. It would increase λ.

61. $? = 1.12$

63. See website.

65. *a.* 100   *b.* 100

67. *a.* blue < yellow < red *b.* red < yellow < blue

69. *a.* higher by 23 Hz   *b.* lower by 23 Hz

71. *a.* 2824 Hz   *b.* 2667 Hz

73. See website.

75. *a.* roughly $x = 12$ cm and $x = 22$ cm *b.* roughly $x = 17$ cm   *c.* roughly $x = 12$ cm and $x = 22$ cm

77. 8.6 m/s

79. Methane

81. 0.225 m

83. See website.

85. 3.1 m/s

87. min = 12 m, max = 14.4 m

89. Stretch the Slinky.

91. 10^5 times larger

93. 3 beats per second

95. 1.04 s

97. 0.42 Hz

99. 1.01×10^6 W/m^2

101. *a.* 10^{11} times as great   *b.* 3.16×10^5 times as great   *c.* 0.316 m   *d.* No.

CHAPTER 15

7. 2.994×10^8 m/s

9. equal to

11. See website.

13. *a.* See website.   *b.* Along center line, p.d. = 0   *c.* At 90°, p.d. = distance from S_1 to S_2 *d.* See website.

15. *c.*

17. *a.* 0   *b.* 0   *c.* −1.20 cm

19. See website.

21. *a.* 3λ *b.* 1.5λ

25. *a.* 0.020 m   *b.* 0.020 m   *c.* 0.040 m *d.* 0.0410 m

27. 6.6° or 20° or 35° or 53°

29. See first two pages of Section 15-5.

31. See last paragraph of Section 15-4.

35. *a.* 7.4×10^{-5} m   *b.* It wouldn't change. *c.* It would decrease.   *d.* 7.5 cm

37. 4.8×10^{-7} m

39. *a.* It only produces certain wavelengths, and not those in between.   *b.* There can be constructive interference for electromagnetic radiation at all wavelengths, not just in the visible range.

41. *Hint:* The panel is like the edge of the slit in Figure 15-24.

43. Expect rings around the center bright spot.

45. See website.

47. *a.* destructive   *b.* constructive
   c. destructive

49. *a.* 2.1×10^{-7} m   *b.* 4.2×10^{-7} m

51. See website.

53. 5.3×10^{-7} m, yellow

55. See website.

57. See website.

59. *a.* $P_1 = P_2 < P_3$ (Path difference varies
   with θ.)   *b.* $P_3 < P_1 < P_2$ (See website.)

61. No. There is oscillation between
   $y = \pm Y$.

63. *a.* 62°   *b.* 84°

65. See website.

67. *a.* The lines in the pattern will gradually shift.   *b.* The pattern will shift.

69. $d_B < d_A$ Show that $d_B = (\lambda_{blue}/\lambda_{red})d_A$

71. *a.* No "pattern" of fringes appears; there
   is constructive interference only along
   the center line.   *b.* The fringes are too
   close together to be seen separately.

73. *a.* 0.16 m   *b.* 0.80 m

75. bob up and down with double the
   frequency. This halves the wavelength,
   as does reducing the propagation speed
   by half.

77. 20°

79. three

81. Yes. The total amplitude varies between
   $\pm(Y_1 + Y_2)$

83. ≈100 m

85. *a.* 2 m, 6 m, 10 m, 14 m   *b.* No. (See
   website for explanation.)

87. Remain the same. The path difference
   depends only on the positions of the
   two sources and P, which don't change.

CHAPTER 16

3. *a.* 4.32 m   *b.* 0   *c.* 4.44 m

5. Remain the same. See website.

7. B < C < A < D (or B < C = A < D
   if you neglect the fact that C is slightly
   further from S_2 than A is from S_1)

9. *a.* 0.33 m   *b.* 0   *c.* 0.30 m

11. *a.* from the mirror surface   *b.* from
   1 m behind the mirror surface

13. See website.

15. *a.* 0.20 m and 0.80 m beyond mirror A
   b. 0.30 m and 0.70 m beyond mirror B
   c. all 0.12 m

17. A would be better. The cone of
   reflected rays from B would diverge too
   much to be easily gathered by an eye
   or lens.

19. See website.

21. *a.* 0.23 m virtual, upright, enlarged
   b. no observable image   *c.* −0.10 m
   real, inverted, same size   *d.* −0.05 m
   real, inverted, reduced

23. *a.* 0.065 m virtual, upright, enlarged
   b. 0.05 m   *c.* 0.033 m   (*a–c* are virtual,
   upright, reduced)   *d.* 0

25. See website.

27. *a.* When $s_o > \frac{1}{2}$, image is inverted.
   When $s_o < \frac{1}{2}$, image is upright.
   b. Upright for all values of s_o.

29. *a.* No. The surface is always convex
   and produces a reduced image   *b.* Yes.
   Everywhere except at the center, where
   the image is the same size as the
   object.

31. *a.* 0.10 m   *b.* 0.30 m   *c.* No. It would
   be behind him.

33. 1.01%

35. *a.* B   *b.* B

37. The larger *n* is, the more the ray
   bends.

39. *a.* 19°   *b.* 30°

41. No. Find *n* and compare it to n_{water}.

43. *a.* ϕ_t is smaller.   *b.* ϕ_r is unchanged.
   c. Transmitted v is slower.   *d.* λ is
   smaller.   *e.* *f* is unchanged.

45. 24.4°

47. *a.* Not usually   *b.* Yes, if it is bent too
   sharply.

49. less than

51. *a.* 0.8 m   *b.* 3 m

53. It will remain the same. (Compare with
   Problem 16-5.)

55. Yes, if the object is between a point
   source and a screen and moves parallel
   to the screen.

57. 3.65 cm

61. *a.* No. (You would have to go to ∞)
   b. At the center.

63. You need more than one ray from each
   object point to determine an image point.

65. A is correct. See website.

67. See website.

69. It will bend towards the horizontal
   upon entering, then bend gradually
   away from the horizontal (upward).
   See website.

71. *a.* 12 m   *b.* 4 m

73. *d*

75. 1.2 m

77. 0.225 m

79. *a.* It will decrease because λ is less in
   glass.   *b.* It will increase because the
   rays bend away from the center line on
   emerging from the glass.

CHAPTER 17

5. *a.* Very little (very slightly dimmer). The
   image of the bulb is still formed by the
   rest of the lens.   *b.* only in the latter
   case   *c.* See website.

7. because the biconvex lens inverts the
   image

9. *a.* All. Light travels in all directions
   from each point on the bulb.   *b.* same

11. remain where the screen had been

13. See website.

15. *a.* 1.0 m   *b.* 0.20 m

17. 3.15 m

19. 0.30 m

21. *a.* −0.47 m   *b.* ∞   *c.* 0.70 m   *d.* 0.35 m

23. 1.67 m

25. Two intersecting lines are sufficient to
   determine a point.

27. *a.* 0.75 m   *b.* 0.083 m

29. 1.14 m

31. *a.* 0.24 m, virtual, upright, inverted
   c. 0.10 m, real, inverted, neither (same
   size)

33. See website.

35. 1.7 cm

37. See website.

39. Think about why the focal length of
   the objective lens must be long.

41. 0.516 cm

43. See website.

45. See website.

47. It will remain in the same position. See
   website for explanation.

49. See website.

51. C < D < A < B

53. 0.067 m

55. Linear magnification less; angular magnification greater. The image is smaller
   than the moon itself, but subtends a
   greater angle.

57. 15.3 cm

59. *a.* −0.78 m   *b.* −1.61 m

65. *a.* real   *b.* inverted   *c.* reduced

67. −6.0 m

69. 29.6 cm

71. Diffuse, so that light will travel in all
   directions from the image point.

73. 30 cm, 120 cm

75. 135 cm

77. 43 cm above the surface

79. *a.* 0.54 cm   *b.* real   *c.* virtual

CHAPTER 18

7. See website.

9. It won't be affected.

11. See website.

13. *a.* Neutral. It was grounded.   *b.* No.
   See website.

15. See website.

17. See website.

19. *a.* You are a projection on an otherwise flat surface. (See Figure 18-6.)
   b. Lie down.   *c.* No. See website.

21. No. They can be ions or any other charged particles.

23. Right to left.

25. See website.

27. See website.

29. *a.* Bulb A has a lower resistance, so the capacitor discharges faster. *b.* The capacitor connected to A was more fully charged.

31. No. You are shifting charge from one end of the battery to the other to produce a surplus of + charge at one end and a deficit of + charge at the other.

33. See website.

35. *a.* Maximum brightness is greater. See website. *b.* It takes longer to dim to half its brightness. The greater stored charge takes longer to "drain" through the resistor just as a larger pot of water would take longer to drain through a coffee filter.

37. *a.* + *b.* neutral *c.* − *d.* neutral *e.* neutral *f.* neutral

39. A = B = C

41. *a.* All positive in one region, all negative in another. No. *b.* See website.

43. attracted to the north pole (not because it is a magnetic pole but because it is a conductor)

45. See website.

47. Insulators. Charge wouldn't build up on conductors.

49. Ground, to draw off any build-up of charge.

51. See website.

53. *a.* Yes. A tiny amount of charge will move until the charge density is the same on the bulb as on the terminal connected to it. *b.* No. There is no sustained flow and the little bit of movement is far too little to heat the bulb filament to glowing.

55. See website.

59. *a.* Same. See website. *b.* The bulb in Figure 18-27 glows longer because it takes the two capacitors longer to become fully charged.

61. See website.

63. They would still be attracted. Their molecules would polarize oppositely.

65. It must flow − to + in order for the current to flow in the same direction through all parts of the circuit.

67. *b*

69. *a.* your heart *b.* the voltage or charge density gradient

CHAPTER 19

3. *a.* 0.050 N *b.* 500 N *c.* No change.

5. *a.* 2.3×10^{-22} N *b.* 4.2×10^{42}

7. *a.* 0.010 m to left of origin *b.* 0.010 m to right of origin

9. 19 800 m/s^2

11. 10.9 N, $\theta = 108.5°$

13. *a.* 1.2×10^8 N/C to left *b.* 1.5×10^8 N/C to right *c.* 9.9×10^7 N/C to left *d.* 5.4×10^7 N/C to right

15. *a.* 1.08×10^7 N/C, $\theta = 109°$ *b.* 1.73×10^{-12} N, $\theta = -71°$

17. 1.25×10^4 N/C, downward

19. *a.* 1.8×10^{-8} N/C *b.* 1.8×10^{-8} N/C *c.* Cannot be answered. No position is given and the field varies with position.

21. *a.* 2.7×10^{-6} N/C, $\theta = 33°$ *b.* 4.74×10^5 m/s^2, $\theta = 213°$ or $-147°$

23. *a.* right *b.* left *c.* left *d.* left, left *e.* left *f.* right *g.* left *h.* left, left *i.* It is away from the + terminal and towards the − terminal.

25. *a.* Less than. The lines are further apart. *b.* To the right of. The tangent at A (field direction) passes to the right of C. The initial acceleration is in this direction.

27. See website.

29. The one moved to the right of B. The field due to the plane is constant, but the field due to a line of charge drops off as $1/r$.

31. It decreases PE_{grav}. The electrons must be moved together against the mutual repulsive force to increase their PE_{elec}. But PE_{grav} increases when they are pulled apart against their mutual gravitational attraction.

33. 2.04×10^{-18} J

35. *a.* 400 m/s *b.* 1.06×10^3 m/s

37. 3.84×10^{-19} J (increase)

39. 1.66×10^4 m/s

41. *a.* -7.2×10^{-15} J *b.* -2.4×10^{-15} J *c.* $+2.4 \times 10^{-15}$ J *d.* -9.6×10^{-15} J

43. 1.5 J/C

45. *a.* 1.63×10^{-18} J *b.* 6.1×10^{17} J

47. 4800 V/m or 4800 N/C

49. No. The student needs to use two different metals for the electrodes.

51. See website.

53. (*c*)

55. *a.* $E_A = E_B = E_C = E_D = E_E$ *b.* $pot_D = pot_E < pot_A = pot_B < pot_C$

57. 7.3×10^{-8} C

59. See website.

61. *b.* $E = kq/x^2$; From that distance, the ring looks like a point charge.

63. See website.

65. See website.

67. See website.

69. 30 N/C towards the + pole

71. 281 base pairs

73. 5.27×10^{13} m/s^2 to left

75. *a.* 6.5×10^6 m/s *b.* Cannot be answered because the separation between plates is not known.

77. $E_A = 30$ V/m, $E_B = 10$ V/m

79. A (Why?)

81. The ratio remains the same because both vary as $1/r$.

83. *a.* The same by both wires. *b.* The same by both wires. *c.* Twice as large along wire 1. *d.* The same by both paths. *e.* Twice as large along path 1. *f.* Twice as large along path 1. *g.* Yes.

CHAPTER 20

3. *a.* 6.4×10^{-18} A *b.* 5.2×10^{-18} A *c.* 4.6×10^{-18} A

5. Less than. The flow is jointly proportional to density ρ and speed v. Verify that $\rho_A v_A = 55\rho_A$ and $\rho_B v_B = 50\rho_A$

7. 1.5 C

9. No. Typical drift velocities are so low that electrons will travel only a few cm in this time.

11. *a.* 1.5 V *b.* 1.5 V

13. *a.* 0.30 C *b.* 1.8 C *c.* No. It depends on the voltage produced by the voltage supply connected across it.

15. See website.

17. 2.1 V

19. *a.* $C_2/C_1 = 4$ *b.* $C_2/C_1 = 2$

21. *a.* 8.7×10^{-11} F *b.* No. Surfaces of dimes aren't flat so the gap width between them would vary.

23. *a.* lattice vibrations and impurities *b.* They would accelerate.

25. 5.0 A

27. *a.* $R_B < R_C < R_E < R_A = R_D$ *b.* $\rho_A = \rho_B = \rho_C = \rho_D = \rho_E$

29. 1.46×10^{-4} m

31. 0.234 V

33. No. $\rho = 1/nec$. If n varies, so does ρ, and so, too, does R.

35. 24 Ω

37. When it is not connected. The meter reads the value of $V = \mathcal{E} - Ir$. This equals $\mathcal{E}$ only when I is zero.

39. *a.* 11.4 Ω *b.* 0.130 A *c.* 1.53 V

41. *a.* 0.24 W *b.* 14.4 J

43. 144 Ω

45. *a.* 0.0126 F *b.* 0.159 C

47. 1.06×10^{-3} A

49. *a.* 1.32×10^{-6} m/s *b.* 1.25×10^{17}

51. *a.* increase *b.* more than. The resistance increases with temperature. *c.* more than. Before the wire is fully heated, the resistance is lower so V^2/R is greater.

53. 1.06×10^{-9} C

55. *a.* ∞ *b.* 0

57. 1.22×10^{-5} C

59. 0.0045 J

61. See website.

63. 52 s

65. See website.

67. D < B = C < A

69. resistance

71. See website.

73. *a.* They are equal because the potential difference is the same across both wires. *b.* It is half as great from wire 2. Since it is twice as long, its resistance is doubled so $P = V^2/R$ is halved.

CHAPTER 21

5. *a.* neither *b.* series *c.* parallel *d.* neither

7. *a.* A_3 *b.* A_2 or A_4 *c.* A_1 *d.* V_2 *e.* V_1 *f.* none

9. *a.* battery *b.* R_1, R_2 *c.* battery *d.* C

11. *a.* 14 Ω *b.* 0.67 Ω

13. 0.05 Ω

15. 28.9 Ω

17. No. The potential difference across each parallel path between the same two endpoints is the same.

19. *a.* 0.065 A through each *b.* 0.30 A through the 5 Ω resistor, 0.19 A through the 8 Ω resistor, 0.15 A through the 10 Ω resistor

21. In parallel. The potential difference across each is then the same, so each draws the same current.

23. (i)

25. 20 Ω

27. 12.9 Ω

29. *a.* 0.17 A *b.* 0.033 A *c.* Yes.

31. *a.* 55 mF *b.* 5 mF

33. *a.* 0.92 mF, less than any of the individual capacitances *b.* 1.4 mC or 0.0014 C *c.* 0.0014 C

35. *a.* A_1 and A_2 *b.* A_1 and A_2 *c.* directly *d.* inversely *e.* They must add. *f.* The inverses must add.

37. *a.* $V_1 = 2$ V, $V_2 = V_3 = 1$ V *b.* $Q_1 = 9$ μC, $Q_2 = 5$ μC, $Q_3 = 4$ μC

39. *a.* 0.0888 A *b.* 0.0296 A *c.* 0

41. 9.0×10^{-3} s

43. *a.* $V_1 = 1.875$ V, $V_2 = V_3 = 1.125$ V *b.* $I_1 = 0.15$ A, $I_2 = 0.0375$ A, $I_3 = 0.1125$ A

45. $I = 0.060$ A counterclockwise through R_1 and R_2, 0.24 A to right through R_3

47. *a.* 0.167 A *b.* No. This isn't possible. You can show that the current through R_2 is 0.33 A, so A_2 must read more than 0.33 A.

51. *a.* 0.15 A *b.* 0.05 A

53. It increases. See website.

55. *a.* A goes out because almost all the current flows by the negligible-resistance path in parallel with it. *b.* B glows brighter because the potential drop across the parallel paths is negligible so all of the potential drop is now across B.

57. *a.* The capacitance of the combination must be zero. *b.* The capacitance of the combination is the capacitance of the other capacitor. *c.* It can hold no charge (like if there are just wire tips with ≈ 0 area instead of plates).

59. Yes. See website.

61. *a.* equivalent conductance = conductance$_1$ + conductance$_2$ *b.* $\frac{1}{\text{equiv.cond.}} = \frac{1}{\text{conductance}_1} + \frac{1}{\text{conductance}_2}$

63. See website.

65. $\frac{1}{4}$

67. *a.* A_1 reads 0.0375 A, A_2 reads 0.1875 A *b.* $R_{AB} = R_{BC} = 60$ Ω

69. *a–d.* All greatest at first. *e* and *f.* See website.

71. *a.* No effect. The potential difference is the same across both. *b.* It would lower it. See website.

73. The changes are additive: Total change = change$_1$ + change$_2$ + change$_3$ + ⋯

75. Yes, because the total current drawn is additive. Drawing too much current blows fuses/circuit breakers or burns out wires.

77. 43 200 J

79. is positive and its value gets smaller

81. *a.* 27 Ω *b.* 60 Ω

83. doesn't dim at all

85. *a.* $P_{40W} = 14.4$ W, $P_{60W} = 9.6$ W *b.* $P_{40W} = 40$ W, $P_{60W} = 60$ W

87. It increases. More total current is drawn, while V remains the same for each, so VI_{total} increases.

89. See website.

91. equal to

93. *a.* 0.37 A *b.* 0.033 W

CHAPTER 22

3. No. The needle is long enough so that the field lines curve over its length, so the field need not be the same in magnitude or direction at its two ends.

5. *a.* 45° below +x axis (assuming positive direction is to right) *b.* same

7. *a.* 0.010 m to left of origin *b.* 0.010 m to right of origin

9. *a.* It doubles. *b.* It drops to zero. *c.* It doesn't change.

11. the source of the magnetic field (or simply the field)

13. *a.* 9.6×10^{-17} N into page *b.* zero *c.* 3.6×10^{-16} N into page

15. *a.* 8.5×10^{-4} T in +z direction (out of page if page is the xy plane) *b.* 6.4×10^{7} m/s^2 radially inward

17. 1920 N/C upward

19. *a.* negative, so that the right hand rule will give a force directed radially inward *b.* 0.57 m *c.* 0.28 m

21. 39°

23. 0.0048 N

25. CD < AB = BC = DE = EF

27. *a.* any direction perpendicular to the plane of the page *b.* Any direction in (or parallel to) the plane of the page

29. *a.* 0 *b.* 0.0055 m · N *c.* 0.0053 m · N

31. *a.* out of page *b.* downward *c.* out of page *d.* It would veer toward the right *e.* It wouldn't change.

33. *a.* toward top of page *b.* out of page *c.* to right

35. *a.* out of page *b.* toward top of page

37. *a.* 0.5 cm *b.* 2 cm in +y direction *c.* 1.33 cm in −y direction *d.* 1.0 cm, $\theta = 71°$ (counterclockwise from +x axis)

39. *a.* 4.8×10^{-5} T *b.* 8.9×10^{-6} T

41. *a.* south *b.* north *c.* the same as before the break—from the original south to the original north *d.* the same *e.* the same

43. *a.* same *b.* spin-down

45. *a.* Yes, current-carrying wires. *b.* Yes, stationary charges. *c.* No, because a single particle cannot have a net moving charge if it has no charge.

47. See website.

49. Magnetic forces never do work because they are always perpendicular to the motion.

51. *a.* $\frac{1}{2}$ *b.* Halfway around. $s = r\theta$. s is the same if the speed is the same, then if r is halved, θ must double.

53. *a.* No. The power is high voltage, low current so the force IlB is small. *b.* 250 A

55. *a.* High. See website. *b.* $\frac{1}{2}$ Place a high resistor in series. See website.

57. *a.* one *b.* an infinite number *c.* none

59. C < D < A = B = E

61. See website.

63. not rotate at all

65. See website.

67. 3.1 T

69. 0.29 T

71. remain the same

CHAPTER 23

3. Yes. by induction

5. *a.* at position 2 *b.* between position 1 and position 2

7. *a.* B *b.* negligible change *c.* θ
   d. none *e.* A

9. 16 cm^2

11. 0.015 A

13. *a.* 1.5×10^{-3} V *b.* 6.4×10^{-4} V

15. A = B < D < C = E

17. *a.* from a to b *b.* from a to b; See
    website for explanations.

19. From a to b See website.

21. From a to b. It must produce a mag-
    netic field toward the right to oppose
    the drop in the given field.

23. counterclockwise; The flux toward the
    north decreases, so the current must
    produce a field toward the north.

25. *a.* It varies sinusoidally. *b.* The induced
    EMF is proportional to $A_\perp$. *c.* (It is
    vertical.) *d.* (It is horizontal.)

27. 0.090 A

29. *a.* 3.7 V *b.* 0

31. *a.* $\omega = 60\pi$ rad/s, $f = 30$ Hz *b.* 50 V
    c. 0.025 s *d.* −29.4 V *e.* 5.3 T

33. *a.* 0 *b.* 1800 A/s *c.* 0 *d.* 9 V

35. *a.* 720 A/s *b.* 5.4 V across resistor,
    3.6 V across coil *c.* 9 V

37. 2.5×10^{-3} s

41. *a.* primary to outlet, secondary to bulb
    b. 50 turns

43. *a.* 2.5×10^5 V *b.* 3300 A

45. See website.

47. its resistance per unit length

49. 0.023 V

51. 8.6 m/s

53. See website.

55. See website.

57. See website.

59. *a.* 0.5 A *b.* 2 A *c.* 20 A *d.* 2400 Ω

61. 1.67 A

63. See website.

65. 8×10^{-5} s

67. 3.72 cm

69. *a.* 0.15 A *b.* −300 A/s *c.* −100 A/s

71. 11.8 A

73. 1-B 2-E 3-A 4-D 5-F 6-A

75. equal to; The EMF doubles but the
    resistance also doubles, so $I = \mathcal{E}/R$
    remains the same.

CHAPTER 24

3. A = B = C

5. See website.

7. No. The magnetic field would also vary
   sinusoidally.

9. A. The path becomes a straight line
   where there are no magnetic or electric
   forces.

11. *a.* 0 *b.* 0

13. the velocity selector and using an
    electric field to accelerate a particle

15. α and β rays are both particles; γ and X
    rays are both electromagnetic radiation.

17. See page 717.

19. *a.* two black lines very close together
    in an otherwise continuous spectrum
    b. See website.

21. The wavelength is halved.

23. It decreases.

25. *a.* No. *b.* Yes. *c.* No. *d.* No. *e.* Yes.
    f. No. *g.* No. *h.* Yes. *i.* Yes. *j.* Yes.
    k. No.

27. The speed of light is the same for all
    observers.

29. *a.* 2.2 m/s *b.* No. *c.* Yes.

31. See website.

33. *a.* 0, 19.6 m, 0 *b.* 0, −80 m, 0
    c. $(\Delta x)_A = (\Delta x)_C = -2$ m,
    $(\Delta x)_B = (\Delta x)_D = -52$ m
    d. $v_A = v_C = -2$ m/s,
    $v_B = v_D = -52$ m/s,

35. **ia.** Yes. **ib.** No. **ic.** No. **id.** No.
    ie. Yes. **iia.** No. **iib.** No. **iic.** Yes.
    iid. Yes. **ie.** No.

37. β rays

39. X rays and γ rays because in both
    there is an oscillating electric field

41. No. It would be deflected out of the
    page but its speed wouldn't change.

43. See website.

45. *c*

47. No. Dense materials and large molecules
    have continuous absorption spectra.

49. No. The wavelengths must be smaller
    than the objects observed.

51. 3.0×10^8 m/s

CHAPTER 25

3. *a.* 105 mi/h *b.* 3.0×10^8 m/s

5. 1. The rocket is traveling away from
   a major gravitational attractor.
   2. The elevator's upward acceleration is
   decreasing.

7. *a.* definitely is not *b.* possibly is
   c. definitely is *d.* definitely is not

9. *b* only

11. *a.* not an event (occurs over a dis-
    tance and over an interval) *b.* an
    event (occurs at one instant at a single
    point) *c.* not an event (occurs over a
    distance and over an interval) *d.* not
    an event (refers to whole length of
    spring, not a single point) *e.* an event
    (occurs at one instant at a single
    point)

13. 1.5×10^4 s

15. depends on your age; It will always be
    double your actual age.

17. *a.* 1.34 s *b.* 29.8 s *c.* 29.8 s

19. measurements of the positions at the
    two ends at the same instant

21. See website.

23. *a.* 0.60 m *b.* 1.00 m

25. E < A = C = D < F < B

27. *a.* 73.3 m *b.* 87.3 m *c.* 14 m
    d. 1.17×10^{-7} s (agrees to within
    rounding-off error)

29. 303 m

31. 110%

33. 2.02×10^{-13} J or 1.26 MeV

35. greater than (See problem 25-14.)

37. *a.* 2.63×10^{-13} J or 1.64 MeV
    b. 9.11×10^{-31} kg

39. 5.42 eV

41. 2.05×10^{-22} kg · m/s

43. No, because there are no repulsive
    gravitational forces.

45. *a*

47. the motions of the Earth and planets
    that were then the basis for time
    measurements

49. See website.

51. A sees it as a rectangle with the shorter
    side in A's direction of motion. B sees
    it as a parallelogram with the shorter
    diagonal in B's direction of motion.

53. The rod is perpendicular to the direc-
    tion of relative motion.

55. No. See website.

57. *a.* Approximately. Collisions occur over
    a tiny Δt, not at a single instant.
    b. Exactly. The origins are points.
    c. Approximately. It occurs over a tiny
    Δt, not at a single instant. *d.* Approxi-
    mately. The light is emitted over a tiny
    Δt. *e.* Not at all. Motion is over an
    interval Δt, not at a single instant.

59. *a.* much less than *b.* 42.5, 9000

CHAPTER 26

3. See website.

5. Reduce the separation between the
   plates.

7. *a.* 2130 N/C *b.* 320 V

9. See website.

11. 1.3×10^6 m/s

13. In principle, yes. Electrons would fol-
    low circular paths. B would have to be
    large enough so that r is less than the
    distance between the plates.

15. *a.* 0.40 W/m^2 *b.* yes *c.* 1.3×10^5 m
    or 130 km

17. less than, so that $1/m^2 > 1/n^2$, which
    is required for λ to be positive

19. *a.* 4.05×10^{-6} m *b.* 7.46×10^{-6} m

21. singly charged deuterium ($_1^2$H) ions.

23. The α particles emitted are helium
    nuclei.

25. See website.

27. *a.* 9.1×10^{-5} T   *b.* Yes. Electric PE is converted into KE.   *c.* No. The KE remains constant because magnetic forces do no work. They only change the electron's direction.

CHAPTER 27

5. ceases to exist

7. *c*

9. *a.* 1.10×10^{15} Hz
   b. 2.55×10^{15} Hz   *c.* 6.0 V

11. No. Find f for the laser light to see if it is below f_o.

13. *a.* -2.18×10^{-19} J   *b.* -13.6 eV

15. none of these

17. *a.* increases more in transition A
   b. in transition B   *c.* Yes   *d.* No

19. No. As values of $1/n^2$ get closer together, values of n^2 get further apart.

21. *a.* $r_2 = 2.11 \times 10^{-10}$ m,
   $r_3 = 4.75 \times 10^{-10}$ m   *b.* further apart

23. 2.48×10^{14} Hz, 2.94×10^{14} Hz

25. *a.* 10.2 eV   *b.* -1.0000×10^{-12}
   c. 0.249999999999, 3.4 eV
   d. No. See website.

27. *a.* It is absorbed and the electron is excited to the $n = 2$ state.   *b.* It is not absorbed.   *c.* It is absorbed and the electron breaks away from the atom (the atom is ionized).

29. If $m = 60$ kg and $v = 5$ m/s (about half of top speeds for the 100-m dash), $\lambda = 2.2 \times 10^{-36}$ m. (Use your own values for m and v.)

31. *a.* 7.04×10^{-7} m/s. No.   *b.* 39 500 m/s; Yes.

33. 1

35. See website.

37. $|A\rangle$ and $|B\rangle$

39. 2.5×10^{-25} kg · m/s

41. *a.* 1.05×10^{-20} kg · m/s
   b. 3.14×10^{-14} J

43. the ability to absorb energies over a broad range of wavelengths

45. *a.* It is unchanged.   *b.* It ceases to exist as a photon.

47. a singly ionized helium atom, a triply ionized carbon atom

49. *a.* $1 \rightarrow 2$: $-\phi + 0 + hf = KE_2$
   $2 \rightarrow 3$: $KE_2 = eV_o$.   *b.* These are the two parts of Equation 27-1, which is in effect the work-energy theorem applied to a photoelectron.

51. No. Only bound state energy levels are quantized. The photoelectron is no longer in a bound state.

53. *a.* 1.99×10^{-24} kg · m/s
   b. 1.15×10^{-24} kg · m/s To one significant figure, this equals Δp_x, in agreement with the comment that $\Delta p_x \approx |p_x|$.

55. *a.* $n \approx 8$ *b.* $n \approx 17$ *c.* < 17
   d. $\approx 5 \times 10^{-8}$ m   *e.* very high temperatures, resulting in large energy input and consequent large energy output

59. It decreases. With increasing orbit size, Δx increases, so the uncertainty principle tells us Δp must decrease.

61. *a.* A photon must have energy hf at least equal to E_I to break an electron away from an atom and thus ionize the atom.   *b.* and *c.* See website.

63. All photons travel at speed 3×10^8 m/s relative to any observer, never at octopus speed.

67. *a.* 14   *b.* none

69. *a.* $|4\,0\,0\,-\frac{1}{2}\rangle$ $|4\,0\,0\,+\frac{1}{2}\rangle$
   b. $|2\,0\,0\,-\frac{1}{2}\rangle$ $|2\,0\,0\,+\frac{1}{2}\rangle$
   $|2\,1\,-1\,-\frac{1}{2}\rangle$ $|2\,1\,-1\,+\frac{1}{2}\rangle$
   $|2\,1\,0\,-\frac{1}{2}\rangle$ $|2\,1\,0\,+\frac{1}{2}\rangle$
   $|2\,1\,1\,-\frac{1}{2}\rangle$ $|2\,1\,1\,+\frac{1}{2}\rangle$

71. 2.6×10^{-11} m

CHAPTER 28

3. 20

5. -2 -2 -4
   1 -1 0
   -1 1 0

7. *a.* $^3_2$He   *b.* $^1_0$n   *c.* $^{15}_7$N + $^0_{+1}$e

9. 1.56×10^{16} nuclei

11. *a.* 8×10^{21} nuclei   *b.* 5.6×10^{22} electrons   *c.* 1.00×10^{-6} s   *d.* 3.8×10^{18}
   e. 3.8×10^{18} *f.* See website.

13. *a.* decreases   *b.* decreases

15. *a.* C-11: $0.0081 N_o$, Fl-18: $0.0015 N_o$
   b. Room B   *c.* Exposure would be the same in both, since both samples will have effectively completely decayed, emitting equal numbers of β^+ particles.

17. *a.* about 200/min   *b.* 180 000

19. 2936

21. *a.* 1.3×10^4 years   *b.* 3.84×10^{-12} s^{-1}
   c. about 200   *d.* about 20   *e.* about 2

23. *a.* $^{32}_{16}$S   *b.* $^{72}_{32}$Ge

25. *a.* $^{214}_{83}$Bi   *b.* $^{214}_{82}$Pb   *c.* $^{218}_{85}$At

27. See website.

29. *a.* Against nuclear forces, to overcome short range nuclear attractions in separating nucleons.   *b.* Against Coulomb forces, to overcome electrostatic repulsions in bringing same-charge nucleons together.

31. See website.

33. See website.

35. See website.

39. 0.7 mGy of α particles. (Use Table 28-3 and compare the dose equivalents.)

41. They are the same. See website.

43. greater (Use Table 28-3 and compare the dose equivalents.)

45. *a.* 208 J   *b.* 208 s   *c.* The radiation treatment because it is ionizing radiation (the QF of the bulb's light would be zero) .

47. See website.

49. 86 min

51. *a.* 8.45×10^{13} J   *b.* 2×10^6 kg
   c. They would have to carry much less fuel weight.

53. *a.* 0.002 J   *b.* 0.4 J   *c.* about 200 times as much   *d.* No. the energy is a risk only if it causes ionization.

55. 10

57. *c.* $\lambda = 0.6931/T_{1/2}$

Photo Credits

Chapter 1 *Opener:* Courtesy NASA/CXC/SAO. Page 3 *(top left):* Sally A. Morgan/Corbis Images. Page 3 *(top right):* Courtesy NASA Langley Research Center. Page 3 *(bottom):* Tom Ives/Corbis Images. Page 4 *(left):* SUSUMU NISHINAGA/Photo Researchers. Page 4 *(center and right):* Courtesy Thomas Eisner, Cornell University. Page 5 *(left):* Stanford Linear Accelerator Center/Photo Researchers. Page 5 *(right):* Courtesy Space Telescope Science Institute/NASA.

Chapter 2 *Opener:* Getty/Brand X Pictures. Page 14: Art Wolfe/Stone/Getty Images. Page 17: Kevin Schafer/Corbis Images.

Chapter 3 *Opener:* David Fleetham/Taxi/Getty Images. Page 51 *(top left):* Bettmann/Corbis Images. Page 51 *(center):* Tek Image/Science Photo Library/Photo Researchers. Page 51 *(top right):* Lester Lefkowitz/Corbis Images. Page 52: Chip Simons/Taxi/Getty Images. Page 54: Peter Gridley/Taxi/Getty Images. Page 55: Richard Megna/Fundamental Photographs. Page 66: NASA/Science Photo Library/Photo Researchers. Page 86: Karl Weatherly/Stone/Getty Images.

Chapter 4 *Opener:* Gandee Vasan/Stone/Getty Images. Page 93: Terri Froelich/Index Stock. Page 94: Fundamental Photographs. Page 99: LWA-Dann Tardif/Corbis Images. Page 107: Marc Vaughn/Masterfile.

Chapter 5 *Opener:* John Kane Photography. Page 127: Marc Vaughn/Masterfile. Page 129: Sandra Behne/BONGARTS/Sports Chrome, Inc. Page 131: Andre Lichtenberg/Taxi/Getty Images.

Chapter 6 *Opener:* Bruce Forster/Stone/Getty Images. Page 157 *(center):* Courtesy Jeffrey Formby Antiques, Moreton in Marsh, United Kingdom. Page 157 *(bottom):* Courtesy J. Touger. Page 158: From U. Ganiel, *The Physics Teacher,* vol. 30, Jan. 1992, pp. 18–19. Page 172: Sandro Vannini/Corbis Images. Page 179: Martin Bond/Science Photo Library/Photo Researchers. Page 180: ML Sinibaldi/Corbis Images.

Chapter 7 *Opener:* and page 190: Andrew Davidhazy. Page 191: Photo by Harold E. Edgerton. ©The Harold and Esther Edgerton Family Trust, courtesy of Palm Press, Inc. Page 193 *(top):* ©AP/Wide World Photos. Page 193 *(bottom):* NASA/Science Photo Library/Photo Researchers. Page 196: Christophe Simon/AFP/Getty Images.

Chapter 8 *Opener:* Image courtesy of NASA Landsat Project Science Office and USGS EROS Data Center. Page 214: Courtesy Brookhaven National Laboratory. Page 216: Aaron Horowitz/Corbis Images. Page 218: David Madison/Stone/Getty Images. Page 221: Image Source/Alamy Images. Page 225: Drawing from Newton's *Principia.* Courtesy AIP Niels Bohr Library, American Institute of Physics. Page 229: Courtesy NASA/Johnson Space Center.

Chapter 9 *Opener:* Courtesy NASA Goddard Space Flight Center. Page 243 *(top):* Courtesy National Optical Astronomy Observatories. Page 243 *(center left):* Lois Greenfield. Page 243 *(center right):* Images created by artist Ed Jones. Courtesy NASA Goddard Space Flight Center. Page 257: Jack Guez/AFTP/Getty Images. Page 258: Mike Powell/Getty Images News and Sport Services.

Chapter 10 *Opener:* Oli Tennent/Stone/Getty Images. Page 274: Juan Manuel Silva/Age Fotostock America, Inc. Page 276: Patrick HERTZOG/AFP PHOTO/Getty Images. Page 288: Adrian Lyon/Taxi/Getty Images. Page 289: Chris Jones/Corbis Images.

Chapter 11 *Opener:* J.A. Kraulis/Masterfile. Page 312: Dr. Arthur Tucker/Science Photo Library/Photo Researchers. Page 322: Courtesy NASA/Johnson Space Center. Page 325 *(center):* K & K Ammann/Taxi/Getty Images. Page 325 *(bottom):* Manoj Shah/Getty/Image Bank.

Chapter 12 *Opener:* PhotoDisc, Inc./Getty Images. Page 337: Tom Brakefield/Corbis Images. Page 342: Dennis Stock/Magnum Photos, Inc. Page 356: Bridgeman Art Library/NY. Page 357 *(center):* Hulton Archive/Getty Images. Page 357 *(bottom):* Courtesy GM Media Archives. Page 360: Hans Wolf/The Image Bank/Getty Images.

Chapter 13 *Opener:* GJLP/Science Photo Library/Photo Researchers. Page 373: Reproduced with permission of Bensussen Deutsch and Assoc., Inc. All rights reserved. Page 378: Kim Karpeles/Alamy Images. Page 389: Jamie Budge/Corbis Images. Page 390: Courtesy The New York Public Library, Astor, Lenox and Tilden Foundation. Page 391: ©Science Museum, London/Topham-HIP/The Image Works. Page 391 *(inset):* Araldo de Luca/Corbis Images. Page 394: Simon Fraser/Royal Victoria Infirmary, Newcastle Upon Tyne/Photo Researchers.

Chapter 14 *Opener:* Philip Long/Stone/Getty Images. Page 406: Courtesy Education Development Center. Page 421: Jeremy Woodhouse/Masterfile. Page 426 *(top far left):* Alvis Upitis/The Image Bank/Getty Images. Page 426 *(top left):* Frank Siteman/Age Fotostock America, Inc. Page 426 *(top and center):* Mark Gamba/Corbis Images. Page 426 *(top right):* Connie Ricca/Corbis Images. Page 426 *(center):* Hugh Burden/Taxi/ Getty Images. Page 426 *(bottom):* Docwhite/Taxi/Getty Images. Page 435: Courtesy National Weather Service.

Chapter 15 *Opener:* Dale Sanders/Masterfile. Page 448 *(top):* Fundamental Photographs. Page 448 *(center):* Courtesy Education Development Center. Page 449: Richard Megna/Fundamental Photographs. Page 464: Photo courtesy Uri Haber-Schaim. Page 466: Carolina Biological/Visuals Unlimited. Page 469 *(left):* Ken Kay/Fundamental Photographs. Page 469 *(center):* From M. Cagnet, et al., *Atlas of Optical Phenomena,* Springer-Verlag, 1962. Page 470 *(top):* Ken Kay/Fundamental Photographs. Page 470 *(bottom):* From Clark, Hurd and Ackerson "Single Colloidal Crystals," *Nature* **281**:57 (1979). Page 471 *(left):* Richard Megna/Fundamental Photographs. Page 471 *(right):* Nelson, Alan G./Animals Animals/Earth Scenes.

Chapter 16 *Opener:* Courtesy J. Touger. Page 484 *(top):* Gail Mooney/Masterfile. Page 484 *(center):* Hubble Space Telescope/Space Telescope Science Institute. Page 501: From A.P. Deschanel, *Natural Philosophy,* 13th ed., NY, 1894. Courtesy The New York Public Library, Astor, Lenox and Tilden

Foundations. Page 502 (*left*): Courtesy John Hopkins School of Medicine. Page 502 (*right*): Science Photo Library/Photo Researchers. Page 503: Courtesy Bausch & Lomb.

Chapter 17 *Opener:* Lester Lefkowitz/Corbis Images. Page 520: Paul Kaye/Corbis/Cordaiy Photo Library, Ltd. Page 525: From Deschanel's *Natural Philosophy,* 13th ed., New York, 1894. Courtesy The New York Public Library, Astor, Lenox and Tilden Foundations. Page 526: Dick, Michael/Animals Animals/Earth Scenes. Page 534: Roger Ressmeyer/Corbis Images.

Chapter 18 *Opener:* Courtesy of International Business Machines Corporation. Page 550 (*bottom left*): From Joseph Priestly, *The History and Present State of Electricity,* 1769. Courtesy The New York Public Library, Astor, Lenox and Tilden Foundations. Page 550 (*bottom right*): Peter Menzel/Peter Menzel Photography. Page 551: Richard Megna/Fundamental Photographs. Page 552: David Phillips/Visuals Unlimited. Page 555: Bruce Ayres/Stone/Getty Images.

Chapter 19 *Opener:* Martyn F. Chillmaid/Science Photo Library/Photo Researchers. Page 579: Eurelios/Phototake. Page 580: Julian Baum/Science Photo Library/Photo Researchers. Page 588: Courtesy Johnson Controls, Inc.

Chapter 20 *Opener:* Lester Lefkowitz/Corbis Images. Page 602: Dennis Kunkel /Phototake. Page 610: Andrew Syred/Photo Researchers. Page 613: Courtesy Adamson Systems Engineering, Ontario, Canada.

Chapter 21 *Opener:* Martin Bond/Science Photo Library/Photo Researchers. Page 623: Michael Dalton/Fundamental Photographs. Page 626: Kevin Fleming/Corbis Images. Page 629: Corbis-Bettmann. Page 637: P.M. Motta, G. Macchiarelli, A. Caggiati and F.M. Magliocca/Science Photo Library/Photo Researchers.

Chapter 22 *Opener:* G. Glatzmaier, Los Alamos National Laboratory, P. Roberts, UCLA/Science Photo Library/Photo Researchers. Page 652 (*left*): Yoav Levy/Phototake. Page 652 (*right*): Courtesy Arbor Scientific. Page 658: Courtesy J. Touger. Page 659: Corbis Images.

Chapter 23 *Opener:* Don Klumpp/Stone/Getty Images. Page 679 (*top*): Lester Lefkowitz/Corbis Images. Page 689 (*left*): Lester Lefwokitz/Corbis Images. Page 689 (*right*): Peter Turnley/Corbis Images. Page 690: Lester Lefkowitz/Corbis Images. Page 696: Courtesy J. Touger.

Chapter 24 *Opener:* Courtesy NASA Kennedy Space Center. Page 711: Photo by Matthew Brady. Courtesy Library of Congress. Page 712: Courtesy Magellan Mission, NASA Jet Propulsion Laboratory. Page 713 (*left*): David Parker/Science Photo Library/Photo Researchers. Page 713 (*right*): NASA/Science Photo Library/Photo Researchers.

Chapter 25 *Opener:* Bill Brooks/Masterfile. Page 731: Corbis-Bettmann. Page 750 (*left*): From *The Times,* London, November 7, 1919; © NI Syndication, London 2005. Page 750 (*right*): Courtesy The New York Times.

Chapter 26 *Opener:* Alfred Pasieka/Science Photo Library/Photo Researchers. Page 761: Peter Fowler/Science Photo Library/Photo Researchers.

Chapter 27 *Opener:* © Andreiko & Theo Kerdemelidis, http://www.innerphase.co.nz/. Page 783: Courtesy Education Development Center. Page 788: From A. Tonomura, J. Endo, T. Matsuda and T. Kawasaki, *Am. J. Phys.* 57(2):117, Feb. 1989. Page 793: Richard Megna/Fundamental Photographs.

Chapter 28 *Opener:* Nathan Benn/CORBIS. Page 805: Courtesy Kamioka Observatory, ICRR (Institute for Cosmic Ray Research), The University of Tokyo. Page 809: Larry Dale Gordon/The Image Bank/Getty Images. Page 813 (*center*): Corbis Images. Page 813 (*bottom*): Richard Cummins/ CORBIS. Page 819: Kai Pfaffenbach/Reuters/Landov.

Index

Four-dimensional space-time, 743
Fraunhofer, Joseph, 717
Fraunhofer lines, 717
Free-body diagram, 116
    of body in free fall, 118f
Free fall, 34–38
    constant acceleration equations of motion
        applied to, 35
    definition of, 39
    of object thrown upward, 37–38
Free space
    permeability of, 664
    permittivity of, 603
Freezing, entropy change during, 354
Frequency, 260, 435. *See also* Angular
    frequency
    allowable, 428, 436
    angular or phase, 382
    angular velocity and, 246–247
    audible, 425
    audio, 436
    definition of, 412
    determining, 383
    fundamental, 430
    low, 426
    matching, 394–395
    in musical sound, 424–425
    resonant. *See* Resonance
    of sound waves, 422–423
    threshold, 764, 767t, 772, 795
    very-high, 426
    of wave, 412–413
    wavelength and, 425
Friction, 94
    in water slide, 128–129
Friction-like forces, 280–281
Frictional forces, 117, 140–143, 143, 149,
    181, 182
    compensating for, 105
    curved roadways and, 221–222
    radial force and, 219–220
    situations involving, 173–174
    sliding, 219
    static, 218–219
Frictionless motion, 130–131
Fringes, 462–463, 475
    separating, 467
Fringing, 579
Frisch, Otto, 813
Fuel rods, 814
Fuels, 159
Functions
    of angle, 378–379
    sinusoidal, 379, 380
Fundamental frequency, 430, 436
Fusion, 810, 811–812, 820
    latent heat of, 318

G

γ rays, 716, 803, 819
Galileo Galilei, 93, 223–224
Galvanic cell, 588
Galvanometer, 661, 669
    calibrating, 661–662
Gases
    densities of, 272t
    diffusion of, 337
    emission spectra of, 716–717

kinetic theory of, 336–348, 714, 759t
macroscopic properties of, 342–344
pressure and flow of, 281
temperature as measure of molecular
    motion in, 353f
volume and temperature of, 337–338
    at constant pressure, 339–340
Gauge pressure, 289
Gauss, Karl Friedrich, 577, 655
Gauss's law for magnetic fields, 577, 580,
    667, 709
Geiger, Hans, 761, 762
Geiger counter, 808, 820
Gel electrophoresis, 579
Gemstone, facets of, 500–501
General relativity theory, 734–735,
    749–751, 751
Generators
    electric, 689–693
    hand-cranked, 692–693
Geometric optics, 483–505, 504
    definition of, 485–486
Germer, L. H., 783
Gilbert, William, 544
Global energy system, 161
Gold foil experiment, 767t
Gold nucleus, radius of, 762–763
Gradient, 558, 559
Gram-molecular mass. *See* Mole
Graphs
    of accelerated motion, 24–27
    definition of, 38
    of instantaneous velocity versus clock
        reading, 22–24
    interpreting slope of, 20–22
    of pin ball motion, 32f
    of position versus clock reading, 22
    representing motion, 19–22
    of uniformly accelerated motion, 27–29
    of x versus t, 22–23
Gravitation
    circular motion and, 214–215
    universal law of, 223–229, 233, 235–236
Gravitational acceleration, 34–38, 39, 227
    pendulum in determining, 392
Gravitational analog to resistance, 607
Gravitational attraction, 53
Gravitational charge, 545
Gravitational circuit, 585–586
Gravitational constant, universal, 227
Gravitational fields, 580
    space-time geometry and, 749
Gravitational force, 94, 118–121, 156f
    direction of, 224–225
    electrical forces and, 652–653
    electrostatic force and, 580–583
    magnitude of, 225
    weight and, 100, 107
Gravitational forces, 143
Gravitational interaction, 225
Gravitational lens, 750–751
Gravitational mass, 225, 558. *See also*
    Gravitational charge
Gravitational potential energy, 157, 181,
    229–230, 233
    in the case of ball thrown upward, 167–168
    in the case of dropping ball, 166–167
    in fluid flow, 284–287

problems on, 236
quantitative definition of, 165–170
on slope, 169
Gravity, center of, 135
Ground, 558
Ground state, 777, 795
Grounding, 548

H

Hahn, Otto, 813
Half-life, 636, 804, 820
Halley's Comet, 243f
Harmonics, 430, 436
    wavelengths and frequencies of, 432
Hearing, threshold of, 427
Heart, pumping function of, 289
Heat, 4, 181, 326. *See also* Specific heat
    allowable and forbidden exchange of, 355
    definition of, 305
    distributed energy and, 157–159
    latent, 318
    light and, 612
    measurement of, 306
    mechanical equivalent of, 309, 326
    of transformation, 317–321
        of common materials, 318t
    units of, 309–310
Heat engines, 357–360, 363
    efficiency of, 358–359, 360, 363
Heat pump, 358, 363
Heat transfer, 326
    by conduction and radiation, 322, 323
    changes of state and, 317–321
    from cooler to warmer object, 353
    energy and, 307–310
    factors affecting, 323–326
    modes of, 321–323
    pumping and, 350–351
    rate of, 323–324, 326, 348–349
    surface area and, 324–325
    temperature differences and, 305–307,
        310–317
    through metals, 347
Heavy nuclei decay, 811
Heisenberg, Werner, 790
Heisenberg uncertainty principle, 790–791, 795
Helical or spiral motion, 51
    in magnetic field, 659
Helium
    discovery of, 717
    doubly charged ions of, 759
    line spectrum of, 780
Hemodynamics
    fluid dynamics in, 287, 289–291
Henry, Joseph, 693
Henry (H), 693
Hertz, Heinrich, 711
High-temperature reservoir, 357
High-voltage transmission lines, 696
Hiroshima bombing, 813, 815
Home electrical service, 697–698
Hooke's law, 174–176, 181, 389, 395
Hoover Dam, 690
Horizontal forces, 130f
Horizontal motion, 50–51
Horizontal motion equations, 72–73
Horizontal oscillator, 374
Horsepower, 179, 182

SI Units

Quantity	Name of Unit	Symbol	Expression in Terms of Other SI Units	Quantity	Name of Unit	Symbol	Expression in Terms of Other SI Units
Length	meter	m	Base unit	Pressure, stress	pascal	Pa	N/m^2
Mass	kilogram	kg	Base unit	Viscosity	—	—	$Pa \cdot s$
Time	second	s	Base unit	Electric charge	coulomb	C	$A \cdot s$
Electric current	ampere	A	Base unit	Electric field	—	—	N/C
Temperature	kelvin	K	Base unit	Electric potential	volt	V	J/C
Amount of substance	mole	mol	Base unit	Resistance	ohm	Ω	V/A
Velocity	—	—	m/s	Capacitance	farad	F	C/V
Acceleration	—	—	m/s^2	Inductance	henry	H	$V \cdot s/A$
Force	newton	N	$kg \cdot m/s^2$	Magnetic field	tesla	T	$N \cdot s/(C \cdot m)$
Work, energy	joule	J	$N \cdot m$	Magnetic flux	weber	Wb	$T \cdot m^2$
Power	watt	W	J/s	Specific heat capacity	—	—	$J/(kg \cdot K)$ or $J/(kg \cdot C°)$
Impulse, momentum	—	—	$kg \cdot m/s$	Thermal conductivity	—	—	$J/(s \cdot m \cdot K)$ or $J/(s \cdot m \cdot C°)$
Plane angle	radian	rad	m/m	Entropy	—	—	J/K
Angular velocity	—	—	rad/s	Radioactive activity	becquerel	Bq	s^{-1}
Angular acceleration	—	—	rad/s^2	Absorbed dose	gray	Gy	J/kg
Torque	—	—	$N \cdot m$	Exposure	—	—	C/kg
Frequency	hertz	Hz	s^{-1}				
Density	—	—	kg/m^3				

The Greek Alphabet

Alpha	A	α	Iota	I	ι	Rho	P	ρ
Beta	B	β	Kappa	K	κ	Sigma	Σ	σ
Gamma	Γ	γ	Lambda	Λ	λ	Tau	T	τ
Delta	Δ	δ	Mu	M	μ	Upsilon	Υ	υ
Epsilon	E	ε	Nu	N	ν	Phi	Φ	ϕ
Zeta	Z	ζ	Xi	Ξ	ξ	Chi	X	χ
Eta	H	η	Omicron	O	o	Psi	Ψ	ψ
Theta	Θ	θ	Pi	Π	π	Omega	Ω	ω